Physics of Strong Fields

NATO ASI Series

Advanced Science Institutes Series

A series presenting the results of activities sponsored by the NATO Science Committee, which aims at the dissemination of advanced scientific and technological knowledge, with a view to strengthening links between scientific communities.

The series is published by an international board of publishers in conjunction with the NATO Scientific Affairs Division

A	**Life Sciences**	Plenum Publishing Corporation
B	**Physics**	New York and London
C	**Mathematical and Physical Sciences**	D. Reidel Publishing Company Dordrecht, Boston, and Lancaster
D	**Behavioral and Social Sciences**	Martinus Nijhoff Publishers
E	**Engineering and Materials Sciences**	The Hague, Boston, Dordrecht, and Lancaster
F	**Computer and Systems Sciences**	Springer-Verlag
G	**Ecological Sciences**	Berlin, Heidelberg, New York, London,
H	**Cell Biology**	Paris, and Tokyo

Recent Volumes in this Series

Volume 150—Particle Physics: *Cargèse 1985*
edited by Maurice Lévy, Jean-Louis Basdevant, Maurice Jacob, David Speiser, Jacques Weyers, and Raymond Gastmans

Volume 151—Giant Resonances in Atoms, Molecules, and Solids
edited by J. P. Connerade, J. M. Esteva, and R. C. Karnatak

Volume 152—Optical Properties of Narrow-Gap Low-Dimensional Structures
edited by C. M. Sotomayor Torres, J. C. Portal, J. C. Mann, and R. A. Stradling

Volume 153—Physics of Strong Fields
edited by Walter Greiner

Volume 154—Strongly Coupled Plasma Physics
edited by Forrest J. Rogers and Hugh E. Dewitt

Volume 155—Low-Dimensional Conductors and Superconductors
edited by D. Jérome and L. G. Caron

Volume 156—Gravitation in Astrophysics: *Cargèse 1986*
edited by B. Carter and J. B. Hartle

Series B: Physics

Physics of Strong Fields

Edited by

Walter Greiner

Johann Wolfgang Goethe University
Frankfurt am Main, Federal Republic of Germany

Springer Science+Business Media, LLC

Proceedings of a NATO Advanced Study Institute on
Physics of Strong Fields,
held June 1-14, 1986,
in Maratea, Italy

Library of Congress Cataloging in Publication Data

NATO Advanced Study Institute on Physics of Strong Fields (1986: Maratea, Italy)
Physics of strong fields.

(NATO ASI series. Series B, Physics; vol. 153)
"Published in cooperation with NATO Scientific Affairs Division."
Includes bibliographical references and index.
1. Heavy ion collisions—Congresses. 2. Quantum chromodynamics—Congresses. 3. Broken symmetry (Physics)—Congresses. 4. Gravitation—Congresses. 5. Astrophysics—Congresses. I. Greiner, Walter, 1935- . II. Title. III. Title: Strong fields. IV. Series: NATO ASI series. Series B, Physics; v. 153.
QC794.8.H4N38 1986 530.1 87-11104
ISBN-13: 978-1-4612-9052-0 e-ISBN-13: 978-1-4613-1889-7
DOI: 10.1007/ 978-1-4613-1889-7

Originally published by Plenum Press, New York in 1987
Softcover reprint of the hardcover 1st edition 1987
A Division of Plenum Publishing Corporation
233 Spring Street, New York, N.Y. 10013

PREFACE

The NATO Advanced Study Institute on *Physics of Strong Fields* was held at Maratea/Italy from 1-14 June, 1986.

The school was devoted to the advances, theoretical and experimental, in physics of strong fields made during the past five years. The topic of the first week was almost exclusively quantum electrodynamics, with discussions of symmetry breaking in the ground state, of the physics of strong fields in heavy ion collisions and of precision tests of perturbative quantum electrodynamics. The famous positron lines found at GSI (Darmstadt) and the related question "new particle versus vacuum decay" - (yes or no or both) - constituted the center of experimental advances. This was followed in the second week by the presentation of a broad range of other areas where strong fields occur, reaching from nuclear physics over quantum chromodynamics to gravitation theory and astrophysics.

We were fortunate to be able to call on a body of lecturers who not only made considerable personal contributions to this research but who are also noted for their lecturing skills. Their enthusiasm and dedication for their work was readily transmitted to the students resulting in a very successful school. This enthusiasm is also reflected in the contributions to these Proceedings which, as I believe, will in time become a standard source of reference for future work on the physics of strong fields and will help to spread the benefits of the school to a larger audience than those who were able to attend. I regret that the East German and Soviet colleagues were unable to participate. They have made important contributions to the field so that their presence would have certainly further contributed to the atmosphere of the meeting.

Special highlights were two distinguished lectures: "The future of nuclear physics" by D.A. Bromley and "Van der Waals forces and zero point energy" by H.B.G. Casimir. Professor Casimir was personally involved in several aspects of the early development of quantum mechanics and, in particular, of the recognition of the vacuum as a special object (Casimir effect). In his lecture he exhibited the roots of the modern developments of these ideas. Professor Bromley is the discoverer of nuclear molecules in light ion-ion scattering. These seem to be of importance also for very heavy nucleus-nucleus systems, particularly in connection with the possibility to observe experimentally the vacuum decay in supercritical fields.

The School was sponsored by and had the support of the NATO Research Council, the Bundesministerium für Forschung und Technologie, Vanderbilt University (Nashville/Tennessee), and the Gesellschaft für Schwerionenforschung (GSI).

The meeting took place at the Hotel "Villa del Mare" at Acquafredda di Maratea which contributed with its beautiful scenery, its facilities and last but not least its perfect management to the success of the school.

Finally I wish to acknowledge the contributions made by all those who took part in organizing and running the school. Above all, my thanks go to my secretaries Mrs. Ruth Lasarzig and Ellen Pfister, who handled the correspondence before, during and after the school, managed the budget and helped to make these Proceedings possible. I am grateful to the Organizing Committee for their help in preparing the school, in particular I wish to thank my colleague Reiner Dreizler and many other members of the Institute of Theoretical Physics, Johann Wolfgang Goethe-University at Frankfurt am Main.

Walter Greiner

CONTENTS

STRONG FIELDS IN PERSPECTIVE

Walter Greiner

Institut für Theoretische Physik
Johann Wolfgang Goethe Universität
Frankfurt am Main, Germany

This is the second NATO Advanced Study Institute on Physics of Strong Fields. In summer 1981 we gathered at Lahnstein/Germany and discussed the exciting developments in Quantum Electrodynamics of Strong Fields. The latter is still the principal topic of this meeting, but it is also supplemented with strong field aspects in other areas, like strong quark-gluon-fields, strong gravitational fields, the various aspects of the vacuum as a physical object (its structure, Casimir effect etc.).

At the time of the Lahnstein NATO ASI the first structures appeared in the positron spectra measured in conjunction with collisions of very heavy ions like U+U, U+Cm. Such stuctures were anticipated theoretically by Rafelski, Müller et al.[1] in 1978 and - more quantitatively - by J.Reinhardt et al.[2] in 1981. The excitement was great, and, in fact it still is. Meanwhile many more facts have been assembled. First the positron lines seem to have been clearly established both by the group working with the Orange spectrometer (Berderman, Bosch, Kienle, Koenig, Kozhuharov) and the group working with the EPOS spectrometer (Backe, Bethge, Bokemeyer, Cowan, Greenberg, Schwalm, Schweppe, Stiebing). Also Krankeleit's group (Oeschler, Senger, Krämer) has orally reported the observation of positron line structures. The latter group has definitively observed the time-delay effect in connection with deep inelastic heavy ion collisions (time delay $\tau \approx (1 \sim 2) \cdot 10^{-21}$sec) both in the positron and in the δ-electron spectra. This important work demonstrates that the idea of an atomic clock in heavy ion reactions, as proposed by Soff et al.[3], is successful. We are at the beginning of a new spectroscopy allowing to measure short delay times in nuclear reactions in a direct and absolute way. The status of strong field research up to 1984 is summarized in the book entitled Quantum Electrodynamics of Strong Fields[4].

Already in 1983 when U+U and U+Cm-positron data were compared, the line structure in the positron spectra surprisingly appeared at the same energy. When data for various systems with different united charge $Z = Z_1 + Z_2$ became available, a (nearly) Z-independent positron line seemed to emerge. It was in 1983 that Andreas Schäfer et al.[5] looked into the possibility of creating new particles in strong fields. The discussion was carried out within the standard model, whose Higgs-sector is still unknown. The idea was suggestive: due to the interaction between the photon field and the Higgs-field $\hat{\phi} = \binom{\phi^{(+)}}{\phi}$ as well the anti-Higgs $\hat{\phi}^* = \binom{\phi^{(-)}}{\phi^*}$, the charged fields are dragged off from their vacuum expectation value

$$\phi_{(o)} = \binom{o}{v} \rightarrow \binom{o + \chi^{(+)}}{v + \chi}, \qquad \phi^*_{(o)} = \binom{o}{v} \rightarrow \binom{o + \chi^{(-)}}{v + \chi^*}.$$

It was assumed that the $\chi^{(-)}$-particles are captured in strongly bound states becoming overcritical and thus forming a condensate, while the $\chi^{(+)}$-particles - conserving charge - are emitted. In particular these $\chi^{(+)}$-Higgs bosons (Goldstone bosons) were thought to decay into

$$\chi^{(+)} \rightarrow e^+ + \nu,$$

thus producing positrons. The same mechanism could happen to the neutral χ-Higgs-boson, which would decay into

$$\chi \rightarrow e^+ + e^-.$$

The idea in conjunction with heavy ion positron experiments is summarized in the following figure taken from the original paper of Schäfer et al.[5]

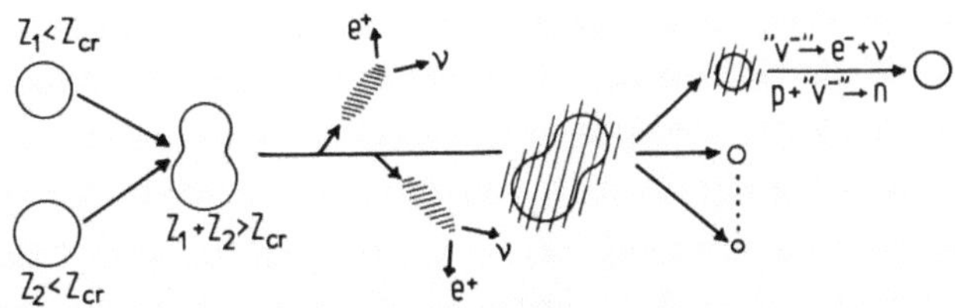

Fig. 1: Illustration of the formation and decay of a Higgs-particle condensate in a heavy ion encounter as suggested by A. Schäfer et al.[5]

This paper already contains the idea that a condensate of the new boson (in this case the Higgs boson) might be formed which could eventually lead to low energy production of these particles in the c. m. system.

After careful investigation of the processes described above (by Matthias Grabiak, A. Schäfer and G. Staadt[6]), it turned out that the production probability is far too small; the reason is that excitations of the charged components $\chi^{(+)}$ and $\chi^{(-)}$ of the Higgs field, which seemed to lead to the process described above, are essentially excitations of the W^+ and W^--field and so are only possible if the available energy is of the order of magnitude of

$$m_W \approx 80 \text{ GeV}$$

which is far out of reach.

Nevertheless, the particle idea had been planted into the brains, and in particular my experimental friend Jack Greenberg pursued it by systematically investigating a whole series of other overcritical systems with different total charge $Z = Z_1 + Z_2$. Paul Kienle investigated the U+U, Th+U and Th+Th systems. The results of the EPOS group are shown in figure 2 and also compared with the theoretically expected scaling of the

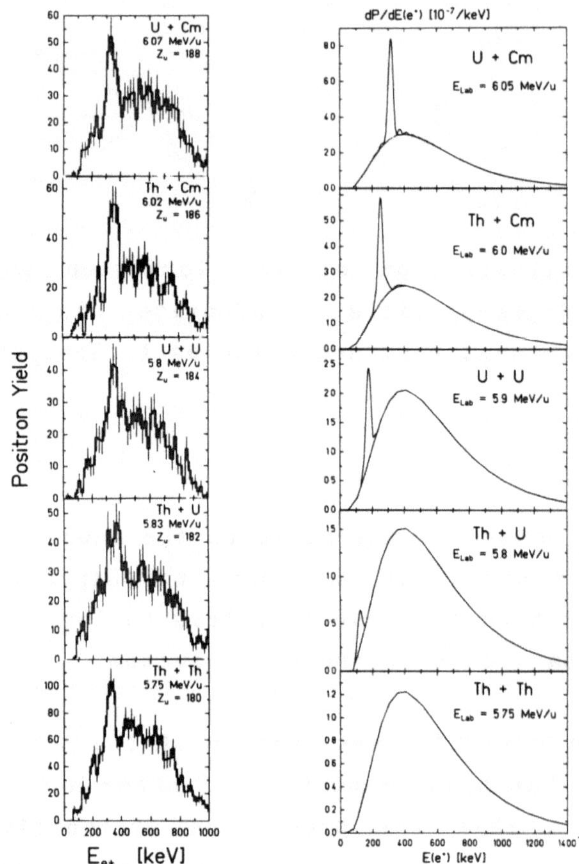

Fig. 2: a) Experimental positron spectra for various systems
b) Theoretical spectra with spontaneous vacuum decay line for nose-to-nose giant nuclear molecules.

line structure, if it is assumed that the positron line is due to spontaneous vacuum decay from a giant nuclear molecule of the simple geometrical nose to nose structure. A giant nuclear molecule is, of course, the most naive nuclear model one can start with, and nuclear structure could be much more involved. Vasak, Neise et al.[7] have therefore investigated the idea whether nuclear structure eventually undergoes a massive phase transition to cold quark matter. Zhang, Derreth et al.[8] did the same for infinite nuclear matter. Since extensive studies of Maruhn, P.G. Reinhardt et al.[9] showed that conventional mass formulas, i.e. ordinary nucleonic nuclear matter extrapolated to the giant nuclei region, do not reveal a strong binding effect for giant nuclear systems (it particularly does not yield the necessary 180 MeV/nucleon additional binding energy for the U+Cm-complex in order to lower the energy of the spherical configuration down to the energy region where the positron lines are found), the fact that cold quarkic giant nuclei could be obtained with a bag constant $B^{1/4}$ = 145 MeV was of great interest. Even though many of the assumptions, especially the non-interacting giant quark bag, are perhaps too naive, the model prediction of rather small, spherical giant quark nuclei has to be considered a most amusing possibility.

An even more important experimental step forward seemed to be the observation of $e^+ e^-$ coincidence spectra by the EPOS group, figure 3.

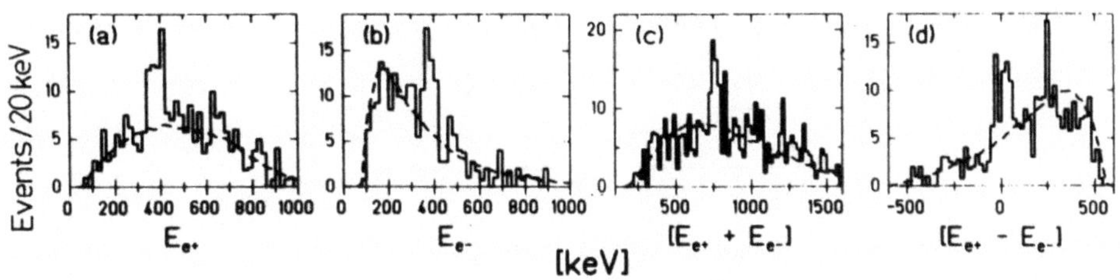

Fig. 3: e^+e = Coincidence Spectra by the Epos group

At first sight, it seemed to confirm that there is indeed a new particle involved. Greenberg gave an interview to Physics Today (B. Schwarzschild, Physics Today, Nov. 1985, p. 17) and the High energy Physics Communitiy as well as a larger fraction of the Nuclear Physics Community became interested. Phenomenological proposals were made (Balantekin et al.[10]) and new axion models were proposed (Peccei et al.[11]), but all the less utopic models could simply be ruled out on theoretical grounds, especially in the work by Schäfer, Reinhardt et al.[12]. Others followed this reasoning (see Reinhardt's talk on this subject at this conference). The essential argument goes like this: Suppose we are dealing with a neutral particle being either scalar, pseudoscalar, vector, pseudovector, etc. Its basic interaction with the electron-positron field Ψ_e and the nuclear (quark) field Ψ_N should essentially be of the form

$$g_e \bar{\Psi}_e \Gamma \Psi_e X + g_N \bar{\Psi}_N \Gamma \Psi_N X$$

where $\Gamma = 1$ (scalar), $\Gamma = \gamma_5$ (pseudoscalar), $\Gamma = \gamma^\mu$ (vector), $\Gamma = \gamma^\mu \gamma^5$ (pseudovector), $\Gamma = \sigma^{\mu\nu}$ (tensor) is the vertex function for the various kinds of particles and g_e, g_N are the coupling constants to the electron-positron and nuclear (quark) fields respectively. g_e and g_N are the two unknowns. They can be fixed from precision experiments. For example, there should be a contribution of the new particle to the (g-2)-experiment (see figure 4) and there should be a new interaction of the electron with a nucleus due to the exchange of a X-particle (see figure 5).

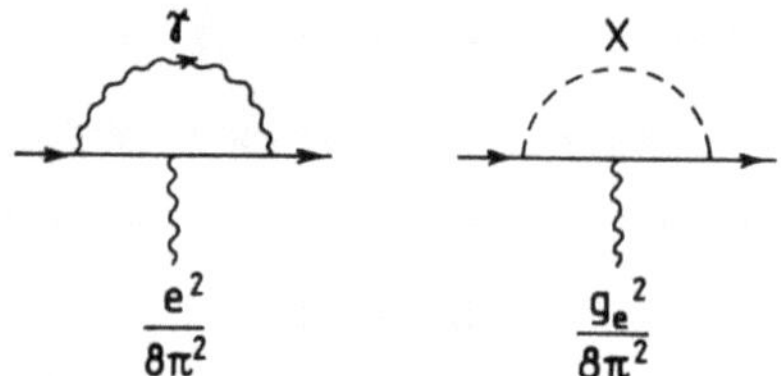

Fig. 4: The dominant contributions to (g-2) from QED (a) and a new particle X (b).

This in turn would modify the Lamb-shift. Hence, from (g-2) and from the Lamb-shift, but also from positronium levels, from neutron-nucleus scattering, from Delbrück scattering and many other precision measurements upper bounds for g_e and g_N can be found. Practically from all cases one gets

$$\frac{g_e^2}{4\pi} < 4 \times 10^{-8}$$

$$\frac{g_N^2}{4\pi^2} < 10^{-9}$$

As Reinhardt et al.[12] have shown (see Reinhardt's talk), this in turn yields a cross section for positron production which is about 4 - 5 orders of magnitude too small compared to the cross section measured unter the

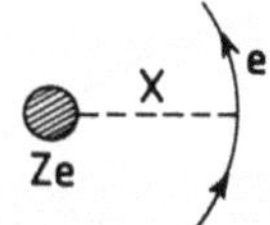

Fig. 5 : A new scalar particle X interacting with the quark-and e^+-e^--fields would cause a new interaction between electrons and nuclei.

positron line seen in the GSI - experiments. This statement assumes that all produced neutral ϕ-particles are miraculously slowed down and decay at rest. This is, in fact, very difficult to believe. More realistically the X-particles will be produced with a bremsstrahlung type spectrum, from which only a tiny few, namely the very slow ones (shadowed domain in the following figure) decay inside the detection volume. If this fraction is taken into account, the above quoted numbers change to an e^+-cross section which is by the factor 10^{-13} too small compared to the measured one. Hence the positron lines cannot stem from a new particle, except for ultra-exotic ones (see A. Schäfer's talk). The latter we are only inclined to believe, if we are driven into a corner out of which there is no other way for escape. Of course, the above quantitative argumentation to a large extent depends on the assumed bremsstrahlung production spectrum. If this were changed, e. g., by a condensation phenomenon as originally anticipated by Schäfer et. al.[5], the low energy components might eventually be enhanced and thus the cross section for a positron line might increase.

A question always bothered us and also many others not involved in the experiments themselves, namely whether the line structure observed is real or a statistical fluctuation triggered by cuts, psychology and other influences. There are many examples of observations at the limit of statistical relevance which could not stand up in

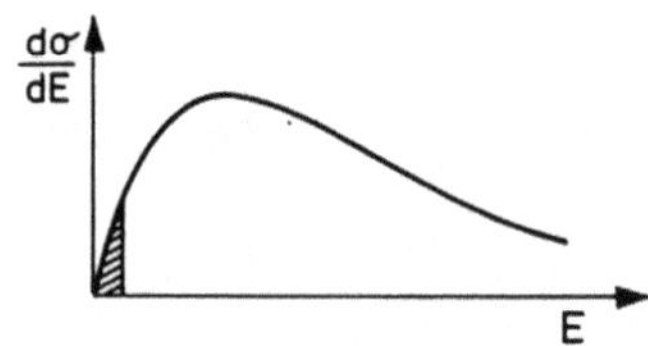

Fig. 6: Typically a particle production spectrum is of bremsstrahlung type. The shadowed region at very low energies, if selected by the experiments, could eventually cause an e^+-e^--positron line after decay. All other particle decays would smear out a line structure.

high statistics data; but - fair enough - there are also counter examples. Let's hope that we are dealing with such a one here. The high statistics data by the EPOS-group, analyzed and presented here by K. Stiebing, seem to indicate that there are possibly several electron lines at different energies in coincidence with one positron line. This would be highly important; therefore it must be clearly established by further measurements and analysis whether this observation is correct or not. Also, the principal positron line of the singles spectra seems now to appear not at 310 KeV as in all the years before, but at about 430 KeV. Perhaps this is now the vacuum decay line? The question arises how it scales with $Z = Z_1 + Z_2$; we do not know yet. But, let's reflect on the new observations of possibly several electron lines in conjunction with one positron line and vice versa, appearing in the coincidence spectra. If this turns out to be correct, the new particle will be dead. Instead, it would suggest what I believe is an equally exciting discovery, that we are dealing with a giant nuclear system, its level structure strongly coupled to spontanous decay of the vacuum. This yields on the one hand Stokes-satellites to the spontaneous vacuum decay, which can be nearly as strong as the vacuum line itself (Stefan Schramm discusses this issue in his contribution.). On the other hand, if one considers a cascade decay of the giant nucleus (figure 7) as P. Schlüter and G. Soff discussed a year ago[13], the first transition might eject an electron, the second one a positron. Since the electronic levels are partially filled and partially empty also two electron and two positron emission might occur and all this in various combinations. A multitude of e^+-e^- lines with various positron and (different) electron energies should then be seen; not only in overcritical but also in undercritical systems. What an exciting spectroscopy would then eventually become possible!? The multitude of e^+-e^--coincidence and of single lines (see P. Kienle's summary talk) could fit into such a picture. What does not fit is that experimentalists up till now claim that the positron energy always equals the electron energy.

This, together with the many lines observed, could eventually also be explained by a complex particle built out of several e^+-e^- pairs, that is $(e^+$-$e^-)^n$ strongly bound to a total mass of about 1600 KeV, as B. Müller and his associates have suggested[14]. Such a complex could only decay into one e^+-e^--pair or two photons; it could have excited states which are expected to be rather high in energy, because of the extremely strong binding (all masses of (n-1) e^+-e^--pairs are bound away!). Fine structure splitting will probably produce low lying excitations of the complex. An illustration of these tightly bound polypositronium clusters is given in fig. 8.

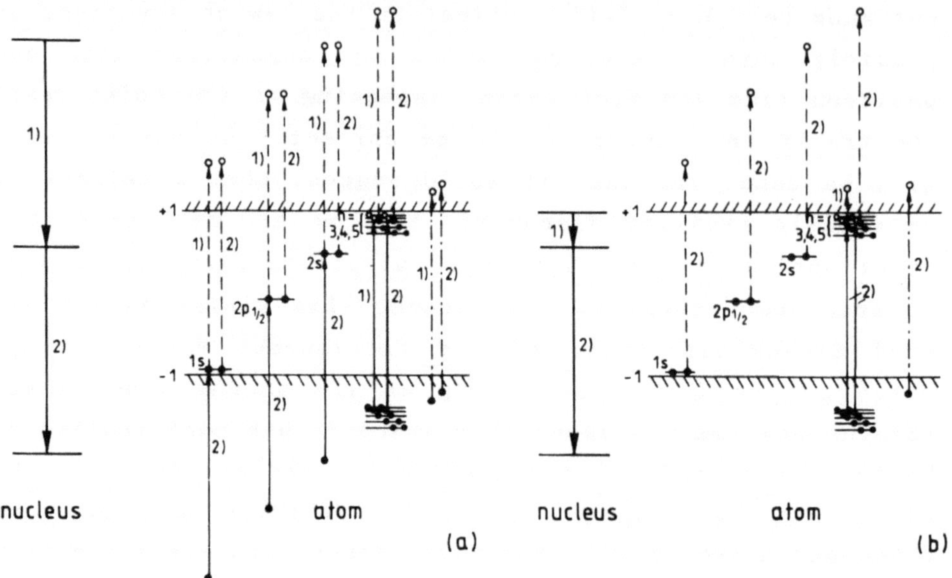

Fig. 7: Cascade conversion. At the left a cascade of the (giant) nucleus is indicated; at the right the conversion electron and the conversion positron transitions are shown. Obviously both electron and positron would be in coincidence, but have different energies and momenta would not be correlated.

Whether the electromagnetic binding alone will be sufficient to hold such clusters together is at the moment quite doubtful. Probably new nonlinear (many-body-)interactions in the e^+-e^--field are needed. A gauge invariant theory must then be constructed, yielding new gauge bosons of high mass etc, in order to insure the short range of the interaction. One has to see whether this idea leads to contradictions somewhere. Equally beautiful is in this connection the thought that the e^--e^+-field itself is a Higgs-field and that the contact terms of the form $(\bar{\Psi}_e\Psi_e)^2$ and $(\bar{\Psi}_e\Psi_e)^4$, needed in the Higgs-sector are doing the job. If such nonlinear interactions exist the vacuum might exhibit cluster structure. A theory for a clustered QED-vacuum does not exist at present. This is an extremely interesting concept which must be pursued.

Indeed, the strong binding of the electrons in giant atoms (molecules) and the corresponding shrinking of their wave functions as well as the hole states in the innermost shells (which can be considered as bound positrons) seem to constitute the ideal formation scenario for such strongly bound $(e^+e^-)^n$-clusters. For a real condensate of them the level structure has too small density, except if due to the time-dependence, magnetic effects and Heisenberg broadening increased the level structure. This seems unlikely, but must be quantitatively clarified. K. Rumrich and G. Soff are presently carrying out this program.

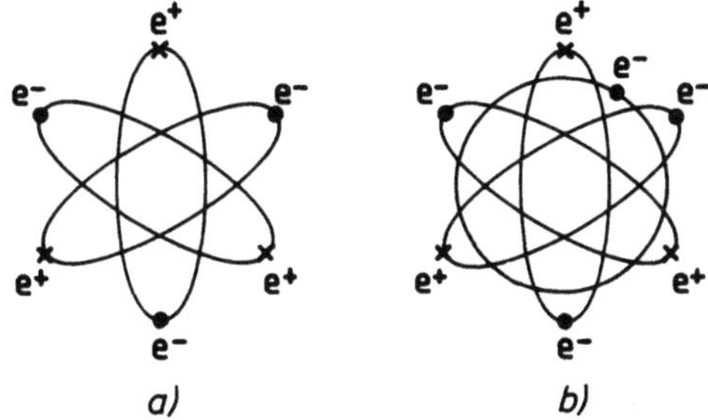

Fig. 8: Illustration of tightly bound polypositronium clusters
a) n = 3-cluster, b) n = 3-cluster with additional electron.
Other types are also thinkable, of course.

There are a number of wrong suggestions (Wong[15]) and rather doubtful calculations floating around. The latter picks up Barut's idea of strongly bound and small sized e^+-e^--configurations due to their interaction via the anomalous magnetic moments (Wong and Becker strangely take the full magnetic moment —— the g=2 part is already contained in the Dirac equation). Up to now Barut has not convincingly demonstrated that such micro-positronium states exist. One can doubt this, because of the very many

QED-corrections coming in at small distances and the strong fields encountered there.

Exotic particle scenarios have been discussed by A. Schäfer et al. (see A. Schäfer's talk). They are utopic enough to be not in contradiction with present day experiments, but also far enough off the main road of today's physics that many more convincing experimental information in their favour has to be available before one can believe any such model.

The question also arises whether the e^+-e^--peaks, if real, might indicate a yet unknown solid state effect. For example, it is well known that positrons of low energy interpenetrate crystals much further than expected from lifetime estimates of positrons in an electron plasma. Warke and Greiner[16] have derived a nonlinear Schrödinger equation for such positrons (the non-linearity stemming from the interaction of e^+ with its surrounding) and shown that soliton solutions exist, which give the positron much longer stability on its way through the crystal. These solitons can be viewed as a kind of quasi-stable positronium states (e^+-e^-)which break up as soon as the soliton leaves the crystal. In other words, the positron neutralizes within the medium, dragging an electron along. Thus positron and electron have essentially the same energy. If such states were also formed with a certain probability at much higher positron energy (i. e. 200 - 1000 KeV) it is thinkable that such solitons could preferably penetrate the crystal in certain energy windows, thus leading eventually to peak structure (think about Lummer-Gehrke plates for photons as an analogy). Such an effect would easily explain most features observed up till now; also the yet unpublished but orally transmitted results of Karl Erb (Oak Ridge) who seems to see related (if not the same) resonances by bombarding a thorium target with the positron spectrum of a natural β^+-emitter. The characteristics of such a spectrum are similar to the dynamically induced positron spectrum of heavy ion encounters. The idea deserves to be studied more closely. If true, one would have found a kind of positron filter.

The fact that some of the observed positron resonances (singles) seem to appear in a regular pattern in under-critical systems at ion energies and impact parameters assuring no nuclear contact, could point towards a "sudden rearrangement" of quasimolecular orbitals (matrix elements) at a certain distance of the ions. This would mean that non-adiabatic and still unknown atomic processes play a role in the dynamics, similar to what was proposed in ref. 17.

We thus see that the positron puzzle is open. Perhaps none of the ideas presented here work. Perhaps also various phenomena are simultanously seen. It is certainly a great challenge to find out the solution! This conference hopefully contributes to its clarification. What is needed are additional systematic experiments; excitation functions of the positron resonance as a function of ion energy; a systematic study of elastic nucleus-nucleus scattering to check out the giant nuclear molecule idea; the positron resonances should be measured in conjunction with the heavy ion residues coming out of the collision; the energy loss of the fragments is important to know; the exact windows (parameters) under which the e^+-resonances appear have to be found; the invariant mass in the e^+-e^- channel has to be determined; etc, etc.

This NATO ASI on Physics of Strong Fields, being intermediate between a school and a topical meeting, addresses itself to most topical research. I am happy that so many distinguished scientists were willing to contribute as lecturers, seminar speakers and discussion leaders.

In retrospect I can state that the very extensive discussions were extremely stimulating and enlightening. It was characteristic of the fine atmosphere that discussions after many lectures were nearly as long as the lectures themselves. Very often the talks were interrupted by enlightening questions and the corresponding answers; thus leading to clarification and a better understanding of the issues discussed. Very many students from all over the world have applied to attend. Unfortunately the size had to be cut from over 180 applicants to about 100 attendants.

Some distinguished speakers from the United States did finally not show up, even though they originally agreed to be present, because of their fears of a possible terrorist attack (Ghaddafi, Camorra, ...) on either the airplanes or the school itself. I am grateful to the many courageous colleagues and friends who finally came despite.of this modern neurosis. Thank God: all those fears were unjustified!

It gives me great pleasure to thank a number of institutions whose support was essential:

1. Nato Scientific program, especially Dr. Craig Sinclair.
2. Deutsches Bundesministerium für Forschung und Technologie, especially Dr. D. Hartwig.
3. Gesellschaft für Schwerionenforschung (GSI) at Darmstadt; especially its director Prof. P.Kienle.
4. Vanderbilt University at Nashville/Tennessee and the Joint Institute for Heavy Ion Research at Oak Ridge/Tennessee, especially its director Prof. J. Hamilton.

The hotel "Villa del Mare" at Aquafredda di Maratea did its best to make our two week stay enjoyable.

Finally my thanks go especially to my personal secretary Mrs. Ruth Lasarzig who helped me tremendously in all practical matters concerning the organization of this Advanced Study Institute. Without her assistance and continuous efforts this school would not have been what it was: an interesting and successful top level meeting on a highly interesting field of research.

REFERENCES

1. J. Rafelski, B. Müller, W. Greiner, Z. Physik A285 (1978) 49
2. J. Reinhardt, U. Müller, B. Müller, W. Greiner, Z. Physik A303 (1981) 173

3. G. Soff, J. Reinhardt, B. Müller, W. Greiner, Phys. Rev. Lett. 43 (1979) 1981
4. W. Greiner, B. Müller, J. Rafelski, Quantum Electrodynamics of Strong Fields, Springer-Verlang, Heidelberg and New York (1985)
5. A. Schäfer, B. Müller, W. Greiner, Phys. Lett. 149B (1984) 455
6. M. Grabiak, A. Schäfer, G. Staadt, to be published
7. D. Vasak, L. Neise, W. Greiner, Do nucleons dissolve in giant nuclei?, preprint LBL 20561
8. Q. Zhang, C. Derreth, A. Schäfer, W. Greiner, J. Phys. G12 (1986) L19
9. M. Seiwert, J.A. Maruhn, W. Greiner, J. Friedrich, Z. Physik 321 (1985) 653 P.G. Reinhardt, M. Rufa, J. Maruhn, W. Greiner, J. Friedrich, Z. Physik 323 (1986) 13
10. A.B. Balantekin, C. Bottcher, M. R. Strayer, S. J. Lee, Phys. Rev. Lett. 55 (1985) 461
11. R.D. Peccei, T. T. Wu, T. Yanagida, Phys. Lett. 172 (1986) 435iW. Bardeen, R. D. Peccei, T. Yanagida, Constraints on variant axion models, preprint Desy 86-054
12. A. Schaefer, J. Reinhardt, B. Müller, W. Greiner, G. Soff, J. Phys. G11 (1985) L69
    A. Schäfer, J. Reinhardt, B. Müller, W. Greiner, Models for new particle production in nuclear collisions, preprint GSI-86-12
13. P. Schlüter, U. Müller, G. Soff, T. de Reus, J. Reinhardt, W. Greiner, Z. Physik A323 (1986) 139
14. B. Müller, J. Reinhardt, W. Greiner, A. Schäfer, J. Phys. G12 (1986) L109
15. C.-Y. Wong, Phys. Rev. Letters,
    C.-Y. Wong, R. L. Becker, Scalar magnetic (e^+e^-) resonance as possible source of anomalous e^+ peak in heavy-ion collisions, ONRL-preprint
16. C.S. Warke and W. Greiner, Phys. Lett. 80A (1980) 399
17. R.K. Smith, B. Müller, W. Greiner, J.S. Greensberg, C.K. Davis, Phys. Rev. Lett. 34 (1975) 117

STATUS OF PRECISION QED

IN LIGHT AND HEAVY ATOMS

Peter J. Mohr

National Science Foundation
Washington, DC 20550
and
National Bureau of Standards
Gaithersburg, MD 20899

INTRODUCTION

This paper presents a review of QED in various systems. Precision tests in weakly bound systems are reviewed, tests of radiative corrections in highly ionized atoms and inner shells of neutral heavy atoms are described, and a more speculative application of QED to quarks bound in a cavity is presented.

PRECISE TESTS OF QED IN LIGHT ATOMS

Electron g-Factor Anomaly

A fundamental test of QED is the comparison of theory and experiment for the magnetic moment of the electron. The moment is written in terms of the anomaly a_e defined by

$$g_e = 2(1 + a_e) \tag{1}$$

that characterizes the deviation of the g factor from the Dirac value of $g_e = 2$. A recent measurement of the anomaly by Van Dyck et al. has yielded the accurate value[1]

$$a_e = 1\ 159\ 652\ 193(4) \times 10^{-12} \tag{2}$$

The theoretical value is calculated as a power series in the fine structure constant α to obtain a series of the form

$$a_e = a_e^{(2)} \frac{\alpha}{\pi} + a_e^{(4)} \left[\frac{\alpha}{\pi}\right]^2 + a_e^{(6)} \left[\frac{\alpha}{\pi}\right]^3 + a_e^{(8)} \left[\frac{\alpha}{\pi}\right]^4 + \cdots \tag{3}$$

The leading term is the well-known Schwinger term

$$a_e^{(2)} = \frac{1}{2} \tag{4}$$

The fourth-order term is also known in analytic form

$$a_e^{(4)} = \frac{197}{144} + \frac{\pi^2}{12} - \frac{\pi^2}{2} \ell n\, 2 + \frac{3}{4}\zeta(3) = -0.328\ 478\ 966 \tag{5}$$

Many diagrams contributing to the sixth-order term have been calculated analytically, and the rest have been calculated numerically[2]

$$a_e^{(6)} = 1.176\ 5(13) \tag{6}$$

The eighth-order coefficient has recently been calculated by Kinoshita and Lindquist who obtain the value[2,3]

$$a_e^{(8)} = -0.8(1.4) \tag{7}$$

This massive calculation consists of the evaluation of 891 Feynman diagrams. In addition to the QED contributions, there are small corrections for muon vacuum polarization loops, hadronic vacuum polarization loops (estimated from e^+-e^- annihilation into hadrons data), and weak interaction contributions

$$a_e(\text{muon vac pol}) = 2.8 \times 10^{-12} \tag{8}$$

$$a_e(\text{had vac pol}) = 1.6(2) \times 10^{-12} \tag{9}$$

$$a_e(\text{weak int}) = 0.05 \times 10^{-12} \tag{10}$$

A recent analysis by Taylor of condensed matter measurements of α gives[4]

$$\alpha^{-1} = 137.035\ 981(12) \tag{11}$$

With this value for α, the total calculated value for the anomaly is

$$a_e = 1\ 159\ 652\ 307(41)(102) \times 10^{-12} \tag{12}$$

where the first number in parentheses is the error due to uncertainty in the theoretical evaluation of the eighth-order terms and the second number in parentheses is the uncertainty associated with the fine structure constant α.

If the theory is assumed to be correct, then the comparison of theory and experiment for the g factor yields a value for the fine structure constant α_g given by

$$\alpha_g^{-1} = 137.035\ 994(5) \tag{13}$$

A similar comparison of theory and experiment for the muonium hyperfine splitting gives

$$\alpha_{\Delta\nu}^{-1} = 137.035\ 991(25) \tag{14}$$

so that there is excellent agreement between these independent determinations of α.

Muon g-Factor Anomaly

The measured value of the muon g-factor anomaly is[5,6]

$$a_\mu = 1\ 165\ 911(11) \times 10^{-9} \tag{15}$$

Recent work by Kinoshita and coworkers has improved the theoretical precision of a_μ.[7] The power series for the QED contributions is

$$a_\mu = a_\mu^{(2)} \frac{\alpha}{\pi} + a_\mu^{(4)} \left(\frac{\alpha}{\pi}\right)^2 + a_\mu^{(6)} \left(\frac{\alpha}{\pi}\right)^3 + a_\mu^{(8)} \left(\frac{\alpha}{\pi}\right)^4 + \cdots \tag{16}$$

where the lowest-order term is the same as the lowest-order electron term

$$a_\mu^{(2)} = \frac{1}{2} \tag{17}$$

Higher-order terms differ from the electron terms, because of the relatively larger effect of electron vacuum polarization corrections. The next three terms are

$$a_\mu^{(4)} = 0.765\ 858\ 10(10) \tag{18}$$

$$a_\mu^{(6)} = 24.073(11) \tag{19}$$

and

$$a_\mu^{(8)} = 140(6) \tag{20}$$

A sizable correction arises from the hadronic vacuum polarization correction. An estimate is obtained from measurements of e^+e^- annihilation into hadrons

$$\sigma_T(e^+e^- \to \text{hadrons}) \longrightarrow a_\mu(\text{had}) = 702(19) \times 10^{-10} \tag{21}$$

Weak interaction effects contribute

$$a_\mu(\text{weak}) = 195(1) \times 10^{-11} \tag{22}$$

The total theoretical value, based on the condensed matter value of α,[4] is[7]

$$a_\mu = 11\ 659\ 200(20) \times 10^{-10} \tag{23}$$

which agrees with experiment. The hadronic vacuum polarization correction is necessary to have this agreement. Further measurements could lead to a test of the theory of weak interactions at the one-loop level.

Hydrogen Lamb Shift

A classic test of bound-state QED is the comparison of theory and experiment for the hydrogen Lamb shift. Recent measurements of the Lamb shift are listed in Table 1.

The theoretical contributions are listed in Table 2. The self-energy value is based on an extrapolation to low Z of a nonperturbative calculation

Table 1. Recent Lamb Shift Measurements

1 057.862(20) MHz	Newton Andrews & Unsworth (1979)[8]
1 057.845(9) MHz	Lundeen & Pipkin (1981)[9]
1 057.8514(19) MHz	Pal'chikov Sokolov & Yakovlev (1984)[10]

that has been confirmed for the 1S case.[11,12] The other corrections such as the higher-order radiative corrections are well established. Reduced mass-corrections include the higher-order (in $Z\alpha$) terms from the self energy. In the finite nuclear size correction, the assumed value for the proton rms charge radius is $R = 0.862(12)$ fm, as measured by Simon et al.[13]

Recent work on the Lamb shift has led to an order-of-magnitude estimate of the effect of the finite size of the nucleus on the self energy.[14] More recently, an explicit calculation of the lowest-order effect of the finite nuclear size on the vacuum polarization (Uehling potential) has been made, with the result[15]

$$\Delta E_{SU} = \alpha(Z\alpha)^2 |\psi(0)|^2 \frac{\pi}{2} (R/\lambda\!\!\!^{-}_e)^2 mc^2 \tag{24}$$

Both of these corrections were found to be negligible. Higher-order recoil contributions to the Lamb shift are being examined by Grotch and coworkers.[16]

Hydrogen Hyperfine Structure

One of the most accurately measured physical quantities is the ground-state hyperfine splitting in hydrogen. The experimental value is[17,18]

Table 2. Theoretical Contributions to the Lamb Shift in Hydrogen

Contribution	Order[mc^2]	Value[MHz]
Self Energy	$\alpha(Z\alpha)^4\cdots$	1 085.812
Vacuum Polarization	$\alpha(Z\alpha)^4\cdots$	-26.897
Fourth Order	$\alpha^2(Z\alpha)^4$	0.101
Reduced Mass	$\alpha(Z\alpha)^4 \frac{m}{M}\cdots$	-1.647
Relativistic Recoil	$(Z\alpha)^5 \frac{m}{M}$	0.359
Nuclear Size	$(Z\alpha)^4(R/\lambda\!\!\!^{-}_e)^2$	0.145
Total		1 057.873(20)

$$\Delta\nu = 1\ 420.405\ 751\ 766\ 7(10)\ \text{MHz} \tag{25}$$

The theory is substantially less well known, with precision limited by uncertainties in the proton size and structure. The known theoretical contributions are summarized in Table 3. Sapirstein has recently made an improvement in the accuracy of the term of order $\alpha(Z\alpha)^2E_F$.[19] The term δ_p represents the unknown proton polarizability correction that depends on the internal structure of the proton. If the theoretical value of the hyperfine splitting with δ_p taken as a free parameter is set equal to the experimental value, then the solution for the fractional level shift due to the proton magnetic polarizability is $\delta_p = 1.6(9)$ ppm. An independent limit imposed by polarized electron-proton scattering data is $|\delta_p| < 4$ ppm.[20]

Muonium Lamb Shift

Recent advances in the production of muonium in vacuum has made possible the first measurements of the Lamb shift in this system. The results of two independent measurements are[21,22]

$$\text{TRIUMF:} \quad S_\mu = 1\ 070\ {}^{+12}_{-15}\ \text{MHz} \tag{26a}$$

$$\text{LAMPF:} \quad S_\mu = 1\ 057\ {}^{+29}_{-25}\ \text{MHz} \tag{26b}$$

Table 4 lists the theoretical contributions to the muonium Lamb shift. The main difference between the Lamb shift in hydrogen and muonium is the fact that the mass of the muon is roughly an order of magnitude less than the mass of the proton. Consequently, the nuclear motion corrections are larger, and give the largest uncertainty to the theoretical value. On the other hand, the muon structure is known, so there is no uncertainty analogous to the proton size effect in hydrogen.

Table 3. Theoretical Contributions to the Hyperfine Splitting in Hydrogen

E_F	1 418.840 8 MHz
a_eE_F	1.645 4
$\frac{3}{2}\alpha^2E_F$	0.113 3
$O(\alpha^2)E_F$	-0.136 5
$O(\alpha^3)E_F$	-0.010 5(1)
$O(m/M)E_F$	-0.049 1(13)
δ_p	?
Total	1 420.403 4 MHz

Table 4. Theoretical Contributions to the Lamb Shift in Muonium

Contribution	Order[mc^2]	Value[MHz]
Self Energy	$\alpha(Z\alpha)^4 \cdots$	1 085.812
Vacuum Polarization	$\alpha(Z\alpha)^4 \cdots$	-26.897
Fourth Order	$\alpha^2(Z\alpha)^4$	0.101
Reduced Mass	$\alpha(Z\alpha)^4 \frac{m}{m_\mu} \cdots$	-14.626
Relativistic Recoil	$(Z\alpha)^5 \frac{m}{m_\mu}$	3.188
Total		1 047.58(30)

Muonium Hyperfine Structure

The measured value of the ground-state hyperfine splitting in muonium is[23]

$$\Delta\nu = 4\ 463.302\ 88(16)\ \text{MHz} \tag{27}$$

Theoretical contributions to the splitting are listed in Table 5. The corrections are similar to the corrections in hydrogen, except that the recoil terms are larger and require more extensive calculation in order to achieve high precision. Recoil calculations have been reviewed by Bodwin, Yennie, and Gregorio.[24] Theory and experiment are sufficiently precise that the comparison of theory and experiment for the muonium hyperfine splitting provides a determination of the fine structure constant that is competitive with other determinations. The value $\alpha_{\Delta\nu}$ obtained by equating the theory value with the experimental value, with α taken as a free parameter, is given in Eq. (14).

Table 5. Theoretical Contributions to the Hyperfine Splitting in Muonium

E_F	4 459.033 4 MHz
$a_e E_F$	5.170 9
$\frac{3}{2}\alpha^2 E_F$	0.356 2
$O(\alpha^2)E_F$	-0.429 0
$O(\alpha^3)E_F$	-0.032 9
$O(\alpha m/m_\mu)E_F$	-0.800 3
$O(\alpha^2 m/m_\mu)E_F$	0.005 2
Total	4 463.303 5 (15) MHz

Table 6. Theoretical Contributions to the $1^3S_1 - 2^3S_1$ Level Difference in Positronium

$\frac{3}{8}$ Ry	1 233 690 730.(1.3) MHz
$O(\alpha^2)$Ry	-82 006.
$O(\alpha^3)$Ry	-1 527.
$\pm\ \alpha^4 \ell n(\alpha^{-2})$Ry	± 46.
Total	1 233 607 197. MHz

Positronium Energy Levels

Positronium is a system that is nearly purely leptonic and so is, at least in principal, amenable to precise calculation. The calculations are difficult, because of the need to include high-order recoil effects.

A high precision measurement of the $1^3S_1 - 2^3S_1$ energy separation in positronium has yielded the value[25]

$$\Delta E = 1\ 233\ 607\ 185(15)\ \text{MHz} \tag{28}$$

Theoretical contributions to the splitting are listed in Table 6.[26] The last line before the total in that table gives an indication of the order of the next uncalculated terms; the coefficient is likely to be somewhat less than 1, so the number is probably an overestimate.

The experimental hyperfine structure splitting in the ground state of positronium is[27]

$$\Delta\nu = 203\ 389.10(74)\ \text{MHz} \tag{29}$$

This value is consistent with theory, which is considerably less accurate.[24] Further work on the theory is in progress.[3]

Table 7. Theoretical Contributions to the Orthopositronium Decay Rate

Γ_0	7.211 μs^{-1}
$O(\alpha)\Gamma_0$	-0.172
$O(\alpha^2 \ell n\ \alpha^{-2})\Gamma_0$	-0.001
$\pm\ \alpha^2\Gamma_0$	± 0.0004
Total	7.038 μs^{-1}

Positronium Decay Rate

The measured value of the orthopositronium decay rate is[28]

$$\Gamma = 7.051(5)\ \mu s^{-1} \tag{30}$$

The theoretical contributions are listed in Table 7.[29-31] In this case, the difference between theory and experiment is larger than a rough estimate of the higher-order uncalculated terms would indicate. Additional work on both theory and experiment will be of value in resolving the reason for the disagreement.

QED OF RELATIVISTIC ELECTRONS

Perturbative vs. Nonperturbative QED

Perturbative QED calculations, as in the case of the anomalous magnetic moment of the electron, yield results as a power series in α as indicated in Eq. (3). In contrast with this expansion, the expressions for the energy levels of strongly-bound electrons are not amenable to expansion in powers of α. Since α in the formulas for the bound-state energies appears in the combination $Z\alpha$, the expansion parameter is not necessarily small. In fact, it is desirable to calculate nonperturbatively to all orders in $Z\alpha$ for certain contributions, because the power series converges slowly. If this is done, one obtains a mixed series

$$E = f_0(Z\alpha) + \frac{\alpha}{\pi} f_2(Z\alpha) + \left[\frac{\alpha}{\pi}\right]^2 f_4(Z\alpha) + \cdots \tag{31}$$

in which α is an expansion parameter and the exact dependence on $Z\alpha$ is retained. With an obvious redefinition of the functions in the series in (31), this series is expressed as an expansion in $1/Z$

$$E = [g_0(Z\alpha) + \frac{1}{Z} g_2(Z\alpha) + \frac{1}{Z^2} g_4(Z\alpha) + \cdots](Z\alpha)^2 mc^2. \tag{32}$$

which more accurately represents the rate of convergence for few-electron atoms.

Furry Picture

A field theoretical framework in which binding effects to all orders in $Z\alpha$ are taken into account is the Furry picture.[32] In this picture, the zeroth-order wave functions are solutions of the Dirac equation in an external potential, which for a few-electron atom can be taken as the Coulomb potential (units in which $\hbar = c = m = 1$ are employed here)

$$[-i\vec{\alpha}\cdot\vec{\nabla} + V(x) + \beta - E_n]\varphi_n(\vec{x}) = 0 \tag{33}$$

The electron-positron field $\psi(x)$ is expanded in terms of a complete set of time-dependent solutions

$$\varphi_n(x) = \varphi_n(\vec{x})\exp(-E_n t) \tag{34}$$

with positive energy (+) solutions multiplying electron annihilation operators and negative energy (-) solutions multiplying positron creation operators as

$$\psi(x) = \sum_{(+)} a_n \varphi_n(x) + \sum_{(-)} b_n^{\dagger} \varphi_n(x) \tag{35}$$

The creation and annihilation operators satisfy the usual anticommutation relations

$$\{a_n, a_{n'}^{\dagger}\} = \delta_{nn'} \qquad \{b_n, b_{n'}^{\dagger}\} = \delta_{nn'}$$

$$\{a_n, a_{n'}\} = \{a_n^{\dagger}, a_{n'}^{\dagger}\} = \{b_n, b_{n'}\} = \{b_n^{\dagger}, b_{n'}^{\dagger}\} = 0 \tag{36}$$

Electron creation operators generate one-electron states from the Furry picture vacuum

$$|n\rangle = a_n^{\dagger}\,|0\rangle \tag{37}$$

where n denotes the complete set of quantum numbers $n \longrightarrow \{n, j, \ell, m\}$, and multielectron states are generated by linear combinations of products of creation operators. The unperturbed electron-positron energy is

$$H_0 = \sum_{(+)} a_n^{\dagger} a_n E_n - \sum_{(-)} b_n^{\dagger} b_n E_n \tag{38}$$

which gives the expected relations

$$H_0|n\rangle = E_n|n\rangle \tag{39}$$

$$H_0|n_1 n_2 n_3 \cdots\rangle = (E_{n_1} + E_{n_2} + E_{n_3} + \cdots)|n_1 n_2 n_3 \cdots\rangle \tag{40}$$

Interactions with electromagnetic radiation are generated by the interaction Hamiltonian

$$H_I(x) = j^{\mu}(x)A_{\mu}(x) - \delta M(x) \tag{41}$$

where the current operator is

$$j^{\mu}(x) = -\frac{e}{2}\,[\bar{\psi}(x)\gamma^{\mu}, \psi(x)] \tag{42}$$

and the mass renormalization operator is

$$\delta M(x) = \frac{\delta m}{2}\,[\bar{\psi}(x), \psi(x)] \tag{43}$$

In (41), $A_\mu(x)$ is the vector potential of the quantized radiation field. Energy level shifts due to interaction with the electromagnetic field are conveniently defined by the expression

$$\Delta E_n = \lim_{\substack{\epsilon\to 0\\ \lambda\to 1}} \frac{\langle n|U_\epsilon(\infty,0)[H_0 + \lambda H_I^\epsilon(t=0) - E_n]U_\epsilon(0,-\infty)|n\rangle}{\langle n|U_\epsilon(\infty,-\infty)|n\rangle} \tag{44}$$

where

$$H_I^\epsilon(t) = e^{-\epsilon|t|}\int d\vec{x}\, H_I(x) \tag{45}$$

which according to the theorem of Gell-Mann and Low, in the symmetric form of Sucher, is[33,34]

$$\Delta E_n = \lim_{\substack{\epsilon\to 0\\ \lambda\to 1}} \frac{i\epsilon}{2} \frac{\frac{\partial}{\partial\lambda}\langle n|S_{\epsilon,\lambda}|n\rangle_c}{\langle n|S_{\epsilon,\lambda}|n\rangle_c} + \text{const} \tag{46}$$

In (46), the subscript c denotes the fact that only matrix elements corresponding to connected Feynman diagrams are retained. $S_{\epsilon,\lambda}$ is the adiabatic S matrix that has the perturbation expansion given by

$$S_{\epsilon,\lambda} = 1 + \sum_{j=1}^{\infty} S_{\epsilon,\lambda}^{(j)} \tag{47}$$

with the definition

$$S_{\epsilon,\lambda}^{(j)} = \frac{(-i\lambda)^j}{j!}\int d^4x_j \cdots \int d^4x_1\, e^{-\epsilon|t_j|}\cdots e^{-\epsilon|t_1|} \times T[H_I(x_j)\cdots H_I(x_1)] \tag{48}$$

Level shifts given by Eq. (46) can be written in a simple form that is independent of the number or state of electrons in the atom, provided that the unperturbed state is restricted to be a linear combination of terms of the form

$$|n\rangle = a_1^\dagger a_2^\dagger \cdots |0\rangle \tag{49}$$

consisting only of electrons with no positrons. With this restriction, the lowest order (in α) level shifts are given by the expression[35]

$$E^{(2)} = -4\pi i\alpha \int d(t_2-t_1)\int d\vec{x}_2 \int d\vec{x}_1\, D_F(x_2-x_1)$$

$$\times \left\{ \frac{1}{2} \sum \bar{\varphi}_n(x_2)\gamma_\mu\varphi_m(x_2)\bar{\varphi}_k(x_1)\gamma^\mu\varphi_\ell(x_1)\ \delta(E_n+E_k,E_\ell+E_m)\ \langle a_n^\dagger\, a_k^\dagger\, a_\ell\, a_m\rangle \right.$$

$$+ \sum \bar{\varphi}_n(x_2)\gamma_\mu S_F(x_2,x_1)\gamma^\mu\varphi_m(x_1)\ \delta(E_n,E_m)\ \langle a_n^\dagger\, a_m\rangle$$

$$\left. - \mathrm{Tr}[\gamma_\mu S_F(x_2,x_2)] \sum \bar{\varphi}_n(x_1)\gamma^\mu\varphi_m(x_1)\ \delta(E_n,E_m)\ \langle a_n^\dagger\, a_m\rangle \right\}$$

$$- \delta m \sum \int d\vec{x}\ \bar{\varphi}_n(x)\varphi_m(x)\ \delta(E_n,E_m)\ \langle a_n^\dagger\, a_m\rangle \tag{50}$$

where $\delta(a,b) = 1$ if $a = b$ or $\delta(a,b) = 0$ if $a \neq b$. This can be written in a more compact form

$$E^{(2)} = \sum B_{nk\ell m}\ \delta(E_n+E_k,E_\ell+E_m)\ \langle a_n^\dagger\, a_k^\dagger\, a_\ell\, a_m\rangle$$

$$+ \sum \Sigma_{nm}\ \delta(E_n,E_m)\ \langle a_n^\dagger\, a_m\rangle$$

$$+ \sum U_{nm}\ \delta(E_n,E_m)\ \langle a_n^\dagger\, a_m\rangle \tag{51}$$

where B, Σ, and U correspond to the exchanged photon interaction, the self energy, and the vacuum polarization corrections. From this expression, it is clear that the self energy and vacuum polarization correction to a many-electron atom is just the sum of the corrections for the individual electrons to lowest order in perturbation theory. Both of these operators are diagonal in the standard basis of Dirac Coulomb wave functions.

Extension to Many Electrons

To take into account the effects of screening in a many-electron atom the perturbation method described above can be modified by including a correction δV to the Coulomb potential in the zero-order basis functions so that the total potential $V + \delta V$ simulates the effective potential for an individual electron in the atom

$$[-i\vec{\alpha}\cdot\vec{\nabla} + V(x) + \delta V(x) + \beta - E_n]\varphi_n(\vec{x}) = 0 \tag{52}$$

The correction term is then included in the interaction Hamiltonian

$$H_I(x) = j^\mu(x)A_\mu(x) - j^0(x)\ \delta A_0(x) - \delta M(x) \tag{53}$$

where

$$\delta V(x) = -\ e\ \delta A_0(x) \tag{54}$$

in order to restore the correct nuclear potential V perturbatively.

Table 8. Comparison of Theory and Experiment for the Lamb Shift in One-Electron Atoms with Z = 15-18 in Units of THz

Z	15	16	17	18
Self energy	21.394(5)	26.838(5)	33.194(6)	40.546(7)
Vacuum polarization	-1.262(0)	-1.629(0)	-2.072(0)	-2.597(1)
Higher order	0.005(7)	0.007(9)	0.008(11)	0.011(14)
Nuclear size	0.112(1)	0.151(1)	0.206(3)	0.276(2)
Relativistic recoil	0.007(6)	0.009(8)	0.010(9)	0.012(11)
Total[36,37]	20.255(11)	25.375(13)	31.347(16)	38.247(19)
Experiment	20.13(20)[38]	25.14(24)[39] 25.27(6)[40]	31.19(22)[41]	37.89(38)[42]

High-Z One-Electron Atoms

The most simple application of this formalism is to a one-electron atom. In this case, the exchanged-photon term vanishes, and the corrections are just the self energy and vacuum polarization effects. There are additional corrections corresponding to higher-order terms in α contained in the formalism, but not discussed here. In addition, the finite size of the nucleus is significant. For a consistent picture, this effect can be included in the zeroth-order potential V(x). Nuclear recoil effects are not contained in the external field formulation, but they are small in high-Z atoms. Table 8 shows a comparison of theory and experiment for the Lamb shift in one-electron atoms for Z in the range 15 to 18. The theoretical values in that table and in Table 9 are based on compilations in Refs. 36 and 37. In Table 9, theory and experiment are compared for the 1S - 2P transitions in hydrogenlike argon (Z = 18). Both comparisons provide a test of the theory at approximately the level of 1%.

Table 9. Theoretical Wavelengths [pm] for Lyman-α Transitions in Ar^{17+} and 1S QED Corrections (eV) Compared to Experimental Values

		$1S-2P_{1/2}$	$1S-2P_{3/2}$	
Self energy				1.2168
Vacuum polarization				-0.0853
Higher order				0.0005
Nuclear size				0.0090
Relativistic recoil				0.0003
Total[36,37]				1.1413(7)
Transition Wavelength		373.6516(1)	373.1101(1)	
Experiment	Ref. 43:	373.6522(19)	373.1105(19)	1.145(16)
	Ref. 44:	373.6514(40)	373.1142(70)	1.139(35)

Inner-Shell Holes in Many-Electron Atoms

For inner-shell holes in high-Z neutral atoms, one-electron radiative corrections provide a first approximation to the actual QED effects. As described by Eq. (51), the lowest-order radiative correction is the sum of the radiative corrections of the individual electrons. The exchanged photon correction is largely taken into account by the electron-electron Coulomb interaction in the many-electron level calculation together with the transverse photon (Breit) correction treated as a perturbation.

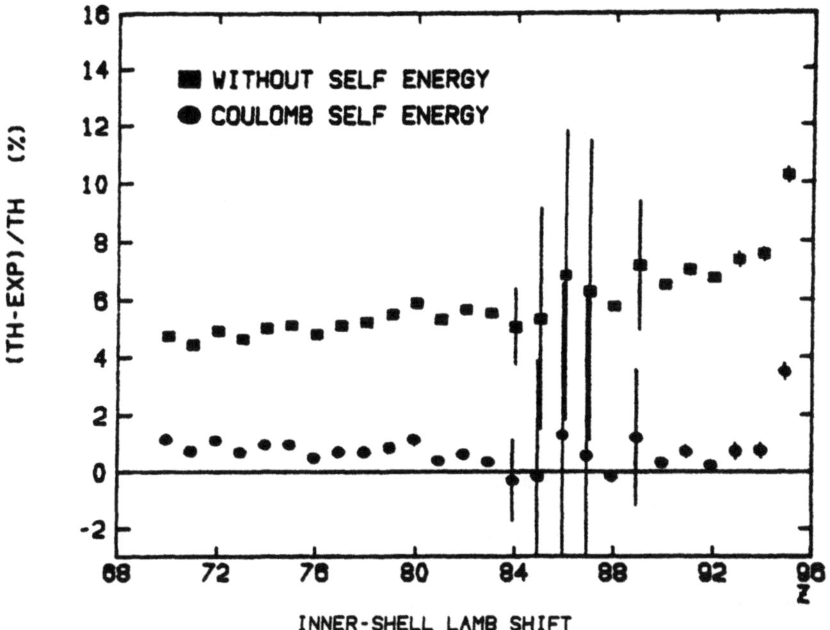

Figure 1. Comparison of Theory[45,46] and Experiment[47] for the Inner-Shell $2S_{1/2} - 2P_{1/2}$ Vacancy Transition in Heavy Atoms

Figures 1 and 2 illustrate the fact that there is a qualitative improvement in the theoretical transition energies if the lowest-order Coulomb self energy is included in the individual orbital energies as suggested by Eq. (51). Fig. 1 shows the fractional difference between theory and experiment for the inner-shell Lamb shift energy difference, and Fig. 2 shows the same comparison for the $1S_{1/2} - 2P_{1/2}$ vacancy energy difference. From the comparisons, it is clear that there is considerable improvement if the self energy is added, and that further work is needed to quantitatively explain the systematic difference that remains.

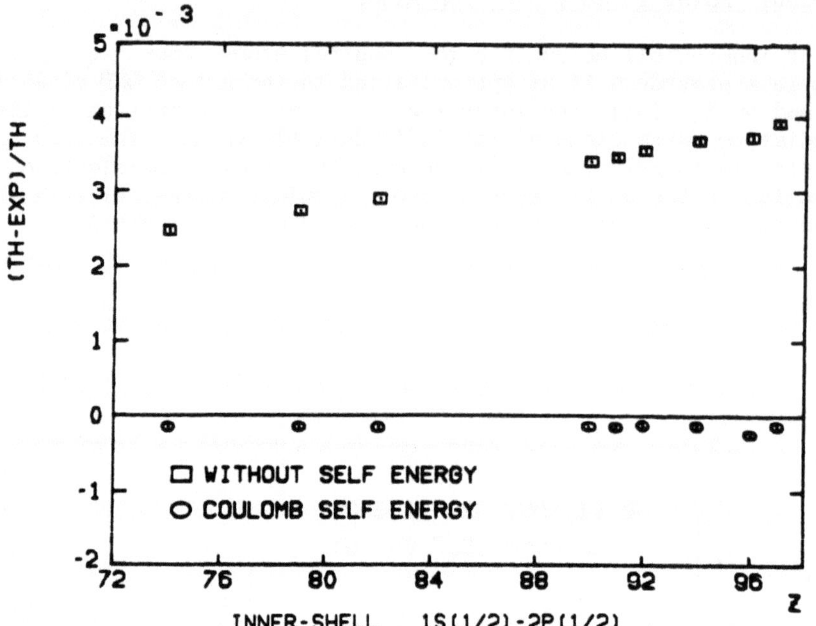

Figure 2. Comparison of Theory[45,46,48] and Experiment[49] for the Inner-Shell $1S_{1/2}$ - $2P_{1/2}$ Vacancy Transition in Heavy Atoms

QED CORRECTIONS FOR QUARKS IN A CAVITY

It is of interest to see to what extent the methods that are highly successful in predicting radiative corrections in atoms are applicable to other bound systems such as nucleons. This section describes work done in collaboration with Jonathan Sapirstein on this problem.[50] Part of the motivation for exploring this question is the well known intriguing fact that the mass difference between the neutron and proton

$$\frac{m_n - m_p}{m_p} = \frac{\alpha}{\pi} 0.59 \tag{55}$$

is of the order of an electromagnetic correction. For example, if the electromagnetic self-energy of a quark in a nucleon has the value ΔE for a unit charge, then due to the different fractional charge

$$q_u = \frac{2}{3} e \qquad\qquad q_d = -\frac{1}{3} e \tag{56}$$

of the up and down quarks, the self energies (proportional to the square of the charge) will be different

$$\delta E_u = \frac{4}{9} \Delta E \qquad\qquad \delta E_d = \frac{1}{9} \Delta E \tag{57}$$

As a result, the differential energy shift between the neutron with odd quarks and the proton with odd quarks is

$$\Delta m_n - \Delta m_p = \left[\frac{4}{9} + \frac{1}{9} + \frac{1}{9}\right] \Delta E - \left[\frac{4}{9} + \frac{4}{9} + \frac{1}{9}\right] \Delta E = -\frac{1}{3} \Delta E \tag{58}$$

To the extent that the mass of a quark is a meaningful concept, the bound-state radiative level shifts of quarks affect the relation between the observed nucleon mass and the hypothetical quark masses

$$m_p = 2m_u + m_d - BE + 2\delta m_u + \delta m_d \tag{59a}$$

$$m_n = m_u + 2m_d - BE + \delta m_u + 2\delta m_d \tag{59b}$$

where BE is the strong-force binding energy, another hypothetical quantity, of the quarks in the nucleons.

Leaving aside the precise physical interpretation of the result, we shall examine the problem of calculating the electromagnetic corrections for the quarks in the idealization that they are Dirac particles bound in an infinitely deep cavity.

Cavity Wave Functions

For a Dirac particle bound in a cavity of radius R, the zeroth-order basis functions are the eigenfunctions of the equation

$$[-i\vec{\alpha}\cdot\vec{\nabla} + V(x) + \beta m - E_n]\varphi_n(\vec{x}) = 0 \tag{60}$$

where the potential is

$$V(x) = \lim_{V_0\to\infty} \begin{cases} 0 & x < R \\ \beta V_0 & x > R \end{cases} \tag{61}$$

This definition of the potential as a limit of a finite potential is useful for avoiding ambiguities in the properties of the solutions. The wave functions are written in the well-known form

$$\varphi(\vec{x}) = \begin{bmatrix} u_1(x)\chi_\kappa^\mu(\hat{x}) \\ iu_2(x)\chi_{-\kappa}^\mu(\hat{x}) \end{bmatrix} \tag{62}$$

with radial eigenfunctions u_1 and u_2 of the equation

$$\begin{bmatrix} m + V_r - z & -\frac{d}{dx} + \frac{\kappa - 1}{x} \\ \frac{d}{dx} + \frac{\kappa + 1}{x} & -m - V_r - z \end{bmatrix} \begin{bmatrix} u_1(x) \\ u_2(x) \end{bmatrix} = 0 \tag{63}$$

where $V_r = \theta(x-R)V_0$. The solution of (63), for $0 < x < R$, regular at $x = 0$ is

$$\mathcal{U} = \begin{bmatrix} u_1(x) \\ u_2(x) \end{bmatrix} \qquad \begin{aligned} u_1(x) &= \sqrt{z+m}\; j_{\kappa+}(cx) \\ u_2(x) &= \sqrt{z-m}\, \frac{\kappa}{|\kappa|}\, j_{\kappa-}(cx) \end{aligned} \tag{64}$$

A general solution in the region $0 < x < R$ is

$$\Psi = a\mathcal{U} + \mathcal{V} = \begin{bmatrix} au_1(x) + v_1(x) \\ au_2(x) + v_1(x) \end{bmatrix} \qquad \begin{aligned} v_1(x) &= \sqrt{z+m}\; h^{(1)}_{\kappa+}(cx) \\ v_2(x) &= \sqrt{z-m}\, \frac{\kappa}{|\kappa|}\, h^{(1)}_{\kappa-}(cx) \end{aligned} \tag{65}$$

In Eqs. (64) and (65), $\kappa\pm = |\kappa \pm \frac{1}{2}| - \frac{1}{2}$ and $c = (z^2 - m^2)^{1/2}$ with Im (c) > 0. In the region $x > R$, for finite V_0, the solution regular as $x \longrightarrow \infty$ is

$$\Psi = \lambda\bar{\mathcal{V}} = \begin{bmatrix} \lambda\bar{v}_1(x) \\ \lambda\bar{v}_2(x) \end{bmatrix} \qquad \bar{\mathcal{V}}(m) = \mathcal{V}(m + V_0) \tag{66}$$

In the limit of large V_0 the exterior solutions have the ratio

$$\lim_{V_0 \to \infty} \frac{\bar{v}_2}{\bar{v}_1} = -1 \tag{67}$$

The eigenvalues are determined by the continuity conditions

$$u_1(R) = \lambda\bar{v}_1(R) \tag{68a}$$

$$u_2(R) = \lambda\bar{v}_2(R) \tag{68b}$$

which, in view of (67), means that

$$u_1(R) + u_2(R) = 0 \tag{69}$$

This is equivalent to the boundary condition

$$[i\vec{\gamma}\cdot\hat{x} + 1]\varphi_n(\vec{x}) \qquad \text{at } |\vec{x}| = R \tag{70}$$

of the MIT bag model in the static approximation.

Cavity Green's Function

Calculation of the QED corrections for the quark is facilitated by having an expression for the Green's function for the radial Hamiltonian

$$H_\kappa(x) = \begin{bmatrix} m + V_r & -\frac{d}{dx} + \frac{\kappa - 1}{x} \\ \frac{d}{dx} + \frac{\kappa + 1}{x} & - m - V_r \end{bmatrix} \tag{71}$$

The eigenfunctions U_n are given by

$$H_\kappa(x)U_n(x) = E_n U_n(x) \tag{72}$$

with normalization

$$\int_0^\infty dx\, x^2\, U_n(x)^T U_n(x) = 1 \tag{73}$$

In terms of the complete set of solutions, the radial Green's function is defined as

$$G_\kappa(x_2, x_1, z) = \sum_n \frac{U_n(x_2)U_n^T(x_1)}{E_n - z} \tag{74}$$

and satisfies the equation

$$[H_\kappa(x_2) - z]G_\kappa(x_2, x_1, z) = \frac{1}{x_2 x_1}\, \delta(x_2 - x_1) \tag{75}$$

An explicit construction of G in terms of the solutions in Eqs. (64) and (65) is given by

$$G_\kappa(x_2, x_1, z) = \frac{1}{J(z)} \left\{ \theta(x_1-x_2)\mathcal{U}(x_2)\mathcal{W}^T(x_1) + \theta(x_2-x_1)\mathcal{W}(x_2)\mathcal{U}^T(x_1) \right\}$$

$$J(z) = x^2(u_2 w_1 - u_1 w_2) = x^2(u_2 v_1 - u_1 v_2) = \frac{1}{ic} \tag{76}$$

which gives

$$G_\kappa(x_2, x_1, z) = ic\left\{ \theta(x_1-x_2)\mathcal{U}(x_2)\mathcal{V}^T(x_1) + \theta(x_2-x_1)\mathcal{V}(x_2)\mathcal{U}^T(x_1) \right\}$$

$$+ ica\left\{ \theta(x_1-x_2)\mathcal{U}(x_2)\mathcal{U}^T(x_1) + \theta(x_2-x_1)\mathcal{U}(x_2)\mathcal{U}^T(x_1) \right\} \tag{77}$$

or

$$G_\kappa(x_2, x_1, z) = F_\kappa(x_2, x_1, z) + ica\,\mathcal{U}(x_2)\mathcal{U}^T(x_1) \tag{78}$$

where F_κ is the free particle Green's function for $V \equiv 0$. The constant a is fixed by the condition of continuity of the general solution at the boundary $x = R$ to be

$$a = -\frac{v_1(R) + v_2(R)}{u_1(R) + u_2(R)} \tag{79}$$

In the remainder of this section, we note a few results that are of interest in the subsequent discussion.

The spectrum of the radial Hamiltonian consists of discrete eigenvalues, since the well is infinitely deep. Hence, there is no branch point in G_κ and

$$G_\kappa(x_2,x_1,z+i\epsilon) = G_\kappa(x_2,x_1,z-i\epsilon) \tag{80}$$

as $\epsilon \longrightarrow 0$ for z real and not an eigenvalue. This might be surprising at first sight in view of Eq. (78) in which the function F_κ is known to have branch points at $z = \pm m$. A closer examination reveals that the branch points are cancelled by corresponding singularities in the second term on the right-hand side of (78).

The normalized radial eigenfunctions are contained in the residues of the poles of the radial Green's functions. According to Eq. (78), the poles are determined by the function

$$f(z) = \frac{1}{ica} = \frac{i}{c}\frac{u_1(R) + u_2(R)}{v_1(R) + v_2(R)} = f(E_n) - f'(E_n)(E_n-z) + \cdots \tag{81}$$

In view of the eigenvalue condition in (69), the first term on the right-hand-side in (81) is zero, and the second term determines the residue. This together with a comparison of Eqs. (74) and (78) gives the normalized radial eigenfunction as

$$\frac{1}{\sqrt{-f'(E_n)}}\,\mathcal{U}(x) \tag{82}$$

Another result of importance is the fact that

$$\langle V\rangle = 0 \tag{83}$$

This can be demonstrated by explicitly constructing the wave functions for finite V_0, and then taking the limit as $V_0 \longrightarrow \infty$.

QED Corrections

The lowest-order QED corrections are given by Eq. (50) modified to deal with more than one type of particle. They consist of the self energy

$$E^{(2)}_{SE} = -4\pi i\tau^2\alpha \int d(t_2-t_1) \int d\vec{x}_2 \int d\vec{x}_1\, D_F(x_2-x_1)$$

$$\times\, \bar{\varphi}_i(x_2)\gamma_\mu S_F(x_2,x_1)\gamma^\mu\varphi_i(x_1) - \delta m \int d\vec{x}\, \bar{\varphi}_i(x)\varphi_i(x) \tag{84}$$

the vacuum polarization

$$E^{(2)}_{VP} = 4\pi i\tau^2\alpha \int d(t_2-t_1) \int d\vec{x}_2 \int d\vec{x}_1\, D_F(x_2-x_1)$$

$$\times\, \mathrm{Tr}[\gamma_\mu S_F(x_2,x_2)]\, \bar{\varphi}_i(x_1)\gamma^\mu\varphi_i(x_1) \tag{85}$$

for a single quark type, of charge $q = \tau e$ with $m = 1$ in (60), in state i, and the photon exchange correction

$$E_{PE}^{(2)} = -4\pi i\tau_i\tau_j\alpha \int d(t_2-t_1) \int d\vec{x}_2 \int d\vec{x}_1\, D_F(x_2-x_1)$$

$$\times\, \bar{\varphi}_i(x_2)\gamma_\mu\varphi_i(x_2)\bar{\varphi}_j(x_1)\gamma^\mu\varphi_j(x_1) \tag{86}$$

for a simple product state.

Photon Exchange

With the aid of the relation

$$\int d(t_2-t_1)\, D_F(x_2-x_1) = \frac{i}{4\pi|\vec{x}_2-\vec{x}_1|} \tag{87}$$

we obtain

$$E_{PE}^{(2)} = \tau_i\tau_j\alpha\int d\vec{x}_2\, d\vec{x}_1\, \frac{1}{|\vec{x}_2-\vec{x}_1|}\, \varphi_i^\dagger(\vec{x}_2)\alpha_\mu\varphi_i(\vec{x}_2)\varphi_j^\dagger(\vec{x}_1)\alpha^\mu\varphi_j(\vec{x}_1) \tag{88}$$

for the photon exchange energy. The evaluation of this energy is elementary, and the results can be parameterized in terms of functions f_1 and f_2 defined by

$$E_{PE}^{(2)} = \tau_i\tau_j\,\frac{\alpha}{\pi R}\,[f_1(mR) + \langle\vec{\sigma}_2\cdot\vec{\sigma}_1\rangle f_2(mR)] \tag{89}$$

with calculated values appearing in Table 10.

Vacuum Polarization

From the definition of the propagation function, we have

$$\mathrm{Tr}[\gamma_0 S_F(x,x)] = \frac{1}{2}\left[\sum_{(+)}|\varphi_n(x)|^2 - \sum_{(-)}|\varphi_n(x)|^2\right] \tag{90}$$

Table 10. Values of the Functions f_1 and f_2

mR	f_1	f_2
0	4.016	-0.428
1	4.299	-0.325
6	5.100	-0.077
10	5.298	-0.035

so that

$$E_{VP}^{(2)} = -\tau^2\alpha \int d\vec{x}_2 \int d\vec{x}_1 \frac{1}{|\vec{x}_2 - \vec{x}_1|} \frac{1}{2} \left[\sum_{(+)} |\varphi_n(x_2)|^2 - \sum_{(-)} |\varphi_n(x_2)|^2 \right]$$

$$\times\ \varphi_i^\dagger(x_1)\varphi_i(x_1) \tag{91}$$

where $\alpha = \frac{e^2}{4\pi}$

This level shift can be interpreted to be the result of interaction of the electron with the potential generated by the vacuum polarization charge density

$$\rho_{VP}(x) = -\tau e \frac{1}{2} \left[\sum_{(+)} |\varphi_n(x)|^2 - \sum_{(-)} |\varphi_n(x)|^2 \right] \tag{92}$$

Inspection of the radial differential equation for the quark wave functions reveals the symmetry relation

$$u_1(-E_n, -\kappa) = u_2(E_n, \kappa) \tag{93}$$

for the solutions. It follows that

$$|\varphi_{E_n,\kappa}|^2 = u_1^2(E_n,\kappa)\chi_\kappa^\dagger\chi_\kappa + u_2^2(E_n,\kappa)\chi_{-\kappa}^\dagger\chi_{-\kappa} = |\varphi_{-E_n,-\kappa}|^2 \tag{94}$$

and so

$$\sum_{(+),\kappa} |\varphi_{E_n,\kappa}|^2 = \sum_{(+),\kappa} |\varphi_{-E_n,-\kappa}|^2 = \sum_{(-),\kappa} |\varphi_{E_n,-\kappa}|^2 = \sum_{(-),\kappa} |\varphi_{E_n,\kappa}|^2 \tag{95}$$

This relation together with the definition of ρ_{VP} in (92) shows that the vacuum polarization effect vanishes.

Self Energy

The self-energy correction requires the most detailed calculation. For finite mass quarks, as contrasted with massless quarks, there is an infinite mass renormalization necessary, which adds a complication to the numerical evaluation. The mass renormalization is done with mass-shell renormalization. This procedure is not entirely satisfactory, since free quarks appear to be unavailable to define an experimental mass. However, the procedure most closely follows the conventional QED calculations for bound electrons where it is known to provide accurate results.

It is useful to reexpress the propagation function that appears in Eq. (84) as

$$S_F(x_2,x_1) = \langle 0|T[\psi(x_2)\bar{\psi}(x_1)]|0\rangle$$

$$= \sum_{(+)} \varphi_n(x_2)\bar{\varphi}_n(x_1)\theta(t_2-t_1) - \sum_{(-)} \varphi_n(x_2)\bar{\varphi}_n(x_1)\theta(t_1-t_2)$$

$$= \frac{1}{2\pi i}\int_{-\infty}^{\infty} dz \sum_n \frac{\varphi_n(\vec{x}_2)\bar{\varphi}_n(\vec{x}_1)}{E_n - z(1+i\delta)} e^{-iz(t_2-t_1)}$$

$$= \frac{1}{2\pi i}\int_{-\infty}^{\infty} dz\, G(\vec{x}_2,\vec{x}_1,z(1+i\delta))\, e^{-iz(t_2-t_1)} \tag{96}$$

where $\theta(0) = \frac{1}{2}$ as in (90). In terms of G, the self energy level shift is

$$E_{SE}^{(2)} = -i\frac{\tau^2\alpha}{2\pi}\int d\vec{x}_2 \int d\vec{x}_1\, \varphi_i^\dagger(\vec{x}_2)\alpha_\mu \int_C dz\, G(\vec{x}_2,\vec{x}_1,z)\alpha^\mu\varphi_i(\vec{x}_1)$$

$$\times \frac{1}{x_{12}}\left[e^{-bx_{21}} - e^{-b'x_{21}}\right] - \delta m \int d\vec{x}\, \varphi_i^\dagger(\vec{x})\beta\varphi_i(\vec{x}) \tag{97}$$

where C is the Feynman contour, which is Wick rotated so that the integration extends from $-i\infty$ to $+i\infty$ in the complex z plane. To make the terms in this expression separately finite, the photon propagation function has been regulated by making the replacement

$$\frac{1}{k^2+i\epsilon} \longrightarrow \frac{1}{k^2+i\epsilon} - \frac{1}{k^2-\Lambda^2+i\epsilon} \tag{98}$$

in momentum space. With this regulator, the mass renormalization term δm is

$$\delta m = \frac{\tau^2\alpha}{\pi}\left[\frac{3}{4}\ell n\, \Lambda^2 + \frac{3}{8}\right] \tag{99}$$

In (97) b and b' are defined by

$$b = -i[(E_n-z)^2 + i\epsilon]^{1/2} \qquad Re(b) > 0$$

$$b' = -i[(E_n-z)^2 - \Lambda^2 + i\epsilon]^{1/2} \qquad Re(b') > 0 \tag{100}$$

Note that this definition of b constrains the integration contour in Eq. (97) to cross the real axis in the complex z plane at the point $z = E_n$. To isolate the divergent parts of the self energy, it is convenient to work with the expression

$$E_{SE}^{(2)} = i\frac{\tau^2\alpha}{4\pi^3}\int_C dz \int d\vec{k}\left[\frac{1}{(E_i-z)^2 - k^2 + i\epsilon} - \frac{1}{(E_i-z)^2 - k^2 - \Lambda^2 + i\epsilon}\right]$$

$$\times \langle \alpha_\mu \frac{1}{\vec{\alpha}\cdot\vec{p} - \vec{\alpha}\cdot\vec{k} + V + \beta - z}\alpha^\mu \rangle \tag{101}$$

The most singular parts of this expression can be isolated in the first few terms of the well-known expansion

$$\frac{1}{A+B} = \frac{1}{A} - \frac{1}{A} B \frac{1}{A} + \frac{1}{A} B \frac{1}{A} B \frac{1}{A} + \cdots \tag{102}$$

that gives

$$\frac{1}{\vec{\alpha}\cdot\vec{p} - \vec{\alpha}\cdot\vec{k} + V + \beta - z} = \frac{1}{\vec{\alpha}\cdot\vec{p} - \vec{\alpha}\cdot\vec{k} + \beta - z}$$

$$- \frac{1}{\vec{\alpha}\cdot\vec{p} - \vec{\alpha}\cdot\vec{k} + \beta - z} V \frac{1}{\vec{\alpha}\cdot\vec{p} - \vec{\alpha}\cdot\vec{k} + \beta - z} + \cdots \tag{103}$$

The singular terms arise from poor convergence of the integration over z and $\vec{k}$ in (101). The first and second terms in the expansion in (103) give the following behavior in the asymptotic region

$$\int_C dz \int d\vec{k} \frac{1}{k^2} \frac{1}{z} \longrightarrow \text{linear div} \longrightarrow \text{log div (symmetric int)}$$

$$\int_C dz \int d\vec{k} \frac{1}{k^2} \frac{1}{z^2} \langle V \rangle \longrightarrow \text{log div} \times \langle V \rangle \longrightarrow 0 \tag{104}$$

The higher-order terms in this expansion give convergent integrals as shown by power counting in the integrand. These estimates indicate that the divergence of the self energy is confined to the first term in the expansion in powers of V. In the general case, the second term would also contain a divergence; however in this particular case, $\langle V \rangle$ happens to vanish and that term is finite. This fact is particularly convenient in view of the natural breakup of the Green's function into the free part, corresponding to the leading term in the expansion in powers of V, and a remainder that has a particularly simple structure, as given in Eq. (78) for the radial Green's functions, or

$$G(\vec{x}_2,\vec{x}_1,z) = F(\vec{x}_2,\vec{x}_1,z) + G_R(\vec{x}_2,\vec{x}_1,z) \tag{105}$$

for the full Green's function.

The contribution of the free Green's function to the self energy is

$$E^{(2)}_{SE,F} = - i \frac{\tau^2\alpha}{2\pi} \int d\vec{x}_2 \int d\vec{x}_1 \, \varphi_i^\dagger(\vec{x}_2)\alpha_\mu \int_C dz \, F(\vec{x}_2,\vec{x}_1,z)\alpha^\mu \varphi_i(\vec{x}_1)$$

$$\times \frac{1}{x_{12}}\left[e^{-bx_{21}} - e^{-b'x_{21}}\right] - \delta m \langle \beta \rangle \tag{106}$$

where F is just

$$F(\vec{x}_2,\vec{x}_1,z) = \left[\left(\frac{d}{x_{21}} + \frac{1}{x_{21}^2}\right) i\vec{\alpha}\cdot\vec{x}_{21} + \beta + z\right] \frac{e^{-dx_{21}}}{4\pi x_{21}} \tag{107}$$

where $d = (1 - z^2)^{1/2}$ with Re (d) > 0. Inspection of (107) reveals that the divergence of the integral in (106) at large z corresponds to the region of coordinate space where $\vec{x}_2 \approx \vec{x}_1$. In order to isolate the divergent parts of the integration in simple expressions, we expand the wave function at point $\vec{x}_1$ about point $\vec{x}_2$ in Eq. (106)

$$\varphi_i(\vec{x}_1) = \varphi_i(\vec{x}_2) + (\vec{x}_1-\vec{x}_2)\cdot\vec{\nabla}_2\varphi_i(\vec{x}_2) + \cdots \tag{108}$$

The two leading terms in this expansion contain the divergent terms, and have a sufficiently simple structure that the coordinate integrations can be evaluated analytically. The final integration over z has the form

$$\begin{aligned} E^{(2)}_{SE,F} &= \int_{C'} dz\ f(z) - \delta m\langle\beta\rangle \\ &= \int_{C'} dz\ [f_0(z) + f_1(z) + f(z) - f_0(z) - f_1(z)] - \delta m\langle\beta\rangle \end{aligned} \tag{109}$$

where f_0 and f_1 denote the functions that correspond to the two terms in the expansion in (108). The first two terms are of order $\ell n\ \Lambda^2$, and the combination of the last three terms gives a finite limit as $\Lambda \longrightarrow \infty$. In particular, we have

$$\begin{aligned} &\int_{C'} dz\ [f_0(z) + f_1(z)] - \delta m\ \langle\beta\rangle \longrightarrow \text{finite analytic result} \\ &\int_{C'} dz\ [f(z) - f_0(z) - f_1(z)] \longrightarrow \text{finite numerical result} \end{aligned} \tag{110}$$

The finite remainder is calculated by partly analytic and partly numerical methods which give accurate results for that part of the level shift.

The remainder of the level shift, corresponding to the second term on the right-hand side of Eq. (105), is numerically evaluated by expanding the Green's function in a series of eigenfunctions of angular momentum. The radial Green's functions have a particularly simple form for this part of the calculation as shown by Eq. (78). The numerical methods employed are similar to those used for strong-field Coulomb self energy calculations.[48,51,52]

Table 11. Values of the Free Contribution F_F, the Remainder Contribution F_R, and the Total F

mR	F_F	F_R	F
0	0.618	2.452	3.070
1	-0.602	2.466	1.864
6	-2.775	3.167	0.392
10	-3.221	3.397	0.176

Conclusion

In Table 11, numerical results are given in terms of the function F defined by

$$E_{SE}^{(2)} = \frac{\tau^2 \alpha}{\pi R} F(mR) \tag{111}$$

The function F(mR) appears to smoothly decrease as mR increases. The value of F for zero mass is

$$F(0) = 3.070 \tag{112}$$

which is consistent with the early result of Chodos and Thorn[53] F(0) = 2.4(3) and with current results.[54,55]

The QED corrections do not explain the neutron-proton mass difference as a purely electromagnetic effect for any value of the quark masses within the model considered, but the electromagnetic shifts should be taken into account in phenomenological analyses of mass splittings of particle multiplets. For such applications, further work is necessary to obtain a quantitative estimate of the model dependence of the corrections.

REFERENCES

1. R. S. Van Dyck, Jr., P. B. Schwinberg, and H. G. Dehmelt, in Atomic Physics 9, ed. by R. S. Van Dyck, Jr. and E. N. Fortson (World Scientific, Singapore, 1984).
2. T. Kinoshita and W. B. Lindquist, Phys. Rev. Lett. 47, 1573 (1981).
3. T. Kinoshita and J. Sapirstein, in Atomic Physics 9, ed. by R. S. Van Dyck, Jr. and E. N. Fortson (World Scientific, Singapore, 1984).
4. B. N. Taylor, J. Res. Natl. Bur. Stand. 90, 91 (1985).
5. J. Bailey, K. Borer, F. Combley, H. Drumm, F. J. M. Farley, J. H. Field, W. Flegel, P. M. Hattersley, F. Krienen, F. Lange, E. Picasso, and W. Von Rueden, Phys. Lett. 68B, 191 (1977).
6. F. J. M. Farley and E. Picasso, Ann Rev. Nucl. Part. Sci. 29, 243 (1979).
7. T. Kinoshita, B. Nižić, and Y. Okamoto, Phys. Rev. Lett. 52, 717 (1984).
8. G. Newton, D. A. Andrews, and P. J. Unsworth, Phil. Trans. Roy. Soc. London A 290, 373 (1979).
9. S. R. Lundeen and F. M. Pipkin, Phys. Rev. Lett. 46, 232 (1981).
10. V. G. Pal'chikov, Yu. L. Sokolov, and V. P. Yakovlev, JETP Lett. 38, 418 (1984).
11. P. J. Mohr, Phys. Rev. Lett. 34, 1050 (1975).
12. J. Sapirstein, Phys. Rev. Lett. 47, 1723 (1981).
13. G. G. Simon, Ch. Schmitt, F. Borkowski, and V. W. Walther, Nucl. Phys. A333, 381 (1980).
14. G. P. Lepage, D. R. Yennie, and G. W. Erickson, Phys. Rev. Lett. 47, 1640 (1981).
15. D. J. Hylton, Phys. Rev. A 32, 1303 (1985).
16. G. Bhatt and H. Grotch, Phys. Rev. A 31, 2794 (1985).
17. H. Hellwig, R. F. C. Vessot, M. W. Levine, P. W. Zitzewitz, D. W. Allan, and D. J. Glaze, IEEE Trans. Instrum. Meas. IM-19, 200 (1970).

18. L. Essen, R. W. Donaldson, M. J. Bangham, and E. G. Hope, Nature 229, 110 (1971).
19. J. Sapirstein, Phys. Rev. Lett. 51, 985 (1983).
20. V. W. Hughes and J. Kuti, Ann. Rev. Nucl. Part. Sci. 33, 611 (1983).
21. C. J. Oram in Atomic Physics 9, ed. by R. S. Van Dyck, Jr. and E. N. Fortson (World Scientific, Singapore, 1984).
22. A. Badertscher, V. W. Hughes, D. C. Lu, M. W. Ritter, K. A. Woodle, M. Gladisch, H. Orth, G. zu Putlitz, M. Eckhause, J. Kane, and F. G. Mariam in Atomic Physics 9, ed. by R. S. Van Dyck, Jr. and E. N. Fortson (World Scientific, Singapore, 1984).
23. F. G. Mariam, W. Beer, P. R. Bolton, P. O. Egan, C. J. Gardner, V. W. Hughes, D. C. Lu, P. A. Souder, H. Orth, J. Vetter, U. Moser, and G. zu Putlitz, Phys. Rev. Lett. 49, 993 (1982).
24. G. T. Bodwin, D. R. Yennie, and M. A. Gregorio, Rev. Mod. Phys. 57, 723 (1985).
25. S. Chu, A. P. Mills, Jr., and J. L. Hall, Phys. Rev. Lett. 52, 1689 (1984).
26. T. Fulton, Phys. Rev. A 26, 1794 (1982).
27. M. W. Ritter, P. O. Egan, V. W. Hughes, and K. A. Woodle, Phys. Rev. A 30, 1331 (1984).
28. D. W. Gidley, A. Rich, E. Sweetman, and D. West, Phys. Rev. Lett. 49, 525 (1982).
29. M. A. Stroscio, Phys. Rev. Lett. 48, 571 (1982).
30. G. S. Adkins, Phys. Rev. A 27, 530 (1983).
31. G. S. Adkins, Ann. Phys. (N.Y.) 146, 78 (1983).
32. W. H. Furry, Phys. Rev. 81, 115 (1951).
33. M. Gell-Mann and F. Low, Phys. Rev. 84, 350 (1951).
34. J. Sucher, Phys. Rev. 107, 1448 (1957).
35. P. J. Mohr, Phys. Rev. A 32, 1949 (1985).
36. P. J. Mohr, At. Data Nucl. Data Tables 29, 453 (1983).
37. W. R. Johnson and G. Soff, At. Data Nucl. Data Tables 33, 405 (1985).
38. P. Pelligrin, Y. El Masri, and L. Palffy, Phys. Rev. A 31, 5 (1985).
39. V. Zacek, H. Bohn, H.Brum, T. Faestermann, F. von Feilitzsch, G. Giorginis, P. Kienle, and S. Schuhbeck, Z. Phys. A 318, 7 (1984).
40. A. P. Georgiadis, D. Müller, H.-D. Sträter, J. Gassen, P. von Brentano, J. C. Sens, and A. Pape, Phys. Lett. A 115, 108 (1986).
41. O. R. Wood II, C. K. N. Patel, D. E. Murnick, E. T. Nelson, M. Leventhal, H. W. Kugel, and Y. Niv, Phys. Rev. Lett. 48, 398 (1982).
42. H. Gould and R. Marrus, Phys. Rev. A 28, 2001 (1983).
43. H. F. Beyer, R. D. Deslattes, F. Folkmann, and R. E. LaVilla, J. Phys. B 18, 207 (1985).
44. E. S. Marmar, J. E. Rice, E. Källne, J. Källne, and R. E. LaVilla, Phys. Rev. A 33, 774 (1986).
45. M. H. Chen, B. Crasemann, M. Aoyagi, K.-N. Huang, and H. Mark, At. Data Nucl. Data Tables 26, 561 (1981).
46. P. J. Mohr, Phys. Rev. A 26, 2338 (1982).
47. J. A. Bearden, A. F. Burr, Rev. Mod. Phys. 39, 125 (1967); Atomic Energy Levels (U. S. Atomic Energy Commission, Oak Ridge 1965).
48. P. J. Mohr, Ann. Phys. (N.Y.) 88, 26, 52 (1974).
49. R. D. Deslattes, E. G. Kessler, Jr., L. Jacobs, and W. Schwitz, Phys. Lett. 71A, 411 (1979).
50. P. J. Mohr and J. R. Sapirstein, Phys. Rev. Lett. 54, 514 (1985).
51. G. E. Brown, J. S. Langer, and G. W. Schaefer, Proc. Roy. Soc. (London) A 251, 92 (1959).
52. A. M. Desiderio and W. R. Johnson, Phys. Rev. A 3, 1267 (1971).
53. A. Chodos and C. B. Thorn, Nucl. Phys. B104, 21 (1976).
54. R. L. Jaffe, private communication.
55. J. Baacke and H. Usler, private communication.

THE MANY FACETS OF THE DIRAC VACUUM

Berndt Müller

Institut für Theoretische Physik
Johann Wolfgang Goethe-Universität, Postfach 111932
D-6000 Frankfurt/Main, West Germany

ABSTRACT

The properties of the Dirac vacuum in strong external fields is reviewed, with particular emphasis on phase transition to a charged vacuum.

1. THE CHARGED VACUUM IN SUPERCRITICAL FIELDS

When atomic structure is extrapolated from the known boundary of chemical elements (nuclear charge Z=109) into the region Z=170-190, one finds that the $1s_{1/2}$-state, the atomic K-shell, gains tremendously in binding energy. As shown in Fig. 1 the $1s_{1/2}$-state - and also the next higher state, the $2p_{1/2}$ level - traverses the gap between the positive and negative energy continuum solutions of the Dirac equation, and is predicted to reach a binding energy of $2m_e$=1.022 MeV at the critical nuclear charge Z_c=173±1. The uncertainty derives from our lack of precise knowledge of the extrapolated nuclear charge distribution and from possible radiative corrections of higher order that are not accounted for in the calculations.*)

What happens at and beyond this critical charge was clarified in the years 1971-72 by our group at Frankfurt (Müller, Peitz, Rafelski, and Greiner, 1972; Müller, Rafelski, and Greiner 1972) and by the Moscow group (Zel'dovich and Popov, 1972). The transition from a just subcritical 1s-state to the supercritical state is most easily understood in the framework of Fano's theory of configuration interaction and autoionizing states. We start with the reduced Hilbert space of a just critical atom spanned by the 1s-state $|\varphi_0\rangle$ and the negative energy continuum of s-wave states $|\varphi_E\rangle$, as shown in Fig. 2 (left part):

$$H_c|\varphi_0\rangle \approx -m_e|\varphi_0\rangle \tag{1a}$$

$$H_c|\varphi_E\rangle \approx E|\varphi_E\rangle \quad , \qquad E<-m_e. \tag{1b}$$

*) For a point-like nuclear charge distribution the Dirac Hamiltonian is not self-adjoint for $Z>\alpha^{-1}\approx 137$. The origin of this is the singular behaviour of the vacuum in the point nucleus limit for $Z\alpha>1$.

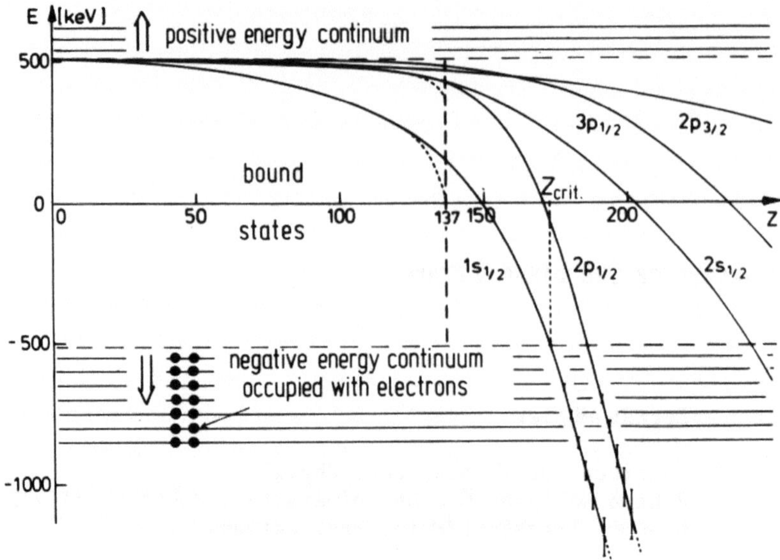

Fig. 1. Atomic binding energies as function of nuclear charge.

When a few, Z', protons are added to the nucleus to render the potential supercritical, the 1s-state is drawn into the continuum and only continuum solutions exist:

$$(H_c+Z'U(r))|\psi_E\rangle = |\psi_E\rangle \tag{2}$$

with $E<-m_e$. The solutions of the supercritical potential, $|\psi_E\rangle$, are expanded in terms of those of eqs. (1):

$$|\psi_E\rangle = a(E)|\varphi_0\rangle + \int_{-\infty}^{-m_e} dE' b_{E'}(E)|\varphi_{E'}\rangle \quad . \tag{3}$$

Elementary methods for solving integral equations yield the following result for the probability $|a(E)|^2$ of admixture of the critical 1s-state to the supercritical continuum state $|\psi_E\rangle$:

$$|a(E)|^2 = \frac{1}{2\pi}\Gamma_E \left[(E-E_\varphi-F(E))^2 + \frac{1}{4}\Gamma_E^2\right]^{-1} \tag{4}$$

where

$$E_\varphi = -m_e + Z'\langle\varphi_0|U(r)|\varphi_0\rangle \ , \tag{5a}$$

$$V_E = Z'\langle\varphi_E|U(r)|\varphi_0\rangle \quad , \qquad \Gamma_E = 2\pi|V_E|^2 \ , \tag{5b}$$

and

$$F(E) = P\int_{-\infty}^{-m_e} dE' \frac{|V_{E'}|^2}{E-E'} \quad . \tag{5c}$$

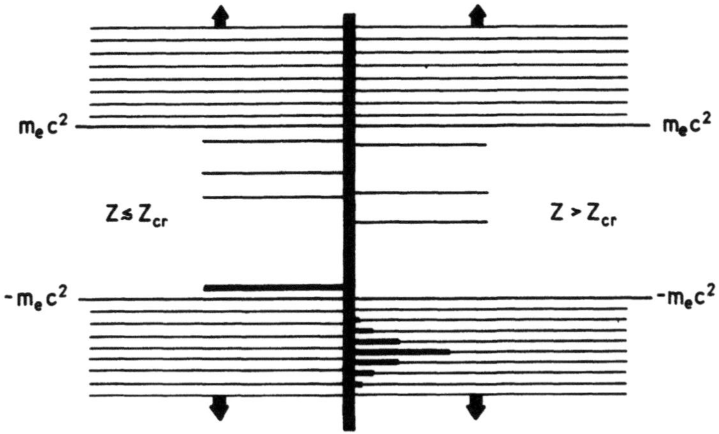

Fig. 2. Transition form Z_c (left) to $Z_c + Z'$ (right).

Obviously, the bound 1s-state turns into a resonance in the negative energy continuum located at E_{res}=E +F(E_{res}). The width of the resonance, Γ_E, is of the order of 1 keV, corresponding to a lifetime in the range 10^{-18}–10^{-19}s. The supercritical situation is shown schematically in Fig. 2 (right part).

The reason why the 1s bound state turns into a resonance is intuitively clear: the vacant K-shell is unstable against pair-decay when the binding energy E_k exceeds twice the electron rest mass. A pair is created, the electron occupying the 1s state while the positron is emitted freely with kinetic energy $E_p = E_k - 2m_e$. When the K-shell is fully occupied by two electrons the spontaneous decay process is stopped by the action of the Pauli principle. The intuitive picture is easily corroborated by arguments based on second quantized field theory (Rafelski, Müller, and Greiner, 1974; Fulcher and Klein 1974).

As the supercritical K-shell resonance is part of the negative energy continuum, i.e. of the Dirac sea, it is customary to consider it as part of the vacuum. Consequently one speaks of the <u>charged vacuum</u> state in supercritical quantum electrodynamics, and the spontaneously occurring process of pair creation involving the vacant 1s state is known as <u>decay of the (neutral) vacuum</u>. That the supercritical vacuum, indeed, contains a nonvanishing charge is illustrated in Fig. 3, showing the vacuum polarisation charge density

$$\rho_{VP}(r) = \frac{1}{2} \sum_{E<-m_e} \psi_E^+(r)\psi_E(r) - \frac{1}{2} \sum_{E\ -m_e} \psi_E^+(r)\psi_E(r) \tag{6}$$

for the supercritical nuclear charge Z=184. It is very similar to the charge distribution contained in the just subcritical 1s-state at Z=172, which is also shown. Note that the space integral over ρ_{VP} for Z=184 does

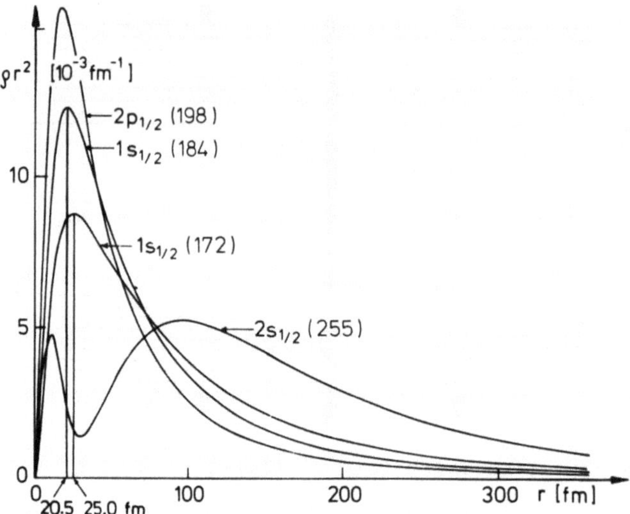

Fig. 3. Real vacuum polarization in comparison with K-shell density at Z_c.

not vanish, indicating that the vacuum charge is real and not only a displacement charge as for $Z<Z_e$.

So far we have restricted our considerations to the single-particle picture. It is a relevant question whether these conclusions survive when higher-order processes of quantum field theory are taken into account. The basic corrections are shown in the Feynman diagrams of Fig. 4. It is clear that the usual perturbation expansion in powers of ($Z\alpha$) cannot be trusted at the critical Z, hence all order must be summed by using exact propagators in the external field (indicated by heavy lines in Fig. 4). Denoting the exact electron propagator by G(x,y) and the free photon propagator by D(x,y), the corrections to the binding energy of the 1s-state can be written as:

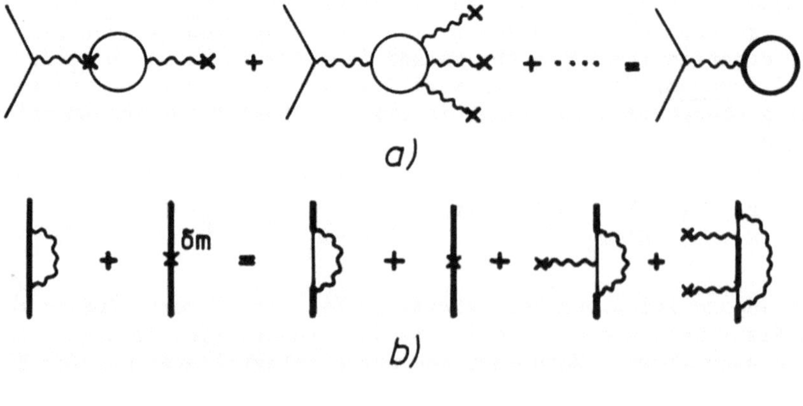

Fig. 4. Vacuum polarization (above) and self-energy corrections (below) to all order in the nuclear charge.

(a) vacuum polarization (Fig. 4a):

$$\Delta E_{1s}^{VP} = -ie^2 \int d^3x\, d^3y\ \bar{\psi}(x)\gamma^0\psi(x)D(x-y)\int\frac{d\omega}{2\pi}\, Tr[\gamma^0 G(y,y';\omega)]_{y'\to y} \tag{7a}$$

(b) self-energy and vertex corrections (Fig. 4b):

$$\Delta E_{1s}^{SE} = ie^2 \int dt\, d^3x\, d^3y\ \bar{\psi}(x)\gamma^\mu G(x,y)\ \gamma_\mu\ \psi(y)\ D(x-y) + \delta m \int d^3x\ \bar{\psi}(x)\psi(x) \quad . \tag{7b}$$

Here $\psi(x)$ denotes the 1s wavefunction. Eq. (7a) was evaluated by Gyulassy (1975) and by Rinker and Wilets (1975), who obtained a shift of

$$\Delta E_{1s}^{VP}\ (Z_c) = -\ 10.68\ \text{keV} \tag{8a}$$

at the critical point, increasing the binding energy. Expression (7b), which is more difficult to evaluate, was calculated by Soff et al. (1982) who found a repulsive contribution

$$\Delta E_{1s}^{SE}\ (Z_c) = +\ 10.99\ \text{keV} \tag{8b}$$

at the diving point. The almost complete cancellation between the two contributions means that the total shift in the K-shell energy due to field theoretic corrections of order α is only +0.31 KeV, less than 10^{-3} of the total binding energy. It would be highly surprising, if higher orders in α (not $Z\alpha$!) would change this picture. It therefore seems clear that the transition to a charged vacuum state must occur at a critical nuclear charge $Z_c \approx 173$.

It is legitimate to ask the question: how far do we have experimental proof that binding energies comparable to the rest mass of the electron, or twice the rest mass, can actually be achieved? This question actually has two aspects: First, how can the strong binding be set up experimentally and second, how can its presence be observed? The answer to the first question was given around 1970 by the Moscow and Frankfurt groups (see Greiner (1983) for a historical perspective). In collisions of two very heavy nuclei the electric field of a nucleus with charge Z_1+Z_2 is simulated temporarily. The solution of the Dirac equation with two Coulomb centers showed that the critical binding of $2m_e$ should be reached in, e.g., U+U collisions of a distance $R_c \approx 30$ fm (Müller, Rafelski, and Greiner, 1973). However, the binding energy varies with time as the nuclei move rapidly on their Rutherford trajectories and it is not so easy to determine the binding energy at a certain internuclear distance experimentally.

An approximate method was, nonetheless, proposed by Soff, Müller, and Greiner (1978), making use of the generalization of Bang and Hansteen's (1959) scaling law for direct ionization. In the case of superheavy collision systems, ionization occurs predominantly at the point of closest approach of the nuclei, R_o. One can then show (Müller et al. 1978,1983; Bosch et al., 1980) that the ionization probability of the 1s-state on a given scattering trajectory depends - to a good approximation - only on the binding energy at distance R_o:

$$P = D\ \exp\left(-\ \frac{\gamma R_o}{v}\ E_B^{1s}\ (R_o)\ \right) \quad , \tag{9}$$

where γ is a numerical constant, v the beam velocity, and D is only a function of (Z_1+Z_2) which can be obtained by comparison with full-scale numerical calculations. In Fig. 5 we have shown binding energies extracted from measurements of 1s-vacancy production in the Pb+Cm (Z_1=82, Z_2=96) system by Liesen et al. (1980), in comparison with results of a two-center Dirac calculation carried out by W. Betz (1980). Although it is not possible to conclude that critical binding can be achieved (the system is too "light"), the existence of binding energies in the range between 500 and 800 keV appears to be fairly well established.

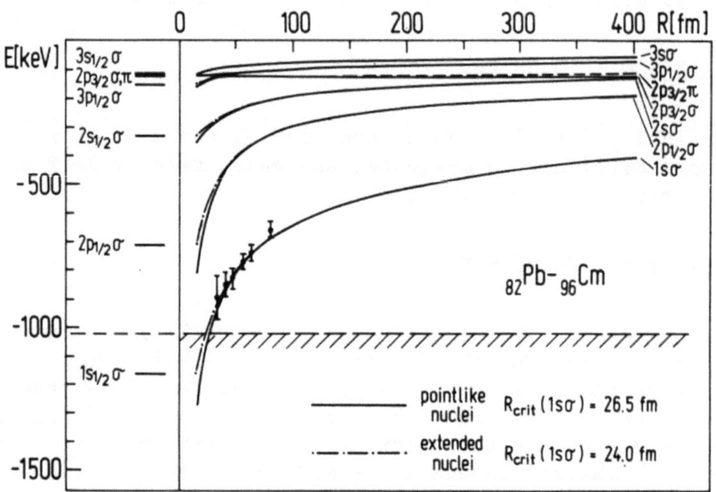

Fig. 5. Binding energy of the 1sσ-state in Pb+Cm, determined from measured Cm K-hole probabilities with help of the scaling law (9).

2. THE VACUUM STATE IN STRONG FIELDS: GENERAL THEORY

2.1 Role of the Vacuum State in Quantum Field Theory

To be specific, let us consider the ground state of quantum electrodynamics in an external field. The Lagrangian

$$L_{QED} = \bar{\psi}\gamma^{\mu}(i\partial_{\mu}-eA_{\mu})\psi-m_e\bar{\psi}\psi-\frac{1}{4}F_{\mu\nu}F^{\mu\nu}-A_{\mu}j^{\mu}_{ext} \tag{10}$$

yields, upon variation with respect to ψ and A_{μ}, the following equations of motion:

$$[\gamma^{\mu}(i\partial_{\mu}-\hat{A}_{\mu}) - m_e]\ \hat{\psi}(x) = 0, \tag{11a}$$

$$\partial_{\mu}\hat{F}^{\mu\nu} = e\hat{\bar{\psi}}\ \gamma^{\nu}\hat{\psi} + j^{\nu}_{ext} \quad . \tag{11b}$$

In the Heisenberg picture, where the state of the quantum field is time-independent, these equations specify the evolution of the field operators.

Whenever the external source j^{ν}_{ext} can be considered to be much larger than the dynamical source term $e\bar{\psi}\gamma^{\mu}\psi$, i.e. the charge current contained in the electron-positron field, eqs. (11) can be decoupled and the external field A^{ext}_{μ} may be substituted in eq. (11a) for the full field operator A_{μ}. The evolution of the electron-positron field can then be solved by expanding the field operator in terms of eigenfunctions:

$$\hat{\psi}(x) = \sum_n \hat{b}_n \varphi_n(x) \tag{12}$$

with

$$[\gamma^{\mu}(i\partial_{\mu}-eA^{ext}_{\mu}) - m_e]\, \varphi_n(x) = 0 \quad . \tag{13}$$

If the field is static, the eigenfunctions will be stationary:

$$\varphi_n(x) = \varphi_n(\vec{x})\exp(-E_n t) \quad ; \tag{14}$$

but the particle mode operators $\hat{b}_n$ are not explicitly time-dependent in any case: $\partial\hat{b}_n/\partial t = o$. The equal-time anticommutation relations for the field operators $\psi,\bar{\psi}$ yield the relations

$$\left\{\hat{b}_n,\hat{b}^+_m\right\} = \delta_{nm} \quad , \quad \left\{\hat{b}_n,\hat{b}_m\right\} = \left\{\hat{b}^+_n,\hat{b}^+_m\right\} = 0 \quad , \tag{15}$$

showing that electrons and positrons are fermions. In terms of the particle mode operators, the Hamiltonian takes the form

$$\hat{H} = \sum_n E_n (\hat{b}^+_n \hat{b}_n - \frac{1}{2}) \quad . \tag{16}$$

The contribution $(\frac{1}{2}E_n)$ is the zero point energy for each mode which is negative for fermions.

The vacuum state is now defined as the Fock state at lowest energy under given conditions, i.e. given the external field and possible boundary conditions. As the number operator $\hat{b}^+_n\hat{b}_n$ has eigenvalues 0 and 1 (this can be shown from eq. (15)), the lowest energy state $|o\rangle$ is found by choosing

$$\hat{b}^+_n \hat{b}_n|o\rangle = o \qquad \text{for} \qquad E_n>o \quad , \tag{17a}$$

$$\hat{b}^+_n \hat{b}_n|o\rangle = |o\rangle \qquad \text{for} \qquad E_n<o \quad , \tag{17b}$$

i.e. by filling all negative energy states with a particle (electron). This is the so-called Dirac sea (see Fig. 6), which was invented by Dirac to keep electrons with positive energy from falling into the negative energy states, emitting an infinite amount of radiation energy. Calculations are facilitated by defining hole operators $\hat{d}_n = \hat{b}^+_n$ for $E_n<o$, casting the Hamiltonian into the convenient form:

$$\hat{H} = \sum_{E_n>o} E_n\hat{b}^+_n\hat{b}_n + \sum_{E_n<o} (-E_n)\hat{d}^+_n\hat{d}_n + E_{vac} \quad , \tag{18}$$

where

$$E_{vac} = -\frac{1}{2}\sum_n |E_n| = \langle o|\hat{H}|o\rangle \tag{19}$$

is the vacuum energy. By an analogous calculation, the electric current density in the vacuum state is obtained as

$$j^\mu_{vac}(x) = \langle o|\hat{j}^\mu(x)|o\rangle = -\frac{1}{2} e \sum_n \mathrm{sgn}(E_n)\bar{\psi}_n(\vec{x})\gamma^\mu\psi_n(\vec{x}) \quad . \tag{20}$$

Both expressions (19) and (20) are symmetric against an exchange of electrons and positrons, i.e. $E_n \to (-E_n)$ and $e \to (-e)$, taking account of charge conjugation invariance of QED.

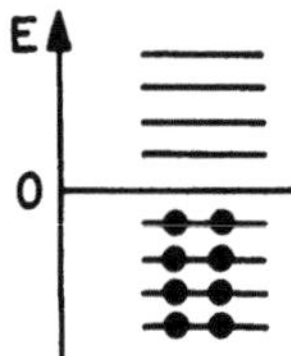

Fig. 6. Dirac Sea

In defining the vacuum as the state of lowest energy under specified conditions we have implied that the state must be physically accessible. There is, however, one criterion of accessibility we have not yet considered: conservation of electric charge, which acts as a kind of superselection rule. Physical transitions can only proceed to states of the same total charge. In order to conserve charge we have to add a Lagrange multiplier times the charge operator to the Hamiltonian when we minimize the energy:

$$\hat{H} \to \hat{H} - \mu\hat{Q}/e \quad . \tag{21}$$

μ, a kind of chemical potential, measures the energy required to supply a unit of charge. When the potential attracts electrons, we have $\mu = -m_e$, because the vacuum electron "supply" is the Dirac sea. In a different way, $(-m_e)$ is the smallest energy emitted by a positron (minus sign because the energy is lost from the system). We thus have to minimize

$$\hat{H} + m_e\hat{Q}/e = \sum_n (E_n+m_e)(\hat{b}^+_n\hat{b}_n-\frac{1}{2}) \tag{22}$$

which is obviously achieved by setting

$$\begin{aligned} \hat{b}^+_n\hat{b}_n|o\rangle &= o \qquad \text{for} \quad E_n > -m_e \\ \hat{b}^+_n\hat{b}_n|o\rangle &= |o\rangle \quad \text{for} \quad E_n < -m_e \quad . \end{aligned} \tag{23}$$

More generally, one can introduce a Fermi surface $E_F = \mu$ separating the vacant from the occupied electron states; the choice $E_F = -m_e$ corresponds to the vacuum state.

2.2 Properties of the Vacuum State

In the context of the external field approximation, all properties of the vacuum of QED can be expressed in terms of the Feynman propagator

$$i\, S_F(x,x'|A) = \langle o|T(\hat{\psi}(x)\hat{\bar{\psi}}(x'))\,|o\rangle =$$

$$= \sum_n \begin{cases} \varphi_n(x)\bar{\varphi}_n(x')\theta(E_n - E_F) & \text{for} \quad t>t' \\ \varphi_n(x)\bar{\varphi}_n(x')\theta(E_F - E_n) & \text{for} \quad t<t' \end{cases} \quad . \tag{24}$$

Here "T" denotes the time-ordered product of the field operators. [In the fully interacting quantum field theory, all time-ordered n-point functions must be known to define the vacuum state.]

In particular, the vacuum polarization current and the vacuum energy have simple expressions in terms of S_F:

$$j^{\mu}_{vac} = \langle o|\hat{j}^{\mu}(x)|o\rangle = -ie\ \mathrm{tr}[\gamma^{\mu}S_F(x,x|A)] \quad , \tag{25}$$

where the limit of equal arguments in the propagator must be symmetric with respect to past and future. Similarly, the vacuum energy is

$$E_{vac}[A] = -i\int d^3x\ \mathrm{tr}[\gamma^0 H_D(x|A)S_F(x,x|A)] \quad , \tag{26}$$

where $H_D(x|A)$ is the Dirac Hamiltonian in the external field A_{μ}. Since E_{vac} does not vanish in the limit $A_{\mu}\to o$ as a sum of positive terms, (see eq. (19)), it is useful to subtract this infinite constant:

$$E'_{vac} = E_{vac}[A] - E_{vac}[o] = \int_0^1 d\lambda \frac{d}{d\lambda} E_{vac}[\lambda A] \quad . \tag{27}$$

With help of the Dirac equation for the propagator, the derivative of E_{vac} can be expressed as

$$\frac{d}{d\lambda} E_{vac}[\lambda A] = E_F \frac{d}{d\lambda} \sum_n \theta(E_F - E_n(\lambda)) + \int d^3x A_{\mu}(x) j^{\mu}_{vac}(x|\lambda A). \tag{28}$$

The first contribution comes from all states penetrating through the Fermi surface, while the second term represents the interaction of the external field with the vacuum polarization (see (Plunien et al.,1986) for details of the derivation).

In a similar manner, the derivative of the total vacuum charge can be expressed in terms of the number of states penetrating the vacuum Fermi surface $E_F=-m_e$:

$$\frac{d}{d\lambda} Q_{vac}[\lambda A] = e \frac{d}{d\lambda} \sum_n \theta(E_F - E_n(\lambda)) \equiv e \frac{d}{d\lambda} N_{vac} \quad . \tag{29}$$

Combining eqs. (28) and (29), the vacuum energy takes on its final form:

$$E'_{vac}[A] = E_F Q_{vac}/e + \int d^3x A_{\mu}(x) \int_0^1 d\lambda j^{\mu}_{vac}(x|\lambda A) \quad . \tag{30}$$

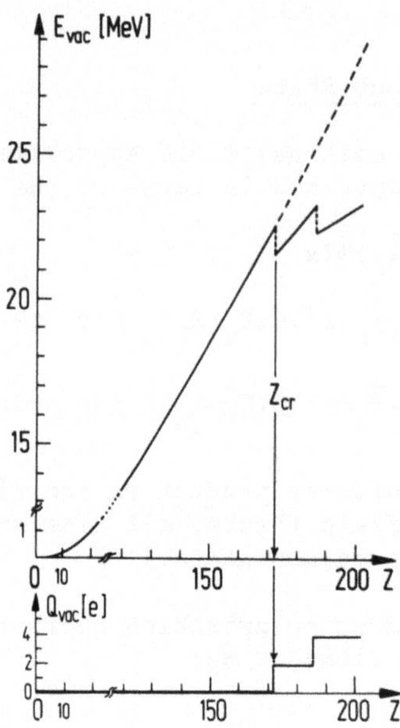

Fig. 7. Vacuum energy and charge as function of Z. E_{vac} and Q_{vac} change discontinuously when a bound state "dives" into the Dirac sea.

The vacuum charge and energy in the Coulomb of nuclei with charge Z are shown in Fig. 7. One clearly sees the sudden transition from neutral to charged vacuum at Z_c=173 with a simultaneous drop in the vacuum energy by $2m_e$. The next transition occurs when the $2p_{1/2}$-state becomes supercritical at Z_c'=185.

If the external potential is so strong that the vacuum state becomes highly charged, a statistical method may be applied to evaluate the vacuum charge density (Müller and Rafelski, 1975). In the framework of a local plane-wave approximation (Thomas-Fermi approximation), the vacuum charge density (20) can be written in the form

$$\rho_{vac.}(x) \approx \frac{e}{2}\left[2\int_0^\infty \frac{d^3k}{(2\pi)^3} + 2\int_0^{k_F(x)} \frac{d^3k}{(2\pi)^3} - 2\int_{k_F(x)}^\infty \frac{d^3k}{(2\pi)^3}\right] =$$

$$= e\cdot 2\int_0^{k_F(x)} \frac{d^3k}{(2\pi)^3} = \frac{e}{3\pi^2}\, k_F(x)^3 \tag{31}$$

where (with $E_F = -m_e$):

$$k_F(x) = \left[(E-A^0(x))^2-m_e^2\right]^{1/2}\theta(E_F-A^0(x)-m_e) \tag{32}$$

is the generalized Fermi momentum. The factor 2 in eq. (31) counts the different spin states. In a systematic analysis one finds that (31) is the first term in a gradient expansion (see lecture by R. Dreizler in this volume). When the expression (31) is added to the external source ρ_{ext}, a self-consistent theory of the vacuum charge is obtained. One important result is that the external potential cannot exceed a limit

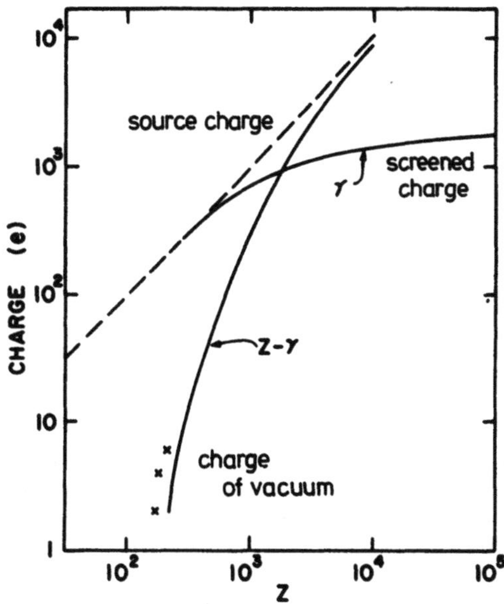

Fig. 8. Screening of very large charges by the real vacuum polarization charge.

$$V_{max} = (\frac{3\pi^2}{e} \rho_{ext})^{1/3} \tag{33}$$

independent of the size of the external source. For nuclear density, one finds $V_{max} \sim 250$ MeV. When the external charge is increased more and more, the vacuum begins to screen the source charge to an ever-growing degree (see Fig. 8). This effect has recently become of some interest in connection with the semi-macroscopic droplets of potentially stable strange quark matter, the so-called nuclearites, that were postulated by Witten (1983). These droplets would consist of a very large number of up- and down-quarks and a somewhat smaller number of strange quarks, and could have a very large bare charge. They would then be surrounded and partially screened by a cloud of vacuum electrons (Farhi and Jaffe, 1984); Pacheco and Sañudo, 1985).

2.3 Fractional Vacuum Charge

The vacuum charge in quantum electrodynamics is always integer,since it is given by the number of states that have crossed the Fermi surface. This is, actually, a somewhat simplified representation of the real situation, for the following reason. The naive evaluation of eq.(25) for the vacuum current density reveals the presence of an infinite contribution, which is strictly proportional to the external source current. Of course, the spatial integral over this charge density does not vanish, rather it is infinitely large even for an arbitrarily small source. However, this infinite contribution to $j^{\mu}_{vac}(x)$ can be absorbed completely, as it is well known, as renormalisation of the unit of electric charge. Only after this renormalisation the vacuum charge vanishes, at least for subcritical fields.

In a very interesting development it has been shown during the past decade that the charge of the Dirac vacuum can take arbitrary real, non-integer multiples of the unit charge in the presence of non-electromagnetic interactions (Jackiw and Rebbi, 1976; Su, Schrieffer and Heeger, 1979; Goldstone and Wilczek, 1981). Instead of presenting the general theory of such phenomena (see e.g., Capri, Ferrari and Picasso, 1984,1985; Ball,1986) we will illustrate the effect on the example of a simple toy model (MacKenzie and Wilczek, 1984).

Let us consider a one-dimensional Dirac particle (a two-spinor!) in a mixed scalar-pseudoscalar potential of the form

$$\phi(x) = m\,(e^{i\gamma_5\alpha(x)} - 1) =$$

$$= m\,(1-\cos\alpha(x) + m\, i\gamma_5 \sin\alpha(x)) \quad , \tag{34}$$

where

$$\alpha(x) = \alpha_o \theta(x) \quad . \tag{35}$$

The potential (34) is not as artificial as it may seem, it would be produced, for instance, by a constant axion field of strength α_o in the region x>o interacting with electrons. Interestingly, the potential can be absorbed in a chiral transformation of the wave-function, so that the Dirac equation becomes apparently free:

$$(i\gamma^\mu\partial_\mu - m)\,\psi' = o \quad , \tag{36}$$

with

$$\psi'(x) = \begin{cases} \psi(x) & \text{for} \quad x<o \\ e^{\frac{i}{2}\gamma_5\alpha_o}\,\psi(x) & \text{for} \quad x>o \;. \end{cases} \tag{37}$$

Thus the solutions are plane waves in the regions x<o, x>o, separately which must be matched such that $\psi(x)$ - not $\psi'(x)$ is continuous at x=o. The result of this procedure are two sets of continuum states (for positive and negative energies, which are given in detail by MacKenzie and Wilczek) and, quite surprisingly, a single bound state, that we derive now. Denoting the bound state energy E_o, the regular solutions of eq.(36) in the two regions are (Dirac spinors in one spatial dimension have only two components!)

$$x<o : \qquad \psi(x) = A \begin{pmatrix} 1 \\ \dfrac{-i\kappa}{E_o+m} \end{pmatrix} e^{\kappa x} \tag{38a}$$

$$x>o : \qquad \psi(x) = B \begin{pmatrix} \cos\frac{\alpha_o}{2} & i\sin\frac{\alpha_o}{2} \\ i\sin\frac{\alpha_o}{2} & -\cos\frac{\alpha_o}{2} \end{pmatrix} \begin{pmatrix} 1 \\ \dfrac{i\kappa}{E_o+m} \end{pmatrix} e^{-\kappa x} \quad , \tag{38b}$$

where $\kappa = (m^2 - E^2)^{1/2}$. Equating the ratios of upper and lower components at x=o, we obtain the eigenvalue equation

$$\tan\left(\tfrac{1}{2}\alpha_o\right) = \kappa/E_o \quad , \tag{39a}$$

or

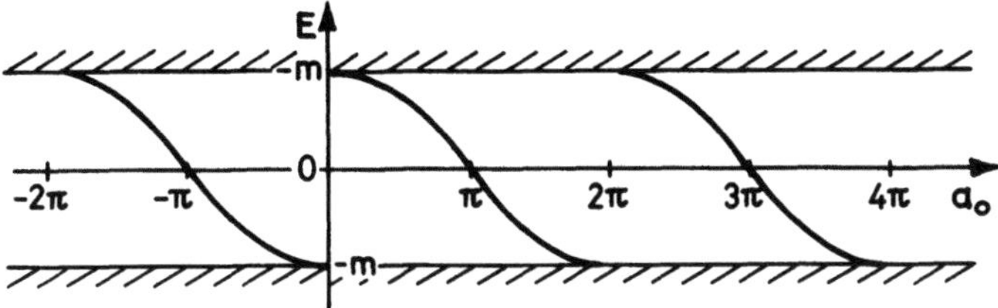

Fig. 9. Bound state energy as function of α_0 in the toy model.

$$E_0 = m \cos(\alpha_0/2) \quad . \tag{39b}$$

The energy of the bound state is shown as function of α_0 in Fig. 9, it is periodic with period $\Delta\alpha = 2\pi$ as is evident from the definition (34). At $\alpha_0 = 0$ a bound state begins to drop from the upper continuum, crosses the $E = 0$ line at $\alpha_0 = \pi$ and joins the lower continuum at $\alpha_0 = 2\pi$, when a second bound state develops at $E = +m$ and starts to approach the Dirac sea, etc.. Naively, we would expect the vacuum charge to be zero at $\alpha_0 = 0$, one at $\alpha_0 = 2\pi$, two at $\alpha_0 = 4\pi$, etc. But what happens in between?

The answer is found by computing the vacuum current density as sum over the wavefunctions, eq. (6). The result of this calculation is (for details see Capri et al.,1984; MacKenzie and Wilczek, 1984):

$$\rho_{vac}(x) = \frac{\kappa}{\sqrt{2}}\, e^{-2\kappa|x|}\, \mathrm{sgn}(\alpha_0) -$$

$$- \frac{m^2}{2\pi} \int_m^\infty dE\, \frac{2iE/p\, \sin(2p|x|)}{E^2 + p^2 \mathrm{ctg}^2(\alpha_0/2)} \quad , \tag{40}$$

where $p = (E^2 - m^2)^{1/2}$.

The integration over x can be done analytically and yields the vacuum charge:

$$Q_{vac} = \frac{1}{2} - \frac{m^2}{2\pi^2} \sin\frac{\alpha_0}{2} \cos\frac{\alpha_0}{2} \int_0^\infty \frac{dp\,(p^2+m^2)^{-1/2}}{(p^2 + m^2 \sin^2\alpha_0/2)} = \frac{\alpha}{2\pi} \quad . \tag{41}$$

The result obviously reproduces the special cases discussed above, but the surprise is that (41) interpolates smoothly between these integer values! At first, one might think that the non-integer part should be absorbed by some renormalization procedure, but this cannot be so, because the potential is a pseudo scalar whereas the vacuum charge is part of a four-vector. Thus the result (41) must be taken seriously.

There is actually one special case, where the logical necessity of this result can be intuitively understood: $\alpha_0 = \pm\pi$. $\alpha_0 = \pi$ corresponds to a situation where the bound state has come down to $E = 0$ from $E = +m$, while in

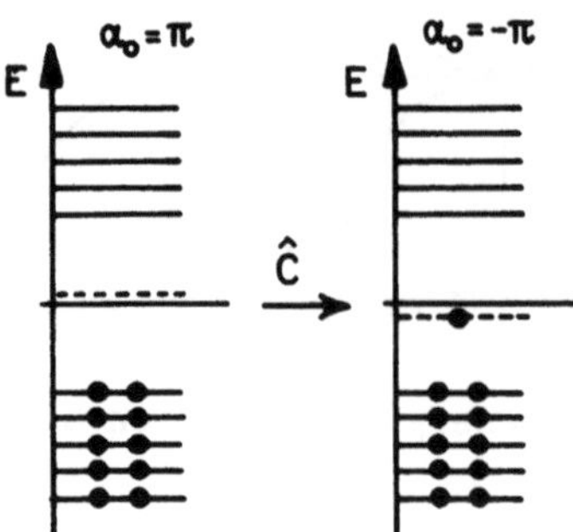

Fig. 10. The cases $\alpha_0=\pi$ and $\alpha_0=-\pi$ are related by charge conjugation.

the case $\alpha_0 = -\pi$ it has risen upward from E = -m (see Fig. 9). Thus, for $\alpha_0 = \pi$ the bound at E=o is empty, whereas that for $\alpha_0 = -\pi$ is occupied in the filled Dirac sea picture. However, the cases α_0 and $(-\alpha_0)$ are generally related by charge conjugation. Using the representation $\gamma^0 = \sigma_1$, $\gamma^1 = i\sigma_2$, we have $\gamma_5 = \gamma^0\gamma^1 = -\sigma_3$, and the charge conjugation transformation is $C = \gamma_5 K$, where K is complex conjugation. Obviously,

$$C \exp(i\gamma_5\alpha) C^{-1} = \exp(-i\gamma_5\alpha) \quad , \tag{42}$$

as asserted above. Now the argument goes as follows: Since the situations for $\alpha_0=\pi$ and $\alpha_0=-\pi$ differ only in the occupation of the bound state (see Fig. 10), we have

$$Q_{vac}(-\pi) = Q_{vac}(\pi) + 1 \quad . \tag{43}$$

On the other hand, owing to the fact that the two cases are mutually charge conjugate, we must have

$$Q_{vac}(-\pi) = -Q_{vac}(\pi) \quad , \tag{44}$$

requiring the assignment of half-integer charge:

$$Q_{vac}(\alpha_0 = \pi) = \frac{1}{2} \quad ! \tag{45}$$

Still, in one dimension the fractional charge is somehow the artefact of an idealisation. For any potential of finite extent, $\alpha(x)$ would return to zero at some, however large value of x. This would correspond to a jump of magnitude $(-\alpha_0)$ at this location, which induces the charge $-\alpha_0/2\pi$ in the vacuum. The combined effect of both jumps would therefore be a net charge Q_{vac}=o. In fact, one can prove that

$$Q_{vac} = \frac{1}{2\pi} [\alpha(+\infty) - \alpha(-\infty)] \quad . \tag{46}$$

Topologically, only asymptotic values $\alpha(\pm\infty) = 2n\pi$ are allowed, because they yield $\phi(\pm\infty)$=o according to eq. (34). In more than one spatial dimension one has to study chiral fields of more complicated symmetry, e.g. SU(2) fields, to obtain similar phenomena. The most widely studied cases are magnetic monopole fields in nonabelian gauge theories and the so-called Skyrmion, a topological soliton solution of the chiral sigma model which induces in the nucleon vacuum a non-vanishing baryon number. (For more details we refer to the lecture of Prof. J. Eisenberg in this volume.)

We finally mention briefly that the expression for the vacuum charge, derived from eq. (20)

$$Q_{vac} = -\frac{e}{2} \sum_{E_n} \operatorname{sgn}(E_n) \quad , \tag{47}$$

is a special case of the generalized Zeta function

$$F(s) = \sum_n |E_n|^{-s} \operatorname{sgn}(E_n) \tag{48}$$

which plays an important role in the mathematical theory of Fock space operators. The value ½F(o) is called the spectral asymmetry, for which powerful theorems were derived by Atiyah, Patodi and Singer (1973), Callias (1978) and Bott and Seeley(1978). These allow to derive expressions like (46) under very general assumptions and, in particular, give the vacuum charge under the influence of the topology of space.

REFERENCES

Atiyah, M.F., Patodi, V.K., and Singer, I.M., 1973, Bull. Lond. Math. Soc., 5:229

Ball, R. D.,1986, Phys. Lett.,171B:435

Bang, J., Hansteen J. H., 1959, K. Dan. Vidensk. Selsk. Mat. Fys. Medd., 31:No.13

Betz, W., Dissertation, Frankfurt 1980 (unpublished)

Bosch, F., 1980, Z. Phys. A296:11

Bott, R, Seeley R., 1978, Comm. Math. Phys. 62:235

Callias, C., 1978, Comm. Math. Phys. 62:213

Capri, A. Z., Ferrari, R., Picasso, L. E., 1984, Phys. Rev. D30:2136;1985, Phys. Rev. D32:2037

Fulcher, L. P., Klein, A., 1974, Ann. Phys. (N.Y.) 84:335

Goldstone, J., Wilczek, F., 1981, Phys. Rev. Lett. 47:968

Greiner, W., Opening Remarks in: Quantum Electrodynamics of Strong Fields, NATO-ASI at Lahnstein, ed. Greiner, W., Plenum, N.Y. 1983)

Gyulassy, M., 1975, Nucl. Phys. A244:497

Jackiw, R, Rebbi, C., 1976, Phys. Rev. D18:3398

Jaffe, R. J., Farhi, E, 1984, Phys. Rev. D30:2379

Liesen, D., et al., 1980, Phys. Rev. Lett. 44:983

MacKenzie, R., Wilczek, F., 1984, Phys. Rev. D30:2194, 2260

Müller, B., Peitz, H., Rafelski, J., Greiner, W., 1972, Phys. Rev. Lett. 28:1235

Müller, B., Rafelski, J., Greiner, W., 1972, Z. Phys. 257:62,183

Müller, B., Rafelski, J., Greiner, W., 1973, Phys. Lett. 47B:5

Müller, B., Rafelski, J., 1975, Phys. Rev. Lett. 34:349

Müller, B., Soff, G., Greiner, W., Ceausescu, V., 1978, Z. Phys. A285:27

Müller, B., Reinhardt, J., Greiner, W., Soff, G., 1983, Z. Phys. A311:151

Pacheco, A. F., Sañudo, J., 1986, Phys. Lett. 169B:1

Rafelski, J., Müller, B., Greiner, W., 1974, Nucl. Phys. B68:585

Rinker, G. A., Wilets, L., 1975, Phys. Rev. A12:748

Soff, G., Müller, B., Greiner, W., 1978, Phys. Rev. Lett. 40:540

Soff, G., Schlüter, P., Müller, B., Greiner, W., 1982, Phys. Rev. Lett. 48:1465

Su. W. P., Schrieffer, J. R., Heeger, A. J., 1979, Phys. Rev. Lett. 42:1698

Witten, E., 1984, Phys. Rev. D30:272

Zel'dovich, Ya. B., Popov, V. S., 1972, Sov. Phys. Usp. 14:673

ARE THE POSITRON PEAKS CAUSED BY INTERNAL ELECTRON-POSITRON PAIR CONVERSION?

G. Soff, P. Schlüter* and W. Greiner

Gesellschaft für Schwerionenforschung (GSI), Planckstraße 1
Postfach 110 541, D-6100 Darmstadt, West Germany

Institut für Theoretische Physik, Johann Wolfgang Goethe-Universität
Postfach 111 932, D-6000 Frankfurt am Main, West Germany

1. INTRODUCTION

In this introduction we first indicate the significance of internal electron-positron pair conversion [1-7] as background process to the positron spectra from giant quasiatoms. The next section briefly deals with the basic theoretical formalism. The performed approximations are emphasized. After that we investigate the possible role of conversion processes a) in the actinide region ($Z \simeq 90$), b) in giant systems ($Z > 150$) and c) in heavy ion fragments ($Z \lesssim 20$). Finally we present a brief conclusion.
In collisions of very heavy ions with $E_{Lab} > 3$ MeV/u both nuclei are Coulomb excited. Transfer reactions or even deep-inelastic nuclear reactions can take place which lead to additional excitations of the nuclei. This internal excitation energy may be carried away by a photon or may be transferred to a bound electron or to an electron of the negative energy continuum, which leads to ionization and electron-positron pair creation, respectively. The latter process requires nuclear transition energies ω larger than twice the electron rest mass. Nuclear E0-transitions are characterized by the absence of single photon emissions, because a photon must carry at least one unit of angular momentum. Such processes form the main source of non-atomic positrons, and they have to be well understood if one wants to draw firm conclusions about the presence or absence of spontaneous pair creation [8-9]. The basic processes under investigation are depicted schematically in fig. 1. The nucleus which makes an E0-transition is labelled by its initial and final state angular momenta J_i, $J_f = J_i$, and eigenenergies E_i, $E_f = E_i - \omega$ ($\hbar = 1$). Process a describes the electron-positron pair creation. An electron of the negative energy continuum ($\epsilon = -E < -m_0c^2$) with Dirac quantum number κ is lifted to the positive energy continuum. The final state energy obviously amounts to $E' = \epsilon + \omega$, whereas the angular momentum quantum number remains unchanged. Since neither the initial state energy nor the final state energy are fixed one expects a continuous energy distribution for the emitted positrons. Process b indicates the conversion of a K-shell electron ($n = 1$, $l = 0$, $j = \frac{1}{2}$, $\kappa = -1$)

with energy eigenvalue $E_{1s_{\frac{1}{2}}}$. Thus bound states with definite energies are involved. Energy conservation then simply causes monoenergetic lepton emission for a fixed nuclear transition energy ω. Process c symbolizes monoenergetic positron production. Here an electron of the negative energy continuum is excited to a bound state, e.g., to the $1s_{\frac{1}{2}}$-state. Thus we focus our attention on: i) The pair conversion coefficient β, defined as the ratio of the pair production probability (process a) compared with that of photon emission for a specific nuclear transition with energy ω. Since the energy of the electron and the positron takes continuous values we may express β also as integral of the differential pair conversion coefficient $d\beta/dE$. The lower bound of the integral is determined by the rest mass of the electron, which corresponds to vanishing kinetic energy, while the upper bound is given by the nuclear transition energy ω minus m_0c^2. ii) The conversion coefficient α, defined as the ratio of the probabilities of inner-shell vacancy formation (process b) and photon emission. In particular, this mechanism is important for low energy nuclear transitions. iii) The ratio η of the two E0-conversion probabilities for electron-positron pair creation and for the ionization of bound state electrons. This ratio is completely determined by the density of the electron wave functions at the nuclear origin, thus being independent of the nuclear wave function.

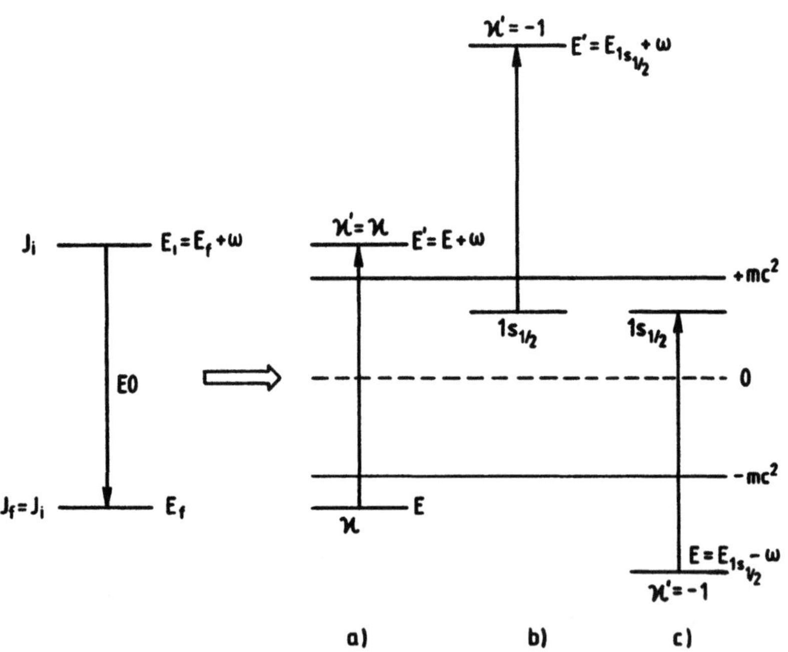

Fig. 1: Schematic representation of electron conversion processes accompanying nuclear E0-transition from a state E_i, J_i to a state $E_f = E_i$-ω, $J_f = J_i$: a) Electron-positron pair production leading to a continuous energy distribution of positrons and electrons. b) Conversion of K-shell electrons - a monoenergetic electron-production mechanism and c) Monoenergetic positron production.

The presence of these conversion processes as background contribution has been verified, e.g., in measurements of δ-electron production. In figure 2 we compare our theoretical results [10] for the double differential cross section for δ-electron emission with the experimental data of Herath-Banda et al [11] for the system I - U. The double differential cross section is shown versus the kinetic electron energy in the center of mass system. The lower curve represents coincidences between electrons and K-holes.

The upper curve indicates our result for the total δ-electron emission rate. Remarkable agreement is achieved for the total and the coincident spectra for the slope as well as for the absolute numbers. Note, that in the calculations no scaling or fitting has been applied. The bumps in the coincidence spectra arise from nuclear Coulomb excitation and subsequent internal conversion in the U-nucleus. The corresponding nuclear transitions are indicated. In summary we may state that our theoretical considerations correctly describe inner-shell excitation and δ-electron production in superheavy systems. The strong relativistic effects are reflected in the rather high production rates of high energy electrons. But internal conversion processes obviously may lead to considerable background contributions.

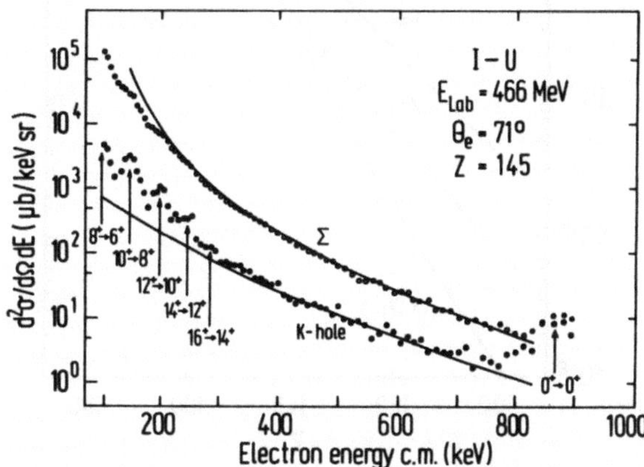

Fig. 2: Experimental double differential cross sections for the δ-electron yield (Σ, open dots) and for the 1s atomic coincident yield (K, full dots) versus kinetic electron energy. The system I - U at 466 MeV bombarding energy is considered. The full lines are the results of our coupled channel calculations [10]. Experimental data by M.A. Herath-Banda and collaborators [11].

At present the positron emission in elastic and deep-inelastic collisions of very heavy ions is subject of widespread experimental activities. Narrow peak structures at a kinetic positron energy of about $E_{e^+} = 300$ keV were observed [12-17] at projectile energies close to the nuclear Coulomb barrier. Currently there exists no convincing theoretical explanation of these puzzling cusps which reproduces all experimental facts. One might suppose that those peaks are of trivial origin, e.g., they might be caused by nuclear background processes. In this contribution we thus will discuss the theoretical framework of conversion processes in single atoms and in supercritical compound systems and the corresponding experimental data.

The competing modes of a nuclear transition are predominantly determined by the transition energy ω, the angular momentum selection rules and the electron density inside the nuclear interior. Hence a first estimate on the importance of conversion processes especially in superheavy systems can be gained by evaluating the electron probability density distribution inside the nucleus. Fig. 3 displays $|\Psi(r=0)|^2$ for the 1s-, 2s- and $2p_{\frac{1}{2}}$-state, respectively versus the nuclear charge number Z. A strong increase is visible below Z = 160, which is most pronounced for the $2p_{\frac{1}{2}}$-state. Between Z = 92 (U) and Z = 184 $|\Psi(r=0)|^2$ changes by more than four orders of magnitude. For Z > 160 a saturation behaviour appears indicating the sharp localization of the electron wave function close to the origin.

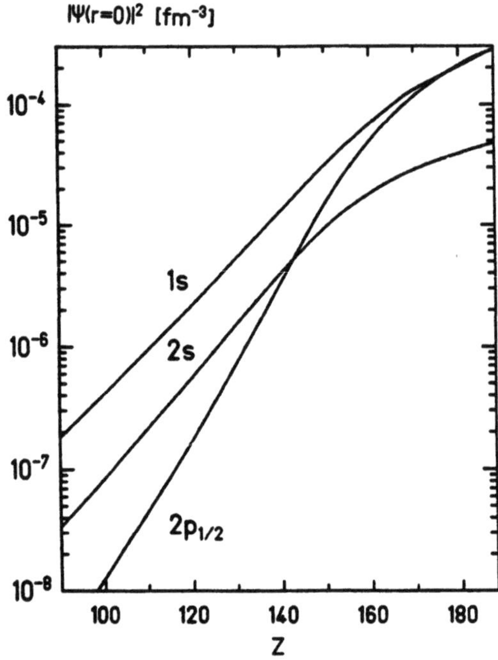

Fig. 3: Radial electron density $|\Psi(r=0)|^2 = f^2(0) + g^2(0)$ at the nuclear origin (r = 0) for 1s-, 2s- and $2p_{\frac{1}{2}}$-electrons as function of the nuclear charge number Z in superheavy atoms. The relative increase is most pronounced for the $2p_{\frac{1}{2}}$-state.

Also in connection with the hypothesis that the puzzling positron peaks result from a two-body decay of a new elementary particle [18-24] electron-positron pair conversion processes play a decisive role. Such a new particle should be detectable in nuclear transitions once the energetical production threshold is exceeded. Thus experimental investigations would provide upper limits for the allowable coupling constant of this particle.

2. BASIC THEORETICAL FORMALISM

The theoretical ingredients concerning internal pair formation are presented in refs. 1 and 2 and need not to be repeated here. We calculated the differential conversion coefficient $d\beta/dE$ for internal e^+, e^--pair creation with respect to the positron energy

E, which is defined by ($\hbar = c = m_e = 1$)

$$\frac{d\beta}{dE} = \frac{dP_{e^+,e^-}}{dE}\frac{1}{P_\gamma} \tag{1}$$

P_{e^+,e^-} is the probability for electron-positron pair formation. P_γ denotes the photon emission probability in a nuclear transition of multipolarity EL or ML (L $\geq$ 1) and transition energy ω. In the following P_{e^-} signifies the complementary probability for ionizing a 1s-electron. The corresponding quantity for nuclear E0-transitions will be denoted by $d\eta/dE$. In this case single photon emission is strictly forbidden.

$$\frac{d\eta}{dE} = \frac{dP_{e^+,e^-}}{dE}\frac{1}{P_{e^-}} \tag{2}$$

The general theoretical pair creation probability is given by [1]

$$P_{e^+,e^-} = \frac{2\pi}{2J_i+1}\sum_{M_i=-J_i}^{J_i}\sum_{M_f=-J_f}^{J_f}\sum_{\substack{\kappa=-\infty\\ \kappa\neq 0}}^{\infty}\sum_{\substack{\kappa'=-\infty\\ \kappa'\neq 0}}^{\infty}\sum_{\mu=-j}^{j}\sum_{\mu'=-j'}^{j'} W_{e^+,e^-} \tag{3}$$

Here J_i and J_f denote the total angular momentum quantum number of the nucleus in the initial and final state, respectively. M_i and M_f indicate the corresponding magnetic substates. κ and κ' signify the Dirac angular momentum quantum number of the electron initial and final state, respectively. Again μ and μ' are quantum numbers of possible magnetic substates of the electron. The probability W is determined by the retarded Coulomb interaction via the matrix element $U_{if}^{(2)}$

$$W_{e^+,e^-} = \int_1^{\omega-1} |U_{if}^{(2)}|^2\, dE \tag{4}$$

with

$$U_{if}^{(2)} = -\alpha\int_0^\infty d\tau_n \int_0^\infty d\tau_e\, j_n(\vec{r}_n)\, j_e(\vec{r}_e)\, \frac{\exp\{-i\omega|\vec{r}_n-\vec{r}_e|\}}{|\vec{r}_n-\vec{r}_e|} \tag{5}$$

The subscripts n and e indicate nucleus and electron coordinates, respectively. $d\tau_n$ and $d\tau_e$ are the corresponding volume elements. E is the total positron energy. After a multipole expansion the second-order matrix element $U_{if}^{(2)}$ may be decomposed into an electric and magnetic part

$$U_{if}^{(2)} = \sum_{L=1}^{\infty}\sum_{M=-L}^{L}\left\{U_{if}^{mag}(L) + U_{if}^{el}(L)\right\} + U_{if}^{L=0} \tag{6}$$

The electric part follows from

$$U_{if}^{el}(L) = 4\pi\, i\alpha\omega\, V_\gamma^{(e)}\left(M_{(s)}^{el} + M_{(d)}^{el}\right) \tag{7}$$

$V_\gamma^{(e)}$ characterizes the associated photon emission amplitude. The static part $M_{(s)}^{el}$ of the matrix element predominantly is determined by the nuclear charge number Z, the nuclear transition energy ω and the multipolarity L of the transition. In contrast to this the dynamic part $M_{(d)}^{el}$ depends explicitly on the specific nuclear structure. This contribution is neglected in all our calculations of internal conversion coefficients except for electric monopole transitions in which the static part $M_{(s)}^{el}$ vanishes.
The calculation of branching ratios in the E0-case is exceptionally simple, if one assumes that the only non-vanishing nuclear matrix element

$$M^{(k)} = \int_0^\infty d\tau_n\, \rho_n(\vec{r}_n)\, r_n^k, \quad k = 2,4,6... \tag{8}$$

is that with $k = 2$. In this case only electron states contribute which possess $j = \frac{1}{2}$. Employing this decisive approximation it follows [2]

$$\frac{d\eta}{dE} = \frac{\sum_{\kappa=\pm 1} |C_{e^+,e^-}|^2}{|C_{K-e^-}|^2} \tag{9}$$

The quantities C are determined by electron transition densities at the origin ($r \to 0$). The nuclear matrix element cancels in the above fraction for $d\eta/dE$. Since we don't know explicitly the nuclear matrix element we express in discussions of conversion processes all quantities through ratios.

Employing electron wave functions for the Coulomb field of a point nucleus one can derive simple analytical expressions for the differential conversion coefficient in nuclear E0-transitions. We obtain [2]

$$\frac{d\eta}{dE} = \frac{(EE' - \gamma^2)\, e^{\pi(B'-B-\bar{B})}}{\pi(Z\alpha)^2(\omega + 2\gamma)\,\Gamma(2\gamma+1)} \left(\frac{pp'}{Z\alpha\bar{p}}\right)^{2\gamma-1} \left|\frac{\Gamma(\gamma+iB)\,\Gamma(\gamma+iB')}{\Gamma(\gamma+i\bar{B})}\right|^2 \tag{10}$$

with

$$B = \frac{Z\alpha E}{p} \tag{11}$$

$$B' = \frac{Z\alpha E'}{p'} \tag{12}$$

$$\bar{B} = \frac{Z\alpha \bar{E}}{\bar{p}} \tag{13}$$

and

$$p = \sqrt{E^2 - 1}, \tag{14}$$

$$\bar{E} = \gamma + \omega, \tag{15}$$

$$\gamma^2 = \kappa^2 - (Z\alpha)^2 \tag{16}$$

3. CONVERSION PROCESSES IN HEAVY ATOMS

We computed the differential conversion ratio $d\eta/dE$ with respect to the positron energy E for E0-transitions. As bound state only the atomic K-shell has been taken into account. The conversion probability of higher bound states is at least an order of magnitude smaller. Figure 4 displays the change of $d\eta/dE$ with nuclear charge Z for the transition energies $\omega = 1423$ keV and $\omega = 2023$ keV, respectively. The spectra are centred around E_{max}^{kin} reflecting the strong Coulomb repulsion of low energetic positrons. With decreasing Z and increasing ω the energy distribution becomes more and more symmetric and therefore equal to the electron spectra. On the other hand, the Coulomb repulsion of positrons is best visible in high Z elements and transition energies close to the pair production threshold.

Calculated total conversion coefficients η for nuclear E0-transitions are compared with available experimental data in table 1 (see also refs. 25 - 27). Fair agreement is achieved.

We also computed the differential conversion coefficient $d\beta/dE$ for nuclear EL and Ml transitions with $L \geq 1$. For the nucleus $_{92}$U the energy distribution of emitted positrons is similar to those presented in fig 4. Total conversion coeficients β are presented in table 2 for various nuclear transitions. The comparison with experimental data shows again fair agreement. β is typically in the order of 10^{-4}.

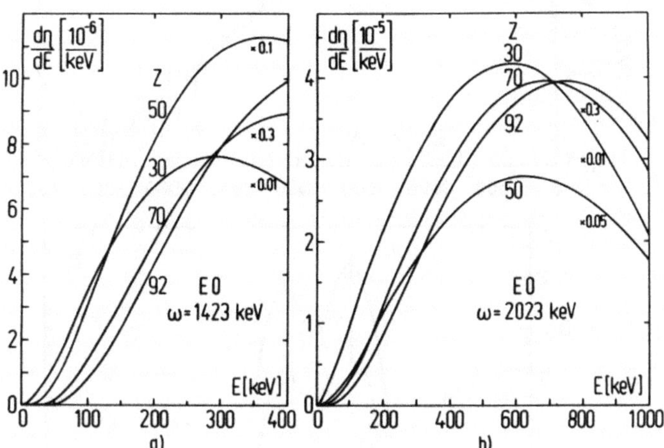

Fig. 4: Z-dependence of the differential conversion probability ratio $d\eta/dE$ versus the kinetic positron energy E for electric monopole transitions with ω = 1423 keV and ω = 2023 keV.

Table 1: Comparison of theoretical and experimental total conversion coefficients $\eta = P_{e^+,e^-}/P_{K-e^-}$. The experimental results for $1/\eta$ are cited in ref. 25.

Nucleus	Transition energy [MeV]	$\frac{1}{\eta}$(exp)	$\frac{1}{\eta}$(this work)
$^{16}_{8}O$	6.06	$(4.0\pm0.46)\cdot10^{-5}$	$3.8\cdot10^{-5}$
$^{40}_{20}Ca$	3.353	$6.9\cdot10^{-3}$	$7.1\cdot10^{-3}$
$^{42}_{20}Ca$	1.836	0.111 ± 0.02	0.138
$^{90}_{40}Zr$	1.752	2.08 ± 0.08	2.465
$^{140}_{58}Ce$	1.903	6.3	6.7
$^{214}_{84}Po$	1.416	440-625	368

Now we can discuss the possibility whether the observed structures in positron spectra may originate from nuclear E0 transitions. One convincing argument against this interpretation is related to the shape of the e^+-energy distribution. According to fig. 4, the halfwidth of the spectra should be at least 150 keV. However, the observed structure is much narrower. The second argument is connected with the energy distribution of emitted γ rays and δ electrons. Nuclear transitions of, e.g., multipolarity E1 or E2 should also be observable in the emitted photon spectra provided that proper

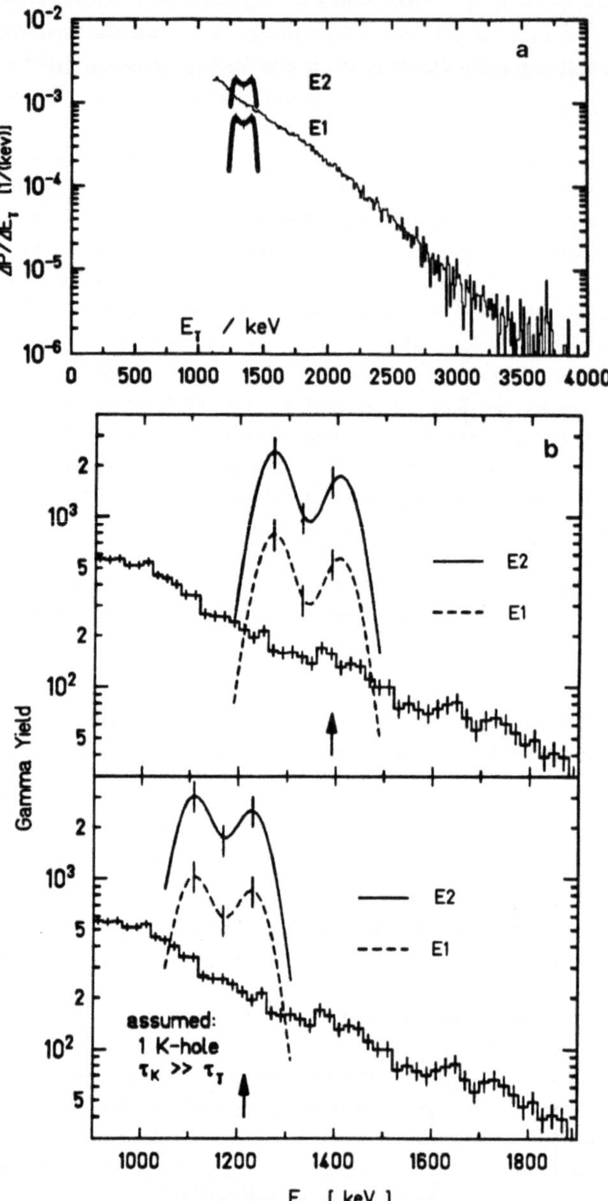

Fig. 5: Part a: The γ-ray spectrum for U+U according to refs. [13,16]. The expected γ-lines are shown, if the positron line structure would be due to conversion. - Part b: γ-ray spectrum [12,14] from 6.05 MeV/u U+Cm collisions. The data shown are coincident with scattered particles with the same kinematic conditions as yield the peak in the positron energy spectrum. The expected γ-ray peak is plotted for the assumption that the positron structure is due to monoenergetic pair conversion of an E1 or E2-transition in the uranium nucleus (lower part). One K-hole at the moment of conversion has been assumed. The calculated γ-peak includes Doppler shift, detector efficiency and resolution.- The upper part shows the same as before, now for the assumption that the positron structure is due to ordinary nuclear pair conversion of an E1 or E2-transition in either one of the nuclei, uranium or curium.

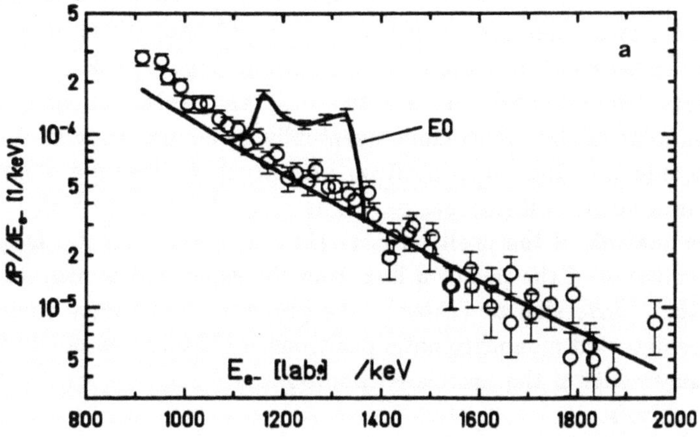

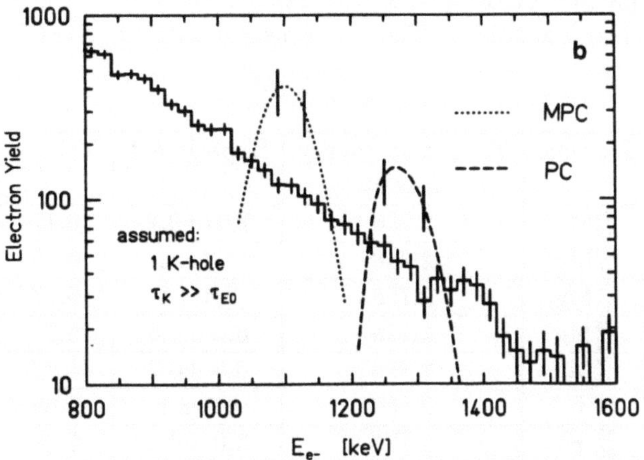

Fig. 6: Part a: δ-electron spectrum for U+U according to refs. [13,16] . The full curve shows our theoretical result. If the observed line structure in the positron spectrum would be due to E0-conversion, the indicated bump in the δ-electron spectrum should be seen. Obviously the δ-electron spectrum is smooth. Part b: Electron spectrum from 6.13 MeV/u U+Cm collisions. The data are coincident with scattered particles in the kinematic region which yields a 316 keV peak in the positron energy spectrum. The expected internal conversion electron peak is plotted over the data assuming that the positron structure is due to pair conversion (PC) or monoenergetic pair conversion (MPC) of an E0-transition in the uranium nucleus. (Note: The positron line width argues against the PC process.) The plotted peaks are calculated for one K-hole at the moment of E0-conversion. In order to explain both the positron peak intensity and the structureless electron spectrum, there must be more than one K-hole at the moment of E0-conversion (95% efficiency and detector resolution).

Doppler shift corrections are performed. This is shown for the systems U + U and U + Cm in figs. 5 and 6. In both cases the expected γ line is indicated if the positron structure would be due to conventional conversion. Obviously, this can be ruled out. If the observed structures are caused by nuclear E0 transitions, one should also observe a distinct peak in the δ-electron distribution. Such a peak does not exist, as the measurements demonstrated. Since both, γ-ray and δ-electron spectra do not show any substantial structure which could eventually be connected with the positron line, we can conclude that the sharp positron line structure does not stem from a trivial conversion in a single collision partner. These conclusions are also substantiated by a line-shape analysis of the positron peak [14]. The experimental facts seem to contradict an emission of the positron line from the separated atoms. In conclusion we may state that conversion processes in the projectile and target nuclei as a possible source of the detected monochromatic positrons could be ruled out experimentally by direct measurements of the associated photon and δ-electron spectra as well as by a line-shape analysis.

Table 2: Comparison of theoretical and experimental total conversion coefficients β for various nuclear transitions. The experimental data are cited in ref. 1.

$\beta * 10^4$

Author	Nucleus	Transition energy [MeV]	Experiment	Theory	Multipolarity
Bloom	$^{24}_{12}$Mg	2.758	7.01±0.2	6.77	E2
Cleland	$^{24}_{12}$Mg	2.758	6.7 ±1	6.77	E2
Mims	$^{24}_{12}$Mg	2.758	7.4 ±1	6.77	E2
Slätis	$^{24}_{12}$Mg	2.758	8.0 ±0.5	6.77	E2
Spring	$^{24}_{12}$Mg	2.758	6.6 ±0.7	6.77	E2
Brunner	$^{144}_{60}$Nd	2.180	6.6 ±1	7.09	E1
Allan	$^{205}_{82}$Pb	1.776	3.0 ±0.6	3.68	E1
Allan	$^{205}_{82}$Pb	1.863	4.5 ±0.8	4.31	E1
Allan	$^{205}_{82}$Pb	1.903	4.3 ±0.8	4.60	E1
Allan	$^{206}_{82}$Pb	1.595	2.5 ±0.5	2.36	E1
Allan	$^{206}_{82}$Pb	1.719	3.7 ±0.6	3.26	E1
Spring	$^{208}_{82}$Pb	2.615	4.3 ±0.7	3.72	E3

4. CONVERSION PROCESSES IN SUPERHEAVY SYSTEMS

Now we will turn to the question, whether observed peak structures in positron spectra may result from a conversion process in a giant nuclear system [6] which exists for a specific time T. There are two distinct positron production modes (see fig. 7). The first one, the continuum-continuum conversion, can be excluded immediately, since, as can be deduced from figures 8 and 9, it yields a more or less triangular shaped spectrum. Therefore this width is by at least a factor of 2 too large to be compatible with the measured width $\Gamma \simeq 70$ keV. The spectral distribution of emitted positrons is illustrated in figs. 8 and 9 for nuclear transitions in the atom Z = 169 with ω = 1323, 1423, 1523, 1623 and 1723 keV, respectively. The dashed lines indicate the end points of the corresponding positron spectra. In each case the width of the energy distribution is much to large to be responsible for the narrow lines observed in the differential e^+-production cross sections. In the second case, the MPC (monoenergetic pair conversion), there are several possibilities depending on the origin of the participating hole. At present the experimental data seem to indicate a peak energy more or less independent of the nuclear charge.

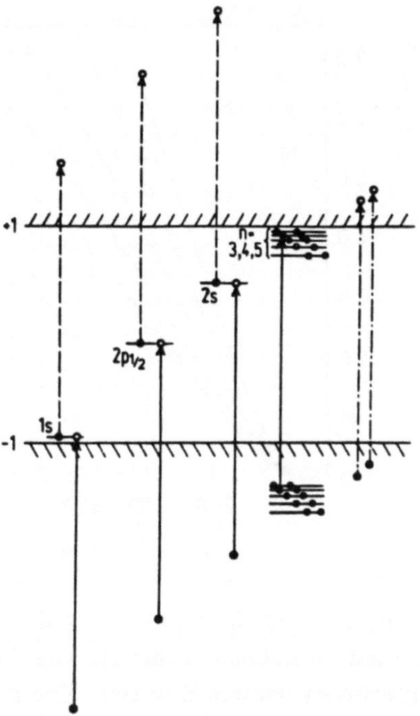

Fig. 7: Various monoenergetic conversion processes in a giant, for reasons of lucidity here subcritical system. Full lines denote processes leading to monoenergetic positron creation by conversion into bound electron states, whereas the dashed lines show conversion of bound electrons into the upper continuum. Electron-positron pair creation is denoted by the dashed-dotted line.

Assuming that the peak originates from monoenergetic pair conversion into the K- or L-shell, this means that the nuclear transition energy must depend in a predetermined manner on the charge of the giant nucleus. But as long as the nuclear transition energy ω cannot be fixed independently, it seems to be more natural to suppose a more or less constant value for the transition energy ω. Then also the binding energy of the final state must be nearly unchanged for different combined nuclear charges Z_u. Following this argument the most probable cause is the conversion into the M-shell. Its binding energy depends only weakly on Z while the conversion coefficients are still rather large. Increasing the principal quantum number leads to strongly decreasing conversion probabilities. The binding energies of the 3s- and 3p-state differ by less than 30 keV for $Z_u \simeq Z_{cr}$. Hence both states may contribute to the peak. This process is depicted in figure 7. In order to discuss this possibility we will assume the most favourable conditions: A fraction q of all detected nuclei originates from a giant nucleus. All these nuclei are excited, so that nuclear transitions during the life-time T of the giant nucleus become possible. At the moment we will restrict ourselves to the case of just one transition.

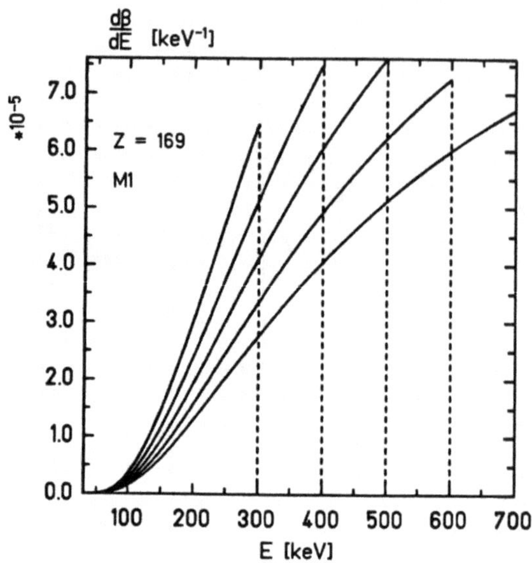

Fig. 8: Differential conversion coefficient for e^+,e^--pair creation versus kinetic positron energy E. $d\beta/dE$ is depicted for nuclear transition energies $\omega = 1323$, 1423, 1523, 1623 and 1723 keV in the superheavy system Z = 169. The multipolarity M1 is considered.

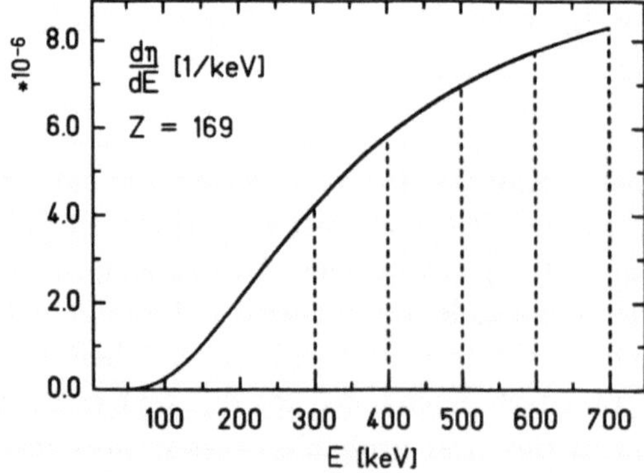

Fig. 9: The differential conversion coefficient $d\eta/dE$ for electric monopole transitions in the superheavy system Z = 169 as function of the kinetic positron energy E. The differential e^+,e^--pair production probability is compared with the ionization probability of 1s-electrons for the nuclear transition energies $\omega = 1323$, 1423, 1523, 1623 and 1723 keV. The dashed lines indicate the end points of the distinct positron spectra.

The most favourable multipolarity is E0. In this case single photon emission is prohibited and a rather high fraction decays by MPC. Photon emission, on the other hand, is dominant for EL-transitions with $L \geq 1$. In this case the fraction q must be about one order of magnitude larger than in the E0-case to get the same number of positrons. As an alternative one might consider magnetic dipole transitions; here only a few percents of the excited nuclei decay by photon emission. Moreover, the branching ratios of the various conversion processes are nearly identical with those for multipolarity E0.

In the supercritical system $Z = 188$ the electron K-shell is imbedded as resonance in the negative energy continuum. A vacant K-shell thus may lead to spontaneous positron production. Pronounced peaks in the positron spectra due to the spontaneous part of the positron production mechanism were predicted thereby indicating the energetical location of the 1s-resonance in the negative energy continuum.

In fig. 10 we show the binding energy of various bound states [28]. The arrows indicate the considered radiative transitions to the 1s-state. Most of the transitions may also take place via internal electron-positron pair conversion since the corresponding energy differences are above the pair production threshold. But we shall not discuss this type of Auger process in more detail.

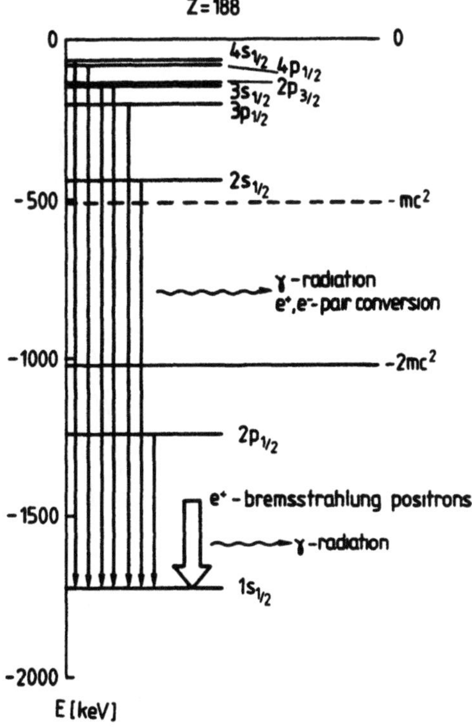

Fig. 10: Binding energies of various electron states in the superheavy atom $Z = 188$. The arrows indicate the considered radiative transition to the vacant 1s-resonance state. Additional radiative transitions from electron states of the Dirac sea to the K-shell are depicted qualitatively by the big arrow.

A supercritical nucleus $Z > Z_{cr}$ which undergoes a transition with $\omega > 2m_0c^2$ during the nuclear reaction period T may transfer this excitation energy to one of the electrons in the negative energy continuum. The remaining hole represents a positron. But also the K-shell electron can be lifted to the upper continuum. If T is longer than the spontaneous decay width of the K-shell resonance the K vacancy will be filled again leading to spontaneous positron emission. This is a sort of 'positron-gun' firing several shots, the energy of which is supported by the excitation energy of the giant nuclear system.
For the nuclear charge distribution of the giant system Z = 184 a homogeneously charged sphere with a radius R_N = 10.88 fm has been assumed. In fig. 11 we show the differential pair conversion coefficient $d\beta/dE$ as function of the positron energy E. The considered nuclear transition energies and multipolarities are indicated. The appearance of the pronounced peak at $E = E_R$ is striking. The resonance shows up only in the κ = -1 channel. Since a giant nuclear compound system very likely is not a static object, its internal dynamics may influence the spectrum of emitted positrons. Clearly, a rigorous treatment of such effects cannot be based on the semiclassical approximation, but requires a fully quantum mechanical reaction theory for the nuclear scattering.

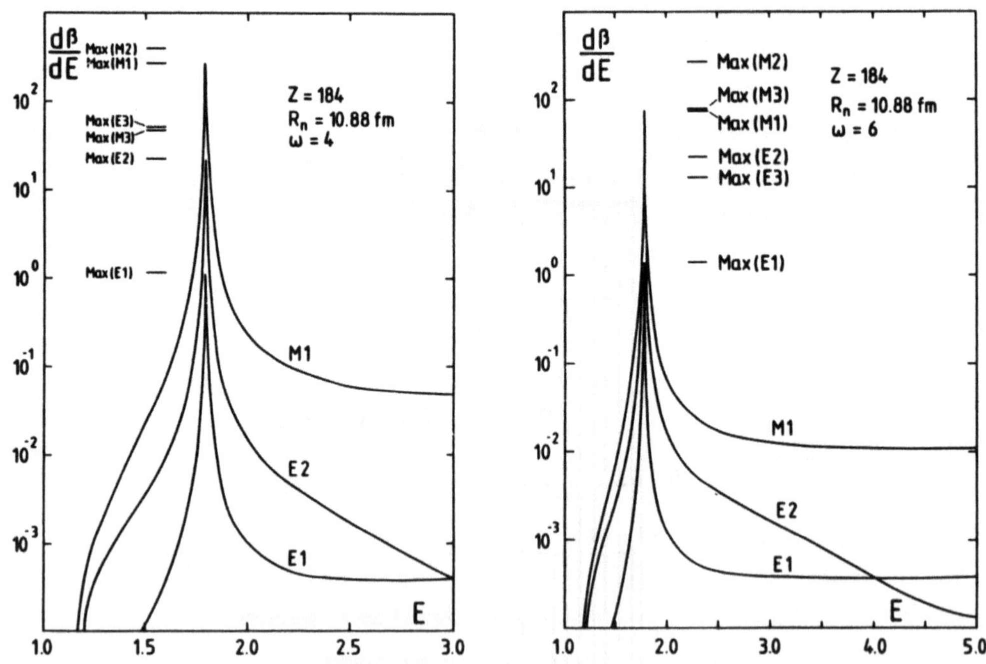

Fig. 11: Differential conversion coefficient for e^+,e^--pair creation versus kinetic positron energy E. $d\beta/dE$ is depicted for nuclear transition energies ω = 2 and 4, respectively, in the supercritical system Z = 184. The considered multipolarities are indicated.

The bound state resonance imbedded in the negative energy continuum is also reflected in the differential conversion coefficient $d\eta/dE$ for nuclear E0-transitions which is presented in fig. 12 for the supercritical system Z = 188 (U + Cm). The cusp in the

spectrum at E $\simeq$ 230 keV signals the $2p_{\frac{1}{2}}$-resonance while the peak at the endpoint of the positron distribution for ω = 1723 keV is caused by the $1s_{\frac{1}{2}}$-resonance.
In our explicit calculations we considered two systems being subcritical and supercritical, respectively. The first one is Pb + Cm (Z_u = 178) at a bombarding energy E_{lab} = 5.9 MeV/u. Here the energy of the 1s-state amounts to -0.994 at the distance of closest approach for central collisions (R_{min} = 17.0 fm). We presume that a giant nucleus is formed at this stage. As we have pointed out, the conversion rates are formed by multiplying the densities of the participating wave functions at the origin. Surprisingly, the electron densities of the states in the upper continuum are nearly independent of central charge ($178 \leq Z \leq 188$), parity (± 1) and energy ($1 \leq E \leq 4$). It varies only slowly in the range $2.2 \cdot 10^{-3}$ - $2.5 \cdot 10^{-3}$. In the lower continuum the density ρ(r=0) depends more sensitively on the physical parameters.

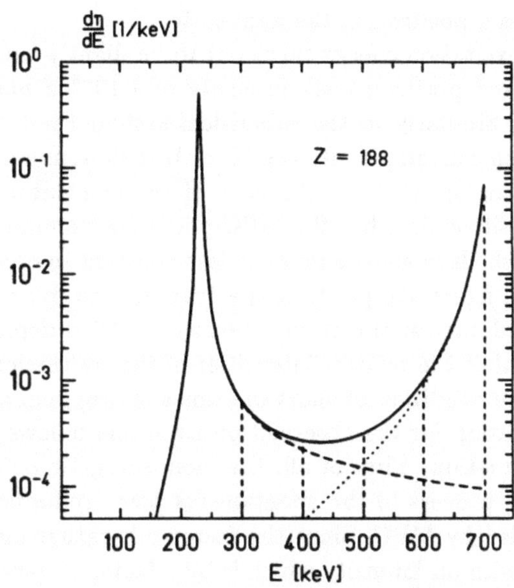

Fig. 12: The same as in fig. 9 for Z = 188. The differential e^+,e^--pair production probability is compared with the ionization probability of 2s-electrons. The dashed and dotted lines give the contributions of the κ = 1 and κ = -1 states, respectively. The cusp at E $\simeq$ 230 keV signals the $2p_{\frac{1}{2}}$-resonance imbedded into the negative energy continuum.

Assuming a peak energy E_+ = 1.63 and conversion into the 3p-state an excitation energy ω = 2.49 follows. Assuming a positron peak intensity of $5 \cdot 10^{-6}$ this leads to q = $2.2 \cdot 10^{-2}$. With decreasing life-time T the normalized intensities remain nearly unchanged. For exceptionally small T the normalization factor q may increase considerably. The rates of the MPC into the K- or L-shell are even larger despite the smaller hole numbers. Consequently the positron spectrum should be dominated by these 'side peaks'. This spectrum is depicted in figure 13. Here Doppler broadening has been simulated by a Gaussian curve with a characteristic width $\Delta E = 0.1 \cdot \sqrt{E^2 - 1}$. If these peaks could be observed, the binding energies of the other bound states could be determined immediately. Thus a spectroscopy of giant atoms would become possible.

Up to now, however, there seems to be no unambiguous experimental evidence for such 'side peaks'.
Most of the holes in the M-shell are accompanied by electrons in the K- or L-shell. Thus the life-time of the giant nucleus need not be exceptionally long. However, only a small fraction of all nuclei decays by MPC into the M-shell.
The other system, which we consider, is U + Cm at a bombarding energy $E_{lab} = 6.05$ MeV/u. This system is supercritical ($Z_u = 188$) and at the distance of closest approach the total 1s-energy amounts to -1.555. The study of this system is complicated by the presence of the resonance, which allows for another positron production mechanism. As we know from the subcritical case, the positron yield due to conversion is rather small. In contrast to this, now every K-hole may decay spontaneously. If we require, that the 'conversion peak' is not only a side effect, the life-time of the giant nucleus must be small compared with the spontaneous decay time. By this it becomes possible to describe a K-hole in the initial state as well as the formation of a K-hole, which does not show up as a positron in the final state.
In this system the excitation energy turns out to be slightly smaller, namely $\omega = 2.45$. Assuming a measured positron peak intensity of $2 \cdot 10^{-5}$ a nuclear formation rate q $= 1.0 \cdot 10^{-2}$ follows. Similarly to the subcritical system Pb + Cm also here the 'side peaks' resulting from conversion into the K- or L-shell dominate the spectrum. These 'side peaks' are changing with the charge Z of the giant atom considerably, as does the spontaneous positron line, but the MPC-positrons stemming from transitions into the M- and higher shells would be more or less constant in energy. This is illustrated in figure 13. In this figure the positron spectrum due to dynamical positron creation and MPC for a fixed nuclear transition energy $\omega = 2.5$ is depicted schematically. We have to emphasize that the relative intensities of the conversion lines were derived by extrapolation. If the condition of short life-times is dropped, an additional argument against this explanation for the observed positron line follows from the possibility of spontaneous pair creation. First of all, the monoenergetic pair conversion can be no effective concurring process to the spontaneous pair production since only 1% of all K-holes may be filled by MPC. Since the factor q is rather large, this gives rise to a spontaneous peak with an intensity which is by a factor 30 larger than the 'conversion peak'.
From the extraordinary large formation rate q it may be concluded that for supercritical, as well as subcritical systems, an explanation of the detected positron peaks by MPC within the discussed scenario seems to be unlikely.
In the following we will discuss some modifications and complications of the explanation previously mentioned. One might think that it is possible to reduce the nuclear excitation rate q if there are multiple transitions in the nucleus. To discuss this we have to distinguish two cases. If the excitation energies of the first transitions in the cascade are larger than about 2, these transitions will lead predominantly to holes in the K- and L-shell. Therefore the final transition with $\omega \simeq 2.5$ will enlarge the positron 'side peaks' by MPC into these shells. In the supercritical case a strong enhancement of the spontaneous positron peak is the necessary consequence. The picture is changed if the first transitions in the cascade are of low energy, so that conversion out of the K- or L-shell are forbidden energetically. In this case the hole number in the M-shell will be increased. However, the previously derived results will be changed only quantitatively since in these calculations already rather large hole numbers have been used.
These 'side peaks' vanish if there are no holes in the K- and L-shell. The holes, which

have been produced during the incoming part of the collision, may be filled by atomic transitions. Clearly this mechanism works only if the life-time of the giant nucleus as well as its internal decay time are larger than the atomic transition times. The life-time of a hole in the K- or L-shell of giant atoms is in the range 10^{-18} s to 10^{-17} s.

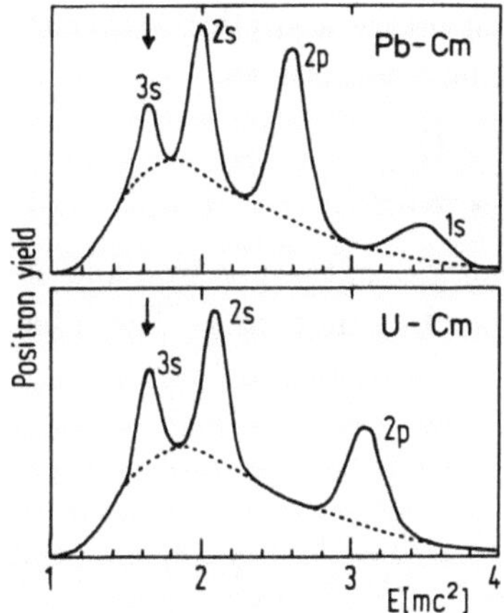

Fig. 13: Schematic positron spectrum for Pb + Cm collisions. The positrons originate from MPC into the K-, L-, and M-shell in a giant atom and from dynamical positron creation processes during Rutherford scattering calculated within the coupled channel formalism (dashed line). The arrow denotes the position of a peak with given intensity. Its cause is MPC into the bound state(s) (a) 3s and 3p, (b) 2s, (c) 2p, and (d) 1s. The Gaussian line shape of width $\Delta E = 0.1\sqrt{E^2 - 1}$ simulates Doppler broadening. The same is shown also for the system U + Cm ($Z_u = 188$). The spontaneous peak is not shown. Its position coincides with the location of the arrow ($E^+ = 1.56$). Its absolute intensity depends on the life-time of the giant nuclear system and is up to a factor of 30 larger than the shown conversion line.

5. CONVERSION PROCESSES IN HEAVY ION FRAGMENTS

In this section we investigate conversion processes in light nuclei with transition energies above the e^+, e^--pair creation threshold within an analytical framework [7]. In particular we evaluate the ratio of electron transition probabilities from the negative energy continuum into the atomic K-shell and into the positive energy continuum, respectively. The possible role of monoenergetic positron conversion with respect to the striking peak structures observed in e^+-spectra from very heavy collision systems is examined.

We study the hypothesis that a neck links the projectile and target nucleus at the distance of closest approach. In the course of separation a light ion fragment could be formed with an excitation energy above the pair production threshold. The nuclear transition may occur by transferring the energy to an electron in the negative energy continuum thus yielding a positron. The final state of the electron could be a vacant bound state, preferably the K-shell, or a positive energy continuum state. The first of these conversion processes is associated with monoenergetic positron emission. As an example the ratio of the probabilities of these two different transition modes is calculated assuming a transition energy of $\omega = 2.6\, m_e c^2$ in a beryllium nucleus (Z = 4). The multipolarities E0, E1, E2, M1 and M2 are considered.

Neglecting nuclear size corrections and nuclear penetration effects one may derive in the limit $Z\alpha \to 0$ an elementary analytical expression for the positron spectrum resulting from electric monopole transitions,

$$\frac{d\eta}{dE} = \frac{pp'(EE'-1)}{2\pi(Z\alpha)^3\sqrt{\omega}(\omega+2)^{3/2}}. \tag{17}$$

Here E and E' denote the energies of the positron and electron (E + E' = ω), whereas p and p' denote the corresponding momenta. The integration over positron energies E yields

$$\eta = \int_1^{\omega-1} dE\, \frac{d\eta}{dE} \tag{18}$$

$$= \frac{\omega+2}{\omega}\left(\frac{\omega-2}{8Z\alpha}\right)^3 [4\,{}_2F_1(-\tfrac{1}{2},\tfrac{1}{2};2;x^2) - x\,{}_2F_1(-\tfrac{1}{2},\tfrac{3}{2};3;x^2)], \tag{19}$$

with the abbreviation

$$x = \frac{\omega-2}{\omega+2}. \tag{20}$$

$\eta_{1s}^{(+)}$ denotes the ratio of the probability for exciting a negative energy continuum electron to a vacant K-shell to the probability for ionizing an electron out of a completely occupied K-shell to the positive energy continuum. We obtain in the limit $Z\alpha \to 0$ the simple result

$$\eta_{1s}^{(+)} = x^{3/2}, \tag{21}$$

independent of Z. For $\omega = 2.6$ and Z = 4 this leads to

$$\frac{\eta}{\eta_{1s}^{(+)}} \simeq \frac{1}{22(Z\alpha)^3} \simeq 2000. \tag{22}$$

In consequence monoenergetic positron E0-conversion is strongly suppressed in light nuclei which is reflected in the factor $(Z\alpha)^{-3}$. In completely stripped low-Z ions nuclear E0-transitions predominantly occur via electron transitions to the positive energy continuum leading to a continuous e^+-spectrum. In principle, two-photon emission is also allowed. The positron distribution $d\eta/dE$ is displayed in figure 14. A broad and flat spectrum (solid line) is obtained being typical for results derived within the Born approximation (Z = 0). For the integrated conversion ratios it follows $\eta \simeq 87$ and $\eta_{1s}^{(+)} \simeq 0.047$. For comparison the dashed line shows the calculated positron spectrum employing relativistic electron wave functions for the Coulomb potential of an extended nucleus. Small Coulomb distortion effects are visible.

Now we turn the discussion to nuclear EL- and ML-transitions with $L \geq 1$. In contrast to electric monopole transitions here single photon emission evidently represents a competing process. The ratio of the electron transition probability from the negative energy continuum into a vacant K-shell and the photon emission probability will be denoted by $\alpha^{(+)}$. Similarly the quantity $\alpha^{(-)}$ determines the ordinary K-shell conversion from the completely occupied bound state to the positive energy continuum. Thus $\alpha^{(+)}$ is correlated with monoenergetic positron production whereas $\alpha^{(-)}$ corresponds to monoenergetic electron emission. Again, in the limit $Z\alpha \to 0$, we can deduce the handy expressions

$$\alpha^{(\pm)}(EL) = \frac{2\alpha(Z\alpha)^3}{L+1}(\omega \mp 2)^{L-1/2}/\omega^{L+5/2}[(L+1)\omega^2 + 4L], \tag{23}$$

$$\alpha^{(\pm)}(ML) = 2\alpha(Z\alpha)^3(\omega \mp 2)^{L+1/2}/\omega^{L+3/2}. \tag{24}$$

$\alpha^{(-)}(ML)$ as well as $\alpha^{(-)}(EL)$, for the limiting cases $\omega \to 0$ and $\omega \to \infty$, agree with the results of ref. 29.

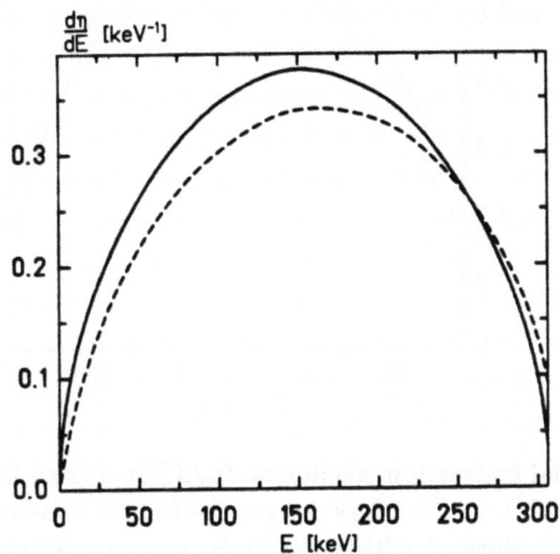

Figure 14: Differential conversion coefficient $d\eta/dE$ in 1/keV for an electric monopole transition in beryllium with $\omega = 1329$ keV versus the kinetic positron energy in keV. Solid line: Analytical result derived in the limit $Z\alpha \to 0$. Dashed line: Numerical result employing relativistic Coulomb wave functions.

Considering continuum-continuum transitions, we evaluated the positron spectrum $d\beta/dE$ resulting from electric dipole transitions. In figure 15 we compare the outcome (solid line) of Born approximation calculations ($Z = 0$) with numerical data employing Coulomb wave functions [30] (dashed line). Again both distributions are very similar and exhibit no striking structures. Table 3 provides the total conversion coefficients β and $\alpha^{(\pm)}$ for various multipolarities.

Table 3: Total conversion coefficients β and $\alpha^{(\pm)}$ for $\omega = 1329$ keV in beryllium and various multipolarities. Asymptotic expressions valid for $Z\alpha \to 0$ are used in the computations.

	E1	E2	M1	M2
β	1.42E-4	3.47E-5	2.34E-5	6.39E-6
$\alpha^{(+)}$	8.69E-8	2.16E-8	1.55E-8	3.58E-9
$\alpha^{(-)}$	2.41E-7	4.58E-7	3.28E-7	5.81E-7

Here the absolute positron production probabilities are suppressed by about four orders of magnitude compared with the allowed photon emission probability. The monoenergetic e^+-production mode is additionally suppressed by about three orders of magnitude. Consequently there should be a strongly enhanced photon peak associated with a possible line structure in the positron distribution.
Based on these intensity arguments we may conclude that the pronounced e^+-peak structure detected in collisions of very heavy ions most likely is not correlated with monoenergetic positron conversion in light ion fragments.

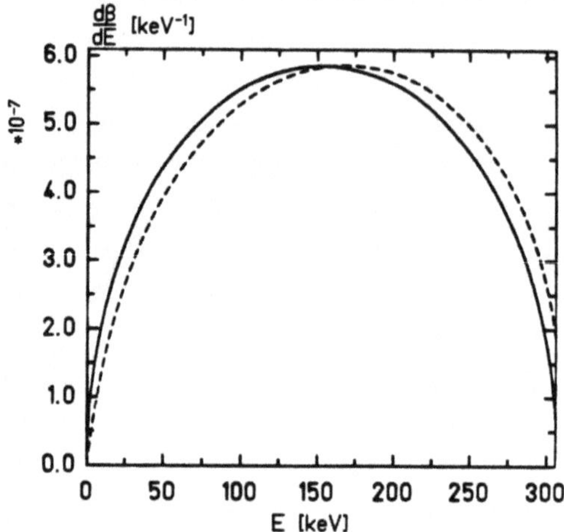

Figure 15: Differential conversion coefficient $d\beta/dE$ in 1/keV for electric dipole transitions in beryllium with $\omega = 1329$ keV versus the kinetic positron energy in keV. Solid line: Analytical result derived within the Born approximation (Z = 0). Dashed line: Result employing relativistic Coulomb wave functions.

6. CONCLUSIONS

We calculated conversion coefficients in the charge range $0 \leq Z \leq 188$. It was demonstrated that measurements of photons and δ-electron spectra serve to determine the role of conversion processes in positron production experiments. Employing the calculated conversion coefficients one may conclude that internal electron-positron pair creation in actinide nuclei cannot account for the narrow peak structures in positron spectra. In addition we can state the monoenergetic pair creation in heavy ion fragments is rather unlikely and, likewise, that conversion processes in giant systems most probable are not responsible for the e^+-peak structures. For the latter conclusion the basic arguments are as follows: a) A rather large rate ($q \simeq 1\%$) for the formation of a giant nuclear system with specific excitation energy compared with elastic scattering events must be assumed to reproduce the measured peak intensity. b) The observed constancy of the e^+-peak energy as function of Z implies the existence of pronounced satellite lines whose energy position depend on Z. c) In supercritical systems we inevitably expect a dominant peak structure related with the spontaneous positron emission. d) There are no explanations for the observed correlated e^+,e^--events.
Future theoretical investigations should be concerned with angular correlations between emitted electrons and positrons in internal pair creation processes. These correlations might contribute as background process in e^+,e^--coincidence measurements.

*Present address: Siemens AG, ZT ZTI SOF 212, Otto-Hahn-Ring 6, D-8000 München 83, West Germany.

REFERENCES

1. P. Schlüter, G. Soff, W. Greiner, Phys. Rep. 75, 327 (1981)
2. G. Soff, P. Schlüter, W. Greiner, Z. Physik A303, 189 (1981)
3. P. Schlüter, G. Soff, W. Greiner, Z. Physik A286, 149 (1978)
4. P. Schlüter, G. Soff, Atomic Data and Nuclear Data Tables 24, 509 (1979)
5. P. Schlüter, T. de Reus, J. Reinhardt, B. Müller, G. Soff, Z. Physik A314, 297 (1983)
6. P. Schlüter, U. Müller, G. Soff, T. de Reus, J. Reinhardt, W. Greiner, Z. Physik A323, 139 (1986)
7. P. Schlüter, G. Soff, W. Greiner, Phys. Rev. C33, 1816 (1986)
8. W. Greiner, B. Müller, J. Rafelski, Quantum Electrodynamics of Strong Fields, (Springer, Berlin, 1985)
9. W. Greiner, ed., Quantum Electrodynamics of Strong Fields, NASI series B80, (Plenum, New York, 1983)
10. G. Mehler, T. de Reus, J. Reinhardt, G. Soff, U. Müller, Z. Physik A320, 355 (1985)
11. M.A. Herath Banda, W. Koenig, B. Martin, F. Güttner, H. Skapa, J. Soltani, K. Dworschak, F. Bosch, Ch. Kozhuharov, A.V. Ramayya, Phys. Rev. A33, 861 (1986)
12. J. Schweppe, A. Gruppe, K. Bethge, H. Bokemeyer, T. Cowan, H. Folger, J.S. Greenberg, H. Grein, S. Ito, R. Schule, D. Schwalm, K.E. Stiebing, N. Trautmann, P. Vincent, M. Waldschmidt, Phys. Rev. Lett. 51, 2261 (1983)
13. M. Clemente, E. Berdermann, P. Kienle, H. Tsertos, W. Wagner, C. Kozhuharov, F. Bosch, and W. Koenig, Phys. Lett. 137B, 41 (1984)
14. T. Cowan, H. Backe, M. Begemann, K. Bethge, H. Bokemeyer, H. Folger, J.S. Greenberg, H. Grein, A. Gruppe, Y. Kido, M. Klüver, D. Schwalm, J. Schweppe, K.E. Stiebing, N. Trautmann, P. Vincent, Phys. Rev. Lett. 54, 1761 (1985)
15. H. Tsertos, E. Berdermann, F. Bosch, M. Clemente, P. Kienle, W. Koenig, C. Kozhuharov, W. Wagner, Phys. Lett. 162B, 273 (1985)
16. P. Kienle, J. Phys. Soc. Jpn. 54 Suppl. II, 549 (1985)
17. T. Cowan, H. Backe, K. Bethge, H. Bokemeyer, H. Folger, J.S. Greenberg, K. Sakaguchi, D. Schwalm, J. Schweppe, K.E. Stiebing, P. Vincent, Phys. Rev. Lett. 56, 444 (1986)

18. A. Schäfer, J. Reinhardt, B. Müller, W. Greiner, G. Soff, J. Phys. G11, L69 (1985)

19. A.B. Balantekin, C. Bottcher, M.R. Strayer, S.J. Lee, Phys. Rev. Lett. 55, 461 (1985)

20. J. Reinhardt, A. Schäfer, B. Müller, W. Greiner, Phys. Rev. C33, 194 (1986)

21. A. Schäfer, W. Greiner, B. Müller, J. Reinhardt, Consequences of a light vector boson in nuclear decays, preprint

22. W.A. Bardeen, R.D. Peccei, T. Yanagida, Constraints on variant axion models, Preprint DESY 86-054

23. R. Eichler, L. Felawka, N. Kraus, C. Niebuhr, H.K. Walter, S. Egli, R. Engfer, Ch. Grab, E.A. Hermes, H.S. Pruys, A. Van der Schaaf, D. Vermeulen, W. Bertl, N. Lordong, U. Bellgardt, G. Otter, T. Kozlowski, J. Martino, Phys. Lett. 175B, 101 (1986)

24. F.W.N. de Boer, K. Abrahams, A. Balande, H. Bokemeyer, R. van Dantzig, J.F.W. Jansen, B. Kotlinski, M.J.A. de Voigt, J. van Klinken, Search for short-lived axions in a nuclear isoscalar transition, preprint 1986

25. R. Lombard, C.F. Perdrisat, J.H. Brunner, Nucl. Phys. A110, 41 (1968)

26. A. Passoja, P. Tikkanen, A. Krasznahorkay, Z. Gácsi, T. Kibédi, T. Fényes, Nucl. Instr. Meth. 223, 96 (1984)

27. A. Passoja, High-resoltion study of E0 internal-pair transitions from excited 0^+ states in ^{58,60,62}Ni, Research Report No. 4/1980, Jyväskylä, Finland

28. G. Soff, W. Greiner, J. Phys. B15, L681 (1982)

29. S.M. Dancoff, P. Morrison, Phys. Rev. 55, 122 (1939)

30. H.-J. Bär, G. Soff, Physica 128C, 225 (1985)

ATOMIC CLOCK PHENOMENA IN COLLISIONS OF VERY HEAVY IONS

G. Soff, T. de Reus, U. Müller, J. Reinhardt, B. Müller, and W. Greiner

Institut für Theoretische Physik, Johann Wolfgang Goethe-Universität, D-6000 Frankfurt am Main, West Germany

Gesellschaft für Schwerionenforschung (GSI), Planckstraße 1 Postfach 110 541, D-6100 Darmstadt, West Germany

1. INTRODUCTION

Peculiar properties of electron and positron states in superheavy systems may be utilized to deduce informations on the time-scale of nuclear reactions[1-17,64-66]. One major subject of this report deals with spectra of electrons which are emitted during heavy-ion collisions, traditionally denoted as δ-electrons. The slope in the kinetic energy distribution of ejected δ-electrons provides a finger-print of the occuring nuclear delay and deceleration times[3,5,64,65] of the underlying deep-inelastic collisions on the time scale of 10^{-21} s. Theoretical expectations will be confronted with experimental observations[14-17]. In addition a theoretical scaling law[12] is derived which may serve as a tool to determine the nuclear stopping time in intermediate-energy collisions. Furthermore we investigate the influence of a time delay caused by a nuclear reaction on the production yield of K-shell vacancies[4,6,10,16,17] and on the quasimolecular X-ray spectrum[7,8,18] in superheavy collision systems. Finally we analyze the advantages of positron production in crossed beams of completely stripped ions.

Over the last decade our knowledge on atomic structure of superheavy quasimolecules in the range $110 \leq Z_{tot} \leq 188$ has increased considerably[13,19-43]. Heavy-ion collisions, in which superheavy quasimolecules are formed for a short period of time, offer us a unique tool to investigate the electronic structure of ultra-high Z-systems, which are not otherwise accessible to experiment. Comparison of K-vacancy formation, δ-electron and positron emission with available experimental data suggests the validity of the quasimolecular picture, which will be taken as the theoretical framework of our calculations.

One of the pivotal points still under discussion in this context is the question, whether in systems with $Z_{tot} \geq 174$ a vacant 1sσ-state dives into the lower continuum and subsequently decays into adjacent positron states[20,27-29]. In case of nuclear time delay this process should cause a pronounced structure in the positron spectra.

Besides this exciting phenomenon of vacuum decay into a new twofold negatively charged stable vacuum ground state, electron excitation in heavy-ion collisions may be employed for the determination of delay and deceleration times on the nuclear time scale, i.e. offering an atomic clock, operating in the range 10^{-21} - 10^{-24} s. In deep-inelastic heavy-ion collisions this provides a test for classical nuclear reaction models[44,45]. In collisions at intermediate energies an independent measurement of the deceleration time is of interest for comparison, e.g., with the results of the pion bremsstrahlung model[46,47].

After a presentation of the underlying theoretical framework we will examine the possibility to use δ-electron emission as clock hand on the nuclear time scale from the early concepts of an atomic clock in deep-inelastic heavy-ion collisions towards recent predictions for intermediate energy collisions.

Our considerations will be restricted solely to the interplay of atomic and nuclear physics in superheavy collision systems. Many exciting experiments were performed also for light-ion collisions. The corresponding physical outcome is summarized in the excellent review articles by W. Meyerhof[1] and U. Heinz[2].

2. THEORETICAL FRAMEWORK

In collisions of very heavy ions, transient quasiatoms of charge number $Z_{tot} = Z_p + Z_t > 1/\alpha$ are formed, whose 1sσ-electron binding energies are as large as or even larger than the electron rest mass. This obviously necessitates a fully relativistic treatment of the electron motion. The colliding nuclei, in contrast, are assumed to follow classical non-relativistic trajectories, i.e. Rutherford hyberbolae, for collisions below the Coulomb barrier of the two nuclei.

At smaller internucler distances the most strongly bound electrons move in the combined field of both nuclei at each instant during the course of the collision. In the adiabatic picture electronic wave functions and binding energies are obtained by the stationary solutions of the two-centre Dirac equation[48,49] with $\hbar = c = 1$:

$$[\vec{\alpha} \cdot \vec{p} + \beta m_e + eV_{TC}(\vec{r},\vec{R}) - E_n(\vec{R}(t))]\ \varphi_n(\vec{r},\vec{R}) = 0. \tag{1}$$

Here eV_{TC} is the combined Coulomb potential of the colliding nuclei acting on the considered electron with label n. In the one-electron approximation eV_{TC} is the potential of two extended nuclei with charge densities $\rho_1(\vec{r_1})$ and $\rho_2(\vec{r_2})$

$$V_{TC}(\vec{r},\vec{R}) = -\int \rho_1(\vec{r}_1{}')/|\vec{r}-\vec{r}_1{}'|\, d^3r_1{}' - \int \rho_2(\vec{r}_2{}')/|\vec{r}-\vec{r}_2{}'|\, d^3r_2{}'. \tag{2}$$

We solve equation (1) by performing a multipole expansion of equation (2) and restricting ourselves to the dominant monopole ($\ell=0$) term. For two homogeneously charged spheres with radii a_i and charges Z_i the monopole potential reads in the centre of charge system[11]

$$V_o(r,R) = \Sigma_{i=1}^{2} -Z_ie/S_i \qquad \text{for } r \le S_i - a_i$$

$$V_o(r,R) = \Sigma_{i=1}^{2} -3Z_ie/(2a_i^3)\,[a_i^3(1/r + 1/S_i)/3 + (S_i^3/r + r^3/S_i)/24 - a_i^2(S_i/r + r/S_i)/4 - (r^2 + S_i^2)/6 + a_i^2/2 + rS_i/4 - a_i^4/(8rS_i)] \qquad \text{for } S_i - a_i < r < S_i + a_i \tag{3}$$

$$V_o(r,R) = \Sigma_{i=1}^{2} -Z_ie/r \qquad \text{for } r \ge S_i + a_i$$

S_i are the distances of the individual nuclei from the centre of charge. Equation (3) describes the potential of two extended bare nuclei separated by the internuclear distance R in the monopole approximation. Since binding energies and matrix elements play a substantial role in all calculations, electron screening must be included in the solutions of the stationary Dirac equation to obtain quantitatively reliable results. In particular this is true for supercritical collisions, where the 1sσ-state becomes bound by more than $2m_ec^2$. Another motivation derives from the availability of improved experimental data. In our present calculations we use a monopole Hartree-Fock-Slater (HFS) potential, both for the determination of the wave functions and their eigenenergies. These wave functions and energies are then used for dynamical calculations. Due to our restriction to the $\ell = 0$ term for the two-centre potential we also restrict the screening potential to be of monopole type[24].

In the framework of the Hartree-Fock approximation, the dynamical evolution of the electrons in the time varying Coulomb field of two colliding nuclei is described by the solutions Φ_j of the time-dependent two-centre Dirac equation:

$$i\,\partial/\partial t\ \Phi_j(R(t)) = H_{TCD}(R(t))\ \Phi_j(R(t)). \tag{4}$$

Since the relativistic two-centre Hamiltonian $H_{TCD}(R(t))$ depends sensitively on the internuclear separation R(t), we expand the total wave function Φ_j into Born-Oppenheimer states, represented by the stationary molecular eigenstates of the Hamiltonian,

$$\Phi_j(t) = \Sigma_k\ a_{jk}(t)\ \varphi_k(R(t))\ \exp\{-i\chi_k(t)\}, \tag{5}$$

with

$$\chi_k(t) = \int^t E_k(t')\, dt'.$$

The sum in equation (5) includes an integration over continuum states with respect to positive and negative frequencies. Inserting the expansion of equation (5) into equation (4) and projecting with stationary eigenfunctions of equation (1), a set of first-order coupled differential equations for the occupation amplitudes $a_{jk}(t)$ is obtained[33].

$$\dot{a}_{ij}(t) = - \Sigma_{k \neq j}\, a_{ik}(t)\, \langle \varphi_j | \partial / \partial t\, | \varphi_k \rangle \, \exp\{i(\chi_j - \chi_k)\}. \tag{6}$$

The initial condition is $a_{ij}(-\infty) = \delta_{ij}$. Dividing the time-derivative operator into a radial part and a rotational coupling and neglecting the latter, the coupled equations (6) are solved by numerical integration. In the independent-particle approximation excitations of many-electron systems are described by incoherent summation over one-electron transition probabilities. After the collision the number of particles p occupying a state above the Fermi level, up to which the quasimolecular levels were initially filled, is given by[33]

$$N_p = \Sigma_{r<F}\, |a_{rp}(\infty)|^2 \qquad (p > F), \tag{7}$$

while the number of holes in a state q below the Fermi level is given by

$$N_q = \Sigma_{r>F}\, |a_{rq}(\infty)|^2 \qquad (q < F). \tag{8}$$

A positron is represented as a hole in the negative energy continuum. For the number of correlated particle-hole pairs N_{pq} one finds

$$N_{pq} = N_p\, N_q + |\Sigma_{r<F}\, a^*_{rp}\, a_{rq}|^2. \tag{9}$$

This formula should be applied to analyse experiments, where coincidences between electrons and 1s-vacancies are measured. To evaluate particle-particle or hole-hole correlations, the sign of the second term in equation (9) has to be inverted. If K-holes are measured in coincidence with emitted electrons a problem arises from accidental coincidences, since no distinction is made between $E_{s\sigma}$ and $E_{p1/2\sigma}$ δ-electrons, i.e., their parity is not experimentally determined. Taking this into account and considering different possible spin directions, which were neglected until now, the total number of particle-hole pairs is found to be

$$P^{tot}_{E,1s} = 2\, |\Sigma_{r<F}\, a^*_{r,Es\sigma}\, a_{r,1s\sigma}|^2 + 4\, (N_{Es\sigma} + N_{Ep1/2\sigma})\, N_{1s\sigma}, \tag{10}$$

where the last term describes accidental coincidences. Equation (10) is valid only for sufficiently asymmetric systems, since otherwise a supplementary term due to vacancy sharing has to be added, describing accidental coincidences between $2p_{1/2}\sigma$-vacancies and electrons. In this case we obtain:

$$P_{pq}^{tot} = P_{E,1s\sigma} + P_{E,2p1/2\sigma} \tag{11}$$

with

$$P_{E,1s\sigma} = 2\,(N_{Es\sigma,\,1s\sigma} + N_{Es\sigma}N_{1s\sigma}) + 4N_{Ep1/2\sigma}N_{1s\sigma}, \tag{12}$$

$$P_{E,2p1/2\sigma} = 2\,(N_{Ep1/2\sigma\,,2p1/2\sigma} + N_{Ep1/2\sigma}N_{2p1/2\sigma}) + 4N_{Es\sigma}N_{2p1/2\sigma}. \tag{13}$$

In this case it becomes difficult to distinguish K-holes from the heavier collision partner from those of the lighter collision partner. Due to vacancy sharing a quasimolecular $2p_{1/2}\sigma$-hole can cause an atomic K-vacancy (K_H) in the heavier collision partner in the asymptotics $R \to \infty$. In the same way a $1s\sigma$-vacancy can cause atomic K-vacancies (K_L) in the lighter collision partner. The subscript pq stands for particle-hole coincidences. For only slightly asymmetric systems it both terms in equation (11) have to be weighted according to:

$$N_{pq_H} = (1 - w)\ N_{E,1s\sigma} + w\ N_{E,2p1/2\sigma} \tag{14}$$

$$N_{pq_L} = w\ N_{E,1s\sigma} + (1 - w)\ N_{E,2p1/2\sigma} \tag{15}$$

The vacancy sharing factor is calculated according to Meyerhof[50]:

$$w/(1 - w) = \exp\{-2x\}, \tag{16}$$

where

$$2x = \sqrt{2}\,\pi\,(\sqrt{E_H} - \sqrt{E_L})/\sqrt{m_e}v_p. \tag{17}$$

E_H and E_L denote the K-shell ionization energies of the heavier and the lighter collision partner, resp., v_p the velocity of the projectile. In the following we deal with the quantities P which include a summation over possible spin orientations instead of N for the number of particles and holes in a particular state. Thus $P_{2p1/2\sigma} = 2N_{2p1/2\sigma}$, e.g., can reach a maximum value of 2 and therefore is no probability in the conventional sense. In order to demonstrate the agreement between theory and experiment in collisions below the Coulomb barrier, we discuss δ-electron emission in I+Pb collisions. The probabilities for $s\sigma$ and $p_{1/2}\sigma$ δ-electrons have been added incoherently to obtain the total spectrum. Figure 1 displays the double differential cross section, with respect to the kinetic energy and solid angle of the electron, for the emission of δ-electrons versus kinetic δ-electron energy in the system I + Pb at E_{lab} = 500 MeV, measured by Herath-Banda et al[42]. The notation F = 3sσ, $4p_{1/2}\sigma$ for the Fermi surface indicates that only the levels 1sσ - 3sσ and $2p_{1/2}\sigma$ - $4p_{1/2}\sigma$ are assumed to be occupied initially. The full lines indicate coupled channel HFS calculations.

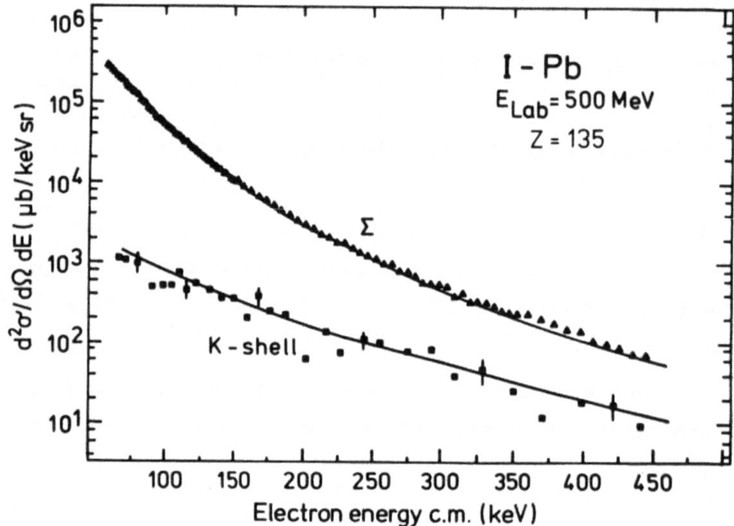

Fig. 1: Delta-electron emission in the system I + Pb at a beam energy of E_{lab} = 500 MeV. The experimental data were measured by Herath-Banda et al[42]. The full line displays coupled-channel HFS calculations for the total electron spectrum as well as for 1sσ-coincident rates (lower line). The Fermi level was chosen to be F = 3sσ, $4p_{1/2}\sigma$.

The overal agreement between the measurements of Herath-Banda and coworkers[42] and coupled channel HFS results is very satisfying, since our calculations yield parameter-free absolute values.

3. THE CONCEPT OF AN ATOMIC CLOCK IN DEEP-INELASTIC COLLISIONS

One of the central questions in deep-inelastic heavy ion scattering is the determination of the reaction time of superheavy nuclear composite systems. Diffusion models and fragmentation theory have come into widespread use to describe mass transfer and energy dissipation in the considered reactions. In these models, the experimentally observed widths of the fragment masses, charges, angular momenta and energy distributions are in first approximation proportional to the available reaction time T. Based on theoretical models, typical values of the order of $T \sim 10^{-21} - 10^{-20}$ s are deduced from the experimental data. For more relaxed collisions the assumptions of a linear correspondence between variance and time becomes questionable. This adds to the wish for having an independent and precise 'clock' that can be used to measure the time scale in deep-inelastic nuclear reactions[3-5]. In the following we discuss the measurement of T by means of the kinetic energy distribution of emitted δ-electrons. The latter is of exponential form, the steepness being a function of the combined nuclear charge, $Z_{tot} = Z_p + Z_t$, and the minimal distance of approach of the two nuclei, R_{min}. For a typical centre-of-mass energy of 3 MeV/u a minimal distance $R_{min} \sim 15$ fm corresponds to a time scale of the order of 10^{-21} s. This indicates that the change of the time structure due to nuclear reactions may produce observable effects in the ionization process.

We will show that the nuclear sticking time leads to a phase shift in the ionization amplitude which finally can produce oscillations in the energy distribution of emitted δ-electrons[3,5]. The principal mechanism, how a nuclear time delay influences the excitation amplitudes becomes most transparent in perturbation theory.

It is well known that this perturbative treatment correctly reproduces the dependence of 1sσ-ionization on kinematical parameters. Only the absolute magnitude of the ionization probability is found to be too small by a factor 3 - 5 compared with results from coupled channel calculations[33] and with experimental data. The high-momentum component of the ionization, which is of interest here, is adequately described by the monopole approximation for the wave functions. In perturbation theory ($a_{ii} \cong 1$) equation (6) is solved explicitly[25] by:

$$a_{ij}(t) = -\int_{-\infty}^{t} dt' \, \langle\varphi_j|\partial/\partial t'|\varphi_i\rangle \, \exp\{\, i(\chi_j(t') - \chi_i(t'))\}. \tag{18}$$

We neglect any effect of the internal nuclear structure on the electrons and also the influence of energy and angular momentum dissipation on the shape of the scattering trajectory, but introduce a time delay T at the distance of closest approach. During this time we assume that the radial nuclear charge distribution remains unchanged, i.e. that the two nuclei form a composite system of fixed shape which may, however, rotate. In first-order perturbation theory we can write the modified excitation amplitudes as[5]

$$a_{ij}(\infty) = a_{ij}(0) - \exp\{-i\Delta\Omega\}\, a_{ij}^{*}(0). \tag{19}$$

Here $a_{ij}(0)$ is the amplitude after the incoming branch of the trajectory at $t = 0$ and $\Delta\Omega = (E_j - E_i)T/\hbar$ denotes the phase shift induced by the nuclear sticking time. The corresponding transition probability is

$$P_{ij} = |a_{ij}(\infty)|^2 = 4|\mathrm{Re}\, a_{ij}(0) \sin(\Delta\Omega/2) + \mathrm{Im}\, a_{ij}(0) \cos(\Delta\Omega/2)|^2. \tag{20}$$

The interference between incoming and outgoing path of the trajectory therefore produces oscillations with a width[3,5]

$$\Delta E = 2\pi\hbar/T. \tag{21}$$

For an explicit calculation we quote results for the system $^{136}Xe + ^{208}Pb$ with E_{Lab} = 7 MeV/u, because it is asymmetric enough to allow for measurements in coincidence with a K-X-ray from the target-like fragment. Our conclusions hold qualitatively for any other superheavy system. The resulting δ-electron spectrum[5] from direct 1sσ-ionization, calculated in first-order perturbation theory, is shown in Fig. 2 for T = 0, $3 \cdot 10^{-21}$ s and $1 \cdot 10^{-20}$ s.

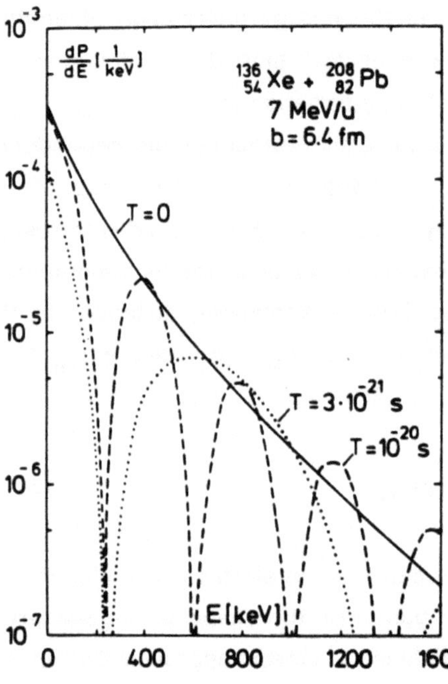

Fig. 2: Differential 1sσ-ionization probability (normalized to 2) in the system Xe + Pb at 7 MeV/u with respect to kinetic δ-electron energy for different nuclear delay times, calculated in first-order perturbation theory[5]. The impact parameter b = 6.4 fm corresponds to a grazing collision, i.e. $b \cong b_{gr}$.

The spectra reflect the one to one correspondence between T and the energy distance ΔE between two neighbouring maxima. The question now arises, how much of the oscillatory structure in the spectrum is smeared out by multi-step excitation processes. A glance at Fig. 3 settles this point[3].

The same system was evaluated using the coupled channel code and assuming a Fermi level of F = 3sσ, 4p½σ. We find that the interference pattern is slightly washed out, but there remains a maximum-minimum ratio of about 5 to 1. Hence we conclude that a measurement of the δ-electron spectra in coincidence with K-vacancy formation might allow the determination of a nuclear time delay with fixed value T.

One should keep in mind, however, that in an actual experiment the nuclear reaction is characterized by a distribution function of reaction times f(T). Oscillations in the δ-electron spectrum will survive averaging over f(T) only if the distribution is centred around a finite mean value and is sufficiently narrow.

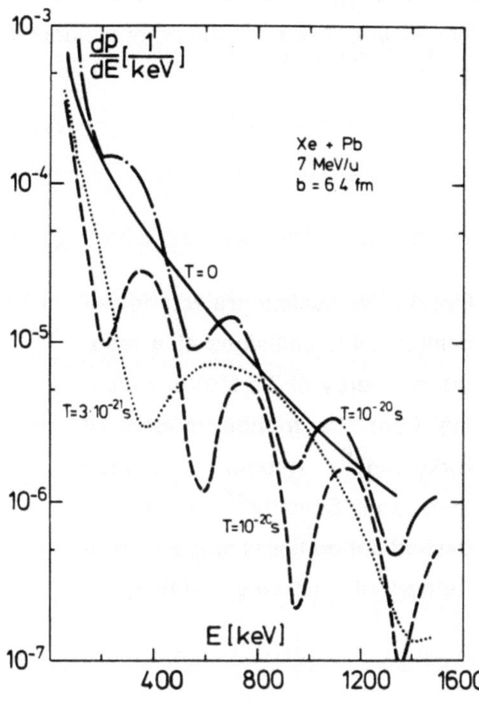

Fig. 3: Delta electron spectrum evaluated using the full coupled channel code[3] for the same system and kinematical parameters as in Fig. 2. The dashed curve which displays δ-electron emission in coincidence with 1sσ-vacancies exhibits more pronounced oscillations in comparison with the dashed-dotted curve without the coincidence condition.

4. COMPARISON WITH EXPERIMENTAL DATA

Instead of using the schematic model discussed in the last section, it is more convincing to adopt trajectories calculated from a nuclear model which is consistent with the elastic and inelastic heavy-ion scattering data. Deep-inelastic reactions have been discussed in terms of many models with different degree of refinement.

In the following we will employ two different macroscopic friction models for the nuclear motion in U+U collisions. The first model, proposed by Birkelund and coworkers[45] (furtheron denoted as model I), is based on the proximity nuclear potential of Blocki and the one-body nuclear friction in the proximity formalism of Randrup. The model of Birkelund incorporates nuclear intrinsic rotation and has a set of dynamical variables $\{R, P, \Theta_t, \ell_t, \Theta_p, \ell_p\}$, i.e. the internuclear distance and the orientation angles of the individual nuclei and their corresponding conjugate momenta, for which the classical equations of motion are solved. As an alternative we employ the nuclear trajectories of Schmidt and coworkers[44] who have proposed a macroscopic friction model (model II), which in a simple way accounts for neck formation in the separating system. Thereby one is able to explain the experimentally observed high energy loss, where up to ~ 30% energy dissipation for b ~ 0 can be achieved.

Strong deviations from a Coulomb trajectory and, at the same time, increased reaction times are found in both models. This is demonstrated in Fig. 4 for the case of head-on collisions (b = 0) of U+U at a bombarding energy of E_{Lab} = 7.5 MeV/u. Here model II predicts delay times up to $\Delta T \sim 1.1 \cdot 10^{-21}$ s defined with respect to the point of nuclear separation.

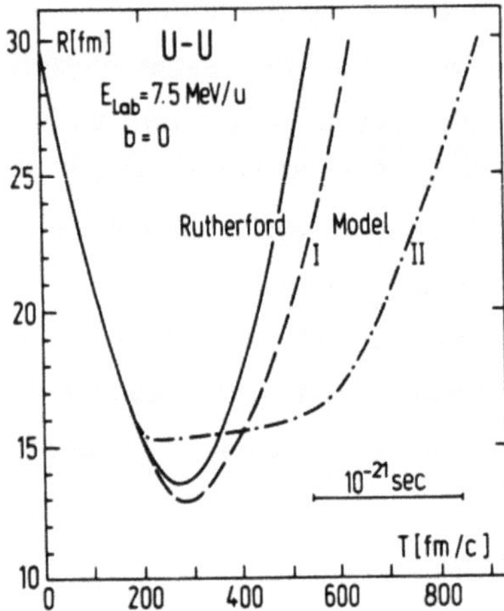

Fig. 4: The nuclear trajectories of central U+U collisions at a laboratory energy of 7.5 MeV/u resulting from the friction models of Birkelund[45] (model I, dashed line) and Schmidt[44] (model II, dashed-dotted line) compared with Rutherford scattering (full line).

The spectra of emitted δ-electrons for model I and II decrease considerably in their high-energy part compared to spectra calculated for Rutherford scattering and their fall-off is steeper[10]. All these effects are established best for the innermost electrons which are extremely sensitive to the nuclear charge configuration and thus can be strongly influenced by the modified nuclear kinetics. Recently experimental data have been published[14,15] concerning the energy spectra of δ-electrons and positrons emitted in U+U and U+Cm collisions above the Coulomb barrier. To obtain a signature for close contact, the atomic excitations have been measured in coincidence with fission fragments, being detected in a laboratory angular window of $\theta_{Lab} = 40^\circ \pm 5^\circ$. For the bombarding energy of E_{Lab} = 5.9 MeV/u an angular window of $\theta_{Lab} = 45^\circ \pm 10^\circ$ was chosen.

For a quantitative comparison with the experiment one has to integrate the theoretical impact parameter-dependent spectra over all values of b leading to a nuclear reaction, weighted by the corresponding probability $w_f(b)$ to induce nuclear fission

$$dP_{e^\pm}/dE_{e^\pm} = \int b\, db\, [\, dP_{e^\pm}(b)/dE_{e^\pm}\,]\; w_f(b) \,/ \int b\, db\, w_f(b). \qquad (22)$$

The experimental analysis yields a quarter point angle of $\Theta_{1/4} = 87.5^{\circ} \pm 2^{\circ}$, which corresponds to a classical distance of closest approach of 16.85 fm, fitting nicely into the systematics of strong absorption radii given by Birkelund and others. Although a procedure to determine $w_f(b)$ from these data is not unique, due to contributions from forward and backward scattering, the final results depend only slightly on the estimates used. In consequence δ-electron and positron spectra change only slightly within the experimentally given boundaries.
Figure 5 shows the experimental data[14] for δ-electron emission in 8.4 MeV/u U+U collisions in comparison with theoretical results including electron shielding[10,11].

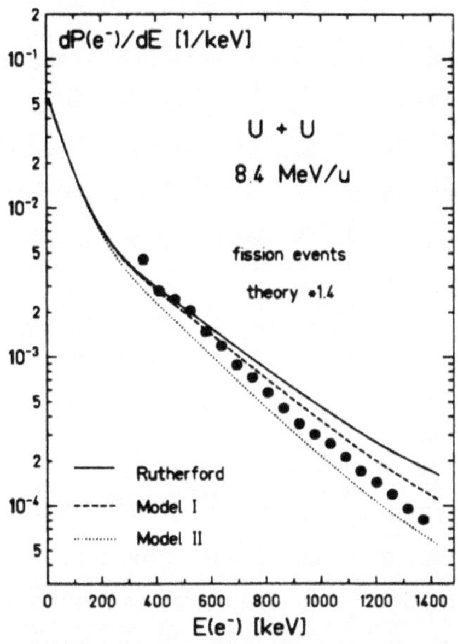

Fig. 5: Spectrum of δ-electrons emitted in 8.4 MeV/u U+U collisions measured by Backe and coworkers[14] in coincidence with nuclear fission residues. The experimental data are compared with calculations assuming Coulomb scattering (full lines) and friction model trajectories. The calculated spectra are shifted up by a factor 1.4.

Delta-electron spectra at 8.4 and also at 10 MeV/u bombarding energy taken in coincidence with fission products (i.e., following a nuclear reaction) fall off notably steeper than the assumption of a Rutherford trajectory would predict. For comparison theoretical probabilities for δ-electron emission are shown where even for small impact parameters pure Rutherford scattering is assumed, thus simulating 'transparent' nuclear matter, followed by a nuclear fission process (solid line). The dashed lines display spectra calculated with the modified trajectories of reaction model I from Birkelund[45], whereas the dotted line represents calculations for model trajectories of Schmidt[44] (model II). The theoretical probabilities have been scaled by a factor 1.4.
In all calculations the trajectories of a binary system have been used, assuming that the fission process, which is delayed by $\sim 10^{-20}$ s, does not severely modify the high-energy δ-electron emission. The slope of δ-electron emission probabilities is reproduced best by calculations based on the trajectories of model II. Analogous results have been obtained for a bombarding energy of $E_{Lab} = 7.5$ MeV/u. Again the δ-electron spectrum calculated for Rutherford scattering has a slope which is too flat to fit the experimental data of Backe and coworkers. Predictions using model I miss both

slope and absolute value. Model II leads to δ-electron spectra having the correct slope, but also a scaling factor is needed to fit the absolute values of the experimental data. However, the steepness of the slope offers an appropriate tool for selecting the most favourable reaction model. The experimental data are best described when using model II of Schmidt and coworkers[44] yielding a delay time of about $2 \bullet 10^{-21}$ s, whereas model I underestimates the time delay ($\sim 4 \bullet 10^{-22}$ s).

5. K-VACANCY FORMATION IN DEEP-INELASTIC HEAVY-ION COLLISIONS

Now we turn the discussion to the influence of a nuclear reaction time on the K-vacancy production yield in superheavy quasimolecules. The corresponding experimental research was motivated by a proposal of R. Anholt[4]. Those electrons, which are excited from the quasimolecular 1sσ-state during a collision of two sufficiently asymmetric nuclei, cause the emergence of K-holes in the heavier collision partner after separation of the nuclei. In case of symmetric collisions as ,e.g., U+U however, it is impossible to attach the resulting vacancies to one of the collision partners due to vacancy sharing[50]. Thus the rates for quasimolecular 1sσ- and $2p_{1/2}\sigma$-vacancies have to be added so that the ionization probabilities are normalized to 4.

Assuming a schematic model[10], where a fixed time delay T at a constant internuclear distance R is sandwiched between the two symmetric branches of a Rutherford hyperbola, vacancy formation P_K is found to oscillate as function of the time delay. An evaluation of K-vacancies using classical friction models for the trajectory as mentioned in the preceding section, however, seems to be more adequate and will be discussed now.

Figure 6 displays the dependence of K-vacancy formation on the impact parameter b ($100\hbar$ = 1.4 fm) for a U+U collision at 7.5 MeV/u bombarding energy.

In the case of Rutherford scattering $P_K(b)$ varies within 10% for impact parameters $b \leq 10$ fm. Here vacancy formation increases with decreasing impact parameter. Using the trajectory of model II a different behaviour shows up: the K-hole yield also increases with decreasing impact parameter as long as $b \geq 6$ fm. At smaller impact parameters, however, a drastic decrease down to $P_K \sim .6$ for $b = 0$ is observed.

The onset of deviations from Rutherford scattering starts earlier in model I of Birkelund and coworkers at $b \sim 7.5$ fm. Here vacancy formation also decreases for $b < 7.5$ fm, but exhibits a minimum around $\ell \sim 300\,\hbar$. At even smaller values for the impact parameter, $P_K(b)$ increases again, similarly to Rutherford scattering.

In Fig. 7 we compare theoretical K-hole production in a 7.5 MeV/u U+U collision with experimental data of Stoller and coworkers[16,17] measured in an angular window $\Theta_{Lab} = 45^{\circ} \pm 13^{\circ}$ as function of the total kinetic energy loss. The Q-value as function of the impact parameter is given theoretically, dependent on the underlying model, so that $P_K(b)$ can be uniquely associated to $P_K(Q)$.

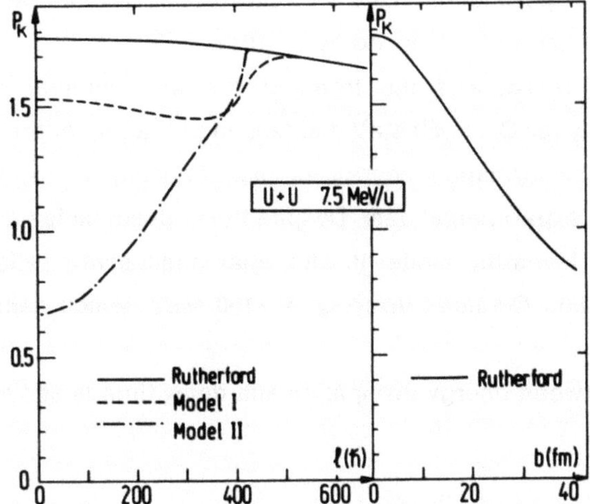

Fig. 6. K-vacancy rates in U+U collisions at E_{Lab} = 7.5 MeV/u versus angular momentum is displayed in the left part of the figure, and as function of the impact parameter on the right ($100\hbar$ = 1.4 fm). The full lines represent vacancy formation obtained in case of Rutherford scattering, the dashed curve results when using the nuclear friction model I of Birkelund and coworkers[45] and the dashed-dotted curve applies for model II of Schmidt and others[44].

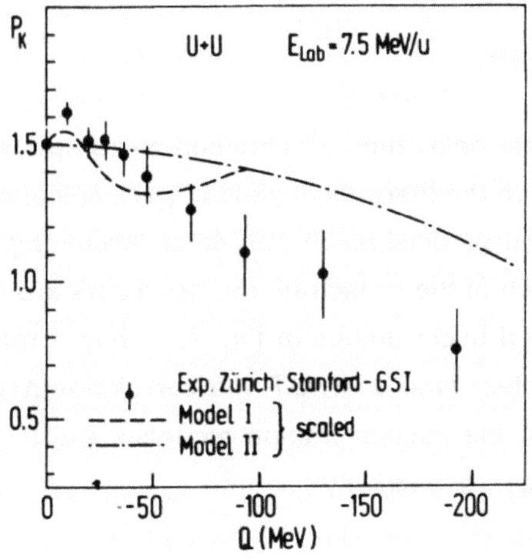

Fig. 7: The dependence of K-vacancy production rates as function of the Q-value in 7.5 MeV/u deep-inelastic U+U collisions. Experimental data[16,17] are compared with coupled channel results[10,11] for model I and model II.

In the experimental setup those events were selected, where deep-inelastic reactions with a definite Q-value still yielded two U-like stable fragments. $P_K(Q=0)$ represents the integrated (from $\Theta_{Lab} = 32^\circ$ to 58°) ionization rates from elastic and quasi-elastic collisions. To fit the experimental value for $P_K(Q=0) \cong 1.5$ the vacancy rates resulting from model I and II had to be reduced by ~ 10%.
Whereas model I reproduces the trend of the experimental data of Stoller and coworkers[16,17] only for Q ≥ -50 MeV and fails in predicting energy dissipation beyond ~ 85 MeV, vacancy formation decreases monotonically in model II, however, still overestimating the experimental data. Despite these deficiencies qualitative agreement is achieved, again favouring model II. The latter qualitatively reflects the decrease of $P_K(Q)$ with increasing Q-values up to Q = -190 MeV, which cannot be described in model I.
The correlation between energy dissipation and delay time is shown in Fig. 8.

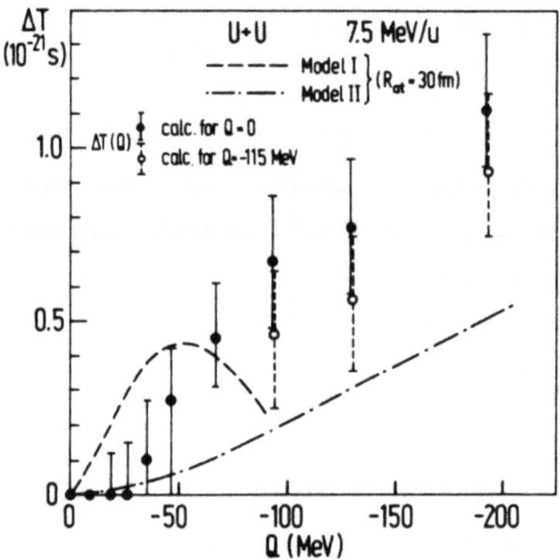

Fig. 8: Correlation of energy loss Q with nuclear delay time ΔT obtained by comparison of the experimental data displayed in Fig. 7 with the theoretical yields $P_K(T)$, calculated within the schematic model and neglecting energy dissipation (full dots). Assuming an energy loss of 115 MeV on the outgoing branch of the trajectory, the open dots are obtained. The error bars reflect the experimental uncertainties in Fig. 7. Using a reference distance $R_{at} = 30$ fm to define the nuclear time delay ΔT, the correlation ΔT(Q) can also be obtained from the calculations for the nuclear friction models I and II, represented by dashed and dashed-dotted curves, respectively.

The experimental data of Stoller and coworkers[16,17] (full dots) of $P_K(Q)$ were correlated to $P_K(T)$ obtained within a schematic model (sharp defined delay time T, during which the internuclear distance R was kept constant) in order to relate Q to a delay time

ΔT. Within this schematic model the incoming and outgoing branches of the Rutherford trajectory are assumed to be symmetric, i.e., no energy loss occurs. Introduction of energy dissipation ($Q = -115$ MeV) on the outgoing branch of the trajectory destroys this symmetry and leads to modified results (open dots).

A procedure yielding $\Delta T(Q)$ for the friction model trajectories demands to choose arbitrarily a reference distance R_{at}, at which ΔT is defined with respect to Rutherford scattering. The energy loss Q is determined uniquely by the model calculations. Choosing $R_{at} = 30$ fm yields the dashed and dashed-dotted curves in Fig. 8 for model I and model II, respectively.

However, the theoretical values for the nuclear time delay ΔT displayed in Fig. 8 should be interpreted with care due to the arbitrariness in the choice of R_{at}. Furthermore the correlation between $P_K(Q)$ and $P_K(T)$ by means of the schematic model certainly is an oversimplification. But at least the trend of increased delay times for increasingly dissipative collisions is clearly visible.

6. QUASIMOLECULAR X-RAY RADIATION IN PB + PB COLLISIONS WITH NUCLEAR CONTACT

Another atomic process in which nuclear reactions are expected to lead to measurable effects is quasimolecular X-radiation[7,8] (MO X-rays). For a review article we refer to ref. 18. In experiments utilizing pre-stripped and decelerated ions and gas targets, Tserruya, Schmidt-Böcking, Schuch and collaborators[51,52] have succeeded in observing the interference between X-ray emission during approach and separation of the two nuclei. Minima in the X-ray spectrum occur whenever the phase

$$\Phi_0(E_x) = \int_{R_{min}}^{R_0(E_x)} (\Delta E(R) - E_x)\, dR \,/\, v(R) \qquad (23)$$

where $\Delta E(R)$ is the transition energy in the quasimolecule, E_x the X-ray energy at R_0 with $E_x = \Delta E(R_0)$ and $v(R)$ the radial velocity of the nuclei, takes on multiples of π. The dependence of Φ_0 on E_x can be used to map out the functional dependence of the quasimolecular transition energy on nuclear separation R.

More interesting for our purpose is the observation that a time-delay T due to a nuclear reaction introduces an additional phase shift

$$\Delta\Phi(E_x,T) = (\Delta E(R_{min}) - E_x)\, T \qquad (24)$$

which causes modifications of the interference pattern.

Several calculations have been performed by J. Kirsch et al[7] for the Pb + Pb system with a projectile energy close to the nuclear Coulomb barrier, i.e. $E_{lab} = 5.8$ MeV/u. At first we compare two calculations where a K-hole is brought into the collision and the

L- and M-shells are initially filled. In one calculation the electronic transition amplitudes are taken to be constant, in the other calculation the amplitudes are treated dynamically. Figure 9 shows the result[7]. Obviously, no major difference can be seen. In particular, the minima at 150 keV and 275 keV are not filled up, i.e. the interference structure is not smeared out by the collision dynamics.

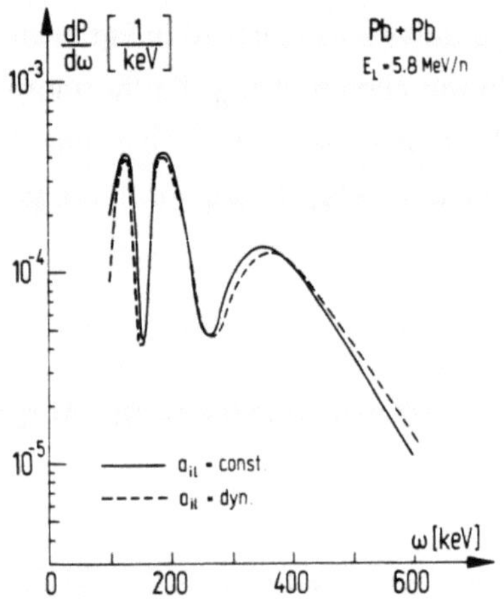

Fig. 9: The differential Pb + Pb radiation spectrum. The full curve represents a calculation[7] with constant amplitudes a_{ij}; dashed curve: the corresponding calculation with dynamic a_{ij}.

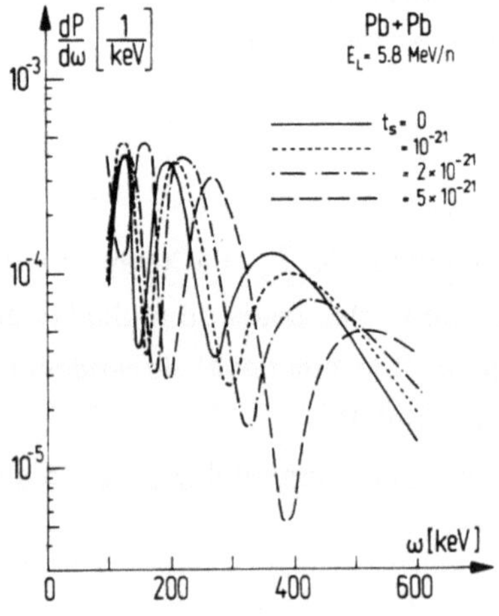

Fig. 10: The differential MO radiation cross-section for several nuclear sticking times[7]. The occupation amplitudes are dynamically calculated with a single K-vacancy present initially.

We now come to the main goal of the calculations, viz. to what happens to the minima if the path of the projectile is delayed due to nuclear reactions with the target. In Fig. 10 the results[7] for several nuclear sticking times are shown. The minimum at 150 keV as well as the minimum at 275 keV are shifted to higher X-ray energies. In first approximation the shift is proportional to the length of the nuclear contact time.

For the Pb + Pb system at a bombarding energy of 5.8 MeV/u it was found that $dE_x^0/dT \sim 25$ keV/10^{-21} s for the interference minimum at the highest energy, and about 10 keV/10^{-21} s for the next lower minimum.

In conclusion we may state that the energy shift of the minima in the differential cross-section $dP/d\omega$ is shown to be proportional to the nuclear sticking time T. Therefore the observation of the shift may serve as a clock for nuclear reaction times. This method is particularly well suited to measure absolute values of the reaction time since the interference phase can be calibrated for the case of Rutherford collisions where the time scale is well known.

7. DELTA ELECTRONS EMITTED IN INTERMEDIATE HEAVY-ION COLLISIONS AS MESSENGERS OF NUCLEAR DECELERATION TIMES

More than 20 years ago a bremsstrahlung experiment was proposed by Eisberg and coworkers[53,54] to measure the nuclear time delay in order to offer an unambiguous separation between compound nucleus reactions and direct reactions by utilizing the fact, that compound nucleus reactions are delayed. The usefulness of employing electromagnetically interacting particles at intermediate energies for this purpose, however, has not been widely discussed yet.

The recently observed subthreshold production[55,56] of neutral pions motivates a determination of the reaction time scale. Since neutral pion production has been measured at bombarding energies as low as 20 MeV/u, it is believed to be of collective origin because the independent nucleon model with Fermi motion fails in predicting pions below 50 MeV/u. Hence the pions should depend strongly on the evolution of the nuclear collision zone, respectively on the deceleration time.

In our treatment[12] of electron emission we will assume that the nuclear charge and current density during the reaction can be described classically. This is probably a reasonable approximation as long as the energy carried away by the particle is a small fraction of the total centre-of-mass energy, and if the measurement averages over many nuclear final states.

As complementary channel to photon and pion emission we propose the investigation of electron spectra in intermediate energy collisions (E_{Lab} = 20 - 100 MeV/u). While in atomic collisions with relativistic heavy ions (E_{Lab} = 82 - 670 MeV/u) K-vacancy production[57] has been successfully explained using atomic models such as plane-wave Born approximation, we retain the molecular model described in the introduction which should still be valid in the considered energy range.

In first-order perturbation theory the differential emission probability of a quasimolecular bound electron with energy E_i into a continuum state with energy E becomes ($A_o \equiv V_o$)

$$dN^{e^-}/dE = \Sigma_{i<F} |\int dt < \varphi_E | (E_i - E)^{-1} \, e\partial V_o/\partial t - ie/\hbar \, \vec{\alpha} \bullet \vec{A} | \varphi_i > \exp\{i(E - E_i)t/\hbar\}|^2. \quad (25)$$

To be more specific we may ask whether the mean nuclear deceleration time τ in central intermediate energy heavy-ion collisions is reflected in the spectra of emitted δ-electrons. In order to deal with this question we use a simplified semiclassical model, assuming that the relevant information on the reaction can be described through a nuclear trajectory R(t).

The central question we aim at[12] in this section is whether the investigation of high-energy electrons emitted in central heavy ion collisions might help to illuminate the underlying space-time evolution of the nuclear reaction. In particular we are interested to determine nuclear stopping times via a measurement of the emitted yield of high-energy δ-electrons. Again we employ a quasimolecular semiclassical theory where the nuclear motion is described classically. Since we use the molecular model and neglect the influence of relativistic kinematics and retardation effects, the results are valid only for bombarding energies $E_{lab} \leq 100$ MeV/u. As a first step a simple trajectory accounting for nuclear deceleration is considered, neglecting Coulomb effects, which is reasonable for bombarding energies above, e.g., 20 MeV/u.

In order to obtain the transition matrix elements and binding energies we solve the stationary problem using molecular basis states. The internuclear two-centre potential is expanded into multipoles, where only the leading monopole term $V_o(r,R)$ is considered. The latter approximation restricts the validity of our calculations to total nuclear charges $Z_1 + Z_2 \geq 110$, where the dynamical coupling matrix elements increase drastically at small internuclear distances R.

To demonstrate the basic features, we restrict ourselves to first-order perturbation theory. Another motivation for using first-order perturbation theory derives from the desire of deducing an analytic expression for the electron emission probability in dependence on the nuclear stopping time τ. In this case the amplitude for the transition of a quasimolecular 1sσ-electron to the continuum state E is approximated by

$$a(E) = -\int_{-\infty}^{\infty} dt \, \dot{R}(t) / (E_{1s\sigma} - E) < \varphi_E | \, e\partial V_o/\partial R | \varphi_{1s\sigma} > \exp\{i(E - E_{1s\sigma})t/\hbar\}. \quad (26)$$

For large continuum energies E we set $E - E_{1s\sigma} \cong$ const. $\equiv \Delta E$, independent of R(t). The matrix element in (26) can be approximated by

$$< \varphi_E | e\partial V_o/\partial R | \varphi_{1s\sigma} > / \Delta E \cong N(E, Z_{tot}) \, R / (R_m^2 + R^2), \quad (27)$$

where R_m indicates the position of the maximum in the matrix element, being in the vicinity of the contact point R_{cont} of the two nuclei. $N(E,Z_{tot})$ indicates a normalization factor being weakly dependent on the involved continuum state with energy E. In a first approximation we consider only central collisions (b = 0), where the relative nuclear motion comes to a complete stop. This may be described by the following trajectory

$$R(t) = v_{\infty}\tau/2 \; \ell n[\exp\{2t/\tau\} /(1 + \exp\{2t/\tau\})], \tag{28}$$

depending on a unique value for the stopping time τ. In order to obtain a closed expression for high-energy electron emission describing also the results for finite stopping times τ, we applied the calculus of residues to equation (26) for complex times t. An approximation of the integrand around those poles having the smallest imaginary part of t was performed. From these considerations we obtained the following scaling law[12].

$$dP_{e^-}/dE_{e^-} \cong N^2(E,Z_{tot}) \; [\; (\pi/2\gamma)^2 \exp\{-\Delta E(R_m/v_{\infty} + \tfrac{1}{2}\pi\tau)/\hbar\} +$$

$$(2\pi^2/\gamma) \; (1 + \tfrac{1}{2}\pi\alpha^2)^2 \exp\{-\Delta E \; (\tfrac{1}{2}R_m/v_{\infty} + \tfrac{3}{4}\pi\tau)/\hbar\}/ \; f(\alpha,\beta) +$$

$$(2\pi)^2(1 + \tfrac{1}{2}\pi\alpha^2)^4\exp\{-\Delta \; E\pi\tau/\hbar\}/f^2(\alpha,\beta) \;], \tag{29}$$

where $\alpha = \Delta E\tau/(2\hbar)$, $\beta = R_m\Delta E/(\hbar v_{\infty})$, $\gamma = 2R_m/(\tau v_{\infty})$ and $f(\alpha,\beta) = (2\beta)^2 + (\alpha\ell n2)^2$.
In equation (29) also the singularities in the complex t-plane originating from the trajectory have been taken into account. To judge the validity of this scaling law we also performed a numerical integration of eq. (26). The results are confronted with the scaling law (29) in Fig. 11, where we have considered a Pb + Pb collision at $E_{lab} = 60$ MeV/u, setting $R_m = 11$ fm and $N^2 = 10^{-3}$ MeV^{-1}.
Both limits for small and large stopping times τ are described by (29). For values of τ inbetween 1 and 12 fm/c the data are in agreement within a factor of about 2. Except for the low energetic part of the spectrum, equation (29) describes the result obtained within first-order perturbation theory in a satisfactory way.
As pointed out earlier the slope of the emitted electrons in the high-energy wing of the δ-electron spectrum may serve to determine nuclear stopping times. The slope is steepest for large values of τ and decreases with smaller values of τ which represent more abrupt decelerations. The novel point is the ΔE^{-4}- decrease of the spectra in the limit $\tau \rightarrow 0$.
The importance of the present consideration is provided by the analytic result (29) for the electron emission probability[12]. This allows for an easy estimate of the electron emission yield in central intermediate-energy heavy ion collisions. Beyond that the nuclear stopping time τ can readily be extracted from the high-energy wing of the measured electron spectra.

8. POSITRON PRODUCTION IN CROSSED BEAMS OF BARE URANIUM NUCLEI

Up to now atomic positron production processes have been studied experimentally[14,15,22,58-63] by bombarding solid targets of, e.g., Th, U, and Cm with highly stripped heavy-ion projectiles in the same region of nuclear charge. Two experimental groups[58-63] detected peaks in positron spectra of those heavy-ion collisions. Since one does not know definitely how to interpret these structures, measurements carrying additional informations seem to be needed. In one type of new experiments positron production in collisions of two bare nuclei, e.g., uranium on uranium, might be investigated. This situation can be realised experimentally with crossed beams of high-energy, fully stripped heavy ions as discussed now in context with the SIS proposal at GSI, Darmstadt.

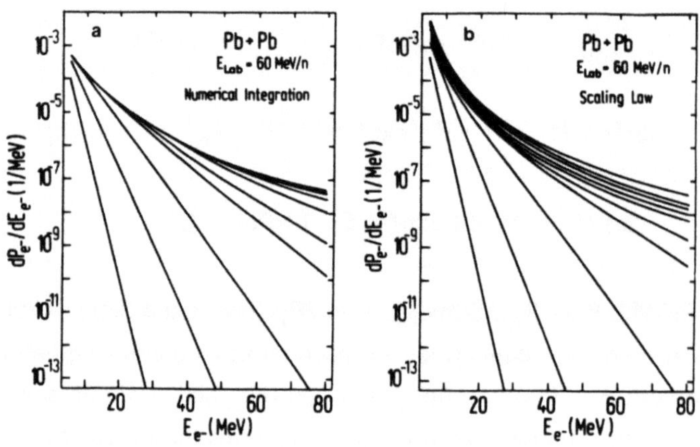

Fig. 11: a) Numerical integration describing electron emission from the quasimolecular 1sσ-state into the upper Dirac continuum in first-order perturbation theory. The result[12] is obtained for a central (b = 0) 60 MeV/u Pb + Pb collision. Different curves belong to different nuclear stopping times τ. With decreasing slope: τ = 96, 48, 24, 12, 9, 6, 4, 3, 2, 1 fm/c. b) Same as in part a), however, obtained by use of the scaling law.

In this section we will discuss the qualitative and quantitative difference between atomic positron production processes, and the corresponding probabilities, in conventional scattering experiments on the one hand and in crossed beams' experiments of bare uranium nuclei on the other hand. The underlying formalism[21] concerning positron production in heavy-ion collisions has to be applied for both types of scattering experiments. For conventional scattering the time dependent atomic wave function $\Phi_i(R(t))$ is expanded in a basis $\varphi_k^{HFS}(r,R)$ of self-consistent adiabatic, molecular wave functions[24] including electron screening in the Hartree-Fock-Slater (HFS) approximation. For these calculations it is assumed that only 50 electrons in the highest molecular states are missing. In contrast, in collisions of bare nuclei, no electron-electron interaction is present, thus leading to an expansion of Φ_i in Coulombic molecular wave functions $\varphi_k^{Cb}(r,R)$. This modification will lead to a stronger diving of the resonant 1sσ state within the negative energy continuum and, at the same time, to an increased resonance decay width Γ_R. E.g., for R = 18 fm, i.e. the distance of closest approach in 5.7 MeV/u U + U head-on collisions, the 1sσ binding energy shifts by more than 100 keV from 1165 keV to 1270 keV. For R = 16.5 fm, corresponding to a bombarding energy of E_{Lab} = 6.2 MeV/u at impact parameter b = 0, we note a binding energy of 1200 keV for HFS-states and of 1310 keV for hydrogen-like states. Looking for the spontaneous decay width Γ_R we notice an increase by a factor of about 3.5 for R = 18 fm and of about 2.7 for R = 16.5 fm, respectively, when switching off the electron-electron interaction. Taking these values by themselves the spontaneous positron creation process will be strongly enhanced with the bare collisions' scenario yielding, for very long nuclear reaction times of the two uranium nuclei, a positron line shifted by 100 keV.

Table 1: Binding energy and spontaneous decay width of the resonant 1sσ electron state for two internuclear distances R in U + U scattering. HFS denotes electron states including electron shielding, whereas Cb characterises Coulombic electron states. R = 18 fm for E_{Lab} = 5.7 MeV/u corresponds to the distance of closest approach for central collisions.

R(fm)	E_{Lab}(MeV/u)	$E_{1s\sigma}^B$(keV)		Γ_R(keV)	
		HFS	Cb	HFS	Cb
18.0	5.7	1165	1270	.31	1.07
16.5	6.2	1200	1310	.63	1.69

Aside from these effects additional modifications in the positron spectra originate from the fact that for bare nuclei a Fermi level of F = 0 has to be considered, i.e., in order to calculate the creation probability for a positron of given energy one has to sum up the transition probabilities into all bound states and, as for conventional scattering, into the whole upper electron continuum. We clearly see the difference: transitions into the vacant 1sσ state, usually suppressed by the small probability for having created a 1sσ hole, now have a chance to dominate the positron spectra. Thus we have to expect the spontaneous positron creation process to be enhanced drastically once more, but this will be true for dynamical positron production as well.

In the following we want to discuss this model more quantitatively, namely for the case of U + U scattering at two relative kinetic energies, E_{rel} = 680 MeV and E_{rel} = 740 MeV. These energies correspond to E_{Lab} = 5.7 MeV/u and 6.2 MeV/u, respectively, when bombarding solid uranium targets with highly stripped uranium atoms. Within the crossed beams' scenario the relative angle between the two uranium rays and the kinetic energy of the high-energy particles have to be carefully adjusted in order to obtain the desired kinematical conditions. For eight impact parameters (b = 0 up to b = 40 fm) transitions between more than 6 bound states, 15 upper continuum and 30 lower continuum states for both, $\kappa = +1$ and $\kappa = -1$, have been calculated.

We first want to discuss the production probabilities of positrons for U + U scattering at various impact parameters b. For pure Rutherford head-on collisions, i.e. b = 0, the maximum differential emission probability $dP^{e^+}/dE_{e^+}|_{max}$ will be enhanced by a factor of about 30, namely from 6.2×10^{-7}/keV for conventional scattering (Fig. 12, part a) up to 1.76×10^{-5}/keV (Fig. 12, part b) for E_{rel} = 740 MeV. To explain this tremendous increase it is important to remember that for bare uranium scattering a vacancy in the diving molecular $1s\sigma$ state is already present, while in conventional scattering events the $1s\sigma$ hole probability even at the distance of closest approach (R = 2a) is only about 5.5% (normalized to 200% due to spin degeneracy). Thus the maximum differential emission probability in the $s_{1/2}$-channel jumps drastically by about two orders of magnitude, whereas the $p_{1/2}$-states contributions increases by a factor of about 10. For the lower bombarding energy of E_{rel} = 680 MeV we find an increase in the positron spectra by a factor of about 40, namely from 4×10^{-7}/keV to 1.6×10^{-5}/keV.

Since the binding energies are considerably larger in collisions of bare nuclei due to the absence of screening as discussed above, the corresponding positron spectra exhibit a shift of about 40 - 50 keV towards higher kinetic positron energies.

The enhancement factor for 740 MeV relative kinetic energy increases from about 30 in central collisions (as discussed for b = 0), to about 65 in medium range collisions (b ≅ 25 fm), and up to a factor of about 130 in peripheral collisions (b ≅ 40 fm). This effect again is caused by the strong suppression of positron creation in peripheral conventional scattering due to tiny hole probabilities for bound electron states. For all impact parameters calculated, the position of the maximum in the energy distribution of induced positrons is shifted towards higher kinetic positron energies.

All features discussed are revealed also by the folded spectra of emitted positrons originating from usual Rutherford scattering of U + U in given angular detection windows: the increase in probability, being most pronounced for distant collisions, and furthermore the energetic shifts of the spectral maximum towards higher kinetic energies. However, in the bare nuclei scattering scenario not only the induced contribution to positron emission will be increased, but also the spontaneous positron production processes are enhanced.

Let us finally discuss the hypothesis that nuclear collisions with a very long delay time are present, which had been suggested as an explanation for the experimentally observed positron line structures. Positron spectra for pure Rutherford scattering, and for an arbitrary nuclear delay time of T = 2×10^{-20} s at the distance of closest approach, are displayed in Fig. 13, calculated within the schematic trajectory model[6,9] for E_{rel} =

740 MeV. Switching from conventional scattering to bare nuclei scattering, for $T \neq 0$ the maximum differential positron emission probability, i.e. the height of the spontaneous positron line, rises by a factor of about 75 to 6.9×10^{-4}/keV. For asymptotically long nuclear reaction times $T \gg \hbar/\Gamma$ the total emission probability in the spontaneous line will increase from 0.055 per collision considered to about 2, the number of holes in the 1sσ state.

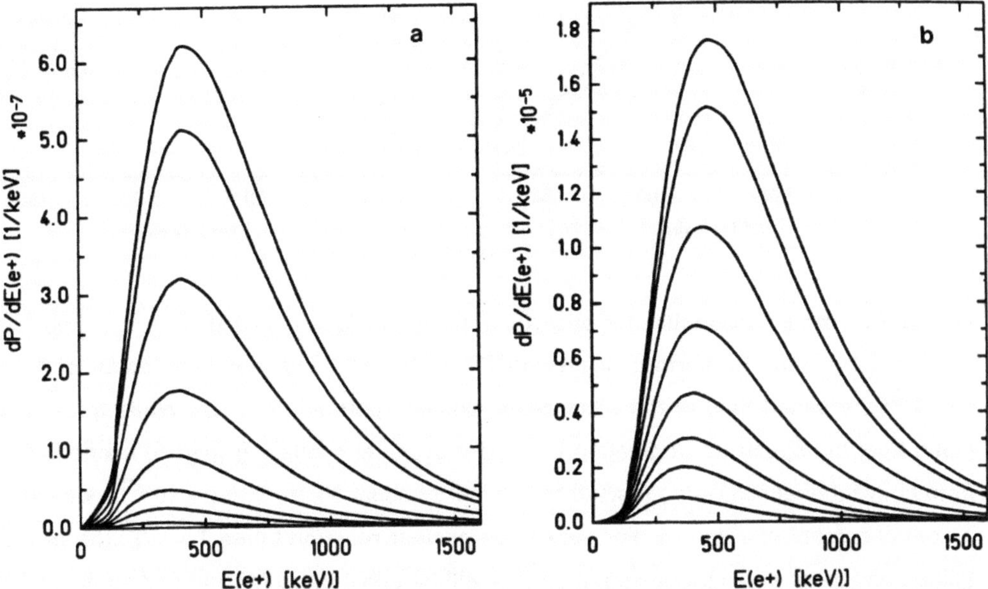

Fig. 12: Energy distribution of positrons emitted in U + U collisions at a relative kinetic energy of E_{rel} = 740 MeV for impact parameters b = 0 to b = 40 fm. The considered impact parameters are 0, 5, 10, 15, 20, 25, 30 and 40 fm, respectively. While an enhancement of a factor of about 30 is found for head-on collisions comparing conventional scattering (part a) to bare nuclei scattering (part b), for peripheral collisions the increase is about 130. The energy shift towards higher kinetic positron energies in part b of about 50 keV (for b = 0) and the steeper decrease of the spectra in the high-energy region are also obvious.

These values are correct unless fast conversion processes in the nuclear compound system do empty the resonant K-shell thus generating new possibilities to create monoenergetic positrons spontaneously.

For higher bombarding (or relative) energies, however, the increase in the maximum differential emission probability as well as in the total line probability will be smaller, due to the higher 1sσ hole probability $P_{1s\sigma}(R=2a)$ for higher bombarding energies in conventional scattering, and vice versa for lower bombarding energies.

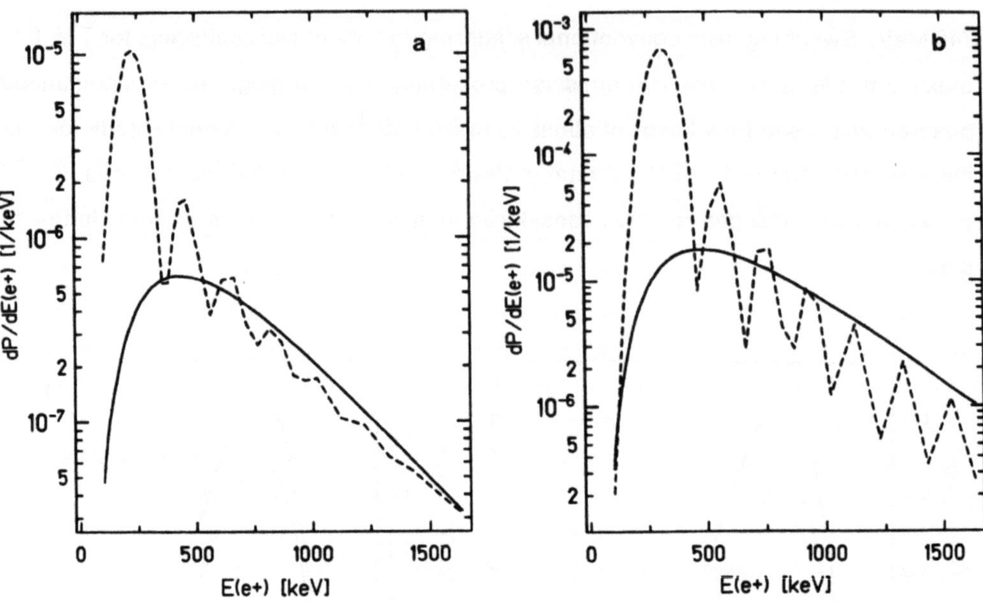

Fig. 13: Differential probability for positron emission in E_{rel} = 740 MeV U + U head-on collisions for pure Rutherford scattering (T = 0) and a nuclear time delay of T = 2×10^{-20} s (dashed line) at the distance of closest approach R = 2a. While for T = 0 (solid line) the spectrum originating from bare uranium scattering (part b) exceeds this from conventional scattering (part a) by a factor of about 30, for T = 2×10^{-20} s we found an enhancement of about 75. For very large nuclear reaction times $T \gg \hbar/\Gamma$, the probabilities under the spontaneous positron line will be about 0.055 (part a) and about 2 (part b), respectively, reflecting the K-hole probabilities at the distance of closest approach, $P_{1s\sigma}(R=2a)$. The position of maximum positron creation probability shifts from about 400 keV to about 220 keV for T = 2×10^{-20} s, with a limit of about 180 keV for $T \gg \hbar/\Gamma$ in part a. For part b the corresponding values are 450 keV, 300 keV, and 285 keV, respectively.

The spectra displayed in Fig. 13 also reveal the energetic shift of the resonant 1sσ state within the lower electron continuum due to the neglect of electron-electron interaction[24]: For increasing nuclear delay times T, the position of the spontaneous positron line shifts from the dynamically caused probability maximum $E_e+|_{max} \cong 400$ keV and 450 keV, respectively, to the position of the resonant state, i.e., $E_e+ = E_{1s\sigma}{}^B(2a) - 2mc^2$, cf. Table 1. This yields 140 keV and 250 keV, respectively, for E_{rel} = 680 MeV, and 180 keV and 290 keV, respectively, for E_{rel} = 740 MeV. This dependence of the positron lines' position on charge state and kinematics may serve to prove the atomic origin of a line structure.

The question, whether spontaneous positron creation can be separated better from dynamical processes when dealing with collisions of bare nuclei, thus depends on kinematical conditions.

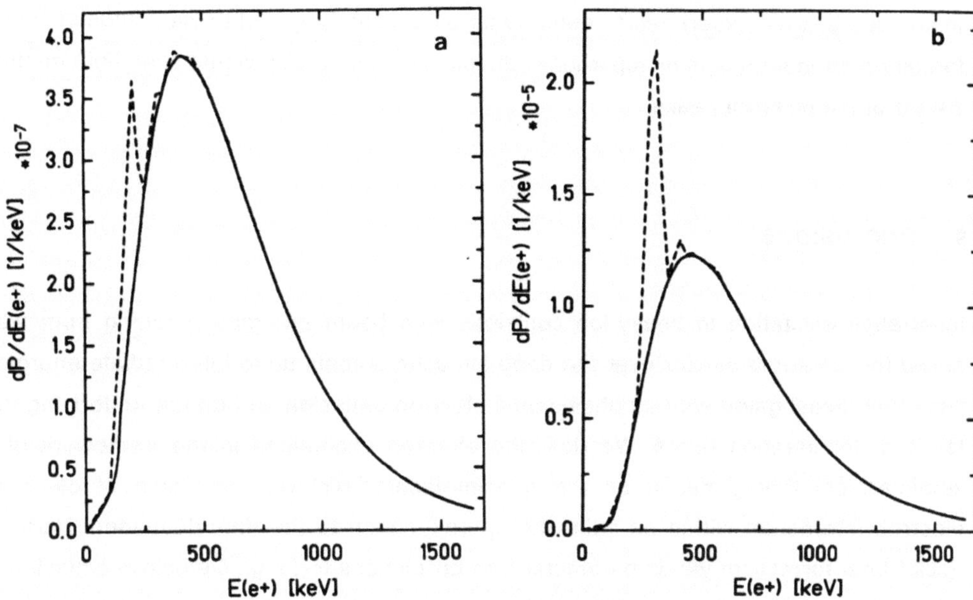

Fig. 14: Spectra of positrons stemming from Rutherford scattering of U + U at E_{rel} = 740 MeV in an angular detection window of $\Theta = 90^\circ \pm 6^\circ$ superposed by a contribution originating from hypothetical long lasting nuclear reactions with sharp delay time T = 5×10^{-20} s and nuclear cross section $\sigma_N(\Theta) = q \times \sigma_R(\Theta) \cong 7$ mb (dashed line). Pure Rutherford scattering is indicated by the solid line. While for conventional scattering experiments (part a) the spontaneous positron line will be hard to detect, in bare nuclei collisions (part b) a line structure at E_{e^+} = 285 keV (for the kinematical assumptions made) is dominating the positron spectrum.

Fig. 14 shows a hypothetical superposition of positrons, originating from ordinary Rutherford scattering of U + U at E_{rel} = 740 MeV in an angular detection window of $\Theta_{cm} = 90^\circ \pm 6^\circ$, plus spontaneous positrons from long lasting nuclear molecule formation of sharp lifetime T with a nuclear reaction cross section $\sigma_N(\Theta) = q \times \sigma_R(\Theta) \cong 7$ mb in the assumed angular window. Even for the longest delay time T shown, in conventional scattering experiments there will be only a small chance to detect the spontaneous positron line. Switching to the bare nuclei scenario, spontaneous positrons will give a nearly equal contribution for $E_{e^+} \cong 250$ - 350 keV. Similar results have been obtained also for angular windows $\Theta_{cm} = 70^\circ \pm 6^\circ$ and $\Theta_{cm} = 50^\circ \pm 6^\circ$. For E_{rel} = 680 MeV, however, the calculations yield less encouraging results for all cases, i.e. for all detection windows and for both types of scattering experiments. One should remember

also that values for the nuclear reaction cross section $\sigma_N(\Theta)$ and for the sharp delay time T as well as the nuclear charge distribution during the reaction (here two spheres with internuclear separation R = 2a), etc., are chosen quite arbitrarily so that quantitative statements for the spontaneous line structure are arbitrary as well.
However, putting aside possible experimental difficulties, the scattering of bare nuclei in crossed beams' experiments seems to be an encouraging tool to yield additional information on inner shells in superheavy atomic systems. It also would shed light on the nature of the positron peaks.

9. CONCLUSIONS

Inner-shell excitation in heavy-ion collisions with beam energies reaching from well below the Coulomb barrier over the deep-inelastic domain up to intermediate energies has been investigated with emphasis on δ-electron emission and on the underlying delay and deceleration times. We describe electron excitations in the framework of a semiclassical theory based on the quasimolecular picture. Inclusion of electron-electron interaction within an adiabatic approximation to the time-dependent Hartree-Fock-Slater formalism yields parameter-free predictions for, e.g., δ-electron emission in the system I+Pb at E_{lab} = 500 MeV which agree well with experimental measurements, both in shape and absolute values.
For bombarding energies allowing the penetration of the Coulomb barrier, we introduced a schematic model for the description of a nuclear compound system with a lifespan T. The resulting δ-electron emission shows oscillatory patterns having a width of ΔE being inversely proportional to T. In deep-inelastic heavy ion collisions, however, where due to energy dissipation the trajectories become asymmetric and only relatively short delay times occur, the schematic model is only of restricted value. Instead we used classical friction models describing the nuclear motion classically and compared our calculations with experimental data for U+U at 8.4 MeV/u bombarding energy. The current experimental status is summarized in ref. 15.
The experimentally measured emission probability[15] of δ-electrons versus kinetic electron energy shows a slope being notably steeper than the one being predicted assuming Rutherford scattering. The slope is best reproduced within the model of Schmidt and coworkers also accounting for neck formation, which yields the longest delay time of about $2 \cdot 10^{-21}$ s. Comparison with the model of Birkelund and others based on the nuclear proximity prescription and yielding shorter delay times of only about $4 \cdot 10^{-22}$ s, leaves discrepancies in the slope thus favouring the former model. Similar results were obtained by considering K-vacancy formation in deep-inelastic U + U collisions.
At intermediate energy collisions of heavy ions, where also subthreshold pion production is observed, we stressed how electron emission depends on the space-time

evolution of the internuclear collision zone. Restricting ourselves to first-order perturbation theory and a simple Fermi-type model for the trajectory, valid only for head-on ($b = 0$) collisions, we predict the emission of electrons with kinetic energies up to 50 MeV in 60 MeV/u Pb+Pb collisions.

The slope of the δ-electron spectra in the high energy wing provides a tool to determine the underlying deceleration time scale: for increasing deceleration times the slope becomes steeper while it decreases in case of more abrupt decelerations, i.e. shorter deceleration times.

In order to obtain quantitative predictions which can be compared with measurements, a full coupled channel calculation also including $p_{1/2}$ states is required which could increase the probabilities by about an order of magnitude. Furthermore an ansatz for the impact parameter dependence of close collisions and the outgoing trajectory is needed. Electron screening effects are negligible. In future calculations a more refined model for the time-evolution of the nuclear charge and current distribution will be studied.

The experimental feasibility depends on the intensity of background effects producing high energy electrons, in particular pair conversions of γ-rays and pion decay $\pi^0 \rightarrow \gamma + e^+ + e^-$. These effects, however, are expected to yield about the same numbers of electrons and positrons. However, the slope of the atomic positron spectra with respect to their kinetic energy is expected to be notably smaller in comparison with the one obtained for atomic δ-electron emission, so that the spectra could be disentangled by studying the ratio $P^{tot}_{e-}/P^{tot}_{e+}$.

Concluding, we note the possibility to use δ-electron emission in deep-inelastic and intermediate energy collisions as a clock-hand to determine delay and deceleration times on the nuclear time-scale, i.e. $T = 10^{-21} - 10^{-24}$ s. The shape of the high-energy wing of the δ-electron spectra provides the underlying nuclear reaction times.

ACKNOWLEDGEMENT

We are grateful for the fruitful collaboration with our experimental colleges H. Backe, P. Senger, R. Krieg, E. Kankeleit, H. Oeschler, E. Bozek, M. Krämer, R. Anholt and W. Meyerhof.

REFERENCES

1. W.E. Meyerhof, J.F. Chemin, in: Advances in Atomic and Molecular Physics, eds. D.R. Bates, B. Bederson, (Academic Press, New York, 1985), Vol. 20, p. 173.
2. U. Heinz, Interplay of nuclear and atomic physics in ion-atom collisions, preprint
3. G. Soff, J. Reinhardt, B. Müller, W. Greiner, Phys. Rev. Lett. 43, 1981 (1979)
4. R. Anholt, Phys. Lett. 88B, 262 (1979)
5. J. Reinhardt, B. Müller, W. Greiner, G. Soff, Z. Physik A292, 211 (1979)
6. U. Müller, J. Reinhardt, G. Soff, B. Müller, W. Greiner, Z. Physik A297, 357 (1980)

7. J. Kirsch, B. Müller, W. Greiner, Z. Physik D1, 47 (1986)
8. J. Kirsch, B. Müller, W. Greiner, Phys. Lett. A94, 151 (1983)
9. U. Müller, G. Soff, T. de Reus, J. Reinhardt, B. Müller, W. Greiner, Z. Physik A313, 263 (1983)
10. U. Müller, G. Soff, J. Reinhardt, T. de Reus, B. Müller, W. Greiner, Phys. Rev. C30, 1199 (1984)
11. T. de Reus, J. Reinhardt, B. Müller, W. Greiner, U. Müller, G. Soff, Progress in Particle and Nuclear Physics Vol. 15, Ed. A. Faessler, Pergamon Press (1985) 57
12. T. de Reus, J. Reinhardt, B. Müller, W. Greiner, G. Soff, Phys. Lett. 169B, 139 (1986), Phys. Lett. 173B, 491 (1986) (E)
13. G. Soff, U. Müller, P. Schlüter, J. Reinhardt, T. de Reus, A. Schäfer, K.-H. Wietschorke, B. Müller, W. Greiner, AIP Conference Proceedings No. 136, p. 204, New York, 1985, Eds.: H.P. Kelly, Y.-K. Kim, Atomic Theory Workshop on Relativistic and QED Effects in Heavy Atoms, National Bureau of Standards, Gaithersburg, USA, 1985.
14. H. Backe, P. Senger, W. Bonin, E. Kankeleit, M. Krämer, R. Krieg, V. Metag, N. Trautmann, J.B. Wilhelmy, Phys. Rev. Lett. 50, 1838 (1983)
15. R. Krieg, E. Bozek, U. Gollerthan, E. Kankeleit, G. Klotz-Engmann, M. Krämer, U. Meyer, H. Oeschler, P. Senger, Phys. Rev. C34, 562 (1986)
16. Ch. Stoller, M. Nessi, E. Morenzoni, W. Wölfli, W.E. Meyerhof, J.D. Molitoris, E. Grosse, Ch. Michel, Phys. Rev. Lett. 53, 1329 (1984)
17. C. Stoller, Nucl. Instr. Meth. B10/11, 432 (1985)
18. R. Anholt, Rev. Mod. Phys. 57, 995 (1985)
19. W. Greiner, ed., Quantum Electrodynamics of Strong Fields, NASI series B80, (Plenum, New York, 1983)
20. W. Greiner, B. Müller, J. Rafelski, Quantum Electrodynamics of Strong Fields, (Springer, Berlin, 1985)
21. J. Reinhardt, B. Müller, W. Greiner, Phys. Rev. A24, 103 (1981)
22. F. Bosch, B. Müller, Prog. Part. Nucl. Phys. 16, 195 (1985)
23. G. Soff, J. Reinhardt, B. Müller, W. Greiner, Phys. Rev. Lett. 38, 592 (1977)
24. T. de Reus, J. Reinhardt, B. Müller, W. Greiner, G. Soff, U. Müller, J. Phys. B17, 615 (1984)
25. W. Betz, B. Müller, G. Soff, W. Greiner, Phys. Rev. Lett. 37, 1046 (1976)
26. J. Kirsch, B. Müller, W. Greiner, Z. Naturforsch. 35a, 579 (1980)
27. B. Müller, J. Rafelski, W. Greiner, Z. Physik 257, 62 and 183 (1972)
28. W. Pieper, W. Greiner, Z. Physik 218, 327 (1969)
29. V.S. Popov, Sov. J. Nucl. Phys. 12, 235 (1971)
30. J. Reinhardt, B. Müller, W. Greiner, G. Soff, Phys. Rev. Lett. 43, 1307 (1979)
31. G. Soff, B. Müller, W. Greiner, Phys. Rev. Lett. 40, 540 (1978)
32. G. Soff, W. Greiner, W. Betz, B. Müller, Phys. Rev. A20, 169 (1979)
33. G. Soff, J. Reinhardt, B. Müller, W. Greiner, Z. Physik A294, 137 (1980)

34. F. Bosch, D. Liesen, P. Armbruster, D. Maor, P.H. Mokler, H. Schmidt-Böcking, R. Schuch, Z. Physik A296, 11 (1980)
35. D. Liesen, P. Armbruster, H.H. Behncke, S. Hagmann , Z. Physik A288, 417 (1978)
36. D. Liesen, P. Armbruster, H.H. Behncke, F. Bosch, S. Hagmann, P.H. Mokler, H. Schmidt-Böcking, R. Schuch, Electronic and Atomic Collisions. N. Oda, K. Takayanagi, Eds., (North-Holland, Amsterdam, 1980), 337
37. D. Liesen, P. Armbruster, F. Bosch, S. Hagmann, P.H. Mokler, H.J. Wollersheim, H. Schmidt-Böcking, R. Schuch, J.B. Wilhelmy, Phys. Rev. Lett. 44, 983 (1980)
38. R. Anholt, W.E. Meyerhof, C. Stoller, Z. Physik A291, 287 (1979)
39. F. Güttner, W. Koenig, B. Martin, B. Povh, H. Skapa, J. Soltani, T. Walcher, F. Bosch, C. Kozhuharov, Z. Physik A304, 207 (1982)
40. M.A. Herath-Banda, A.V. Ramayya, C.F. Maguire, F. Güttner, W. Koenig, B. Martin, B. Povh, H. Skapa, J. Soltani, Phys. Rev. A29, 2429 (1984)
41. M.A. Herath-Banda, A.V. Ramayya, W. Koenig, B. Martin, H. Skapa, J. Soltani, Phys. Rev. Lett. 53, 1646 (1984)
42. M.A. Herath-Banda, W. Koenig, B. Martin, F. Güttner, H. Skapa, J. Soltani, K. Dworschak, F. Bosch, C. Kozhuharov, A.V. Ramayya, Phys. Rev. A33, 861 (1986)
43. K.E. Stiebing, H. Schmidt-Böcking, W. Schadt, K. Bethge, R. Schuch, P.H. Mokler, F. Bosch, D. Liesen, S. Hagmann, P. Vincent, Z. Physik A319, 239 (1984)
44. R. Schmidt, V.D. Toneev, G. Wolschin, Nucl. Phys. A311, 247 (1978)
45. J.R. Birkelund, L.E. Tubbs, J.R. Huizenga, J.N. De, D. Sperber, Phys. Rep. 56, 107 (1979)
46. D. Vasak, B. Müller, W. Greiner, Phys. Scr. 22, 25 (1980)
47. D. Vasak, W. Greiner, B. Müller, T. Stahl, M. Uhlig, Nucl. Phys. A428, 291 (1984)
48. B. Müller, J. Rafelski, W. Greiner, Phys. Lett. 47B, 5 (1973)
49. B. Müller, W. Greiner, Z. Naturforsch. 31a, 1 (1976)
50. W.E. Meyerhof, Phys. Rev. Lett 31, 1341 (1973)
51. I. Tserruya, R. Schuch, H. Schmidt-Böcking, J. Barette, W. Da-Hai, B.M. Johnson, M. Meron, K.W. Jones, Phys. Rev. Lett. 50, 30 (1983)
52. R. Schuch, H. Schmidt-Böcking, I. Tserruya, B.M. Johnson, K.W. Jones, M. Meron, Z. Physik A320, 185 (1985)
53. R.M. Eisberg, D.R. Yennie, D.H. Wilkinson, Nucl. Phys. 18, 338 (1960)
54. H. Feshbach, D.R. Yennie, Nucl. Phys. 37, 150 (1962)
55. H. Noll, E. Grosse, P. Braun-Munzinger, H. Dabrowski, H. Heckwolf, O. Klepper, C. Michel, W.F.J. Müller, H. Stelzer, C. Brendel, W. Rösch, Phys. Rev. Lett. 52, 1284 (1984)
56. H. Heckwolf, E. Grosse, H. Dabrowski, O. Klepper, C. Michel, W.F.J. Müller, H. Noll, C. Brendel, W. Rösch, J. Julien, G.S. Pappalardo, G. Bizard, J.L. Laville, A.C. Müller, Z. Physik A315, 243 (1984)

57. R. Anholt, W.E. Meyerhof, Ch. Stoller, E. Morenzoni, S.A. Andriamonje, J.D. Molitoris, O.K. Baker, D.H.H. Hoffmann, H. Bowman, J.S. Xu, Z.Z. Xu, K. Frankel, D. Murphy, K. Crowe, J.O. Rasmussen, Phys. Rev. A30, 2234 (1984)
58. J. Schweppe, A. Gruppe, K. Bethge, H. Bokemeyer, T. Cowan, H. Folger, J.S. Greenberg, H. Grein, S. Ito, R. Schule, D. Schwalm, K.E. Stiebing, N. Trautmann, P. Vincent, M. Waldschmidt, Phys. Rev. Lett. 51, 2261 (1983)
59. M. Clemente, E. Berdermann, P. Kienle, H. Tsertos, W. Wagner, C. Kozhuharov, F. Bosch, W. Koenig, Phys. Lett. 137B, 41 (1984)
60. T. Cowan, H. Backe, M. Begemann, K. Bethge, H. Bokemeyer, H. Folger, J.S. Greenberg, H. Grein, A. Gruppe, Y. Kido, M. Klüver, D. Schwalm, J. Schweppe, K.E. Stiebing, N. Trautmann, P. Vincent, Phys. Rev. Lett. 54, 1761 (1985)
61. H. Tsertos, E. Berdermann, F. Bosch, M. Clemente, P. Kienle, W. Koenig, C. Kozhuharov, W. Wagner, Phys. Lett. 162B, 273 (1985)
62. P. Kienle, J. Phys. Soc. Jpn. 54 Suppl. II, 549 (1985)
63. T. Cowan, H. Backe, K. Bethge, H. Bokemeyer, H. Folger, J.S. Greenberg, K. Sakaguchi, D. Schwalm, J. Schweppe, K.E. Stiebing, P. Vincent, Phys. Rev. Lett. 56, 444 (1986)
64. J. Reinhardt, B. Müller, W. Greiner, Coherent production of positrons in heavy ion collisions, Proc. of the International Workshop on Coherence and Correlations in Atomic Collisions, London, 1978, Eds.: H. Kleinpoppen, J.F. Williams, (Plenum, New York, 1980), p. 331
65. E. Kankeleit, Nukleonika 25, 253 (1980)
66. R. Krieg, E. Bozek, U. Gollerthan, E. Kankeleit, G. Klotz, M. Krämer, U. Meyer, H. Oeschler, P. Senger, Nucl. Instr. Meth. B9, 762 (1985)

NARROW CORRELATED POSITRON-ELECTRON PEAKS FROM SUPERHEAVY COLLISION SYSTEMS

Thomas E. Cowan and Jack S. Greenberg

A.W. Wright Nuclear Structure Laboratory
Yale University
New Haven, Connecticut

EPOS Collaboration

H. Backe[5], K. Bethge[3], H. Bokemeyer[1], T. Cowan[2], H. Folger[1], J.S. Greenberg[2], K. Sakaguchi[3], P. Salabura[3], D. Schwalm[4], J. Schweppe[2], K.E. Stiebing[3], and P. Vincent[6]

[1]GSI, [2]Yale Univ., [3]Univ. Frankfurt
[4]Univ. Heidelberg, [5]Univ. Mainz, [6]BNL

I. INTRODUCTION

Positron production from superheavy collision systems [1-8] has been under investigation for a decade at the Gesellschaft für Schwerionenforschung in Darmstadt, West Germany. This discussion will concentrate on the recent studies of correlated emission of electrons with positrons [7] carried out by the EPOS collaboration. Although these are the experiments we are currently pursuing, they have obviously evolved from this decade of work marked by progress in both experiment and theory. Guided by the ongoing developments, immediate investigations have changed from the original goals, but the past has very much influenced the present experimental approach. It is helpful, therefore, towards understanding the directions that have been adopted in the most recent work, to recount briefly the developments that have led to the present interesting juncture in the experiments.

As is well known, investigation of positron production in heavy-ion collisions was first motivated by the search for the spontaneous decay of the QED vacuum state which is predicted to occur in the presence of supercritical electric fields [9,10]. Such fields can be generated with a combined charge $Z_u = 173$. The signal for this unique process is the emission of a monoenergetic positron with a kinetic energy which reflects the capture of the electron of the spontaneously created pair into the previously ionized 1s state of the quasi-atom.

The early experiments [1,2,8,11-16] with superheavy collision systems verified that superheavy quasimolecules with deeply bound 1s states are indeed formed in such collisions, indicating that strong fields could be available transiently. In addition they showed [11,12] that a sufficiently large probability for K-hole creation exists on the ingoing path of the collision so that spontaneous positron emission could occur at the turning point of the heavy-ion trajectory. However, the collisions are expected to involve a short time scale of $\sim 10^{-21}$ sec. This inherent feature leads to several consequences which compound the difficulty of detecting spontaneous positron emission. Some of the more significant ones are 1) the short time available during the collision is in competition with the much longer intrinsic time scale of spontaneous positron emission $\sim 10^{-19}$ s, 2) the natural line width of a few keV is broadened beyond ~300 keV, and 3) the amplitude for spontaneous positron emission only adds coherently to a dominating continuous background due to dynamic production of positrons by the rapidly evolving Coulomb field [17-19].

Expecting such continuous distributions, the first experiments on positron production naturally focussed on the broad features of the spectra [1,2]. Together with more recent studies [20-30], they largely confirm the characteristics of positron production predicted by theory. But in addition to this anticipated behavior, there was an unexpected observation by the EPOS collaboration of an enhancement of low-energy positron production for close collisions between uranium ions and curium atoms at energies near the Coulomb barrier [20-24,31]. Subsequent measurements with these systems revealed a narrow peak in the positron energy distribution at 316 ± 10 keV, having a width of ~80 keV [3].

A number of interesting features are connected with the observation of this line. It is associated with quasi-elastic scattering events which differ slightly in their ion angle-angle correlations from elastic Rutherford collisions. An investigation of γ-ray and electron energy distributions accumulated simultaneously under identical conditions as the positron peak, reveal no competing γ-ray or electron internal conversion (IC) lines [3,5] which would be predicted [32] to be readily detectable if the positron peak were due to transitions in the final-state nuclei. The mere narrow width of the line, together with the observation that this width is independent of HI scattering angle, further precludes a nuclear origin; the magnitude of the width indicates a velocity of the emitting source which is consistent with the velocity of the CM [3,5]. Similar lines were observed in U+U collisions by the EPOS apparatus [16,22-27] and by the "Orange" spectrometer in U+U and U+Th [4,6,28].

It was particularly provocative to find that the positron peak energy associated with the U+Cm system corresponded to the 1s binding energy (minus $2m_ec^2$) predicted for a configuration where the nuclear centers are separated by 17 fm (the distance of closest approach for head-on collisions at 6.05 MeV/u). This apparent coincidence encouraged speculation that the observed structure may be the signature of spontaneous positron emission. Indeed, it was demonstrated that the narrow line width, its intensity, and the dependence on projectile energy could all be incorporated into the theory of spontaneous positron emission if, in a small fraction (10^{-3}) of the collisions, a di-nuclear complex were formed whose lifetime exceeds 10^{-19} s [33-37], an order of magnitude longer than the typical scattering time where supercritical binding can be maintained. However, this attractive explanation was challenged by subsequent measurements which probed the very distinctive feature that for spontaneous e^+ emission the energy of the positron has an unusual $\sim Z_u^{20}$ dependence on the nuclear charge.

To pursue this point, available supercritical collision systems were systematically studied with the EPOS spectrometer, ranging from U+Cm (Z_u = 188) where the spontaneous positron peak is predicted to occur at $E_{e+} \approx 300$ keV, to Th+Th (Z_u = 180) where $E_{e+} \approx 80$ keV [5]. In addition, the Th+Ta (Z_u = 163) collision system [38,39] was investigated as an example where the nulcear charge is clearly below the supercritical charge threshold so that spontaneous positron emission is not expected. As mentioned above, data was also available in the U+U and U+Th systems in experiments by the "Orange" spectrometer group [4,6,28]. As shown in Fig. 1 taken from the EPOS data [5], narrow positron peaks between 300 and 400 keV are observed in all these systems. The spectra in Fig. 1 correspond to kinematic conditions chosen empirically to enhance the peak over the continuous backgrounds by taking advantage of the apparent difference in scattering angle correlations between the peak and the continuous distributions of dynamic and nuclear positrons which are associated with elastic scattering kinematics. In each case, the line is narrow (typically ~80 keV), is observed with a cross-section of $d\sigma_{e+}/d\Omega_{HI} \sim 10$ μb/sr, and no competing γ-ray or electron lines are observed which have sufficient intensity to explain the data in terms of a nuclear conversion process. The positron line energies for the

U+U and U+Th systems observed by the "Orange" spectrometer group also share a nearly common energy [6].

Viewed in any simple way, this apparent independence of peak energy on combined nuclear charge is in contradiction to the predictions by QED for spontaneous e^+ emission. Attempts to accommodate the measured line energies within spontaneous positron theory must involve a radical variation of the nuclear charge distribution, or of the ionization state of the complex, as a function of Z_u so as to lead to nearly constant binding of the $1s\sigma$ orbital [5,40]. The possibility that one common giant nucleus is formed by nucleon evaporation in all of these collisions, leading to a single positron peak energy, can be ruled out by the observed HI kinematics. For example, the fast (10^{-19} s) emission of an α particle from the compound system would require a kinetic energy of ≈30 MeV [5,41]. Assuming that a $Z_u = 180$ complex is formed in U+Cm collisions would imply that more than 100 MeV would be lost from the CM of the remaining nuclides. The sum of the HI scattering angles is sensitive to total CM energy loss and the near elasticity of events associated with the e^+ peak sets a total kinetic energy loss limit of ≤20 MeV, excluding this possibility. It is also difficult to understand how the thorium nuclei (colliding at the Coulomb barrier) can form the compressed spherical complex necessary for $Z_u \leq 180$ and subsequently

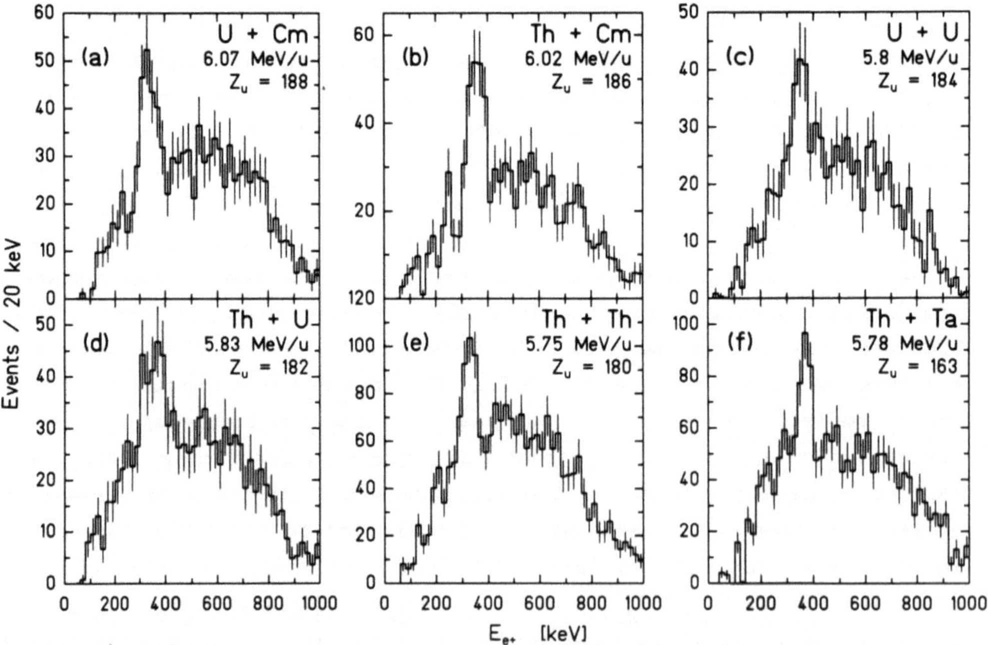

Figure 1. Positron energy spectra for six collision systems and bombarding energies as indicated. Kinematic constraints chosen as discussed in text.

break up into a nearly binary, quasi-elastic exit channel. Finally, the combined nuclear charge of the Th+Ta system is clearly too small to produce critical binding of the 1s orbital required for spontaneous positron emission [see also Ref. 42].

A difficult puzzle is therefore presented by these experimental observations: narrow lines are observed in several collision systems ranging over $163 \leq Z_u \leq 188$, all of which cannot presently be explained by conventional mechanisms involving nuclear and QED processes. The nearly common peak energy (a simple average of the data gives $\langle E_{e+} \rangle$ = 336 ± 10 keV) suggests a common source. The uniqueness of a sharp positron line suggests an unorthodox phenomenon. An obvious speculation [5,43-46] for such a source is the two-body decay of a previously undetected object. If such an object exists, some of its properties follow from the available experimental data. For example, a neutral object X^0 decaying into e^+e^- would have a mass of ~1.7 MeV/c^2. The data seem to favor emission from a system moving with about the CM velocity. The lifetime of the object is bounded on the one hand by the uncertainty principle ($\tau > \Delta E/h \simeq 10^{-19}$ s), and on the other by the fiducial volume of the EPOS spectrometer ($\tau < \ell/v_X \sim 10^{-9}$ s). For many reasons, an object with these properties could have gone undetected to date.

Not to forego an obvious possibility, a search was carried out for the other member of a two-body decay. The conventional candidate is an electron, and the EPOS apparatus was modified to search for a monoenergetic electron associated with the positron [7]. It led to the detection of a narrow peak in the energy distribution of electrons emitted in coincidence with the monoenergetic positron line from U+Th collisions at 5.83 MeV/u. As shown in Fig. 2, the energy (375 ± 10 keV) and width (70 ± 15 keV) of the electron peak are equal within statistical error to the energy, 380 ± 10 keV, and width 80 ± 15 keV of the positron line. The structure in the sum of electron and positron energies (Fig. 2c) is as narrow (80 ± 20 keV) as the individual lines, suggesting the correlated cancellation of the positron and electron Doppler shifts. Such a cancellation would be expected for the back-to-back emission of positron and electron from the two-body decay of a neutral object.

Monte-Carlo simulations of the coincident positron-electron peak shapes and intensities expected for the decay of a 1.8 MeV object show a marked resemblance to the measured data [7]. It has been assumed in Fig. 2e-h that the object is created at rest in the CM system of the HI collision with sufficient probability (~10^{-5}) to account for 3‰ of the total observed positron yield. As will be discussed in more detail below, an extensive Monte-Carlo study appears to exclude all known nuclear conversion processes as well as a wide variety of hypothetical positron producing mechanisms as an explanation for these data.

A follow-up series of measurements on U+Th have provided more details on the coincident electron-positron structure. Particularly, with considerably more data and an improvement of the resolution of the electron detector over that of the initial measurement, two central features have emerged clearly. It was established that the sum-energy peak is indeed considerably narrower (<40 keV) than the individual positron or electron lines. Moreover, the narrow sum-energy peak, which appears to be free of Doppler broadening associated with the positron peaks of Fig. 1, in turn provides the needed energy resolution to unambiguously demonstrate that more than one set of correlated positron-electron peaks is present. These features are illustrated in the spectra shown in Fig. 3, taken from the recent measurements, which exhibit two prominent lines at sum energies of ~620 and ~810 keV. There are many details of these data which will be presented in the discussion that follows, but at this point it bears emphasis that the mere appearance of the earlier mentioned very narrow peak in the sum-energy spectrum can have important implications regarding the possible source of our anomalous positron lines.

As previously stated our concentration will be on the later aspects of the decade's work. A detailed interpretation of these data is intimately linked with the characteristics of the EPOS spectrometer. Many of its features can be used to advantage to extract information on the kinematic characteristics associated with the positron and electron emission. This contribution develops these points. In particular, it discusses several of the experimental aspects of our research for coincident positron-electron

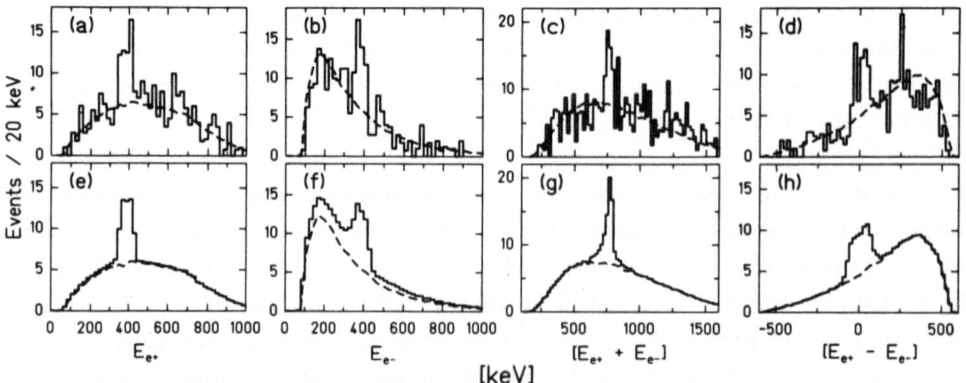

Figure 2. Projections of the coincident e+ and e- intensity distributions collected for 5.83 MeV/u U+Th collisions (a-d), compared to Monte-Carlo simulations of the two-body decay of a 1.8 MeV/c^2 neutral object (e-h). The columns correspond to projections onto the E_{e+}, E_{e-}, $(E_{e+}+E_{e-})$, and $(E_{e+}-E_{e-})$ axes.

emission (including a detailed discussion of the performance of the apparatus and the results of Monte-Carlo calculations which model its operation), presents an evaluation of several hypotheses for its origin, and arrives at model-independent conclusions about the characteristics of the source. For a more complete discussion of these topics, the reader is referred to Ref. 41, of which this paper is an abbreviated subset.

II. PRELIMINARY CONSIDERATIONS

Compatibility of Two-Body Decay with Narrow Positron Peaks

Since the positron-electron coincidence experiments are primarily motivated by the search for a two-body decay mechanism, an essential first question to be addressed is whether the decay of neutral particles could at all in principle account for the narrow positron lines presented in Fig. 1.

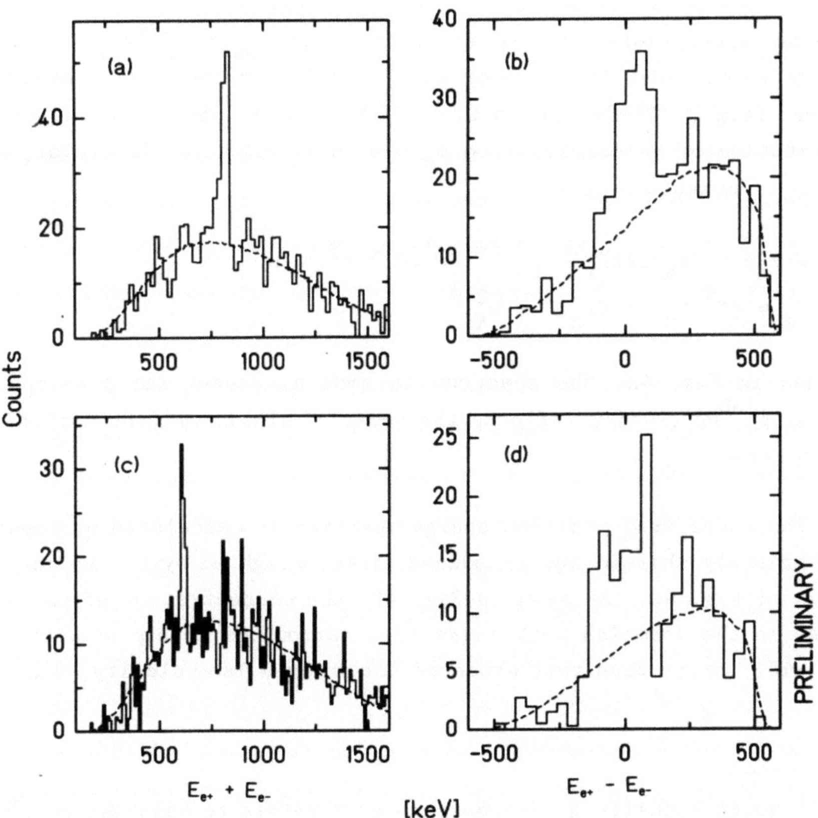

__Figure 3__. Results of a preliminary analysis of U+Th collisions near 5.87 MeV/u (Feb, 1986). $(E_{e+}+E_{e-})$ and $(E_{e+}-E_{e-})$ projections for two subsets of data gated on beam energy, heavy-ion scattering angle and e+ or e- TOF chosen to enhance the prominent sum lines at ~810 keV and ~620 keV, respectively.

The narrow width of the peaks implies that the mean laboratory velocity of the emitter is of order v_{CM}. Due to the dynamic, time-dependent nature of the heavy-ion collision, however, one expects that a neutral particle should be created with a broad range of energies [44,45,47,48]. An important consideration within the framework of a particle-decay scenario, therefore, is whether a broad range of emission velocities can produce to a narrow positron peak. The alternative would again involve a source linked to the formation of a long-lived di-nuclear system in order to provide the quasi-static system for the production of the particle-like object.

In addressing this question, we first note that because of the collection of positrons over nearly 4π sr in EPOS, the laboratory distribution of monoenergetic positrons emitted with total energy E^*_{e+} from a system moving with velocity β (and Lorentz factor γ),

$$E^{lab}_{e+} = \gamma(E^*_{e+} + \beta P^*_{e+}\cos\theta^*_{e+}) \tag{1}$$

reflects the maximum Doppler-broadening,[1] forming a rectangle of width $2\gamma\beta P^*$, centered at an energy $E = \gamma E^*$. For creation and decay of an object X^0, the appropriate emitter velocity is $(\gamma\beta)_{em} = P^{lab}_x/M_x c$. For illustration we consider the case of an X^0 created in the CM of a heavy-ion collision (e.g., Th+Cm at 6.02 MeV/u, $v_{CM} = .056c$), with a constant distribution of velocities from $\beta_x = 0$ to $\beta_x = 1$. The laboratory distribution of X^0 momenta is given by

$$w(P_x) = (P_x/2\gamma\beta E_x) \int_{P-}^{P+} w^*(P^*)/P^* \cdot dP^*, \tag{2}$$
$$P_\pm = [\gamma^2(E_x \pm \beta P_x)^2 - M_x^2]^{\frac{1}{2}}.$$

As seen in Fig. 4a, the spectrum is smooth, except for a sharp, Jacobian peak at $\beta^{lab}_x = \beta_{CM}$ which reflects the boost of slowly emitted particles up to the CM velocity.

The associated positron energy spectrum is calculated by superimposing appropriately shifted and broadened lines, weighted by the momentum distribution of Fig. 4a. As seen in Fig. 4b, the concentration of particle velocities in the Jacobian peak leads to a narrow positron peak with a width of only ~10% larger than that expected for a fixed lab velocity, v_{CM} [5]. Neu-

---

1 Although actually a misnomer since it refers to massless phenomena, the Doppler effect is used throughout this work to refer to relativistic kinematic shifts of positrons and electrons from a moving emitter.

tral objects created with larger velocities produce much wider positron distributions, shifted to higher energy, forming a broad continuum contribution of positron energies beneath the narrow peak.

This example demonstrates schematically that the peaks in the positron spectrum of heavy-ion collisions could be consistent with a X^0-creation scenario, if sufficient probability exists for the source to be created with low P_x. The tailing to high energy from fast objects would be hard to differentiate from the contribution of nuclear and dynamic positrons. Moreover, if the lifetime of the X^0 is greater than 10^{-10} s, many of the high-momentum objects escape the ~2 cm fiducial volume of the EPOS spectrometer before decaying, reducing the high energy positron component and enhancing the appearance of a narrow peak [5,7,41]. It therefore is evident that, in principle at least, the emission of a monoenergetic positron in the two-body decay could be consistent with observing narrow positron peaks.

Expected Signature

The experimental signature for the two-body decay of a neutral object which lives long enough to move away from the influence of the Coulomb field of the heavy ion, would be the back-to-back emission of equal energy positron and electron in the rest frame of the emitter. In the laboratory system, the kinetic energy distributions of positron and electron will be broadened, and the opening angle between their momentum vectors reduced from 180°. The ideal experiment would obviously involve measuring the energy and angle of the coincidently emitted positrons and electrons, recon-

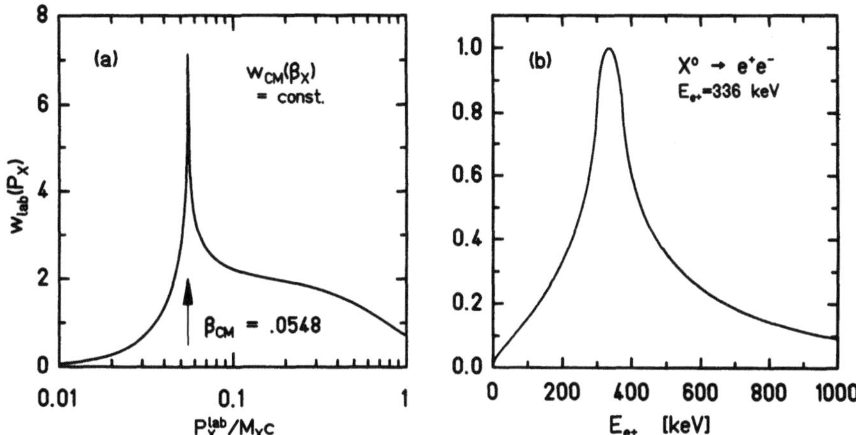

Figure 4. (a) Laboratory momentum distribution of a 1.8 MeV/c² object emitted isotropically from the CM system with a constant velocity distribution from 0 to c. The Jacobian peak at v_{cm} produces a narrow line (b) in the e^+ energy spectrum having a width ≤100 keV.

structing their center-of-mass, and searching for resonances in the invariant e^+e^- mass spectrum.

The salient aspects of such an experiment can be carried out with a modified version of the EPOS spectrometer based on the premise that we are dealing with a slowly moving source. The relatively low source emission velocities ($v_{em} \leq 0.1c$), deduced from the narrow positron peaks, implies that the positron-electron opening angle in the laboratory remains large, $\theta_{e+e-} > 165°$. This property can be exploited by redistributing the solenoid magnetic field to separate the EPOS spectrometer into two solenoid detection systems, transporting positrons to one side and electrons to the other. Detection of back-to-back, relative to isotropically emitted positrons and electrons, is thus enhanced. A search is then made for monoenergetic electron peaks which are coincident with the previously observed narrow positron structures.

Although only very approximate measurement of the positron and electron emission angles is provided by their time-of-flight (TOF), high resolution e^+ and e^- energy measurement substantially determines the positron-electron kinematics. In addition to Eqn. 1, the electron has energy,

$$E^{lab}_{e-} = \gamma(E^*_{e-} + \beta P^*_{e-} \cos\theta^*_{e-}). \tag{3}$$

By adding the total energy of the coincident positron and electron, their Doppler shifts cancel to first order (since $P^*_{e+} = P^*_{e-}$ and $\theta^*_{e-} = \pi - \theta^*_{e+}$), yielding

$$E^{lab}_{e+} + E^{lab}_{e-} = \gamma(E^*_{e+} + E^*_{e-}) = \gamma M_X. \tag{4}$$

Since β and γ refer here to the laboratory velocity of the source, Eqn. 4 gives the total laboratory energy of the neutral object which is very nearly its rest mass for a considerable range of small X^0 momenta. A characteristic signature of a two-body final state is then a sharp peak in the distribution of the sum of positron and electron energies, whose width (determined primarily by intrinsic detector resolution) is smaller than the Doppler-broadened energy distribution of the individual coincident positron and electron peaks.

III. EXPERIMENTAL ARRANGEMENT

EPOS Solenoid Spectromenter

The rather harsh nature of the heavy-ion collision environment places stringent demands on an experiment to measure positron production in HI collisions. Large detection efficiency for positrons is required because of their small peak production rates (10^{-5} per collision), while at the same time, excellent discrimination must be maintained against backgrounds of δ-electrons, γ-rays, X-rays, and scattered nuclei which are produced with 10^4 to 10^6 greater probability. The sought-for monoenergetic electron signal must be extracted from the hugh flux of low-energy δ-electrons. An investigation of the role of the nuclear interaction in positron production requires a determination of the scattering kinematics. Simultaneously, γ-rays and conversion electrons must be measured in order to evaluate backgrounds involving nuclear pair conversion processes.

The EPOS (Electron POsitron Solenoid) Spectrometer (shown in Fig. 5) addresses these requirements by employing a magnetic solenoid field to transport positrons[2] (with high efficiency over a wide energy range) away from the intense flux of γ-rays, X-rays, and heavy-ions near the target to a well-shielded region of reduced background where they can be identified and their energies measured with high resolution. Electrons are simultaneously transported to the opposite end of the spectrometer to another set of detectors. The target chamber is instrumented with position sensitive heavy-ion detectors to determine the scattering kinematics event-by-event in coincidence with detected positrons and electrons, and γ-ray detectors monitor nuclear radiations.

The principle features of the EPOS spectrometer involving detection of positrons, γ-rays, and scattered heavy-ions remain essentially unchanged in the current arrangement, modified to detect positrons in coincidence with electrons, and have been reported in detail elsewhere. An understanding of the electron detection system and certain aspects of the positron detection are important for the present work and are presented here in abbreviated form. The reader is referred to Refs. 23, 16, 20, 21, 41, 49, and 50 for further details.

Positron Detection. Positrons emitted from the target region spiral along the 1.8 kG solenoidal magnetic field with an orbital radius $\rho = (P_{e+}c/eB)\sin\theta_{e+}$ (ρ = 1.25 cm for T_{e+} = 350 keV emitted at $\theta_{e+} = 90°$) to a pencil-like Si(Li) detector, positioned on the symmetry axis 83 cm from the target, where their energies are measured with ~12 keV resolution. Unambiguous identification of the e^+ is ensured by the coincident detection of at least one 511 keV annihilation photon in an eight-fold cylindrical

---

[2] Emission and transport of positrons and electrons is described by the polar coordinate system indicated in Fig. 5. The +Z axis is in the direction of the e^+ Si(Li) detector, +X is in the beam direction, and +Y is the vertical axis.

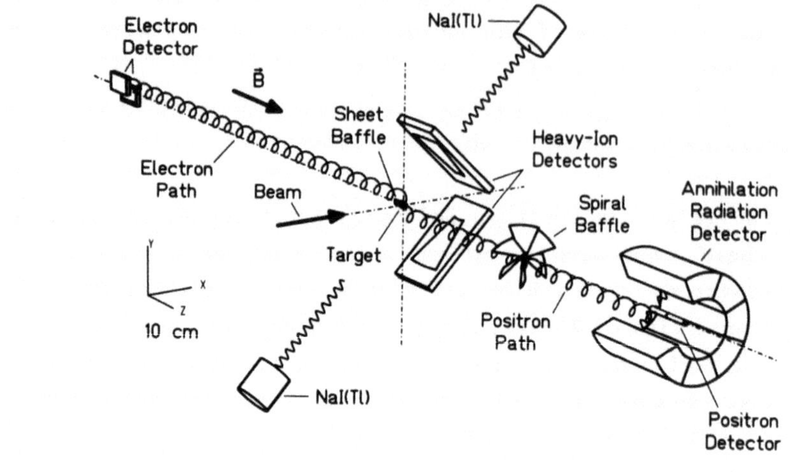

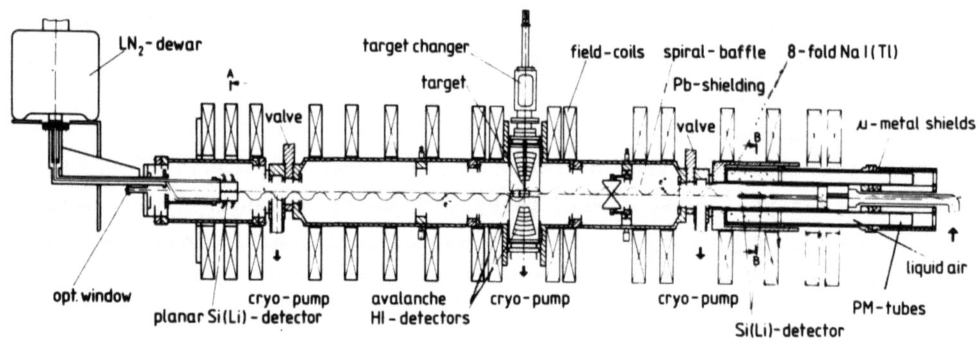

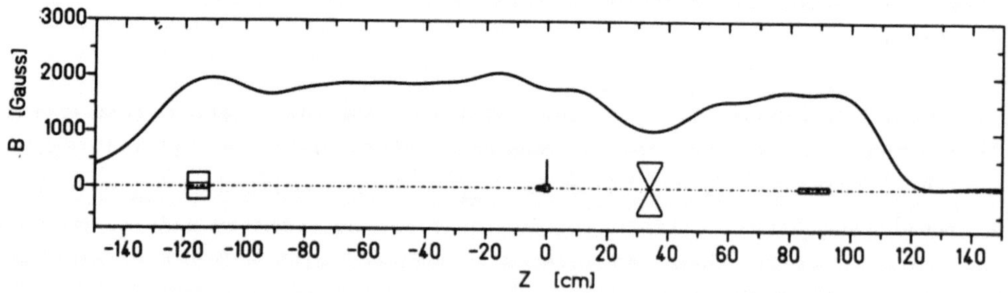

Figure 5. A perspective drawing of the EPOS spectrometer (upper panel), a schematic view (middle), and the magnetic field configuration (lower panel). Polar coordinates, with +Z along the solenoid axis and +X in the beam direction (noted in upper panel), are used to describe transport efficiency.

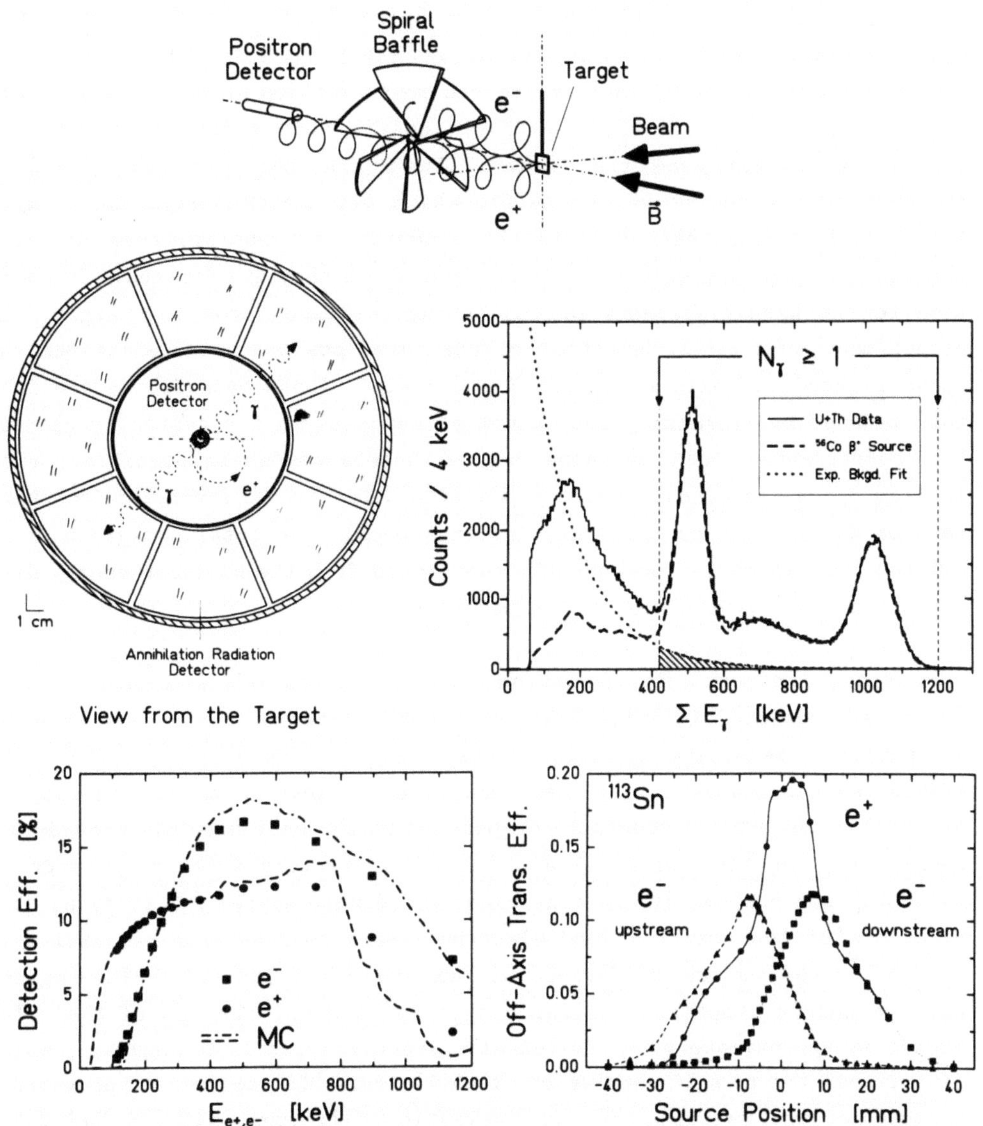

Figure 6. Upper panel: Spiral baffle mechanically stops electrons while transmitting positrons ($\varepsilon_{e-}/\varepsilon_{e+}\approx10^{-3}$). Middle panel: 511 keV annihilation radiation detected in 8-segment NaI crystal array. Sum of γ-ray energies exhibits one and two photon full-absorption peaks. Spectrum for U+Th data (solid histogram) includes exponential background (dotted line) in excess of shape from ^{56}Co β^+ source (dashed curve); shaded background is ≈5.3% of the yield in $N_\gamma\geq1$ analysis gate. Lower left: Energy dependence of e+ and e- detection efficiencies (MC calculations compared to source measurements). ε_{e+} includes 60% detection efficiency of NaI array. Lower right: Measured transport efficiency for e+ or e- emission off-axis. Upstream and downstream e- detectors plotted separately.

array of NaI(Tl) crystals surrounding the Si(Li) detector (Fig. 6b). Delta electrons, which are also transported by the solenoid field, are mechanically blocked by a six-bladed fan-like spiral baffle, which exploits the opposite orbital helicity of positrons and electrons (Fig. 6a). The Si(Li) detector projects a very small geometric cross-section perpendicular to the solenoid axis, and therefore presents only a very small solid angle to γ-rays and X-rays from the target. It is further shadowed by a disk attached to the center of the baffle which attenuates γ-rays and X-rays. This axial arrangement exploits the property that positrons spiral back through the same magnetic field-line from which they are emitted. In addition to the high efficiency of collecting positrons from the target, the arrangement suppresses detection of scattered positrons and electrons and those created off-axis by external conversion. These backgrounds are further reduced by minimizing the amount of material in the central region of the spectrometer, and avoiding high-Z substances where possible. Under In-Beam operating conditions, the positron detector has a very low counting rate of $R_{Si(Li)}$~500 Hz (2500 Hz, instantaneous). The total background from all sources, which is subsequently subtracted from the positron energy distribution, constitutes ≈10% of the positron yield.

Heavy-Ion and γ-Ray Detection. Positrons (as well as γ-rays and electrons) are measured in coincidence with both scattered heavy-ions detected in position sensitive Parallel-Plate Avalanche Counters [21,50]. The PPAC's use continuous delay-line readouts to achieve an angular resolution as small as $\delta\theta = 0.5°$. Contacts evaporated on the entrance foil provide a ϕ sensitivity in ≈10° bins. The detected laboratory scattering range is $20°<\theta<70°$ and $-28°<\phi<28°$. A convenient coordinate system for studying correlations of the positron (and electron) peaks to heavy-ion kinematics is defined by the sum and difference of the coincident ion scattering-angles. For the nearly symmetric collision systems studied, the sum of the polar angles is approximately 90° for elastic scattering, and the angular resolution translates to a $\Sigma\theta$ resolution of ~0.7°. This "$\Sigma\theta$" coordinate provides a measure of the reaction "Q" for inelastic collisions, with 25 MeV total energy lost from the binary system corresponding to a 1° shift.

In the presence of the solenoid magnetic field, the scattered ions are deflected by up to 2° along their trajectories. There is an indication that differences from the velocity-dependent equilibrium ionic charge distributions [51] appear to be associated with the positron-peak producing collisions. This interesting possibility is suggested by the observation that the pattern of scattering-angle correlation associated with the peak is rotated slightly away from elastic scattering kinematics. It is *this* feature which has been exploited in the kinematic cuts used to produce the spectra of Fig. 1. This point is discussed in greater detail by K. Stiebing in another contribution to this conference.

γ-rays emitted from the individual nuclei excited in the collision are detected by two NaI(Tl) detectors, positioned in the vertical plane at 45° and 225° with respect to the beam direction. γ-Ray energy distributions measured in coincidence with scattered heavy-ions, under conditions identical to those used for the positrons and electrons, provide the opportunity to investigate a variety of nuclear hypotheses which could explain the observed positron and coincident electron structures.

Electron Detection. A very high flux of low-energy delta electrons is one of the salient features of superheavy collision systems. Dealing with this situation presents the principle difficulty in carrying out coincidence measurements of electrons with positrons. We therefore dwell on this aspect of of the apparatus in some detail.

In addition to the usual problems of measuring with high counting rates while maintaining good energy resolution, the presence of a high multiplicity of delta electrons, together with high efficiency, compounds the problem of detecting delta electrons simultaneously with the interesting monoenergetic electron signal. Because of the very steep exponential character of the δ-electron energy distribution, this problem was addressed by maintaining high detection efficiency above the interesting energy range of E_{e^-}>300 keV, while striving for a very sharp cut-off in detection efficiency below this threshold. The method chosen employed a geometry similar to that used previously by Backe et al. [52]. It adds a further feature to exclude the low energy part of the spectrum. Two planar Si(Li) detectors (32×65 mm^2 $\times$ 3 mm thick) are oriented vertically, parallel to the solenoid axis at a distance 112 cm from the target. As shown in Fig 7c, the counters face each other and are 9 mm from the axis, one displaced upward and the other downward from the axis. Delta electrons having a small orbital radius, $\rho \propto P_{e^-} \sin\theta_{e^-}$, may then spiral between the counters undetected. Detection is thereby suppressed for low electron energies (small P_{e^-}) and flat emission angles, θ_{e^-} near 180°, with respect to the solenoid axis. Figure 7d illustrates the detection efficiency near threshold, showing that emission for $P_{e^-} < \ell(eB/2c)/\sin\theta_{e^-}$ is excluded.

Increasing the inter-detector distance to cut off all undesired electron energies leads to a smooth decrease in efficiency (i.e., for steeper emission angle) for the interesting energy range as well. A complementary, second-stage, low-energy suppression, consists of a small sheet of aluminum (10 mm $\times$ 20 mm $\times$ 1 mm) placed beside the target. Low energy electrons with emission angles steep enough such that their orbital radius is sufficient to reach the off-axis Si(Li)'s have correspondingly shorter pitch-length (distance traveled along the axis in one orbit) and thus strike the sheet baffle. The pitch-length is given by $Z_o = 2\pi(P_{e^-}c/eB)\cos\theta_{e^-}$, leading to a suppression for $P_{e^-} < \eta(eB/2\pi c)/\cos\theta_{e^-}$ (see Fig. 7d), which complements the

$1/\sin\theta_{e-}$ dependence provided by detector displacement. A small magnetic mirror field on the electron side of the spectrometer limits detectable emission angles to $\theta_{e-} > [\pi - \sin^{-1}(B_o/B_{max})] = 112°$, so the sheet-baffle length can be safely extended to the pitch-length of a 300 keV electrons emitted at 112° without reducing detection efficiency for interesting events. The electron-energy/emission-angle acceptance was optimized by

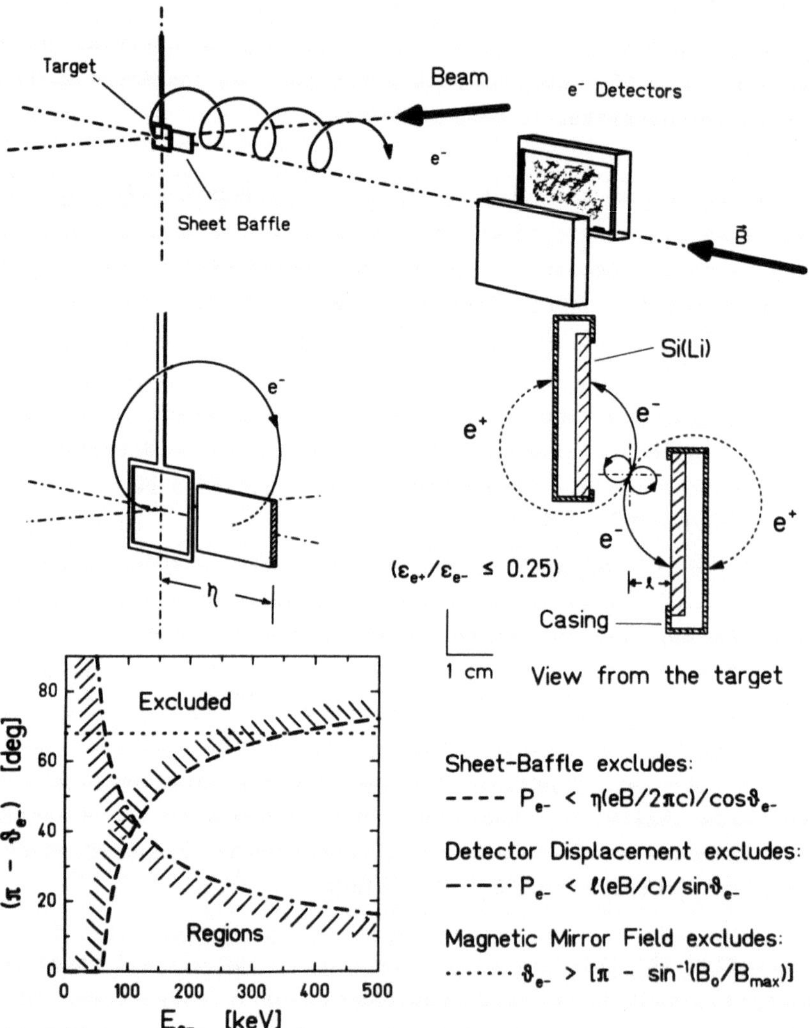

Figure 7. Suppression of intense low-energy δ-electron flux by: 1) Off-axis arrangement of the 32×65 mm² (3 mm thick) Si(Li) detectors (middle-right). Low-energy electrons emitted at flat angles spiral between the counters. Positrons of opposite orbital helicity have a reduced probability (×0.25) for detection. 2) A sheet baffle (middle-left) absorbs low-energy electrons emitted at steep angles (i.e., short pitch lengths).

choosing the sheet baffle dimensions and detector geometry which best suppress detection for E_{e^-}<300 keV, while simultaneously enhancing detection for E_{e^-}>300 keV, as calculated by an adiabatic Monte-Carlo ray-tracing technique [41].

The Si(Li) detectors were each covered with a light-tight 3 $\mu g/cm^2$ (10 μm) Al foil, and cooled to -30° C in the first experiments providing $\delta E \approx 30$ keV. For subsequent measurements, energy resolution was improved to ≈10 keV with LN_2 cooling. Electron backscattering from the detectors leads to a lineshape peak-to-total ratio of 59% at 350 keV. The "in-beam" performance of the chosen detector system is as follows. Under typical conditions (one particle nA beam on ~300 $\mu g/cm^2$ targets), the instantaneous counting rate in the individual detector segments was 100 kHz, with an accidental coincidence rate of ≤10%. A measurement of the K-conversion lines of a ^{207}Bi source accumulated simultaneously with beam hitting a 300 $\mu g/cm^2$ gold target, showed that for instantaneous rates as high as 250 kHz, no shift in gain or degradation in energy resolution occurred. For the particular geometry used, positron-electron coincidences constituted ≈20% of the positron rate. The average multiplicity of electrons emitted per collision is ≈4 for E_{e^-}>100 keV. The pile-up rate, i.e., the probability of observing more than one electron in one detector, was measured by investigating the coincident rate between the two electron detectors. Pileup, from multiplicity alone, is ≈10%, rising to ≈20% due to chance events. The net result is an ~20% decrease in the electron peak detection efficiency, yielding a net lineshape peak-to-total ratio at 350 keV, including outscattering, of ≈47% [41].

Detection Efficiency

Figure 6d summarizes the EPOS efficiency for transporting and detecting positrons and electrons integrated over 4π sr. The curves are calculated using a recent version of the ray-tracing program ECPI [53] which follows the trajectory of positrons and electrons emitted from the target, through the spectrometer, to the detectors, by pointwise integration of their equations of motion in the solenoid field. The data points represent source measurements using ^{113}Sn conversion electrons (simulating a range of energies by varying the magnetic field strength). Both the shape and the absolute magnitude of both positron and electron efficiency are well reproduced.

The transport efficiency, integrated over azimuth angles for energies between 300 and 400 keV, is displayed in Fig. 8a as a function of the e^+/e^- polar emission angle (with respect to the solenoid axis), illustrating the back-to-back orientation of positron and electron acceptance regions. Figure 8b shows the ϕ dependence of detection, integrated over polar angle. These distributions are fairly uniform, with the principle regions of positron and electron acceptance oriented colinearly back-to-back with one another. It is interesting to note that the two electron detectors sample complementary regions of emission angle: the backward counter detects electrons emitted upward, and the forward counter, those emitted downward.

To visualize the EPOS acceptance, it is useful to picture these calculations in three dimensions. The regions of solid angle sampled resemble two thick, hollow, coaxial cones, joined at their common vertex. While obviously enhancing the observation of back-to-back emitted positrons and electrons, a broad range of opening-angles will be detected with significant efficiency down to $\theta_{e+e-} \sim 20°$.

The full effect of the spiral baffle, sheet baffle, target frame, and detector geometry is presented in the two-dimensional θ-ϕ dependence of transport efficiency illustrated in Fig 9 [41]. The dark (and light) regions indicated directions of a 350 keV positron or electron, parameterized by cosθ and ϕ w.r.t the solenoid, in which a hit (or miss) is recorded in the corresponding Si(Li) detector. Several details of the EPOS transport system are readily apparent. For example, for positron detection

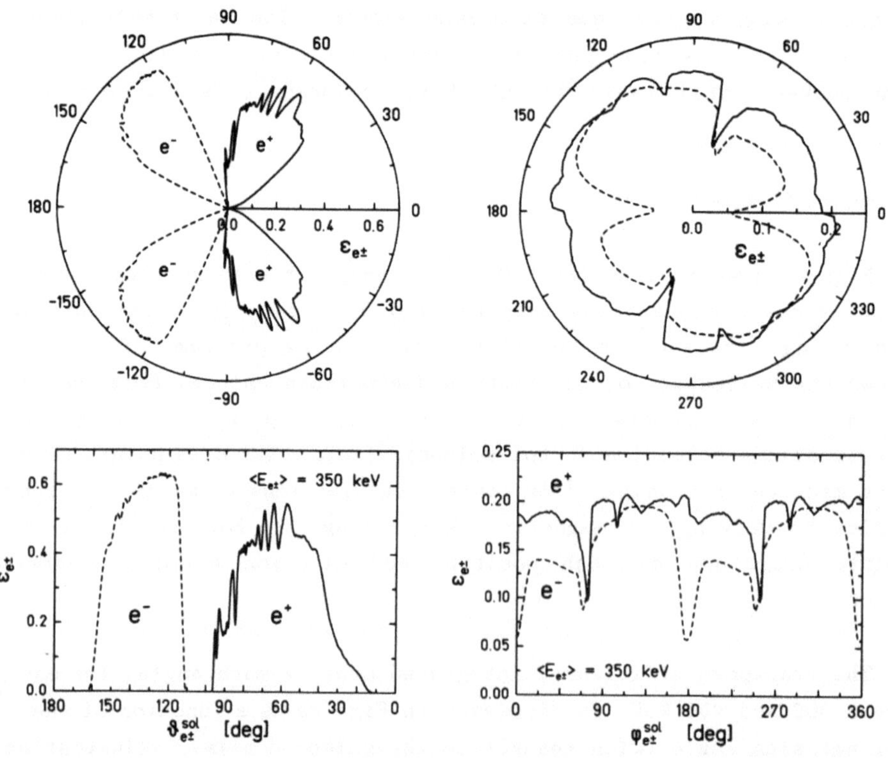

Figure 8. EPOS transport probability for E_{e+} and E_{e-} between 300 and 400 keV, shown as a function of: (left) polar emission angle, θ, w.r.t. the solenoid axis (integrated over ϕ), and (right) ϕ (integrated over θ).

($\cos\theta > -.2$), the six-bladed spiral baffle produces a six-fold periodicity of efficiency with azimuth angle. Gaps occur in detection as a function of θ (vertical bands) when after an integral number of orbits, the positron returns to the axis at the position of the spiral-baffle central disk (34 cm from the target). Similar oscillations arise when electrons strike the casing of the planar-Si(Li) detectors. Also visible are horizontal bands at $\phi = 80^\circ$ and 260° corresponding to vertically emitted positrons and electrons which hit the target frame.

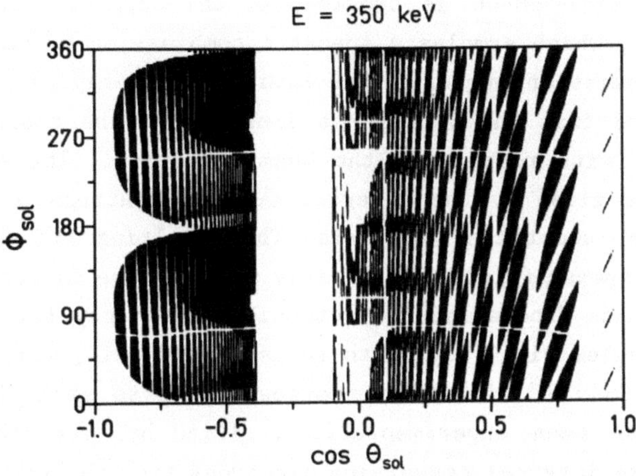

Figure 9. Transport efficiency for 350 keV positrons and electrons emitted from a point target with angles θ, ϕ w.r.t. the solenoid axis. Dark (light) regions denote hits (misses) in corresponding detector. Evident features include the six-fold periodicity of e+ detection in ϕ from the spiral baffle, gaps in efficiency caused by vertically oriented target frame at $\phi \approx 80^\circ$ and 260°, and the oscillatory structure in $\cos\theta$ when, after an integral number of orbits, the positron hits the central disk of the spiral baffle or an electron strikes the Si(Li) detector casing.

Although the combination of magnetic transport system and baffles produces an efficiency response function with considerable fine-structure, these fluctuations do not generally lead to structure in the observed distribution of isotropically emitted positrons or electrons. Their periodicity is small compared to the total region of acceptance, and they change smoothly with positron and electron energy. Upon integration of the detection efficiency over physically reasonable angular ranges, many "oscillations" are averaged over, almost eradicating any fine-structure from the resulting energy distributions. Even simply integrating the position of emission over the size of the beam spot (dia. ~5 mm) significantly smooths the detection efficiency [41] because, compared to the typical orbit size ~3 cm, this represents a smearing of the phase of the e^+ or e^- trajectory by a substantial fraction of an orbit. A trivial example of the smoothness of detection efficiency when integrating over 4π is the total energy-dependent detection efficiency of Fig. 6d, as verified by direct source measurements. Figure 10 shows the β^+ energy distribution measured with a ^{56}Co source, again illustrating the absence of oscillatory structure in the detection of continuous distributions.

A differential check is provided by the Doppler-broadened positron lineshape [3,5], which samples a specific combination of emission energies correlated to emission angles. The laboratory energy of a monoenergetic positron emitted from the CM depends linearly on the cosine of the polar emission angle with respect to the beam direction. The efficiency at a fixed energy is given by the integral over the pattern of Fig. 9 on the appropriate cone about the beam axis. The resulting lineshape, corrected for detector response (Fig. 11) is fairly smooth. The characteristic dip in the center results from a reduced detection efficiency for positrons emitted at small angles with respect to the solenoid axis, which travel nearly perpendicular to the CM motion, corresponding to small Doppler shifts. This result has been experimentally verified by reversing the solenoid field to measure internal conversion electrons from an excited ^{126}Ba nucleus formed in $^{12}C(^{118}Sn,4n)$ fusion reactions at 5.86 MeV/u. Figure 11a shows the good agreement for the $2^+ \rightarrow 0^+$ K-line, between the measured width

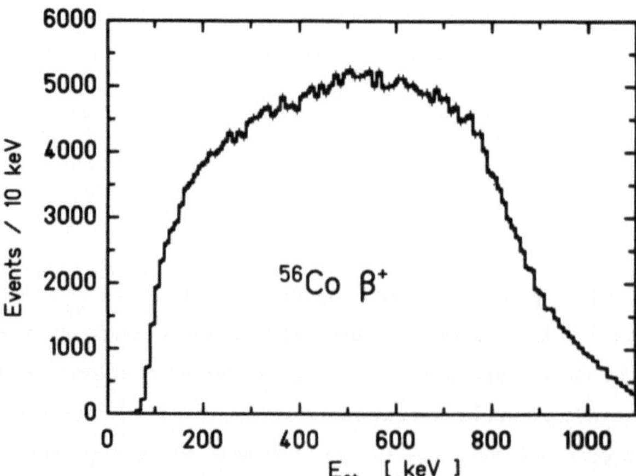

Figure 10. Measured energy distribution of positrons from $^{56}Co\ \beta^+$ source.

(107 ± 5 keV) and that calculated (109 keV) for the emitter velocity, v_{CM} = .101c. Only a slight (~10%) reduction of the linewidth (also understood) is observed for emission away from the solenoid axis as indicated in Fig. 11c, where a 10 mm displacement of the target and the 188 ps 2^+ lifetime results in an average ≈15 mm off-axis emission.

While verifying the smooth nature of the EPOS detection efficiency individually for positron and electron emission, these arguments are not immediately applicable to coincident detection. Quantitative calculations of the detected energy distribution of coincident positrons and electrons are necessary to investigate the possibility that narrow lines may emerge from broad e^+/e^- distributions. In addition, although the efficiency for isotropically emitted positrons and electrons is simply the product of the curves in Fig. 6d -- $\varepsilon_{e+e-}(E_{e+},E_{e-}) = \varepsilon_{e+}(E_{e+})\times\varepsilon_{e-}(E_{e-})$ -- for two-body decays the laboratory positron-electron angular correlation depends on the velocity of the emitter, so coincidence detection efficiency can only be calculated in a model-dependent way for a variety of source velocity distributions. Equal attention must also be paid to background processes since their signature must be understood as obscuring backgrounds, as well as eventual explanations of the structure in the coincidence spectra. These points have been addressed with Monte-Carlo simulations discussed next.

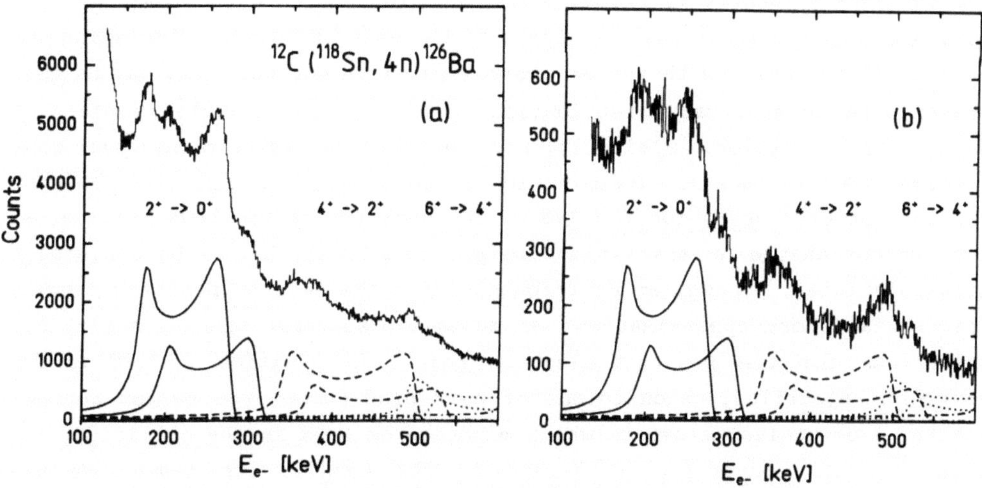

Figure 11. Energy distribution of electrons (transported to the e+ detector with reversed magnetic field) from 5.86 MeV/u $^{118}Sn\rightarrow^{12}C$ collisions, measured in coincidence with γ-rays, compared to calculated Doppler-broadened K and L internal conversion lines from ^{126}Ba (v_{CM} = .101c). (a) The measured $2^+\rightarrow0^+$ K-line width (107 ± 5 keV) agrees with calculated width of 109 keV. (b) With target positioned downstream, off-axis emission (≈15 mm) causes <10% reduction in linewidth.

IV. MONTE-CARLO SIMULATIONS OF POSITRON-ELECTRON COINCIDENCES

The computer code MCSPEC (Monte-Carlo Simulation of Positron Electron Coincidences) [41,54] was developed to generate a wide range of event scenarios and to model the response of the EPOS spectrometer in detecting positrons and coincident electrons. Several sources of coincident positrons and electrons are modeled, e.g., the production and two-body decay of neutral or charged objects, multi-body decays, various types of nuclear internal pair conversions, dynamic positron and δ-electron emission, or any generalized emission of positrons and electrons from the target. The generated positrons and electrons are then transformed to the lab system, where the positron and electron are tracked through the EPOS spectrometer to determine whether or not each is detected. The pulse-height responses of the Si(Li) counters are taken into account, pileup from the δ-electron flux is added, and electronic pulse-shaping and trigger-logic efficiencies are simulated. Finally, the generated events are analyzed with a logical duplicate of the experimental data analysis program. Projections of the simulated events may then be compared with identical cuts applied to the experimental data.

Event Generators. Event generation for each specific scenario is flexible to allow a detailed study of various aspects of the simulated positron-electron source. For example, the production and decay of a neutral particle-like object includes its angular and momentum distributions w.r.t. the emitting frame (CM of the heavy-ion collision, the scattered nuclei, or the stationary laboratory frame). The positron and electron momentum vectors are constructed in the rest frame of the emitting source, including the angular distribution with respect to the particle motion. They may be generated with an arbitrary distribution in e^+/e^- opening-angle in place of back-to-back emission. Similarly, studies of of nuclear internal pair conversion (IPC) include 1) determination of the scattering angle distribution for the detected heavy-ion (including the geometry of the EPOS PPAC's), 2) any nuclear charge or transition energy, 3) E0, E1, E2, or M1 conversion multipolarities, 4) angular distributions for the emitted positron, and 5) theoretical (Born approximation) or mutually isotropic opening-angle distributions for the positron-electron pair. In addition to IPC in the heavy-ion ejectiles, calculations of pair production from target contaminants, from nuclear clusters which might be emitted in the collision, and from external conversion of γ-rays in the target, its frame, or the sheet-baffle material, have been investigated.

Efficiency Determination. Calculation of the large quantity of MCSPEC events necessary for studying a wide variety of production processes requires a computationally efficient evaluation of the EPOS Spectrometer response. The standard technique employed involves the interpolation of a look-up table of detection probabilities calculated for a large range of positron and electron emission with the ECPI ray-tracing code [53]. The mesh size (the pixel size in Fig. 9) is small compared to the periodicity of detection efficiency to ensure high accuracy (corresponding to steps of 10 keV in $E_{e+,e-}$, 1.5° in ϕ, and .002 in $\cos\theta$). On the average, 60% of positron "hits" are assumed to be detected to account for the mean efficiency for detecting 511 keV annihilation quanta in the NaI crystal array.

Detector Response. Correction for the response of the solid state positron and electron detectors includes finite energy resolution, tailing due to energy-loss, pile-up due to the rescattering (≈8% probability) of an undetected 511 keV annihilation photon in the positron detector, the high multiplicity of δ-electrons striking the electron detector (≈20% probability), and the finite Si(Li) resolution for the fast TOF pickoff. Pulse-height corrections (i.e., energy resolution, tailing, and pileup) are made for positrons or electrons by the standard method of sampling at random the inverse, F^{-1}, of the cumulative probability of observing an energy E, given incident energy E_o,

$$F(E,E_o) = [\int_0^E R(E',E_o)\, dE']/[\int_0^\infty R(E',E_o)\, dE'], \quad (5)$$

where $R(E,E_o)$ is the experimental response function determined from measurements with conversion electron line sources, by iterative unfolding of β-decay source spectra, or from measurments of pileup as described above.

The time-of-flight of the positrons and electrons to their respective detectors, which depends on their velocity parallel to the solenoid axis, is evaluated by interpolating a smooth function of energy and polar emission angle, calculated with ECPI. The fast timing resolution of the Si(Li) detectors is ≈4 ns FWHM for electrons and ≈7.5 ns FWHM for positrons, compared to average flight times to the counters from 4 to 15 ns, with a mean of ~8 ns. Low-energy electronic discriminator thresholds are measured separately for each detector. No attempt is made to evaluate the detector response in a "microscopic" Monte-Carlo fashion tracking the penetration of a positron or electron into the silicon crystal. To the extent that energy-loss, backscattering, and time resolution are independent of the angle and position of incidence, or more realistically, that a broad range of Doppler-shifted emission momenta and incidence angles map onto closely spaced detected positron and electron energies, this pseudo-Monte-Carlo approach is extremely accurate.

Event Analysis. Detected positrons and electrons are analyzed by a logical duplicate of the experimental data analysis program [55]. Adjustments are made to the times-of-flight to correct, in an average way, for positron or electron energy and PPAC timing. (PPAC's provide the experimental "start" signal, which depends on heavy-ion TOF as a function of the scattering angles θ and ϕ). Resulting time distributions reflect, in principle, only the positron or electron emission angle with respect to the solenoid axis, and real time delays (e.g., lifetimes). Positrons are experimentally identified by a clean three-fold coincidence signal: 1) prompt kinematic coincidence between both PPACs indicating a quasi-elastic binary heavy-ion scattering event, 2) prompt time and energy signal in the positron Si(Li) detector, and 3) coincident detection of $\geq$1 511 keV photon in the NaI(Tl) annihilation radiation detector array. As a subset of positrons, we look for a fourth condition, 4) a prompt time and energy signal in one of the electron detectors. If both electron counters register, the event is discarded to avoid double-counting of the positron, or piling-up of the electron signals which distort the positron-electron coincidence distributions. For each sequential pair of detector elements in the four-fold coincidence pattern, accidental background is evaluated by subtracting the event distributions averaged over several neighboring beam-pulses from the data of the prompt coincidence beam-pulse. The energies of the positron and electron signals are combined to give sum and difference energies as described below. Event sorting is performed first for detected positrons, of which coincidence events are a subset, thus retaining an absolute normalization for comparison to experiment.

Two-Body Decay Coincidence Detection

The above computations to evaluate the acceptance and response of the EPOS spectrometer can be used to quantitatively determine the efficiency for detection of back-to-back events, and, in particular, the expected signature of the creation and two-body decay of neutral particle-like objects. Figure 12 presents expected coincident positron-electron yields for the decay of a neutral object, of mass 1.8 MeV/c^2, created at rest in the CM of the HI collision system (v_{CM} = .056c). The energy spectrum of all positrons detected for 250000 detected positrons (2×10^6 generated X°s) is shown in Fig. 12a, taking into account all detection efficiencies and detector response. For clarity, contribution from continuous atomic and nuclear processes is neglected. Since the velocity of the neutral object coincides with the CM motion, this spectrum gives the Doppler-broadened laboratory lineshape of a monoenergetic positron peak emitted from the CM, comparable to previous calculations (Refs. 3,5,23 and Fig. 11). Although slight oscillatory structure remains in the emitted distribution, it is small compared to the Si(Li) energy resolution so that the observed peak is quite smooth.

Figure 12b plots the positron energy versus coincident electron energy, and Figs. 12c,d present the positron energy distribution when gating on the coincident electron peak ($340 < E_{e-} < 420$ keV), and the electron distribution for $340 < E_{e+} < 420$ keV, respectively. The diagonal orientation of the positron-electron energy correlation pattern illustrates the correlated cancellation of Doppler shifts characteristic of two-body decay. While the individual positron and electron lines exhibit the full Doppler broadening appropriate for the emitter velocity, their first-order Doppler shifts ($\gamma\beta P^*_{e+}\cos\theta^*_{e+}$ and $\gamma\beta P^*_{e-}\cos\theta^*_{e-}$) are equal and opposite, resulting in the diagonal correlation pattern.

Sum-Energy Distribution. Adding the positron and electron laboratory energies leads to the narrow peak of Fig. 12e whose ≈50 keV width, due entirely to finite energy resolution of the detectors, is smaller than either the positron or electron peaks of Figs. 12c,d. As noted above, improvement of electron energy resolution in later experiments reduced the minimum expected sum-energy peak width to ≈25 keV.

The rationale behind the wedge-shaped contour (labeled C in Fig. 12b) used to select the data included in the sum-energy projection of Fig. 12e is to account for the change in kinematic broadening of the individual positron and electron distributions as a function of energy. For example, the two-body decay of a 2 MeV/c^2 object, created at rest in the CM, would lead to positron and electron lines centered at 490 keV, with FWHM of about 96 keV, while a mass of 1.4 MeV/c^2 would produce a mean e^+ or e^- energy of 190 keV with FWHM of only 53 keV. A wider range of positron and electron energies must therefore be added when searching for higher-energy structures, and a smaller range for low energy sum-lines [7]. The wedge-shaped cut is easily parameterized by $(E_{e+}-E_{e-})/(E_{e+}+E_{e-})$ = constant, and roughly follows the momentum dependence of the positron or electron linewidth. It effectively excludes as much background intensity (for which $E_{e+} \neq E_{e-}$) as possible from the sum-energy projection.

Difference-Energy Plot. Complementary to the sum-energy projection is a plot of the the difference of positron and electron energies, for a given

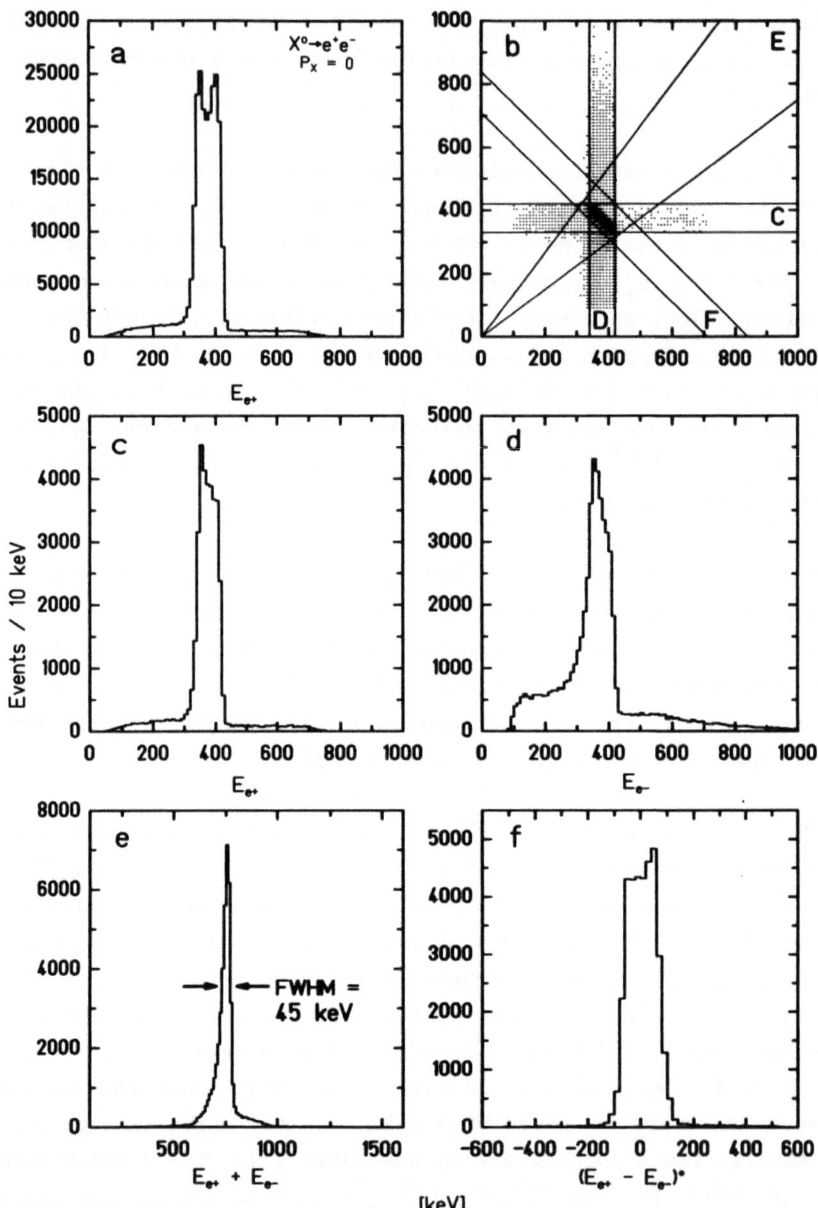

Figure 12. MCSPEC simulation of $X^0 \rightarrow e^+e^-$ decays for a 1.8 MeV/c² object created at rest in CM. (a) Total detected e+ distribution. (b) Plot of E_{e+} vs. E_{e-} illustrates the correlated cancellation of Doppler shifts. (c)-(f) Projections on the E_{e+}, E_{e-}, $(E_{e+}+E_{e-})$, and $(E_{e+}-E_{e-})$ axes, respectively, for data within gates indicated in (b). The ratio of detected coincidence positron-electron peak events to total observed peak positrons is 16%. Sum-peak width of ~50 keV reflects detector resolution (1985).

sum energy. Figure 12f presents events within the diagonal contour labeled D in Fig. 12b, plotted against $\Delta E_{\pm} = E_{e+} - E_{e-}$, for $710 < (E_{e+}+E_{e-}) < 840$ keV. This projection provides information regarding the symmetry of the positron and electron peak energies and, for correlated (back-to-back) emission, is the best measure of the Doppler broadening and source velocity. As will be discussed below, the shape of the difference-energy spectrum is central in discriminating against conventional origins of coincident positron-electron emission.

Detection Efficiency. The detection efficiency for back-to-back emission of positrons and electrons from a source moving with the CM is directly given by the yields of the simulated events. If detector response is neglected, the ratio of observed coincident e^+/e^- pairs to total detected positrons is $N_{e+e-}/N_{e+} = 0.32$ [41]. Since the positron detection efficiency is approximately 12% (Fig. 6), the absolute coincidence detection efficiency is $\varepsilon_{e+e-} = 4\%$, at $E_{e+,e-} = 380$ keV. Lineshape tailing and pileup in the Si(Li) detectors reduce the fraction of total events that can be identified in the positron, electron, or coincidence peaks. The lineshape peak-to-total ratio (P/T) for positrons is ≈80%, while for electrons P/T ≈ 50%. The relative detection ratio of coincident peak events is $(N_{e+e-}/N_{e+})_{pk} = 0.16$. Including "pile-up rejection" of double e^- hits, this falls to 14%. The absolute coincident peak efficiency is $\varepsilon^{pk}_{e+e-} \equiv 1.4\%$. The larger low-energy tailing associated with electron detection not only decreases the coincident peak detection efficiency, but shifts the mean detected electron distribution to slightly lower energies than for positrons. In parts c,d of the figure, the mean observed positron energy is 380 keV while $E_{e-} = 375$ keV. The difference-energy peak is similarly shifted to slightly positive values [7].

Non-Correlated Coincidence Detection

The enhancement of detection efficiency for back-to-back compared to spacially uncorrelated emission is demonstrated by the MCSPEC simulation shown in Fig. 13. Here it was assumed that the positron and electron are emitted isotropically with respect to each other from the CM system with equal energy = 380 keV. For comparision, the same number of total events is generated, and identical projections are plotted on the same absolute scale as those of Fig. 12. The detection efficiency for no spacial correlation is much smaller than for back-to-back emission. The total N_{e+e-}/N_{e+} ratio, excluding detector response, is 0.155 (compared to 0.32), and for peak events $(N_{e+e-}/N_{e+})_{pk} = .07$ (=6% including pile-up rejection), a factor of two smaller than for back-to-back emission [7].

As is apparent by the two-dimensional energy correlation plot in Fig. 13b, the pattern of events for isotropic emission differs substantially from the diagonal structure of two-body decay. Since the direction of the electron and positron are not relatively back-to-back, the Doppler shift of one lepton is unrelated to, and thus does not cancel, the Doppler shift of

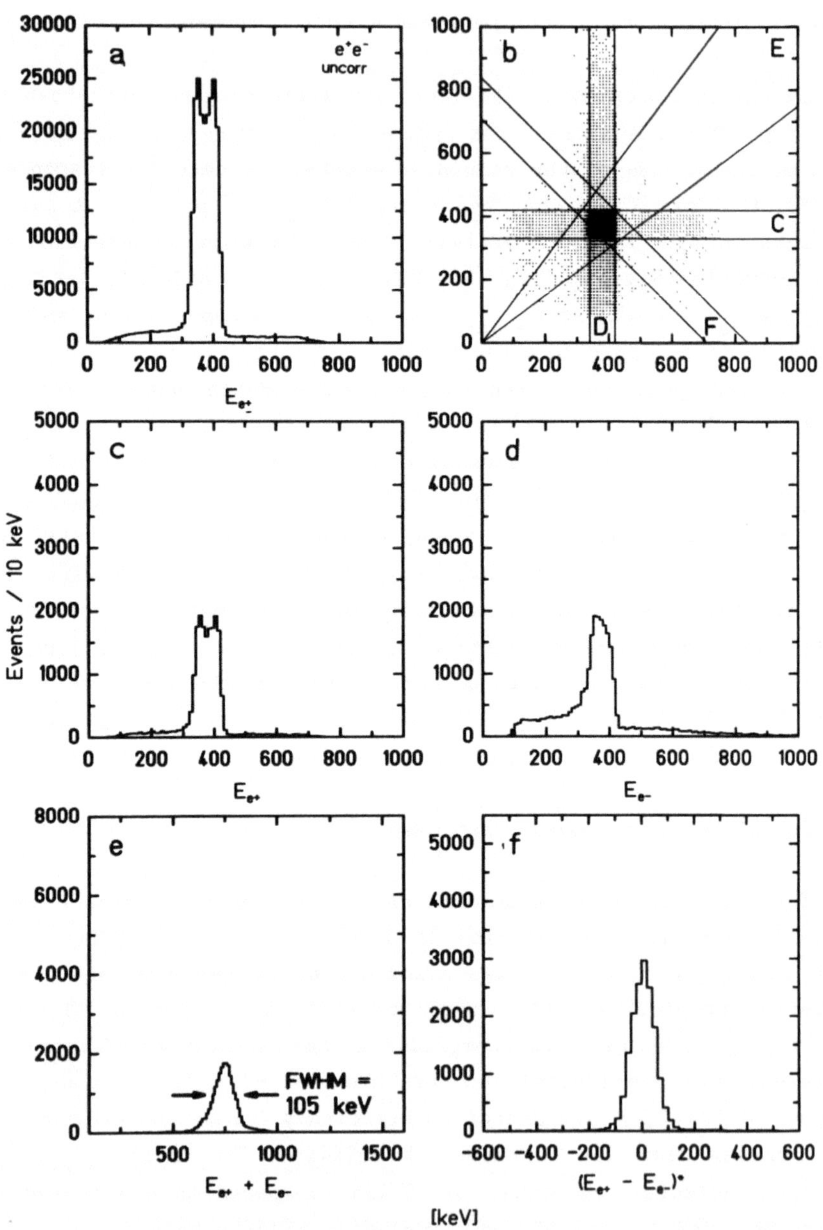

Figure 13. MCSPEC simulation of monoenergetic (380 keV) e+ and e- isotropically emitted from the CM with no relative angular correlation. Parts (a)-(f) same as Fig. 12. (e) Sum-energy peak width, ~105 keV, is approximately quadrature sum of individual e+ and e- Doppler-broadened linewidths (v_{CM} = .056c). The coincidence peak intensity relative to total observed e+ peak counts is 7%.

the other. The sum-energy peak FWHM for isotropic emission is approximately the quadrature sum of the individual lepton linewidths, about 105 keV for a mean source velocity of v_{CM} (Fig. 13e). The difference peak exhibits a similar width, in comparison to back-to-back emission for which the Doppler correlation leads to a width of 160 keV (Fig. 12f). For monoenergetic positron and electron lines having different energies, a similar rectangular pattern in the energy-energy plot can occur anywhere in the correlation plane, with suitably modified widths. The wedge-cut may not necessarily overlap such intensity patterns, complicating identification, and the appropriate difference peak will not be centered near $E_{e+}-E_{e-} = 0$.

Continuous Pair Creation Detection

The preceeding calculations show that detection of positrons and electrons resulting from a two-body decay is enhanced compared to uncorrelated, isotropically emitted monoenergetic lines and leads to a peak in the sum-energy spectrum, identifying such events as candidates for representing the two-body decay of a neutral object. In addition, the positron Doppler-broadened lineshape, verified in Fig. 11a,b, remains intact when detected in coincidence with electrons (Fig. 12c), i.e., no fragmenting of the distribution occurs from the inclusion of the oscillating electron efficiency pattern, leading to additional structure. MCSPEC simulations of continuous dynamic positron and delta-electron emission similarly reveal no structure in the coincident intensity distributions. An important further check is a simulation of nuclear IPC, the only conventional source of energy-correlated positron-electron pairs in HI collisions, which accounts for ≥20% of the total detected positron yield.

Figure 14 presents the expected energy distributions [56] for a 1.8 MeV E0 transition in one of the scattered nuclei (Z≈92). In this three-body process, the heavy recoiling nucleus removes momentum but very little energy from the decay positrons and electrons, so that over their individually broad range, the positron and electron energies sum to a constant, $E_{e+}+E_{e-} = W_o-2m_ec^2$. Part (a) presents the nearly triangular positron energy distribution, shifted by the Coulomb suppression of low-energy positrons (and high-energy electrons) to higher E_{e+}. A small fraction of the total intensity falls within the windows C,D of Fig. 14b, leading to peak-like structure in the projections on E_{e+} and E_{e-} axes (Fig. 14c,d). The dashed histogram is calculated assuming that the opening-angle distribution between positron and electron is given by the Born approximation (valid for Z→0) [57-58]:

$$dN(E0)/d(\cos\theta_{e+e-}) \propto P_{e+}P_{e-}(E_{e+}E_{e-} - m_e^2 + P_{e+}P_{e-}\cos\theta_{e+e-}) \qquad (6)$$

which exhibits a preference for the positron and electron to be emitted in the same direction, peaking at $\theta_{e+e-} = 0°$. Maximum anisotropy occurs for

$E_{e+} \simeq E_{e-}$ while as E_{e+} or E_{e-} approaches zero the distribution becomes more spherically symmetric. Since an exact calculation for Z≈92 is not yet available, for comparison a very conservative limiting case assuming iso-

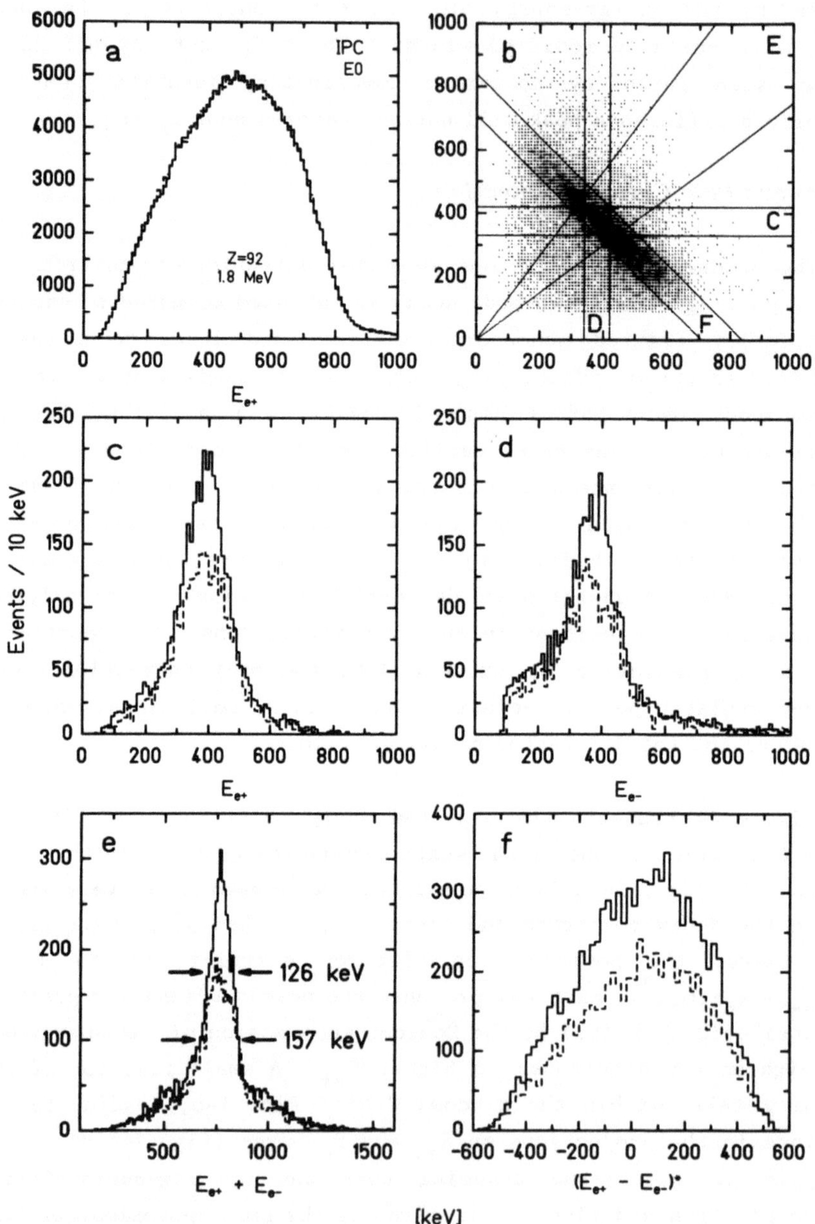

Figure 14. Simulation of IPC of a 1 8 MeV E0 transition in the scattered heavy-ions (Z=92) Parts (a)-(f) as in Fig. 12. Solid histograms assume relative isotropy between emitted e+ and e-. The dashed curves include $w(\cos\theta_{e+e-})$ calculated in Born approximation (Eqn. 6).

tropy between the positron and electron is shown by the solid histrogram. The intensity of coincident positrons and electrons near 380 keV relative to total detected positrons is small: ≈0.82‰ in the Born approximation, or ≈1.02% assuming isotropy. The broad angular distribution and rather large scattered heavy-ion velocities ($.05 < v_{nuc} < .12c$) combine to form a wide sum-energy peak (FWHM = 126 or 157 keV) within the wedge-cut (Fig. 14e). The difference-energy distribution (Fig. 14f) reflects the broad range in positron and electron energy, in contrast to the well-defined peak of two-body decay (Fig. 12f). No fine-structure or narrow lines emerge from an interference pattern between the positron and electron detection efficiencies.

This very important result has been verified directly by measurements of IPC from the 1.76 MeV E0 transition in ^{90}Zr. A ^{90}Sr source was employed which β^- decays to the ground-state of ^{90}Y, in turn β-decaying 99.9+‰ to the 0^+ ground-state of ^{90}Zr, and 0.011% to the 1.76 MeV 0^+ first excited state. About 30‰ of $0^+ \rightarrow 0^+$ transitions emit positron-electron pairs, and ≈70‰ internally convert atomic electrons. The only source of positrons in the decay scheme is the E0 transition, whose long lifetime (61 ns) implies that only the correlated electrons are in prompt time coincidence, providing a very clean e^+e^- signal. β-decay electrons are emitted a factor of 6×10^4 more frequently, simulating the adverse conditions of huge δ-electron fluxes encountered experimentally in the heavy-ion collision.

Figure 15 presents results of a two day measurement in which ca. 70,000 total positrons and 4500 coincident pairs were detected. Parts a and b give the total positron energy distribution and the positron-electron energy correlation diagram of the coincidence events. Figures 15c,d present the total coincident positron and electron yields, respectively. For comparison, the dashed lines represent the results of an MCSPEC calculation (generating ten times more data) normalized to the total singles positron yield. The ratio of coincidence to positron singles events is not adjusted. The energy spectrum of the emitted positrons is adequately described (Z = 40) by [57-58],

$$dN/dE_{e+} = P_{e-}E_{e-}P_{e+}E_{e+} \bullet F(-Z,E_{e+})\ F(Z,E_{e-}), \qquad (7)$$

the phase space distribution times a product of Fermi functions

$$F(Z,E) = 2\pi n/(e^{2\pi n}-1); \qquad n = Z\alpha E/P, \qquad (8)$$

which account for effects of the Coulomb field. The Born approximation result for the positron-electron opening-angle correlation has been assumed.

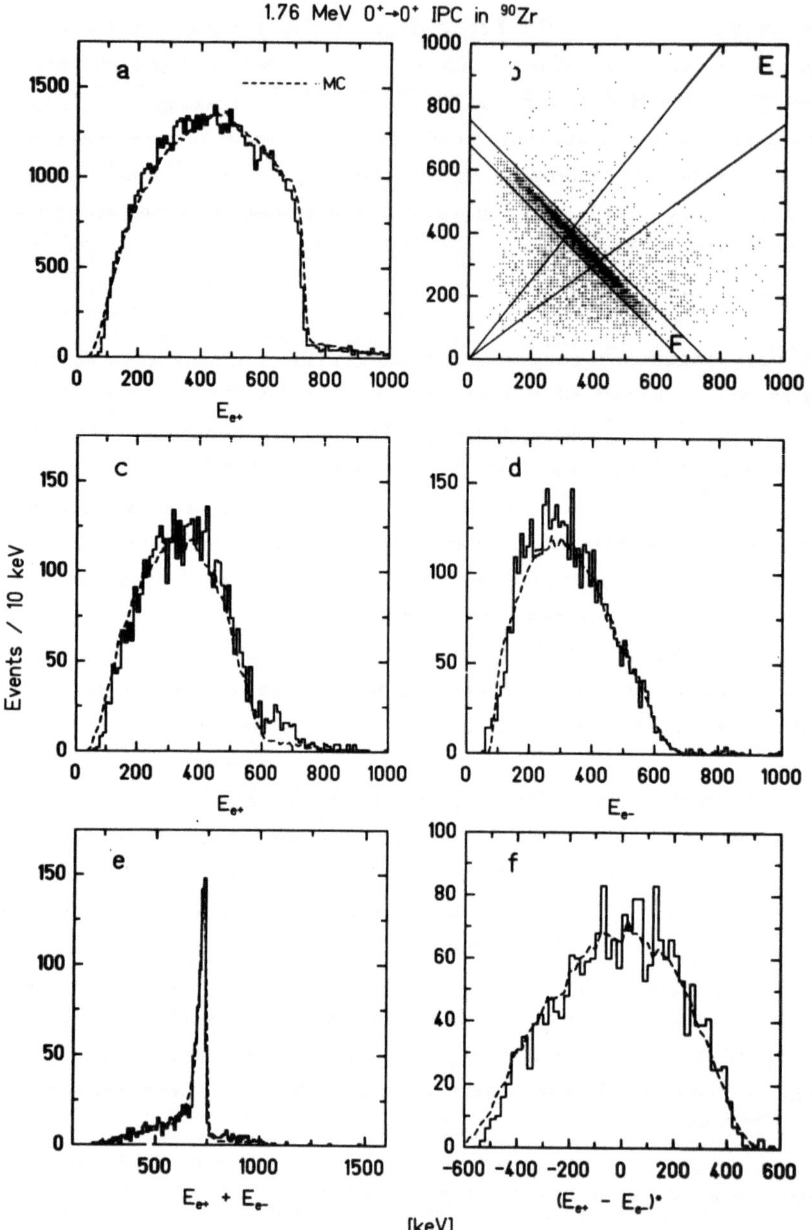

Figure 15 IPC positrons and electrons from the 1 76 MeV $0^+\to0^+$ transition in ^{90}Zr (3×10^{-5} branching ratio from a ^{90}Sr decay) (a) Of 72,000 total e^+, (b) ~4500 are in coincidence with e^- (c)-(d) Total coincident e^+ and e^- yields (e)-(f) Projections onto $(E_{e+}+E_{e-})$ and $(E_{e+}-E_{e-})$ axes for events within the corresponding contours in (b) Dashed curves MCSPEC simulation of 700,000 detected positrons (normalized to the detected yield) The narrow sum-energy peak width from the stationary source reflects Si(Li) resolution only (≈25 keV, 1986)

The good agreement within statistical uncertainty between measured and calculated coincidence ^{90}Zr E0 pair production demonstrates the accuracy with which spectral shapes and coincidence efficiencies can be calculated with MCSPEC. The experimental ratio of coincidence events to total detected positrons is (N_{e+e-}/N_{e+}) = 6.47(10)% compared to a calculated 6.12(3)% - a relative discrepancy of only 5.4(1.6)%. The shapes of the coincident energy distributions, in particular the difference spectrum, are well reproduced. It should be noted that the narrow sum-energy peak (FWHM≈25 keV), which reflects the resolution of the improved Si(Li) detectors, is not Doppler broadened as in Fig. 14 since the ^{90}Sr source is at rest in the lab. In addition to clearly being able to explain IPC for nuclear charges up to Z=40, theory successfully describes pair conversion in ^{206}Pb [59], extending the applicability of these calculations to the region of interest around Z≈90 (employing relativistic calculations as for Fig. 14).

No significant fine-structure is observed in the ^{90}Zr measurement, even though positrons and electrons are emitted with correlations over a range of energies and opening angles. For a given decay kinematic (fixed positron energy, electron energy, and opening angle), the broad acceptance of the EPOS spectrometer (resembling two back-to-back coaxial cones) requires an integration over a range of orientations, which in turn averages over a large number of the rather finely spaced efficiency oscillations (Fig. 9). Unless emission of positron and/or electron is extremely anisotropic, focussed in a small region comparable to the periodicty of detection, no observable structure should arise. An MCSPEC study of anisotropic emission for two-body decay, IPC, and dynamic positron/δ-electron emission revealed no significant efficiency-related structure for distributions of the form $|Y_{\ell m}|^2$ for $\ell \leq 2$ [41]. The calculations and measurements presented here are valid for emission near (< few mm) the solenoid axis, covering a source lifetime range of $\leq 10^{-10}$ s. Preliminary calculations and source measurements of off-axis pair conversion do not indicate the formation of narrow structure. Calculations of IPC in which the direction of the scattered ion is specified also produce no structure. It seems very difficult to construct a situation in which the interplay between emission kinematics and positron and electron detection efficiencies quantitatively leads to narrow lines or a significant distortion of the expected coincident yields, assuming smooth distributions in kinetic energy, or broad distributions in emission angle.

The preceeding simulations also show that the sum-energy and difference-energy projections provide physically natural coordinates for characterizing both two-body decay and the major correlated positron-electron background, nuclear IPC. Quantitative comparison to the coincident peak data of Figs. 2 and 3 requires an understanding of the continuous backgrounds of dynamic positron, δ-electron, and nuclear pair production within these projections.

Figure 16 presents a high statistics sample of ≃150000 detected positrons and 32500 coincident pairs from recent measurements of U+Th colli-

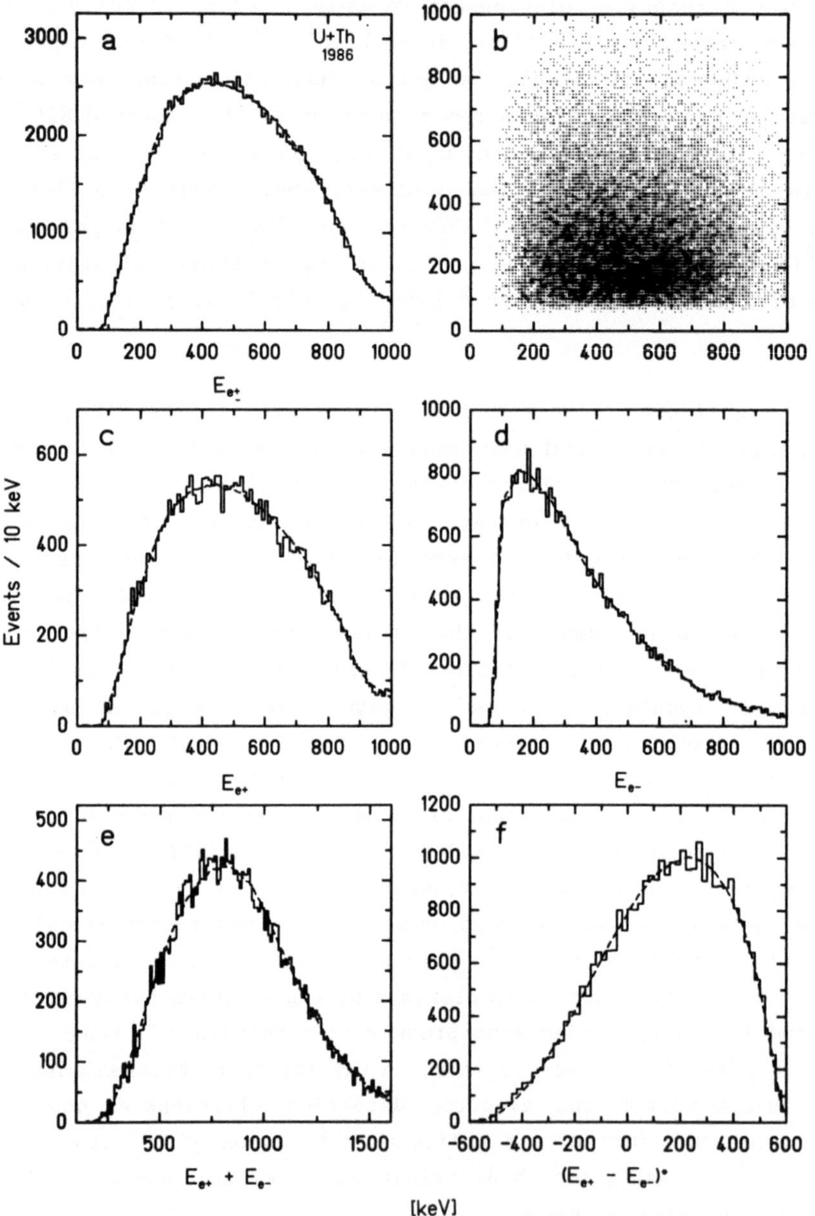

Figure 16 Total data for 5 81 to 5 90 MeV/u U+Th (Feb/Mar 1986) (a) 154,000 total detected e^+ (b) 32,500 coincidence events (c-f) Projections of (b) onto the E_{e+}, E_{e-}, $(E_{e+}+E_{e-})$, and $(E_{e+}-E_{e-})$ axes Dashed lines are MCSPEC simulations of continuous dynamic and nuclear contributions expected from theory and adapted to fit spectra (c) and (d) (e) Sum- and (f) difference-energy spectra are well described assuming independent e^+ and e^- energy distributions (Eqn 9)

sions between 5.8 and 5.9 MeV/u. The dashed curves running through the total coincidence e^+ and e^- spectra of Fig. 16c,d represent the expected yields calculated from the theoretical shape of the dynamic and nuclear positron contributions, and the theoretical exponential slopes of the δ- and nuclear IPC electron yields [41,54]. A 30% contribution to the total e^+ yield is assumed from pair conversion in the scattered nuclei, and after correction for EPOS transport efficiency and detector response, the form agrees with the data except for an ~5% excess of observed intensity between 250 and 450 keV. The ratio of the exponential δ-ray spectrum to the flatter nuclear pair electron distribution was determined by fitting a linear combination of these two components, after correction for efficiency and detector response, to the data of Fig. 16d. One must assume ≈21% of the observed coincident electrons are of nuclear IPC origin. The absolute production probability of positrons and electrons agrees with theoretical expectations to within <20%. The curves of Fig. 16c,d have been adapted slightly to the exact shape of the coincident e^+ and e^- yields in order to more accurately determine the expected background shapes for sum-energy and difference-energy cuts.

The spectral shape of the electron distribution is found not to depend on the coincident positron energy, and vice versa, as illustrated in Fig. 17, which presents the coincident positron (electron) spectra when gating on different electron (positron) energy intervals (first and second columns). The bulk of the coincidence production is atomic in origin where an interdependence arises only indirectly via HI impact parameter, whose effect on the shape of the positron and electron energy distributions is quite weak. A good approximation to the double-differential production probability is a simple product of the individual positron and electron production probabilities:

$$d^2P_{e+e-}/dE_{e+}dE_{e-} = (dP_{e+}/dE_{e+})(dP_{e-}/dE_{e-}). \tag{9}$$

The dashed lines of Fig. 17 show a high statistics (10^7 generated e^+) MCSPEC simulation, reflecting Eqn. 9 by drawing on the spectral forms for positron and electron energy distribution (after EPOS response) and normalizing to the entire event yield. The total sum-energy and difference-energy distributions, shown in Fig. 16e,f, are accurately described by Eqn. 9. The third column of Fig. 17, shows the sum-energy distributions for different wedge-shaped cuts denoted in Fig. 17a. The vertical scale for each remains

---

[3] To facilitate the comparison, these data are displayed as a function of $\Delta E^*_{\pm} = (E_{e+}-E_{e-})/(E_{e+}+E_{e-})\times 700$ keV. With this parameterization, the difference-energy spectral shape does not change substantially as a function of sum-energy.

constant to illustrate the fall-off of coincidence intensity as the contour swings toward lower E_{e+} and higher E_{e-}. In the last column, the difference-energy spectra[3] are plotted for a variety of sum-energies. In both cases, the shape and magnitude of the expected continuum distributions (Eqn. 9) are in very good agreement with the observed data. No fit of the expected intensity to the data within each cut is performed.

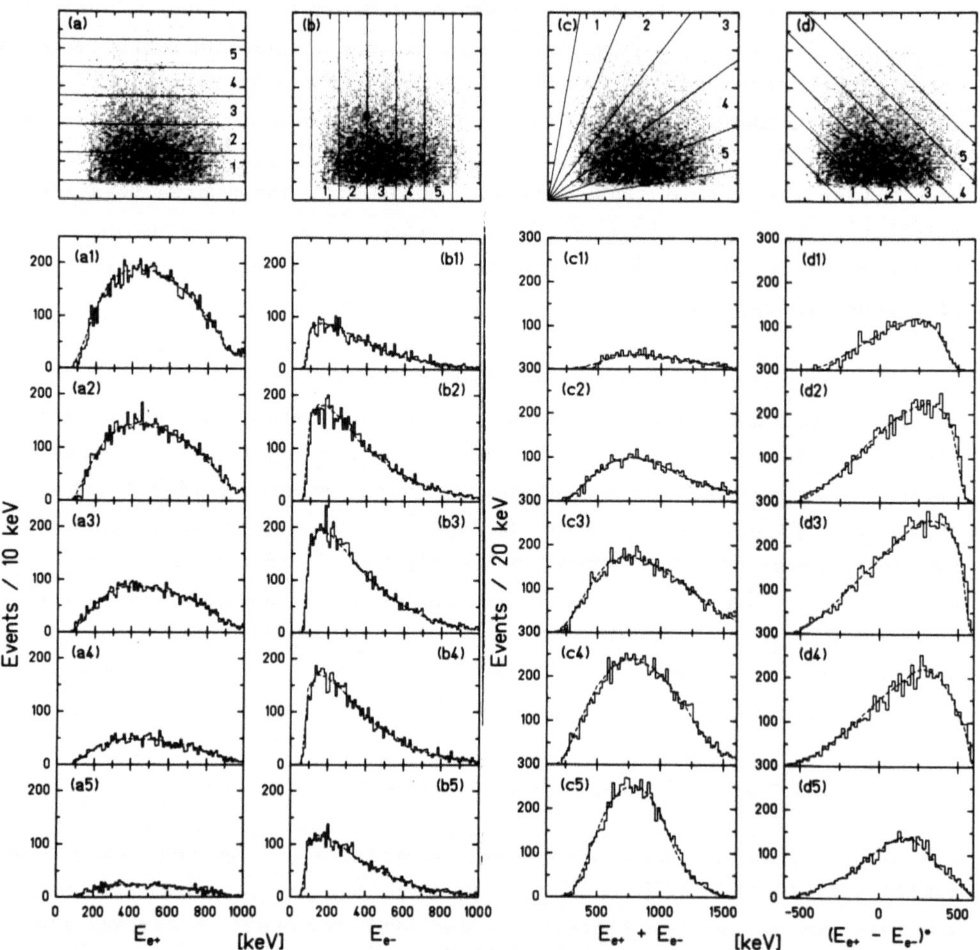

Figure 17. Projections of data of Fig. 16: (a) Positron energy distribution, gating on E_{e-} as marked; (b) e^- spectra for various E_{e+}; (c) E_{e+}+E_{e-} distributions for various wedge cuts; and (d) E_{e+}-E_{e-} spectra vs. sum-energy. Positron and electron spectral shapes appear independent of one another. Dashed lines: identical cuts applied to MCSPEC calculation (Eqn. 9), normalized to total coincidence yield.

In the central region of the event correlation plane (i.e., $E_{e+} \approx E_{e-}$), Fig. 17 shows that the MCSPEC calculation over-estimates the data by ~1.5%. In the boundary regions (E_{e+} or $E_{e-} \approx 0$), the data is underestimated. This slight (but statistically significant) redistribution of intensity from the central to outer regions of correlation energies indicates that, while for the bulk of coincidence events it is valid to assume an effective independence between coincident positron and electron energies, some degree of positron-electron correlation remains. The ≈20% nuclear IPC coincidences may account for this effect since for $E_{e+} \sim E_{e-}$, the opening-angle distribution between positron and electron exhibits maximum anisotropy, hence reduced detection efficiency, while for $E_{e+} \neq E_{e-}$, it approaches isotropy, with relatively larger detection probability. Observation of IPC positrons and electrons should therefore be less probable in the central region of the event plane as compared to the outer regions.

Importance of Various Projections. The relative sensitivity of the E_{e+}, E_{e-}, (E_{e+}+E_{e-}), and (E_{e+}-E_{e-}) projections as indicators of back-to-back positron-electron emission has been quantitively evaluated with a Monte-Carlo technique. A series of simulations [41] of 2000 detected coincidences each were generated in which it was assumed that ~4% of the total positron yield arose from the two-body decay of a 1.8 MeV/c^2 neutral object, X^0, created at rest in the CM. After making the four projections of Fig. 12c-f, it is found that the sum-energy and difference-energy spectra most clearly exhibit evidence for the coincidence peak. The positron energy distribution shows less evidence. The electron energy spectrum is the least sensitive indicator of the existence of structure, principally due to the steep exponential δ-electron distribution which results for low statistics in an imbalanced clustering of events at low energies, making the identification of a coincident peak difficult. As per its design, the sum-energy peak has better energy resolution, having eliminated the 1st order Doppler shifts, leading to the emergence of more prominent structure. If a sum-energy peak is observed, then the difference spectrum becomes sensitive because by gating on sum-energies tightly around the peak, the maximum background, especially at energies near the positron and electron peaks, are excluded from the resulting projection.

The sensitivity of the sum-spectrum, and in combination, the difference-spectrum, suggests a new approach for analyzing experimental data, different from that previously employed in searching for positron peak structures. For the data of Fig. 1, the peaks were enhanced by cutting away the continuous dynamic and nuclear backgrounds with heavy-ion scattering angle cuts, which exploited the good PPAC resolution and apparent difference of the peak-related scattering events from Rutherford elastic kinematics [3,5]. This was accomplished at the cost of total peak intensity. In the present EPOS configuration, coincidence detection efficiency leads to at most 14% of the intensity of a positron singles peak appearing in the corresponding coincident line. Instead of ~50 e^+ peak counts, less than 10 coincident counts should be accumulated in a typical experiment, insufficient for a meaningful evaluation of the existence of structure. An alternate strategy could involve a direct search in the total data sample for structure in the sum-energy distribution. If corresponding difference-energy, positron, and electron peaks are found in conjunction with a sum-energy line, the increased intensity should be advantageous in extracting valuable information, despite a smaller peak-to-background ratio.

V. EXPERIMENTAL DATA AND DISCUSSION

Figure 18 presents a plot of the coincident positrons and electrons for U+Th collisions accumulated in the first series of coincidence experiments performed in June and July of 1985 [7]. The beam energy for the displayed data ranged from 5.80 to 5.85 with a mean of 5.83 MeV/u. Targets were 250 $\mu g/cm^2$ layers of ThF_4, evaporated on a 40 $\mu g/cm^2$ carbon foil, covered with an additional 20 $\mu g/cm^2$ of carbon to reduce sputtering. The entire scattering angle range of the PPAC's between 20° and 70° is included for a total of 9600 detected positrons, with 1690 coincident positron-electron events. The shapes of the individual coincident positron and electron energy distributions are in good agreement with that expected from the theoretical atomic and nuclear continua as described above.

Although very little structure is readily apparent in the total coincident yield, the sum-energy spectrum derived from plotting the events contained within the wedge-shaped contour (labeled C) versus $E_{e+} + E_{e-}$ reveals a narrow peak centered at 760 ± 20 keV, as shown in Fig. 19c, which has a width of 80 ± 20 keV. Correlated with this peak, are narrow lines in the positron and electron energy distributions. Plotting positron energy for electrons with kinetic energy between 340 and 420 keV reveals a peak centered at 380 ± 15 keV with a width of 80 ± 15 keV (Fig. 19a). Gating on positrons between 340 and 420 keV yields an electron line at 375 ± 15 keV with width 75 ± 15 keV (Fig. 19c). Finally, by gating on the sum of positron and electron energies between 710 and 830 keV, the projection of the data on the difference axis exhibits structure centered near $E_{e+} - E_{e-} = 0$ (Fig. 19d).

In each of the spectra in Fig 19a-d, the dashed curves represent the distribution expected assuming that the entire 1690 coincidence events arise from the continuous atomic and nuclear processes. A high statistics MCSPEC calculation has been normalized to the coincident event yield, after which the identical projections as taken for the data in each spectrum are made, providing a model-independent evaluation of the continuous background. Although these four projections are correlated, the apparent intensity of each structure varies according to the extent that the different cuts exclude background events in the energy region surrounding the peak. Fits of the background shapes individually to the spectra give the following estimates of the coincident line intensity: $N_{e+} = 26.7 \pm 7.7$, $N_{e-} = 31.7 \pm 7.6$, $N_{\Sigma} = 35.3 \pm 9.4$, and $N_{\Delta} = 38.5 \pm 9.7$ counts.

For comparison, Figs. 19i-j present the results of a MCSPEC calculation involving the production (at rest in the CM) and isotropic two-body decay of a neutral object. In order to reproduce the observed data, a mass of $M_x = 1.8$ MeV/c^2 and an intensity such that only 3% of the total detected positron yield arise from the decay of this object, have been assumed. The shape and magnitudes of the positron, electron, sum-energy, and difference-energy peaks are remarkably well reproduced within this model. The sum-energy peak width of 80 ± 20 keV is in agreemeent with that expected

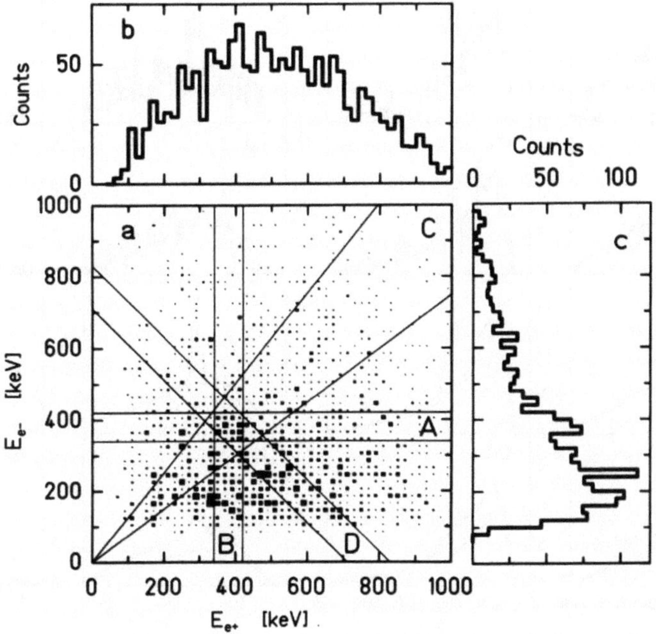

Figure 18. (a) Intensity distribution of coïncidence events as a function of E_{e+} and E_{e-} for 5.83 MeV/u U+Th collisions (July, 1985). (b),(c) Projections of these data on E_{e+} and E_{e-} axes.

solely from detector resolution of ~50 keV. Even the slightly lower mean electron peak energy is understood in terms of the larger low-energy tailing in the electron Si(Li) detector response.

The coincidence peak intensity is confined to a small region centered around $E_{e+} \approx E_{e-} = 380$ keV. Figures 19e-h show the positron, electron, sum-energy, and difference-energy projections for gates taken on either side of the corresponding cuts in Fig. 18a. The absence of significant structure in these spectra illustrates the correlated nature of the positron and electron peak events, and is in full agreement with the predictions from a two-body decay origin.

The relative intensity of the coincident sum-energy signal at 760 keV can be obtained by comparing it to the peak in the total positron yield enhanced by heavy-ion kinematic cuts. For laboratory scattering angles between 20° and 40°, a positron enhancement similar to those of Fig. 1 emerges from the data with a mean energy of 380 keV (Fig. 20a) [41]. The associated coincident electron peak is enhanced accordingly over the continuous delta-electron background. The ratio of coincidence to total posi-

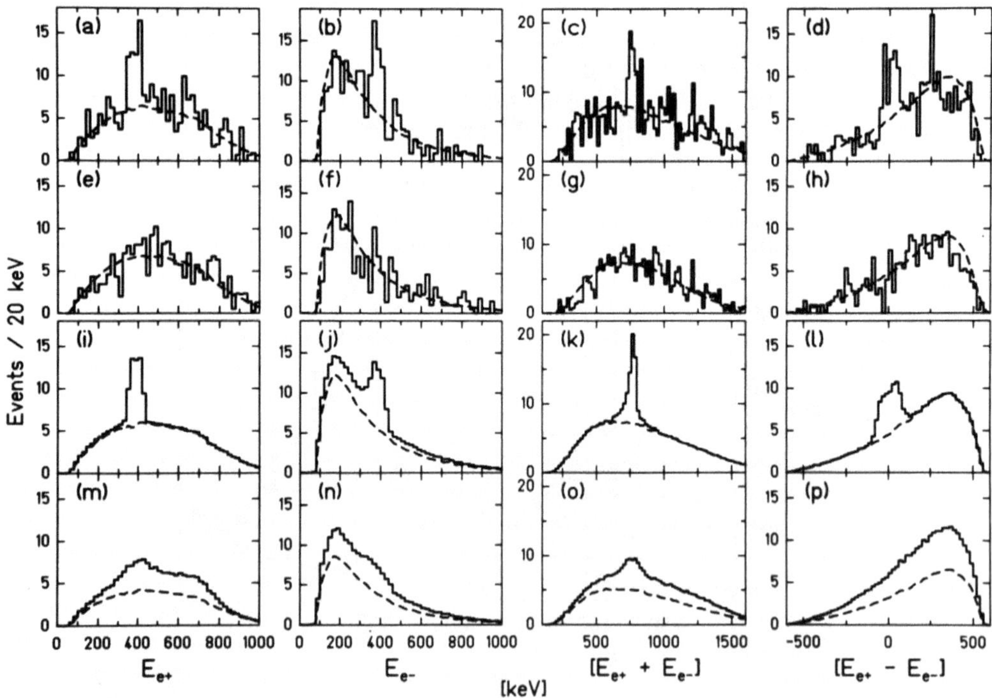

Figure 19. Projections of U+Th coincidence data from Fig. 18 (a-d), and MCSPEC simulations of the two-body decay of a 1.8 MeV/c² neutral object (i-ℓ) and IPC of a 1.8 MeV E1 transition in the scattered nuclei (m-p). Columns correspond to projections onto the E_{e+}, E_{e-}, $(E_{e+}+E_{e-})$, $(E_{e+}-E_{e-})$ axes. Parts (e)-(h) are the average of similar gates adjacent to either side of the gates A-D of Fig. 18.

tron peak events is 19 ± 8%, which is consistent with 14% expected for the two-body decay of neutral objects produced at rest in the CM. Because of the rather weak statistical significance of this ratio due to the limited peak intensities available within this angular region, this comparison is not inconsistent with a ratio of 6% expected for spacially uncorrelated emission.

Statistical Significance

The size of the peak, an estimated net intensity above background of 32 ± 8 counts, represents a 4σ statistical significance. One may alternatively assess the probability that no coincidence peak exists, i.e., that the entire coincident event yield is due to continuous atomic and nuclear processes (dashed curves in Fig. 16) from which the observed structure is merely a statistical fluctuation. In the region $340 < E_{e+} < 420$ and $340 < E_{e-} < 420$ keV, 24 ± 4.9 counts are expected from continuous backgrounds. The observed 54 events in this interval therefore represents a 30 count, 6σ excursion above the expected intensity, strongly arguing against the null-hypothesis that continuum production explains the data [60].

The data presented here were accumulated with targets identical to those employed in the Th+Th measurement of Fig. 1, and represent ≈2/3 of the total data collected on the U+Th system. An additional ~4500 positrons were measured at somewhat lower beam energies (5.75 to 5.80 MeV/u) on targets fabricated with a different ThF_4 evaporation procedure on different carbon backing foils [61]. Viewed independently, these targets exhibit no statis-

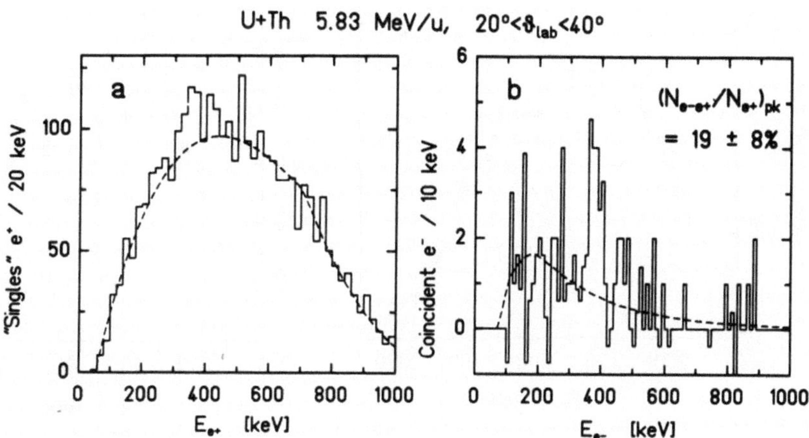

Figure 20. Total positron, and coincident electron ($340 < E_{e+} < 420$ keV) spectra for U+Th collisions (Fig. 18) with heavy-ions scattered between 20° and 40° in lab. The relative coincident e^- peak intensity to the e+ peak is 19 ± 8%, consistent with 14% expected for back-to-back emission.

tical evidence for the appearance of the peak (N^{pk}_{e+e-} = 5 ± 4). While this could reflect a reduced production cross-section at lower projectile energies, independent evidence alternatively suggests that these targets exhibited a rapid deterioration in the heavy-ion beam. If resonant processes contribute to the production of the sum-energy peak, then when non-uniformities set in the thick portion of the target may lead to an overwhelming contribution of the continuous positron backgrounds which obscure the peak. With this in mind, in a subsequent experiment using thorium beam we compared data collected for only short irradiation times (≤1 hour) to that including longer heavy-ion bombardment of the targets. Figure 21 shows that a peak in the total positron distribution emerges at ≈325 keV (using no kinematic selection), which seems to be correlated with "fresh" targets [62]. In any case, adding the remaining 1/3 of the U+Th data measured in 1985 does not significantly dilute the results presented in Fig 19a-d [41]. The coincidence structure remains an ~6σ deviation from the smooth background, and the net peak intensity is 3.3σ.

Positron/Electron Identification and Instrumental Backgrounds

Various aspects of the EPOS spectrometer preclude the possibility that the positron and electron signals are produced by trivial instrumental effects or secondary scattering processes. First, the spiral baffle attenuates the flux of 350 keV electrons from reaching the Si(Li) by a factor of ~700 (see Fig. 6), ruling out, for example, the possibility that a conver-

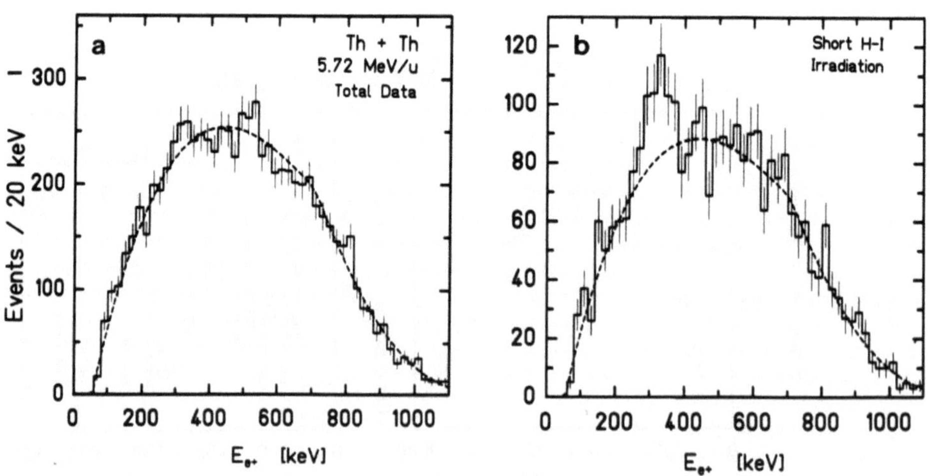

Figure 21. (a) Total positron energy distribution (20° < θ_{HI} < 70°) for Th+Th collisions at 5.75 and 5.70 MeV/u, July 1985. (b) Subset of data selecting approximately the first hour of heavy-ion irradiation on each target.

sion electron is mistaken for a positron. Second, unambiguous positron identification is provided by the 8-fold NaI crystal array which detects the coincident characteristic annihilation radiation. Gating on the positron-electron coincidence line reveals two clean γ-ray peaks at 511 and 1022 keV [41] (corresponding to the detection of one or both annihilation quanta, respectively), clearly signaling a positron event. The two-photon peak is particularly noteworthy, as its occurrence precludes the possibility that the "positron" peak is in fact the Compton edge of one 511 keV photon scattered in the Si(Li) in conjuction with the coincident detection of the second 511 photon in the NaI array. In experimental arrangements where the annihilation quanta are not measured, this effect is particularly dangerous as it may trivially lead to narrow structure near the Compton-edge energy of 341 keV.

The time difference between Si(Li) detector and heavy-ion PPAC further distinguishes between positrons emitted from the target, or created in the vicinity of the positron detector by γ-rays or evaporation neutrons. A γ-ray would require 3 ns, and a typical evaporation neutron (E_n~3 MeV) ~35 ns transit time from target to Si(Li). Neither is observed. Rather, a mean flight time of 8 ns is measured, equal to that observed for the continuous dynamic and nuclear positron distributions.

Applied to electron detection, time-of-flight arguments similarly imply that the observed event is a lepton emitted from the target. Together with its small geometric cross-section, 1.3×10^{-4} sr, this rules out direct γ-rays and neutrons from the target, as well as the possibility that one 511 keV photon from the detected positron travels two meters across the solenoid to Compton scatter in the electron detector 6.5 ns later. The small positron production probability, $P_{e+} \sim 10^{-4}$, further argues against the possibility that a second positron is created in the collision which annihilates near the electron detector in which a 511 keV photon scatters.

It is also very unlikely that the electron signal could be a mis-identified monoenergetic positron incident in the counter, because the efficiency of detecting positrons in the planar-pair of Si(Li) crystals is a factor of ≃4 less than the electron detection efficiency (Fig. 7). To achieve the observed coincidence peak intensity, one must then assume an average multiplicity of about ten monoenergetic positrons (E_{e+}≃380 keV) emitted from the collision. Due to double and triple counting in the positron detector, peaks at 760 keV (two positrons) 20% larger than, and at 1140 keV (three positrons) still half as intense as, the 380 keV peak would be observed in the positron energy spectrum. In addition, for these multiple-e^+ structures, 25% to 40% of the annihilation radiation detector energy spectrum would exceed 1022 keV due to detection of more than two 511 keV photons [41]. Neither of these effects are observed, and scenarios based on emission of several monoenergetic positrons (e.g., the Spontaneous Positron "machine-gun" [63], in which conversions in the di-nuclear complex repeatedly empty the K-shell which is then filled by Spontaneous Positron Emission) are ruled out by the data. MCSPEC simulations also rule out the trivial possibility that a peak-like structure in the electron spectrum at 300 keV could be produced by the summing of two δ-electrons (whose detected energy distribution exhibits a maximum at ~150 keV). We conclude that the observed signals in fact represent coincidences between monoenergetic positrons and electrons which are emitted from the target region within a few nanoseconds of a heavy-ion collision.

Correlation Between Positron and Electron Peaks

A next step in evaluating these data is to determine whether the emission of the monoenergetic positron is correlated to the emission of the monoenergetic electron signal. It is possible that in addition to narrow positron peaks, unrelated internal conversion electron lines, from a particularly strong nuclear transition, are detected in these collisions, leading to the coincident signal.[4] Figure 22b illustrates the positron-energy vs. electron-energy diagram appropriate for this class of origins involving independent production of positron and electron peaks. The vertical band illustrates the presence of positron peak intensity at 380 keV for all electron energies, and the horizontal stripe indicates the electron line intensity independent of positron energy. If projections on either side of the positron peak are taken, the electron line should be clearly evident, and vice versa. In fact, the total coincident positron and electron yields should provide the highest statistical indication of the presence of independent peaks. Figures 19e-h show projections analagous to those of Figs. 19a-d, averaging regions on both sides of the corresponding cuts. The coincidence structure is absent from these neighboring regions of the event plane.

Gating on either side of the wedge cut should similarly produce a sum-energy spectrum with two "peaks" where the cuts cross the banded contour. The difference-energy distribution for adjacent sum-energies should exhibit two peaks, one on either side of $\Delta E_{\pm} = 0$. Such features are absent in Figs. 19g,h. Figure 23 expands this analysis, presenting positron or electron spectra for 100 keV wide gates on electron or positron energies around the peak region. In all cases, the dashed lines represent the

[4] Hypothesizing a narrow conversion line to explain the coincidence electron peak does not address the fundamental issue of the origin of the positron peak. It should be noted that no satisfactory explanation of the narrow positron lines of Fig. 1 has been found.

expected distributions assuming the entire coincident yield is attributable to the smooth dynamic and nuclear distributions The absence of peak structure in regions adjacent to the coincident signal rules out the possibility that the positron and electron peaks are physically unrelated

Correlated Positron/Electron Pair Creation

Internal Pair Conversion The most obvious source of correlated positrons and electrons from heavy-ion collisions is the IPC of transitions in the final-state nuclei As discussed above, this leads to a broad distribution of positron and electron energies for which $E_{e+}+E_{e-} = W_o-2m_ec^2$, with

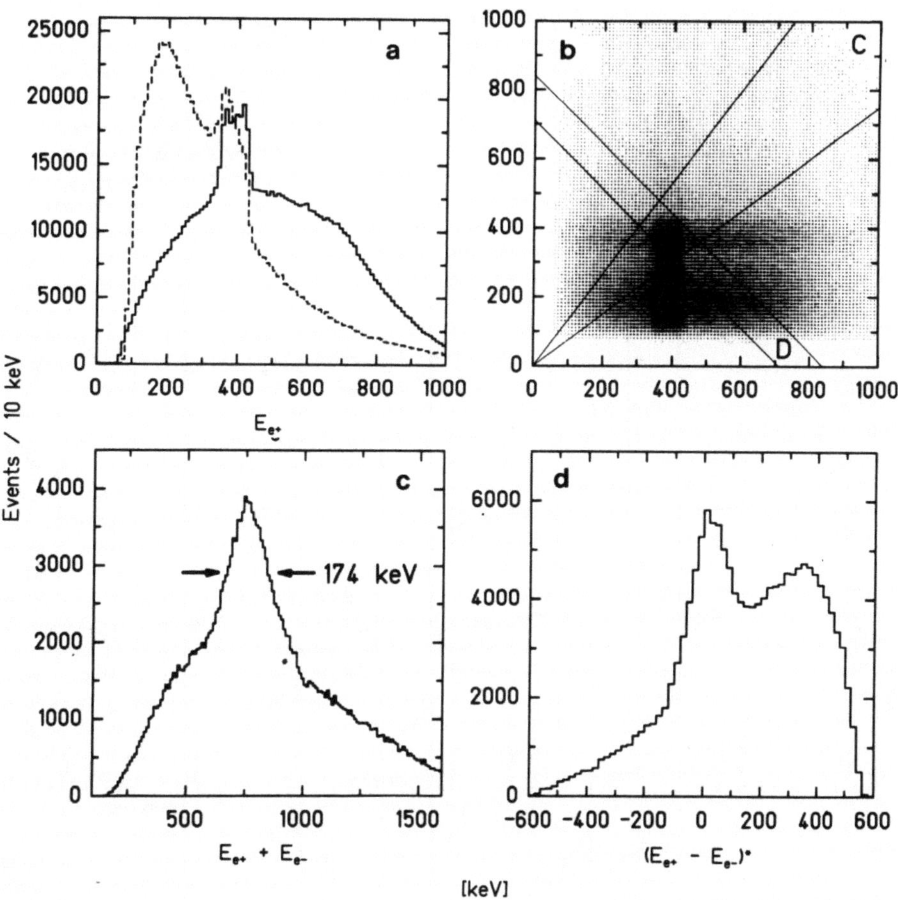

Figure 22 Simulation of independent monoenergetic e^+ and e^- emission (a) Total coincident e^+ (solid) and e^- (dashed) spectra (b) Energy correlation pattern exhibits horizontal and vertical bands superimposed on the continuous background Independent emission is characterized by wide sum-energy distribution (c) and structure in adjacent cuts (as evident from b)

the possibility that the central region of the IPC event yield, for $E_{e+} \sim E_{e-}$, simulates the observed coincidence signal.

Figure 24 presents an MCSPEC simulation of IPC of a 1.8 MeV E1 transition, typical for high energy (>1 MeV) excitations in the scattered nuclei (Z≈92). The dashed histogram corresponds to the expectations of mutually isotropic positron-electron emission. Unlike E0 conversions, the E1 Born approximation angular distribution [57]

$$dN(E1)/d(\cos\theta_{e+e-}) \propto P_{e+}P_{e-}[1 + W_0^2 m_e^2/F^2 + (E_{e+}^2 + E_{e-}^2)/F], \qquad (10)$$
$$F = (m_e^2 + E_{e+}E_{e-} - P_{e+}P_{e-}\cos\theta_{e+e-}),$$

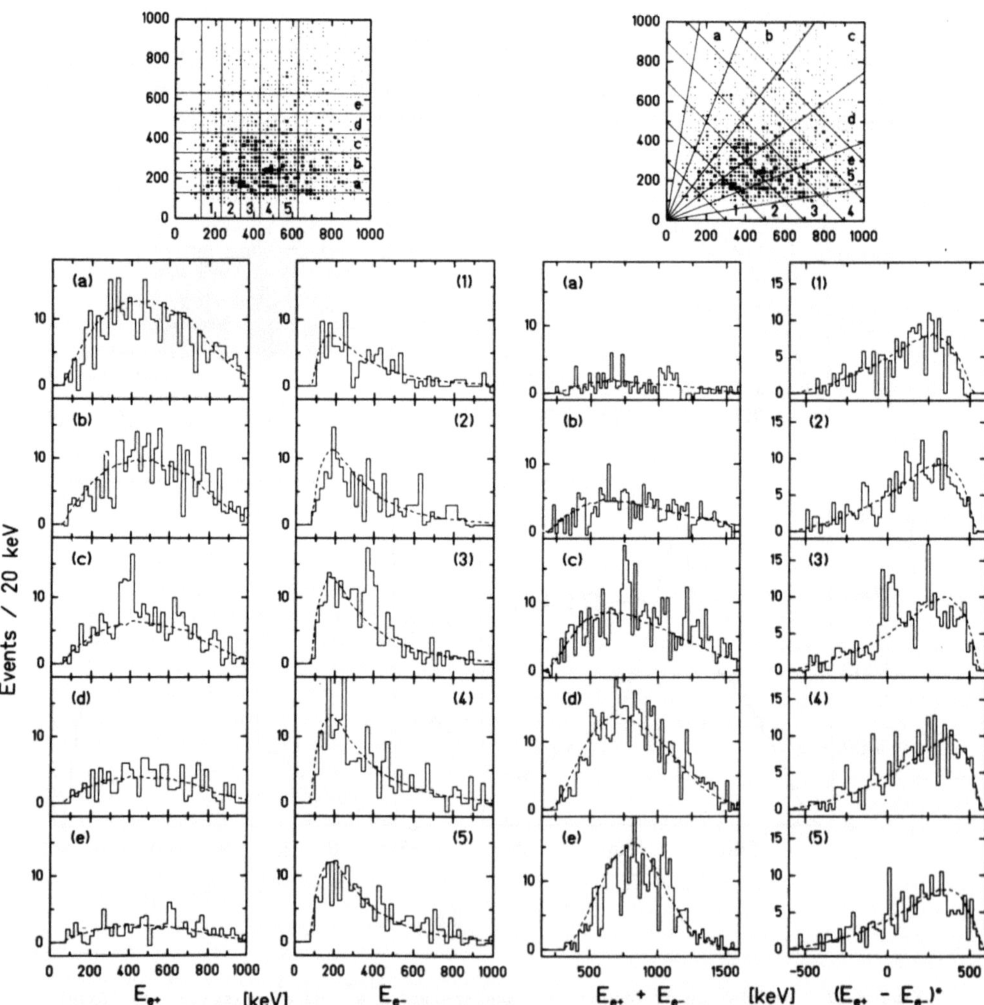

Figure 23. (Left) Coincident e+ and e- spectra projected from the U+Th data of Fig. 18 for conditions as marked. (Right) Sum- and difference-energy distributions for gates as shown. Peak structure appears to be confined to $E_{e+} \approx E_{e-}$.

does not vanish for $\theta_{e+e-} = 180°$ emission. When integrated over the EPOS acceptance, it leads to only a ≈20% reduction in detected intensity compared to relatively isotropic e^+e^- emission.

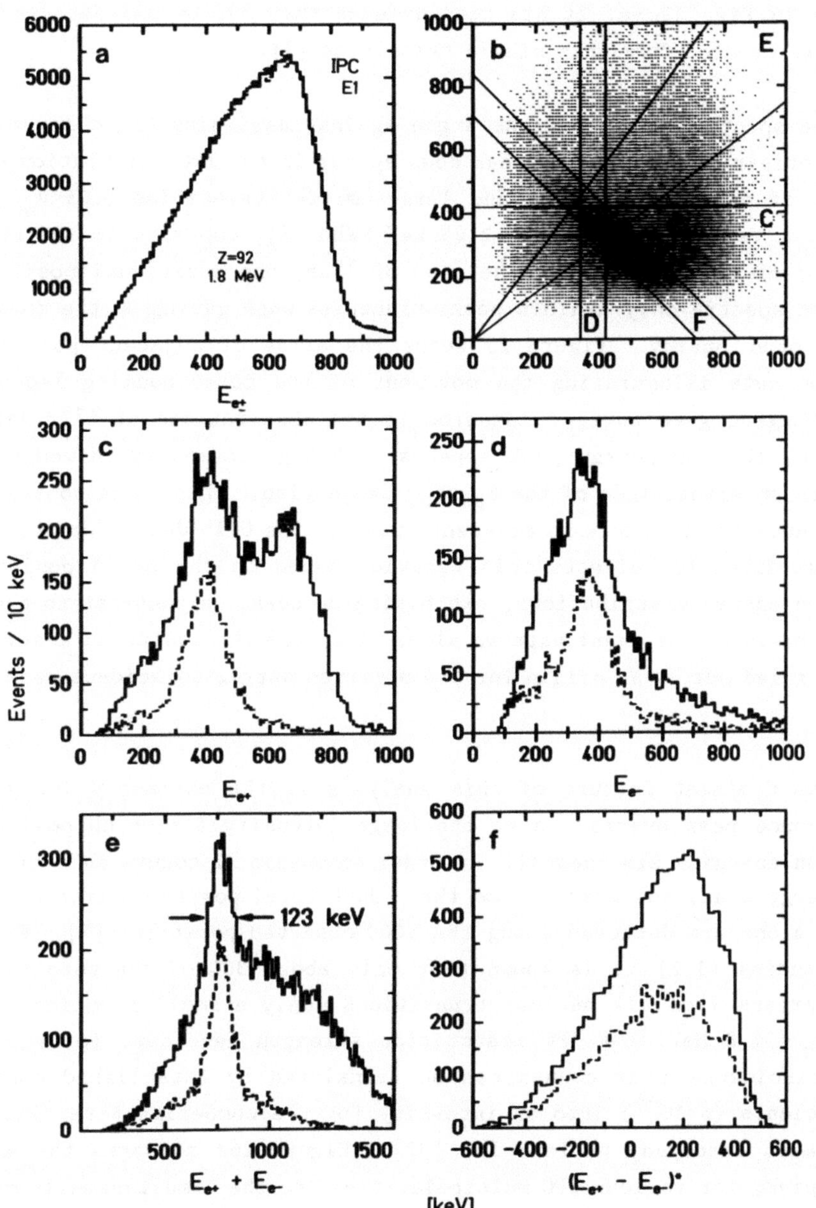

Figure 24. MCSPEC simulation of IPC of an 1.8 MeV E1 transition in the scattered heavy-ions (Z=92). (a)-(f) Same as Fig. 12. Dashed histogram assumes isotropic θ_{e+e-}. Born approximation angular distribution decreases the coincidence yield by ~20%. Solid histogram includes detection of δ-electrons with IPC positrons when correlated e^- partner is not observed. (e) Sum-energy peak intensity represents 0.75% of total e^+ yield.

The solid histogram presents the distortion of the pair yields from the additional contribution of δ-electrons coincident with 20% of the IPS positrons when the correlated e^- is not detected. In addition to the e^+ peak near 380 keV (for 340 $\leq E_{e-}$ <420 keV), Fig. 24c reflects the shape of the total e^+ energy distribution. For comparison to the data, the expected IPC yield (assuming isotropic e^+e^-) has been added to the continuous backgrounds in Fig. 19m-p. It has been assumed that 30% of all detected positrons are created from this single transition [7].

The spectral distributions argue against assigning the observed peaks to IPC origin. A sum-energy peak does appear in the IPC calculation but the absence of angle-correlation and large emitter velocities ($.05<v_{nuc}<.12c$) lead to a very wide structure ≥125 keV FWHM [7], compared to the observed width of 80 ± 20 keV. As in the case of ^{90}Zr, the individual positron and electron spectra only exhibit peak structures when gating on the complementary lepton energy. Figure 25 shows the array of adjacent positron and electron cuts illustrating the movement of the corresponding lepton peak with changing gate energy, compared to the observations of ^{90}Zr data. In addition, the sum-energy peak appears with high intensity in wedge-shaped contours on either side of the $E_{e+} \approx E_{e-}$ projection. This is in contrast with the absence of structure in adjacent cuts for the U+Th data. The difference spectrum directly reflects this behavior based on the broad positron and electron energy distributions, exhibiting no peak, in contrast to the data. On the basis of the event pattern alone, IPC from the scattered heavy nuclei can be ruled out as an origin for the observed narrow coincident peak structures.

The dominant feature of this analysis is the extremely low yield of coincidence peak events due to the large intensity for which positron and electron energies are unequal. In order to obtain 32 counts in a coincident sum-energy peak, one must assume that ≈5000 total positrons from this transition alone are detected among the 9600 observed positrons [7]. From earlier studies [1,2] it is known that only about 30% of the total positron yield arises from all nuclear transitions, only a small fraction of which have W_0 = 1.8 MeV. The IPC transtition strength necessary for explaining the coincidence peak counts can be translated by established conversion coefficients ($\beta \approx 10^{-4}$) into an intensity for the competing γ-ray line ($\lambda \geq 1$) which must accompany nuclear IPC [32]. Figure 26a compares the expected γ-ray peak, for various IPC multipolarities, to the simultaneously measured γ-ray yield, conservatively assuming isotropic positron-electron emission. If the transition is E0, the simulation of Fig.14 is appropriate, requiring 3500 total detected positrons from this IPC transition. Figure 26b presents the expected size of the competing IC electron line, compared to the simultaneously measured high-energy electron spectrum. If the Born Aprroximation for the θ_{e+e-} distribution is correct, the expected γ-ray and e^- peak intensities are increased by 20% (E1) or a factor of ~2 (E0,E2). In either case, the measured yields are smaller than the required peak intensities by an order of magnitude.

These intensity considerations, independent of arguments based on the broad energy distributions encountered, rule out conventional IPC in the separated nuclei as a source for the observed coincident positron-electron structure. MCSPEC simulations indicated that the coincidence distributions are not altered when the heavy-ions are scattered into a narrow angular range (e.g., to large angles with lower velocities, or to specific values of θ and ϕ), nor do they substantially change if anisotropic emission of the IPC positron is assumed with respect to the HI motion [41]. Even if a fragmenting of the broad IPC distribution by the overlap of the oscillatory positron and electron transport efficiencies *were* to occur through some

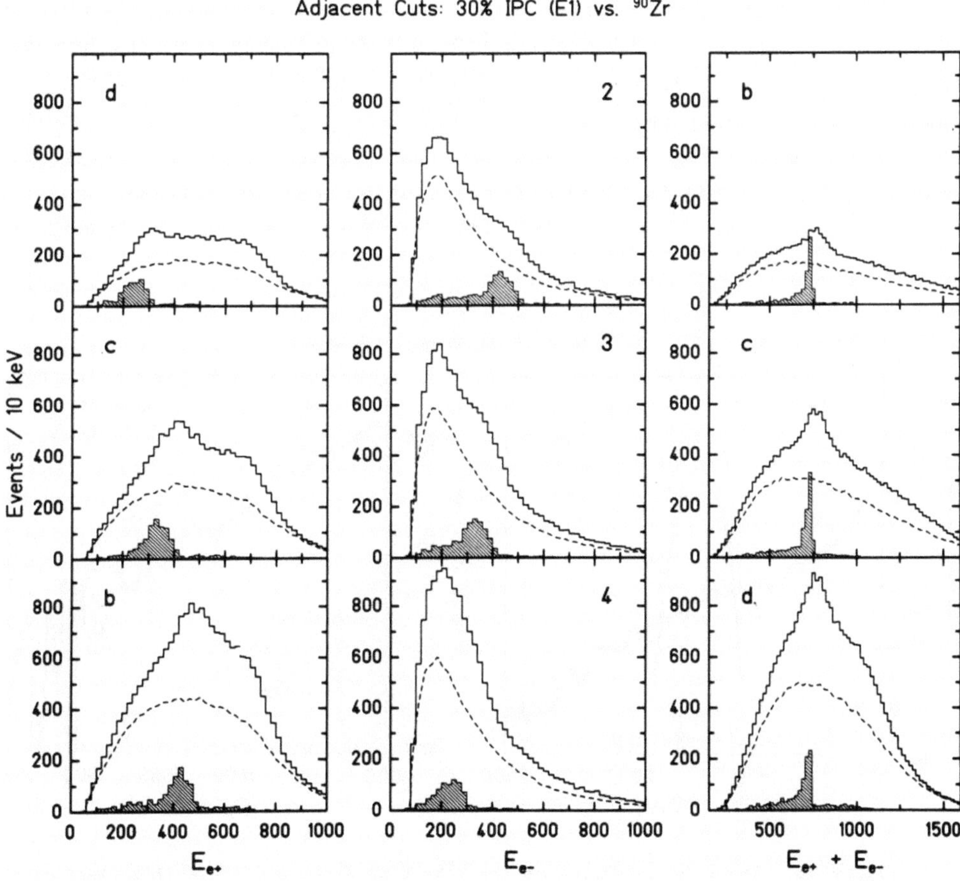

Figure 25. Adjacent cuts for the E1 simulation of Fig. 19m-p compared to ^{90}Zr measurement (hatched). Projections onto E_{e+}, E_{e-}, and $(E_{e+}+E_{e-})$ axes are identical to those of Fig. 23 as marked. Structure is observed, for IPC, in the neighboring cuts, in contrast with U+Th data of Fig. 23.

exotic combination of kinematic variables, theoretical conversion coefficients would have to be in error by orders of magnitude in order for IPC *not* to be excluded on intensity grounds alone.

Pair Creation in Nuclear Fragments or Clusters. A further possibility is that a low mass (Z~0) nuclear fragment is formed in these collisions, in which IPC of an E0 transition leads to the observed structure. This consideration is motivated by the fact that the small nuclear charge leads to a more symmetric energy distribution, determined almost purely by phase space ($\propto P_{e+}E_{e+}P_{e-}E_{e-}$), somewhat reducing the total singles positron yield which must accompany the coincident IPC signal. A cluster might be formed with no atomic electrons, rendering a search for the competing IC e^- line useless.

An MCSPEC simulation of this scenario is shown in Fig. 27. It closely resembles the low-Z ^{90}Zr measurement, exhibiting a very broad difference-energy spectrum, incompatible with the data. Adjacent cuts also must exhibit significant structure in the positron, electron, and sum-energy spectra. Creating the cluster without noticeably perturbing the observed quasi-elastic scattering kinematics (ΔQ<20 MeV) probably requires formation from the neck of the di-nuclear complex [64]. The resulting velocity, $\sim v_{CM}$, leads to a sum-energy peak width ≥105 keV, larger than the ≃40 keV width of the recent data (Fig. 3). Finally, the symmetric energy distribution results in only a marginal reduction in singles positron intensity relative to coincidence yield, since for Z≃0 the Born approximation result for

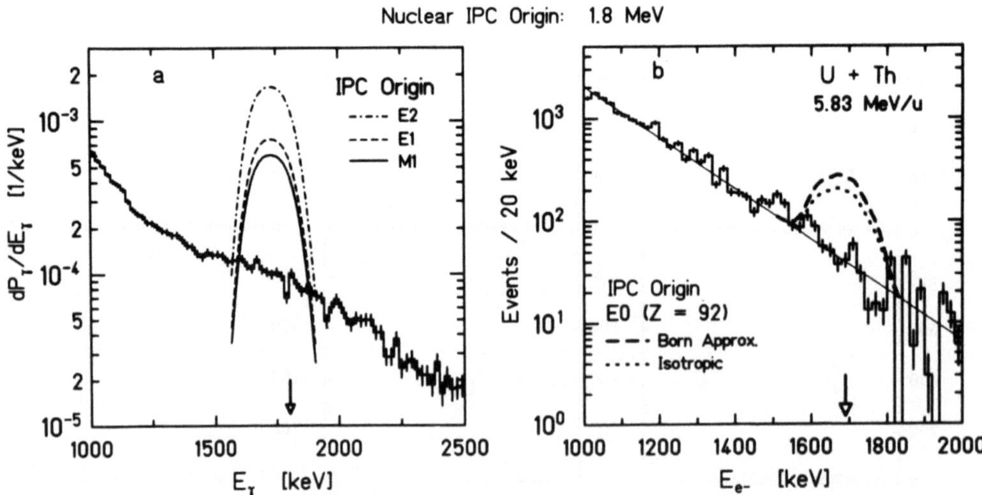

Figure 26. Required intensity of the competing γ-ray and conversion electron lines, compared to the γ, e^- distributions measured simultaneously with data of Fig. 18. (1.8 MeV IPC in scattered nucleus assumed). (a) Expected γ-ray peaks for isotropic θ_{e+e-}. Born-approx. θ_{e+e-} dependence requires increased γ-ray peak intensity. (b) Expected IC electron line (for E0) is shown added to an exponential fit through the data.

$dN/d(\cos\theta_{e+e-})$ applies. One must still assume that ≥35% of the measured positron spectrum arises from this single transition to explain the 32 ± 8 sum peak counts.

One possibility to accommodate the observed data within this model assumes that the fragment converts several cm away from the target (τ_{E0} >

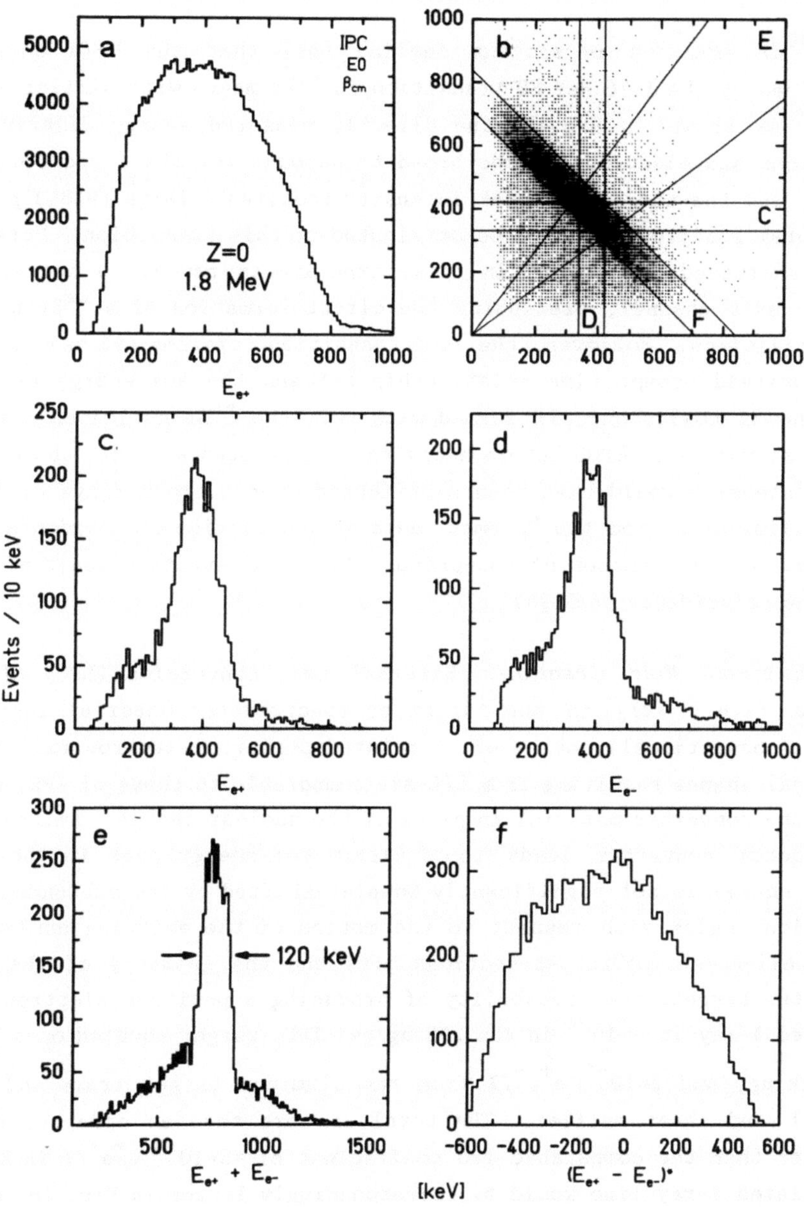

Figure 27 Simulation of a 1 8 MeV E0 IPC in a nuclear fragment (Z ≈ 0) moving with v_{cm} (a)-(f) As in Fig 12 Despite the smaller spread of e^+ and e^- energies (compared to Z=92), IPC must account for ≥35% of total poistron yield No narrow difference-energy peak emerges

10^{-9} s), perhaps leading to narrow structure by a suppression of some range of detected coincident positron and electron energies and opening angles. However, positron detection efficiency decreases more slowly off-axis than coincidence detection probability, so the ratio of positrons to coincidence events should, if anything, significantly increase. Considerably more than 30% of the detected e^+ yield must therefore arise from that particular transition. This intensity argument holds for any other mechanisms involving pair production away from the target region.

^{90}Zr IPC. A persistent feeling [65] that the U+Th data may be explained by the 1.76 MeV E0 transition in ^{90}Zr meets with similar difficulties. As shown in Fig.15, the directly measured energy distribution of positrons and electrons is too broad to account for the narrow coincidence peaks, and the sum-energy peak intensity requires a large (≈38%) portion of the total positron yield to be attributed to this transition. Detection of both scattered heavy-ions in these experiments reveals a nearly elastic binary exit channel, precluding the direct formation of a ^{90}Zr nuclide in the collisions. Moreover, the long transition lifetime (61 ns) contradicts the observed prompt time relationship between the sum-energy peak events and the HI collision. If formed with v>0, the ^{90}Zr nuclei exit the ≈2 cm fiducial volume of EPOS before decaying. If created at rest, the coincident pair intensity would have been distributed over a broad range of times and then eliminated from the "prompt" data by subtraction of accidental coincidences. An examination of non-prompt (Δt>10 ns) events reveals no evidence for the coincidence peak [41].

External Pair Creation. External pair conversion (EPC) of nuclear γ-rays (i.e., $\lambda \geq 1$) in the target or spectrometer material represents a final (conventional) source of correlated positron-electron emission. The spectral shapes resulting from EPC are comparable to those of IPC, using the Z of the converter material in place of the nuclear charge. Emission from a stationary converter leads to a narrow sum-energy peak if the incident γ-ray energy is not significantly Doppler-shifted by the subtended range of emission angles with respect to the motion of the emitting nucleus. From the well-known EPC cross-sections [66] and the geometry of the material near the target, the probability of producing a positron-electron pair per emitted γ-ray is 5×10^{-6} in the 200 μg/cm² ThF_4 target mounted on a 50 μg/cm² C backing, and 2×10^{-5} e^+e^-/γ from the aluminum target frame and both the spiral and sheet baffles. The total is more than an order of magnitude smaller than the comparable IPC coefficient of $\approx5\times10^{-4}$ e^+e^-/γ in Z≈92. The associated γ-ray line would be correspondingly larger in Fig. 26, exceeding the observed γ-ray intensity by more than a factor of ~100.

It is apparent that no known internal or external pair conversion processes lead to narrow peaks in the energy distributions of coincident positrons and electrons which could reproduce the measured data. Attempts to accommodate pair conversion processes to the observed coincidence structure through a variety of unusual mechanisms have been modeled with MCSPEC simulations including 1) specifying the exact direction, or range of directions, for the scattered nucleus or for the emitted cluster, 2) assuming various broad or extremely narrow distributions for the positron-electron opening-angle, and 3) determining the direction of emission, or distribution of emission directions, of the positron with respect to the motion of the nuclide or fragment. No combination of the above leads to a significant change in the broad difference-energy distribution or in the relative magnitude of detected pairs to total observed positrons [41]. Over the normal range of nuclear velocities, the broad θ_{e+e-} distribution leads to wide sum-energy peaks (between 100 and 200 keV), in addition to the large spread in individual positron and electron energies. Greater than at least 30% to 60% of the total observed positron yield must arise from the single transition at 1.8 MeV, in order to explain the observed coincident peak area alone. The large required positron yields and the absence of competing γ-ray and electron conversion lines appear, on intensity grounds alone, to rule out conventional and exotic IPC and EPC processes as a possible explanation for the coincidence peak data.

Sequential Positron/Electron Emission

A third general class of models may contribute to production of coincident positrons and electrons. Instead of the emission of both e^+ and e^- from from a single decay, monoenergetic positrons and electrons might be produced sequentially in a cascade-like process [7]. If a monoenergetic electron is emitted every time that the positron is created, this would avoid the difficulties associated with independent e^+e^- emission (i.e., that structures would have been easily observable in adjacent cuts), as well as the problems with pair conversion processes (i.e., the very broad distribution of positron and electron energies). One simple example is the sequential decay of a series of nuclear transitions in which a positron creating conversion is uniquely preceeded or followed by a separate internal electron conversion.

Figure 28 illustrates the coincidence yield expected if the peak positron were due to IPC of a 1.4 MeV transition in U or Th, followed, with branching ratio of unity, by IC of an atomic K electron by a ~500 keV transition. The coincident positron peak is ≈200 keV wide, reflecting the triangular IPC energy distribution, much larger than the observed FWHM = 80 ± 20 keV. The sum-energy structure also exhibits a very large width of ~150 keV because of the absence of correlation between the emission

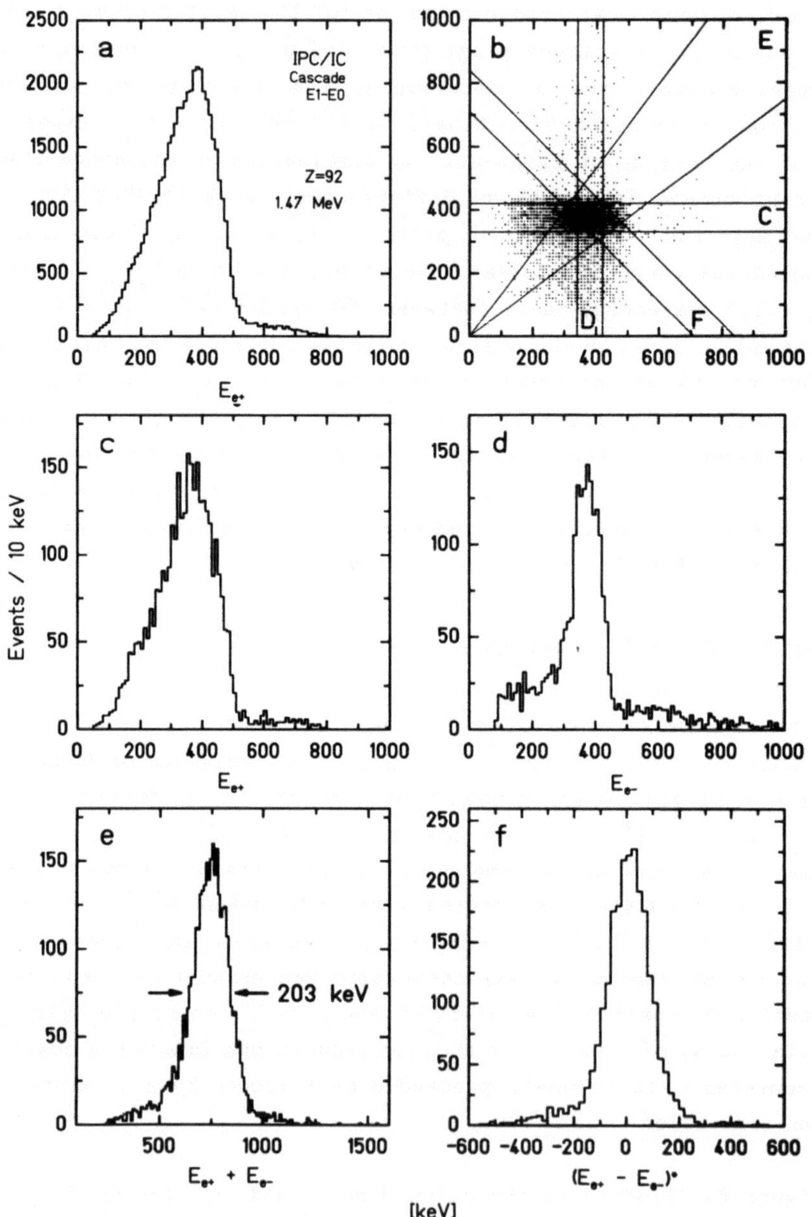

Figure 28. Simulation of a nuclear cascade process: (a),(c) Positrons are produced by 1.4 MeV IPC, and (d) coincident e^- from a 500 keV IC in the scattered nuclei. Positons exhibit a triangular IPC energy distribution with width >180 keV. Assuming an e^-/e^+ branching ratio of unity, ≈6% of the e+ peak counts are associated with coincidence events in the very wide sum-peak (e).

angles of positron and electron. Nuclear IPC origins have previously been shown unable to account for the narrow positron peak widths, or required transition intensities, for the data of Fig. 1. Figure 29 compares the expected competing γ-ray or IC electron line from the 1.4 MeV positron emitting transition to the simultaneously measured yields of Fig. 26 [41].

The intensity and positron linewidth arguments could be mitigated if one assumed Monoenergetic Positron Creation (MPC) [32,67] in an E0 conversion as the origin of the coincidence positron peak, enhanced by a preceeding depletion of the K-shell by e^- IC, accounting for the electron line. If the K-shell is empty with 95% probability at the moment of MPC, no competing K-conversion e^- lines (Fig. 29b) would be observed [3,5]. Production of monoenergetic positron and electron lines which exhibit no angular correlation leads to an E_{e+}-E_{e-} diagram such as Fig. 13. The widths of the individual positron and electron peaks depends on their mean energy and the nuclear velocity, leading to sum-energy peak widths ranging from 100 to 200 keV, depending on HI scattering angle.

Cascade-like decays could also, in principle, occur for particularly fast transitions in a giant di-nuclear complex, where the strong localization of the electronic wavefunctions increases the probability of MPC conversion. Assuming a branching ratio of unity and relative isotropy between e^+ and e^-, Fig. 13 describes the expected coincidence yield appropriate for an emitter velocity of v_{CM}. Aside from questions based on nuclear physics concerning the plausibility of this scenario, the formation of intrinsically narrow e^+ and e^- lines requires a long compound system lifetime (10^{-19} s), in which case spontaneous positron emission should occur with probability which dominates over MPC transitions [68]. Associating the coincidence positron line with spontaneous emission, rather than MPC, is difficult in light of the large measured peak energy (E_{e+} = 380 keV) compared to the prediction of QED ($E_{e+}\simeq$140 keV for U+Th), as well as the near constancy of peak energies in Fig. 1 over a large range of combined charge ($163 \leq Z_u \leq 188$).

The creation of monoenergetic positrons and electrons in cascade or sequential processes is not limited to nuclear conversions. Within the context of two-body decay origins, for example, a neutral object could be produced which decays sequentially through an intermediate state, $X^0 \to e^- \chi^+$, followed by $\chi^+ \to e^+ \nu$. A charged object formed directly in the collision and decaying by $\psi^+ \to e^+ \nu$ [69] or $(e^+e^-e^+) \to e^+\gamma$ [70,71] would produce monoenergetic positrons. Creation and decay of the charge conjugate, ψ^- or $(e^-e^+e^-)$, could in principle account for the coincident monoenergetic electron line.

The necessary feature of all these sequential processes is a one-to-one correspondence between monoenergetic positron and electron production. The absence of correlation between the e^+ and e^- emission angles results in an intensity pattern typifed by Fig. 13, exhibiting a wide sum-energy peak width (from 105 keV for v_{CM} to ~160 keV for $\langle v_{nuc} \rangle$), and a ratio of coincidence events to total detected positrons of ≤6%. In addition, the production of positrons and electrons with equal energy does not arise naturally in cascade models and must be assumed ad hoc by adjusting the nuclear transition energies or X^0, X^+ masses. While a pair of conjugate charged objects may naturally explain nearly equal positron and electron emission energies, production must occur at large distances (~10^4 fm) from the compound system in order to avoid an asymmetric shift in the detected positron and electron kinetic energies ($\langle E_{e+} - E_{e-} \rangle_{obs} \leq 20$ keV) by the strong Coulomb field. If those objects are produced with a distribution of momenta, the e^+ and e^- would be Doppler-shifted by different source velocities, substantially smearing out the correlations in the energy-energy plane.

VI. FURTHER RESULTS

Several further experimental results may have particular bearing on the interpretation of the data, particularly in the context of a two-body decay. In addition to the U+Th data presented above, data were accumulated on the Th+Cm and Th+Th systems. In Th+Cm around 5.85 MeV/u, and in Th+Th at 5.75 MeV/u, the data seem to indicate the presence of a coincident sum-energy peak near 750 keV [7], which added to the U+Th data discussed above, yields ≈45 coincident peak events, as shown in Fig. 30. Positron, electron, and difference-energy peaks are comparable to those presented in Fig. 19,

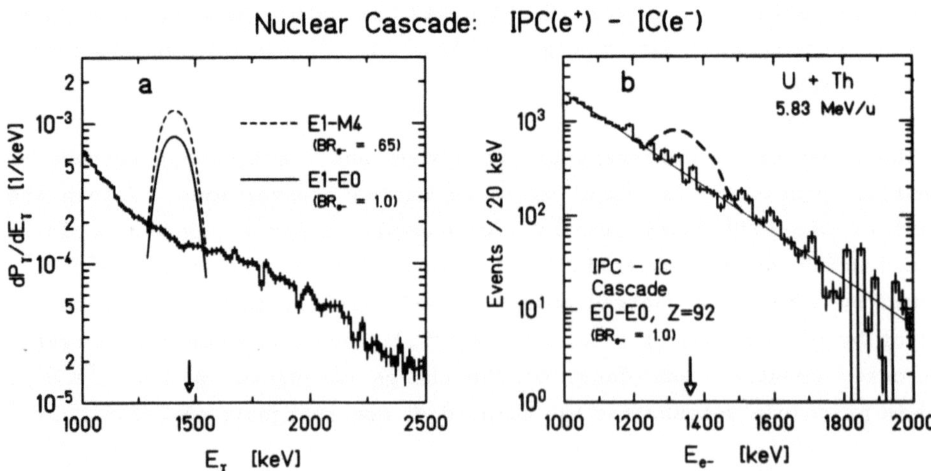

Figure 29. Data of Fig. 26 compared to expected intensity of (a) γ-ray, and (b) e^- IC lines competing with the 1.4 MeV IPC branch of the nuclear cascade of Fig. 28.

but the increased statistical significance of the sum-energy spectrum suggests a narrower sum-line with width 60 ± 15 keV, centered at 750 ± 15 keV.

At lower beam energies, around 5.70 MeV/u in Th+Th and 5.75 MeV/u in U+Th, a second interesting feature in the coincident event plane is observed. As shown in Fig. 31, there appears to be evidence that a positron peak at 310 keV is observed in coincidence with an electron peak at ~295 keV. Although the statistical evidence is weak here, there is an indication that more than one coincidence structure may be associated with positron peak energies between ~300 and ~400 keV [7].

These features have been confirmed in a second series of coincidence measurements undertaken in February and March 1986. Electron energy resolution was improved to <15 keV, so that the minimum expected width of the sum-energy peak is ≈25 keV, if due entirely to detector response. New heavy-ion PPAC detectors with a factor three larger solid angle (at a cost

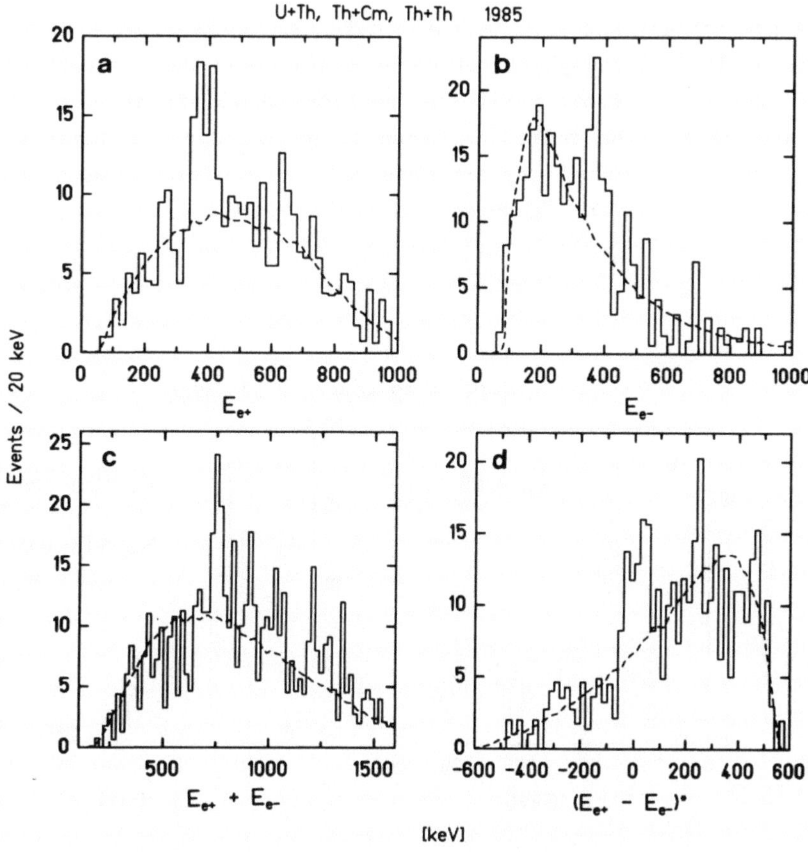

Figure 30. Data from 5.88 MeV/u Th+Cm and 5.75 MeV/u Th+Th collisions are shown added to those of Fig. 19. The coincident e^+/e^- peak intensity is increased to ≈45 counts, with a sum-line width of 60 ± 15 keV.

of decreasing the scattering-angle resolution from .5° to ~1.2°) were employed in order to acquire greater statistical accuracy. This change seemed justified in light of the shift of our analysis approach away from exploiting heavy-ion scattering-angle correlations to suppress continuous backgrounds, towards exploiting the sensitivity and selectivity of the EPOS coincidence spectrometer in detecting correlated positron-electron emission.

The study concentrated on the U+Th system at energies between 5.82 to 5.90 MeV/u. A .03 MeV/u recalibration of the UNILAC energy places the preceeding U+Th measurement at ~5.86 MeV/u. The targets were ~250 μg/cm² ThF_4 evaporated on 20 μg/cm² carbon backings, covered with an evaporated C layer of 5 μg/cm². As mentioned previously (see also Fig. 21), in the Th+Th measurement in July 1985 it was found that the positron peak near 320 keV seemed to be enhanced by selecting only data collected for short irradiations of targets, i.e., during the first hour of heavy-ion beam hitting each target. During the subsequent run, we therefore changed targets every 1½ hours, using a total of over 200 targets in 20 days of beam time.

A preliminary analysis of these data has revealed several very interesting results. Although target deterioration and the possibility of sharp beam-energy resonances clearly are questions which affect the collection of data and have to be addressed, even if we include all the data at this point, sum-energy structures are observed. In addition to excess intensity

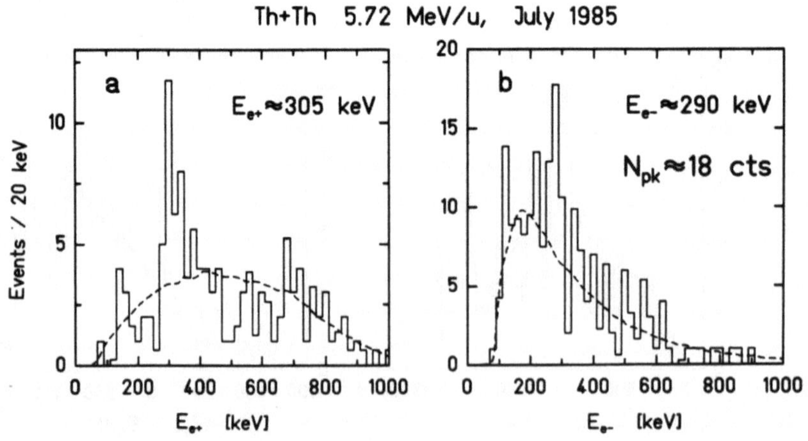

Figure 31. Th+Th collisions at 5.72 MeV/u. (a) Coincident peaks at $E_{e+} \approx 305$ keV and (b) $E_{e-} \approx 290$ keV suggest the presence of additional correlated structure.

near 750 keV, prominent narrow lines are evident near ~600 and ~800 keV. The higher of these, already at a 3σ level in the total data seems to become prominently visible by selecting the higher beam energies sampled, as seen in Fig. 32 a,b. Here the sum-energy distribution and corresponding difference-energy spectrum are plotted for data accumulated for beam energies between 5.87 and 5.90 MeV/u and prompt positron-electron timing. In contrast to the data of Fig. 3, no HI scattering angle region is selected. A narrow sum-energy peak is observed centered at 810 ± 10 keV, which is associated with narrow positron and electron lines. Its intensity of 100 ± 15 counts represents >6σ level of confidence. The narrow width of 40 ± 10 keV is clearly smaller than the individual positron and electron peak widths.

Figure 32 c,d presents data on the lower energy sum peak at 620 ± 10 keV, for beam energies between 5.85 and 5.90 MeV/u, with a further selection on positron arrival times slightly delayed (~3 ns) from the mean dynamic positron TOF. The sum-energy peak intensity is 55 ± 12 counts, and the FWHM is ≈25 keV, significantly smaller than the associated difference-energy distribution width of ≈180 keV. Possible interpretations of the apparent TOF delay will be discussed in more detail below.

The simulation studies presented above of various possible origins for the coincident structures apply equally well to these new data. Additionally, the class of cascade-like models involving monoenergetic, isotropic emission from the CM system, leading to sum-energy peak widths of 105 keV, can now be quantitatively ruled out based on the extremely narrow observed widths of ≈25 and 40 keV. Taken together, these data convincingly preclude explanations based on known nuclear conversions, as well as a variety of other processes described above. The characteristic patterns of correlated positron and electron energies appear to emulate those expected for the two-body decay of a neutral source away from the influence of the Coulomb field.

VII. TWO-BODY DECAY ORIGIN

Theoretical attempts to explain these data in terms of one or several new elementary particles [43-48,65,73-76,83-101] have met with difficulties in light of the negative results of recent experimental particle searches [102-117]. In this connection it may be worthwhile to note that the environment in which these coincident peaks are observed is very different than usually encountered in most other realms of physics. The heavy-ion quasi-molecular collision is a unique "laboratory" characterized by large transient electric and magnetic fields, formed on the time scale of 10^{-21} s, with strengths exceeding 10^{16} V/cm and 10^{13} Gauss (at 100 fm) [73]. Because of the strong fields, non-linear effects may be important. The presence of

~500 nucleons and ~100 electrons in a confined region of space may further complicate matters. It has been suggested that nuclear effects [45-47,73,74], collective coupling to quarks [75], or many-body interactions among the positrons and electrons [76] may all play a role in enhancing a particular production mechanism, and investigations along these lines are discussed by many authors. Proposals range from quasi-atomic effects [77-80], to the formation of composite objects with an internal structure [70,71,76,81,82], to the creation of condensates of elementary bosons [73,83-85], to ideas involving a restructuring of the vacuum [69,73,86]. As the body of evidence against a simple interpretation in terms of a single new elementary particle grows, it is increasingly important to understand the particular role of the heavy-ion collision environment in producing these coincidence signals.

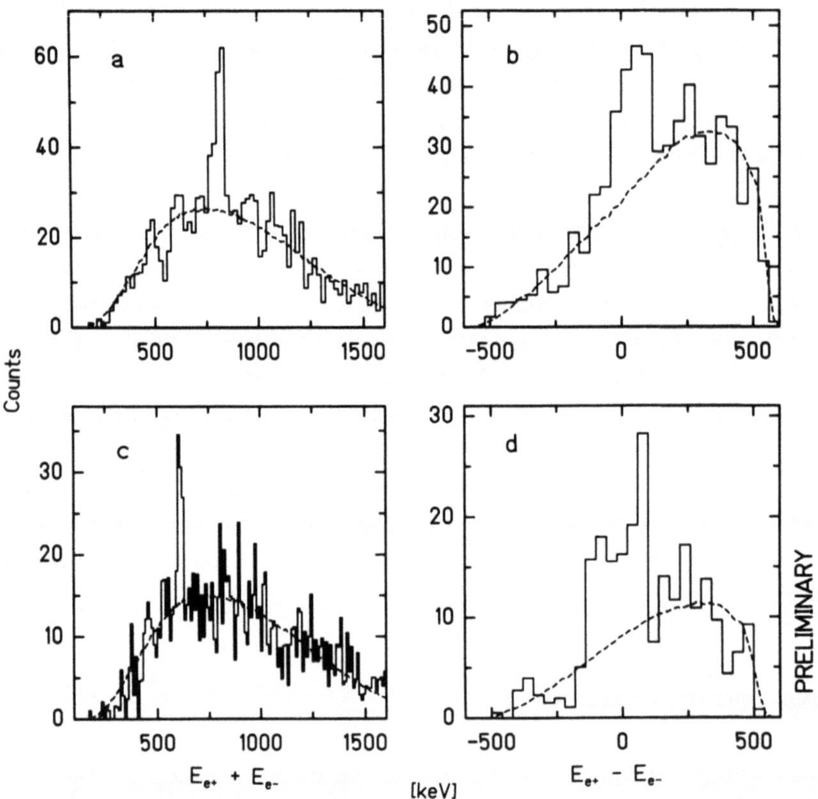

Figure 32. Results of a preliminary analysis of U+Th collisions near 5.87 MeV/u (Feb, 1986). (a) Sum-energy and (b) Difference-energy projections for U+Th collisions between 5.87 and 5.90 MeV/u, selecting events for prompt e+ and e- TOF. The sum-energy line at 810 keV contains 100±15 counts, and has a width of 40±10 keV. Data of (c) and (d) are selected for 5.85 to 5.90 MeV/u collisions with prompt electron TOF but slightly delayed e+ TOF. The ≈620 keV sum-energy line has 55±12 counts with a width of 25±10 keV.

Up to this point our discussion has been largely concerned with ruling out classes of certain well-defined model-dependent origins for the coincidence peaks. One can extract experimental model-independent information from the data, as well as address questions such as the existence of multiple structures and the required small momentum for a created object, within a more detailed investigation of the kinematics of the emitted positron and electron.

Positron/Electron Lines from a Moving Source

As mentioned above, monoenergetic positrons or electrons emitted from a system (denoted by *) which moves with velocity $v = \beta c$ with respect to the laboratory frame, will be detected with Lorentz transformed kinetic energies and emission angles. The magnitude of these shifts depends only on the momentum and direction of the emitted positron or electron:

$$E_{e+} = \gamma \,(E^*_{e+} + \beta P^*_{e+} \cos\theta^*_{e+}) \tag{11a}$$

$$E_{e-} = \gamma \,(E^*_{e-} + \beta P^*_{e-} \cos\theta^*_{e-}) \tag{11b}$$

where $\theta^*_{e+,e-}$ is the lepton direction with respect to the system's motion and E^*, P^* are the emitted lepton energy and momentum. The individual positron (or electron) peak, when integrating over 4π sr, can exhibit a full width of $2\gamma\beta P^*_{e+}$ (corresponding to emission parallel and antiparallel to the emitter motion), centered at a lab energy of $E_{e+} = E^*_{e+} + (\gamma-1)E^*_{e+}$. For a 350 keV positron, this amounts to: FWHM = 77.7 keV, δE = 1.35 keV.

The sum-energy and difference-energy coordinates introduced previously are defined as:

$$\Sigma E_{\pm} + 2m_e c^2 = \gamma(E^*_{e+} + E^*_{e-}) + \gamma\beta(P^*_{e+}\cos\theta^*_{e+} + P^*_{e-}\cos\theta^*_{e-}), \tag{12a}$$

$$\Delta E_{\pm} = \gamma(E^*_{e+} - E^*_{e-}) + \gamma\beta(P^*_{e+}\cos\theta^*_{e+} - P^*_{e-}\cos\theta^*_{e-}). \tag{12b}$$

The position and width of the resulting sum and difference peaks depend not only on the emitted energy, but also on the direction the positron and electron are ejected. For the special case where the electron and positron energies are equal ($E^*_{e+} = E^*_{e-} \equiv E^*$), Eqns. 12a,b simplify to:

$$\Sigma E_{\pm} = 2\gamma(E^* - m_e c^2) + \beta\gamma P^*(\cos\theta^*_{e+} + \cos\theta^*_{e-}), \tag{13a}$$

$$\Delta E_{\pm} = \beta\gamma P^*(\cos\theta^*_{e+} - \cos\theta^*_{e-}). \tag{13b}$$

The kinematic "Doppler-shift" now depends only on the relative emission directions of positron and electron. For example, if both positron and electron are emitted always in the same direction, then $\cos\theta^*_{e+} = \cos\theta^*_{e-}$ and the difference-energy is always zero, while the sum distribution exhibits a full width of $4\gamma\beta P^*$. For back-to-back emission, this situation is reversed yielding a narrow sum peak and a $4\gamma\beta P^*$ keV wide difference peak. For mutually isotropic emission, after averaging over direction, the sum *and* difference peaks exhibit a width of $2\sqrt{2}\gamma\beta P^*$, $\sqrt{2}$ larger than that of the individual positron or electron line alone (see Fig. 13). Figure 33 illustrates this point by presenting calculated sum- and difference-energy

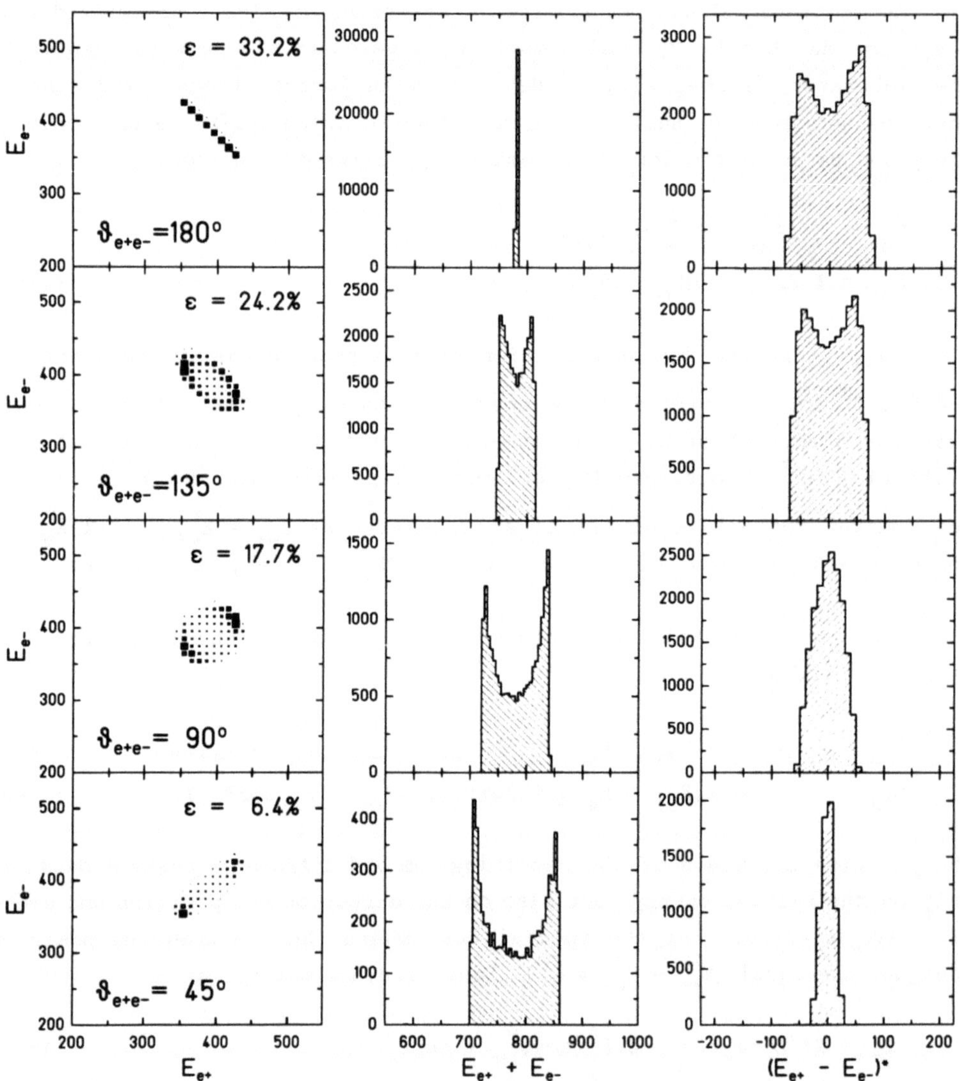

Figure 33. Monoenergetic e^+ and e^- (380 keV) emitted isotropically from CM with fixed relative opening angle. (Left) Energy correlation plot, and the widths of the sum-energy peak (center) and difference-energy peak (right) illustrate the effect of angular correlation on kinematic broadening. Detector response and backgrounds have been omitted for clarity. ε denotes relative coincident/total e^+ detection efficiency.

peaks, assuming monoenergetic positrons and electrons emitted with fixed opening angle from a frame moving with v_{CM}. For clarity, detector resolution has been omitted. The choice of velocity does not affect the salient feature that only for large opening angles -- $\theta_{e+e-} \gg 90^\circ$, approaching back-to-back emission -- do a narrow sum and wide difference peak emerge, which resemble the observed data.

Multi-Body Final States

The occurrence of monoenergetic positrons and electrons confined to a small range of nearly equal energy, arises most naturally in the two-body decay of a neutral state. For a three-(or more)-body final state, each positron or electron takes on a range of energies from zero to some maximum, which is determined by the particular combination of decay-product masses. In general, the positron and electron energy correlation has a broad oval pattern, several hundred keV wide, as indicated in Fig. 34a,b for third body masses of 0 and 511 keV/c^2, respectively. Nuclear IPC represents a limiting case of multi-body decays in which the remaining participant(s) have much larger mass than the positron and electron, so that they remove very little kinetic energy from the decaying system. The positron-electron correlation pattern is then, of course, the familiar diagonal band for which $E_{e+} + E_{e-} =$ constant (Fig. 34c). However, the momentum taken up by the heavy partners in a multi-body final state erases any back-to-back tendency for the remaining positrons or electrons. A broad distribution in positron-electron opening angle results which, from Eqn. 13a, leads to a broad sum-energy distribution, as we have seen for specific calculations of IPC scenarios. The expectedly wide sum-line and the particularly broad range of positron and electron energies are in contradiction to the pattern of the observed coincidence data. Even if the emission is from rest in the lab, so that the sum-energy peak remains narrow, as we have discussed for external pair creation, the broad positron and electron and difference-energy distributions cannot be accommodated by the well-defined difference-energy peak and absence of structure in adjacent cuts observed in the heavy-ion data. Multi-body (greater than two) decay kinematics of a state of fixed total energy appear to be ruled out.

Two-body Decay from Rest in the Lab

Within the context of a two-body final state, the small width of the sum-energy peak and the wider difference-energy peak arise naturally for back-to-back emission from a *moving* source. The magnitude of the broadening depends on the motion of the emitting state. If the emitter is at rest in the lab (as shown in Fig. 35a) no kinematic broadening occurs for any of the lines -- the various lineshapes are determined solely by the intrinsic width of the state and detector response. Since the sum-energy peak measures the total kinetic energy of the positron-electron combination, for a two-body final state from an object at rest in the lab, the sum-line exhibits the intrinsic width of the parent. This energy, however, will be divided equally between the positron and electron. The individual positron and electron lines therefore exhibit only half the sum-energy peak width, and the difference-energy will be identically zero (neglecting, for the moment, detector resolution). This pattern (Fig. 35b) is the exact reverse of that predicted for a moving source for which the kinematic broadening dominates the intrinsic width of the state (see Fig. 12b). For a particle lifetime greater than 10^{-18} s ($\delta E < 5$ keV), the linewidths are determined by detector resolution: ~15 keV for the positron and electron, and ~25 keV for the sum- and difference-energy structures. We conclude that the appearance of a narrow sum-energy peak in the data, accompanied by much wider positron, electron, and difference-energy peaks, rules out the possibility that we are observing the two-body decay of an object at rest in the lab.

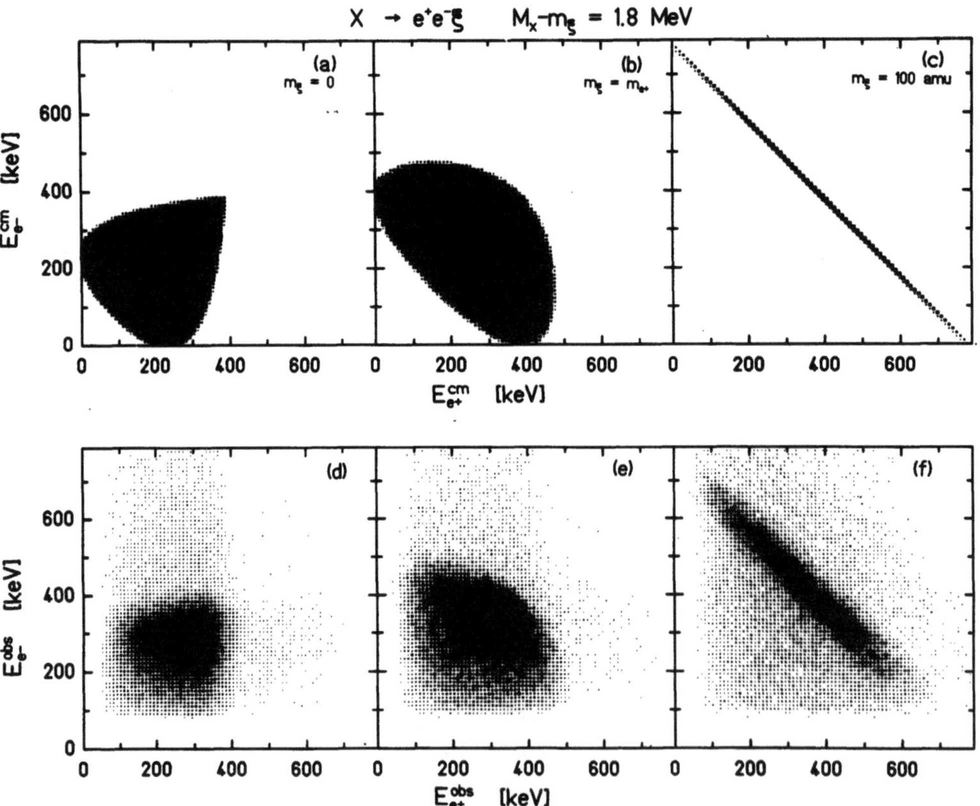

Figure 34. MCSPEC simulation of 3-body decay. (a-c) E_{e+} vs. E_{e-} in X rest frame, given by phase space distributions only. M_ξ = 0, m_{e+}, and 100 amu, as marked. (d-f) Corresponding detected coincidence distributions assuming creation of the object at rest in heavy-ion CM system.

While two-body decay of a neutral particle-like object provides a natural explanation for the observed peak widths and the near equality of positron and electron energy, it should be noted that the EPOS experiment does not reconstruct the CM system or yield the invariant mass of the positron-electron system. An analysis of the kinematic energy correlations, however, provides extensive, if indirect, information regarding the emitter. In the considerations above, the multi-body (≥3) decay of a particular state, and the two-body final state of an object at rest in the lab, have been shown to contradict the observations. One is left with the very general picture that a nearly monoenergetic positron and electron are emitted from some system which is moving with respect to the laboratory frame. The data exhibit wide difference-energy peaks centered around zero, $\langle E_{e+}-E_{e-}\rangle \sim 0$, associated with much narrower sum-energy peaks. The ratio of their statistical widths, $RMS_{\Sigma}/RMS_{\Delta}$ (~25 keV/180 keV for the ~620 keV sum-line, and ~40 keV/140 keV for the ~810 keV structure) is largely independent of source velocity. Within this context, Eqn. 13 and Fig. 36 indicate that the positron and electron must be ejected from the emitting source with large mean relative opening angles larger than 150° which is compatible with two-body decay in which the positron and electron are emitted back-to-back, $\theta_{e+e-} = 180°$.

In this analysis, it is assumed only that the positron and electron are each nearly monoenergetic and that they are emitted from the same inertial

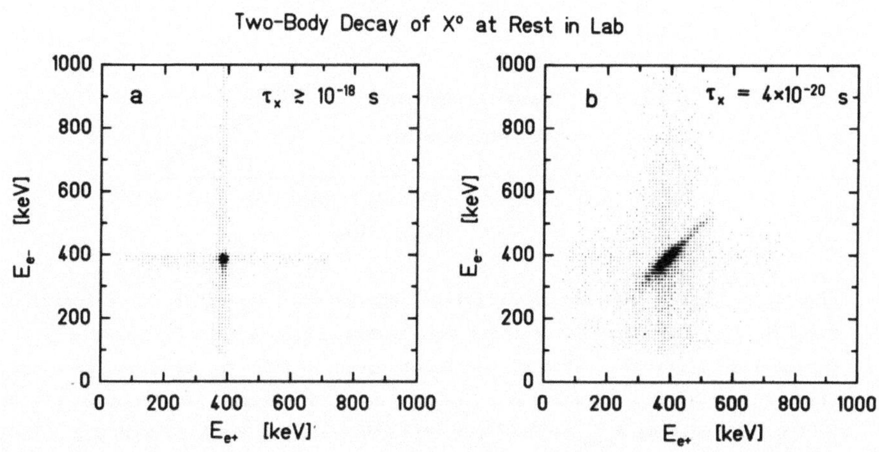

Figure 35 Expected pattern for two-body decay of an object at rest in Lab, if (a) its intrinsic width is smaller than detector resolution (i e , $\tau_x \geq 10^{-18}$ s), or (b) $\Delta E_x = 150$ keV ($\tau_x \approx 4\times10^{-21}$ s) Continuous background omitted for clarity.

frame over a broad range of angles. The dashed line in Fig. 36 presents the ratio of sum-energy to difference-energy peak width for emission from the CM system calculated in Fig. 33. A detailed analysis of various combinations of anisotropic e^+ emission, source momentum and direction, and distributions as well as discrete values of θ_{e+e-}, indicates that the difference-energy peak width decreases more quickly than the sum-energy line width from their maximum values, given by the source velocity times the rms width of $(\cos\theta^*_{e+} - \cos\theta^*_{e-})$ or $(\cos\theta^*_{e+} + \cos\theta^*_{e-})$, respectively [41]. The solid curve in Fig. 36 plots this limiting case, corresponding to uniform detection over 4π sr,

$$rms_\Sigma/rms_\Delta = [1 + \cos\theta_{e+e-}]^{\frac{1}{2}}/[1 - \cos\theta_{e+e-}]^{\frac{1}{2}}, \quad (14)$$

representing the smallest opening angle for a given ratio of rms widths. The dotted curves include detector resolution. $\theta_{e+e-} > 140°$ for the ~810 keV sum-energy line and $\theta_{e+e-} > 165°$ for the ~620 keV peak therefore provide fairly model-independent limits (90% C.L.) of the minimum opening angle between the ejected positron and electron, in the rest frame of their source. Back-to-back emission, and the subsequent identification of the signal with the two-body decay of a neutral object, is clearly suggested by this result.

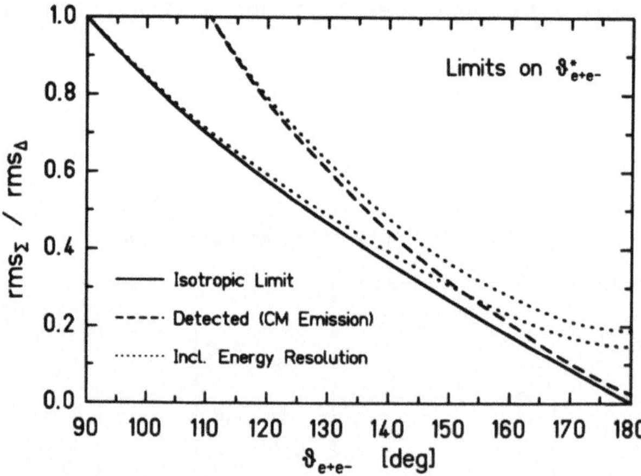

Figure 36. Ratio of sum- to difference-energy RMS peak widths as a function of fixed θ_{e+e-}. The solid curve shows the kinematic broadening of emitted e^+ and e^-; dashed line plots the ratio of widths observed in EPOS for emission from the CM (Fig. 33). The dotted and lines include detector energy resolution. $R \leq 0.40$ (90% C.L.) implying $\theta_{e+e-} > 140°$ for ~810 keV line, and $R \leq 0.20$ (90% C.L.) implying $\theta_{e+e-} > 165°$ for ~620 keV peak.

However, the two-body nature is not proven. One may hypothesize a modification to the Cascade-like processes discussed above, namely, that in some way the emission directions of the positron and electron are distributed predominantly opposite to each other such that the mean opening angle is greater than ~140° of 165°.[5] Then, both the shape of the coincident event pattern, and the relative coincidence-to-total positron peak detection efficiency will simulate the expectations of two-body decay. This class of models still retains the difficulty that transitions in the separated nuclei, as well as spontaneous decay of the QED vacuum, are not viable positron production mechanisms. More exotic schemes for producing monoenergetic positrons and electrons at the same energy which are spacially correlated such that $\langle\theta_{e+e-}\rangle \geq 140°$ or 165°, cannot be ruled out from the coincidence data alone. Physically viable mechanisms must be evaluated on an individual basis in light of the additional observed properties of the U+Th coincidence data.

Emitter Velocity

From the preceeding kinematic arguments, we conclude that both the equal energy of the observed positron and electron signals, and their angular correlation, is explained by a process that closely resembles the two-body final state of a decaying neutral object. The pattern of positron-electron energy correlation becomes more complicated if that neutral object is not created nearly at rest in the heavy-ion collision. Its total lab energy is kinematically shifted by its motion relative to the emitting frame. Consider, for example, a particle emitted from the CM of the HI collision with momentum P_x^* and total energy $E_x^* = (P_x^{*2} + M_x^2)^{\frac{1}{2}}$. In the lab,

$$E_x = \gamma_{cm}[(P_x^{*2} + M_x^2)^{\frac{1}{2}} + \beta_{cm}P_x^*\cos\theta_{x-cm}^*]. \tag{15}$$

The sum-energy line at $(E_x-2m_ec^2)$ will therefore be shifted from $(M_x-2m_ec^2)$ and broadened. For isotropic creation, the shift is $(\gamma_{cm}E_x^*-M_x)$, i.e., just the additional mean laboratory kinetic energy of the object, and the width is $FWHM_\Sigma = 2(\gamma\beta)_{cm}P_x^*$. Figure 37 plots the positron-electron energy correlation pattern (left panel) for a neutral object of mass 1.8 MeV/c^2 emitted isotropically from the CM system, with kinetic energies T_x = 0, 10, 50, 200 keV, as marked. Even for very small energies, the width of the indi-

---

[5] A rather narrow angular distribution is required. For example, a $|Y_{10}|^2$ distribution in θ_{e+e-} already appears to be excluded by the present experimental data.

vidual positron and electron peaks and the difference distribution (right panel) broaden very rapidly. The sum-energy peak (center panel) width is summarized as a function of the emitted kinetic energy in Fig. 38b.

An important consequence of the motion of the e^+e^- source in the CM is the subsequent decrease in the opening angle between the positron and electron. As the velocity of the object increases, the back-to-back correlated positron and electron are swept forward in the lab system. The positron or electron lab angle is

$$\theta_{e+,e-} = \tan^{-1}[\sin\theta_{e+}/\gamma_x(\cos\theta_{e+} + \beta_x/\beta_{e+})], \tag{16}$$

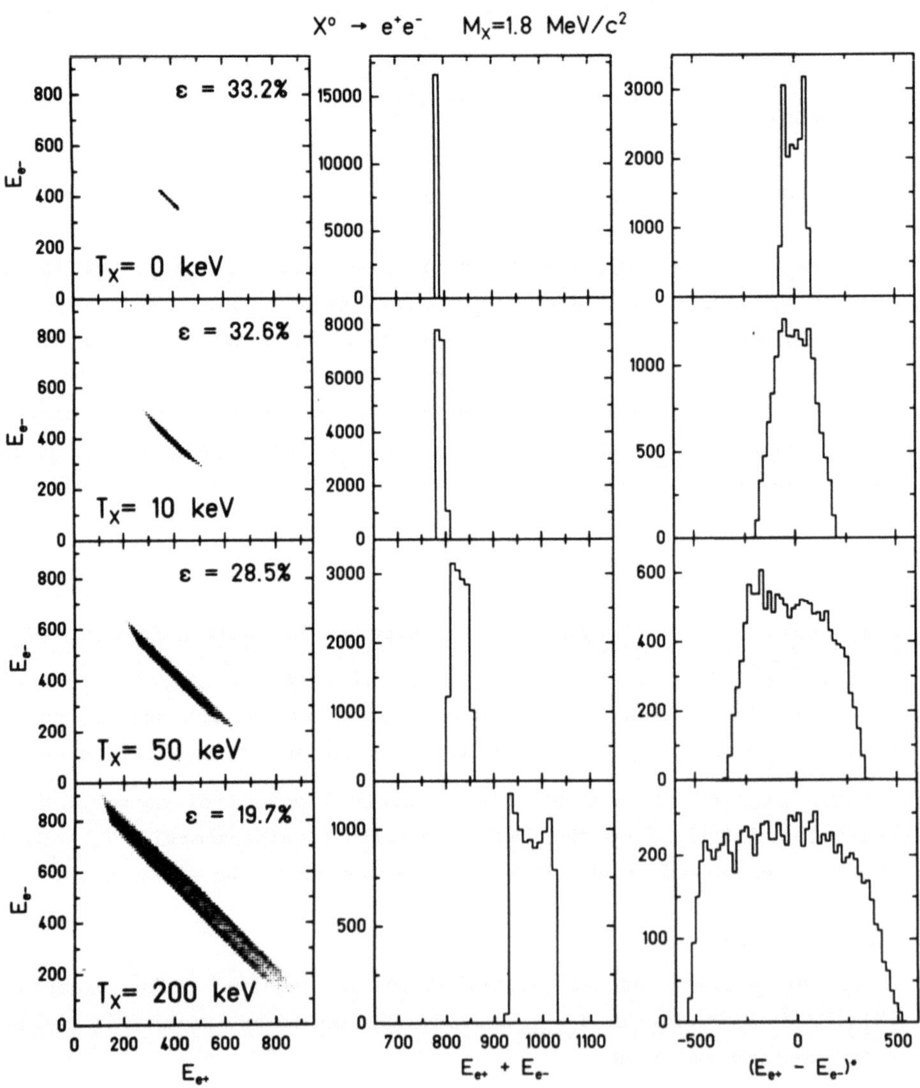

Figure 37. Simulation of 1.8 MeV/c² X° created isotropically in CM with fixed kinetic energy T_x as explained in text (detector resolution omitted).

where γ_x and β_x denote the X^0 velocity in the lab. The opening angle, assuming back-to-back emission, is about $2 \cdot \tan^{-1}(\beta_{e+}/\beta_x\gamma_x)$. Figure 38a presents an exact calculation of the mean laboratory opening angle, averaged isotropically over all emission directions, as a function of the CM kinetic energy, for a neutral state of 1.6 or 1.8 MeV/c^2. For creation at rest in the CM, v_{lab} = .056c, and the opening angle is already reduced from 180° to ~172°.

The decrease in opening angle naturally leads to a decrease in detection efficiency as the relative positron-electron distribution deviates farther from back-to-back emission. The solid line in Fig. 38c plots the absolute EPOS acceptance for the energetic neutral object. From the spreading of events in the energy correlation plane evident in Fig. 37, the portion of the data within the wedge-shaped sum-energy projection also decreases with increasing energy, producing a decrease in analysis efficiency (shown by the dashed line in Fig. 38c).

For objects created with large energies in the collision, coincident events are distributed over a wide area in the correlation plane, and tend to merge with the bulk of continuous dynamic positron, δ-electron, and nuclear IPC pair backgrounds. It is for this reason that even the creation of a particle with a broad range of momenta can lead to the appearance of

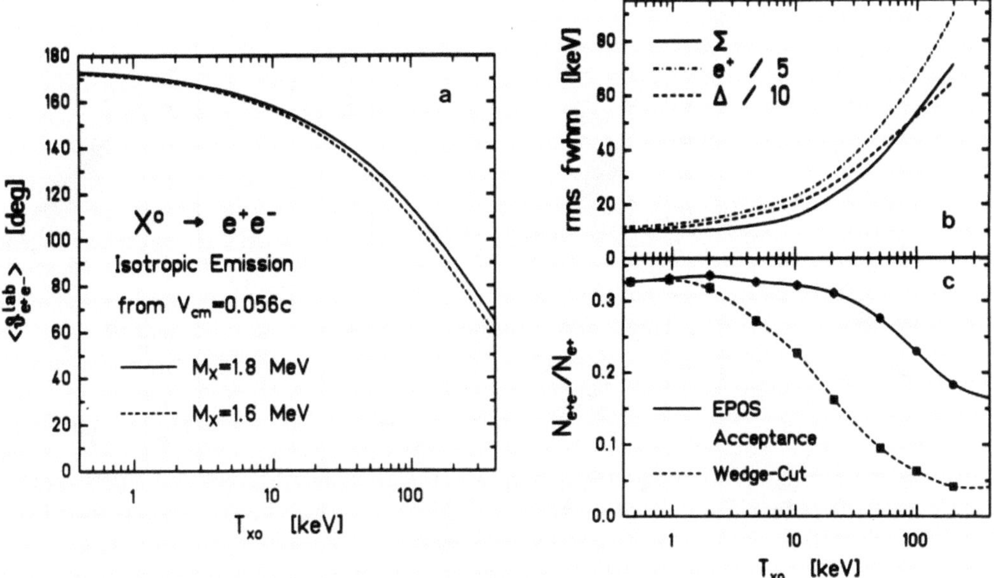

Figure 38. (a) Mean laboratory e+/e- opening angle for the two-body decay of X^0 emitted from CM system with fixed kinetic energy T_x. For creation at rest, θ_{e+e-} = 172° from CM motion. (b) Widths of sum-energy, difference-energy (÷10), and individual positron peaks (÷5) vs. T_x. (c) Detection efficiency of coincidence relative to positron events. Solid-line indicates EPOS acceptance, dashed curve is portion of events in wedge-cut.

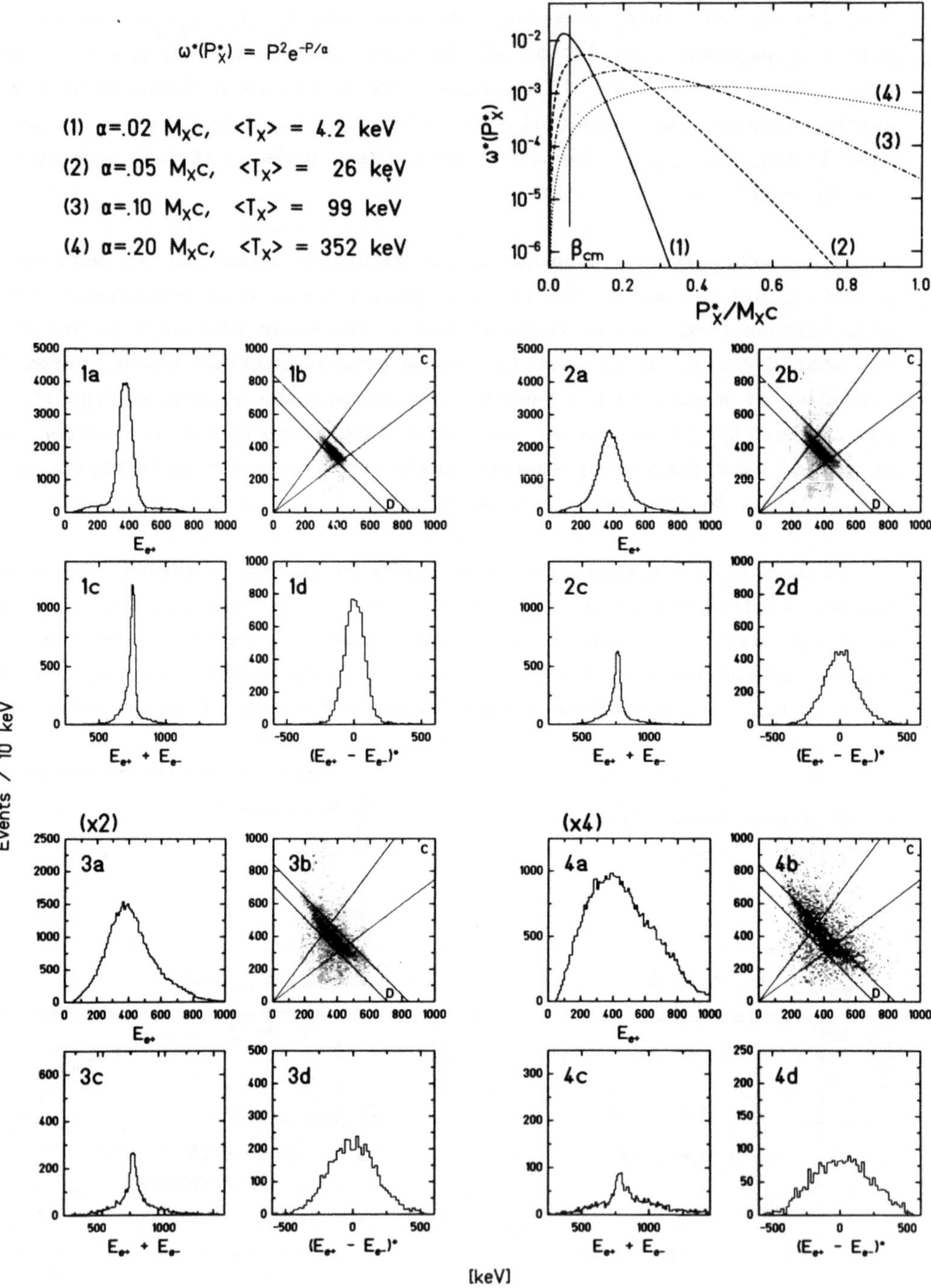

<u>Figure 39</u>. MCSPEC simulation of the two-body decay of 1.8 MeV/c² objects created isotropically in the CM system with a distribution of momenta. (a) Total detected e^+, (b) coincident e^+-e^-, (c) sum-energy, and (d) difference-energy distributions. Quadrants correspond to $w(P) = P^2e^{-P/\alpha}$, α = .02,.05,.1, and .2 M_xc, as marked. 50000 detected e+ generated for each case. Note expanded scale for α = .10, .20 M_xc. (Detector response included.)

narrow positron structures as discussed in Fig. 4. To place this on a quantitative footing, Fig. 39 presents the expected coincidence yields for various distributions of X^0 velocity. In order to account for the P^2 phase space volume element, and the anticipated exponential decrease in production at large momentum, we have assumed a momentum distribution of the form $w^*(P_x^*) = P^2 e^{-P/\alpha}$. The four curves for different values of α = .02, .05, .1, and .2 $M_x c$ lead to mean emitted particle kinetic energies of 4.2, 26, 99, and 352 keV (assuming M_x = 1.8 MeV). For increasingly broad distributions, we observe that the total positron energy spectrum becomes quite wide over a range in which the sum-energy line remains fairly narrow.

The superposition of low momentum components which suffer little kinematic shift, together with larger momenta particles, leads to peaks in the positron, electron, sum-energy, and difference-energy spectra which are at nearly the same position as expected for creation at rest. Coincidence intensity is focussed within the region $E_{e+} \simeq E_{e-}$, as hinted at by the triangular shape of the difference spectrum, centered at $\Delta E_{\pm} = 0$. Up to rather large mean emission energies, ~100 keV, very little significant structure emerges in cuts adjacent to the positron and electron peak regions. This surprising result, in the context of the broad positron energy distribution, underlines the difference between kinematic correlations here and those observed for IPC, where the much broader difference-energy spectrum and adjacent cuts were sufficient to rule out a nuclear origin. The angular correlation is also different. For $\alpha \leq .1\ M_x c$, the positron and electron are typically emitted with opening angles $\theta_{e+e-} > 90°$. The detection efficiency of accelerated and energy broadened particle-decay products is therefore still much larger than for IPC, whose angular distribution peaks at $\theta_{e+e-} = 0°$. One requires only a moderate increase in the production cross-section, beyond the 3% of total positron yield appropriate for creation at rest, in order to adequately describe the observed data. A good fit to our data is for $\alpha = 0.07\ M_x$, and ~10% of the total positron yield.

Theories involving the dynamic production [43-45,47,48,74] of a neutral particle electromagnetically from the Coulomb field or by strong interactions from the assembly of quarks, predict a suppression of low-momentum creation, beyond phase space volume, by an extra factor of P^3 to P^5. From the above analysis it seems unlikely that these scenarios can describe the observed data. A necessary condition for understanding the narrow positron, electron, and difference-energy peak widths is the presence of a sufficient intensity of objects created with low momentum (P_x/M_x < .1). As demonstrated in Fig. 39, the high-P_x components tend to smear out becoming indistinguishable from the continuous backgrounds. At some point, however, the narrow peak becomes overwhelmed by the increased continuum intensity.

Detection of positrons or electrons is suppressed if the neutral state has a long lifetime, so that the fast objects (v_x>.1c) escape the EPOS fiducial volume. Figure 6 presents the dependence of positron and electron detection efficiency on the perpendicular distance of the source from the solenoid axis. Coincidence detection efficiency should decrease even more quickly. Lifetimes of $\tau > \ell/v = 10^{-9}$ s will be affected by the velocity-filter behavior [5,7,41]. The presence of detected or undetected high-P_x contributions, however, requires an increase in the absolute X^0 production cross-section. Other proposed mechanisms [85,86] can, in principle, accommodate creation of a neutral object at rest in the CM of the collision.

This analysis seems to indicate that a viable description of the data in terms of the two-body decay of a neutral state need not require creation of the object only at rest in the CM system of the heavy-ion collision. The experimental observation that neutral particle-like objects must be formed with sufficient intensity at low velocities is consistent with creation with rather broad momentum distributions, due to the preferential detection of slowly moving objects in the EPOS spectrometer. Caution must therefore be exercised in evaluating theoretical models based on the expected broad range of created momentum. For example, an empirical distribution suggested by Balantekin et al. [44],

$$dN_x/dP_x = \Sigma\ \delta(\vec{P}_x - \vec{k}_{e+} - \vec{k}_{e-})\ (dN_{e+}/d\vec{k}_{e+})\ (dN_{e-}/d\vec{k}_{e-}), \tag{17}$$

extends to only slightly higher mean momenta than our acceptable estimate, $w^*(P_x^*) = P^2 e^{-P/0.07Mc}$, and is thus nearly consistent with the coincidence measurements. The comparison is improved if the neutral particle lives for $\geq 10^{-9}$ s.

Multiple Structure

The observation of at least three narrow sum lines raises the unattractive prospect of hypothesizing a whole family of neutral objects to explain the data, instead of just one. An alternate possibility is that a single object is created with different discrete values of kinetic energy. Chodos et al. [45, see also 47,73,74]] suggested a process in which the multiple positron peak energies arise from a single object receiving increments of kinetic energy from the nuclear complex. In order to retain a small Doppler-broadened linewidth, however, the velocity of the decay positron and electron, in the rest frame of the source, must be small compared to v_{CM}, thus $T_{e+,e-} \leq 1$ keV. Only very small particle masses of $M_x \leq 1023$ keV, can therefore be accommodated, and the small available lepton momentum implies that the positron and electron travel in nearly the same direction ($\theta_{e+e-} \leq 4°$), and are not detected in coincidence in the EPOS spectrometer.

Alternatively, one may assume that a single object of mass ≈1.65 MeV/c^2, explaining the lower observed sum-energy peak, is boosted by ~100 to ~200 keV in order to fit the higher lines. From Fig. 38b we note, however, that the broadening of the positron and electron energy distributions is very large, inconsistent with the data. The observed narrow sum-energy peak widths of 25 to 40 keV set a limit on the kinetic energy of the object, from Fig. 38b, of $T_x < 60$ keV. This is at the limit of associating the 760 keV sum-energy peak with the 810 keV sum-line observed at slightly higher beam energies in the recent experiments. Figure 37 (1st and 3rd panels) evaluates the possibility that these are the same structure, assuming a single object of mass 1.8 MeV/c^2 accelerated by either 0 or 50 keV. The expected broad difference-energy distribution, in light of the observed narrow positron and electron peak associated with this line seems to argue against this interpretation.

The data suggest that the different observed sum-energy peaks are not related simply by the kinematic boost of a single object. We conclude that more than one, and possibly several objects, must be formed in the HI collision, with different values of total energy ranging from ~1.6 to ~1.9 MeV. We note in passing that several recent theoretical models which avoid the assumption of a new family of elementary bosons describe the multiple structures as various excited states of a single composite object.

Correspondence of Positron Peaks to e^+e^- Coincidences

The positions of the different sum-energy peaks correlate closely to the range of positron peak energies observed in previous heavy-ion experiments. Sum energies of 620, 760 and 810 keV appear to correspond to narrow positron peaks observed at ~320, 380, and 410 keV, respectively. The presence of many structures in the data also provides an understanding, in hindsight, of the comparative difficulty encountered in the positron peak searches over the past five years. Whereas the present intrinsic sum-energy resolution of ≈25 keV is small compared to the ~100 keV separation of these sum-energy lines, the associated positron peaks are separated by only ~50 keV, and each experiences a Doppler-broadening of ~80 keV. The individual positron lines are not individually resolved, but overlap, forming a broad excess of intensity in the positron energy distribution. The presence of this excess in the positron spectrum has, for the EPOS group, been an on-line indicator of the presence of structure. Interestingly, the measured total coincident positron yield, when compared to the shape of the continuous dynamic and nuclear positron distribution (corrected for efficiency and detector resolution), shows a similar excess of events between

~250 and ~450 keV. A distribution of particle emission velocities, which smears out the individual lines, may, together with the presence of multiple structures, contribute to the typical absence of narrow peaks in the total positron distribution.

Correlations with Kinematic Parameters

The possibility of enhancing one of the several prominent sum-lines, to the exclusion of the others, closely mirrors our experience with "singles" positron peaks in previous experiments. The coincidence peak data of Fig. 32 are selected by beam-energy and, for the 620 keV line, delayed positron arrival times. Similarly, the spectra of Fig. 1 are selected by beam-energy and scattering angle regions for which the positron line appears most prominently in each data sample. In addition to corroborating the correspondence between the previous and present experiments, this behavior seems to suggest that *real* correlations exist between the production of each coincidence structure and some particular aspect of the heavy-ion collision environment.

Projectile energy and/or target condition seem to play an essential role in determining the production of the sum-energy peaks. Following an initial indication that the ~320 keV peak in U+Cm collisions was produced over a narrow range of bombarding energies [3], we have concentrated on improving energy definition and reducing target deterioration to investigate this further [62]. In the 1986 experiment, over a three day period of constant beam energy (5.87 MeV/u), stable UNILAC tune, and for a homogeneous batch of ThF_4 targets, only the 810 keV sum-line emerged as shown in Fig. 40. No cuts in ion scattering angle have been taken to produce this spectrum. At lower beam energies (≤5.85 MeV/u), this structure appears, if at all, with significantly lower intensity.

One striking example of similar behavior in the positron singles data was presented in Fig. 21. For Th+Th collisions at 5.70 MeV/u, the total positron energy spectrum is displayed, taking approximately only the first hour of data from each irradiated target. Whereas the total sample again exhibits only a broad excess of intensity, these "fresh-target" data exhibit a marked positron peak at ~320 keV, again with no selection of ion scattering angles or positron or electron flight-times. It appears that target condition plays an important role in these experiments.

The simplest physical interpretation of these data is that peak production is associated with one or several very narrow resonances, or perhaps threshold behavior, in the heavy-ion scattering system. It should be noted that the mean target thickness was ~0.05 MeV/u, and the energy resolution of the UNILAC beam ~0.02 MeV/u, in these experiments. Most of the EPOS data were accumulated at beam-energies chosen to keep the nuclear overlap for head-on collisions constant for the particular projectile-target nuclide combination [41].

The correlation of positron and sum-energy peak production on heavy-ion scattering angle appears of secondary importance in revealing the structures. In our original experiments [3,5], ion-angle cuts were employed to "cut away" the continuous background of dynamic and nuclear positron production by exploiting the very good PPAC angular resolution in order to select regions slightly separated from the elastic scattering kinematics. The effectiveness of this approach seemed to be facilitated by an apparent rotation in the angle-angle plane of scattered heavy-ions associated with the positron peak. This behavior could be qualitatively understood if the ionic charge-states of the ejectiles associated with peak production differed from the velocity-dependent charge-states characteristic of Rutherford scattering [51]. Therefore, the kinematic cuts do not directly reflect a dependence of peak production on the heavy-ion angular distribution. Rather, all the cuts of Fig. 1 are over very large ranges of ion scattering angle, and the data suggest a rather broad distribution of the positron peak production with scattering angle.

These characteristics seem also to be reflected in the coincidence data. Although heavy-ion angular cuts may be employed to slightly enhance

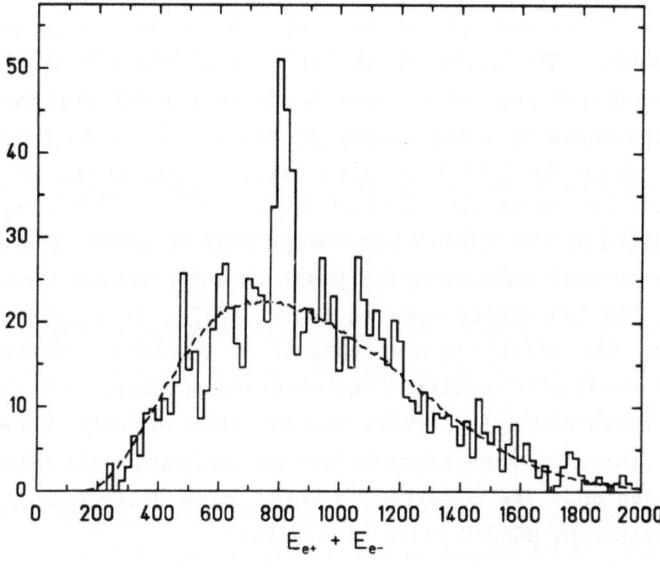

Figure 40. Sum-energy spectrum for U+Th collisions at 5.87 MeV/u. No heavy-ion angular cuts applied.

the peak over background (compare Fig. 3 with HI cuts to Fig. 32 without), they are obviously not necessary for extracting the bulk of the coincidence peak intensity. Sorting the data first by beam-energy and time-of-flight yields a more effective trigger for the peak events, and the additional enhancement by angle cuts is, in comparison, small. Again, the peak data are distributed over a very broad range of ion-scattering angles, and the different sum-lines, and hence positron peaks, appear in overlapping angular regions. With the increased sensitivity of the coincidence detection method, however, there appears to be some additional evidence of a dependence of the sum-line intensity on the scattering angles (θ and ϕ) of the heavy-ions [41].

An investigation of the correlation of peak production to the flight-time of the positrons and electrons to their respective detectors is a rather new approach which arose from the observation in July 1985 that the ~320 keV positron peak in Th+Th collisions is associated with longer flight-times to the Si(Li) detector. Figure 41 presents a comparision of the total positron energy distribution (without ion-angle cuts) for flight-times delayed ($-5 < \Delta t < -1$ ns) or prompt ($1 < \Delta t < 5$ ns) with respect to the average arrival time for the total continuous dynamic positron distribution (Note: $\Delta t_{e+} \equiv t_{HI} - t_{e+}$). Similar results are found for the associated 620 keV sum-energy line in U+Th collisions, as shown in Fig. 32, for which a comparably delayed window is selected. On the other hand, the 810 keV sum-energy peak is associated with prompt arrival times, centered closely about the mean of the TOF distribution.

The positron or electron TOF depends on its velocity parallel to the solenoid axis, proportional to $\cos\theta_{e+}$, and varies between ~4 and ~15 ns, with a mean of ≈8 ns and FWHM ~5 ns, for the continuous positron and electron backgrounds. The marginal Si(Li) time resolution of ~4 ns for electrons and ~7 ns for positrons, provides only a rough determination of the their emission angle as either steep or fairly flat with respect to to the spectrometer axis. Arrival time also reflects any physical delay for the emission, such as source lifetimes of order $\geq 10^{-9}$ s. The prompt TOF distribution exhibited by the 810 keV sum-energy peak is actually what one expects for the back-to-back emission of a positron and electron from a slowly moving source. The laboratory opening angle, θ^{lab}_{e+e-}, is larger than ~140°. In this case, the coincidence acceptance of the EPOS spectrometer, determined by the overlap of positron and electron efficiencies, selects a range of moderate emission angles. Very steeply emitted positrons have a reduced probability for the associated electron to enter the electron angular acceptance region. The observed "prompt" time distribution is quantitatively understood by MCSPEC calculations [41].

In light of this, the delayed positron TOF for the 620 keV sum-line and corresponding positron peak may reflect a long source lifetime. However, the coincident electron peak does not appear to exhibit as large a delay, suggesting that the positrons may be preferentially emitted perpendicular to the solenoid axis, i.e., in the plane of the HI scattering. The coincident electron can still be detected if θ^{lab}_{e+e-} is smaller than ~160°, that is, if the emitter is not at rest in the CM system of the heavy ions. A somewhat larger velocity is in any case indicated for the 620 keV line by the rather broad difference peak shown in Fig. 32d.

The association of the production of the narrow coincidence positron- electron signals with beam-energy, target thickness or condition, heavy-ion scattering angle, and target-detector TOF seems to suggest a basic correlation of the various structures with certain aspects of the physics of the heavy-ion collision. For the first time, with the coincident detection of both positron and electron, we are able to individually resolve the multiple structures in the data, and these correlations can now be studied in more detail.

VIII. SUMMARY AND OUTLOOK

The interesting puzzle presented by the unexpected appearance of sharp

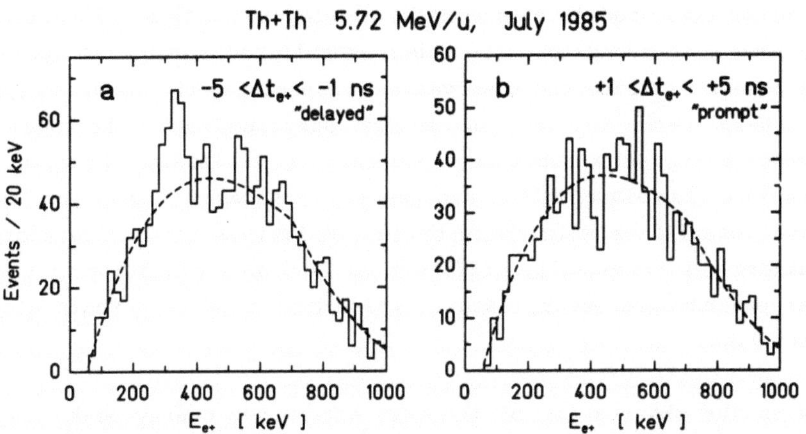

Figure 41. Energy spectrum of positrons collected in Th+Th collisions at 5.70 MeV/u (20° < θ_{HI} < 70°) which arrive at the Si(Li) detector (a) slightly delayed, or (b) prompt, with respect to the mean positron TOF. Similar behavior is observed for the 620 keV sum-peak (see Fig.3).

peaks in the positron spectra from superheavy collision systems and, in particular, the striking feature that a near degeneracy of peak energies occurs for all these systems, led us to explore the possibility that the source of these peaks could be a two-body decay process involving the electron as a member of a pair of particles in the final state. This venture into the positron-electron coincidence experiments has proven to be rewarding from several points of view. From the standpoint of the physics involved, these experiments have raised new and interesting issues, pointed out new directions to be explored both experimentally and theoretically, and conclusively dismissed previously viable options for explaining the peaks. Technically, they have introduced a more sensitive probe to clarify and remove previous ambiguities in interpreting the data and provided a more discriminating technique for pursuing new studies.

Of course, one of the salient results emerging from the coincidence studies was the observation of simultaneously emitted positrons and electrons in narrow peak structures with similar energies and widths. As our detailed discussion of the functioning of the apparatus has demonstrated, the appearance of the coincidence peak structures cannot be attributed to a pathological response of the detector system. They seem to be real and produced in the three collision systems studied, U+Th, Th+Th, and Th+Cm, although most of the evidence has been obtained with the U+Th system. The mere occurrence of a sharp electron peak in coincidence with the positron peak rules out the possibility that the source of the peak can be spontaneous positron emission, Monoenergetic Pair Conversion, or annihilation of some structures involving clusters of electrons and positrons. None of these mechanisms produce an accompanying electron with a well-defined energy. We have also found that the other established atomic and nuclear mechanisms cannot simulate the observation. Especially the combined appearance of a sharp peak in the sum-energy spectrum, and the peak in the difference-energy distribution, presents a difficult obstacle for any explanations involving IPC processes. Arguments based on comparing observed intensities with Monte-Carlo simulations further reinforce these conclusions. It is equally difficult to find an explanation of the data in nuclear cascade processes, each of which must produce a sharp electron or positron line.

With the observation of the very narrow sum-energy peak, whose width is clearly narrower than the individual electron and positron peaks and is consistent with the width expected from the energy resolution of the detectors, it is difficult to avoid the conclusion -- based solely on measurements -- that there is a high degree of spacial correlation between the electron and positron. Comparing the widths of the sum-energy and difference-energy peaks implies that the angle between the electron and positron is greater than 140° for the ~810 keV peak or 165° for the ~620 line, which is consistent with back-to-back emission. This model-independent observation, together with almost equal energies for the electron and positron, of course, suggests a two-body decay of an object which lives long enough to escape the influence of the Coulomb fields of the nuclear system. A comparison of the widths of the individual peaks, the sum-energy peak, and the difference-energy peak, also implies that the source of the pair cannot be at rest in the laboratory, and places limits on its mean laboratory velocity between v_{CM} = .05c and ~.2c.

The previously-suspected possibility that there is more than one peak structure, has now been confirmed. In fact, it is the sharp sum-energy peak which provides a new sensitivity and a new probe to resolve the neighboring structures in the positron spectra in a reliable way. In hindsight, it can now be understood that the previous confusion regarding the appearance of the peaks was due to the production of overlapping neighboring structures, which could not be resolved due to Doppler broadening, and therefore sometimes only appeared as an enhancement of the continuous spectrum.

In addition to being faced with interesting conclusions, we are also challenged by a number of equally interesting questions that have emerged from the coincidence studies. 1) If the two-body decay of a neutral parent is involved, then a mechanism has to be found to produce this object with an appreciable contribution of velocities not very different from the CM velocity. As our discussion has brought out, this general conclusion can be mitigated somewhat by a long lifetime of the source ($\tau \geq 10^{-10}$ s) which, together with the finite fiducial volume of the apparatus, can act as a velocity filter. But in essence, the data is consistent with a low velocity distribution, and any explanation of the observations in terms of the two-body decay of a new object must reproduce such a velocity distribution. 2) The observation of more than one structure clearly complicates any explanation in terms of a new object. Therefore a key question concerns the relationship between these structures. There are indications of a correlation of the individual peaks to beam energy, time of detection, and scattered ion kinematics -- some of these suggesting underlying connections with nuclear properties. These must be explored further. 3) The question of source lifetime is particularly interesting. The limits are currently set by the peak width corresponding to 10^{-19} s and the fiducial volume of the apparatus and measured coincidence timing corresponding to ~10^{-9} s. 4) The possible role of the very large Z_u in production is still not understood. Whether these superheavy systems a unique vehicle for this source of positrons can, of course, be answered by further experiments with lower Z systems. 5) In this connection an additional question involves the role of nuclear contact. Are the peaks related to producing strong electric fields for long lifetimes? This point is obviously also related to the correlations referred to above.

These questions are presently being addressed in new experiments with heavy-ion scattering at GSI. Concurrently, some aspects of this puzzle are being explored in related experiments involving electron-positron scattering, production of related objects in nuclear transitions, and high-energy experiments. With all of these together, we may anticipate an answer to our puzzle in the near future.

ACKNOWLEDGEMENTS

We would like to take this opportunity to thank the members of the EPOS group, past and present, whose combined efforts made this a very challenging and successful collaboration. We especially acknowledge H. Folger and his staff in the GSI target-lab for their extensive efforts in the fabrication of actinide targets, and Werner Kreuzer, without whose technical assistance in designing, constructing, and maintaining the apparatus, these experiments would have been much more difficult. The dedication of the UNILAC operating staff in pushing the machine to its limits of resolution and stability contributed greatly to these experiments and is deeply appreciated.

The role of W. Greiner and his colleagues of the Univ. Frankfurt Theory group in originally motivating these experiments and their active support over the years is warmly acknowledged. We have benefitted greatly from discussions with W. Greiner, B. Müller, J. Reinhardt, G. Soff, U. Müller, A. Schäfer, P. Schlüter, and T. de Reus. T.C. wishes to personally thank our host, Professor Greiner, for his very kind invitation to speak at Maratea, and for rearranging the conference schedule to allow a thorough discussion of these data.

T. C. gratefully acknowledges the Kernphysik I and II divisions and the administrative staff of GSI for kind hospitality during his stay in Germany. We also thank R. Cowan for assistance in preparing this manuscript. This work was supported in part by the U.S. Department of Energy Contract

No. DE-AC02-76ER03074, and by the Bundesministerium für Forschung und Technologie of the Federal Republic of Germany.

REFERENCES

1. H. Backe, L. Handschug, F. Hessberger, E. Kankeleit, L. Richter, F. Weik, R. Willwater, H. Bokemeyer, P. Vincent, Y. Nakayama, and J.S. Greenberg, Phys Rev Lett 40:1443 (1978).
2. C. Kozhuharov, P. Kienle, E. Berdermann, H. Bokemeyer, J.S. Greenberg, P. Vincent, H. Backe, L. Handschug, and E. Kankeleit, Phys Rev Lett 42:376 (1979).
3. J. Schweppe, A. Gruppe, K. Bethge, H. Bokemeyer, T. Cowan, H. Folger, J.S. Greenberg, H. Grein, S. Ito, R. Schulé, D. Schwalm, K.E. Stiebing, N. Trautmann, P. Vincent, and M. Waldschmidt, Phys Rev Lett 51:2261 (1983).
4. M. Clemente, E. Berdermann, P. Kienle, H. Tsertos, W. Wagner, C. Kozhuharov, F. Bosch, and W. König, Phys Lett 137B:41 (1984).
5. T. Cowan, H. Backe, M. Begemann, K. Bethge, H. Bokemeyer, H. Folger, J.S. Greenberg, H. Grein, A. Gruppe, Y. Kido, M. Klüver, D. Schwalm, J. Schweppe, K.E. Stiebing, N. Trautmann, and P. Vincent, Phys Rev Lett 54:1761 (1985).
6. H. Tsertos, E. Berdermann, F. Bosch, M. Clemente, P. Kienle, W. König, C. Kozhuharov, and W. Wagner, Phys Lett 162B:273 (1985).
7. T. Cowan, H. Backe, K. Bethge, H. Bokemeyer, H. Folger, J.S. Greenberg, K. Sakaguchi, D. Schwalm, J. Schweppe, K.E. Stiebing, and P. Vincent, Phys Rev Lett 56:444 (1986).
8. For additional references relevant to these phenomena see Quantum Electrodynamics of Strong Fields, W. Greiner, ed., Plenum Press, New York (1983).
9. W. Pieper and W. Greiner, Z Phys 218:327 (1969);
   B. Müller, J. Rafelski, and W. Greiner, Z Phys 257:62, 183 (1972).
10. S.S. Gershtein and Ya.B. Zel'dovich, Lett Nuovo Cimento 1:835 (1969).
    Ya.B. Zel'dovich and V.S. Popov, Sov Phys Usp 14:673 (1972).
11. J.S. Greenberg, H. Bokemeyer, H. Emling, E. Grosse, D. Schwalm, and F. Bosch, Phys Rev Lett 39:1404 (1977).
12. H.H. Behncke, P. Armbruster, F. Folkmann, S. Hagmann, J.R. MacDonald, and P. Molkler, Z Phys A289:333 (1979).
13. C. Kozhuharov, in: Electronic and Atomic Collisions, S. Datz, ed., North-Holland, Amsterdam (1982), pp. 179-193.
14. J.S. Greenberg, in: Electronic and Atomic Collisions, proc. of 11th ICPEAC, Kyoto (1979), p. 351.
15. H. Bokemeyer, in: Selected Topics in Nuclear Structure, J. Styczen and R. Kulessa, eds., Jag. Univ. Krakow (1978).
16. For a review of many of the early experiments, see J.S. Greenberg and P. Vincent, in: Treatise on Heavy-Ion Science, Vol. 5: High Energy Atomic Physics, D.A. Bromley, ed., Plenum, New York (1985), pp. 138-421.
17. J. Reinhardt, B. Müller, and W. Greiner, Phys Rev A24:103 (1981).

18. T. Tomoda, and H.A. Weidenmüller, Phys Rev A26:103 (1982); and, T. Tomoda, Phys Rev A26:174 (1982).
19. For other, more recent calculations see,
C. Bottcher and M.R. Strayer, Phys Rev Lett 54:669 (1985); and,
Y. Hirata and H. Minakata, Phys Rev D34:2493 (1986).
20. J. Schweppe, PhD Thesis, Yale University (1985).
21. A. Gruppe, PhD Thesis, Univ. Frankfurt (1985).
22. J.S. Greenberg, in: Ref. 8, p. 853.
23. H. Bokemeyer, K. Bethge, H. Folger, J.S. Greenberg, H. Grein, A. Gruppe, S. Ito, R. Schulé, D. Schwalm, J. Schweppe, N. Trautmann, P. Vincent, M. Waldschmidt, in Ref. 8, p. 173.
24. J.S. Greenberg, in: X-Ray and Atomic Inner Shell Physics, B. Crasemann, ed., Plenum, New York (1982), p. 173.
25. J.S. Greenberg, in: Nuclear Physics with Heavy Ions, P. Braun-Munzinger, ed., Harwood Academic Publishers, New York (1983), p. 419.
26. D. Schwalm, in: Electronic and Atomic Collisions, J. Eichler, I.V. Hertels, and N. Stolterfoht, eds., Elsevier Science Publishers B.V. (1984), p. 295.
27. H. Bokemeyer, in: Selected Topics in Nuclear Structure, J. Stachura, ed., Jag. Univ., Krakow (1984), p. 335.
28. P. Kienle, in: Ref. 8, p. 293.
29. M. Clemente, PhD Thesis, Tech. Univ. München (1984).
30. H. Tsertos, PhD Thesis, Tech. Univ. München (1985).
31. H. Bokemeyer, H. Folger, H. Grein, S. Ito, D. Schwalm, P. Vincent, K. Bethge, A. Gruppe, R. Schule, M. Waldschmidt, J.S. Greenberg, J. Schweppe, and N. Trautmann, GSI Scientific Report 1980, GSI-81-2:127 (1981).
32. P. Schlüter et al., Z Phys A314:297 (1983).
33. J. Rafelski, B. Müller, and W. Greiner, Z Phys A285:49 (1978).
34. J. Reinhardt, U. Müller, B. Müller, and W. Greiner, Z Phys A303:173 (1981).
35. U. Müller et al., Z Phys A313:263 (1983).
36. U. Heinz, J. Reinhardt, B. Müller, W. Greiner, and W.T. Pinktston, Z Phys A316:341 (1984); and,
U. Heinz et al., Ann Phys 158:476 (1984).
37. Nuclear scattering potentials have been discussed in:
M.J. Rhoades-Brown, V.E. Oberacker, M. Seiwert, and W. Greiner, Z Phys A310:287 (1983);
M.R. Stayer, R.Y. Cusson, H. Stöcker, J.A. Maruhn, and W. Greiner, Phys Rev C28:228 (1983);
M. Seiwert, W. Greiner, V. Oberacker, and M.J. Rhoades-Brown, Phys Rev C29:477 (1984);
M. Seiwert, J.A. Maruhn, W. Greiner, J. Friedrich, Z Phys A321:653 (1985);
N. Malhorta and R.K. Gupta, Phys Rev C31:1179 (1985);
M.W. Katoot, V.E. Oberacker, and W.T. Pinkston, Phys Lett 172B:292 (1986); and,
H.M.M. Mansour et al., Phys Rev C34:1278 (1986).
38. J. Schweppe, in: Electronic and Atomic Collisions, D.C. Lorents, W.E. Meyerhof, and J.R. Peterson, eds., Elsevier Science Publishers B.V. (1986), p. 405.

39. H. Bokemeyer, H. Folger, H. Grein, T. Cowan, J.S. Greenberg, J. Schweppe, A. Balanda, K. Bethge, A. Gruppe, K. Sakaguchi, K. Stiebing, D. Schwalm, P. Vincent, H. Backe, M. Begemann, M. Klüver, and N. Trautmann, GSI Scientific Report 1984, GSI 85-1:177 (1985).
40. U. Müller et al., Z Phys A323:261 (1986).
41. T. Cowan, PhD Thesis, Yale University (1987). A number of detailed calculations which only appear in this thesis are referred to throughout this paper.
42. W. König, contribution to this conference.
43. A. Schäfer, J. Reinhardt, B. Müller, W. Greiner, and G. Soff, J Phys G11:L69 (1985).
44. A.B. Balantekin, C. Bottcher, M.R. Strayer, and S.J. Lee, Phys Rev Lett 55:461 (1985).
45. A. Chodos and L.C.R. Wijewardhana, Phys Rev Lett 56:302 (1986).
46. N. Mukhopadhyay and A. Zehnder, Phys Rev Lett 56:206 (1986).
47. B. Müller and J. Reinhardt, Phys Rev Lett 56:2108 (1986).
48. J. Reinhardt et al., Phys Rev C33:194 (1986).
49. H. Bokemeyer, contribution to this conference.
50. K.E. Stiebing, contribution to this conference.
51. V.S. Nikolaev and I.S. Dmitriev, Phys Lett 28A:277 (1968).
52. H. Backe et al., Phys Rev Lett 50:1838 (1983).
53. E. Liarokapis, R. Schule, and T. Cowan, "Efficiency Calculation by Path Integration," ECPI computer program, unpublished.
54. T. Cowan, "Monte Carlo Simulations of Positron-Electron Coincidences," MCSPEC computer program and Yale Report, unpublished.
55. Coincident positron-electron analysis techniques and computer programs developed by T.Cowan and J.S.Greenberg, June 1985. For more information regarding the analysis of positron spectra, see Refs. 20, 21, and 41.
56. G. Soff, P. Schlüter, and W. Greiner, Z Phys A303:189 (1981).
57. M.E. Rose and G.E. Uhlenbeck, Phys Rev 48:211 (1935).
58. R. Wilson, in: "α-, β-, and γ-Ray Spectroscopy," K. Siegbahn, ed., North-Holland, Amsterdam (1968), p. 1557.
59. P. Schlüter, G. Soff, and W. Greiner, Phys Rep 75:327 (1981).
60. Eadie et al., Statistical Methods in Experimental Physics, North-Holland, Amsterdam (1971), p. 272.
61. H. Folger, private communication.
62. For more information regarding measurements of target thickness and in-beam deterioration see Ref. 50 and:
M. Klüver, M. Begemann, K.E. Stiebing, (1983) unpublished;
D. Kraft, K. Stiebing, K. Bethge, M. Begemann, M. Klüver, and H. Bokemeyer, Scientific Report 1985, IKF, Univ. Frankfurt, (1986); and
D. Kraft, Dipl. Thesis, IKF, Univ. Frankfurt (1987).
63. G. Soff, contribution to this conference.
64. P. Schlüter, G. Soff, and W. Greiner, Phys Rev C33:1816 (1986).
65. L.M. Krauss and M. Zeller, Phys Rev D34:3385 (1986).
66. E. Storm and H.I. Israel, Nuclear Data Tables A7:565 (1970).
67. L.A. Sliv, Dokl Akad SSSR 64:521 (1949) and JETP 25:7 (1953).
68. P. Schlüter et al., Z Phys A323:139 (1986).
69. A. Schäfer, B. Müller, and W. Greiner, Phys Lett 149B:455 (1984).

70. C.Y. Wong, Phys Rev Lett 56:1047 (1986).
71. M.-C. Chu and V. Pönisch, Phys Rev C33:2222 (1986).
72. J. Rafelski and B. Müller, Phys Rev Lett 36:517 (1976); and G. Soff, J. Reinhardt, and W. Greiner, Phys Rev A23:701 (1981).
73. A. Schäfer, J. Reinhardt, B. Müller, and W. Greiner, Z Phys A324:243 (1986).
74. D. Carrier, A. Chodos, and L.C.R. Wijewardhana, Phys Rev D34:1332 (1986).
75. Bing-An Li and H.T. Nieh, preprint ITP-SB-86-23 (1986).
76. B. Müller, J. Reinhardt, W. Greiner, and A. Schäfer, J Phys G12:L109 (1986).
77. W. Lichten and A. Robatino, Phys Rev Lett 54:781 (1985).
78. J. Reinhardt, B. Müller, W. Greiner, Phys Rev Lett 55:134 (1985).
79. R.H. Lemmer and W. Greiner, Phys Lett 162B:247 (1985).
80. G. Scharf, Phys Lett 177B:429 (1986).
81. C.Y. Wong and R.L. Becker, ORNL preprint (1986).
82. A.O. Barut, contribution to this conference.
83. L.M. Krauss and F. Wilczek, Phys Lett 173B:189 (1986).
84. B. Müller and J. Rafelski, Phys Rev D34:2896 (1986).
85. B. Müller and J. Rafelski, preprint UFTP 170/1986.
86. L.S. Celenza et al., Phys Rev Lett 57:55 (1986).
87. K. Lane, Phys Lett 169B:97 (1986).
88. E. Ma, Phys Rev D34:293 (1986).
89. A. Zee, Phys Lett 172B:377, (1986).
90. R.D. Peccei, T.T. Wu, and T. Yanagida, Phys Lett 172B:435 (1986).
91. M. Suzuki, Phys Lett 175B:364 (1986).
92. A. Schäfer et al., Mod Phys Lett A1:1 (1986).
93. U.E. Schröder, Mod Phys Lett A1:157 (1986).
94. S.J. Brodsky et al., Phys Rev Lett 56:1763 (1986).
95. L.M. Krauss and M.B. Wise, Phys Lett 176B:483 (1986).
96. Y.S. Tsai, Phys Rev D34:1326 (1986).
97. C.M. Hoffman, Phys Rev D34:2167 (1986).
98. E. Masso, Phys Lett 181B:388 (1986).
99. N. Mukhopadhyay and A. Zehnder, preprint SIN-PR-86-02 (1986).
100. W.A. Bardeen, R.D. Peccei, and T. Yanagida, preprint DESY 86-054.
101. S. Barshay, preprint TH Aachen (1986).
102. F. Bergsma et al., Phys Lett 157B:458 (1985).
103. F.P. Calaprice et al., Phys Rev Lett 56:302 (1986).
104. G. Mageras et al., Phys Rev Lett 56:2672 (1986).
105. T. Bowcock et al., Phys Rev Lett 56:2676 (1986).
106. M.J. Savage et al., Phys Rev Lett 57:178 (1986).
107. R. Eichler et al., Phys Lett 175B:101 (1986).
108. A. Konaka et al., Phys Rev Lett 57:659 (1986).
109. C.N. Brown et al., Phys Rev Lett 57:2101 (1986).
110. A.L. Hallin et al., Phys Rev Lett 57:2105 (1986).
111. W.E. Meyerhof et al., Phys Rev Lett 57:2139 (1986).
112. H. Albrecht et al., Phys Lett 179B:403 (1986).
113. F.W.N. de Boer et al., Phys Lett 180B:178 (1986).
114. M. Davier, J. Jeanjean, and H.N. Ngoc, Phys Lett 180B:295 (1986).
115. C.V.K. Baba et al., Phys Lett 180B:406 (1986).
116. K.A. Erb, I.Y. Lee, and W.T. Milner, Phys Lett 181B:52 (1986).
117. D.A. Bryman and E.T.H. Clifford, Phys Rev Lett 57:2787 (1986).

INVESTIGATION OF POSITRON LINE EMISSION IN HEAVY ION COLLISIONS WITH THE EPOS-SPECTROMETER

H. Bokemeyer
Gesellschaft für Schwerionenforschung
D-6100 Darmstadt, Germany

EPOS Collaboration
H. Backe[4], K. Bethge[3], H. Bokemeyer[1], T. Cowan[2], H. Folger[1], J.S. Greenberg[2], K. Sakaguchi[3], P. Salabura[3], D. Schwalm[5], J. Schweppe[2], K.E. Stiebing[3], N. Trautmann[4], and P. Vincent[6]

[1]GSI, [2]Yale Univ., [3]Univ. Frankfurt, [4]Univ. Mainz
[5]Univ. Heidelberg, [6]BNL

I. INTRODUCTION

Positron production in nuclear collisions is well known from internal pair conversion processes of nuclear transitions with energies exceeding $2m_0c^2$. The pair production is related to the competing γ-decay of the nuclear transitions through conversion coefficients and the resulting positron yield is essentially proportional to the γ-ray intensity. An additional strong source of positrons, however, has been detected in very heavy ion collisions already in the first year of the UNILAC operation.[1,2] Compared to the positron yield expected from nuclear processes as inferred from the γ-ray spectral intensity ($E_\gamma > 1$ MeV), the new source dominates the positron production for the high-Z collision systems with the combined charge $Z_{ua} = Z_1 + Z_2 > 160$ (Figure 1 on page 196). From its steep increase with either Z_{ua} (see c) or with the decreasing internuclear distance R_{min} (see b), the new contribution was identified as quasiatomic positron production[3,4]. Indeed, the deep ($V_c > 20$ MeV) Coulomb potential temporarily formed during the collision for some 10^{-20}sec changes fast. Mostly when the logarithmic derivative $\dot{R}/R$ of the internuclear distance R(t) (part (a) of Figure 2 on page 197) is maximized, this time variation gives rise to induced Coulomb ionization type[6]) transitions from the negative energy continuum states to bound quasiatomic or continuum electronic states (Figure 2 on page 197).

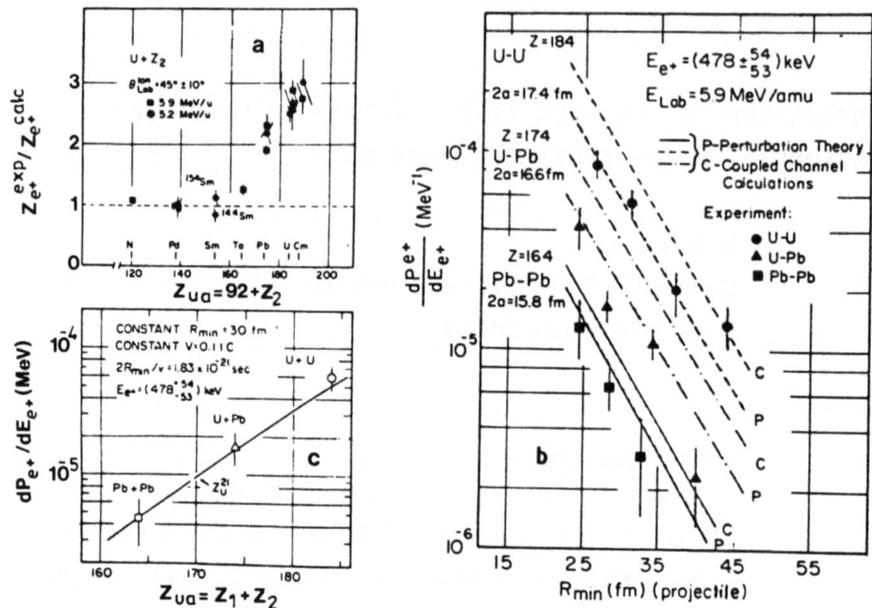

Figure 1. Early experimental evidence for quasiatomic positron production[1,2])
(a) Number of totally observed positrons in collisions with U as projectile relative to the number of positrons calculated from the γ-ray spectrum versus the nuclear charge of the united system.
(b) Quasiatomic production probability for positrons of ~500 keV as a function of R_{min} the distance of approach for various high-Z systems. The straight lines represent quasiatomic theory[3]) at different levels of completeness.
(c) Quasiatomic production probability for positrons of ~500 keV versus the combined nuclear charge Z_{ua} selected for trajectories of same internuclear distance and same relative velocity.

Together with accompanying quasiatomic processes as δ-electron and K-hole production or MO-transitions, the dynamically induced positron production is well described within the theory[3,5]) of the two center Dirac equation including multistep processes. The relativistically enhanced compression of the electronic wave functions in the strong quasiatomic Coulomb potential results in both an enhancement and a localization of the induced transitions to regions close to the turning point of the trajectory. Evaluation of the transition matrix elements, appropriately restricted for these small internuclear separations to the monopole term and radial coupling, reveals scaling prescriptions with the minimum momentum transfer q_{min}[6,7]) or alternatively a critical interaction time $\hat{t}$[8])as the relevant parameters. This mechanism quite naturally predicted[3]) the above mentioned strong Z_{ua}-dependence and the steep increase for closest collisions.

The extensive theoretical and experimental work was initially motivated from the search for spontaneous positron emission, a resonant electronic transition predicted[9,10]) to occur in very heavy ion collisions when in systems with $Z_{ua} > 172$ the electronic

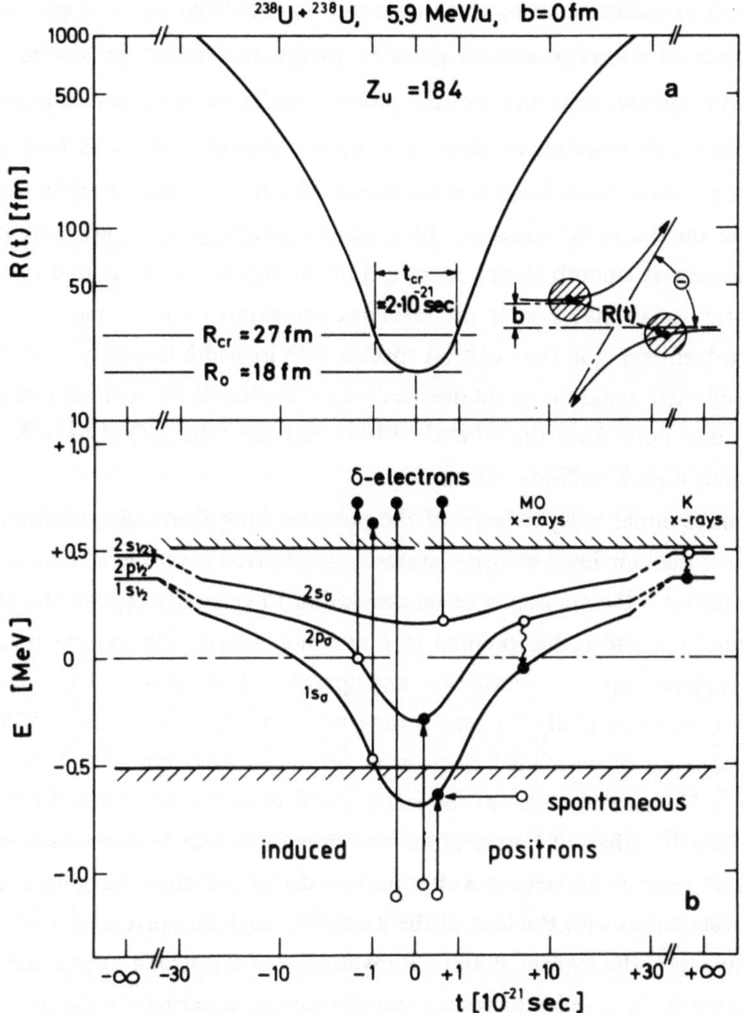

Figure 2. Internuclear distance R(t) (a) and binding energies (b) of three quasiatomic levels as a function of time, when assuming Rutherford scattering. The insert in (a) schematically defines the impact parameter b, the distance of approach R(t) (R_{min} = R(t=0)) and the CM scattering angle Θ. Typical quasiatomic transitions are indicated in (b).

binding energy of the quasiatomic 1sσ-state exceeds $2m_0c^2$ and the level dives into the negative energy continuum. Indeed, the negative energy continuum completely filled with electrons is the principal physical QED-vacuum state and spontaneous positron emission signals the basic decay of a neutral into a charged vacuum state and thus these investigations became of great theoretical interest. Soon it became clear, however, that the identification of spontaneous positron emission was hindered by the quasimolecular induced processes all revealing smooth positron energy and impact parameter distributions; moreover, spontaneous and induced positron emission are coherent processes[5)] and there exists no prescription to single out the spontaneous events as long as only elastic scattering is concerned. But it has been pointed out[11,12)] that considerable delay of the collision on the order of $\Delta t > 10^{-20}$sec by strong nuclear interactions may structure the otherwise smooth energy spectra. For the supercritically bound systems ($Z > Z_c \sim 172$) this delay may favour the resonant transition over the induced part. A pronounced peak emerging in the positron spectra with its width inversely related to the sticking time will then signal the spontaneous decay. Candidates for such structures have indeed been found both from the EPOS[13,14,15)] and the ORANGE[16,17,18)] collaboration in various high-Z collision systems.

However, an essential independence of the positron lines from all quasiatomic and nuclear details of the individual collision systems was derived from the EPOS-data. This fact virtually ruled out the spontaneous interpretation to generally explain the origin of narrow positron lines and rather pointed to a common source. Having excluded more conventional explanations, this source was anticipated with the decay of a so far undetected light particle.[15)] Actually the search for coincident electrons in the EPOS set-up revealed the existence of correlated electron positron line-emission with both lines of same energy.[19)] The narrow width of the respective sum-energy line hints for a cancellation of the Doppler effect of both leptons and thus for a near back-to-back emission characteristic, all being in agreement with a particle decay. Attempts have been made to connect this observation with the idea of the axion[20)], originally invented[21,22)] to fundamentally explain[23)] the evident P and CP invariance of the strong interaction. In the meanwhile, however, in a series of recent complementary experiments the axion again is convincingly ruled out[25)] also in its variant form[24)], same as it already was excluded[26)] in its standard form. Moreover, it became difficult to imagine a production mechanism for a particle in these heavy ion collisions to explain consistently the positron data and at the same time not to disagree with the present knowledge of the respective coupling-mechanism.[27)] In very recent experiments with the EPOS-spectrometer, in addition, it was found that there may be more than one set of correlated lines in the positron and electron spectra with both leptons of comparable energy.

An overview of the field as concerns the quasiatomic aspects the reader may find in ref. 4,28-30) and is reflected in various conference proceedings[31-36)]. This report covers the development of the experiments of the EPOS-collaboration. It first describes the experimental set-up in its various states of completion and then reviews the experimental results starting from the detection of positron lines, their systematic investigation, and finally the electron-positron correlation. The article is accompanied by the presentations of K. Stiebing on special aspects of the heavy ion detection in the EPOS-spectrometer and of T. Cowan on details of the positron-electron correlation measurement.

II. THE POSITRON SPECTROMETER SYSTEM EPOS

The Electron Positron Solenoidal spectrometer system (EPOS)[37)] was designed to flexibly account for the experimental demands resulting from the various aspects of heavy ion positron spectroscopy. A list of these demands contains

- the optimization of the experiment for the measure of positron parameters actually of interest as there are the total positron cross-sections, the global and detailed shapes of the positron spectra and impact-parameter distributions and recently the electron-positron correlations and eventually the positron angular distributions,
- the clean detection of positron emission with cross sections in the order of only several 100 μb, accompanied from strong processes of nuclear and atomic origins competing during the heavy ion collision,
- the determination of all collision parameters both to distinguish elastic and inelastic reaction channels, and to verify different formation of the quasiatom from the measure of the internuclear distance R_{min} (see Figure 2 on page 197) particularly for a range near the minimuum separation,
- and the measure of nuclear excitations by detecting nuclear γ-radiation and internal conversion electrons in order to determine the contributions from nuclear internal pair conversion to the positron yield.

The principal arrangement of the EPOS-spectrometer as installed at the beam line Z2 of the UNILAC accelerator of GSI, Darmstadt, is schematically shown in Figure 3 on page 200. Positrons and electrons are both transported simultaneously in the solenoidal field to the appropriately shaped Si(Li)-detectors for their individual detection. Thus the set-up combines all the advantages of wide momentum-band characteristics, largest efficiency, and maximum energy resolution. Positrons, electrons, and γ-rays are recorded event-by-event in coincidence with the scattered heavy ions, which are detected in position sensitive parallel plate heavy ion avalanche counters. The essential parts of the spectrometer

- the magnetic solenoid,
- the positron detection system,
- the electron detection part (shown is a version optimized for conversion electron spectroscopy),
- the heavy ion detectors, and
- the γ-ray detctors,

can easily be identified in the figure. To a certain extent, these have partly been discussed elsewhere[13,28,33)], but for completeness the different parts will all together be described in the following. The interested reader is also referred to ref. 38,39.

Positrons and electrons proceed in a homogenous magnetic field of a solenoid of induction B with opposite helicity but otherwise identical spiralling trajectories. In polar

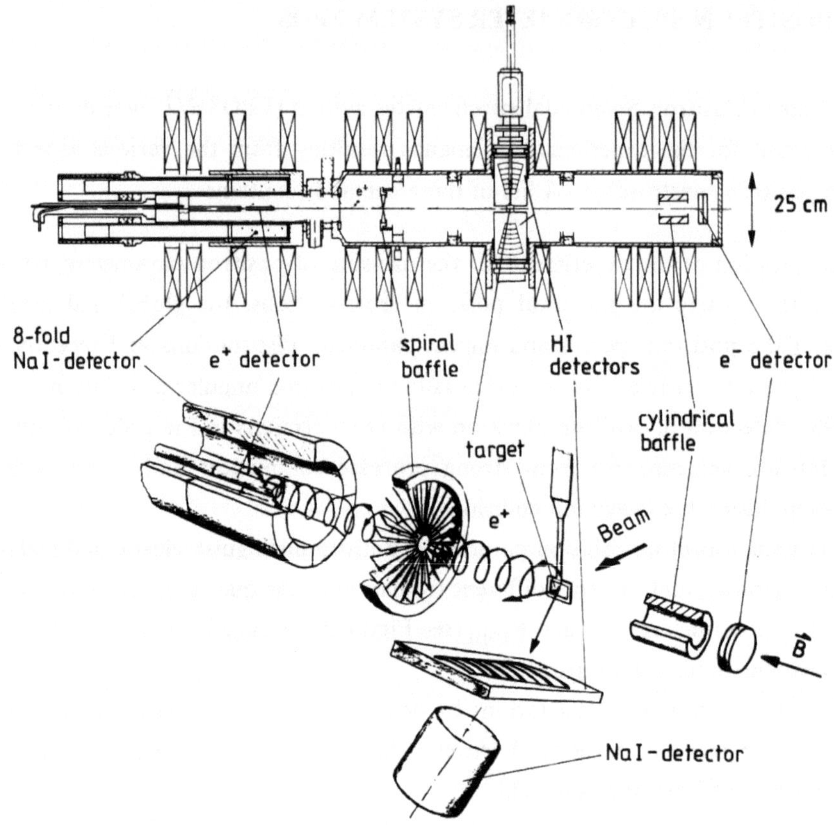

Figure 3. Schematic view of the positron spectrometer system EPOS in a version optimized in addition for conversion electron spectroscopy together with a perspective drawing of the main components (only one out of two heavy ion detectors are shown for reasons of clarity).

coordinates r, φ and z oriented parallel to B, the trajectories are described together with the propagation time t according

$$r = 2\rho_B \sin\varphi \; ; \; z = 2\rho_B \varphi \; tg\theta_0 \; ; \; \varphi = \tfrac{1}{2}\omega_B t \; \mathrm{sign}(q)$$

with E, p, θ_0 the total energy, momentum and emission angle and q the charge of the lepton. Here the gyration freguency ω_B and the spiral radius ρ_B are defined as

$$\omega_B = |eB/(\gamma m_0 c)| = 8.947 \times 10^6 \; B(\text{Gauß})/E(\text{MeV}) \; [1/\text{sec}]$$

$$\rho_B = |(pc \sin\theta_0)/(eB)| = 0.33 \times 10^4 \; p(\text{MeV/c})\sin\theta_0/B(\text{Gauß}) \; [\text{cm}]$$

The spiral's guiding center at a distance ρ_B from the field line of emission crosses this field line repetitively with the pitch length $z_p = 2\pi cp \cos\theta_0/(eB) = 2\pi v_0 \cos\theta_0/\omega_B = 1.04 \times 10^4 \; p(\text{MeV/c})/B(\text{Gauß})$ [cm]. Thus the electron/positron trajectory follows the field, and this basic feature of solenoidal transport - in the adiabatic approximation - also holds for slightly inhomogenous magnetic fields. The magnetic field of the EPOS-system of 0.1 - 0.5 T is produced from a flexible arrangement of a series of commercially available pan-cake coils of 260 mm inner diameter. The detailed shape of the field can easily be adjusted for various demands. It ideally transports electrons and positrons with energies ranging from ~0.1 MeV to ~1 MeV emitted from a target placed on the axis of the solenoid with practically constant maximum efficiency to a region far-off the highly radioactive target-zone. It belongs to the basic concepts of solenoidal transport, that all trajectories independent of their emission angle θ_0 and their energy are again completely collected anywhere along the z(solenoid)-axis within a distance equal to the pitch length z_p evaluated for maximum positron/electron energy.[40] For pratical reasons the two arms of the solenoid on both sides of the target are used in a way that one arm is optimized for positron detection, the other for electron detection. In addition a magnetic mirror with the magnetic field increasing along the z-axis on that side of the target opposite the detection of the respective particle is typically installed. For either positrons or electrons a solid angle substantially exceeding 2π for the transport into one arm of the solenoid can be reached. The maximum angle θ_{max} of emission still accepted is given to $\sin(\pi-\theta_{max}) = (B_0/B_{max})^{1/2}$ with B_0 the field strength at the target site. Note that this mirror effect can also lead to substantial loss of solid angle, when the field between target and detector happens to exceed the value at the target point. The solid angle of 4π is in most cases shared between the two leptons such that the positron has the larger acceptance. Together with the energy range accepted this allows for the effective and simultaneous measurement of the complete spectrum of both positrons and electrons at the same time, which turns out to be a big advantage of solenoidal devices. In very heavy ion collisions, however, it is the high intensity of the multiplely and promptly emitted ($\sim 10^{-21}$sec) δ-electrons which limits the use of this solenoid technique simply from counting rate considerations and special δ-electron suppression techniques had to be developed. The respective techniques provided for the electron measurements in the EPOS set-up are discussed further below. For the spectroscopy of the extremely weak quasiatomic positron production with a probability of $P_{e+} \sim 10^{-7}$ per keV and collision which are also emitted promptly ($\sim 10^{-21}$sec) during the collision, an even more effective suppression technique is introduced. This makes use of the helicity of the positron's trajectory in the field and will be discussed first.

The obvious solution to utilize the helicity of the positron's trajectory is to simply block the electron trajectories of opposite helicity with a properly adjusted spiral baffle. In case of a homogenous field along the z-axis the steepness of the spiralling trajectory is derived to $d\varphi/dz = \frac{1}{2}\mathrm{sign}(q)\mathrm{tg}\theta_0/\rho_B$; the steepness is a constant of the motion, but depends on the sign of the charge q. The use of a helical baffle centered on the solenoid axis with radial blades properly inclined with the angle θ_b relative to the solenoid axis therefore is the natural choice for being transparent for positrons and blocked for electrons. Extended helical baffles are well known from magnetic lens-spectrometers and typically are strongly dispersive. Using blades with a short width or at least in the order of the pitch length z_p, baffles can be designed as only slightly dispersive elements in order to retain the wide-band characteristics of the transport solenoid. Principally the transmission through such a baffle, also for the correct sign of the charge, is limited from geometrical considerations to emission angles of $\theta_o < \theta_{o,max}$ with $tg\theta_{o,max} \cong \pi tg\theta_B$. This reduction of the solid angle is partly compensated when the baffle is operated in a region of reduced magnetic field strength of a value $B = B_b$ where the positron trajectories, opposite as in the magnetic mirror, are bent forward according the equation $\sin\theta_b/\sin\theta_o = (B_b/B_o)^{1/2}$ with the index o or b referring to either target or baffle position, respectivley.[41] The EPOS-baffle is carefully shaped using trajectory codes in order to suppress δ-electrons strongest for the high-intensity low-energy part of their spectral distribution which exponentially decreases for the higher energies and, on the same time, to remain maximally transparent for positrons throughout the complete energy range of up to ~800 keV. The resulting transport efficiency levels at values of ~45% and is a smooth function of the energy with $\varepsilon > 0.5\varepsilon_{max}$ between 150 keV and ~1 MeV. The efficiency is reduced in the size in order to account for the additional measurement of the annihilation radiation performed to identify the positron (see below), and is shown together with the suppression factor in Figure 4 on page 203. The size and shape of the efficiency is determined with calibrated electron line sources and is compared with calculations based on the trajectory code; the smooth character of the shape in this energy region is moreover prooved with the measurement of the continous spectra of β^+ and β^- sources. Electrons scattered at the blades to a small extent may penetrate through the baffle, then leading to serious problems in cases of extreme electron to positron ratios as is typical in these experiments. Thus the suppression is only complete, when the detector with its sensitive volume is limited to a region close to the axis. As mentioned above, it is a basic feature of the solenoid, that the axis, on one hand, is the optimal place to detect positrons emitted from the target. On the other hand, it is a location of geometrically reduced efficiency for particles emitted off this axis as it is the case for the electrons scattered from the baffle. Moreover, the choice of such a detector geometry to a great extent also excludes the eventual detection of positrons which stem from external pair conversion outside the target like in the walls of the vacuum vessel.

With this technique δ-electron counting rates can be sufficiently reduced to allow for the operation of high resolution solid-state detectors for the positron detection and to avoid electron-positron sum-coincidences at a level of considerably less than 1%. However, the above mentioned special requirements for both maximum positron efficiency and scattered-electron suppression demanded the unconventional construction of a thin but very elongated, pencil-like shaped Si(Li)-detector mounted along the z-axis at a distance of ~83 cm from the target position (Figure 5 on page 205). For practical reasons the Si(Li)-crystal is subdivided into two pieces of 5 cm length each, both coupled to two separate preamplifiers and voltage supplies. In the presently used version

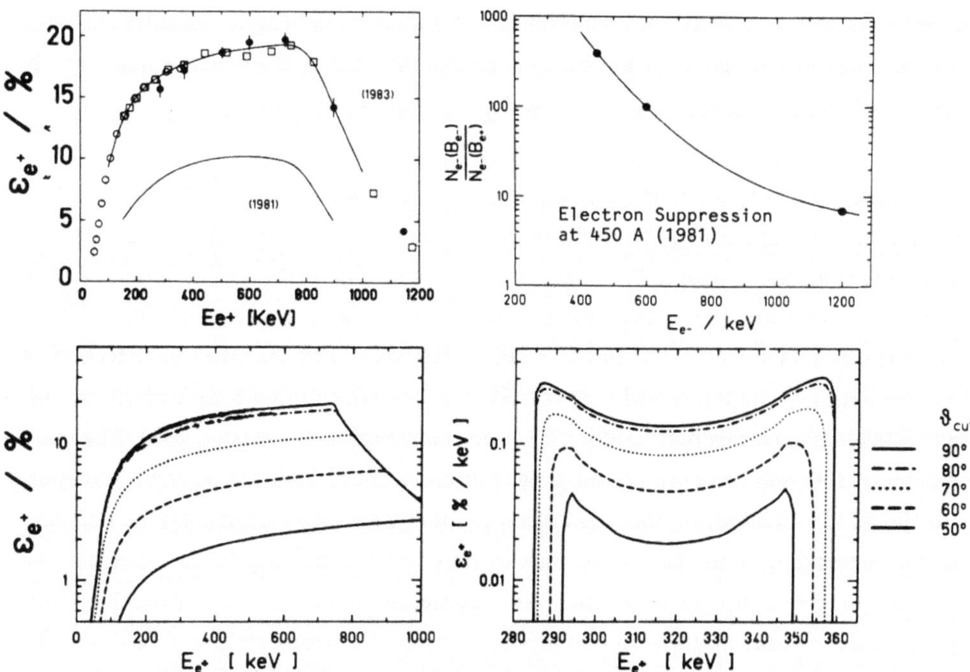

Figure 4. Various aspects of the EPOS positron detection efficiency. (left up): Positron full-energy-peak detection efficiency as a function of the positron kinetic energy including the NaI-crystal efficiency for the identification of the annihilation radiation and the detector response together with the results from calibration measurements. (right up): e^--suppression factor (ratio of total e^--transmission for opposite magnetic field-settings) versus kinetic energy together with the values determined from source measurements. (left down): Various calculated efficiency curves obtained by assuming that positrons with $\theta_{cut} \leq \theta_{e+} \leq \pi - \theta_{cut}$ defined relative to the solenoid axis are excluded from the transport to the positron counter. Only cut-off angles $\theta_{cut} > 80^0$ are consistent with the calibration measurements. (right down): Doppler profile of an intrinsically narrow positron line at 320 keV calculated for different cut-off angles as indicated and an emitter velocity of $\beta = 0.056$.

the front piece is moreover sensitive on its front end. The counters are coaxially drifted with a sensitive thickness of 2.5 mm; this thickness already corresponds to an energy range of fully stopped positrons of up to ~1.4 MeV, but the limit of complete energy loss is still extended to higher energies due to the inclined impact and spiralling trajectory of positrons in the magnetic field. Both detector pieces are mounted on a 5 mm ∅ Al-rod and are cooled together with the FET's from a small automatically filled liquid-nitrogen reservoir. In order to avoid backscattering, all mechanical parts are placed far away from the Si(Li)-crystal at a position where the magnetic field has already been reduced, thus forming a magnetic mirror for particles eventually scattered back towards the Si(Li)-detector. The intrinsic resolution is limited mainly by the large capacitance of the detectors. Nevertheless, the resolution obtained of better than 10 keV is considerably smaller than the line width originating from the Doppler effect when the positron is emitted in flight into a solid angle of almost 4π (see below); the time resolution obtained with the detectors is < 7 ns. Special care has been taken to achieve a constant detection efficiency as well as a constant resolution along the detector length. Besides the electrical contacts in the middle of the detector of together 2 mm, the detector has a nearly constant efficiency within $< 3\%$ and no φ-dependence has been observed.

To finally identify positrons unambigously, their characteristic annihilation radiation can be used. Predominantly, the positron in a solid annihilates through the $1S_0$-state of the positronium atom within 10^{-10}sec by emission of two 511 keV collinear γ-rays. The annihilation through the $3S_1$-state leading to 3 γ-rays as well as annihilation in flight is retarded by a factor of ~ 370 and of ~ 10, respectively. For the effective detection of the annihilation γ-quanta, a NaI-ring crystal of 2″ thickness and a length of 20 cm has been installed symmetrically around the Si(Li)-detector (Figure 5 on page 205). The inner diameter of 100 mm has been chosen largely in order not to limit the positron transport onto the Si(Li). In addition, the ring crystal is subdivided into 8 segments; the collinear angular correlation of the two γ-quanta can thus be used to even further increase the selectivity of the set-up for positrons. As a consequence of the use of extended light-pipes in order to operate the photomultipliers outside the strong magnetic field and the loss of light output due to the unfavourable and elongated shape of the NaI-crystals, the energy resolution of each segment amounts to $\sim 13\%$ at 511 keV.

Mostly, it is sufficient to detect the full energy of only one of the two γ-rays. Thus the NaI's efficiency in principle can be doubled but than has to be corrected for double-counting. The probability of observing just one out of the two γ-rays in the photopeak for the ring crystal in this case reads:

$$P^{ph}_{\gamma\gamma} = 16\Omega_S(1-\varepsilon_\gamma(\Omega_C/\Omega_S))f\varepsilon_\gamma$$

and the probability of observing both γ-rays each in the photopeak is

$$P^{phph}_{\gamma\gamma} = 8\Omega_S f^2\varepsilon_\gamma^2(\Omega_C/\Omega_S).$$

The solid angle of the whole crystal amounts to $8\Omega_S \cong 78\%$ and $f\,\varepsilon_\gamma$ describes the γ-ray photopeak efficiency of the crystal per unit solid angle; the ratio (Ω_C/Ω_S) reflects the geometrical reduction of the solid angle for the second γ-ray when the annihilation occurs outside the center of the Si(Li)-crystal. Double-counting is easily prevented by linear summing of the energy signals of each crystal during the analysis which also improves the peak-to-Compton ratio. The resulting sum-spectrum directly exhibits a unique pattern for one- and two-photon detection. Still remaining background due to scattered X- and γ-rays contributes increasingly at low γ-energies and was found to be mostly correlated with low positron energies (originating from misidentified δ-electrons). This background was found to be typically less than 10% and can be further reduced without loss of efficiency, when the collinearity condition is activated for only those γ-events of incomplete energy deposition. Moreover, it can be quantitatively subtracted by comparing the sum-spectrum measured in-beam with that obtained from a β^+-source measurement. The efficiency now becomes maximal for the mixed identification of at least one γ-ray detected. This corresponds to a sum-energy window

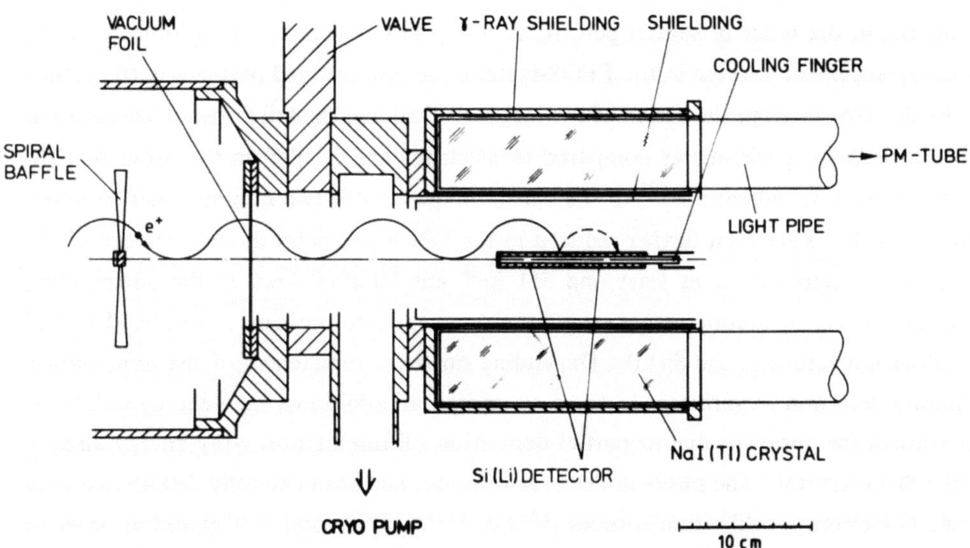

Figure 5. Enlarged view of the Si(Li)-detector arrangement used to measure the kinetic energy of the positrons. The Si(Li)-detector is cooled close to liquid-air temperatur by a temperature controlled feed-through cryostat (not shown in the figure) of very slim shape in order to fit into the surrounding 8-fold NaI detector arrangement. A face-sensitive and non-face-sensitive Si(Li)-detector has been used throughout different periods.

ranging from 450 keV to 1050 keV and the probability to detect the annihilation is given to

$$P_{\gamma\gamma}^{>450} = 16\Omega_S(1-\tfrac{1}{2}\varepsilon_\gamma(\Omega_C/\Omega_S)(2-1/f))f\,\varepsilon_\gamma$$

and amounts to 58%; this exceeds values which could be obtained with the same crystal if only one γ-ray is emitted per event. The two-γ detection probability $P_{\gamma\gamma}^{phph}$ associated with a sum-energy window of 900 keV to 1050 keV is extremely selective and can always be activated in the analysis, but since it reaches an efficiency of only ~16% appears too small for effective in-beam positron spectroscopy. Moreover, this high degree of selectivity turned out to be unnecessary in most of the measurements performed so far. In fact, the sum-crystal efficiency depends on the individual place of annihilation along the Si(Li)-detector and thus could modify the shape of the positron spectrum. This effect is contained in the ratio (Ω_C/Ω_S) and the reduction was found to be of ~3% for the $E_\gamma > 450$ keV window and of ~12% for the two-γ detection for positrons of ~150 keV kinetic energy and considerably decreases to values of <1% and 5%, respectively, for positrons of ~500 keV.

Si(Li)-detectors generally suffer from low-energy tails in the single line response due to outscattering of electrons/positrons and due to bremsstrahlung. The outscattering-ratio for Si is practically energy independent for the energy range presently of interest and amounts to values of >25% for perpendicular incidence and is still increasing for tangential penetration.[42] The bremsstrahlungs loss grows with energy and contributes to the tail in the order of several percent at ~800 keV. In the special geometry of the Si(Li)-arrangement selected in the EPOS-system, the outscattered particles partly return to the detector through the action of the magnetic field and the tail portions are reduced. Tail portions for positrons as compared to electrons are smaller on the order of 10% from principal considerations[43)] of the scattering-process of two non-identical particles. The tail portions are even further reduced in the EPOS-geometry in consequence of the need of the detection of at least one 511 keV annihilation γ-ray in the surrounding sum-crystal. This condition rather excludes an outscattered positron to still be identified if it does not return to the Si(Li). Depending on the completeness of the annihilation γ-quanta detection requirement in the ring crystal, an additional high-energy tail, however, arises for positrons due to partial deposition of annihilation γ-ray energy already in the Si(Li)-crystal. The positron detector response has been carefully determined with single line electron calibration sources (^{109}Cd, ^{113}Sn, ^{207}Bi and ^{137}Cs) and in order to obtain the high-energy tail with β^+-sources (^{22}Na, ^{56}Co, ^{68}Ge). With the aim to minimize the influence of additional energy-loss tailing from the backing-material of the source on the result, target equivalent sources only have been utilized. A functional dependence of the response similar as it is discussed in ref. 44 was used with typical low-energy tail portions ranging from ~30% at 300 keV to ~60% at 900 keV and ~6% for the high-energy tail of the total intensity. Fig. 4 contains the positron peak-detection efficiency including the sum-crystal response (E_γ > 450 keV) and the Si(Li)-detector full-energy-peak response and amounts to ≤20%.

In in-beam experiments with heavy projectiles the energy resolution for positrons/electrons is moreover limited from kinematic effects because they are emitted from a moving source with $\beta = v/c$ and $\gamma^2 = (1-\beta^2)^{-1}$. The total laboratory energy is gi ven to $E = \gamma E^*(1+(p^*c/E^*)\beta\cos\theta^*)$ with E^*,θ^* the total energy and the emission angle in the CM-frame with respect to the beam direction. Isotropically emitted positrons/electrons with an intrinsically sharp energy distribution centered at $E^* = E_0^*$ then appears in the lab-frame with a broadened rectangularly shaped line centered practically unshifted at the energy $E_0 = \gamma E_0^*$ and with upper/lower cut-off limits of $E_{1/2} = \gamma(E_0^* \pm \beta p_0^*)$ corresponding to forward and backward emission, respectively. The solid angle of the solenoid arranged perpendicular to the beam direction is symmetric with respect to forward and backward emission. Thus in the EPOS-system a positron line always appears at the practically unshifted at the position intrinsic energy $E = E_0^*$ of the ($\gamma \cong 1$) but is maximally broadened with the width given by $\Delta E_D = 2\gamma\beta p_0^*$. This amounts to a value of $\Delta E_D \cong 70$ keV for $\beta = 0.05$ at ~300 keV which considerably exceeds the intrinsic resolution of the positron counter of ~10 keV. It is obvious, that this Doppler width represents an excellent and direct measurement of the emitter velocity as long as essentially isotropic emission can be assumed. The detailed shape of positron lines finally measured in the EPOS-spectrometer depends on the details of the angular dependent transport efficiency and is determined with the transport code mentioned above again incorporating Monte Carlo techniques. Note, that a meaningful experimental reduction of the kinematic width is always accompanied by a considerable loss of efficiency. This can be seen from a variation of an artificially introduced cut-off for large emission angles changing both the efficiency and the line shape as demonstrated in Figure 4 on page 203. The size of the efficiency of the solenoid was found to be in excellent agreement between measurement and calculations; thus any eventual experimental restriction of the solid angle of values large enough to otherwise influence the line width can safely be excluded.

But in view of the importance of this feature an experiment to directly measure the Doppler profile in case of emission from a fast moving system was performed. Excellent agreement between the calculated and measured kinematic line profile was found from the measurement of the conversion electrons from the nuclear decay of the 2^+(188 ps)[45]-level which is populated promptly in the compound reaction ^{12}C(^{118}Sn,4n)^{126}Ba at 5.86 MeV/u. The spectrum (Figure 6 on page 208) was obtained in the original EPOS set-up only modified for electron detection by simply inverting the magnetic field. One should note, however, that efficiency and kinematic line profile may be distorted, when emission takes place from long-living systems propagating fast, which then may escape from the region of the maximum acceptance of the solenoid. Considerable reduction of the width was found only (see part (b) of Figure 6 on page 208) when the place of emission is drastically removed to distances > 1 cm in agreement with the fall-off of the efficiency at a comparable distance; for an emitter velocity of $\beta \cong 0.1$ in the beam direction this corresponds to an emission life-time in the order of ~ 100 ps.

Even when optimized for positron detection, the EPOS-spectrometer in the opposite arm bears the advantage for an easy simultaneous measurement of electrons in order to obtain either information on possible nuclear transitions - perferentially of EO-type, on δ-electron spectral shapes, or on principal electron-positron coincidences. In all cases the highly intense δ-electrons of low energies (approximately below 200 keV) of high multiplicity[46] have to be prevented from detection in order to avoid sum coincidences or even operational failure of the detectors. Principally this can be obtained by the use of absorber foils, but this technique is applicable only for the measure of high-energy electrons (≥ 1 MeV) with a comparably negligible energy loss in the absorber foil. All experimental techniques used so far therefore, in addition, utilize the action of the solenoidal magnetic field to selectively reduce δ-electron intensity by a restriction to angular and/or momentum windows. The actual technique employed significantly differs for the various applications.

The identification of electron conversion lines from nuclear transitions in the presence of the δ-electron background is hindered from the strong broadening of the lines when the electrons are emitted from the fast moving nuclei. Certainly, for such ideal low-Z

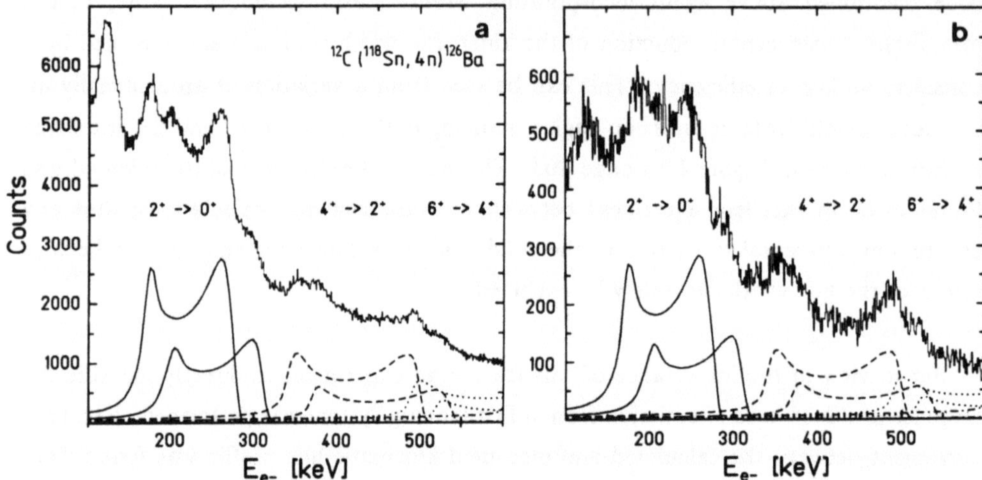

Figure 6. (a) Electron conversion spectrum of the reaction $^{12}C(^{118}Sn,4n)^{126}Ba$ at 5.86 MeV/u as measured in the EPOS set-up with the positron detector at inverted but otherwise identical magnetic field together with the instrumental Doppler-effected line profiles ($\beta = 0.1$) of the K- and L-conversion lines of arbitrary intensities. Although the most prominent 2^+-decay has an effective decay-length of 0.56 cm ($\tau^{2+} = 188$ ps[45]) the measured FWHM of the K-line of $\Delta E = (107 \pm 5)$ keV agrees very well with the calculated value of $\Delta E = 109$ keV.
(b) Same as (a) but with the target removed 1 cm down-stream. Only the long-living 2^+-state decays sufficiently off-axis to result in a considerable reduction of the characteristic line width.

situations as in the case of ^{126}Ba mentioned above, also maximally broadened conversion lines can be identified. For high-Z systems, however, with increased δ-ray background a considerable reduction of the Doppler-broadening is typically necessary. To preserve the advantage of a simultaneous measurement, i.e. not to reduce the solid angle for positron detection, only electrons which escape the magnetic mirror at small angles with respect to the axis are used for the conversion spectroscopy which typically exploits cross sections of large values as compared to the positrons. Independent of their energy this limits the transport of electrons already to values of the emission angles θ_0 (with respect to the solenoid axis) of $145^0 < \theta_0 < 180^0$. In order to limit the Doppler width to $\Delta E_D < 80$ keV for energies of ~ 1 MeV - the intrinsic resolution of a planar Si(Li)-counter for these energies is in the order of ~ 30 keV - the detection of the electrons has still to be restricted further to only those trajectories associated with emission angles of $170^0 < \theta_0 < 180^0$. Within the energy range of interest of 800 keV $< E_{e-} < 1.5$ MeV this is achieved by means of an appropriate hollow cylindical baffle of ~ 10 cm length installed along the solenoid axis. The principle action of this baffle is easily understood from Figure 7 on page 209 where electron trajectories of different emittance angle θ_0 ($\theta_{e+} = 180^0 - \theta_{e-}$)are shown. In order to minimize the backscattering of electrons, which may otherwise seriously influence the clean detection of positrons, the cylinder is machined from a low-Z material (Al). Moreover, it is positioned at a region of already reduced magnetic field strength thus forming a magnetic mirror for the backscattered electrons. The remaining low-energetic δ-electron back-

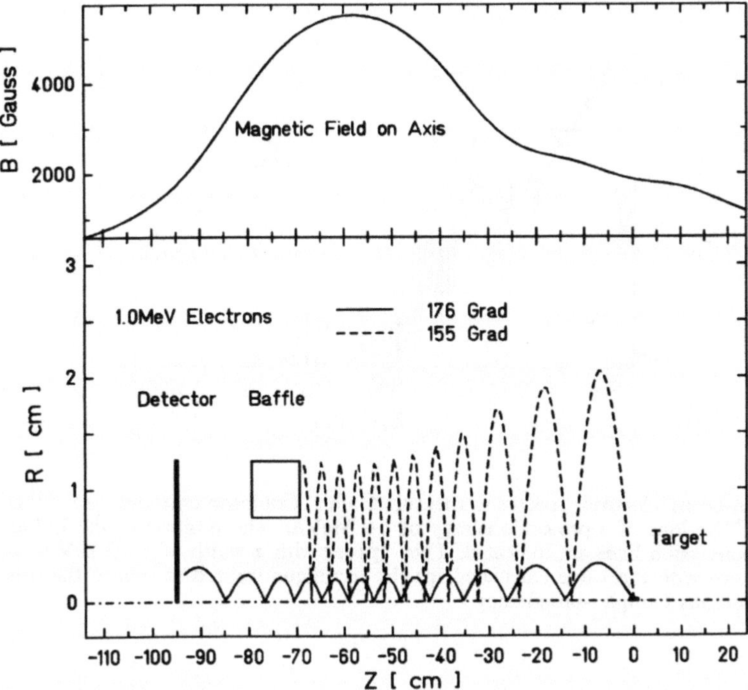

Figure 7. Mechanical baffles in the solenoid field can be used to limit the emission angle of the electrons and thus the respective Doppler-broadening: electrons with an emission angle $\theta_0 > 170^0$ (i.e. $\theta_{e-} > 10^0$) (both with respect to the solenoid axis) can pass the cylindrical baffle to the detector. The Doppler-broadening is thus limited to $\Delta E_D \sim 80$ keV at 1 MeV kinetic energy of the electrons and $\beta \sim 0.1$.
(a) Magnetic field distribution along the solenoid axis
(b) Spiralling trajectories plotted along the solenoid axis of electrons of 1 MeV kinetic energy emitted at $\theta_0 = 176^0$ and $155.^0$.

ground at the electron detector was suppressed in these experiments by the absorption in a Cu-foil of 31 mg/cm^2 thickness mounted in front of the cooled (-20°C) planar Si(Li)-detector. An overall resolution of ~30 keV was obtained with this assembly for the 660 keV conversion line of a ^{137}Cs-source at the target position. Figure 8 on page 210 shows an in-beam electron spectrum measured with this arrangement after Coulomb excitation of the second 0^+-state of ^{238}U with ^{136}Xe projectiles at 3.6 MeV/u. The line obtained from the 990 keV K-conversion shows the expected width of ~70 keV and obviously such weak E0-transitions can be identified in this assembly, although it should be well noted that the quasiatomic δ-electron intensity will still increase by nearly one order of magnitude[29)] for the very high-Z systems.

Recently the EPOS-spectrometer was modified for positron-electron correlation measurements. This demand was to test the assumption of a two-body decay of an object of ~1.8 MeV/c^2 restmass into a positron electron pair with both of same energy of ~350 keV and near back-to-back emission. For the measurement of such low energies a complete cut-off of the δ-electron background below ~150 keV is crucial. But still remaining quasiatomic electrons with energies exceeding 200 keV establish a coincidence

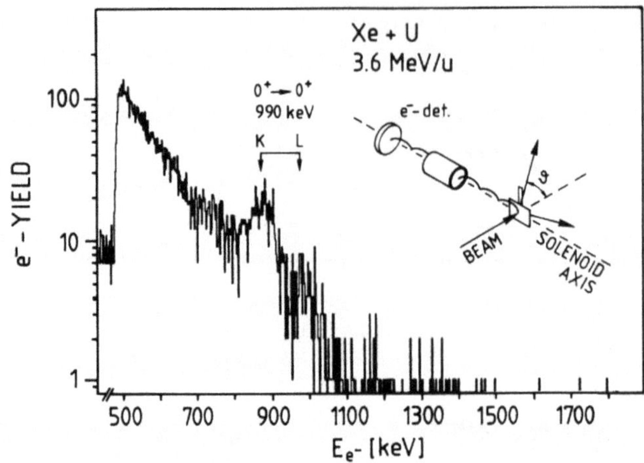

Figure 8. In-beam electron spectrum measured after Coulomb-excitation of ^{238}U with ^{136}Xe-ions of a projectile energy of 3.6 MeV/u. The position of the EO electron conversion lines are indicated. They appear with a width of ~70 keV in accordance with the value calculated for the technique utilized to reduce the otherwise ruinous Doppler-broadening.

probability together with the quasiatomic positrons of $P_{e+e-} \sim 0.7$ per collision.[46] This necessarily limits the useful electron detection efficiency to values of ~20% in order not to cover up the searched process of a correlated positron electron pair decay with a probability in the order of or less than 1 % of the positron yield.

The low-energy cut-off for the electrons is achieved from the combined action of a small sheet baffle and the radial dislocation of the electron detectors from the solenoid axis (Figure 9 on page 211). The sheet baffle of 2 cm length and 1 cm height is installed along the solenoid axis closely neighbouring the target. The pair of planar Si(li)-counters is arranged parallel to the solenoid axis at a position of 112 cm from the target with a radial distance of 18 mm. The low-energy electrons, especially when emitted steep with respect to the axis, will then be stopped when they hit the sheet baffle during the early part of their trajectories. Alternatively they simply pass between the two Si(Li)-detectors, when their trajectories' spiral diameters do not exceed the distance of the detectors from the axis (0.9 mm each). Due to the built-in 'circular sense' (see cross-section in Figure 9 on page 211) of the electron detector-arrangement with the entrance windows facing the solenoid axis, the detectors are differently accessible for either electrons or positrons depending on their respective helicity at a given magnetic field direction. Originally a similar assembly was used in ref. 46 to maximally suppress electrons with respect to positrons. The geometry of this set-up is rather optimized for the highest electron efficiency but still supresses the positron detection in the electron counter by a factor of ~4. The rectangular detector area of each of the pair is subdivided into two equal parts with a sensitive area of 32 x 64 mm², the depletion depth amounts to 2.9 mm. The size of the detectors is selected largely in order to collect electrons of up

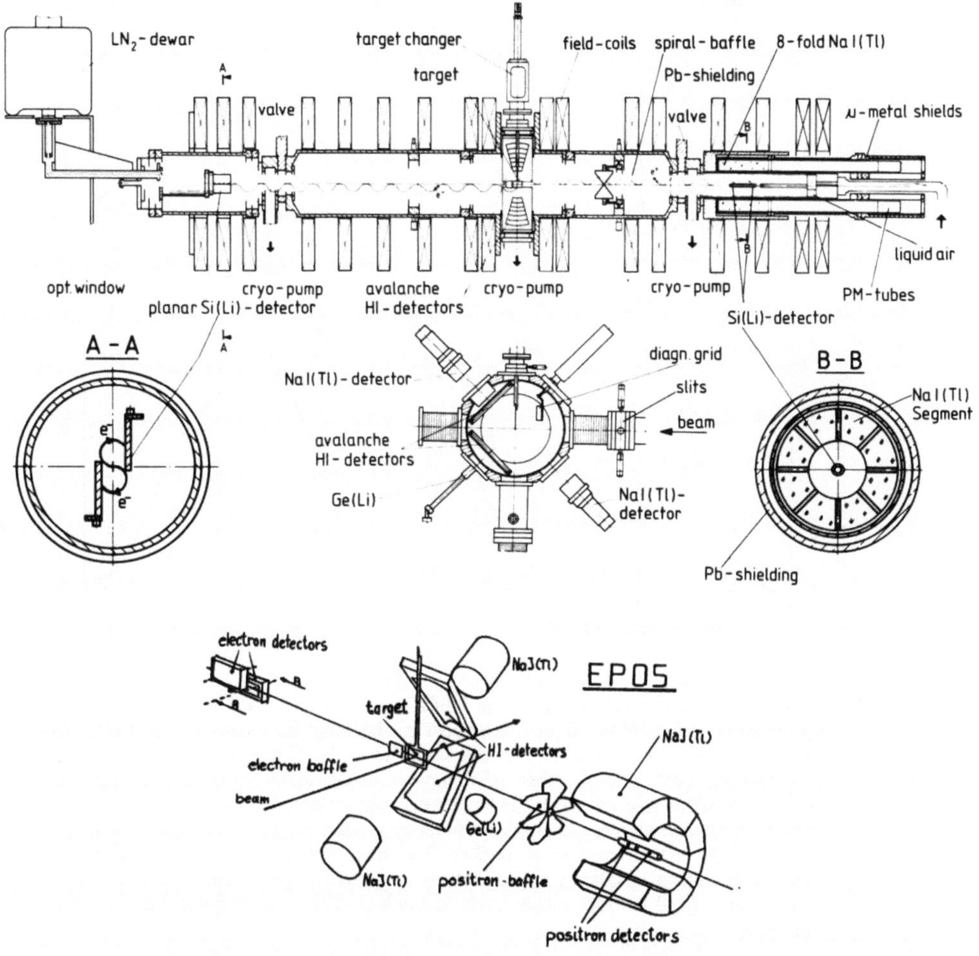

Figure 9. Schematic view of the EPOS-spectrometer in a version as recently used for the positron-electron correlation experiments together with (middle row) cross sections of the electron detector area (enlarged scale), the target area and the positron detector area and (down) perspective drawing of the main components.

to ~900 keV at practically maximum efficiency at the present magnetic field setting. The magnetic field for this experiment is modified such that positrons are accepted with an emittance angle θ_0 between ~20° and 120° and electrons between 120° and 170°. Electrons of ~660 keV are measured with a resolution of better than 10 keV when the detectors are cooled with liquid nitrogen and the intrinsic time resolution is ~5 ns.

The resulting positron singles and electron-positron coincidence efficiency is shown in Figure 10 on page 212 as calculated with the trajectory code and Monte Carlo techniques. The coincidence efficiency is computed for comparison either for back-to-back emission (correlated efficiency) and for a random opening-angle distribution assuming for the electron and the positron the same energy. A back-to-back emitted pair has a ~15 % probability to actually be identified in the set-up as a pair, once the positron has been observed. The correlated efficiency is roughly twice as large than the efficiency for

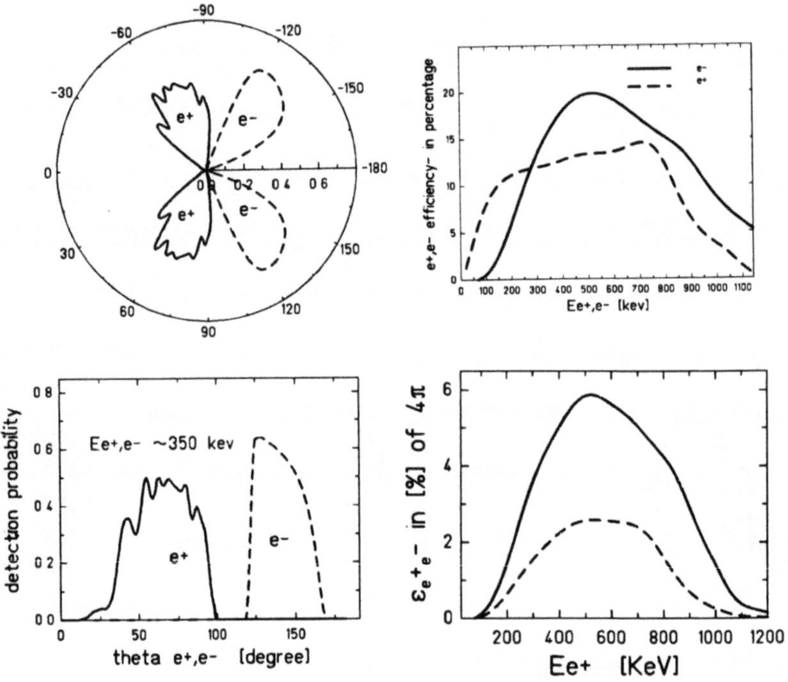

Figure 10. The total detection efficiency of the EPOS-spectrometer in the electron-positron coincidence mode. The singles detection probability for positrons and electrons of a kinetic energy of 350 keV is displayed in polar (left up) and cartesian representations (left down) (the efficiency is rotational symmetric around the solenoid axis) as a function of the electron/positron emission-angle θ_0. Integrated over all emission angles the singles efficiency is shown as a function of the kinetic energy (right up) for positrons (dashed line) and electrons (solid line). The coincidence detection efficiency (right down) is given for positrons and electrons of same energy (380 kev) for either back-to-back (full line) or angularly uncorrelated emission.

uncorrelated isotropic emission. Uncorrelated emission can approximately be assumed for coincident quasiatomic electrons and positrons and also for E0 pair conversion[47]; for the internal pair conversion processes of higher multipolarities with an opening-angle distribution[48] strongly enhanced towards 0°, the efficiency is even more reduced. The efficiencies quoted in the figure are total detection efficiencies. The tails contain on the order of ~30 % of the total intensity when realistic detector response functions are considered. The detailed coincidence response of the device has been systematically studied with Monte Carlo techniques together with the solenoid transport code and realistic detector response-functions obtained from measurements with various electron line- and β^+-sources. The calculations concerning the pair conversion scenario have been found in excellent agreement with the measurement of the particularly relevant 1.76 MeV E0 pair conversion decay of a ^{90}Sr-source prepared to optimally simulate the target conditions (Figure 11 on page 214). The individual spectral shapes are very well reproduced, an agreement which is sensitively dependent on the details of the detector response functions. Also the calculated overall singles-to-coincidence detection efficiency ratio of ~6 % for nearly uncorrelated events excellently fits to the data. The electron-positron coincidence set-up together with the Monte Carlo calculations is elaborated in detail in ref. 49.

The instrumental solutions for the detection of the scattered heavy ions in solenoidal systems is to a certain extent influenced by the only limited available space in the target region which, on the other hand, is needed for the installation of large area counters. The need of positioning the target in a region of maximum magnetic field strength (besides the mirror) as mentioned above, demands for a close arrangement of the field producing coils in this area. Nevertheless, the installation of particle counters in order to incorporate kinematic coincidence techniques well known from nuclear reaction experiments, was performed in the EPOS set-up already from the very beginning. In view of the large counting rate and the high charge density typically induced from heavy ions when passing through matter, gas counters are mostly used in such experiments instead of expensive solid state detectors which moreover suffer from the irradiation.

Two parallel plate avalanche gas counters appropriately constructed for the close geometry have been installed symmetrically with respect to the beam direction (Figure 3 on page 200 and Figure 9 on page 211) inside the solenoid coils and thus have to operate in a magnetic field of ~ 0.2 T. The measure of θ_{ion} defined with respect to the beam direction is performed with elliptically shaped, continuous meander delay-line[50,39] of ~ 200 ns total delay incorporated as a thin foil into the back of each of the counters and is operated as anode (Figure 12 on page 215). The φ-angle is measured simultaneously with a segmented second anode foil positioned as entrance window opposite to the common cathode. Connected with a discrete delay network the counter can reach a resolution of $\Delta\varphi < 1^0$ but in most cases was used with broad φ-stripes of $10^0 < \Delta\varphi < 20^0$. Typically each of the counters cover an angular range of $\sim 15^0 < \theta_{ion} < 75^0$ and $|\varphi_{ion}| < 30^0$ which very recently has been enlarged to $|\varphi_{ion}| < 60^0$. The detectors each are positioned in a way to completely overlap with the corresponding angular ranges in the case of elastic scattering. With the anodes each separated by < 2 mm from the common cathode foil these counters reach a time resolution of $\delta t < 600$ ps when operated with a continuous gas flow of n-Heptan at ~ 10 Torr and ~ 600 V cathode voltage with both anodes grounded. No count-rate limitations - being most critical for the continuous θ_{ion} delay-line read-out - have been observed up to 200 kHz instantanous counting. The gas counters experience in addition to their own electric field the perpendicularly oriented magnetic field of the solenoid. This situation of crossed electric and magnetic field does not critically effect the operation of the counters as long as the mean gyration radius ρ_d of the secondary electrons exceeds the mean free path λ_d in the counter gas or, equivalently, as long as the Lorentz-angle α defined between the direction of the drift velocity and the electric field of $\tan\alpha = \lambda_d/\rho_d$ remains sufficiently small.[51] A small offset due to the resulting drift of the secondary electrons perpendicular to both the electric and the magnetic field is corrected in the calibration. The parallel plate avalanche counters and their operation are described in more detail in ref. 39 and 52.

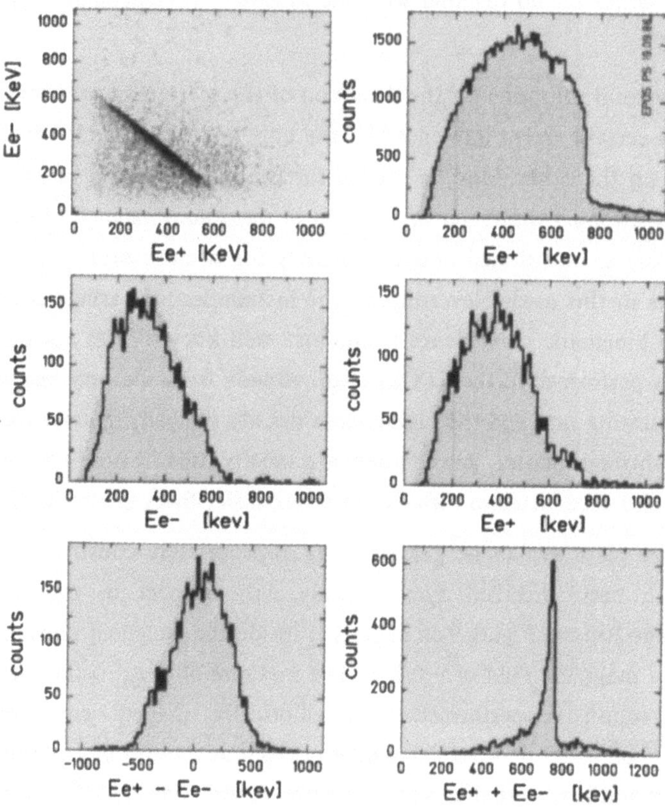

Figure 11. Experimental (solid line) and Monte Carlo simulated spectra (hatched region) of the nuclear internal pair conversion of the 1.76 MeV transition in ^{90}Zr. The spectral distributions are shown for the singles detection of the positron (right up), as a two-dimensional intensity distribution (left up), and as projections of this distribution onto either the positron energy, the electron energy (second row, respectively), the sum and difference of the positron and electron energies (down row). The calculations contain the detection efficiencies of Figure 7 on page 209 and realistic detector response functions and are normalized to the total experimental yield.

The angular resolution obtained for a pair of elastically scattered ions with both detectors each at a distance of 10 cm from the target and inclined under 45° to the beam axis is $\delta\theta \leq 0.5°$ (see (b) in Figure 13 on page 216) and is presently limited by target thickness effects and the beam geometry. The two branches in part (b) of Figure 13 on page 216 represent a typical elastic scattering pattern being twofold because the target- and projectile-like nuclei are not individually identified in the detectors. For elastic scattering all relevant kinematic CM-parameters including the CM ion scattering-angles θ_1 and θ_2, the impact parameter b, or the distance of approach R_{min} can unambigously be determined all over the measured angular region. This needs a simultanous measure of the two individual heavy ion scattering-angles θ_1 and θ_2 and the difference Δt_{12} of the ion's time-of-flight. Using this kinematic coincidence technique the impact parameter dependence of the positron production cross-section can easily be followed up to the closest collisions. Moreover, this technique allows for an effective separation against

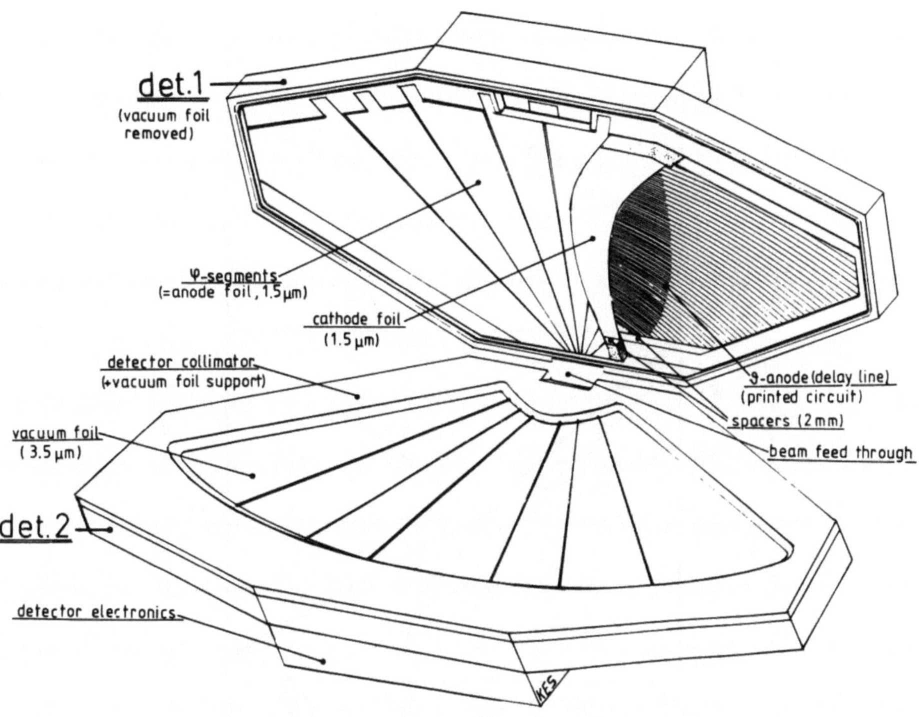

Figure 12. An exploded view of the EPOS parallel plate avalanche counters for the detection of the scattered ions.

positron background originating from inelastic collisions with low-Z target-impurities or -compounds and backing- or surface-layers, all leading to substantially different ion angular correlations with the heavy projectile-like scattering-partner limited to forward lab scattering-angles.

Clearly, the Ta-targets are most favourable as concerns the target quality and typically the resolution is reduced to $\delta\theta \sim 0.8^0$ for targets of increased thickness or less homogenity as it was the case for the molecular plated ^{248}Cm-targets. But even with the decreased resolution the close and distant collision branches can still be separated in the set-up for the only slightly assymetric collision system ^{232}Th + ^{248}Cm at angles outside the $\theta = 45^0$ region, or the two branches appear to be partially separated for the nearly symmetric ^{238}U + ^{248}Cm system. For the detection of only small deviations from pure Rutherford-scattering due to a slight inelasticity (Q-value) including mass transfer, the composed coordinate $\Sigma\theta = \theta_1 + \theta_2$ appears to be most sensitive. Therefore two-dimensional event distributions versus $\Sigma\theta$ and the orthogonal coordinate $\Delta\theta = \theta_1 - \theta_2$ (see Figure 13 on page 216) are commonly utilized during the analysis. Actually the Q-value is directly related to the $\Sigma\theta$-coordinate. In particular, when no mass transfer is assumed, the Q-value at $\Delta\theta = 0$ $(\theta_1 = \theta_2 = 45^0)$ is analytically given to $Q = \frac{1}{2} E_1 (tg^2(45^0 - \Sigma\theta/\sqrt{2}) - 1)$ for symmetric collisions $(A_1 = A_2)$. This results in a value of $\Delta Q \sim 50$ MeV for a shift of $\delta(\Sigma\theta) \sim 1.5^0$ in the case of U + U-collisions.

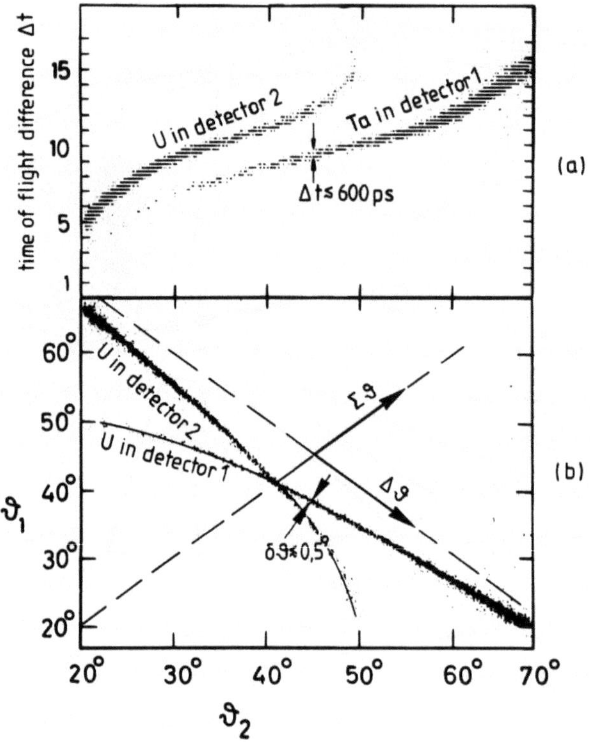

Figure 13. Two-dimensional ion scattering-event distributions for elastic $^{238}U + ^{181}Ta$ collisions measured at a beam energy of 5.9 MeV/u in (b) versus θ_1 and θ_2, the lab ion scattering-angles with the composed coordinates $\Sigma\theta = \theta_1 + \theta_2$ and $\Delta\theta = \theta_1 - \theta_2$ indicated and in (a) versus Δt_{12}, the difference in the time-of-flight of the two ions (with arbitrary off-set) and θ_2.

Note, that the angular scattering pattern necessarily has to be symmetric around $\Delta\theta = 0$ and this symmetry principally holds for all cases of symmetric, asymmetric, elastic or inelastic collisions. Certainly, this 'up-down' symmetry is not necessarily displayed from the intensities of the various scattering or production processes, but rigorously holds for the production probabilties obtained when normalized to the scattered particle intensity. However, the symmetry is partially released by the action of the magnetic field directed perpendicular to the ion scattering plane. To a certain extent this additional deflection is compensated in the calibration procedure of the particle counters. The calibration parameters are determined from the cross-over angles $\theta = \arccos(1 + 0.5(A_1/A_2)^{1/2})$ and cut-off angles $\theta_{max} = \arcsin(A_2/A_1)$ measured for elastic scattering of asymmetric projectile masses A_1 and target masses A_2 ($A_1 > A_2$). The measured values contain the action of the magnetic field but they have to be scaled between the various collision systems assuming equilibrated charge states[53] which vary with the velocity of the ion. To date it is known, that the target thickness of ~300 μg/cm² of heavy elements typically used in the experiments, is insufficient to equilibrate strongly enhanced charge states which actually are expected from large-angle scattering.[53] First measurements to determine the target thickness dependence of the ionic charge state have been performed.[55,56] For the details and also for the related variation of the ion angular scattering-distributions due to the magnetic field the reader is referred ref. 39 and 52. However, it is clear that the 'up-down' symmetry contained in the set-up may partially be broken for non-equilibrated charge states and also for long-living compound systems separating outside the target.

The set-up is supplemented (Figure 9 on page 211) with NaI(Tl)- and Ge-detectors for γ- and X-ray detection. These counters all are arranged in the scattering-plane at some 40 cm distance from the target and form angles of 45^0 either backward or forward with the beam direction. The two NaI(Tl)-crystals are positioned opposite to each other and the spectrum of their energy sum may well be used to compensate the Doppler effect to first order. The contribution of neutrons mostly present in the γ-ray detectors placed in the forward hemisphere is identified by means of their time-of-flight delayed as compared to the γ-rays.

The deterioration of the targets during the heavy ion bombardment is monitored with surface barrier detectors installed at 45^0 sideward to the scattering-plane and cooled to -5^0C. Serious deterioration was observed already at the very early measurements. This can be concluded from the strongly increased low-energy tailing of the projectile or target lines in the monitor spectrum together with drastic changes in the intensity ratios derived for carbon, oxygen and/or other light target elements. Enormous effort has been put into the development of more stable targets. In addition, a target wobbler appropriately constructed for the solenoid's special geometry has been introduced throughout some periods of the experiments. Moreover, targets are microscopicaly probed before and after the irradiation with proton recoil techniques.[57,58] Strongest inhomogenities of the target material distribution ranging on the order of factors 2 to 3 and respective change of the energy-loss thickness was detected with the proton recoil technique for the ^{248}Cm-targets after irradiation of only some hours with heavy ion beam currents of ~1 pnAmp. These rare targets were obtained by molecular plating from a liquid solution, a technique[59] developed to highest performance for a most efficient target production from only mg-substrates. But also targets produced with more conventional techniques and materials suffer from growing inhomogenities in the order of up to 30% when irradiated with 1-2 pnAmp currents already after ~2 hours. Presently the targets were evaporated as fluorides and sandwiched between ~20 µg/cm² C-backing and ~5 µg/cm² C-surface layers. The results of target-probing with proton recoil techniques is presented in ref. 52.

During the course of the experiments the exact measurement of the projectile energy, its reproducibility, and its long-term constancy became of increased importance. With an accuracy ΔE_b of the beam energy needed to values better than 0.05 MeV/u, the sufficiently exact determination is already limited from target-thickness uncertainties: a layer of 100 μg/cm² of the heavy elements results in an energy loss of ~ 0.02 MeV/u. Moreover, such condition also exceeds the accuracy typically achieved throughout longer periods during the operation of the UNILAC although its conceptional beam-energy accuracy amounts to values of better than $2 \times 10^{-3} E_b$, i.e. 0.01 MeV/u. Presently the beam energy is absolutely determined simultaneously from a measure in the UNILAC precision time-of-flight pass in a separate branch of its beam transport system with the beam pulsewise switched into this line. In addition, the beam energy is continuously monitored from a time-of-flight set-up installed directly in front of the EPOS-experiment.[60] Using non-destructive beam pick-up probes mounted with a separation of 8 m, a time resolution was derived from the zero-crossing of the time pick-off signals of better than 600 ps with discriminator levels set to accept beam intensities exceeding ~15 nA. An effective down-scaling of the time-of-flight signals originally running with the UNILAC-frequency of 127 MHz is absolutely necessary in order to obtain this accuracy with standard spectroscopy electronics and is automatically performed when the positron events are used as a stochastic hardware trigger. The values of the time-of-flight difference are added to the parameter list of each positron event when stored in the event-by-event mode from the GSI SATAN/GOLDA[61] computer network and can be used as individual trigger during the analysis.

III. EXPERIMENTS AND RESULTS

It is the aim of this article to follow the development of the EPOS-experiments on heavy ion positron production since the preceeding 'QED-conference' in Lahnstein 1982[4]. The most exciting phenomena of these experiments clearly was the detection of pronounced line structures in the positron spectra whose origin in spite of the enormous effort put into their investigation remained an open question till today. This subject will be the main part of this section. On the other hand, the quasiatomic positron yield is dominated by the dynamic production[1,2,3] and any search for the details of the structures of the positron spectra has to be preceeded by a systematic investigation of the gross properties of both nuclear and quasiatomic positron production including subcritical and supercritical collision systems. More details on this section can also be found in ref. 38 and 39.

Nuclear positron production (Figure 1 on page 196) originates from internal pair conversion of nuclear deexcitations with transition energies exceeding $2m_0c^2$. The process is excellently described theoretically[47,48,62,63] when the electromagnetic properties of the nuclear transitions are known. The branching with respect to the competing γ-ray emission is characterized from the pair conversion coefficient $\beta(Z,E_\gamma,EM)$ (EM represents the electromagnetic type and multipolarity of the transition). For a nuclear transition energy range of only several MeV, $\beta(Z,E_\gamma,EM)$ depends only weakly on the nuclear charge number and is of the order of 10^{-4} to 10^{-3} but decreases considerably for electromagnetic transitions of higher multipolarities other than E1 or M1. For E0-transitions the nuclear internal pair conversion competes only with electron conversion and the ratio $\eta(Z,E,EM)$ of the pair to the electron conversion is of the order of 10^{-3}. In nuclear spectroscopy of light nuclei, where the electron conversion is weak, the pair conversion coefficients are used to establish multipolarity and type of the electromagnetic nuclear transitions. A single nuclear transition of a nucleus with Z = 1 equivalently exploits for both electrons and positrons continuous, bell-shaped energy distributions as it is simply derived from phase-space arguments. But for increasing charge number the Coulomb potential modifies the wave functions near the nucleus and the Born approximation generally used in the calcualtions iń no longer valid. This Coulomb distortion results in triangularly shaped energy spectra with the centroid progressively shifted towards the highest available energy of $E_{e+,max} = E_\gamma\text{-}2m_oc^2$ for the positrons and oppositely for the electrons. For light nuclei the shape change of the spectra can be accounted for with correction functions[47)], but for the heavy elements the process had to be completely recalculated[64,65)] using relativistically correct Coulomb wave functions.

A direct experimental separation of nuclear and quasiatomic positron production for the heaviest systems is presently not possible. An indirect method[66)] is based on the observation that the nuclear source of positrons dominates in low-Z collisions ($Z_{ua} < 160$, see part (a) in Figure 1 on page 196) and on the assumption that the extrapolation to high-Z systems can be described according ref. 2) to $N_{e+} = N_{e+}(\text{quasiatomic}) + N_{e+}(\text{nuclear}) = P_{e+}N_p + C_{e+}N_\gamma$. The quasiatomic contribution $P_{e+}N_p$ is proportional to the number of scattered particles N_p and P_{e+} describes the quasiatomic positron production probability; P_{e+} depends most strongly on the combined nuclear charge $Z_{ua} = Z_1 + Z_2$ and minimum separation R_{min} during the collision. The nuclear positron production $C_{e+}N_\gamma$ is proportional to the γ-ray intensity with energies of $E_\gamma > 1$ MeV; C_{e+} is an effective internal pair conversion coefficient for heavy ion collision systems that has been found experimentally[2)] to be relatively independent of Z_{ua} (compare (a) in Figure 1 on page 196) and the details of the γ-ray production. Generally the determination of C_{e+} from low-Z systems assumes, that the global features of nuclear deexcitation spectra (Figure 14 on page 220) and the admixture of transition multipolarities are approximately the same for all deformed heavy ions. This was found in ref. 2 and in more detail[38)] is examined in the EPOS-experiments for several low- and high-Z systems.

The effective internal pair conversion coefficient C_{e+}, or equivalently an empirical admixture of nuclear transition multipolarities, is fixed by a comparison of the measured positron production in lower-Z collision systems with that calculated from the γ-ray spectra using state of the art internal pair conversion coefficients[63-65)]. A systematic

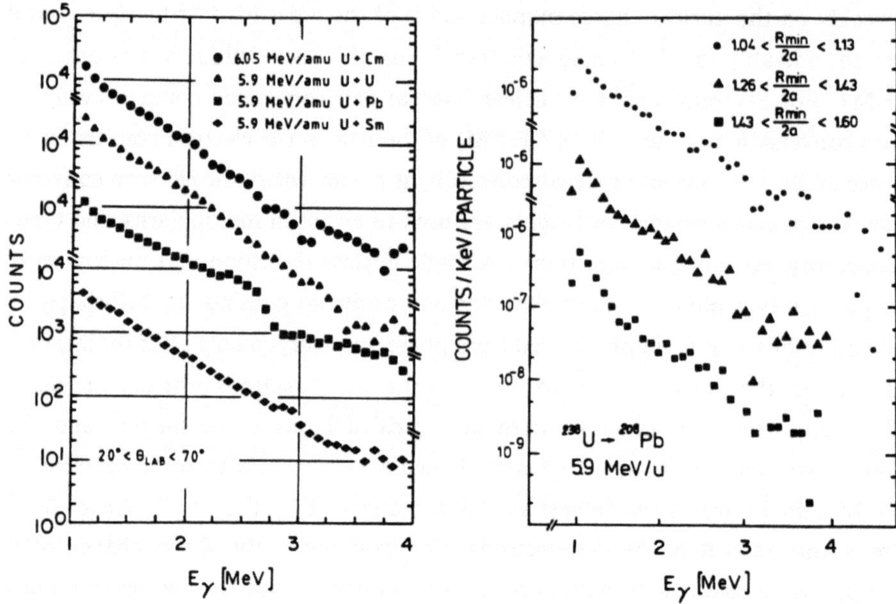

Figure 14. γ-ray spectra for various collision systems and ion-angular regions: (left) spectra of the measured γ-ray yield per particle scattered into an lab ion-angular region $20^0 < \theta_{ion} < 70^0$ for the collision systems and beam energies indicated. (right) spectra of the measured γ-ray yield per scattered particle for $^{238}U + ^{208}Pb$ collisions for three regions of distance of approach R_{min}, as indicated ($2a = R_{min}(b=0)$).

study of low-Z collision systems ($Z_{ua} \leq 160$) such as ^{238}U + $^{144,154}Sm$ and ^{238}U + ^{165}Ho shows, that the positron production from nuclear processes is consistent with a mixture of E1 and E2 transitions such that solely E1 is assumed for $E_{e+} > 800$ keV and solely E2 for $E_{e+} < 600$ keV with a smooth transition in between. Part (a) of Figure 15 on page 222 and of Figure 16 on page 223 compares the positron spectra and impact parameter dependence for the system ^{238}U + ^{154}Sm at 5.9 MeV/u corrected for all experimental efficiencies with those calculated for nuclear positron production from the simultaneously measured γ-ray spectra. The compared probabilities per scattered ion are determined absolutely and only the empirical E1/E2 admixture is adjusted.

The shapes of the γ-ray spectra of the deformed nuclei as shown in part (a) of Figure 14 appear to be essentially independent of the collision system and, moreover, are practically unchanged within the impact parameter or ion angular region accepted in the EPOS-system (see (b) in Figure 14) The extrapolation to the high-Z systems therefore is just a recalculation of the nuclear positron production probability with mean pair conversion coefficients β(Z) relevant for the individual collision system and is then scaled to the actual γ-ray intensity obtained for the window of ion scattering-angles of interest. For asymmetric systems a mean pair conversion coefficient averaged between the coefficients of the individual nuclei is incorporated to these calculations. Only ^{208}Pb exploits γ-ray spectra which distinctively deviate from the typical exponential character observed for the deformed nuclei (part (b) in Figure 14). Therefore, if ^{208}Pb is one of the collision partners a separate calculation is performed.

The global aspects of the quasiatomic positron production, as concerns the positron production probability differential either with respect to the positron energy or to the ion scattering angle, are displayed in the parts (b), (c), and (d) of the figures 15 and 16. Here the experimental probability is defined as

$$(dP_{e+}/dE_{e+})^{exp} \equiv \int_{\Delta\theta}(d^2\sigma_{e+}/(d\Omega dE_{e+}))d\Omega)/\int_{\Delta\theta}(d\sigma_R/d\Omega)d\Omega,$$

with $d\sigma_R$ the Rutherford cross section, which has to be adequately symmetrized if forward and backward collisions are not separable because of experimental or principal considerations. The ion angular integration range $\Delta\theta$ for the total spectra covers the experimental particle counter acceptance of typically $25^0 < \theta < 65^0$ defined relative to the beam direction. The impact parameter dependence then is evaluated according

$$P_{e+}^{exp}(R_{min}) \equiv \iint_{\delta\theta\Delta E}(d^2\sigma_{e+}/(d\Omega dE_{e+}))dE_{e+}\, d\Omega/\int_{\delta\theta}(d\sigma_R/d\Omega)d\Omega$$

with the positron energy range ΔE selected in this case to 100 keV $< E_{e+} <$ 1 MeV. P_{e+} is typically averaged over an ion angular bin of $\delta\theta = 10^0$ as indicated from the binsize in Figure 16 on page 223. For the asymmetric system $^{238}U + ^{208}Pb$ the positron production could be followed up to the closest near head-on collisions where both quasiatomic and nuclear contributions are largest. For the other heavier systems the data are directly plotted versus the lab scattering-angle because in these cases the distinction between the two kinematic branches corresponding to backward or forward scattering in the CM-frame is principally impossible or was not unique with the resolution accessible at that time. With global windows positioned in the $(\Sigma\theta,\Delta\theta)$-scattering plane the plotted data in both figures are confined to elastic or quasielastic scattering-events. The data have been completely corrected for the experimental response, in particular, for the Si(Li)-detector response function and the efficiency of the positron detection system, and are normalized to the scattered particle rate. The errors indicated reflect the statistical uncertainty in the originally measured data as well as the propagation of these errors through the detector line-shape deconvolution procedure. For an absolute comparison with measurements obtained in other experiments or with the theory an additional error of ~20% mostly due to uncertainties in the absolute efficiency determination has to be taken into account in the scale. The rather fair agreement of the shapes and the absolute size of both the experimental energy spectra and angular distributions with the quasiatomic theory[3] employing Coulomb trajectories together with the nuclear positron contribution is obvious.

This detailed consistency over a wide range of Z_{ua} establishes the quasiatomic dynamic mechanism as the process essentially responsable for the very heavy ion positron production. Agreement with the quasiatomic theory was also reported from other experiments including the absolute cross sections when renormalization of the calculated yield of up to ~20% is applied between theory and experiment.[17,18,46] In addition

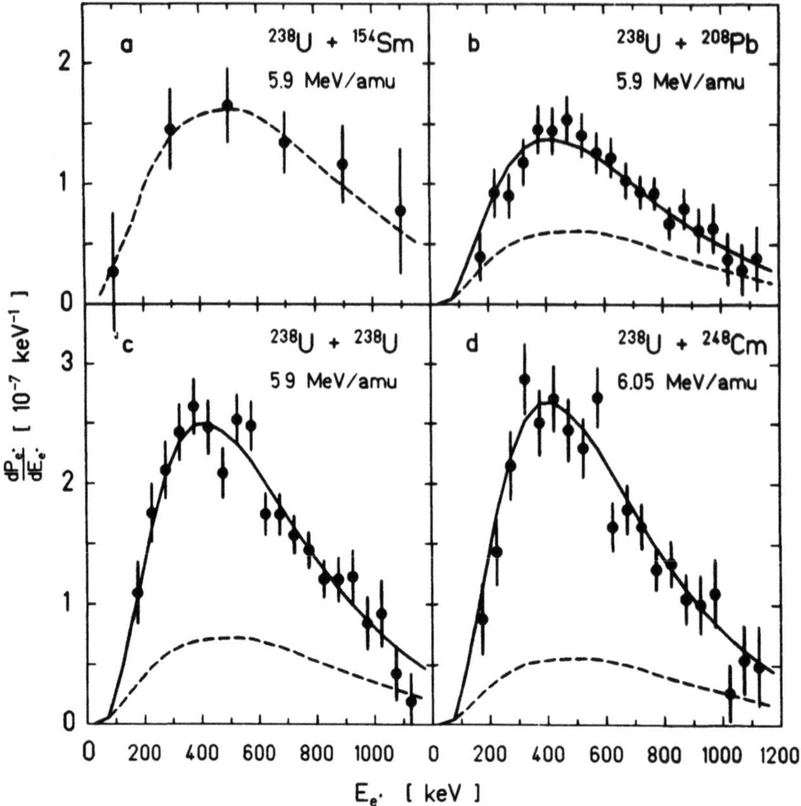

Figure 15. Kinetik energy distributions of positron production probabilities averaged over a lab ion angular region $25^0 < \theta < 65^0$, measured for four different scattering systems and projectile energies as indicated. The solid lines represent quasiatomic theory[3)] assuming elastic scattering added onto the nuclear positron production probability (dashed lines).

to the spectral distributions, also the slope of the experimental impact parameter dependence is consistent with the dynamical aspects of the quasiatomic positron emission including the closest collisions (see (b) in Figure 16 on page 223), as determined in the U + Pb system. Note, that for the closest collisions also the nuclear positron production contributes maximally and obviously is consistently described as well. The exponential slope agrees with expressions attributed[67,68)] to induced quasiatomic processes of $P_{e+}(R_{min}) \propto \exp(-\alpha\ R_{min})$ with $\alpha\ =\ 0.15\ fm^{-1}$ [33)]. For CM forward scattering (distant collisions) the slope also follows the similar formulation $P_{e+}(b) \propto \exp(-b/a)$ originally used to identify[6,16,69)] the quasiatomic positron production as a Coulomb ionization process. The dynamical aspect of the process is also well parameterized with the critical time $\hat{t}$ defined as the time interval between the two extrema of $\dot{R}/R(t)$, the logarithmic time derivative of the distance R(t) of the two nuclei (see Figure 2 on page 197), which determines the size of the quasiatomic transition amplitude. This results in the expression $P_{e+} \propto \exp\ (-2\hat{t}/(c\lambda_c))$ with $2\hat{t}\ =\ (2a/v_1)^{1/2}(\varepsilon + 1.16 + 0.45/\varepsilon)$, and $v_1\ =\ (2E_1/M_1)^{1/2}$ the projectile velocity at infinity, ε the excentricity, and $\lambda_c\ =\ \hbar/m_0c\ =\ 386$ fm the Compton wavelength of the electron.[8,29)]

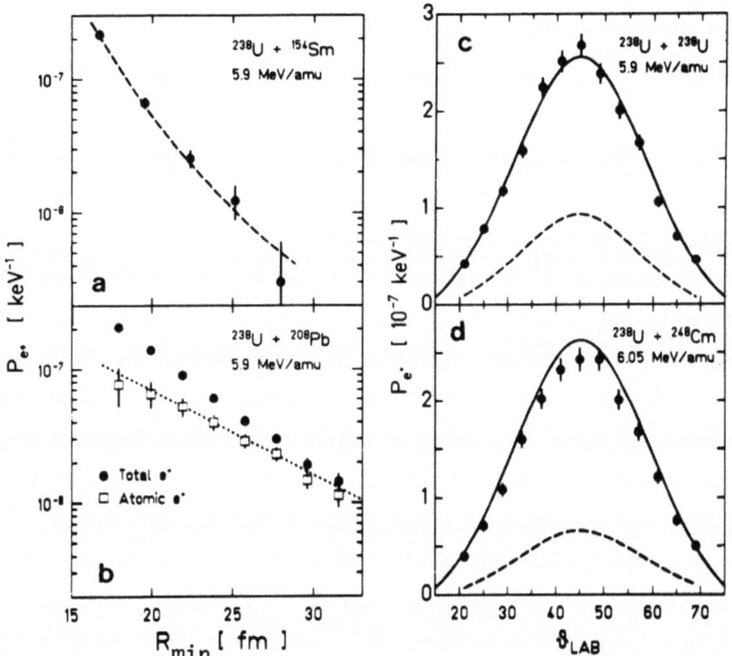

Figure 16. Positron production probabilities averaged over kinetic positron energies from 100 keV to 1 MeV versus the distance of closest approach R_{min} or the lab ion scattering-angle for the systems same as in Figure 15 on page 222. Open symbols: measured positron production probability after subtraction of the nuclear background. Dotted line: Quasiatomic theory[3] assuming Rutherford trajectories. Solid and dashed lines same as in Figure 15 on page 222.

It has been shown that the positron spectra as well as the ion angular distributions are consistently discribed in their overall shapes from the quasiatomic theory. But this may not necessarily hold for all details of the positron production of the very heavy collision systems. First deviations have already been reported independently during the Lahnstein conference[4] from both the EPOS[13] and the ORANGE[16] group. Indeed, a first indication in the EPOS-experiments for the later identified line structures was found in the impact parameter dependence of the $^{238}U + ^{248}Cm$ collision system (Figure 7 in ref. 13): when gated on positron energies $E_{e+} < 450$ keV the respective slope was found to be considerably steeper then for $E_{e+} > 500$ keV. This observation is clearly in disagreement with that what is expected from the dynamic theory which rather predicts the opposite behaviour. These early measurements have still been performed with a scintillator as the positron detector of rather poor energy resolution. When exchanged for the high resolution Si(Li)-detectors, the above observation soon led to the identification of line-like structures in the spectra of the $^{238}U + ^{238}U$ collisions near the Coulomb barrier energy.[13] Certainly the structure observed at ~ 320 keV which was reported during the Lahnstein conference[13] with a yield of 45 ± 11 events above the smooth spectrum specifies a confidence level corresponding to $\sim 4\sigma$ and is of limited statistical relevance. But the appearance of the structure in addition to the background at two regions of the lab ion scattering-angles symmetric with respect to $\Delta\theta = 0$ and the near loss of the structure at adjacent angular regions argued against only stochastic or instrumental origin.

It became clear at that time, that spontaneous positron production motivating these experiments can not at all be distinguished from the dynamic mechanism as long as only elastic scattering is concerned. It was proposed[11], however, that the spontaneous channel might be enhanced, if through nuclear attractive interaction the quasiatomic life-time can be increased beyond $\sim 10^{-20}$ sec. For supercritically bound systems a pronounced peak in the positron spectra with a width in the order of $\Delta E = 2\hbar/T = 1.32 \times 10^{-21}/T(\mathrm{sec})$ [MeV] inversely related to the sticking-time T will signal the spontaneous decay. Scattering at certain internuclear distances through an intermediate compound-like nuclear system may lead to target- and projectile-like products with ion-angular distributions rather determined from the compound-nuclear break-up than from elastic scattering. Sticking of these very heavy nuclei, however, appears even less probable for deep inelastic collisions[70] although such conclusion certainly depends on a detailed study of all reaction parameters. With beam energies then selected to differ only marginally from the Coulomb-barrier energy, the resulting deviations from the elastic scattering-pattern might actually be small, but eventually can still be detected in a measurement of the collision kinematics. The incorporation of the kinematic coincidence technique in order to determine the complete scattering-kinematics was actually a special aspect of the EPOS-experiments already during the first experiments performed. The selection of only binary scattering-events through the two-dimensional windows in the $(\Sigma\theta,\Delta\theta)$-coordinate plane globally restricts the positron data to elastic and quasielastic events. But indeed, the width of the two-dimensional distribution for the high-Z collisions at beam energies of 5.9 MeV/u close to the Coulomb barrier exceeds the typical resolution obtained from scattering of lower-Z ions. Besides the trivial contribution from less homogenous targets, this broadening gives room for a slight inelasticity or additionally broadened ionic charge-state distributions incorporated in the scattering process. Following the ideas of sticking it appears natural to now optimally use the excellent kinematic resolution of the set-up and to scan ion angular regions close to the elastic domain for interesting events. In view of the quasiatomic character of the phenomena searched for it appears even more natural also to concentrate the experimental efforts to collision systems of maximum combined charge Z_{ua}.

Figure 17 on page 225 displays two spectra of a proceding experiment[14] with ^{238}U-beam at 6.05 MeV/u on ^{248}Cm-targets. The positron spectra shown here and in most cases later on, are not corrected for the various features of the experimental efficiency in order to safely avoid any distortion of the spectra due to deconvolution procedures. Although a normalization of the spectra to the ion scattering-yield in the selected (θ_1,θ_2)-window is always available, the spectra are typically displayed with the

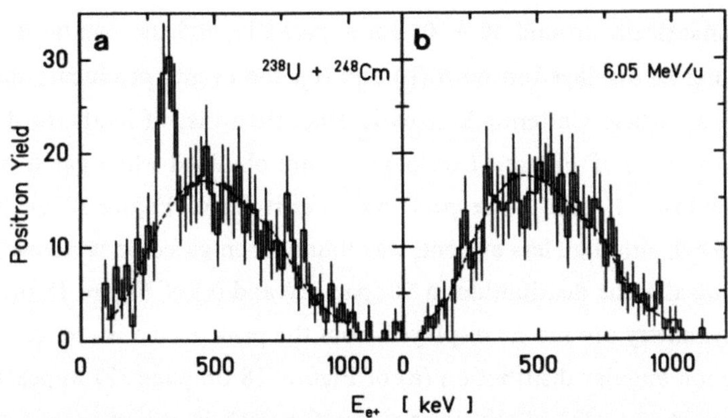

Figure 17. Positron energy spectra as observed for $^{238}U + ^{248}Cm$ collisions at an incident beam energy of 6.05 MeV/u. The ion angular regions corresponding to (a) and (b) are neighbouring regions preferentially overlapping with elastic scattering angles of $50^0 < \theta_{CM} < 80^0$ in the case of (a) and for (b) $100^0 < \theta_{CM} < 130^0$ together with slightly inelastic events. Dashed lines: dynamic theory assuming pure elastic scattering added onto the calculated nuclear pair conversion background, convoluted with the positron detection efficiency and normalized to the spectral region outside the peak.

original yield in order to preserve the access to the original counting-statistics. The solid lines in Figure 17 on page 225 are obtained from calculations of the nuclear and the dynamic positron production both modified by the experimental efficiency. The spectra were obtained for closely neighbouring ion angular regions (actually identical $\Delta\theta$ window, see below) and the line-producing spectrum corresponds to scattering events again slightly outside the angular region where elastic scattering dominates. The one-dimensional lab ion scattering-angle window quoted in the figure caption corresponds to the area actually selected in the two-dimensional $(\Sigma\theta,\Delta\theta)$-coordinate frame. The ion-angular distribution (a) of Figure 18 on page 227 already exhibits structure when gated with the prominent line in the energy spectrum (a) of Figure 17. The selection of only one out of the two lab ion scattering-angles (θ_1,θ_2) is apparently sufficient to already raise the interesting events slightly over the background which typically exhibits a smooth shape for the unbiased angular distribution. But in order to obtain the prominant line spectrum (a) of Figure 18 on page 227 θ_2 as the other scattering-angle coordinate has additionally to be optimized. This final selection may well depend on the actual resolution of the individual experiment and is equivalent to a two-dimensional gate in the $(\Sigma\theta,\Delta\theta)$-coordinate frame. Indeed, part (a) and (b) of Figure 17 are obtained from the same window in $\Delta\theta$ and correspond to adjacent cuts in $\Sigma\theta$. More precise information on this ion angular behaviour can be obtained from the different plots of the two-dimensional ion angular distribution shown in (a), (b), and (c) of Figure 18 on page 227. To a certain degree the ratio of peak positrons to dynamic positrons (see (a)) is distributed symmetrically around $\Delta\theta = 0$. But apparently, and this can be seen from the distribution along $\Sigma\theta$ displayed in both (b) and (c), the events producing the peak are associated with a particle scattering-kinematic other than that of Rutherford scattering. The spectra (a) and (b) of Figure 17 on page 225 are obtained when gated on the two-dimensional windows '1' and '2', respectively in part (c) of Figure 18 on page 227; a structure for $\Delta\theta > 0$, although less evident, was found when gated on window '3'. Clearly, when considering also the distribution in $\Sigma\theta$ (part (b) and (c) of Figure 18 on page 227) the degree of symmetry around $\Delta\theta$ displayed from the peak events is reduced. One word of caution, the ion angular distribution (a) of Figure 18 on page 227 appears very suggestive. But according to (b) of the same figure the peak in the spectra may also well exist all over the experimental range of $\Delta\theta$. The angular distribution (a) in this case is then less defined from a distinct individual ion angular distribution for the peak production along θ, but rather reflects a gain in the peak-to-background ratio in a region where the two-dimensional ion angular distribution for the peak events in the experiment sufficiently deviates from the elastic scattering pattern with dynamically produced positrons dominanting.

To date, the details of the distribution of these favourable regions in the $(\Delta\theta,\Sigma\theta)$-ion scattering plane are still not fully understood. As discussed above, the proposed sticking of the nuclei may well allow for ion angular distributions in variance to elastic scattering but the only poor reproduction of the inherent symmetry of the instrumental set-up remains an open question. Moreover, proceding measurements discussed below again revealed asymmetric distributions (Figure 19 on page 228) in the $(\Delta\theta,\Sigma\theta)$-plane for the peak producing events which all essentially appear to be slightly clockwise rotated in this coordinate frame as compared to the elastic pattern. The latter is straightened out in this plot for reasons of better comparison. Also a close inspection of the $^{238}U + ^{238}U$-data introduced in ref. 13 revealed a certain structure which indeed appeared to be enhanced along such a rotated kinematic pattern (Figure 20 on page 230). (It might be important to realize, that the indicated structures in addition to the rather smooth spectrum in (b) could well be reproduced from a series of three lines at energies of $E_{e+} \sim 400$ keV, ~ 550 keV, and ~ 690 keV with the Doppler-effected instrumental line width properly increasing with the positron energy). On the other hand, as explained in the preceding section II, ionic charge state distributions for the peak-producing events different from the velocity-dependent ionic equilibrium charge together with the action of the magnetic field may well account for these deviations from the inherent symmetry. In fact, already a constant velocity-independent ionic charge of $\sim 60^+$ attributed to both scattered ions when associated with the peak producing events, is enough to account for a clockwise rotation of the kinematic pattern relative to the elastic scattering-distribution. For the latter the ions are charged according the velocity-dependent equilibrium charge state distribution. The calculated[39)] deviations from the elastic pattern with a value $\Delta(\Sigma\theta) \sim -1^0$ at $\Delta\theta \sim 20^0$ qualitatively reproduce what has been observed. It appears an

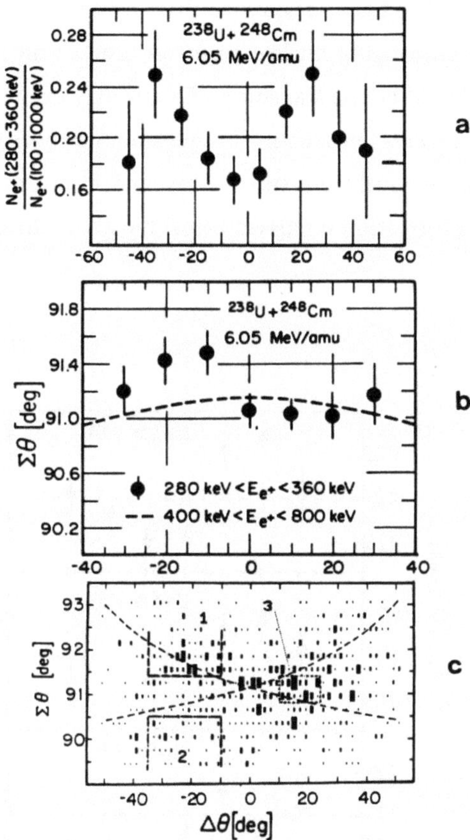

Figure 18. Two-dimensional ion angular correlation selected for the positron peak energy as obtained from ^{238}U + ^{248}Cm collisions at a beam energy of 6.05 MeV/u: (a) Ratio of the yield of positrons from energies 280 keV < E_{e+} < 360 keV to that of the total spectrum versus the composed Δθ ion scattering-coordinate. (b) The centroid of the positron ion angular distribution parallel to the ΣΘ-axis is plotted as a function of Δθ. The circles correspond to a coincidence with positrons of 280 keV < E_{e+} < 360 keV and the dashed line to positrons of 400 keV < E_{e+} < 800 keV. (Both distributions are equivalently corrected for obvious nonlinearities in the read-out) (c)Scatter plot of the positron events for positron kinetic energies of 280 keV < E_{e+} < 360 keV.

interesting coincidence, that high ionic charge states are actually expected[68,71] for closest collisions where also sticking may occur and such charge states, indeed, can survive for the presently used target thicknesses[54,55,56] or alternatively could be produced from a separation outside the target (compare section II and ref. 52). However, if these observations can eventually be attributed to some exclusive compound-elastic scattering-process causally associated with the appearance of the positron lines remains to be solved.

So far it is obvious from the experimental and theoretical work that positron production based on Rutherford trajectories does not allow for pronounced peak structures and it was tried to connect the line at ~320 keV obtained with the $^{238}U + ^{248}Cm$ collisions (Figure 17 on page 225) with the long-sought process of spontaneous positron production.[12] On the other hand, deviations from a typically smooth character could also stem from the accompanying nuclear positron production and an obvious source for line-like structures could be the transition of a strongly excited isolated nuclear state in one of the separated reaction products. This question has been addressed by a detailed inspection of the simultaneously measured γ-ray and electron conversion spectra.[14] However, the body of information discussed below mostly excludes such an explanation.

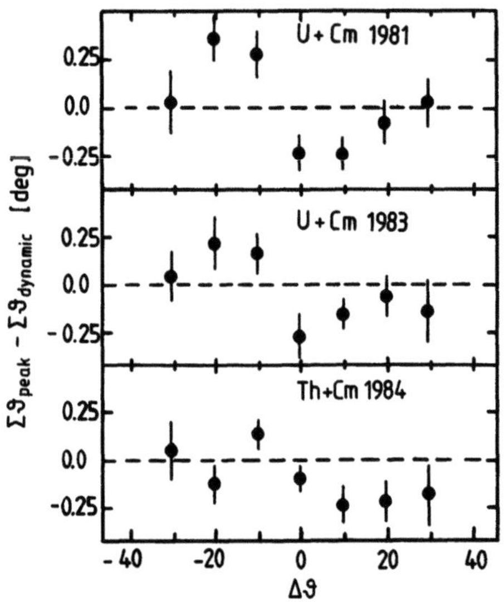

Figure 19. Two-dimensional ion angular correlations of events associated with positron energies inside the peak region of 280 keV < E_{e^+} < 360 keV for the ^{238}Th + ^{248}Cm collisions (closed circles). The centroid of the positron ion angular distribution parallel to the $\Sigma\theta$-axis is plotted relativ to the unbiased distribution (dashed line) versus $\Sigma\theta$ and $\Delta\theta$.

The two mechanisms that may produce peak structures are normal internal pair conversion and internal pair conversion involving the capture of the electron into an empty inner shell orbital. Irrespective the actual number of vacancies, the latter process is by one order of magnitude less intense than the competing normal version. The simple inspection of the peak width already substantially eliminates normal pair conversion as the source of the peak - but not necessarily the monoenergetic version of the pair decay. The observed width of ~80 keV of the line is much too narrow to fit the significantly broader pair conversion spectrum which for these heavy nuclei and transition energies generally displays an intrinsic width of already ~150 keV and no considerable reduction through an unfavourable shape of the efficiency is in sight. A comparison of the positron line with the internal pair conversion spectrum is displayed in part (a) of Figure 21 on page 231.

No corresponding γ-ray or electron conversion line could be detected either in the γ-ray or electron spectra at the energetic positions determined from the positron peak energy. These spectra all are measured event-by-event in coincidence with the scattered ions simultaneously with the positrons, and are selected for the identical angular regions in the $(\Delta\theta,\Sigma\theta)$-scattering plane found advantageous for the identification of the positron structure. The comparison of the absolute yields between the γ-ray or electron spectra and the positron spectra therefore is simply based on a normalization to the scattered-ion event rate and does not suffer from instrumental uncertainties as it is the case when different experiments have to be compared. Figure 22 on page 232 compares the observed γ-ray spectrum and the γ-ray line profiles as expected for an isolated nuclear E1 and E2 transition. The intensities of the γ-ray lines are calculated[72] for both normal and monoenergetic pair conversion from the yield of the positron line according to the equations $P_\gamma = P_{e+}/\beta$ and $P_\gamma = (2/x)P_{e+}/\alpha_{e+K}$, respectively. Here β and α_{e+K} are the internal pair conversion coefficients[73] for the normal and the monoenergetic version, respectively, and for the latter process capture into the K-shell with x the number of vacancies $(0 \leq x \leq 2)$ is assumed to be most probable. The calculation is performed for the U-nucleus but as far as the normal conversion process concerns, the result is essentially identical for the Cm-nucleus because of the weak dependence of $\beta(Z)$ on the nuclear charge. The calculated line profile includes the effects of Doppler shifts, the resolution, and the efficiency of the NaI-counters. The double-humped form results from the different Doppler shifts obtained for forward and backward-scattered uranium detected either in the upper or lower heavy ion counter. Higher multipolarities than E2 require even more γ-ray intensity because $\beta(Z,EM)$ decreases with increasing multipolarity. The difference in the line position between (a) and (b) results from both the additional resource of available transition energy when the binding of the electron in the K-shell is included and the unlike transition energies derived from the different adjustment of either a monoenergetic or a triangularly shaped spectrum to the positron peak. The calculations were done under the assumption that one K-shell vacancy $(x=1)$ is still present at the time of the nuclear transition. Less K-vacancies as well as capture into other shells instead of the 1s-shell result in even larger γ-ray line intensities necessary to explain the positron intensity. The absence of large enough structure in the measured γ-ray spectrum virtually rules out a nuclear origin for the ~320 keV peak involving nuclear transitions of multipolarity E1 and higher.

For E0-transitions, on the other hand, the conversion electron spectra have to be inspected and the comparison is made in Figure 23 on page 233 for the calculated line intensities now corrected for the efficiency and the electron spectrum shown with the original yield. The measurement has been performed using the Doppler shift reduction

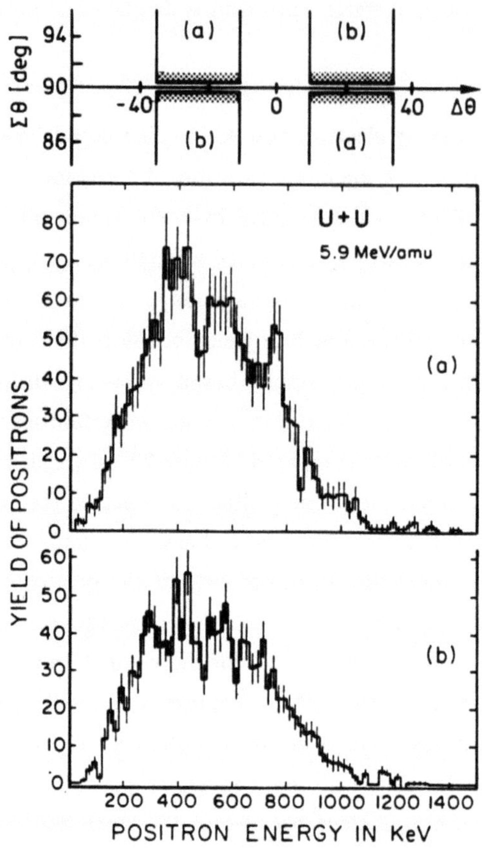

Figure 20. Measured yield of positrons versus the positron kinetic energy of the $^{238}U + ^{238}U$ collision system at a beam energy of 5.9 MeV/u. (a) Positrons in coincidence with particles scattered into the angular regions 'a' indicated in the schematic $(\Delta\theta, \Sigma\theta)$-frame on top, (b) positrons in coincidence with regions 'b'.

technique explained in section II and was checked for consistency with the observation of the 910 keV E0-transition of ^{238}U in a Coulomb ionization experiment (Figure 8 on page 210); Gaussian lines are assumed to approximately describe the remaining Doppler-broadening. For E0-transitions the number of associated electrons is given[65,72] for normal pair conversion by $P_{e-} = (2x-1)(1/\eta)P_{e+}$ and for energetic conversion by $P_{e-} = -(1-2/x)(1/\zeta)P_{e+}$ with η and ζ the respective ratio of the pair conversion to the electron conversion process. The intensity of the calculated lines clearly exceeds the yield of the smooth electron spectrum. Because of the common but opposite dependence of the monoenergetic pair conversion and the electron conversion on the degree of the ionization of the K-shell, monoenergetic pair conversion increases when at the same time K-conversion is reduced. The calculation is performed for the K-vacancy lifetime τ_K exceeding the lifetime τ_{E0} of a fictive E0-transition. Shorter vacancy lifetimes τ_K result in even larger intensities for the conversion lines. On the other hand, the K-electron conversion line will completely disappear for totally empty K-shells $(x = 2)$ during the time the nuclear transition takes place.

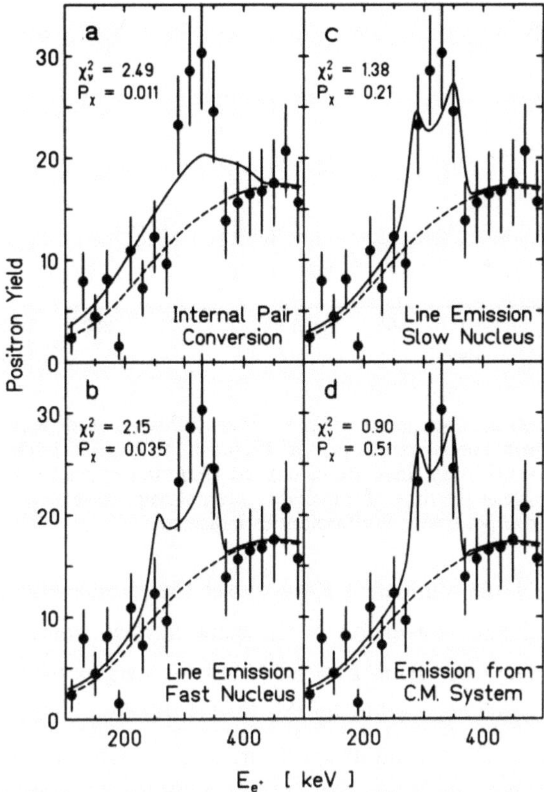

Figure 21. Maximum likelihood fits to the selected positron spectrum of the $^{238}U + ^{248}Cm$ collision system of part (a) in Figure 17 on page 225 based on different possible origins for the peak as indicated. The smooth dashed curve represents dynamic positron production for elastic scattering and is added onto the nuclear background. The fit involves varying the peak intensity, its position, and the intensity of the continous distribution and is performed for the entire spectrum with χ^2 stated for a 200 keV region around the peak.

The γ-ray and electron measurements virtually exclude nuclear processes to provide an explanation for the peak of ~320 keV found in the $^{238}U + ^{248}Cm$ system. But the peculiar dependence of monoenergetic pair and K-electron conversion on the number of K-vacancies opens a small possibility that a singular nuclear transition may still produce a well observed positron structure and, on the same time, escape from the detection in both γ-ray and electron conversion spectra. However, a K-shell vacancy in these heavy ions typically lives $\sim 10^{-17}$sec, and it is thus difficult to imagine that almost two K-vacancies ($x > 1.85$ is consistent with the data) survive for times comparable to lifetimes of 10^{-12} to 10^{-14} sec typical for E0-transitions. Note again, enhanced vacancy production is indeed expected for closest heavy ion collisions to provide initially large numbers of vacancies and the peculiar behaviour of the ion angular event correlation attributed to the action of the magnetic field may, in addition, even support the existence of long-living high charge states. On the other hand, it remains very unprobable that the K-electron conversion from an E0-transition is not at all detected in the electron spectra, although it should be vitally present for some less extreme atomic configurations.

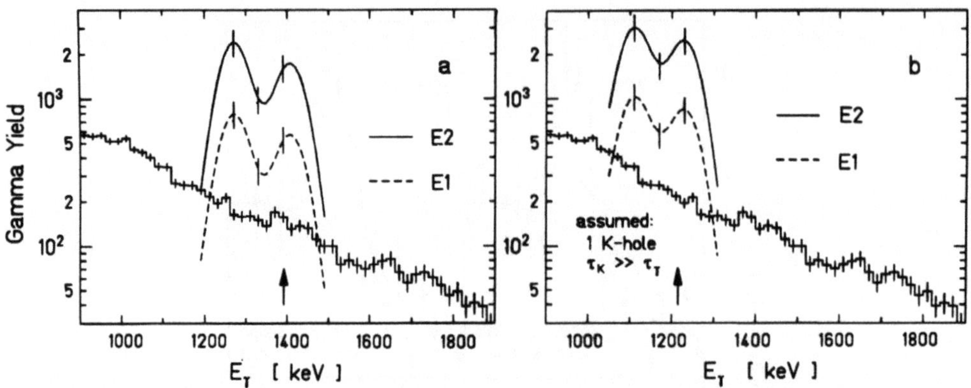

Figure 22. γ-ray spectrum obtained in $^{238}U + ^{248}Cm$ collisions at identical conditions as for the positron line spectrum (a) in Figure 17 on page 225. The intensities of the superimposed γ-ray lines are calculated assuming normal (a) or monoenergetic (one K-hole at the time of transition) internal pair conversion (b) as the origin for the positron peak with multipolarities indicated.

Moreover, even for a large number of K-vacancies the monoenergetic pair conversion competes with normal pair conversion of the same nuclear transition. The continous spectrum of normal pair conversion now is shifted to energies below the positron line of the monoenergetic pair conversion by the K-shell binding energy. For the extreme case of $x = 2$ the yield of the continous spectrum is maximally suppressed and the ratio of ζ/η amounts to ~ 6.7 in the case of the U-ion. To date, no such contribution in the positron spectra has been identified.

Interesting information on the origin of the line emission could be drawn from the shape of the positron lines itself.[15] In particular, the width provides a direct measure of the emitter system velocity as long as positron emission is isotropic in the CM-frame and the intrinsic width can be neglected (compare instrumental section and Figure 4 on page 203 and Figure 6 on page 208). It was already concluded from part (a) of Figure 21 on page 231, that the broad pair conversion spectrum is inconsistent with the observed width of the positron line. Part (b) and (c) of Figure 21 on page 231 now compares the monoenergetic version of the pair conversion with the same data. The calculated profiles contain all details of the instrumental response and monoenergetic emission is assumed isotropically from either the fast (b) or the slow nucleus (c). Because these data stem from ion scattering angles outside the 45°-region the velocities of the separated ions differ considerably. Whereas emission from the slow ion is still consistent with the data, emission from the fast ion (most probably Cm-like when considering the selection of the two-dimensional $(\Delta\theta, \Sigma\theta)$-gate of part (c) of Figure 18 on page 227 in terms of elastic scattering) provides a very poor fit to the data. Additional information can be obtained from a second independent measurement of the $^{238}U + ^{248}Cm$ collision system. It should be noted, that this second experiment was performed in a modified set-up. In particular, the spiral baffle was exchanged to the presently used construction with a reduced number ($n = 6$) but enlarged area of the blades, in order to improve both the absolute size

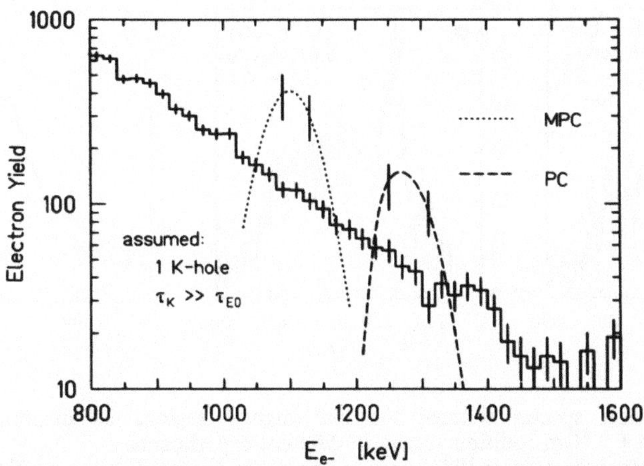

Figure 23. Electron spectrum as obtained in $^{238}U + ^{248}Cm$ collisions at identical conditions as for the positron line spectrum (a) in Figure 17 on page 225. The electron structure associated with the positron peak assuming E0-transitions in the U-nucleus is shown for normal (dashed line) and monoenergetic (dotted line) pair conversion. The calculated line intensities are modified for the efficiency of the electron detection.

of the efficiency and the sensitivity at positron energies below ~200 keV. The latter now leads to a different shape of the smooth background in the positron spectra when displayed with the original experimental yield. Very similar to the previous case a narrow line is identified again in this measurement and the two line spectra of the two different $^{238}U + ^{248}Cm$ runs are compared in Figure 24 on page 234. The two-dimensional ion angular distribution of the peak events labeled with '1983' in Figure 19 on page 228 appears to be rather similar as compared to the old experiment, but in this measurement the peak is most pronouced at an ion angular region around 45^0 ($\Delta\theta = 0$). At this angular region both separated nuclei move with comparable velocity and the fit to the line with a Doppler profile assuming monoenergetic pair conversion appears to be inconsistent with the data (see part (c) in Figure 24 on page 234).

The complementary information on the line width measured at two different ion angular regions in these two experiments is summarized in Figure 25 on page 234. There the two experimental values of the width (among other measurements, see below) as a function of the ion lab scattering-angle θ_{ion} are shown together with the calculated line width when emission from the fast or slow nucleus or the CM-system is assumed. The emitter velocity resulting from the two practically equal values of the line width is too low to agree well with the kinematic scattering-angle dependent velocity of either of the two separated nuclei. But the values rather fit to a velocity close to the value of the CM-velocity. The well agreement of the Doppler profile assuming CM-velocity is displayed in parts (d) and (c) of Figure 21 on page 231. Clearly, nuclear internal pair conversion in all variants originating from the separated target- or projectile-like nuclei is excluded to be the source of the narrow positron line. A common velocity with values

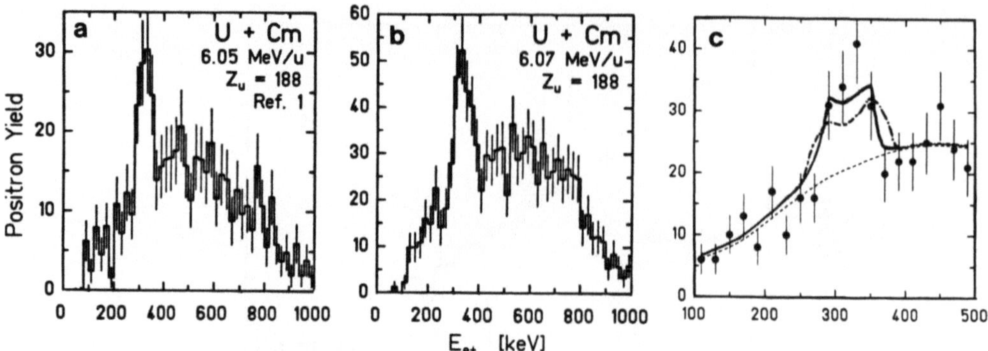

Figure 24. Positron spectra selected for ion angular regions as measured with the $^{238}U+^{248}Cm$ collision system in different experiments.
(a) Measurements in 1981, same as part (a) in Figure 17 on page 225.
(b) Measurement in 1983 with improved efficiency, gated with a lab ion angular window around 45° (compare (b) of Figure 19 on page 228). (c) same as (b) together with maximum likelihood fits, carried out similarly as for Figure 21 on page 231 assuming an intrinsically narrow positron line emitted from the CM (solid line) and the separated nuclei (dotted line).

deduced from the experiments close to the velocity of the CM rather points for a quasiatomic origin. In this respect, also the apparently limited choice of projectile energies so far associated with the identification of the structure may be considered. Indeed it was found that the appearance of the structure in the $^{238}U+^{248}Cm$ collision is confined to a narrow beam energy interval close to 6 MeV/u. More quantitative measurements of the excitation function are limited by the extent of target inhomogenities and deteriorations during the irradiation with the heavy ion beam, which lead to differences of the projectile's energy-loss on the order of 0.1 MeV/u. This experimentally introduced energy loss straggling spoils the initial projectile-energy accuracy which, on the other hand, is needed to conclusively detect an eventually underlying resonance- or threshold-like behaviour. The efforts spent in this experiment to quantitatively control the target conditions have been described above in the instrumental section and in ref. 52.

All these observations found in connection with the detection of the narrow positron line in $^{238}U+^{248}Cm$ seem to suggest that quasiatomic spontanous positron emission could be identified with the line phenomena.[12,14,18] Moreover, the line energy favourably coincides with the excess of the electronic $1s\sigma$ binding-energy over $2m_0c^2$ as actually calculated for a reasonable configuration of the U+Cm quasiatom. However, following this suggestion the intrinsic line-broadening, as simply inferred from the uncertainty relation because of the only temporary existence of the quasiatom, has to be limited at least to the value of the experimental line width of ~70 keV, or even to a value of less than ~40 keV. The latter value is determined from the experiments when the experimental Doppler broadening is additionally taken into account. Consequently, a rather long-living quasiatom providing for the supercritical charge ($Z_{ua} > Z_{cr}$) has to

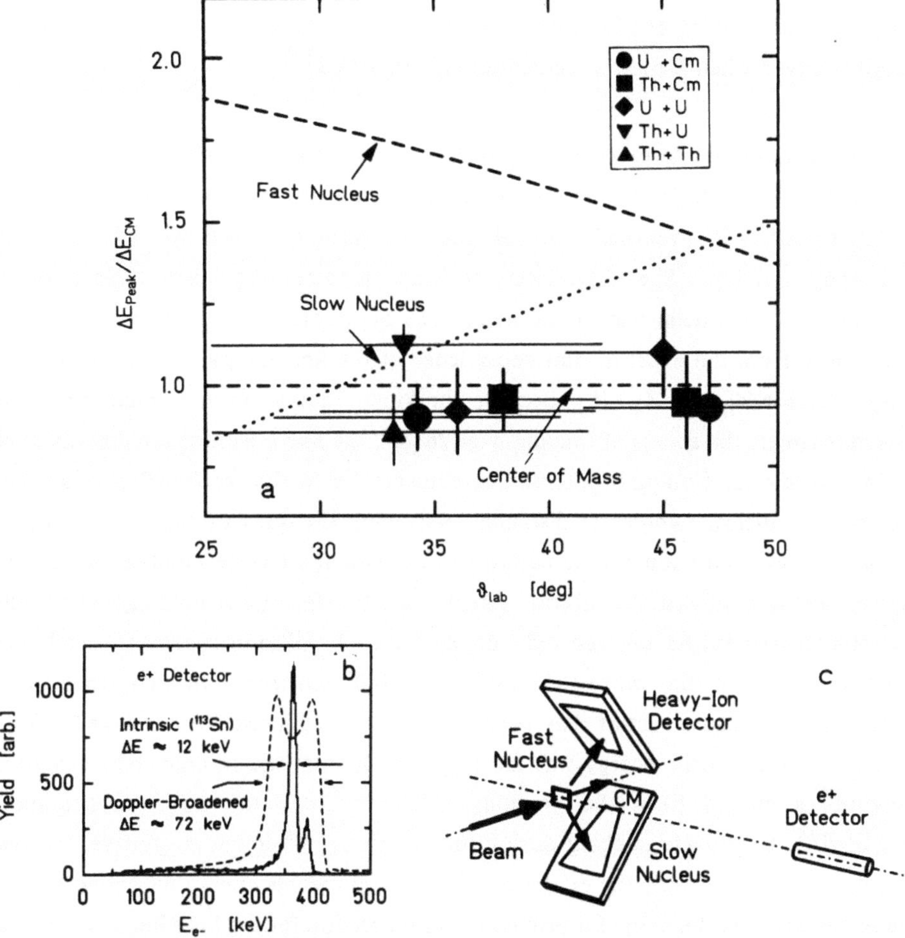

Figure 25. (a) Doppler-broadening of positron lines as expected for target, projectile, and CM-emission for elastic scattering of ^{238}U on ^{248}Cm plotted as a function of the laboratory scattering angle. Broadening is expressed as a fraction of the calculated width for CM-emission. The calculations are compared with the measured positron line-width of the systems $^{238}U+^{248}Cm$, $^{232}Th+^{248}Cm$, $^{238}U+^{238}U$, $^{232}Th+^{238}U$ and $^{232}Th+^{232}Th$, as obtained from the spectra in Figure 26 on page 237.
(b) Comparison of the positron detector resolution (measured with a ^{113}Sn-source) with the Doppler broadening for positrons emitted from the CM-system ($\beta = 0.05$). The experimental Doppler-broadened line shape has been verified with conversion electrons from the compound reaction $^{12}C(^{118}Sn,4n)^{126}Ba^*$.
(c) Schematic view of the positron and particle detection geometry.

be formed during the collisions with its decomposition delayed on the order of $\sim 10^{-19}$ sec through internuclear interaction of a di-nuclear system.[11,12,14,74] Although delay times of up to $\sim 10^{-21}$ sec have recently been deduced in dissipative heavy ion collisions from δ-electron measurements[76,77], no experimental evidence for such long-living di-nuclear systems is presently available for the regime of the very heavy collision partners. On the other hand, irrespective of supporting experimental information, quasiatomic theory to a certain degree can consistently explain both intensities and width of the narrow structures observed in $^{238}U+^{248}Cm$ collisions, if a mean nuclear delay time of $\sim 10^{-19}$ sec is either schematically introduced[11,12,78] or more profoundly derived when S-matrix techniques are employed[79].

Theoretical efforts to explore the possibility of such giant nuclear systems revealed models[74,75] of di-nuclear compounds weakly bound in rotational states within small potential pockets[80] eventually formed when the surfaces of the two deformed nuclei sufficiently overlap. The discussion of such models was also subject of this conference.[81] A crucial test of the hypothesis of spontaneous positron emission can be obtained from the experimental dependence of the line energy on the united system charge number $Z_{ua} = Z_1 + Z_2$. Because the kinetic energy of spontaneously emitted positrons equals the access of binding over $2m_0c^2$, the line positions are directly related to the quasiatomic Coulomb potential or equivalently to the combined nuclear charge and the internuclear two-center distance. Assuming the quasiatomic model is correct, the line energy then should systematically scale with the charge number. A variety of supercritical and subcritial collision systems has therefore been investigated at actual projectile energy values selected individually for each collision system to result in an equal separation of the two nuclear surfaces at 180^0 scattering. In all systems investigated a pronounced structure has been identified[15] (Figure 26 on page 237). The scaling of the projectile energy was motivated both from the resonance-like excitation function as ensued from the previous $^{238}U + ^{248}Cm$ data and from the models suggested[74] to provide sticking during the collisions. Although suggestive, the result does not necessarily imply the existence of a sharp projectile energy dependence for the line phenomena; in the case of a production cross section for the lines being independent from the projectile energy the above scaling is meaningless.

Similarly as described above in detail together with the $^{238}U + ^{248}Cm$ system, favourable regions in the $(\Delta\theta,\Sigma\theta)$-ion scattering plane are again selected to optimize the line intensity with respect to the smooth background spectrum. For the $^{232}Th + ^{248}Cm$ collision system the existence of favourable regions are obvious from part (c) of Figure 19 on page 228. Again, the reasons for this behaviour are not completely understood, but the proposed weak sticking of the nuclei for the interesting events may well allow for the conservation of a target- and projectile-like character of the scattered ions and at the same time exhibit ion angular distributions only slightly different from elastic scattering. Irrespective of the individual $(\Delta\theta,\Sigma\theta)$-gates selected, it was found in general that the pronounced positron line production is related to binary scattering into target- and projectile-like events of apparently only slight inelasticity of < 20 MeV.

Also the electron and γ-ray spectra again all measured simultaneously with the positrons have been carefully inspected for eventual nuclear deexcitations of corresponding energy and intensity. Already on this basis with no electron or γ-lines found, nuclear internal pair conversion can virtually be ruled out for all measured systems to

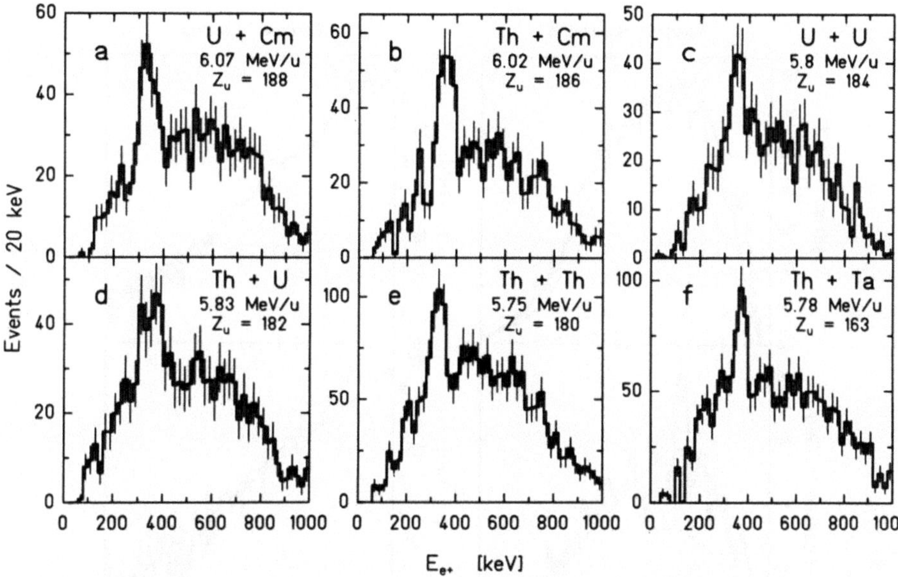

Figure 26. Positron energy spectra for six subcritical and supercritical collision systems and bombarding energies as indicated. Kinematic constraints have been chosen as discussed in the text and in ref. 14 and 15.

explain the narrow positron lines. Only the ^{232}Th$+^{181}$Ta system exhibits an enhanced γ-ray intensity at the corresponding γ-ray energy. Neither the shape nor the intensity, however, of the observed positron line can be accounted for from this γ-ray bump, as can be seen from a careful comparison of the measured and calculated positron spectra (Figure 27 on page 238). The nuclear contributon are composed from a weak dynamic[82] (~20%) and the internal pair conversion positron spectrum. The latter is dominant for this low-Z system and is derived from the measured γ-ray spectrum similarly as described above. The continuous part of the γ-ray distribution is assumed to be of mixed E2/E1 character same as above. The structure at ~1.3 MeV in the γ-ray spectrum is tentatively attributed to a transfer produced target (Ta)-like nucleus in accordance with a Doppler shift analysis of the γ-ray structure and is taken to be of E1-type leading to maximum positron intensity.

All the observed positron lines exhibit a width of ~70 keV. Individually normalized to the experimental width at CM-velocity the values are plotted in Figure 25 on page 235 as a function of the projectile laboratory scattering-angle. Compared with the values calculated for emission from the separated either fast or slow nuclei, the experimental values typically appear to be smaller and rather independent of the scattering-angle. Indeed, only limited information on the ion angular distribution of the line width is presently available for the individual systems, but the whole body of these data collectively argues against line emission from a separated nucleus. Same as it already was concluded from the line width analysis based only on the data of the previously measured ^{238}U$+^{248}$Cm system, these data as a whole point to an origin exhibiting a common

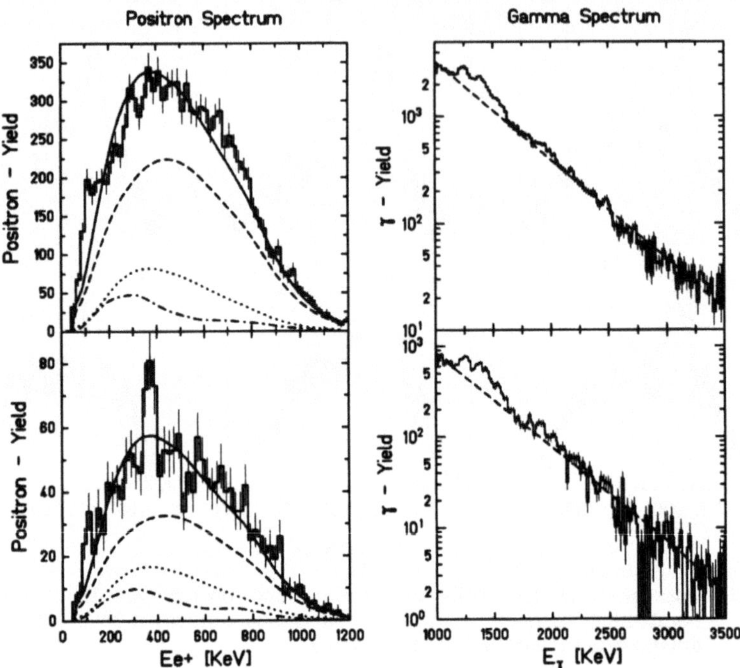

Figure 27. Positron und γ-ray spectra as measured in $^{232}Th + ^{181}Ta$ collisions at bombarding energies of 5.75 MeV/u and 5.80 MeV/u integrated over all lab ion scattering-angles from $20^0 < \theta_{Ion} < 70^0$ (upper row) gated with kinematic constraints as described in the text and in ref. 34. The solid curves in the positron spectra are calculations composed from dynamic positron production[82] (dotted) and nuclear pair conversion as calculated from the corresponding γ-ray spectrum assuming different multipolarities for the smooth part of the γ-ray spectrum (E2/E1 mixture) (dashed lines) and for the structures (E1) (dashed dotted).

velocity of a value close to or slightly less than the CM-velocity. Note that a substantial anisotropy in the intrinsic emission characteristic may weaken these arguments and also the direction of the velocity not necesserily has to coincide with the beam direction.

Clearly, however, the positron energy spectra of Figure 26 on page 237 do not show a systematic decrease of the individual line energies when going from the heavier to the lighter systems. On the contrary, with the line energies essentially concentrated around a mean energy of ~340 keV and individual deviations only in the order of or less than the experimental line width, they rather display (Figure 28 on page 239) a remarkable constancy with Z_{ua}. Such behaviour can hardly be reproduced within the spontaneous positron production scenario. A strong positron-energy dependence proportional to a high power of Z_{ua} in the order of 20 is expected if the nuclear and ionic configuration is kept constant throughout the various systems.[12] A modification of the quasiatomic Coulomb potential in order to compensate its trivial charge number dependence is certainly possible. Such modifications contain the variation of the compound nuclear shapes ranging from spherical to just-touching configurations, the decrease of the compound nuclear radius as a result from droplet model calculations with an appropriately

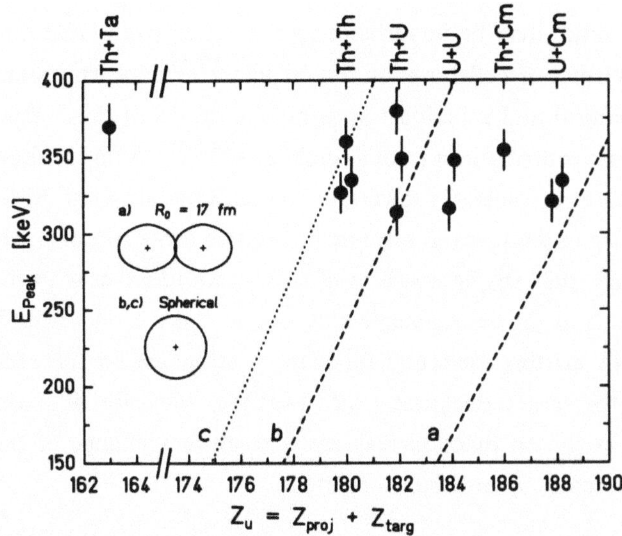

Figure 28. Mean energies of the positron peaks of the spectra shown in Figure 26 on page 237 and of proceding measurements as a function of Z_{ua}. Calculations[12] are peak energies for spontaneous positron emission from static nuclear complexes with either (a) deformed shapes consisting of the two nuclear centers separated by 17 fm in a head-on collision of major axis or (b) spherical nuclear shapes and normal density, both with fifty-fold ionization. Bare nuclei are presented for the spherical shape (c).

modified compressibility which is still consistent with the body of the present knowledge throughout the stable elements,[83] and the variation of the quasiatomic charge including the completely stripped situation with no electronic shielding. With all these changes incorporated, the Z_{ua}-dependence to a certain extent can be adjusted. But it is difficult to imagine why for the individual systems various configurations partly extreme to each other should just favourably appear in order to result in a near constancy of the $1s\sigma$ binding energy. Thus, although spontaneous positron emission in heavy-ion collisions as an explanation may still fit for the heaviest systems[75] as $^{238}U + ^{248}Cm$, this process appears to be inappropriate to generally interprete these pronounced line structures. Moreover, these lines all appear with a comparable cross section of ~ 10 μb/sr both in the critical and subcritical collision systems, which again is difficult to explain with the quasiatomic theory. However, it should be noted, that the quoted cross section is related to the size of the Rutherford cross section associated with the selected $\Delta\theta$-window irrespective of the choice of $\Sigma\theta$. Clearly, the favourable events may stem from collisions with values of the impact parameter different to those associated with the $\Delta\theta$-window when elastic scattering is assumed. The angular distribution of the cross section is presently unknown, although a rather constant yield of the lines is anticipated from the data including these angular regions where because of enhanced background the lines could not be identified. Therefore, the line intensity of spectra, obtained in a procedure of selecting favourable angular regions to experimentally improve the peak-to-background ratio, may be in variance to the actual size of the cross section. The ORANGE-spectrometer collaboration also reports[17,18] the appearance of line structures in both critical and subcritical systems; the details of these experiments are discussed in separate contributions to this conference[84,85]. Some of these structures they report are found at considerably lower positron energies as in the EPOS-measurements and, in particular in the lower-Z systems, at reduced cross sections. However, it is important to realize, that any comparison of cross sections either with the theory or with other experiments might be meaningless as long as there is only insufficient knowledge on an eventually existing excitation function. A steepness on the order of the energy equivalent of the target thickness (≤ 20 MeV) in the case of a threshold or even resonance-like excitation function may easily result in a change of the experimentally observed cross section by orders of magnitude.

The near constancy of the line position, line width, and to a certain extent also the constancy of the production cross-section rather signals independence of the pronounced line production from all nuclear and quasiatomic details of the collision system and points for a common source. An obvious speculation is that such a source is the two-body decay of a previously undetected light particle into an electron positron pair.[19] Along these lines various speculations for such a source have been proposed in a series of papers and is also discussed during this conference. Up to the time of writing (September '86) the list of hypothetical explanations contains the eventual detection of the axion,[86,87,88] of pairwise charged Higgs-particles,[89] of an electron-positron quasiparticle strongly bound through magnetic interaction at subatomic distances,[90,91] of a poly-electron ($e^+e^+e^-$) complex,[92] and effects of non-linear force equations resulting in either poly-positronium droplets[93] tightly bound through a many-body short-range force or in solitons formed from heavy quasi-electrons and -positrons in a new form of the vacuum state[94]. These candidates decay into electron-positron pairs but the coincident electrons and positrons are differently correlated depending on the phase space available when decaying. The two-body decay of a particle (or a strongly bound object) leads to a back-to-back emission with electrons and positrons of comparable energy and thus exhibits a characteristic pattern. Concentrating on such a light particle produced in the field of the two colliding nuclei, then it is clear that when it decays it has to be sufficiently unbound and dislocated from the nuclei in order to both at least approximately preserve the characteristic free-particle two-body decay kinematics and to eliminate final state interactions in the Coulomb field. Note, that uneffected from further experimental evidence, the hypothetical particle has to move slowly with respect to the CM-system in order to still match with this scenario the experimental evidence of narrow positron lines; depending on the production mechanism and the corresponding energetic distribution, this limits the mean kinetic energy of the particle in the CM-frame to considerably less than 100 keV, unless very special angular distributions are considered.

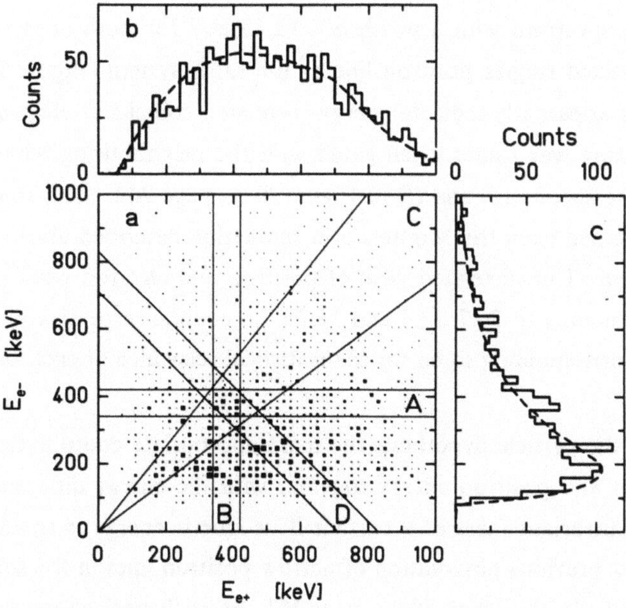

Figure 29. Intensity distributions of coincidence events for 5.83 MeV/u $^{238}U + ^{232}Th$ collisions, displayed two-dimensionally (a) as a function of E_{e+} and E_{e-} and projected onto the E_{e+} (b) and E_{e-} -axis (c). The calculated spectra result from Monte Carlo calculations for internal pair conversion and quasiatomic positron-electron production and are normalized to the total yield (dashed lines).

Consequently an electron-positron correlation experiment has been performed[19)] with the EPOS-spectrometer appropriately modified as explained in section II in order to allow for the coincident detection of low energetic electrons. For more details, including the various Monte Carlo calculations which will be shortly mentioned below, see ref. 49. Figure 29 on page 241 shows the two-dimensional positron-electron coincidence spectrum of a first experiment[19)] in June '85 for 5.83 MeV/u $^{238}U + ^{232}Th$ collisions, which includes ~2/3 of the available data selected because of severe target deterioration problems discussed in the preceding section. Again the positron-electron coincidence events are globally restricted to those originating from binary, mostly elastic ion scattering-events as discussed above, but no additional ion angular regions are now selected for reasons of pure statistics. The parts (b) and (c) of Figure 29 on page 241 show projections onto the positron and electron energy coordinates. The spectra are well reproduced by Monte Carlo calculations of nuclear internal pair conversion and dynamic quasiatomic processes when adjusted to the total yield.

A striking correlation between electrons and positrons, however, became apparent when the projections are gated on certain energy windows. Parts (a) and (b) of Figure 30 on page 243 display such projections on the positron and electron energies gated with the two-dimensional windows A and B, respectively, shown in Figure 29 on page 241. The structures at 380 ± 15 keV in the gated positron and at 375 ± 15 keV in the gated electron spectrum with a width of 75±15 keV for both lines favourably fit the previously detected singles positron line of the same system (Figure 26 on page 237). The structures apparently indicate energy-correlated coincident electron-positron emission; no structure was found when gated with the neighbouring windows of A and B displayed in the spectra (e) and (f) of Figure 30 on page 243. The smooth background curve was obtained from the Monte Carlo simulation described above without any further adjustment. The correlated yield of positron and electron peak events above this background amounts to 26.7±7.7 and 31.7±7.6 counts, respectively, specifying a confidence level corresponding to 6σ for the statistical relevance of each of the peaks.

In view of the particle hypothesis the more appropriate coordinates are the sum ΣE of the electron and positron energy together with the energy difference ΔE; ΣE corresponds to the invariant mass of a particle if its kinetic energy is small, as it is actually required by the previous observation of narrow positron lines in the singles spectra. The spectra (c) and (d) of Figure 30 on page 243 are such projections onto the sum and difference energy of the same data now gated with the two-dimensional windows C and D, respectively, of Figure 29. A wedge-shaped window C is selected in order to approximately consider the enlarged Doppler-broadening when the electron/positron energy increases. A sharp line appears in the sum-spectrum at 760±20 keV in agreement with twice the individual electron or positron line energies, together with a clear structure in the difference-spectrum around zero energy with 35.3±9.4 and 38.5±9.7 events, respectively, above the calculated background. Again no structures are found in the windows neighbouring C and D (see (g) and (h) in Figure 30 on page 243).

In the limiting case of a particle decaying at rest in the heavy ion CM-frame, the Doppler shifts of the positron and the electron emitted back-to-back at CM-angles Θ and π+Θ defined relative to the direction of the CM-velocity, are of equal size but opposite in sign. Consequently the Doppler shift exactly cancels in the sum of both energies. The Doppler width of the singles positron line and of the line in the difference-spectrum is determined from the CM-velocity only. In this case the line position in ΣE becomes an exact measure of the invariant mass. For an additional intrinsic particle velocity the cancellation is no longer complete. But for values of the intrinsic velocity of the particle less or comparable to the velocity of the CM, the feature of the narrow sum-energy line essentially remains together with a still clear appearance of both the singles and the difference-energy lines. For larger values of the particle velocity this feature is diluted and in particular, the singles positron line and the line in the difference-spectrum are smeared out with the additional broadening resulting from the particle's intrinsic velocity. From the line-shape analysis of the narrow singles positron lines of the previous measurements a common emitter velocity with values not exceeding the CM-velocity was derived. In order to match this result, the intrinsic velocity of a hypothetical particle with respect to the CM on the average is indeed limited to consid-

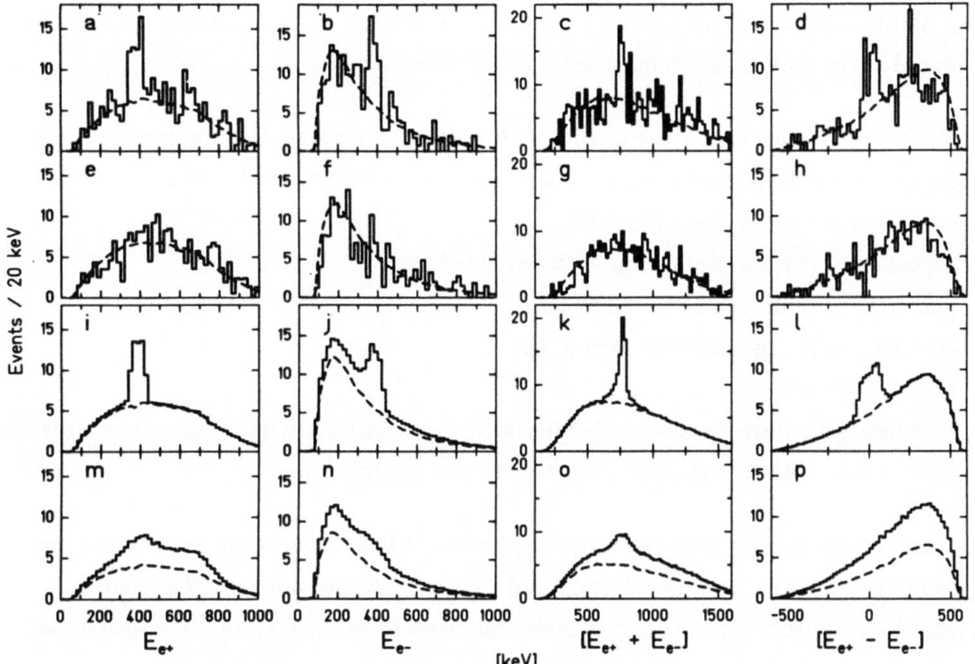

Figure 30. (a) -(d) Projections of the positron-electron coincidence intensity distribution of Figure 29 gated with the two-dimensional windows A through D of Figure 29 onto either the positron energy, the electron energy, the sum or the difference of positron and electron energies.
(e) - (h) same as (a) - (d) but with two-dimensional similar windows symmetrically neighbouring the windows A through D. For comparison are also shown the results of Monte Carlo calculations for the two-body decay of a particle (scaled by 6×10^{-4}) ((i) - (l) and for internal pair decay of a nuclear excited state (scaled by 6×10^{-3}) ((m) - (p)).

erably less than the velocity of the CM (~0,05c). Thus if the experiment exploits a sharp sum-energy line, the actual existance of a near back-to-back emitting process leading to the Doppler shift cancellation is most probable and then the energy of the line directly determines the invariant mass. However, the assumption of near-isotropic positron emission introduced in the emitter-velocity determination from the shape of the singles positron lines may weaken this argumentation. Then a future measure of the full momentum correlation, more direct as it can be performed in this historically grown-up experiment, then eventually has to be considered. In fact, the width of 80 ± 20 keV observed in this measurement for the sum-energy line is apparently less than what is expected from a convolution of the Doppler-effected individual electron and positron lines. This observation signals a cancellation and argues for a back-to-back emission, but the statistical relevance of these data does not allow to draw any final conclusions. The third row of Figure 30 represents the results of a Monte Carlo simulation of the hypothetical particle scenario. Containing all details of the experimental response, a weak particle pair decay rate sufficient to contribute an amount of 3% to the total singles positron yield fairly reproduces the experimental data; a particle of 1.8 MeV/c^2 rest mass produced at rest in the CM-frame is assumed for this calculation.

Various processes which eventually contribute to correlated electron-positron line production have been systematically investigated with Monte Carlo simulation techniques. These scenarios range from the

- hypothetical particle production with correlated and
- uncorrelated (although of less physical relevance) electron-positron emission,
- internal pair production from high-Z and
- low-Z fragments, to
- monoenergetic internal pair production with the electron obtained from an internal electron conversion process of a subsequent cascade transition.

Internal pair production from the target- or projectile-like nuclei is the most prominent conventional process. Then the electron and positron energies are again correlated to a constant energy-sum as determined from the transition energy, but the individual electron and positron energies are distributed continuously: the internal pair conversion process is solely distributed parallel to the ΔE-axis and (ΣE,ΔE) are again the most appropriate coordinates. Projections according the window C of Figure 29 on page 241 but also of neighbouring windows onto the sum-energy coordinate for the pair conversion process all yield lines in the sum-energy spectra. Because the Doppler effect is no longer canceled in the sum-energy as it is the case for pronounced back-to-back emission, these lines now are considerably broadened. But as a most significant feature, the pair conversion process as a three-body-decay never reveals structures in the ΔE-spectrum. The fourth row of Figure 30 on page 243 now shows the Monte Carlo simulation of the internal pair conversion of a 1.8 MeV E1 nuclear transition assumed in a Z=91 nucleus gated with the identical two-dimensional energy windows as the spectra in row (a)-(d) of this figure. Most strikingly, the smooth spectrum in projection (p) for the ΔE-spectrum is in clear contradiction to the corresponding experimental data in (d); this to a great extent excludes nuclear internal pair conversion as a conceivable explanation now also for correlated positron-electron emission. Moreover, this simulation assumes already 30% of the singles positron yield to originate solely from this transition. The strong discrepencies of the calculations based on this percentage to the experimental data as concerns the individual line shapes both in the positron singles spectrum (not shown) and the projections of the coincidence yield are obvious. But considerably more than 30% is needed to at least match the observed intensity of the sum-energy line. Note that the response of the present set-up on an internal pair conversion process, namely the E0 pair conversion from the 1.76 MeV transition of ^{90}Zr was extensively studied with a ^{90}Sr-source and was found in excellent agreement with the corresponding Monte Carlo simulation (compare Figure 11 on page 214 and accompanying text). Moreover, also a contamination with ^{90}Zr built up from U-fragmentation can be ruled out to account for the observed correlated line emission because of the apparent difference of the in-beam data from both the ^{90}Zr-source measurement and the internal pair conversion simulation.

Presently all attempts but the two-body decay scenario fail to reproduce the experiment. On the other hand, the production of a so far undetected particle of ~ 1.8 MeV/c^2 rest mass has been shown in the meanwhile to be extremely unlikely. Any production mechanism using linear electromagnetic, leptonic, or quark coupling in order to match the positron results appears to be inconsistent with precision measurements involving the respective coupling-constant and/or fails to sufficiently enhance the production cross section at low kinetic energies of the decaying particle.[95] Also the existence of the axion, which has been mostly considered as a candidate for such a particle, was already ruled out in its standard form.[26] Partly motivated by the unexplained positron lines a variant axion model with the coupling of the axion non-trivially different to the various quarks has been invented[20,24] to circumvent the existing experimental bounds. With a mass of ~ 1.8 MeV as anticipated now from the positron lines, the axion life time when decaying either into two γ-rays or into an electron-positron pair amounts[96] to $\sim 5 \times 10^{-12}$sec. This life time is much shorter than it was determined before when axion masses below 1 MeV have been assumed. Then previous experiments may no longer apply because the pseudoscalar particle could simply have escaped from the detection (this mostly concerns beam dump and nuclear decay[97] experiments). Very recently, however, a series of respective experiments have been performed and more stringent experimental bounds have been derived in this mass and life time domain from either the radiative decay of the Y vector meson,[98,99] the multilepton decay of π^+ and μ^+,[100,101] a high-energy electron beam dump experiment,[102] or the branching of the nuclear decays of isovector (^{14}N)[103] and isoscalar (^{10}B)[104] transitions. No evidence for a short lived light neutral particle was reported and the axion model-variant is ruled out as well.

The simplifying assumption of a common mean energy of all positron lines observed in the past certainly led to the exciting detection of the correlated electron-positron line emission. However, the individual deviations from the mean value of the line energies exceed the experimental uncertainties and have to be taken seriously. Moreover, it is presently unclear how the positron lines reported[17,18] from the ORANGE-group partly observed at considerably lower but recently also comparable[85] positron energies correlate with the EPOS-results. The analysis of the ^{232}Th$+^{232}$Th collision system measured together with the ^{238}U$+^{232}$Th system in the same EPOS correlation experiment[19] indeed already indicates that there exist correlated electron-positron lines (and corresponding sum-energy lines) of different energies. Positron line energies of ~ 310 keV and ~ 370 keV (together with the respective electron lines) have been found in two different

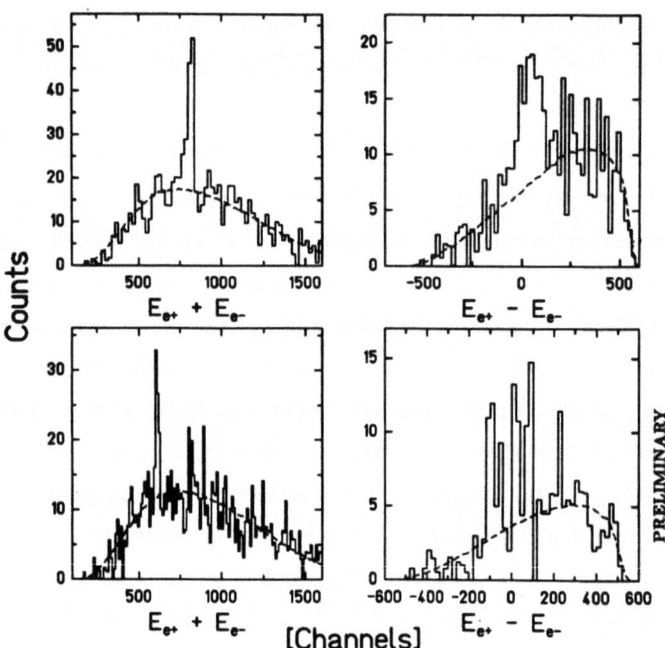

Figure 31. A preliminary analysis of a measurement in February '86 of the EPOS-collaboration of the $^{238}U + ^{232}Th$ collision system at beam energies around 5.87 MeV/u reveals two prominent sum-energy peaks significantly narrower than the individual positron and electron lines. The lower and upper panels show sum and difference-energy spectra for two subsets of the data gated on beam energy, heavy-ion scattering angle, and positron/electron time-of-flight chosen to enhance these two sum-energy lines, respectively.

sets of the data corresponding to projectile energies of 5.70 and 5.75 MeV/u, respectively.

Very recent experiments with the EPOS-spectrometer in 1986 still being analysed reproduced the existence of electron-positron line correlations. Clearly two pairs of correlated positron/electron lines have been detected in the $^{238}U + ^{232}Th$ collision system at projectile energies ranging from 5.81 to 5.90 MeV/u. Each pair leads to pronounced sum-energy lines of ~ 620 keV and ~ 810 keV, respectively, both significantly narrower than the individual positron and electron lines, and to clear line structures at zero energy in the respective difference-energy spectra (Figure 31). The data allow for the additional presence of the previously observed 760 keV sum-energy line, but its rigorous identification appears to be difficult. It was found, on the other hand, that the peak-to-background ratio of the sum-energy lines can be alternatively enhanced with additional gates on the positron or electron time-of-flight, the projectile energy, and the ion scattering angular regions. It also was observed mostly for the 810 keV sum-energy line, although with less evidence, that a choice of still more stringent angular windows to a similar extent enhances the peak-to-background ratio of both the sum-energy line and the corresponding singles positron line. Apparently this observation relates the correlated electron-positron line emission to the singles positron lines detected in the previous measurements. Along this context it should also be realized, that the energy range of the pronounced singles positron lines to a large extent overlaps with the corresponding energy range which contains the sum-energy lines observed in the recent experiments. This also points to a common underlying process for both singles and correlated line emission. Positron and electron time-of-flight in these experiments is first of all determined from the path length in the solenoidal field. The deviations from the typical time-of-flight distribution mostly observed for the 620 keV sum-energy line therefore may also reflect non-isotropic positron emission oriented preferantially perpendicular to the solenoid axis, although a non-trivial delay cannot be excluded presently.

IV. Final Remarks

Clearly, the positron puzzle is not solved and there is no quick solution in sight. Starting from the mere existence of the positron lines, the body of experimental information was considerably enlarged mainly from the general exclusion of nuclear transitions to produce the lines, from the systematic variation of Z_{ua} resulting in an essential independence of the line position and the line width from the individual collision systems, and finally from the finding of a correlated positron-electron line emission. Conventional and exotic processes including spontaneous positron emission and the pair decay of a hypothetical axion have been proposed to explain the lines, and also partly motivated the experimental research. But there is presently no process proved to consistently describe the existing data, although the fictive assumption of a near back-to-back decay of a light object slowly moving with the CM of the colliding heavy ions still is in agreement with the experiments. Certainly the experiments will continue. It appears important to clarify the long-standing question of an eventually existing threshold- or resonance-like behaviour of the line-production cross section and also its relation to the heavy-ion kinematics. The sharp sum-energy line with the Doppler effect apparently canceled now offers a much better measure of the line energy than the strongly Doppler broadened singles positron lines did so far. Therefore the question of a systematic variation of the line energies has again to be addressed with an extension of the positron-electron coincidence technique to considerably different (lower) projectile energies and to other (lighter) projectile-target combinations. This will, in particular, clarify the influence of the nuclear interaction on the line production and probe the actual need of the strong electric and magnetic fields maximally provided in these heaviest collision systems. The eventual existence of an emitter-system lifetime in the ns-range is particularly interesting and argues for an independent and appropriately optimized measurement using various recoil techniques; however, the alternative and more probable possibility of an anisotropy in the positron-electron emission is as important also because of its implication on the line width analysis of the singles positron lines. But moreover, the study of this behaviour which is apparently different for the individual sum-energy lines may provide us with more insight into the underlying process.

ACKNOWLEDGMENT

Positron spectroscopy has now been persued at GSI for about one decade and much help has been obtained from people inside and from many places outside GSI by personal assistance and contributions, detailed discussions, or through permanent motivating interest in this continously developing field. Let me representatively for all those persons express my thanks to the members of the ORANGE[18] and TORI[76] groups, to the staff members of GSI, in particular the UNILAC crew, and the Frankfurt-GSI theory group, in particular, W. Greiner, B. Müller, U. Müller, J. Reinhardt, Th. deReus, A. Schäfer and G. Soff. Especially, however, I want to thank my colleagues from the EPOS-group[105] for the intensive and fruitful collaborations we persued together throughout many years and through all stages of the experiment. I particularly thank M. Begemann-Blaich, T. Cowan, K. Sakaguchi, and P. Salabura for supplying me with various details of the EPOS-experiments, which are part of their PhD-theses. This report also gained considerably from details I found in the previously completed PhD-theses of A. Gruppe and J. Schweppe.

Finally I want to thank Walter Greiner and his 'team' for the pleasant organization of this conference.

References

1. H. Backe, L. Handschug, F. Hessberger, E. Kankeleit, L. Richter, F. Weik, R. Willwater, H. Bokemeyer, P. Vincent, Y. Nakayama, and J.S. Greenberg, Phys.Rev.Lett. **40**(1978)1443
2. C. Kozhuharov, P. Kienle, E. Berdermann, H. Bokemeyer, J.S. Greenberg, P. Vincent, H. Backe, L. Handschug, and E. Kankeleit, Phys.Rev.Lett. **42**(1979)376
3. J. Reinhardt, B. Müller, and W. Greiner, Phys.Rev. **A24**(1981)103
4. Quantum Electrodynamics of Strong Fields, edited by W. Greiner, Plenum, New York (1983)
5. T. Tomoda, and H.A. Weidenmüller, Phys.Rev. **A26**(1982)103
   T. Tomoda, Phys.Rev. **A26**(1982)174
6. P. Armbruster and P. Kienle, Z. Physik **A291**(1979)399
7. B. Müller, G. Soff, W. Greiner, and V. Ceausescu, Z.Phys. **A285**(1978)27
8. E. Kankeleit, Nukleonika **25**(1980)253
9. W. Pieper and W. Greiner, Z.Physik **218**(1969)327;
   B. Müller, J. Rafelski, and W. Greiner, Z.Physik **257**(1972)62,183
10. S.S. Gershtein and Ya.B. Zel'dovich, Lett.Nuovo.Cimento 1(1969)835;
    Ya.B. Zeldovich and V.S. Popov, Sov.Phys.Usp. **14**(1972)663
11. J. Rafelski, B. Müller, and W. Greiner, Z.Physik **A285**(1978)49
12. J. Reinhardt, U. Müller, B. Müller, and W. Greiner, Z.Physik **A303**(1981)173
13. H. Bokemeyer, K. Bethge, H. Folger, J.S. Greenberg, H. Grein, A. Gruppe, S. Ito, R. Schul&aca.e, D. Schwalm, J. Schweppe, N. Trautmann, P. Vincent, M. Waldschmidt, in ref. 4 p. 273
14. J. Schweppe, A. Gruppe, K. Bethge, H. Bokemeyer, T. Cowan, H. Folger, J.S. Greenberg, H. Grein, S. Ito, R. Schul&aca.e, D. Schwalm, K.E. Stiebing, N. Trautmann, P. Vincent, and M. Waldschmidt, Phys.Rev.Lett. **51**(1983)2261

15. T. Cowan, H. Backe, M. Begemann, K. Bethge, H. Bokemeyer, H. Folger, J.S. Greenberg, H. Grein, A. Gruppe, Y. Kido, M. Klüver, D. Schwalm, J. Schweppe, K.E. Stiebing, N. Trautmann, and P. Vincent, Phys.Rev.Lett. **54**(1985)1761
16. P. Kienle in ref. 4 p. 293
17. M. Clemente, E. Berdermann, P. Kienle, H. Tsertos, W. Wagner, C. Kozhuharov, F. Bosch, and W. König, Phys.Lett. **B137**(1984)41
18. H. Tsertos, E. Berdermann, F. Bosch, M. Clemente, P. Kienle, W. König, C. Kozhuharov, and W. Wagner, Phys.Lett. **B162**(1985)273
19. T. Cowan, H. Backe, K. Bethge, H. Bokemeyer, H. Folger, J.S. Greenberg, K. Sakaguchi, D. Schwalm, J. Schweppe, K.E. Stiebing, and P. Vincent, Phys.Rev.Lett. **56**(1986)444
20. R.D. Peccei, T.T. Wu, and T. Yanagida, Phys.Lett. **B172**(1986)435
21. S. Weinberg, Phys.Rev.Lett. **40**(1978)223
22. F. Wilczek, Phys.Rev.Lett. **40**(1978)229
23. R.D. Peccei and H.R. Quinn, Phys.Rev.Lett. **38**(1977)1440
24. W.A. Bardeen, R.D. Peccei, and T. Yanagida, preprint 86/054, DESY, Hamburg, Germany andPAE/PTH 27-86, MPI, Heidelberg, Germany, 1986
25. A.B. Balantekin, contribution to this conference
26. A. Zehnder, in Fundamental Interactions in Low Energy Systems, Plenum, New York (1985)
27. J. Reinhardt, A. Schäfer, B. Müller, W. Greiner, Phys.Rev. **C33**(1986)194
28. Treatise on Heavy Ion Science, Vol. 5: High Energy Atomic Physics, edited by D.A. Bromley, Plenum Press, New York (1985)
29. H. Backe and C. Kozhuharov in Progress in Atomic Spectroscopy, Part C, edited by H. Beyer and H. Kleinpoppen, Plenum Press, New York (1984), p. 28
30. F. Bosch, B. Müller, Progress in Part. and Nucl. Phys. **16**(1985)195
31. J.S. Greenberg in Proceedings of the X-82 Int. Conf. on X-Ray and Atomic Inner Shell Physics 1982, University of Oregon, Eugene, Oregon, USA (1982)
J.S. Greenberg in Proceedings on the Atomic Theory Workshop on Relativistic and QED Effects in Heavy Atoms, NBS, Gaithersburg, Maryland, USA (1985)
32. H. Bokemeyer in Selected Topics in Nucl. Structure, XIX. Winter School of Physics, Zakopane, 1984, edited by J. Stachura, Inst. of Nucl. Phys., Jag. Univ., Krakow (1984) p. 335
H. Bokemeyer in Selected Topics in Nucl. Structure, XXI. Winter School of Physics, Zakopane, 1986, edited by R. Broda and J. Stachura, Inst. of Nucl. Phys., Jag. Univ., Krakow (1986)
33. D. Schwalm in Electronic and Atomic Collisions, edited by J. Eichler, I.V. Hertels, and N. Stolterfoht, Elsevier Science Publishers B.V. (1984) p. 295
34. J. Schweppe in Electronic and Atomic Collisions, edited by D.C. Lorentz, W.E. Meyerhof, and J.R. Peterson, Elsevier Science Publishers B.V., 1986, p. 405
35. J. Reinhardt, T. deReus, W. Greiner, B. Müller, U. Müller, A. Schäfer, P. Schlüter, S. Schramm, and G. Soff in Electronic and Atomic Collisions, edited by D.C. Lorentz, W.E. Meyerhof, and J.R. Peterson, Elsevier Science Publishers B.V., 1986
36. P. Kienle in Nucleus Nucleus Collisions from the Coulomb Barrier up to the Quark Gluon Plasma, Erice, Italy (1985) and report preprint 85-31, GSI, Darmstadt, Germany (1985)
37. H. Bokemeyer, H. Folger, H. Grein, T. Cowen, J.S. Greenberg, J. Schweppe,K. Bethge,A. Gruppe, K.E. Stiebing, R. Schule, M. Waldschmidt, D. Schwalm, S. Ito, S. Matsuki, Y. Nakayama, P. Vincent, to be published
38. J. Schweppe, PhD-thesis, Yale Univ. New Haven, USA, 1985, unpublished
39. A. Gruppe, PhD-thesis, J.W. Goethe Univ., Frankfurt, Germany, 1985 and report-85-4, GSI, Darmstadt, Germany, 1985
40. G.A. Burginyon and J.S. Greenberg, Nucl. Inst. Meth. **41**(1966)109
41. H. Backe, H. Bokemeyer, E. Kankeleit, E. Kuphal, Y. Nakayama, L. Richter, R. Willwater, Laborbericht Nr. 66, Inst. für Kernphysik, Techn. Hochschule, Darmstadt, Germany, 1975
42. M. Waldschmidt and S. Wittig, Nucl. Inst. Meth. **64**(1968)189
43. α-, β- and γ-ray-spectroscopy, edited by K. Siegbahn, North Holland Publ. Co., Amsterdam, 1979
44. Damkjaer, Nucl. Inst. Meth. **200**(1982)377
45. G. Seiler-Clark, D. Husar, R. Nowotny, H. Graf, and D. Pelte, Phys. Lett. **80B** (1979)345
46. H. Backe, P. Senger, W. Bonin, E. Kankeleit, M. Krämer, R. Krieg, V. Metag, N. Trautmann, J.B. Wilhelmy, Phys. Rev. Lett. **50**(1983)1838
47. R. Wilson in ref. 43, p. 1557
48. M.E. Rose, Phys. Rev. **76**(1948)668 and errata Phys. Rev. **78**(1950)184
49. T. Cowan, contribution to this conference
50. P. Fuchs, H. Emling, E. Grosse, D. Schwalm, H.J. Wollersheim, R. Schulze, Annual Report-77-1, GSI, Darmstadt, Germany, 1977, p. 195
51. H. Kleinknecht, Detektoren für Teilchenstrahlung, B.G. Teulner, Stuttgart, 1984
52. K.E. Stiebing, contribution to this conference

53. N.S. Nikolaev, I.S. Dimitriev, Phys. Lett. **28A**(1968)277
54. D. Maor, P.H. Mokler, D. Schüll, and Z. Stachura, Nucl. Instr. Meth. **194**(1982)377
55. D. Schüll, W.F.W. Schneider, H. Bokemeyer, R. Bock, H. Freiesleben, F. Pühlhofer, D. Bangert, B. Kohlmeyer, M. Marinescu, Scientific Report 1982 83-1, GSI, Darmstadt, 1983, p. 14
56. K. Stiebing, H. Bokemeyer, A. Bösser, H. Folger, R. Künkel, W. Schneider, and D. Schüll, to be published and contribution to this conference
57. D. Kraft, K. Stiebing, K. Bethge, M. Begemann, M. Klüver, and H. Bokemeyer, Scientific Report 1985, Inst. für Kernphysik, J.W. Goethe Univ., Frankfurt, (1986) p. ... and to be published
58. D. Kraft, master thesis, Inst. für Kernphysik, J.W. Goethe Univ., Frankfurt, 1987
59. N. Trautmann, M. Weber, D. Gombalies-Datz, R. Heinmann, H. Folger, and W. Hartmann, Preprint-82-40, GSI, Darmstadt, 1982 and Proc. 11th World Conf. Int. Nucl. Target Dev. Soc., University of Washington, Seattle, 1983, p. 120
60. P. Strehl, J. Klabunde, V. Schaa, H. Vilhjamsson, D. Wilms, report-79-13, GSI, Darmstadt, Germany, 1979
61. F. Busch, D. Croome, H. Göringer, V. Hartmann, J. Lowsky, D. Marinescu, M. Richter, K. Winkelmann, EDAS-Report 83-4, GSI, Darmstadt, Germany, 1983
62. G.K. Horton, Proc. Phys. Soc. **60**(1943)457
63. P. Schlüter, G. Soff, W. Greiner, Phys. Rep. **75**(1981)327
64. P. Schlüter, G. Soff, and W. Greiner, Z. Phys. **A286**(1978)149
65. G. Soff, P. Schlüter, and W. Greiner, Z. Phys. **A303**(1981)189
66. W.E. Meyerhof, R. Anholt, Y. El Masri, I.Y. Lee, D. Cline, F.S. Stephans, R. Diamond, Phys. Lett. **69B**(1977)
67. B. Müller, G. Soff, W. Greiner, V. Ceausescau, Z. Physik **A285**(1978)27
68. D. Liesen, P. Armbruster, F. Bosch, S. Hagmann, P.H. Mokler, H.J. Wollersheim, H. Schmidt-Böcking, R. Schuch, and J.B. Wilhelmy, Phys. Rev. Lett. **44**(1980)983
69. J. Bang and J.M. Hansdeen, Phys. Lett. **72A**(1979)218; Pysica Scripta **22**(1981)609
70. T. Tanabe, R. Bock, M. Dakowski, A. Gobbi, H. Stelzer, U. Lynen, and A. Olmi, Nucl. Phys. **A324** (1980)194
71. J.S. Greenberg, H. Bokemeyer, H. Emling, E. Grosse, D. Schwalm, and F. Bosch, Phys. Rev. Lett. **39** (1977)1404
72. P. Schlüter, Th. deReus, J. Reinhardt, B. Müller, and G. Soff, Z. Physik **A314** (1983)297
73. P. Schlüter, and G. Soff, At. Data Nucl. Data Tables **24** (b): 509 (1979)
74. U. Heinz, J. Reinhardt, B. Müller, W. Greiner, and W.T. Pinktston, Z. Phys. **A316** (1984)341
75. S. Schramm, J. Reinhardt, U. Müller, and W. Greiner, Z.Phys **A323** (1986)275
76. B. Blank, E. Bozek, U. Gollerthan, H. Jäger, E. Kankeleit, G. Klotz-Engmann, M. Krämer, R. Krieg, U. Meyer, H. Oeschler, M. Rhein, and P. Senger, Scientific Report 1985 86-1, GSI, Darmstadt, Germany, 1986, p. 182
77. H. Backe, M. Begemann-Blaich, M. Klüver, W. Konen, P. Senger, P. Glässel, D.V. Harrach, K. Wallenwein, K. Stiebing, H. Bokemeyer, and K. Poppensieker, Scientific Report 1985 86-1, GSI, Darmstadt, Germany, 1986, p. 184 and P. Senger contribution to this conference
78. U. Müller, G. Soff, Th. deReus, J. Reinhardt, B. Müller, and W. Greiner, Z. Phys. **A313**(1983)263
79. T. Tomoda and H.A. Weidenmüller, Phys.Rev. **C28**(1983)739
80. M.J. Rhoades-Brown, V.E. Oberacker, M. Seiwert, and W. Greiner, Z.Phys. **A310**(1983)
81. V. Oberacker, contribution to this conference and ref. therin
82. Th. deReus, J. Reinhardt, U. Müller, W. Greiner, G. Soff, Preprint, Inst. für Theoret. Physik, J.W. Goethe Univ., Frankfurt, Germany, 1986 and GSI-86-5, subm. to Physica C
83. M. Seiwert, J.A. Maruhn, W. Greiner, J. Friedrich, Z.Physik **A321**(1985)653
84. C. Kozhuharov, contribution to this conference
85. W. König, contribution to this conference
86. A. Schäfer, J. Reinhardt, B. Müller, W. Greiner, and G. Soff, J. Phys. **C11**(1985)169
87. A.B. Balantekin, C. Bottcher, H.R. Strayer, and S.G. Loe, Phys.Rev.Lett. **55**(1985)461
88. A. Chodos and L.C.R. Wijewardhana, Phys.Rev.Lett. **56**(1986)302
89. A. Schäfer, B. Müller, and W. Greiner, Phys.Lett. **149B**(1984)455
90. Chenk-Yin Wong and R.L. Becker, preprint
91. A. Barut, contribution to this conference
92. Chenk-Yin Wong, Phys.Rev.Lett. **56**(1986)1047
93. B. Müller, J. Reinhardt, W. Greiner, and A. Schäfer, J. Phys. **G12**(1986)L109
94. L.S. Celensor, V.K. Mishra, and C.M. Skakein, Phys.Rev.Lett. **57**(1986)55
95. A. Schäfer, J. Reinhardt, W. Greiner, and B. Müller, Mod.Phys.Lett. **A1**(1986)1 and ref. therin
96. T.W. Donnelly, S.Y. Freedman, R.S. Lytel, R.D. Peccei, and M. Schwartz, Phys.Rev. **D18**(1978)1607 (note the error in equ. (15): the first m_a^2 to be exchanged to m_e^2)
97. F.P. Calaprice, R.W. Dunford, R.T. Kouzes, M. Müller, A. Hallin, M. Schneider, and D. Schreiber, Phys.Rev. **D20**(1979)2708

98. G. Mageras, P. Franzini, P.M. Tuts, S. Youssef, T. Zhao, J. Lee-Franzini, and R.D. Schamberger, Phys.Rev.Lett. **56**(1986)2672
99. T. Bowcock et al., Phys.Rev.Lett. **56**(1986)2676
100. R. Eichler, L. Felawka, N. Kraus, C. Niebuhe, H.K. Walter, S. Egli, R. Engfer, Ch. Grab, E.A. Hermes, H.S. Pruys, A. von der Schaaf, W. Berth, N. Nordong, U. Bellgardt, G. Otter, T. Kozlowski, J. Martino, Phys.Lett. **175**(1986)101
101. L.M. Krauss, M.B. Wise, Phys.Lett. **B176**(1986)483
102. A. Konaka, K. Imai, H. Kobayashi, M. Masaika, K. Miyake, T. Nakamara, N. Nagamine, N. Sasao, A. Enomoto, F. Fukishima, E. Kikutami, H. Koiso, H. Matsumoto, K. Nakahara, S. Ohsawa, T. Taniguchi, I. Sato, and J. Urakawa, Phys.Rev.Lett. **57**(1986)659
103. M.J. Savage, R.D. McKeown, B.M. Fillipone, and L.W. Mitchell, Phys.Rev.Lett. **57**(1986)178
104. F.W.N. de Boer,K. Abrahams, A. Balanda, H. Bokemeyer, R. van Dantzig, J.F.W. Jansen, B. Kotlinski, M.J.A. de Voigt, and J. van Klinken, Preprint, KVI, Groningen, The Netherlands, 1986, acc. for publ. in Phys.Lett. (1986)
105. The previous and present members of the EPOS group are: H. Backe, A. Balanda, M.L. Begemann, K. Bethge, H. Bokemeyer, T. Cowan, H. Folger, J.S. Greenberg, H. Grein, A. Gruppe, Y. Kido, M. Klüver, S. Matsuki, K. Sakaguchi, P. Salabura D. Schwalm, R. Schul&aca.e, K.E. Stiebing, N. Trautmann, and P. Vincent

SPECIAL ASPECTS OF THE EPOS EXPERIMENTS ON POSITRON EMISSION FROM SUPERHEAVY COLLISION SYSTEMS*

K. E. Stiebing
Institut für Kernphysik
der Johann Wolfgang Goethe - Universität
Frankfurt am Main, FRG (IKF)

The first observation of monoenergetic positrons emerging from collisions of very heavy nuclei (e.g. U+U, U+Cm etc) in experiments performed at GSI[1,2], initiated substantial experimental and theoretical effort to understand the mechanisms that produce these positrons. Part of these conference proceedings reflects this endeavor to solve the "positron puzzle". Two experimental results found by the EPOS collaboration[3] have been matter of extensive discussions in this context :

a) the meaning of the projection windows on the kinematic heavy-ion correlation ("kinematic cuts"), which were set to achieve an optimal peak-to-background ratio of the positron lines above the spectrum of dynamical and nuclear positrons.

b) The apparent "resonance-" or "threshold-" like dependence of the positron line production on the impact energy.

This article is intended to reconsider the above conditions in the view of additional information which has been obtained in specially designed experiments performed at GSI and IKF. In a first experiment the charge-state distributions of ejectiles from a similar heavy ion collision system (Pb on Au) at an impact energy of 5.9 MeV/u as a function of target thickness have been measured to judge the influence of the magnetic field of the EPOS-apparatus on the particle kinematics. In the second experiment the method of proton backscattering has been employed to investigate the target structure and its changes by the heavy-ion bombardement.

*For a description of these experiments see ref.3

A) Kinematic coincidence

In order to clearly separate true events (elastic or quasi-elastic heavy-ion scattering) from background events like electron pile up, recoils from light target components or fission events, it is a useful technique to measure the angular and time correlations of projectile and recoil particle by coincidence technique. A very useful coordinate system to discuss angular correlations of two particles 1 and 2 of a binary event is then (in the laboratory frame):

$$\Delta\vartheta = \vartheta_1 - \vartheta_2$$
$$\text{and } \Sigma\vartheta = \vartheta_1 + \vartheta_2 .$$

In this representation, the elastic scattering of a completely symmetric system results in a straight line parallel to the $\Delta\vartheta$-axis at $\Sigma\vartheta=90°$. An asymmetric system separates into two branches both intersecting the $\Sigma\vartheta$ axis at a value $\Sigma\vartheta>90°$ for $A_{proj}<A_{recoil}$ or $\Sigma\vartheta<90°$ for $A_{proj}>A_{recoil}$ (see fig. 1). In the EPOS apparatus parallel-plate-avalanche counters are used to obtain time and angular information. For the geometry given, a separation of close and distant collisions sets in at mass differences for heavy nuclei of $\Delta A>16u$. In these counters no determination of the total energy of the respective ejectiles can be performed and thus only for elastic scattering, the kinematic correlation determines the system unambigously.

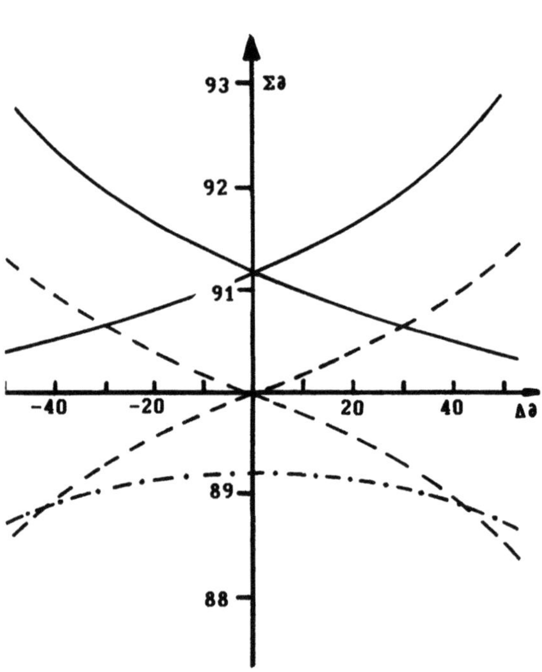

fig.1: Kinematic correlations for: a)elastic scattering of $^{238}U+^{248}Cm$ (solid lines), b)inelastic scattering, no mass transfer, $\Delta Q=-30$ MeV (dashed lines) c)inelastic scattering, emission of 2α-particles ($\Delta Q=-60$ MeV), mass transfer into symmetric exit channels ^{239}A (dashed dotted lines)

To illustrate the influence of inelasticity in fig.1 two examples for the case of U+Cm are shown. In the first case, only a loss of the total kinetic energy of TKEL=30MeV is assumed which can clearly be distinguished from the elastic correlation within the experimental $\Sigma\vartheta$-resolution of $\simeq\pm.8°$. In the second case the assumption has been made that a compound nucleus has been formed which emits two α-particles and then separates symmetrically into two nuclei of mass A=239, disregarding the influence of recoil-momentum transfer from the α-particle on the compound nucleus. For the emission of α-particles from the compound nucleus an energy loss of TKEL=30MeV/α has been taken into account. The examples demonstrate that inelasticity changes the kinematic correlations quite drastically. For reactions where mass transfer occurs but no loss (e.g. by particle emission) a maximum TKEL of ±16MeV can be deduced from the experimentally observed FWHM of the $\Sigma\vartheta$-distribution at $\Delta\vartheta$=0. If light particles are emitted like p, n, α, etc, the kinematic correlation only coincides with the elastic one in the limits given above for those clusters that are formed with an extremly low TKEL. For example the symmetric fission of the compound nucleus into masses A=233 (i.e. a total loss of 20u=5α) which is especially interesting for the assumption of a common source of the positron peaks, would be only in accordance with the observed width of the $\Sigma\vartheta$- distribution for a TKEL in the range of 10-50 MeV, in clear discrepancy with the emission of α-particles. From fig.1 it becomes evident that there exists a mirror symmetry at the $\Sigma\vartheta$-axis which should be conserved for all events.

By setting windows on the kinematic correlation it was possible to bring out the monoenergetic positron lines very prominently above the background of dynamical and nuclear positrons. Two facts, however, introduce problems into a straightforward relation of these cuts to a certain type of scattering events. Firstly, the mirror symmetry at the $\Sigma\vartheta$-axis is no longer conserved for these cuts, i.e. although being close to the "elastic" distribution associated with dynamic and nuclear positrons, the cuts lie slightly at higher $\Sigma\vartheta$-values for negative $\Delta\vartheta$ and at smaller $\Sigma\vartheta$ for positive $\Delta\vartheta$. Secondly, no definite $\Delta\vartheta$-dependences can be deduced from these cuts to allow an interpretation in terms of far or distant collisions. Adding information from different measurements suggest rather that the events are observed at all angular regions not coinciding, however, with the kinematic correlation for dynamical and nuclear events, which conserves mirror symmetry, when a correction for the declination for the particle in the solenoid field is applied. For this correction equilibrium charge-state distributions[5] are assumed. Relative to this correlation the cuts are slightly rotated as can be seen from fig.2.

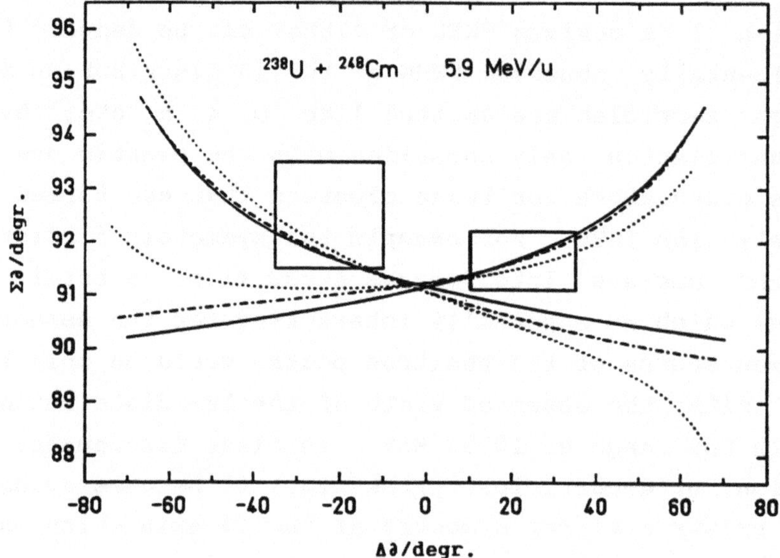

fig.2:Influence of the solenoid field onto the kinematic correlation[4]: a) elastic U+Cm kinematic, no magnetic field(solid line),b) same kinematics with magnetic field assuming equilibrium charge-state distribution[5](dashed-dotted line), c) a constant charge state of 60^+ is assumed (dotted line).

If one takes into account the magnetic field of the apparatus, such a "rotation" indeed takes place if the ejectiles associated with monoenergetic positrons would be emitted with an angular-independent charge-state distribution which is sufficiently different from the charge-state distributions for "elastic"-events.

B)Measurement of the target-thickness dependence of charge-state distributions emerging from superheavy collision systems

in collaboration with: H. Bokemeyer, H. Folger, D. Schüll, W. Schneider, GSI
R. Künkel, Hahn-Meitner Inst., Berlin, FRG,
A. Bösser, Univ. Marburg, FRG

Charge-state distributions of projectiles at $\vartheta_{lab}=0°$ and scattered into small angles (up to 5°) have been subject to extended studies at GSI[5,6,7]. Here, however, only symmetric distributions of a small width of $\Delta q \simeq 5\ e^+$ have been observed. For the regime of very close encounters, which means scattering angles $\vartheta_{LAB}>30°$, only few data of poor statistical significance exist, suggesting that these distributions indeed change drastically and become asymmetric towards high charge states[8]. In order to provide more systematic data for this case, we investigated the target-thickness dependence of charge-state distributions of projectile and recoil particles scattered into $\vartheta_{lab}=35°$ for 5.9 MeV/u Pb on Au collisions at the GSI magnetic spectrometer. Gold has been chosen as target because it can be produced as self-supporting, very uniform foils as thin as 45μg/cm² and generally suffers only little deterioration under heavy-ion bombardement of moderate flux (i.e. 1 to 2 part.nA). For comparison with small-angle-scattering data, distributions at $\vartheta_{lab}=5°$ and $\vartheta_{lab}=15°$ have also been measured.

In fig.3 the projectile charge-state distributions observed at $\vartheta_{lab}=5°$ and $\vartheta_{lab}=35°$ are shown for the various target thicknesses measured. In accordance with the above-mentioned measurements, the distributions at 5° show no marked contributions at high charge states and can be fitted by a single Gaussian with a width of $\Delta q \simeq 5e^+$ (dashed lines). The centroid shift of 1.5 e^+ indicates that at least in target thicknesses of $\Delta x \simeq 100\mu g/cm^2$ no equilibration is achieved, which is again in good agreement with the above experiments.

For the case of $\vartheta_{lab}=35°$, however, the picture changes quite drastically. A substantial centroid shift over 10 units takes place when going from 20μg/cm² (on carbon backing) to 370μg/cm². The low charge-state sides of the 370μg/cm² and 870μg/cm² distributions coincide, indicating that for these components equilibrium is reached at thicknesses around

300μg/cm². A strong asymmetry towards high charge states is observed especially for the distributions from the two thickest targets. For 370μg/cm², charge state components differing from the central charge state by approx. 10 units are still contributing with a relative weight of 10%. (The dashed lines here are drawn to guide the eye and do not represent fit functions.) Since in ^{208}Pb little nuclear excitation is expected that could lead to substantial post-foil ionization, this asymmetry is a measure of the "equilibration" thickness or the net electron-capture cross section in the target material. In the EPOS experiments, however, the nuclear-excitation channel cannot be excluded, and the distributions shown in fig.3 are a lower limit for the actual case of the positron measurements, where similar target thicknesses (250 to 400μg/cm²) are used. Especially, if the monoenergetic positrons are associated with charge states substantially higher than for the elastic scattering events, this difference will be preserved to a considerable degree. On the other hand, the data show that changes in the target thickness by heavy-ion bombardement may strongly influence the relative contribution of high charge states.

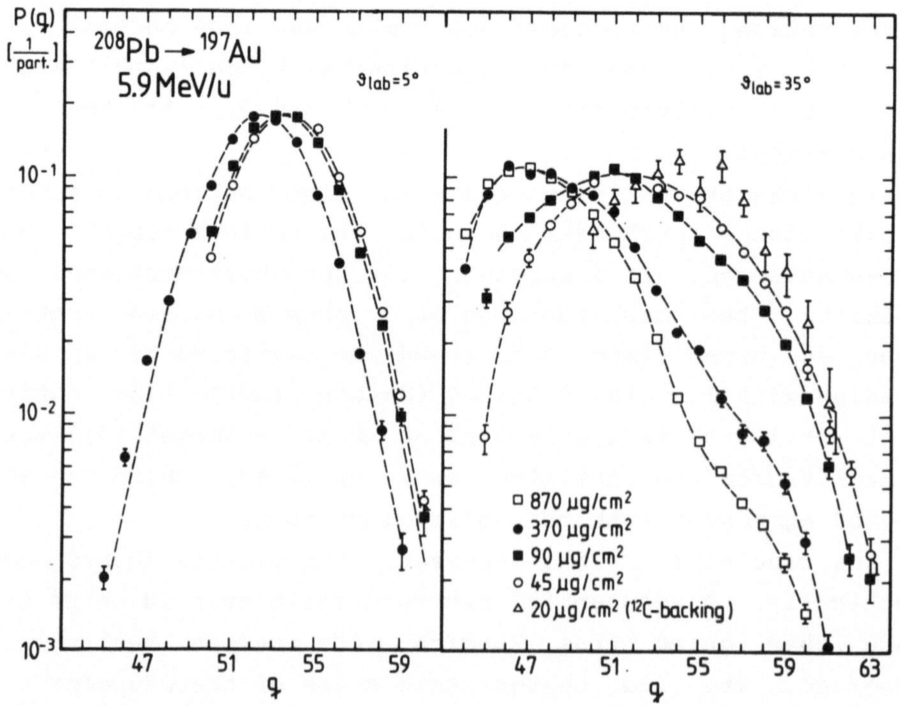

fig.3:Pb-Charge-state distributions at ϑ_{lab}=5° and ϑ_{lab}=35°

C)Investigations of the target structure by proton-backscattering spectroscopy

in collaboration with: D. Kraft, K. Bethge, IKF
H. Bokemeyer, H. Folger, GSI
M. Klüver, M. Begemann-Blaich,
H. Backe, Univ. Mainz, FRG

Generally, very heavy ions will change the target structure and composition drastically, even at small doses. Besides a larger energy deposition inside the target material, which is in the order of 100-130 MeV/mg/cm^2 for C-targets and for U-targets 40-50 MeV/mg/cm^2, one has to take into account their capability to transfer large linear momenta. This will lead to stable dislocations. To avoid sputtering of target material from the carbon backing, all targets used at EPOS are additionally covered with a thin carbon foil.

Any experiment that depends on an accurate knowledge of the projectile-energy loss inside the target has to take into account this fact. This is usually done by monitoring the scattering yield with a surface-barrier detector at an angle where the energy dependence is a smoothly varying function over the opening angle of the detector and where for nearly symmetric systems projectile- and recoil-ion intensities are almost equal but separated within the energy resolution of the detector. The information contained in these spectra, however, is a convolution of target shape and beam-energy stability which has to be determined independently e.g. by time of flight technique. Besides fairly poor resolution for heavy-ion detection, solid state detectors themselves are subject to substantial damage by the beam. As these problems turned out to be an essential point of consideration for the EPOS experiments[3)] a second independent way was used to obtain information about the target structure alone.

The targets have been investigated at the 2.5 MV van de Graaff accellerator of the IKF by proton-backscattering spectroscopy[9)]. The incident energy was 1.6 MeV. Collimation of the beamspot to a size of .5x.5mm^2 on the target with fluxes of 10-30nA allowed a lateral scanning of the target to test microstructure and large scale homogeneity of the target layer, which was evaporated onto the carbon backing as spot of ϕ5mm diameter(see e.g. the inset in fig.6). Protons of 1.6MeV, besides having a good emittance, have the advantage of boosting the signal for light elements (C, O, N, F) by contributions from elastic nuclear scattering. At scattering angles of $\vartheta_{lab}=175°$ the intensity of protons scattered from carbon is approximately ten times higher than that expected for pure Rutherford scattering. In this way spectra of good statistics for all target constituents could be taken in 5-10 min/per point.The shape of the peaks is a direct measure of the energy-loss profile at the target spot investigated, provided that FWHM>ΔE (ΔE is the experimental resolution). Besides this, the scattering yield is an independent measure of the total amount of target material on the spot. The information, of course, is not restricted by the detector resolution and can be applied also for layers of very small thickness. A third independent information can be obtained for the case of sandwich targets from the separation of front- and backlayer of the target. In targets, where a well separated layer of target material is enclosed between two foils this separation must be just the FWHM of the peak for scattering from the target material itself. Backscattering is also a useful technique to test the purity of the target. As the total energy in the backscattering spectrum depends on the position of an impurity inside the target and the resolving power decreases for heavy elements an identification of the element may become ambigious. A combination with another methode like x-ray analysis should be employed here, if necessary.

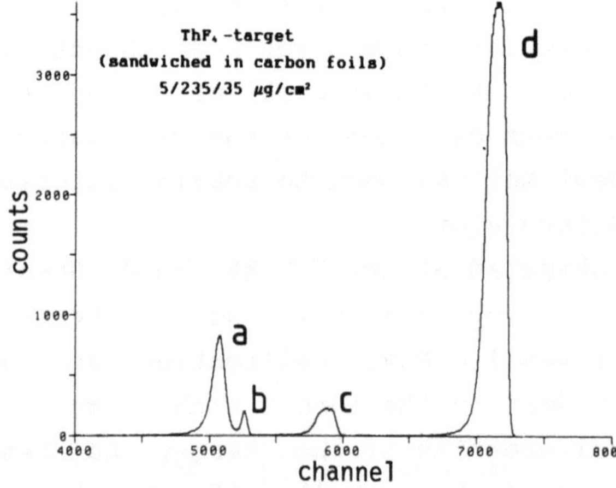

fig.4: Proton backscattering spectrum from a typical non irradiated ThF_4-target. The peaks are: protons scattered from carbon backing (a) from carbon front layer (b) from F(c) and from Th (d)

In fig.4 an example is given for a non irradiated target of ThF_4 (235 µg/cm²) on a carbon backing of 35 µg/cm² covered with a carbon foil of 5 µg/cm². The peaks of protons scattered from Th and F have identical widths much broader than the experimental energy resolution of $\Delta E \simeq 11$ keV. They are in very good agreement with an energy loss calculated for the target thickness measured by the GSI-target laboratory, if ThF_4 compound is assumed. Fig.4 therefore is a demonstration of a target which has a very homogeneous microstructure and where no marked contaminations are observed. It shows that protons are an adequate probe due to the advantages described above. Heavier particles like α-particles,at the cost of a much longer measuring time would allow an improved depth resolution which is not necessary for the purpose of the present investigation.

To illustrate the possible influence of an irradiation, in fig.5 two spectra for scattering from a 1.1mg/cm² metallic Th foil are shown. The upper spectrum was measured at a point which had not been hit by the heavy-ion beam and shows an excellent microscopic homogeneity. The lower spectrum was taken at a point close to a hole burnt into the foil by the U-beam. Here one has to assume that the structure has turned into a "lumpy" shape manifested by a wide variation of energy losses reaching from half the energy loss in the homogeneous target to many times of this value on this sharply located point.

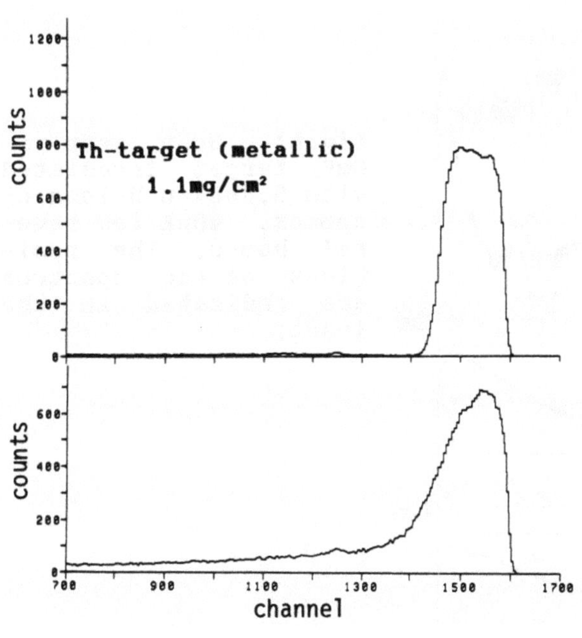

fig.5: Proton backscattering spectra from a 1.1 mg/cm² metallic Fh target. Upper spectrum of a positron which was not hit by the heavy ion beam, lower spectrum close to a hole burnt by the U-beam

Consequences for the EPOS-measurements:

It becomes clear that such a change during a positron measurement besides strongly influencing the magnetic field effect will prevent an exact determination of the projectile energy (at a level of several percent(≃.1MeV/u)). Therefore all targets employed in EPOS-measurements have been investigated by this method after a beam time. Additionally, samples are scanned prior to irradiation in order to have a standard. As one example of the typical shape of irradiated CmF_3 targets evaporated on carbon layers at ORNL, USA, in fig.6 a scan is shown. The positions are indicated on the inset. The second spectrum, which is taken at a position outside the beam spot may be used as a standard. The target shows drastic changes into a lumpy shape,in spectrum 3 the 10% level has an energy loss of 4 times, in spectrum 4 of 6 to 7 times the homogeneous value (see the energy scale in the picture). Besides these drastic deteriorations of this target in the irradiated region, two rather strong impurities are seen.

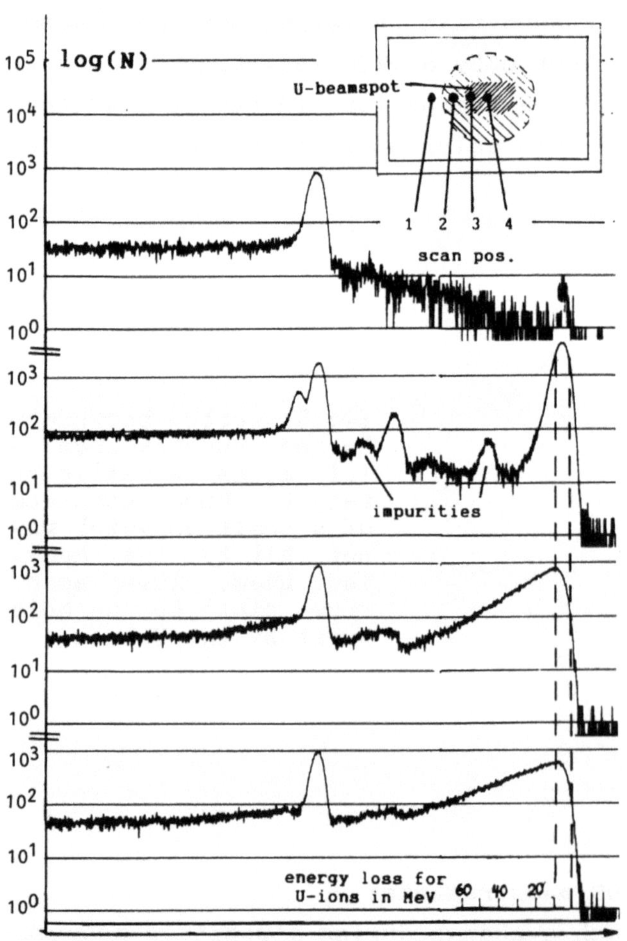

fig.6: Scan over a CmF_3 target irradiated with 5.9MeV/u U-ions of approx. 40nA for several hours. The positions of each spectrum are indicated in the inset.

Similar deteriorations have been observed for all Cm- targets[10], each of which, due to the lack of Cm-material, remained in the beam for many hours.As a consequence of this fact, ThF_4 targets are now used for experiments performed at GSI. Besides the excellent initial homogeneity demonstrated in fig.3, these targets can be produced free of impurities and in large number. Every two hours the target is changed to minimize the effects of radiation damage. In a "warm-up phase" the beam intensity is increased from zero to maximum within 10 - 15 min. First results from the new targets show that indeed the radiation damage can be kept small by this procedure. A final scanning in the proton beam, however, is still useful to provide a basis to eliminate datasets from targets showing major damage, which for instance may be due to a sudden warm up.

The results given above have been worked out in close cooperation with my colleagues from the EPOS-group: H. Backe, K. Bethge, H. Bokemeyer, T. Cowan, H. Folger, J.S. Greenberg, K. Sakaguchi, P. Salabura, D. Schwalm, J. Schweppe, N. Trautmann and P.Vincent.I would like to thank H.Schmidt-Böcking for valuable discussions.

The experiments have been supported by Bundesministerium für Forschung und Technologie (BMFT) and GSI.

References

1) H. Bokemeyer, K. Bethge, H. Folger, J.S. Greenberg, H. Grein, A. Gruppe, S. Ito, R. Schule, D. Schwalm, J. Schweppe, N. Trautmann, P. Vincent, M. Waldschmidt; in Quantum Electrodynamics of Strong Fields, edited by W. Greiner, Plenum, New York (1983), page 273
2) P. Kienle; in the proceedings cit. in ref. 1, page 293
3) H. Bokemeyer in these proceedings
4) A. Gruppe, PhD-thesis, J.W. Goethe-Univ., Frankfurt(1985), and GSI-85-4
5) V.S. Nikolaev and I.S. Dimitriev, Phys.Lett.28A, (1968) 277
6) Franzke, unpublished data

7) D. Maor, P.H. Mokler, D. Schüll and Z. Stachura, Nucl.Instr. and Meth. 194 (1982) 377-380

8) D. Schüll, W. F. W Schneider, H. Bokemeyer, R. Bock, H. Freiesleben, F. Pühlhofer, D. Bangert, B. Kohlmeyer, M. Marinescu; GSI scientific report; GSI 83-1

9) See e.g. W. -K. Chu, J.W. Mayer, M. -A. Nicolet, Backscattering spectrometrie, Academic Press, New York (1978)

10)K. Bethge et al.IKF Annual Report (1983), page 60

SPECTROSCOPY OF POSITRONS FROM HEAVY ION - ATOM COLLISIONS

C. Kozhuharov

Gesellschaft für Schwerionenforschung (GSI)
Postfach 110 451
D-6100 Darmstadt 11, West Germany

INTRODUCTION

In this lecture I would like to present some results of our experimental investigations on pair creation in heavy ion collisions, particularly emphasizing our studies of collision systems with combined charge numbers greater than 173. I would like to discuss both the gross features of positron production as well as the occurrence of a narrow structure in the spectra of positrons, emitted in U+U and U+Th collisions. A discussion of recent experiments with subcritical collision systems - i.e., for which the combined charge is smaller than 173 - is contained in the article by W. Koenig in this volume [1]. The experiments I shall review, were carried out by a GSI Darmstadt-Technische Universität München collaboration *.

The basic motivation of our studies originated from the possibility to study superheavy atoms, formed transiently during heavy ion collisions. Fig. 1 shows schematically the formation of the quasiatom Z=184 in collisions of uranium ions with uranium atoms at bombarding energies close to the Coulomb barrier. In the upper part the internuclear distance is shown as a function of the collision time. The lower part of Fig. 1 displays the evolution of the binding energies of the innermost electronic shells during the collision. Their tremendous increase with decreasing internuclear distance should lead to a "diving" of the 1sσ-state into the negative energy continuum. On the other hand, due to the time dependence of the two center Coulomb field, energy and momentum can be transferred to the bound and to the continuum electrons, which are then excited into higher lying empty states or into the positive energy continuum, as also sketched in the lower part of the Fig. 1. As a consequence, vacancies are created both in the bound states as well as in the negative energy continuum, and the latter could be detected as the so called *dynamically induced positrons*. Also indicated is a process, where a preionized 1sσ-level dives into the continuum, becomes a resonance embedded there, and might eventually decay via *spontaneous positron emission*. Note however, that as long as the collision partners follow pure Rutherford trajectories, the diving time is as short as $2 \cdot 10^{-21}$ sec, whereas the spontaneous decay width corresponds to $\tau \sim 10^{-19}$ sec [2]. Thus, the measured spectrum is expected to be totally dominated by the broad distribution of the dynamically induced positrons.

---

* Members of the group: E. Berdermann, F. Bosch, M. Clemente, M. Franosch, S. Huchler, P. Kienle, W. Koenig, C. Kozhuharov, H. Tsertos, and W. Wagner

But already in 1980, at the very beginning of our investigations, the measured spectra of positrons, emitted in U+U and U+Th collisions, exhibited an unexpected structure, superimposed over the dynamically induced positrons and over the - also broad and structureless - background due to nuclear excitation processes (Fig. 2) [3]. The estimated line width of ~ 70 keV leads to a time scale, which is approximately an order of magnitude longer than the collision time. It stands to reason that this surprising result gave a very strong impetus for further series of measurements, which are the subject of this lecture.

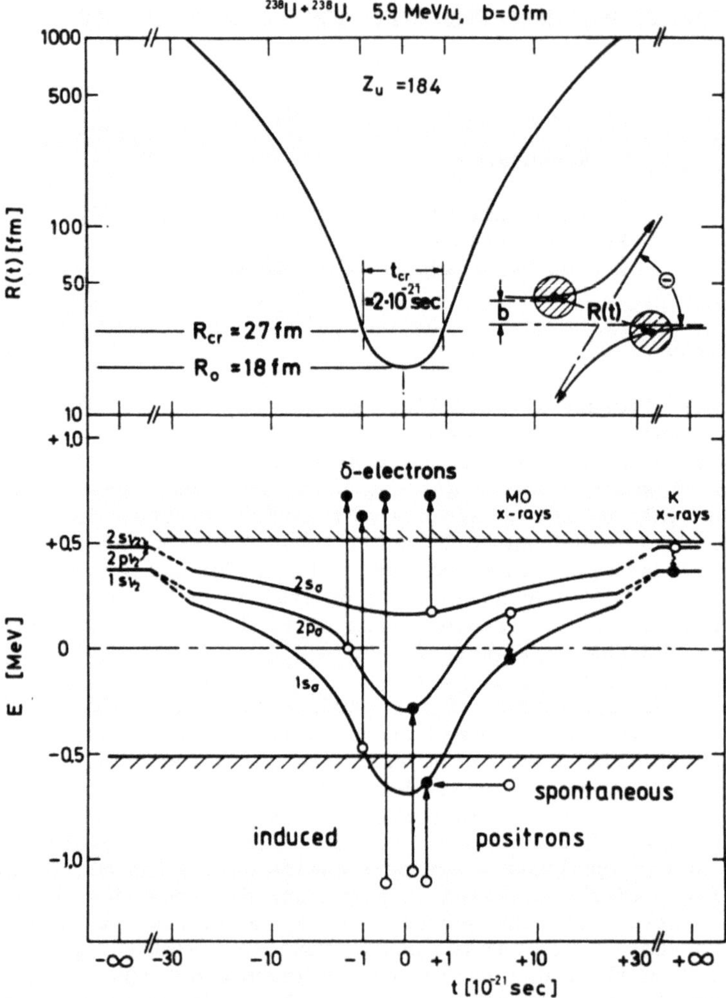

Fig. 1: Schematic presentation of a supercritical collision of uranium atoms at 5.9 MeV/u bombarding energy.
Upper part: Dependence of the internuclear distance R(t) on the collision time in a head on collision following classical Rutherford trajectory.
Lower Part: Evolution of the binding energies E of the innermost electronic shells. Also indicated are the pair production processes, induced by the collision dynamics, and the spontaneous positron creation.

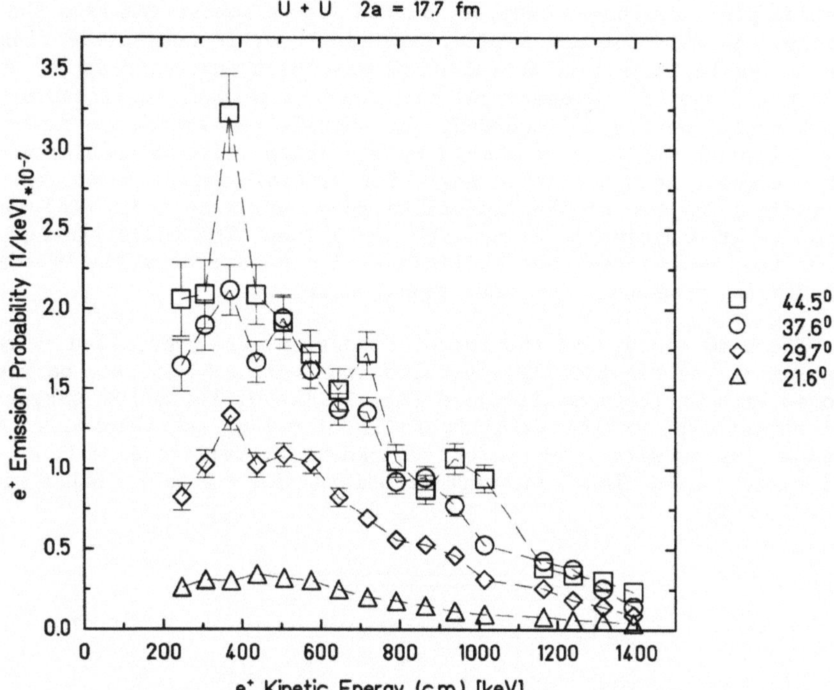

Fig. 2: **Spectra of positrons associated with particles scattered at angles (Lab.) as indicated [3]. Also indicated is the collision diameter 2a for U+U collisions at 5.9 MeV/u.**

EXPERIMENTAL METHOD

As already shown in Fig. 2, the probabilities for pair creation in heavy ion collisions are rather small: ~10^{-7} per collision and keV. The γ and δ-ray backround, produced in the in-beam measurements, is several orders of magnitude higher. Thus, an experimental setup with high background suppression and high detection efficiency - for measurements of small cross sections and for coincidence measurements - is needed. Fig. 3 shows the experimental arrangement used [4,5,6,7]. The target is in the object point of an iron-free orange-type β-spectrometer. Positrons, emitted from the target between 30° and 70° with respect to the beam direction, are momentum analyzed by the toroidal magnetic field produced by 60 coils and focused onto the positron counters placed in the focal area. They pass through a position sensitive proportional gas counter, which subdivides the accepted momentum range of ($\Delta p/p$)=14% into six bins and suppresses the background from γ-rays. This counter surrounds a 12 cm long, conically shaped plastic scintillator, where the positrons are stopped and their energy is recorded additionally to the momentum determination. Within the focused momentum band the overall detection efficiency is 14% of 4π. Electrons, emitted from the target, are defocused by the magnetic field, and cannot reach the positron counters. γ-rays are shielded by a lead absorber between the target and the detectors. The total positron spectrum was measured by scanning the magnetic field repeatedly up and down, and the duration of each particular setting of the magnetic field was defined by a constant number of scattered particles. With this multiscaling method no unfolding of positron pulse height spectra is necessary, and this could certainly be an advantage when dealing with structured spectra.

The scattered particles were detected with a position sensitive, annular parallel plate avalanche counter, placed 7.8 cm downstream from the target. The anode of the detector was subdivided into 12 or 16 concentric rings, allowing thus an angular sensitive detection of particles scattered in the angular range between 13° and 52°. Kinematical coincidences between scattered projectile and target nuclei were also recorded. The angular resolution was ±2.5° on an average. (Although trivial, it should be mentioned that the annular geometry leads to the highest possible solid angle for particle detection at certain scattering angle.) Because of the individual read out, each anode ring was capable of operating at counting rates as high as 10^6/sec. The modest energy resolution, 25% of the energy loss, was sufficient to separate elastically scattered ions from fission fragments and light reaction products.

The beam energy and the target condition were controlled by recording the spectrum of the elastically scattered ions with a Si surface barrier detector, denoted with BM (beam monitor) in Fig. 3. The U, Th, and Pd targets used, were self supporting, rolled metallic foils of ~ 1 mg/cm² thickness. Additionally uranium targets with ~ 500 μg/cm² thickness (metallic) as well as ~ 300 μg/cm² (oxide) were used. The foils were irradiated for ~ 2 to ~ 4 hours with a beam of

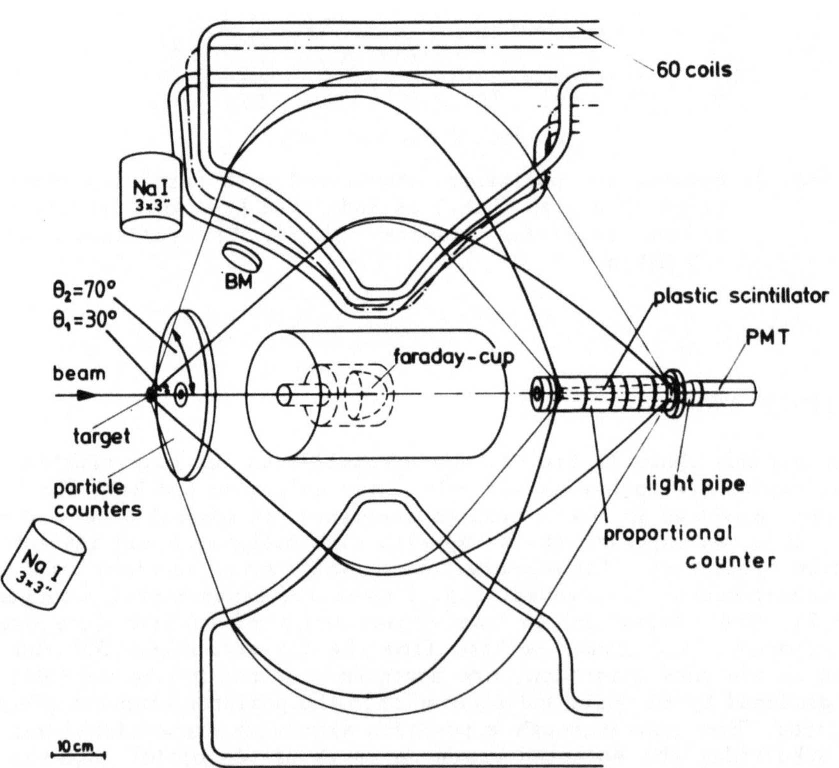

Fig. 3: Schematic view of the experimental setup in the orange-type β-spectrometer. Positrons are focused by the toroidal magnetic field onto a position sensitive proportional gas counter which surrounds a plastic scintillator. Scattered particles are detected with an annular parallel plate avalanche counter with 12 anode rings.

~ $3 \cdot 10^9$/sec and replaced a with fresh one, after low energy tails in the scattered particle spectrum signalled an advancing target deterioration.

The spectra of γ-rays were recorded simultaneously with the positrons, using two 3"×3" NaI scintillator counters placed at 90° and 125° relative to the beam direction at distances from the target of 19 cm and 23 cm, resp. The coincidence conditions with respect to scattered particles were the same as the requirements for positron particles coincidences.

The entire detection system was adjusted and calibrated (in beam) with electrons in the same momentum range by simply reversing the magnetic field of the β spectrometer.

Physical quantities in positron spectroscopy

The objective of positron spectroscopy in heavy ion collisions is in principle to investigate the triple differential cross section for positron creation:

$$\frac{d^3\sigma_{e^+}\ (E_{e^+},\Theta_{e^+},\Phi_{e^+},E_1,\Theta_1,Z_1,M_1,\Theta_2,M_2)}{dE_{e^+}\ d\Omega_{e^+}\ d\Omega_p} \qquad (1)$$

which is differential with respect to the positron kinetic energy E_{e^+}, to the solid angle element $d\Omega_{e^+}$ of the emitted positron, and to the solid angle element $d\Omega_p$ of the scattered particle. It has been already assumed that the scattering process is quasi elastic, and this is indeed the case in the experiments described below. The quantity Θ_{e^+} describes the emission angle of the positron with respect to the beam axis, Φ_{e^+} - with respect to the scattering plane, whereas Θ_1 and Θ_2 are the scattering angle relative to the beam direction of the detected ejectiles and recoils. The cross section is furthermore a function of the charge numbers Z_1,Z_2, and masses M_1,M_2 of the projectile and target nuclei, resp. as well as of the bombarding energy E_1.

From an experimental point of view it is very convenient to measure the energy and solid angle differential probability for positron production:

$$\frac{d^2P_{e^+}}{dE_{e^+}\ d\Omega_{e^+}} = \frac{d^3\sigma_{e^+}/(dE_{e^+}\ d\Omega_{e^+}\ d\Omega_p)}{(d\sigma_R/d\Omega_p)} \approx \frac{(\Delta^2N_{e^+}/\varepsilon_{e^+})}{\Delta E_{e^+}\ \Delta\Omega_{e^+}N_p} \qquad (2)$$

which is obtained by normalizing the cross section (1) with respect to the Rutherford cross section. Experimentally, the number $\Delta^2N_{e^+}$ of positrons in the energy interval ΔE_{e^+} and solid angle element $\Delta\Omega_{e^+}$ is measured in coincidence with the scattered particles, detected under certain scattering angle. The single counting rate of the latters is recorded simultaneously, hence the total number of particles N_p is also known. Thus, knowing the positron detection efficiency ε_{e^+}, the probabilities for positron emission can be easily calculated. It is obvious that using this method, the beam intensity, the target thickness, and the efficiency for particle detection cancel in Eq.(2) - i.e. - they cannot influence the results.

It should be noted that the positron emission probabilities and cross sections have been transformed from the laboratory system to the c.m. system prior to the presentation in this lecture. Furthermore, assuming isotropic emission in the c.m. system, these quantities were integrated over $d\Omega_{e^+}$.

RESULTS AND DISCUSSION

The early studies [4,8,9] on positron creation in heavy ion collisions already identified two sources of positrons: One is the internal pair creation following nuclear excitation processes and the other is the pair creation due to the collision dynamics.

Positrons of nuclear origin

In the collision of two heavy ions at energies close to the Coulomb barrier, nuclear levels of the collision partners may be excited which decay by transitions having energies greater then $2m_ec^2$ The use of uranium beams results in the excitation of a large number of such nuclear levels and the observed γ-ray spectra exhibit a smooth exponential fall-off toward higher energies. Thus, the superimposed energy distributions of the positrons due to internal pair conversion of all these transitions would also exhibit a smooth spectrum which cannot be distinguished from the contributions due to the collision dynamics. However, the measured γ-ray spectra can be converted into positron spectra using theoretical conversion coefficients [10], as demonstrated first by Meyerhof et al.[11]. This procedure was successfully tested for several light collision systems, where practically no contributions from dynamically induced pair creation is expected. Indeed, it turned out that for collision of heavy ions with a combined charge number smaller then 160, all observed positrons could be attributed to internal pair creation [4,8,9]. Fig. 4 shows, as an illustrative example, the spectrum of positrons, associated with particles scattered between 22° and 38° for U+Pd collisions at 5.9 MeV/u bombarding energy [6,7]. The solid line represents the positron spectrum calculated from the

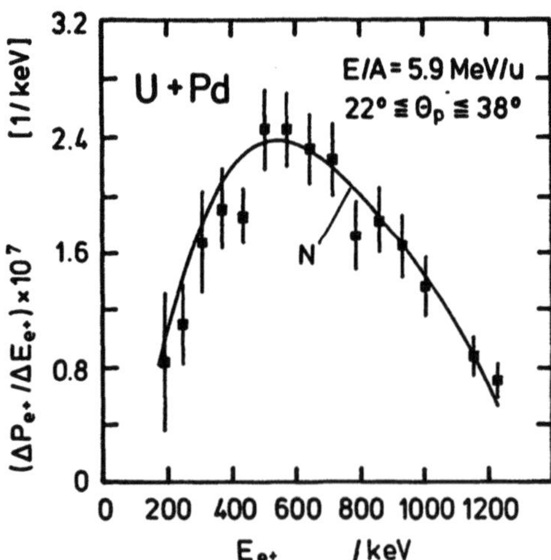

Fig. 4: Spectrum of positrons associated with scattered particles in the angular range as indicated for U+Pd collisions at 5.9 MeV/u. The solid line represents the positron spectrum obtained by converting the simultaneously measured γ ray spectrum [6,7].

γ-ray spectrum, measured simultaneously under the same kinematical conditions. The high energy γ-ray originate predominantly from statistical E1 transitions, hence E1 conversion coefficients have been used. As can be seen, the data are well reproduced within the limits of error.

Positrons of atomic origin

As already mentioned, in collisions of very heavy ions, transitions from the negative energy continuum into preionized bound states or even into the positive energy continuum can be induced (cf. Fig. 1). The remaining holes are detectable as positrons and have been indeed detected in a series of experiments [4,8,9] for collisions systems with united charge number higher than 160, where their contribution is no longer obscured by the background due to nuclear processes. This new source of positrons exhibits a remarkably steep increase with the united atom charge (Z^{20}), as predicted by the theory [2],

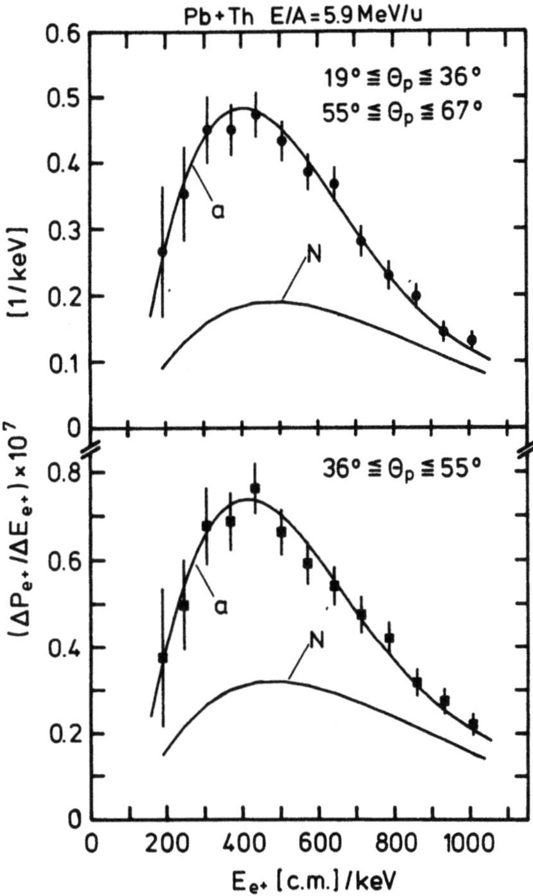

Fig. 5: Positron spectra from Pb+Th collisions at 5.9 MeV/u, associated with particles scattered in the angular ranges as indicated [7]. The nuclear background (N) is subtracted from the data. Curves (a) present results of theoretical calculations [12].

which demonstrates its electromagnetic origin. Fig. 5 shows spectra of positrons, emitted in collisions of Pb+Th at 5.9 MeV/u and associated with particles scattered in two different angular ranges as indicated [7]. The background due to nuclear processes, shown as a solid line denoted by N, is already subtracted from the data points. The smooth solid curve, denoted by 'a', repres-

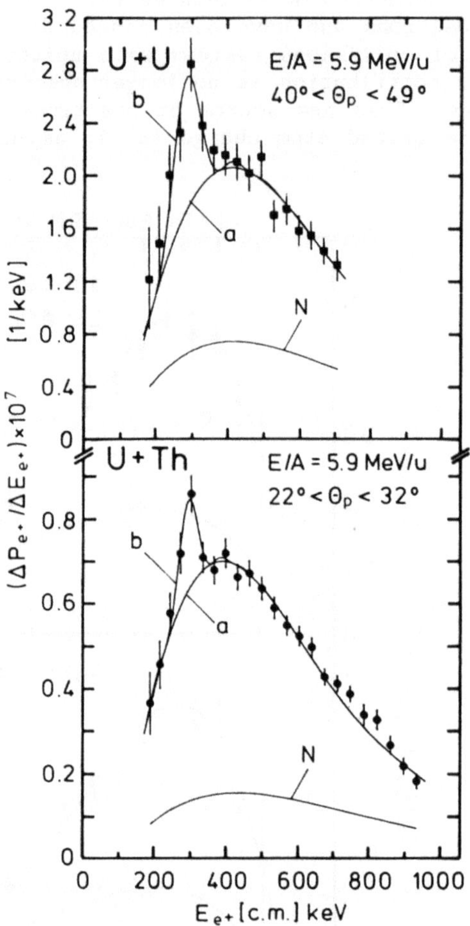

Fig. 6: Spectra of positrons emitted in U+U and U+Th collisions at 5.9 MeV/u for the angular range of scattered particles as indicated [7]. N - background due to nuclear processes (subtracted). Curves (a) - results of theoretical calculations [12,13], curves (b)- line fit to the data.

ents results of theoretical calculations [12] and is in good agreement with the measured spectral distribution. Here, and in all further data, the theoretical values were scaled down by a common factor of 0.8 - i.e. the theory overestimates the data slightly. Nonetheless - as we will also see in the discussion of the U+U and U+Th collision systems - the theory reproduces very well the impact parameter dependence, the dependence on the united atom charge, and the spectral distribution of the data for a broad range of these parameters.

Observation of a narrow structure

Fig. 6 displays spectra of positrons, emitted in collisions of U+U (upper part) and U+Th (lower part) at 5.9 MeV/u for the selected angles of scattered particles as indicated [7]. The procedures for nuclear background determination (curves N, subtracted from the data) and for the comparison with the theoretical results (curves a) are as described in the previous paragraphs. Note the good agreement between theory [12,13] and the smooth, high energy part of the measured spectra. Superimposed over the expected smooth distributions a pronounced, statistically significant peak, centered at appr. 300 keV , can be observed. From the width of appr. 70 keV a live time of the emitting source of appr. 10^{-20} (or longer) can be deduced, solely based on the uncertainty relation. The spectra, however, do not change if an additional kinematic coincidence is required. Hence, practically all observed positrons originate from binary collisions, where the transferred mass and energy ought to be very small, and where no fission occurs.

The observed symmetric line shape differ from the typical saw-tooth like shape for a internal pair creation following nuclear excitation. Also the fwhm of the latter of appr. 120 keV in its rest frame, together with the additional Doppler broadening of appr. 55 keV, is significantly larger than the observed small width. Monochromatic positron emission in nuclear transition could occur if the electron of the pair is captured into a preionized atomic state, predominantly into the K-shell [14]. Since the live time of a K vacancy is with appr. 10^{-17} much shorter then that of a 1.3 MeV excited nuclear state, the probability for such a process should be reduced by several orders of magnitude.

Nevertheless the possibility that the line is of nuclear origin was examined carefully by investigating electron and γ-ray spectra in search for lines with energies and intensities corresponding to the observed positron line. The lower part of Fig. 7 displays the measured γ-ray spectrum from U+U collisions [6]. Inserted are the hypothetical lines from E1 and E2-transitions, drawn in an absolute scale, Doppler broadened, and folded with the detector resolution. Even for the most unfavorable case of E1-transitions, the expected line intensity is of the same order of magnitude as the measured continuous background and should have been detected. On the other hand, if the observed positron line is due to contributions from a pair converted E0 transition, this transition should also manifest itself as a K-conversion line in the electron spectrum. The measured electron emission probabilities [6] are shown in the upper part of the Fig. 7 and are well reproduced by theoretical predictions for δ-ray emission (curve a) [15] scaled by a factor of 1.25. The intensity of the conversion line from the hypothetical E0-transition is calculated using the theoretical ratio of internal pair creation to internal conversion coefficients [10]. This hypothetical line was Doppler broadened, folded with the response function of the spectrometer, and superimposed over the curve (a). Again, as can be seen, no such line was detected.

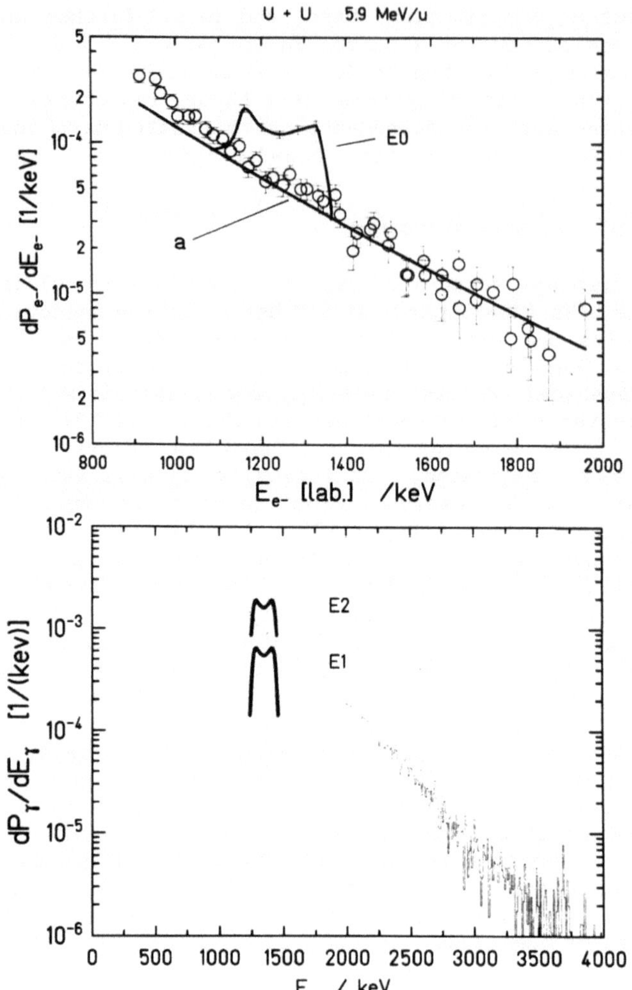

Fig. 7: Electron (top) and γ-emission probabilities per collision for U+U encounters at 5.9 MeV/u [6] The double humped, Doppler broadened lines would be expected, if the observed U+U positron line from Fig. 6 would originate from nuclear pair conversion decays of multipolarities as indicated. The hypothetical electron K-conversion line is superimposed on a calculated δ-ray spectrum [15] scaled by 1.25 (curve a), whereas the lines in the γ-spectrum are drown in an absolute scale.

I now turn to discuss some of the properties of the positron line: energy, width, and intensity and their dependence on the scattering angle. The dependence on the united atom charge will be discussed in the lecture of W. Koenig [1], together with the results from subcritical collision systems.

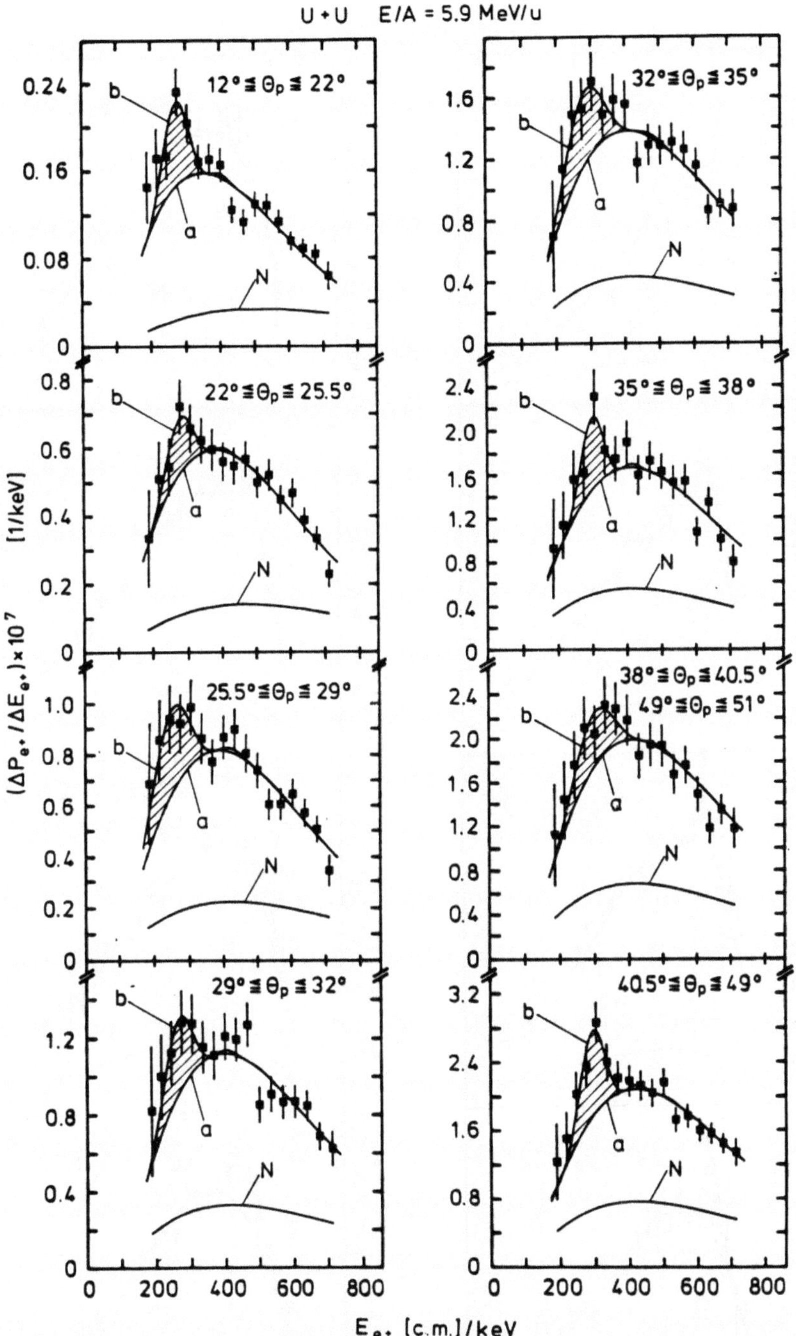

Fig. 8: Positron creation probabilities per energy interval and per scattered ion as a function of the positron kinetic energy for U+U collisions at 5.9 MeV/u are shown for 8 angular ranges as indicated [7]. The nuclear background (N) is subtracted from the data. Curves denoted by (a) represent results of theoretical calculations for dynamically induced positron emission [12,13]. Curves denoted by (b) are line fits to the data.

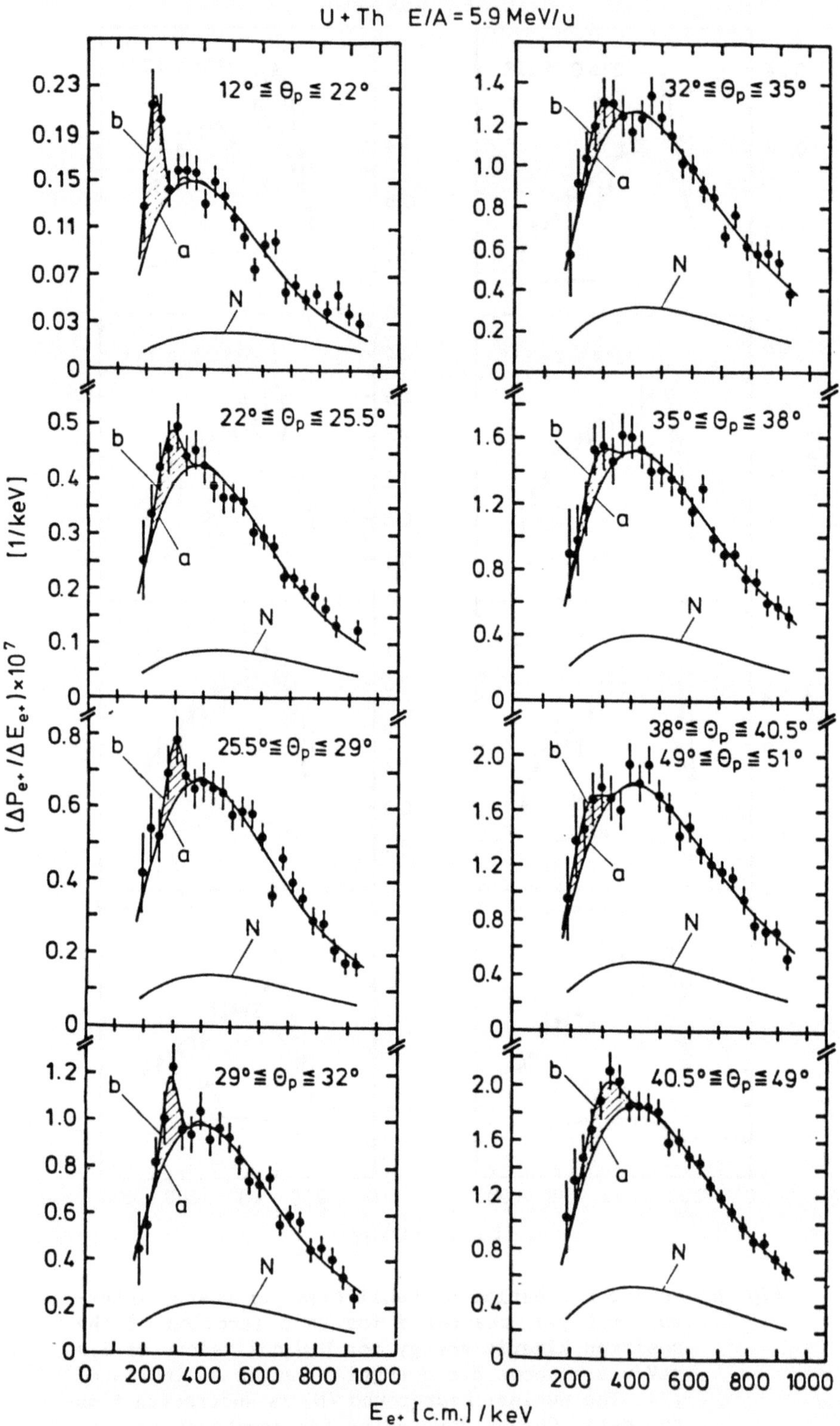

Fig. 9: **Positron creation probabilities as in Fig. 8 for U+Th collisions. Data from Ref. 7.**

Fig. 8 and Fig. 9 show the energy differential probabilities for positron emission per scattered ion as a function of the positron kinetic energy for U+U and U+Th collisions at 5.9 MeV/u [7]. The positrons are associated with particles, scattered under 8 different angles ranging from 12° to 51°. Again, the background due to nuclear processes has been determined as described previously and was subtracted from the data (curve N). Note that the smooth part of the spectra at energies higher than appr. 400 keV is well reproduced by the results of theoretical calculations for the dynamically induced positron emission [12,13]. The already mentioned common normalization factor was used consistently for all angles and for both collision systems. It should be emphasized again, that the same theory was used to describe the spectra for the Pb+Th system, presented in Fig. 5. As we shall see in the lecture of W. Koenig [1], the same theory also fits the data in the U+Ta, U+Au, and Pb+Pb collision systems. Also for the sake of consistent data analysis, a line superimposed on the smooth distribution was fitted to the data. As can be seen, all spectra exhibit lines centered around 300 keV. Note that the ordinate of the drawings for every particular scattering angle has its different scale - i.e. the probabilities for positron emission increase by more than an order of magnitude when going from smaller to larger scattering angles.

Fig. 10 summarizes the line energies and widths - i.e. the centroids and the FWHM from the line fit - as a function of the (lab.) scattering angle of the particles. Within the statistical fluctuations both the energy and the width do not show any significant dependence on the scattering angle. This can be interpreted as an indication that the line is emitted from the c.m. system. Under this assumption the energies were Doppler corrected and fitted by a constant. The average values for the energies amount to (280±6) keV for U+U and (277±6) keV for U+Th and the corresponding widths are (80±8) keV (U+U) and (75±6) keV (U+Th), respectively. The almost constant line energy for both collision systems is difficult to understand in the framework of a model, suggesting that the line could originate from spontaneous positron creation in the Coulomb field of a long lived di nuclear system, formed in central collision [16,17], because in the latter case a strong dependence of the energy on the united atom charge is expected.

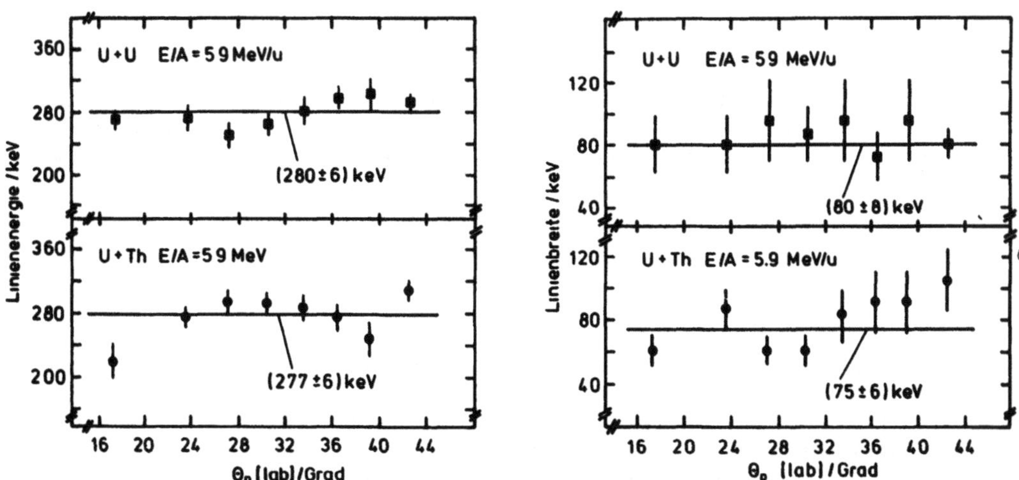

Fig. 10: Line energy (left part) and line width (FWHM, right part) of the positron line as a function of the particle scattering angle (Lab.) for U+U and U+Th collisions at 5.9 MeV/u. The solid line represents a constant, fitted to the data points [7].

The differential cross sections for line emission are shown in in Fig. 11 as a function of the (c.m.) scattering angle of the particles Θ_p [7]. They are best reproduced by a $1/\sin\Theta_p$ dependence for both collision systems. From this, one is lead to conclude that the emission is isotropic in the reaction plane. A fit with a constant to the data points is - albeit slightly less probable - yet not unreasonable. One should also keep in mind that, in the case of symmetric collisions, the picture could be obscured by the fact that contribution from scattering angle Θ_p cannot be distinguished from those associated with $(\pi - \Theta_p)$. The total cross sections for line emission are (104±8) μbarn in the case of U+U and (51±4) μbarn for the U+Th collision system - i.e. the cross section drops down by a factor of two when the united atom charge decreases by merely two protons, as can be also seen in Fig. 11. If this strong Z dependence holds, it is obvious that hypothetical effects in the subcritical region would be very weak - comparable with the error bars in the spectrum for the Pb+Th collisions (cf. Fig. 5). Therefore further measurement, with significantly better statistics, are needed (and were carried out - cf. e.g. Ref. 1) in order to investigate the subcritical systems in the search for structures.

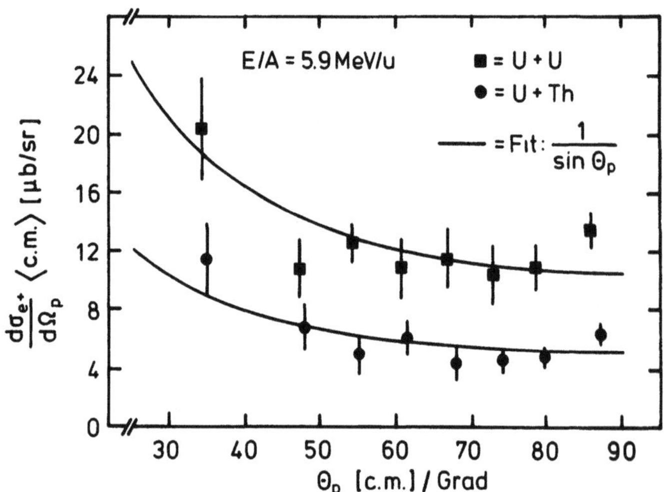

Fig. 11: Differential cross section for line emission as a function of the (c.m.) particle scattering angle for U+U and U+Th encounters [7]. The data points are fitted by a $(1/\sin\Theta_p)$ angular distribution.

Let us now turn to the dependence of the positron line on the bombarding energy. This problem has been studied for U+U collisions at 5.7, 5.9, and 6.2 MeV/u. Fig. 12 shows the measured positron spectra [7], associated with particles scattered in the angular region between ~ 40° and ~ 50°. The data are analysed in the same consistent way as described previously: curves denoted by (a) represent the theory which is in good agreement with the measurement, reflecting the increase of the dynamically induced positrons with the increasing bombarding energy. The information about the peak structure is extracted by applying the same line fit procedure to the data points. Its differential cross section amounts to (15±3) μbarn/srad at 5.7 MeV/u, to (13.5±1.3) μbarn/srad at

5.9 MeV/u, and to (18±4) μbarn/srad at 6.2 MeV/u. This leads to a rather weak dependence on the bombarding energy within a 120 MeV energy interval. There is a certain discrepancy between this finding and measurement of the EPOS collaboration [18], where preliminary results indicate a resonance like behavior of the cross section. It should be mentioned, however that this type of measurements are rather difficult, because of the very fragile targets. As described previously, the target condition was controlled and the targets were replaced quite often. The beam intensity was kept also rather low which was certainly not favorable for the statistics. Further measurements, with significantly improved statistics, are therefore needed and planed, utilizing a beam pulse synchronized target wheel.

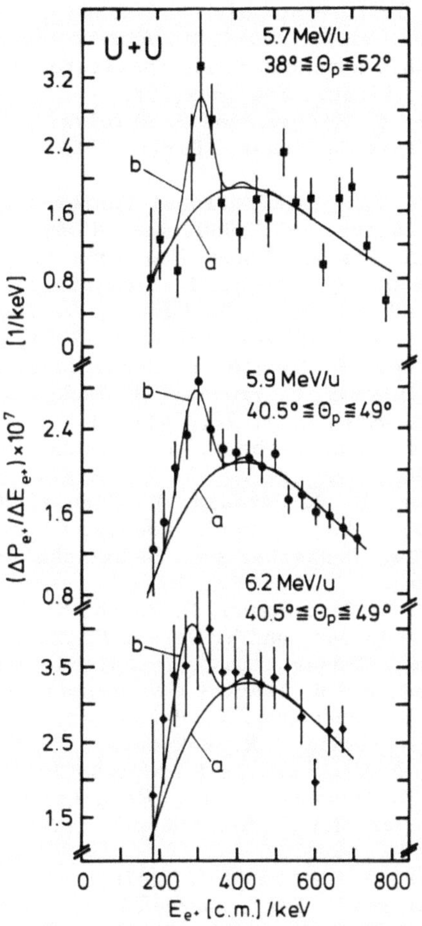

Fig. 12 Energy differential positron production probabilities for U+U collision system at 5.7, 5.9 and 6.2 MeV/u for the angular ranges as indicated [7]. Curve a - theory for induced positron emission [12,13], b - line fit to the data.

For a further discussion of these topics I would like to refer again to the lecture given by W. Koenig at this conference [1].

ACKNOWLEDGMENTS

My special thanks go to my friends and collaborators, who have carried out the experiments I discussed. I would also like to thank P. Armbruster, W. Greiner, B. Müller, and J. Reinhardt for many helpful discussions. I am especially indebted to U. Müller, T. de Reus, and G. Soff for making their results available to us prior to their publication. This work was supported by the Bundesministerium für Forschung und Technologie.

REFERENCES

1. W. Koenig, lecture at this conference
2. W. Greiner (ed.), *Quantum Electrodynamics of Strong Fields*, Proceedings NATO Advanced Study Institute, June 15-26,1981, Lahnstein, FR Germany, Plenum, New York(1983)
   W. Greiner, B. Müller, and J. Rafelski: *Quantum Electrodynamics of Strong Fields* Springer, Berlin(1985), and the references cited there
3. P. Kienle, 7th International Conference on Atomic Physics, MIT, Cambridge, August 4-8, 1980, in: *Atomic Physics 7*, (D. Kleppner and F.M. Pipkin, eds.), Plenum, New York(1981), p. 1
   E. Berdermann, F. Bosch, M. Clemente, F. Güttner, P. Kienle, W. Koenig, C. Kozhuharov, B. Martin, B. Povh, H.Tsertos, W. Wagner, Th. Walcher, *GSI Scientific Report*(1980), p. 128
4. C. Kozhuharov, P. Kienle, E. Berdermann, H. Bokemeyer, J.S. Greenberg, Y. Nakayama, P. Vincent, H. Backe, L. Handschug, and E. Kankeleit, *Phys. Rev. Lett. 42*,376(1979)
5. H. Tsertos, Diplomarbeit, Technische Universität München(1980), unpublished
6. M. Clemente, E. Berdermann, P. Kienle, H. Tsertos, W. Wagner, C. Kozhuharov, F. Bosch, and W. Koenig, *Phys. Lett. 137B*,41(1984)
   M. Clemente, Dissertation, Technische Universität München(1983), unpublished
7. H. Tsertos, E. Berdermann, F. Bosch, M. Clemente, P. Kienle, W. Koenig, C. Kozhuharov, and W. Wagner *Phys. Lett. 162B*,273(1985)
   H. Tsertos, Dissertation, Technische Universität München(1985), unpublished, and H. Tsertos *GSI Report*, 85-13, unpublished
8. H. Backe, L. Handschug, F. Heßberger, E. Kankeleit, L. Richter, F. Weik, R. Willwater, H. Bokemeyer, P. Vincent, Y. Nakayama, J.S. Greenberg, *Phys. Rev. Lett. 40*,1443(1978)
9. H. Backe, C. Kozhuharov, in: *Progress in Atomic Spectroscopy*, Part C (H.J. Beyer and H. Kleinpoppen, eds.), Plenum, New York(1984), p.459
10. P. Schlüter, G. Soff, and W. Greiner, *Z.Phys.A286*,149(1978)
    P. Schlüter and G. Soff, *ADANDT 24*,509(1979)
    P. Schlüter, G. Soff, and W. Greiner, *Phys.Rep.75(6)*,329(1981)
11. W.E. Meyerhof, R. Anholt, Y. El Masri, D. Cline, F.S. Stephens, and R. Diamond, *Phys.Lett.B69*,41(1977)
12. U. Müller, G. Soff, T. de Reus, J. Reinhardt, B. Müller, W. Greiner, *Z.Phys.A313*,263(1983), see also Ref. 2
13. T. de Reus, U. Müller, private communication, see also Ref. 2
14. L.A. Sliv, *Dokl.Akad.SSSR,64*,521(1949)
15. G.Soff, private communication
16. J. Reinhardt, U. Müller, B. Müller, and W. Greiner, *Z.Phys.A303*,173(1981), see also Ref. 2
17. T. Tomoda and H.A. Weidenmüller, *Phys.Rev.C28*,173(1981)
18. H. Bokemeyer, lecture at this conference

POSITRON EMISSION FROM SUBCRITICAL SYSTEMS*

E. Berdermann[1], F. Bosch[2], M. Clemente[1], M. Franosch[1], S. Huchler[1], J. Kemmer[1], P. Kienle[2], W. Koenig[2] , C. Kozhuharov[2], H. Tsertos[1] and W. Wagner[1]

[1] Physik Dept., Techn. Univ. Muenchen, D-8046 Garching West Germany
[2] Ges. f. Schwerionenforschung mbH, D-6100 Darmstadt 11 West Germany

* talk presented by W. Koenig

INTRODUCTION

This contribution is a continuation of the talk presented by C. Kozhuharov during this conference [1], thus a detailed introduction is omitted.

The data presented in this contribution differ in so far from the data shown during the conference and presented partially in the summary talk by P. Kienle [2] as most of the preliminary results are now completely analyzed, i.e. the efficiency of the apparatus was taken into account and absolute positron-emission probabilities are given.

The structures observed in positron spectra obtained for overcritical heavy-ion collisions are far from being understood. In order to reduce the number of speculations and models to explain these structures and to get additional hints with respect to the possible origin of the observed positron lines several subcritical heavy-ion systems were investigated by the ORANGE-collaboration.

EXPERIMENTAL SET-UP

The spectral distribution of positrons, emitted during a heavy ion

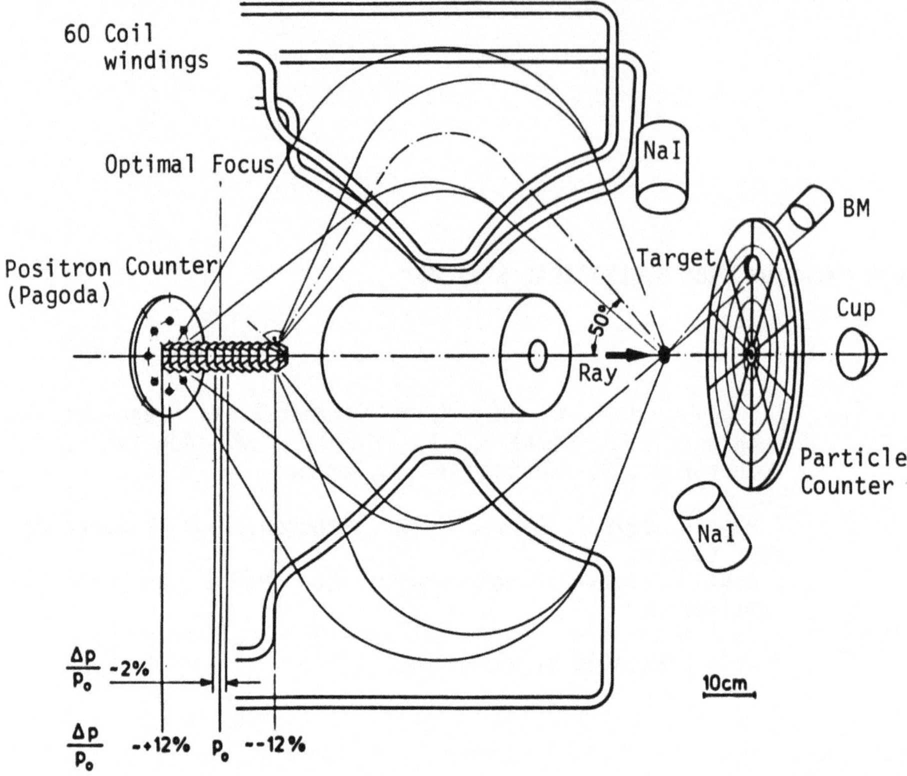

Fig. 1: Experimental set-up, consisting of an ORANGE-spectrometer combined with a multi-detector system ('PAGODA') for the positron detection and a Θ- and Φ-sensitive parallel plate avalanche counter used for the heavy ion detection. The ORANGE-spectrometer consists of 60 coils arranged cylindrically around the beam axis. The 'PAGODA' consists of 10 planes ('roofs') each containing 6 cooled high resolution Si-detectors.

collision, were obtained with the experimental set-up shown in Fig. 1. The main component is an iron-free 'orange-type' spectrometer consisting of 60 coils arranged radially around the beam axis. Compared to the set-up discussed by C. Kozhuharov [1], several improvements were made in order to obtain a higher energy resolution, a better signal to noise ratio and a significantly higher efficiency:

Positron Detector

Positrons emerging from the target area are focussed in backward direction onto a so-called PAGODA detector. It consists of 10 'pagoda roofs' mounted along the beam axis. Each roof is of hexagonal shape and contains 6 high resolution Si-detectors. The Si-detectors are cooled by alcohol down to -30 °C. A matrix readout (rows and columns) is used to get the information about energy and arrival time of particles separately for each detector. The energy and time resolution obtained in-beam for a 300 keV positron amount to 10 keV and 10 ns, respectively.

The division of the positron detector into 10 roofs allows to make use of the dispersive properties of the Orange-spectrometer. For a given spectrometer current a momentum byte of $\sim$ 2 % is focussed onto each roof corresponding to an energy resolution of also 10 keV at 300 keV due to the spectrometer dispersion. A total momentum byte of $\sim$ 20 % is analyzed simultaneously at a given spectrometer current.

Heavy Ion Detector

Projectile- and target-recoil ions are detected by a large area, position sensitive parallel-plate avalanche detector. It has annular geometry thus measuring all ions independently of the orientation of the scattering plane with respect to the emerging positrons. This detector is mounted 100 mm downstream of the target and covers an angular range Θ from $\sim$ 15° to 55°. A delay line readout performed separately for the left and right hand side of the detector gives an angular resolution of better than 1°.

The arrangement mentioned above allows to investigate kinematical coincidences between the two outgoing heavy ions, i.e. the Q-value of the collision and the mass drift can be obtained assuming a binary reaction mechanism. In addition, the detector is divided into 12 Φ-segments with separate readout. This allows together with the segmentation of the pagoda roofs to determine the orientation of the scattering plane relative to the emitted positron with a resolution of $\sim$ 60°.

Target Wheel

For the study of structures in positron spectra data with good statistics and high resolution are crucial. Additionally, the energy straggling of the incident ion should be limited in order to obtain information about the dependence of these structures from the beam energy. This requires homogeneous targets limited in thickness. Because a rapid target deterioration takes place for most of the heavy targets (U, Th, Pb) if the beam intensity is increased above $\sim$ $^1/_2$ particle nA (pnA) the latter requirement is the limiting factor with respect to the high count rate necessary to get good statistics within a limited beam time of typical several days.

In order to take full advantage of the count rate capability of the experimental set-up and the beam intensities available at GSI, a rotating target wheel was developed. It allows to increase the beam intensity by roughly a factor of 10 compared to a stationary target without any significant sign of target deterioration over a period of several days.

The rotating target wheel takes advantage of the time-structure of the heavy-ion beam which has a macro-pulse structure of ~ 4 - 5 ms beam followed by a ~ 16 - 15 ms pause. Nine targets each ~ 70 mm long, are mounted on a circle with ~ 400 mm diameter. The wheel rotation is synchronized with respect to the beam pulse such that one ion pulse moves across the whole target whereas within the beam-pause interval the next plus one target is put into place. Every second turn the same target is used again and can cool down in between. Thus, melting effects which result in the occurence of drop-like structures on the target can be avoided completely. Even target deterioration being proportional to the total dose of heavy ions are reduced because the total target area covered by the beamspot for one target load is ~ 200 times larger than in case of a stationary target.

In addition to the higher beam intensities acceptable due to this construction this wheel has further advantages. Using two target sets with different thicknesses and thus different energy losses of the incident heavy ion beam, a direct comparison of positron spectra measured at different beam energies can be obtained. Because one can switch between the two target sets every 20 ms this comparison is performed under exactly the same experimental conditions.

EXPERIMENTS AND DATA ANALYSIS

The following three subcritical systems were investigated with the experimental set-up described above.

System	Z_u	E_{ion}/A MeV/u	$d_{targ.}$ μg/cm²
U on Ta	165	5.9	1000
U on Au	171	5.9	950-1000
Pb on Pb	164	5.7	400 / 800

Z_u symbolizes the sum of the projectile- and target-atomic number, whereas E_{ion}/A specifies the incident beam energy and d_{targ} the target thickness.

All three systems are characterized by an atomic number of the united system low enough to prevent the K-shell from diving into the negative continuum even if a completely stripped compound nucleus is formed [3].

In case of Ta and Au stationary targets were used because the development of the target wheel was not finished at that time. These elements are practically the only ones in this regime of atomic numbers which can withstand high beam currents without deterioration. On the other hand the contribution of nuclear positrons emitted after Coulomb excitation of the target nuclei is large compared to f.e. Pb due to the different nuclear structure. As will be shown later, this is a major disadvantage of these two systems. In order to maintain the advantage of the Pb+Pb system - namely the scattering of two double magic typical shell model nuclei - the incident beam energy for this system was below the value chosen for the other two systems. According to Ref. 4 no significant contribution from nuclear positrons emitted after nuclear reactions is expected at this energy.

Before the experimental results will be discussed, a short description of the data analysis will be given. In order to avoid a lengthy discussion of the ∿ 500 spectra created from the 70 parameters stored event by event this discussion will focus on the response of the apparatus to a positron.

The redundance of the energy determination via the pulse height of the Si-detector and via the magnetic field of the spectrometer is used to obtain a clean signature of a positron emitted from the target. Whereas the spectrometer analyzes the momentum and charge of a particle emitted the Si-detector gives a signal proportional directly to the energy. Thus, by plotting the count rate versus the energy differences given by the energy obtained from the pulse height minus the energy calculated from the spectrometer current (magnetic field) a peak at zero-energy difference will occur only for positively charged particles steming from the target and having the mass of a positron. Other particles like electrons or gamma-rays scattered from the walls will show up as a continuous background. Besides a unique signature for positrons this method delivers in addition an accurate and quantitative determination of the remaining background contribution. It shows further the energy resolution in beam at any time during the experiment.

Energy difference spectra as described above are shown in Fig. 2. The upper part shows the result of a calibration measurement obtained with opposite magnetic field for δ-electrons and a single 'pagoda roof' (6 Si-detectors). The spectrum is integrated over all kinetic energies of the electrons. In heavy ion collisions the emission probability for δ-electrons is several orders of magnitude higher than for positrons. Thus, they are ideally suited to check the apparatus in beam within a short time. The measured FWHM of 13 keV is in good agreement with the expected resolution of 10 keV for each independent energy determination. The small background contribution visible in the specturm is due to backscattering of electrons from the detector, a well known process [5].

The lower part shows an energy-difference spectrum in case of positrons obtained for the system U+Au at a rather low kinetic energy of the positrons of 230 keV. At these low energies the background due to multiply scattered electrons is not negligible. However, it can be separated clearly from the peak produced by positrons emitted from the target. At 300 keV kinetic energy this background contributes to about 5 % of the total intensity taken within the FWHM of the peak.

The lower part of Fig. 2 shows the result for all 10 'pagoda roofs' added together and summed up over the whole beam time.

This resulting resolution for the energy difference amounts to 20 keV. Thus, the resolution with respect to the kinetic energy of the positrons which is taken as the average of the energies obtained via the magnetic field and the pulse height of the 'pagoda' detectors amounts to $^1/_2$ of this value, i.e. 10 keV. In this context it should be mentioned that the Doppler broadening due to the acceptance angle of the Orange spectrometer also amounts to 10 keV at $\sim$ 300 keV.

The energy-difference spectra contain another interesting information namely about the life-time of the system which emits the positrons. If the emitter is not at rest in the lab-system a finite life-time leads to an emission behind the target which results in a shift of the peak shown in Fig. 2b towards positive energies. A flight path of $\sim$ 1 mm gives rise to a shift of the centroid by 1 keV. Assuming a speed of the emitter

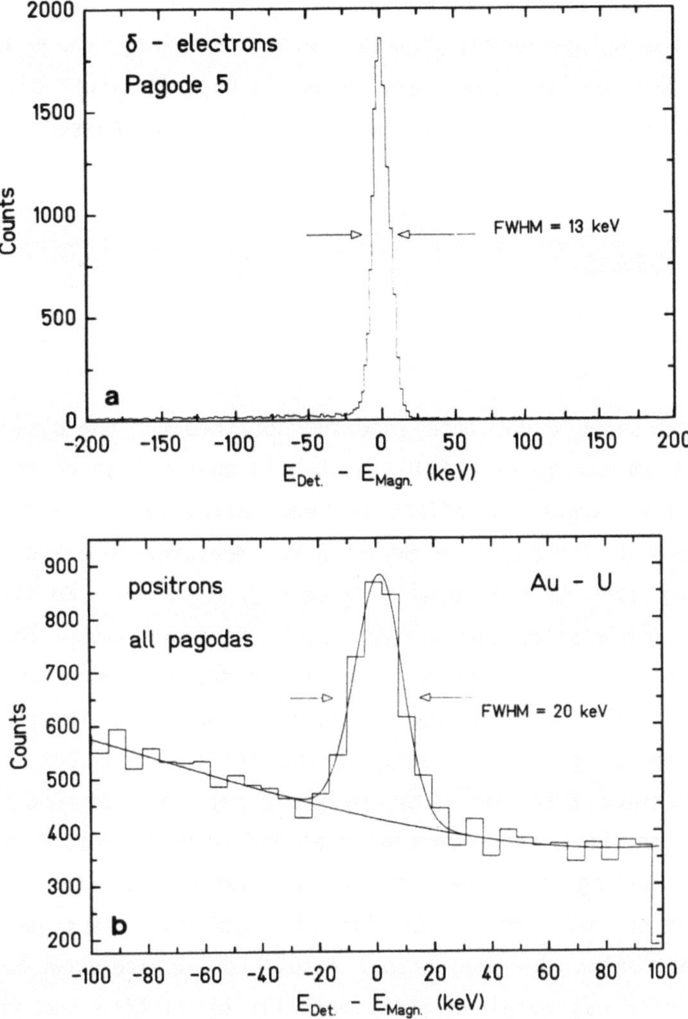

Fig. 2: Energy difference spectra obtained for δ-electrons (upper part, 2a) and positrons (lower part, 2b). The x-axis shows the difference between the energies obtained from the pulse height of the Si-detectors and the energy calculated from the magnetic field of the spectrometer. Fig. 2a shows a single pagoda roof, the spectrum is integrated over all electron energies (135 keV < $E(e^-)$ < 450 keV).
Fig. 2b shows a spectrum obtained at an energy of 230 ± 6 keV, summed over all pagoda roofs (cf. page 4).

equal to the center-of-mass velocity of the heavy ions this corresponds to a life-time τ = 60 ps. Whereas the dynamically induced positron emission takes place within 10^{-21} s and nuclear positrons are emitted within a few ps or less recent results obtained by the Tori collaboration suggested a life-time of $\tau \sim$ 100 ps of the system which causes the narrow lines observed in positron spectra [6].

Besides a correction with respect to the detection efficiency of the apparatus which is nearly constant within an energy range in between 150 keV and 500 keV the peak area shown in Fig. 2b gives directly one point in the energy spectrum for the system U+Au, discussed below (Fig. 4a).

EXPERIMENTAL RESULTS

1. U + Ta

Figure 3 shows a positron spectrum obtained for the system U+Ta at an incident beam energy of 5.9 MeV/u. Due to energy loss of the incident ions within the target the effective beam energy ranges from 5.9 MeV/u down to 5.7 MeV/u. The positron emission was measured in coincidence with outgoing heavy ions in the angular range $\theta_{lab} = 15^\circ-50^\circ$. No differentiation between projectiles and target-recoil ions was made. The emission probability per keV is given on an absolute scale. Figure 3a shows the experimental results without any suggestive lines drawn through the data points. Figure 3b gives in addition the contribution due to nuclear positrons obtained from the measured gamma-ray spectrum (dotted line). The dashed line shows the expected total positron-production probability obtained by adding the theoretical calculation for the dynamically induced positron emission [7]. As for all systems investigated by the ORANGE collaboration the theoretical result was scaled down by a factor 0.8. This factor was obtained experimentally by fitting the high energy parts of the positron spectra ($E(e^+)$ = 400-800 keV) measured for the U+U system [1,8]. Whereas the relative efficiency of the NaI-gamma-ray detector used to measure the gamma-ray spectrum can be determined quite accurately with respect to the positron-detection set-up the absolute efficiency of the total apparatus might be affected by systematic errors of up to 10% - 20%. The measured positron spectrum is dominated by nuclear positrons, the contribution of dynamically induced positrons amounts to ∿25% only. Because of the different velocities of the emitters of dynamically induced and nuclear positrons the results are specified in the lab-system.

Besides the smooth contribution expected for dynamically induced as well as for nuclear positrons the measured spectrum shows two additional structures. Both are statistically significant by more than 6 standard deviations if the dashed line is taken as a reference. The probability that **all** data points can be described by a smooth function is below 4 %.

A fourth order polynomial using the logarithm of the x- and y-values was fitted to the data points. This type of a function describes perfectly well both the calculated spectrum of dynamically induced positrons and the background due to nuclear positron emission.

Taking the dashed curve as a reference, the energy and width (FWHM) of the lower lying line amount to 230±5 and 35±11 keV, respectively. The total probability per collision normalized to the measured elastic scattering amounts to (3.2±0.8) x 10^{-7}. These results were obtained by

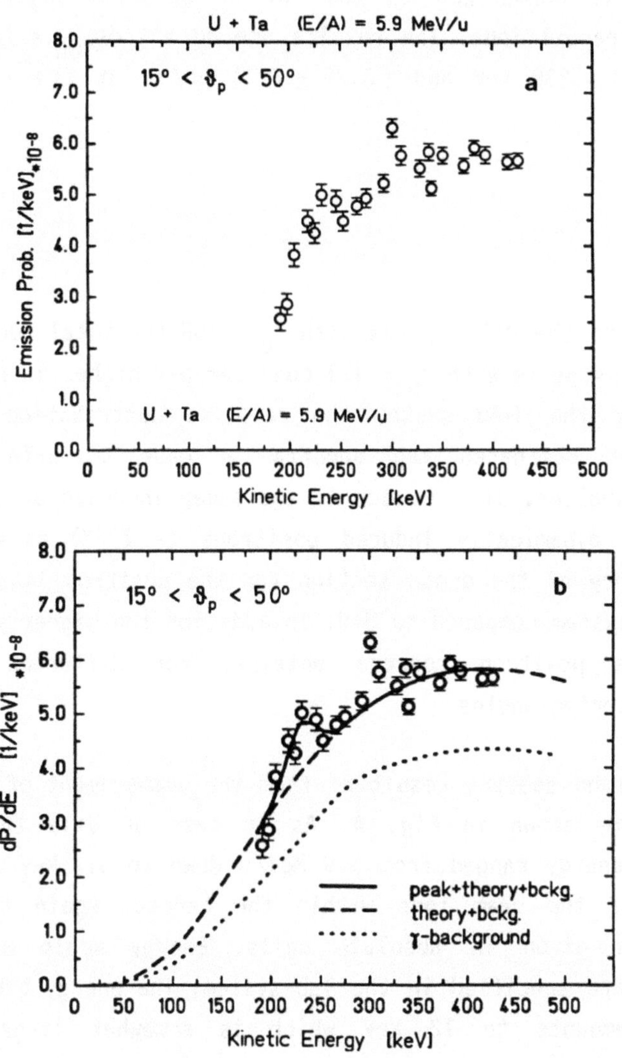

Fig. 3: Positron spectra measured for the system U+Ta (Z_u = 165) at an incident beam energy of 5.9 MeV/u.
The upper part shows the measured differential emission probability only. The lower part includes a comparison with the contribution from nuclear positrons and the sum of the nuclear and dynamically induced emission. The latter was calculated by T. de Reus et al. [7] and scaled down by a factor 0.8 (see text).

fitting a Gauss function to the data points as shown in Fig. 3b. The second structure at 310 keV is extraordinarily small. The width of ~ 14 keV is just consistent with an emission from the center-of-mass system. Due to the uncertainty concerning the width the total probability can only be estimated to be of the order of 2×10^{-7}. The total probability of the lines can be converted into a cross section by multiplying it with the corresponding Rutherford cross section for the projectile and target ions. In order to allow a direct comparison with results obtained for other systems at somewhat different ranges of cm-scattering angles the cross section is normalized to the sum of cm-solid angles for both projectile and recoil ions. The results are $d\sigma(e^+)/d\Omega_{ion} = (0.40 \pm 0.1)$ μb/ sr at $E(e^+) = 230$ keV and (0.25 ± 0.1) μb/ sr at $E(e^+) = 310$ keV, respectively.

2. U + Au

Compared to the U+Ta system with $Z_u = 165$ the total nuclear charge of the united system is with $Z_u = 171$ considerably higher in case of U-Au collisions. For the U+Au system the relative contribution of nuclear positrons which dominated the spectra in case of U+Ta should be significantly smaller. This is due to the steep increase of the emission proability for dynamically induced positrons ($\propto Z_u^{20}$) as well as the strong dependence of the cross section for the positron lines estimated from the U+Ta system compared to U+U. In addition the higher yield allows to investigate positron spectra obtained for different ranges of heavy-ion scattering angles.

Two positron spectra resulting from the bombardment of a Au-target with U-ions are shown in Fig. 4. As in case of U+Ta the effective incident beam energy ranged from 5.9 MeV/u down to 5.7 MeV/u due to the energy loss of the beam ions within the target. Again the emission probability is given in absolute units. Energy scale and emission probabilities are specified in the lab-system. The energy binning of the data points amounts to 12 keV which is somewhat larger than the resolution of the system. The full line shows the sum of the contribution due to nuclear positrons obtained from the measured gamma-ray spectrum and the theoretically calculated contribution from the dynamically induced positron emission [7]. The latter was multiplied by the factor 0.8 as already described for the U+Ta system. The dotted line gives the contribution from nuclear positrons separately. The spectral distribution obtained in coincidence with all heavy ions registered by the parallel plate avalanche counter is shown in the upper part of Fig. 4 whereas the lower part shows the positron spectrum obtained for heavy ion scattering angles in between 31.5° and 50° in the lab-system.

Similar to the U+Ta case statistically significant structures show up in the energy range in between 200 and 400 keV. These structures are lying on top of the smooth distribution given by theory and nuclear background. They are more pronounced in case of kinematical constraints to the heavy ions (Fig. 4b), i.e. by selecting the more central collisions. Due to the strong increase of the Rutherford-cross section the upper spectrum in Fig. 4 is dominated by the forward scattered particles in between Θ_{lab} = 16.5° and 30°. Fig. 4b shows two lines at a position of 245 $\pm$ 5 keV and 313 $\pm$ 5 keV, respectively. If again the theoretically calculated production probability together with the nuclear contribution is used as a reference (upper full curve in Fig. 4) the total probabilities per elastic collision amount to (11.3 $\pm$ 2.2) x 10^{-7} and (13.7 $\pm$ 2.1) x 10^{-7}, respectively. The peak widths are 40 $\pm$ 6 keV for the lower lying and 35 $\pm$ 5 keV for the higher lying line. The same evaluation for the lower lying line in Fig. 4a gives a probability of (4.0 $\pm$ 0.8) x 10^{-7} keV and a FWHM of 29 $\pm$ 5 keV. This reduction of the probability in case of the larger angular range of scattered heavy ions is caused by the increase of the Rutherford cross section at small angles. The total line cross section normalized to the solid angle of both the scattered projectiles and the target recoil ions seems to be nearly independent of the heavy-ion scattering angle. These cross sections amount to $d\sigma(e^+)/d\Omega_{ion}$ = (0.72 $\pm$ 0.14) μb/sr for the restricted angular range of Θ_{ion} = 31.5° - 50° and (0.62 $\pm$ 0.12) μb/sr for the larger range of Θ_{ion} = 16.5° - 50°. These cross sections are ~ 50 % larger than the corresponding values obtained for the lighter U+Ta system. It should be mentioned in this context that the choice of the upper full curves in Fig. 4 as reference lines is justified by the fact that the low energy range ($E(e^+)$ < 210 keV) as well as the high energy range ($E(e^+)$ > 400 keV) is perfectly well described by the sum of dynamically induced and nuclear positrons. In these energy ranges the mean deviation between the full line and the data points amount to (0.5 $\pm$ 6.8) % ($E(e^+)$ < 210 keV) and (0.6 $\pm$ 1.0) % ($E(e^+)$ > 400 keV) in case of Fig. 4a and (7 $\pm$ 9) %, respectively (0.1 ± 1.6) % in case of Fig. 4b. Although this perfect agreement must be considered accidental due to the uncertainties in the absolute calibration of the apparatus it makes an adjustment of the smooth background function with respect to the data points unnecessary.

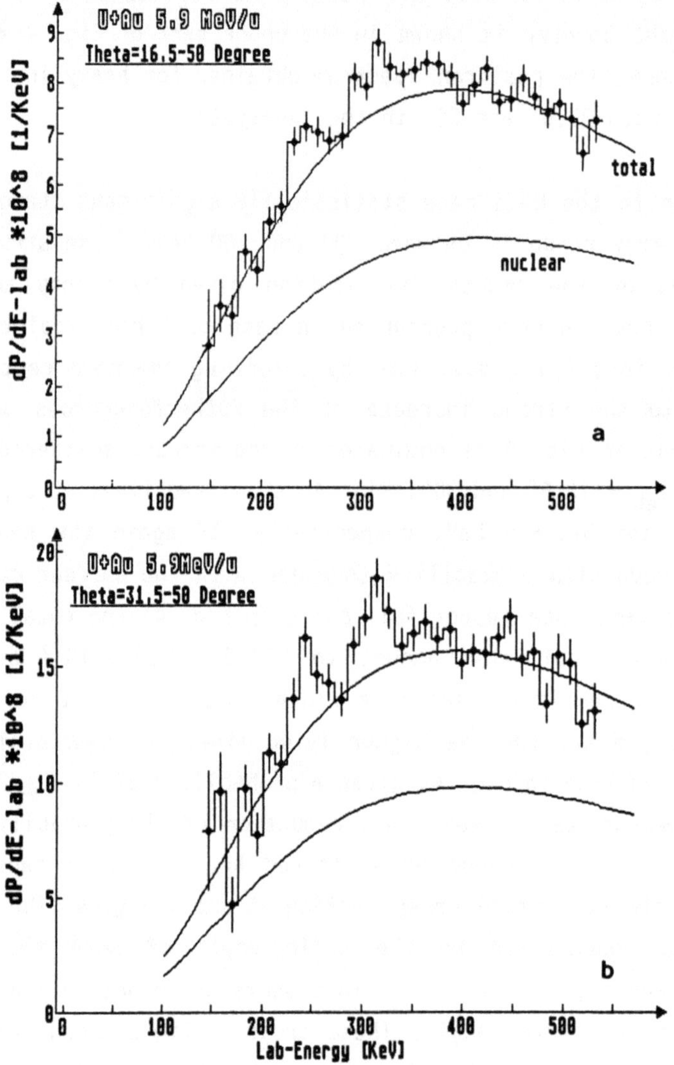

Fig. 4: Positron spectra measured for the system U+Au (Z_u = 171) at an incident beam energy of 5.9 MeV/u.
The upper part shows the positron emission probability per collision for heavy ions scattered within the angular range Θ_{lab} = 16.5° - 50°, whereas the lower part shows the spectrum obtained for a restricted range of heavy-ion scattering angles Θ_{lab} = 31.5° - 50°. The upper lines denote the sum of the nuclear and dynamically induced positron emission, the lower lines specify the nuclear contribution to the positron spectrum separately. The dynamical part, calculated by T. de Reus, was scaled down by a factor 0.8 (see text).

The asymmetry of the U+Au system allows to separate the scattered projectiles from the target-recoil ions by the method of kinematical coincidences. Furthermore this method was used to calculate the total kinetic energy of the two outgoing particles. The results are shown in Fig. 5a and b for heavy ions without the requirement of a coincidence with positrons. Using the time-of-flight difference between the heavy ions and the two scattering angles the total kinetic energy was calcula-

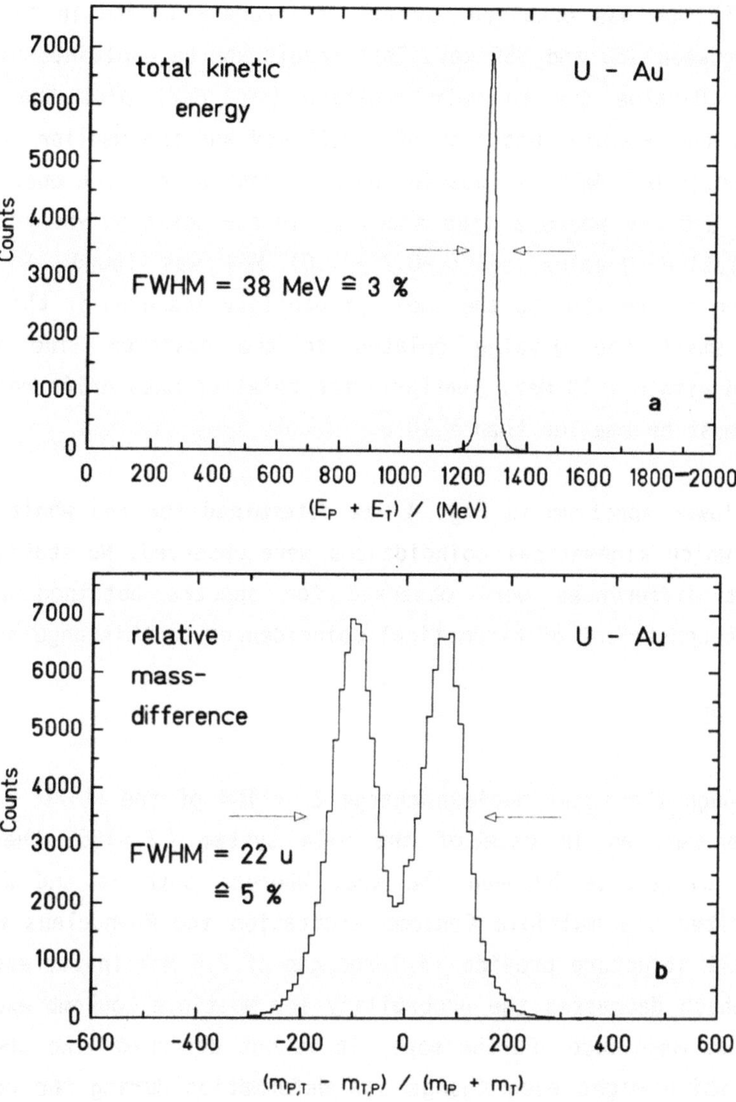

Fig. 5: Total kinetic energy (upper part) and relative mass difference (lower part) of the outgoing heavy ions measured for U-Au collisions at 5.9 MeV/ u. The data were obtained using the method of kinematical coincidences as described in the text.

ted without any assumption about the relative masses, i.e. any possible mass transfer. In the same way the relative mass difference was obtained without any constraints with respect to the reaction Q-value. However, a binary reaction was assumed. This method limits the resolution to ΔE = 38 MeV corresponding to $\Delta E/A$ = 0.16 MeV/u and 22 mass units, respectively. The calibration uncertainties for this method do not allow an accurate calculation of the total kinetic energy on an absolute scale. However, a very accurate comparison can be made for the centroids of the total kinetic energies obtained for elastically scattered heavy ions and those measured in coincidence with positrons. By this way a Q-value of

(-1.5 ± 0.5) MeV was obtained for all positrons detected in the energy range in between 150 and 550 keV. This result can be explained nicely by the atomic Q-value due to pair creation (∿ 1 MeV) plus the kinetic energies of the measured positron of ∿ 0.35 MeV and the smaller energy of the electron (∿ 0.1 MeV). Triggering on positrons within the energy range of 210 and 270 keV where a line shows up in the positron spectrum (Fig. 4b) a relative Q-value of (-0.7 ± 1.0) MeV was found. Since the contribution of the line to the whole probability measured in this energy range is small the Q-value related to the positron line is only constrained within ± 10 MeV. Similarly the relative mass drift related to this line must be smaller than ∿ 10 u.

The lower spectrum in Fig. 4 was determined for the whole angular range for which kinematical coincidences were observed. No statistically significant differences were observed for spectra obtained with and without the constraint of kinematical coincidences in this angular range.

3. Pb + Pb

Although the total nuclear charge Z_u = 164 of the Pb+Pb system is nearly the same as in case of the U+Ta system (Z_u=165) there is a remarkable difference between the two. Whereas both Ta and U can be easily excited via multiple Coulomb excitation the Pb-nucleus with its closed shell structure provides a large gap of 2.6 MeV in its excitation spectrum which decreases the probability for multiple Coulomb excitation by orders of magnitude. Furthermore, it is not deformed like the 'soft' U-nucleus which might even change its deformation during the collision [10]. A sub-Coulomb collision of two Pb-ions can be described reasonably well as the scattering of two billard balls, i.e. it represents a very clean system with respect to atomic processes. This is in particular important with respect to the speculation of mono-energetic pair creation with the electron filling a vacancy in the K-shell of the outgoing highly ionized ion. Such a process requires a nuclear transition at ∿ 1.2 MeV which should be observed in the gamma-ray spectrum. Whereas in case of U+Ta and U+Au the gamma-ray spectra increase exponentially to lower energies [8], the spectrum measured for the Pb+Pb system is rather clean in particular at lower gamma energies. This is shown in Fig. 6. Practically the whole intensity up to 2.6 MeV can be explained by the photopeak of the 2.6 MeV Pb-line together with the 1st and 2nd escape and compton scattering within the 3'' x 3'' NaI-detector. Above 2.6 MeV a small contribution due to the transition from the 2^+ excited state at 4.1 MeV

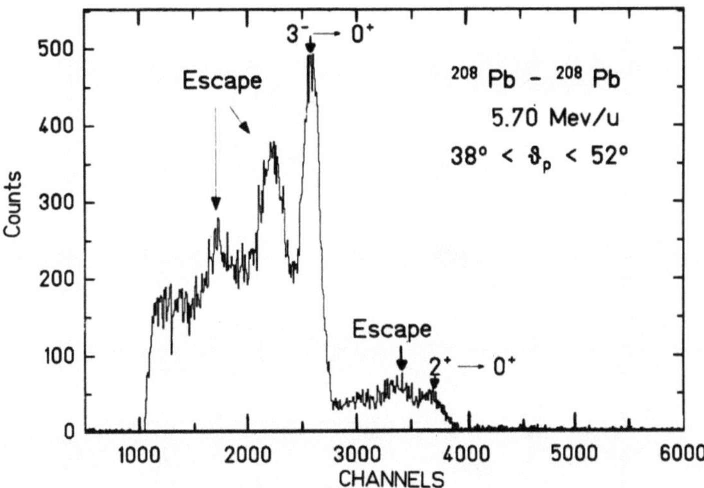

Fig. 6: Gamma-ray spectrum obtained for the system Pb+Pb at 5.7 MeV/u. The spectrum was measured in coincidence with heavy ions scattered within the angular range $\Theta_{lab} = 38^0 - 52^0$. Beside the two transitions at 2.6 MeV ($3^- \rightarrow 0^+$) and 4.1 MeV ($2^+ \rightarrow 0^+$) and the corresponding response of the NaI detector (escape lines, compton scattering) no additional lines are visible.

to the ground state is visible. Any nuclear reactions - f.e. sticking of the two nuclei which would lead to a few MeV excitation of the outgoing fragments - should be clearly observable in this spectrum as an additional source of gamma rays if the intensity of the corresponding positrons stemming from pair creation would exceed the error bars of the measured positron spectra discussed below.

Figure 7 shows three positron spectra measured in coincidence with heavy ions scattered into different angular ranges. As expected the contribution of nuclear positrons is reduced by more than a factor of 8 with respect to the U+Ta system.

It should be mentioned in this context that the absolute calibration of the contribution of nuclear positrons is preliminary and uncertain by ± 15 %. (The NaI detector was mounted in a different position in this experiment requiring a new calibration.) A discrepancy of ∿ 10 % between two source measurements with respect to the e^+ -detection efficiency in the high energy regime above ∿ 440 keV might lead to a corresponding reduction of the data points. (An additional check is performed momentarily.)

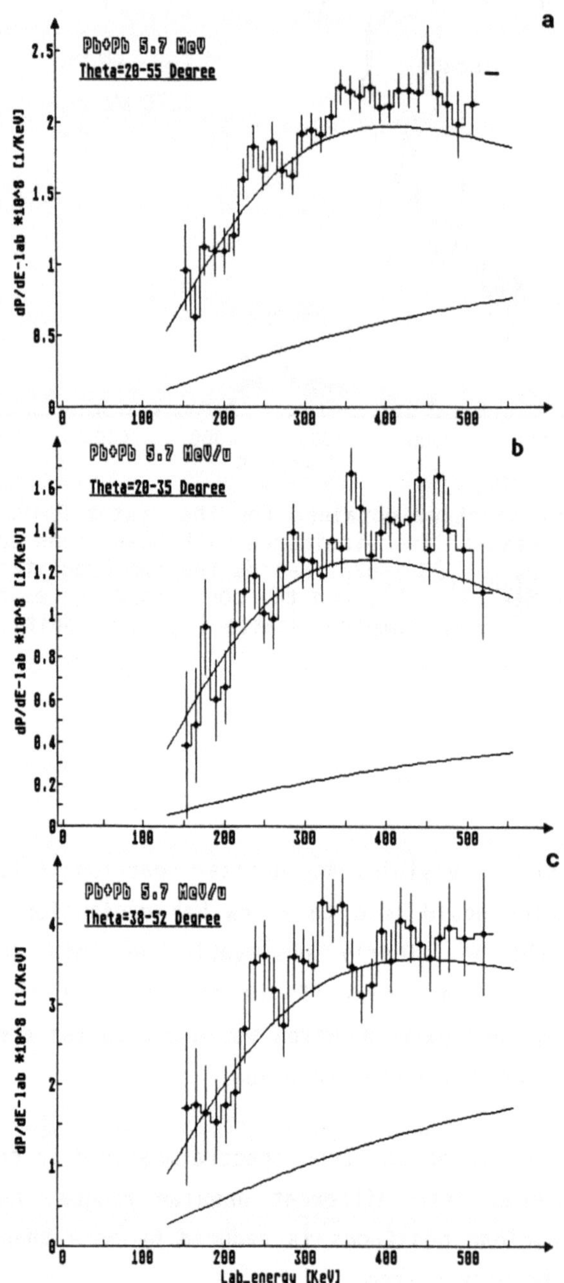

Fig. 7: Positron spectra obtained for the system Pb+Pb (Z_u = 164) at an incident beam energy of 5.7 MeV/u.
The uppermost part shows the positron emission probability per collision for heavy ions scattered within the angular range of θ_{lab} = 20° - 55°, the middle part and the lower part show spectra obtained for restricted ranges of heavy-ion scattering angles θ_{lab} = 20° - 35° and 38° - 52°, respectively.
The upper lines denote the sum of the nuclear and dynamically induced positron emission, the lower lines specify the nuclear contribution to the measured positron spectrum separately. The dynamically induced emission was calculated by T. de Reus et al. and scaled down by a factor of 0.8 (see text).

In Fig. 7a no pronounced structures show up at the energies where lines were observed in case of the U+Ta and U+Au systems. Only an indication of the line observed at ∿ 240 keV seems to be visible in this spectrum. For the spectrum obtained at small scattering angles θ_{ion} = 20°-30° the situation is similar (Fig. 7b). Around 360 keV two points show a deviation from the full line of together nearly 4 standard deviations. However, because of the above mentioned uncertainties a more detailed statistical analysis of the spectra will be performed after these uncertainties are settled. The spectrum obtained for more central heavy ion collisions, i.e. scattering angles in between 38° and 52° is on the other hand quite similar to the corresponding spectrum measured for the other two subcritical systems (cf. Fig. 9). It shows again two statistically significant lines at energies of 244 ± 5 keV and 330 ± 6 keV. The question whether or not the small bump in between is an indication of a third line can be discussed more precisely by considering the analysis shown in Fig. 8. The binning of 12 keV steps used in Fig. 7 for the positron energies is already larger than the energy resolution of the detection system. Thus a more precise information can be obtained by performing the same analysis of the energy difference spectra - mentioned at the beginning - by using the same binsize but shifting these bins in energy by 6 keV i.e. half the binwidth. The resulting two spectra denoted by different symbols are shown in Fig. 8. The higher resolution obtained this way shows more clearly that the line at 240 keV is enclosed in between two distinct minima and the shoulder below the peak at 330 keV becomes more pronounced.

The deviation of the data points from the upper full line in Fig. 7c and 8 is statistically not significant in neither the low- ($E(e^+) < 215$ keV) nor the high-energy range ($E(e^+) > 365$ keV). The probability that the full curve is consistent with the data points amounts to ∿ 50 % in both cases whereas it drops below 9 % in between. Thus the same procedure as for the other two subcritical systems can be used to extract a line width and intensity at least for the two structures at 244 keV and 330 keV, respectively. The resulting total emission probabilities amount to $(3.2 \pm 1.0) \times 10^{-7}$ for the lower lying line and $(4.0 \pm 1.1) \times 10^{-7}$, respectively. The structure in between at ∿ 292 keV gives an emission probability of $(1.7 \pm 0.6) \times 10^{-7}$. The corresponding line widths of (25±5) keV at 244 keV, (20 ± 5) keV at 292 keV and (30 + 6) keV at 330 keV are comparable to the values obtained for the U+Ta and U+Au system. In the same way as described before cross sections can be obtained which are also similar to the values measured for the U+Ta system. The numbers are $d\sigma(e^+)/d\Omega_{ion}$ = (0.34±0.11) μb/sr, (0.18±0.06) μb/sr and (0.42±0.12) μb/sr, respectively.

Finally, it should be mentioned that both target sets - the thin and thick targets - show the same behaviour. Although the statistical errors become larger by this subdivision a relative comparison can be made. Within statistical errors the two sets give identical results. This supports the conclusion that the observed structures do not result from a sharp resonance with respect to the beam energy. The latter would mean that either the set of thin targets would show no structure at all or the structures would be damped by about a factor of 2 for the set of thick targets because the 2nd half of the target would contribute to the smooth background only.

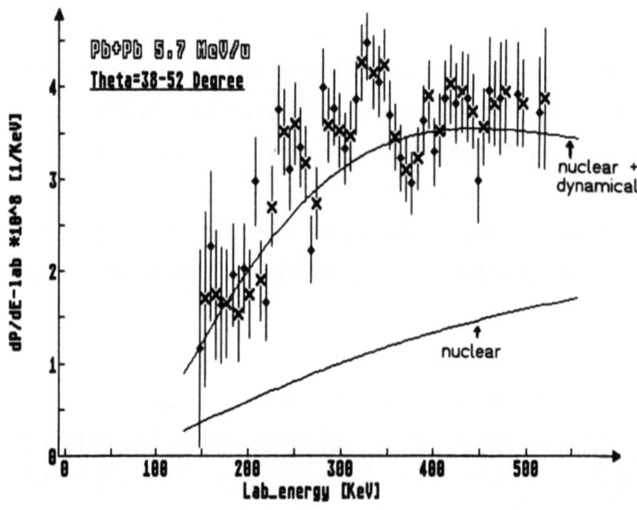

Fig. 8: Positron spectrum measured for the system Pb+Pb at an incident beam energy of 5.7 MeV/u. The crosses and squares denote different bins of the positron kinetic energy. Whereas the bin width amounted to 12 keV in both cases the bins are shifted by 6 keV with respect to each other (cf. page 17).

SUMMARY AND CONCLUSIONS

The results obtained for the three subcritical systems U+Ta, U+Au and Pb+Pb are summarized in Figs. 9 to 11. As seen in Fig. 9 the positron spectra of all 3 systems show a striking similarity. Even if each single spectrum might not be convincing, all three taken together lead to the conclusion that the positron spectra contain lines lying on top of the expected smooth contributions. The width of these lines is unexpectedly small, around 35 keV or even less. This number must be compared to an experimental resolution of 10 keV and a Doppler broadening of also 10 keV if the emitter has the velocity of the cm-system of the colliding ions.

In Fig. 10 the line positions are shown together with the results obtained by the ORANGE collaboration for overcritical systems (open circles) and the results of the EPOS collaboration [11,12,13] (full symbols). The open circles at Z_u = 182 and 184 denote results obtained earlier for the U+U and U+Th systems with the spectrometer turned by 180°, i.e. for positrons emitted in the forward direction [1,8]. The resolution of the system amounted to ~ 60 keV at that time. The question whether or not the measured peak at 280 keV with a width of ~ 80 keV results from a superposition of 2 or more lines is currently under investigation.

The cross sections for the positron lines are shown in Fig. 11 as a function of the combined nuclear charge Z_u. In order to compare the low resolution results obtained for the systems U+U and U+Th with the high resolution measurements performed at the subcritical systems the lines found in between 230 keV and 340 keV were added together (full circles in Fig. 11). The cross section for the line at ~ 240 keV (lab-system) is given separately (open circles). The steep decrease of the cross section with decreasing Z_u is obvious. Using a power law, a fit to the data points gives a dependence of $d\sigma(e^+)/d\Omega_{ion} \propto Z_u^{26}$. This dependence of Z_u is even more pronounced than the Z_u^{20} dependence of the production cross section for dynamically induced positrons.

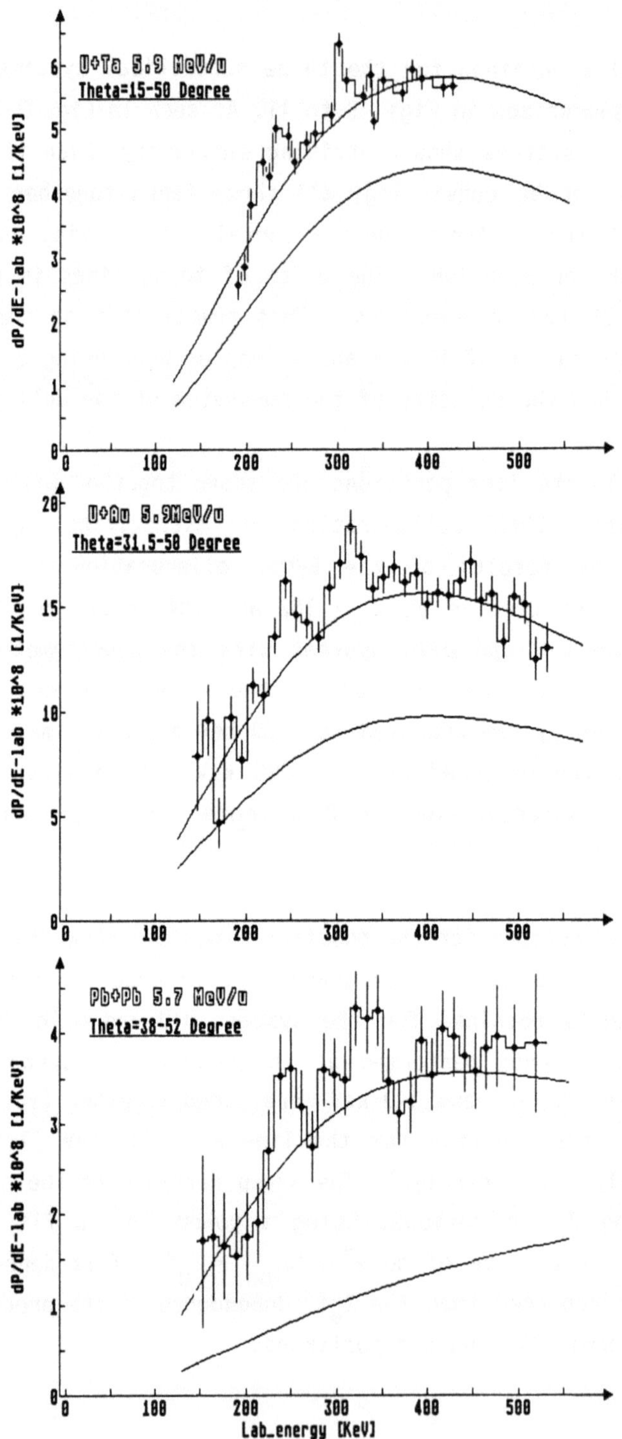

Fig. 9: Positron spectra obtained for the three systems U+Ta, U+Au, and Pb+Pb. Shown are differential positron emission probabilities as a function of the kinetic energy given in the lab-system.
The upper full lines describe the total contribution due to nuclear pair conversion and dynamically induced positrons whereas the lower lines specify the nuclear contribution separately.

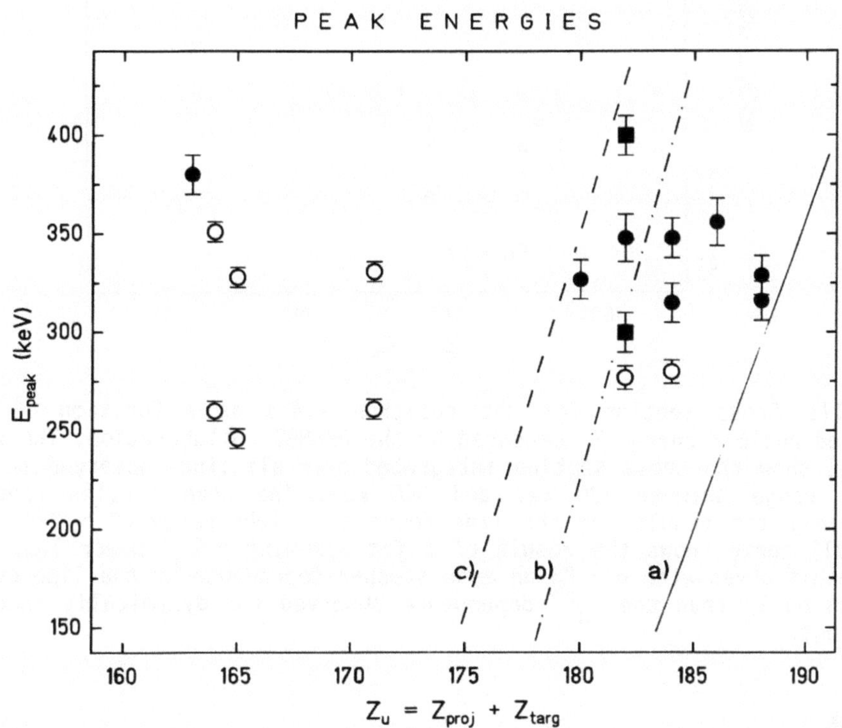

Fig. 10: Energies at which positron lines were observed in various systems. For better comparison all values are transformed into the cm-system of the heavy ions.

Full symbols:
Line energies found by the EPOS collaboration [11,12,13]

Open circles:
Line energies observed by the ORANGE collaboration [1,8].

The full line labeled a) denotes the diving depth of the 1s electrons into the negative energy continuum if elastical scattering is assumed and a mean charge state of q = 50. The dashed-dotted line labeled b) gives the same result for completely ionized ions and the dashed line labeled c) assumes in addition complete fusion of the two ions.

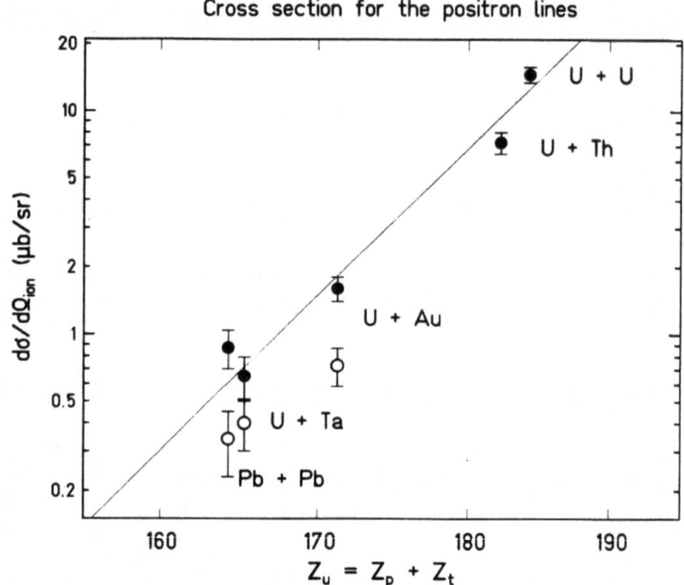

Fig. 11: Cross section for the positron lines as a function of the combined nuclear charge Z_u measured by the ORANGE collaboration. The full circles show the cross section integrated over all lines observed in the energy range between 230 keV and 340 keV. The open circles specify separately the result for the line found at a lab-energy of ~ 240 keV. The full curve shows the result of a fit assuming a Z_u^n power law. The fit result gives with n = 26 an even steeper dependence of the line cross section on Z_u than the Z_u^{20} dependence observed for dynamically induced positrons.

OUTLOOK

Although efficiency and resolution of the experimental set-up have been improved significantly during the last years, the data analysis is still affected by the rather large statistical errors. This is caused by the extraordinary small cross sections found for the positron lines in subcritical systems as well as by the fact that multiple line structures were found. Furthermore the still open question about the velocity of the emitting source requires an experimental answer. Last but not least the first electron-positron coincidence measurements performed by the EPOS collaboration [21,13] have revealed a number of open questions. In this context a more precise determination of the relative angle between the e^+ and e^- is of special interest in order to differentiate between pairs of nuclear origin and a source emitting positrons and electrons back to back, i.e. an intermediate state or particle which decays into e^+ and e^- without an additional heavy particle being involved.

All these considerations have lead us to the construction of a double ORANGE spectrometer called POSEIDON (Fig. 12). This new set-up consists of two identical ORANGE-type spectrometers facing each other. Each spectrometer is equipped with a 'Pagoda'-type high resolution, multi-detector system.

This set-up allows f.e. to study simultaneously the positron emission in the forward and backward direction. Such an experiment should allow to determine the velocity of the emitting sources with good accuracy by comparing the Doppler shifts for the emission in and opposite to the beam direction. F.e. a source moving with the velocity of the cm-system of the heavy ions yields a relative shift of $\sim$ 40 keV which is well above the resolution of the system.

In case of e^+-e^- coincidences all emission angles θ and Φ of all particles involved can be obtained with an accuracy in between $\pm$ 15° and

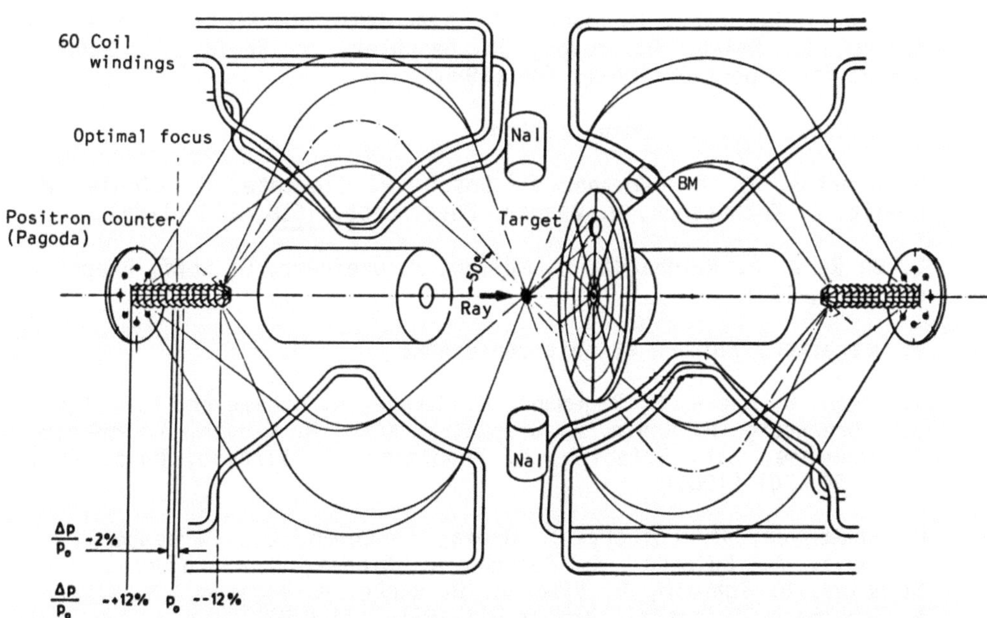

Fig. 12: The double ORANGE spectrometer 'POSEIDON' momentarily under construction at GSI. It consists of two ORANGE-spectrometers facing each other and allows to measure e^+ - e^- coincidences as well as detecting positrons differentially in angle within a large angular range covering forward and backward emission simultaneously.

$\pm$ 30°. This information can be obtained in a straightforward way due to the segmentation of both the heavy-ion and the lepton detectors with separate read-out of the segments.

We hope that this new set-up will shed some light on the positron puzzle. However, it might well happen that the number of open questions will be increased further.

REFERENCES

1. C. Kozhuharov, lecture at this conference and references therein
2. P. Kienle, experimental summary of this conference
3. B. Fricke, G. Soff, At. Data Nucl. Data Tables 19, 83 (1977)
4. R. Bass, Nuclear Reactions with Heavy Ions, Springer Berlin, Heidelberg, New York (1980) and A. Gobbi, private communication
5. K. Siegbahn, ed., Alpha-, Beta- and Gamma-Ray Spectroscopy, Vol. 1
6. B. Blank, E. Bozek, H. Jäger, E. Kankeleit, G. Klotz-Engmann, M. Krämer, R. Krieg, U. Meyer, H. Oeschler, M. Rhein, P. Senger, Scientific Report GSI-86-1, 180 (1986)
7. T. de Reus, private communication
8. H. Tsertos, E. Berdermann, F. Bosch, M. Clemente, P. Kienle, W. Koenig, C. Kozhuharov, W. Wagner, Phys. Lett. 162B, 273 (1985)
9. T. de Reus, J. Reinhardt, U. Müller, W. Greinerr, G. Soff, Preprint GSI-86-5, 1986
10. V. Überacker, lecture at this conference
11. T. Cowan, H. Backe, M. Begemann, K. Bethge, H. Bokemeyer, H. Folger, J.S. Greenberg, H. Grein, A. Gruppe, Y. Kido, M. Klüver, D. Schwalm, J. Schweppe, K.E. Stiebing, N. Trautmann, P. Vincent, Phys. Rev. Lett. 54 1761 (1985)
12. H. Bokemeyer, H. Folger, H. Grein, T. Cowan, J.S. Greenberg, J. Schweppe, A. Balanda, K. Bethge, A. Gruppe, K. Sakaguchi, E.K. Stiebing, D. Schwalm, P. Vincent, H. Backe, M. Begemann, M. Klüver, N. Trautmann, Scientific Report GSI-85-1, 177 (1985)
13. T. Cowan, lecture at this conference

THE CONSEQUENCES OF SUDDEN REARRANGEMENTS OF ELECTRONIC SHELLS

T. de Reus, G. Soff, O. Graf and W. Greiner

Gesellschaft für Schwerionenforschung (GSI), Planckstraße 1
Postfach 110 541, D-6100 Darmstadt, West Germany

Institut für Theoretische Physik, J.W. von Goethe- Universität
Postfach 111 932, D-6000 Frankfurt am Main, West Germany

INTRODUCTION

We investigate the influence of sudden rearrangement processes in the electron distribution of the two- centre potential on electron and positron spectra emitted in heavy ion collisions. Using a schematic ansatz for the matrix elements, a scaling law is deduced, which predicts oscillations in the emission spectra. Coincident electron-positron emission in the system Pb + Pb at $E_{Lab} = 5.7$ MeV/n was evaluated using a full coupled channel code under the assumption that the rearrangement effects may be simulated by adding a Gaussian to the matrix elements at position R_s. Pronounced structures are found for certain cuts in the plane spanned by the kinetic electron and positron energies. Surprisingly, electron and positron production cross sections peak at the same energies. Hence there is great resemblance of this phenomenon with experimental observations on $e^+ - e^-$ spectra in heavy ion collisions.
The original prediction of spontaneous positron emission, due to vacuum decay[1,2] in overcritical electric fields in the vicinity of long living giant nuclear systems, causes a pronounced structure in the positron spectra but applies only for systems with total nuclear charge Z > 174. Taking this prediction at face value, the positron peak position $T_{e^+} = E_{e^+}^{peak}$ depends sensitively on the charge number Z as long as the nuclear (molecular type) structure is the same for various giant systems. Experimental findings of the EPOS- and ORANGE- collaborations[3–12] at GSI, however, hint at an independence[3,4,5] or only a weak dependence of T_{e^+} on the combined nuclear charge Z. Earlier measurements for the U + U system show two structures in the positron spectra (see the experimental reports in ref. 2) which agrees with recent data for various systems, also indicating the existence of two or more structures[9–12]. Even more decisive are measurements of positron emission in so called subcritical collision systems ($Z \leq 174$), such as Th + Ta (Z = 163) or U + Ta (Z = 165), in which narrow structures are also present[7,9–12]. For the latter systems, however, spontaneous positron emission as the source of these structures is unlikely.

These deficencies led to the search for alternative explanations of the mechanism which causes the line structure, such as the $e^+ - e^-$ decay of an up to now unknown neutral particle[13] by pair creation. According to this idea the positrons generating the structures are correlated with electron spectra exhibiting similar structures. Indeed, $e^+ - e^-$ coincidences yield narrow structures[8,9,11] in the difference and sum spectra, suggesting the possible creation of new light elementary bosons[13–16] which would, however, contradict precision measurements in QED[16,17]. Many proposals following these early ideas, except for the very utopic ones, can be ruled out similarly because they lead to strong contradictions with well established physics[17,18].

In this paper we investigate the effect of sudden rearrangement processes in the electron cloud surrounding the quasimolecule created by the projectile and target. Such a mechanism has been considered earlier as an explanation of oscillatory structures observed in quasimolecular X- ray spectra from medium heavy ion collisions[19]. It is different from a recently published suggestion[20], where a special type of coupling was assumed to cause structures in positron spectra, which could not be affirmed by our considerations[21,22], even though it bears some similarities with it. This scenario would cause a discontinuity in the two-centre potential at a certain distance R_s, changing the potential from $V^{out}(r,R)$ to $V^{in}(r,R)$. At large distances $V^{out}(r,R)$ could consist of the fully screened potential of the target atom, whereas the projectile particle displays a certain degree of ionisation. In contrast $V^{in}(r,R)$ may reveal only the presence of the two nuclear centres and the innermost electrons. This can be simulated by constructing the potential according to

$$V(r,R) \quad = \quad \Theta(R-R_s)\, V^{out}(r,R) + \Theta(R_s - R)\, V^{in}(r,R). \tag{1}$$

In order to investigate the effect of such a sudden rearrangement potential on the emitted electron and positron spectra we briefly mention the underlying formalism without going into details.

The stationary two-centre Dirac equation is solved for different internuclear distances R, considering only the monopole approximation $V_0(r,R)$ of the full two-centre potential,

$$\hat{H}_{TCD}\, \phi_i(r,R) = [\vec{\alpha}\cdot\vec{p} + \beta m + V_0(r,R)]\, \phi(r,R) \quad = \quad E_i\, \phi_i(r,R). \tag{2}$$

This provides an optimal basis set for expanding the time dependent wavefunction $\Psi_i(t)$ by means of the molecular eigenstates $\phi_i(r,R)$

$$\Psi_i(t) \quad = \quad \sum_j a_{ij}(t)\, \phi_j\, e^{-\chi_j(t)}\,, \tag{3}$$

with

$$\chi_j(t) \quad = \quad \int^t E_j\, dt'\,.$$

Inserting eq. (3) into the time-dependent two-centre Dirac equation yields the following set of coupled channel equations[23]

$$\dot{a}_{ij}(t) \quad = \quad -\sum_{k\neq j} a_{ik}(t) < \phi_j|\frac{\partial}{\partial t} + i\hat{H}_{TCD}|\phi_k > \, e^{-i(\chi_k - \chi_j)}. \tag{4}$$

The number of emitted particles p or holes q is obtained by

$$N_p \quad = \quad \sum_{r<F} |a_{rp}|^2\,, \qquad \text{for } p > F\,, \tag{5}$$

$$N_q \quad = \quad \sum_{r>F} |a_{rq}|^2\,, \qquad \text{for } q < F\,, \tag{6}$$

where F indicates the Fermi level. The number of correlated particle hole pairs reads

$$N_{pq} \;=\; N_p \cdot N_q + |\sum_{r<F} a^*_{rp}\, a_{rq}|^2. \tag{7}$$

Since in the present experiments no distinction is made between positrons or electrons with $s\sigma$- or $p_{\frac{1}{2}}\sigma$-parity entering the calculation, expression (7) has to be modified. Taking also into account spin degeneracy, we obtain the following expression for positrons emitted in coincidence with electrons and vice versa

$$\begin{aligned}\frac{d^2P_{e+,e-}}{dE_{e+}\,dE_{e-}} \;=\; & 4\left(\sum_{r<F}|a^s_{r,E_{e-}}|^2+\sum_{r<F}|a^p_{r,E_{e-}}|^2\right)\cdot\left(\sum_{r>F}|a^s_{r,E_{e+}}|^2+\sum_{r>F}|a^p_{r,E_{e+}}|^2\right)\\ & +2\left(\left|\sum_{r<F}a^{*s}_{r,E_{e-}}\;a^s_{r,E_{e+}}\right|^2+\left|\sum_{r<F}a^{*p}_{r,E_{e-}}\;a^p_{r,E_{e+}}\right|^2\right).\end{aligned} \tag{8}$$

The superscripts s and p indicate amplitudes of $s\sigma$- or $p_{\frac{1}{2}}\sigma$-parity, respectively. Restricting ourselves to subcritical systems ($Z < 174$) the potential coupling term in equation (4) vanishes and we are left with

$$\dot a_{ij}(t) \;=\; -\sum_{k\neq j} a_{ik}(t)\,\dot R\; <\phi_j|\frac{\partial V_0(r,R)}{\partial R}|\phi_k>\; e^{-i(\chi_k-\chi_j)}. \tag{9}$$

Inserting the potential from equation (1) we obtain the following expression

$$\begin{aligned}\dot a_{ij}(t) \;=\; & -\sum_{k\neq j} a_{ik}(t)\dot R\;\{<\phi_j|\Theta(R-R_s)\,\frac{\partial V^{out}(r,R)}{\partial R}+\Theta(R_s-R)\,\frac{\partial V^{in}(r,R)}{\partial R}|\phi_k>\\ & +<\phi_j|\delta(R-R_s)\,V^{out}(r,R)-\delta(R_s-R)\,V^{in}(r,R)|\phi_k>\}\;e^{-i(\chi_k-\chi_j)}.\end{aligned} \tag{10}$$

Thus, in addition to the radial matrix elements also potential couplings have to be evaluated.

A thorough calculation requires the replacement of the theta function by e.g. a Fermi-type step function with a finite width, so that the delta function in equation (10) is also replaced by a finite sized expression. Beyond that the potential coupling matrix elements are needed.

In a first step we will elude this procedure and simulate the effect of the second term in equation (10) by adding a spatially localized function to the radial matrix elements at the position R_s. The effect of a structure in the matrix elements at position R_s with a width ΔR can be estimated in first-order perturbation theory, when the matrix element is approximated according to

$$\frac{1}{\Delta E} < f|\frac{\partial V}{\partial R}|i> \;\sim\; \frac{N_1}{R}+\frac{N_2}{R}\,\frac{R_s}{\pi\Delta R}\,\frac{1}{1+(\frac{R-R_s}{\Delta R})^2}, \tag{11}$$

where the second term is a representation of the δ-function in the limit $\Delta R \to 0$. We obtain

$$\begin{aligned}\frac{dP_{e+}}{dE_{e+}} \;\sim\; & \frac{4N_2^2n^2e^{-2\delta\beta}}{(1+n^2)^2}(n\sin\delta\alpha-\cos\delta\alpha)^2+\left(N_1\pi+\frac{N_2n}{1+n^2}\right)^2 e^{-2\delta\gamma}\\ & +\;\frac{4N_2ne^{-\delta\,(\beta+\gamma)}}{1+n^2}\left(N_1\pi+\frac{N_2n}{1+n^2}\right)\,(n\sin\delta\alpha-\cos\delta\alpha).\end{aligned} \tag{12}$$

with the abbreviations:

$$\begin{aligned}
\delta &= \frac{a\Delta E}{\hbar v_\infty}, \qquad \gamma = \pi + \frac{b}{a} - \arccos\left(\frac{a}{R_{min}-a}\right), \\
\mathcal{R} &= \frac{R_s - a}{a\varepsilon}, \qquad n = \frac{R_s}{\Delta R}, \qquad \mathcal{I} = \frac{\Delta R}{a\varepsilon}, \\
z &= \frac{1}{2}[\mathcal{R}^2 + \mathcal{I}^2 - 1 + \sqrt{(\mathcal{R}^2 + \mathcal{I}^2 - 1)^2 + 4\mathcal{I}^2}], \\
\alpha &= \varepsilon\sqrt{z - \mathcal{I}^2} + \operatorname{arsinh}\sqrt{z}, \\
\beta &= \varepsilon\mathcal{I}\sqrt{1 + \frac{1}{z}} + \arcsin\left(\frac{\mathcal{I}}{\sqrt{z}}\right).
\end{aligned} \tag{13}$$

In the limit $\Delta R \to 0$ equation (12) simplifies to

$$\begin{aligned}
\frac{dP_{e^+}}{dE_{e^+}} &\sim (N_1\pi)^2 \exp\left\{-2\frac{a\Delta E}{\hbar v_\infty}\left(\pi + \frac{b}{a} - \arccos\left(\frac{a}{R_{min}-a}\right)\right)\right\} \\
&+ 4N_2^2 \sin^2\left\{\left(\frac{a\Delta E}{\hbar v_\infty}\right)\varepsilon\sqrt{\left(\frac{R_s - a}{a\varepsilon}\right)^2 - 1} + \operatorname{arsinh}\left(\varepsilon\sqrt{\left(\frac{R_s - a}{a\varepsilon}\right)^2 - 1}\right)\right\} \\
&+ 4\pi N_1 N_2 \sin\left\{\left(\frac{a\Delta E}{\hbar v_\infty}\right)\varepsilon\sqrt{\left(\frac{R_s - a}{a\varepsilon}\right)^2 - 1} + \operatorname{arsinh}\left(\varepsilon\sqrt{\left(\frac{R_s - a}{a\varepsilon}\right)^2 - 1}\right)\right\} \\
&\times \exp\left\{-\frac{a\Delta E}{\hbar v_\infty}\left(\pi + \frac{b}{a} - \arccos\left(\frac{a}{R_{min}-a}\right)\right)\right\}.
\end{aligned} \tag{14}$$

The first term represents the regular result based on the first term in equation (11). The second term results from interference of the amplitudes at R_s on the incoming and outgoing branch of the trajectory. The third expression in (14) evolves by interference of the amplitudes associated with the central region around R_{min} with those at R_s, with half of the frequency of the second term. For a width $\Delta R \leq 70$ keV oscillations are observable in the spectra with a period ΔE associated to the third term in equation (14). The oscillation frequencies have approximately the values one readily estimates by applying Heisenbergs' uncertainty principle for the mentioned distances.

Very recent measurements of the ORANGE- collaboration[7,10,12] dealt with the system Pb + Pb at a beam energy of 5.7 MeV/n. Structures in the positron spectra at different positions $T_{e^+} \approx 230$, 310 and 420 keV have been detected, which may hint towards an oscillation pattern. Stimulated by these results we performed a coupled channel calculation including 8 or 6 bound states for the $s\sigma$ or $p_{\frac{1}{2}}\sigma$ channel, respectively, 50 electron and 50 positron states in order to resolve the oscillations. Similar to equation (11) a Gaussian with a width of 50 fm was added to all matrix elements at a two-centre distance of $R_s = 1200$ fm. The height of the structure was assumed to be 20% of the corresponding value of the matrix element at the distance of closest approach. Beside a sudden change in the potential due to equation (1) also crossings of bound states[24] in the complex two-centre correlation diagrams or electron compression waves[25,26] might cause structures in the matrix elements. Similarly explicit electron- positron correlation effects could generate strongly enhanced coupling strength.

Figure 1 shows the result of a coupled channel calculation with the mentioned assumptions for an impact parameter b = a = 8.17 fm, corresponding to a scattering angle $\Theta_{Lab} = 45^o$. The double differential emission probability for positrons or electrons according to equation (8) with respect to kinetic positron and electron energies is depicted in a contour plot.

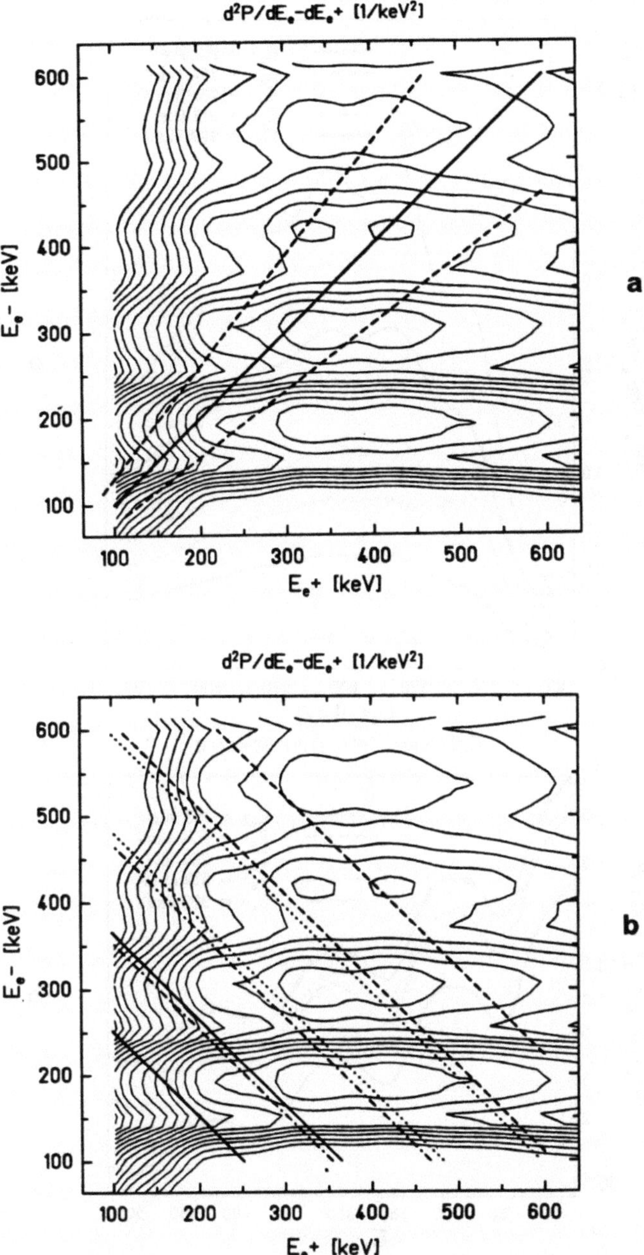

Figure 1: a) Double differential emission probability of coincident electron- positron emission in Pb + Pb at $E_{Lab} = 5.7$ MeV/n for an impact parameter b = a. A Gaussian at $R_s = 1200$ fm with a width of $\Delta R = 50$ fm was added to all matrix elements. The height of the Gaussian amounts to 20% of the considered matrix element at the distance of closest approach. The emission probability is presented in a contour plot versus kinetic electron and positron energies. Each line corresponds to a fixed value for the emission probability, increasing from 10^{-11} keV^{-2} to $3 \cdot 10^{-10}$ keV^{-2} by a factor of 1.185 in logarithmic representation. Absolute numbers are displayed in figure 2). The cuts indicate how the data in fig. 3 a) are obtained. Part b) displayes the cuts associated with figure 3 b).

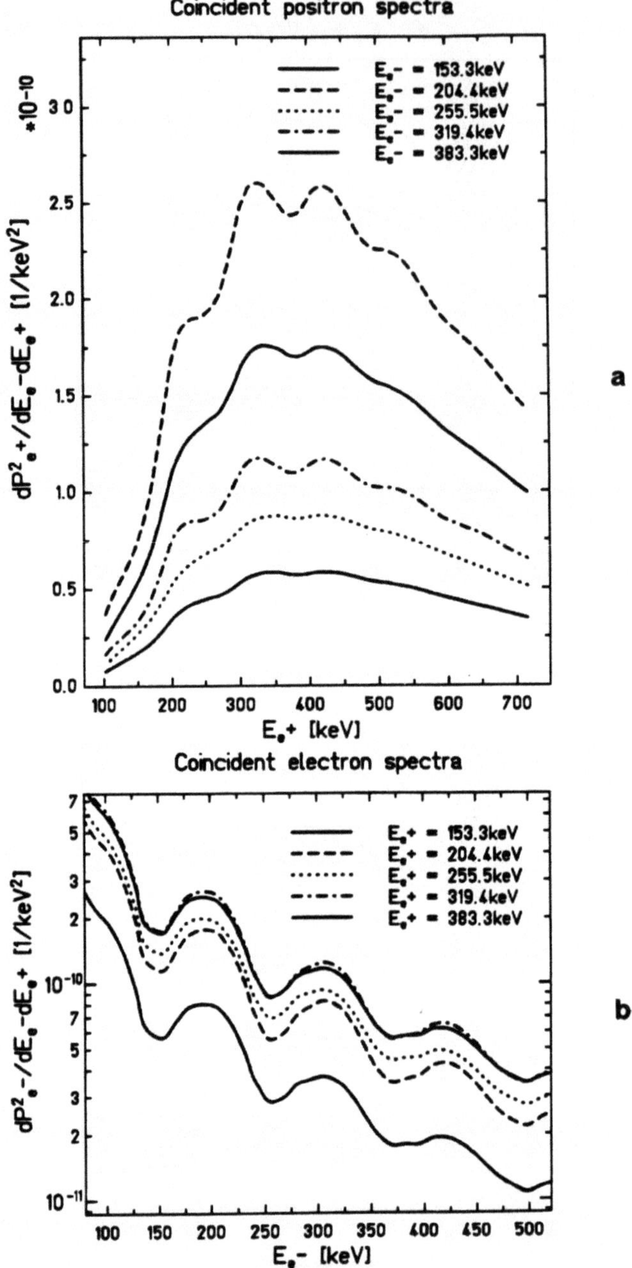

Figure 2: a) Positron emission in coincidence with different kinetic electron energies is displayed versus the positron energy, corresponding to different cuts parallel to the abscissa of figure 1. The lower full line belongs to $E_{e^-} = 383.3$ keV. The structures in the spectra for the electron energies $E_{e^-} = 204.4$ keV and 319.4 keV are most pronounced and largest in amplitude, since at these energies the δ- electron spectra exhibit maxima. The other electron energies are associated with minima in the electron spectra, with the result that the oscillations are smeared out. b) In contrast the complementary cuts parallel to the ordinate of figure 1, which represent electron spectra, retain their structures for different positron energies E_{e^+}. The lower full line belongs to $E_{e^+} = 383.3$ keV.

In contrast to coincidences between δ-electrons and K-vacancies, where the accidental term exceeded the coherent sum in equation (8) by factors 10 to 100, the coherent sum for correlated electron- positron emission has the same magnitude or even exceeds the accidental term, which is a small number. Quite surprisingly both, electron and positron spectra exhibit maxima at about the same kinetic energies, as the cuts displayed in figure 2 demonstrate. The oscillations in the electron spectra, depicted in part b) of the figure, essentially retain their strength for different positron energies, whereas the oscillations in the positron spectra are most pronounced for electron energies associated with a maximum in the coincident electron spectra.

Performing a cut along the full line of figure 1 a) therefore enhances the structures. With the condition of equal kinetic energies $E_{e^+} = E_{e^-}$ coincident electron- positron emission is plotted as a function of the sum energy $E_{e^+} + E_{e^-}$ in figure 3 a). Pronounced structures at the sum energies around 410, 610 and 850 keV are obtained with a width of $\Delta E \sim 100$ keV. The dashed curve indicates results for projections perpendicular and symmetric to the diagonal in figure 1 a) as indicated by the dashed lines for a width $\Delta E = 0.18\,(E_{e^+} + E_{e^-})$, which according to ref. 5 allows for kinematical broadening. The structure at 850 keV vanishes, whereas the peaks at 410 and 610 keV are retained. Averaging over adjacent cuts, excluding the latter region, these structures vanish.

Keeping the sum energy fixed, we also performed cuts perpendicular to the diagonal as indicated in fig. 1 b), with a width of $\Delta E = 80$ keV. The results are displayed in figure 3 b). For sum energies $E_{sum} = 408.8$ keV and $E_{sum} = 638.8$ keV "hills" in the contour plot in figure 1 b) are crossed, so that a structure at $E_{e^+} \approx E_{e^-}$ evolves. Crossing "valeys" as for the dashed and dashed dotted curves, in contrast generates minima at $E_{e^+} \approx E_{e^-}$ and structures symetrically arranged around this value. For different impact parameters b the oscillations will be shifted, but the period ΔE should modify only negligibly. Therefore we expect that the oscillations in fig. 2 might be washed out when performing an integration over differnt impact parameters.

As discussed above, the appearance of structures in the matrix elements due to sudden rearrangement effects in the electron cloud or crossings between bound or continuum states may cause structures in the positron and electron spectra. This has been demonstrated for the system Pb + Pb at a bombarding energy of 5.7 MeV/n for an impact parameter b = a, corresponding to a lab angle of 45°. Considering the double differential data $d^2P_{e^+,e^-}/dE_{e^+}dE_{e^-}$, these structures are even enhanced if a cut along $E_{e^+} = E_{e^-}$ is performed, since the position of the structures appears at about the same kinetic energies, both in the positron and electron spectra. The structures displayed in figure 2 show only a weak dependence on the bombarding energy. Their position is shifted by ~ 30 keV when reducing the beam energy from 5.7 to 5.0 MeV/n. An experimental investigation of the beam energy dependence and the accompanying electron spectra could readily clarify our conjecture.

We mention that the peak structures in the various cuts become narrower if more than one sudden rearrangement is considered. The idea of sudden rearrangements has been introduced ad hoc. The surprising consequences of this idea bears close similarity to many observations in the $e^+ - e^-$ spectra of heavy ion collisions. If true, the observed interference structures should appear in practically all collision systems. Then a detailed microscopic justification of our assumptions is called for. It may reveal rather complicated (collective ?) structure in the atomic shells. The ejection mechanism due to sudden rearrangements introduces strong quasi- positronium like correlations into $e^+ - e^-$ pairs.

Concluding, we propose to study the effects of sudden rearrangement processes during

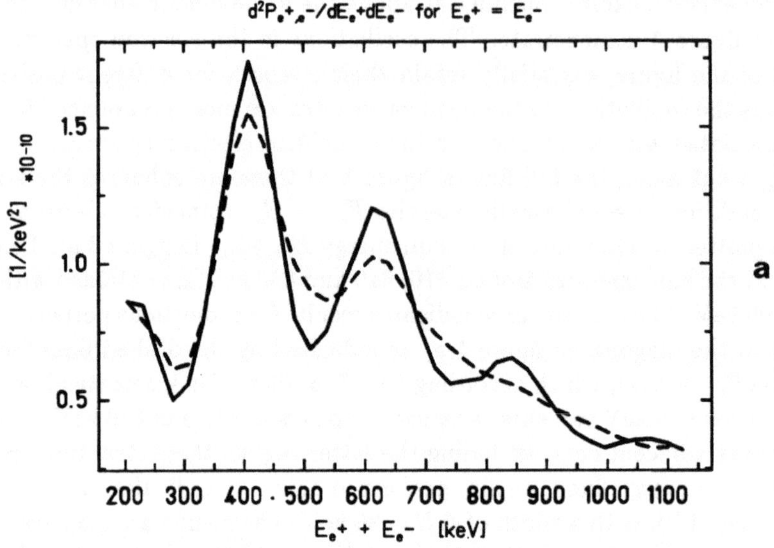

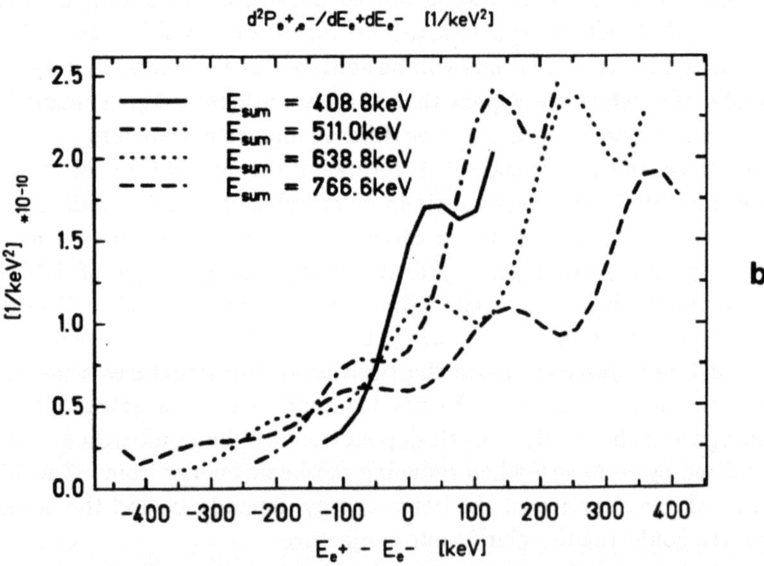

Figure 3: a) A coincident electron- positron spectrum obtained along the cut $E_{e^+} = E_{e^-}$ indicated in figure 1 a) is displayed by the full line. Pronounced structures evolve around $E_{e^+} + E_{e^-} \approx 410$, 610 and 850 keV. Averaging over $\Delta E = 0.18 \cdot (E_{e^+} + E_{e^-})$ as indicated by the dashed lines in fig. 1 a), yields the dashed curve. b) Cuts perpendicular to the diagonal as shown in figure 1 b) with a width of $\Delta E = 80$ keV are performed for different sum energies $E_{sum} = E_{e^+} + E_{e^-}$.

heavy ion collisions, which might be responsible for the appearance of multiple structures in the $e^+ - e^-$ spectra measured in sub- and supercritical collisions.

ACKNOWLEDGEMENT

The authors are indebted to J. Reinhardt and B. Müller for valuable suggestions and critical reading of the manuscript. We acknowledge fruitful discussions with the members of the EPOS- and ORANGE collaborations.

REFERENCES

1. W. Greiner, B. Müller, J. Rafelski, Quantum Electrodynamics of Strong Fields, (Springer, Berlin, 1985), and further references therein
2. W. Greiner, ed., Quantum Electrodynamics of Strong Fields, NASI series B80, (Plenum, New York, 1983)
3. J. Schweppe, A. Gruppe, K. Bethge, H. Bokemeyer, T. Cowan, H. Folger, J.S. Greenberg, H. Grein, S. Ito, R. Schule, D. Schwalm, K.E. Stiebing, N. Trautmann, P. Vincent, M. Waldschmidt, Phys. Rev. Lett. **51**, 2261 (1983)
4. M. Clemente, E. Berdermann, P. Kienle, H. Tsertos, W. Wagner, C. Kozhuharov, F. Bosch, and W. Koenig, Phys. Lett. **137B**, 41 (1984)
5. T. Cowan, H. Backe, M. Begemann, K. Bethge, H. Bokemeyer, H. Folger, J.S. Greenberg, H. Grein, A. Gruppe, Y. Kido, M. Klüver, D. Schwalm, J. Schweppe, K.E. Stiebing, N. Trautmann, P. Vincent, Phys. Rev. Lett. **54**, 1761 (1985)
6. H. Tsertos, E. Berdermann, F. Bosch, M. Clemente, P. Kienle, W. Koenig, C. Kozhuharov, W. Wagner, Phys. Lett. **162B**, 273 (1985)
7. P. Kienle, J. Phys. Soc. Jpn. **54** Suppl. II, 549 (1985)
8. T. Cowan, H. Backe, K. Bethge, H. Bokemeyer, H. Folger, J.S. Greenberg, K. Sakaguchi, D. Schwalm, J. Schweppe, K.E. Stiebing, P. Vincent, Phys. Rev. Lett. **56**, 444 (1986)
9. T. Cowan, in: Physics of Strong Fields, Ed.: W. Greiner, (Plenum, New York)
10. W. Koenig, ibid.
11. H. Bokemeyer, ibid.
12. Ch. Kozhuharov, ibid.
13. A. Schäfer, B. Müller, W. Greiner, Phys. Lett. **149B**, 455 (1984)
14. A. Schäfer, J. Reinhardt, B. Müller, W. Greiner, G. Soff, J. Phys. **G11**, L69 (1985)

15. A.B. Balantekin, C. Bottcher, M.R. Strayer, S.J. Lee, Phys. Rev. Lett. **55**, 461 (1985)

16. J. Reinhardt, A. Schäfer, B. Müller, W. Greiner, Phys. Rev. **C33**, 194 (1986)

17. J. Reinhardt, contribution to: Physics of Strong Fields, Ed.: W. Greiner, (Plenum, New York)

18. A. Schäfer, ibid.

19. R.K. Smith, B. Müller, W. Greiner, J.S. Greenberg, C.K. Davis, Phys. Rev. Lett. **34**, 117 (1975)

20. W. Lichten, A. Robatino, Phys. Rev. Lett. **54**, 781 (1985)

21. J. Reinhardt, B. Müller, W. Greiner, Phys. Rev. Lett. 55, 134 (1985)

22. W. Lichten, A. Robatino, Phys. Rev. Lett. 55, 135 (1985)

23. T. de Reus, J. Reinhardt, B. Müller, W. Greiner, G. Soff, U. Müller, J. Phys. **B17**, 615 (1984)

24. G. Soff, W. Greiner, W. Betz, B. Müller, Phys. Rev. **A20**, 169 (1979)

25. J.F. Hofmann, H. Stöcker, W. Scheid, W. Greiner, V. Ceausescu, E. Badralexe, Z. Phys. **A280**, 131 (1977)

26. W. Schäfer, H. Stöcker, B. Müller, W. Greiner, Z. Phys. **A288**, 349 (1978)

ARE THE GSI EVENTS CAUSED BY PARTICLE DECAY?

Joachim Reinhardt, Berndt Müller and Walter Greiner

Institut für Theoretische Physik
Johann Wolfgang Goethe-Universität, Postfach 111932
D-6000 Frankfurt/Main, West Germany

Andreas Schäfer

Gesellschaft für Schwerionenforschung
Postfach 110 541
D-6100 Darmstadt, West Germany

1. INTRODUCTION

From the beginning in 1977 the main goal of the positron experiments[1] at GSI was finding experimental evidence that the spontaneous decay of the neutral QED vacuum into a charged vacuum exists in strong electric fields, as it had been predicted by theory[2-5]. When the presence of a narrow, peak-like structure[6,7] in the positron spectra of U+U and U+Cm collisions was discovered and, step by step, confirmed in the years 1981-83, this was quite naturally taken as indication of spontaneous positron emission in the strong field created by the combined action of these nuclei. Moreover, theoretical models developed by our group[8,9] and by others[10] showed that the existing data were compatible with the assumption that a giant nuclear complex[11] was formed in these collisions with a lifetime in the range 10^{-20}-10^{-19}s. The observed positron line position agreed reasonably well with the predicted position of the supercritical K-shell resonance in the positron continuum.

This interpretation, however, became less convincing when the EPOS collaboration[12] began to discover virtually identical structures in all investigated combinations of nuclei from U+Cm ($Z_1+Z_2=188$) down to Th+Th ($Z_1+Z_2=180$). Within the experimental uncertainty, caused by limitations in statistics and Doppler broadening, the line position appeared to be (nearly) the same in all cases. This was in gross contrast to the most natural expectations derived from the theoretical models mentioned above. To wit, based on the assumption that a molecule-like nuclear configuration was formed in each case, the line should have moved from about 330 keV in U+Cm down to around 100 keV in Th+Th where it would be so weak as to be practically undetectable[13] (see Fig. 1). This was very clearly not what was being observed. To the contrary, the near Z-independence of the line position pointed to a common source of the positrons in these peak-like structures, at least it did so for an unbiased observer. One possibility certainly was that always the same giant nuclear system was formed in every

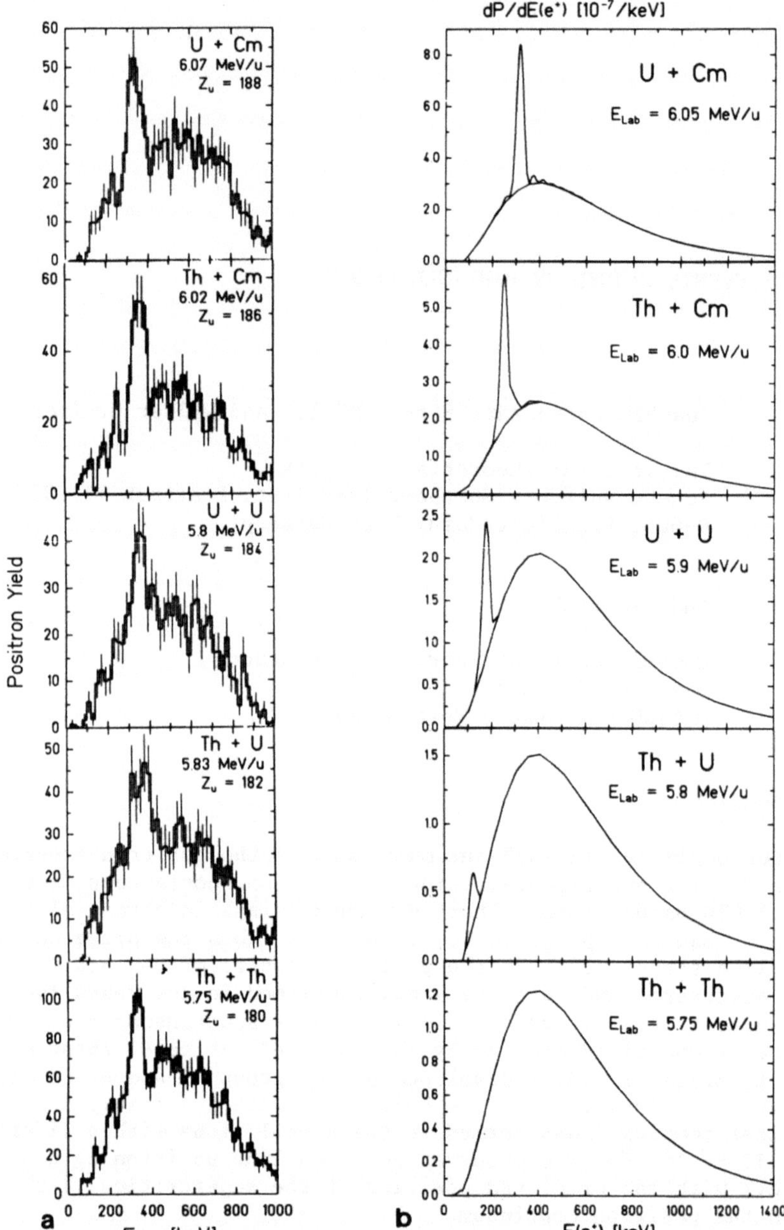

Fig. 1. a. Positron energy spectra for the collision systems U+Cm (Z_u=188), Th+Cm (186), U+U (184), Th+U (182), and Th+Th (180) at bombarding energies close to the Coulomb barrier measured by the EPOS group at GSI.

b. Positron spectra corresponding to part (a) calculated under the assumption that in all systems elongated giant nuclear molecules of similar type are formed[13].

collision, which would have to be a highly stripped, spherical nucleus with Z=180. When the initial combined charge was larger than 180, the surplus charge would have to be emitted, most likely in the form of alpha-particles. Two facts contradicted this explanation rendering it unviable. The experimental counterargument was that emission of four alpha-particles in the U+Cm system would by necessity be associated with a reaction Q-value far outside the observational limits[14]. The theoretical flaw was that the spherical giant nuclear configuration required for the explanation was predicted to be as much as 150 MeV higher in energy than what was available in the experiment. All subsequent attempts to explain a lowering of the energy in the framework of conventional nuclear models failed[15].

Another obvious source of monochromatic positrons would be the decay of a light, previously unknown particle, either $X^+ \to e^+ + \nu_e$ or $X^0 \to e^+ + e^-$, which we began to consider earnestly in the summer of 1984. This interpretation of the data encounters two main difficulties that were immediately apparent, and which are still the crucial stumbling blocks for the veritable flood of models proposed over the past year. Difficulty number one was why such a light particle with mass somewhere between 1.5 and 2 MeV had not been detected before, either in nuclear reactions and decays or in precision spectroscopy. Problem number two was to explain the narrowness of the observed line-width which implied that the particles would be decaying essentially at rest. This result, in fact, rules out the majority of models proposed so far, since the Fourier spectrum available in the heavy-ion collision has a characteristic width of several hundred keV implying that the particles would be produced moving quite rapidly. We will discuss these problems in detail in the following sections.

The model[16] proposed by us in 1984 was based on the realisation that both these difficulties could, in principle, be solved at the same time by involving Bose condensation as mechanism for particle production. Firstly, condensate formation is an entirely nonperturbative process, and it would therefore not be surprising if it had not occurred in any other physical process studied before. Secondly, if the condensate wave function would correspond to a bound state in the strong field of the two nuclei, that is adiabatically dissolved at the end of the collision, it would be quite natural that the particles emerge almost at rest in the center-of-mass system. We still feel that, so far, no other mechanism has been proposed which provides a credible explanation for these two basic problems encountered by the particle interpretation.

Our original model[16] made use of the observation that the critical nuclear charge for a charged, spin-less boson $X^\pm$ with mass around 1.6 MeV would be only $Z \approx 130$, much less than the value 173 required for the electron-positron field. To avoid the possibility of perturbative $(X^+ + X^-)$-pair production we tried to identify the X-particle with the charged component of a Higgs field - perhaps the Higgs iso-doublet of Weinberg and Salam - which can be gauged away in all orders of perturbation theory. Closer study of this model has revealed that this mechanism does not work for any reasonable value of nuclear charge Z because of the large value of the vacuum expectation value of the Higgs field implied by the great masses of the intermediate vector bosons[17]. Also in the meantime the discovery of a correlated electron line[18] has rendered the assumption of a decay chain $X^+ \to e^+ + \nu_e$ untenable. If the particle interpretation is correct, the new boson must be electrically neutral suggesting that condensation would have to occur due to the action of a nonelectromagnetic force.

Before we begin to discuss various models for new particle production in heavy-ion collisions and the limits imposed on them by other data[19-23], it is useful to briefly review the experimental situation. Our experimental colleagues will hopefully forgive us, if this review is overly simplified in their eyes. In our view, the situation presents itself as follows[6,7,12,18,24,25]:

(a) Narrow, peak-like structures have been observed in positron spectra from all investigated collision systems, ranging from Th+Ta(Z=163) up to U+Cm(188). The width of these structures is invariably about 80 keV; the structures that were found to be reproducible all fall into the energy range 250-400 keV (laboratory energy).

(b) The observed cross-section of events in the peak is of the order of 10 μb/sr.

(c) The EPOS collaboration has reported[24] an extremely pronounced dependence of the intensity and location of the peak on beam energy, which may be even stronger than permitted by naive estimates based on the target thickness.

(d) The peak appears to be associated with collisions that differ slightly from the kinematics of elastic scattering.

(e) A correlated peak[18] in the electron spectrum was discovered by the EPOS collaboration in the systems Th+U, Th+Th and Th+Cm. The precise position of the peak seems to differ with system and beam energy, but the energy of the electron peak agrees with that of the positron peak to within about 10 keV in every case. The intensity of the coincident events appears to exhaust the singles peak event rate. The fact that the sum-energy spectrum reveals a much narrower peak (width as small as 30 keV) indicates back-to-back emission of electron and positron in the c.m. and laboratory systems.

If the observed events are attributed to production and decay of a new particle, a number of conclusions can be drawn immediately:

(1) The particle must be a neutral boson with mass around 1.8 MeV and non-negligible branching ratio for the decay into an e^+e^--pair.

(2) A large fraction of the produced particles must decay while moving very slowly (with velocity not larger than $v_{cm} \sim \frac{1}{20} c$). Unless a mechanism can be found that explains how the neutral particles can be stopped, this implies production of the particles almost at rest in the nuclear c.m. system.

(3) The lifetime of the particle cannot much exceed 1 ns.

(4) The observed equality of e^+ and e^- peak energies, $|E_{e^+}-E_{e^-}| \lesssim 10$ keV , requires that the decay must occur far away from the nuclear Coulomb field (more than 20 000 fm). In combination with the low velocity this sets a lower limit for the lifetime of 10^{-18} s.

(5) The remarkably strong beam-energy dependence indicates that the (slow) production of the decaying particles may be associated with a special type of collisions occurring preferentially at the Coulomb barriers, e.g. with collisions involving an extended nuclear contact ("sticking").

(6) Existence of several possible peak positions requires the existence of several particles within a narrow mass range or, alternatively, a single particle with a dense excitation spectrum pointing toward an internal structure or compositeness of the particle.

We finally remind the reader of the definition of a particle state in relativistic quantum physics:

A particle is a state that transforms under a definite irreducible representation of the Lorentz group.

This means that any state characterized by an invariant mass m and spin and parity J^π is called a particle, if it satisfies the relativistic energy-momentum relation $E^2 = m^2c^4+\vec{p}^2c^2$. It may by elementary (like e^+, e^-, photon γ, quark q, etc.) or composite (like the nucleon, the hydrogen atom in its ground state, a molecule, etc.).

Bearing these things in mind, we now turn to a more detailed survey of attempts to interpret the GSI data by a new particle.

2. KINEMATICAL REQUIREMENTS ON THE DECAY $X \rightarrow e^+ + e^-$

Before discussing in any detail the nature and properties of a new particle, X, created in heavy ion collisions we have to ascertain under which conditions this hypothesis can explain the experimental observations, in particular the narrow positron line width $\Gamma \lesssim 80$ keV. The two-body decay of a particle into an electron-positron pair will lead to a sharp (monochromatic) positron energy only in the particle rest frame. The latter does not agree with the laboratory frame since realistically the particles will be created with a finite energy spread and (presumably) in the moving heavy ion center-of-mass system. The resulting proper momentum $\vec{P}_x$ of the particle is added to the momentum $\vec{P}'$ released by the decay. Since the relative angle between these vectors is not fixed the resulting positron energy spectrum will be broadened. This effect would not be manifest in a complete measurement of the e^+-e^- two-body kinematics (i.e. the invariant mass), which has not yet been performed, but line structures have been found already in the singles positron spectra.

Thus the mere fact that particles with an appropriate mass $m_x = 2(m_e+E_{e^+}(kin))$ are created does not suffice to guarantee the emergence of a line in the positron spectrum. The particles have to be produced with low velocity. To give a quantitative discussion let us assume for simplicity that the production mechanism is isotropic in the heavy ion center-of-mass frame. If also the decay is isotropic (which is obvious for spin-0 particles) the following simple expression for the energy spectrum of positrons (and electrons) applies

$$\frac{dw}{dE'} = \frac{m_x}{\sqrt{m_x^2 - 4m_e^2}} \int_{E_-}^{E_+} dE_x \frac{1}{\sqrt{E_x^2 - m_x^2}} \frac{dw}{dE_x} \tag{1}$$

where

$$E_\pm = \frac{m_x}{2m_e^2} \left(E'm_x \pm \sqrt{(E'^2-m_e^2)(m_x^2-4m_e^2)}\right) .$$

E' is the c.m. positron energy. The distribution (1) still has to be transformed to the laboratory frame which entails an additional Doppler broadening.

We will discuss two types of models for the particle spectrum $\frac{dw}{dE_x}$.

(1) If the particles are radiated during the course of the heavy ion collision, their energy distribution dw/dE_x must be rather broad, reflecting the available "Fourier frequencies" of the nuclear motion or, put differently, the time-energy uncertainty relation. At energies close to the Coulomb barrier the characteristic, unavoidable energy spread is of the order of several hundred keV. This essentially is true for nuclear and atomic Coulomb excitation and for x-ray emission processes.

As it turns out, the possible emergence of a line structure in the spectrum of the decay products depends sensitively not only on the falloff constant of the particle spectrum but even more so on its behavior at small kinetic energies. This is readily understood from Eq.(1) since the decay of fast particles leads to a large motional Doppler broadening.

The number of neutral bosons dynamically produced will be determined by a squared transition matrix element, the structure of which will depend on the detailed production mechanism, multiplied by a phase space factor

$$dw = |M_{fi}|^2 \frac{d^3p_x}{(2\pi)^3E_x} = |M_{fi}|^2 \frac{p_x dp_x d\Omega_x}{(2\pi)^3} \quad . \tag{2}$$

Normally M_{fi} will be bounded at small momentum p_x, it may even vanish with a power of p_x, e.g. if selection rules call for a higher multipolarity. Thus the energy spectrum of X particles will be suppressed at small values of the momentum.

To investigate the shape of the resulting positron spectrum let us model the particle energy distribution by

$$\frac{dw}{dE_x} \simeq \frac{p_x}{m_x} \frac{1}{\Gamma} e^{-(E_x-m_x)/\Gamma} \quad , \tag{3}$$

i.e. introducing an exponential cutoff at high energies E_x. The resulting positron spectrum is shown in Fig. 2 for three values of the falloff constant Γ = 100, 300, and 1000 keV. We have assumed a mass of m_x=1.68 MeV corresponding to E_{e^+}=330 keV. It is obvious that the shape of the spectra comes nowhere near the narrow line observed in the experiments.

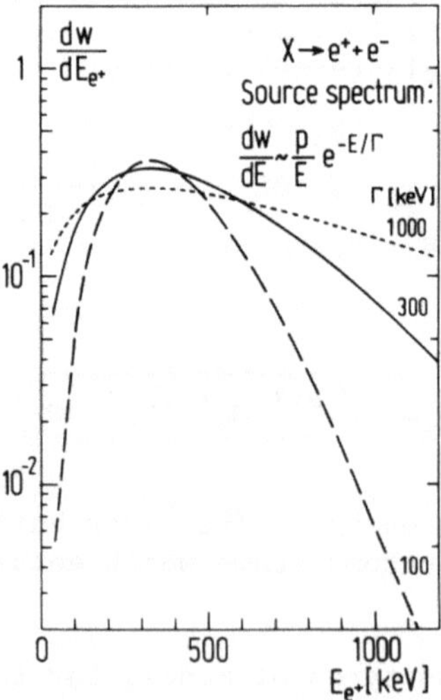

Fig. 2. Positron spectrum resulting from the decay $X \rightarrow e^+e^-$, if the energy spectrum assumed for the X particles is of the form $dw/dE_x \simeq (p_x/E_x)\cdot 1/\Gamma \exp[-(E_x-m_x)/\Gamma]$. Three different values of the falloff constant Γ have been assumed.

As already discussed qualitatively this situation would change if the low-velocity component of the particle spectrum was enhanced. This can be described, e.g.,by making an ansatz similar to (3) for the momentum distribution

$$\frac{dw}{dp_x} \simeq \frac{1}{\Gamma} e^{-p_x/\Gamma} \quad . \tag{4}$$

The expression (4) differs from (3) mainly through the factor $1/p_x \cdot dp_x/dE_x = E_x/p_x^2$ which heavily weights the low momentum part of the spectrum. The resulting positron spectra, shown in Fig. 3, are 70-130 keV wide, taking the same values for Γ as above. The curves have a flattened maximum due to the Doppler broadening of the c.m. motion. In a given experiment this effect will depend on the detector geometry.

With the ansatz (4) sufficiently narrow positron lines can be produced for reasonable values of the falloff constant Γ. However, the enhancement of low momentum emission implied in (4) is not justified by perturbative models for the particle creation process.

To complete the discussion of energy distributions, a further effect has to be mentioned which can lead to a narrow positron line[19,12]. If the lifetime τ of the X particle is comparable to the time needed to leave the finite sensitive volume of the detector then mainly the decay products of slow particles will be detected. This again will enhance the central region of the energy spectrum compared with the wings. Neglecting any details of the experimental set-up, the effect can be taken into account by multiplying the particle spectrum with the (velocity dependent) decay probability

$$P(E_x) = 1 - \exp(-\Delta R m_x/\tau p_x) \quad , \tag{5}$$

where ΔR is a measure for the extension of the sensitive region of the detector. This condition has to be imposed in the laboratory frame.

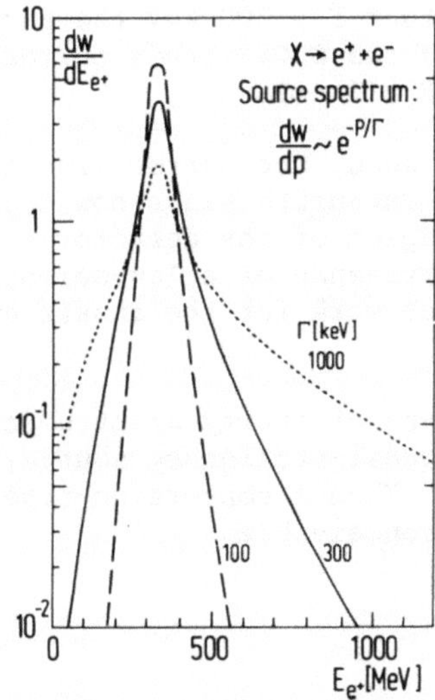

Fig. 3. Same as Fig. 2 if the spectrum assumed for the X particles is of the form $dw/dp_x \simeq 1/\Gamma \exp(-p_x/\Gamma)$.

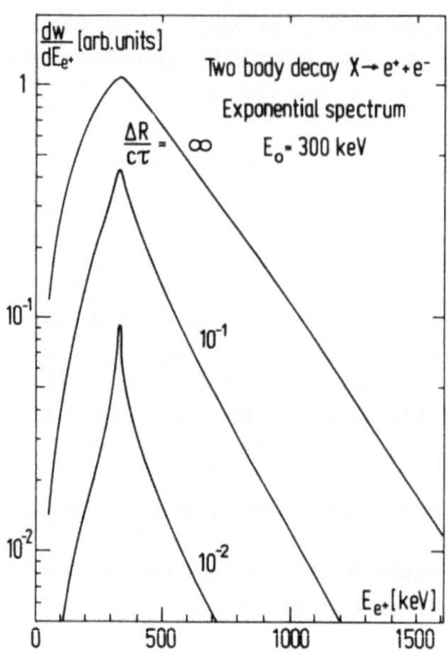

Fig. 4. Positron spectrum as in Fig. 2, assuming Γ=300 keV. The two lower curves demonstrate the effect of a finite lifetime allowing the escape of fast X particles out of the sensitive region of the detector. Two values of the ratio $\Delta R/\tau$ have been assumed, cf. Eq. (5).

In Fig. 4 we have taken an energy spectrum leaving out the factor p_x, i.e. being intermediate between the cases (3) and (4). Then the lifetime effect indeed can be used to obtain a sufficient narrow positron spectrum. Assuming a falloff constant Γ = 300 keV the parameter $\Delta R/c\tau$ was chosen as ∞ (as in Fig. 2), 0.1 and 0.01 (this corresponds to lifetimes of 0 s, 10^{-9}s, and 10^{-8}s, if $\Delta R \simeq 3$ cm).

However, as can be seen, the narrow linewidth in Fig. 4 is bought at the expense of emission intensity since now most of the particles decay outside the sensitive region of the detector. Furthermore, since the mechanism depends on the presence of a low-momentum component in the particle spectrum, it will not work for the ansatz of Eq. (3) and similar realistic models.

(2) A different type of energy spectrum can be expected if the emission originates from a quasi-stationary source, e.g. a long lived ($\gtrsim 10^{-20}$s) "giant" nuclear system. Then a conversion-type process leading to monoenergetic particles is conceivable

$$\frac{dw}{dE_x} \simeq \delta\,(E_x - E_o)\ . \tag{6}$$

Then from (1) it follows that the resulting positron spectrum is box-shaped centered around $E_x = E_o$ with a width

$$\Gamma = \sqrt{1 - 4m_e{}^2/m_x{}^2}\ \sqrt{E_o{}^2 - m_x{}^2} \tag{7}$$

To obtain a narrow width Γ, as required by the experiments, we obviously have two choices[22]: (i) $E_o \simeq m_x$ or (ii) $m_x \approx 2m_e$. This result has a simple interpretation: In the first case the X particles move with a low velocity while in the second case the energy released in the decay is small. Then the smearing in positron energy resulting from the superposition of the two velocities will be small. This is depicted schematically in Fig. 5. One should note, however, that both explanations call for quite coincidental values of the particle mass m_x in order to produce a narrow line width $\Gamma \lesssim 80$ keV. Solving (7) for m_x we obtain with E_o = 1.68 MeV

$$\text{(i)}\quad m_x = 1.67696 \text{ MeV} = E_o - 3.0 \text{ keV} \qquad \text{or}$$

$$\text{(ii)}\quad m_x = 1.02385 \text{ MeV} = 2m_e + 1.9 \text{ keV}\ .$$

In the first case the kinetic energy of the produced particle is limited to a value of 3 keV (which corresponds to a velocity $\beta_x \simeq .06$). For the second solution the mass of the particle only barely may exceed the threshold value of twice the electron mass. We should note that the quoted allowed excess energies will be reduced further if Doppler broadening due to the center of mass motion is taken into account.

From Fig. 5 we see that it is easy to distinguish experimentally between the two possibilities. In the second case the emitted e^+e^- pair moves nearly in parallel direction in the laboratory frame. Already from present coincidence data this seems to be excluded. We will come back to the assumption of monoenergetic X-production in Section 4.

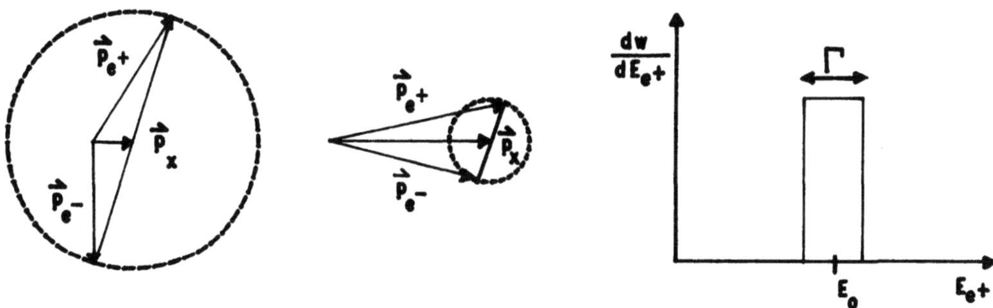

Fig. 5. The pair decay of particles emitted with a sharp energy E_o leads to a box shaped positron spectrum. The width Γ of this spectrum will be narrow if either the particle momentum (left) or the kinetic energy released in the decay (right) is small.

3. PHENOMENOLOGICAL COUPLINGS AND THEIR BOUNDS

The "new particle" hypothesis was entirely motivated by the GSI observations of positron lines [6,7,12], and more recently e^+e^- correlations. Trying to fit it into a theoretical framework at present is completely speculative. Therefore in the following we will give a phenomenological discussion [19,21,26] trying to avoid the limitation of specific models. We will investigate the coupling of X to various other fields and derive bounds on the coupling strengths from the fact that no traces of the new particle were observed in any other experiment in atomic, nuclear or particle physics. In the next section these bounds will be used to estimate the expected number of created particles in comparison with experiment.

Throughout the analysis we will assume (i) that "X" is an elementary object having no discernible internal structure, (ii) couples linearly to other fields. Going beyond these assumptions of course will considerably widen the room for speculations. Some ideas of this type will be discussed in sections 5,6.

The coupling of X to the leptons (ℓ=e,μ,τ) will be given by the interaction

$$L_{x\ell} = \sum_{\ell} g^{i}_{x\ell}\, \bar{\psi}_{\ell}\, \Gamma_i\, \psi_{\ell}\, \phi_x \tag{8}$$

where the index i denotes the Lorentz structure of the interaction which may be of scalar, pseudoscalar, vector, or axial vector type: $\Gamma_S = 1$, $\Gamma_P = i\gamma_5$, $\Gamma_V = \gamma_\mu$, $\Gamma_A = \gamma_\mu\gamma_5$. In a similar fashion the coupling to the quarks can be written as

$$L_{xq} = \sum_{q} g^{i}_{xq}\, \bar{\psi}_q\, \Gamma_i\, \psi_q\, \phi_x \tag{9}$$

where q = (u,d), (c,s),(t,b). The transition from (9) to the coupling to physical hadrons is not trivial, involving, e.g., for i = P current algebra techniques. One may introduce an effective interaction, again with the structure

$$L_{xN} = \sum_{N} g^{i}_{xN}\, \bar{\psi}_N\, \Gamma_i\, \psi_N\, \phi_x \tag{10}$$

where now N = neutron or proton.

In addition to the coupling to quarks and leptons the particle will also interact with the photon[22,33,34]. This is of particular interest in view of the strong electromagnetic fields present in heavy ion collisions. If we denote by $F_{\mu\nu} = \partial_\mu A_\nu - \partial_\nu A_\mu$ the electromagnetic field strength tensor, by $\tilde{F}_{\mu\nu}$ its dual, and by $G_{\mu\nu}$ the analogous tensor for the X-field the simplest ansatz for such an interaction reads

$$L_{x\gamma} = \frac{1}{2}\, g^{S}_{x\gamma}\, F_{\mu\nu}F^{\mu\nu}\phi_x = g^{S}_{x\gamma}\, (\vec{E}^2-\vec{B}^2)\phi_x \qquad \text{for scalar X}$$

$$L_{x\gamma} = \frac{1}{4}\, g^{P}_{x\gamma}\, F_{\mu\nu}\tilde{F}^{\mu\nu}\phi_x = g^{P}_{x\gamma}\, \vec{E}\cdot\vec{B}\, \phi_x \qquad \text{for pseudoscalar X}$$

$$L_{x\gamma} = \frac{1}{4}\, g^{V}_{x\gamma}\, F_{\mu\nu}F^{\mu\nu}F_{\tau\sigma}G^{\tau\sigma} + \frac{1}{4}\, g'^{V}_{x\gamma}\, F_{\mu\nu}\tilde{F}^{\mu\nu}\tilde{F}_{\tau\sigma}G^{\tau\sigma} \qquad \text{for vector X}$$

$$L_{x\gamma} = \frac{1}{4}\, g^{A}_{x\gamma}\, F_{\mu\nu}\tilde{F}^{\mu\nu}F_{\tau\sigma}G^{\tau\sigma} + \frac{1}{4}\, g'^{A}_{x\gamma}\, F_{\mu\nu}F^{\mu\nu}\tilde{F}_{\tau\sigma}G^{\tau\sigma} \qquad \text{for axial vector X}$$

(11a-d)

According to these interaction terms spin-0-particles will couple to two photons and spin-1 particles to three photons. In the vector case also a term

$$L_{x\gamma} = \frac{1}{2}\, g''^{V}_{x\gamma}\, \partial_\nu F_{\mu\nu}\, \partial^\lambda G^{\mu\nu} \tag{11e}$$

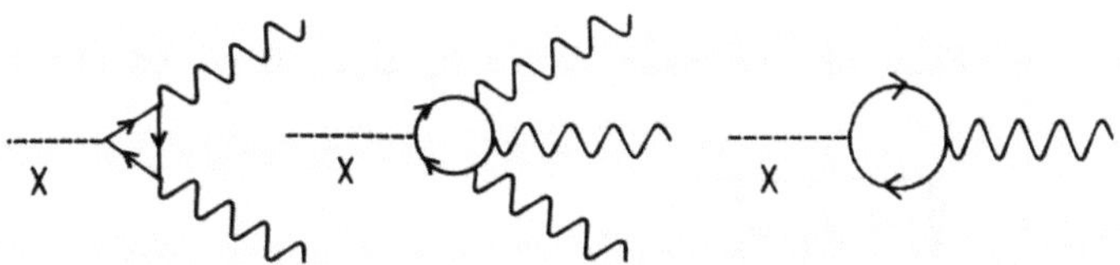

Fig. 6. The lowest order graphs leading to the effective couplings between X and the electromagnetic field as described by eqs. (11a,b), (11c,d), and (11e), respectively.

is conceivable which would transform (off-shell) photons into X-bosons. The interaction (11c) has the advantage that it will operate preferentially in the presence of strong electromagnetic fields, as they are encountered in heavy ion collisions.

The X-γ coupling can be viewed as an effective interaction resulting from intermediate loops of charged fermions. The corresponding graphs are shown in Fig. 6. For the scalar and pseudoscalar case the two-photon vertex is quite simple to calculate in the one-loop approximation. For a pseudoscalar particle, e.g., one can adopt the standard procedure known from the decay $\pi^0 \to \gamma + \gamma$. The result is

$$g^P_{x\gamma} = \sum_i \frac{e^2}{2\pi^2} \frac{g^P_{xi}}{m_i} \qquad \text{with } i = e, \mu, \ldots, u, d, \ldots \tag{12a}$$

and

$$g^S_{x\gamma} = \sum_i \frac{e^2}{12\pi^2} \frac{g^S_{xi}}{m_i} \quad . \tag{12b}$$

The expressions were derived neglecting the X-momentum compared to the mass of the fermion. In general the contributions are momentum dependent, approaching zero for very large X-momenta P_x. The effective coupling constant peaks for $P_x \sim 2m_i$. For the momenta in the order of a few MeV the approximation (12) is justified except for the electron loop. But even there it should only be wrong by perhaps a factor of two.

vector

A similar calculation for the three-photon vector coupling gives

$$g^V_{x\gamma} = \sum_i \frac{2}{45} \frac{e^3}{(4\pi)^2} \frac{g^V_{xi}}{m_i 4} \quad . \tag{12c}$$

From (12) the X-γ coupling constant can be estimated if information on the X-quark and X-lepton coupling is available.

To close the general discussion of coupling types let us give the <u>lifetime</u> of X. The partial width for the decay $X \to e^+ + e^-$ is found to be

$$\Gamma_{e^+e} = \frac{1}{2} m_x \alpha^i_{xe} F_i(m_x/m_e) \quad . \tag{13}$$

Here $\alpha^i_{x_e} = (g^i_{xe})^2/4\pi$ is the analogue of the Sommerfeld constant and F_i is a slowly varying function of the mass ratio m_x/m_e. E.g. in the pseudoscalar case we have

$$F_P\ (m_x/m_e) = \sqrt{1-4(m_e^2/m_x^2)}\quad . \tag{14}$$

For a particle mass of m_x = 1.68 MeV the functions F_i take on a value of F_S=0.5, F_P=0.79, F_V=0.63 and F_A=0.33. The lifetime corresponding to (13) is

$$\tau = \hbar/\Gamma \simeq 10^{-21}\,\mathrm{s}/\alpha^i_{xe} \tag{15}$$

Competing with this decay is the channel $X\to 2\gamma$ (for J=0) or $X\to 3\gamma$ (for J=1). The partial decay width into two photons is [23]

$$\Gamma_{2\gamma} = \frac{1}{16\pi}\,(g^S_{x\gamma}m_x)^2\ m_x \qquad \text{(scalar)}\quad , \tag{16a}$$

$$\Gamma_{2\gamma} = \frac{1}{64\pi}\,(g^P_{x\gamma}m_x)^2\ m_x \qquad \text{(pseudoscalar)}\ , \tag{16b}$$

and in the vector coupling case we have[28]

$$\Gamma_{3\gamma} = \frac{1}{64\pi^3}\ \frac{1}{960}\,(g^V_{x\gamma}m^4_x\,)^2\ m_x \qquad \text{(vector)}\ . \tag{16c}$$

Judging from the theoretical estimates for the X-γ coupling constant, eq. (12), the pair decay $X\to e^+ + e^-$ should be the dominant decay channel.

If one chooses a particular model for X its coupling type and certain relations between the various coupling constants entering (8), (9) and (11) are predicted. According to the presently accepted concepts of elementary particle physics it is suggestive to assume that a new boson X will be related to the Higgs sector of the standard SU(3)xSU(2)xU(1) model. Indeed the existence of one such particle, the axion[29,30] has been predicted by theory. The axion is a light pseudoscalar boson which was postulated in order to explain the absence of CP violation in the strong interaction. In its original form the "standard" axion model made quite definite predictions on the coupling constants, namely

$$g_{xq} = \frac{m_q}{F}\cdot\begin{cases} x & \text{for}\quad q = u,c,t \\ 1/x & \text{for}\quad q = d,s,b \end{cases}\ ,$$

$$g_{x\ell} = \frac{m_\ell}{F}\quad 1/x \quad \text{for}\quad \ell = e,\mu\ \tau\ \ .$$

Here F was thought to be the symmetry breaking scale of the weak interaction, F=250 GeV. x is a free parameter, the ratio between the vacuum expectation value of two Higgs fields. The axion mass is predicted to be m_a=75(x+$1/x$) keV which in our case would call for x≈20 (the alternative value x≈$1/20$ would lead to a much too long lifetime of X).

The axion has been searched for in a considerable number of experiments involving the decay of heavy quarkonium systems (J/ψ and γ) and of other elementary particles (κ,π,μ). high energy beam dump experiments[73] and decays of excited nuclear states. By now the standard axion model has been definitely ruled out through these experiments[31]. In view of this development "variant" axion models have been constructed showing a reduced coupling to the heavy quarks and a short lifetime (since some of the experiments were not sensitive to particles decaying within less than $\sim 10^{-13}$s). These new models, however, also are in conflict with experiment. We will not review this subject in detail, referring the reader to refs. 34-36. In the following discussion we will not adopt a particular model for the nature of the new particle, keeping the discussion as general as possible.

The Coupling to the Electron

Bounds on the coupling of a light neutral boson to the electron can be deduced from various high precision atomic physics data for which excellent agreement with the predictions of QED has been achieved. This agreement would be spoilt by the introduction of an additional interaction. We will discuss two such effects: the anomalous magnetic moment of the leptons[19] and the hyperfine splitting in positronium [37].

The existence of a light particle coupling to leptons will make itself felt not only if it is created on mass shell but also through virtual processes, i.e. vacuum fluctuation effects. Most notably, the g factor of the electron (and muon) will be affected. Since these values are known experimentally, and understood theoretically, to an exceedingly high accuracy, this will lead us to stringent upper limits for the X-particle-lepton coupling constant g^i_{xe} (and $g^i_{x\mu}$).

The main contribution to the anomalous magnetic moment of a Dirac particle is determined by the first-order radiative correction to the photon-lepton vertex (cf. Fig. 7b). This graph can be calculated using standard methods. It leads to an expression for the lepton vertex function which can be decomposed into electric and magnetic form factors. The magnetic moment is determined by the static limit (momentum transfer $q^2 \to o$) of the form factor $F_2(q^2)$. We define the lepton anomaly

$$a = \frac{1}{2}(g-2) = F_2(o) \quad . \tag{17}$$

This can be split into the known QED part plus an additional contribution due to the graph of Fig. 7b:

$$a = a_{QED} + \Delta a \quad . \tag{18}$$

Doing the calculation one finds

$$\Delta a = \frac{1}{2\pi}\alpha^i_{xe} K_i(m_x/m_e) \quad , \tag{19}$$

where the function K_i can be given in closed form[21] for the various types of coupling (i=S,P,V,A). Since the expressions for the coefficients K_i

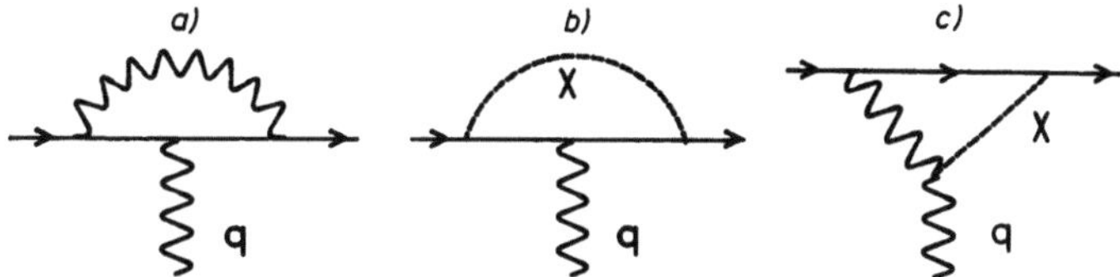

Fig. 7. Contribution to the anomalous magnetic moment. (a) lowest order QED vertex correction, (b) correction due to a particle X coupling to the lepton, (c) the same for X coupling to the photon.

are not very illuminating we have plotted their values against the mass ratio m_x/m_e in Fig. 8. Also indicated in the figure are the values of m_x/m for the case of electrons and of muons if one assumes m_x=1.68 MeV.

From the known level of agreement between theory and experiment for the anomalous magnetic moment we deduce limits for the coupling constant which are given in Table I. We used the recent values[38,39] $\Delta a<2\cdot 10^{-10}$ for the electron and $\Delta a<1\cdot 10^{-8}$ for the muon.

Table I. Upper bounds on the coupling constant between X and the electron or muon as deduced from the anomalous magnetic moment

i	α^i_{xe}	$\alpha^i_{x\mu}$	$(m_e/m_\mu)^2\alpha^i_{x\mu}$
S	$7\cdot 10^{-9}$	$4\cdot 10^{-8}$	10^{-12}
P	10^{-8}	10^{-7}	$2\cdot 10^{-12}$
V	$3\cdot 10^{-9}$	$7\cdot 10^{-8}$	
A	$5\cdot 10^{-9}$	$3\cdot 10^{-8}$	

From these results and eq. (15) a lower limit for the lifetime of X against pair decay can be deduced, e.g. $\tau_{e^+e^-}>10^{-13}$s for the pseudoscalar case. The last column of Table I shows the X-μ coupling constant scaled down by the mass ratio $(m_e/m_\mu)^2$. Assuming electron-muon universality this number should equal the X-e coupling constant provided that the X-particle is one of the Higgs fields generating the lepton masses. Due to the large value of the muon mass, this prescription obviously would lead to much stronger bounds although the muon anomaly is less precisely measured. The resulting lifetime limit is $\tau_{e^+e^-}>5.10^{-10}$s. This comes close to the border of the allowable region defined by the requirement that in the GSI experiment

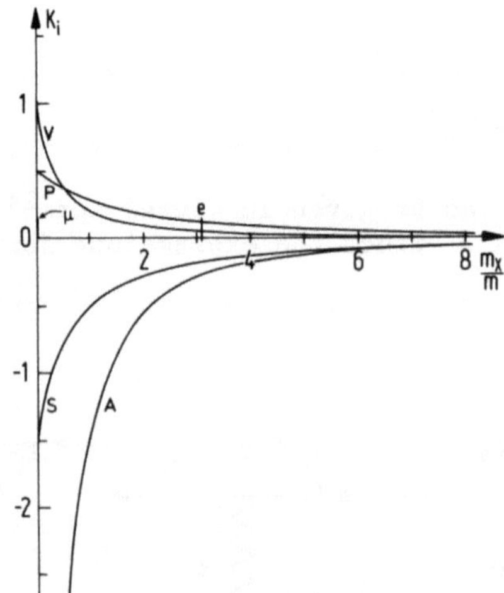

Fig. 8. The coefficients $K_i(m_x/m)$ describing the contribution of the postulated new particle to the anomalous magnetic moment of a lepton of mass m, cf. Eq. (19).

a sizable fraction of the particles has to decay within the detector. Let us note that there is a possible loophole in the (g-2)-argument: If there exist several new particles of different coupling type (e.g. scalar and pseudoscalar), their contributions to the anomalous magnetic moment may cancel each other, thus allowing for larger values of the coupling constant. Even though this possibility seems quite remote an independent bound on α_{xe} can be obtained from the hyperfine splitting of positronium[37]. The energy of positronium bound states will be shifted due to the annihilation ($q\simeq 2m_e$) and scattering ($q\simeq o$) interaction involving X instead of the photon. The energy shift is approximated by the density of the wavefunction at zero distance multiplied by a matrix element

$$\Delta E_n = M|\varphi_n(o)|^2$$

where

$$M = -\bar{u}\Gamma_i v \frac{g_{xe}^2}{(2m_e)^2 - m_x^2} \bar{v}\Gamma_i u + \bar{u}\,\Gamma_i u \frac{g_{xe}^2}{-m_x^2} \bar{v}\Gamma_i v$$

$$\equiv \frac{g_{xe}^2}{m_x^2 - 4m_e^2} A_i - \frac{g_{xe}^2}{m_x^2} B_i \,, \qquad i=S,P,V,A \,. \tag{20}$$

u and v denote the electron and positron spinors. The coefficients A and B depend on the relative spin orientation of e^+ and e^-. Therefore a contribution to the splitting between the singlet and triplet states is predicted. One finds $\Delta A(^3S_0 - ^3S_1) = -2$ for i = P,V,A and $\Delta B\ (^3S_0 - ^3S_1) = 4$ for i=A, the other contributions vanish.

The agreement between theory[40] and experiment[41] for the hyperfine splitting in the positronium groundstate is about 50 ppm or $|\Delta E| < 10$ MHz. From this follows a bound on the coupling constant

$$\alpha_{xe}^i < 10^{-6} \qquad \text{for } i = P,V,A. \tag{21}$$

This is less sensitive than the g-2 bound but it is not subject to possible cancellations.

The Coupling to the Photon

Apart from the theoretical estimates already discussed also experimental information is available on the coupling between a light neutral boson X and the electromagnetic field.

We first discuss the bounds imposed by Delbrück scattering, i.e. elastic photon scattering off the Coulomb field of a nucleus. This process has been analyzed[37] for the two-photon interaction of a spin-0 particle X, i.e. the couplings of eq. (11a,b). The resulting graphs which have to be added to the well-known box-shaped QED diagram are shown in Fig. 9. The unpolarized scattering cross section is given by

$$\frac{d\sigma}{d\Omega} = \frac{1}{2} \sum_{\vec{\varepsilon},\vec{\varepsilon}'} |2\pi\omega \langle \vec{k},\vec{\varepsilon}'|M|\vec{k},\vec{\varepsilon}\rangle|^2 \quad . \tag{22}$$

In the case of pseudoscalar coupling the matrix element entering (22) resulting from the first graph of Fig. 9a can be written as

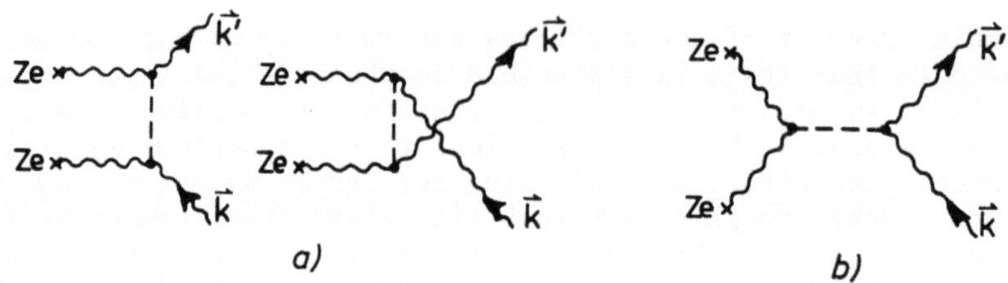

Fig. 9. Graphs describing the contribution of a scalar or pseudoscalar particle X to Delbrück scattering.

$$<\vec{k}';\vec{\varepsilon}'|M|\vec{k},\vec{\varepsilon}>=\frac{1}{2}\omega(g^{P}_{X\gamma})^{2}\int d^{3}q\int d^{3}q'\delta^{3}(\vec{k}-\vec{k}'-\vec{q}-\vec{q}')\ \vec{E}(\vec{q})\cdot\vec{\varepsilon}_{1}E(\vec{q}')\cdot\vec{\varepsilon}_{1}\left((k+q)^{2}-m_{X}^{2}\right)^{-1}, \tag{23}$$

where $\vec{\varepsilon}_1, \vec{\varepsilon}_1'$ are the polarization vectors perpendicular on those, $\vec{\varepsilon}$ and $\vec{\varepsilon}'$, of the incoming and outgoing photon. Eq. (22) contains an integration over the Fourier components of the nuclear Coulomb field strength $\vec{E}(\vec{q})$ in momentum space. The graph of Fig. 9b does not contribute in the pseudoscalar case since $\vec{E}\cdot\vec{B}=0$ for the electrostatic Coulomb field. The result of an evaluation of (22), (23) is shown in Fig. 10 (solid line) together with experimental data[42] for Delbrück scattering off uranium. Comparing these results one may conclude

$$g^{P}_{X\gamma} \quad < \quad .5 \quad \text{GeV}^{-1}. \tag{24}$$

This is a very conservative bound as no coherent addition to the (dominant) QED contribution has been performed.

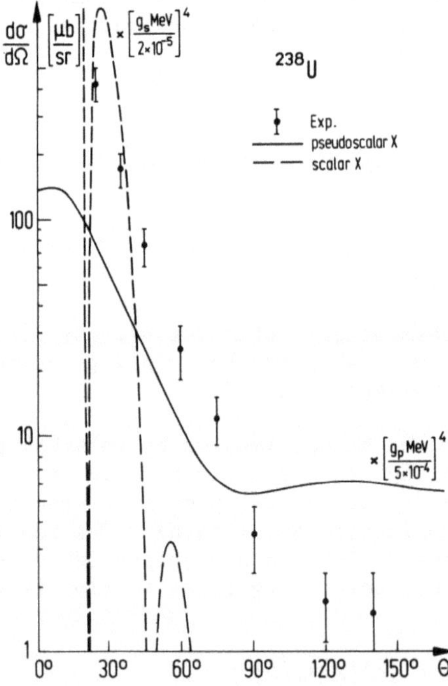

Fig. 10. Delbrück scattering cross section due to the presence of a scalar (dashed line) or pseudoscalar (full line) light neutral boson.

For a scalar particle the graph of Fig. 9b turns out to be dominant. It is given by

$$<\vec{k};\vec{\varepsilon}'|M|\vec{k},\vec{\varepsilon}>=\frac{1}{2}\omega(g_{x\gamma}^{S})^2 \int d^3q\int d^3q'\vec{E}(\vec{q})\cdot\vec{E}(\vec{q}')\vec{\varepsilon}\cdot\vec{\varepsilon}'((q+q')^2-m_x^2)^{-1} . \quad (25)$$

To make the integral finite the electric field strength $\vec{E}(\vec{q})$ for an extended nucleus has to be used. The result is given by the dashed line in Fig. 10 which now leads to an upper limit

$$g_{x\gamma}^{S} < 0.2 \text{ GeV}^{-1}. \quad (26)$$

Let us stress that the results (24) and (25) apply to the X-γ coupling in the presence of the strong electric field of a uranium nucleus, i.e. in a situation which bears some resemblance to heavy ion collisions.

For completeness we mention that the electron g-factor can be exploited to get information also on the X-γ coupling. This has been investigated for the pseudoscalar case[43]. The corresponding vertex correction graph, Fig. 7c) has a logarithmic ultraviolet divergence. Introducing a momentum cutoff Λ the contribution to the electron anomaly in the limit $m_e \ll m_x$ is

$$\Delta a = \frac{1}{8\pi^2} m_e \, g_{xe}^{P} \, g_{x\gamma}^{P} \ln(\frac{\Lambda^2}{m_x^2}) \quad , \quad (27)$$

which must be smaller than $2\cdot10^{-10}$. This leads to a bound for the product of coupling constants to the electron and to the photon. To draw some conclusion on the latter we take $\Lambda = 1$ GeV and $g_{xe}^{P}>10^{-6}$ (in order to have a lifetime $\tau<10^{-8}$s) which gives

$$g_{x\gamma}^{P} \lesssim 2 \text{ GeV}^{-1} \quad . \quad (28)$$

This is comparable to the Delbrück result (24).

Finally, a bound can be found for the three photon coupling, eq.(11c), of a vector boson[44]. In an external electric field this coupling leads to a mixing between the photon and the vector boson. Therefore the particle can be produced through the conversion of electromagnetic transitions in heavy nuclei where it would compete with the process of ordinary pair conversion. From the measure of agreement between experimental and theoretical[45] pair conversion coefficients in, e.g., $3^-\rightarrow0^+$ of ^{208}Pb, a bound of

$$g_{x\gamma}^{V} < 3.10^{-9} \text{ MeV}^{-4} \quad (29)$$

is obtained[44].

The Coupling to Hadrons

Not surprisingly when the domain of strong interaction is involved, it is less simple to find model independent bounds on the coupling of X to the hadrons. What first comes into mind are low energy processes involving a nucleus which as a whole couples to the new field. The structure of the effective coupling (10) depends on the coupling type i. In the nonrelativistic limit X couples either to the nuclear density (i=S,V) or to the nuclear spin density (i=A). The pseudoscalar coupling in this limit reduces to zero.

In the domain of atomic physics the exchange of X between the electron and the nuclear source leads to an additional Yukawa-type interaction potential of range $1/m_x \sim 120$fm. This new interaction will influence the binding energies of atomic electrons, depending on the product of the X-N and X-e coupling constants. The most stringent constraint of this type comes from the Lamb shift in hydrogen[19]. The experimental value $\Delta E = 1057.845 \pm .009$MHz agrees with theoretical predictions to better than .03MHz. From this a limit for the combined coupling constants

$$g^i_{xp}\, g^i_{xe} < 2.10^{-8} \tag{30}$$

is deduced for the coupling types i=S,V,A. Due to the vanishing source term pseudoscalar particles, however, will not influence the Lambshift in first order.

Further useful bounds on the coupling of light scalar (and also vector) particles can be gained from low energy (~1keV) neutron-nucleus scattering[46]. From the measured nearly isotropic angular distribution of the scattering cross section one concludes

$$g^s_{xn} < 6.10^{-5} \; . \tag{31}$$

In contrast to (30) this stringent bound does not depend on the undetermined value of g_{xe}.

In addition to processes where a nucleus as a whole acts as static external source for the X-field, nuclear transitions of sufficient energy should lead to the emission of real X-quanta. Searches for this process[47,48], which competes with the "background" of nuclear pair conversion if X is shortlived, have been performed mainly in view of the axion hypothesis. The interpretation of the result has to rely on model assumptions on the coupling and on the nuclear structure involved[49]. As an example, in a recent experiment which was particularly sensitive for short X-lifetime the X-conversion of the 9.17MeV $2^+ \to 1^+$ isovector transition in ^{14}N was searched for. From the observed null effect an upper limit g^p_{xN}(isovector) $<1.4 \cdot 10^{-2}$ was deduced.

A further field on which one may hunt for traces of X are "exotic" decays of elementary particles [34-36,50-52]. E.g. analyzing experimental data on the branching ratios of $\Sigma^+ \to pe^+e^-$, $K^+ \to \pi^+\gamma\gamma$ Suzuki[43] deduces an upper limit for the coupling of X to light quarks

$$g^p_{xq} < 6.10^{-4} \; . \tag{32}$$

Particularly sensitive for X-production is a recent experiment[36] on the decay of the charged pion, where a very small branching ratio for the channel $\pi^+ \to e^+e^-e^+\nu$ was found. Although the theoretical analysis of this process is quite involved, the measurement stands in clear contrast to the predictions of the various axion models proposed up to now[34].

4. CONVENTIONAL MODELS FOR PARTICLE PRODUCTION

Mechanisms for the production of X-bosons based on the linear couplings discussed in the last section fail to reproduce the observed GSI positron lines both on grounds of intensity and energy distribution. In the following we will demonstrate this by quoting several estimates and model calculations for the production process.

The Leptonic Production

As the hypothetical particle X is known to couple to the electron field - it is observed through its leptonic decay channel - one might conjecture that it is produced via this interaction[20]. It could be emitted from the rapidly changing electron cloud in a heavy ion collision just like a bremsstrahlung photon or, if atomic inner shells are involved, like a quasimolecular X-ray[53]. The theoretical description of this process can be taken over from the photon case. Essentially the number of particles emitted per energy interval dp_o and solid angle element $d\Omega$ in the semiclassical approximation for the nuclear motion is

$$\frac{dw}{dp_o d\Omega} = \frac{p}{16\pi^3}\left(g^i_{xe}\right)^2 \sum_{\substack{n>F\\ m<F}} \left|\int_{-\infty}^{+\infty} dt\, e^{ip_o t} \langle\phi_n|\Gamma_i e^{-i\vec{p}.\vec{x}}|\phi_m\rangle\right|^2 . \qquad (33)$$

Here the summation goes over the filled initial states m (including the lower continuum) and the empty final states n. The wavefunctions ϕ_k satisfy the time-dependent Dirac equation and thus describe the dynamics of the collision. The production cross section is obtained by multiplying (33) by the nuclear scattering cross section and integrating over impact parameter b. No detailed calculation of X-production based on this formalism has been performed. Nevertheless an estimate can be extracted from the existing calculations[54] and measurements of X-ray emission in heavy ion collisions. In high-Z systems the electronic transition matrix elements for the various types of coupling entering (33) are about equal in magnitude. Therefore, very roughly, the cross section for X-boson creation should correspond to the cross section for the emission of photons with energy $E_\gamma > m_x$ multiplied by the ratio of coupling constants $\alpha^i_{x\gamma}/\alpha$. For the collisions under consideration the integrated X-ray cross section is of the order of several hundred μb. Using an upper limit of $\alpha^i_{x\gamma} < 10^{-8}$ (cf. Sect. 3) this leads to the estimate $\sigma_x < 5.10^{-10}$b. This falls short of the measured intensity [6,7] by 5 orders of magnitude. Therefore X-boson creation from the atomic electron shell can not account for the observed positron lines.

Nuclear Bremsstrahlung of Light(pseudo-) scalar Bosons

Let us assume that the particle X is produced by coupling to the nuclei. At low collision energies, at or below the Coulomb barrier, one expects that the emission is mainly caused by the collective deceleration of the colliding nuclei, i.e. by a mechanism of bremsstrahlung type. A calculation of this process based on the semi-classical approximation was given in ref. 21. The number of emitted particles is

$$\frac{dw}{dp_o d\Omega} = \frac{p}{16\pi^3} \left| \int d^4x\, e^{i(p_o t - \vec{p}.\vec{x})} \rho_i(\vec{x},t) \right|^2 . \qquad (34)$$

It depends on the Fourier transform of the external source term $\rho_i(\vec{x},t)$ which is generated by the localized nuclear density distributions moving on classical trajectories. It is proportional to the nucleon density in the scalar case or to the nuclear spin-density for pseudoscalar coupling, each multiplied by an effective X-nucleon coupling constant. For simplicity we consider point-like nuclei with nucleon numbers A_1, A_2 moving on Rutherford hyperbolae $\vec{R}_i(t)$. Then the source term is

$$\rho_s(\vec{x},t) = g^s_{xN} \sum_{i=1}^{2} A_i\, \delta^3(\vec{x}-\vec{R}_i(t)) \qquad (35a)$$

$$\rho_p(\vec{x},t) \simeq g^p_{xN} \frac{1}{2M} \sum_{i=1}^{2} A_i \; \vec{p}\cdot\vec{s}_i \; \delta^3(\vec{x}-\vec{R}_i(t)) \; . \tag{35b}$$

In the latter expression M denotes the nucleon mass and $\vec{s}_i$ is the average spin per nucleon. To arrive at (35b) a self-interaction term has been neglected and in transforming to the derivative coupling the nonrelativistic approximation was used.

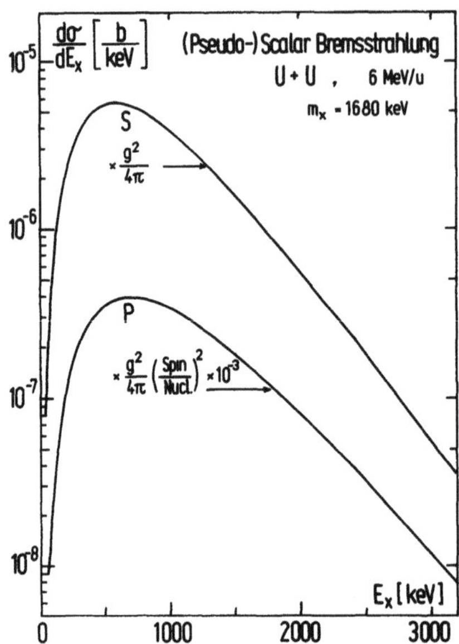

Fig. 11. Differential cross section for scalar (upper curve) and pseudoscalar (lower curve) particles produced by nuclear bremsstrahlung in a 6MeV/nucleon U + U collision calculated within the semiclassical approximation.

With (35) eq. (34) reduces to a Fourier integral in the time variable. It can be solved in closed form if one performs a series expansion of the exponential function $\exp(-i\vec{p}\cdot\vec{R}_i)$ which corresponds to a multipole expansion of the radiation field[55]. It is sufficient to keep only the lowest nonvanishing order since the wavelength 1/p is much larger than the characteristic length scale of the collision. Due to cancellations the first contributing term is of quadrupole type. In the pseudoscalar case (35b) we assume that the nuclear spins $\vec{s}_i$ are approximately constant in time. One gets maximum emission if they are aligned antiparallel.

Fig. 11 shows the resulting differential emission cross sections $d\sigma/dp_o$ calculated for the example U+U at 6MeV/u bombarding energy and assuming a particle mass m_x=1.68MeV. The spectra exhibit broad maxima at p_o~600keV. At low kinetic energies the intensity is strongly suppressed by a factor $(p/p_o)^5$. According to the discussion of section 2 such a spectrum will certainly not lead to a narrow positron line, even in the hypothetical case of long particle lifetime when most of the fast emitted particles decay outside the detector. Leaving aside this problem the argument against nuclear bremsstrahlung production of X can also be based on the intensity. The expected total emission cross section is

$$\sigma^S_x = 6.4 \cdot 10^{-3} b \cdot (g^S_{xN})^2/4\pi \; , \tag{36a}$$

$$\sigma_x^P = 5.3\ 10^{-7}b\cdot(g_{xN}^P)^2/4\pi * (\text{spin/nucleon})^2 \tag{36b}$$

Equating the product of σ_x times the fraction of particles decaying within the detector (as determined by the lifetime) with the measured positron cross section we arrive at a lower bound for the coupling constants $\alpha_{xN}^1\cdot\alpha > 10^{-13}$. In the scalar case this violates the constraint deduced from the Lambshift, cf. Sect. 3. For pseudoscalar particles the production cross section is so small that it calls for values of the coupling constant which are completely unrealistic, $\alpha_{xN}^p > 10^4$. Thus the nuclear bremsstrahlung production of X particles can be ruled out [21,56].

Electromagnetic Production

For completeness also the coupling of X to the electromagnetic field has to be considered as the basis of the X-production mechanism[27,22,23]. We are justified to treat the rapidly varying strong electromagnetic fields $\vec{E}$ and $\vec{B}$ produced by the colliding nuclei as external fields. The general expression for the number of created particles is the same as in the bremsstrahlung case, eq. (34). The source terms for the scalar and pseudoscalar case now read

$$\rho_s(\vec{x},t) \simeq g_{x\gamma}^s \vec{E}^2 = g_{x\gamma}^s \left(\frac{Z_1 e(\vec{x}-\vec{R}_1)}{|\vec{x}-\vec{R}_1|^3} - \frac{Z_2 e(\vec{x}-\vec{R}_2)}{|\vec{x}-\vec{R}_2|^3} \right)^2 \tag{37a}$$

or

$$\rho_p(\vec{x},t) = g_{x\gamma}^p \vec{E}\cdot\vec{B} = -g_{x\gamma}^p \frac{Z_1 Z_2 e^2}{|\vec{x}-\vec{R}_1|^3 |\vec{x}-\vec{R}_2|^3} \frac{\vec{L}\cdot\vec{x}}{\mu} \tag{37b}$$

$\vec{L}$ is the orbital angular momentum and μ the reduced mass. Two point nuclei in nonrelativistic motion were assumed for simplicity. The effect of the finite nuclear extension will only further reduce the results.

For the pseudoscalar case an evaluation of the X-production cross section was given in ref. 27. The final result

$$\sigma_x^p \simeq 10^{-7}(g_{x\gamma}^p)^2\text{GeV}^2 b \tag{38}$$

falls short of the experimental value $\sigma_{e^+} \simeq 10^{-4}$b by about three orders of magnitude if one inserts the upper bound for the coupling constant, eq. (24). Of course, also the energy spectrum is much too broad to produce a positron line.

Monoenergetic Production

The stumbling block for most models of X-particle production that strive to explain the GSI experiments is the narrow width of the singles positron lines. As discussed in section 2 (eq. (6) and below) this could be explained under very special circumstances if the energy distribution of the created particles is essentially monochromatic. To avoid the customary dynamical broadening of the emission spectrum a nuclear physics mechanism might be called into action. Indeed, if long-lived "giant" nuclear systems were formed in the collision the process of spontaneous pair production[8] and of pair conversion in the united nucleus[57] should lead to narrow lines in the positron spectrum. To explain the e^+e^- correlations one would have to combine this concept with the idea that a neutral particle is produced. Although such a combination of speculations seems to be highly construed let us consider its consequences.

In a crude semiclassical picture the two nuclei may be assumed to rotate and/or vibrate about an intermediate distance 2a during a prolonged time interval of contact. A simple ansatz is

$$\vec{R}(t) = 2a(1+\varepsilon \sin \Omega' t)\{\cos\Omega t,\ \sin \Omega t,\ 0\} \text{ for } 0\leq t\leq T \tag{39}$$

where Ω and Ω' are the rotational and vibrational frequencies and ε is the vibrational amplitude. Via the Fourier transform implied in (34) periodic motion in $\vec{R}(t)$ will lead to narrow (width $\Delta E \sim 2\pi/T$) lines in the energy spectrum of emitted particles. Such a model was proposed in ref. 22 for the conversion of nuclear rotations into pseudoscalar particles, mediated by the coupling to the electromagnetic field. The resulting expression for the number of created particles is [22,26]

$$W_P^{rot} = (g_{x\gamma}^P)^2 \frac{\pi}{945} (Z_1 Z_2 e^2 a\Omega)^2 p^7 a^4 T \tag{40a}$$

at the energy of twice the rotational frequency $p_o = 2\Omega$. The result is severely suppressed for small X-momenta by the combined dynamical and phase-space factor p^7. Furthermore a rotational frequency of $\Omega \sim 850$keV is unrealistic as it would call for a nuclear angular momentum of $L \simeq 1000\hbar$. Somewhat more promising is the conversion of a nuclear vibration (combined with a rotation to provide the $\vec{B}$-field). This leads to [26]

$$W_P^{vib} = (g_{x\gamma}^P)^2 \frac{2\pi}{3} (Z_1 Z_2 e^2 a\Omega)^2 p^3 \varepsilon^2 T \tag{40b}$$

at the energy $p_o = \Omega'$, the vibrational energy. (40b) is much less suppressed at small p and in addition more realistic values for the frequency Ω can be used.

For the production of scalar particles the calculation has to be repeated with the coupling (37a). As point-charges would lead to a divergence a cutoff must be introduced. A typical result for the conversion of pure vibration is[23]

$$W_S^{vib} = (g_{x\gamma}^S)^2 \frac{\pi^3}{10} (Z_1 Z_2 e^2 a)^2 p^5 \varepsilon^2 T \tag{40c}$$

at $p_o = \Omega'$.

When we insert reasonable values into eqs. (40a-c), e.g. $Z_1=Z_2=90$, a=8fm, $\varepsilon<0.3$, $T<10^{-19}$s, and $m_x = 1.7$MeV, the predicted particle numbers are roughly $W_P^{rot} < 10^{-20}(g_{x\gamma}^P)^2$ GeV², $W_P^{vib} < 3.10^{-10}(g_{x\gamma}^P)^2$ GeV², $W_S^{vib} < 2\cdot 10^{-10}(g_{x\gamma}^S)^2$ GeV². These numbers still have to be multiplied by the supposedly small probability to produce a sufficiently long-lived "giant" nucleus. With the bounds on the coupling constants $g_{x\gamma}^i$ that have been found in section 3 there is no chance to come anywhere near the experimental production cross sections. This remains true for the light particle scenario, $m_x \approx 2m$, although the numbers are considerably larger due to the higher momentum p.

Even if the effective coupling constant in the vicinity of the giant nucleus were increased by many orders of magnitude over its value in ordinary atoms other severe problems would remain. Since for a giant nucleus one expects a dense level spectrum a multitude of X-conversion lines should appear (which have to compete with the deexcitation channels via photon emission and ordinary conversion), a fact that may be overlooked in the simple semiclassical model discussed above. Furthermore, for the required large values of the reaction time $T\sim10^{-19}$s lines due to spontaneous positron creation with their characteristic strong Z-dependence should be dominantly visible[67].

Resonant X-Production in e^+e^--Scattering

Since the conjectured object X decays into electron-positron pairs it should be also be visible as a resonance in e^+e^- - scattering. Here we will collect a few results on the production cross sections that can be expected[21].

A weakly coupling neutral particle of mass m_x will appear as a narrow resonance in the s-channel (the annihilation graph) of Bhabha scattering at a total center of mass energy $E_{cm}=m_x$. Since colliding beam experiments presently are not feasible, a typical experiment will scatter an e^+ beam on the electrons of an atomic target fixed in the laboratory frame[58]. Let us first assume that the target electrons are essentially free and at rest. Then due to energy-momentum conservation the resonance will appear at a total positron energy $E_R=m_x{}^2/2m_e-m_e$. If we take, e.g., $m_x=1.7$MeV this corresponds to a kinetic positron energy $E_{kin}=1.81$MeV. The produced particle will move with a velocity $\beta_x=\sqrt{1-4/\rho}$, where $\rho=m_x{}^2/m^2{}_e$, in the forward direction, $\theta_x=0$.

An elementary calculation gives the production cross section for a pseudoscalar particle

$$\sigma^P_x = \pi^2 \alpha^P_{xe} (1-4/\rho)^{-1/2} \frac{1}{m_e} \delta(E-E_R) . \qquad (41a)$$

The scalar cross section is smaller by a factor $1-4/\rho$, i.e.

$$\sigma^S_x = \pi^2 \alpha^S_{xe}(1-4/\rho)^{1/2} \frac{1}{m_e} \delta(E-E_R) . \qquad (41b)$$

If one takes into account the finite lifetime of the particle, the δ-function will be broadened to a Breit-Wigner curve with the width Γ_i taken from (13). This leads to

$$\sigma^P_x = \frac{1}{m^2} (\alpha_{xe}{}^P)^2 \frac{\pi}{4}\rho^2 \frac{m^2}{(E-E_R)^2+(m_x\Gamma_p/2m)^2} \qquad (42)$$

and a similar expression for the scalar case, reduced by a factor $(1-4/\rho)^2$. The X-production cross section translates into the spectrum of electrons and positrons after the decay $X\to e^++e^-$. In the following we will only quote the more favourable pseudoscalar case:

$$\frac{d\sigma^P}{d\Omega_1} = \frac{1}{m^2} (\alpha^P_{xe})^2 \frac{\cos\theta_1}{(1-b^2\cos^2\theta_1)^2} \qquad \frac{m^2}{(E-E_R)^2+(m_x\Gamma_p/2m)^2} \qquad (43)$$

where $b^2 = (E-m)/(E+m)$ which at resonance coincides with the squared center of mass velocity $\beta_x{}^2$. Two-body kinematics gives a unique relation between the positron (electron) angle θ_1 and its energy, namely

$$E_1 = m \frac{1+b^2\cos^2\theta_1}{1-b^2\cos^2\theta_1} . \qquad (44)$$

The cross section (43) has to be compared with ordinary Bhabha scattering[59]. In principle there will be interference terms[21] since the amplitudes for X- and photon exchange have to be added coherently (both the s- and t-channels). This, however, can be neglected if the resonance is narrow.

Whether the resonance can be detected experimentally depends on the magnitude of the coupling constant α^1_{xe}. We recall that if X is an elementary object the width Γ is limited to be $\Gamma \approx \alpha_{xe} m < 10^{-8}$MeV which is very small indeed. Exactly at its peak value the resonance cross section would exceed Bhabha background by many orders of magnitude. To be realistic, however, the cross section must be averaged over the energy spread ΔE of the beam (and the momentum distribution of the electrons, see below).

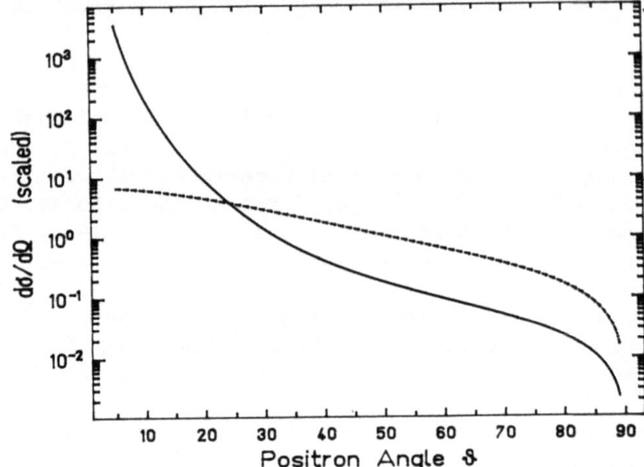

Fig. 12. Comparison of ordinary Bhabha scattering (full line) with the resonant production and subsequent decay of a pseudoscalar particle of mass m_x=1.7MeV (dashed line). The incident kinetic energy is E_{kin} = 1.81MeV. The coupling constants have been factored out.

Assuming for simplicity that the energy distribution has a Lorentzian shape the resonance factor in (42), (43) is replaced by $(\Delta Em/\Gamma_i m_x).m^2/[(E-E_R)^2+(\Delta E/2)^2]$. The maximum value of the averaged positron cross section then is

$$< \frac{d\sigma^p}{d\Omega_1} >_{max} = \frac{1}{m^2} \alpha^p_{xe} \frac{\cos\theta_1}{(1-b^2\cos^2\theta_1)^2} \frac{8}{\sqrt{\rho(\rho-4)}} \frac{m}{\Delta E} . \qquad (45)$$

Integrated over the positron angle we get

$$< \sigma^p_x >_{max} = \frac{1}{m^2} \alpha^p_{xe} \; 2\pi \sqrt{\rho/(\rho-4)} \; \frac{m}{\Delta E} . \qquad (46)$$

The ordinary Bhabha cross section is of the order α^2/m^2 which now has to be compared with $(\alpha^1_{xe}/m^2).(m/\Delta E)$ for the resonance scattering. The angle dependent functions multiplying these factors are plotted in Fig. 12. (To set the scale, note that $1/m^2$=1490 barn). For the resonance and Bhabha differential cross sections to be equal, the beam energy spread ΔE must be very small, e.g. ΔE=.2keV at $\theta_1=30^\circ$ if we take $\alpha^p_{xe}=10^{-8}$. With sufficient counting statistics, of course, a smaller signal would suffice to indicate the resonance so that ΔE may be larger. If the energy spectrum of the incident positron beam is continuous, e.g. originating from a β^+-emitting radioactive source, the value of ΔE is essentially replaced by the detector resolution used to identify the resonance.

The discussion up to now was based on the simplifying assumption that the target electrons are at rest. Then the X-particles are produced only at the sharp energy $E_i=E_R$ and move exactly in forward direction. In reality the target contains bound electrons which (in the inner shells of high-Z atoms) can carry a fairly large momentum, thus relaxing the kinematic conditions required to enter the resonance.

Let us assume that the electron is bound to an infinitely heavy nucleus. Then the created particle will have a fixed energy $E_X = E_i + E_n$ where E_n is the bound state energy. A simple expression for the X-production cross section by annihilation of an incident positron with the bound electron can be given

$$\frac{d\sigma}{dE_X d\Omega_X} = \pi\delta(E_X - E_i - E_n)\,\frac{p_X m}{p_i}\,(g^P_{xe})^2\;\mathcal{F}(\vec{q},\vec{p}_i) \tag{47}$$

with the spin-averaged electronic form factor

$$\mathcal{F}(\vec{q},\vec{p}_i) = \frac{1}{2}\sum_{\pm s_i} |\bar{v}(\vec{p}_i,\vec{s}_i)\gamma_5\,\varphi_n(\vec{q})|^2 \,. \tag{48}$$

Here $\varphi_n(\vec{q})$ is the electron wavefunction in momentum space

$$\varphi_n(\vec{q}) = \int \frac{d^3x}{(2\pi)^{3/2}}\, e^{-i\vec{q}.\vec{x}}\, \varphi_n(\vec{x}) \tag{49}$$

which is to be evaluated at the momentum transfer $\vec{q}=\vec{p}_X-\vec{p}_i$ absorbed by the nucleus. $v(\vec{p}_i,\vec{s}_i)$ denotes the normalized spinor of the incoming positron. To arrive at this simple result the Coulomb wavefunction of the positron has been replaced by the plane wave $\sqrt{m/E_i}\; v(\vec{p}_i,\vec{s}_i)\exp(-i\vec{p}_i.\vec{x})$. This approximation will overestimate the true cross section due to the neglect of Coulomb repulsion.

The total production cross section can be expressed as an integral over momentum transfer

$$\sigma_X = 2\pi^2 \frac{m}{p_i^2}\left(g^P_{xe}\right)^2 \theta(E_i+E_n-m_X)\int dq\; q\,\mathcal{F}(\vec{q},\vec{p}_i) \tag{50}$$

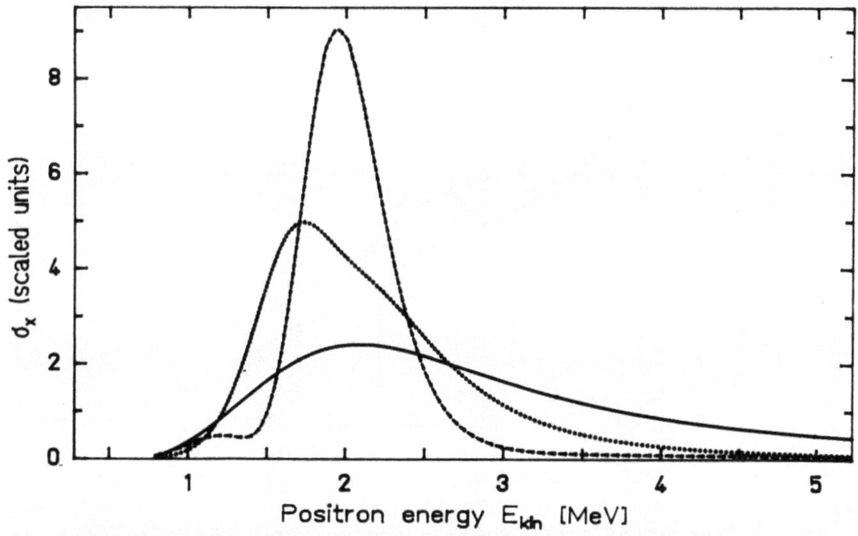

Fig. 13. Cross section for X-production (m_X=1.7MeV) by positrons impinging on inner shell electrons of a thorium atom (Z=90) in dependence of the positron kinetic energy E_{kin}. The states considered are $1s_{1/2}$ (full line), $2s_{1/2}$ (dashed), $2p_{1/2}$ (dotted).

between the limits $q_{min}=|p_i-p_x|$ and $q_{max}=p_i+p_x$, where $p_x^2=(E_i+E_n)^2-m_x^2$. The angle between the vectors q and p_i is fixed by the relation $2qp_i\cos\theta_q = p_x^2 - p_i^2 - q^2$.

Neglecting screening the hydrogen-like Dirac bound state wavefunctions in momentum space, $\varphi_n(\vec{q})$, can be calculated analytically. Then the form factor (48) follows at once. Fig. 13 shows the resulting resonance cross section as a function of the positron kinetic energy E_{kin} for a thorium target (Z=90). The electrons of the states $1s_{1/2}$ (full line), $2s_{1/2}$ (dashed line) and $2p_{1/2}$ (dotted line) are considered. The mass of the resonance was taken as m_x=1.7MeV. The numbers have to be multiplied by the factor α^p_{xe}/m^2. We note that, due to the momentum spread of the bound wavefunctions, the resonance can be excited in a large range of incident energies, in contrast to eq. (41). When averaging over a wide enough energy window, the magnitude of the cross section is nearly independent of the bound state and also agrees with the value for free electrons at rest. Hence resonant X-production will be best visible if the electrons are as weakly bound as possible.

The momentum distribution of the bound electron also will lead to a finite angular window of the emission around the forward direction. For the discussed example the angular spreading is of the order 10°. Finally also the distribution of positrons (electrons) following from the decay of X can be calculated. The doubly differential cross section is

$$\frac{d\sigma}{dE_1 d\Omega_1 dE_2 d\Omega_2} = \frac{1}{(2\pi)^2}\left(g^p_{xe}\right)^4 \frac{m p_1 p_2}{p_i}\,\delta(E_1+E_2+E_i-E_n)(E_1E_2-\vec{p}_1\cdot\vec{p}_2+m^2)$$

$$\cdot\frac{1}{(2m^2-m_x^2+2E_1E_2-2\vec{p}_1\cdot\vec{p}_2)^2+m_x^2\Gamma_p^2}\,\mathcal{F}(\vec{q},\vec{p}_i) \qquad (51)$$

where $\vec{q} = \vec{p}_1+\vec{p}_2-\vec{p}_i$.

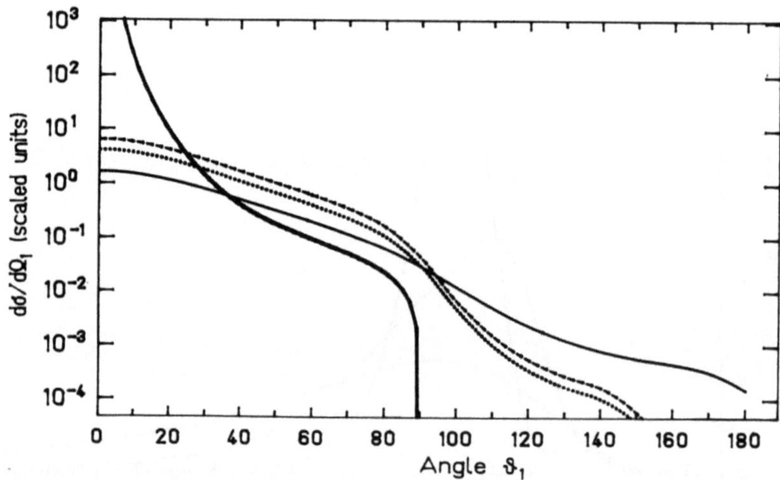

Fig. 14. Differential cross section for positron emission following the resonant production of a 1.7MeV pseudoscalar particle by positrons of E_{kin}=1.81MeV incident on the inner shell electrons of a thorium target. For comparison the Bhabha cross section for free target electrons is included (thick line). The coupling constants α^p_{xe} and α have been factored out.

In the limit of small resonance width (on-shell production of X) a simplified expression for the positron (electron) differential cross section can be written

$$\frac{d\sigma}{dE_1 d\Omega_1} = \frac{1}{2}\frac{m}{p_i}\left(g^p_{xe}\right)^2 (1-4/\rho)^{-1/2} \int_0^{2\pi} d\tilde{\varphi} \;\; \mathcal{F}(\vec{q}, \vec{p}_i) \tag{52}$$

where the integration variable $\tilde{\varphi}$ is the azimuthal angle of $\vec{p}_2$ with respect to $\vec{p}_1$. The momentum $\vec{q}$ now depends on the angle θ_1 in a rather complicated way.

Due to the nuclear recoil the fixed relation between angle and energy, eq. (44), now is smeared out. Fig. 14 shows the resulting angular distribution $d\sigma/d\Omega_1$, integrated over E_1 at an energy E_{kin}=1.81MeV, the resonance energy for free electrons. Also plotted in the figure is the corresponding Bhabha cross section (thick line). The results have to be multiplied by α^p_{xe}/m^2 and α^2/m^2, respectively. Since $\alpha_{xe} < 10^{-6}\alpha$ Bhabha scattering clearly will be dominant.

5. MODELS INVOLVING SEVERAL X-PARTICLES

All phenomenological models discussed so far assumed the existence of a single X-particle, which can be treated as elementary at energy scales of the order of a few MeV. The observation of several line-structures however forces one to assume the existence of either several particles or of a composite object with several excited states. Let us start by analyzing the first possibililty. There is a marked difference whether the X-particles are of scalar-/pseudoscalar- or of vector-/axial vector - nature. In the first case they would certainly belong to the Higgs-sector of the theory, whereas in the second case they would have to be identified with gauge-fields for purposes of renormalizability. As long as one does not introduce a new interaction, the only fields with which they could be identified are the gluons. Although gluons are generally assumed to be massless a soft symmetry-breaking leading to small gluon masses (of the order of a few MeV might be acceptable[60,61]). The massive gluons could then be deconfined. Such a soft symmetry breaking could even lead to different masses for different gluons. In ref. 62 a scheme was discussed, which made only one gluon (equivalent to the G^8 gluon) massive, whereas the others stayed massless. This scheme could easily be generalized to give rise to a whole mass spectrum. Even without referring to any specific model the decay channel of such massive gluons can be analyzed. For the masses of interest (namely 1.4 to 1.8 MeV) the dominant decay channel can be either $X \to e^+e^-$ or $X \to 3\gamma$. In both cases the symmetry violation leads to a coupling of the X (which has a color-octet change) to a color-singlet $q\bar{q}$ -pair, which then annihilates (Figure 15). Note that the 3γ coupling is again bounded by e.g. the X-γ coupling in the presence of strong electric fields.

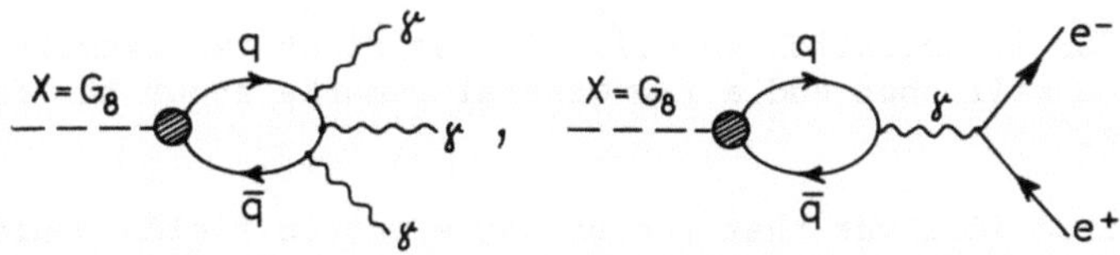

Fig. 15. Possible decay modes of massive, unconfined gluons.

If the X particles are scalar a.d/or pseudoscalar they will probably be identified with components of the Higgs-fields. As the Higgs-sector is completely untested experimentally (and in fact might even be a mathematical artefact) one is free to introduce as many particles as required. In fact, it is mainly a question of skill and patience to find the best suited combination of initial group representation and symmetry-breaking potential. It is exactly this arbitrariness which is used in the axion models[29-33], where one starts from two Higgs-doublets instead of one.

Instead of postulating several low-mass elementary particles with similar properties one can interpret the different mass-eigenstates as the excitation spectrum of a composite object, just as the J/ψ resonances are understood as $c\bar{c}$-states.

Candidates for such a composite object were discussed in refs [63-65]. The main idea is that a new electron-electron interaction, e.g. a nonlinear coupling, could lead to bound states of an equal number of electrons and positrons. If the total energy of such a 'poly-positronium' were about 1.7MeV it would have rather interesting properties. The only possible decay channels would be:

a) $X \to e^+ + e^- + m\gamma \quad m \geq o$

b) $X \to n\gamma \quad n \geq 2$ or 3, depending on the spin of X

Each of these decays requires a multiple pair-annihilation among the constituents. Thus the decay time could very well be of the order of 10^{-9}sec. One has, however, to make sure that the e^+e^- decay probability (without any additional photons) is not smaller than a few percent. A major problem for such a model is the following. The nonlinear effects must become large in strong electric fields to explain why the 'polypositronium ' is only produced in high-Z nuclear collisions. This fact can then be understood as being due to the high electron densities occuring in very strong fields. On the other hand atomic effects like K-hole production are described correctly be standard QED and it is not clear why these would not be substantially modified by the new interaction, leading to disagreement between theory and experiment. Until now it was not possible to construct a toy model showing at least a few of the required properties. Therefore at the time being the polypositronium hypothesis must be regarded as purely speculative[68].

6. EXOTIC PRODUCTION MECHANISMS

We have just discussed that the nature of the hypothetical new particle (s) is completely unclear. At the same time it has become obvious that the production mechanism must be very special and highly selective. To find candidates for the production mechanism it is necessary to check all conceivable processes for every type of particle. Clearly such a search can never be exhaustive. The fact that, up to now, nobody has succeeded in finding any valid candidate for the production process may not mean that particle scenario is wrong.

Just for illustration we will discuss first two examples[26] which do not work and will then add a few general remarks about the formation of condensates.

The first idea was that the strong electric fields could render decays possible which are usually forbidden. As an example we discussed the decay of a nuclear excitation with pionic quantum numbers (i.e. an isospin-one, unnatural parity state) into an iso-singlet pseudoscalar particle (see figure 16).

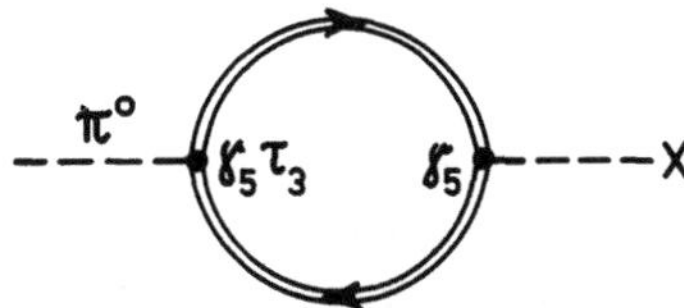

Fig. 16. Isospin-violating π^0-decay in a strong electric field. The double-line stands for the pion-propagator in an electric field.

With a few approximations this process can be calculated using the Euler-Heisenberg-Lagrangian. The final result is that the decay probability is in principle a complicated function of the nuclear charge Z but the non-linear terms are still unimportant for $1<Z\alpha<2$. Furthermore the spectrum of the X particles produced is suppressed for small momenta and is very broad such that no positron-line could emerge.

The second attempt started from the observation that the strong electro-magnetic fields induce a local CP-violating term for the gluon field[69]:

$$\mathcal{L}_{int} = \text{const.}\ \vec{E}\ .\ \vec{B}\ G^a_{\mu\nu}\ \tilde{G}^{a\mu\nu} = \frac{\alpha_s\,\theta(r)}{8\pi}\ G^a_{\mu\nu}\ \tilde{G}^{a\mu\nu} \tag{53}$$

This term could lead to axion production. However a detailed calculation shows that this mechanism effectively reduces to a X-2γ-coupling which has already been excluded[26], since $\theta(r)=\Sigma_q(\alpha/m_q{}^4)\vec{E}\cdot\vec{B}$.

With the failure of these and other attempts the only promising idea one is left with is the formation or modification of some kind of Higgs-condensate. This idea is roughly the following.

The Higgs fields ϕ^a couple to the fermions by

$$\mathcal{L}_{int} = \sum_{ij} \Gamma^a{}_{ij}\ \psi_i\ \phi^a\ \psi_j\ . \tag{54}$$

The effect of the fermion density on spontaneous symmetry breaking can probably be amplified by allowing for a non-linear fermion Higgs coupling[62]. Another suggestion along these lines is derived from nonlinear terms in the Peccei-Quinn model[29,70]. By inserting the definition of the axion field into equation (54) one gets:

$$\begin{aligned}\mathcal{L}_{int} &= \text{const.}\ \{\ \bar{q}_L\ q_R\ \exp(\ iax/f\ )+\ \bar{q}_R\ q_L\ \exp(\ -iax/f\ )\ \}\\ &= \text{const.}\ \{\ \bar{q}\ \ q\ \cos(\ ax/f\ )\ -\ \bar{q}\ i\gamma_5\ q\ \sin(\ ax/f\ )\ \}\end{aligned} \tag{55}$$

Due to the non-linear term $\bar{q}\,q\cos(ax/f)$ the axion field a could in principle condensate in the presence of a large nucleus.

In general there are several possibilities for the modifications of spontaneous symmetry breaking in the presence of fermion fields:

1. The vacuum expectation value can be shifted. Including the term (54) the Higgs potential reads (for just one Higgs field)

$$V_{\phi} = -\mu^2\phi^2 + \lambda\phi^4 + \alpha(x)\phi \tag{56}$$

with $$\alpha(x) = \langle vac| \sum_i \Gamma_i \bar{\Phi}_i \phi_i |vac\rangle \tag{57}$$

Here $|vac\rangle$ denotes the physical vacuum state including the local quark densities. Neglecting the kinetic contributions it is obvious that the vacuum expectation value is shifted locally

$$\langle vac|\phi|vac\rangle = \phi_o \left(1- (1/4\mu^2)\, \alpha(x)/\phi_o + O[(\alpha/\phi_o)^2]\right) \tag{58}$$

with $$\phi_o{}^2 = \mu^2/2\lambda \quad . \tag{59}$$

2. For a sufficiently large value of a(x) spontaneous symmetry breaking can be suppressed altogether.

3. If μ itself depends on the fermion density, a Higgs field component that does not condensate for vanishing fermion density, can acquire a non-zero vacuum expectation value in the presence of fermions.

In any of these cases a copious production of particles would be associated with the corresponding vacuum change. However all these processes share the following general problems:

(a) A fine-tuning of parameters is needed to obtain an effect in a nuclear complex with the total baryon number $A \simeq 400$ and avoid any visible effect for ordinary nuclei $A \leq 250$.

(b) It is rather unclear how the condensating particles should be emitted and if so why they should be emitted with sufficiently small momentum. In principle, formation of a bound state during the nuclear collision could help resolve this problem[66].

(c) One has to find a convincing explanation as to why no drastic effect is observed for reactions above the Coulomb barrier, like U+U at E/A=10MeV/nucleon.

(d) Again, no contradiction with a huge body of data in nuclear and particle physics is permitted to occur. No general study of the compatibility of particle models based on nonlinear dynamics or condensate formation has been performed, so far.

7. CONCLUSIONS

After two years of speculation that the positron peaks and e^+-e^- coincidence event discovered at GSI might be caused by elementary particle decay, no definite conclusion has been reached. In fact, theoretical and experimental aspects of this fascinating hypothesis have - to a large extent - developed in opposite directions. On the side of theory the hypothesis has lost a good deal of attractiveness because every single model which was quantitatively analyzed has turned out untenable. Either there was a clear contradiction to well-founded experimental data from atomic or nuclear physics, or the models fell far short of describing the GSI events.

On the side of the GSI experiments, on the other hand, all evidence gathered during the past year is compatible with the hypothesis of particle decay, albeit of several distinct particle states, and many pieces of data clearly seem to suggest a two-body decay. It is true that all recent attempts to search for new light particles in nuclear or high-energy physics processes have been futile, and a first search for 2γ-coincidences in U+Th collisions has been unsuccessful[72]. However, these findings have limited direct bearing on the search for the source of the GSI events, except through theoretical models that may well be too narrow-minded.

One lesson from this experience is that an unambiguous refutation or confirmation of the particle hypothesis most likely must come from the experiments conducted at GSI. Obviously, a measurement of the invariant mass of the observed e^+-e^- correlated pairs is required; alternatively, the problem would be solved if a different source of the coincidence events could be demonstrated. Given the difficulty of these tasks, a final resolution is not likely to come soon. Other experimental approaches may, therefore, be helpful: especially so electron-positron scattering with high energy resolution, which could - at least in principle - rule out the particle hypothesis. However, also in this case the required extreme energy resolution poses formidable experimental problems.

Returning to theoretical considerations, it is noteworthy that the disillusioning experience is almost exclusively based on simple scenarios. The conclusions reached so far may be summarized as follows: Explanation of the GSI events requires a highly nonlinear particle production mechanism and must involve several discrete particle states. On the basis of present status of the analysis, the most likely candidates are particles with internal structure and a fairly rich excitation spectrum. The production mechanism probably must involve the temporary formation of a bound state around the nuclei or, even better, in the center-of-mass system, and possibly an as yet unknown change in the vacuum state in the presence of strong fields. Finally, let us note that - irrespective of the outcome - the studies initiated by the "particle hypothesis" will have helped to close an existing but previously unnoticed gap[71] in our knowledge of the particle spectrum in nature : that of a weakly interacting neutral particle in the MeV mass range with lifetime below 1ns.

ACKNOWLEGDEMENTS

We like to thank many friends and colleagues for their helpful discussions, in particular all members of the experimental groups at GSI and W. Meyerhof and his collaborators.

REFERENCES

1. H. Backe, L. Handschug, F. Hessberger, E. Kankeleit, L. Richter, F. Weik, R. Willwater, H. Bokemeyer, P. Vincent, Y. Nakayama, and J. S. Greenberg, Phys. Rev. Lett. 40:1443 (1978); C. Kozhuharov, P. Kienle, E. Berdermann, H. Bokemeyer, J. S. Greenberg, Y. Nakayama, P. Vincent, H. Backe, L. Handschug, and E. Kankeleit, Phys. Rev. Lett. 42:376 (1979).
2. W. Greiner, B. Müller, J. Rafelski, Quantum Electrodynamics of Strong Fields, Springer, (1985).
3. B. Müller, J. Rafelski, and W. Greiner, Z. Phys. 257:62 (1972); 257:183 (1972); Ya. B. Zeldovich and V. S. Popov, Sov. Phys. Usp. 14:673 (1972).
4. J. Rafelski, B. Müller, and W. Greiner, Z. Phys. 49:A285 (1978).
5. J. Reinhardt, B. Müller, and W. Greiner, Phys. Rev. A24:103 (1981).

6. J. Schweppe, A. Gruppe, K. Bethge, H. Bokemeyer, T. Cowan, H. Folger, J. S. Greenberg, H. Grein, S. Ito, R. Schulê, D. Schwalm, K. E. Stiebing, N. Trautmann, P. Vincent, and M. Waldschmidt, Phys. Rev. Lett. 51:2261 (1983).
7. M. Clemente, E. Berdermann, P. Kienle, H. Tsertos, W. Wagner, C. Kozhuharov, F. Bosch, and W. Koenig, Phys. Lett. 137B:41 (1984).
8. J. Reinhardt, U. Müller, B. Müller, and W. Greiner, Z. Phys. A303:173 (1981).
9. U. Heinz, U. Müller, J. Reinhardt, U. Müller, B. Müller, and W. Greiner, Ann. Phys. (N.Y.) 158:476 (1984); S. Schramm, J. Reinhardt, U. Müller, B. Müller, and W. Greiner, Z. Physik A323:275 (1986).
10. T. Tomoda and H. A. Weidenmüller, Phys. Rev. C28:739 (1983).
11. W. Greiner, Proc. Intl. Conf. on Nucl. Phys., P. Blasi and R. A. Ricci, eds., 11:635,(Bologna, 1983); M. Seiwert, W. Greiner, W. T. Pinkston, J. Phys. G11:L21 (1985).
12. T. Cowan, H. Backe, M. Begemann, K. Bethge, H. Bokemeyer, H. Folger, J. S. Greenberg, H. Grein, A. Gruppe, Y. Kido, M. Klüver, D. Schwalm, J. Schweppe, K. E. Stiebing, N. Trautmann, and P. Vincent, Phys. Rev. Lett. 54:1761 (1985).
13. U. Müller, T. de Reus, J. Reinhardt, B. Müller, and W. Greiner, Z. Physik A323:261 (1986).
14. H. Tsertos, E. Berdermann, F. Bosch, M. Clemente, P. Kienle, W. Koenig, C. Kozhuharov, and W. Wagner, Phys. Lett. 162B:372 (1985).
15. L. W. Neise, J. Maruhn, and W. Greiner, Z. Phys. A323:195 (1986); J. Fink, U. Heinz, J. A. Maruhn, and W. Greiner, Z. Phys. A323:189 (1986).
16. A. Schäfer, B. Müller, and W. Greiner, Phys. Lett. 149B:455 (1984).
17. M. Grabiak, contribution to this conference.
18. T. Cowan, H. Backe, K. Bethge, H. Bokemeyer, H. Folger, J. S. Greenberg, K. Sakaguchi, D. Schwalm, J. Schweppe, K. E. Stiebing, P. Vincent, Phys. Rev. Lett. 56:444 (1986).
19. A. Schäfer, J. Reinhardt, B. Müller, W. Greiner, G. Soff, J. Phys. G11:L69 (1985).
20. A. B. Balantekin, C. Bottcher, M. R. Strayer, and S. J. Lee, Phys. Rev. Lett. 55:461 (1985).
21. J. Reinhardt, A. Schäfer, B. Müller, W. Greiner, Phys. Rev. C33:194 (1986).
22. A. Chodos and L. C. R. Wijewardhana, Phys. Rev. Lett. 56:302 (1986).
23. K. Lane, Phys. Lett. 169B:97 (1986).
24. T. Cowan et al., contribution to this conference.
25. W. Koenig et al., contribution to this conference.
26. A. Schäfer, J. Reinhardt, B. Müller, and W. Greiner, Z. Phys. A324:243 (1986).
27. B. Müller and J. Rafelski, Phys. Rev. D34:in print (1986).
28. S. Barshay, preprint, TH Aachen (1986).
29. R. D. Peccei and H. R. Quinn, Phys. Rev. Lett. 38:1440 (1977).
30. S. Weinberg, Phys. Rev. Lett. 40:223 (1978); F. Wilczek, ibid. 40:279 (1978).
31. A. Zehnder, Proc. of the 1982 Gif-sur-Yvette Summer School.
32. L. M. Krauss and F. Wilczek, Phys. Lett. 173B:189 (1986).
33. R. D. Peccei, T. T. Wu, and T. Yanagida, Phys. Lett. 172:435 (1986).
34. W. A. Bardeen, R. D. Peccei, T. Yanagida, DESY 86-054:preprint.
35. B. Balantekin, contribution to this conference.
36. R. Eichler et al., Phys. Lett. 175:101 (1986).
37. A. Schäfer, J. Reinhardt, W. Greiner, and B. Müller, Mod. Phys. Lett. 1:1 (1986).
38. P. Mohr, contribution to this conference.
39. T. Kinoshita and J. Sapirstein, New Developments in QED, Atomic Physics 9 (World-Scientific, Singapore, 1984).
40. G. T. Bodwin and D. R. Yennie, Phys. Rep. 43:267 (1978).

41. M. W. Ritter, P. O. Egan, V. W. Hughes, and K. A. Woodle; Phys. Rev. 30:1331 (1984).
42. S. Kahane and R. Moreh, Phys. Lett. 47B:351 (1973).
43. M. Suzuki, Phys. Lett. 175:364 (1986).
44. A. Schäfer, B. Müller, J. Reinhardt, preprint (1986).
45. P. Schlüter, G. Soff, and W. Greiner, Phys. Rep. 75:327 (1981).
46. R. Barbieri and T. E. O. Ericson, Phys. Lett. 57B:270 (1975); U. E. Schröder, Mod. Phys. Lett. A1:157 (1986).
47. F. P. Calaprice, R. W. Dunford, R. T. Kouzes, M. Miller, A. Hallin, M. Schneider, and D. Schrieber, Phys. Rev. 20:2708 (1979).
48. M. J. Savage, R. D. McKeon, B. W. Fillippone, and L. W. Mitchell, Phys. Rev. D34:1332 (1986); F. W. N. deBoer et al., Groningen preprint (1986).
49. T. W. Donnelly, S. J. Freedman, R. S. Lytel, R. D. Peccei, and M. Schwarz, Phys. Rev. D18:1607 (1978).
50. E. Ma, Phys. Rev. D34:293 (1986).
51. L. M. Krauss and M. Zeller, YTP86-08,preprint.
52. G. Mageras et al., Phys. Rev. Lett. 56:2672 (1986); T. Bowcock et al., Phys. Rev. Lett. 56:2676 (1986); H. Albrecht et al., DESY86-077, preprint.
53. J. Reinhardt and W. Greiner in: Treatise on heavy ion science, ed. D. A. Bromley, vol.5:p.3, (Plenum 1980).
54. J. Kirsch, B. Müller, and W. Greiner, Z. Naturforsch. 35a:151 (1983).
55. J. Reinhardt, G. Soff, and W. Greiner, Z. Phys. A276:285 (1976).
56. D. Y. Kim and M. S. Zahir, URTP-86-03:preprint.
57. P. Schlüter, U. Müller, G. Soff, Th. de Reus, J. Reinhardt, and W. Greiner, Z. Physik A323:139 (1986).
58. K. A. Erb, I.Y. Lee, and W. T. Milner, to be published.
59. J. M. Jauch and F. Rohrlich: The Theory of Photons and Electrons, Springer (1970).
60. E. W. Kolb, G. Steigman, and M. S. Turner, Phys. Rev. Lett. 47:1357 (1981).
61. A. deRújula, R. C. Giles, and R. L. Jaffe, Phys. Rev. D17:285 (1978).
62. A. Schäfer, B. Müller, W. Greiner, Phys. Rev. Lett. 50:2047 (1983).
63. B. Müller, J. Reinhardt, W. Greiner, and A. Schäfer, J. Phys. G12:2109 (1986).
64. L. S. Celenza, V. K. Mishra, C. M. Shakin, and K. F. Liu, Phys. Rev. Lett. 57:55 (1986).
65. C.-Y. Wong and R. L. Becker, Oak Ridge preprint (1986).
66. S. Brodsky and M. Karliner, private communication; B. Müller and J. Rafelski, UFTP170,preprint (1986).
67. B. Müller and J. Reinhardt, Phys. Rev. Lett. 56:2108 (1986).
68. Models for composite X-particles are being investigated by K. Geiger, C. Ionescu and A. Scherdin, Frankfurt am Main.
69. B. Müller and J. Rafelski, Proc. of the 2nd Intl. Conf. on Nucleus Nucleus Collisions Visby, Sweden (1985).
70. R. Peccei, Invited talk present at the DPG annual meeting, Heidelberg 1986.
71. N. C. Mukhopadhyay and A. Zehnder, Phys. Rev. Lett. 56:206 (1986).
72. W. E. Meyerhof, J. Molitoris, K. Danzmann, D. Spooner, F. S. Stephens, R. M. Diamond, E. M. Beck, A. Schäfer, and B. Müller, Phys. Rev. Lett., in print.
73. A. Konaka et al., Phys. Rev. Lett. 57, 659 (1986).

ON THE POSSIBILITY OF NEW PARTICLE PRODUCTION IN HEAVY-ION COLLISIONS*

A. B. Balantekin**

Physics Division
Oak Ridge National Laboratory
Oak Ridge, Tennessee 37831

INTRODUCTION

Experimental identification of narrow peaks in the positron spectra emitted in the low-energy collisions of heavy ions[1-4] has brought many surprises. Such structures were predicted to be present when the collision leads to a long-lived nuclear complex.[5] However, the observed Z-dependence of the peak energy and the cross section does not seem to be consistent with the hypothesis that quasimolecular nuclear configurations are formed. The 1s binding energy for those systems is expected to vary strongly with the total nuclear charge Z.[6] Hence, when this binding energy exceeds twice the electron mass,[7] the spontaneously created positron would have an energy strongly dependent on Z. Initial data[1-3] indicated that there is at least one constant peak energy for all the systems investigated, supercritical or subcritical. More recent measurements[8,9] seem to unveil the existence of a multiple structure in the positron emission spectra, again with little Z-dependence.

In this talk, I consider an alternative hypothesis, namely two-body decay of a particle,[10,11] which would explain the constancy of the peak energy. I also discuss the consequences of such a hypothesis and recent theoretical and experimental work motivated by it.

PHENOMENOLOGY OF PARTICLE PRODUCTION

I assume a pseudoscalar interaction Lagrangian of the form[10]

$$\mathscr{L}_{int} = g_e \bar{\psi}_e \gamma_5 \psi_e \phi_a \tag{1}$$

*Research sponsored by the U.S. Department of Energy under contract DE-AC05-84OR21400 with Martin Marietta Energy Systems, Inc.

**Present address: Department of Physics, University of Wisconsin-Madison, Madison, Wisconsin 53706 where the interaction strength g_e, and the mass of the particle, m_a, are free parameters to be fitted by the data. One can repeat similar analyses for other types of the interactions,[11] but a pseudoscalar interaction suffices to illustrate the basic principles. For the purposes of the analysis which follows, the Lagrangian in Eq. (1) can be either elementary or effective. The mass of the particle is assumed to be greater than twice the electron mass, so that its dominant decay mechanism is into electron-positron pairs. Since these positrons are products of a two-body decay, they would yield sharp peaks in the experimental spectrum. This hypothesis has a built-in assumption about the lifetime of the particle. It must live longer than heavy-ion reaction time ($\sim 10^{-20}$ sec), but shorter than the time of flight to the positron detectors ($\sim 10^{-9}$ sec). So if a subsequent analysis yields a lifetime outside the above limits, this hypothesis should be discarded.

The peak at 300 keV positron kinetic energy has three pieces of information: its energy, width (<70 keV), and the cross section (~200 μb).[3] The mass of the particle is adjusted to fit the peak energy, the production multiplicity to fit the integrated cross section, and the momentum to fit the width. One gets[10] $m_a \approx 1.6$–1.8 MeV momenta with less than 0.1 $m_e c$ with approximately 10^{-5} neutral particles being produced per reaction.

If we make a model for production, one can deduce the interaction strength g_e from the multiplicity quoted above. Given the lack of definitive models, we can extract limits on the interaction strength from the absence of this particle from other experimental observations. A particle interacting with electrons and positrons via the Lagrangian in Eq. (1) would contribute to various quantum electrodynamics processes. Here I consider the limits coming from the most thoroughly tested prediction of Q.E.D., the anomalous magnetic moment of electron.[10] A similar analysis has been carried out for other Q.E.D. processes.[12,13] For a particle with mass 1.6–1.8 MeV, the contribution to the anomalous magnetic moment is $\sim 5 \times 10^{-3}\, g_e{}^2$.[10] Thus, combining very accurate recent measurements[14] and theoretical work on higher order Q.E.D. corrections,[15] one concludes that this contribution would not have been observed if $g_e \sim 10^{-4}$.[10] Since this particle predominantly decays into electron-positron pairs, its lifetime can then be calculated to be $\sim 10^{-13}$ sec[10] which is well within the previously mentioned bounds.

Light particles, even if they couple to matter very weakly, could carry away large amounts of energy from the stars and could have significant effects on other astrophysical processes. Now, using the constraints from red giant stars as an example, I illustrate that a particle, which would explain the heavy-ion data, would not contribute to the astrophysical processes. Light particles, if they directly couple to fermions as in Eq. (1), can be photoproduced in red giant stars and carry away energy. The observational bounds on energy loss is 100 ergs/gr-sec, and a significant portion of this could be due to the production of a particle if its mass is less than 5×10^2 keV.[16] Obviously, this is much lighter than 1.8 MeV.

A crucial experiment to confirm the particle interpretation of heavy-ion data is to measure the yield versus invariant mass by detecting electrons and positrons in coincidence. This is a very difficult experiment. However, it has been observed that the positrons associated with the narrow peak are correlated with the simultaneous emission of electrons whose energy spectrum also contains a narrow peak at approximately the same energy and with the same width.[8,17] For a detailed discussion of these experiments, I refer the reader to the other contributions to these Proceedings.[8,9]

THE AXION HYPOTHESIS

The standard theory of strong interactions, quantum chromodynamics, does not necessarily have CP invariance, since gauge principle allows CP-violating terms. But there is, at present, no experimental evidence for CP violation in strong interaction physics. Sometime ago, it was observed[18] that a global U(1) symmetry would suppress CP-violating instanton effects. As a consequence of breaking this U(1) invariance, one can introduce a neutral pseudoscalar particle, called an axion.[19] So far, all experiments trying to observe axions have failed. One then might be tempted to rule out the particle interpretation of heavy-ion data using the previous negative results from the axion search experiments. However, essentially all axion searches performed prior to 1986 are not sensitive to a particle decaying as fast as 10^{-13} sec.[10,20,21] A summary of axion search experiments as of mid-1985 is presented in Fig. 1 which displays the possible window.

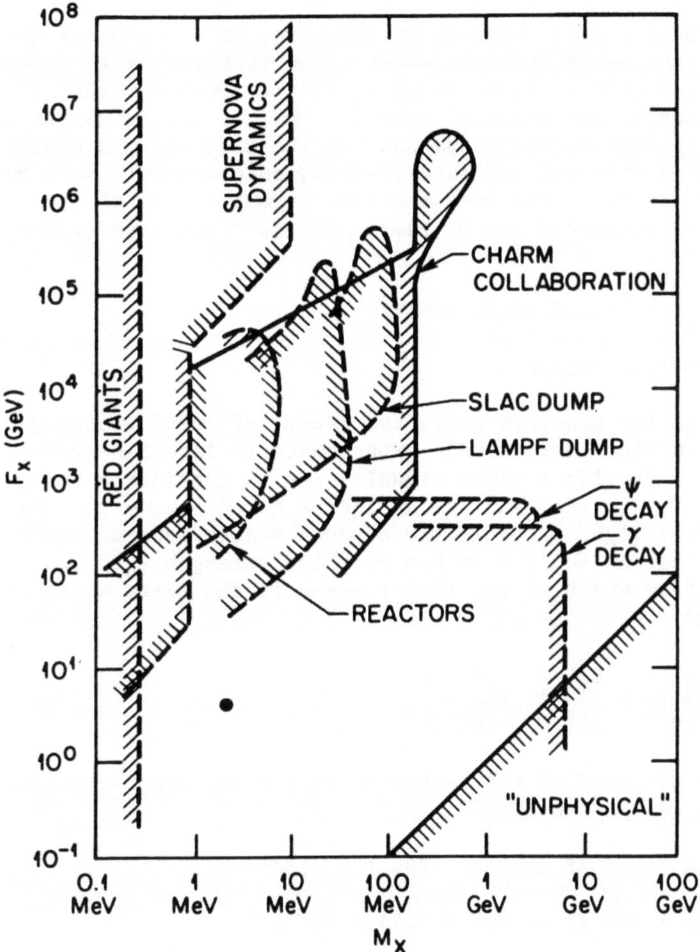

Fig. 1. A summary of world data for axion searches up to 1985 as presented by CHARM collaboration in Ref. 22. F_x = 250 X GeV and M_x is the axion mass. The shaded areas in the parameter space are excluded by the specified considerations. The CHARM collaboration data are from the 400-GeV proton beam dump experiment reported in Ref. 22. The black dot denotes the parameters of the particle which would explain the heavy-ion data.

In the standard axion model[19] one introduces two Higgs doublets. The ratio of the vacuum expectation values of these two fields is denoted by X. If the generic Lagrangian in Eq. (1) represents the coupling of axion to electron, the interaction strength g_e can be related to X

$$X = m_e\left(G_F\sqrt{2}\right)^{1/2}/g_e \tag{2}$$

where G_F is the Fermi constant. For an axion of mass 1.65 MeV, X is either 4.4×10^{-2} or 22 assuming three generations of quarks and taking the ratio of u and d quark current algebra masses to be 0.56. The latter X value yields too long a lifetime ($\sim 10^{-6}$ sec) to be compatible with experiments,[10] while the former value gives a lifetime of about 10^{-12} sec, in agreement with the electron anomalous magnetic moment considerations.

As I enumerate in the next section, recent experiments and reanalyses of old data to accommodate the short axion lifetime still rules out the standard axion model. In particular, it was found out that if there is a neutral, pseudoscalar boson of mass 1.6 MeV, it should couple to light quarks more strongly than to heavy quarks. In the past year, a number of new models, taking this universality breaking into account, have been proposed.[23-26] Again in the next section, I compare two of those models, those of Krauss and Wilczek[23] and of Peccei et al.[24] with the recent experiments.

RECENT PARTICLE SEARCH EXPERIMENTS

Radiative Upsilon Decays

A search for possible radiative decays of heavy vector mesons into light pseudoscalar particles is experimentally feasible since the monochromatic photons provide a clear signal. Let us consider a heavy vector meson composed of the quark q and its antiquark $\bar{q}$ and assume that this quark couples to the light particle via a pseudoscalar interaction Lagrangian similar to that in Eq. (1) with a coupling strength g_q. The ratio of the radiative decay width of the vector meson to the decay width into $\mu^+\mu^-$ pairs is given by[27]

$$\frac{\Gamma(V\to a+\gamma)}{\Gamma\left(V\to\mu^+ + \mu^-\right)} = \frac{g_q^{\,2}}{2\pi\alpha}\left(1 - \frac{m_a^{\,2}}{m_V^{\,2}}\right) \tag{3}$$

where m_V is the mass of the vector meson and the quantum chromodynamics corrections are neglected.

For the decay of upsilon, Eq. (3) predicts

$$B[\Upsilon(1s) \to \gamma + a] \sim (2.7 \pm 0.7) \times 10^{-4}\ X^2 \tag{4}$$

in the standard axion model.[19] Previous experimental searches looking for this decay reported null results,[28] but they have assumed that the axion escapes the detector region before decaying. However, axions with a lifetime of $\sim 10^{-13}$ sec would decay dominantly inside the detector. For $X = 0.04$, Eq. (4) gives a branching ratio of 0.16 ± 0.04. Two searches for axions decaying into e^+e^- inside the detector were recently performed by groups using the CUSB[29] and CLEO[30] detectors. Both groups report an upper limit of $\sim 2 \times 10^{-3}$ for the branching ratio which conclusively excludes the possibility of short-lived axions within the original model.

One can, nevertheless, use Eq. (3) in conjunction with the above experimental results to obtain an upper limit on g_b.[21] Those limits would be consistent with the new axion models[23,24] since in these models the couplings of the axion to heavy quarks are suppressed.

Electron Beam Dump Experiments

A recent search for neutral penetrating particles has been performed using a 2.5-GeV electron beam at the National Laboratory for High Energy Physics (KEK) in Japan.[31] After the beam hits a tungsten target, the axions were supposed to penetrate a dump and to decay into e^+e^- pairs in a decay volume. Neutral particles are assumed to be produced by the Primakoff process by photons

$$\mathscr{L} = \frac{g_\gamma}{m_e} F_{\mu\nu} \tilde{F}^{\mu\nu} \phi_a \tag{5}$$

as well as by bremsstrahlung from electrons (cf. Eq. (1)). A portion of the two-dimensional parameter space of $\alpha_e = g_e{}^2/4\pi$ and $\alpha_\gamma = g_\gamma{}^2/4\pi$ is excluded at the 90% confidence level[31] assuming $m_a = 1.8$ MeV. In particular, neutral particles living longer than 10^{-12} sec are excluded for all values of g_e, when $\alpha_\gamma \gtrsim 10^{-21}$. Also, when $g_\gamma = 0$, the interval $0.022 < X < 0.074$ is excluded for the standard axion model. Let me reiterate that a standard axion would require $X \sim 0.04$ to explain the heavy-ion data.

Nuclear M1 Transition Experiments

If there is enough energy, a neutral particle coupling to up and down quarks can be produced in nuclear transitions.[32] This is an ideal situation in which to use perturbation theory since the relevant couplings are small. Furthermore, since a neutral, pseudoscalar particle acts like a "magnetic" photon,[19] the emission of such a particle would follow the same selection rules as the magnetic transitions in nuclei.[33,34] Consequently, one could look for the decay of a nuclear state by neutral particle emission in competition with, say, M1 transitions.

The decay of the 9.17-MeV $J^\pi=2^+$, T=1 state in ^{14}N has been recently investigated for this purpose.[35] This state decays into the $J^\pi=1^+$, T=0 ground state by an M1 transition with a branch of 85%.[36] Savage et al. measured the branching ratio of neutral particle emission decay to that of M1 transition by studying the angular correlation of e^+e^- pairs.[35] Assuming $m_a = 1.7$ MeV and a lifetime less than 10^{-11} sec, they obtain an upper limit of 4×10^{-4} for this branching ratio, as opposed to the standard axion model prediction of 0.55. Since this transition is isovector, the above result can also be used to provide an upper limit of 1.4×10^{-2} to the neutral particle-nucleon isovector coupling constant.

Other Limits from Nuclear e^+e^- Pair Measurements

Calaprice and his collaborators reanalyzed[37] early Brookhaven data[38] determining multipolarities of various nuclear electromagnetic transitions. In particular, they studied the decays of the 3.56-MeV 0^+ state in ^{6}Li, and the 7.03-MeV 2^+ state in ^{14}N by internal pair conversion to the ground states of these nuclei. They deduced the branching ratio of neutral particle emission decay to the e^+e^- pair emission decay to be less than 0.14 for ^{6}Li and less than 5 for ^{14}N. The standard axion,[19] Krauss-Wilczek,[23] and Peccei et al.[24] models predict this ratio to be 7.2(300), 10(0), and 10(0) respectively for ^{6}Li(^{14}N). Thus, these two new axion models seem to be also ruled out.

NEUTRAL PARTICLE PRODUCTION MECHANISMS IN HEAVY-ION COLLISIONS

During the past year, considerable theoretical activity has been devoted to the investigation of various possible mechanisms to produce light, neutral particles in heavy-ion collisions.[10,39-42] Generally, one assumes that the heavy-ion system generates a current $j(\vec{x},t)$ which is the source of the neutral particles. The resulting interaction Lagrangian

$$\mathscr{L}_{int} = \phi_\alpha \, j(\vec{x},t) \tag{6}$$

leads to an exactly solvable quantum field theory problem.[43] The momentum distribution, $F(\vec{k})$, of the produced particles is given by

$$F(\vec{k}) \sim e^{-\bar{n}} \left| j(\vec{k},\omega_k) \right|^2 / \omega_k \tag{7}$$

where $\omega_k = \sqrt{\vec{k}^2 + m_a^2}$, $\bar{n}$ is the average number of the particles produced, and

$$j(\vec{k},\omega_k) = \int d^4\vec{x} \; j(\vec{x},t) \exp\left[i\left(\vec{k}\cdot x - \omega_k t\right)\right]. \tag{8}$$

The function j was modeled in several ways. One possibility is to take[10]

$$j(k) \sim \sum_{\vec{p},\vec{q}} \delta_{\vec{k},\vec{p}+\vec{q}} \left(\frac{dN_+}{d\vec{p}}\right)\left(\frac{dN_-}{d\vec{q}}\right) \tag{9}$$

where $\left(dN_+/d\vec{p}\right)$ and $\left(dN_-/d\vec{q}\right)$ are the differential number of positrons and electrons respectively computed using the techniques of Ref. 44. Another possibility is to consider[39-42] effective pseudoscalar

$$j(\vec{x},t) \sim \vec{E}\cdot\vec{B} \tag{10a}$$

or scalar

$$j(\vec{x},t) \sim \vec{E}^2 - \vec{B}^2 \tag{10b}$$

couplings to the classical electromagnetic field of heavy ions.

One can unfold the experimental data directly to obtain the momentum distribution given in Eq. (7). This had been done in Ref. 10, and the distribution shown in Fig. 2b (solid line) was obtained. Probably the most salient feature of this figure is that the momentum spread is very small $\left(\lesssim 0.1\; m_e c\right)$. On the other hand, F(k) obtained by substituting Eq. (9) into Eq. (7) is much broader (Fig. 2b, dashed line). Other authors[39-42] using the source terms given in Eq. (10), modeled along very different lines, arrive at a similar conclusion. Thus, the theory seems to fail in describing the production mechanism if indeed such particles are produced in heavy-ion collisions. However, at this conference it was reported[8] that one can fold the broad distribution in Fig. 2 by the kinematical constraints of the experiment and obtain a rather sharp momentum spread.

COMPOSITE AND EXOTIC PARTICLES

So far, I have confined my discussion to the particles which are either elementary or with some possible structure which cannot be probed at these energies. A number of other possibilities were also considered concerning the creation of composite particles or exotic states which I will summarize very briefly.

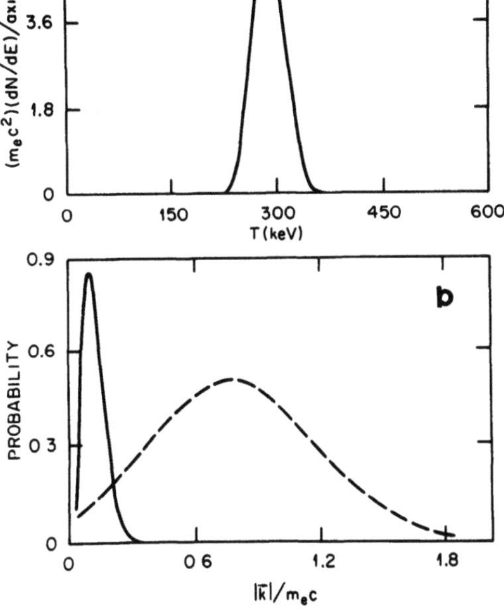

Fig. 2. a) Differential number of positrons emitted in the heavy-ion collisions. b) Momentum distribution of the neutral particles required to fit the data (solid line) and obtained from Eqs. (7) and (9) (dashed line).

Wong suggested that a polyelectron complex $(e^+e^+e^-)$ could be produced in heavy-ion collisions.[45] The decay of this complex into a positron and a photon would create a sharp positron line. However, the decay of such a complex is dominated by the two-photon mode $e^+e^+e^- \to e^+ + 2\gamma$. Indeed the branching ratio $\Gamma_{1\gamma}/\Gamma_{2\gamma}$ is calculated to be $\sim 10^{-11}$.[46] Consequently, if the positron peak with a cross section of 200 μb is due to the one-photon decay mode, most of the polyelectrons produced in heavy-ion collisions would decay through the two-photon mode yielding a total positron cross section of ~200 $\times 10^{11}$ μb = 2×10^7 barns. Such a situation is manifestly not realized in the experiments.

Celenza et al. suggested that these peaks could originate from the decay of nontopological solitons.[47] Other possible mechanisms might be due to a change in the vacuum structure of a Higgs field[48] or due to the existence of systems with highly localized and tightly bound states of several electron-positron pairs.[49] A detailed discussion of these subjects is given elsewhere in this volume.[42,50]

CONCLUSIONS

Experimental data still seem to allow a neutral particle with mass ~1.6 - 1.8 MeV and a lifetime of $\sim 10^{-12} - 10^{-13}$ sec, but the situation might soon change. This particle is not the standard axion; the recent models proposed by Krauss-Wilczek and Peccei et al. also seem to be ruled out. Furthermore, if this particle is being produced in heavy-ion collisions, we do not know its production mechanism where the main difficulty appears to be obtaining the correct momentum spread.

REFERENCES

1. J. Schweppe et al., Phys. Rev. Lett. 51:2261 (1983).
2. M. Clemente et al., Phys. Lett. 137B:41 (1984).
3. T. Cowan et al., Phys. Rev. Lett. 54:1761 (1985).
4. H. Tsertos et al., Phys. Lett. 162B:372 (1985).
5. J. Rafelski, B. Müller, and W. Greiner, Z. Phys. A285:49 (1978).
6. W. Pieper and W. Greiner, Z. Phys. A218:327 (1969); S. S. Gershtein and Y. B. Zeldovich, Lett. Nuovo Cim. 1:835 (1969); J. Rafelski, L. P. Fulcher, and W. Greiner, Phys. Rev. Lett. 27:958 (1971); V. S. Popov, Sov. Phys. JETP 32:526 (1971); see also W. Greiner, ed., "Quantum Electrodynamics of Strong Fields," Plenum, New York (1983).
7. B. Müller, W. Peitz, J. Rafelski, and W. Greiner, Phys. Rev. Lett. 28:1245 (1972).
8. T. Cowan, contribution to these proceedings.
9. H. Bokemeyer, contribution to these proceedings.
10. A. B. Balantekin, C. Bottcher, M. R. Strayer, and S. J. Lee, Phys. Rev. Lett. 55:461 (1985).
11. A. Schafer et al., J. Phys. G11:L69 (1985).
12. J. Reinhardt, A. Schafer, B. Müller, and W. Greiner, Phys. Rev. C33:194 (1986).
13. J. Reinhardt, contribution to these proceedings.
14. R. S. Van Dyck, P. B. Schwinberg, and H. G. Dehmelt, Phys. Rev. Lett. 38:310 (1977).
15. T. Kinoshita and W. B. Lindquist, Phys. Rev. Lett. 47:1573 (1981).
16. A. Barroso and G. C. Branco, Phys. Lett. 116B:247 (1982).
17. T. Cowan et al., Phys. Rev. Lett. 56:444 (1986).
18. R. D. Peccei and H. R. Quinn, Phys. Rev. Lett. 38:1440 (1977).
19. S. Weinberg, Phys. Rev. Lett. 40:223 (1978); F. Wilczek, ibid., 40:279 (1978).
20. N. C. Mukhopadhyay and A. Zehnder, Phys. Rev. Lett. 56:206 (1986).
21. A. B. Balantekin, C. Bottcher, M. R. Strayer, and S. J. Lee, in: "Proceedings of Atomic Theory Workshop on Relativistic and Q.E.D. Effects in Heavy Atoms," H. P. Kelly and Y.-K. Kim, eds., American Institute of Physics, New York, p. 302 (1985).
22. F. Bergsma et al., Phys. Lett. 157B:458 (1975).
23. L. M. Krauss and F. Wilczek, Phys. Lett. 173B:189 (1986).
24. R. D. Peccei, T. T. Wu, and T. Yanagida, Phys. Lett. 172B:435 (1986).
25. D. Y. Kim and M. S. Zahir, U. of Regina preprint URTP-86-03 (1986).
26. E. Ma, Phys. Rev. D34:293 (1986).
27. F. Wilczek, Phys. Rev. Lett. 39:1305 (1977).
28. M. Sivertz et al., Phys. Rev. D26:717 (1982).
29. G. Mageras et al., Phys. Rev. Lett. 56:2672 (1986).
30. T. Bowcock et al., Phys. Rev. Lett. 56:2676 (1986).
31. A. Konaka et al., University of Kyoto preprint (1986).
32. S. B. Treiman and F. Wilczek, Phys. Lett. 74B:381 (1978).
33. T. W. Donnelly et al., Phys. Rev. D18:1607 (1978).
34. A. Barroso and N. C. Mukhopadhyay, Phys. Rev. C24:2382 (1981).
35. M. J. Savage, R. D. McKeown, B. W. Filippone, and L. W. Mitchell, Phys. Rev. Lett. 57:178 (1986).
36. F. Ajzenberg-Selove, Nucl. Phys. A360:1 (1981).
37. F. P. Calaprice, in: "Proceedings of Second Conference on the Intersections Between Particle and Nuclear Physics," D. F. Geesaman, ed., American Institute of Physics, New York (1986).
38. E. K. Warburton et al., Phys. Rev. 133B:42 (1964).
39. A. Chodos and L.C.R. Wijewardhana, Phys. Rev. Lett. 56:302 (1986); D. Carrier, A. Chodos, and L.C.R. Wijewardhana, Yale preprint YTP85-33 (1986).
40. K. Lane, Phys. Lett. 169B:97 (1986).
41. B. Müller and J. Rafelski, U. of Cape Town preprint UCT-TP 40/1986.
42. B. Müller, contribution to these proceedings.

43. C. Itzykson and J.-B. Zuber, "Quantum Field Theory," McGraw-Hill, New York, p. 163 (1980).
44. C. Bottcher and M. R. Strayer, Phys. Rev. Lett. 54:669 (1985).
45. C. Y. Wong, Phys. Rev. Lett. 56:1047 (1986).
46. M. C. Chu and V. Pönisch, Phys. Rev. C33:2222 (1986).
47. L. S. Celenza, V. K. Mishra, C. M. Shakin, and K.-F. Liu, Phys. Rev. Lett. 57:55 (1986).
48. A. Schäfer, B. Müller, and W. Greiner, Phys. Lett. 149B:455 (1984).
49. B. Müller, J. Reinhardt, W. Greiner, and A. Schäfer, J. Phys. G12:L109 (1986).
50. A. O. Barut, contribution to these proceedings.

QUASIATOMIC SPECTROSCOPY AS A TOOL FOR DEEP INELASTIC COLLISIONS

Egbert Kankeleit

Institut für Kernphysik der Technischen Hochschule Darmstadt

ABSTRACT

The $\hat{t}$ scaling model is extended for application to deep inelastic collisions with time varying shapes of nuclear charge distributions. The quasiatomic spectroscopy for very heavy systems is determined by the second moment of the charge distribution and is nearly independent of the position of the bound state energies. The sensitivity of this spectroscopy to the details of the evolution of charge distributions in deep inelastic collisions are examplified.

1. INTRODUCTION

The purpose of this note is to review the $\hat{t}$ scaling model, often mentioned during this conference in connection with spectral shapes of electron and positron spectra (Oeschler et al.,Senger et al). This model, introduced in 1978 (Ka 1978), is based on the theoretical approach of the Frankfurt group conducted by Prof. Greiner. It certainly can not achieve the accuracy of their rather involved calculations but its presentation may be justified by its simplicity by which the most relevant features and parameter dependencies of the quasiatomic spectroscopy are pointed out and by its use of experimentalists for numerically fast least square fitting procedures.

In this connection the treatment of deep inelastic collisions and the study of the dynamical evolution of the reaction is of particular interest. This new field of research has been presented by extensive contributions at this conference and in earlier publications (Backe et al. 1983, Krieg et al. 1986). The question to be answered is to what parameters is the quasiatomic spectrocopy sensitive and what can with what accuracy be measured with regard to the study of the time evolution of reactions in which two nuclei may form complicated shapes and eventually split into two or more clusters in the exit channel.

2. BASIC FEATURES OF THE $\hat{t}$ SCALING MODEL

In the semiclassical treatment the distance R between projectile of massnumber A1 and charge Z1 hitting a target nucleus (A2,Z2) the minimum distance in a coulombic collision is given by

$$2\,a= \frac{e^2\ Z1\ Z2}{\mu\ E1}$$

with e^2=1.44 MeV/fm, μ the reduced mass in the entrance channel, E1 the energy per nucleon of the projectile. The velocity of the projectile long before the collsion is

$$v=\sqrt{2*E1/(A1*931.5)}\ *c.$$

With a scattering angle δ_{cm} we obtain the excentricity

$$\epsilon=1/\sin(\delta_{cm}/2),$$

and the impact parameter

$$b = a \sqrt{\epsilon^2 -1}\ ,$$

the distance of minimum approach at this angle:

$$R_m = a\ (\ \epsilon^2-1\ ).$$

There is no closed expression for the time dependence of the distance as a function of time, but vica versa with $x=(R/a -1\ )/\epsilon$

$$t=a/v\ [\ \epsilon \sqrt{x^2 -1} +\ln\ (x + \sqrt{x^2 -1}\ )\ ].$$

We will see that the quantity

$$\frac{\dot R}{R} = \frac{v}{R} \sqrt{1- \frac{2\ a}{R} - \frac{a^2\ (\epsilon^2-1)}{R^2}}$$

will be of particular importance. This function is now very well approximated by the decisive formula:

$$\frac{\dot R}{R} = \kappa \frac{t}{t^2\ + \hat t^2}$$

with $\kappa\simeq 1+0.174/\epsilon$ and

$$\hat t \simeq \frac{a}{v}\ (\ \epsilon +1.60 + \frac{0.50}{\epsilon}\ )$$

the characteristic time constant of the coulombic motion. In the following we will use natural units of the electron ($\hbar$=1,c=1,mc^2=1), which are most useful in this context, where the lepton energies are experimentally of the order mc^2 and the times involved $\hbar/mc^2$ = 1.29 10^{-21} s. The electromagnetic interaction induces transitions between electrons in bound states or in states of the upper continuum, the so-called δ-electrons, or electrons in states in the lower continuum in the Dirac sea, the holes produced representing the positrons. In first order the amplitude for transitions between eigenstates E_i and E_f is:

$$a(E_i,E_f)=-\int_{-\infty}^{\infty} dt\ \langle f|\ \frac{\partial}{\partial t}\ |i\rangle\ e^{\ i \int_{-\infty}^{t} dt'(E_f-E_i)}$$

.Using commutator relations and neglecting rotational coupling we get

$$a(E_i,E_f)= \int_{\infty}^{\infty} dt\ R\ \langle f|\ \frac{\partial V}{\partial R}\ |i\rangle\ \frac{1}{(E_f-E_i)} * \frac{\dot R}{R}\ e^{\ i \int_{-\infty}^{t} dt'(E_f-E_i)}\ .$$

The potential V(R,r) depends on the distance R between the two nuclei and their charge distribution. The derivative of V(R,r) with respect to the distance R exhibits in r a very particular behaviour. In greater distances of r the potential will be in monopole approximation ~1/r and not depend on R. The derivative is, therefore, zero. In closer distances the situation becomes in general complicated in the two center potentials. But it has been shown by Soff et al. (1979) that it is sufficient to take the monopole part in the expansion of the potential. The wave length is large compared to the extension of $\partial V/\partial R$. As a consequence for pointlike nuclei the potential of the two charges $Z=Z1+Z2$ will be constant $V=-\alpha\, Z/R$ for $r<R/2$ and drop off according to $V=-\alpha\, Z/r$ for $r>R/2$ as depicted in Fig.1. A change $\partial V/\partial R$ would, therefore, occur

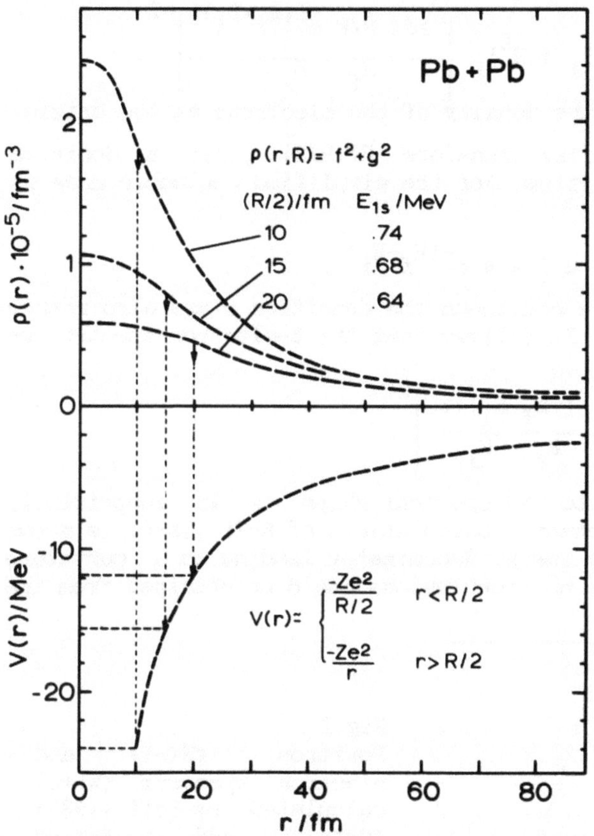

Fig.1
Density of 1s electron and monopole potential for two pointlike Pb nuclei separated by R=20, 30 and 40 fm as a function of distance r. $\partial V(r,R)/\partial R$ contributes only from r=0 to the vertical dashed lines.

only within the two charges. This simplifies the calculation of the matrix element considerably and also offers a very transparent picture for the excitation process. The main contribution comes from the moving bottom of the potential and is proportional to the probability of the electron to be within this domain,- very much like in a flipper game. As we see from Fig.1 the active domain is small compared to the wave-length and in good approximation we may just consider the wave-function nearly constant in this domain and use its values at the origin. In that case the wave-functions $\Psi_{i,f}(0)$ at the origin can be brought out of the space integration and we are left with the integration over the active volume. In this we follow Church and Weneser (1956) in their treatment of the monopole conversion, which is the very same process. We thus arrive at

$$a(E_i,E_f)=-\frac{4\pi Z\alpha}{12}\,(\Psi R)_i\,(\Psi R)_f\,\frac{1}{E_f-E_i}\int_{-\infty}^{\infty}dt\,\frac{\dot{R}}{R}\,e^{i\,(E_f-E_i)\,t}\,.$$

In the curly bracket {} the wave-function is taken at the origin and we have included one R each. It will be shown below that these will be nearly time independent. We have also used an important result of Müller et al. (1978) that the time varying energies $E_{i,f}$ can be savely taken to be constant at the value of minimum approach. The consequence is that the amplitude factorizes into a time independent quantity depending only on the binding properties of the electronic system and a Fourier integral of the dynamic quantity $\dot{R}/R$.

The transition probability between bound state or continuum states, with P beeing then the probability per energy interval is

$$P(E_i,E_f)=\left(\frac{4\pi Z\alpha}{12}\right)^2 \{\rho r^2\}_i \{\rho R^2\}_f \left|\frac{\int dt\ \dot{R}/R\ e^{i(E_f-E_i)t}}{E_f - E_i}\right|^2 .$$

The brackets {} include now the density of the electrons at the origin.

It is by now apparent that Fourier transform $FT\{\dot{R}/R\}$ plays a decisive role for the dynamics of the system. For the simplified coulombic case as shown above we can write

$$FT\{ \frac{\dot{R}}{R} \} \cong i\ \pi\ \kappa\ e^{-(E_f-E_i)\hat{t}} .$$

For processes in the positive continuum the densities ρ are also rather independent of the energies E. It follows that the δ-electron spectra are proportional to

$$\left[\frac{e^{-(E_f-E_i)*\hat{t}}}{E_f - E_i} \right]^2 .$$

That this result with respect to the spectral shape is in surprisingly good agreement with coupled channel calculations of Soff (1985) is shown in Fig.2. If there would be no energy denominator leading to a curvature in the exponential slopes, no conclusions could be obtained from the

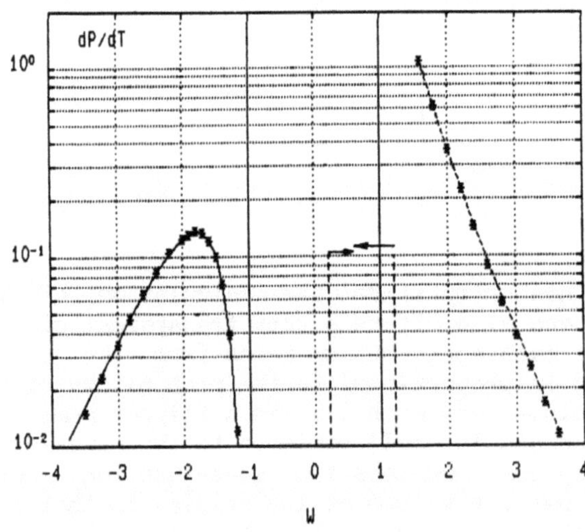

Fig.2
Positron (W<-1) and electron spectrum (W>+1) calculated by Soff (1985) (dots) and fitted according to the $\hat{t}$ scaling law. Indicated are the corresponding effective energies E_f=1.2 and E_i=.22. Z'=184*0.76.

spectral shape with regard to the initial energy E_i from which the transition into the continuum originates. We would just have the exponential slope $\exp(-E_\delta*2*\hat{t})$, which is indeed reached at high delta electron energies E_δ. A similar fit has been done for the positron spectrum in Fig 2. In this case the $\{\rho R^2\}$ for the positron is not constant in energy (Ka 1980) but includes the repulsion in the coulomb field which was taken to be $e^{-2\pi Z'\alpha W/p}$ according to Rose (1961) with an effective charge Z' as fitting parameter. In Fig 2 the following formula was used in the fitting procedure:

$$a(1)* \frac{e^{-2\pi\alpha\ 184\ W/p\ a(2)-2\hat{t}|W-a(3)|}}{|W-a(3)|^2}$$

and the following variables a(1-3)

δ-electrons:	12.34	0.0	0.22
positrons :	1.40	0.76	1.20.

The effective energies a(3) show that in the coupled channel calculation the δ-electrons come essentially from about 400 keV binding energy and the positrons from a transfer into the upper continuuem at 100 keV.

3. THE MONOPOLE POTENTIAL

So far we have considered the monopole potential of two point charges. In the calculations of the wave-functions we have to consider extended charge distributions not only of separated nuclei but also those of complicated shapes during the evolution of a deep inelastic collision. In Fig.3 the monopole part of the charge distribution of two U 238 nuclei, here taken as spherical, and the monopole potentials they are generating are depicted. The density of these charge distributions are given by:

$$\rho_N(r) = \frac{2\ Z}{4\ \pi\ R_N^3\ R} \frac{3}{4} \left[\frac{(R/2-R_N)(R/2+R_N)}{r} - R + r \right].$$

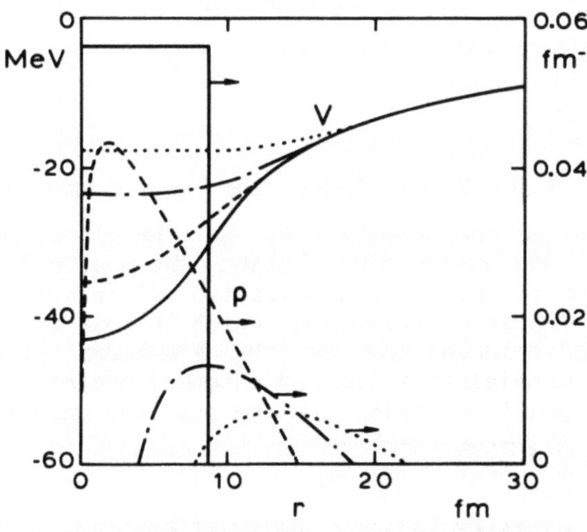

Fig.3
Monopole charge and monopole potential for 'spherical' U nuclei separated by R=15,22.5 and 30 fm. R_N=7.3 fm.

The monopole potential is obtained from a simple integration

$$V(r)/(-4\pi\ Z\alpha) = \int_0^r r' \frac{\rho(r')}{r} r'^2 + \int_r^\infty dr' \frac{\rho(r')}{r'} r'^2 \} ,$$

$$= \frac{(R/2-R_N)^3 (R/2+3\ R_N)}{16\quad r} - \frac{(R/2+3\ R_N)^2 (r/2-2\ R_N)}{4} + $$

$$+ \frac{3\ (R^2/4-R_N)\ r}{8} - \frac{R\ r^2}{8} + \frac{r^3}{16} .$$

For an exact treatment, the electronic eigenstates have to be calculated for each of the time varying potentials during the course of a reaction. With complicated shapes this would be quite a formidable job. Fortunately this is not necessary. Because the wave-functions, compared to muonic 1s states in heavy nuclei, are still rather constant in the domain of the nuclear charge distribution, it turns out that only the second moment of the charge distribution is to be known,- a fact very familiar to hyperfine spectroscopists for low Z atoms. But it is also known that for heavier atomic and particular muonic 1s atomic states a shift from the second to the first moment occurs. As shown in a simple numerical calculation, there is a backward trend to the second moment for super heavy quasiatomic systems. In Fig.4 two extreme potentials due to a point charge and due to a homogeneous distribution have been scaled to

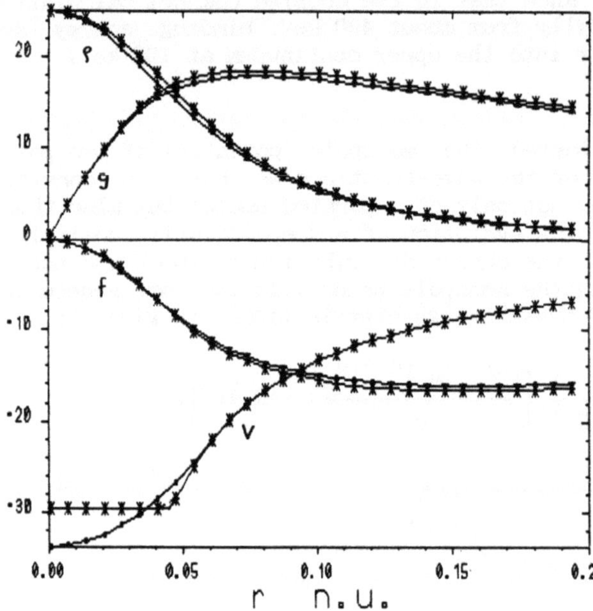

Fig.4
Potentials of 184 protons on a charged shell at r=.045 n.u.=17.5 fm and constant density distribution for r<.060 n.u.=23 fm both leading to a 1s binding energy at -1. Only small differences show up in the density and the great and small components of the Dirac wavefunctions.

produce a binding energy of E_{1s}=-1 for Z=184. Indeed the second moments turn out to be nearly the same as the densities do. We, therefore, can perform the calculation for just one charge distribution, the most simple pointlike for instance, and refer to this with complicated distributions by just calculating the rms charge distribution, which is most easily done. This coincides with the calculations by the Frankfurt group comparing their eigenstates for unshielded point and extended nuclei. The bound state energies for a point distribution are just the ones for extended nuclei at distance $R= 2\sqrt{\langle r^2\rangle - 3/5\, R_N^2}$

whereby $\sqrt{\langle r^2\rangle}$ is here half the distance between the point charges.

The matrix element in the transition amplitude has to be reformulated if in the case of complicated charge distribution the notion of a distance R is not any more given. We rewrite the matrix element in the following way:

$$\langle f|\frac{\partial}{\partial t}|i\rangle\ (E_f - E_i) = \int \Psi_f \Psi_i\ \frac{\partial V}{\partial t}\ d\vec{r}\ ,$$

with $-4\pi\ \rho_{if}(r) = -4\pi\Psi_f\Psi_i = \Delta V_{if}$ this becomes

$$= \int \Delta V_{if}/(-4\pi)\ \dot{V}\ d^3\vec{r}\ .$$

Using Green's second Theorem we obtain

$$=\int V_{if}\ \Delta\dot{V}/(-4\pi)\ d^3\vec{r}\ .$$

The potential generated by the electrons

$$V_{if} \simeq \text{const} - 4\pi/6\ \rho_{if}(0)\ r^2 + O(r^4)$$

is again dominated by the density at the origin. Using this approximation the above expression becomes:

$$=Z\alpha\ 4\pi/6\ \rho_{if}(0)\ \frac{\partial}{\partial t}\int r^2\ \rho_N\ d^3\vec{r}$$

with the nuclear charge distribution ρ_N. Now the integral is just the second moment again. Putting every thing together we obtain the same expression as above for the transition amplitude and probability if we replace R by $\bar{R} = 2\sqrt{\langle r^2\rangle}$, as may be expected from the foregoing.

4. THE ELECTRONIC WAVEFUNCTIONS

As we have seen above the transition probability is governed by the density of the electrons at the origin in their respective states. Let us first consider the densities $\{\rho * R^2\} = 4\pi(f(0)^2+g(0)^2)R^2$, with f,g the two components of the Dirac wave-function as a function of Z in Fig.5. The uprising pair of curves, belonging to 40 fm and to 20 fm nuclear separation, respectively, in the middle of the picture are the 1s (κ=-1) bound state electrons. For low Z there is just a factor $(40/20)^2=4$ difference between these curves. But with increasing charge, the isotope or isomer effect becomes more important. Indeed the densities become independent of R when Z approaches the critical charge.

If we want to calculate the probability for a one step 1s photo excitation into the continuum at 1.5 mc^2 (.25MeV kinetic energy), we have to multiply the corresponding $\{\rho R^2\}$ values for the 1s state and the one of the continuum, giving the static part and multiplying this with the

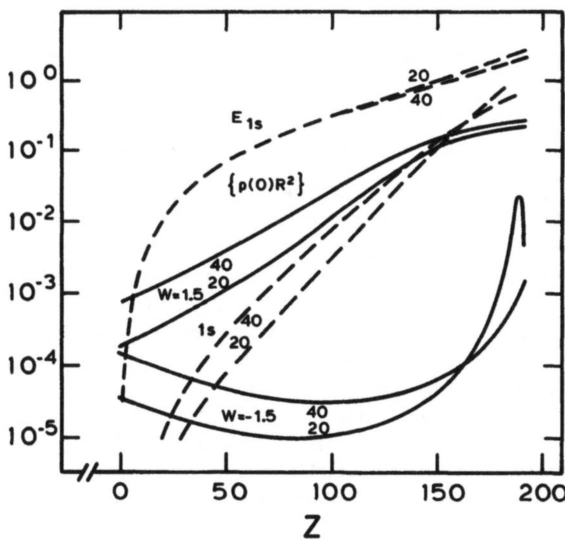

Fig.5
1s binding energy and $\{\rho(0)R^2\}$-values for 1s bound states and s-wave continuum states for W=+1.5 and -1.5 mc^2 and R=40 and 20 fm separation between identical nuclei in the point charge limit.

dynamic part :

$$\pi \frac{e^{-2\hat{t}(E-E_{1s})}}{(E-E_{1s})^2},$$

which then determines via $\hat{t}$ the dependence on projectile energy and scattering angle. Similarly, the vacancy production is given by the integral over the continuum states, leading to an incomplete gamma function, as discussed in detail in (Ka 1978). The nearly exponential dependence has been nicely demonstrated by the experiments of Liesen et al. (1980).

The positron wave-functions are similarly only little dependent on the separation R, as long we deal with the fast processes as we do, at which even the diving 1s states behave essentially as bound states due to poor overlap with the continuum or long tunneling times. The uprising densities of the positrons with Z should be taken only in extrapolation of the under critical situation. The resonance in Fig.5 at $Z \simeq 180$ and R= 20 fm, which is due to the overlap of the diving state with the W=-1.5 continuum state, should be ignored. If we now consider the positron production the main process occurs by filling the low bound states with a probability proportional to the corresponding densities $\{\rho R^2\}$. The total production probability can thus be obtained by multiplying with the above vacancy production probability, a product of four densities giving rise to the observed strong Z dependence

We have so far ignored the fact that in the domain of very heavy systems, the nondiagonal matrix elements for electrons above the lower continuum are of the order one. A strong coupling results between all bound and continuum states and a coupled channel calculation is required. It was shown in (Ka 1978) that even for multistep processes the general dynamic dependences still exists as for one step processes. This,indeed,is the explanation for the good agreement in shape between the coupled channel calculations and the one step formula as shown above in Fig.2.

One more aspect in this connection is of importance, which will be dealt with in more detail elsewhere. For a heavy system, like U-U, the densities above the lower continuum are not only independent of the distance R but also independent as a function of E if we define now the density per energy interval $\rho/dE/dn$ as done for the continuum states. For nonrelativistic atoms the densities are $\rho(0)_n = (Z*\alpha/n)^3/\pi$, and the energies $E_n = (Z\alpha/n)^2/2$. For large n the density per energy interval is $\rho/(\partial E/\partial n)=Z\alpha/\pi$, i.e. independent of E_n and identical with the density of the wave-function in the upper continuum. The same seems to be nearly true for heavy relativistic systems. The densities thus defined are about 0.5 (n.u.) for $Z\cong180$ from the 1s state up to high continuum energies. This connection can be derived from Fig.6 where we have plotted the densities and energies normalized to the one for nonrelativistic atoms. It is seen that for the 1s,2s and 3s states both the normalized densities and energies are nearly the same. But this implies that the quasiatomic processes are rather insensitive to the position of the bound states, with the exception of the Fermi level which hardly can go below the 1s state. For the experiments performed so far the Fermi level is taken to be at about n=3. We also recognize from Fig.6 that the normalized binding energies increase only by a factor two to three for the heaviest systems

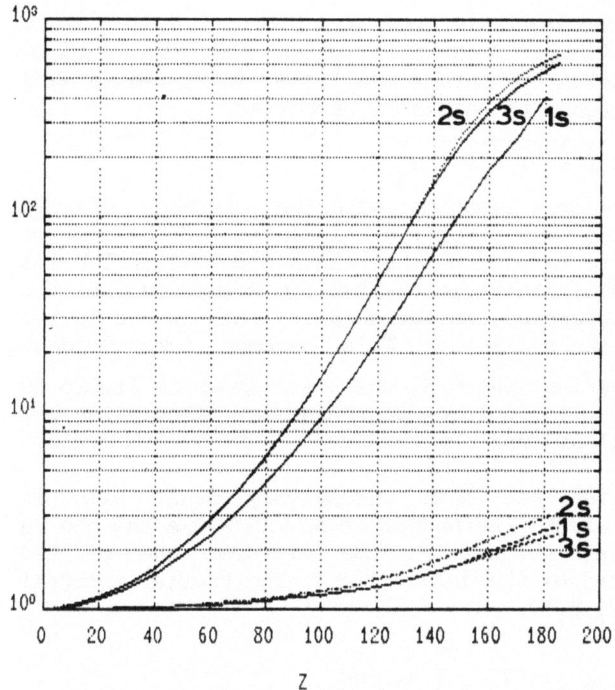

Fig.6
Densities at origin per energy interval of 1s,2s,3s electrons (upper curves) divided by $(Z\alpha/n)^3/\pi$ and binding energies (lower curves) divided by $(Z\alpha/n)^2/2$.

while the normalized densities rise by three orders of magnitude - about a factor hundred beyond the factor ten which is known from hyperfine interaction in heavy elements. This so-called relativistic enhancement is actually due to a slowing down of the electrons to less than speed of light. The electrons spend longer times in the deep potential as compared to the nonrelativistic ones with unlimited velocity.

5. TRAJECTORIES IN DIC

We now turn to the aim of this contribution, the problem of evolution of nuclear charge distribution in a deep inelastic collision and its effect on electron and positron quasiatomic spectra. The first realistic calculations in this regard were presented in (Ka 1978) using the model of two spherical nuclei with frictional forces (Wolschin 1977, Schmidt et al. 1978). For superheavy systems, like U+U, it became clear and it was experimentally shown by Backe et al. (1983) that the slowing down of the nuclear motion leads essentially to a reduction in the high energy or frequency part of the positron and δ-electron spectra. Many detailed calculations have been done, in the meantime, as referred to in the contribution by Oeschler. There are serious aspects missing in this model, which are quite essential for the very heavy systems. According to our knowledge the two nuclei actually seem to combine to a highly deformed nucleus in an intermediate phase and then in forming a long neck split apart such that nuclear forces keep the two fragments for a long time together before coulomb forces overwhelm. This explains the low observed kinetic energy of the fragments. A model by Feldmeier (1984) which allows for a large variety of shapes including a neck, according to a prescription by Blocki and Swiatecki (1981), describes very succesfully many heavy inelastic collisions. But the notion of radial distance from which the monopole potential can be generated is not existing for the shapes occuring during the collision. Instead we use the above result and calculate the rms of the charge distributions. This will be examplified on a central collision reaction U+U at 7.5 MeV . This is very easily done in the Feldmeier program in which for the folding potential a related expression is anyhow available. In Fig.7a we have plotted as a matter of convenience the distance parameter:

$$R'=2\sqrt{\langle r^2\rangle - 3/5\, R_N^2}$$

as a function of time which becomes identical with the distance of two separated spheres. As mentioned above for the calculation of the nuclear potential the rms radius has to be used as the relevant parameter for its extension. The curve also shown, symmetric to t=0, corresponds to the coulomb trajectory. The difference in the outgoing path is due to the friction and sticking during the collision. This becomes more evident considering the logarithmic derivatives $\dot{R}/R$. which are shown in Fig.7b in more detail. The figure not only includes the Feldmeier calculation but also three other cases we will discuss.

The most simple case which has been extensively dealt with in connection with the 'atomic clock' (Reinhardt et al. 1979) is the one in which the derivative in the exit channel is the mirrored one of the entrance channel but shifted by the delay time T. The Fourier integral splits up into two parts ($\omega=E_f-E_i$):

$$\int_{-\infty}^{0} \frac{\dot{R}}{R}\, e^{i\omega t} dt + e^{i\omega T} * \int_{0}^{\infty} \frac{\dot{R}}{R}\, e^{i\omega t} dt.$$

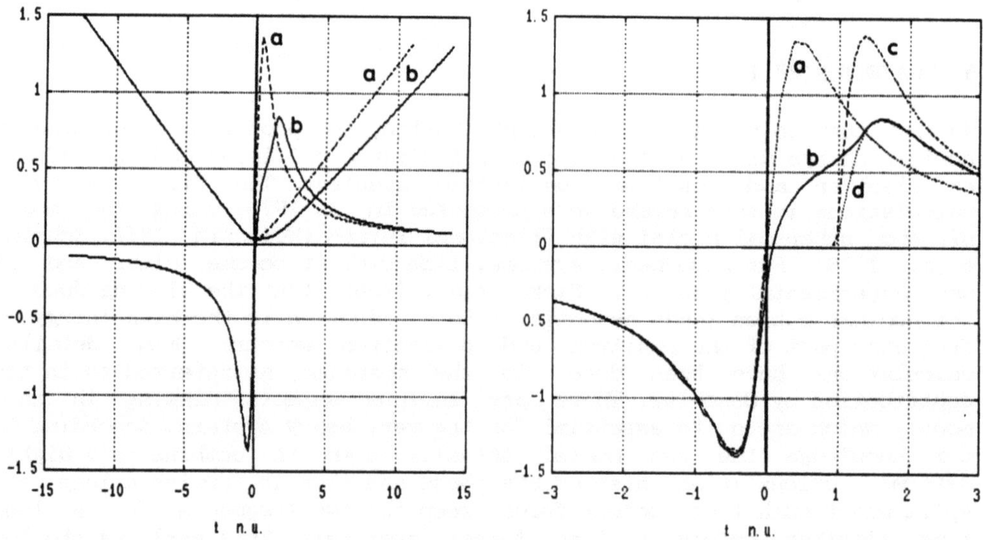

Fig 7a,b
R'(t) and its logarithmic derivative for a central 7.5 MeV/u U-U for a coulombic (a) and DIC according to the Feldmeier model (b). Fig. 7b includes in addition the 'atomic clock' fit (c) to the entrance part of curve (b) ($\hat{t}_i$=0.43) delayed by time T=0.96, and curve (d) which takes care of the damping : $\hat{t}_f$=0.64.

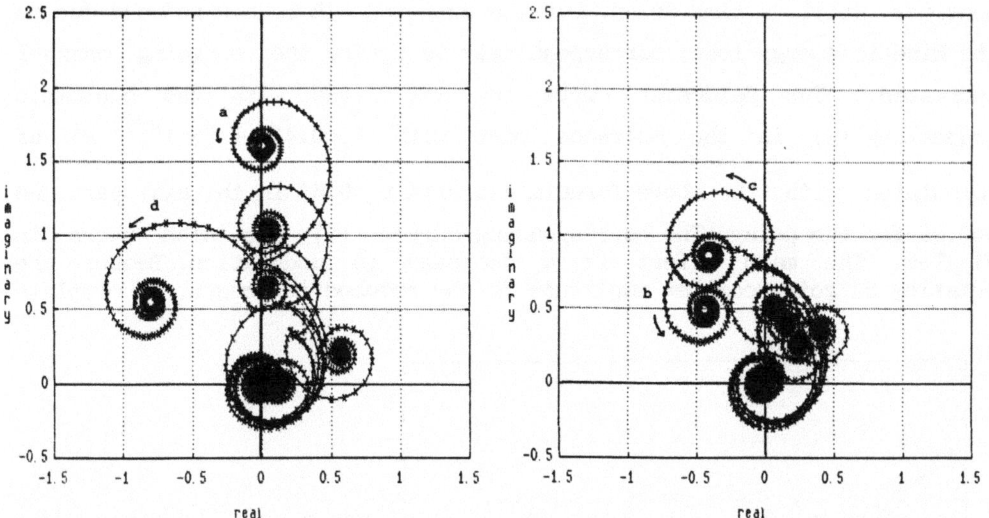

Fig.8a,b

Time integral of $\dot{R}'/R'\ e^{i\omega t}$ for the curves (a-d) of Fig 7 for $\omega=E_f-E_i=2$, 3 and 4

Fig.8 shows the integral as a function of time. If we tacitly assume that till time t=0 at minimum approach we have nearly coulomb trajectories this can be most easily calculated by fast Fourier transform (actually by first differentiating $\dot{R}/R$ and then dividing the FT by $i\omega$ to reduce the periodicity problem) or with the $\hat{t}$ model using the relation:

$$\int_0^{\infty} \frac{t}{t^2+\hat{t}^2} e^{i\omega t}\,dt = 0.5\, e^{-\omega\hat{t}}\{\, E_1(\omega\hat{t})\, e^{2\omega\hat{t}} - E^*(\omega\hat{t})\, i\pi \,\}$$

$$\simeq 0.5\pi\, e^{-(\omega\hat{t})}\{0.107/(\omega\hat{t})+0.275-0.327(\omega\hat{t})-0.097(\omega\hat{t})^2+i\,\}$$

which gives a sufficient good fit. In this case $\dot{R}/R$ is characterized by the two times $\hat{t}_{i,f}$, the distance of the extrema from the zero point and the delay time T. In Fig.8 we see the time evolution of the amplitude in the complex plane with three values for $\omega\hat{t}$ as parameter. The time delay produces like a technical delay line a phase rotation of the outgoing amplitude around the incomig one. The total amplitude, $a(\infty)-a(-\infty)$, cancels to zero with a periodicity of $\omega = 2\pi/T$. For T=1 [n.u.] = 1.288 10^{-21} s this corresponds to a change in energy transfer of about 3 MeV. As was pointed out by Oeschler and Senger at this conference the delay times involved are of this order for the very heavy systems and for these it is an experimental challenge to search for an oszillatory modulation of the spectra over the broad dynamic range of 3 MeV.

The picture just discussed is highly unrealistic for these heavy systems because of the missing energy loss or damping. As a next step we consider still a time delay. But the outgoing $\dot{R}/R$ is now reduced due to the kinetic energy loss, correspondingly is $\hat{t}_f$ for the outgoing channel increased. The Feldmeier curve (b) is fitted with the coulombic trajectory (a) in the entrance part with $\hat{t}_i=0.43$ (.55 10^{-21} s) as calculated with the above formula and with $\hat{t}_f=0.64$ in the exit part and T=0.96 for the delay. The corresponding curves (c) are also shown in Fig.7-9. The main effect is a decrease in modulation. Because the rotating outgoing complex amplitude is now reduced in length, a complete

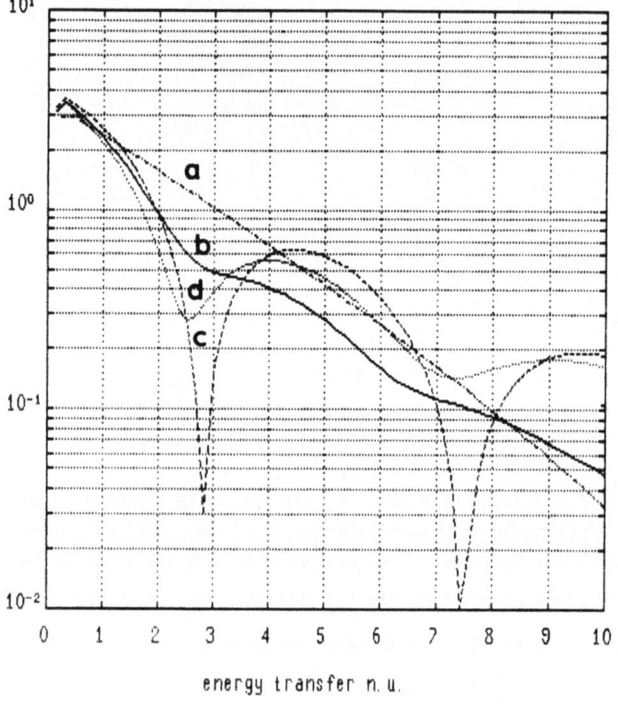

Fig.9
Absolute square of the Fourier transform of $\dot{R}'/R'$ (the distance between the end points of curves in Fig.8), as a function of the energy transfer ω.
(a) coulomb trajectory,
(b) Feldmeier calculation
(c) 'atomic clock',
(d) damping in the outgoing channel due to friction.

cancellation is excluded. It is this model which has been used by the experimental groups and it was shown by Oeschler and Senger that many trajectories have to be summed over to include the experimental and physical distributions of impact parameters for a given kinematical condition. Finally we see from that the Feldmeier curve (b) deviates quite strongly from zero within the domain of the 'delay time' $\dot{R}/R$. Obviously after contact of the nuclei there is still a strong motion particular in the nuclear surface which turns out to be quite essential. This is contrary to the model of two sticking spherical nuclei. Actually, as illustrated in the figures there is a further reduction in modulation which should be visible experimentally if progress goes on in many respects as discussed by Oeschler and Senger.

We thus see that this spectroscopic technique might be a most suitable instrument to study details of the monopole charge evolution during a reaction. It seems that experimentally we a starting of an oszillatory structure. With some optimism this reminds of the early period of elastic electron scattering in which the rms radius was the only derivable quantity in the beginning like it is the delay time in this field. Only with improved techniques the very detailed nuclear charge information could be obtained. The same may be expected for the dynamics of deep inelastic collisions.

ACKNOWLEDGEMENT

I am most grateful to W.Greiner for organizing this conference with its many interesting discussions not only about physics but also about those topics related to problems of NATO which supported this conference.

REFERENCES

Backe H., Senger P.,Bonin W., Kankeleit E., Kraemer M, Krieg R,Metag V, Trautmann N and Wilhelmy J.B., Phys.Rev. L. 50,1838 (1983)
Blocki J and Swiatecki W.J.,preprint LBL-12811 (1981)
Church E.L. and Weneser J., Phys. Rev. 103,1035 (1956)
Feldmeier H., in Nuclear Structure and Heavy-Ion Dynamics, Course LXXXVII, Societa Italiana di Fisica, Bologna, edited by L. Moretto and R. Ricci (North/Holland, Amsterdam, 1984).
Kankeleit E., Nukleonika 25,253 (1980)
Kankeleit E., in Gross Properties of Nuclei and Nuclear Excitations, International Workshop XIII, Hirschegg, Austria, January 1985, edited by H Feldmeier.
Krieg R., Bozek E., Gollerthan U., Kankeleit E., Klotz-Engmann G, Kraemer M.,Meyer U, Oeschler H. and Senger P., Phys.Rev.C, Aug.86.
Liesen D., Armbruster P.,Bosch F., Hagmann S., Schmidt-Boecking H., Mokler P.H., Wollersheim H.J., Schuch R., Wilhelmy J.B., Phys Rev.L. 44, 983 (1980)
Reinhardt J., Mueller B., Greiner W. and Soff G. Z. Physik A292 (1979)
Soff G., Greiner W., Betz W. and Mueller B., Phys. Rev.A, July 1979.
Rose M.E. Relativistic Electron Theory, John Wiley & sons (1961)
Schmidt R, Toneev V., Wolschin G., Nucl. Phys. A311, 247 (1978)
Soff G., Reinhardt J., Betz W.,J.Rafelski Physica Scripta 17 (1978)

charge information could be obtained. The same may be expected for the dynamics of deep inelastic collisions.

POSITRONS AND ELECTRONS EMITTED IN ELASTIC AND DISSIPATIVE HEAVY ION COLLISIONS

H. Oeschler, B. Blank, E. Bożek[a], U. Gollerthan, H. Jäger, E. Kankeleit, G. Klotz-Engmann, M. Krämer, R. Krieg[b], U. Meyer[c], C. Müntz, M. Rhein, and P. Senger[d]

Institut für Kernphysik, Technische Hochschule Darmstadt, Germany

INTRODUCTION

The main research line of the Tori group is the study of the reaction dynamics of dissipative collisions between heavy ions via positron and electron spectroscopy. The last five years since the Lahnstein-Conference[1] are marked for our group by the installation of a new experimental device for detecting positrons and electrons emitted in these collisions, the so-called Tori spectrometer[2]. The first part of this report is devoted therefore to describe the main characteristics of this apparatus.

The introduction and the chapter on theory will be rather short as in the preceeding expose by G. Soff the general motivation and the coupled-channels calculation[3] have been described. A very transparent approach, the so-called scaling model[4], has been introduced by E. Kankeleit and is also presented during this meeting. Both approaches are used in the comparison with the data. A short chapter deals with the problem how to separate positrons from atomic processes - the ones we are interested in - from those originating from nuclear pair conversion.

In section 3 positron and δ-electron spectra from elastic collisions are compared with the calculations. This comparison serves as test of the theoretical description since all components are known. A systematic study of U + U collisions from the Coulomb barrier up to 10 MeV/nucleon is given in Ref. 5 and the main aspects will be presented. In the subsequent section positrons and δ-electrons from dissipative collisions are compared with calculations based on trajectories from various friction models. In these models the interaction time turns out to be the essential quantity determining the spectral shape. For a fixed interaction time a oscillatory pattern of the spectra is expected[3]. By dividing the dissipative collisions into subclasses that they represent a narrow range of impact paramaters and hence a small variance in the nuclear interaction time we have the chance to observe experimentally these oscillations. This technique has been also applied in more detail to the reaction Pb + U and preliminary results are shown at the end of section 4.

1. The Tori Spectrometer

The so-called Tori spectrometer[2] consists essentially of two quarter toroids forming an "S-shaped" solenoid. Here only the essential features of the instrument are explained. A detailed description of the apparatus is given in Ref. 2.

The main point of this device is its ability to separate positrons and electrons in space due to a drift motion acting in inhomogeneous magnetic fields. While in homogeneous magnetic fields charged particles spiral around the field lines, in the Tori spectrometer the particles experience different magnetic field strengths during one turn which leads to a drift perpendicular to the S plane of the Tori spectrometer in opposite directions according to the sign of their charges.

Electrons and positrons can be detected simultaneously without influencing each other. This possibility is used for electron-positron coincidence measurements as discussed in the talk by M. Krämer.

The main components of the apparatus are depicted in Fig. 1. In the first quarter toroid, positrons and electrons drift in opposite directions on their way to the middle plane where they are spatially completely separated. Here the electrons are absorbed by a semicircular diaphragm above which a detector array of usually six Si(Li) diodes is installed. Positrons pass unhindered through the open part of the middle plane and drift back to the central line in the second quarter where they are focussed on a Si(Li) diode with a diameter of 5 cm and a depletion depth of 5 mm. Only a few of the electrons scattered at the diaphragm or on the walls of the tube can reach the second quarter toroid. Measurements with beta and electron conversion sources show that the suppression is about $1{:}10^4$.

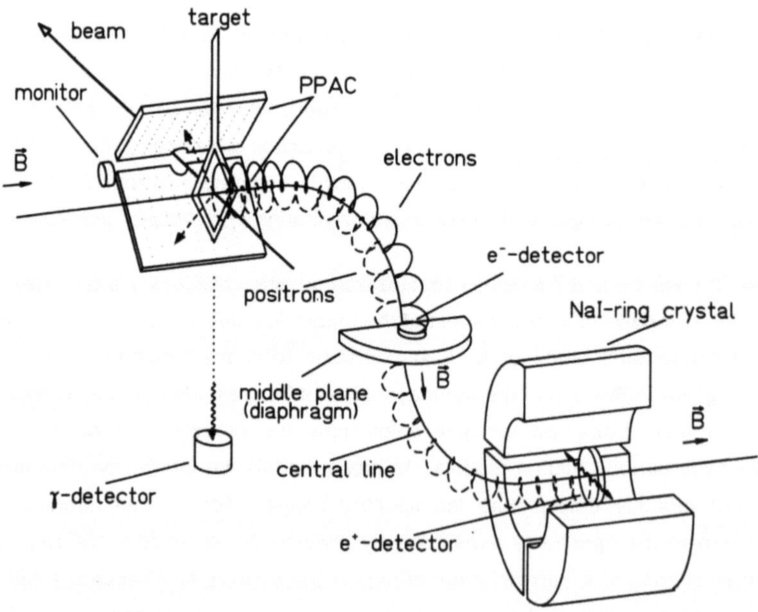

Fig.1: The Tori spectrometer and its detection devices.

In order to discriminate the positrons from scattered electrons, their annihilation radiation is identified by a four-sector NaI-ring crystal. Various conditions concerning the multiplicity and the energies of the required NaI signals are used[2]: The strongest one requires two photo-peak signals of 511 keV in opposite NaI segments, but then only 5% of the emitted positrons are accepted (Fig. 2 line a). Usually the sum of energy spectra of all four NaI crystals has to be between 511 keV and 1022 keV. Then an efficiency of 40 % is achieved (Fig. 2 line b). Due to the high suppression ability for electrons this condition is sufficient to obtain positron spectra free of pile up or summing with electrons. Figure 2 shows also the high transmission of 70 %, i. e. the probability that an emitted positron reaches the e^+-detector without considering the detection probability of the annihilation radiation.

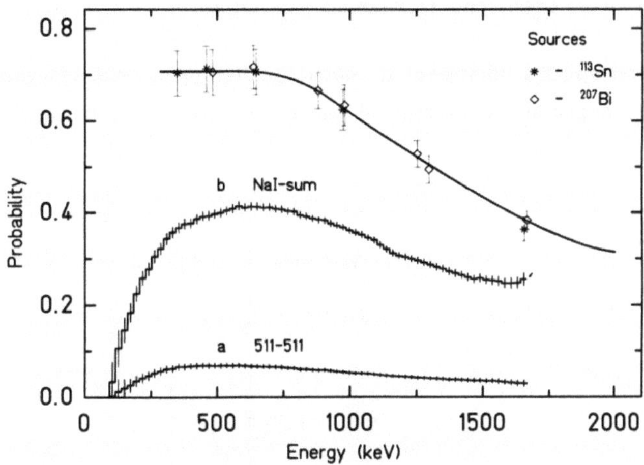

Fig.2: Transmission of the spectrometer onto a 5 cm-diode and detection probability for positrons requiring additionally the observation of the annihilation radiation.

The scattered heavy ions are detected by a pair of position-sensitive parallel-plate avalanche counters (PPAC) covering the scattering angles from 18 to 70 degrees. In the latest version the detectors have an x- and y-position read out with 1 mm spacing. The wires and stripes are connected by delay-line chips of 2 ns per intervall. By reading out both ends of the delay-line chain, double events in one counter can be detected and the fission of one or both partners can thus be observed. Additionally, scattered heavy ions can be distinguished from fission fragments by their energy-loss signal.

In order to detect γ-rays originating from nuclear reactions, a 7.5 × 7.5 cm NaI and recently a large BaF_2 crystal is mounted at an adjustable distance of 30 to 80 cm from the target. The counting rate and the spectral distribution of the γ-rays allow an estimation of the contribution of positrons created by internal pair conversion as described in the next chapter.

2. Positrons from Nuclear Pair Conversion

The determination of the positrons from nuclear pair conversion is one of the main problems in positron spectroscopy.

From the measured gamma spectrum $dN(E_\gamma)/dE_\gamma$ a spectrum of "nuclear" positrons has been determined according to

$$dN(E_{e+}) = dN(E_\gamma) \times d\beta/dE_{e+}(E_\gamma, E_{e+}, Z, M\lambda)$$

where $d\beta/dE_{e+}$ is the pair conversion coefficient depending on the γ energy E_γ, on the nuclear charge Z and the multipolarity of the transition $M\lambda$. The pair conversion coefficients $d\beta/dE_{e+}$ of Schlüter et al.[6] have been used. (For details see Ref. 5).

The positron spectra coincident to elastic scattering in case of systems where atomic contributions are negligible (undercritical) can be reproduced by converting the measured γ

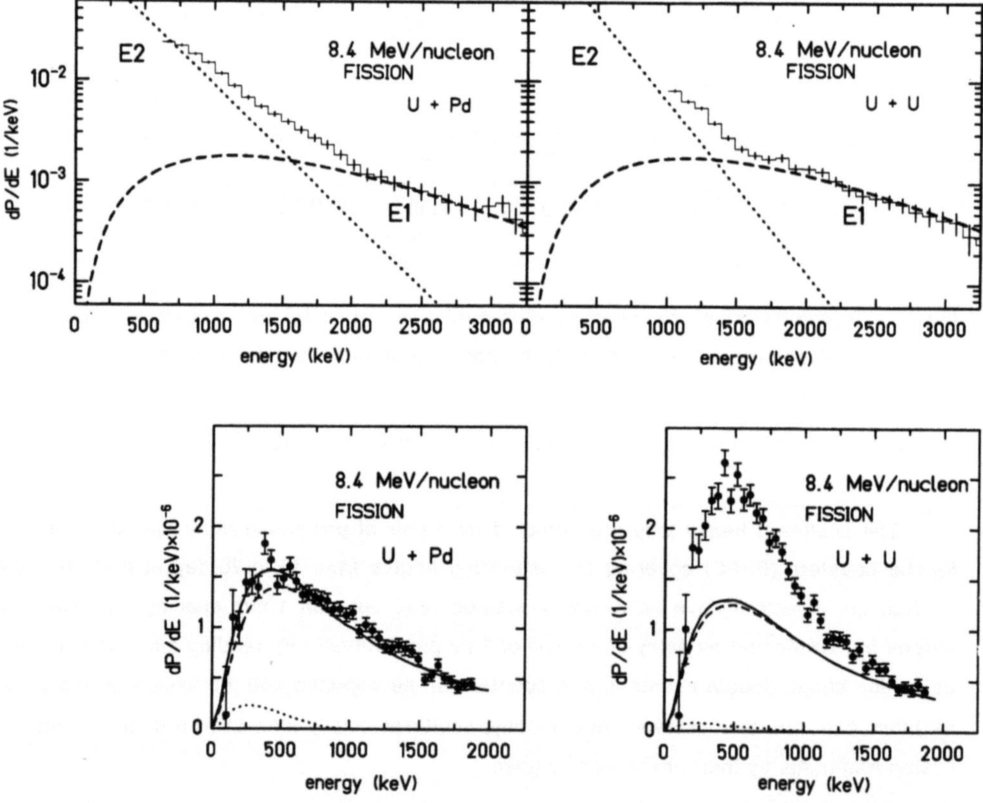

Fig.3: upper part: Decomposition of gamma-spectra in an exponential part and one representing statistical transitions.
lower part: Corresponding contributions to the calculated positron spectra from internal pair conversion.

spectra assuming E1 transitions in the energy region above 1 MeV. These transitions originate from quasi-elastic processes which cannot be separated from pure elastic scattering by our particle counters. Since an E1 multipolarity is valid for a variety of collision systems at different kinematical conditions, the same multipolarity can be supposed for elastic collisions of U + U.

Dissipative collisions lead to high nuclear excitations. In analogy to (HI,xn) reactions, the γ spectra have been decomposed into a low-energy part with an exponential slope (collective E2 transitions) and a part reflecting statistical transitions. Figure 3 shows the spectral distributions of γ rays and positrons from dissipative ^{92}U + ^{46}Pd collisions at 8.4 MeV/nucleon. Assuming an E1 multipolarity for the high-energy transitions and an E2 multipolarity for the low-energy part the positron spectra (lower part of Fig. 3) can be reproduced by converting the measured γ spectra according to the two contributions. Neither a single multipolarity (E1 yields too high and E2 too low calculated production probabilities), nor a simple mixture can reproduce the measured shape. As the outgoing products are the same in the dissipative U + U collisions (uranium fissions), the same multipolarity can be supposed.

Nuclear excitations grow strongly with bombarding energy, yet the surplus of energy is taken away by neutrons and it turns out, that the production probability for atomic positrons increases faster than the positron production by nuclear conversion (see Ref. 5).

3. Positrons and Electrons from Elastic Collisions

In this chapter positrons and electrons emitted in elastic collisions from the Coulomb barrier up to 10 MeV/nucleon are discussed. The positron spectra shown in Fig. 4 mostly exhibit structureless shapes which are well described by coupled-channels calculations[3] using Rutherford trajectories with normalization factors of 0.7 to 1.0. The analogous electron spectra (Fig. 5) are also well reproduced by the same theoretical framework; however, normalization factors of 1.35 and 1.65 are needed. Note that the difference in production probability between positrons and electrons amounts to four orders of magnitude.

Figures 4 a and b show atomic positron spectra from U + U collision systems at the Coulomb barrier, where only elastic scattering is expected. The spectral distribution of positrons from peripheral collisions (a) agrees well with the theory of dynamically induced (atomic) positrons normalized with a factor of 0.7. The spectra in Figure 4 b represents the more central collisions (6.5 fm $<$ b $<$ 11.5 fm). Compared with theoretical calculations scaled with the same factors as above, the positron spectrum shows an enhancement of 20 % around 300 keV. This might be correlated with the observation of peak structures, one of the main subjects of this meeting.[7]

At 7.5 MeV/nucleon the positron spectra are presented for two impact parameter regimes defined by the laboratory scattering angles $25^0 < \theta < 37^0$ and $37^0 < \theta < 45^0$ (Fig. 4 c,d). The experimental points measured close to the grazing angle (45^0) deviate from a smooth distribution showing an unexpected bumplike structure. This structure could have a different origin and as it occurs just at the grazing angle nuclear processes might be its origin.

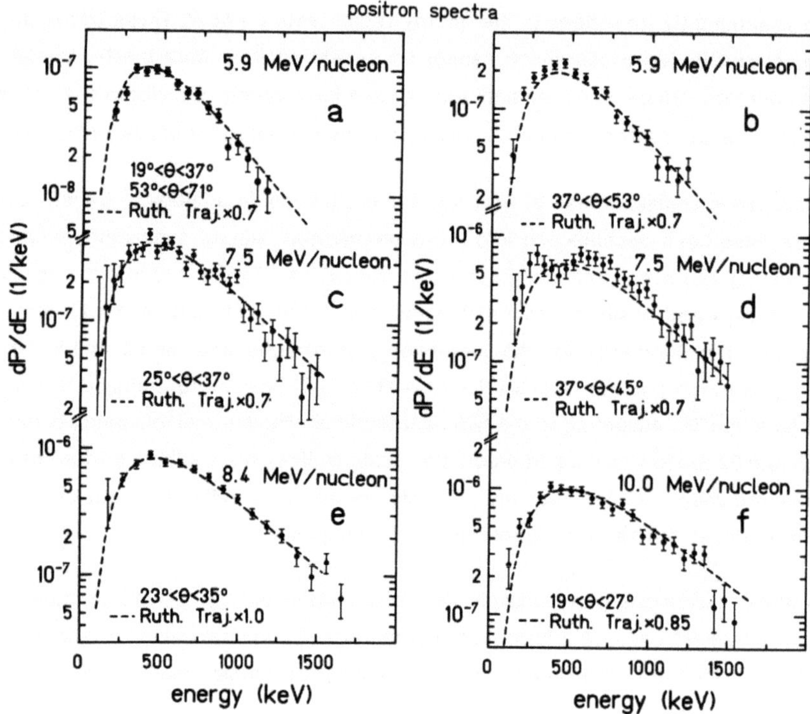

Fig.4: Positron spectra from elastic U + U collisions at various incident energies. The indicated angles are laboratory scattering angles. The dashed lines are calculations by U. Müller, T. de Reus et al.[3]

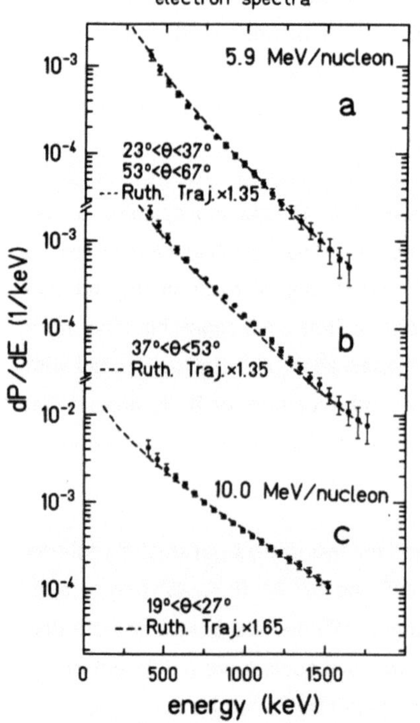

Fig.5: Electron spectra from U + U collisions at 5.9 MeV/u and 10.0 MeV/u beam energy.

The positron spectra obtained at 8.4 MeV/nucleon and 10 MeV/nucleon shown in Fig. 4 e and f, respectively, exhibit smooth curves which agree well with the theory. These data have been obtained with a high statistical quality as firstly the production probabilitiy increases with incident energy and secondly the nuclear contribution is relatively smaller. The δ-electron spectrum measured at 10 MeV/nucleon under the same kinematical condition as positrons is shown in Fig.5c. To describe the data the theoretical values have to be multiplied by a factor 1.65.

In Fig. 6 the normalization factors needed to multiply the coupled-channels calculation in or-

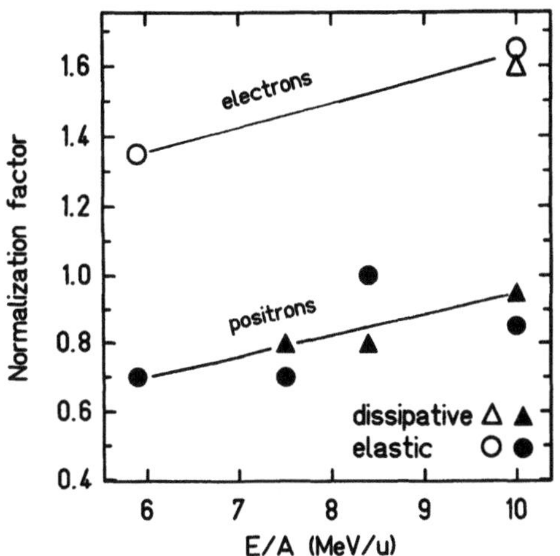

Fig.6: Summary of normalization factors needed to multiply the calculations in order to fit the experimental results.

der to agree with the experimental data are summarized. Two general trends can be remarked: (i) While the positron production is slightly overestimated, the electron production is underestimated by about 50 %. (ii) Both for the positron and the electron production a clear trend can be remarked showing normalization factors increasing with bombarding energy. The second statement is not weakened by the absolute normalization uncertainty of about 20 % as the 5.9 and 10 MeV/nucleon measurements were performed in the same experimental series.

4. Positrons and Electrons Associated with Dissipative Collisions

4.1. Theoretical Remarks

The data of this work are compared either with coupled-channels calculations[3] (they yield absolute probabilities) or with the result of the scaling model[4] (describing only the spectral shape). Both approaches have been presented during this conference and, here, only a few aspects are given in the frame work of the adiabatic first order perturbation theory[4].

For direct excitations of an electron-positron pair the transition amplitude can be written as

$$a_{if} = - \int_{-\infty}^{\infty} dt \, \langle f|\partial/\partial t|i\rangle \times \exp[\, i \int_{-\infty}^{t} dt' \, (E_f - E_i) / \hbar \,]$$

with E_f and E_i the total energy of initial and final state, respectively.

The time derivative of the Coulomb potential contains as essential quantity the term $\dot{R}/R$. This function is shown in Fig. 7 together with the corresponding trajectories. The time between the extrema of the $\dot{R}/R$ curve is used to define the collision time $2\hat{t}$. For dissipative collisions the additional time is then defined as nuclear interaction time t_{int}. The production probability is mainly determined by the Fourier transform of $\dot{R}/R$. Using the approximations introduced in Ref. 4

$$\dot{R}/R \cong t/(t^2 + \hat{t}^2)$$

where $\hat{t}$ depends only on kinematical parameters, it reads

$$dP/dE_{e^+} \cong h(E_{e^+},\hat{t})/(E_{e^+} + 2m_0c^2)^2 \times \exp[-2\hat{t}\,(E_{e^+} + 2m_0c^2)/\hbar]$$

where E_{e^+} is the kinetic energy of the positron. The function $h(E_{e^+},\hat{t})$ describes the decrease of the positron intensity towards low energies similar to the Fermi function in the beta decay.

Figure 8 shows some examples of calculated spectra. Elastic scattering yields a nearly exponential decrease towards higher energies. In the case of dissipative collisions a fixed nuclear interaction time has been used which results in an oscillatory spectral shape (dotted and dash-dotted lines). A distribution of interaction times smears out these oscillations and yields a steeper decline of the spectra for energies up to 1 MeV (full line). First experimental evidence has been reported in Ref. 8.

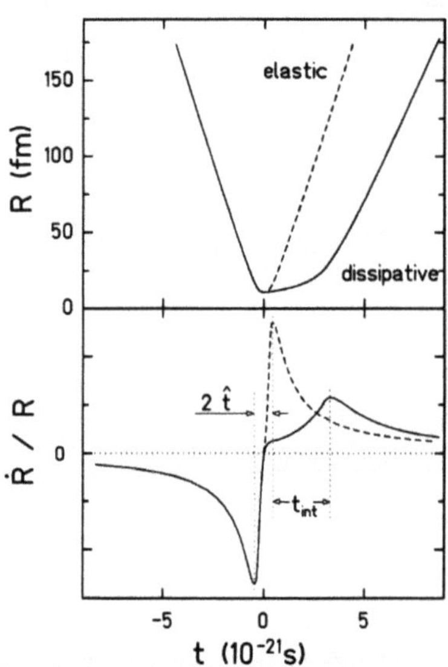

Fig.7: Time dependence of R (distance between the two nuclei) and $\dot{R}/R$ ($\dot{R}$ the relative velocity).

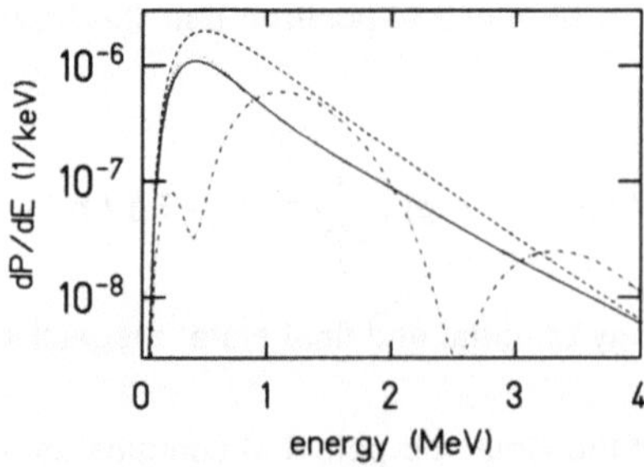

Fig.8: Calculated positron spectra for elastic (dashed line) and disspative collisions (dotted and dash-dotted line: fixed t_{int}, full line: distribution of t_{int}).

4.2. Test of Reaction Models

Dissipative collisions in the U + U system can be well distinguished from elastic and quasi-elastic processes since at least one and usually both reaction products fission. This fact leads to a broad distribution of the outgoing fragments in the angular correlation measured by the two PPAC, while for elastic collisions the sum of the two scattering angles is always 90°

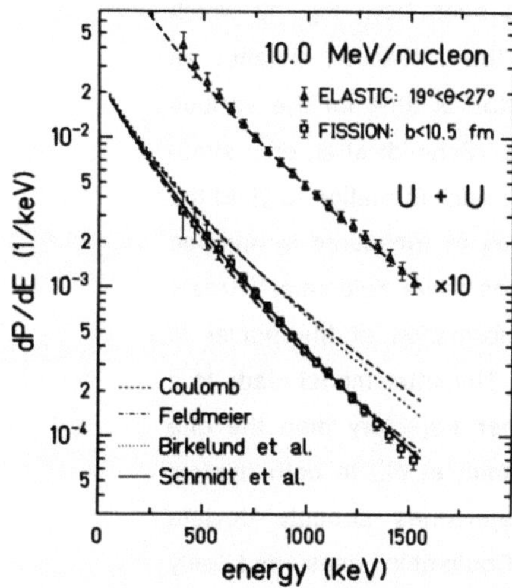

Fig.9: Electron spectra from elastic scattering and dissipative collisions compared with scaling-model[4] calculations based on different reaction models.

The electron spectra obtained in coincidence with elastic and dissipative collisions are shown in Fig. 9. The upper part shows the data from elastic scattering exhibiting good agreement with the calculation. Electrons originating from dissipative collisions (lower part) exhibit a steeper slope; their shapes cannot be reproduced using Rutherford trajectories. One has to take into account the retardation in the collision process due to friction forces. Among many different models of deep inelastic reactions three were chosen to calculate the trajectories.

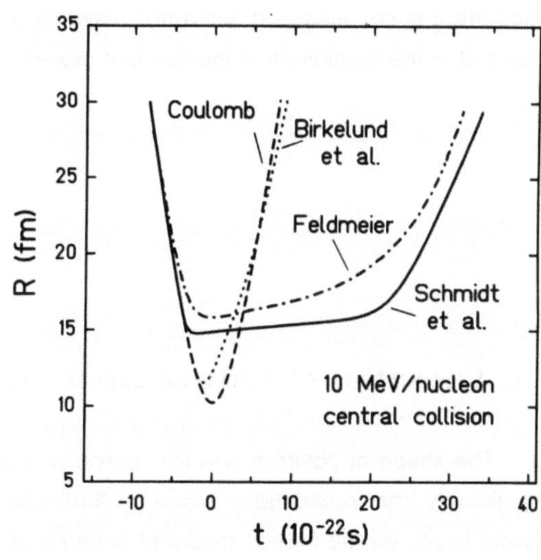

Fig.10: Time dependence of the internuclear distance according to various reaction models.

Two macroscopic friction models use classical equations of motion and a Coulomb and proximity potential. In the one proposed by Birkelund et al.[9], the friction coefficients, which are the essential quantity, are mainly fitted to fusion reactions where heavy-on-heavy systems are hardly considered. Schmidt et al.[10] apply a dominant radial friction factor which leads to a fast deceleration when the two nuclei approach. This can be seen from Fig. 10 which shows the interaction distance as a function of time for the various models. Schmidt et al. only simulate the neck formation to yield the Q values as measured in damped collisions while Feldmeier[11] treats the deformation of the nuclei in detail. The latter model leads to a smoother trajectory than the one of Schmidt et al. In both models the trajectories strongly deviate from Coulombic ones and only those lead to an agreement with the measured electron spectra, while the trajectories from the model of Birkelund et al. fail (Fig. 9). The nuclear interaction times according to these models are shown in Fig. 11. The definition of the quantity t_{int} can be seen from Fig. 7.

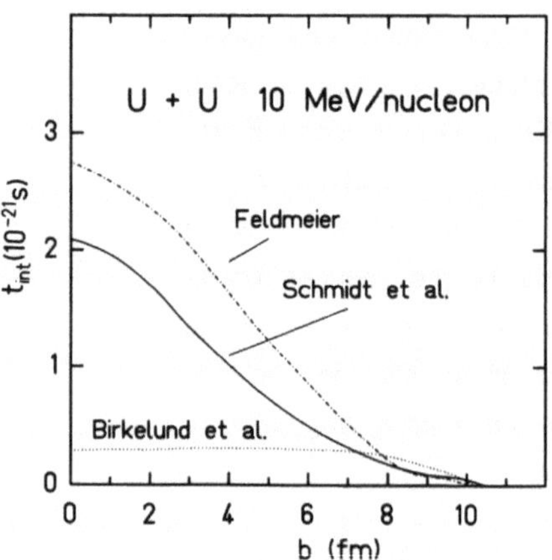

Fig.11: Nuclear interaction times versus impact parameter calculated from the various reaction models.

The interaction times are about 2×10^{-21} s for the most central collisions (see Fig. 11). Since there is no significant difference between the calculated results the positron spectra are analysed in the framework of the Schmidt model only.

4.3. Systematics of Positron Spectra from Dissipative Collisions

The shape of positron spectra associated with deep inelastic collisions of U + U differ significantly from calculations based on Rutherford trajectories but agree well with theoretical results based on the friction model of Schmidt et al. (Fig. 12). Multiplication factors of 0.8, 0.8 and 0.95 for the various energies are used. The δ-electron spectra have to be multiplied by 1.6 to fit the data. The trends of the normalization factors were already discussed along with Fig. 6.

The model of Schmidt et al. yields nuclear interaction times reaching for central collisions values of 1.2×10^{-21} s, 1.7×10^{-21} s and 2.1×10^{-21} s at incident energies of 7.5, 8.4 and 10 MeV/nucleon, respectively.

The information concerning the nuclear interaction time extracted from the positron and δ-electron spectra is consistent. This rules out the discrepancy reported in Ref. 8. Further studies of dissipative collisions will therefore use δ-electrons which have about 10^4 higher production rate. In principle the study of the reaction dynamics can be extended to lower Z values. However, for light atomic numbers one encounters the problem to separate δ-electrons from conversion electrons.

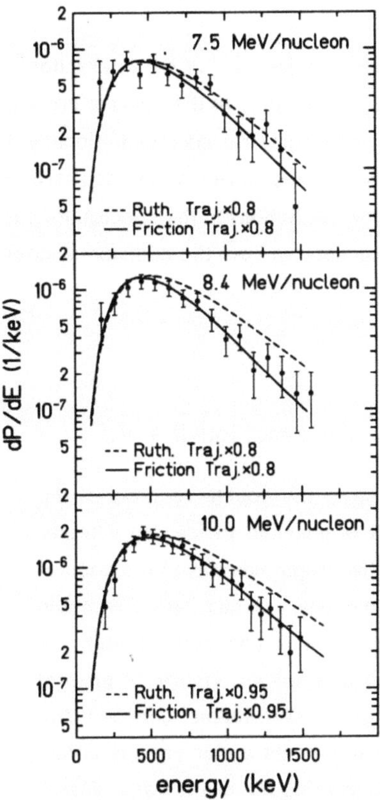

Fig.12: Comparison of measured positron spectra for dissipative collisions at 7.5, 8.4 and 10.0 MeV/u with calculations of the Frankfurt group.[3]

A different method to study nuclear contact times in U + U collisions at 7.5 MeV/nucleon incident energy was used by Stoller et al.[12] In their analysis only the very rare events with Q-values up to -200 MeV and still yielding two U-like stable nuclei were selected. From the corresponding K-shell ionisatión probability, nuclear reaction times of about 1×10^{-21} s were extracted.

Nuclear contact times can also be extracted using simple classical assumptions to deduce the deflection function from the experimental cross sections $d^2\sigma/dEd\Theta$. In Ref. 13 this method is applied to the results[14] of the reaction U + U at 7.42 MeV/nucleon. The results are shown in Fig. 13. Assuming a non-sticking situation between the two nuclei values of t_{int} up to 2×10^{-21} s are obtained. A sticking condition is unlikely and would yield times twice as long.

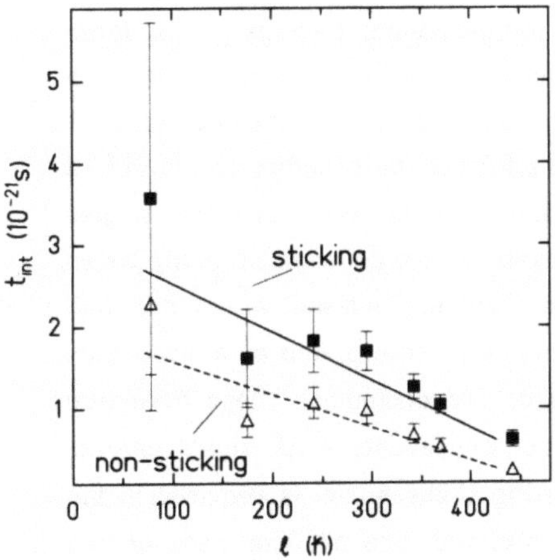

Fig.13: Relation between the nuclear interaction time and impact parameter as obtained by the analysis of the data of Ref. 14 using a simple geometrical model.

4.4. Subgroups of Dissipative Collisions and Impact Parameter Selection

From the analysis of the interaction time versus impact parameter it is evident that the predicted oscillations in the spectra are smeared out due to the broad distribution of t_{int} when studying all dissipative collisions globally. We will now discuss subgroups of identified three- and four-body events which may be ascribed to selected ranges of impact parameter in the hope to observe oscillations in the spectra. Positron spectra from the reaction U + U and preliminary δ-electron spectra from the collision system Pb + U are presented.

4.4.1. The Reaction U + U

Due to the double read-out of the PPAC it is possible to identify whether one or two fragments have hit a counter simultaneously. When two fragments are measured in one counter both angles are determined. Figure 14 present the angular correlation between the emitted fragments under various conditions concerning the event class. Figure 14b shows the subgroup of four-body events, where each PPAC has detected two fragments. They are given as a function of the averages of the two measured angles which corresponds roughly to the primary scattering angle of the fissioning nucleus. The large opening of the fission cone leads to a decreasing detection efficiency towards the edges of the counters and therefore mainly events with primary angles around 45° are detected. In many cases one fragment misses a counter and could be misinterpreted as a three-body event. To distinguish these from true three-body events a ΔE signal characterizing U-like nuclei is required in the counter with one hit. The class of true three-body events is shown in Fig. 14c. Two distinct groups can be seen:

(I) Collisions with high Q values and

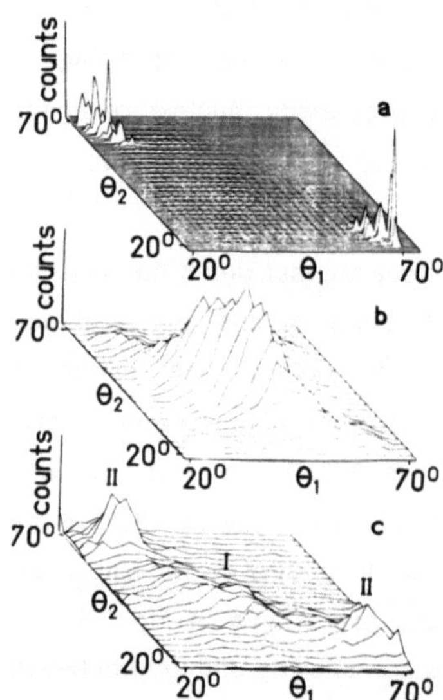

Fig.14: θ_1-θ_2 correlation between the emitted fragments in the collision U + U at 10.0 MeV/u.

a) All events exhibiting two classes: elastic scattering with $\theta_1 + \theta_2 = 90^0$ and sequential fission observed by their broad distribution.

b) Only identified four-body events.

c) Only identified three-body events.

laboratory scattering angles around 30° - 50°. These events are most likely associated with mass transfer yielding one "light" nucleus which survives.
(II) Collisions with low energy loss and scattering angles close to the grazing angle. In these peripheral collisions the nuclei are excited very little and have a chance that one survives

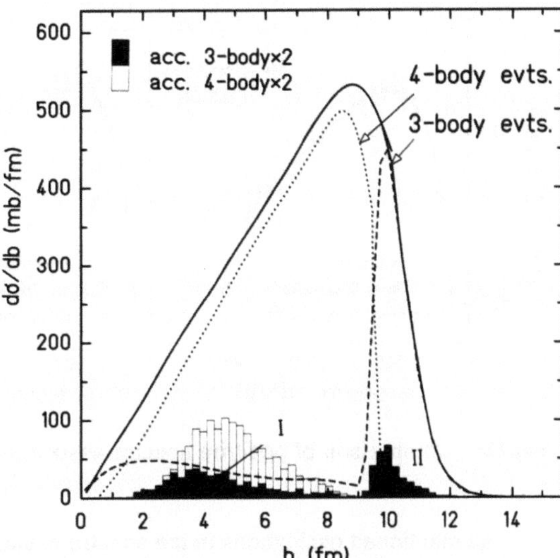

Fig.15: Calculated impact parameter distributions of various dissipative processes of U + U at 10 MeV/nucleon (dashed and dotted lines) and the fraction detected by the particle counters (histograms).

From the primary angles Q values have been calculated neglecting mass transfer. As we replace the unknown primary angle by the average of the two measured angles we assume symmetric fission. This procedure is correct on average but not for each event. In order to estimate systematic errors and the precision of this procedure a Monte-Carlo program was developed to simulate sequential fission. Impact-parameter-dependent Q values and a deflection function proposed by Schmidt et al.[10] were used. The mass transfer of the primary reaction was extrapolated from Ref. 14. The sequential fission is calculated using Q values from the Viola systematics[15] and fission fragment mass distributions suggested by Ref. 16. The width of the mass distribution affects the results very little. Using these Monte-Carlo simulations a mean Q value of four-body events of -395 MeV with an error of ±70 MeV is obtained. The two classes of three-body reactions correspond to Q values of -50 MeV and -450 MeV, respectively, with an error of ±70 MeV.

By this procedure we obtain the impact parameter regimes leading to the various processes. The dotted and dashed lines in Fig. 15 represent the distributions of three- and four-body reactions. Their shapes are adjusted to fit the measured angular correlation shown in Fig. 14. The histograms in Fig. 15 represent the fraction of events detected by the counters. Again, the two classes of three-body events are seen: peripheral (II) and central (I) collisions.

In Fig. 16 the measured positron spectra corresponding to all dissipative collisions, to identified four-body and to class I three-body events are shown together with coupled-channels calculations which are based on the impact parameter selection given in Fig. 15. Within the limited statistics the data agree with the theory The absolute yield increases by 10 % when selecting the more central collisions from all reactions to the four-body events and again by 20 % from four-body to three-body central-collision events. This agrees with the theory and proves that smaller and smaller impact parameters are selected.

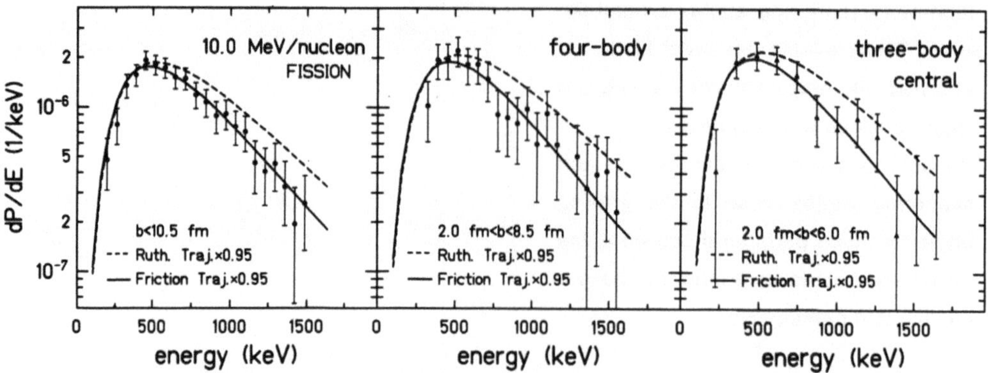

Fig.16: Comparison of positron spectra associated with different impact parameter regimes.

As mentioned oscillations in the spectra could be observed if the range of interaction times is sufficiently narrow. As impact parameters and interaction time are correlated (see Fig. 11) the presented selection is a first step towards a impact parameter selection. The limited statistics does not allow for a finer analysis, e.g. further restriction of the impact parameter range via additional Q-value cuts. This is studied in more detail in the next chapter discussing the reaction Pb + U.

4.4.2. The Reaction Pb + U

As already mentioned, oscillations in the δ-electron spectra are expected for a narrow distribution of nuclear interaction times. These oscillations could eventually be seen if collisions of restricted range of Q values or impact parameter can be selected. This study has been continued with the reaction Pb+U at 8.4 MeV/nucleon . The different reaction channels of binary, three- and four-body events represent a favorable division: Binary events originate mainly from elastic scattering and serve to test the model calculations using Rutherford trajectories. Three-body events represent the dominant class as mostly the Pb-like fragment survives while the U-like partner undergoes fission. This class extends from peripheral to central collisions and has further been divided into subgroups according to Q values and scattering angles. Four-body events are correlated with high Q values. Furthermore, mass transfer from the Pb-like nucleus to the heavier one is typical for this class as only then both nuclei have a chance to fission as was shown in Ref. 17.

Figure 17 exhibits the angular correlation for the three event classes. Figure 18 shows the δ-electron spectra corresponding to the chosen regions indicated in Fig. 17 together with calculations using first order perturbation theory[4]. For binary reactions the shape is well reproduced by the calculation; the height is adjusted to match the experimental values. The spectra belonging to reactions with high Q values both from three- and four-body events show a steep descent up to ≅ 2 MeV followed by a flatter slope towards higher energies. This shape can be interpreted as a first indication of oscillations in the δ-electron spectra. The spectra from three- and four-body events do not differ from each other and both classes represent similar

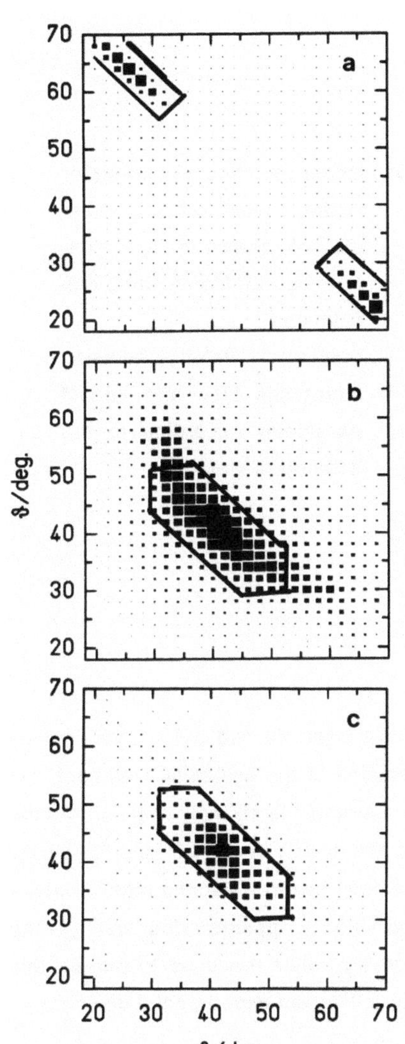

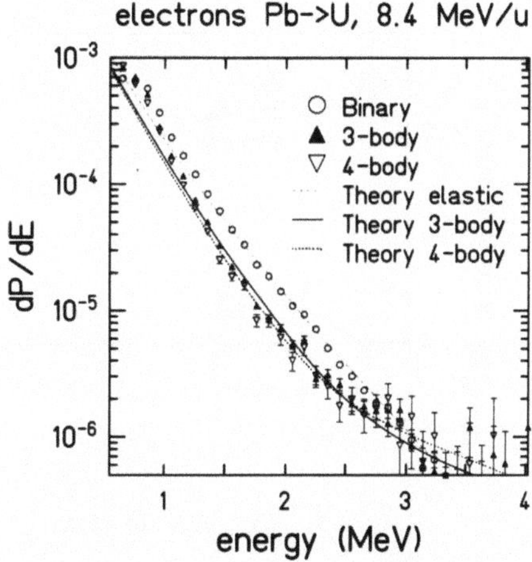

Fig.18: δ-electron spectra originating from the event classes with the limits given in Fig 17. The calculations are performed using the scaling model.[4]

Fig.17: The various event classes of the reaction Pb + U given in θ_1-θ_2 representation. a) binary, b) three-body c) four-body events.

impact parameter regimes as shown in Fig. 19. These distributions have been obtained by a Monte Carlo calculation. Furthermore, the two classes correspond to similar Q values, but the mean mass transfer differs by about 10 mass units as can be deduced from the data of Ref. 17. Within the limits of these data no influence of the mass transfer upon the interaction time can be seen.

The bottom part of Fig. 19 shows the relation between the interaction time and the impact parameter according to the model of Schmidt et al. The presented data correspond to interaction times around $t_{int} = 0.7 \times 10^{-21}$ s. The variance is about 0.7×10^{-21} s and apparently too large to observe oscillations.

A contribution of electrons from nuclear pair conversion amount to about 10 %. No structure is seen in the γ-spectra and no subtraction was performed. E0 contributions cannot be excluded. They are unlikely as both three- and four-body events yield similar electron spectra.

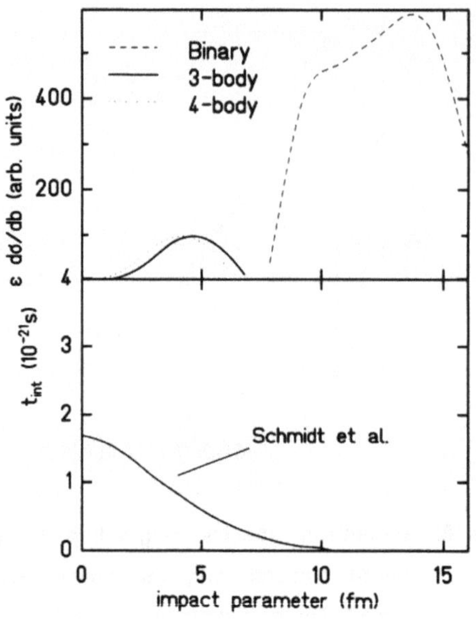

Fig.19: Upper part: Impact parameter regimes leading to the event classes shown in Fig. 17 determined by Monte Carlo calculations.
Lower part: Relation between interaction time and impact parameter according to the Schmidt model.

Recently, a more refined analysis of these data have been carried out. Chosing only events where the fission axis is about parallel to the direction of the fissioning fragment, the error of the Q-value reconstruction caused by the simple averaging is strongly reduced. As example of such an analysis preliminary δ-electron spectra are shown in Fig. 20; one for binary events similar to the spectrum of Fig. 18 and one of three-body events selecting high Q values and a small range of scattering angles. A minimum around 1.7 MeV appears rather well; yet the statistics is limited. New experiments are under way studying other reactions to pursue this question. A similar study is presented by P. Senger using a different experimental device.

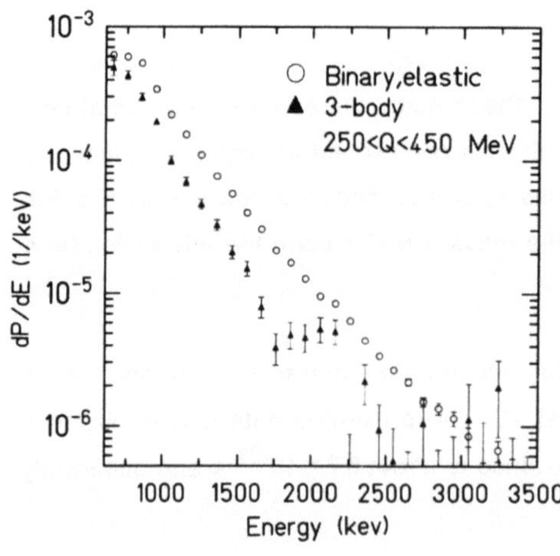

Fig.20: Preliminary δ-electron spectra obtained under special conditions (see text).

A model-independent interaction time can be extracted if two minima are seen. Then the interaction time is given by h divided by the distance of the two minima. To extend the energy range for electrons up to 8 MeV a phoswich-plastic scintillator has been installed in the Tori spectrometer.

Summary and Conclusions

- The Tori spectrometer is well suited for the coincident detection of positrons and δ-electrons emitted in heavy ion collisions. It has a positron-detection efficiency of about 40 % and a high suppression ability for electrons of $1{:}10^4$. Electrons and positrons can be detected simultaneously.

- The collision system U + U was studied at various incident energies from the Coulomb barrier at 5.9 MeV/nucleon up to 10 MeV/nucleon. Measured positron and δ-electron spectra associated with **elastic collisions** are well reproduced by the theory assuming pure Rutherford trajectories.

- Positrons emitted under selected kinematical conditions at bombarding energies 5.9 MeV/nucleon and 7.5 MeV/nucleon exhibit an approximately 20 % enhancement for energies around 200 - 500 keV. This observation might be correlated with the measured line structures reported elsewhere[7].

- The positron and δ-electron spectra obtained in coincidence with **dissipative collisions** exhibit much steeper slopes in the measured energy range than expected from calculations based on Rutherford trajectories. This effect can be explained assuming interaction times up to 2×10^{-21}s. As an example one δ-electron spectrum was used to demonstrate the possibility to test reaction models. It is compared to calculations using trajectories from three different friction models. The shape is reproduced best by the models of Schmidt et al.[10] and the one of Feldmeier[11].

- The shape of the positron and electron spectra both for elastic and dissipative collisions are well described by coupled-channels calculations[3]. Their absolute yields have to be normalized by factors ranging from 0.7 to 1.00 for positrons while the electron emission is underestimated requiring normalization factors of 1.35 to 1.65. The normalization factors seem to increase slightly with incident energy.

- Dissipative collisions had been separated into three- and four-body events. The associated positron spectra show that different ranges of impact parameters are selected. The technique of impact parameter selection has been applied in studying the reaction Pb + U at 8.4 MeV/nucleon. The δ-electron spectra show a clear indication for oscillations when selecting three-body events with high Q values.

ACKNOWLEDGEMENTS

These results are part of the Ph. D. Thesis of R. Krieg and M. Krämer. For performing the coupled-channels calculations and for many very fruitful discussions we would like to acknowledge U. Müller, T. de Reus and G. Soff. We would like to thank J. Foh, Institut für Kernphysik, for building dedicated electronics and H. Folger and the GSI target laboratory for the preparation of the excellent targets. This work was supported by the Bundesministerium für Forschung und Technologie and by the Gesellschaft für Schwerionenforschung, Darmstadt.

REFERENCES

[a] from Institute of Nuclear Physics, Cracow, Poland
[b] now Lurgi, Frankfurt
[c] now GSI, Darmstadt
[d] also University of Mainz

1. Quantum Electrodynamics of Strong Fields, edited by W. Greiner, Nato Advanced Study Institute Series B: Vol. 80, Plenum Press 1983.
2. E. Kankeleit, U. Gollerthan, G. Klotz, M. Kollatz, M. Krämer, R. Krieg, U. Meyer, H. Oeschler, P. Senger, Nucl. Instr. Meth. **A234**, 81 (1985).
3. G. Soff, J. Reinhardt, B. Müller, W. Greiner, Phys. Rev. Lett. **43**, 1981 (1979);
   T. de Reus, J. Reinhardt, B. Müller, W. Greiner, G. Soff, U. Müller, J. Phys. **B17**, 615 (1984);
   U. Müller, G. Soff, J. Reinhardt, T. de Reus, B. Müller, W. Greiner, Phys. Rev. **C30**, 1199 (1984);
   T. de Reus, J. Reinhardt, B. Müller, W. Greiner, U. Müller, G. Soff, Progress in Particle and Nuclear Physics, Ed. A. Faessler, Pergamon Press, Vol. **15**, 57 (1985).
4. E. Kankeleit, Nukleonika **25**, 253 (1980).
5. R. Krieg, E. Bozek, U. Gollerthan, E. Kankeleit, G. Klotz-Engmann, M. Krämer, U. Meyer, H. Oeschler, P. Senger, Phys. Rev. **C43**, 562 (1986).
6. P. Schlüter, G. Soff, W. Greiner, Phys. Rep. **75**, 327 (1981).
7. J. Schweppe, A. Gruppe, K. Bethge, H. Bokemeyer, T. Cowan, H. Folger, J. Greenberg, H. Grein, S. Ito, R. Schule, D. Schwalm, K. E. Stiebing, N. Trautmann, P. Vincent, M. Waldschmidt, Phys. Rev. Lett. **51**, 2261 (1983);
   M. Clemente, E. Berdermann, P. Kienle, H. Tsertos, W. Wagner, C. Kozhuharov, F. Bosch, W. Koenig, Phys. Lett. **B137**, 41 (1984);
   T. Cowan, H. Backe, M. Begemann, K. Bethge, H. Bokemeyer, H. Folger, J. S. Greenberg, H. Grein, A. Gruppe, Y. Kido, M. Klüver, D. Schwalm, J. Schweppe, K. E. Stiebing, N. Trautmann, P. Vincent, Phys. Rev. Lett. **54**, 1761 (1985);
   H. Tsertos, E. Berdermann, F. Bosch, M. Clemente, P. Kienle, W. Koenig, C. Kozhuharov, W. Wagner, Phys. Lett. **B162**, 273 (1985).
8. H. Backe, P. Senger, W. Bonin, E. Kankeleit, M. Krämer, R. Krieg, V. Metag, N. Trautmann, J. Wilhelmy, Phys. Rev. Lett. **50** , 1838 (1983).
9. J. Birkelund, L. Tubbs, J. Hui, J. De, D. Sperber, Phys. Rep. **56**, 107 (1979).
10. R. Schmidt, V. Toneev, G. Wolschin, Nucl. Phys. **A311**, 247 (1978).

11. H. Feldmeier in: Nuclear Structure and Heavy-Ion Dynamics, S.274, LXXXVII Corso, Soc. Italiana di Fisica, Bologna, ed. by L. Moretto and R. Ricci, North Holland (1984). In this model the distance R is calculated via the rms charge distribution of two penetrating nuclei.
12. Ch. Stoller, M. Nessi, E. Morenzoni, W. Wölfli, W. E. Meyerhof, J. D. Molitoris, E. Grosse, Ch. Michel, Phys. Rev. Lett. **53**, 1329 (1984).
13. R. Krieg, Ph. D. thesis, Technische Hochschule Darmstadt, 1985.
14. H. Freiesleben, K. D. Hildenbrand, F. Pühlhofer, W. F. Schneider, R. Bock, D. v. Harrach, H. J. Specht, Z. Physik **A292**, 171 (1979).
15. V. E. Viola,Jr., Nucl. Data **A1**, 391 (1966).
16. R. Vandenbosch and J. R. Huizenga, "Nuclear Fission", Academic Press, New York and London, 1973.
17. T. Tanabe, R. Bock, M. Dakowski, A. Gobbi, H. Stelzer, H. Sann, U. Lynen, A. Olmi, D. Pelte, Nucl. Phys. **A342**, 194 (1980).

DO NUCLEONS DISSOLVE IN GIANT NUCLEI?

D. Vasak, Ch. Derreth, Q. Zhang, A. Schäffer
M. Grabiak, S. Schramm, L. Neise and W. Greiner

Institut für Theoretische Physik
Johann Wolfgang Goethe Universität
Postfach 111932
D-6000 Frankfurt/Main, West Germany

ABSTRACT

An interesting and until now unanswered question is how quarks, confined within a bag, interact when their number is very large. Can their residual interaction be understood in terms of perturbation theory or is it basically non-perturbative ? We have studied two models which are as remote as any other possibility. In the first model we assume the $3A$ quarks of a nucleus to be essentially free in a giant quarkbag. Of course in usual nuclei the structure of individual nucleons is well established but this picture might change if large baryon numbers $A \gg 1$ are considered. The second model preserves the picture of individual nucleons (bags) but allows the quarks to leak out through windows in the bag surface. Each bag is fixed on a lattice site of a simple cubic lattice and the ground state band of this MIT bag crystal is calculated.

I. MOTIVATION

If one takes the MIT bag [1,2] radius of approximately $R \approx 1.1 fm$ seriously there is some difficulty in explaining the nuclear equilibrium density $\rho_0 \approx 0.17 fm^{-3}$ without considerable overlap of the bags. The mean distance of two nucleons can be estimated by $d \approx (volume/nucleon)^{\frac{1}{3}} = (\frac{1}{\rho_0})^3$ which is therefore in the order of $d \approx 1.8 fm$ (see figure 1). Even if the effective bag radius becomes smaller when the pressure of the pion cloud is taken into account this might lead to a probability for the quarks to leak out of their confining bag. On the other hand there are some explanations of the well

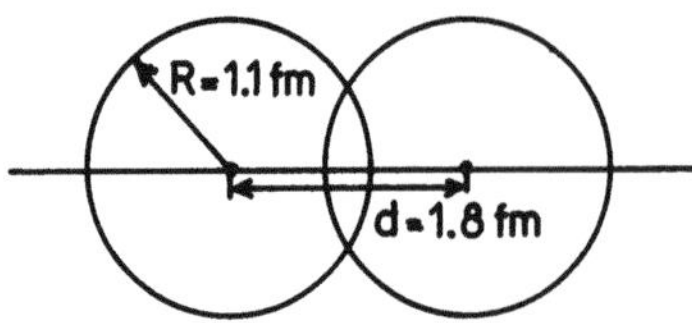

Figure 1: Overlap of nucleons at equilibrium density ρ_0.

known EMC effect which predict that the effective bag radius becomes larger with increasing baryon number [3] . The onset of colour conductivity through tunneling in connection with the EMC effect has been discussed by Pirner [4]. For short times larger quark bags with six or nine or even twelve [5] quarks might be formed in nuclei. If this picture is extrapolated to large baryon numbers it seems to be valuable to study quark bags with arbitrarily large numbers of quarks in it.

The analogy to solid state physics also supports our argument. In a gas of single metal atoms (e.g. sodium vapor) there is no conductivity (as long as the atoms are not ionized) and the electrons are bound to a single nucleus. On the other hand the electrons are forming conduction bands in metallic crystals with a large number of atoms. In this case they are not bound to a single atom. The wavefunction is delocalized through the whole crystal with maxima at the lattice sites.

Conductivity requires at least $10^3 - 10^4$ atoms in the crystal to ge t sufficient level density. If this picture is carried over to the nucleons in a nucleus it might be that bands of colour conductivity are formed in nuclei with large baryon number. The picture is not affected by the substitution of the coulomb potential by a confinement potential $V(r) \propto -\frac{\alpha_s}{r} + \kappa r$. Whereas in the case of metallic crystals the deepest bound levels are not affected very much it might be that in the case of a 'bag crystal' even these levels become bands. In fact our calculations show that this is the case. We point out that conductivity is not bound to a crystaline structure as the existence of liquid metals shows. It is an interesting question if colour conductivity which seems to be not present in usual nuclei up to $A \approx 250$ becomes important for baryon numbers in the order of $A \approx 500$ which can be reached in very heavy ion collisions. This would imply a completely new look on recent experiments at GSI with very heavy ions (U-U, U-Cm etc.) even if colour conductivity takes place only on a very short time scale. Another field of application might be the structure of neutron stars where additionally high densities are involved.

If the picture of overlapping MIT bags is taken seriously and extrapolated to very large baryon numbers it might be that the inhomogenities of 'true vacuum' between the nucleons (i.e. the σ-field in the language of the soliton bag model) become smaller and smaller leading to a cold quark gas in a giant bag. These objects, which we call giant quark nuclei(GQN), are investigated in the next section of this talk. The model of a giant quark bag filled with $3A$ quarks should give at least an estimate of the binding energies involved and how they compare with usual liquid drop formulae. The last part of this talk is then concerned with the more complicated model of a lattice of MIT bags where the quarks can move freely through windows in the surface of overlaping bags.

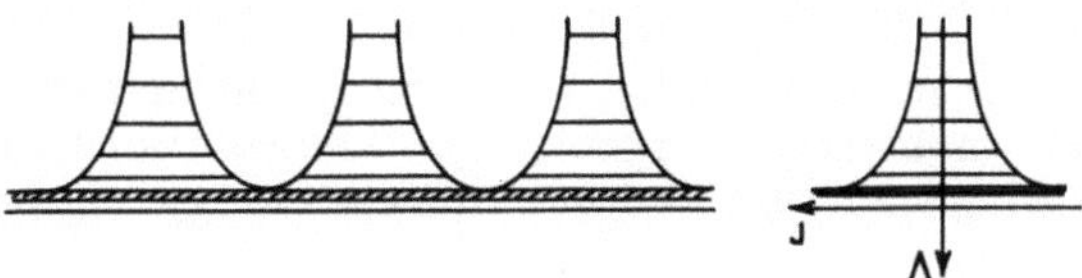

Figure 2: Formation of conduction bands in metallic crystals (schematic).

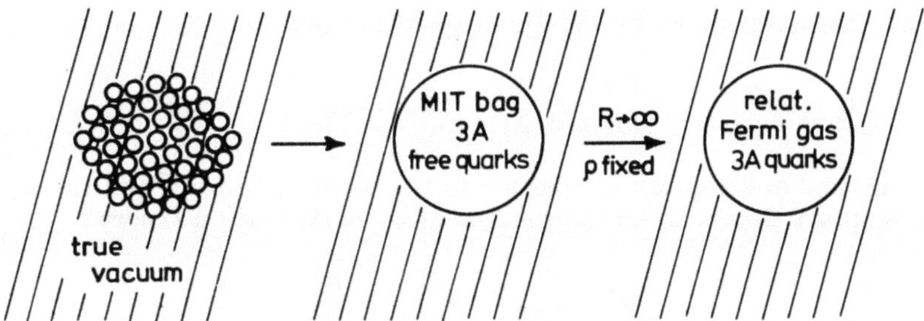

Figure 3: Stages of modeling: bag cluster, giant bag and Fermi gas.

II. MIT BAG AND RELATIVISTIC FERMI GAS FOR 3A QUARKS

We consider now a nucleus with the mass number A, in which quarks are not clustered into nucleons, but move freely throughout the interior of the whole nucleus, i.e. a MIT bag filled with $3A$ quarks. For very large A this model should go over into a free relativistic Fermi gas.

The MIT bag model is defined by the free Dirac equation inside the spherical bag with radius R and two boundary conditions for the quark wavefunctions $q(x)$ at the surface

$$\begin{aligned} &a)\quad (i\partial\!\!\!/ - m)q = 0 \qquad r \le R \\ &b)\quad i n\!\!\!/\, q = q, \qquad r = R \qquad \text{linear boundary condition} \\ &c)\quad -\frac{1}{2}n^{\mu}\partial_{\mu}(\bar{q}q) = B, \qquad r = R \qquad \text{nonlinear b. c.} \end{aligned} \tag{1}$$

where B is the bag constant (vacuum pressure) and n^{μ} is the unit vector normal to the bag surface.

The static spherical solutions of a) are

$$\Psi_{n\kappa r}(\vec{r}) = A_{\kappa n}\begin{pmatrix} j_l(\frac{\omega_{\kappa n} r}{R})\chi_{\kappa r}(\Omega) \\ -\frac{i\omega_{\kappa n} sgn(\kappa)}{\epsilon_{\kappa n}+\mu} j_{\bar{l}}(\frac{\omega_{\kappa n} r}{R})\chi_{-\kappa r}(\Omega) \end{pmatrix} \tag{2}$$

The quantum numbers n, κ, r correspond to the radial nodes of the spherical Bessel functions j_l for angular momentum l_κ , the Dirac quantum number and the projection of the total angular momentum respectively. The solutions can be scaled with the radius of the bag. This leads to the dimensionless energy eigenvalues

$$\epsilon_{\kappa n} = E_{\kappa n}/R = [\omega_{\kappa n}^2 + \mu^2]^{1/2} \tag{3}$$

where $\mu = mR$ is the mass of the quarks and the $\omega_{\kappa n}$ are determined by the l.b.c. b):

$$j_l(\omega_{\kappa n}) = -\frac{\omega_{\kappa n} sgn(\kappa)}{\epsilon_{\kappa n} + \mu} j_{\bar{l}}(\omega_{\kappa n}) \tag{4}$$

The l_κ and $\bar{l}_\kappa$ are determined by the Dirac quantum number

$$l_\kappa = \begin{cases} \kappa & \text{for } \kappa > 0 \\ -\kappa - 1 & \text{for } \kappa < 0 \end{cases} \qquad \bar{l}_\kappa = l_{-\kappa} \quad , \quad j = |\kappa| - \frac{1}{2} \tag{5}$$

Usually the up and down quark masses are in the order of 10 MeV and the $\omega_{\kappa n}(\mu)$ for nonvanishing quark mass can be approximated by perturbation theory

$$\omega_{\kappa n}(\mu) = \omega_{\kappa n}(0) + \frac{1}{2} \frac{\mu}{\omega_{\kappa n}(0) + \kappa} \tag{6}$$

The total energy of a GQN is now given by a sum over the occupied quark levels plus the volume energy

$$E_{tot}(R) = \frac{C}{R} + \frac{4\pi}{3} R^3 B, \qquad C = \sum_i \epsilon_i \tag{7}$$

where i runs over all occupied quantum numbers.

As it is well known the nonlinear boundary condition c) for $j = \frac{1}{2}$ states is equivalent to a minimization of the total energy with respect to the bag radius. This corresponds to a balance of the Fermi pressure of the quarks and the vacuum pressure B. Therefore we replace condition c) by

$$\left.\frac{\partial E_{tot}(R)}{\partial R}\right|_{R_0} = 0 \quad \Longrightarrow \quad R_0 = \left[\frac{C}{4\pi B}\right]^{1/4} \tag{8}$$

whence it follows for the total energy of the GQN in equilibrium

$$E_{tot} = \frac{16\pi}{3} R_0^3 B = \frac{4}{3} (4\pi B)^{1/4} C^{3/4} \tag{9}$$

Two corrections to the total energy (9) can be included easily. The Coulomb energy of the $3A$ quarks can be approximated by the energy of a homogeneously charged sphere

$$E_{Coul} = \frac{C_{Coul}}{R}, \qquad C_{Coul} = \frac{3}{5} \alpha Z^2 \tag{10}$$

where $Z = \frac{2}{3} Z_u - \frac{1}{3} Z_d$ and Z_u, Z_d are the numbers of up and down quarks, respectively. As well a correction due to the spurious center of mass motion of the total wave function can be included via

$$\sum_i \epsilon_i \longrightarrow \left[\left(\sum_i \epsilon_i \right)^2 - \sum_i \omega_i^2 \right]^{1/2} \tag{11}$$

because the total mass of the GQN can be approximated by

$$M = [E_{kin}^2 - < \vec{p}^2 >]^{1/2} \approx \left[E_{kin}^2 - \sum_i < \vec{p}_i^2 >\right]^{1/2}$$

The two corrections result in a replacement of C by

$$C = \left[(\sum_i \epsilon_i)^2 - \sum_i \omega_i^2\right]^{1/2} + C_{Coul} \tag{12}$$

and formula (9) remains unaffected.

Not included are corrections due to colour magnetic or Colour electric interactions. This problem is discussed to some extent in the talk of M. Grabiak (see this proceedings).

For large A the results can be approximated very well by a relativistic Fermi gas (massless quarks). In this case the number of quarks of one flavour is given by

$$N = \frac{g}{(2\pi)^3} \int d^3x \int d^3k, \quad g = 6 = 2_{spin} \times 3_{colour} \tag{13}$$

$$N = g\frac{Vk_f^3}{6\pi^2} = \frac{4}{3\pi}(k_f R)^3 \tag{14}$$

k_f is the Fermi momentum and $V = \frac{4\pi}{3}R^3$ the volume of the bag. From (12) the dimensionless Fermi energy $\omega_f = k_f R$ for massless quarks can be determined

$$\omega_f = \left(\frac{3\pi}{4}N\right)^{1/3} \tag{15}$$

Figure 4 shows how this Fermi energy compares to that one derived from the occupation of cavity modes in a bag (one flavour, massless quarks).

As it is expected the Fermi gas reproduces the smooth dependence of ω_f on the number of quarks and differences arise from shell effects.

In the same way the total kinetic energy of the massless quarks can be expressed by the Fermi energy ω_f

$$E_{kin} = \frac{g}{(2\pi)^3} \int d^3x \int d^3|\vec{k}| = g\frac{Vk_f^4}{8\pi^2} = \frac{1}{\pi}(R^3 k_f^4) = \frac{1}{\pi}\frac{\omega_f^4}{R} \tag{16}$$

If ω_f is eliminated by (15) equation (6) reads

$$C = E_{kin}R = \frac{1}{\pi}\left(\frac{3\pi}{4}N\right)^{4/3} \tag{17}$$

For two flavours the sum of C_u and C_d has to be taken instead. The total energy of a sphere with radius R filled with a Fermi gas of massless up and down quarks

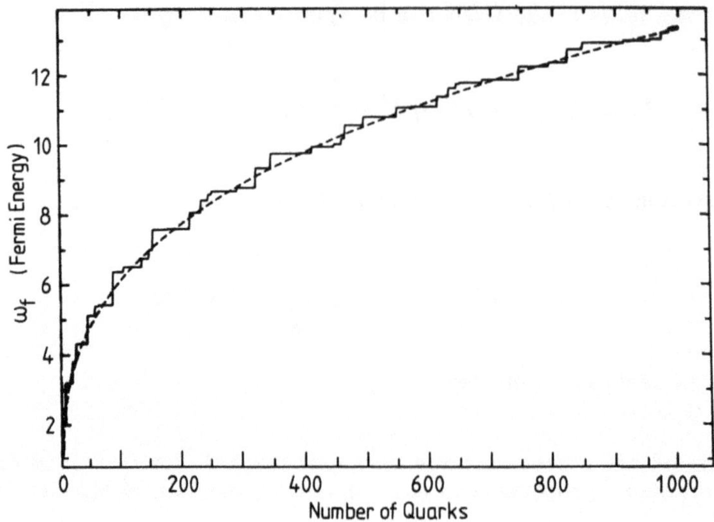

Figure 4: Fermi energy in a relativistic Fermi gas (dashed curve) and in a spherical bag (solid curve).

has now the same form as (7) and formulae (8) and (9) hold as well. Of course the Coulomb correction (10) can be included in the Fermi gas case in the same way as for the MIT bag. If the total energy per baryon is calculated for symmetric nuclear matter $Z_u + Z_d(Z = A/2)$ in the Fermi gas model the result is

$$\frac{E}{A}(\text{Fermi gas}) = \frac{3\pi}{2}\left(\frac{2}{\pi}\right)^{3/4}(4\pi B)^{1/4} \approx 6.323B^{1/4} \tag{18}$$

whereas the liquid drop model yields in this case

$$\frac{E}{A}(LDM) = -15MeV + 938MeV = 923MeV \tag{19}$$

Surprisingly enough the formulae (18), (19) give the same value just for a bag constant $B^{1/4} = 145.9MeV$!

In figure 5 the charge number Z with minimal total energy for a given baryon number A is displayed.

If the Coulomb energy is neglected it is $Z = A/2$, whereas the inclusion of the Coulomb energy leads to a valley of energetically favoured GQN analogous to the valley of stability in the LDM. The staggering represents shell effects while the smooth behaviour is well approximated by the Fermi gas model. The GQN seem to have considerable more neutrons than ordinary nuclei do have but this might be a precipitate conclusion.

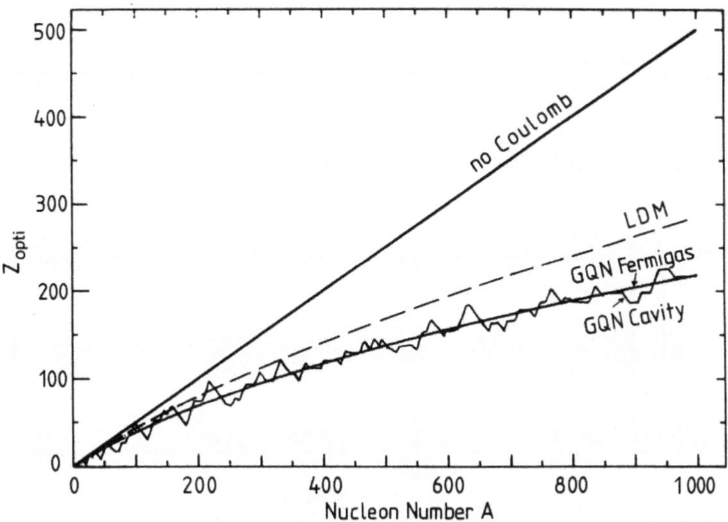

Figure 5: The optimal charge number Z in various models.

It is well known that the symmetry energy in the liquid drop formulae calculated form the Fermi gas model is too small compared to experimental fits and the same might be true for the GQN. In the case of the LDM the additional binding energy of a neutron and a proton with parallel spins is missing and an analogous energy might exist for up and down quarks as well.

Figure 6 shows binding energies

$$B.E. = -\left(\frac{M_{total}}{A} - m_{nucl}\right), \qquad m_{nucl} = 938MeV \tag{20}$$

of the different models (Fermi gas, GQN, LDM) for massless quarks ($m_u, m_d = 0$) and one curve for GQN with down quark mass $m_d = 5MeV$. The bag constant has been chosen to be $B^{1/4} = 145MeV$. Clearly the results are strongly depending on the value of B, but it is very interesting that the binding energy in the liquid drop model has an intersection with the corresponding GQN energies. This means it is possible that up to a certain mass number A_0 usual nuclei (quarks clustered into nucleons) are favorable, but that for very heavy systems the GQN phase is preferred. The transition point A_0 depends very strongly on the bag constant. If $B^{1/4}$ becomes larger than the critical value 145.9 MeV no intersection exists and all GQN and Fermi gas curves move below the LDM. The value of $B^{1/4} = 145MeV$ corresponds to a transition point $A \approx 200$. Of course no definite conclusions can be drawn from this crude model. At least the colour magnetic interaction has to be estimated which is necessary for a correct description of the properties of the nucleon.

The model clearly exhibits the very interesting feature that with increasing mass number A something drastical might happen to the structure of nuclei.

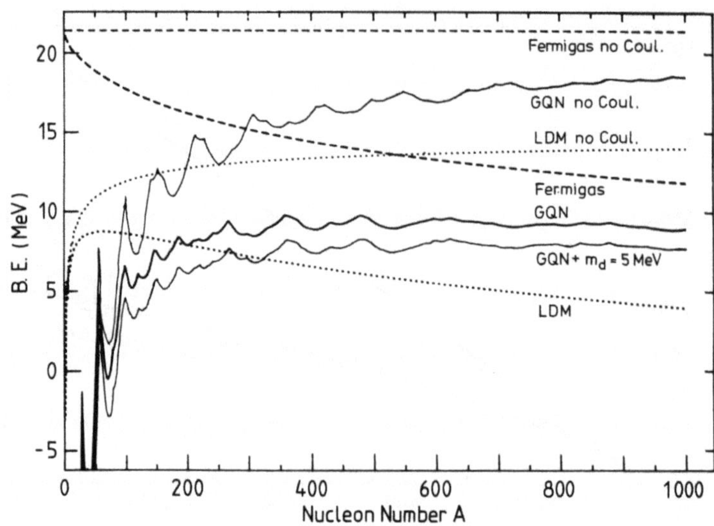

Figure 6: The binding energy as a function of nucleon number for the discussed models

III. THE MIT BAG CRYSTAL MODEL

On the previous pages a deconfinement model for giant nuclei assuming a giant spherical MIT bag has been discussed. The following considerations present a model with a less degree of deconfinement. In this approach infinite nuclear matter is modelled by a crystal of partially overlapping spherical MIT bags representing nucleons [6]. The crystal ansatz has been employed by several authors when calculating properties of nuclear matter using models of strongly interacting matter such as skyrmions and soliton bags. Most authors use the Wigner-Seitz approximation which gives only the ground state of the energy band at pseudomomentum $k = 0$. We do not use this approximation but calculate the energy band of a simple cubic crystal taking the anisotropy of the lattice into account. A somewhat similar work has been done by Achtzehnter [7] about the soliton lattice.

Let us introduce our model in greater detail now. We assume that quarks move freely in nuclear matter at normal nuclear density already. Nuclear matter is composed of spherical MIT bags that overlap partially, thus forming a multiconnected bag. The bag centers are localised on the lattice sites of the sc lattice. Massless quarks move inside this giant bag. This picture is similar to a solid, however in a solid the electrons are influenced by the ion core's charge distribution whereas quarks are entirely free inside of a MIT bag. The regions of true QCD vacuum between the bags constitute a lattice that stabilizes the crystal. The model resembles the muffin-tin approximation of solid state physics but differs from it by the infinitely high potential walls.

To calculate the quark energy band the infinite crystal is reduced to a single nucleon bag with periodic boundary conditions. Figure 7 gives an impression of a bag in this model. Because of the cubic symmetry it has six circular windows connecting it to its neighbours. In the sc lattice the windows are centered on the coordinate axes and are perpendicular to them. On the window planes the quarks wavefunctions must

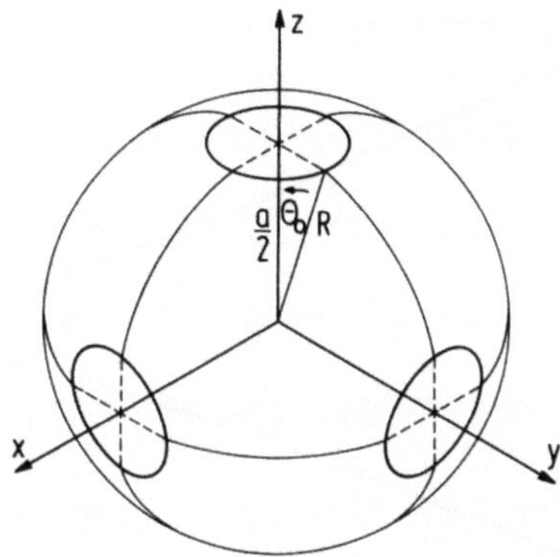

Fig. 7: The shape of a single bag with overlap in the crystal model.

satisfy the periodic boundary condition containing the translational symmetry of a perfect infinite crystal

$$\psi_{+j} = \exp(ik_j a)\psi_{-j} \tag{1}$$

where j denotes the coordinate axis label, $\vec{k}$ is the pseudomomentum vector and a is the lattice constant.

On the remaining bag surface the linear MIT boundary condition is applied (eq. (1b) of the previous contribution (II) decomposes into two equivalent parts):

$$\psi_u = -i(\vec{\sigma}\cdot\vec{n})\psi_d \quad , \tag{2}$$

ψ_u and ψ_d being the upper and lower Pauli spinor component of the quark wavefunctions.

The general form of the massless quark Bloch wavefunction with energy E is given in spherical polar coordinates by

$$\psi_{\vec{k},E}(r,\theta,\phi) = \sum_{\kappa\mu} a_{\kappa\mu}(\vec{k},E)\Phi^E_{\kappa\mu}(r,\theta,\phi) \tag{3}$$

with the Dirac spinor (eq. (2) of (II) for massless particles)

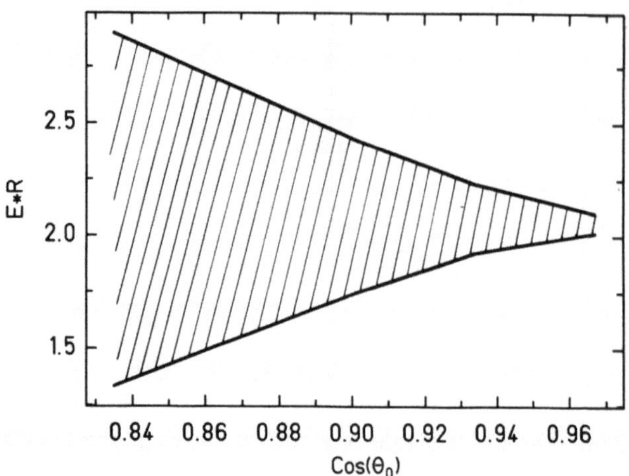

Fig. 8: The energy band of quarks as a function of overlap.

$$\Phi^E_{\kappa\mu}(r,\theta,\phi) = A^E_\kappa \begin{pmatrix} j_{l_\kappa}(Er)\chi_{\kappa\mu}(\theta,\phi) \\ i sign(\kappa E) j_{l_{-\kappa}}(Er)\chi_{-\kappa\mu}(\theta,\phi) \end{pmatrix} \quad , \qquad (4)$$

where A^E_κ is a normalization constant, $j_l(x)$ is a spherical Bessel function and $\chi_{\kappa\mu}$ a spinor spherical harmonic.

Inserting eq. (3) into eqs. (1) and (2) and denoting the results by $G_{\kappa\mu}(\vec{k}, E; \theta, \phi)$, we get an equation for the coefficients $a_{\kappa\mu}$:

$$\sum_{\kappa\mu} G_{\kappa\mu}(\vec{k}, E; \theta, \phi) a_{\kappa\mu}(\vec{k}, E) = 0. \qquad (5)$$

The boundary condition that applies on a boundary point is selected by the polar angle (θ, ϕ). We transform eq. (5) into a matrix eigenvalue problem by multiplying it by its Hermitian conjugate and integrating over (θ, ϕ). The energy at which the matrix is singular is the solution for given pseudomomentum $\vec{k}$. By this procedure we lose some information as only the overall fullfillment of the conditions integrated over the surface is investigated.

In order to obtain the energy band we calculate E for 4 different values of $\vec{k}$ including the center of the Brillouin zone ($k = 0$) and its corner ($\vec{k} = (1, 1, 1)$). The four values of E determine the first Fourier coefficients of the band shape that will be needed to calculate the kinetic energy of the quarks.

The behaviour of the resulting energy band as a function of overlap, expressed

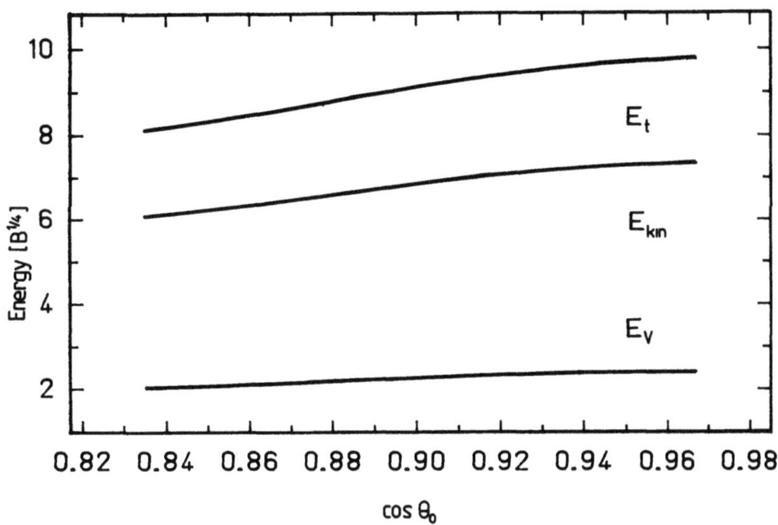

Fig. 9: The total energy and its contributions as a function of overlap.

in terms of the angle θ_0 explained in fig. 7, is shown in fig. 8. The lower limit of the band is its edge at $\vec{k} = \vec{0}$ and the upper limit its corner at $\vec{p} = \vec{1}$. As $\theta_0 \to 0$, the band shrinks into the MIT level $E = 2.04$. With increasing overlap, the upper limit rises above the closed bag value whereas the lower limit sinks below it. The total kinetic energy is obtained by folding the band with the Fermi distribution. The band is filled up to the Fermi energy which is determined by the condition that one nucleon bag contains three quarks. To this energy we have to add the volume energy $E_V = BV$ and minimise the sum with respect to the lattice constant a to get a stable configuration. The result is shown in fig. 9 in units of the bag constant $B^{1/4}$.

Clearly, the infinite crystal favours the free quark gas. This result may be modified quantitatively by the inclusion of the Casimir energy which, unfortunately, cannot be calculated in the crystal geometry. The Coulomb energy has been neglected as for individual or quasi-individual nucleons it is small compared to other contributions, e.g. the N-Δ-splitting. The recoil energy due to the spurious center of mass motion is zero in the case of infinitely extended matter. As in (II) the colour magnetic interaction has been neglected; thus our model describes a state of matter in which N and Δ are degenerate. In any case the quarks are expected to float around freely in infinite nuclear matter.

REFERENCES

[1] A. Chodos, R.L. Jaffe, K. Johnson, C.B. Thorn and V.F. Weisskopf, Phys. Rev. D9 (1974), 3471

[2] A. Chodos, R.L. Jaffe, K. Johnson and C.B. Thorn, Phys. Rev. D10 (1974), 2599

[3] A.W. Hendry et al., Phys. Lett. 136B (1984) 433

[4] H.J. Pirner, Deep inelastic lepton nucleus scattering, International review of nuclear physics vol. 3

[5] H. Faissner and B.R. Kim, Phys. Lett. 130B (1983) 321

[6] Q. Zhang, Ch. Derreth, A. Schäfer, W. Greiner, J. Phys. G 12 (1986) L19

[7] J. Achtzehnter, Diploma thesis, 1985 Universität Giessen

VACUUM VIBRATIONS

R.H. Lemmer* and Walter Greiner

Institut für Theoretische Physik
Johann Wolfgang Goethe-Universität
D-6000 Frankfurt/Main, Germany

ABSTRACT

The possibility of vacuum vibrations of the QED vacuum in the presence of the strong electric field of a heavy ion is examined.

INTRODUCTION

We examine the possibility of collective vacuum vibrations in the presence of the strong electric field of a heavy ion of charge $Z \lesssim Z_c$, that is just subcritical with respect to the spontaneous breakdown of the QED vacuum. The electron levels in such Coulomb fields lie close to the negative energy continuum of the Dirac single particle Hamiltonian. If these levels are rendered empty by collisional ionization, then electron excitation into them from the filled negative energy electron sea allows the system to undergo density oscillations similar to those encountered in the closed outer shells of heavy atoms[1].

DENSITY OSCILLATION IN HEAVY ATOMS

In order to obtain insight into the physical description of such vacuum vibrations, let us first discuss the well known case of collective vibrations of the outer shells in heavy atoms. A transparent description of such collective behaviour is provided by time dependent Hartree-Fock (TDFA) theory. For an arbitrary basis set of single particle states,which we denote by x, x' etc., the time-dependent Hamiltonian in the presence of an external yield W(t) is given as a functional of the single particle density operator $\rho(t)$ by

$$\langle x|H\,[\rho]\,|x'\rangle = \langle x|T+V|x'\rangle + \int dy\,dy'\langle xy|v|x'y'\rangle\langle y'|\rho|y\rangle + \langle x|W(t)|x'\rangle \tag{1}$$

Here V is the nuclear Coulomb potential, and v the electron-electron interaction. Then $\rho(t)$ obeys the equation of motion

* On leave from the Physics Department and Wits-CSIR Schonland Research Center for Nuclear Science, University of the Witwatersrand, Johannesburg

$$[H[\rho],\ \rho] = i\frac{\partial}{\partial t}\rho \tag{2}$$

Two situations can arise: (i) W=o, $\rho=\rho_o$ a constant in time. Then the right hand side of eq. (2) vanishes and ρ_o is determined self-consistently by the condition

$$[H[\rho_o],\rho_o] = o \tag{3}$$

that is completely equivalent to time-independent Hartree-Fock theory, $H\ [\rho_o]|\alpha\rangle = E_\alpha|\alpha\rangle$. (ii) $W(t) \neq o$, $\rho(t) \neq$ constant due to the first instance to the presence of the driving field. However, in view of the ρ-dependence on the right hand side of eq. (1), self-sustaining oscillations in the absence of W(t) are also possible. These are the collective modes of vibration. To study them we follow the standard wisdom and write $\rho(t) = \rho_o+\rho_1(t)$ where $\rho_1(t)$ will be taken as small. Working to first order in the amplitudes $\rho_1(t)$ in the basis set of Hartree-Fock states $|\alpha\rangle$ as given above, one finds the standard random phase approximation (RPA) for the matrix elements of $\rho_1(t)$,

$$[i\frac{\partial}{\partial t} - E_\alpha + E_{\alpha'}]\langle\alpha|\rho_1(t)|\alpha'\rangle =$$

$$= [n_{\alpha'}-n_\alpha]\ [\sum_{\beta\beta'} \langle\alpha\beta|v|\alpha'\beta'\rangle\ \langle\beta'|\rho_1(t)|\beta\rangle$$

$$+ \langle\alpha|W(t)|\alpha'\rangle\ ] \tag{4}$$

Here the occupation numbers $n_\alpha = 1$ or o according as the state $|\alpha\rangle$ is full or empty.

We now examine how $\rho_1(t)$ responds to a periodic external field

$$W(t) = V(x)\ e^{-i\omega t} + h.c \tag{5}$$

in eq. (4), noting that the response also includes a change in the electron-electron interaction as described by the first term under the sum. It is this term that makes free collective oscillation of the density possible in the absence of the driving field.

In the presence of the periodic field (5) the solutions of eq. (4) are also periodic. If we set

$$\langle m|\rho_1(t)|i\rangle = x_{mi}\ e^{-i\omega t} + y_{mi}\ e^{+i\omega t} \tag{6}$$

after noting that only unoccupied (m) and occupied (i) states are coupled by the right hand side of eq. (5) , then the usual RPA equations result. We will however not use the RPA equations in this usual form because of the following difficulty. In contrast to nuclei, important contributions to the solution for ρ_1 come from unoccupied states m that lie in the continuum of the Hartree-Fock spectrum. It is therefore useful to consider instead the matrix elements of $\rho_1(t)$ in the "mixed" representation

$$\langle x|\rho_1(t)|i\rangle = \sum_m \psi_m(x)\ \langle m|\rho_1(t)|i\rangle$$

$$= \eta_{i+}(x)\ e^{-i\omega t} + \eta_{i-}(x)\ e^{+i\omega t} \tag{7}$$

where the sum is over unoccupied states only. Thus the amplitudes $\eta_{i\pm}(x)$

must be orthogonal to all Hartree-Fock orbitals $\psi_m(x)$:

$$\int dx\ \eta^{+}_{i\pm}(x)\ \psi_m(x) = 0 \tag{8}$$

The unoccupied states are now described in a coordinate representation, leading to coupled integro-differential equations for the amplitudes $\eta_{i\pm}(x)$. One finds [2]

$$(\pm\omega - H + E_i)\ \eta_{i\pm}(x) = V_{\pm}(x)\psi_i(x) + \sum_j \int dy \langle xj|v|iy\rangle \eta_{j\pm}(y) + \sum_j \int dy \langle xy|v|ij\rangle\ \eta_{j\mp}(y) - \sum_j \lambda_{ij}\psi_j(x) \tag{9}$$

In these equations H is the Hartree-Fock Hamiltonian, including the nuclear Coulomb Fields $\psi_i(x)$ are occupied Hartree-Fock orbitals, v is the two-electron interaction, and the λ_{ij} are Lagrange multipliers that enforce condition (8). When the $V_{\pm}(x)$ are set to zero one obtains the eigenmodes ω_λ of the system from

$$(\pm\ \omega_\lambda - H + E_i)\eta_{\lambda i\pm}(x) = \text{RHS of eq.(9) with } V_{\pm} = 0 \tag{10}$$

Notice that the eigenmode spectrum ω_λ can be both discrete and continuous.

VACUUM VIBRATIONS

W. Johnson et al[2] have used a relativistic generalization of eq.(9) to study the photo-absorption in heavy atoms with impressive success. Here we wish to apply the relativistic version of eq.(9) in yet another way to study vacuum vibrations in the nearly critical field of a heavy atom[3]. In order to do so, we assume that (i) the nucleus of charge $Z<Z_c$, where $Z_c \simeq 170$ is the critical charge, is completely ionized. (ii) all self-energy effects are neglected (iii) the electron-electron interaction is approximated by a static Coulomb field. We then compute the "photo-absorption" of the vacuum by considering interacting electron-electron hole (i.e. positron) excitations out of the negative energy sea.

The calculation has been carried out in the dipole approximation for which the matrix elements of the interaction V(x) of the electrons with the photon field (of polarization $\vec{\varepsilon}$) read

$$V_{\lambda o} = -i\sqrt{2\pi\omega}\,(\vec{D}_{\lambda o}\cdot\vec{\varepsilon}) \tag{11}$$

where the dipole moment matrix is

$$\vec{D}_{\lambda o} = \sum_i \int dx[\ \eta_{\lambda i_+}(x)\ \vec{d}\ \psi_i(x) + \psi^{+}_i(x)\vec{d}\eta_{\lambda i-}(x)\ ] \tag{12}$$

One very important feature of the relativistic RPA is that it renders the calculation gauge invariant[2]. The expressions above have therefore been given in the very convenient length gauge[4]. A knowledge of $D_{\lambda o}$ leads immediately to the photo-absorption cross section with the ejection of a positron,

$$\phi = 4\pi^2\omega \sum_\lambda |\vec{D}_{\lambda o}\cdot\vec{\varepsilon}|^2_{\omega_\lambda=\omega} \tag{13}$$

This cross-section is shown in Fig. 1 for two values of Z, viz 160 and 169. For comparison the pair production cross section, as adapted from

Heitler[5], is also shown. These calculations have been carried out using the methods described in ref.2, together with the assumption of a pure Coulomb form for the electron interaction, and a Fermi shape of half-width t = 2.5 fm and radius R = 10.39 fm for the nuclear charge distribution. As can be seen in Fig. 1, the cross section increases drastically in magnitude as the lowest discrete $1s_{1/2}$ level approaches the lower continuum when Z = 169. At this value of Z the cross section exhibits a broad peak (of width ~ 600 keV) centered at about 1.3 MeV ~ 2.5 m, where m is the electron mass. One can understand the location of such peaks very roughly by observing that the bound-to-continum interaction matrix elements are enhanced for those continuum wave functions that carry wave numbers $k \sim Z\alpha m$ reciprocal to the localization in space of the bound states. For $Z\alpha \sim 1$, this places the peak cross section near $\omega = (k^2+m^2)^{1/2} = 2.2m$.

One can anticipate a similar increase in positron production rates during the collision of two heavy ions. For that case one also observes one (or more) peak structures on a background of well-understood dynamical positron production processes[6]. The process discussed above will also contribute to the dynamical positron production background but the resulting structure is too wide to be considered as a candidate for the observed positron peaks in heavy-ion collisions.

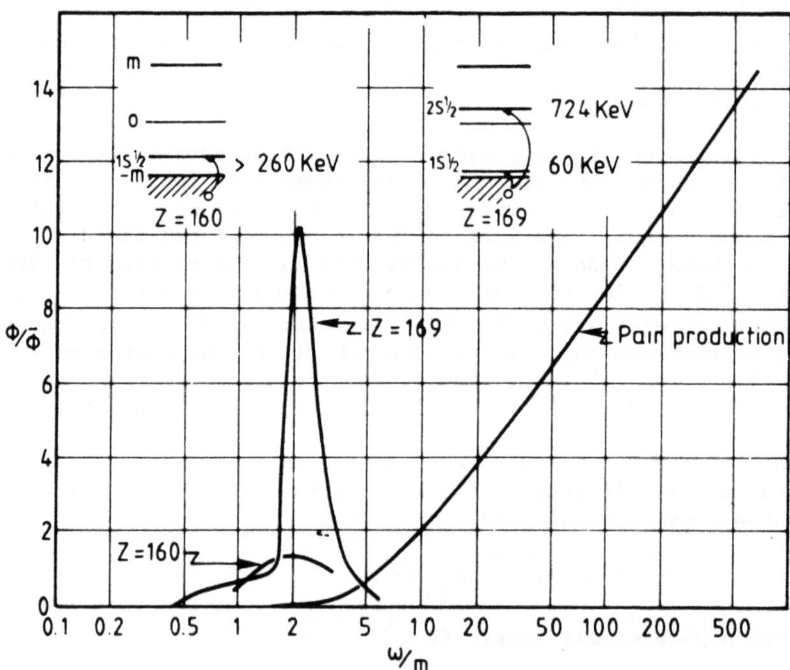

Fig. 1. The calculated photoproduction cross section for positrons in the field of a heavy ion Z, in units of $\bar{\phi} = \alpha Z^2 r_e^2$ where r_e is the classical radius and α the fine structure constant. The pair cross section, as calculated in ref.5, is also shown for comparison. The cross sections near the two thresholds (0.12m for Z = 160, 051m for Z = 169) are too small to be plotted on the scale used. The insets show the two classes of electron-positron excitations that were taken into account for Z = 160 and 169. Both calculations assume a Fermi shape for the nuclear charge density distribution, with half-width and radius parameters[6] t = 2.5 fm and R = 10.39 fm.

REFERENCES

1. G. Wendin, Phys. Rev. B6:42 (1972)
2. W. R. Johnson, C. D. Lin, K. T. Cheng, and C. M. Lee, Phys. Scr. 21: 409 (1980)
3. R. H. Lemmer and W. Greiner, Phys. Lett 162B:247 (1985)
4. V. B. Berestetskii, E. M. Lifshitz, and L. P. Pitaevskii, Quantum-Electrodynamics, Course of Theoretical Physics, Vol. 4, 2nd Ed (Pergamon, New York, 1982)
5. W. Heitler, The Quantum Theory of Radiation, 3rd Ed (Oxford U.P., London, 1954) p. 262
6. J. Rafelski, L. P. Fulcher, and A. Klein, Phys. Rep. 38:227 (1978)

QUANTUM MECHANICAL TREATMENT OF HEAVY-ION COLLISIONS

S. Schramm, J. Reinhardt, U. Müller, B. Müller and W. Greiner

Institut für Theoretische Physik der Johann Wolfgang Goethe-Universität, Postfach 111932, D-6000 Frankfurt, West Germany

A quantum mechanical description of nuclear motion in heavy-ion collisions is developed. Positron spectra of supercritical collision systems are calculated exhibiting one or more line structures if resonances of the nuclear motion are excited.

INTRODUCTION

In the following we develop a theoretical framework to calculate electron and positron spectra in heavy-ion collisions. Especially we have a look at time-delayed scattering presumably occuring at energies close to the Coulomb barrier. The deeper interest in studying electron and positron excitations is the desire to contribute to the explanation of sharp line structures in the positron spectra. These line structures were measured by the ORANGE[1] and EPOS[2] group at GSI in the last years. Here we will look at the possibility that e^+-lines are created by atomic excitation mechanisms during the collision.

One possible mechanism that can produce positron lines is the effect of spontaneous positron production. This effect was predicted for very strong electrical fields nearly twenty years ago[3]. The origin of that effect is shown in figure 1. During the collision of two heavy ions the energies of the bound states of the combined system of target and projectile depend strongly on the varying internuclear distance R. For superheavy collision systems with a combined charge $Z_T \geq 173$ the deepest bound molecular state dives into the lower continuum at some critical internuclear distance R_{cr}. That supercritical state then is an unstable state in the continuum of positrons. Therefore a hole in that state will decay within a time of $\approx 10^{-19}$s becoming a free spontaneously emitted positron which can be detected. For this mechanism to work, however, the "dived" state has to stay supercritical for a period of time comparable to the spontaneous decay time. Unfortunately the time scale of usual Rutherford scattering gives a period of supercriticality of the order of just 10^{-21}s. That difference of two orders in time would make spontaneous positron production unobservable due to kinematical broadening and the background of positrons produced by dynamical mechanisms.

To re-establish the feasability of observing spontaneous positrons one

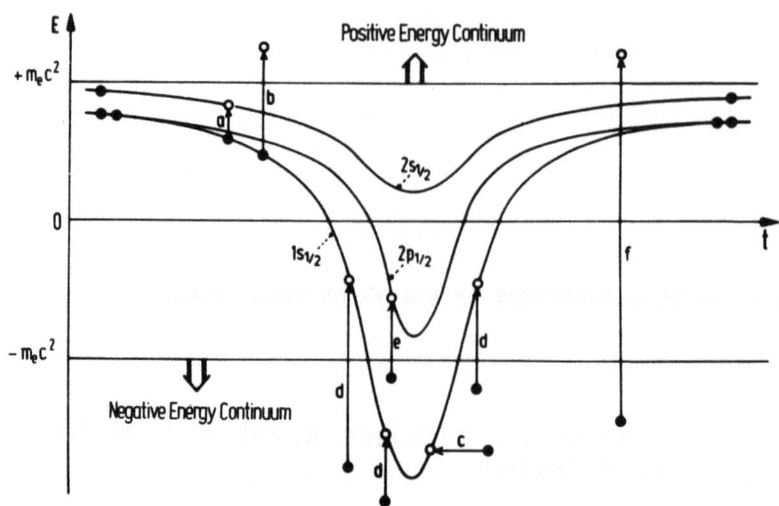

Fig.1: Time-evolution of the energy of bound electron states during a supercritical heavy-ion collision with possible creation of spontaneous positrons (process c).

may think of a delay mechanism holding the two nuclei together for some time to increase spontaneous positron production[4].

Calculations[5] of the scattering potential of various heavy-ion systems show pocket-like structures in the potential at energies near to the Coulomb barrier. These pockets support quasi-bound (resonant) states. If the nuclei are getting trapped in one of these states inside the pocket a quasi-stable nuclear molecule is formed. Due the large lifetime of that system ($\geq 10^{-20}$s) spontaneous positron production is enhanced considerably and may be observable as a line structure in the positron spectrum. Although the existence of a potential pocket is quite model dependent[6] in the following we will assume that it exists.

The problem of a theoretical description of the collision then is to include resonance effects of the pocket. One possibility to solve that problem is to calculate electronic excitation probabilities adopting a classical Rutherford trajectory (or some more realistic trajectory) to treat the nuclear motion[7]. Resonance effects may be simulated by introducing a sticking time - or a distribution of sticking times[8] - at the point of closest approach of the nuclei. To overcome the restrictions of this classical ansatz one has to formulate a quantum mechanical formalism to describe the combined electronic and nuclear dynamics as it is done in the following part.

THEORERTICAL FORMULATION

The Schrödinger equation of the combined electronic and nuclear dynamics may be written as (CM frame)

$$\hat{H}\,|\Phi\rangle = [\hat{P}^2/2\mu + V(R) + H_{TCD}(R,r)\ (+\ \ldots)\ ]\ |\Phi\rangle \tag{1}$$

$V(R)$ being the scattering potential of the nuclei supporting one or several pockets. H_{TCD} is the two-centre Dirac hamiltonian of the electrons depending on the internuclear distance R.
This ansatz presumes that nuclear and electronic dynamics take place in different spatial regions. The (...) indicate that in principle one may add additional hamiltonians, e.g. H_{nuc} to treat inner degrees of freedom of the nuclei. In the following we will restrict our investigation to the simple

form (1). We expand the wavefunction Φ with respect to eigenfunctions of the electrons' hamiltonian:

$$|\Phi\rangle \;=\; \Sigma\, F_n(R)/R\; |\phi_n\rangle$$
$$\text{with} \quad \hat{H}_{TCD}\, |\phi_n\rangle \;=\; \varepsilon_n(R)\; |\phi_n\rangle \tag{2}$$

The prefactors F(R) have to be determined. They will show up to be the wave-functions describing the nuclear motion. The calculation of F(R) is done by solving (1) for the angular part of the internuclear co-ordinate R with the help of rotation matrices. Then multiplying with $\phi_m^*(r,R)$ and integrating over the electron's co-ordinate leads to a differential equation for the radial part of F(R):

$$[d^2/dR^2 + 2\mu/\hbar^2(E-V_J-\varepsilon_n)]\; F^J{}_n(R) \;=\; \Sigma\; [G_{nm}\; d\, F^J{}_m(R)/dR + F_{nm}\; F^J{}_m(R)]$$

$$\text{with} \quad G_{nm} = -2\langle n|\partial/\partial R|m\rangle \quad \text{and} \quad F_{nm} = 2\mu/h^2\langle n|\hat{H}_{TCD}|m\rangle \tag{3}$$

The left-hand side of the upper equation describes the motion of the two nuclei moving in a potential given by the scattering potential (which may contain one or more pockets) and the centrifugal barrier depending on the angular momentum of the partial wave. The total energy E is decreased by $\varepsilon(R)$ taking into account that the nuclear motion is E minus the electronic energy. Since nuclear motion and electronic system are dynamically connected transitions between the electronic channels may occur. That can be seen on the right-hand side of (3). Here we look at the most important couplings $\langle|\partial/\partial R|\rangle$ and $\langle|\hat{H}_{TCD}|\rangle$. $\langle|\partial/\partial R|\rangle$ comes in due to the variation of the electronic wavefunctions with varying internuclear distance R. It is usually called dynamical coupling. $\langle|\hat{H}_{TCD}|\rangle$ should vanish due to (2), but for the supercritical case a special non-diagonal basis of electronic wavefunctions is chosen projecting the supercritical state out of the lower continuum[9]. For that case $\langle|\hat{H}_{el}|\rangle$ mediates the creation of spontaneous positrons.
The solution of the system of coupled differential equations (3) may be done by expanding the wavefunction F(R) with respect to solutions of the elastic scattering problem, i.e. by adopting the wavefunctions $\Omega(R)$ which solve:

$$[d^2/dR^2 + 2\mu/\hbar^2(E-V_J-\varepsilon_n)]\; \Omega^J{}_n(R) \;=\; 0 \tag{4}$$

Assuming that we have already calculated the asymptotically in- and outgoing wavefunctions $\Omega^{\pm}(R)$ the complete radial wavefunction F(R) may be expanded by:

$$F^J{}_n(R) \;=\; a^{J+}{}_n(R)\; \Omega^{J+}{}_n(R) \;+\; a^{J-}{}_n(R)\; \Omega^{J-}{}_n(R) \tag{5}$$

Inserting (5) into (3) leads to coupled first-order differential equations for the excitation amplitudes $a^{\pm}{}_n(R)$[10]:

$$\begin{aligned} da^{\pm}{}_m/dR \;=\; \mp\, i/2\; \Sigma\; [&(G_{mn}\; \Omega^{\mp}{}_m\; d/dR(\Omega^+{}_n) + F_{mn}\; \Omega^{\mp}{}_m\; \Omega^+{}_n)\; a^+{}_n \;+ \\ &+ (G_{mn}\; \Omega^{\mp}{}_m\; d/dR(\Omega^-{}_n) + F_{mn}\; \Omega^{\mp}{}_m\; \Omega^-{}_n)\; a^-{}_n\;] \end{aligned} \tag{6}$$

As $\Omega^{\pm}$ are asymptotically in- and outgoing wavefunctions for large R the product terms $\Omega^+\,\Omega^-$ have to be of the form $\approx e^{2ikR}$. The contributions to $a^{\pm}$ proportional to these terms should average to zero in a good approximation. The resulting differential equations for large R have the following form:

$$da^{\pm}/dR \;=\; \mp\, i/2\; \Sigma\; [G_{mn}\; \Omega^{\mp}{}_m\; d/dR(\Omega^{\pm}{}_n) + F_{mn}\; \Omega^{\mp}{}_m\; \Omega^{\pm}{}_n]\; a^{\pm}{}_n \tag{7}$$

$a^{\pm}$ decouple and their asymptotic values can be identified with the initial and final values of the electronic excitation amplitudes.

To solve (6) or (7) the knowledge of $\Omega^{\pm}$ is required. One has to find the two independent solutions of the differential equation (4). The main difficulty of solving (4) lies in the structure of the potential V(R) including one or even several pockets. A closed exact analytical solution of (8) is not possible. We choose a WKB-approximation of the wavefunction. The procedure to construct the regular solution is developed in ref. 11. The resulting wavefunction is of the form

$$\Omega_{reg} = 1/\sqrt{\pi k} \begin{cases} A_{reg} \sin(\int k dr + \Pi/4) & \text{inside} \\ \sin(\int k dr + \Pi/4 + \alpha) & \text{outside} \end{cases} \text{ the pocket} \tag{8}$$

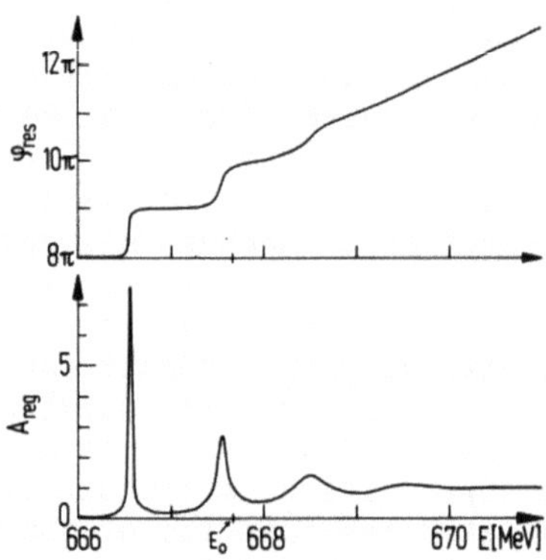

Fig.2: Energy-dependence of the scattering phase shift (upper part) and the amplitude of the regular wavefunction in the pocket (lower part). E_0 marks the energy of the top of the potential barrier.

Apart from the typical WKB-structure there are two significant values A_{reg} and α. Both show resonant behaviour by varying the total energy as it can be seen in fig. 2. Furthermore it can be recognized that the resonances supported by the pocket become very narrow a few MeV below the barrier belonging to the potential pocket. On the other hand the resonant structure is washed out above the barrier so there is only a small effective range of energies contributing to resonant scattering (this argument has to be taken separately for each partial wave, of course).
The irregular solution can be calculated by applying the same method. The asymptotically in- and outgoing wavefunctions needed for the solution of (6) are constructed in the usual way by

$$\Omega^{\pm} = \Omega_{reg} \mp i\, \Omega_{ir} \tag{9}$$

yielding

$$\Omega^{\pm} = 1/\sqrt{\hbar k} \begin{cases} C^{-} \exp[\mp i(\int k dr - \Pi/4)] + C^{+} \exp[\pm i(\int k dr - \Pi/4)] & \text{inside} \\ \exp[\pm i(\int k dr + \Pi/4 + \alpha)] & \text{outside} \end{cases} \tag{10}$$

The whole procedure of calculating solutions of (8) can be extended analytically to handle potentials with an arbitrary number of pockets[12]. For instance assuming two pockets the resulting regular wavefunction is

$$\Omega_{reg} = 1/\sqrt{\hbar k} \begin{cases} A^{(1)}_{reg} \sin(\int k dr + \Pi/4) & \text{, inner pocket} \\ A^{(2)}_{reg} \sin(\int k dr + \Pi/4 + \alpha^{(1)}) & \text{, outer pocket} \\ \sin(\int k dr + \Pi/4 + \alpha^{(2)}) & \text{, outside of pockets} \end{cases} \tag{11}$$

Here there are two amplitudes A and phase shifts α with resonance behaviour to be considered.

The solutions (10) or (11) respectively can be inserted into (6) to yield the complete coupled differential equations for the electronic excitation amplitudes (for the interior of the pocket):

$$\begin{aligned} da^{\pm}_{m}/dR = \pm 1/(8i\sqrt{k_m}) \, \Sigma \, 1/\sqrt{k_n} \{ & [B_m^{\mp} B_n^{\mp *} (F_{mn} - ik_n G_{mn}) \exp\{i(\beta_m - \beta_n)\} + \\ & + B_m^{\pm *} B_n^{\pm} (F_{mn} + ik_n G_{mn}) \exp\{-i(\beta_m - \beta_n)\}] \, a^{\pm}_{n} + \\ & + [B_m^{\mp} B_n^{\pm *} (F_{mn} - ik_n G_{mn}) \exp\{i(\beta_m - \beta_n)\} + \\ & + B_m^{\pm *} B_n^{\mp} (F_{mn} + ik_n G_{mn}) \exp\{-i(\beta_m - \beta_n)\}] \, a^{\mp}_{n} + \\ & + [B_m^{\mp} B_n^{\pm} (F_{mn} + ik_n G_{mn}) \exp\{i(\beta_m + \beta_n)\} + \\ & + B_m^{\pm *} B_n^{\mp *} (F_{mn} - ik_n G_{mn}) \exp\{-i(\beta_m + \beta_n)\}] \, a^{\pm}_{n} + \\ & + [B_m^{\mp} B_n^{\mp} (F_{mn} - ik_n G_{mn}) \exp\{i(\beta_m + \beta_n)\} + \\ & + B_m^{\pm *} B_n^{\pm *} (F_{mn} + ik_n G_{mn}) \exp\{-i(\beta_m + \beta_n)\}] \, a^{\mp}_{n} \} \end{aligned} \tag{12}$$

with $B^{\pm} = A_{reg} \pm A_{ir} \exp(i\tau)$ and $\beta = \int k dr + \Pi/4$.
A_{ir} and τ are the amplitude and phase of the irregular wavefunction in the pocket.
The differential equation outside of the pocket by neglecting fast oscillating terms is given by:

$$da^{\pm}_{m}/dR = \pm 1/(2i\sqrt{k_m}) \, \Sigma_n \, 1/\sqrt{k_n} \, (F_{mn} \pm ik_n G_{mn}) \exp\{\mp i[(\beta_m + \alpha_m) - (\beta_n + \alpha_n)]\} \, a^{\pm}_{n} \tag{13}$$

Performing a few transformations a more compact - but less instructive - system of equations may be derived[12]. We may have a look at two special cases. If there is no resonance in the vicinity of the energy considered, the amplitude A_{reg} is approximately zero and the phase shift is a multiple of Π. The equation (12) inside the pocket vanishes and it remains equation (13) with the resonance phase $\alpha \approx 0$ describing just non-resonant scattering. If the energy E is a few MeV above the barrier there is no resonance structure left as remarked above. Then

$$A_{ir} \approx A_{reg} \approx 1 \quad , \quad \tau \approx 0$$

following

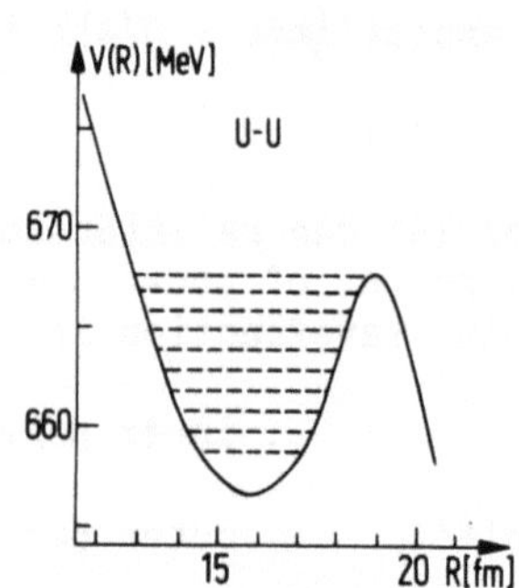

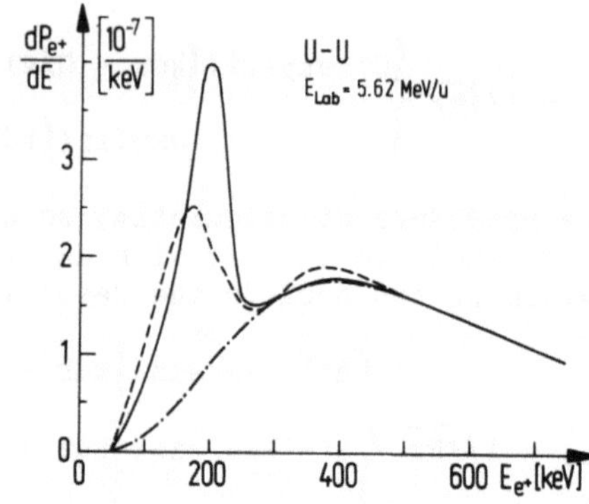

Fig.3 (left): Part of a U-U scattering potential exhibiting a pocket-like structure as used for the calculation of positron spectra. The dashed lines indicate resonant states.

Fig.4 (right): Positron spectra of a U-U collision for an energy of E_{lab}/u = 5.62 MeV. The dash-dotted curve was calculated for a collision without time delay. The spectra were calculated for angular momenta $J=0\hbar$ (dashed lines) and $J=50\hbar$ (full line).

$$da^{\pm}_m/dR = \pm 1/(2i\sqrt{k_m})\ \Sigma_n\ 1/\sqrt{k_m}\ (F_{mn} \pm ik_n G_{mn})\ \exp\{\mp i(\beta_m-\beta_n)\}\ a^{\pm}_n \qquad (14)$$

These equations include non-resonant scattering outside of the pocket and in the region above the pocket. By solving (12) exactly these special cases are of course included automatically.

NUMERICAL RESULTS

The actual numerical calculation has been done in the following way. The electronic excitations outside of the pocket were calculated by applying a classical Rutherford trajectory for the motion of the nuclei. That can be done separately for the incoming and outgoing branch. In the interior of the pocket the coupled equations (12) were solved as a boundary value problem. Taking the resulting excitation amplitudes positron spectra were calculated. To include the energy spread of the ion beam the spectra were averaged over a beam energy interval of $(5\ \text{MeV})_{CM}$. Using a model potential (fig. 3) the resulting e^+-spectrum of a U-U collision for a beam energy of E_{lab}/u = 5.68 MeV is shown in figure 4. In addition to the background of dynamically produced positrons there is a line structure at E=195 keV with a width Γ of ≈80 keV due to spontaneously created positrons. By including in the numerical calculation extra positron channels at higher energies one can see another structure in the positron spectrum as it is shown in figure 5 for E_{e+} = 1200keV. The physical reason for that structure is illustrated in figure 6. A quasimolecule in a higher resonance of the pocket may be deexcited into a lower resonance by transferring energy from the di-nuclear vibration to an electron in the lower continuum. This electron may be excited into the supercritical state. Since the transferred energy is fixed by the energy difference of the resonances in the pocket a line structure due to monoenergetic conversion appears in the positron spectrum. The position of the line is shifted with respect to the spontaneous positron line by the energy difference of the two resonances (here ≈1 MeV).

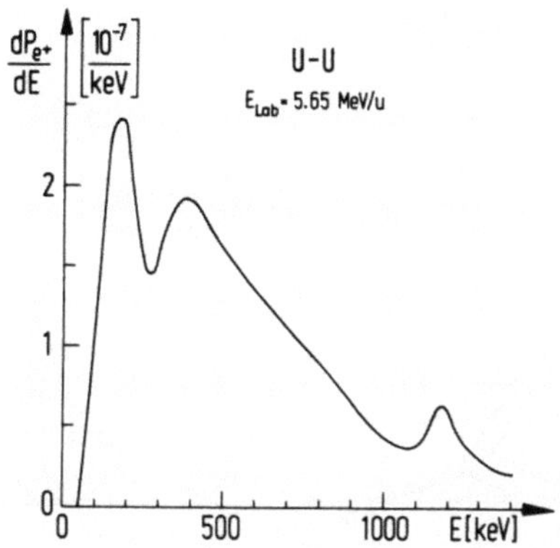

Fig.5: Positron spectrum of a U-U collision showing a second line structure at E = 1200 keV due to a conversion process of the giant nuclear molecule.

Figure 7 illustrates the broadening of the conversion peak. For a fixed sharp beam energy exciting the upper resonance in the pocket (fig. 6, dashed lines) the width of the conversion peak roughly would be given by the narrow width of the lower quasibound state inside the pocket (around 10 keV). For small variation of the total energy the nuclear dynamics tries to stay in the resonance (i.e. at the same energy) so the positron lines are shifted with the total energy as it can be seen in fig. 7. Since both line structures are generated by first exciting the upper resonance in the pocket a variation of the beam energy leads to an averaging over the width of that resonance. Thus

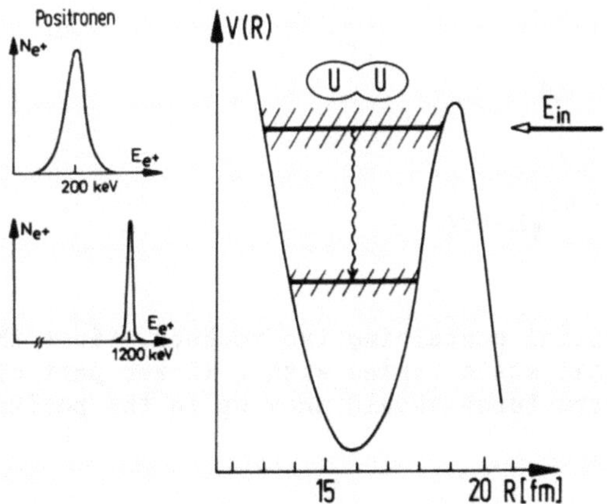

Fig.6: Pocket of the scattering potential. The nuclear quasimolecule is deexcited into a lower resonance, thereby transferring the excitation energy to an electron in the Dirac sea. The resulting positron lines are shown in the left part of the figure.

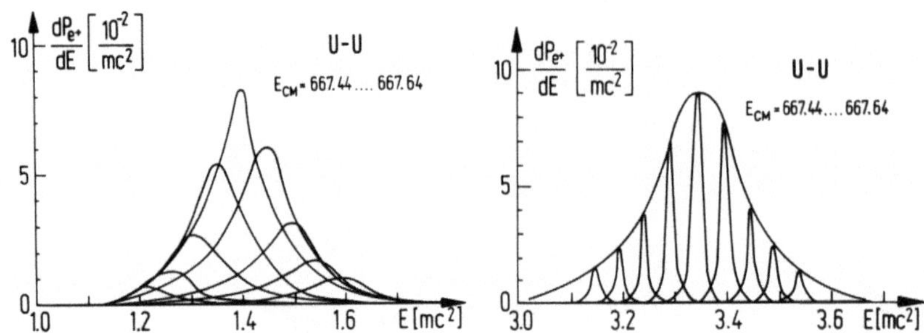

Fig.7: The dependence of position and shape of the spontaneous positron line (left part) and conversion line (right part) upon variation of the beam energy is shown. E_{CM} was varied by 25 keV steps. The lines were calculated by computing only excitations inside the pocket.

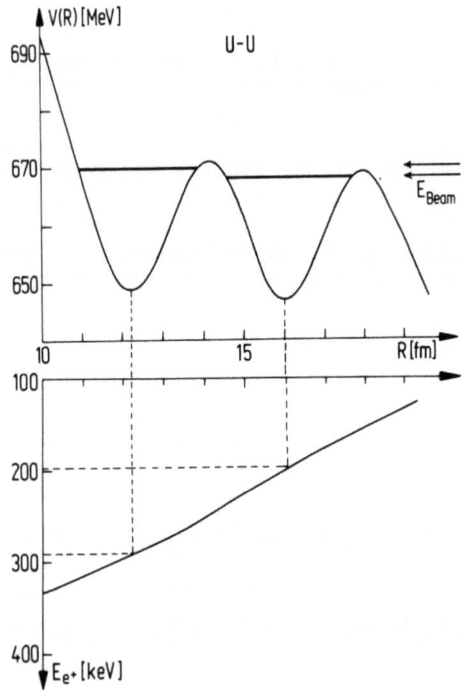

Fig.8: Model potential containing two pockets. Since the energy of the supercritical state varies with R (lower part of the figure), two line structures should show up in the positron spectrum.

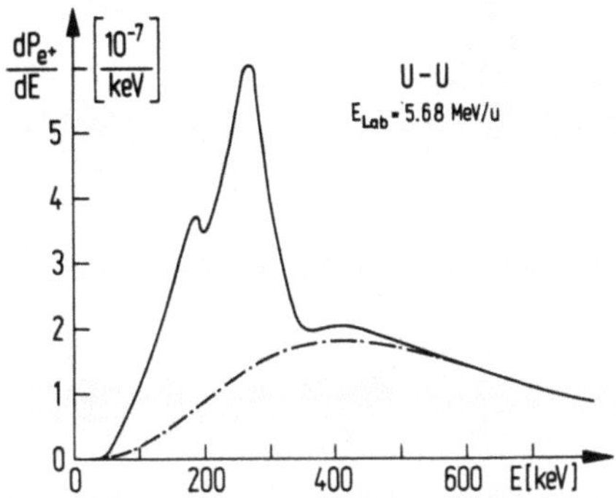

Fig.9: Positron spectrum (full line) of a U-U collision using the potential shown in figure 9. The second spectrum is calculated by neglecting excitations inside the pockets (cf. fig.4).

both lines as seen in fig. 5 have about the same line width of ≈ 100 keV after averaging.

As shown in the theoretical part the discussion directly can be extended to more complex structured potentials. One may speculate that there are more than one pocket-like structures in the scattering potential of superheavy collision systems at the energy of the Coulomb barrier. Without here discussing the implications of these considerations on nuclear physics we just look at the possible influence of multiple pockets on the positron spectra. One quite arbitrarily constructed model potential including two pockets is shown in figure 8. Both pockets support quasibound states (the two states mainly contributing to positron creation are marked by dashed lines). Thus dependent upon which resonances will be excited during scattering two distinct isomeric nuclear quasimolecules can be formed. The effect of isomeric configurations with respect to the positron emission is indicated in the lower part of fig. 8. Since the energy of the supercritical electron state varies strongly with the internuclear distance R, both possible nuclear quasimolecules support a supercritical state at a different energy. Thus the spontaneous positron line induced by the enhanced lifetime of the quasimolecules shows up at different positron energies. The result of a complete calculation is shown in fig. 9. As just discussed there are two positron line structures at E_1=190keV and E_2= 270keV. Although the mainly contributing resonances of both pockets have around the same width $\Gamma \approx$60keV the line at E_1 belonging to the outer pocket is nearly covered by the second line. That is the effect of the increasing coupling matrix elements of spontaneous positron emission for decreasing internuclear distance. Of course the general resulting positron spectrum for a scattering potential including multiple pockets depends on the details of the position, width, and energy of the several potential barriers.
The symmetrized differential cross section of positron creation can be calculated by

$$d^2\sigma/d\Omega dE_{e^+}|_\theta = \langle | \sum_{J,i>F} (2J+1)\, S^J{}_{i\to f} [P_J(\cos\theta) + P_J(-\cos\theta)] |^2 \rangle \qquad (15)$$

'⟨⟩' indicates averaging of the beam energy. $S^J{}_{i\to f}$ is the S-matrix element of

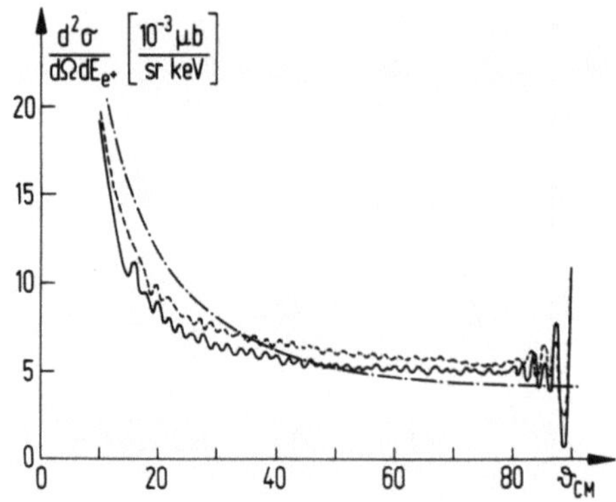

Fig.10: Differential cross section of coincidentally measuring a scattered nucleus an an angle Θ_{CM} and a positron with an energy of 190 keV. Calculations were done by taking account partial waves with a range of angular momenta J = 0-80 (dashed lines) and J = 50-92 (full line). In comparison the dash-dotted graph shows the cross section of an isotropic emission of positrons in the reaction plane.

the transition from electronic channel i to f. F denotes the Fermi level fixed by the chosen initial condition (here F=3) of the collision. Applying that formula with some rough approximations the resulting cross section is shown in fig. 9. The two intervals 0 - 80 $\hbar$ and 50 - 92 $\hbar$ of angular momentum are given by the range of angular momenta contributing to resonance scattering. The overall behaviour of differential cross section is very close to isotropic emission (see figure). The total line cross section is $\sigma \approx 1.6\ \mu b$. Compared to experimental results[13] of around 100 µb a factor of 60 is missing. A possible explanation of the difference may be that in reality the nuclei lose energy during the collision thereby falling into deeper long-lived resonances in the potential pocket. That may increase positron emission considerably.

SUMMARY

A theoretical framework was developed to treat resonant heavy-ion scattering and electron dynamics quantum mechanically. Coupled differential equations for the electronic excitation amplitudes were derived analytically for potentials with one or multiple pocket-like structures. The resulting equations were integrated numerically and positron spectra for U-U collisions at energies near to the Coulomb barrier were calculated. The spectra show line structures due to spontaneous positron emission and mono- energetic pair conversion. The possible existence of various long-lived isomeric nuclear quasimolecules can also lead to various line structures in the positron spectrum as shown above. Severe difficulties remain if one tries to associate these theoretically computed positron lines with the e^+ structures measured at GSI. The electron and positron coincidences as measured by the EPOS group are hardly to be explained by that theoretical framework. Another critical point of explaining the measured positron lines is the Z-dependence of the line position in the various collision systems. In principle that could be cured by assuming a scattering potential varying with the scattering system. There could be multiple pocket-like structures effecting several lines at different energies. Of course the assumed behaviour of the potential would have to be explained by considerations concerning nuclear physics. The possible creation of positron lines in subcritical systems may be explained by the monoenergetic conversion process discussed above, yet further calculations have to confirm that point.

As conclusion it may be stated that these calculations are unlikely to explain all the positron line structures measured so far. On the other hand it is not clear if there is really a single effect generating all the observed lines and if all positron lines have a correlated counterpart in the electron spectrum. Furthermore in experiments planned for the new storage ring SIS to be built at GSI with totally stripped projectiles as $^{92+}$U a strong enhancement of spontaneous positron production can be expected increasing the effects as discussed here considerably.

REFERENCES

1) M. Clemente, E. Berdermann, P. Kienle, H. Tsertos, W. Wagner, C. Kozhuharov, F. Bosch, W. König: Phys. Lett. 137B, 41 (1984)
   H. Tsertos, E. Berdermann, F. Bosch, M. Clemente, P. Kienle, W. König, C. Kozhuharov, W. Wagner: Phys. Lett. 162 B, 372 (1985)
2) J. Schweppe, A. Gruppe, K. Bethge, H. Bokemeyer, T. Cowan, H. Folger, J.S. Greenberg, H. Grein, S. Ito, R. Schule, D. Schwalm, K.E. Stiebing, N. Trautmann, P. Vincent, M. Waldschmidt: Phys. Rev. Lett. 51, 2261 (1983)
   T. Cowan, H. Backe, M. Begemann, K. Bethge, H. Bokemeyer, H. Folger, J.S. Greenberg, H. Grein, A. Gruppe, Y. Kido, M. Klüver, D. Schwalm, J. Schweppe, K.E. Stiebing, N. Trautmann, P. Vincent: Phys. Rev. Lett. 54, 1761 (1985)
3) W. Pieper and W. Greiner: Z. Phys. 218, 126 (1969)
   B. Müller, J. Rafelski, W. Greiner: Z. Phys. 257, 62 and 183 (1972)
   Ya.B. Zeldovich, V.S. Popov: Sov. Phys. Usp. 14, 673 (1972)
4) J. Rafelski, B. Müller, W. Greiner: Z. Phys. A 285, 49 (1978)
5) M. Seiwert, W. Greiner, W.T. Pinkston: J.Phys. G11, L21 (1985)
6) V.E. Oberacker, lecture at this conference
7) U. Müller, G. Soff, Th. de Reus, J. Reinhardt, B. Müller, W.Greiner: Z. Phys. A 313, 263 (1983)
8) J. Reinhardt, B. Müller, U. Müller, W. Greiner, Phys.Rev.A 28, 2558 (1983)
9) J. Reinhardt, B. Müller, W. Greiner, Phys. Rev. A 24, 103 (1981)
10) U. Heinz, B. Müller, W. Greiner: Ann. Phys. 151, 227 (1983)
    U. Heinz, U. Müller, J. Reinhardt, B. Müller, W. Greiner: Ann. Phys. 158, 476 (1984)
11) M.S. Child, 'Molecular Collision Theory', Academic Press, NY 1974, pp. 43-56
12) S. Schramm, J. Reinhardt, U. Müller, B. Müller, W. Greiner: Z. Phys. A 323, 275 (1986)
13) S. Schramm: diploma thesis, Frankfurt (1985)

NUCLEAR CONTACT TIMES IN DISSIPATIVE HEAVY ION COLLSIONS MEASURED VIA δ-RAY SPECTROSCOPY

P.Senger, H.Backe, M.Begemann-Blaich, H.Bokemeyer[2], P.Glässel[1], D.v.Harrach[1], M.Klüver, W.Konen, K.Poppensieker[2], K.Stiebing[3], J.Stroth and K.Wallenwein[1]

Institut für Physik der Universität Mainz
D-6500 Mainz, Federal Republic of Germany

1. II. Physikalisches Institut der Universität Heidelberg
D-6900 Heidelberg, Federal Republic of Germany
2. Gesellschaft für Schwerionenforschung
D-6100 Darmstadt, Federal Republic of Germany
3. Institut für Kernphysik der Universität Frankfurt
D-6000 Frankfurt/Main, Federal Republic of Germany

ABSTRACT

Electron spectra have been measured for elastic and dissipative U + Au collisions at 8.6 MeV/u and analysed within a simple schematic model which describes δ-ray emission in the presence of a nuclear contact time and a total kinetic energy loss ($TKEL$). A nearly linear dependence of the mean nuclear contact time τ and $TKEL$ was found, reaching $\tau = 1.1 * 10^{-21}$ s with a variance $\sigma = \pm 0.4 * 10^{-21}$ s for a $TKEL$ of (400 ± 50) MeV.

INTRODUCTION

This contribution is addressed to the question whether δ-ray spectroscopy can be employed to study nuclear reaction dynamics in dissipative heavy ion collisions. As pointed out by G. Soff [1)] and E. Kankeleit [2)] in their contributions, oscillations are expected in the δ-ray spectra with a period ΔE related to the nuclear contact time T, i.e. the time, for which the radial velocity of the colliding nuclei vanishes. In Fig.1, calculated oscillating δ-ray spectra are shown for discrete nuclear contact times. However, in dissipative nuclear collisions, T is distributed around a mean value, resulting in a damping of the oscillation. In principle, information on the frequency distribution of T, e.g. the mean nuclear contact time τ and the variance σ, can be obtained

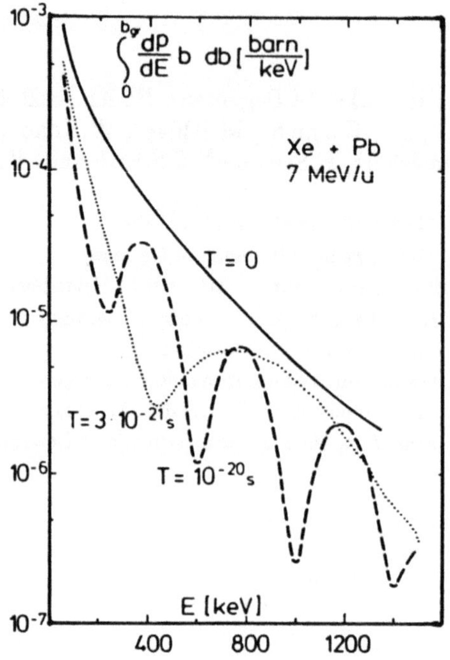

Figure 1: ***Energy differential emission probability for δ-electrons as a function of the electron kinetic energy integrated over all impact parameters $b \leq b_{grazing}$ that lead to a nuclear reaction (taken from ref. 3). T is the nuclear contact time.***

from the measured δ-ray spectra. Furthermore, these quantities can be studied as a function of relevant parameters which characterize the dissipative nuclear reaction: the total kinetic energy loss ($TKEL$), the CM-scattering angle Θ and the mass drift. First intimations of such effects have been observed in the positron and δ-ray spectra taken in coincidence with fission fragments in dissipative U+U and U+Cm collisions[4]. Since that time, the reaction dynamics has been studied systematically by the TORI spectrometer group[5,6] using improved experimental techniques.

We want to report preliminary results of a recent experiment performed at the UNILAC at GSI in Darmstadt. In our new experimental set-up we use a particle detection technique enabling the complete determination of the reaction kinematics with as many as four reaction products in the exit channel. This offers the possibility of studing δ-electrons emitted in deep inelastic collisions of very heavy ions such as U+U in which one or both of the collision partners undergo fission after the first binary reaction stage.

We decided to carry out a first exploratory experiment on the U+Au collision system at an U-beam energy of 8.6 MeV/u. First of all, this is the heaviest system which can be studied without any target problems. Actually, 200 $\mu g/cm^2$ Au-targets, which can easily be prepared, are very stable in a well focused 1p-nA ^{238}U beam. On the other hand , at the beam energy chosen, a large amount of energy can be dissipated in a collision resulting in fission of the uranium projectile. In comparison to binary deep inelastic reactions, these ternary reactions have the advantage that the emission of conversion electrons is considerably reduced, since the conversion coefficients[8)] exhibit a strong Z^3-dependence. From the experimentally determined δ-ray emission probability as a function of Z_u [9)] and from the γ-ray spectra for heavy collision systems [10)], we estimated that in ternary dissipative reactions the conversion electrons should contribute less than 20% to the total electron yield.

As already mentioned, the mean nuclear contact time τ and its variance can be studied as a function of several reaction channels, e.g. also of special contours in the Wilczynski-plot. However, for this conference contribution we restricted the data analysis to the investigation of correlations between τ and $TKEL$.

EXPERIMENT AND DATA ANALYSIS

Our experimental set-up is schematically shown in fig.2. The velocity vectors of up to two heavy reaction products can be measured with each of the two 91×91 cm^2 large area particle-counters [11)]. This is achieved by a position- and a time-of-flight measurement using the UNILAC micro beam pulse as a start signal. The position resolution amounts to $\Delta x = \Delta y \simeq 2$ mm (FWHM). The time-of-flight resolution is limited by the quality of the UNILAC micro bunches being typically 400-500 ps wide. The backward emitted electrons were analysed by a magnetic orange-type β-spectrometer [12)] equipped with an electron plastic-counter combination consisting of 10 independent rings (see Fig.3). This spectrometer accepts a total momentum band $\Delta p/p = 0.2$. The transmission is approximately 0.15. Electrons up to an energy of 2.8 MeV can be detected, limited by the maximum current of the spectrometer power supply of 1600 A.

The dissipative U+Au reaction at 8.6 MeV/u beam energy results in two, three or four particles in the exit channel. However, due to both the high fissility of the uranium-like ejectile and the relatively low fissility of the gold-like recoil, most of the reaction cross section is represented by ternary reactions. Therefore, we will describe in the following only the analysis of these events. We assume that the

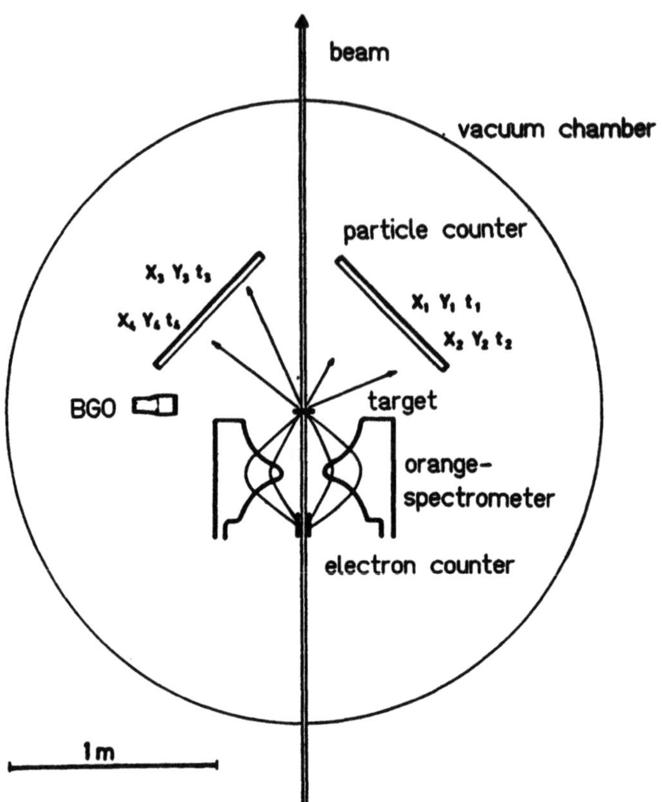

Figure 2: *Experimental set-up*

dissipative U+Au reaction at 8.6 MeV/u proceeds sequentially. That is, after the first binary reaction stage in which the incoming kinetic energy is dissipated into intrinsic excitation and deformation, the uranium like nucleus fissions in a second step at a large distance from the gold-like nucleus with negligible Coulomb final state interactions. This two-step reaction model was found to be valid for dissipative U+U and U+Cm collisions[13].

From the measured velocity vectors $\vec{v}_1$, $\vec{v}_2$ and $\vec{v}_3$ of the reaction products and the conservation laws for momentum

$$\vec{p}_P = M_1\vec{v}_1 + M_2\vec{v}_2 + M_3\vec{v}_3 \tag{1}$$

and energy

$$E_P - TKEL = \frac{(M_1\vec{v}_1 + M_2\vec{v}_2)^2}{2\,(M_1 + M_2)} + \frac{1}{2}M_3\vec{v}_3^2 \tag{2}$$

we can calculate $TKEL$ and the unknown masses M_1, M_2 and M_3. In this notation we assume that the particles 1 and 2 originate from the uranium-like nucleus and particle 3 from the gold-like nucleus. The quantities $\vec{p}_P$, E_P, M_P and M_T represent

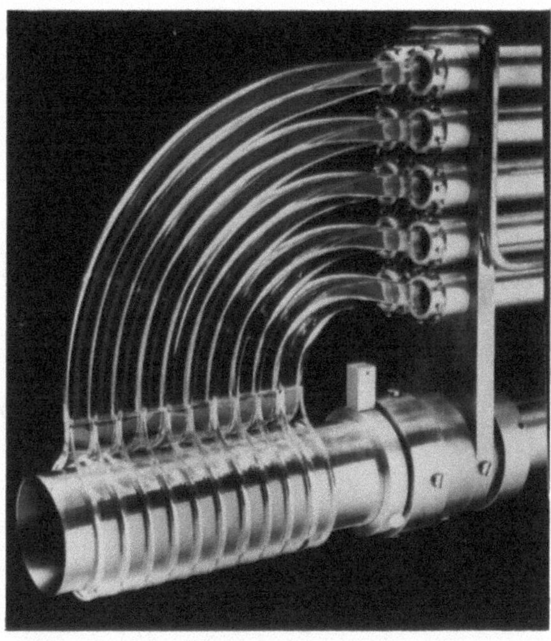

Figure 3: *Plastic-ring counter assembly in the focal region of the orange spectrometer. The NE 104 plastic rings have an inner diameter of 50 mm and are 10 mm wide and 5 mm thick. The plexiglas light guides are coupled to 13 mm diameter photomultiplier tubes (Hamamatsu R 647-01).*

the momentum, energy, mass of the projectile and the mass of the target nucleus, respectively. The mass conservation

$$M_P + M_T = M_1 + M_2 + M_3 \tag{3}$$

is not required in the calculation but this additional equation can be employed to suppress background events since, in the analysis of the three body reactions, we have to discriminate against incomplete four body events in which one fragment misses the particle counters. Most of these events can already be rejected by a coplanarity-check of the velocity vectors. In a true ternary reaction, the CM-velocities of the three particles $\vec{v}_1^{CM}$, $\vec{v}_2^{CM}$, and $\vec{v}_3^{CM}$ span a plane which includes the CM-origin (see Fig.4). Deviations from this coplanarity condition can be characterized by a velocity vector component Δv from the origin in the direction on the normal to this plane (see also ref. 14). The distribution of Δv for ternary events is shown in Fig.5a. The low background below the peak at $\Delta v = 0$ indicates that most of the analysed events actually originate from ternary reactions.

In order to indentify the two particles which originate from the fission reaction

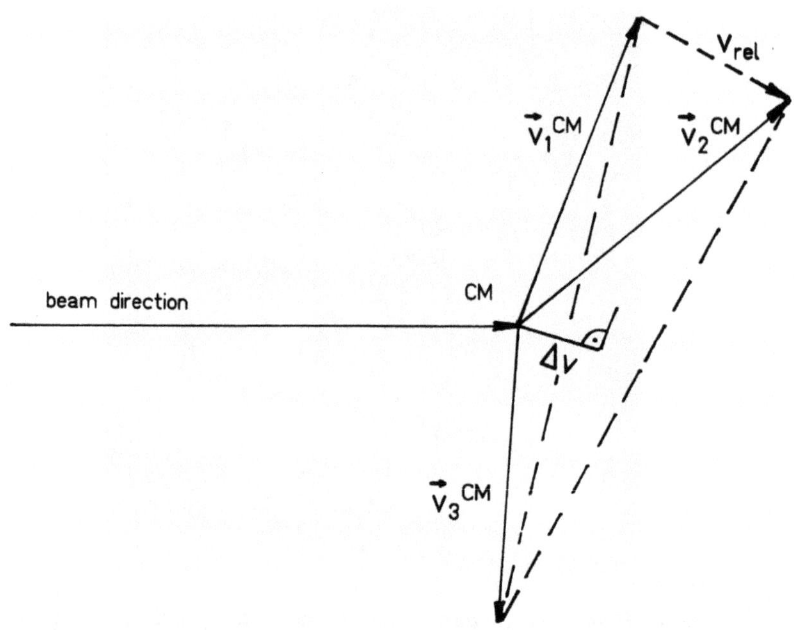

Figure 4: ***Velocity Δv characterizing the deviations from exact momentum conservation for the observed fragments in three body reactions.***

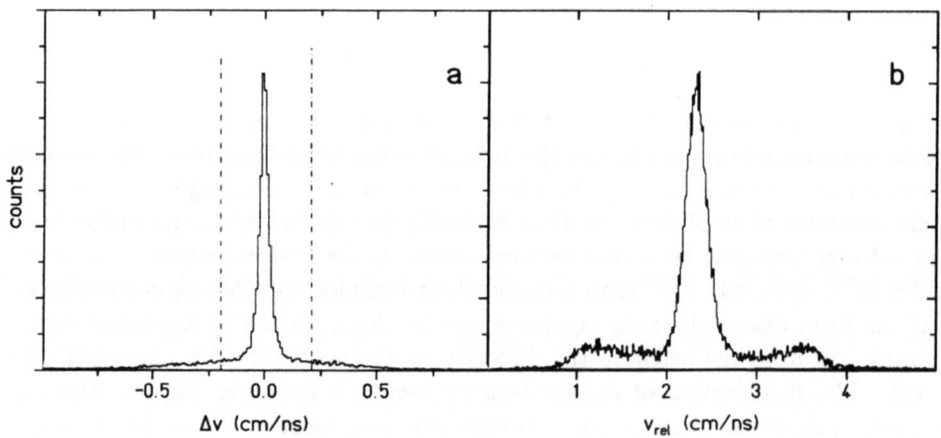

Figure 5: ***(a) Experimental distribution of the CM errors Δv for ternary events. (b) Experimental distribution of the relative velocities of the fission fragments.***

of the uranium-like nucleus, we take advantage of the assumption that the second step is truely decoupled from the first, i.e. that the intermediate fissioning system passes through a compound nucleus state. According to the Viola-systematics [15] the relative velocities of the fission products should be approximately 2.4 cm/ns. Actually a peak appears at this velocity in the event distribution (see Fig. 5b). In the further analysis, all events outside the peak-region around 2.4 cm/ns were rejected. Now the scattering angle of the first binary reaction stage can be determined and the double differential cross section $d^2\sigma/d\Theta_{CM}dE$ can be plotted as a contour map (Wilczinsky plot) in a TKE versus Θ_{CM} diagram (see Fig.6).

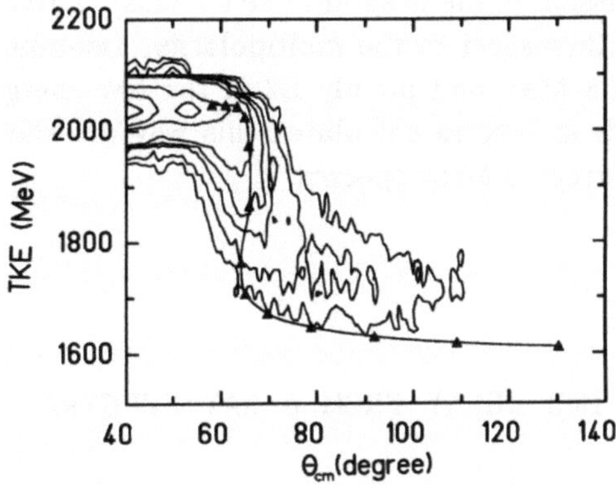

Figure 6: *Wilczynski-plot for the U+Au system at 8.6 MeV/u (preliminary). The line marked with full triangles represents calculations of the expectation values* $< TKE >$ *and* $< \Theta_{CM} >$ *according to the Feldmeier model* [18] .

A few remarks will be added concerning the analysis of the electron spectra. At first, the counts detected in the plastic-ring detector in coincidence with elastic or three-body events were corrected for the γ-ray response of the counter. In dissipative collisions with the highest $TKEL$ of 400 MeV, which also have the highest γ-multiplicity, the background contributes 60% to the counting rate. However, this is true only for the highest electron energies where the electron yield is lowest. The remaining electron counts were corrected for the spectrometer transmission and normalized with respect to the energy band accepted by the ring counter as well as to the particle events of the corresponding reaction class. The electron energy calculated from the electron momentum $p = 2.0506 \left(\frac{keV}{c}\right)\left(\frac{I_0}{A}\right)$, with I_0 being the current of the spectrometer, was first corrected for the 3 cm displacement of the plastic-ring counter in the direction to the target, this displacement being necessary to optimize the focusing of electrons emitted in the CM-system. In addition we corrected for the kinematic Dopplershift of the electron energy, assuming isotropic emission in the CM-system, and for the high voltage of 20 kV at the target, which is needed to protect the particle counters against low energy δ-electrons resulting from Coulomb projectile-target interactions.

Finally, the electron spectra were corrected for the conversion electron contribution. The procedure for estimating this background is based on converting the observed γ-ray spectra with known conversion coefficients into electron spectra. Therefore, the experiment is equipped with a $5.08^{\oslash} \times 3.0$ cm BGO-γ-ray counter (cf. Fig.2) positioned at 90° with respect to the beam direction. The measured γ-ray spectra were unfolded from the detector response function and converted into electron spectra under the assumption that, in elastic collisions, half of the intensity originates from the gold ion with $Z = 79$ and the other half from the uranium ion with $Z = 92$. For ternary reactions we assume that $\frac{1}{3}$ of the intensity originates from $Z = 79$ and $\frac{2}{3}$ from $Z = 46$ (after fission of the uranium-like nucleus). Furthermore, we decomposed the γ-ray spectra with respect to the multipolarity, assuming mainly E1 in the high energy part above 1.5 MeV and mainly E2 in the low energy part below 1 MeV [5]. The conversion electron spectra calculated this way are shown in Fig.7 and Fig.11, together with the corrected δ-ray spectra.

EVALUATION OF TIME DELAY FROM δ-RAY SPECTRA

For Rutherford-scattering the δ-electron ionization amplitude can be calculated in first order time dependent adiabatic perturbation theory according to

$$a(\omega) = \left(\frac{k_1}{\omega}\right) \int_{-\infty}^{\infty} dt \, \frac{\dot{R}(t)}{R(t)} \, exp\{i\omega t\} \tag{4}$$

with k_1 a constant, $\omega = (E + E_b) / \hbar$, E the kinetic energy, $E_b(b)$ an impact parameter dependent mean binding energy of the electron in the quasi atom with $Z_u = Z_P + Z_T$, $\dot{R}(t)$ the relative velocity and R(t) the relative distance of the two nuclei. After a parametrization of $\dot{R}(t)/R(t)$ according to ref.16

$$\frac{\dot{R}(t)}{R(t)} = \kappa \frac{t}{t^2 + \hat{t}^2} \tag{5}$$

the Fourier-integral (4) can be solved analytically and the energy-differential δ-ray production probability reads

$$\frac{dP_{e^-}}{dE_{e^-}} = |a(\omega)|^2 = \left(\frac{k_2}{\omega}\right)^2 exp\{-2\omega\hat{t}\} \tag{6}$$

with [2,16])

$$2\hat{t} = \left(\frac{2a}{v}\right)\left(\epsilon + 1.6 + \frac{0.5}{\epsilon}\right), \tag{7}$$

where 2a is the minimum distance of closest approach in a head-on collision, v the projectile velocity at infinity and $\epsilon = 1/\sin\frac{\Theta_{CM}}{2}$ the excentricity. It is important to note that this perturbative treatment correctly describes the energy dependence of the ionization probability[17)], obtained from exact coupled channel calculations, except for an overall normalization factor. The mean binding energy $E_b(b)$ and the parameter k_2 were determined by fitting dP_{e^-}/dE_{e^-} (Eq. 6) to the experimental δ-ray spectra for Coulomb trajectories (see Fig.7). For the calculation of the δ-ray spectra emitted in dissipative collisions, the parameter k_2 was kept constant.

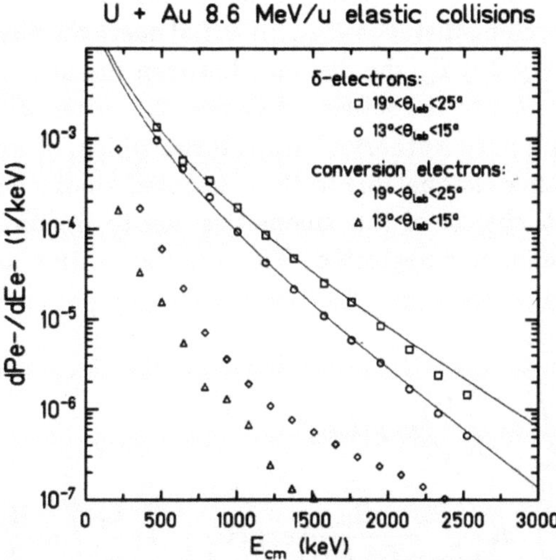

Figure 7: *Preliminary energy differential δ-electron production probability dP_{e^-}/dE_{e^-} as a function of the CM electron energy for elastic scattering angle regions as indicated. The conversion electron contribution is already subtracted, the full lines representing fits according to equation (6) with the normalization factor k_2 and the mean binding energy E_b as free parameters. $E_b = 200$ keV.*

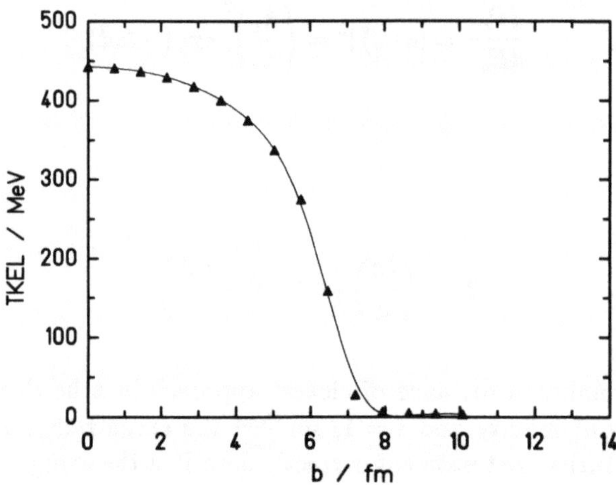

Figure 8: *Expectation value of TKEL as a function of the impact parameter b for U + Au collisions at* 8.6 *MeV/u. Calculations are according to ref. 18.*

In principle, the dissipative collision can be described by a model which is closely related to the one described in ref. 20. For $-\infty < t \leq -T_1$ the colliding particles follow pure Rutherford trajectories. At the time T_1, i.e. the time for which the distance of the nuclei is equal to the touching distance R_{int}, all radial energy is dissipated suddenly. In the sticking period $-T_1 \leq t \leq T_2$, the distance between the nuclei changes linearly as a function of time from R_{int} to the scission distance R_{sc}. After scission at the time T_2, the particles follow again pure Rutherford trajectories which, of course, differ from the trajectories in the entrance channel due to the dissipated kinetic energy and dissipated angular momentum. A change of the kinematics due to fission of the uranium-like projectile in the exit channel is neglected. The δ-ray ionization amplitude reads:

$$
\begin{aligned}
a_T(\omega) &= \frac{k_3}{\omega} \int_{-\infty}^{-T_1} dt \, \frac{\dot{R}_i(t)}{R_i(t)} exp\{i\omega t\} \\
&+ \frac{k_3}{\omega} \int_{-T_1}^{T_2} dt \, \frac{R_{sc} - R_{int}}{(T_2 - T_1)\, R(t)} exp\{i\omega t\} + \frac{k_3}{\omega} \int_{T_2}^{\infty} dt \frac{\dot{R}_0(t)}{R_0(t)} exp\{i\omega t\} \qquad (8)
\end{aligned}
$$

The coherent superposition of the three contributions in equation (8) causes interference effects, and hence oscillations in the energy-differential electron production probability dP_{e^-}/dE_{e^-}. The function $\dot{R}(t)/R(t)$ in equation (8) can also be parametrized according to equation (5), permitting a relatively simple numerical treatment. In addition, $a_T(\omega)$ depends also on the impact parameter b and the dissipated angular momentum. However, b is not an observable in dissipative collisions and a relation between $TKEL$ and b must be assumed. We used the b-dependence of $< TKEL >$

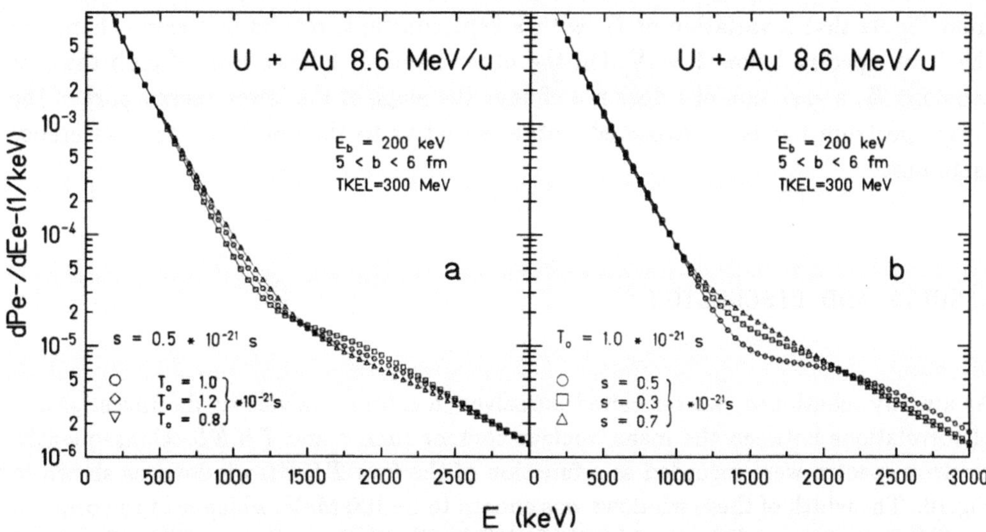

Figure 9: *Calculated electron spectra with a distribution $f(T)$ of the nuclear time delay according to eq. (9). (a) The variance $s = 0.5 * 10^{-21}$ s was kept constant and the most probable delay time T_0 was varied as indicated. (b) $T_0 = 1 * 10^{-21}$ s is kept constant and s varied.*

according to Feldmeiers model, as shown in Fig.8. The dissipated angular momentum can be taken into account in the sticking limit approximation [20]. With these assumptions the kinematics of the dissipative collision process is defined and the δ-ray transition amplitudes (8) can be calculated with the nuclear sticking time $T = T_2 - T_1$ as the only free parameter. Finally we assume that for a fixed $TKEL$ the distribution $f(T)$ of the nuclear sticking time T can be represented by a truncated gaussian:

$$f(T) = \frac{exp\left(-\frac{1}{2}\left[\frac{(T-T_0)}{s}\right]^2\right)}{\int_0^{2s} dT\, exp\left(-\frac{1}{2}\left[\frac{(T-T_0)}{s}\right]^2\right)} \tag{9}$$

defined in the intervall $0 \leq T \leq 2s$.

However, the preliminary results presented here have been obtained with a simplified schematic model. In equation (8) $T_1 = 0$ was chosen and the second integral was neglected. In addition, the scattering angle in the exit channel was assumed to be the same as in the entrance channel. The most probable delay time T_0 and s in equation (9) were obtained by a fit to the measured δ-ray spectra. The sensivity of these parameters to the shape of the δ-ray spectra is demonstrated in Fig.9. We see from Fig.9a that a variation of T_0, with s kept constant, results in a slope change of the δ-ray spectra below 1 MeV. On the other hand, it follows from Fig.9b that, at constant T_0, a variation of s does not change the slope of the lower energy part of the δ-ray spectrum but is, as expected, rather sensitive to the depth of the interference minimum.

RESULTS AND DISCUSSION

As already mentioned, we restricted ourselves in a first analysis to the investigation of correlations between the mean nuclear contact time τ and $TKEL$. Consequently, electron spectra were recorded as a function of the four $TKEL$-windows as shown in Fig.10. The width of these windows was chosen to be 100 MeV, which is of the order of the TKE resolution of about 85 MeV (FWHM). Fig.11 shows the corresponding δ-ray spectra (circles), which are corrected for conversion electrons. In order to demonstrate the drastic effect of the nuclear time delay the data are compared with δ-ray spectra (dashed lines) calculated for trajectories without time delay, but assuming a $TKEL$ and the corresponding impact parameter according to the Feldmeier model[18] (see Fig. 8). The solid lines represent calculations with the same impact parameters and $TKEL$ but taking additionally into account the distribution of the nuclear contact time $f(T)$ as shown in the insert of the figures. With the adopted parameters T_0 and s of the truncated gaussian $f(T)$ of equation (9), the mean value of the delay time τ

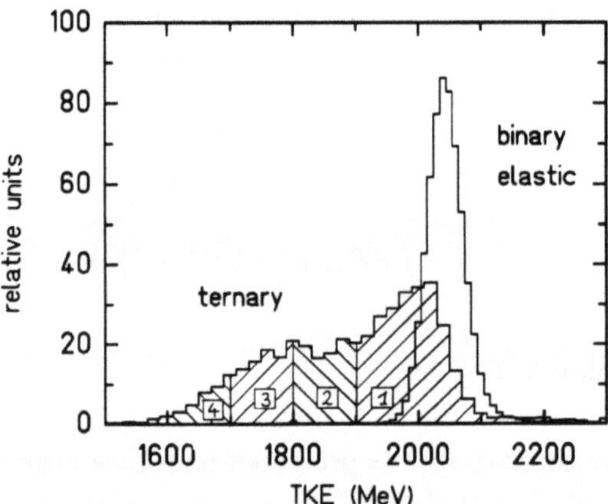

Figure 10: *Experimental distribution of the total kinetic energy for binary elastic and ternary events after the first reaction stage. Displayed data are taken for the U + Au reaction at* 8.6 *MeV/u.*

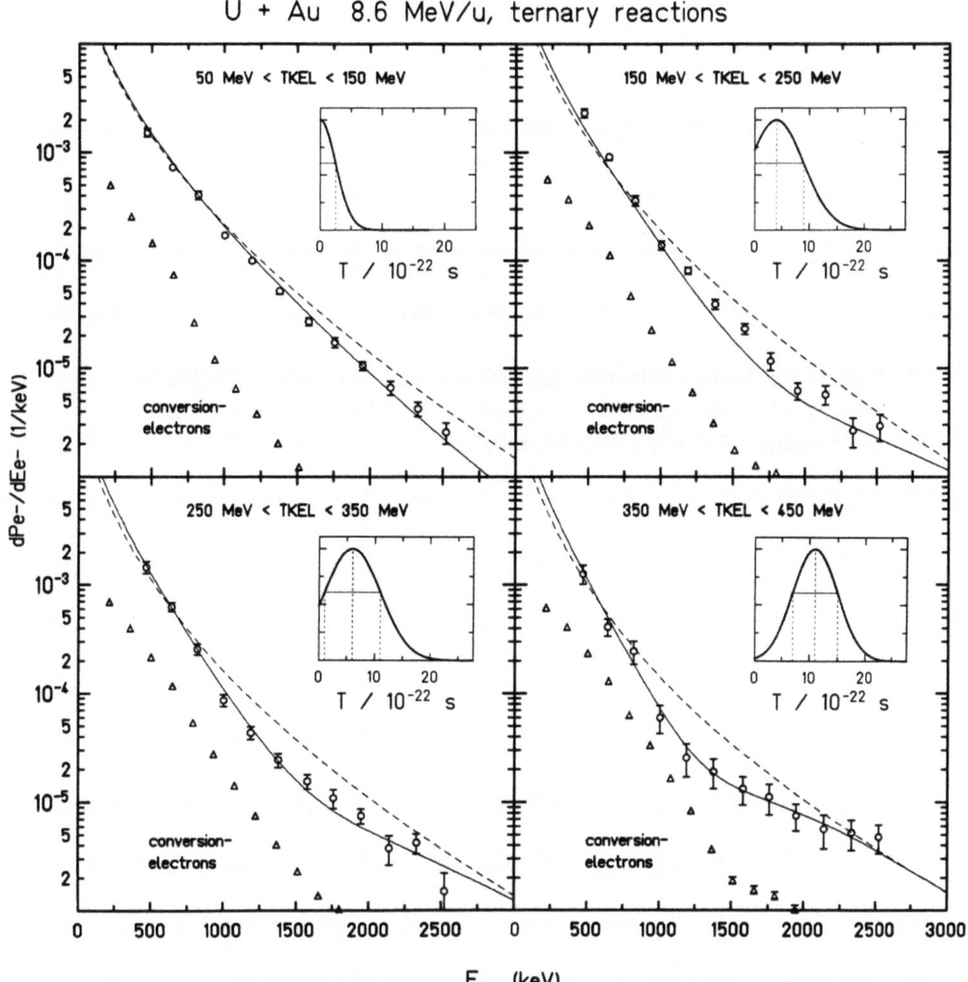

Figure 11: *(o) δ-ray spectra recorded in coincidence with the four TKEL regions indicated in Fig. 10. The conversion electron contribution (Δ) is already subtracted. Dashed lines represent calculated δ-ray spectra assuming trajectories without time delay but with a TKEL as indicated. Full lines represent fits with dissipative trajectories for which the nuclear contact time distributions as shown in the inserts were additionally taken into account (preliminary).*

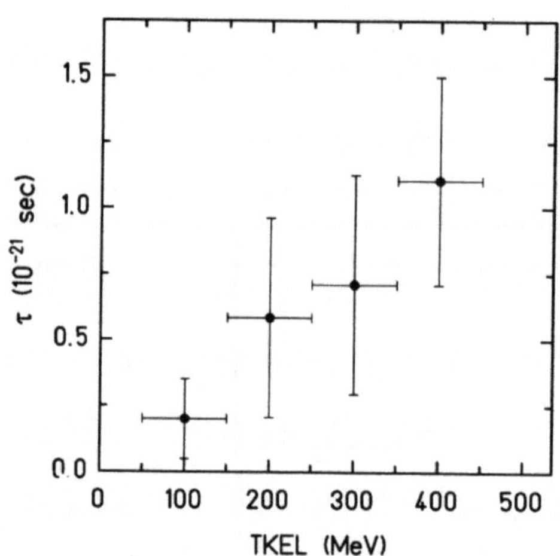

Figure 12: *Mean nuclear contact time τ as a function of TKEL for ternary U + Au reactions at 8.6 MeV/u. The error bars indicate the variance $\pm\sigma$ of τ (preliminary).*

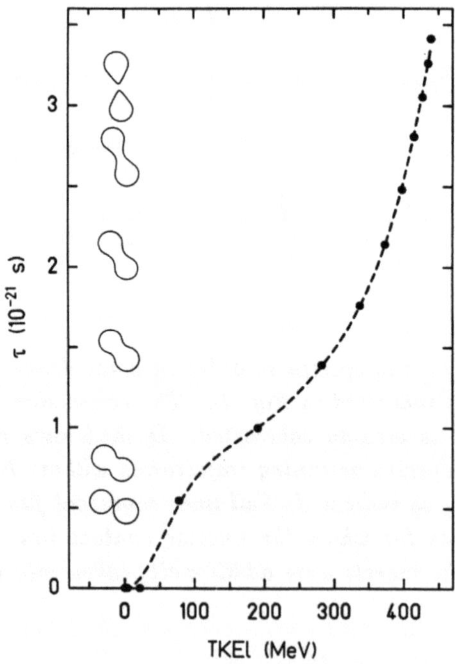

Figure 13: *Calculated nuclear contact time for the U + Au reaction at 8.6 MeV/u [18]. The left hand part shows the time evolution of the nuclear shapes for $TKEL$ = 400 MeV. Note that the nuclear contact time is defined from the first touch of the nuclei to scission.*

and the variance σ were calculated according to

$$\tau = \int_0^\infty dT\, T\, f(T) \qquad (10)$$

$$\sigma^2 = \int_0^\infty dT\, (T - \tau)^2 f(T) \qquad (11)$$

The results are shown in Fig. 12 as a function of the different $TKEL$ windows. The "error bars" represent the variance $\pm\sigma$. It is interesting to note that τ increases nearly linearly with increasing $TKEL$, reaching $1.1 * 10^{-21}$ s for $TKEL = 400$ MeV. On the other hand large fluctuations in the nuclear delay time, which actually increases as a function of $TKEL$, prevent the observation of a pronounced minimum in the δ-ray spectrum taken for $TKEL = 400$ MeV. However, we would like to emphasize that it remains to be investigated whether the physical quantities τ and σ depend on the various model-assumptions made in the analysis of the data as described above. Finally, we would like to mention that nuclear delay times of the same order of magnitude were deduced for deep inelastic U+U collisions at 7.5 MeV/u from the K-shell ionization probability[21)].

The contact time according to the Feldmeier model is shown in Fig. 13. This time cannot be compared directly with our experimental results. Within the Feldmeier model, energy dissipation is connected with a deformation of the nuclear shapes and the exchange of nucleons which takes place after a neck has been formed between the two nuclei. The time evolution of the nuclear shapes during a U + Au collision with a TKEL of 400 MeV is illustrated in the left part of Fig. 13. The nuclear interaction time is defined just as the time period for which a neck exists. The δ-electrons, however, reflect the dynamics of the nuclear charge distribution. As indicated in Fig. 13, the shape of the dinuclear system does not change very much in the first stage of the collision and the dynamics seems to be damped strongly. It is essentially this early time of the nuclear encounter which gives rise to the phase shift between incoming and outgoing ionization amplitudes and hence dominates the measured time delay. In the second stage, the two nuclei start to separate again by expanding the neck. The duration of this stage will contribute much less to the time delay measured with the δ-electrons.

This argument may help to understand qualitatively why our experimental delay time cannot be compared directly with theoretical model calculations. On the other hand, a conclusive test of reaction models can be performed by comparing the measured electron spectra with coupled channel calculations based on the evolution of the charge distribution during a collision as predicted by the reaction models.

OUTLOOK

As we have seen, large fluctuations of the nuclear contact time prevent the observation of pronounced minima in the δ-ray spectra. These fluctuations probably will completely smear out the second minimum which should be present at a δ-ray energy of about 5.5 MeV. Nevertheless, the observation of the second minimum would be of particular importance since the nuclear reaction time could be determined from the energy difference between the first and second minimum and therefore do not suffer from the uncertainty in the mean binding energy E_b of the electrons in the quasi-molecule. In the ongoing analysis we therefore try to find analysing conditions in the Wilczynski-plot which prolong the contact time and minimize its fluctuations. Alternatively, we also will investigate whether the mass flow might be a trigger for long contact times with small fluctuations. Long nuclear contact times in very heavy scattering systems are still of particular interest for the observation of the predicted spontaneous positron peak [7].

Fig. 12 indicates that the mean nuclear contact time τ increases faster than the variance σ as a function of $TKEL$. Therefore, the U + Au collision system should be investigated at higher beam energies for which $TKEL$ in central collisions increases. On the other hand, large nuclear contact times have been found in, for e.g., Pb + Pd collisions only slightly above the Coulomb barrier [19]. The main problem with these low Z_u scattering systems results from the increasing conversion electron contribution.

ACKNOWLEDGEMENT

We are grateful to Hans Feldmeier for fruitful discussions and for supplying us with his computer code.

The experiment has been supported by GSI.

REFERENCES

1. G.Soff, contribution to this conference and references cited therein
2. E.Kankeleit, contribution to this conference and references cited therein
3. G.Soff, J.Reinhardt, B.Müller, W.Greiner, Phys. Rev. Lett. 43 (1979) 1981.
4. H.Backe, P.Senger, W.Bonin, E,Kankeleit, M.Krämer, R.Krieg, V.Metag, N.Trautmann, J.B.Wilhelmy: Phys.Rev.Lett. 50 (1983) 1838.
5. R.Krieg, E.Bozek, U.Gollerthan, E.Kankeleit, G.Klotz-Engmann, M.Krämer, U.Meyer, H.Oeschler, P.Senger: Phys.Rev. C34 (1986) 562.
6. H.Oeschler et al.; contribution to this conference
7. U.Müller, T.de.Reus, J.Reinhardt, B.Müller, W.Greiner; Z.Physik A323 (1986) 261.

8. F.Rösel, H.M.Fries, K.Alder, H.C.Pauli: Atomic Data and Nuclear Data Tables 21 (1978) 291.

9. C.Kozhuharov, in: Proc. NATO Advanced Studies Institute, Quantum Electrodynamics of Strong Fields, Lahnstein 1981 (Plenum Press, New York - London) p.317.

10. P.Senger; thesis, Technische Hochschule Darmstadt 1983, unpublished

11. D.v.Harrach, H.J.Specht; Nucl. Instr. Meth. 164 (1979) 477.

12. E.Moll, E.Kankeleit; Nukleonika 7 (1965) 180.

13. P.Glässel, D.v.Harrach, Y.Civelekoglu, R.Männer, H.J.Specht, J.B.Wilhelmy , H.Freiesleben, K.D.Hildenbrand: Phys.Rev.Lett 43 (1979) 1483.

14. P.Glässel, D.v.Harrach, H.J.Specht, L.Grodzins: Z.Physik A310 (1983) 189.

15. E.E.Viola: Nucl. Data Sect *A*1 (1966) 391.

16. E.Kankeleit: Nukleonika 25 (1980) 253.

17. G.Soff, W.Greiner, W.Betz, B.Müller: Phys.Rev. A20 (1979) 169.

18. H.Feldmeier: Dynamics of heavy ion reactions, in Nucl. Structure and Heavy Ion Dynamics; ed. L.Moretto and R.A.Ricci (North Holland 1984), p.274.

19. K.E.Rehm, H.Essel, P.Sperr, K.Hartel, P.Kienle, H.J.Körner, R.E.Segel, H.Wagner: Nucl.Phys. A366 (1981) 477.

20. H.J. Wollersheim, W.W. Wilcke, J.R. Birkelund, J.R. Huizenga: Phys.Rev. C25 (1982) 338.

21. C.Stoller, M.Nessi, E.Morenzoni, W.Wölfli, W.E.Meyerhof, J.D.Molitoris, E.Grosse, C.Michel: Phys.Rev.Lett. 53 (1984) 1329.

POSITRON-ELECTRON ANGULAR CORRELATIONS IN HEAVY ION COLLISIONS

M. Krämer, B. Blank, E. Bozek[1] , H. Jäger, E. Kankeleit, G. Klotz-Engmann, R. Krieg[2] , C. Müntz, H. Oeschler, M. Rhein, and P. Senger[3]

Institut für Kernphysik, Technische Hochschule Darmstadt, Germany

INTRODUCTION

Recent measurements of positrons in coincidence with electrons emitted in U+Th collisions near the Coulomb barrier show sharp line structures in the sum and difference energy spectra of the two leptons (ref. 1). A possible method to clarify the origin of these lines is to study the angular correlation of the leptons. In this short contribution we want to report on a positron-electron coincidence test measurement performed with the TORI spectrometer. This apparatus offers the possibility to measure positrons in coincidence with electrons emitted into the same and opposite hemisphere as well.

I. Experimental setup

The basic design principles of the TORI spectrometer are described in ref. 2 and in the talk given by H. Oeschler. The important aspects will be summarized here briefly. The TORI spectrometer (Figure 1) is basically an S-shaped solenoidal magnetic transport system for electrons and positrons. Due to the inhomogeneous magnetic field charged particles experience a spatial drift in opposite directions according to the sign of their charges while being guided along the tube. The drift has its maximum value in the middle plane, where the huge amount of electrons is completely separated in space from the positrons. The electrons are stopped at this position by a diaphragm and are detected by a Si(Li) counter. For the positrons the drift is reversed in the second quarter torus and they are detected by a Si(Li) counter as well. The

---

1 from Institute of Nuclear Physics, Cracow, Poland

2 now Lurgi, Frankfurt

3 also University of Mainz

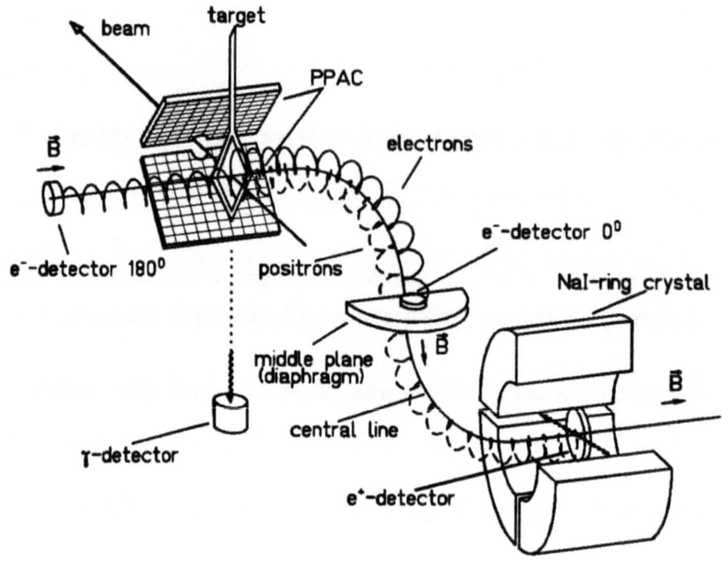

Figure 1. Schematic view of the TORI spectrometer and its detectors

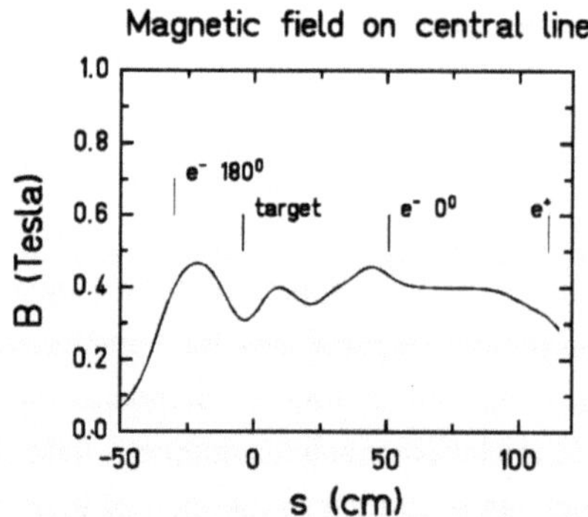

Figure 2. Magnetic field configuration chosen for $e^+ e^-$ coincidence measurement
The position of target and detectors on the central line are indicated.

positron counter is surrounded by a four-fold NaI ring where the detection of the annihilation radiation is requested to identify an event as a positron.

Due to the spatial separation it is possible to measure simultaneously electrons and positrons without interfering each other.

In contrast to the usual setup (ref. 2) the magnetic field strength along the central line was chosen to yield a symmetric acceptance for charged particles in both branches of the spectrometer. On either side of the target there is a maximum of the field strength amounting to 0.45 Tesla which the particles have to penetrate in order to reach the detectors. The resulting acceptance angle is 55^0 with respect to the central line which corresponds to a spectrometer acceptance of 20% of 4π for each hemisphere. As a consequence of this symmetric setup positrons and electrons emitted in the same hemisphere can be measured as coincidence between the positron and the middle plane 0^0 electron counter while antiparallel emission is observable only as coincidence between the positron and the 180^0 electron counter.

2. Experimental results

Coincidence data were taken using a uranium beam of 5.85 MeV per nucleon on thorium fluoride (ThF_4) targets of 200-300 μg cm^{-2} thickness in an approximately 1-2 days test run.

It is convenient to plot the results (spectral distribution of positron energy E_{e^+} and electron energy E_{e^-}) in two-dimensional form. We have chosen the representation of difference $\Delta = E_{e^+} - E_{e^-}$ versus sum $\Sigma = E_{e^+} + E_{e^-}$. In this picture the original E_{e^+} and E_{e^-} axes appear rotated by 45^0.

The two processes of interest to explain line structures have the following signatures:

- If a single strong nuclear transition contributes the sum spectra should exhibit a Doppler broadened peak while the difference spectra are continuous. In the two-dimensional plot it should appear as a ridge parallel to the difference axis.

- If a two-body decay is involved both the sum and difference spectra will show a peak which is sharp in the sum energy spectrum and Doppler-broadened in the difference spectrum.

The one-dimensional spectra of interest can easily be obtained from the two-dimensional ones by projection onto the axes.

In order to check whether the leptons are emitted isotropically or with an angular correlation the following method was applied. The singles energy spectra of electrons and positrons taken in the same run are multiplied to form a tensor product thereby assuming no angular correlation between the directions of the leptons. The two-dimensional result of this procedure after normalization to the number of scattered heavy ions can be treated in the same way as the measured coincidences concerning representation, cuts and projections. If the measured lepton coincidences are not correlated in emission angle their energy distribution would be equal to the tensor product of the singles spectra. It should be mentioned that the projection of the tensor product onto the sum or difference axis essentially means a one-dimensional folding of the energy spectra.

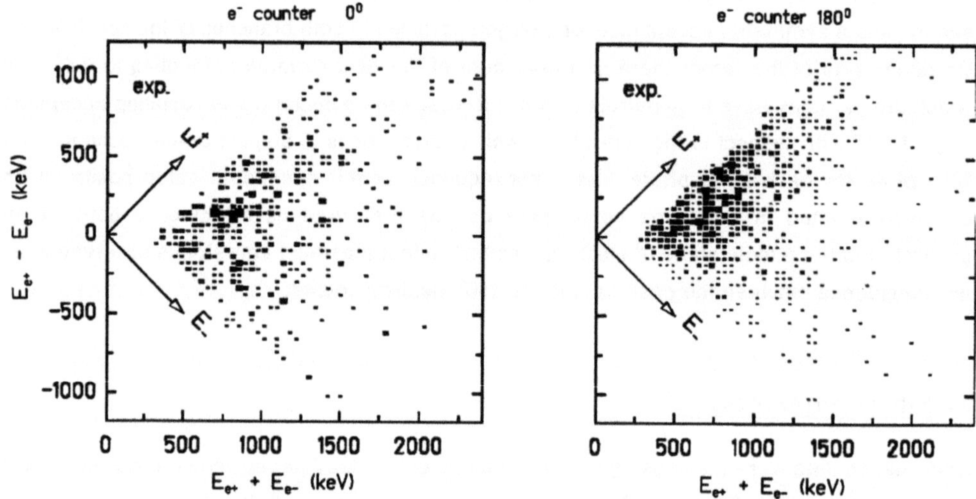

Figure 3. Two-dimensional plot of measured electron-positron coincidences

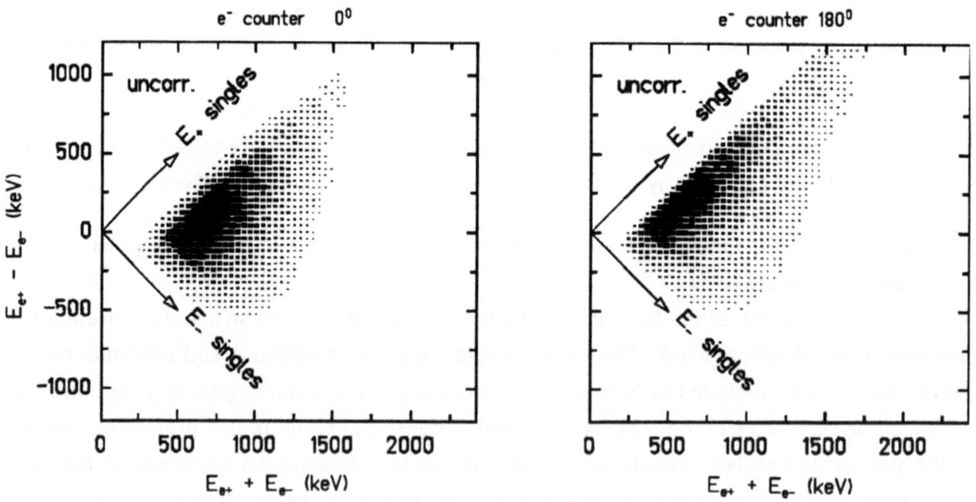

Figure 4. Tensor product of measured singles spectra

Uncorrelated emission angles of electrons and positrons are assumed.

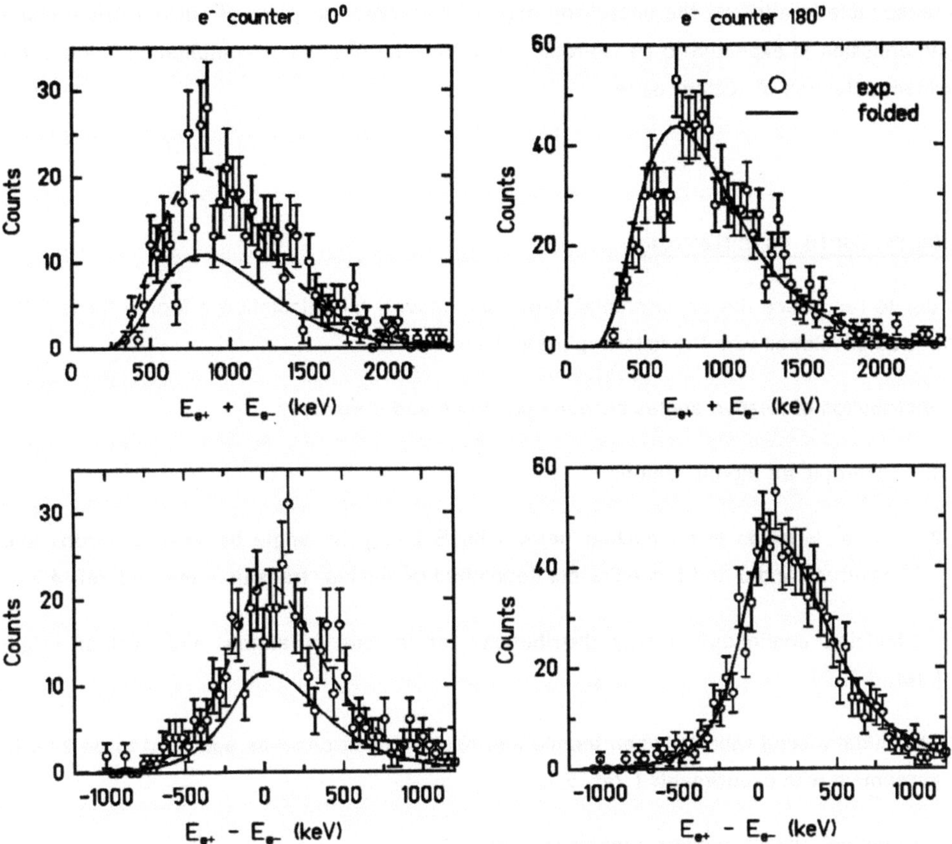

Figure 5. Sum and difference spectra of electron-positron coincidences:
left hand side for 0°, right hand side for 180°,
Symbols are measured coincidence data, the solid curves show the uncorrelated combination of measured singles spectra.
The dashed curve results from scaling of the solid curve by a factor of 1.9 to match the absolute height.

Figure 5 shows the projections onto the the sum and difference energy axes for the two detector combinations together with the results from our check procedure. Within the limits of the present statistics the spectral shape as well as the absolute height of the experimental data for the 180° detector combination is reproduced. We want to emphasize at this point that the solid curves result from experimental energy spectra, neither scaling factors nor efficiency corrections have been used.

The situation is somewhat different for the 0° counter combination. Here the shapes are well reproduced by the folding procedure but a factor of about 1.9 is necessary to match the absolute height. There are more coincidences measured than expected from uncorrelated emission of electron-positron pairs.

This experimental finding shows the existence of a forward peaked angular correlation.

It is reasonable to attribute the underlying angular correlation to pairs of nuclear origin since atomic emission is expected to be isotropic. Therefore the 0^0-counter combination favours the coincident detection of nuclear pairs.

3. Monte Carlo calculations

In order to reproduce the experimental data quantitatively by calculation a Monte Carlo (MC) model was developed with the following ingredients:

- distribution of relative angles between positrons and electrons :
    1. isotropic for atomic leptons,
    2. $\propto a + b \cos \Theta$ for nuclear pairs, with Θ being the angle between electrons and positrons and a and b coefficients dependent on nuclear transition energy (ref. 4),
- calculated continuous energy distributions for leptons of atomic and nuclear origin (ref. 3),
- the initial mixing ratio between atomic and nuclear positrons was assumed to be 2 : 1 in agreement with experiments (ref. 5),
- E1 multipolarity for nuclear transitions (ref. 5),
- multiplicity of atomic electrons $>$ 1 per event to allow for inclusion of summing effects,
- inclusion of the properties of the apparatus such as angle and energy dependent detection efficiency, peak efficiency etc.

MC calculations were performed with high statistics to verify the agreement in shape with the measured data. Figure 6 shows the experimental sum and difference spectra together with the MC results for both counter combinations. Both shape and absolute height of the MC spectra are in good agreement with the data. The MC model calculation verifies quantitatively that the mentioned enhancement in the 0^0 coincidence rate can be ascribed to the non-isotropic nuclear pair creation.

SUMMARY AND CONCLUSIONS

The present statistics obtained with this short test run does not allow to conclude upon sharp

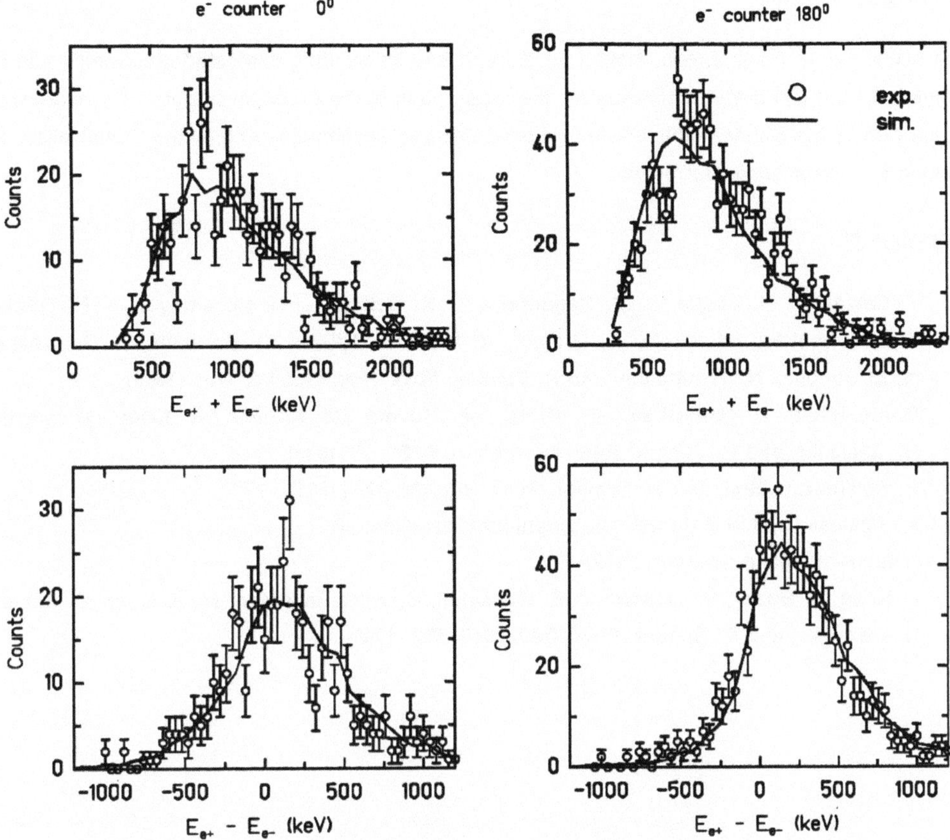

Figure 6. Sum and difference spectra in comparison with MC calculations:
Symbols are measured data points, curves are MC calculations.

line structures in coincidence spectra. The method of observing angular sensitive coincidences of positron-electron pairs offers a possibility to test a possible nuclear origin of the observed line structures.

Since our test run shows that the 0^0 counter combination is sensitive to nuclear pairs any structure appearing in the 0^0 coincidence spectra can be ascribed with high probability to nuclear processes.

On the other hand, structures appearing only in 180^0 coincidences would confirm the assumption of a 2-body decay as origin of the observed line structures.

ACKNOWLEDGEMENTS

We would like to thank J. Foh, Institut für Kernphysik, for building dedicated electronics and H. Folger and the GSI target laboratory for the preparation of the excellent targets. This work was supported by the Bundesministerium für Forschung und Technologie and by the Gesellschaft für Schwerionenforschung, Darmstadt.

REFERENCES

1. T. Cowan, H. Backe, M. Begemann, K. Bethge, H. Bokemeyer, H. Folger, J. S. Greenberg, H. Grein, A. Gruppe, Y. Kido, M. Klüver, D. Schwalm, J. Schweppe, K. E. Stiebing, N. Trautmann, and P. Vincent, Phys. Rev. Lett. **54**, 1761 (1985)
2. E. Kankeleit, U. Gollerthan, G. Klotz, M. Kollatz, M. Krämer, R. Krieg, U. Meyer, H. Oeschler, and P. Senger, Nucl. Instr. Meth. **A234**, 81 (1985).
3. P. Schlüter, G. Soff, and W. Greiner, Phys. Rep. **75**, 327 (1981).
4. A.I.Achieser and W.B.Berestezki, Quantenelektrodynamik, (Teubner,Leipzig) (Moscow,1959)
5. R Krieg, E. Bozek, U. Gollerthan, E. Kankeleit, G. Klotz-Engmann, M. Krämer, U. Meyer, H. Oeschler, and P. Senger, Phys. Rev. **C43**, 562 (1986).

EXOTIC NUCLEAR STRUCTURES AND DECAYS: NEW NUCLEAR COLLECTIVE PHENOMENA

J.H. Hamilton and C.F. Maguire

Physics and Astronomy Department
Vanderbilt University
Nashville, TN 37235, USA

INTRODUCTION

Studies of the properties of exotic nuclei have revealed a surprising richness and diversity in their shapes, structures, and decay modes far exceeding our understandings and expectations of even a decay ago.[1,2,3] By exotic is meant nuclei under extreme conditions not found in nuclei naturally in the earth; that is, nuclei far from stabiity with neutron numbers much larger or much smaller than those of the stable isotopes of a given element, or with Z well beyond 92, or with very high angular momenta even approaching fission limits. In this same decade, studies of heavy ion collisions as reported at this conference have been extended up to uranium on curium with possible evidence for the formations of giant nuclear molecules with combined Z up to 188, which may exist for periods very long compared to their collision times.

From studies of far-off-stability exotic nuclei have come evidence for the coexistence of different nuclear shapes in the same nucleus, new regions of unusually large deformation, new ground-state phase transitions from one shape to another, new magic numbers but now for deformed shapes, and for the importance of reinforcing shell gaps (see ref. 2). New exotic decay modes include a wide variety of beta delayed particle emission (see ref. 1) and heavy cluster emissions such as ^{14}C and ^{24}Ne (see refs. 4, 5). The new deformed magic numbers of 38 and 60 (refs. 2, 3) seen far off stability clearly support that there are likely other "magic" numbers for protons and neutrons which give stability to different deformed shapes. Perhaps these other new magic shell gap numbers at large deformation could influence the sticking of two very heavy nuclei in collisions such as U on Cm. Finally, another area which could have a bearing on the formation, motions, and structures of giant nuclear systems involves the recent observation of very energetic, light particle (proton, alpha) emission with up to 50% and more of the total incoming energy in a collision, for example in 300 MeV ^{32}S on Ta (refs. 6, 7). This is another completely unexpected new collective phenomena where a large fraction of the total energy is transfered to a single particle. Examples of the new insights being gained and their promise for the future will be given. The wide variety of new collective modes being observed in nuclei and in nuclear collisions sugests that indeed we may see other new exotic collective motions in the collision of two very heavy nuclei.

NEW SHAPES AND STRUCTURES

Studies of nuclei far from stability in recent years have brought about major changes in our understandings of the structures of nuclei.[1-3] No longer does a nucleus have to have a single, "permanent" shape which characterizes its low-lying levels. Now throughout the periodic table one finds nuclear shape coexistence where two or more different shapes can coexist with energy levels characteristic of each shape overlapping in energy. These coexisting shapes include small oblate (near spherical)-large prolate; spherical-prolate; triaxial-prolate and other combinations. In these cases sudden ground state phase transitions from one configuration lying lowest to the other being lowest or to even totally different shapes also are observed.

In addition, new "magic" numbers which play an important role in shape coexistence and superdeformations observed in certain regions have been identified. In Fig. 1 are shown the magic numbers (2, 8, 20, 28,...) for the spherical shell model (deformation parameter, $\beta = 0$). Note that 40 is shown as magic for a spherical shape as taken from the review of Baranger and Sorensen.[8] It was the double closed shell structure of ${}^{90}_{40}Zr_{50}$ which originally provided the evidence for the spherical magic character of 40. Subsequently, the double-closed-shell-like structure of ${}^{96}_{40}Zr_{56}$ and more recently well off stability ${}^{68}_{28}Ni_{40}$ provide support for both Z and N of 40 being magic for a spherical shape. Likewise, ${}^{88}_{38}Sr_{50}$ is often taken as the inert double closed shell core in many shell model calculations. Thus, it was quite surprising to discover[9] in ${}^{74,76}_{36}Kr_{38,40}$, where N = 38 and 40, a new region of unusually large deformation with $\beta \sim 0.35$. This new region of very strong deformation is found to be centered around N = Z = 38. It was independently predicted by the calculations of the ground state shapes and masses of over 4000 nuclei by Möller and Nix.[10] Without going through all the experimental evidence, what emerges can be understood by looking at Fig. 2 which shows the single particle levels as a function of deformation, as taken from Bengtsson et al.,[11] and at a summary of the data as given in Table 1. When 40 or the 38 spherical ($\beta = 0$) shell gaps get reinforced by a strong spherical magic number like 28 and 50 and even the weaker subshell gap at 56, then the reinforcing push of the protons and neutrons for the same shape, here spherical, makes nuclei like ${}^{90,96}_{40}Zr_{50,56}$, ${}^{68}_{28}Ni_{40}$ and ${}^{88}_{38}Sr_{50}$ look like spherical double magic nuclei. However, there are competing shell gaps at large prolate deformation ($\beta \sim 0.35$) for N,Z = 38 and 40. These deformed shell gaps are somewhat deeper than their competing spherical gaps at 38 and 40 so that when both N and Z approach 38 and 40, the stronger push for a deformed shape by both the protons and neutrons drives these nuclei to unusually large deformation as first observed in ${}^{74,76}_{36}Kr_{36,40}$ (ref. 9). So, the reinforcing effect of the protons and neutrons can cause a switch in which shell gap is "magic" in this region in stabilizing the nuclear shape to be spherical or deformed.

The reinforcing of the proton and neutron shape driving forces, as seen by the gaps in the single particle spectrum in Fig. 2, also explains the sudden appearance of another new region of equally strong deformation ($\beta \sim 0.35$) in ${}^{100,102}_{40}Zr_{60,62}$ and ${}^{98,100}_{38}Sr_{60,62}$ (as described in detail in ref. 2). As discussed there[2] and seen by looking at the 2^+ energies in Fig. 3, it is the reinforcing of the neutron shell gap at N = 60 at large deformation ($\beta \sim 0.35$) by the proton shell gaps at 38, 40 at the same large deformation ($\beta \sim 0.35$) that leads to this region of very strong deformation and not simply the N = 60 shell gap at large deformation as first proposed. For the N = 60, 62 nuclei as Z increases from 38, the large and suddenness of the onset of deformation (as indicated in Fig. 3 by the 2^+ level energies) are both gone by Z = 42 and clearly by

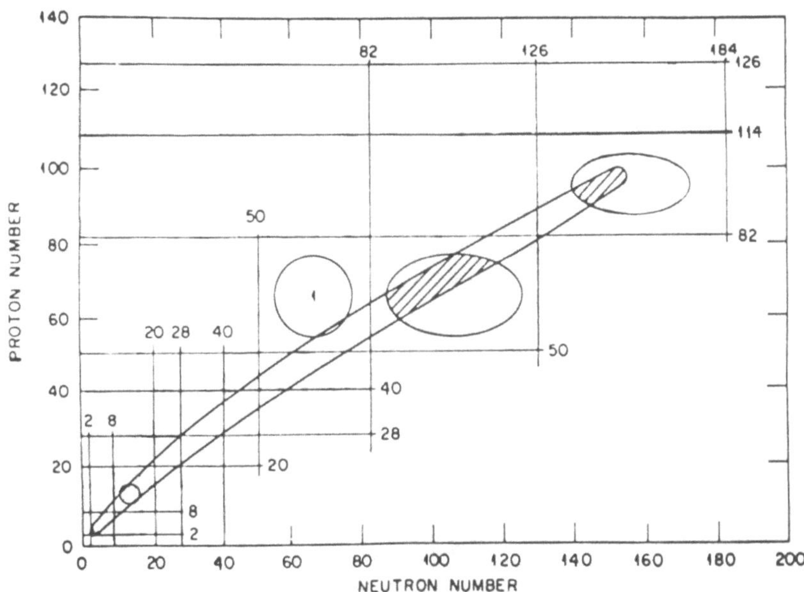

Fig. 1. Chart of the nuclides as a function of N and Z from ref. 8 with the spherical closed-shell magic numbers shown by verical and horizontal lines and deformed regions by ovals.

Z = 44. The suddenness of the onset of deformation in the $_{40}$Zr and $_{38}$Sr nuclei between N = 58 and 60 is also explained by the reinforcing proton and neutron shell gaps. The N = 56 spherical shell gap reinforces the spherical gaps at Z = 38 and 40 to keep these nuclei spherical further out from N = 50 than they would otherwise be. The double-closed-shell-like level structure seen in $^{96}_{40}Zr_{56}$ but not $^{94}_{38}Sr_{56}$ indicates that the Z = 40 and N = 56 spherical shell gap reinforcement is stronger than for 38-58. However, the Z = 38 and N = 60, 62 deformed shell gap reinforcement is stronger to drive the N = 60,62 $_{38}$Sr nuclei to have even larger deformation than the N = 60,62 $_{40}$Zr nuclei. Thus, we have clear evidence for new "magic" numbers, 38 for N and Z, (and perhaps 40 for N and Z) and 60 for N but now for deformed shapes. These "deformed magic" numbers or shell gaps confirm the longstanding prediction of Brack et al.[12] that there should be such magic numbers for different deformations which play the same role at large deformation as the well-known spherical magic number derived by Mayer and Jensen from the spherical shell model. The importance of reinforcing shell gaps on nuclear deformation and of the new "deformed magic numbers" are additional examples of new phenomena which could only be obtained by studying nuclei far from stabilty.

As one looks at Fig 2, one sees that in addition to 38 and 60 there are other shell gaps at different deformations. Theorists have emphasized that the shell gaps at equally large deformation but for an oblate shape ($\beta \sim -0.35$) at Z = N = 36 should be as important as the Z = N = 38 shell gaps. Thus, a new region of the strongest oblate deformation ever observed should exist around N = Z = 36 or 35. The Vanderbilt group initiated studies to seek to identify this new region of very strong oblate deformation and the phase transition from very strong prolate to very strong oblate deformation. All ground state oblate shapes previously observed have had small deformation ($\beta \lesssim -0.15$). Good candidates to observe the sudden transition are the bromine isotopes

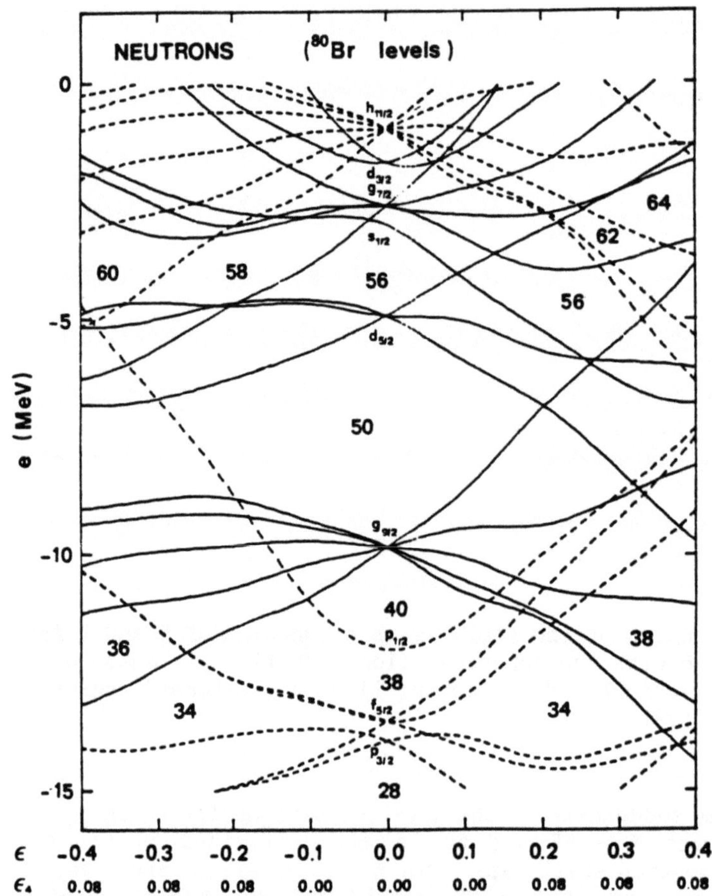

Fig. 2. Single particle levels (ref. 11).

Table 1. Summary of Experimental Data for N = 38 and 40 Nuclei

$^{66,68}_{28}Ni_{38,40}$	- ^{68}Ni spherical double magic energy levels ^{66}Ni spherical, large $E_{2_1^+} = 1.42$	
$^{70,72}_{32}Ge_{38,40}$	- near-spherical ground states and low-lying deformed 0_2^+ states	N=42-50 no deformed bands seen
$^{72,74}_{34}Se_{38,40}$	- near-spherical ground states and low-lying well deformed 0_2^+ states and deformed bands, clear shape coexistence	N=42-50 no deformed bands seen
$^{74,76}_{36}Kr_{38,40}$	- strongly deformed ground states and near spherical 0_2^+ states	N=42,44 γ-soft to N=50 spherical
$^{76,78}_{38}Sr_{38,40}$	- strongly deformed ground states, but no near spherical 0_2^+ states observed	Smooth decrease in deformation as N increases toward 50

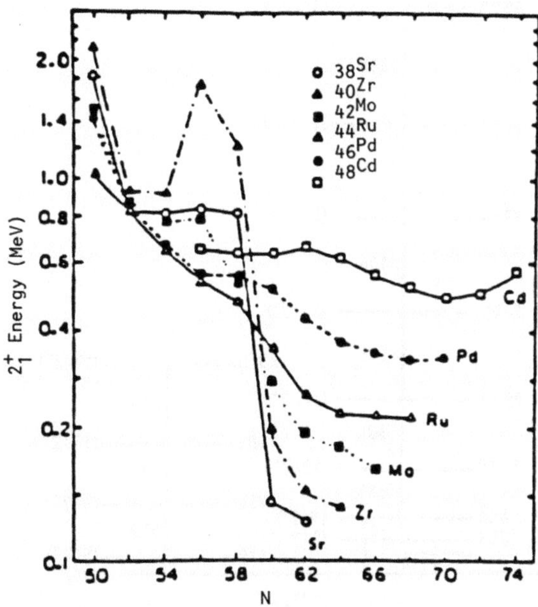

Fig. 3. The 2^+_1 energies for Sr to Cd nuclei with A = 50-70.

$^{75,73,71}_{35}Br_{40,38,36}$ since its Z = 35 is in the center of the expected new region and the N values span the prolate-oblate phase transition region. Unfortunately, the lightest of these isotopes has less than 1% of the total cross section in a heavy ion reaction. Thus, special techniques had to be employed in order to identify the very low cross section products in a heavy ion reaction.

First a five sector neutron detector[13] was built for in-beam n-γ, n-n-γ, and n-γ-γ coincidence studies which have been carried out at the Holifield Heavy Ion Research Facility and the University of Notre Dame. Then this detector was used with the recoil mass spectrometer at the University of Rochester for recoil-mass-n-γ studies. By combining these data, the energy levels in ^{73}Br were identified for the first time (see Fig. 4) and new high spin states in ^{75}Br (ref. 14). The strongest band seen to high spin in both ^{73}Br (Fig. 4) and ^{75}Br is the one built on the $g_{9/2}$ orbital. The transition energies in this band in ^{75}Br, beginning with the $13/2^+ \rightarrow 9/2^+$ transition are 563, 830, 1045, 1209, and 1325, respectively. By comparing these with Fig. 4 one sees that these two bands are nearly identical. (Recently some additional very low energy transitions have been placed[15] below the $9/2^+$ band in ^{73}Br compared to Fig. 4, but the $9/2^+$ bands are identical). The $\Delta I = 2$ sequence of levels in the $9/2^+$ bands in $^{73,75}Br$ are characteristic of a prolate rotor with very large deformation ($\beta \sim 0.35$).

A new five separated sector neutron detector and four large solid angle NaI detectors for light charged particles[6] were built. These detectors were used with the Rochester recoil mass spectrometer in the reaction $^{16}_{8}O + ^{58}_{28}Ni \rightarrow (^{74}Kr)^*$ to search for $^{71}_{35}Br_{36}$. From a comparison of recoil-mass-n-γ and recoil-mass-p-γ coincidences, we identified several transitions as belonging to ^{71}Br (ref. 16). All have energies in the range of 200-400 keV. Since the strongest cascade transitions in $^{73,75}Br$ are in the $9/2^+$ band, we assume this will be the case for ^{71}Br. If this is so, then the low energy of the transitions observed there would indicate a $\Delta I = 1$ sequence in contrast to the $\Delta I = 2$ sequence

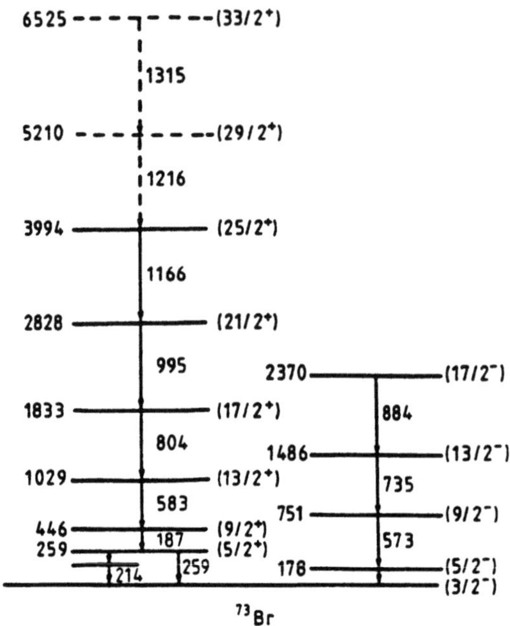

Fig. 4. Energy levels in ^{73}Br from n-γ and recoil mass-n-γ coincidence data[14]. Additional transitions have been placed below the $9/2^+$ level.[15]

in ^{73,75}Br. Based on the single particle orbitals available, a $\Delta I = 1$ sequence is expected for an oblate shape. Thus, these data may be an indication of a sudden phase shift from a very large prolate ground state in ^{73,75}Br to a very large oblate shape for the ground state of ^{71}Br. Analysis of n-n-γ and n-γ-γ data are in progress to establish the bands and spin sequence in ^{71}Br.

As described in more detail in refs. 1, 2, the concept of reinforcing proton and neutron shell gaps illuminates a variety of other regions such as the reinforcing of the spherical subshell gaps at Z = 64 by the strong spherical shell gap at 82 to make $^{146}_{64}Gd_{82}$ a double closed shell nucleus. Indeed, the spherical subshell gap at Z = 64 plays the same role that the N = 56 spherical gap does in the Sr-Zr region in the sudden onset of deformation as discussed above. It is the influence of the spherical Z = 64 gap that keeps the $_{62}$Sm, $_{64}$Gd and $_{66}$Dy nuclei spherical further out from N = 82 than they would otherwise be, so there is a sudden onset of deformation between N = 88 and 90 for these three elements. However, the suddenness of the onset of deformation between N = 88 and 90 disappears for 62 > Z > 66.

New regions and types of nuclear shape coexistence are continuely being found throughout the periodic table. For example, a new and different type of shape coexistence was found this year in the very light Pt nuclei ^{176,178}Pt (refs. 17, 18). Dracoulis et al.[17] studied the yrast cascades in $^{176,178}_{78}$Pt. In Fig. 5 (from ref. 18) their data are compared with similar data for $^{172,174}_{74}$W and $^{174,176}_{76}$Os (refs. 19, 20). The behaviors of the high spins ($\gtrsim 6^+$) states in ^{172}W, ^{174}Os, and ^{176}Pt, as shown in Fig. 5, are remarkably smilar and characteristic of a well-deformed rotor. However, there is a strong deviation in the 2^+ level of ^{176}Pt. Quite similar behavior is seen for the yrast energy levels in the N = 100 isotones except only a small perturbation is seen at 2^+ in ^{178}Pt. Dracoulis et al.[17] interpreted the ^{176}Pt low spin difference to the coexistence of two shapes with quite different deformations. Independently and simultaneously, potential energy surface

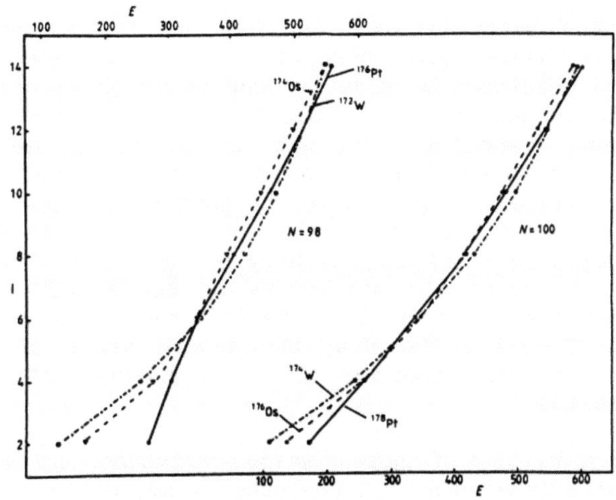

Fig. 5. Yrast level energies vs. nuclear spin in ^{176,178}Pt, ^{174,176}W and ^{172,174}Os.

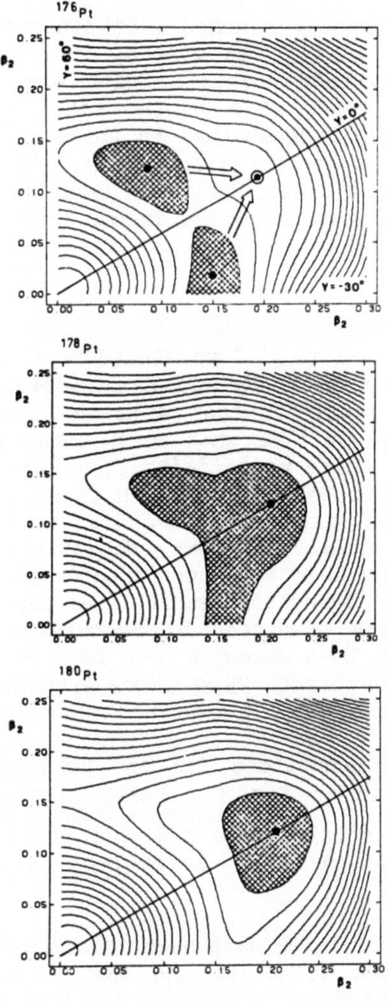

Fig. 6. Potential energy surfaces for $^{176-180}$Pt (ref. 18).

calculations were being carried out for the Pt isotopes.[18] These surfaces (Fig. 6) independently predicted the coexistence of two shapes in ^{178,176}Pt and predicted a sudden ground state phase transition from a strong prolate ground state ($\beta \sim 0.23$) in ^{178}Pt to a weakly deformed ($\beta \sim 0.12$) ground state for ^{176}Pt with a coexisting, nearby, excited prolate deformed shape which had been the ground state of ^{178}Pt. As ^{176}Pt begins to rotate, the minima in the potential energy quickly goes to the prolate shape (flate space in Fig. 6). The weakly deformed band is the excited band in ^{178}Pt and slightly perturbs the 2^+ energy there. Dracoulis et al.[17] noted the similarity of the shape coexistence in ^{176}Pt to that in ^{184}Hg. However, the potential enrgy surface calculations point to a significant difference; the ground state of ^{184}Hg has a small oblate deformation while for ^{176}Pt the ground state has a small, triaxial deformation.

In summary, varieties of nuclear shape coexistence are seen throughout the periodic table (see refs. 1,2 for other examples). In addition, we have firm evidence for new "deformed" magic numbers, 38 for N and Z and 60 for N, which manifest themselves when the proton and neutrons have shell gaps at the same deformation so both the protons and neutrons drive the nucleus toward the same large, prolate deformation and tentative evidence for the $N = Z = 36$ shell gaps at large, oblate deformation. As seen in Fig. 2, there are numbers of other shell gaps at different deformations. The importance of many of these gaps will undoubtedly be seen when nuclei far off stability with the right reinforcement of proton and neutron shell gaps at the same deformation are studied. These data all clearly suggest that in the collision of U on U or U on Cm one could expect to see the stabilizing influence of other new shell gaps, and certainly we expect to see a much wider range of collective phenomena associated with these collisions than we have seen.

EXOTIC DECAY MODES

In 1980 Sandulescu et al.[4] predicted from their calculations a new type of radioactivity for heavy nuclei intermediate between fission and alpha decay. They predicted that a heavy nucleus could emit a heavy cluster such as ^{14}C or ^{24}Ne when such emissions yielded a daughter nucleus at or near double magic $^{208}_{82}Pb_{126}$. Heavy cluster emission is a new collective type of decay mode which is an additional manifestation of the strong nuclear shell structure associated with spherical double magic ^{208}Pb. The first heavy cluster radioactivity, the ^{14}C radioactivity of ^{223}Ra, was discovered by Rose and Jones[21] who were apparently unaware of the predictions of Sandulescu et al.,[4] as noted in ref. 22. As shown in Fig. 7, ^{223}Ra was predicted by Sandulescu et al.[20] to have the largest ratio of ^{14}C to α decay of any isotope. An earlier review of this process is found in ref. 23 and more extensive theoretical analysis and predictions in the recent paper of Poenaru et al.[5]

Basically, the conditions for the splitting of nucleus A_Z into A_{Z_1} + $A2_{Z_2}$ are shown in Fig. 8. Heavy cluster radioactivity occurs when the potential is like curve 2 with a positive Q value and $E_i > 0$. For a potential like curve 1, the nucleus is stable and for curve 3 completely unstable.

Now ^{14}C radioactivities have been observed for $^{222-226}$Ra (refs. 21, 24-28). The expected lifetimes and the ^{14}C/α branching ratios are compared with the original theoretical calculations of Sandulescu et al.[4] and recent calculations from Frankfurt[29] in Fig. 9. While the original calculations do not reproduce the ^{222,224}Ra results, by increasing

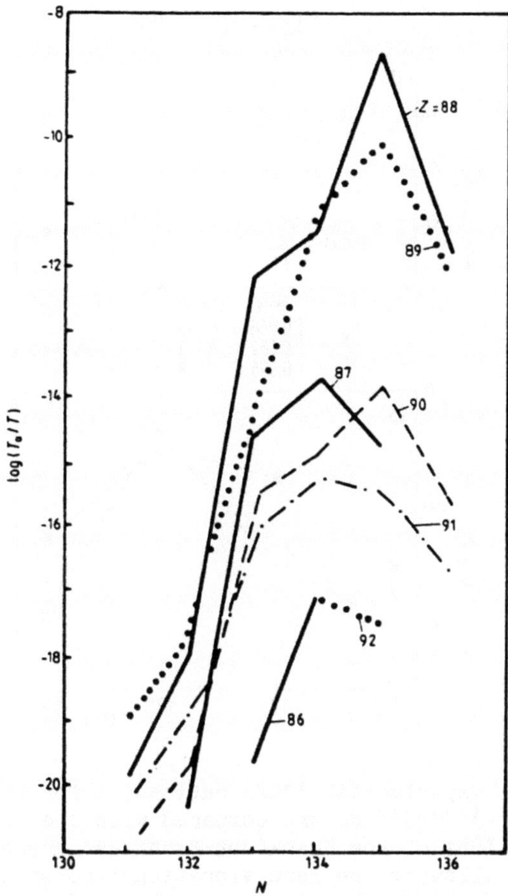

Fig. 7. The $^{14}C/\alpha$ ratios and α partial lifetimes from ref. 4.

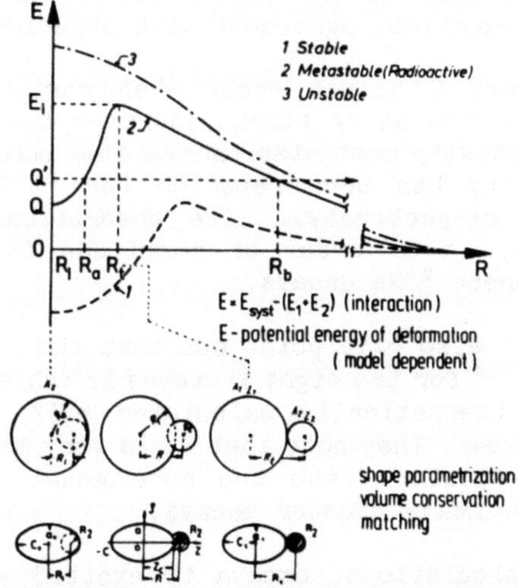

Fig. 8. Possible potential energies for a heavy nucleus. Curve 1 would give a stable nucleus; curve 2 where the Q value is greater than zero would give a radioactive nucleus, and; curve 3 is an unstable nucleus.

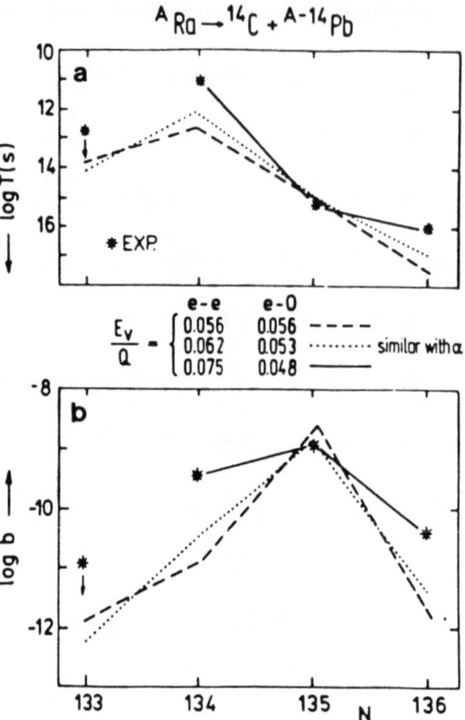

Fig. 9. The experimental $^{14}C/\alpha$ ratios (b) for the decays of ^{222,223}and^{224}Ra are compared with the recent calculations of the Frankfurt-Bucharest group.[29] Note by allowing the zero vibrational point energy to increase for the e-e and decrease for the e-o cases, excellent agreement between theory and experiment is obtained.

the zero point vibrational energy for the even-even and reducing it for the even-odd cases, excellent agreement with experiment is achieved.[29]

Since the largest Q values occur when one daughter product is ^{208}Pb, as the mass of the heavy elment increase the mass and Z of the heavy cluster radioactivity must also increase as illustrated in Fig. 10. Now ^{24}Ne radioactivity has been reported for ^{231}Pa, ^{232}U, ^{233}U and ^{230}Th (refs. 30-33, respectively). The theoretical calculations[5] of the lifetimes are up to an order of magnitude faster than observed experimentally for these ^{24}Ne decays.

Barwick et al.[31] also have point out that the reported spontaneous fission half-lives[34-40] for the eight isotopes in Table 2 are in reasonable agreement with the theoretically calculated half lives[4,5] for heavy cluster radioactivities. They note that it is very likely that what were called spontaneous fission (into two more equal size fragments) in these eight cases are heavy cluster decays.

In the first calculations, decays to excited states of the heavy daughter were neglected[4,5] and this could introduce a serious correction. Very recently Greiner and Scheid[41] have calculated the corrections to the partial half lives for heavy cluster decays to the ground states[4,5] by calculating the total transition rates including those to excited states of the heavy daughter fragments. They find that this increases

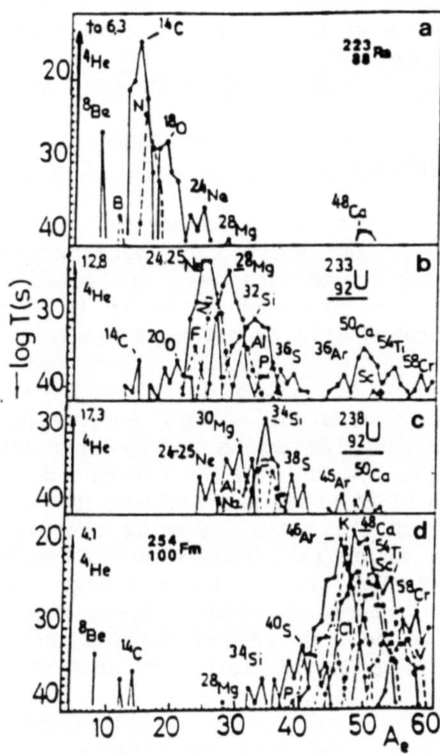

Fig. 10. The heavy cluster radioactivities predicted[5] for heavy elements are shown.

Table 2. Half-Lives Predicted[4,5] for Heavy Ion Emission Compared With Measured Spontaneous Fission Half-Lives[36-42].

Decay Mode	Predicted τ_X	Measured τ_{SF}	Ref.
$^{232}U \rightarrow {}^{24}Ne$	1×10^{13}	6×10^{13}	36
$^{232}Th \rightarrow {}^{26}Ne$	1×10^{22}	$\geq 1 \times 10^{21}$	37
$^{231}Pa \rightarrow {}^{24}Ne$	4×10^{14}	$\geq 1.1 \times 10^{16}$	38
$^{230}Th \rightarrow {}^{24}Ne$	3×10^{17}	$\geq 1.5 \times 10^{17}$	38
$^{233}U \rightarrow {}^{25}Ne$	9×10^{15}	1.2×10^{17}	39
$^{234}U \rightarrow {}^{28}Mg$	2×10^{17}	1.6×10^{16}	40
$^{237}Np \rightarrow {}^{30}Mg$	3×10^{18}	$\geq 1 \times 10^{18}$	41
$^{241}Am \rightarrow {}^{34}Si$	2×10^{15}	2.3×10^{14}	42

the decay rates even in unfavorable cases by at most not more than half an order of magnitude and so is not a serious correction to the original calculations.[4,5] This amount is within the overall accuracy of the calculations. The effect is essentially zero when one daughter is ^{208}Pb whose excited states are very high in energy.

Poenaru et al.[42] have now calculated the possibility of heavy cluster emission from lighter nuclei. They find that all stable nuclei with Z $\gtrsim$ 40 are radioactive with respect to heavy cluster emission but with lifetimes in the range $\sim 10^{40}$ to 10^{50}s.

A NEW NUCLEAR COLLECTIVE PHENOMENA: VERY ENERGETIC PROTON AND ALPHA EMISSION IN HEAVY ION REACTION

Recently Maguire (at Vanderbilt) as part of a collaboration with Argonne, Michigan, Kansas, and Notre Dame constructed several large solid angle NaI detectors to measure very energetic light ions with energies up to 10 times the incident MeV/u in heavy-ion reactions.[6] The detectors were calibrated with 80 MeV α particles on CH_2.

The reactions 300 MeV ^{32}S (9.38 MeV/u) + Ta and 600 MeV ^{58}Ni (10.34 MeV/u) + Ta were studied.[7] Very surprisingly, significant numbers of protons with energies up to and greater than 100 MeV and alpha particles with up to 150-200 MeV were observed. These particles are coming out with up to 10-20 times the incident MeV/u and are carrying off up to 50% and more of the total incoming energy. There is some strong collective effect which is giving rise to the concentration of such a large fraction of the total incident energy into one particle. At present there is no theory to predict such energetic particles. There is much research to be done to investigate when and where these particles are emitted during the collision and what other fragments are in coincidence with them.

Even though one does not have any theoretical understanding, these exciting results generate numbers of interesting speculations. In relation to this conference, one can speculate whether in collisions of U on U or U on Cm at energies in the range of 5-10 MeV/u such fast protons or alpha particles may be emitted at such times and locations as to cool the reaction and lead to longer sticking times for the two heavy nuclei. Also, it is known that there is enhanced alpha emission along the long axis of a prolate deformed nucleus in radioactive decay. If the most favored orientations for two heavy nuclei like U to collide and stick is end-to-end, as has been suggested,[43] there could be added enhancement to such fast alpha or fast proton emission along this axis which, in turn, could increase their sticking time. There could be other consequences as well.

SUMMARY

The thrust of this paper was to illustrate that we are continuing to find in numerous, diverse ways throughout the periodic table ($Z \lesssim 92$) new manifestations of collective nuclear behavior. These include various types of nuclear shape coexistence for low-lying levels, new "magic" numbers for deformed shapes which give stability, for examples, to very large prolate and very large oblate shapes, new collective decay modes such as heavy cluster emission, and collective concentration into a single outgoing proton or alpha particle a large fraction (up to 50% and more) of the incident energy in a heavy ion collision. The wealth and diversity of these collective phenomena strongly suggest that in the collision of two heavy nuclei, like uranium and uranium or curium, that there should be a variety of new collective phenomena including previously inaccessible, exotic collective phenomena.

Already, Greiner[43] has suggested that if two such heavy nuclei collide and stick for even 10^{-19}s that one could see new collective excitation such as a "butterfly" type motion where the opposite ends of the two touching nuclei oscillate up and down about their touching

Fig. 11. Possible new collective excitations associated with two heavy ions which have stuck together as shown (from ref. 43).

point or when the angle between the two symmetry axes is not 180° when at the touching point, they can rotate about a common center of gravity (see Fig. 11). These are but two new easy to visualize collective motions and considerations of other collective behaviors should be made. Some of these phenomena may have strong bearings on the positron line production. Already, at this conference, Raefelski proposed a new way of producing e^+e^- pairs. Also at this conference, Oberacker suggested that the excitation of even one uranium nucleus into its second minima, which has a much large deformation than its ground state, would significantly change the Coulomb energy and so alter possible potential pockets which could lead to sticking. We should look at the possibility that there are new shell gaps at large deformation including very large deformations associated with second minima in the potentials for Z in the range 180-190 and N of 290-298 which could help stabilize two colliding heavy nuclei. One should explore whether the new observed emission of a very energetic proton or alpha particle could, if present in these heavy ion collisions, significantly cool the system so as to influence this sticking of the two nuclei. Finally, it is possible that there is more than one origin for the positron peaks being observed. So, in different colliding nuclear systems or even in the same system, positron lines from more than one origin may be confusing the theoretical interpretations.

This work was supported in part by a the US Department of Energy, Division of Basic Sciences under contract No. DE-AS05-76ER-05034.

REFERENCES

1. J. H. Hamilton, P. G. Hansen, and E. F. Zganjar, in: Reports on Progress in Physics 48: 631 (1985).
2. J. H. Hamilton, in: Nucleus-Nucleus Collisions from the Coulomb Barrier Up to the Quark-Gluon Plasma, (Proc. Int. School of Nuclear Physics), in "Progress in Particle and Nuclear Physics," Vol. 15, A. Faessler, ed., Pergamon Press, NY (1985), p. 107.
3. J. H. Hamilton and J. Maruhn, Scientific American (July, 1986), p. 80.
4. A. Sandulescu, D. N. Poenaru, and W. Greiner, Sov. J. Part. Nucl. 11:528 (1980).
5. D. N. Poenaru, M. Ivascu, A. Sandulescu, and W. Greiner, Phys. Rev. C 32: 572 (1985).
6. C. F. Maguire, D. G. Kovar, C. Beck, C. N. Davids, D. Henderson, M. F. Vineyard, F. W. Prosser, S. V. Reinart, J. J. Kolata, Bull. Am. Phys. Soc. 30:1769 (1985).
7. P. Shulman, F. Becchetti, J. Janecke, R. Stern, D. Kovar, C. Davids, M. Vineyard, C. Beck, and C. F. Maguire, Bull. Am. Phys. Soc. 31:840 (1986).

8. M. Baranger and R. A. Sorenson, Scientific American (1969), p. 58.
9. J. H. Hamilton, R. B. Piercey, R. Soundranayagam, A. V. Ramayya, C. F. Maguire, X.-J. Sun, Z. Z. Zhao, J. Roth, L. Cleemann, J. Eberth, T. Heck, W. Neumann, M. Nolte, R. L. Robinson, H. J. Kim, S. Frauendorf, J. Döring, L. Funke, G. Winter, J. C. Wells, J. Lin, A. C. Rester, and H. K. Carter, in Proc. IV Int. Conf. On Nuclei Far From Stability, CERN81-09 (1981), p. 391; R. B. Piercey, J.H. Hamilton, R. Soundranayagam, A. V. Ramayya, C. F. Maguire, X.-J. Sun, Z. Z. Zhao, R. L. Robinson, H. J. Kim, S. Frauendorf, J. Döring, L. Funke, G. Winter, J. Roth, L. Cleemann, J. Eberth, W. Neuman, J. C. Wells, J. Lin, A. C. Rester, and H. K. Carter, Phys. Rev. Lett. 47: 1514 (1981).
10. P. Möller and R. A. Nix, At. and Nucl. Data Tables 26: 165 (1981).
11. R. Bengtsson, P. Möller, R. A. Nix, and J.-Y. Zhang, Physica Scripta 29:402 (1984).
12. M. Brack, J. Damgaard, A. S. Jensen, H. C. Pauli, V. M. Strutinsky, and C. Y. Wong, Rev. of Mod. Phys. 44:320 (1972).
13. R. B. Piercey, F. E. Dunnam, M. L. Muga, A. C. Rester, A. V. Ramayya, J. H. Hamilton, J. Eberth, and E. F. Zganjar, Int. Conf. on Instrumentation for Heavy-Ion Nuclear Research, Oak Ridge, Oak Ridge National Laboratory, CONF-841005, (1984), p. 27.
14. S. Wen, A. V. Ramayya, S. J. Robinson, C. F. Maguire, W. C. Ma, X. Zhao, T. M. Cormier, P. M. Stwertka, J. D. Cole, E. F. Zganjar, R. B. Piercey, M. A. Herath-Banda, J. Eberth, and M. Wiosna, J. Phys. G Lett. 11:L173 (1985).
15. B. Wörmann, J. Heese, K. P. Lieb, L. Lühmann, F. Raether, D. Alber, H. Grawe, and B. Spellmeyer, Z. Phys. A322:171 (1985).
16. X. Zhao, A. V. Ramayya, C. F. Maguire, J. Kormicki, W. C. Ma, S. Wen, J. H. Hamilton, Z-M. Chen, L. Chaturvedi, T. C. Cormier, M. Statteson, E. F. Zganjar, M. Kortelhathi, J. D. Cole, J. Eberth, N. Schmal, R. B. Piercey, and M. A. Herath-Banda, Bull. Am. Phys. Soc. 31:772 (1986).
17. G. D. Dracoulis, A. E. Stuchbery, A. P. Byrne, A. R. Poletti, S. J. Poletti, J. Gerl, and R. A. Bark, J. Phys. G:Nucl. Phys. 12:L97 (1986).
18. R. Bengtsson, J.-y. Zhang, J. H. Hamilton, and Leon K. Peker, J. Phys. G Letts. (1986), in press.
19. G. D. Dracoulis, P. M. Walker, and A. Johnson, J. Phys. G 4:714 (1978).
20. J. L. Durell, G. D. Dracoulis, C. Fahlander, and A. P. Byrne, Phys. Lett. 115B:367 (1982).
21. H. J. Rose and G. A. Jones, Nature 307:245 (1984).
22. A. Sandulescu, D. N. Poenaru, W. Greiner, and J. H. Hamilton, Phys. Rev. Lett. 54:490 (1985).
23. J. H. Hamilton, Proc. Fifth Adriatic Int. Conf. On Nuclear Physics, in: "Fundamental Problems in Heavy Ion Collisions," N. Cindro and W. Greiner, eds., Singapore, World Scientific, (1984), p. 111.
24. D. V. Alexandrov, A. F. Beliatsky, Yu. A. Gluhov, E. Yu. Nikolsky, B. G. Novatsky, A. A. Ogloblin, and D. N. Stepanov, Pis'ma Zh. Eksp. Teor. Fiz. 40:152 (1984).
25. S. Gales, E. Hourani, M. Hussonnois, J. P. Shapira, L. Stab, and M. Vergner, Phys. Rev. Lett. 53:759 (1984).
26. P. B. Price, J. D. Stevenson, and S. W. Barwick, Phys. Rev. Lett. 54:297 (1985).
27. W. Kutschera, J. Ahmad, S. G. Armato, III, A. M. Friedman, J. E. Gindler, W. Henning, T. Ishii, M. Paul, and K. E. Rehm, Phys. Rev. C 32:2036 (1985).
28. E. Hourani, M. Houssonnois, L. Stab, L. Brillard, S. Gales, and J. P. Shapira, Phys. Lett. 160B:375 (1985).

29. W. Greiner, private communication (1985).
30. A. Sandulescu, Yu. S. Zamyatnin, J. A. Lebedev, B. F. Myasoedov, S. P. Tretyakova, and D. Hasegan, JINR Rapid Communications 5:5 (1984).
31. S. W. Barwick, P. B. Price, and J. D. Stevenson, Phys. Rev. C 31:1984 (1985).
32. S. P. Tretyakova, A. Sandulescu, Yu. S. Zamyatnin, Yu. S. Korotkin, and V. L. Micheev, JINR Rapid Communications 7:23 (1985).
33. S. P. Tretyakova, A. Sandulescu, V. L. Micheev, D. Hasegan, J. A. Lebedev, Yu. S. Zamyatnin, Yu. S. Korotkin, and B. F. Myasoedov, JINR Rapid Communications 13:34 (1985).
34. A. H. Jaffey and A. Hirsch, in Nuclear Fission, by R. Vandenbosch and J. R. Huizenga, Academic Press, NY (1973).
35. G. N. Flerov et al., Dokl. Akad. Nauk SSSR 3:69 (1958) [Sov. Phys. Dokl. 3:79 (1958)].
36. E. Segre, Phys. Rev. 86:21 (1951).
37. B. M. Alexandrov, L. S. Krivokhatski, L. Z. Malkin, and K. A. Petrzhak, Sov. J. At. Energy 20:352 (1966).
38. A. Ghiorso, G. H. Higgins, A. E. Larsh, G. T. Seaborg, and S. G. Thompson, Phys. Rev. 87:163 (1952).
39. V. A. Druin, V. P. Perelygin, and G. I. Khlebnikov, Zh. Eksp. Teor. Fiz. 40:1296 (1961) [Sov. Phys. JETP 13:913 (1961)].
40. V. A. Druin, V. L. Mikheev, and N. K. Skobelev, Zh. Eksp. Teor. Fiz. 40:1261 (1961)[Sov. Phys. JETP 13:889 (1961)].
41. M. Greiner and W. Scheid, J. Phys. G Lett. (in press).
42. D. N. Poenaru, W. Greiner, M. Ivascu, and A. Sandulescu, Phys. Rev. C 32:2198 (1986).
43. W. Greiner, Nucl. Phys. A447:271c (1985).

IONISATION AND TUNNELING IN A STRONG ELECTRIC FIELD

W. Elberfeld and M. Kleber

Physik-Department, Technische Universität Münchën
D-8046 Garching, West Germany

ABSTRACT

This lecture can be of use to those physicists who are interested in the following questions:

i) What is the dynamics of laser-induced ionisation in the strong-field limit?

ii) How does an electric field help an electron to tunnel through a barrier?

iii) What is the traversal time for tunneling?

LASER-INDUCED IONISATION

The dynamics of a bound charged particle becomes quite puzzling when a strong electromagnetic field is present. For example, it takes a computer an almost infinite number of logical operations to find out what happens to a hydrogen atom when it is exposed for a nanosecond to a strong laser pulse. The reason is that such a pulse lasts about 10^8 times longer than the orbiting time of a 1s-electron. ($\hbar^3/(me^4) = 2.4 \cdot 10^{-17}$ sec). Therefore, the time evolution of the wave function has to be calculated over millions of electronic revolutions. The situation is less dramatic for weakly bound electrons. Numerical studies have been done,[1,2] but we believe that ultimately physics will not require calculations which can be carried out only on a big computer.

Let us forget about the long-range Coulomb field,[3] and let us study the effect of a strong oscillating electric field acting on an electron which is bound by a one-dimensional delta potential. For this problem the Hamiltonian is (in atomic units)

$$H = -\frac{1}{2}\,\partial_x^2 - B\,\delta(x) + E(t)\cdot x \quad . \tag{1.1}$$

In the following we shall assume that a sinusoidal classical field is switched on at t = 0:

$$E(t) = F_0 \cos(\Omega t + \varphi) \quad , \quad t > 0 \quad . \tag{1.2}$$

In the absence of the electric field, the one-dimensional atom consisting of one electron in the delta potential, has a single bound state of energy $\epsilon = -\frac{1}{2}B^2$. At t = 0, the initial wavefunction of the electron is therefore given by

$$\psi(x,\ t = 0) = B^{1/2}\, e^{-B|x|} \quad . \tag{1.3}$$

An intense electric field means that the field strength F_0 is large compared to typical intrinsic atomic field strengths which are of the order B^3. In other words, if f is the relative field strength, then

$$f \equiv \frac{F_0}{B^3} \gg 1 \quad . \tag{1.4}$$

In the strong-field limit, $f \gg 1$, perturbation theory does not apply. We can then, however, neglect the binding potential in (1.1). This simplifies the problem enormously. In fact, it is easy to calculate the time evolution of the wavefunction (1.3) with the strong-field Hamiltonian

$$H' = -\frac{1}{2}\,\partial_x^2 + E(t)\cdot x \quad . \tag{1.5}$$

The result is (t > 0)

$$\psi(x,t) = \sqrt{B}\, \exp[-i\, S_{c\ell}(t) + i\, p_{c\ell}(t)\cdot x] \tag{1.6}$$

$$* \; [M(x - x_{c\ell},\ -iB, t) + M(x_{c\ell} - x,\ -iB, t)] \quad .$$

The phase on the right hand side of (1.6) contains the (kinetic) momentum of a classical free particle,

$$p_{c\ell}(t) = p_{c\ell}(0) - \int_0^t dt'\ E(t') \quad , \tag{1.7}$$

driven by the electric field E(t). The other classical quantities are $x_{c\ell}$ (position) and $S_{c\ell}$ (action), with $\dot{x}_{c\ell} = p_{c\ell}$ and $\dot{S}_{c\ell} = \frac{1}{2}p_{c\ell}^2$. In our problem, $x_{c\ell}(0) = p_{c\ell}(0) = S_{c\ell}(0) = 0$.

The last bracket on the rhs of (1.6) consists of two Moshinsky functions, defined by

$$M(x,k,t) = \frac{1}{2}\, e^{i(kx - k^2 t/2)}\ \mathrm{erfc}\left[\frac{x - kt}{\sqrt{2it}}\right] \quad , \tag{1.8}$$

where erfc(y) denotes the complementary error function[4]. We would like to remark that M(x,k,t) is the solution[5] of the one-dimensional free-particle Schrödinger equation for an initial wavefunction

$$M(x,k,0) = \Theta(-x)\, e^{ikx} \quad . \tag{1.9}$$

In other words, if at $t = 0$ a plane wave is confined to the negative x-axis, then $M(x,k,t)$ is the corresponding wavefunction for $t > 0$. Note that M depends on two dimensionless variables only. For example, we could write $M(x,k,t) = \tilde{M}(x/t^{1/2}, kt^{1/2})$.

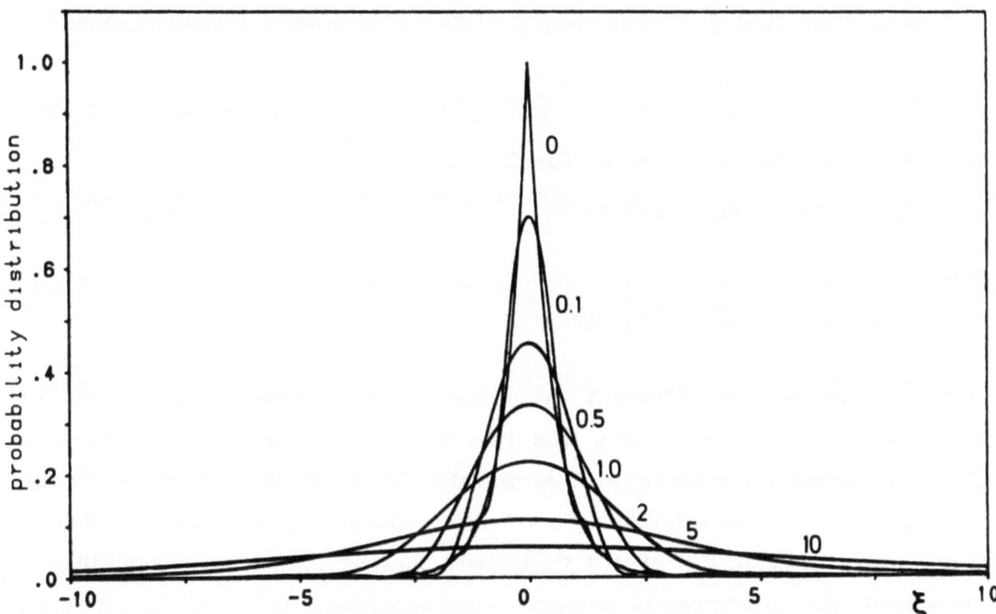

Fig. 1. Probability distribution $|M(\xi,-i,t) + M(-\xi,-i,t)|^2$ for $t = 0$, 0.1, 0.5, 1, 2, 5 and 10.

In fig. 1 we have plotted $|\psi(x,t)|^2$ as a function of $\xi = x - x_{c\ell}(t)$ for various values of t. If we measure time in units of B^{-2} and lengths in units of B^{-1}, then we may choose $B = 1$ without loosing generality. We should mention that the propagator U (time evolution operator), defined by

$$\psi(x,t) = \int dx'\, U(x,t\, ;\, x',0)\, \psi(x',0) \quad , \tag{1.10}$$

can be easily written down:

$$U = (2\pi it)^{-1/2} \exp[-i\, S_{c\ell}(t) + i\, p_{c\ell}(t)\, x]\, \exp[i(x - x_{c\ell}(t) - x')^2/2t] \quad . \tag{1.11}$$

A relativistic version of the so-called Volkov problem (1.5) can be found in Mitter's contribution[6] to the Lahnstein conference.

The calculation of the ionisation probability is somewhat tricky. One can do this job in two different ways, and one can get two different answers for the ionisation probability. We have chosen the $\underline{E}\cdot\underline{r}$ interaction Hamiltonian in (1.1). There is, however, another choice: the $\underline{p}\cdot\underline{A}$ interaction Hamiltonian. In this case, problem (1.1) reads

$$i\,\partial_t\tilde{\psi}(x,t) = [-\frac{1}{2}(\partial_x + \frac{i}{c}A(t))^2 - B\delta(x)]\tilde{\psi}(x,t) \quad . \qquad (1.12)$$

To be consistent with (1.1), we have used the dipole approximation which means that both the electric field and the vector potential are functions of t only. Note that $\tilde{\psi}$ differs from ψ (1.6) by a unitary transformation

$$\tilde{\psi}(x,t) = \psi(x,t)\ \exp[-\frac{i}{c}A(t)\cdot x] \quad . \qquad (1.13)$$

Therefore, the two expressions for the ionisation probability,

$$P(\tau) = 1 - |\langle\psi(0)|\psi(\tau)\rangle|^2 \qquad (1.14)$$

and

$$\tilde{P}(\tau) = 1 - |\langle\tilde{\psi}(0)|\tilde{\psi}(\tau)\rangle|^2 \qquad (1.15)$$

will in general be different from each other, even if $A(0) = 0$, i.e. $\psi(x,0) = \tilde{\psi}(x,0)$. At first sight this result looks like a paradox. There is, however, a simple answer to this puzzle. In (1.14) an experiment would be carried out in the following way: One switches on the electric field at $t = 0$ and turns it off at $t = \tau$; afterwards the ionisation probability is measured. An experimental measurement according to (1.15) would mean that one turns off the vector potential A(t) instead of E(t). Since

$$A(t) = -c\int dt\ E(t) \qquad (1.16)$$

it appears unphysical to think of an experiment[7] where A(t) is turned off. Let us therefore calculate P(t) according to (1.14). By making use of (1.6) it is possible to express P(t) in terms of Moshinsky functions. The reader who wants to know more about gauge invariance in the context of interaction Hamiltonians in quantum optics should consult the paper of Schlicher et al.[8].

The ionisation probability is depicted in figs. 2 and 3. Since $\varphi = 0$ has been studied by Geltman, we use this value of φ in (1.2). Keeping in mind that typical atomic frequencies are of the order B^2 we show one set of results for $\Omega/B^2 \gg 1$ and one for $\Omega/B^2 \ll 1$. It is interesting to observe that the ionisation probability is quasiperiodic in t only for large Ω. This result can be understood from the following argument. In the strong-field approximation the initial wave packet (1.3) feels the electric field but apart from this, it is essentially free. Quantum dispersion of the free wavepacket (1.3) occurs during a spreading time t_s which is of the order B^{-2}. On the other hand, the wave packet oscillates about its initial position with the field frequency Ω.

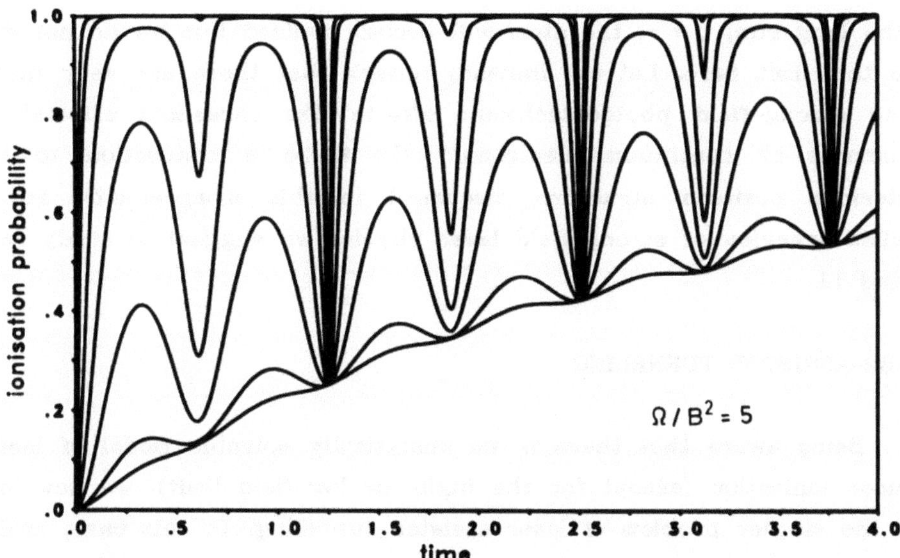

Fig. 2. Ionisation probability as a function of time for f = 1, 2, 5, 10, 20, 50 and 100. The reader will easily identify a curve with fixed value of f, because oscillatory structures become more pronounced when $f = F_0/B^3$ increases. Curves were derived under the assumption f >> 1.

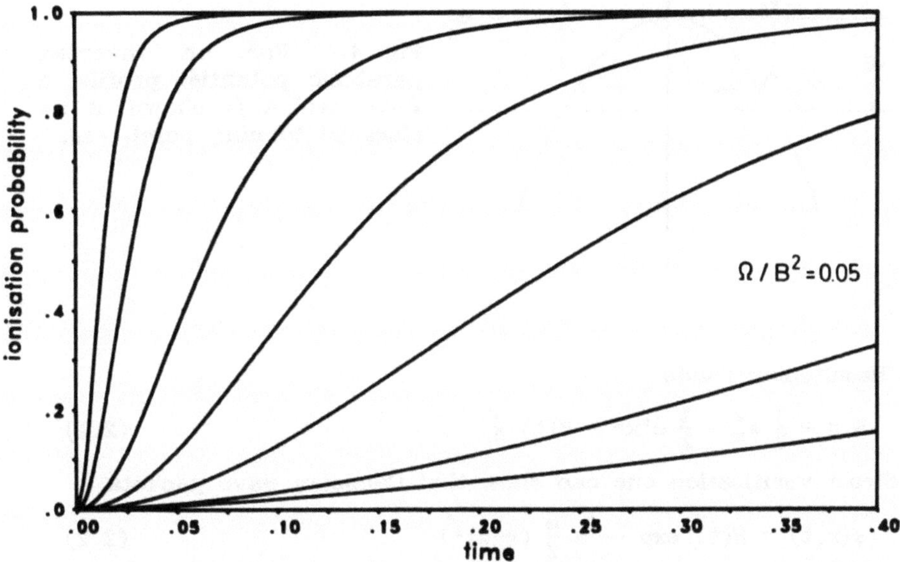

Fig. 3. Same as in fig. 2, except that now $\Omega/B^2 = 0.05$. In analogy to fig. 2, the ionisation probability is relatively flat for f = 1 and rises sharply for f = 100.

Whenever the wave packet comes back to its initial position, the ionisation probability goes through a minimum. However, this effect will only show up if there are many oscillations during the spreading time, i.e.

$$\Omega t_s \approx \Omega/B^2 \gg 1 \quad .$$

In the long run, $t \to \infty$, the atom will become ionized but we do not discuss this limit here. Let us, instead, remark that there are very interesting strong-field photodetachment effects like threshold effects[9] or resonances of continuum electrons[10] (Is there a connection to the mysterious positron structure discussed in this symposium?). For a detailed overview of strong-field laser physics we suggest to study refs. 11 and 12.

LASER-ASSISTED TUNNELING

Being aware that there is no analytically solvable model of laser-induced ionisation (except for the high- or low-field limit), we now look into the simpler problem of laser-assisted tunneling. In this case, analytic results can be obtained for the penetration of a wave packet through an inverted parabolic barrier (fig. 4) in the presence of an arbitrarily strong laser field (1.2).

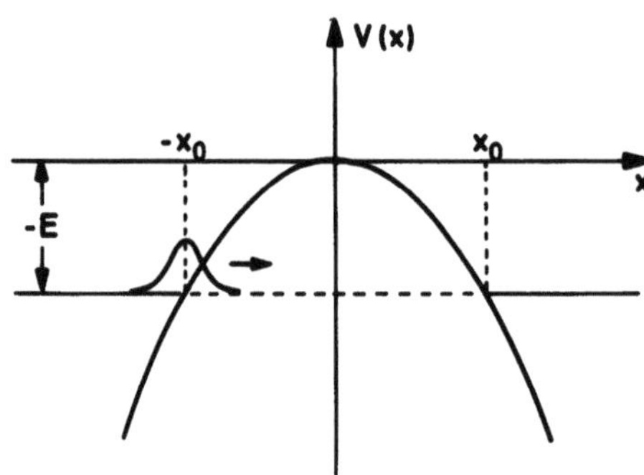

Fig. 4. For an inverted parabolic potential profile, a wave packet is shown at its classical turning point, $-x_0$.

The Hamiltonian reads

$$H = -\frac{1}{2}\partial_x^2 - \frac{1}{2}\omega^2x^2 + E(t)\cdot x \quad . \tag{2.1}$$

By direct verification one can show that Gaussian wave packets

$$\psi(x,t) = N(t)\ \exp\{-\lambda\frac{\omega}{2}(x-\alpha)^2\} \tag{2.2}$$

are solutions of the time-dependent problem

$$(H - i\partial_t)\ \psi(x,t) = 0 \quad , \tag{2.3}$$

if the complex parameters α and λ satisfy the two differential equations

$$i\ \dot{\alpha}\lambda\omega = E(t) - \alpha\omega^2 \quad , \tag{2.4}$$

$$i\ \dot{\lambda}\omega = (\lambda\omega)^2 + \omega^2 \quad . \tag{2.5}$$

It is straightforward to calculate the tunneling probability P of such a wave packet. For a minimum uncertainty wave packet one obtains[13]

$$P = \frac{1}{2}\,\text{erfc}\,(\sqrt{\epsilon} - \rho\,\sin(\varphi-\gamma)) - \frac{1}{2}\,\text{erfc}\,(\sqrt{8\epsilon}) \qquad (2.6)$$

with $\epsilon = - E/\omega$, the field parameter $\rho = F_0\,\omega^{1/2}/[2(\omega^2 + \Omega^2)]^{1/2}$, and $\tan\gamma = \omega/\Omega$. In deriving (2.6) we assumed that the centroid $\langle x\rangle$ of the wave packet reaches the classical turning point, $-x_0$, with $\langle p\rangle = 0$ and minimum uncertainty $\Delta p\Delta x = 1/2$. Furthermore, the electric field (1.2) is switched on when the particle reaches the turning point at $t = 0$. Therefore the tunneling probability P depends on the initial phase φ of the field. For $\sin(\varphi-\gamma) = 1$, laser-assisted-tunneling reaches a maximum value of P; for $\sin(\varphi-\gamma) = -1$ one has minimum P. Since we do not know the initial phase in general, we should average (2.6) over φ. This has been done in fig. 5 for various values of ϵ and ρ. For the maximum value of P see ref. 13.

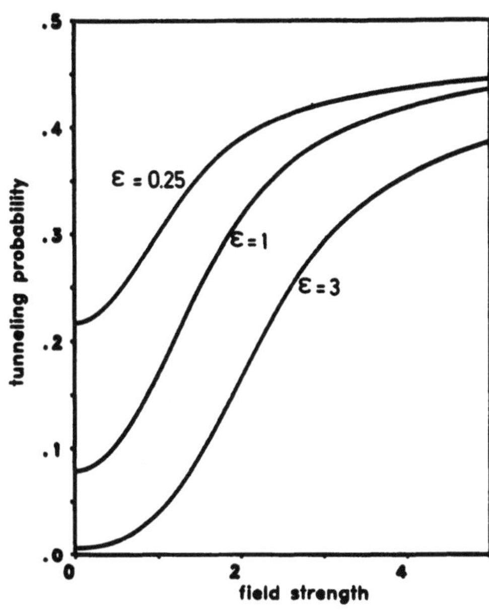

Fig. 5. Averaged tunneling probability as a function of the field strength ρ (see (2.6)) for different values of $\epsilon = - E/\omega$.

WAVE PACKETS VS. INFINITE WAVE

It is legal to ask how long it takes a particle to traverse the barrier. The answer will, of course, depend on how the wavefunction is prepared at some initial time t_i. Suppose, we know the initial preparation. We then gain information about the tunneling dynamics if we measure the tunneling current behind the barrier. A discussion of this current in the presence of an electric field would be too lengthy here. Let us consider only the zero-field limit. The current right behind the parabolic barrier ($x = x_0$, see fig. 4) can be calculated analytically for the minimum uncertainty wave packet (2.2) with $\lambda(t = 0) = 1$ and

$\alpha(t = 0) = - x_0$. The result is

$$j(x_0,t) = \omega \sqrt{\frac{2\epsilon}{\pi \cosh(\omega t)}} \, \frac{\sinh(2\omega t) + \sinh(\omega t)}{\cosh(2\omega t)}$$

$$* \exp\left[- 2\epsilon \frac{(1 + \cosh(\omega t))^2}{\cosh(2\omega t)}\right] \quad . \tag{3.1}$$

At this point of the discussion we may feel uneasy about a pathological feature of the parabolic barrier: it has infinite range. Let us, therefore, discuss tunneling through a localised barrier. If we choose a delta potential barrier, then the Hamiltonian reads

$$H_\delta = - \frac{1}{2} \partial_x^2 + V_0 \, \delta(x) \quad . \tag{3.2}$$

It is very interesting to see that, for this problem, the propagator is

$$U_\delta(x,t;x',0) = U_0(x-x',t;0,0) - V_0 \, M(|x|+|x'|, - i \, V_0,t) \quad . \tag{3.3}$$

In (3.3), U_0 denotes the free propagator and M the already discussed Moshinsky function. By making use of (3.3) one can calculate analytically[14] the time evolution of a certain class of wave packets, including a semi-infinite wave train.

Without going into details we present the results for two cases:

i) a Gaussian wave packet runs against the barrier (figs. 6,7),

ii) a semi-infinite wave train reaches the barrier at $t = 0$ from the left (figs. 8,9).

As one can see from the tunneling current, there is no appreciable time-delay in the transmission of the packet through the barrier (compared to a free passage). Such a behaviour was found more than 50 years ago by Mac Coll[15] who used semiquantitative methods. In order to understand the result, we must keep in mind that a wave packet contains different Fourier components; those with $E > 0$ allow the particle to avoid tunneling by going over the barrier. For a long wave train, on the other hand, there is time-delay as can be seen from fig. 9.

By analyzing the tunneling current one finds that the time a particle needs to go through the barrier is not sharp. One has a distribution of passage times which depends on the initial preparation of the particle's wavefunction. If, instead of tunneling, a particle undergoes total reflection or resonant scattering, then this particle suffers a well-defined time-delay. In passing we mention other interesting problems of quantum tunneling, which we did not consider here: tunneling in a double-well potential[16], resonance scattering[17] and the problem of non-adiabaticity due to time-delay[18].

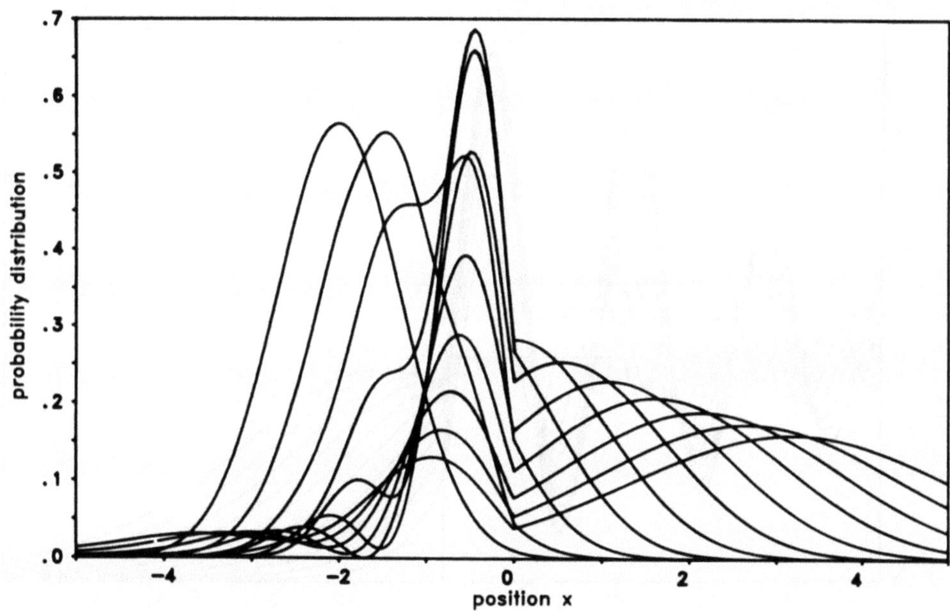

Fig. 6. A Gaussian wave packet starts at $\langle x \rangle = -2$ with $\langle p \rangle = 2$ and $\Delta x^2 = 0.5$. At $x = 0$ there is a delta barrier of unit height ($V_0 = 1$). The profile of the packet is shown for equidistant time steps $\Delta t = 0.25$. As time goes on the wave packet extends further to the right.

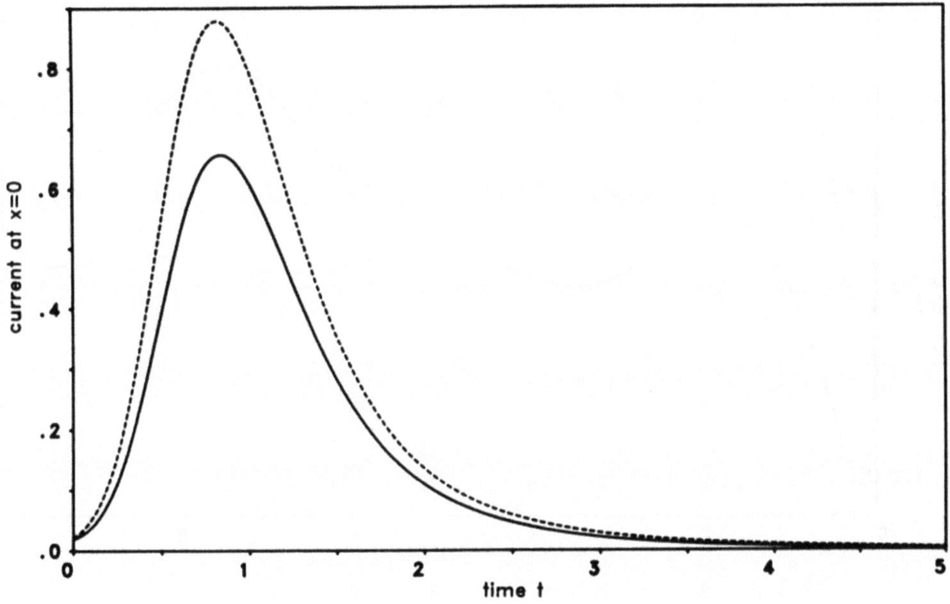

Fig. 7. Particle number current for the Gaussian wave packet (fig. 6) just behind the barrier at $x = 0^+$. Dotted line is the result for a free passage ($V_0 = 0$) of the packet.

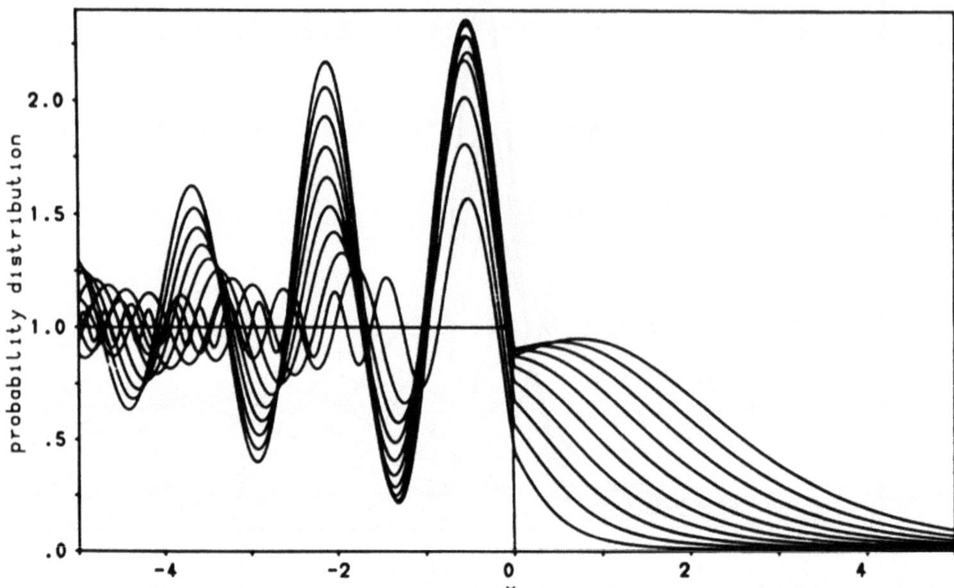

Fig. 8. A semi-infinite wave train $\psi(x,0) = \Theta(-x)\ \exp(ipx)$ with $p = 2$, hits the delta barrier of unit height at $x = 0$ and $t = 0$. Time steps $\Delta t = 0.25$ - as in fig. 6. The oscillatory structure is caused by the interference between incoming and reflected wave.

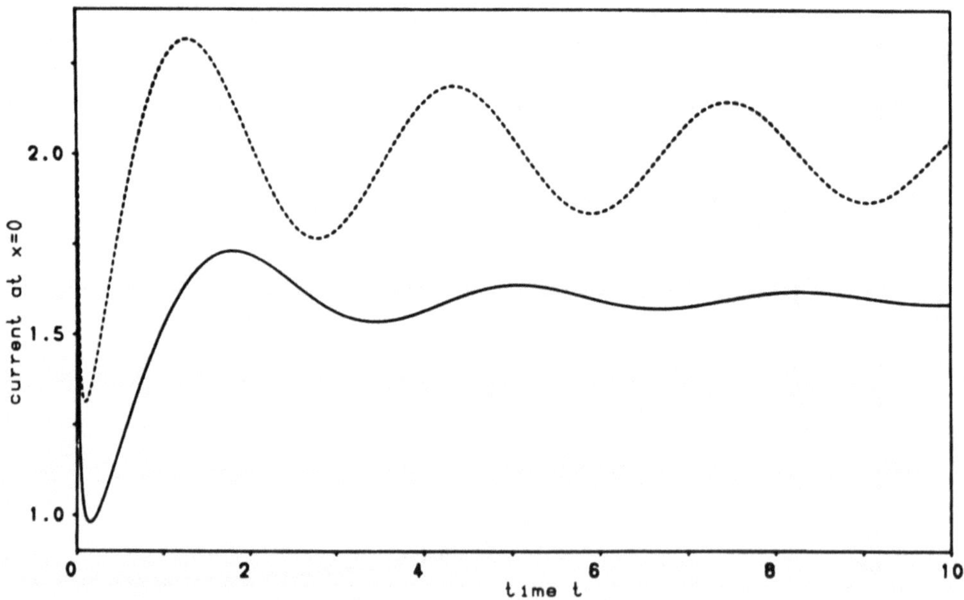

Fig. 9. Particle number current for the semi-infinite wave train (fig. 8) just behind the barrier at $x = 0^{+}$. Compare with fig. 7 for other details and note that now there is a time-delay for tunneling.

ACKNOWLEDGEMENT

This work was supported by the Deutsche Forschungsgemeinschaft and by the Bundesministerium für Forschung und Technologie.

REFERENCES

1. S. Geltman, J. Phys. B 10: 831 (1977)
2. E.J. Austin, J. Phys. B 12: 4045 (1979)
3. F.H.M. Faisal, J. Phys. B 9: 1453 (1973)
4. M. Abramowitz and I.A. Stegun, 1970, "Handbook of Mathematical Functions", Dover Publications, New York
5. M. Moshinsky, Phys. Rev. 88: 625 (1952)
6. H. Mitter, 1983, Particles in Strong Periodic Fields, in: "Quantum Electrodynamics of Strong Fields", W. Greiner, ed., Plenum Press, New York and London
7. H.S. Antunes Neto and L. Davidovich, Phys. Rev. Lett 53: 2238 (1984)
8. R.R. Schlicher, W. Becker, J. Bergou and M.O. Scully, 1984, Interaction Hamiltonian in Quantum Optics, in: "Quantum Electrodynamics and Quantum Optics", A.O. Barut, ed., Plenum Press, New York and London
9. K. Rzazewski, M. Lewenstein and J.H. Eberly, J. Phys. B 15: L661 (1982)
10. A. Tip, 1984, Resonances in Atomic Photo-Ionization, in: "Resonances - Models and Phenomena", S. Albeverio, L.S. Ferreira and L. Streit, eds., Springer, Berlin, Heidelberg, New York
11. M.H. Mittleman, 1982, Introduction to the Theory of Laser-Atom Interaction, Plenum Press, New York and London
12. N.B. Delone and V.P. Krainov, 1985, Atoms in Strong Light Fields, Springer, Berlin, Heidelberg, New York
13. W. Elberfeld, M. Kleber, W. Becker and M.O. Scully, J. Phys. B 19: in print (1986)
14. W. Elberfeld and M. Kleber, to be published
15. L.A. Mac Coll, Phys. Rev. 40: 621 (1932)
16. E. Merzbacher, 1970, Quantum Mechanics, 2nd edn., Wiley, New York
17. E.P. Wigner, Phys. Rev. 98: 145 (1955)
18. B. Müller and N. Takigawa, preprint

ACKNOWLEDGEMENT

QUANTUM MECHANICAL THEORY OF POSITRON PRODUCTION

IN HEAVY ION COLLISIONS WITH NUCLEAR CONTACT

Ulrich Heinz

Physics Department
Brookhaven National Laboratory
Upton, New York 11973, USA

ABSTRACT

The interplay between atomic and nuclear interactions in heavy ion collisions with nuclear contact is studied. The general theoretical description is outlined and analyzed in a number of different limits (semiclassical approximation, DWBA, fully quantal description). The two most important physical mechanisms for generating atomic-nuclear interference, i.e., energy conservation and the introduction of additional phase shifts by nuclear reactions, are extracted. The resulting typical coupling matrix elements are analyzed for their relative importance in atomic and nuclear excitations. The description of nuclear influence on atomic excitations in terms of a classical time delay caused by nuclear reactions is reviewed, and its relationship to the underlying quantal character of the nuclear reaction is discussed.

The theory is applied to spontaneous positron emission in supercritical heavy-ion collisions ($Z_{tot} \gtrsim 173$). It is shown that nuclear contact can lead to line structures in the positron energy spectra if the probability distribution for nuclear delay times caused by the contact has contributions for $T \gtrsim 10^{-19}$ sec. We explicitly evaluate a model where a pocket in the internuclear potential near the touching configuration leads to formation of nuclear molecules, and predict a resonance-like excitation function for the positron peak.

INTRODUCTION

Heavy ion collisions with nuclear contact recently have developed into an increasingly effective tool to study phenomena in a regime where nuclear and atomic excitations occur simultaneously and therefore interfere with each other. One example for this type of phenomena which has recently been the subject of particularly intensive investigations is the emission of spontaneously created positrons in supercritical ($Z_1 + Z_2 > 173$) heavy ion collisions at energies close to the Coulomb barrier. When the first peak structures were found in positron spectra from U+U, U+Th and U+Cm collisions (Schweppe et al. 1983, Clemente et al. 1984), it was attempted to interpret them as spontaneous positrons emitted from a longlived nuclear composite formed in the nuclear collision as a consequence of nuclear interactions at contact (Rafelski et al. 1978, Reinhardt et al. 1981b, U. Müller et al. 1983, Heinz et al. 1983a,1984). Although this explanation has subsequently been excluded by the non-observance of the predicted shift with the combined nuclear charge of the energy at which the positron line occurs (Cowan et al. 1985, Tsertos et al. 1985), there is still some evidence that the appearence of the line structures (which now have been observed in many more, even subcritical systems, see talks at this conference by Bokemeyer, Cowan, Kienle, König, and Kozhuharov) is connected in a sensitive way with the ion beam energy which has to be chosen close to the Coulomb barrier to see the peaks. A likely explanation is that nuclear contact or nuclear interactions play at least some role in the excitation mechanism for these peak structures.

This has implications for the theoretical description, because the usual picture of two ions moving on classical trajectories, and thereby providing a time-dependent Coulomb field leading to atomic excitations, is only valid for well separated nuclei; it generally breaks down if the nuclei touch, allowing the short-range nuclear interactions to modify the nuclear scattering process. In the semiclassical framework their effects may be accounted for heuristically by introducing by hand a classical time delay supposed to be caused by the nuclear interaction; but the validity of this procedure is not obvious, and the value of the time delay to be chosen is an essentially free parameter, where, in principle, it should be determined by the character of the nuclear interactions.

For this reason I will review in this lecture the general quantum mechanical treatment of heavy-ion collisions and the relation to the semiclassical picture. In a strict sense this is the formalism underlying all theoretical calculations shown at this conference. I will show that under a factorization assumption for the atomic and nuclear excitations, which in many practically relevant cases can be justified, the simple semiclassical theory can be used even when nuclear contact occurs; one only has to ensure that a probability distribution for classical nuclear delay times is used which is consistent with the underlying quantum mechanical nuclear scattering amplitude. I will show how to generally derive this classical delay time distribution from the nuclear S-matrix, and present a specific model where this relationship between quantum mechanics and the semiclassical approximation can be worked out semi-analytically. The model describes nuclear scattering through molecular resonances which are supported by a pocket in the internuclear potential at touching distance; the resulting probability distribution of classical delay times has contributions for large times $T \gtrsim 10^{-19}$ sec which lead to formation of a pronounced peak in the positron energy spectrum in supercritical collisions due to spontaneous positron emission. Although I do no longer propose this model as an explanation for the observed line structures, it is instructive to see the basic mechanisms at work, particularly since nuclear interactions may still have some role in the occurrence of these experimentally discovered peaks.

In my presentation I will mostly follow the treatment given in Heinz et al. (1983a,1984). The field is also reviewed in Heinz (1986), where a more complete set of references can be found.

THE THEORETICAL FRAMEWORK

This chapter lays out the general theoretical description of ion-atom or ion-ion scattering events, starting from a completely quantum mechanical treatment of both nuclear and atomic degrees of freedom. Subsequently, different approximation schemes to the full theory are discussed, recovering the known results for several limiting cases, particularly the semiclassical theory which uses classical time-dependent nuclear trajectories.

Nuclear and atomic excitations will be treated on the same footing as far as possible. However, the nuclear dynamics will always be treated nonrelativistically, and the theory is not second quantized, i.e., we work with wavefunctions instead of Fock states. The second quantized description, although conceptually preferable and possible in the semiclassical theory (Reinhardt and Greiner 1977), has not yet been extended to include nuclear excitations.

The Scattering Problem

We face the following problem of stationary scattering theory: our total system is an eigenstate with eigenvalue E of the total energy:

$$H_{tot}\Psi_E(\vec{R},\vec{r},\vec{x},\vec{\xi}) = E\,\Psi_E(\vec{R},\vec{r},\vec{x},\vec{\xi}) \;. \tag{1}$$

$\vec{R}$ is the internuclear distance, $\vec{r}$ ($\vec{x}$) is the set of electronic (photon) coordinates in the nuclear center of mass (CMN) system, and $\vec{\xi}$ are internal nuclear coordinates (possibly collective coordinates) (Fig.1). The Hamiltonian of the total energy is given by

$$\begin{aligned} H_{tot} &= \frac{\vec{P}^2}{2\mu} + V_C(R) + V_N(R) \\ &+ \sum_{i=1}^{Z} H_{TCD}^{(i)}(\vec{r},\vec{R}) + \frac{1}{2M_N}\Big(\sum_{i=1}^{Z}\vec{p}_i\Big)^2 + \frac{1}{2}\sum_{i\neq j}^{Z} V^{ee}(r_{ij}) \\ &+ \sum_{m=1}^{A} H_{int}^{(m)}(\vec{\xi},\vec{R}) + H_{rad}(\vec{x},\vec{\xi},\vec{r},\vec{R}) \;. \end{aligned} \tag{2}$$

$\vec{P}$ is the relative momentum of the two nuclei; μ (M_N) is their reduced (combined) mass; $V_C(R)$ is the Coulomb potential between the nuclei; $V_N(R)$ is some nuclear optical potential to be specified later; $H_{TCD}^{(i)}$ is the two-center Dirac Hamiltonian for a single electron i containing the electron kinetic energy and the interaction with the two nuclei; the next term is the energy contained in a possible collective motion of the complete electron cloud with respect to the CMN; $V^{ee}(r_{ij})$ is the interaction between electron i and j ($r_{ij} = |\vec{r}_i - \vec{r}_j|$). The second to last term is the internal nuclear Hamiltonian, summed over all degrees of freedom for internal motion. The last term, H_{rad}, is the Hamiltonian for electromagnetic radiation; it depends on nuclear and electronic coordinates since it contains coupling terms of the electromagnetic potential to the nuclear and electronic currents. H_{rad} has to be taken into account for the calculation of X-ray spectra, nuclear bremsstrahlung and internal conversion of nuclear into electron excitations.

In (1), (2) the electron and photon coordinates are measured in the nuclear center of mass frame. For increasing mass asymmetry between the colliding nuclei the CMN moves closer and closer to the heavier collision partner ("target"). Therefore in this case often a coordinate system with origin in the target is used to define the electron coordinates. Since the target frame is not an inertial frame, the transformation from the CMN to this frame gives rise to an additional coupling term describing the target recoil, which for non-relativistic electrons takes the form

$$H_R = \frac{m}{M_T}\,\vec{r}\cdot\vec{\nabla}_R\big(V_C(R) + V_N(R)\big) \;. \tag{3}$$

The generalization to relativistic electrons is discussed by Amundsen (1978).

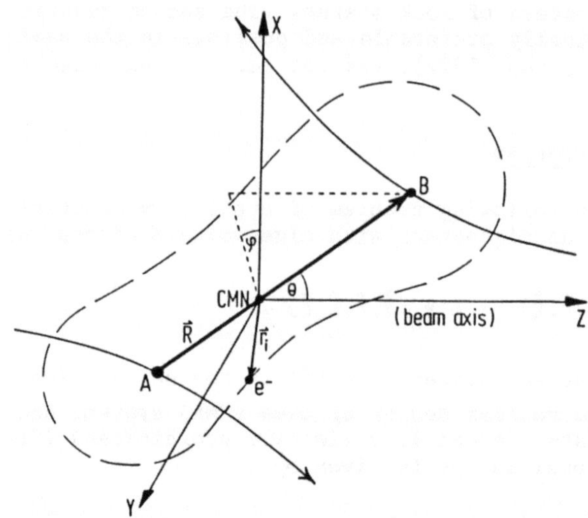

FIG. 1. Definition of the coordinates used in (2). A,B denote the two nuclei, CMN their combined center of mass. (X,Y,Z) denote the laboratory-fixed coordinate system. The distance vector $\vec{R}$ between the two nuclei defines the z-axis of the rotating coordinate system. It orientation with respect to the laboratory system is given by angles θ,ϕ. $\vec{r}_i$ denotes the location of electron i with respect to the nuclear center of mass CMN. The photon coordinates $\vec{x}$ and the internal coordinates $\vec{\xi}$ of the nuclei are not shown in this figure.

Basis Expansion

Of course, nobody wants to numerically integrate the multidimensional Schrödinger equation (1) directly. By choosing basis sets for the photons, electrons, internal nuclear degrees of freedom, and the angular part of the nuclear relative motion, and projecting equation (1) into this basis, a set of coupled differential equations in the internuclear distance R only can be derived. The trick lies in selecting the basis states in such a way that the resulting set of coupled channel equations can be truncated, and that a good approximation is obtained by taking into account only a tractable number of different channels. This implies that for different physical situations different basis sets will prove convenient. This will now be discussed in some detail.

Since photons, once emitted by the system, generally escape without final state interaction, a natural basis for them is provided by plane waves for the electromagnetic vector potential:

$$A_\mu(x) = \sum_\lambda \int \frac{d^3k}{(2\pi)^3} \frac{1}{\sqrt{2\omega}} \varepsilon_{\mu\lambda}(\vec{k})\, e^{-ikx} \equiv \sum_\lambda \int \frac{d^3k}{(2\pi)^3} a_{\mu\lambda}(k) \;. \tag{4}$$

Here $x = (t,\vec{x})$ and $k = (\omega,\vec{k})$. $\lambda = 1,2$ denotes the two transverse polarization directions, and $\varepsilon_{\mu\lambda}(\vec{k})$ is the polarization unit vector.

For the <u>nuclear internal degrees of freedom</u> it seems natural to use eigenstates of $\sum_{m=1}^{A} H_{int}^{(m)}(\vec{\xi},\vec{R})$. Except for a trivial rotation defined by the orientation of the nuclear distance vector $\vec{R} = (R,\theta,\phi)$, these states only depend on the magnitude of R. The trivial rotation is conveniently disposed of by first rotating the nuclear Hamiltonian by the angle (θ,ϕ). Let us call the Hamiltonian in the rotating frame $H_{nuc}(\vec{\xi},\vec{R})$; a set of nuclear basis states is then defined by

$$H_{nuc}(\vec{\xi},\vec{R})\ \chi_N(\vec{\xi},R) = \varepsilon_N(R)\ \chi_N(\vec{\xi},R). \tag{5}$$

Examples for possible choices for H_{nuc} are, among many others, the two-center shell model or collective models. The index N of the wavefunction summarizes all the good quantum numbers of the respective model; among these is the projection of the nuclear intrinsic angular momentum on the vector $\vec{R}$; the total intrinsic angular momentum, however, is not a good quantum number for finite separation R.

The choice of the <u>electron basis</u> depends on the mass asymmetry of the scattering system and on the beam energy. For very asymmetric systems (e.g., proton-induced collisions) an atomic basis centered at the target or a superposition of atomic bases centered at both the target and the projectile have been used.

Such an approach is expected to work if the projectile can be considered as a (time-dependent) perturbation on the target and vice versa, and if sharing of electrons between target and projectile can be assumed to be small. This is particularly true at high collision energies. For larger, more symmetric systems and/or low collision energies (i.e. below the Coulomb barrier), the collision happens adiabatically enough that molecular orbitals can be established, at least by the fast moving inner shell electrons, and a molecular basis will be more appropriate.

It is defined by using eigenfunctions of the two-center Hamiltonian given in the second line of (2). If heavy ions ($Z_1 + Z_2 > 100$) are used, relativistic effects are important, and one solves the two-center Dirac equation (Müller et al. 1973, Müller and Greiner 1976). (The small term corresponding to a collective motion of the electron cloud relative to the CMN has so far always been neglected.) Again, like for the nuclear eigenstates, the basis is most conveniently determined in the rotating frame where the states (let us call them $\phi_n(\vec{r},R)$) only depend on the absolute value of the nuclear distance.

For overcritical systems ($Z_1 + Z_2 > 173$) the negative energy continuum contains one or more resonances from dived electronic bound states. These make the use of exact two-center eigenstates awkward for practical computations. It is then more convenient (Reinhardt et al. 1981a) to project a quasibound state $\tilde{\phi}_R(\vec{r},R)$, which approximates the resonance due to the dived state, out of the positron energy continuum and to orthogonalize this continuum with respect to this quasibound state, yielding modified positron states $\tilde{\phi}_{E_p}(\vec{r},R)$. Instead of the two-center eigenstates, one then uses this projected basis for negative electron energies. The coupling matrix element $\langle \tilde{\phi}_R | H_{el} | \tilde{\phi}_{E_p} \rangle$, which is now non-zero since the projected basis states are not eigenstates of H_{el}, is interpreted as the matrix element for the spontaneous decay of a hole in the quasibound state into a continuum positron with energy E_p (Reinhardt et al. 1981a).

Finally, we may also separate the dependence on the direction of $\vec{R}$ by finding the eigenstates of the angular part of $\vec{P}^2/2\mu$ in the rotating frame. They are found to be given by

$$\mathscr{H}^{J}_{M\Lambda}(\theta,\phi) = (-i)^{\Lambda}\sqrt{\frac{2J+1}{4\pi}}\, d^{j}_{\Lambda M}(\theta)\, e^{iM\phi} = (-i)^{\Lambda}\sqrt{\frac{2J+1}{4\pi}}\, D^{J}_{\Lambda M}(0,\theta,\phi) \,, \quad (6)$$

where θ and ϕ are the angles of $\vec{R}$ in the laboratory frame. The rotation matrices in (6) are interpreted as those responsible for a rotation of a state of total angular momentum J and projection Λ onto the internuclear axis $\vec{R}$ (the latter is given by the sum of the projections on $\vec{R}$ of the intrinsic electron, photon, and nuclear angular momenta) into the laboratory frame, where the total angular momentum has projection M onto the beam axis.

Consider now having chosen such a set of basis states:

$$|\alpha> \equiv \Big|JMnN\mu k\Lambda) \equiv \mathscr{H}^{J}_{M\Lambda}(\theta,\phi)\, [\phi_n(\vec{r},R)\; \chi_N(\vec{\xi},R)\; a_\mu(k)]_\Lambda \,. \quad (7)$$

[Λ is the total projection of all intrinsic angular momenta (nuclei, electrons, and photons) on R.]

The total wavefunction Ψ_E in (1) is then expanded into this basis as

$$\Psi_E(\vec{R},\vec{r},\vec{x},\vec{\xi}) = \sum_\beta C_\beta \frac{F_\beta(R)}{R}\, |\beta> \,. \quad (8)$$

The functions $F_\beta(R)$ describe the dynamics of the internuclear distance R in the different channels β. The constants C_β reflect the initial boundary conditions for the relative motion: a detailed derivation of C_β is given in Blair and Anholt (1982) for an atomic basis and in Heinz et al. (1983a) for a molecular electron basis.

Coupled Channel Equations for the Nuclear Relative Motion

Inserting the expansion (8) into (1) and projecting onto the basis states $<\alpha|$, a set of coupled differential equations for the radial functions $R_\alpha(r) \equiv C_\alpha F_\alpha(R)$ is derived:

$$\left\{\frac{d^2}{dR^2} - \frac{J(J+1)-\Lambda^2}{R^2} + \frac{2\mu}{\hbar^2}\,[E - V_C(R) - V_N(R) - \varepsilon_\alpha(R)]\right\} R_\alpha(R) =$$

$$= - 2 \sum_{\beta\neq\alpha} <\alpha\big|\partial/\partial R\big|\beta> \frac{dR_\beta}{dR}(R) + \frac{2\mu}{\hbar^2} \sum_{\beta\neq\alpha} <\alpha\big|H_{el} + H_{nuc} + H_{rad}\big|\beta>\, R_\beta(R)$$

$$+ \frac{2\mu}{\hbar^2} \sum_{\beta\neq\alpha} Q_{\alpha\beta}\, R_\beta(R). \quad (9)$$

Since the diagonal contribution

$$\varepsilon_\alpha(R) \equiv <\alpha\big|H_{el} + H_{nuc} + H_{rad}\big|\alpha> \equiv \varepsilon_n(R) + \varepsilon_N(R) + \varepsilon_{rad}(R) \quad (10)$$

has been included as an additional, channel- and R-dependent potential on the left hand side, the sum on the right hand side is over all states $|\beta>$, for which at least one of the quantum numbers is different from $|\alpha>$ (see 7). The total angular momentum J and its projection M on the beam axis are conserved quantum numbers, hence the equations are diagonal in J,M. The $\varepsilon_\alpha(R)$ are computed in the rotating frame (H_{el} and $H_{nuc} + H_{rad}$, respectively, are given by the second and third line of (2) expressed in the frame rotating with R); hence they depend only on the length and not on the orientation of R. The channel-dependent contribution of the intrinsic angular momenta to the total angular momentum barrier (which arises when the electron, photon and nuclear wavefunctions are rotated back into the lab frame) is recognized in the term $\sim\Lambda^2/R^2$ on the left side of eq. (9).

The coupling terms on the r.h.s. have the following origins:

$D_{\alpha\beta} \equiv \langle\alpha|\partial/\partial R|\beta\rangle$ is the so-called "radial coupling." It acts between any kind of molecular-type wavefunctions, and it is the most important mechanism for inner shell ionization in relativistic systems.

$H_{\alpha\beta} \equiv \langle\alpha|H_{el}|\beta\rangle$ acts between the electronic basis states whenever they are not eigenstates to H_{el}. This applies if, e.g., in supercritical systems ($Z_1 + Z_2 > 173$) a modified two-center basis is employed as discussed above; $H_{\alpha\beta}$ then describes the spontaneous positron emission.

$U_{\alpha\beta} = \langle\alpha|H_{nuc}|\beta\rangle$ describes the intrinsic nuclear excitations. $V_{\alpha\beta} \equiv \langle\alpha|H_{rad}|\beta\rangle$ describes radiative transitions and internal conversion of nuclear transitions. $Q_{\alpha\beta}$ finally summarizes all further effects, like Coriolis coupling (stemming from the use of a rotating coordinate system), and nucleon and electron translation effects (if a molecular basis is used). The corresponding matrix elements are known and can be easily included (Heinz et al. 1981).

The set of equations (9) is the practical basis of a theoretical description of heavy-ion collisions. No approximation is involved as long as there is no restriction on the number of basis states. Its practical usefulness relies on a convenient selection of basis states and on the availability of approximation schemes to further simplify the structure of the equations (9), such that the number of channels to be retained can be reduced as much as possible. The question of basis states was already discussed. In the remainder of this chapter we will occupy ourselves with the different available approximation schemes to solve equations (9).

Starting from (9), a major theoretical branching point involves a decision whether to continue with a quantum mechanical description of the nuclear relative motion, or whether to replace it by a classical trajectory. In the latter case, one substitutes the nuclear distance R by a time parameter which automatically distinguishes the ingoing ($t<0$) and outgoing ($t>0$) part of the collision, thereby considerably simplifying the formalism. The full quantum theory, however, is necessary for a deeper understanding, e.g., of the real meaning of the delay time T introduced in the semiclassical picture in order to describe nuclear contact (see below). After this point, the theoretical development is quite parallel in both approaches, and one can derive coupled channel equations for excitation amplitudes, solve them in perturbation theory or nonperturbatively, etc., as we will show below.

Most results obtained from (9) one way or the other involve the assumption that atomic and nuclear excitations occur in different spatial regions, $R>R_m$ and $R<R_m$ respectively. [Only recently has progress been made in describing on the basis of (9) simultaneous nuclear and atomic transitions during the nuclear sticking process (see Schramm's talk and Schramm et al. 1986).] If spontaneous decay processes like united atom X-ray emission or spontaneous positron production are involved, the corresponding matrix elements are assumed to be constant in the nuclear region $R<R_m$; this allows to factorize the atomic processes from those nuclear ones leading to sticking.

In this simplified picture, atomic-nuclear interference occurs because a given atomic excitation $\alpha \to \beta$ may happen before or after the nuclear scattering, which happens at different effective energies in the two cases. For an isolated nuclear resonance, the nuclear scattering thus may be on or off resonance, depending on whether the atomic excitation occurs before or afterwards. This results in different contributions to the cross section from the in- and outgoing parts of the collision. The difference between the two nuclear scattering amplitudes can be interpreted as a relative phase which is proportional to the energy derivative of the nuclear scattering phase shift or, classically, to the nuclear sticking time. This is the physics contained in the Blair-Anholt formula and its semiclassical analog, the Ciocchetti-Molinari formula (see below). As the beam energy scans the nuclear resonance, this relative phase between the in- and outgoing transition amplitudes changes, often leading to a characteristic interference pattern in the cross section. If the nuclear resonance is too wide, the change in phase is too small, and the interference is lost. If the nuclear resonance is very narrow compared to the atomic excitation energy, the in- and outgoing amplitudes become incoherent; then one obtains effectively two nuclear resonances, one corresponding to atomic excitations before and one to excitations after the nuclear scattering (see Heinz 1986).

For a nuclear reaction proceeding through a compound nucleus, the same principles apply, but the nuclear scattering amplitude can no longer be parametrized by a single resonance. Rather, the cross section is determined by the nuclear autocorrelation function, i.e. the beam energy average of the product of two scattering amplitudes at different energies. Still, it can be related to a distribution of classical delay times, thus allowing match-up with the semiclassical picture as I will show. These considerations play a role for the process of spontaneous positron production in very heavy ion collisions.

The Semiclassical Approximation - Collisions with Classical Time Delay

It is instructive to rewrite (9) in a more compact form:

$$\left\{ \frac{d^2}{dR^2} - U_\alpha(R) \right\} R_\alpha(R) = X_\alpha(R) \ . \tag{11}$$

U_α comprises all nonderivative terms on the l.h.s. of (9), and $X_\alpha(R)$ summarizes all interchannel couplings. Asymptotically (for $R \to \infty$) the latter vanish, and $U_\alpha(R) \to (2\mu/\hbar^2)(E - \varepsilon_\alpha(\infty))$ becomes very large due to the huge kinetic energy contained in the nuclear relative motion. Therefore, at large R the functions R_α have to be rapidly oscillating (with wavelengths of order 0.1 fm). We can make use of the asymptotic dominance of the total energy (plus corrections from the long-range Coulomb repulsion between the two nuclei and from the angular momentum barrier) to guess and extract the rapidly oscillating part of the functions R_α, leaving a set of differential equations for the remaining "modulation amplitudes" which has much more slowly varying solutions and is easier to handle numerically.

In this subsection we discuss the semiclassical approach, in which the rapidly oscillating part of the R_α is assumed to be given by JWKB-solutions of (11) neglecting the couplings X_α between channels. Within this scheme it is possible to define a classical nuclear trajectory R(t) parametrized by a time variable $-\infty < t < \infty$. This approximation usually breaks down if nuclear reactions are involved; however, one can simulate nuclear reactions by introducing a classical contact or delay time and by appropriately modifying the classical trajectory R(t) during this time interval, as you have seen in other talks at this meeting.

Derivation of Time-dependent Coupled Channel Equations

We consider the equation

$$\left\{\frac{d^2}{dR^2} - \frac{J(J+1)}{R^2} + \frac{2\mu}{\hbar^2}\left[E - V_C(R) - V_N(R) - \bar{\varepsilon}(R)\right]\right\} r_J(R) = 0 \ . \tag{12}$$

It is obtained from (11) by neglecting the coupling terms $X_\alpha(R)$ and the channel dependent contribution to the angular momentum barrier, and by replacing the channel dependent binding energies $\varepsilon_\alpha(R)$ by a channel independent "average" $\bar{\varepsilon}(R)$. Different choices of $\bar{\varepsilon}(R)$ will lead to slightly different nuclear trajectories, but in general (within the domain of applicability of the JWKB approximation) these effects are small and can be neglected. The purpose of these manipulations is that

$$\bar{P}_J^2(R) = 2\mu\left(E - V_N(R) - V_C(R) - \bar{\varepsilon}(R)\right) - \frac{J(J+1)\hbar^2}{R^2} \tag{13}$$

defines a <u>channel independent</u> classical nuclear trajectory (see below) which only depends on the total energy and the total angular momentum (classically related to the impact parameter). Its zero defines a channel independent classical turning point R_o. (There may be more than one R_o, e.g. in a potential with pockets.) The in- and outgoing JWKB-solutions to (12) are given by

$$r_J^{\pm}(R) = [\bar{P}_J(R)]^{-1/2} \exp\left(\pm i\, S_J(R)/\hbar \pm i\, \pi/4\right) \tag{14}$$

with

$$S_J(R) = \int_{R_o}^{R} \bar{P}_J(R')\, dR' . \tag{15}$$

The radial functions R_α in (11) are now expanded into these semiclassical solutions:

$$R_\alpha(R) = a_\alpha^{+}(R)\, r_J^{+}(R)\, e^{i\gamma_\alpha(R)/\hbar} - a_\alpha^{-}(R)\, r_J^{-}(R)\, e^{-i\gamma_\alpha(R)/\hbar} \ . \tag{16}$$

$\gamma_\alpha(R)$ is the additional channel dependent phase shift due to the binding energies $\varepsilon_\alpha(R)$ and the term $\sim\Lambda_\alpha^2/R^2$:

$$\gamma_\alpha(R) = \int_{R_o}^{R} P_\alpha(R')dR' - S_J(R) \simeq \int_{R_o}^{R}\left[\bar{\varepsilon}(R') - \varepsilon_\alpha(R') + \Lambda_\alpha^2\hbar^2/2\mu R'^2\right] \frac{\mu\, dR'}{\bar{P}_J(R')} \tag{17}$$

with

$$P_\alpha^2(R) = 2\mu\left[E - V_N(R) - V_C(R) - \varepsilon_\alpha(R)\right] - \frac{[J(J+1) - \Lambda_\alpha^2]\hbar^2}{R^2} \ . \tag{18}$$

Extracting these phases from the expansion amplitudes $a_\alpha^{\pm}$ in (16) has the advantage of removing all diagonal coupling terms from the equations governing their dynamics. Furthermore all second derivatives $a_\alpha^{\pm}$ can be eliminated by the constraint on (16),

$$r_J^{+}(R)\, \frac{da_\alpha^{+}}{dR} = r_J^{-}(R)\, \frac{da_\alpha^{-}}{dR} \ , \tag{19}$$

which leaves us with the proper number of independent functions, namely one independent modulation amplitude for every channel wavefunction $R_\alpha(R)$.

One now proceeds to insert the ansatz (16) into (11) and to expand in powers of h. The next step involves neglecting all terms containing strongly oscillating factors; it is crucial for defining a classical nuclear trajectory (see below) since it decouples the ingoing amplitudes a_α^- from the outgoing ones, a_α^+. It is only allowed outside of the largest classical turning point R_0; in the case where the internuclear "potential" U_α in (11) contains pockets, resonance scattering is possible, invalidating the concept of a channel independent classical trajectory (different channels may be on or off resonance, depending on $\varepsilon_\alpha(R)$). Still, it may be possible to use JWKB wavefunctions inside the pocket; in matching them to the outside solution, coupling of in- and outgoing amplitudes will, however, be unavoidable (Schramm et al. 1986).

In the final step one defines a time variable through

$$dt = \pm \frac{\mu}{\bar{P}_J(R)} dR \tag{20}$$

with the + (-) sign for the outgoing (ingoing) amplitudes. Matching these amplitudes at the classical turning point ($t = t_o$) by the regularity condition,

$$a_\alpha^+(R_o) = a_\alpha^-(R_o) \ , \tag{21}$$

one finally obtains the time dependent coupled channel equations

$$\frac{da_\alpha}{dt} = - \sum_{\beta\neq\alpha} [\dot{R}(t)\, D_{\alpha\beta}(t) + \frac{i}{\hbar}(H_{\alpha\beta} + U_{\alpha\beta} + V_{\alpha\beta} + Q_{\alpha\beta})] e^{i\chi_{\alpha\beta}(t)}\, a_\beta(t). \tag{22}$$

The phases $\chi_{\alpha\beta}(t)$ are given by

$$\chi_{\alpha\beta}(t) = \frac{1}{\hbar} \int_{t_o}^{t} (\varepsilon_\alpha - \varepsilon_\beta - \frac{(\Lambda_\alpha^2 - \Lambda_\beta^2)\hbar^2}{2\mu R^2})\, dt' \equiv \chi_\alpha(t) - \chi_\beta(t). \tag{23}$$

(The last term in the integral is usually neglected as being small.)

Fully Quantum Mechanical Treatment

A quantum mechanical set of coupled channel equations for the occupation amplitudes can be derived very much along the same lines as in the semiclassical case. The basic improvement lies in a more accurate choice for the rapidly oscillating part of the nuclear relative motion wavefunction which allows one also to go outside the region where the JWKB approximation is applicable. On account of the dominance of the internuclear Coulomb potential at large distances, a natural candidate for the rapidly oscillating part of the nuclear relative wavefunctions R_α is given by in- and outgoing Coulomb waves, i.e. solutions of

$$\{ \frac{d^2}{dR^2} - \frac{J(J+I)-\Lambda_\alpha^2}{R^2} + \frac{2\mu}{\hbar^2} [E - V_C(R) - \varepsilon_\alpha(\infty)]\} H_{L_\alpha}^{\pm}(K_\alpha R) = 0 \ . \tag{24}$$

Here $K_\alpha^2\hbar^2 \equiv 2\mu[E - \varepsilon_\alpha(\infty)]$, and L_α is defined by $L_\alpha(L_\alpha+1) = J(J+1)-\Lambda_\alpha^2$. The analog of the semiclassical expansion (16) reads:

$$R_\alpha(R) = b_\alpha^+(R)\, H_\alpha^+(R) - b_\alpha^-(R)\, H_\alpha^-(R), \tag{25}$$

where $H_\alpha^\pm(R) \equiv H_{L\alpha}^\pm(K_\alpha R)/\sqrt{K_\alpha}$, and the auxiliary condition (19) is replaced by

$$H_\alpha^-\, b_\alpha^{-\prime} = H_\alpha^+\, b_\alpha^{+\prime} \tag{26}$$

(primes denote d/dR). The wavefunction (25) is regular at the origin if

$$b_\alpha^+(R{=}0) = b_\alpha^-(R{=}0) \ . \tag{27}$$

After insertion into (11) one obtains without approximations the coupled channel equations

$$b_\alpha^{\pm\prime} = \pm\, i \sum_\beta \left\{ \left(D_{\alpha\beta}\, H_\alpha^\mp\, H_\beta^{\pm\prime} - \tilde{H}_{\alpha\beta}\, H_\alpha^\mp\, H_\beta^\pm\right) b_\beta^\pm - \left(D_{\alpha\beta}\, H_\alpha^\mp\, H_\beta^\mp - \tilde{H}_{\alpha\beta}\, H_\alpha^\mp\, H_\beta^\mp\right) b_\beta^\mp \right\} \tag{28}$$

Here we defined for $\alpha \neq \beta$

$$\tilde{H}_{\alpha\beta} = \frac{\mu}{\hbar^2}\left(H_{\alpha\beta} + U_{\alpha\beta} + V_{\alpha\beta} + Q_{\alpha\beta}\right) , \tag{29}$$

whereas the diagonal matrix element

$$\tilde{H}_{\alpha\alpha} \equiv \Delta V_\alpha = \frac{\mu}{\hbar^2}\left[V_N(R) + \varepsilon_\alpha(R) - \varepsilon_\alpha(\infty)\right] \tag{30}$$

contains all potential couplings not taken into account in the Coulomb wavefunctions.

In contrast to the semiclassical coupled channel equations (22), in (28) in- and outgoing amplitudes are coupled to each other. The coupling terms contain products of two in- or outgoing Coulomb functions, which outside the nuclear Coulomb barrier are rapidly oscillating and whose semiclassical analog is argued to be negligible. If, however, important physics happens near or inside the Coulomb barrier, where these coupling terms between in- and outgoing amplitudes are no longer rapidly oscillating, they become essential. For example, they generate the proper resonance behavior of the nuclear relative motion wavefunction in collisions which proceed through, say, longlived molecular or compound states, thus giving rise to a large probability density at small internuclear distances. In such a scenario, the occupation amplitudes $b_\alpha^+(\infty)$ are also rapidly dependent on the total angular momentum J, because channels with nearby values of J may be on or off resonance. This invalidates a saddle point approximation for the partial wave sum in the expression for the cross section, and one has to explicitly evaluate

$$\frac{d\sigma_{\alpha\to\beta}}{d\Omega}(\theta) = \frac{1}{4K_\alpha^2}\left|\sum_J (2J+1)\, e^{i(\sigma_J^\alpha + \sigma_J^\beta)}\, a^J_{\alpha\to\beta}\, d^J_{\Lambda_\alpha\Lambda_\beta}(\theta)\right|^2 . \tag{31}$$

rather than being able to use the simple semiclassical formula

$$\frac{d\sigma_{\alpha\to\beta}}{d\Omega}(\theta) = \left|a^{J_s}_{\alpha\to\beta}\right|^2 \cdot \frac{d\sigma_R}{d\Omega}(\theta) . \tag{32}$$

In (31) $K_\alpha = \frac{1}{\hbar}\sqrt{2\mu\left(E - \varepsilon_\alpha(\infty)\right)}$, and the asymptotic phase shifts σ_J^α are semiclassically defined by

$$\sigma_J^\alpha = \frac{1}{\hbar} \lim_{R\to\infty} \left(\int_{R_0}^{R} P_\alpha(R')dR' - P_\alpha(R)\cdot R + \frac{J\hbar\pi}{2} \right) . \tag{33}$$

The saddle point J_s in (32) in the case of Rutherford scattering is related to the impact parameter and the scattering angle by ($P = \mu v$ is the asymptotic relative momentum)

$$J_s = bP/\hbar = \frac{Z_1 Z_2 e^2}{hv} \cot \theta/2 . \tag{34}$$

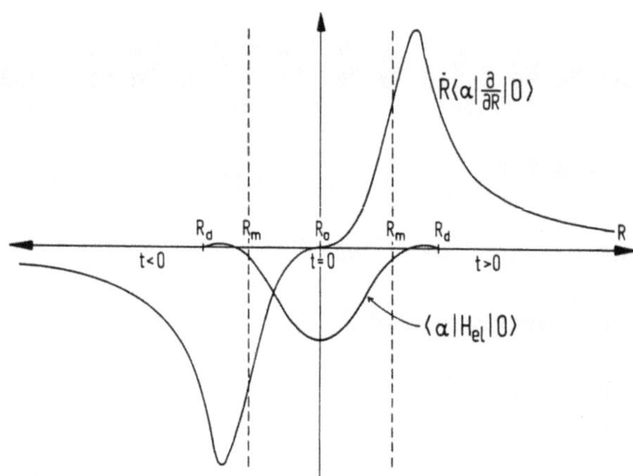

Fig. 2. A schematic picture of radial vs. spontaneous coupling in a semiclassical model. R is the two-center distance $R > R_0$, where R_0 is the classical turning point (semiclassically associated with time t=0). R_d is the diving point, R_m the matching radius defined in the text. For the radial excitation the semiclassically relevant combination $\dot{R}\,\langle\partial/\partial R\rangle$ is drawn which shows explicitly that, semiclassically at least, most of the radial coupling occurs at distances $R \gg R_0$. This semiclassical picture is the motivation for our choice of R_m in the text.

The Factorization Ansatz

As we discussed, sufficiently far away from the nuclear interaction region the collision can be treated semiclassically, i.e., the rapidly oscillating terms in (28) can be neglected. Since the semiclassical equations (22) are so much easier to solve than (28), due to the decoupling of in- and outgoing solutions, we would like to define a matching distance R_m as close as possible to the nuclear interaction regime, such that for $R > R_m$ it is sufficient to solve (22). (Fig. 2) Inside R_m, we will generally have to solve the full set of equations (28) with boundary conditions at R_m determined by the semiclassical solution outside R_m. It is obvious that the solution of (28) would be trivial if inside R_m all couplings would vanish. This is in general not achievable. But even if for $R < R_m$ only the atomic excitations vanish, things simplify because now the solution of (28) inside R_m just determines the nuclear transition S-matrix, and the total excitation amplitude factorizes into a nuclear part from inside R_m and an atomic part from outside R_m. Such an approximation is often justified: the radial coupling vanishes near the point of closest approach due to the vanishing nuclear relative velocity; spontaneous transitions, if their intensity is not very sensitive on the exact nuclear configuration after contact but can be reasonably assumed to be constant inside R_m, can be eliminated by transforming at the matching point R_m to an eigenstate basis in which the spontaneous coupling $H_{\alpha\beta}$ is absent by construction. Under these conditions the asymptotic transition amplitude in angular momentum channel β can be written as:

$$a^{\ell}_{iI\to fF}(\infty,\infty) = -\sum_{n_1 n_2 j} a^{-\ell}_{i\to n_1}(\infty,R_m)\, U_{jn_1}(R_m)\, S^{\ell}_{I\to F}\big(E-\varepsilon_j(R_m)\big)\, U^*_{jn_2}(R_m)\, a^{+\ell}_{n_2\to f}(R_m\infty) \tag{35}$$

This formula was first given by Tomoda and Weidenmüller (1983); its semiclassical analogue had been derived earlier by Müller and Oberacker (1980). $a^{\pm}$ denote the in- and outgoing semiclassical amplitudes from (22) outside R_m; U denotes the transformation matrix to eigenstates at R_m; $S^{\ell}_{I\to F}$ is the nuclear transition matrix element in particle wave ℓ, evaluated at the available scattering energy at R_m, $E-\varepsilon_j(R_m)$. For elastic nuclear scattering this simplifies to

$$a^{\ell}_{i\to f}(\infty,\infty) = -\sum A^{-\ell}_{i\to j}(\infty,R_m)\ \exp\big(2i\delta_{\ell}(E-\varepsilon_j)\big)\ A^{+\ell}_{j\to f}(R_m,\infty)\ , \tag{36}$$

where $A^- \equiv a^- \cdot U$ etc., and the nuclear S-matrix has been parametrized by a phase shift depending on the available nuclear scattering energy.

Perturbative Expansion

If all the transition amplitudes are small, one can solve (22) or (28) in perturbation theory, i.e., neglecting multistep processes $i\to n_1\to\ldots\to f$. Based on (28), this results in replacing the outside amplitudes $a^-_{\alpha\to\beta}(\infty,R_m)$ by

$$B_{\beta\alpha} = \sqrt{\frac{K_\alpha}{K_\beta}}\int_{R_m}^{\infty}\left(D_{\beta\alpha} + i\tilde{H}_{\beta\alpha}/K_\alpha\right)\exp\big(i\bar{q}(R)\cdot R\big)\, dR \tag{37}$$

where

$$\bar{q}(R)\cdot R \simeq (K_\alpha - K_\beta)R + \frac{Z_1Z_2e^2}{2E}(K_\alpha - K_\beta)\ \ln 2K_\alpha R\ . \tag{38}$$

The contribution from the (constant) spontaneous coupling $H^{spont}_{\beta\alpha}$ inside R_m can be evaluated analytically (Blair and Anholt 1982, Heinz et al. 1983a). One obtains the following expression for the cross section to excite channel β:

$$\frac{d\sigma_\beta}{d\Omega}(\theta) \simeq \left|f(\theta,E-\varepsilon_\beta)\right|^2 \sum_\alpha\left|B_{\beta\alpha} - \frac{K_\alpha}{K_\beta}\frac{f(\theta,E-\varepsilon_\alpha)}{f(\theta,E-\varepsilon_\beta)}B^*_{\beta\alpha} - \frac{K_\alpha+K_\beta}{2\sqrt{K_\alpha K_\beta}}\frac{H^{spont}_{\beta\alpha}}{\varepsilon_{\beta\alpha}}\left(\frac{f(\theta,E-\varepsilon_\alpha)}{f(\theta,E-\varepsilon_\beta)}-1\right)\right|^2. \tag{39}$$

The factor $|f(\theta,E-\varepsilon_\beta|^2$ is the nuclear cross section in channel β. The ratios of nuclear scattering amplitudes under the sum over initial channels α can be brought into a more intuitive form by using the quantum mechanical definition of a nuclear delay time as the energy derivative of the scattering phase shift:

$$\frac{f(\theta,E-\varepsilon_\alpha)}{f(\theta,E-\varepsilon_\beta)} \simeq 1 + \varepsilon_{\beta\alpha}\frac{\partial}{\partial E}\ln f(\theta,E-\varepsilon_\alpha) = 1 + i\varepsilon_{\beta\alpha}\tau^{QM}_\alpha/\hbar \simeq \exp(i\varepsilon_{\beta\alpha}\tau^{QM}_\alpha/\hbar)\ . \tag{40}$$

Inserting this into (39) reproduces the semiclassical formula in terms of a classical nuclear delay time τ, as it was derived years ago by Ciocchetti et al. (1963) and used by Gerd Soff in his talk at this meeting.

THE GENERAL RELATIONSHIP BETWEEN THE NUCLEAR S-MATRIX AND A CLASSICAL DELAY TIME DISTRIBUTION

Let us return to the cross section formula (31); in most cases the influence on the intrinsic spins $\Lambda_\alpha, \Lambda_\beta$ is small, and we can replace the d-matrix by $P_\ell(\cos\theta)$, summing over relative orbital momenta ℓ instead of J. Since the amplitudes $a^\ell_{\alpha\to\beta}$ are defined by expanding the relative motion into Coulomb waves, the σ_ℓ in are the Coulomb phaseshifts. Their channel dependence is through the Sommerfeld parameter and small enough to be neglected. In most cases which involve nuclear contact and compound nucleus formation, the nuclear S-matrix hidden in the amplitudes $a^\ell_{\alpha\to\beta}$ (see 35/36) is a very rapidly varying function of energy, due to a huge number of closely spaced compound states. An experiment with a finite beam energy resolution therefore can only measure S averaged over some finite energy interval, rather than resolve individual compound nucleus states. Let us therefore perform an energy average over the cross section, using the fact that compared to the nuclear S-matrix the energy variation of the atomic amplitudes $a^\pm_{\alpha\to\beta}$ and of the Coulomb phases is very slow. To further simplify the calculation, we realise that also the angular momentum dependence of the atomic amplitudes outside the matching radius R_m is slow compared to the Legendre polynomials; they can be extracted from the sum over partial waves and are replaced by their average value evaluated at an appropriate "effective angular momentum". One finds

$$
\begin{aligned}
\left\langle \frac{d\sigma_{iI\to fF}}{d\Omega} \right\rangle \simeq \frac{1}{4K^2} \sum_{nm} a^-_{i\to n}\, a^+_{n\to f}\, a^{-*}_{i\to m}\, a^{+*}_{m\to f} \,\cdot \\
\cdot \sum_{\ell\ell'} (2\ell+1)(2\ell'+1)\, P_\ell P_{\ell'}\, e^{2i(\sigma_{\ell'}-\sigma_\ell)} \\
\cdot \left\langle S_{\ell',I\to F}(E-\varepsilon_n)\, S^*_{\ell,I\to F}(E-\varepsilon_m) \right\rangle ,
\end{aligned}
\tag{41}
$$

The brackets < ... > denote the beam energy average over an interval ΔE as defined, e.g., by the prescription

$$
\langle f(E) \rangle = \frac{\Delta E}{\pi} \int_{-\infty}^{\infty} \frac{f(\varepsilon)d\varepsilon}{(\varepsilon-E)^2 + \Delta E^2} \,. \tag{42}
$$

Equation (41) contains the nuclear autocorrelation function

$$
A^{I\to F}_{\ell\ell'}(E_{nm},\omega) \equiv \langle S_{\ell',I\to F}(E_{nm}-\hbar\omega/2)\, S^*_{\ell,I\to F}(E_{nm}+\hbar\omega/2) \rangle , \tag{43}
$$

where

$$
E_{nm} \equiv E - \frac{1}{2}(\varepsilon_n+\varepsilon_m) \quad ; \quad \hbar\omega \equiv \varepsilon_n-\varepsilon_m \,. \tag{44}
$$

Neglecting the slight channel dependence of $E_{nm} \simeq E$, the Fourier transform $f^{I\to F}_{E,\ell\ell'}(T)$ of the autocorrelation function,

$$
A^{I\to F}_{\ell\ell'}(E,\omega) = \int_{-\infty}^{\infty} dT\, e^{-i\omega T} f^{I\to F}_{E,\ell\ell'}(T) , \tag{45}
$$

can be used to bring the cross section into the following form (Reinhardt et al. 1983):

$$< \frac{d\sigma_{iI\to fF}}{d\Omega}(\theta)> = \int_{-\infty}^{\infty} dT \left| \sum_n a^-_{i\to n} \, e^{-i\varepsilon_n T/\hbar} \, a^+_{n\to f} \right|^2 \cdot$$

$$\cdot \frac{1}{4K^2} \sum_{\ell\ell'} (2\ell+1)(2\ell'+1) \, P_\ell P_{\ell'} \, e^{2i(\sigma_{\ell'}-\sigma_\ell)} \, f^{I\to F}_{E,\ell\ell'}(T) \, . \tag{46}$$

The sum over n is recognized as the semiclassical atomic excitation amplitude for a collision with a classical time delay T. This suggests the interpretation of the variable T as a nuclear delay time. The partial wave sum then defines a different nuclear cross section for every delay time T; the probability of the latter to occur in a collision where the nuclei go from state I to F is given by $f^{I\to F}_{E,\ell\ell'}(T)$:

$$< \frac{d\sigma_{iI\to fF}}{d\Omega} > = \int_{-\infty}^{\infty} dT \left| a_{i\to f,T} \right|^2 \frac{d\sigma^{nuc}_{I\to F}}{d\Omega}(\theta,T) \tag{47}$$

with

$$\frac{d\sigma^{nuc}_{I\to F}}{d\Omega}(\theta,T) = \frac{1}{4K^2} \sum_{\ell\ell'} (2\ell+1)(2\ell'+1) \, P_\ell P_{\ell'} \, e^{2i(\sigma_{\ell'}-\sigma_\ell)} \, f^{I\to F}_{E,\ell\ell'}(T) \, . \tag{48}$$

In the models so far discussed in the literature (see below), the nuclear autocorrelation function (43) does not have any poles in the lower half of the complex ω-plane; in that case, the probability distribution $f_{\ell\ell'}(T)$ in (45) vanishes identically for negative T, and the cross section (47) receives contributions only from positive (and real) delay times. This provides additional support for the interpretation of the variable T as a classical delay time.

The formula (47) is very important in practice because it essentially extends the range where semiclassical methods can be used into a domain where one naively would have assumed it to break down. We realize that even for collisions with nuclear contact where complicated nuclear processes can happen the use of classical trajectories and classical delay times can be possible. Of course, quantum mechanics enters through the back door in that the distribution function of classical delay times $f_{\ell\ell'}(T)$ (which determines the nuclear cross section for the delayed events) has to be chosen consistent with the properties of the underlying quantum mechanical nuclear scattering amplitude; i.e., it has to be the Fourier transform of the appropriate nuclear autocorrelation function. This can become a technically very hard task. Two rather simple examples will be discussed in the following section.

Equations (45), (47) and (48) can, however, also be viewed differently: detailed measurement of atomic excitation cross sections in coincidence with the scattered nuclei (which in general will be in an excited state or have gained or lost a few particles) can be used in an attempt to draw conclusions on the shape of the probability distribution $f_{\ell\ell'}(T)$ for nuclear delay times, and hence to infer characteristic properties of the nuclear S-matrix. In particular, it should be possible to decide whether the distribution of delay times is peaked at T=0 with rather small contributions from finite T, or has a nonvanishing mean value T larger than the typical Rutherford collision time of about 10^{-21} sec (see the talks by Soff and Senger). The implications for the nuclear autocorrelation function would be quite drastic (see below), pointing to a more stochastic nuclear S-matrix with rather short-range correlations in energy and angular momentum in the first case, versus long-range nuclear correlations in energy and/or angular momentum in the second case (Reinhardt et al. 1983).

Let me remind the reader that the basis of this very elegant formulation is the factorization ansatz (35,36) for the excitation amplitudes where all atomic excitations happen outside R_m (and are treated semiclassically, and all nuclear effects are confined to distances less than R_m and are contained in the S-matrix $\tilde{S}_{\beta\to\alpha}(R_m)$. For processes where such a separation is impossible eq. (47) is not applicable. In some sense these would be genuine quantum mechanical nuclear-atomic interference phenomena, because there semiclassical intuition appears to fail completely. Unfortunately, the computational complexities avalanche in that case and have only recently been tackled; Schramm will show in his talk for the case of nuclear sticking caused by a pocket in the internuclear potential that by a proper matching of WKB solutions in- and outside the pocket the interplay of nuclear and atomic transitions <u>during</u> nuclear contact can be described theoretically.

AN EXAMPLE: ELASTIC SCATTERING OFF NUCLEAR MOLECULAR RESONANCES

To illustrate the procedure presented in the last section let us study the elastic scattering off nuclear molecular resonances supported by a pocket near the touching distance in the internuclear potential (see Fig. 3). The pocket is assumed to be sufficiently deep to contain one or more quasibound vibrational states (corresponding to radial excitations of the nuclei around the touching configuration) with energies E_n, n=1, ..., N.

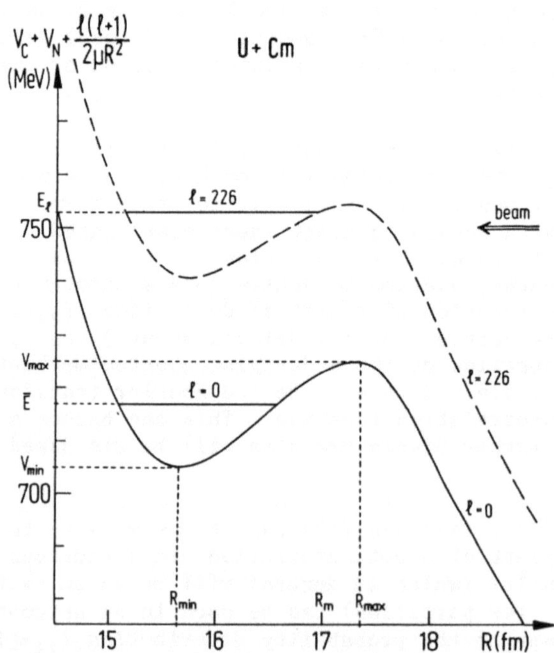

Fig. 3. Schematic picture of the internuclear potential. The parameters indicated are those used in the calculations of this talk: V_{max} = 725 MeV; $V_{max} - V_{min}$ = 20 MeV; $R_{max} - R_{min}$ = 0.1; R_m = 17 fm; $\bar{E} = V_{max}$ - 8 MeV; $\gamma = \hbar^2/(2\mu R^2_{min})$ = 0.7 keV; beam energy = 6.2 MeV/n (E_{cm} = 750 MeV).

On top of each vibrational state one assumes a rotational band with energies

$$E_{n\ell} = E_n + \gamma\, \ell(\ell+1) \ , \tag{49}$$

corresponding to a rotation of the nuclear molecule around its center of mass. ℓ is the angular momentum of the rotation, and $\gamma = \hbar^2/(2\mu R_{min}^2)$ is the rotational constant (inverse moment of inertia) of the molecule, defined by the nuclear separation R_{min} at the minimum of the potential pocket. For large nuclei γ is rather small (~0.7 keV for a U-U molecule (Hess et al. 1984)) which means that rotational excitations with very large angular momenta can be supported by even a rather shallow pocket $\ell_{crit}^{U-U} \sim 300\hbar$ if the band head state is bound by 10 MeV).

It is assumed that each molecular resonance contributes to the elastic nuclear phase shift with a Breit-Wigner form:

$$S_\ell(E) = \prod_{n=1}^{N} (E-E_{n\ell}-i\Gamma_{n\ell}/2)/(E-E_{n\ell}+i\Gamma_{n\ell}/2) \ . \tag{50}$$

The resonance widths are determined from a simple barrier penetration formula:

$$\Gamma_{n\ell} = \frac{\hbar\omega_{min}}{2\pi} \exp\left\{-\frac{2\pi}{\hbar\omega_{max}} \left(V_{max} + \frac{\ell(\ell+1)\hbar^2}{2\mu R_{max}} - E_{n\ell}\right)\right\} \ . \tag{51}$$

ω_{min} (ω_{max}) are the curvature of the potential at the minimum (maximum), and $V_{max}^{\ell} \equiv V_{max} + \ell(\ell+1)\hbar^2/2\mu R_{max}$ is the energy of the effective potential barrier in partial wave ℓ. Obviously the resonances are very narrow and long-lived far below the barrier, and become wide very rapidly close to the barrier top and above.This implies that delayed nuclear scattering will be dominated by a few resonances slightly below the barrier top, because the higher states decay too fast to yield appreciable delay, and the very narrow deep-lying states have hardly a chance to be hit by the incoming nuclei because the tunneling probability through the barrier is too small.

Using the prescription (42) for the energy average, the autocorrelation function and delay time distribution can be obtained analytically from the S-matrix (50). The technical details are given in Heinz et al. (1984a). For N=1 (one rotational band) one finds

$$\left\langle \frac{d\sigma_{i\to f}}{d\Omega}(\theta) \right\rangle = \left|a_{i\to f,T=0}\right|^2 \frac{d\sigma_R}{d\Omega}(\theta)$$

$$+ \int_0^\infty dT \left|a_{i\to f,T}\right|^2 \frac{d\sigma_{delayed}^{nuc}}{d\Omega}(\theta,T) + \text{interference term} \ , \tag{52}$$

with

$$\frac{d\sigma_{delayed}^{nuc}}{d\Omega}(\theta,T) = \frac{1}{4K^2} \left|\sum_\ell (2\ell+1)\, e^{2i\sigma_\ell}\, P_\ell(\cos\theta)\, \alpha_\ell^E(T)\right|^2 . \tag{53}$$

The functions $\alpha_\ell^E(T)$ are given by

$$\alpha_\ell^E(T) = \frac{\Gamma_\ell\sqrt{2\Delta E}}{E-E_\ell+i\Gamma_\ell/2+i\Delta E}\, e^{-iE_\ell T-\Gamma_\ell T/2} \ . \tag{54}$$

The first term in (52) is a "background" of pure Coulomb scattering events, and contains the undelayed transition probability times the nuclear Rutherford cross section. The second term describes the additional contribution from delayed nuclear scattering events (i.e. events which proceed through the molecular resonances). The interference term is rapidly damped out for increasing T; as $\Delta E \to \infty$, its time structure $\Delta E e^{-2\Delta E|T|} \to \delta(T)$ approaches that of the prompt contribution. It is therefore to be interpreted as the flux taken out of the prompt scattering channel due to the presence of delayed scattering events. Typically the delayed term is by three orders of magnitude smaller than the prompt contribution; hence this flux correction to the direct scattering can safely be neglected unless delay times $T < O(1/\Delta E)$ are of interest. (Typically $\Delta E \sim 10$ MeV, and all events with delay times more than 10^{-22} sec can be treated incoherently from the prompt scattering contribution.)

In Fig. 4 we show the nuclear cross section for the delayed events, eq. (53), as a function of delay time T, for the case of a potential pocket which supports ten rotational bands of molecular resonances. Although this "distribution of delay times" looks quite chaotic, it has a distinctive feature: although peaked at small T, it contains appreciable probability for large delay times $T \gtrsim 10^{-19}$ sec. The fall-off is much weaker than an exponential corresponding to stochastic decay of a compound nucleus; the slope of the envelope of Fig. 4 is nearly proportional to 1/T. This means that the long-range energy correlations, introduced into the nuclear autocorrelation function through the law (49) for the energy eigenvalues of the nuclear molecule, are reflected in a delay time distribution with significantly stronger relative importance of large delay times than one obtains from the stochastic compound nucleus theory (Reinhardt et al. 1983). This illustrates very nicely the relationship between the statistical properties of the eigenstate spectrum of the combined nuclear system and the expected time delay in collisions with nuclear contact.

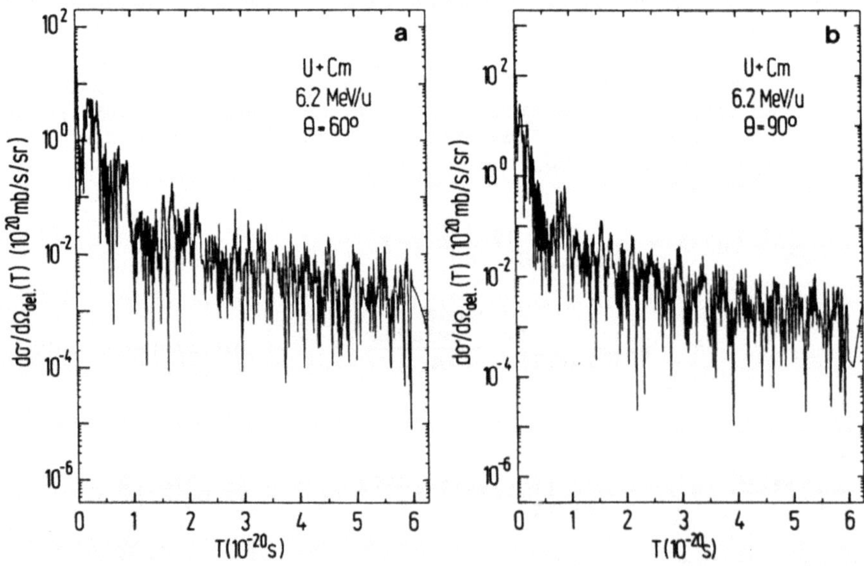

Fig. 4. Time distribution of the delayed nuclear cross section for two nuclear scattering angles, θ = 60° and 90°, employing ten rotational bands of molecular resonances with band heads at 1,2,...,10 MeV below the top of the potential barrier.

POSITRON SPECTRA FROM DELAYED NUCLEAR COLLISIONS

In this section we use these results for the delayed nuclear cross section to compute positron spectra in U + Cm collisions with nuclear contact. We will focus on the second term in (52) describing the positrons coming from genuinely delayed collisions. The reason for doing so is that in our model even at the position of the spontaneous positron line this last term is generally 2 to 3 orders of magnitude smaller than the "dynamical" background of positrons coming from undelayed Rutherford collisions and could be hardly recognized in the complete spectrum. Of course, this is in strong disagreement with the experimental peaks which stick out from the dynamical background by about a factor of 2. On the other hand, this discrepancy was to be expected from our restriction to elastic nuclear scattering: in order to hit nuclear resonances with sufficiently long lifetimes to produce a clear and narrow positron line, in elastic scattering the nuclei first have to penetrate the potential barrier in Fig. 3. The tunnelling probability ("entrance width") in elastic scattering is the same as the subsequent decay probability of the resonance ("exit width") and is correspondingly small for the long-lived nuclear resonances we require, leading to a cross section for collisions with sufficiently long delay times that is much too small.

A possible way out of this dilemma is to let the two nuclei come in above the barrier, thereby increasing the entrance width by orders of magnitude, and subsequently let them lose relative kinetic energy due to internal nuclear excitation, thereby trapping them in one of the long-lived resonances with small exit width below the barrier. This mechanism is known in light nuclear molecules like ^{12}C-^{12}C, ^{16}O-^{16}O, etc., as the double resonance mechanism. Of course, the giant nuclear system, being captured in the pocket, may also decay through other channels like light-particle emission, multiple fission, etc. With our formula (51) for the widths, we estimate that in order to gain a factor 10^3 to 10^4 in the entrance width relative to the exit width, the nuclei would have to lose (with probability close to 1) 2 to 3 MeV of relative energy due to excitations. This does not appear to be a problem in any realistic model for intrinsic nuclear excitations (e.g., already nuclear Coulomb excitation along Rutherford trajectories yields an average nuclear excitation energy of 4 to 6 MeV), and such a factor would be ample to boost the absolute intensity of our spontaneous positron peak into a regime compatible with the experimental peaks. (For a cautionary comment how to <u>not</u> implement this idea see Tom Pinkston's talk at this meeting; with his ansatz for the nuclear S-matrix he actually finds a reduction of the peak cross section due to the presence of inelastic nuclear processes.)

As has become clear in Berndt Müller's talk, the fact that the positron lines fail to shift appreciably in energy as the combined nuclear charge is varied has excluded the spontaneous vacuum decay as a possible explanation of (all) the existing data. Therefore, we do not think that this model provides a viable explanation of the GSI experiments, and we have not performed such more elaborate calculations. We will show results based on the elastic scattering model from the last section simply to demonstrate for the case of spontaneous positron emission the basic mechanisms of nuclear contact on atomic excitations spectra.

In Fig. 5 we show positron emission probabilities from delayed nuclear collisions (53), normalized to the Rutherford cross section, for the form of the delayed nuclear cross section shown in Fig. 4. One clearly sees the line due to spontaneous positron emission; its energy is determined by the matching radius R_m through the 1s binding energy at this value for the nuclear separation (because at this point we transform to the eigenstate basis in order to eliminate the spontaneous coupling inside R_m). More

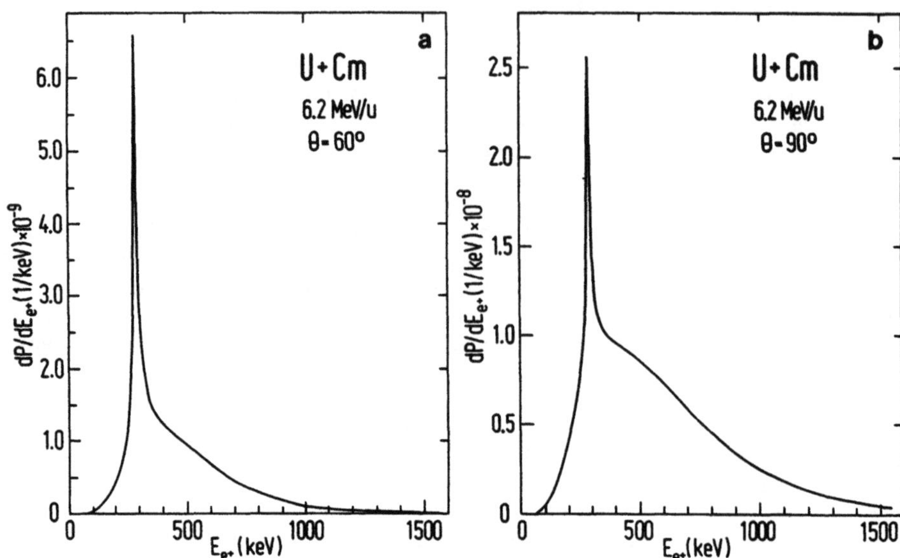

Fig. 5. Positron spectra from delayed nuclear collisions, computed from the nuclear delay time distributions shown in Fig. 4, for two nuclear scattering angles, $\theta = 60°$ and $90°$. The spectra are normalized to the nuclear Rutherford cross section. The "dynamical" positrons from Rutherford events are not added.

properly, the spontaneous coupling should be taken at the location of the minimum of the potential pocket, thereby increasing the peak energy by an amount depending on the location of this pocket in your model. This is clearly seen in the calculations to be presented by Stefan Schramm.

The sharpness of the line is due to its origin in the tail of the delay time distribution (Fig.4) at large delay times $T \gtrsim 10^{-19}$ sec; the broad bump at its base reflects the maximum of this delay time distribution at small values of T. That the peak shows so clearly above the bump demonstrates the "magnifying power" of spontaneous positron emission, allowing to see even small fractions of strongly delayed nuclear events above a large background of rather prompt nuclear collisions (Reinhardt et al. 1981b).

We conclude our study of this model with an investigation of the beam energy dependence of the positron spectra. This beam energy dependence is determined by the fact that only those resonances $E_{n\ell}$, which lie within an energy interval $\pm\Delta E$ aroung the mean incident energy $\bar{E}$, contribute to the delayed cross section. Therefore, at energies far below the potential barrier, resonances are too narrow to be excited, and the cross section for delayed scattering events (and thus the chance to see spontaneous positron emission) is negligible. Far above the barrier, the beam can only hit the high-lying, short-lived resonances, again leading to a very low chance for spontaneous positron emission. Only if the beam energy is barely below the barrier is there a reasonable probability to be captured by a resonance with sufficiently long lifetime to lead to enhanced production of spontaneous positrons (Heinz et al. 1983b).

This is shown in Fig. 6. Since the atomic amplitudes are only weakly energy dependent, the relevant physical parameter is $E - V_{max}$, the c.m. energy relative to the potential energy. In our calculation we varied this parameter by varying V_{max}, i.e., by shifting the whole internuclear

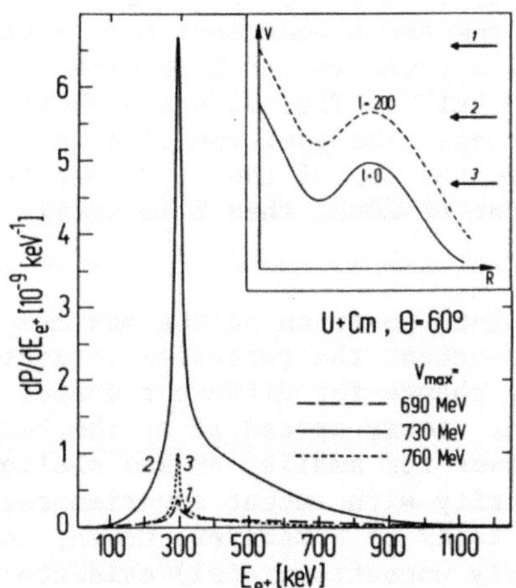

Fig. 6. Positron spectra for three values of $E - V_{max}$, at $\theta = 60°$. Only the contribution from scattering events with delay times $0 < T \leq 6 \cdot 10^{-19}$ sec normalized to the Rutherford cross section is shown. The "background" from Rutherford events is not added.

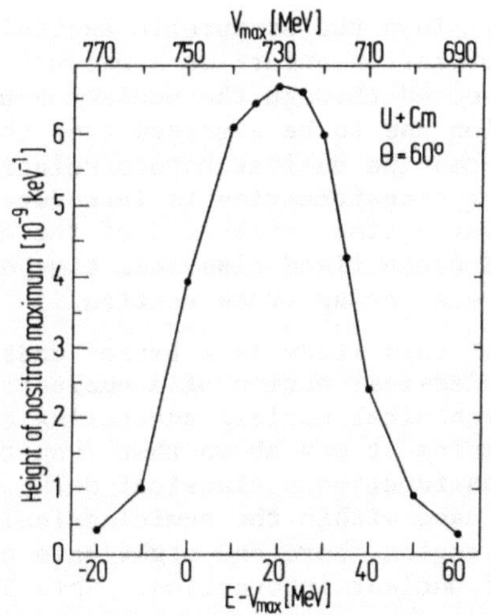

Fig. 7. The height of the maximum in the positron spectra as in Fig. 6 is shown as a function of c.m. energy for $\theta = 60°$.

potential. This certainly is a good first approximation and saves us the computation of the atomic amplitudes for several energies.

Figure 6 demonstrates that the spontaneous positron line at $E_{e^+} \approx$ 300 keV shows a clear resonance behavior: barely visible far below and above the barrier, it is quite prominent if E is close to the barrier. This is demonstrated in more detail in Fig. 7, which shows the "excitation function" for spontaneous positrons. The positron line is strongest if the c.m. energy is ~20 MeV above the top of the (ℓ=0)-barrier; since typical angular momenta are of the order of 200h, then E is barely below the effective potential barrier.

Although the absolute position of the maximum in Fig. 7 as well as its width reflects to some extent the potential parameters used in the calculation, and may slightly change for different shapes of the pocket and for different values of the energy spread ΔE of the beam (we expect the peak in Fig. 7 to become narrower for smaller ΔE and shallower potential pockets), the qualitative similarity with recent experimental data is striking: as stressed again in the talks by Bokemeyer, Cowan, and Kienle, there is some (although not completely uncontroversial) evidence for a resonance-like sensitivity of the experimentally found positron peaks on the beam energy. While such a feature falls out naturally from the assumption of a nuclear potential pocket, no other model has so far given such a feature. This may be indication for some nuclear origin along the lines discussed in this talk in the excitaton mechanism for the measured positron peaks.

SUMMARY

In this talk, the formal framework for the descripton of atomic excitations in heavy ion collisions was based on quantum mechanical scattering theory. This resulted in a unified treatment of nuclear and atomic dynamics and excitations. Outside the nuclear reaction region the nuclear motion may be accurately described using the semiclassical approximation. Neglecting electronic transitions inside the reaction region (except for spontaneous positron creation), the nuclear interaction may be treated independently from atomic processes. Then the asymptotic excitation amplitudes can be represented by a factorization ansatz as a product of incoming and outgoing atomic amplitudes connected through the nuclear S-matrix. The resulting excitation cross section has to be averaged over the spread in beam energy. This procedure introduces the nuclear autocorrelation function. If, in a further step, a Fourier transformation is introduced, the result takes the form of an integral over a time variable T of the atomic transition probability calculated for a fixed classical time delay, multiplied by the corresponding delayed scattering cross section.

The main result of this study is a better understanding of the relationship between the classical notion of a nuclear delay time and the structure of the quantum mechanical nuclear scattering amplitude. At least for elastic nuclear scattering it was shown that from the quantum mechanical nuclear autocorrelation function a classical delay time distribution can be derived, which can be used within the semiclassical framework to consistently extend it into the region where one might have assumed it to break down due to the presence of nuclear interaction. This is of enormous practical significance.

We also pointed out the limits of applicability of this procedure which relies on the mentioned factorization of nuclear and atomic excitation processes. In absence of this property the quantum mechanical scattering framework leads to more complicated coupled channel equations for the transition amplitudes which, however, with present computer technology can be tackled and, as shown by Schramm, have been used to obtain quantitative results.

ACKNOWLEDGEMENTS

I would like to thank Prof. W. Greiner for the warm hospitality extended to me for several weeks before this conference, and him and Dr. J. Reinhardt for the interesting discussions on possible and impossible origins of these positron peaks which puzzle us all. This work was supported under contract DE-AC02-76CH00016 with the U.S. Department of Energy.

REFERENCES

Amundsen P A 1978 J. Phys. B 11:3197.
Blair J S and Anholt R 1982 Phys. Rev. A 25:907.
Clemente M, Berdermann E, Kienle P, Tsertos H, Wagner W, Kozhuharov C, Bosch F, and Koenig W 1984 Phys. Lett. B 137:41.
Cowan T, Backe H, Begemann M, Bethge K, Bokemeyer H, Folger H, Greenberg J S, Grein H, Gruppe A, Kido Y, Klüver M, Schwalm D, Schweppe J, Stiebing K E, Trautmann N, and Vincent P 1985 Phys. Rev. Lett. 54:1761.
Cowan T, Backe H, Bethge K, Bokemeyer H, Folger H, Greenberg J S, Sakaguchi K, Schwalm D, Schweppe J, Stiebing K E, and Vincent P 1986 Phys. Rev. Lett. 56:444.
Ciocchetti G, Molinari A, and Galvano R 1963 Nuovo Cim. 29:1262.
Heinz U 1986 Rep. Prog. Phys., in print
Heinz U, Greiner W, and Müller B 1981 Phys. Rev. A 23:562.
Heinz U, Müller B, and Greiner W 1983a Ann. Phys (N.Y.) 151:227.
Heinz U, Reinhardt J, Müller B, Greiner W, and Müller U 1983b Z. Phys. A 314:125.
Heinz U, Müller U, Reinhardt J, Müller B, and Greiner W 1984a Ann. Phys. (N.Y.) 158:476.
Hess P O, Greiner W, and Pinkston W T 1984 Phys. Rev. Lett. 53:1535.
Müller B and Greiner W 1976 Z. Naturforsch. 31A:1.
Müller B and Oberacker V 1980 Phys. Rev. C 22:1909.
Müller B, Rafelski J, and Greiner W 1973 Phys. Lett. B 47:5.
Müller U, Soff G, deReus T, Reinhardt J, Müller B, and Greiner W 1983 Z. Phys. A 313:263.
Rafelski J, Müller B, and Greiner W 1978a Z. Phys. A 285:49.
Reinhardt J and Greiner W 1977 Rep. Prog. Phys. 40:219.
Reinhardt J, Müller B, and Greiner W 1981a Phys. Rev. A 24:103.
Reinhardt J, Müller B, Müller U, and Greiner W 1981b Z. Phys. A 303:173.
Reinhardt J, Müller B, Greiner W, and Müller U 1983 Phys. Rev. A 28:2558.
Schramm S, Reinhardt J, Müller U, Müller B, and Greiner W 1986 Z.Phys. A 323:275
Schweppe J, Gruppe A, Bethge K, Bokemeyer H, Cowan T, Folger H, Greenberg J S, Grein H, Ito S, Schulé R, Schwalm D, Stiebing K E, Trautmann N, Vincent P, and Waldschmidt M 1983 Phys. Rev. Lett. 51:2261.
Tomoda T and Weidenmüller H A 1983 Phys. Rev. C 28:739.
Tsertos H, Berdermann E, Bosch F, Clemente M, Kienle P, Koenig W, Kozhuharov Ch, and Wagner W 1985 Phys. Lett. B 162:273.

STUDY OF THE CONSEQUENCES OF HYPOTHESIZED POTENTIAL POCKETS USING SIMPLE MODELS

W. T. Pinkston, D. P. Russell, and V. E. Oberacker

Physics Department, Vanderbilt University
Nashville, TN 37235 USA

INTRODUCTION

Two problems are discussed, both related to the suggestion that HI interaction potentials contain minima or pockets which cause "sticking" and time delays in sub-Coulomb collisions.

CAN QUASIMOLECULAR RESONANCES PRODUCE STRONG STICKING?

These remarks are related to the theoretical results on spontaneous e^+ production presented in the contributions by Heinz[1] and Schramm.[2] (The reader is referred to those papers, especially to that of Heinz, for additional references.) Although the observation of sharp positron lines in sub-critical systems seems to rule out the vacuum-decay mechanism, some of the other mechanisms proposed also require time delays long compared the Rutherford collision time. Presumably these time delays result from nuclear contact and "sticking". Sticking is also suggested by experimental results indicating that the positrons seem to emanate from the c.m. of the colliding ions. Thus, the existence and physical origin of sticking and its quantum mechanical description are still open questions.

A detailed discussion of how to include the quantum effects of nuclear scattering in a semiclassical treatment of atomic processes is given in Heinz's contribution.[1] The starting point is the formula of Tomoda.[1,3] The amplitude for a transition from atomic states i to f, and from nuclear states 0 (ground state) to α is given by

$$\mathrm{AMP}^L(i \to f,\ 0 \to \alpha) = -\sum_n a^{(-)L}_{i\to n}(\infty,R)S^L_{o\alpha}(E_n)a^{(+)}_{n\to f}(R,\infty) \tag{1}$$

In (1) L is the orbital angular momentum of relative nuclear motion. The amplitude $a^{(-)L}_{i\to n}$ ($a^{(+)L}_{n\to f}$) is the electronic transition amplitude along the ingoing (outgoing) trajectory. R is a matching radius. Nuclear forces act only for $r < R$; the coupling between electronic states can be neglected in this region. The S-matrix connects ingoing and outgoing Coulomb waves at $r = R$. It depends on the "available energy" at R, $E_n = E - \varepsilon_n(R)$, where ε_n is the energy of the adiabatic electronic state. The cross section for positron production has the form

$$\begin{aligned} \left\langle \frac{d\sigma_{i\to f}}{d\Omega} \right\rangle &= \frac{1}{4K^2} \sum_{n,n'} a^{(-)}_{i\to n'}\, a^{(+)}_{n'\to f}\, a^{(-)*}_{i\to n}\, a^{(+)*}_{n\to f} \\ &\times \sum_{L,L'} (2L+1)(2L'+1)P_L(\cos\theta)P_{L'}(\cos\theta)e^{2i(\sigma_L-\sigma_{L'})} A_{L'L}(E_{n'},E_n)\ . \end{aligned} \tag{2}$$

The time delay arises from the last factor in (2), the autocorrelation function of the nuclear S-matrix,

$$A_{L'L}(E',E) = \sum_\alpha \langle S^{L'}_{o\alpha}(E')\ S^L_{o\alpha}(E)^*\rangle \tag{3}$$

The angular brackets imply an average over bombarding energy ($\Delta E \sim 10$ MeV). Since individual nuclear final states are not resolved in the experiments, there is an incoherent sum over channel states. In the calculations of Heinz et al.,[4] based on the pocket hypothesis, nuclear excitations were neglected, and S was chosen to be of the simple Breit-Wigner form,

$$S^L(E) = \frac{E - E_L - i\Gamma_{L/2}}{E - E_L + i\Gamma_{L/2}} \tag{4}$$

This is essentially a potential scattering model, in which the pocket producing the resonant behavior has no dependence on internal degrees of freedom, such as nuclear orientations. The resonant energies were assumed to be given by the rotational formula, $E_L = \bar{E} + aL(L+1)$; a barrier penetration model was used for calculating the widths. The resulting line intensity[4] was much too small compared to experiment. Similar results were reported here at Maratea by Schramm,[2] who performed a much more sophisticated and realistic calculation. It was suggested that including inelastic couplings and many bands in the calculations should increase the e^+ yield. My purpose is to argue against this assertion.

It was pointed out[5] earlier that making the model more realistic by including several bands and inelastic excitations may even decrease, rather than increase, the positron yield.

The appropriate generalization of (4) to the more general situation in which there are inelastic couplings and overlapping bands is the well-known, many-level Breit-Wigner formula,

$$S^{L}_{\alpha\beta}(E) = \delta_{\alpha\beta} - i\sum_{\lambda} \frac{\gamma_{\alpha L}(\lambda)\gamma_{\beta L}(\lambda)}{E - E_{\lambda L} + i\Gamma_{\lambda L/2}} ,$$

$$\gamma_{\alpha L}(\lambda)^2 = \Gamma_{\alpha L}(\lambda) \quad , \quad \sum \gamma_{\alpha L}(\lambda)\gamma_{\alpha L}(\lambda') = \delta_{\lambda\lambda'}\Gamma_{\lambda L} \tag{5}$$

In order to stress the physics and avoid the distraction of complicated angular momentum algebra, the angular momenta of internal nuclear states have been neglected in (5), so that L is the only angular momentum quantum number. For a single band in which the widths vary slowly with L over the energy interval ΔE, the autocorrelation function becomes

$$A_{L'L}(E',E) \approx (1 - \Gamma_o/\Gamma) + \frac{\Gamma_o}{\Gamma} < \left(1 - i\,\frac{\Gamma}{D_{L'}}\right)\left(1 + i\,\frac{\Gamma}{D^*_L}\right) > \tag{6}$$

The denominators in (6) are $D_L = E - E_L + i\Gamma_L/2$. The first term in (6) is a constant; it results in a contribution with zero time delay. The second term is the same as one would get using the simple formula (4), except for the branching ratio $\Gamma_o/\Gamma < 1$. Such factors are well known in reaction theory; Γ_o/Γ gives the probability that a projectile "knocking on the door" will get into the interior region and form the resonant state. It can be interpreted as the probability, discussed by Oberacker,[6] that deformed nuclei in their ground states approach in orientations favorable to "sticking".

The formalism can also be extended to include the dynamical orientation effect calculated semiclassically by Oberacker.[6] Assuming that both electronic excitations and nuclear Coulomb excitation can be calculated semiclassically, and that they proceed independently for $r > R$, with the Coulex amplitudes being given by $b^{(-)}_{o\alpha}$, then (6) remains valid, except that Γ_o is replaced

$$\Gamma'_o = |\sum_{\alpha} b^{(-)}_{o\alpha}\ \gamma_\alpha|^2 \quad . \tag{7}$$

Oberacker's calculations suggest that $\Gamma'_o/\Gamma < \Gamma_o/\Gamma$ and that $\Gamma'_o/\Gamma \ll 1$. Adding a number of rotational bands increases this branching factor, which becomes $\Sigma_\lambda \Gamma_o(\lambda)/\Gamma_\lambda$; however this sum is not expected to exceed unity. Therefore, it is very likely that the very simple resonance model of ref. 4 provides an upper limit on the positron yield. It is important to note that the microscopic mechanisms are buried in the parameters of the S-matrix, so that any microscopic model resulting in a resonant S-matrix should reproduce these negative results.

NEUTRON TRANSFER IN ^{238}U-^{238}U COLLISIONS

The assumption of potential pockets is somewhat radical. That they may be needed to explain the positron lines is a weak argument for their existence; it must be justified theoretically and also verified in nuclear physics experiments. We have examined neutron transfer for evidence for pockets. Nucleon transfer is a sensitive probe of the nuclear surface region where pockets have been postulated. Data on one and two nucleon transfer in ^{238}U-^{238}U collisions exists.[8] Transfer is a well-studied process, at least in lighter systems, and it seems clear intuitively that such a drastic modification of HI potentials from the conventional Coulomb plus Woods-Saxon forms should give results quite different from those obtained from standard direct reaction theory.

A proper theoretical treatment of transfer reactions between very heavy nuclei such as U would be extremely difficult. Both projectile and target are strongly deformed, so that large numbers of intermediate states should contribute as a result of Coulomb and nuclear inelastic excitation. Also, nothing is known about the optical model parameters in such heavy systems. In addition, as we shall see shortly, the quality of the data does not justify a detailed, realistic treatment. As a result, for these preliminary studies, we resorted to an oversimplified two-state model. In this model, two single-particle states are available to a valence neutron, one on each of two identical, spherical cores, as shown in Fig. 1. The model Hamiltonian is given by

$$H = T(r) + t + V(r) + U(x) + U(y) \quad . \tag{8}$$

T and t are the kinetic energies of the relative motion of the cores and of the neutron, respectively. The mutual interaction of the cores is given by V, the neutron-core interaction by V. A trial function is

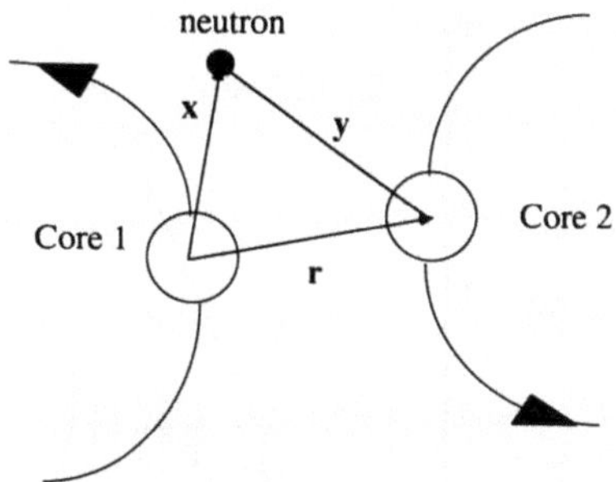

Fig. 1. Coordinates used in two-state model.

substituted in the Schroedinger equation, i.e.

$$H\Psi = E\Psi \ ,$$

$$\Psi = \Phi_i(r)\phi_i(x) + \Phi_f(r)\phi_f(y) \ , \tag{9}$$

$$(t + U(x))\phi_i = \varepsilon_i\phi_i \ , \text{ etc.}$$

The functions ϕ are bound state functions for the neutron. A pair of coupled channel equations results,

$$\begin{aligned} \left[T + V - (E - \varepsilon_i)\right] \Phi_i &= G\Phi_f \ , \\ \left[T + V - (E - \varepsilon_f)\right] \Phi_f &= G\Phi_i \ , \end{aligned} \tag{10}$$

in which

$$G(r) = \lambda\int d^3x \ \phi_i(x) \ U(g) \ \phi_f(y)$$

For simplicity we choose the ϕ to be 3s functions of a Woods-Saxon potential well with conventional geometric parameters and a depth chosen so that the binding on each core is 6 MeV. The adjustable constant λ is included to fit the data. Since $\varepsilon_1 = \varepsilon_2 = \varepsilon$, $Q = 0$, the equations readily decouple by adding and subtracting them. The result is that we only have to solve potential scattering Schroedinger equations with effective potentials, U + G. Many things are neglected in this treatment -- including inelastic channels, recoil effects, non-orthogonality, non-zero angular momentum.

The experimental data[8] is shown in Fig. 2. These are radiochemical data, actually sums of cross sections of all processes which eventually

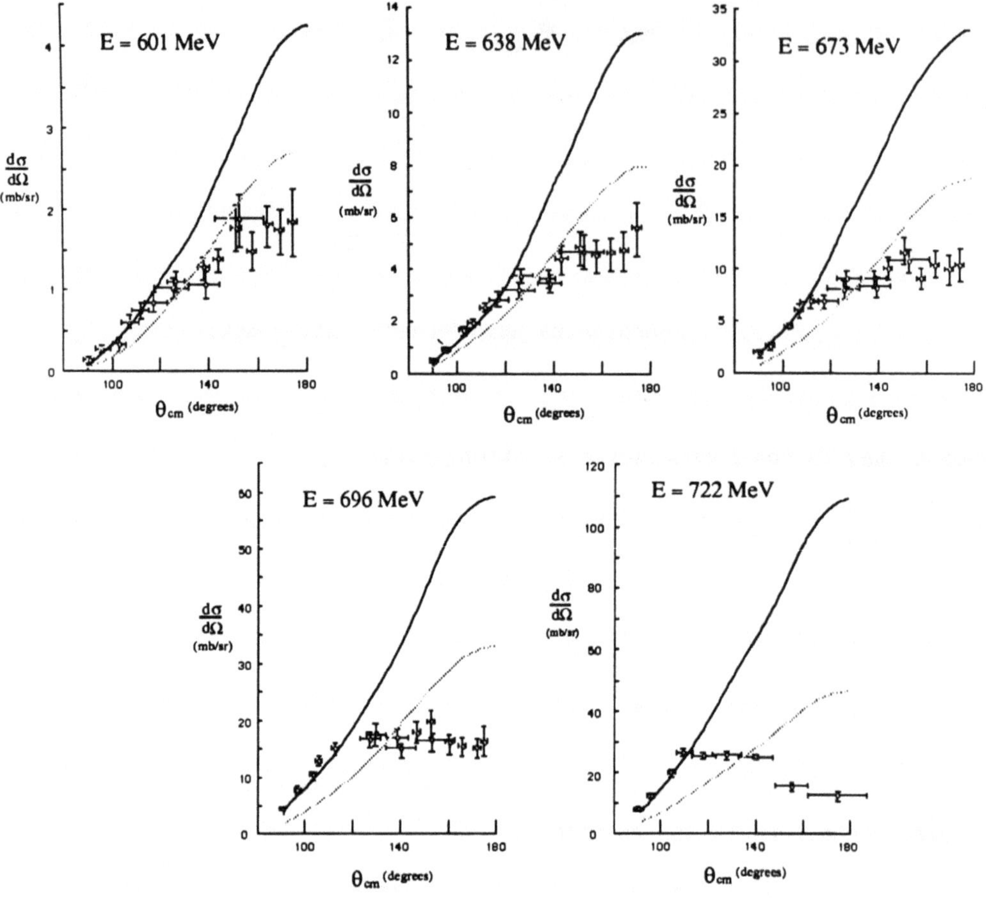

Fig. 2. Experimental data for $^{238}U + ^{238}U$ leading to ^{239}U.

lead to the ground state of ^{239}U. A large amount of information on intermediate states is therefore lost -- information not included in the simple model. The bombarding energies are sufficiently low that one would expect the extreme, sub-Coulomb analytical formula,

$$\frac{d\sigma}{d\Omega} = C(E)e^{-2kD(\theta)} \;,\; k = \sqrt{\frac{2m\varepsilon}{\hbar^2}} \;,\; D(\theta) = \frac{RV_c}{E}\left(I + \csc\frac{\theta}{2}\right) \;, \tag{11}$$

to work reasonably well. This results from solving eqs. (10) in lowest order DWBA, with purely Coulomb distorted waves and the bound-state functions ϕ replaced by their Hankel tails. The authors of ref. 8 fitted eq. (11) to the data in two ways. The dashed lines show their fits to the total cross sections; the solid lines are their fits to the experimental points at smaller angles, on the assumption that the approximate treatment should be most accurate for large impact parameters. Clearly, neither is very good except at the very lowest energies. The "reasonableness" of this sub-Coulomb treatment is illustrated in Fig. 3. The steepest potential curve is the point-Coulomb interaction of the cores. The effect of including a nuclear Woods-Saxon potential, with parameters taken from studies[10] of lighter systems, is also shown. The two lower curves show a "shoulder" potential and one with a pocket, used in our studied. At the higher energies, the data show a dip at large angles, suggesting the onset of absorption of some kind, perhaps the opening of a competing reaction channel. There is also a peak in the neighborhood of 130°, which might signal an increased transfer probability due to orbiting phenomena produced by a barrier top.

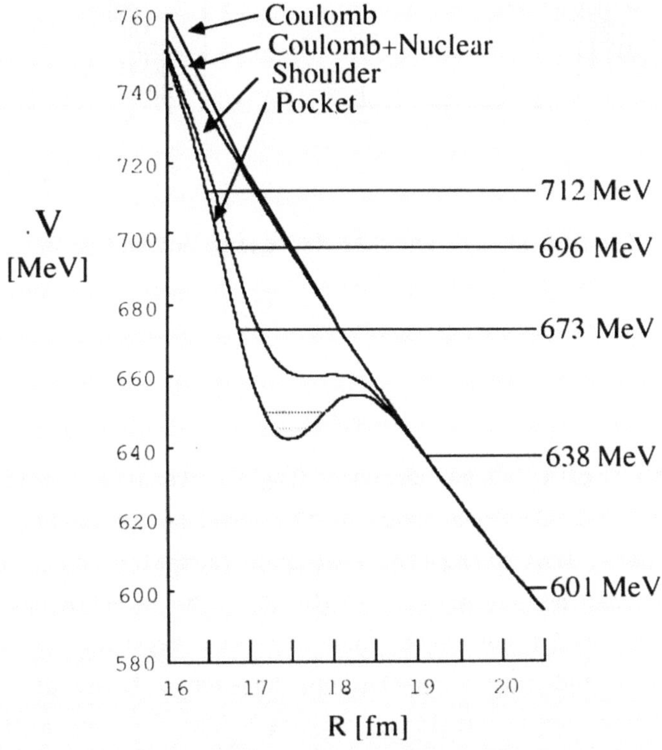

Fig. 3. HI potentials used in the calculations.

The details of our calculations will be given in a later publication. Only a few results will be given here. The coupled channels eqs. (10) were solved using the potential V with the pocket, shown in Fig. 3. Cross sections resulting from these calculations are shown in Fig. 4 for bombarding energies of 696 and 772 MeV. The resulting cross sections are flat for large angles, somewhat like the data, but have large oscillatory

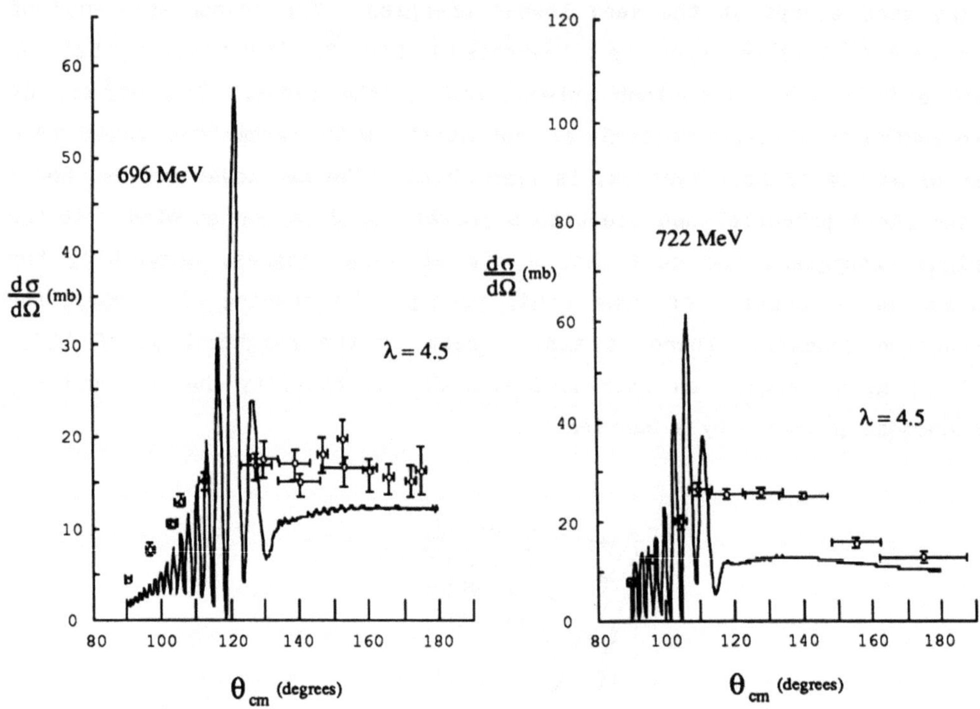

Fig. 4. Results at 696 and 722 MeV using potential with a pocket.

peaks in the angular range corresponding to classical orbiting about the barrier top. Calculations assuming the shoulder potential[11] led to very similar results, indicating that resonance formation had little to do with the overall angular patterns. (Obviously, the oscillatory structure in these figures would not have been observed experimentally, because of finite angular and energy resolution; however a broad peak remains when we angle average these results.) Figure 5 shows a very different kind of fit to the data. A point-Coulomb V was employed, and the coupled equations were solved with a fairly large value of the coupling parameter λ, which was adjusted for a best fit to the 673-MeV data. We consider the agreement to be surprisingly good. In addition, two-neutron transfer is predicted rather well. By assuming that the two-neutron transfer process is dominated by the sequential transfer of particles, a third equation can be added to (10), which can be solved with little additional effort. The total cross

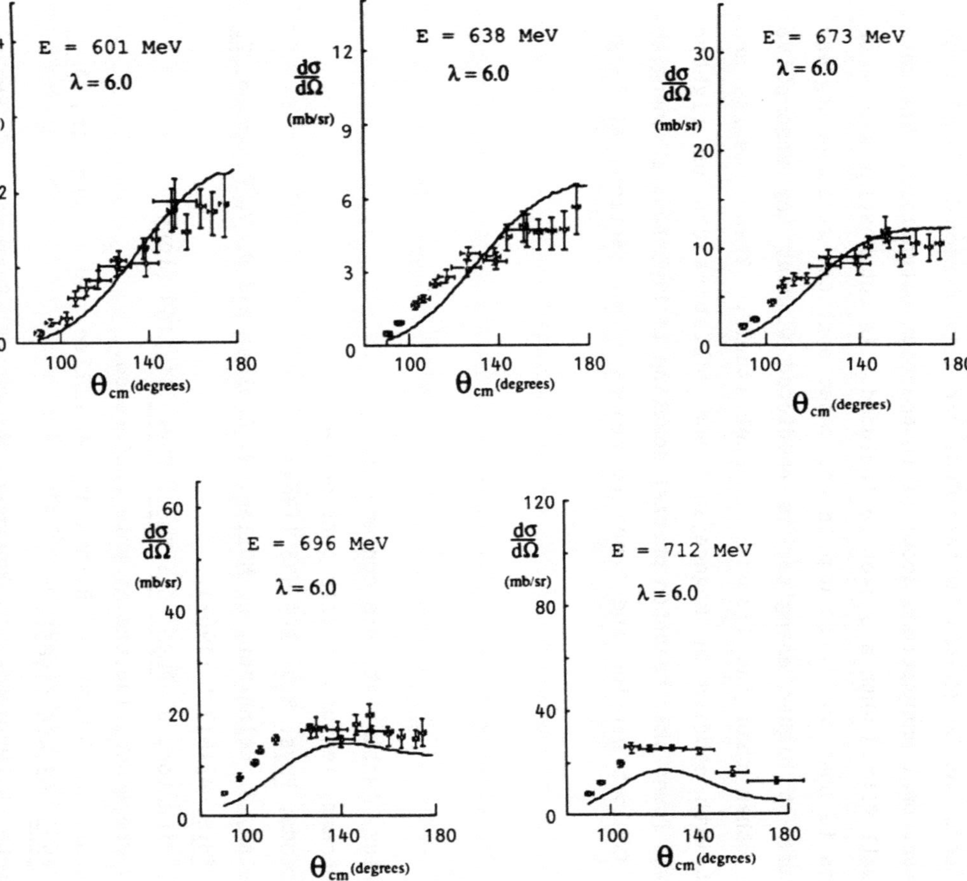

Fig. 5. Coupled channels results (without pocket in potential) compared to experiment.

section for two-neutron transfer was measured[8] to be about 6% of the one-neutron cross section at 673 MeV. In the calculations reported, here, it is predicted to be 10%.

Based on scientific conservatism, we feel obliged to conclude that there is little or no suggestion in the transfer data of anything exotic. Granted that the model is oversimplified and contains adjustable parameters, the parameters can be chosen to reproduce the data reasonably well with quite conventional assumptions about HI interaction potentials. The data are less well fitted when a pocket is included. An interesting aspect of the physics is that the large dip in the cross sections at large angles, which occurs at higher energies, is predicted without any absorptive mechanism other than the transfer channel itself. These results are consistent with studies by Nagarajan[12] and collaborators on lighter systems, who found that reaction channel coupling is important at energies below the Coulomb barrier and tends to reduce cross sections at large angles.

REFERENCES

1. U. Heinz, lecture at this conference.
2. S. Schramm, lecture at this conference.
3. T. Tomoda, Phys. Rev. A29:536 (1984).
4. U. Heinz, J. Reinhardt, B. Müller, W. Greiner and W. T. Pinkston, Z. Phys. A316:341 (1984).
5. W. T. Pinkston, J. Phys. G.: Nucl. Phys. 11:L169 (1985).
6. V. E. Oberacker, lecture at this conference.
7. U. Heinz, U. Müller, J. Reinhardt, B. Müller and W. Greiner, Ann. Phys. 58:476 (1985).
8. G. Wirth, W. Bruechle, M. Bruegger, Wo Fan, K. Suemmerer, F. Funke, V. Kratz, M. Lerche, N. Trautmann, to be published.
9. R. Bass, Ch. 5 in Nuclear Reactions With Heavy Ions, (Springer-Verlag, Berlin, Heidelberg, NY, 1980).
10. M. W. Guidry, R. E. Neese, C. R. Bingham, L. L. Riedinger, J. A. Vrba, I. Y. Lee, N. R. Johnson, G. R. Satchler, P. A. Butler, R. Donangelo, J. O. Rasmussen, D. L. Hillis, and H. H. Kluge, Nucl. Phys. A430:485 (1984).
11. D. P. Russell, W. T. Pinkston, and V. E. Oberacker, Phys. Letts. 158B:201 (1985).
12. M. A. Nagarajan, private communication.

THEORIES OF HEAVY-ION INTERACTION POTENTIALS
FOR GIANT DINUCLEAR SYSTEMS

V.E. Oberacker, M.W. Katoot and W.T. Pinkston

Department of Physics & Astronomy
Vanderbilt University
Nashville, Tennessee 37235, U.S.A.

INTRODUCTION

In this lecture, I would like to discuss the nuclear aspects of the physics of strong electromagnetic fields. In particular, I shall address the following questions:

a) Is it possible to form giant nuclear molecules (e.g. $^{238}U+^{238}U$) with a lifetime of about 10^{-19} s in a heavy-ion reaction at bombarding energies near the Coulomb barrier?
b) What are the implications of dynamical nuclear alignment (due to Coulomb excitation) for nuclear "sticking"?

The formation of very heavy intermediate nuclear molecules plays a vital role not only for spontaneous positron creation, but also for the possible production of light neutral elementary particles (axion) or composite particles (poly-positronium).

Originally, the study of giant nuclear molecules has been motivated by the theory of the spontaneous vacuum decay. Rafelski, Müller and Greiner[1] pointed out that the formation of a rather long-lived nuclear composite system would lead to a clear signature for the "sparking of the vacuum": the emergence of a narrow line structure in the positron energy spectra measured in coincidence with the scattered heavy ions. An abnormally long nuclear reaction time is required because the mean lifetime of a K-shell vacancy entering the negative energy continuum is about 10^{-19} s. In ref.1 it was suggested that deep-inelastic reactions between actinides might lead to prolonged nuclear reaction times. So far, however, all experimental attempts in this direction[2] have failed: delta-electron spectra measured in U+U collisions at 7.5 MeV/u and 8.4 MeV/u suggest reaction times of order $2x10^{-21}$ s. In complete agreement with this result, no positron line structures could be identified in these reactions.

To everybody's surprise, line structures were first observed[3,4] in rather low-energy collisions of several actinide ions (at energies slightly below the Coulomb barrier) Greiner et al.[5] suggested a possible explanation: because of the large static deformation of actinide nuclei, the relative importance of the short-ranged nuclear interaction would increase and thus generate a minimum in the heavy-ion interaction potential. This "pocket" could cause the formation of a rather long-lived nuclear quasimolecule similar to those observed in light-ion reactions[6]. We will see below that the possible formation of intermediate giant nuclear molecules at low beam energies imposes a great challenge for nuclear theory.

This lecture is organized as follows: First, I will review several phenomenological heavy-ion potentials, in particular the proximity and the double-folding approach. In the second part of my talk, I will describe a new microscopic theory for heavy-ion interaction potentials which is based on nonrelativistic many-body theory and two-center shell model wavefunctions. The third part deals with the dynamical alignment of deformed heavy nuclei near the distance of closest approach. Finally, I will summarize the main results and indicate some directions that our research might take in the future.

PHENOMENOLOGICAL HEAVY-ION POTENTIALS - A REVIEW

From the lectures given by Müller, Heinz and Schramm during this conference we have learned that it is possible to interpret the observed positron line structures in terms of the vacuum decay -- at least qualitatively -- provided that one assumes the existence of a metastable nuclear quasimolecule. But how could such a molecule be formed? The <u>proximity model</u> developed by Blocki and Randrup[7,8] helps us to understand this in rather simple terms. The nucleus-nucleus potential is assumed to be a function of the minimum distance between the two nuclear surfaces, s, only; hence, it has the form

$$V_N = V_N(s) = F(\text{geometry}) \cdot \Phi(s), \qquad (1)$$

where $\Phi(s)$ is a universal function which can be derived from the specific N-N interaction. Blocki et al. parametrized their numerical result by a simple analytical expression ("pocket formula")

$$\begin{aligned} \Phi(s) &= -3.437 \exp(-s/0.75) && ; s>1.25 \\ \Phi(s) &= -0.5\,(s-2.54)^2 - 0.0852\,(s-2.54)^3 && ; s<1.25\,. \end{aligned} \qquad (2)$$

For spherical nuclei, s is given by $s = r - R_1 - R_2$, where r is the distance between the nuclear centers of mass and R_i denote the nuclear radii. Seiwert et al.[9] extended the proximity model to deformed nuclei. In this case, the minimum distance between the surfaces, s, depends on the relative orientation of the nuclei and has to be determined numerically. It is easy to see that <u>for a fixed value of r</u>, the potential V_N between deformed nuclei can become much stronger than the potential between spherical nuclei of the same mass. As is evident from the inset in fig.1, for certain relative nuclear orientations the minimum distance between the deformed nuclear surfaces decreases in comparison to the spherical case and, consequently, the nuclear potential increases according to eq.(2). In the proximity model, the optimum orientation (leading to the strongest nuclear potential) depends on the nuclear shape. For actinides with a strong prolate quadrupole deformation and sizable negative hexadecapole deformation, the calculations by Seiwert et al.[9] generally predict an optimum orientation angle of about 50° relative to the internuclear

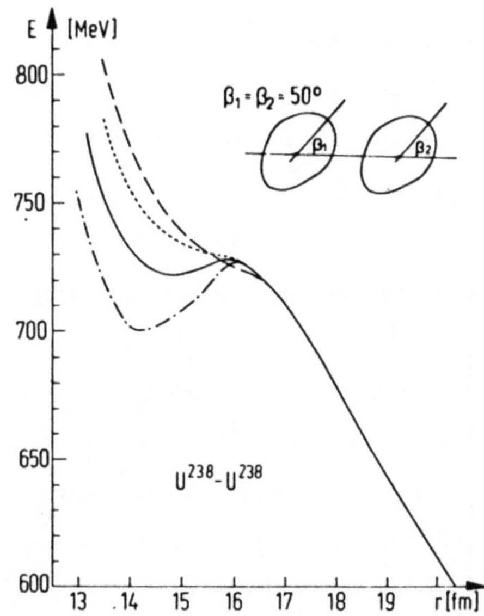

Fig. 1 Proximity potential for $^{238}U+^{238}U$, calculated with different deformation parameters (ref. 9).

axis. Fig.1 shows the heavy-ion potential for $^{238}U+^{238}U$ at this optimum angle for different sets of deformation parameters. The solid line is probably the most reliable result; it is based on high-precision inelastic electron scattering data[10]. Note that a quasimolecular minimum emerges at an internuclear distance r = 15 fm; the depth of this pocket is about 10 MeV, enough to support several nuclear resonances and hundreds of rotational states built on top of these[11].

Another phenomenological model for calculating heavy-ion interaction potentials is the double-folding method. This approach involves averaging a N-N interaction over the static matter distribution of the two nuclei

$$V(\vec{r}) = \int d^3r_1 \int d^3r_2\, \rho_1(\vec{r}_1) v(\vec{r} + \vec{r}_2 - \vec{r}_1) \rho_2(\vec{r}_2). \tag{3}$$

The double-folding model is very appealing because it emerges naturally from a nuclear many-body approach (see next chapter); in fact, at large internuclear distances (negligible overlap of the nuclear densities) it is the correct heavy-ion potential! Satchler and Love[12] used this approach to calculate the real part of the nuclear potential for spherical systems. The effective N-N interaction ("M3Y") was based on G-matrix elements constructed from the Reid soft-core potential[13]. In this way, a large number of elastic and inelastic scattering data could be explained. Rhoades-Brown, Oberacker, Seiwert and Greiner[14] extended the double-folding method to deformed nuclei. For the nuclear densities, deformed Fermi distributions were utilized with deformation parameters extracted from electron scattering data. There are no adjustable parameters in this theory. Fig. 2 shows the total (Coulomb + nuclear) heavy-ion potential for the system $^{238}U+^{238}U$ at three different orientation angles with

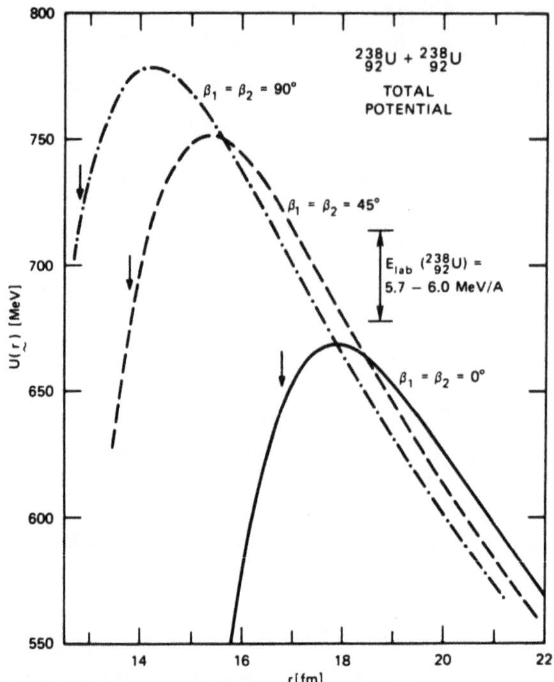

Fig. 2 Total (Coulomb + nuclear) heavy-ion potential for U+U within the double-folding model (ref. 14).

respect to the internuclear axis: 0°, 45° and 90°. Due to the strong nuclear interaction, a potential barrier emerges at distances between 18 and 14 fm, depending on orientation. In contrast to the proximity model, the lowest potential barrier is obtained if the nuclei approach "with their ends touching", i.e. at 0°. This is easily understood: In the double- folding method, the interaction potential is proportinal to the density overlap of the nuclei (cf. eq.(3)) which is largest for this orientation. Note that rotating the nuclei produces a dramatic change in the barrier height: it increases from 669 MeV at 0° to 750 MeV at 45° and reaches its maximum value of 780 MeV at 90° orientation ("equators touching").

At distances smaller than the sum of the half-density radii (marked by an arrow in fig.2) the folding model gives rise to compression of the nuclear density which is unphysical at energies in the vicinity of the Coulomb barrier. This leads to an unrealistically large nuclear interaction (-11000 MeV at r=0 fm for U+U); no "pocket" is formed like in the proximity model; rather, a very deep potential well exists. Shortcomings like these can, ultimately, only be avoided in a completely microscopic description which will be given later. However, the folding model is certainly correct at larger separations. The main problem is that one does not know for sure where it will break down: is it at 5%, 10% or 50% density overlap? Only a quantum-mechanical many-body theory can tell!

Also shown in fig.2 is the range of beam energies (5.7 - 6.0 MeV/u) at which positron line structures have been found in the GSI data[3,4]. This means that only a certain fraction of relative orientations can give rise to "nuclear sticking". If the angle between the nuclear symmetry axis and the distance vector $\vec{r}$ exceeds a critical value the potential resonances (assuming there is a "pocket" behind the Coulomb barrier) are located too far below the barrier and have an extremely small probability of being excited, because of their small width. Hence, based on the double-folding model, we estimate that the nuclei must approach with orientation angles between 0° and about 35° at 6.0 MeV/u in order to form long-lived nuclear molecules. It is clear that only a small fraction of the nuclei will be oriented in the "correct" manner. In the last chapter, we will describe a dynamical treatment of the nuclear alignment.

Without going into too much detail, we would like to mention a few other macroscopic theories of heavy-ion potentials. Münchow, Hahn and Scheid[15] applied also the double-folding formalism. In contrast to our work[14], they used an exactly ellipsoidal nuclear shape (rather than a β_2 and β_4 deformation); in order to get analytical expressions, a Gaussian interaction was utilized; its parameters were fitted such that the resulting heavy-ion potential reproduced the well-known "Bass potential"[16] at larger distances. Their numerical results are similar to ours, but the Coulomb barriers are about 40 MeV higher.

Another interesting theory is the "Yukawa plus exponential" model of Krappe, Nix and Sierk[17]. Obtained by generalizing the modified liquid-drop model so that two semi-infinite slabs of nuclear matter have minimum energy at zero separation, this potential is given in terms of a double-volume integral of a Yukawa and an exponential folding function. For heavy nuclear systems such as $^{84}Kr+^{208}Pb$, the interaction potential is rather similar to the proximity potential. In general, the model reproduces experimental data for elastic scattering, fusion, fission and ground-state masses. When applied to the U+U system[18], one finds a very smoothly increasing potential at distances between 20 and 13 fm: there is neither a potential barrier nor a pocket!

Seiwert, Greiner and Pinkston[19] developed another phenomenological model; it is equivalent to the double-folding model at large separations and to the liquid drop model in the limit of spherical composite systems. When applied to Pb+Pb, Pb+U and U+U, the potential energy shows a quasimolecular minimum corresponding to surface contact of the nuclei.

NUCLEAR MATTER CALCULATION OF THE REAL AND IMAGINARY PART OF THE HEAVY-ION POTENTIAL

Faessler et al.[20,21] proposed a very interesting method for calculating both the real and the imaginary part of the heavy-ion potential. One starts from a realistic N-N interaction (Reid soft-core potential) and considers the collision of two infinite nuclear matter distributions flowing through each other. For this situation, the Bethe-Goldstone equation

$$G = v + v\frac{Q_F}{E - H_0}G. \tag{4}$$

is solved numerically; Q_F denotes the Pauli operator. In momentum space, the system can be characterized by two overlapping Fermi spheres, separated by the relative momentum of the two heavy ions. The resulting reaction matrix G is complex, because the energy denominator can have a pole due to the non-sphericity of the two Fermi spheres (energy-conserving

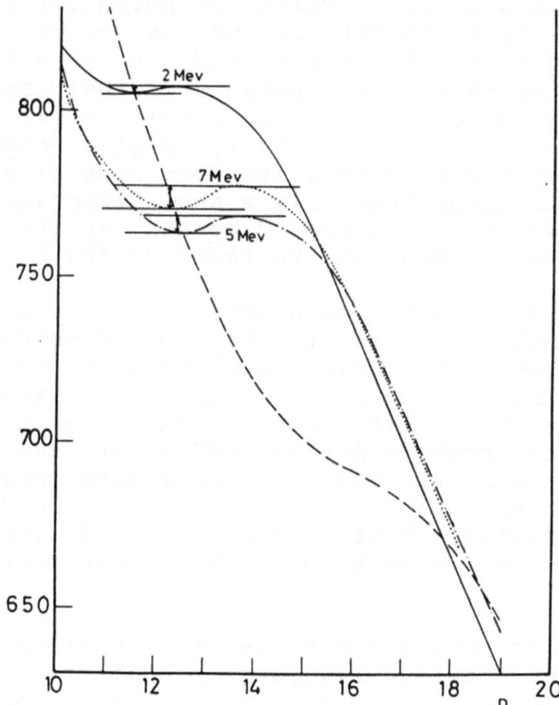

Fig. 3 Real part of the heavy-ion potential for U+U, according to Faessler and collaborators (ref. 21). Small pockets occur at beam energies E=6.4-6.8 MeV/u.

2p- 2h excitations can now occur). The heavy-ion potential can be calculated from the energy-density formalism in conjunction with the "frozen density" approximation. In this picture, one assumes that locally the density distribution of the two ions simply add up to give the total density.

In fig.3 we display the real part of the U+U interaction potential calculated by Ismail et al.[21] for different orientation angles (dashed line: 0°; dot-dashed line: 45°; dotted line: 50°; solid line: 90°). Note that potential pockets of a few MeV depth occur at beam energies between 6.4 and 6.8 MeV/u, but it must be emphasized that <u>there are no quasimolecular minima in the energy range E = 5.7-6.0 MeV/u</u> where positron line structures have been observed. As we said earlier, it is a challenge to produce pockets at low beam energies, not at high beam energies!

The theory of Faessler et al. has several very attractive features: the calculation of the effective N-N interaction from the Bethe-Goldstone equation is "state-of-the-art". Secondly, the method yields a real and imaginary potential at the same time. Furthermore, the heavy-ion potential depends on the beam energy. The main disadvantage is the frozen-density approximation (same as in the double-folding approach). In the following section we will outline a microscopic many-body theory which yields a more realistic density distribution in the nuclear overlap region. It is based on two-center shell model wavefunctions. We shall see that the frozen density approximation appears to be rather suspect at low beam energies and tends to overestimate the nuclear attraction. It would be interesting to combine Faessler's method with our approach.

As we have seen, there are many phenomenological ("macroscopic") approaches for calculating heavy-ion potentials. The first important step towards a more fundamental theory was made by Strutinsky in 1967 ("shell correction" or "macroscopic-microscopic" method). The basic idea was to calculate the smooth part of the collective potential energy from the liquid drop model and to add to that a microscopic correction which arises from fluctuations in the shell structure as a function of the collective parameter. Rapid advances in computer technology finally allowed fully microscopic calculations of nuclear properties (density distributions and binding energies of single nuclei) by the Hartree-Fock (HF) method[22] and of potential energy surfaces by means of the Constrained Hartree-Fock (CHF) method[23]. Both of these rely on nonrelativistic many-body theory and a variational principle.

However, the HF methods, even though most appealing, are not without problems. Constrained HF theories[23] have been highly successful in describing potential energy surfaces of a single nucleus, including nuclear fission. The study of two interacting nuclei is, however, not so straightforward. The main problem is to find a suitable constraint at larger separations. (The quadrupole moment is often used in fission, but this works only up to the saddle point and becomes questionable beyond that). The time-dependent Hartree-Fock method (TDHF) [24,25] gives a detailed description of the time-evolution of the nuclear reaction, but there is no unambiguous way to extract a nuclear potential from this method. This problem is overcome by the more recent quantized adiabatic time-dependent Hartree-Fock method[26] (QATDHF). Although the name suggests that ATDHF is an approximation to TDHF, this is in fact not so: TDHF is restricted to time-dependent Slater determinants and provides a semiclassical picture including dissipation. Quantized ATDHF, on the other hand, deals with a superposition of time independent Slater determinants and leads to a quantum mechanical formulation without dissipation. Its objective is to derive, for a given effective N-N interaction, the lowest large-amplitude collective mode and the corresponding collective Hamiltonian. The TDHF and QATDHF methods are very computer-time consuming and have not been applied successfully to very heavy systems.

We therefore decided to pursue an alternative fully microscopic method which is particularly suited for heavier systems. Our many-body two center theory (MBTC)[27-29] is also based on the nonrelativistic quantum theory of many-body systems and is formulated in the language of second quantization. In contrast to the HF-theories, it utilizes two-center shell model (molecular) wavefunctions rather than selfconsistent wavefunctions. These basis states depend on two collective parameters: internuclear distance and fragment deformation.

Our approach overcomes all the unsatisfactory features of the double-folding method described earlier: the nuclear densities are not kept frozen which avoids unphysical compression of nuclear matter at low beam energies. Furthermore, using second quantization, the Pauli exclusion principle is taken into account at every step of the calculation; this means, some correlations among the nucleons are included. Finally, the use of two-center shell model wavefunctions guarantees that shelleffects are treated in a consistent manner; they might give rise to additional structure in the heavy-ion potentials.

We present now some mathematical details of the theory. We start from a Hamiltonian of the form[29]

$$H(R) = \sum_{i\ell} < i(R)|(-\hbar^2/2m\nabla^2)|\ell(R) > a_i^+ a_\ell$$
$$+1/2 \sum_{ij\ell m} < ij(R)|v_2^N + v_2^C|\ell m(R) > a_i^+ a_j^+ a_m a_\ell$$
$$+1/6 \sum_{ijk\ell mn} < ijk(R)|v_3^N|\ell mn(R) > a_i^+ a_j^+ a_k^+ a_n a_m a_\ell, \quad (5)$$

which consists of the kinetic energy of all nucleons and their mutual Coulomb and nuclear interaction. The latter consists of a two- and three-body term (Skyrme interaction). The creation and annihilation operators a^+_k, a_k refer to a two-center shell model basis. At the moment, the density distribution is restricted to axially symmetric shapes. The s.p. potential consists of two harmonic oscillators in symmetry (z) direction which are cut off at z=0. In the radial direction (ρ) a single oscillator potential is utilized. The frequencies ω_z and ω_ρ of the shell model potential depend upon the collective parameters (internuclear distance and fragment deformation) and are determined numerically from the requirement of surface potential and volume conservation. Fig.4 displays the charge density distribution in symmetry direction for $^{238}U+^{238}U$ at several internuclear distances. Note that at large distances R, we obtain the densities of the individual uranium nuclei and in the limit R-->0 the density distribution of the U+U compound nucleus (A=476). It is apparent that the central density remains essentially constant as a function of R, i.e. there is no compression in the overlap region.

We approximate the ground state of the system by

$$|\Phi_0(R) >= (\prod_{k=1}^{F} a_k^+(R))|0>, \quad (6)$$

which is equivalent to a Slater determinant in configuration space. At every distance R the nucleons occupy the lowest possible two-center energy levels. This adiabaticity assumption appears to be justified at energies near the Coulomb barrier. However, for E >> E_C, this is no longer realistic. (In this case, one could introduce a finite temperature T; the occupation probabilities of the single-particle energy levels $E_k(R)$ are then given by the partition function. The temperature will smear out the sharp Fermi distribution which one obtains at T=0).

From the Hamiltonian (5) and the ground state (6) we can calculate the binding energy as a function of the collective shape parameters R:

$$E(R) =< \Phi_0(R)|H(R)|\Phi_0(R) >= \sum_{i=1}^{F} < i(R)| - \hbar^2/2m\nabla^2|i(R) >$$
$$+1/2 \sum_{i,j=1}^{F} < ij(R)|v_2^C + v_2^N|ij(R) >_{sym}$$
$$+1/6 \sum_{i,j,k=1}^{F} < ijk(R)|v_3^N|ijk(R) >_{sym}, \quad (7)$$

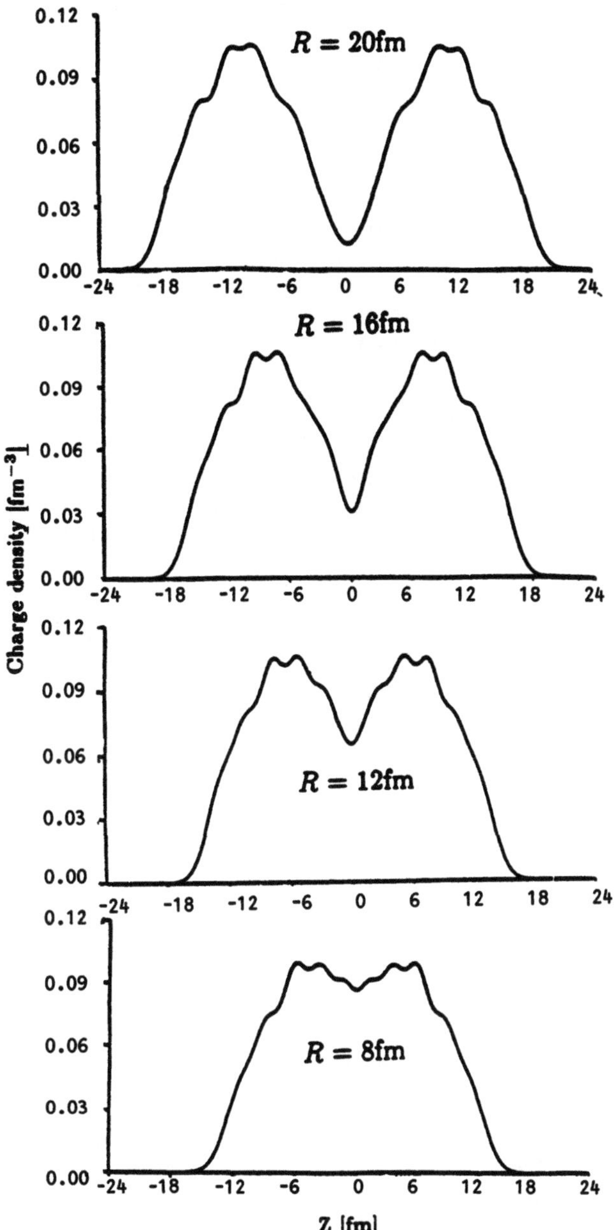

Fig. 4 Charge density for $^{238}U+^{238}U$ in the two-center shell model, for different internuclear separations R.

where the subscript "sym" denotes all additional exchange terms. The heavy-ion potential is the difference between the binding energy of the dinuclear system at distance R and the binding energy of the two separated nuclei (R-->∞)

$$V(R) = E(R) - E(\infty), \tag{8}$$

As in most Hartree-Fock calculations, we utilize a Skyrme force for the effective N-N interaction[30]. For numerical calculations, this is very convenient because the delta-functions in the spatial variable reduce all double integrals to single integrals. Note, however, that due to its momentum dependence, the Skyrme interaction simulates some finite-range effects. For simplicity, we have neglected the spin-orbit term. It is not possible to use one of the standard Skyrme force-parameter sets since we employ two-center shell model wavefunctions instead of selfconsistent wavefunctions. We have therefore determined the five parameters of the N-N interaction from a least-square fit to experimental binding energies throughout the periodic table[29]. At this point we want to stress that we compute the Coulomb potential exactly from the density distribution given by the two-center single-particle wavefunctions; the Poisson equation is solved numerically on a two-dimensional spatial grid with appropriate boundary conditions[31]. While some self-consistent calculations produce a 14% error in the binding energies of heavy nuclei[25] (which amounts to 250 MeV for ^{238}U), our binding energies generally agree with the measured values within about 1% .

With the method outlined above, we have calculated heavy-ion potentials for several spherical and deformed nuclear systems, including $^{28}Si+^{28}Si$, $^{40}Ca+^{40}Ca$ and $^{90}Zr+^{90}Zr$. The resulting potentials have been published[29] and shall not be repeated here, for lack of space. We wish to comment, however, on the reaction Zr+Zr which is of considerable interest because it is the heaviest symmetric system for which complete fusion has been observed. Our theory predicts a double-humped barrier caused by shell effects (very similar to the fission barriers found in actinides). The potential also exhibits two minima: the inner minimum corresponds to the deformed compound nucleus ^{180}Hg and the outer pocket to a very elongated system which could be an indication of the formation of a nuclear quasimolecule. If found experimentally, it would be by far the heaviest nuclear molecule ever observed. The ESTU accelerator currently under construction at Yale University is ideally suited for such experiments[33].

In Fig.5, we display the interaction potential for $^{238}U+^{238}U$. The calculation[31] was carried out for the optimum orientation angle of 0°, and the nuclear deformation parameter β was kept fixed at the value corresponding to the deformation of a single U nucleus. The shape of the configuration is shown by the inset of fig.5. The most important result of our fully microscopic calculation is that <u>there are no potential pockets</u>. Nevertheless, we see a remarkable broad "shoulder" in the region where some of the macroscopic theories predicted secondary minima. Let us try to understand this result. It is illuminating to compare the MBTC-theory to the double-folding model. If we consider a local two-body interaction such as the M3Y-force discussed earlier, the binding energy has the following structure (compare to eq.(7))

$$E(R) = 1/2 \int d^3x \int d^3x' \; v(x,x')\rho(x;R)\rho(x';R)$$

$$-1/2 \int d^3x \int d^3x' \; v(x,x') \sum_s |\rho_s(x,x';R)|^2$$

$$+\sum_{k=1}^{F} < \phi_k(x;R)| - \hbar^2/2m\nabla_x^2|\phi_k(x;R)> . \qquad (9)$$

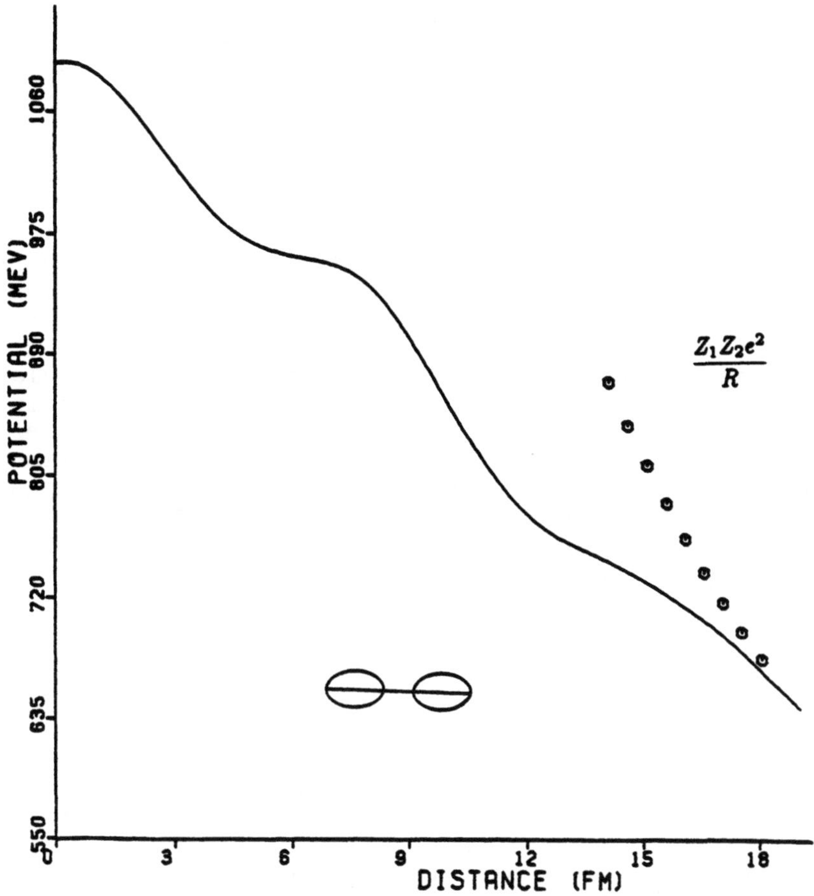

Fig. 5 Fully microscopic calculation of the interaction potential for $^{238}U+^{238}U$. The nuclear orientation and shape is shown by the inset (ref. 31).

The first term in eq.(9) is just the double-folding integral; it is the only term that survives at large internuclear distance. The second term depends on the square of the one-body density matrix; it arises from the Pauli principle and describes the corresponding exchange energy. The third term accounts for the variation of the kinetic energy with R. Our numerical calculations reveal that the magnitude of the exchange energy (which is neglected in the folding models) is crucial for the existence or nonexistence of potential pockets, because it has the opposite sign. As soon as the nuclei overlap, this term becomes sizable. In systems with a very large Coulomb repulsion (e.g. U+U) the exchange energy apparently destroys any minima produced by the direct energy term.

However, it might still be possible to create quasimolecular minima in the U+U interaction potential. Remember, that we have computed only a potential line for a fixed value of the deformation parameter β. There could be minima at other values of β. This requires the calculation of the full two-dimensional potential energy surface $V(R,\beta)$. It is clear that at deformation values corresponding to the shape-isomeric minimum in ^{238}U (β=0.6), the attractive nuclear force would be much stronger than in the normal (ground state) minimum which could produce a "superdeformed" quasimolecular pocket. We plan to investigate this fascinating possibility in the near future.

During this lecture, Wilets pointed out that one should also compute the mass parameters corresponding to the collective coordinates used (separation R and deformation β). This point is well taken and will be studied; the collective masses could be calculated from the two-center shell model within the cranking formalism. One might speculate that after a transformation from the coordinate-dependent masses to a set of constant mass parameters, structures (e.g. quasimolecular minima) might appear in the potential energy. In principle, this is possible, of course, but we believe that this effect cannot be responsible for the formation of a U+U quasimolecule: the experimental data[3,4] clearly require quasi-elastic scattering (Q-values of less than 16 MeV). This seems to imply rather large nuclear separations R where the collective masses should be similar to their asymptotic values (which we have utilized so far).

DYNAMICAL NUCLEAR ALIGNMENT CAUSED BY COULOMB EXCITATION

In connection with fig.2, we mentioned already an important point which was often overlooked in the theoretical analysis of the spontaneous vacuum decay: the nuclei have to approach within a relatively small cone of favorable orientation angles. This "angular window" ranges from 0° to about 35° (with respect to the internuclear axis) at 6.0 MeV/u, and is even smaller at 5.7 MeV/u, the lowest beam energy where positron line structures have been observed. The question is: what is the probability that the nuclei arrive at the distance of closest approach within this cone of orientation angles?

Let us first assume, for simplicity, that the two ^{238}U nuclei remain in their 0^+ ground state during the collision. In this (academic) case, the orientation probability $dP(\Phi)/d\Omega$ is independent of the orientation angle Φ. From simple geometric arguments one finds a probability of 0.18 that one of the nuclei is aligned favorably. However, since both nuclei must be aligned properly, only a fraction $f=0.18^2=0.033$ of all collisions will result in a favorable orientation within the angular window. As I have shown in ref.32, this number is further reduced by dynamical effects (nuclear Coulomb excitation) to a fraction f=0.01 of all collisions at 6.0 MeV/u. This can be understood by qualitative arguments: at large distance, the ^{238}U nuclei are in their 0^+ ground states, and there is no preferential alignment. As time progresses, higher spin states of the g.s. rotational band will be excited which leads to an alignment caused by the uncertainty principle $\Delta J \cdot \Delta \Phi \geq \hbar/2$. The dynamical calculations show that the nuclei are more likely to be oriented at Φ=90° (equators touching) than at Φ=0° (ends touching). Apparently, the nuclei try to minimize their Coulomb interaction energy; at a fixed distance between the centers, the Coulomb energy reaches its minimum value for the 90° orientation. A detailed discussion of the theoretical treatment is given in ref.32.

SUMMARY AND CONCLUSIONS

We have reviewed a variety of phenomenological interaction potentials and found that some of these predict the existence of quasimolecular minima for giant dinuclear systems (U+U, U+Cm), while others do not. This is unsatisfactory. An analysis of the situation shows that at large internuclear distances, the interaction potential should be described by the double-folding model. It thus seems natural to extend this model into the nuclear overlap region; to do this properly requires a nuclear many-body approach. We have formulated a fully microscopic theory which is particularly suited for heavy systems where Hartree-Fock calculations become prohibitively expensive. This theory overcomes all the problems associated with the frozen-density approximation of the double-folding model and takes shell effects into account. At a fixed value of the deformation parameter $\beta=0.26$, our MBTC-theory[29,31] shows no potential pocket for the U+U system; this is essentially due to the strong exchange energy term which has been neglected in the phenomenological models. However, it is still possible that quasimolecular minima exist at larger deformations (corresponding to the shape isomeric state of ^{238}U). This is currently under investigation.

A second important theoretical result is that the nuclei must approach with a well-defined orientation. Because of the dynamic orientation of the nuclear quadrupole moments, the probability for "favorable" alignment is calculated to be only about one percent.

Finally, we would like to suggest to our experimental colleagues to measure positrons at very low beam energy, say 5.0 MeV/u for U+U. We feel that such an experiment could help to resolve the current puzzle about the origin of the positron line structures: since it is impossible to form giant nuclear molecules at 5.0 MeV/u (the nuclear interaction is much too weak to produce potential resonances at this low beam energy) the observation of positron lines would clearly imply a e^+ production mechanism which does not require long delay times (nuclear quasimolecule). If, on the other hand, the experimental data rule out the existence of a positron peak at 5.0 MeV/u, but confirm its existence at higher beam energies, this would be the first indirect fingerprint of the formation of a giant nuclear molecule!

REFERENCES

1. J. Rafelski, B. Müller and W. Greiner, Spontaneous Vacuum Decay of Supercritical Nuclear Composites, Z. Phys. A285 49:52 (1978)
2. H. Backe, P. Senger, W. Bonin, E. Kankeleit, M. Kramer, R. Krieg, V. Metag, N. Trautmann and J.B. Wilhelmy, Estimates of the Nuclear Time Delay in Dissipative U+U and U+Cm Collisions Derived from the Shape of Positron and δ-Ray Spectra, Phys. Rev. Lett. 50 1838:1841 (1983)
3. J. Schweppe, A. Gruppe, K. Bethge, H. Bokemeyer, T. Cowan, H. Folger, J.S. Greenberg, H. Grein, S. Ito, R. Schule, D. Schwalm, K.E. Stiebing, N. Trautmann, P. Vincent and M. Waldschmidt, Observation of a Peak Structure in Positron Spectra from U+Cm Collisions, Phys. Rev. Lett. 51 2261:2264 (1983)
4. M. Clemente, E. Berdermann, P. Kienle, H. Tsertos, W. Wagner, C. Kozhuharov, F. Bosch and W. Koenig, Narrow

Positron Lines from U-U and U-Th Collisions, Phys. Lett. 137B 41:46 (1984)
5. J. Reinhardt, U. Müller, B. Müller and W. Greiner, The Decay of the Vacuum in the Field of Superheavy Nuclear Systems, Z. Phys. A303 173:188 (1981)
6. K.A. Erb and D.A. Bromley, Heavy Ion Resonances, in: "Treatise on Heavy-Ion Science " , Vol.3, D.A. Bromley, ed., Plenum , New York (1984)
7. J. Blocki, J. Randrup, W.J. Swiatecki and C.F. Tsang, Proximity forces, Ann. Phys. 105 427:462 (1977)
8. J. Randrup and J.S. Vaagen, On the Proximity Treatment of the Interaction Between Deformed Nuclei, Phys. Lett. 77B 170:173 (1978)
9. M. Seiwert, N. Abul-Naga, V. Oberacker, J.A. Maruhn and W. Greiner, On the Possibility of Superheavy Nuclear Quasi-molecules in U+U Collisions, in: "GSI Scientific Report 1981", GSI Darmstadt (1982); see also:
M. Seiwert, W. Greiner, V. Oberacker and M.J. Rhoades-Brown, Test of the Proximity Theorem for Deformed Nuclei, Phys. Rev. C29 477:485 (1984)
10. T. Cooper et al., Shapes of Deformed Nuclei as Determined by Electron Scattering, Phys. Rev. C13 1083:1094 (1976)
11. P.O. Hess, W. Greiner and W.T. Pinkston, Structure of Giant Nuclear Molecules, Phys. Rev. Lett. 53 1535:1538 (1984)
12. G.R. Satchler and W.G. Love, Folding Model Potentials from Realistic Interactions for Heavy-Ion Scattering, Phys. Rep. 55 183:254 (1979)
13. G. Bertsch, J. Borysowicz, H. McManus and W.G. Love, Interactions for Inelastic Scattering Derived from Realistic Potentials, Nucl. Phys. A284 399:419 (1977)
14. M.J. Rhoades-Brown, V.E. Oberacker, M. Seiwert and W. Greiner, Potential Pockets in the $^{238}U+^{238}U$ System and Their Possible Consequences, Z. Phys. A310 287:294 (1983)
15. M. Münchow, D. Hahn and W. Scheid, Heavy-Ion Potentials for Ellipsoidally Deformed Nuclei and Application to the System $^{238}U+^{238}U$, Nucl. Phys. A388 381:401 (1982)
16. R. Bass, Nucleus-Nucleus Potential Deduced From Experimental Fusion Cross Sections, Phys. Rev. Lett. 39 265:268 (1977)
17. H.J. Krappe, J.R. Nix and A.J. Sierk, Unified Nuclear Potential for Heavy-Ion Elastic Scattering, Fusion, Fission, and Ground-State Masses and Deformations, Phys. Rev. C.20 992:1013 (1979)
18. M. Seiwert, Ph.D. thesis, Insitut fuer Theoretische Physik der Universitaet Frankfurt, West Germany (1985)
19. M. Seiwert, W. Greiner and W.T. Pinkston, Do Heavy-Ion Potentials have Pockets?, J. Phys. G: Nucl. Phys. 11 L21:L26 (1985)
20. A. Faessler, W.H. Dickhoff, M. Trefz and M. Rhoades-Brown, Microscopic Approach to Real and Imaginary Part of the Heavy Ion Potential, Nucl. Phys. A428 271c:284c (1984)
21. M. Ismail, M. Rashdan, A. Faessler, M. Trefz and H.M.M. Mansour, The Effect of Deformation on the Nucleus- Nucleus Optical Model Potential and How It Produces Pockets, preprint, Institut fuer Theoretische Physik der Universitaet Tuebingen, West Germany (1986)
22. D. Vautherin and D.M. Brink, Hartree-Fock Calculations With Skyrme's Interaction. I. Spherical Nuclei, Phys. Rev. C5 626:647 (1972)
23. H. Flocard, P. Quentin, A.K. Kerman and D. Vautherin,

Nuclear Deformation Energy Curves With The Constrained Hartree-Fock Method, Nucl. Phys. A203 433:472 (1973)
24. P. Bonche, S. Koonin and J.W. Negele, One-dimensional Nuclear Dynamics In The Time-Dependent Hartree-Fock Approximation, Phys. Rev. C13 1226:1258 (1976)
25. M.R. Strayer, R.Y. Cusson, H. Stoecker, J.A. Maruhn and W. Greiner, Time-Dependent Hartree-Fock Studies of Superheavy Molecules, Phys. Rev. C28 228:236 (1983)
26. K. Goeke, F. Gruemmer and P.G. Reinhard, Three-Dimensional Nuclear Dynamics in the Quantized ATDHF Approach, Ann. Phys. 150 504:551 (1983)
27. V.E. Oberacker, Heavy-Ion Potentials For Strongly Deformed Nuclei in: "Fusion Reactions Below The Coulomb Barrier," Proceedings of the Int. Conf. held at MIT, Cambridge, Massachusetts (1984), S.G. Steadman, ed., Springer Verlag, New York (1984)
28. V.E. Oberacker, M.W. Katoot and W.T. Pinkston, Heavy-Ion Interaction Potentials and Static Deformation Effects In Sub-Barrier Fusion, in: "Proc. 23rd Int. Winter Meeting On Nuclear Physics", Bormio, Italy (1985), I. Iori, ed., Istituto di Fisica dell'Universita'·Milano, Italy (1985)
29. M.W. Katoot, V.E. Oberacker and W.T. Pinkston, Microscopic Theory of Heavy-Ion Potentials, Phys. Lett. 172B 292:296 (1986)
30. P. Ring and P. Schuck, "The Nuclear Many-Body Problem", Chapter 4, Springer Verlag, New York (1980)
31. M.W. Katoot, "Microscopic Theory of Heavy-Ion Interaction Potentials", Ph.D. Thesis, Dept. of Physics & Astronomy, Vanderbilt University, Nashville, Tenn. 37235, U.S.A., (1986)
32. V.E. Oberacker, Nuclear alignment: Implications For The Decay Of The QED Vacuum, Phys. Rev. C32 1793:1795 (1985)
33. D.A. Bromley, Yale University, private communication

FUTURE ASPECTS OF POSITRON SPECTROSCOPY

H. Backe

Institut für Physik der Universität Mainz

Postfach 39 80, 6500 Mainz, Fed. Rep. of Germany

INTRODUCTION

In the first week of this conference experimentalists made us familiar with the features of the positron peaks observed in the collision of very heavy ions with atoms. We learned that these peaks are not understandable in terms of known positron production mechanisms like monoenergetic pair decay of excited nuclear levels or the proposed spontaneous positron production in superheavy atoms. In addition, theorists told us that, at present, also the interpretation that the positron peak structures originate from the decay of a previously not observed neutral particle is in contradiction to well-known experimental facts. Some of you may have, in addition, become confused by the results of the three experimental groups which, at first glance, sometimes may look contradictory. Therefore, I will try to summarize and to discuss in this contribution the most important experimental observations of the EPOS, ORANGE and TORI groups and to formulate, using this as a starting point, the open questions for future experimental investigations.

However, I would like to emphasize that what is presented in the following was written under the impression of my present knowledge of the experimental facts and their theoretical interpretations. New insights may be gained in this rapidly developing field overnight which could make the proposed future experiments obsolete already at the time of publication.

OBSERVATION OF POSITRON PEAKS AND DISCUSSION OF GROSS FEATURES

Let me first recall briefly the basic principles of the positron experiments. In Fig. 1 a schematic sketch of an experimental set-up is shown. A target with charge and mass number Z_2, A_2 is bombarded by a heavy ion beam having charge and mass number Z_1, A_1 and a specific energy E_1/M_1. Positrons are detected by spectrometers described elsewhere in detail. In all recent experiments, energy-differential positron-production probabilities $dP_{e^+}^{exp}/dE_{e^+}$ (positron spectra) have been

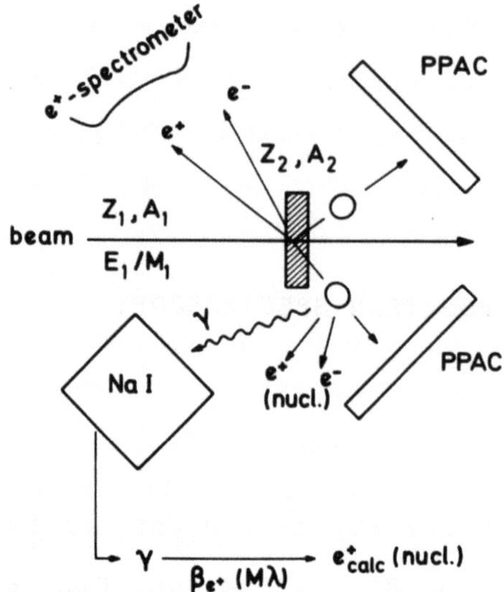

Fig. 1. Schematic sketch of a positron experiment.

investigated. These are derived from the number of positrons $\Delta N_{e^+}^{exp}$ detected at kinetic energy E_{e^+} in the interval between E_{e^+} and $E_{e^+}+\Delta E_{e^+}$ in coincidence with reaction events detected by the particle counters.

After correcting for the positron detection efficiency $\varepsilon(E_{e^+})$ and normalizing to the number of reaction events N_p we get

$$dP_{e^+}^{exp}/dE_{e^+} \simeq \Delta N_{e^+}^{exp}/[\varepsilon(E_{e^+})\ N_p\ \Delta E_{e^+}]. \qquad (1)$$

The particle counter may be able to subdivide the events into various reaction classes, as e.g. into pure Rutherford scattering, deep inelastic collisions with subsequent fission of one or both reaction partners or binary reactions with a well-defined total kinetic energy loss (TKEL). With this coincidence detection technique, background events originating from reactions with the target backing or light target contaminations like oxygen are rejected completely.

A major difficulty in performing positron experiments is that nuclear levels of the collision partners, which could decay by transitions with energies greater than $2m_ec^2$, may be excited by Coulomb interaction in combination with transfer reactions. Positrons resulting from internal pair conversion of these transitions cannot be easily distinguished from the atomic contribution because the time delay is only of the order of 10^{-16} to 10^{-13}s, and therefore timing or recoil techniques cannot be applied. The procedure for estimating the nuclear positron background is based on converting the observed γ-ray spectra with known pair conversion coefficients into positron spectra. Therefore, all experiments are equiped with γ-ray counters.

Spectrometers for use in in-beam positron spectroscopy must have a large positron collection efficiency and a broad energy acceptance band, while at the same time effectively suppressing other background radiation such as electrons, γ-rays, and neutrons. These requirements are fairly well met by two different types of spectrometers, namely the magnetic ORANGE type ß-spectrometer and the positron transport systems EPOS and TORI. In all these experiments silicon detectors are used for determination of the positron energy, yielding redundant information for the ORANGE spectrometer set-up. In summarizing this part I would like to stress that all three groups are equiped with highly sophisticated, state-of-the-art experimental set-ups.

Typical positron spectra are shown in Fig. 2 and Fig. 3. The smooth part of the spectra can be decomposed into a spectrum from pair decay of excited nuclear levels and a dynamically induced part originating from the time-varying Coulomb field of the two colliding nuclei. The dynamically induced positron spectrum in Fig. 3 agrees well with theoretical calculations scaled with a normalization factor of 0.75 ± 0.03. A scaling factor of 0.7 was also found by the TORI group[5,32] for the U+U collision system at a beam energy of 5.9 MeV/u and scattering angles between $19° \leq \Theta_{lab}^{ion} \leq 71°$ while the EPOS collaboration reports agreement again for the same collision system at scat-

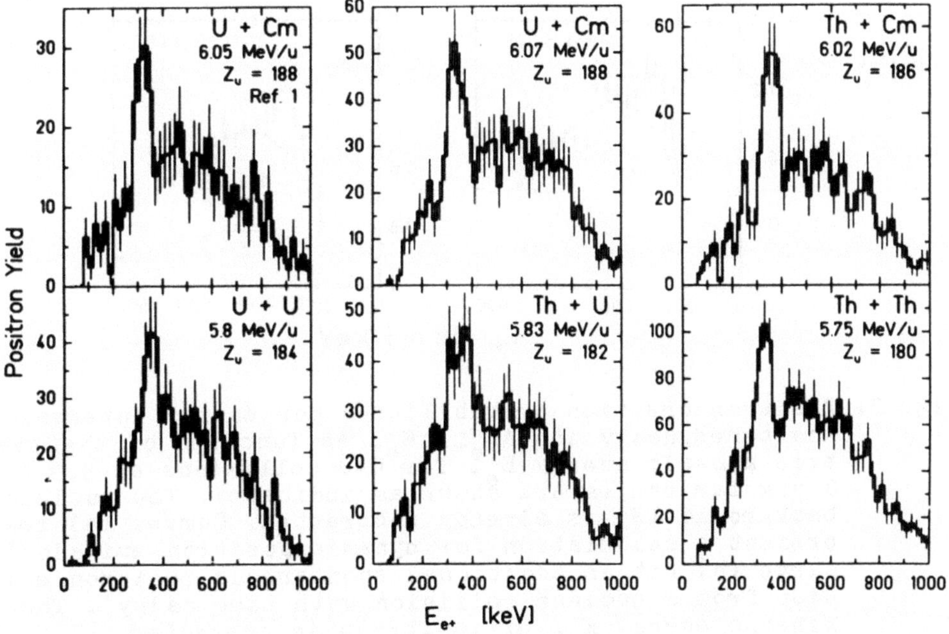

Fig. 2. Positron energy spectra for the five collision systems and bombarding energies indicated. The peaks were enhanced relative to the dynamic background by the choice of kinematic cuts which take advantage of differences in the kinematic behaviour exhibited by the peak events and elastic scattering (EPOS collaboration, ref. 1).

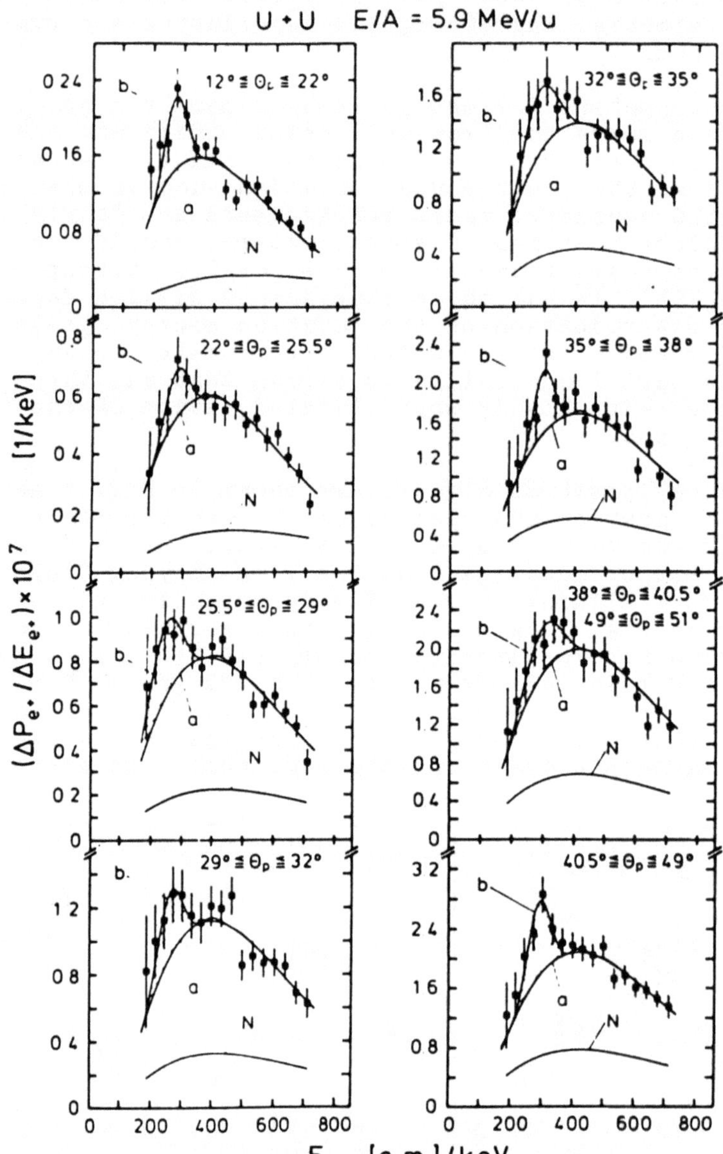

Fig. 3. Positron creation probabilities per energy interval and scattered heavy ion $\Delta P_{e^+}/\Delta E_{e^+}$ as function of the positron kinetic energy E_{e^+} for U+U collisions at 5.9 MeV/u. 8 angular ranges are shown as indicated. The nuclear background (N) is already subtracted. Curves (a) represent a calculation for dynamic positron emission[3], curve (b) fit an additional spontaneous positron emission from a nuclear collision with time delay[4]. The kinetic energy E_{e^+} of positrons is corrected for kinematic shift, assuming the CM system as the positron-emitting source (ORANGE collaboration, ref. 2).

tering angles between $25° \leq \Theta_{lab}^{ion} \leq 65°$[6]. Although the dynamical induced positron production mechanism is, at present, not anymore the focal point of interest, it might be of importance to understand this variance. It could well be that it is somehow connected to the positron peaks seen in all the spectra displayed in Fig. 2 and 3. All positron peak energies which are known to date are presented in Fig. 4 as a function of the united nuclear charge number $Z_u = Z_1 + Z_2$ of the corresponding scattering system.

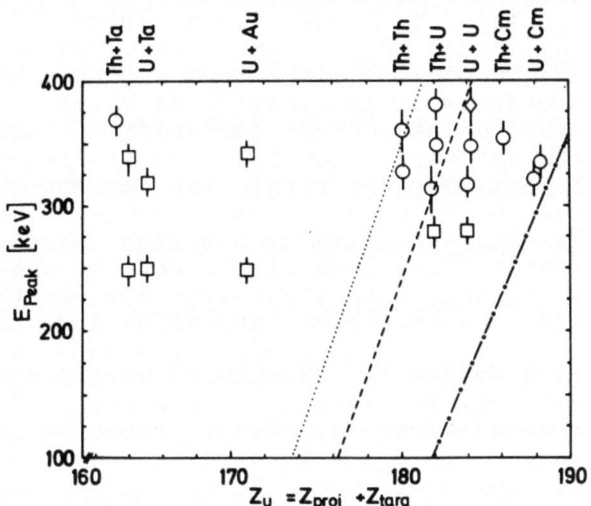

Fig. 4. Summary of positron peak energies observed by the EPOS[30] (o), ORANGE[31] (□ assuming emission from a source moving with CM-velocity) and TORI[23] (◇) collaboration as a function of the united nuclear charge Z_u. The dashed-dotted line shows the theoretical dependence of the resonance energy assuming two sticking, deformed nuclei which overlap in the most elongated configuration at a distance of 17 fm (atomic ionization state q = 50^+). The two other lines correspond to a spherical nuclear shape. Broken line: spherical configuration with q=50^+, dotted line: spherical configuration with bare nuclei.

First intimations of the positron peak structures were discovered about 8 years ago. Already in the forerunner of this conference, held in Lahnstein five years ago[7], they were thoroughly discussed in terms of spontaneous positron emission. It was assumed that a dinuclear system is formed in nearly central collisions of the heavy ions which then breaks apart after about 10^{-19}s into nuclei very similar to those present in the entrance channel. This time is long enough to allow a 1sσ vacancy created in the in-going channel of the collision to decay by spontaneous positron emission. While for the U+Cm scattering system the line energy agrees with this picture, according to the same theoretical calculations the spontaneous positron line should be found in the U+U system at an energy of about 210 keV, assuming in both cases a separation of the nuclei in the quasi-molecule of 16 fm. Experimentally, in this scattering system the line was observed at 320 keV[8] and 280 keV[2]. This clearly implies that in the picture of spontaneous positron creation, the line observed in the U+U system must be emitted from an intermediate nuclear complex of far less elongated shape as in the U+Cm system. A systematic study of the Z_u-dependence of the positron peak structure in super-heavy collision systems has been performed by the EPOS collaboration[1]. As shown in Fig. 2, narrow positron peaks are observed in five supercritical collision systems with combined nuclear charge $180 \leq Z_u \leq 188$. For all systems the positron peak was found at nearly the same energy, while for constant configurations of nuclear charge and electron ionization a strong Z_u-dependence is predicted[9]. As shown in Fig. 4, radically different nuclear charge configuration and ionization states for the nuclear compound system are required to explain the measured peak energies within the context of spontaneous positron emission. Because of these findings the subcritical $^{232}Th + ^{181}Ta$ ($Z_u = 163$) system was also investigated by the EPOS collaboration[10]. A line at 370 keV was observed which could not be attributed to structures seen in the γ-ray spectrum. Recently, the ORANGE collaboration investigated the subcritical U+Ta ($Z_u = 165$), U+Au ($Z_u = 171$) and Pb+Pb ($Z_u = 164$) scattering systems[11]. Structures were found in all these scattering systems. The energies are shown in Fig. 4. Thus, the positron peak energy dependence predicted for spontaneous positron creation could not be verified experimentally (see Fig. 4).

As we learned from the various contributions at this conference, it seems that we are presently far away from understanding the experimental observations and you may indeed ask why this problem has not been solved already. In attempting to answer this question, I would like to mention first of all that in-beam positron spectroscopy is a new field in experimental heavy-ion physics. Only a few other attempts have been undertaken previously in measuring positrons emitted in ion-atoms collisions[12,13,14]. In addition, we should keep in mind that the GSI experiments were initiated with the investigation of the most fascinating but unfortunately also the most complicated scattering systems. Even if we restrict ourselves to the inner shell electrons of the quasi-molecule formed, e.g. in the collision of uranium ions with uranium atoms at beam energies close to the Coulomb barrier, quite a large number of electrons will be subject to enormous time changes of the electromagnetic field. Also nuclear degrees of freedom can be excited or the emission of nuclear clusters might be possible. Therefore, it could well be that certain many-particle aspects of atomic and nuclear excitation in the quasimolecule or the separated highly ionized collision partners moving through the target matter may have not been fully understood. Careful investigations of the positron production mechanism even in light scattering systems with very good statistics, good kinematics, good energy resolution and well-defined target situations (e.g. gas targets) might help to understand the strange observations made in the super-heavy collision systems.

In addition, we should keep in mind that the counting rate for peak events is in the order of only about 0.04/s and long running times of about 3 days are required to record spectra with sufficient statistics. The UNILAC beam time devoted to the investigation of positron peaks since 1982 amounts to about 100 days (see also Fig. 10). Thus, only about 30 spectra were recorded in this period. It also became clear that the positron peaks must be investigated as a function of many parameters like the charge number and the energy of the beam, the charge number of the target, the target thickness, the time of e^+-emission and so on. In addition, it can not be excluded that there are parameters which have not been recognized up to now to be of importance. Thus, only very little is known in this multi-dimensional parameter space and it requires a lot of intuition to make the right experiments or to find a scheme to reduce the many parameters to the relevant ones.

At first glance the results of the three experiments may sometimes look contradictory. However, as long as the positron production mechanism is not understood, it is necessary to judge carefully the similarity of the experimental details before making such statements. For instance, differences in the geometrical arrangement may have a big impact on the observed results. To illustrate this, let us assume that positrons originate from delayed decays of neutral particles which are produced with a special angular distribution so that they move perpendicular to the beam direction. This results in a large discus-like source which we will assume to have a mean radius of say 10 centimeters. For simplicity we neglect effects of kinematic broadening caused by in-flight emission. With an ORANGE spectrometer without energy analyzing Si-detectors a broad structure would be observed since the spectrometer was designed for a point source. However, in the transport systems, part of the positrons which originate from the decay of neutral particles near the symmetry axis will be detected with good energy resolution by the Si-detectors and thus may give rise to a pronounced narrow peak structure. Another scenario is also conceivable in which the situation is just reversed.

FEATURES OF THE OBSERVED POSITRON STRUCTURES

As already mentioned, the positron peaks can be studied as a function of many parameters; see Fig. 5. There are parameters which can easily be controlled. These are e.g. the charge and mass numbers (Z_1,A_1) and (Z_2,A_2) of the beam and the target, respectively, the beam ionization state q_1 and the scattering angles Θ_3, Θ_4. However, whether the collision partners retain their identity in the collision (e.g. no nucleon transfer) is presently an open question. In none of the experiments are the ionization states q_3, q_4 measured. Nevertheless, the experiments equiped with magnetic transport systems are sensitive to them, which complicates the interpretation of the scattering kinematics.

In addition, the positron peak intensity and energy may depend on the beam energy E_{beam} and the energy spread ΔE_{beam}, the target thickness, the level of target deterioration, and perhaps the target-backing. Also the elements of the chemical composition of the target like oxygen or fluorine may be of importance. Only very little is known about the effect of these various parameters on the intensity and energy of the positron lines. In the following I would like to discuss a few features which in my opinion may be of particular importance to understanding the origin of the positron structures.

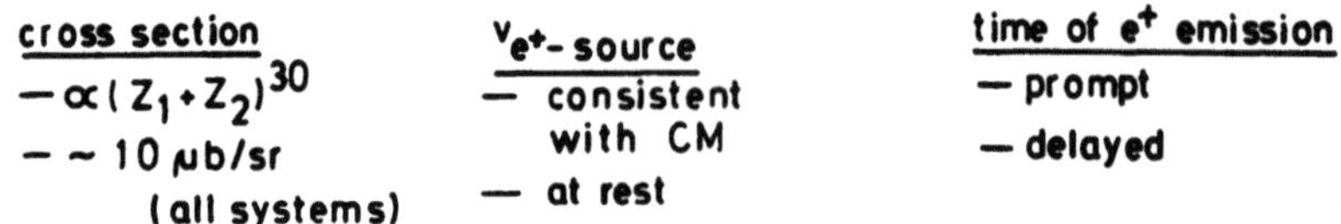

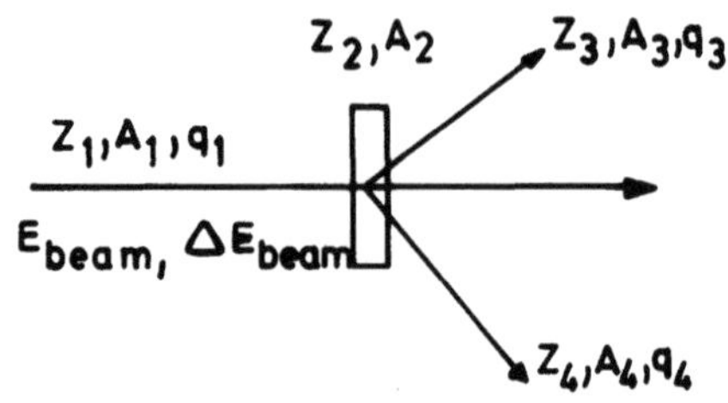

excitation function	target effects	kinematics
— broad	200-1000 $\mu g/cm^2$	— binary
— resonance like		— broad Θ-depend.
— threshold ?	— deterioration	— TKEL ≲ 20 MeV
		— magn. field effect
	— C, O, F	— cluster ?
		— mass drift ?

Fig. 5. Summary of positron line features.

The energy dependence of positron cross-sections has been measured by the ORANGE collaboration for the U+U scattering system. Differential cross-sections for monochromatic positron creation of (15±3) μb/sr at 5.7 MeV/u, 14 μb/sr at 5.9 MeV/u and (18±4) μb/sr at 6.2 MeV/u were found[2] for ion scattering angles between 40° and 50° in the laboratory system. No dependence of the data on the target thickness in the range between 0.3 and 1 mg/cm^2 has been observed. These findings indicate a broad excitation function. The EPOS collaboration, on the other hand, reports that the positron peak is most pronounced for apparently narrow intervals of projectile energy which correspond in each system to a marginal overlap of nuclear surfaces in a head-on collision[1]. Therefore, the beam energy and the spread of the beam energy were carefully controlled. In addition, it was observed by the EPOS collaboration that the positron peak intensity apparently strongly depends on the degree of target deterioration. In a recent measurement the ThF_4 targets with a thickness of about 300 $\mu g/cm^2$ were changed routinely after 2 hours of irradiation with a beam current of 1 p-nA. These findings could indicate that nuclear forces may play an important role in the positron production mechanism. One open question for the future is therefore certainly to investigate carefully the excitation function.

The cross-section $d\sigma_{e^+}/d\Omega_p$ of the positron peaks with energies between 300 and 400 keV observed by the EPOS collaboration is about 10 μb/sr for all systems with $163 \leq Z_u \leq 188$ after having chosen the proper beam energy. The ORANGE spectrometer group reports for the positron structures at CM-energies between 250 ... 280 keV a strong $(Z_1+Z_2)^{30}$-dependence for $U+Z_2$ collisions, $73 \leq Z_2 \leq 92$ ($165 \leq Z_u \leq 184$) with a U-beam energy of 5.9 MeV and a laboratory scattering angle $40 \leq \Theta_{lab}^{ion} \leq 50°$. This might be no contradiction since the lines observed in the low Z_u-systems by the EPOS and ORANGE collaborations differ in energy by more than 140 keV and likely do not have the same physical origin.

I come back to the discussion of the positron peak energies collected in Fig. 4. In different measurements various positron peak energies (center of gravity) have been observed situated for $Z_u \geq 180$ in the energy interval between 280 and 380 keV. Note that sometimes in repeating the experiment with virtually the same parameters the positron line shifted in energy. This could indicate that there is a narrow structure, the energy of which is strongly dependent on parameters which can not easily be controlled, as e.g. beam energy or target energy loss. Still another possibility would be that there are several narrow positron structures which are excited at specific beam energies and/or scattering angles.

Some of these features, particularly the constant positron peak energy as a function of Z_u, are consistent with the speculation that the positron line might be caused by the decay of a previously unobserved neutral particle which decays into a positron-electron pair. (Actually, to explain the various positron peak energies a whole particle family or particles with internal structure must be assumed.) Therefore, the EPOS collaboration performed an experiment in which electron-positron correlations were explored.

The solenoid spectrometer was modified to allow the detection of electrons in coincidence with positrons. This set-up is shown in Fig. 6. Two 32 mm x 65 mm planar Si(Li) electron detectors were installed on the opposite side of the target from the existing cylindrical, positron detector. Using a geometrical configuration similar to that employed in ref. 15 they were oriented with their sensitive surfaces facing each other and parallel to the solenoid axis. They were also removed symmetrically from the target-positron-detector-axis to avoid the high flux of low energy δ-electrons. This arrangement particularly permits the detection with high efficiency of electron-positron pairs with opposing momenta. For 5.83 MeV/u U+Th collisions a narrow peak structure was found at 760±20 keV in the distribution of the sum of the electron and positron energies whose width is consistent with the instrumental distribution. As also explained in detail in ref. 16, a peak structure also appears in the coincidence intensity plotted as a function of the positron and electron energy difference for a constant sum of their energies between 710 keV and 830 keV.

The EPOS collaboration reported at this conference that, in addition to this observation, two other electron-positron sum peaks of about 600 and 800 keV were observed in the U+Th collision system with similar features, see Fig. 7.

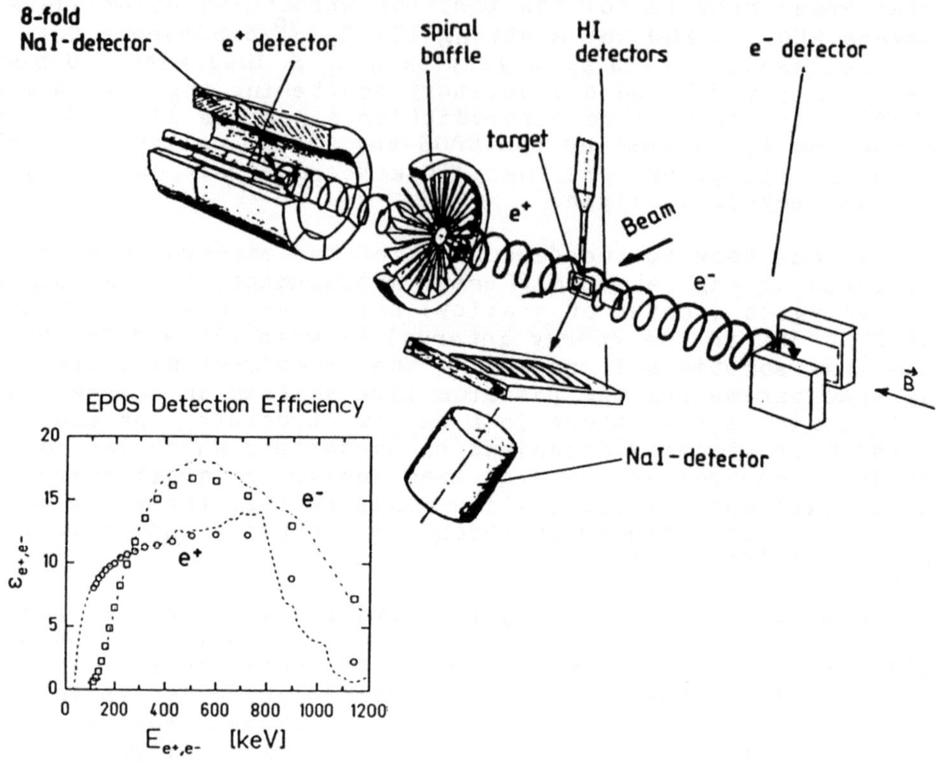

Fig. 6. Schematic view of the modified solenoidal spectrometer EPOS. Two planar 32 mm x 65 mm Si(Li) electron detectors were installed on the opposite side of the target from the existing cylindrical positron detector. Also shown is the detection efficiency for positrons and electrons as a function of the kinetic energy[20,30].

DISCUSSION

Treating all this experimental results seriously, we learn from the ORANGE data that the object which emits positrons with an energy of about 280 keV may be produced in the Coulomb field of the two colliding nuclei. This can be concluded from the experimental indication of a broad excitation function. In addition, the Z_u^{30} - dependence of the cross-section could indicate the importance of the relativistic shrinkage of the electron/positron wave function in the quasi-molecule for the production mechanism. The constancy of the positron cross-section at proper chosen beam energies close to the Coulomb barrier, as reported by the EPOS collaboration, could be interpreted in such a way that nuclear forces may enhance significantly the cross-section. The reason for this could be brought into connection with the formation of a giant dinuclear complex resulting in a significantly prolonged time the high Coulomb field exists. Furthermore, it follows from the EPOS coincidence experiment that the positron-emitting objects somehow resemble neutral particles which decay into electron-positron pairs. A whole particle family or composite particles with internal

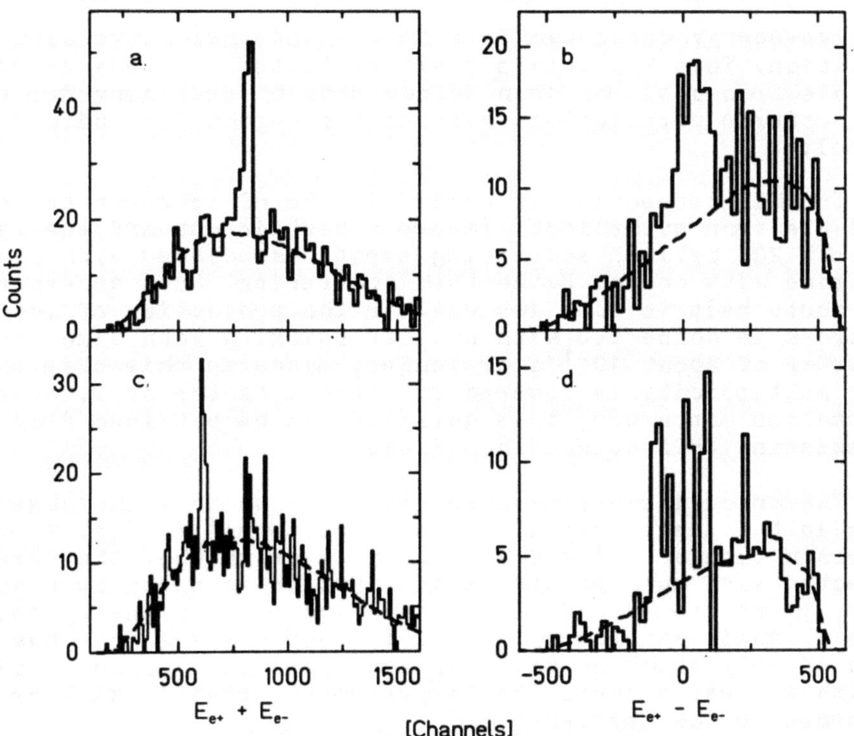

Fig. 7. A preliminary analysis of a measurement in February, 1986, by the EPOS collaboration of the U+Th collision system at beam energies around 5.87 MeV/u reveals two prominent sum peaks significantly narrower than the individual positron and electron lines. The lower and upper panels show sum and difference spectra for two subsets of the data gated on beam energy, heavy-ion scattering angle, and positron/electron time of flight chosen to enhance these two sum lines, respectively[20,30]

structure (responsible for internal states of excitation) must be assumed to explain the various observed positron line energies as well as the indication for the different electron-positron sum correlation energies. The problems associated with the large production cross-section, which is of variance with predictions of QED, as well as the problem with the creation of such particles with very low velocity in the CM system have been discussed in several contributions during this conference [17,18,19,20]. It would certainly be helpful for the solution of this puzzle to gather more experimental facts. In the following section I would like to describe a few ideas for future experiments. In doing this I will orient myself on other known features of the positron peak structures.

WHAT SHOULD BE DONE IN THE FUTURE?

As proposed in ref. 21 nuclear internal conversion transitions in superheavy giant atoms may also yield electron-positron correlations. A conceivable process could be that in a first internal conversion transition, an inner-shell vacancy is formed which is filled by an electron from the negative-energy continuum in a second internal-conversion transition. This hypothesis could be tested with the existing EPOS electron-positron coincidence data by searching for electron-positron correlations outside the region for which $E_{e^+} \simeq E_{e^-}$ holds.

In this connection it would also be of interest to compare the δ-electron multiplicity (above a certain cut-off energy of about 200 keV) of scattering events associated with peak positrons with normal Rutherford scattering. Such an experiment could help to find out whether the production of peak positrons is connected with nuclear sticking with times on the order of about 10^{-19}s or longer, since in this case the δ-ray multiplicity is lowered by about a factor of 2. Some information concerning this question can be obtained from the existing EPOS coincidence data.

The crucial experiment to determine whether the observed peaks in the electron-positron sum spectrum (see Fig. 7) resemble the production of a (neutral) particle or whether they are connected with nuclear and atomic physics going on in a superheavy, giant atom would be to investigate again the Z_u-dependence of their sum energy. Since the 600 keV sum peak has a width of only about 30 keV such an experiment should provide us with a clear answer. This experiment certainly will be performed in the near future.

Detailed insight, concerning also this question, could be gained from a careful study of the interrelation between positron emission and scattering kinematics. At present, all groups agree that positron lines are observed in binary scattering with little TKE loss ($\lesssim$ 20 MeV) and little or no mass drift. The Θ-angle region at which lines were observed is broad. The ORANGE collaboration observes the line at laboratory scattering angles between 12° and 49°[2]. The EPOS collaboration found that the peaks can be enhanced relative to the dynamic background by the choice of kinematic cuts which take advantage of differences in the kinematic behaviour exhibited by the peak events and elastic scattering[22,1]. Also, peak events might be connected to large non-equilibrium ionization states q_3, q_4 of the ejectile and recoil ions which due to the solenoidal magnetic field are bent differently in comparison to normal elastic scattering. This finding was corroborated recently by the TORI group[23] which found that peak events are connected to out-of-plane scattering. An alternative explanation for this observation could be that a third particle (nuclear cluster) is emitted. It would be of particular interest to find out whether the positron peak events are connected to a mass drift between the colliding nuclei. Such a mass drift would be expected if a giant dinuclear complex is formed in the collision.

Another important feature is the time of emission of the peak positrons after the scattering process. From the EPOS experiment it can be stated that the emission must be prompt ($\lesssim$ 1 ns) since otherwise a source moving with CM velocity would escape the sensitive region of the detector. However, also long lifetimes in the order of $\lesssim$ 10 ns are conceivable in case the emitting system moves along the symmetry axis of the solenoid. Such long lifetimes could be detected by timing techniques. (The time resolution of the solenoidal transport system is rather poor because of the large differences of the positron paths as a function of the emission angle). Recently the TORI group reported evidence of delayed emission of the peak positrons[23]. The recoil-shadow technique[24] was applied to positrons. However, this result could not be corroborated in a later measurement with an improved shadow system[25]. With this method the lifetime of the positron emitting object could be measured down to about 10^{-11}s provided its velocity component in the beam direction amounts approximately to 10 % of the velocity of light and the magnetic field is high enough.

The ORANGE collaboration proposed to employ the energy shift of the positrons in an approximately 2 mg/cm^2 low Z target-backing in order to determine whether the time of positron emission is below about 10^{-13}s[26]. In the lifetime range between about 10^{-20} to 10^{-19}s information can be obtained from the Lorentz width being in the range between about 70 and 7 keV, respectively.

In the following the velocity of the peak-emitting source will be discussed. The EPOS collaboration reported that the line shapes are all consistent with the emission of a narrow line by a system moving with the center-of-mass velocity. (This does not mean that the emitting source is the CM system). In particular, the line shape distinguishs between emission from a source moving with the CM velocity and a source with velocities associated with the individual nuclei[1].

The ORANGE spectrometer group[27] investigated the emission probability of the line as function of the azimuthal angle relative to the scattering plane using azimuthal-angle-sensitive particle and positron detectors. No statistically significant deviation from the expected isotopic emission for a source moving in the beam direction was found. On the other hand, the source velocity component in the beam direction can be measured by observing the positron line energy at angles Θ_{e^+} and $\pi-\Theta_{e^+}$ relative to the beam using the up- and down-stream position of the spectrometer. From this measurement the source velocity component in the beam direction was determined to be somewhat smaller than the CM velocity. This experiment will be repeated with the planned double ORANGE spectrometer set-up shown in Fig. 8, yielding a much higher accuracy.

If one believes in the production of an previously unobserved, neutral particle in heavy-ion collisions, this hypothesis should be tested with improved experimental tools. One step in this direction can be made with the TORI spectrometer set-up shown in Fig. 9. Positron-electron coincidences can be recorded with the electron counter shown at the left of the picture and the other one in the middle plane. Back-to-back decay of a neutral particle would result only in coincidences with the electron detector shown at the left hand side but not with the middle plane detector. On the other hand, uncorrelated electron-positron pair production results in coincidences in both electron counters[33].

In addition, the experiments should be improved so that the measurement of the invariant mass is possible. It could well be that the observed peak structures in the positron spectra reflect only the peak of an iceberg. Many such neutral particles might have been produced with high kinetic energies.

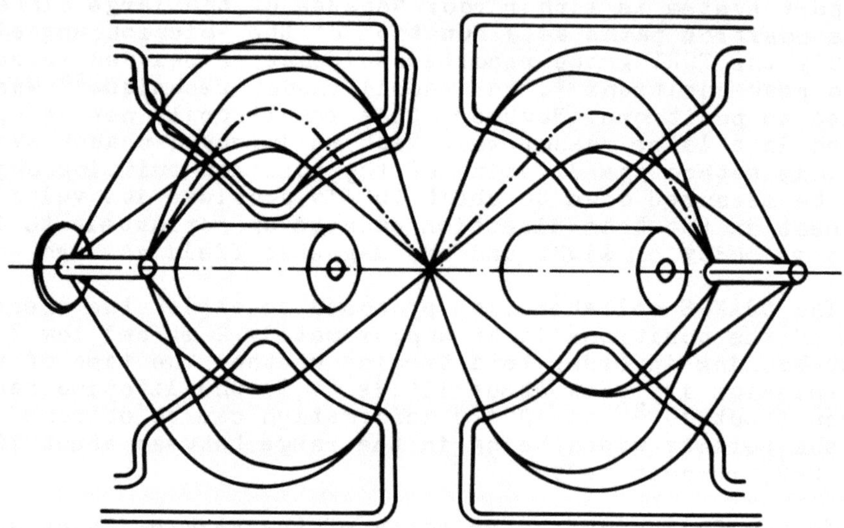

Fig. 8. The double ORANGE spectrometer set-up. Both spectrometers will be equiped with Si-chip detectors mounted on a pagoda-like roof. Positrons will be detected with the left hand side orange spectrometer and electrons with the right hand side one. This spectrometer combination is sensitive to the back-to-back decay into an electron-positron pair of a neutral particle at rest.

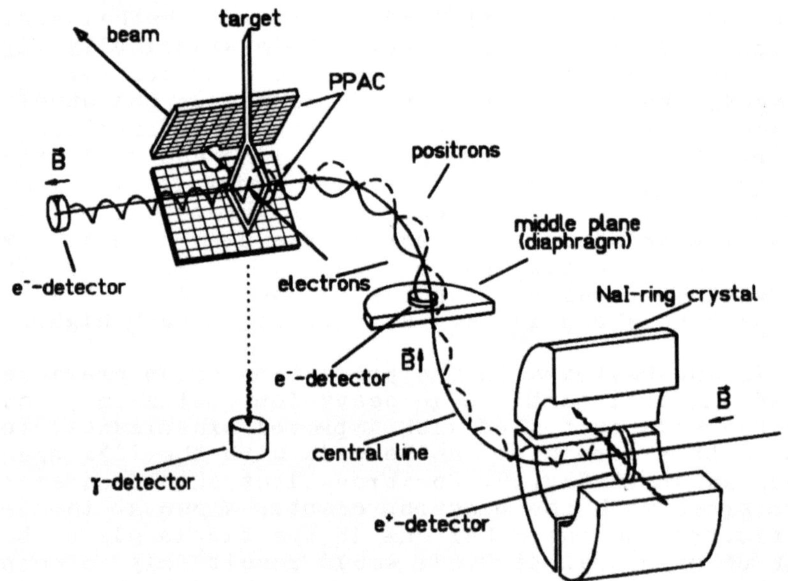

Fig. 9. TORI spectrometer with two electron detectors. Positron-electron coincidences from the back-to-back decay of a neutral particle at rest in the CM system will be detected only with the electron counter shown on the left hand side but not with that at the middle plane[33].

If this is true, even in the electron-positron sum energy spectrum a broad structure will be observed which is indistinguishable from the dynamically induced background. The measurement of the invariant mass requires position sensitive detectors which are able to measure the vector components of the electron- and positron-momentum in more or less a 4π geometry. Perhaps one step in this direction could be the double ORANGE spectrometer set-up shown in Fig. 8.

Finally, another possible experiment which emerged from the particle hypothesis will be discussed. Let us consider the decay of a neutral particle into an electron-positron pair under time reversal. Then, the existance of such a particle should be noticed as a resonance in the Bhabha scattering cross-section at that energy which corresponds to the invariant mass of the particle. Ideally an electron-positron colliding-beam experiment should be performed. Since the decay width of the neutral particle may be rather small (a few meV if the lifetime is on the order of 10^{-13}s), high quality electron and positron beams with kinetic energies of about 400 keV (head on collision) are required. In a much simpler experiment, 2.15 MeV positrons could be scattered on electrons of a solid state target. A high quality positron beam can be provided by a Van-de-Graaff-accelerator equiped with a slow positron source. Considerable progress has been achieved recently in the construction of very intense, low-energy positron sources. The presently most intense eV-positron beam can be made by taking advantage of the bremsstrahlung shower of a high current (10 μA) 300 MeV electron beam. The high-energy positrons are moderated in a specially prepared 25 μm tungsten foil having a negative work function for thermalized positrons. With a 8 μA electron beam, a flux of 10^7 slow positrons per second was already measured[28] which probably can be optimized considerably. Using basically a similar technique, lower intensity beams can be produced by using radioactive β^+-sources in the terminal of the Van-de-Graaff-accelerator[29].

ACKNOWLEDGEMENTS

The experimental work extensively reported from different speakers at this conference benefited greatly by the theoretical work done in Frankfurt since 1967. The impact of the group around W. Greiner, B. Müller, J. Reinhard, J. Rafelski and G. Soff was so strong that actually with the first UNILAC beam in 1976, experimentalists were ready for data taking. The beamtime devoted since then to the positron experiments is shown in Fig. 10 as a function of time. The little beamtime used in 1978 and 1982 coincides with the transitional period between the first, second and third generation experiments. In the first generation experiments, it was demonstrated that the separation of atomic and nuclear positrons is possible. In the second generation experiments, positron spectra were explored and the peak structures discovered, while in the third generation experiments, a detailed investigation of these peak structures is in progress.

I am indebted to the members of the EPOS-, ORANGE- and TORI-group for their cooperation in preparing this talk.

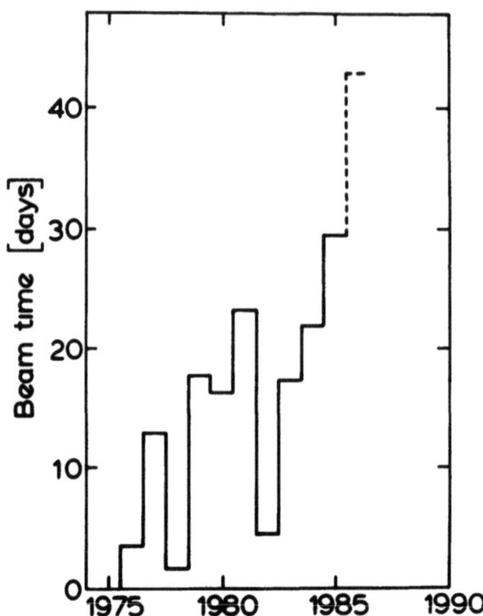

Fig. 10. Target beamtime in days/years devoted to the exploration of positron emission in heavy-ion collisions as a function of time. Data are taken from the GSI annual reports. Also quoted is the expected beamtime for 1986.

REFERENCES

1. T. Cowan, H. Backe, M. Begemann, K. Bethge, H. Bokemeyer, H. Folger, J.S. Greenberg, H. Grein, A. Gruppe, K. Kido, M. Klüver, D. Schwalm, J. Schweppe, K.E. Stiebing, N. Trautmann and P. Vincent, Phys. Rev. Lett. 54 (1985) 1761.
2. H. Tsertos, E. Berdermann, F. Bosch, M. Clemente, P. Kienle, W. Koenig, C. Kozhuharov and W. Wagner, Phys. Lett. 162B (1985) 273.
3. J. Reinhardt, B. Müller and W. Greiner, Phys. Rev. A24 (1981) 103.
4. J. Reinhardt, U. Müller, B. Müller and W. Greiner, Z. Physik A303 (1981) 173.
5. R. Krieg, E. Bozek, U. Gollerthan, E. Kankeleit, G. Klotz-Engmann, M. Krämer, U. Meyer, H. Oeschler and P. Senger, Phys. Rev. C34 (1986) 562.
6. D. Schwalm, in: Electronic and Atomic Collisions, XIII. ICPEAC, Invited Papers, eds: J. Eichler, I.V. Hertel, N. Stolterfoht, Amsterdam, Elsevir Sc. Publ., North Holland 1984, p. 295.
7. see articles in Proc. NATO Advanced Studies Institute, Quantum Electrodynamics of Strong Fields, Lahnstein 1981, ed: W. Greiner (Plenum Press, New York 1983).
8. H. Bokemeyer, K. Bethge, H. Folger, J.S. Greenberg, H. Grein, A. Gruppe, S. Ito, R. Schule, J. Schweppe,

N. Trautmann, P. Vincent and M. Waldschmidt, in: Proc. NATO Advanced Studies Institute, Quantum Electrodynamics of Strong Fields, Lahnstein 1981 (Plenum Press, New York) p. 273.

9. U. Müller, T. de Reus, J. Reinhardt, B. Müller and W. Greiner, Z. Physik A323 (1986) 261.
10. H. Bokemeyer, H. Folger, H. Grein, T. Cowan, J.S. Greenberg, J. Schweppe, A. Balanda, K. Bethge, A. Gruppe, K. Sakaguchi, K.E. Stiebing, D. Schwalm, P. Vincent, H. Backe, M. Begemann, M. Klüver, N. Trautmann, GSI Scient. Rep. 1984, GSI 85-1 (1985) 177.
11. F. Bosch, P. Kienle, W. Koenig, C. Kozhuharov, E. Berdermann, M. Franosch, S. Huchler, J. Kemmer, H. Tsertos and W. Wagner, GSI Scient. Rep. 1985, GSI 86-1 (1986) 179.
12. W.E. Stephens, H. Staub, Phys. Rev. 109 (1958) 1196.
13. W.E. Meyerhof, R. Anholt, Y. El Masri, D. Cline, F.S. Stephens and R. Diamonds, Phys. Lett. B69 (1977) 41.
14. M. Heykants, M. Niecke, Nucl. Phys. A332 (1979) 22.
15. H. Backe, P. Senger, W. Bonin, E. Kankeleit, M. Krämer, R. Krieg, V.. Metag, N. Trautmann and J.B. Wilhelmy, Phys. Rev. Lett. 50 (1983) 1838.
16. T. Cowan, H. Backe, K. Bethge, H. Bokemeyer, H. Folger, J.S. Greenberg, K. Sakaguchi, D. Schwalm, J. Schweppe, K.E. Stiebing and P. Vincent, Phys. Rev. Lett. 56 (1986) 444.
17. A.B. Balentekin, contribution at this conference.
18. J. Reinhardt, contribution at this conference.
19. B. Müller, contribution at this conference.
20. T. Cowan, contribution at this conference.
21. P. Schlüter, U. Müller, G. Soff, Th. de Reus, J. Reinhardt, W. Greiner, Z. Physik 323 (1986) 139.
22. J. Schweppe, A. Gruppe, K. Bethge, H. Bokemeyer, T. Cowan, H. Folger, J.S. Greenberg, H. Grein, S. Ito, R. Schulé, D. Schwalm, N. Trautmann, P. Vincent, M. Waldschmidt, Phys. Rev. Lett. 51 (1983) 2261.
23. B. Blank, E. Bozek, H. Jäger, E. Kankeleit, G. Klotz-Engmann, M. Krämer, R. Krieg, U. Meyer, H. Oeschler, M. Rhein, P. Senger, GSI Scient. Rep. 1985, GSI 86-1 (1986) 180.
24. H. Backe, L. Richter, R. Willwater, E. Kankeleit, E. Kuphal, Y. Nakayama and B. Martin, Z. Physik A285 (1977) 159.
25. E. Kankeleit, H. Oeschler, private communication.
26. F. Bosch, discussion remark.
27. P. Kienle, Talk presented at "Nucleus-Nucleus Collisions from the Coulomb Barrier up to the Quark-Gluon-Plasma", Erice, Italy, April 10-22, 1985, GSI-85-31 PREPRINT, August 1985.
28. A. Picard, diploma thesis, Universität Mainz 1986, unpublished and G. Werth, private communication; see also: M. Begemann, G. Gräff, H. Herminghaus, H. Kalinowsky, R. Ley, Nucl. Instr. Meth. 201 (1982) 287.
29. A. Seeger, MPI für Metallforschung, Stuttgart, private communication.
30. H. Bokemeyer, et al., contribution at this conference.
31. W. Koenig, et al., contribution at this conference.
32. H. Oeschler, et al., contribution at this conference.
33. M. Krämer, et al., contribution at this conference.

INTERFERENCE EFFECTS IN QUASIMOLECULAR RADIATION AND A CLOCK FOR HEAVY ION NUCLEAR REACTIONS*

Itzhak Tserruya

Department of Nuclear Physics
The Weizmann Institute of Science
Rehovot, Israel

ABSTRACT

Quasimolecular radiation has been extensively studied both experimentally and theoretically during the last decade. The use of H-like ions at low velocities brings a new dimension to the study of quasimolecular K x-rays. The most important consequence is the observation of interference structures in the quasimolecular K x-ray spectra which result from the coherent sum of x-ray emission in the incoming and outgoing half of the trajectory. Using very simple assumptions it is possible to obtain, directly from the data, the quasimolecular orbital transition energy as a function of internuclear distance, thus opening the way for spectroscopic studies of quasimolecular orbitals in general. As an extension of this work we propose to use these interference structures as a clock to measure nuclear sticking times in heavy ion nuclear reactions. From calculations we expect the method to be sensitive to sticking times as short as 10^{-20}s for collisions of Sn+Sn.

I. INTRODUCTION

In heavy-ion-atom collisions, when the projectile velocity is small compared to the orbital velocity of the inner-shell electrons, these electrons have time to adjust their orbits to the changing Coulomb field of the projectile-target nuclear system, and their binding energy increases with decreasing internuclear distance. The electrons follow quasimolecular orbitals characteristic of the quasimolecule transiently formed during the collision. If a vacancy exists in one of these orbitals it can decay during the collision emitting a quasimolecular x-ray. In a static picture, the energy of the quasimolecular K x-rays, which are emitted when a vacancy in the innermost orbital – the $1s\sigma$ orbital – decays during the collision, depend on the internuclear distance. Since the latter constantly changes during the collision, the quasimolecular K x-rays form a continuous band extending from the characteristic K x-rays of the separated atoms at large distances to the characteristic K x-rays of the united atom at small distances (see Fig. 1).

Here we will concentrate on quasimolecular K x-rays. They have attracted much attention since the first experimental observation[1], not only because of the intrinsic interest as a new phenomenon, but mainly because quasimolecular emission offers very interesting applications[2]. For example, Greiner and co-workers proposed already in 1969 the possibility of spontaneous positron emission[3] when a vacancy in the 1sσ orbital dives into the Dirac negative energy sea in collisions of, e.g., U+U. It could also be possible to study electronic binding energies of superheavy elements. These suggestions strongly rely on the quasimolecular picture and on the correlation of the quasimolecular orbital binding energy with internuclear distance. The determination of this spectroscopic information has been one of the main experimental goals.

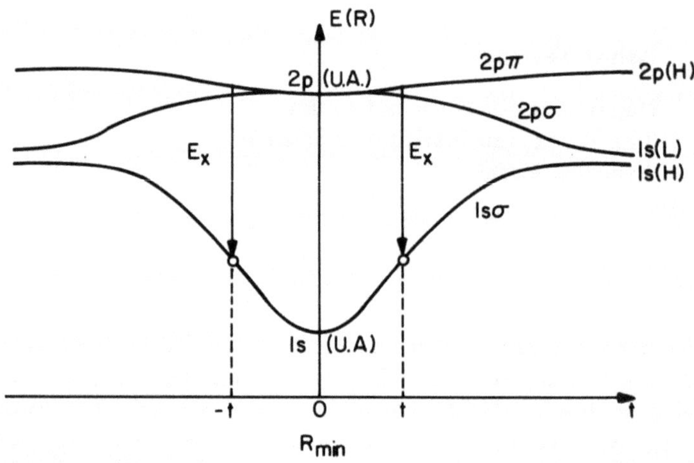

Fig. 1. Schematic correlation diagram

After a brief review of some general features of quasimolecular K x-ray spectra, we will describe in detail a recent study of quasimolecular K x-rays emitted in slow collisions of Cl+Ar using hydrogen-like Cl^{16+} projectiles[4]. This method of using H-like projectiles brings a new dimension to the study of quasimolecular x-rays and opens the way for spectroscopic studies of quasimolecular orbitals. The most important consequence is that it produces interference structure in the x-ray spectra which result from the coherent sum of x-ray emission in the incoming and the outgoing half of the trajectory. Using simple assumptions based on the stationary phase approximation and the uniform approximation, it is possible to obtain directly from the data, the quasimolecular orbital transition energy as a function of internuclear distance[5,6].

As a extension of this work we propose to use these interference structures a a clock to measure nuclear sticking times in heavy ion nuclear reactions. This is a fundamental quantity which plays a basic role in theories of nuclear reactions, in particular in deep inelastic collisions and in fission, and of which we have very little experimental information. Recently several attempts have been made to measure nuclear sticking times using atomic clocks. Krieg et al.[7] use the changes in the slope

Table 1 Relevant quantities involved in the quasimolecular processes of Cl+Ar and Sn+Sn collisions.

	Cl-Ar	Sn-Sn
R_K(fm)	2300	780
E_K(keV)	11.9	120.6
E_p (MeV)	23	715
E_B (MeV)	84	550
τ_c (sec)	$4x10^{-19}$	$4.7\ 10^{-20}$
τ_K (sec)	$4.4\ 10^{-16}$	$5\ 10^{-18}$

R_K, E_K, τ_K are the radius, binding energy and radiative half-life of the united atom K shell. E_p is the projectile energy for which $v_p = \frac{1}{3}v_K$, E_B is the Coulomb barrier and τ_c is the collision time $\tau_c = 2R_K/v_P$.

of positron and δ-electron spectra as the clock. Stoller et al.[8] deduced nuclear reaction times in deep inelastic collisions of U+U from K shell ionization probabilities. All those methods exploit essentially the same feature; a nuclear reaction with a sticking time T causes a phase shift $\Delta\phi = \omega T$ (where ω is the quasimolecular orbital transition frequency at the closest distance of approach) between the incoming and outgoing amplitude. In the method discussed here, this phase shift changes the interference pattern of the quasimolecular K x-ray spectra and, therefore the observation of these changes may serve as a clock for nuclear sticking times. From calculations we expect the method to be sensitive to sticking times as short as 10^{-20}s for collisions of Sn+Sn. The sensitivity increases in heavier systems, reaching $\sim 10^{-21}$s in Pb+Pb collisions[9].

II. CHARACTERISTIC FEATURES OF QUASIMOLECULAR K X-RAYS

Some of the relevant distances, energies and times involved in the quasimolecular processes are listed in Table I for Cl+Ar and Sn+Sn systems. R_K and E_K denote the radius and binding energy of the united atom K shell. E_p is the projectile energy for which $v_p \approx \frac{1}{3}v_K$, where v_p and v_K are the projectile and K shell electron velocity respectively. E_B gives the value of the Coulomb barrier. τ_c and τ_K are the collision time, defined as the time it takes the projectile to cross the K shell ($\tau_c = 2R_K/v_p$ and the K-shell radiative half-life respectively. It is seen that for a light system like Cl+Ar, the adiabaticity condition $v_p \ll v_K$ necessary for the justification of the quasimolecular picture is not fulfilled at energies near or above the Coulomb barriers. However, for heavier systems, like e.g. Sn+Sn, the quasimolecular picture applies at incident energies near or above the Coulomb barrier.

In a very crude picture, the quasimolecular K x-ray production probability is given by the 1sσ vacancy production probability multiplied by the 1sσ decay probability during the collision (which is $\sim \tau_c/\tau_K$). These two factors are very small and it thus appears that quasimolecular K x-ray emission is a very rare process. However, this radiation has been observed and studied in a large variety of colliding systems covering both light and heavy systems.

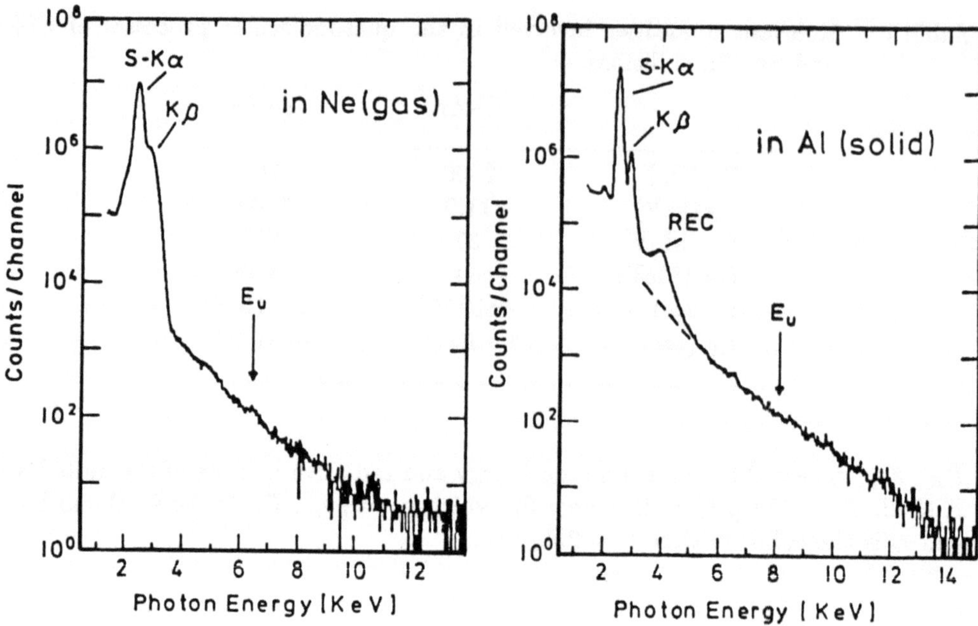

Fig. 2. X-ray spectrum from collisions of 48 MeV S^{7+} on Ne gas target (left panel) and 55 MeV S ion on Al (right panel). E_u denotes the united atom transition energy. (From ref. 10)

As an example we show in Fig. 2 the x-ray spectrum measured for 48 MeV S^{7+} projectiles incident on a Ne gas target under single collision conditions (left panel) and the spectrum for 55 MeV S projectiles on an Al solid target (right panel)[10]. The spectra show the dominant characteristic K x-rays of S and at higher energies the continuous quasimolecular K x-ray band with several orders of magnitude lower intensity. In addition the S+Al spectrum show the radiative electron capture (REC) line which is produced here in a two collision process (in the first collision a K vacancy is produced in the projectile and in the second one a loosely bound target electron is captured into the K shell). It is clear that the quasimolecular K x-ray spectrum exhibits no sharp end-point at the maximum transition energy (E_u) of the united atom expected in a static picture. Instead, due to the broadening effects produced by the dynamics of the collision, the spectra extend beyond this limit. They exihibit a structureless shape with a nearly exponential fall-off which depends only on the projectile velocity[11].

Other types of measurements were also performed. The impact parameter dependence of 1sσ quasimolecular x-rays was first measured for Cl+Cl[12)] and subsequently for heavier systems[13]. Fig. 3 shows as an example the results obtained in Cl+Cl at 35 MeV. One sees that the coincidence spectra extend well beyond the united atom limit with a nearly exponential fall-off. They are very similar in shape to the

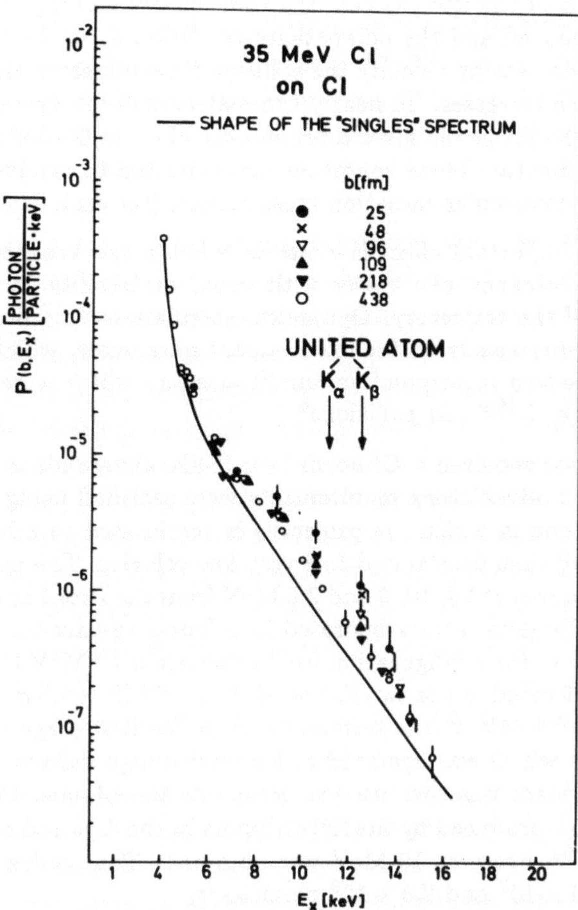

Fig. 3. Quasimolecular K x-ray production probability as a function of transition energy in 35 MeV Cl+Cl collisions for various impact parameters. The solid line indicates the shape of the "singles" x-ray spectrum.

"singles" spectrum, indicated in Fig. 3 by the solid line, and show almost no dependence on the impact parameter. The results illustrated in Figs. 2,3 demonstrate that spectroscopic information could not be obtained from "singles" spectra or from impact parameter dependent measurements. In both cases the radiation exhibits a structureless shape with nearly exponential fall-off[14].

III. THE USE OF LOW-VELOCITY H-LIKE PROJECTILES IN Cl-Ar COLLISIONS

Here we will describe a recent study[4] of the Cl+Ar system in which the experimental conditions used for the study of quasimolecular K x-rays have been radically improved. These improvements have resulted in a pronounced structure in the x-ray spectra from which detailed spectroscopic information about the quasimolecular orbitals has been obtained[5,6]

The essential improvement that has been made is the use of low velocity hydrogenlike ions which made it possible for the first time to bring a K vacancy into the collision. As a result of the low velocity, the collision broadening, which depends on the velocity[11], is reduced and the adiabaticity condition $v_p \ll v_K$ is better fulfilled. Furthermore, with decreasing velocity the collision time increases and thus the decay rate of $1s\sigma$ vacancies increases. In nearly symmetric collision systems, the presence of a K vacancy in the projectile gives a probability close to 0.5 for the presence of a vacancy in the $1s\sigma$ orbital. These conditions are expected to produce a considerable increase in the quasimolecular radiation cross section (see sect. II and ref. 15).

Finally a very important effect is expected when a $1s\sigma$ vacancy is brought into the collision. The vacancy can decay with equal probability in the incoming or the outgoing half of the trajectory. Dynamical calculations[16,17] predict interference structures in the x-ray spectrum for fixed impact parameter, which result from the coherent sum of the two corresponding amplitudes and which were observed for the first time in 10 MeV Cl^{16+}+Ar collisions[4].

The experiments required a Cl beam in a H-like charge state (q=16+) at very low energy. These contradictory requirements were satisfied using the acceleration-deceleration technique in which the projectile is accelerated to a high energy where $v_p \gtrsim v_K$, stripped and then decelerated to a very low velocity. The measurements were done using Cl^{16+} beams at 20, 10, 5 and 2.5 MeV from the Brookhaven National Laboratory dual MP Tandem facility operated in a four-stage acceleration-deceleration mode[18]. Fig. 4 shows the configuration used to obtain a 10 MeV Cl^{16+} beam. Chlorine ions were accelerated in the two stages of the first MP tandem accelerator which was operated at +8.4 MV at the terminal and in the first stage of the second MP tandem accelerator which was operated at a negative high voltage of -8.2 MV at the terminal. The last stage was then used to decelerate the Cl ions. By selecting charge states of 8^+ and 16^+ produced by the foil strippers in the first and second accelerator, respectively, a Cl^{16+} beam at 10 MeV was obtained. The analyzed beam intensity varied between 3.9×10^8 and 2.3×10^9 particles/s.

The experiments used an Ar gas target cell with a three-stage differential pumping system in order to ensure single collision conditions and to preserve the charge state of the incoming beam. The x-rays were detected at 90° relative to the beam direction with a Si(Li) detector which covered a solid angle of 4.5%. A 38μm-thick Al absorber was mounted in front of the Si(Li) detector to absorb the huge yield of projectile K x-rays. The scattered projectiles were detected simultaneously at sixteen angles (covering the impact parameter range between 350 and 3200 fm) with a position-sensitive parallel-plate avalanche counter, with a 2π azimuthal geometry[19]. Coincidences between scattered particles and x-rays were detected event by event. More details about the experimental set-up are given in ref. 4.

IV. INTERFERENCE STRUCTURE IN Cl-Ar X-RAY SPECTRA

We present here the qualitative features of the experimental results obtained in Cl^{16+}+Ar collisions at 20, 10, 5 and 2.5 MeV. The x-ray production probabilities at 10 MeV are shown as function of photon energy in Fig. 5 for various impact parameters. The spectra are plotted for E_x >4 keV. Below that energy the spectra are dominated by the strong yield of the REC line and the characteristic lines. The striking feature of these data is that for impact parameters smaller than ~2250 fm,

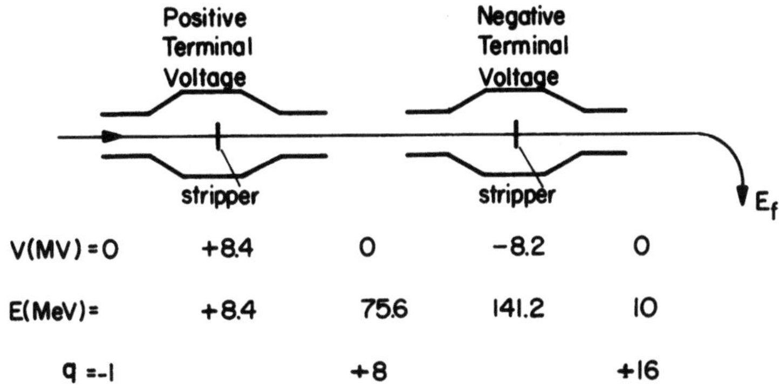

Fig. 4. Operational conditions of the dual MP Tandem accelerator facility at Brookhaven National Laboratory to produce a 10 MeV Cl^{16+} beam.

the spectra show a pronounced and broad maximum which moves toward higher x-ray energy as the impact parameter decreases. The peak moves from ~6 keV at 2600 fm to ~9 keV at 500 fm. Fig. 6 shows the quasimolecular x-ray emission probabilities measured at the four different incident energies at about the same impact parameter b≈1300 fm. The structure is clearly seen in all spectra. It moves towards higher x-ray energies and becomes more pronounced as the bombarding energy decreases. While at 20 and 10 MeV only one oscillation is observed, the data at 2.5 MeV clearly shows two oscillations. The results shown in Figs. 5,6 differ markedly from those obtained in previous studies[12,13] of quasimolecular K x-ray production probability which used projectiles with incident charge state $q>Z-2$. In those studies the coincidence spectra were similar in shape to the single x-ray spectra and showed no structure and almost no dependence on the impact parameter (see section II).

The solid lines in Fig. 5 represent the results of dynamical calculations[20] which include transitions only from the $2p\sigma$ and $2p\pi$ orbitals to the $1s\sigma$ orbital. The calculations were normalized to the experimental results; however, the same normalization factor was used for all the spectra of Fig. 5. The calculations reproduce the measured shapes, and in particular the position and width of the peak are well reproduced. The structure in the calculated spectra arises from the interference between the $1s\sigma$ decay amplitude in the incoming part and the outgoing part of the trajectory. These interference structures were expected to occur when a $1s\sigma$ vacancy is brought into the collision and were observed for the first time in ref. 4. More recently similar interference structures were observed in other collision systems using H-like projectiles S^{15+} and Ge^{31+} [21].

V. SPECTROSCOPY OF Cl-Ar QUASIMOLECULAR ORBITALS

The structures seen in Figs. 5,6 vary in a regular and smooth way both with incident energy and with impact parameter. This is illustrated in Fig. 7 which shows the energies of maxima and minima as observed in all x-ray spectra as function of impact parameter. This regularity has been exploited in order to extract the quasimolecular orbital transition energy as function of internuclear distance[5,6]. This spectroscopic information was obtained using simple and general assumptions in two different and complementary ways based on the stationary phase approximation and on the uniform approximation[22].

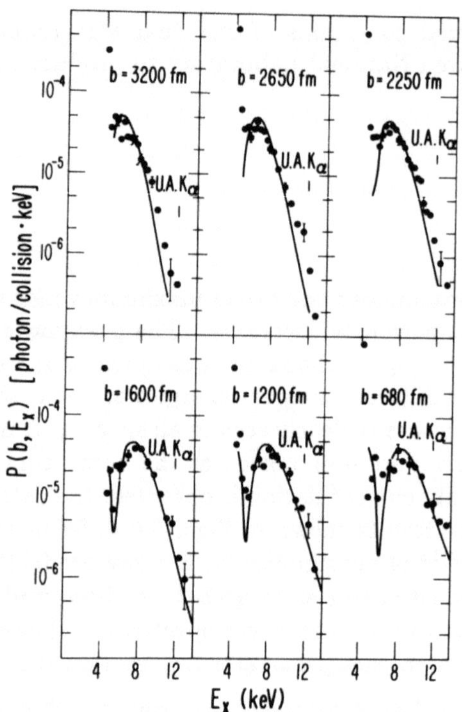

Fig. 5 X-ray production probability as a function of photon energy in Cl^{16+}-Ar collisions at 10 MeV for various impact parameters.

A. Stationary Phase Approximation

The dipole transition matrix element is given by[2,17]

$$D(\omega) = \frac{1}{\sqrt{2\pi}} \int_{-\infty}^{\infty} dt\, D(R(t))\, exp \left[i \int^{t} (\omega - \omega(t'))dt' \right] \quad (1)$$

where D(R(t)) is the time-dependent dipole matrix element and $\omega(t)$ is the time-dependent transition frequency between the quasimolecular orbitals. The typical

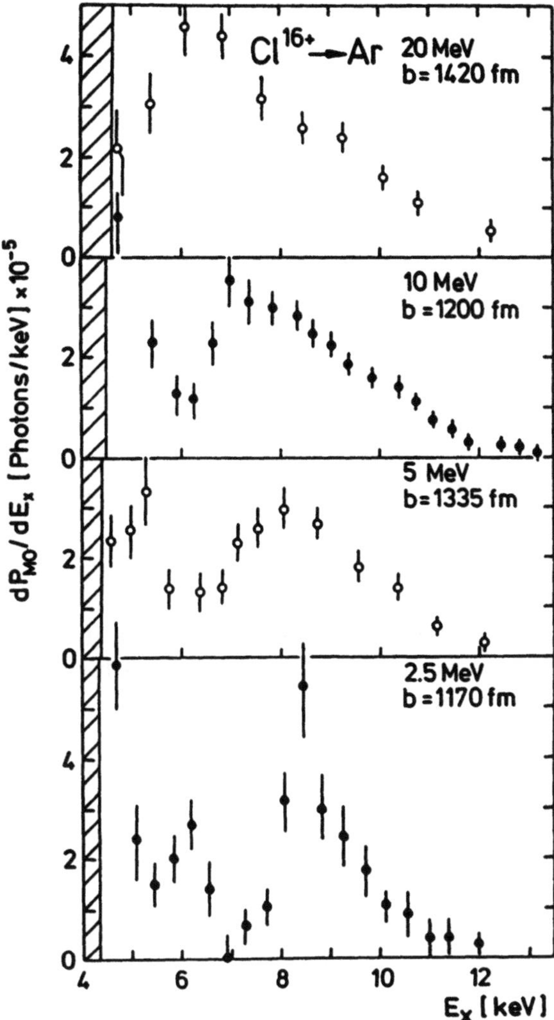

Fig. 6. X-ray production probability as a function of photon energy in Cl^{16+}-Ar collisions at various incident energies and at nearly the same impact parameter.

shape of $\omega(t)$ is shown in Fig. 8a. In the stationary phase approximation one assumes that the phase factor

$$f(\omega,t) = \int_{o}^{t} [\omega - \omega(t')]dt' \tag{2}$$

varies rapidly with time so that the main contribution to the integral (1) comes from the neighborhood of the stationary points defined by

$$\frac{df}{dt} = 0 .$$

Fig. 8a shows that for every frequency ω there are two stationary points ($-t_o$ corresponding to a point on the incoming trajectory and $+t_o$ on the outgoing trajectory)

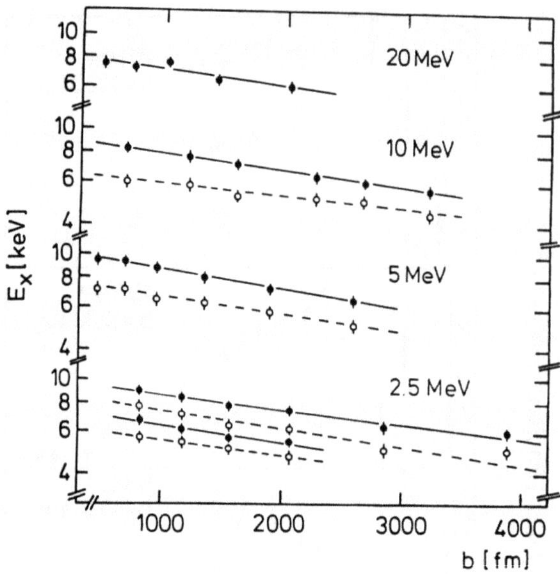

Fig. 7. Energy positions of maxima (full dots) and minima (open dots) as function of impact parameter b.

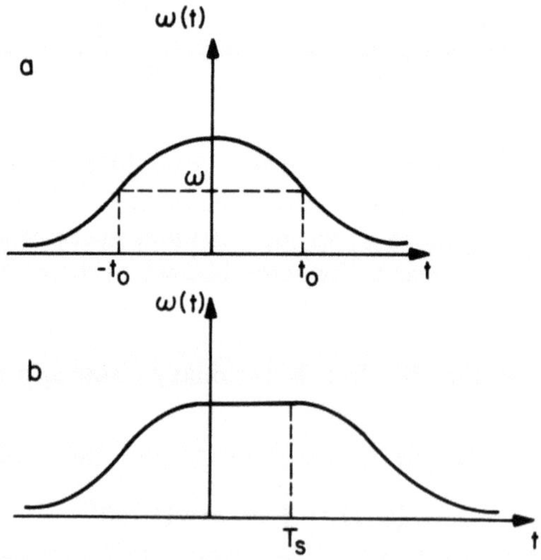

Fig. 8. Typical shape of the time dependent transition frequency without (a) and with (b) nuclear sticking time T_s.

given by $\omega = \omega(t_o) = \omega(-t_o)$. Expanding $f(\omega,t)$ up to second order around t_o,

$$f(\omega,t) \approx f(\omega,t_o) - \frac{1}{2}\left(\frac{d\omega}{dt}\right)_{t_o}(t-t_o)^2 \tag{3}$$

and substituting in (1) gives the contribution of the stationary point to the transition amplitude:

$$D(\omega) \approx D(R(t_o))\left|\left(\frac{d\omega}{dt}\right)_{t_o}\right|^{-\frac{1}{2}} exp\left[if(\omega,t_o) + \frac{\pi}{4}sgn\, f''(\omega,t_o)\right]$$

From the symmetry properties of D(R) and $f(\omega,t)$, it follows that the contributions from $+t_o$ and $-t_o$ are complex conjugate. Summing both contributions gives:

$$D(\omega) \approx 2\,D(R(t_o)\left|\left(\frac{d\omega}{dt}\right)_{t_o}\right|^{-\frac{1}{2}} cos\Phi \tag{4}$$

where

$$\Phi = \int_0^{t_o}[\omega - \omega(t)]dt + \frac{\pi}{4} = f(\omega,t_o) + \frac{\pi}{4} \tag{5}$$

and the photon intensity is given by:

$$I(\omega) \propto \omega\left[D(R(t_o))\left|\left(\frac{d\omega}{dt}\right)_{t_o}\right|^{-\frac{1}{2}}\right]^2 cos^2\Phi\,. \tag{6}$$

The structure in the x-ray spectra is due to the $\cos^2\Phi$ term. Since $\frac{d\omega}{dt}$ and D(R(t)) are slowly varying functions of t, the phase difference between two consecutive extremes, a maximum at ω_{max} and a minimum at ω_{min}, is approximately equal to $\frac{\pi}{2}$ and according to (5) is approximately given by:

$$\Delta\Phi = \frac{\pi}{2} \approx \frac{1}{2}(\omega_{max} - \omega_{min})t \tag{7}$$

where t is the mean collision time $t = \frac{1}{2}(t_{max} + t_{min})$ corresponding to the mean transition frequency $\omega = \frac{1}{2}(\omega_{max} + \omega_{min})$. This result assumes a linear dependence of $\omega(t)$ in the small energy interval (1-2 keV) between consecutive extrema. The transition energy as function of internuclear distance can then be easily obtained in the following way: the collision time is calculated from (7) using experimental values of ω_{max} and ω_{min} and then converted into internuclear distance R by integrating the Coulomb trajectory.

The results are presented in Fig. 9. The points with different symbols represent the results obtained at the different incident energies, they overlap nicely and form a common transition energy curve. The experimental results are compared to the $2p\pi$-$1s\sigma$ (dashed line) and the $2p\sigma$-$1s\sigma$ (dot-dashed line) transition energies obtained[5] by scaling the energies of the H^+-H system according to the procedure given in ref.

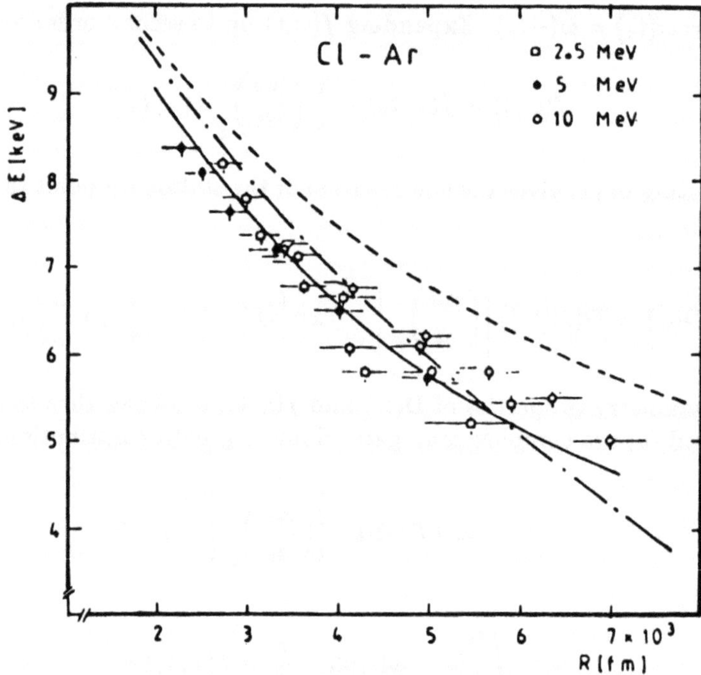

Fig. 9. Quasimolecular transition energy as function of internuclear distance: symbols represent experimental values obtained using the stationary phase approximation; dashed curve (dot-dashed curve) are calculated $2p\pi$-$1s\sigma$ ($2p\sigma$-$1s\sigma$) transition energies by scaling transition energies of the H^+-H system. The solid line is the results of calculations from ref. 24.

23. From general arguments based on the magnitude of the dipole transition matrix elements and the relative population of the orbitals one can predominantly assign the experimental points to $2p\pi$-$1s\sigma$ transitions[5]. It is seen in Fig. 9 that the simple scaling procedure gives transition energies which are systematically higher than the experimental ones. The solid line shows the results[24] of may-electron relativistic molecular calculations of the $2p\pi$-$1s\sigma$ transition energies. These calculations are in very good agreement with the experimental results.

B. The Uniform Approximation

The stationary phase approximation breaks down for $\omega = \omega(o)$ i.e. at the closest distance of approach at t=0 or near the classical maximum transition frequency. At this point the second derivative of $f(\omega,t)$ vanishes and the expansion (3) breaks down. A modification of the stationary phase approximation can be applied which involves a cubic expansion of $f(\omega,t)$ around t_o but which will be only valid for a narrow range of ω. Another approximation scheme exists, the uniform approximation[22], which allows to evaluate $D(\omega)$ both close and far from $\omega(o)$. In this approach a new variable z is introduced:

$$f(\omega,t) = \frac{1}{3}z^3 - \xi \cdot z + A \tag{8}$$

where ξ and A are constants which are chosen so as to make for

$$\begin{aligned} t = t_o &\leftrightarrow z = \sqrt{\xi} \\ t = -t_o &\leftrightarrow z = -\sqrt{\xi} \end{aligned} \tag{9}$$

substituting (9) into (8) gives:

$$A = 0 \qquad \xi = \left[-\frac{3}{2}f(\omega, t_o)\right]^{\frac{2}{3}} \tag{10}$$

substituting (8) into (1) and using the results of (10) yields:

$$D(\omega) = \frac{1}{\sqrt{2\pi}} \int_{-\infty}^{\infty} D(R(t)) \cdot \frac{dt}{dz} \cdot exp\left[i\left(\frac{1}{3}z^3 - \xi z\right)\right] dz. \tag{11}$$

After expanding $D \cdot \frac{dt}{dz}$ in powers of z one obtains finally:

$$\begin{aligned} D(\omega) &= \frac{1}{\sqrt{\pi}} D(R(t_o)) \xi^{\frac{1}{4}} \left|\left(\frac{d\omega}{dt}\right)_{t_o}\right|^{-\frac{1}{2}} \int_{-\infty}^{\infty} exp\left[i\left(\frac{1}{3}z^3 - \xi z\right)\right] dz \\ &= 2\sqrt{\pi} D(R(t_o)) \xi^{\frac{1}{4}} \left|\left(\frac{d\omega}{dt}\right)_{t_o}\right|^{-\frac{1}{2}} Ai(-\xi) \end{aligned} \tag{12}$$

where Ai is the Airy function. The photon intensity is then given by:

$$\begin{aligned} I(\omega) &\propto \omega |D(\omega)|^2 \\ &\propto \omega \cdot Ai^2(-\xi) \end{aligned} \tag{13}$$

From this it follows that the value ω_m for which $I(\omega)/\omega$ gets its first maximum is to a good approximation also the maximum of the Airy function. From eqs. (2) and (10) it also follows that t=0 corresponds to $\omega = \omega(o)$ and to ξ=0. We therefore have:

$$\frac{I(\omega(o))/\omega(o)}{I(\omega_m)/\omega_m} = \left[\frac{Ai(o)}{Ai(-\xi_1)}\right]^2 = 0.44 \tag{14}$$

where ξ_1 is the first maximum of Ai($-\xi$).

Using (14) it is possible to determine $\omega(o)$ simply by inspecting a plot of $I(\omega)/\omega$ and finding the location, in the region above ω_m, at which $I(\omega)/\omega$ drops to 0.44 of its value at the first maximum ω_m, as shown in Fig. 10.

Fig. 11 shows the results from the uniform approximation (full symbols) and from the stationary phase approximation (open symbols). The uniform approximation yields results for small values of R; however, there is a region of R values (between 2500 and 3000 fm covered by two approaches, in which there is good agreement between the extracted values of the transition energies. This agreement between results obtained using different approximations and derived from different portions of the

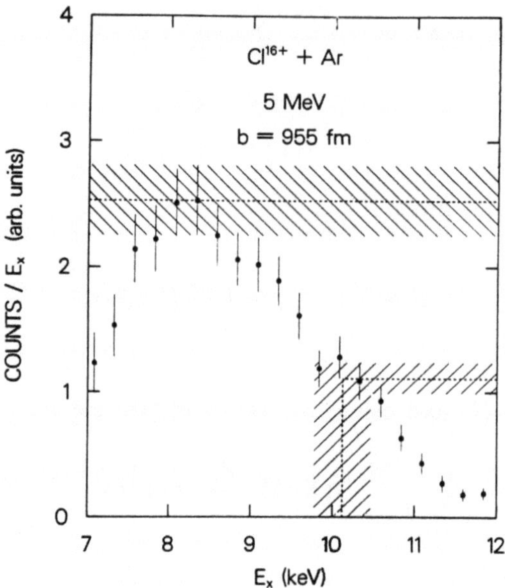

Fig. 10. Quasimolecular radiation yield divided by the x-ray transition energy and application of the Airy analysis method as discussed in the text. The shaded areas represent experimental uncertainties.

spectra indicates the reliability of both schemes, each in its own useful range. It also shows that the two methods complement each other; they allow to extract the quasimolecular orbital transition energies based on simple considerations about the interference pattern.

VI. NUCLEAR REACTION TIMES FROM QUASIMOLECULAR K X-RAY SPECTRA

The interference structure in quasimolecular K x-ray spectra can be used as a clock for nuclear reaction times. As mentioned in the introduction several other attempts have recently been made[2,7,8,25] to use atomic processes in order to determine nuclear reaction times. The principle of the clock is similar in all the proposed methods and will be presented here for the quasimolecular case. The time integral of eq. (1) can be split as

$$D(\omega) = \int_{-\infty}^{0} + \int_{0}^{\infty} = D_{in} + D_{out} \qquad (15)$$

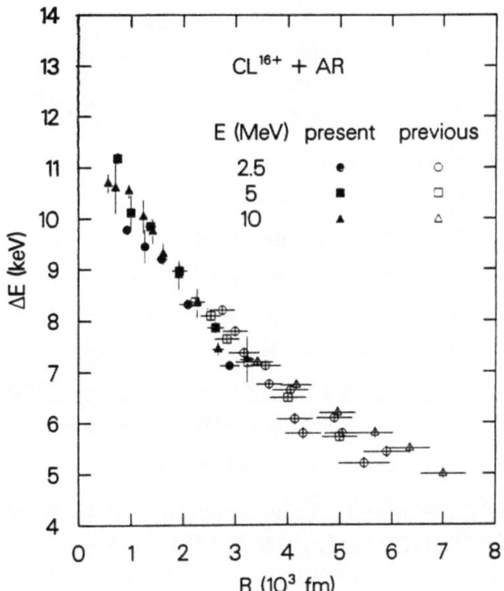

Fig. 11. **Quasimolecular transition energy ΔE as function of internuclear distance R obtained using the uniform approximation (full symbols) and the stationary phase approximation (open symbols).**

where D_{in} is the incoming and D_{out} the outgoing emission amplitudes.

If the incident energy is above the Coulomb barrier, nuclear reactions will take place in collisions at small impact parameters $b \leq R_P + R_T$, where R_P and R_T are the projectile and target nuclear radii, respectively. For the atomic processes one can consider that $b \approx 0$ when a nuclear reaction takes place. Fig. 8b shows the time-dependent transition frequency for a nuclear collision in which the projectile and target nuclei stick together for a time T_s. The corresponding emission amplitude will be given by:

$$
\begin{aligned}
D(\omega) &= \int_{-\infty}^{o} + \int_{o}^{T_s} + \int_{T_S}^{\infty} \\
&= D_{in} + \frac{D(R(o))}{i(\omega - \omega_o)} exp[i(\omega - \omega_o)T_S - 1] + \\
&\quad exp[i(\omega - \omega_o)T_S] D_{out}
\end{aligned}
\tag{16}
$$

The second term gives the decay amplitude of the $1s\sigma$ vacancy during the nuclear reaction time. For a long-lived composite nuclear system this term gives rise to sharp united atom K x-ray line at ω_o. For short reaction times the essential effect

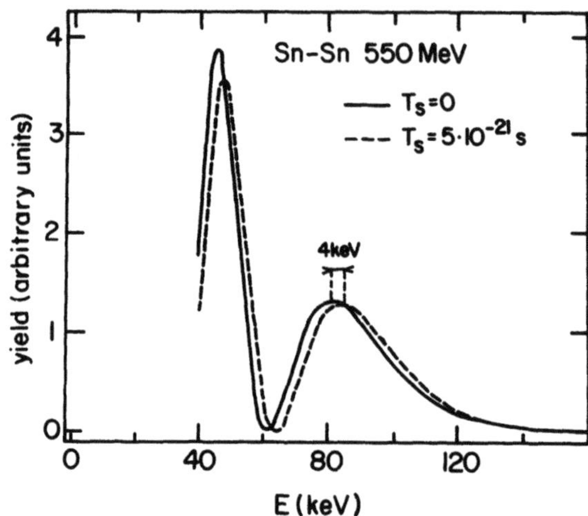

Fig. 12. The quasimolecular K x-ray spectrum at b=0 for 550 MeV Sn+Sn collisions with $T_S=0$ and $T_S=5.10^{-21}$s.

is the introduction of an additional phase between D_{in} and D_{out}. This phase factor produces a change in the interference structure which can be used as the clock to determine the nuclear reaction time.

The sensitivity of the clock is illustrated in Fig. 12 which shows the spectral shape of the quasimolecular radiation from 550 MeV Sn+Sn collisions at b=0 without sticking time, $T_S = 0$, and with a sticking time of $5 \times 10^{-21}s$. The calculations were done for $2p\pi$-$1s\sigma$ transitions using dipole matrix elements as given by Anholt[26] and transition energies scaled from the H^+-H system. It is seen that the sticking time produces a shift of the maximum of ~4 keV. The sensitivity of the clock depends on the product $\omega_o T_S$; it therefore increases for heavier systems and becomes $\sim 10^{-21}s$ for Pb on Pb[9].

For comparison we quote some results of recent calculations about nuclear sticking times. For symmetric systems with $A_1 = A_2 \approx 100$ the calculated sticking times are of the order of $10^{-20}s$ or even more and they decrease rapidly for heavier systems to the more typical values of a few times $10^{-21}s$ [27]. The mean time from the saddle to the fission point of the fissioning nucleus ^{286}Yb is also predicted to be rather long of the order of $10^{-20}s$ or more[28].

Fig. 13 shows the evolution of the interference pattern for various sticking times in the Sn+Sn system. For short nuclear sticking times one observes a shift of the maximum towards higher photon energies; as the sticking time increases more oscillations appear and for long enough sticking times the united atom (U.A.) transition energy becomes the dominant feature of the spectrum.

Several comments of caution should be made about the implementation of the

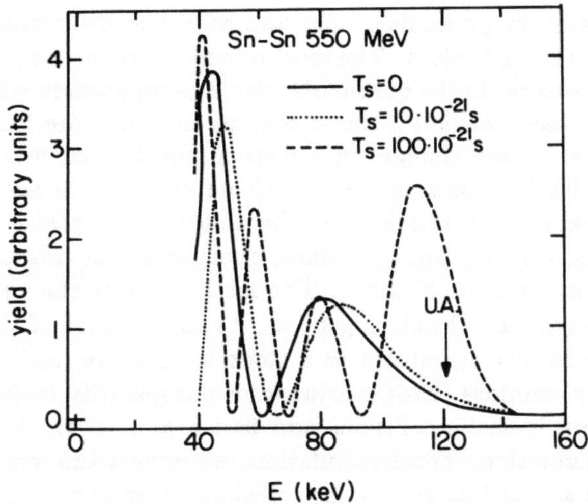

Fig. 13. The quasimolecular K x-ray spectrum at b=0 for 550 MeV Sn+Sn collisions and several nuclear sticking times.

method. *i*) The experiment requires an H-like projectile and therefore an accelerator with deceleration capabilities. Furthermore, the projectile velocity must be low enough to satisfy the adiabicity condition necessary for the validity of the quasimolecular picture and high enough to overcome the Coulomb barrier (see Table 1). These constrains on the projectile limit today the method, for symmetric systems, to $A_1 = A_2 \approx 70 - 100$. *ii*) Due to yield considerations, the experiment must be performed with a solid target. This, however, does not seem to be a serious difficulty. Preliminary measurements[29] indicate that the projectile K vacancy can survive a target thickness of $\sim 20\ \mu g/cm^2$ before being filled. *iii*) The calculations of Figs. 12,13 were done with a fixed nuclear sticking time. In nuclear reactions, there will be a distribution of collision times and thus the sensitivity of the clock will be limited by the experimental ability to select a narrow collision time interval. *iv*) The most serious difficulty is the need to detect quasimolecular x-ray which are produced with a low probability ($\sim 10^{-2}$ per event in Sn+Sn, see Table 1) in the presence of nuclear gamma rays which may have a multiplicity of 10 or more. Fortunately, the quasimolecular K x-rays are in a narrow energy band (40-140 keV in the Sn+Sn system) whereas the nuclear gamma rays are spread in a much wider range, up to several MeV. A very efficient anti-Compton scattering system is needed in order to reduce this problem.

VII. SUMMARY

Quasimolecular radiation has been extensively studied during the last decade. The structureless shape of these spectra with a nearly exponential fall-off has voided the possibility of spectroscopic studies of quasimolecular orbitals in general and of superheavy elements in particular. The use of H-like projectiles at low velocities drastically changes this situation. The most important consequence is the observation of interference structures in the quasimolecular K x-ray spectra which result from the coherent sum of x-ray emission in the incoming and outgoing half of the trajectory. The interference structure changes in a smooth and regular way both with impact parameter and with bombarding energy. Using very simple assumptions based on the stationary phase approximation and the uniform asymptotic approximation it is possible to obtain, directly from the data, quasimolecular orbital transition energy as a function of internuclear distance. The method opens the way for spectroscopic studies of quasimolecular orbitals in general. As an extension of this work we propose to use these interference structures as a clock to measure nuclear sticking times in heavy ion nuclear reactions. This is a fundamental quantity which plays a basic role in theories of deep inelastic collisions and fission and of which we have very little experimental information. From calculations we expect the method to be sensitive to sticking times as short as 10^{-20}s for collisions of Sn+Sn. From the experimental point of view the measurements seem feasible although they are not simple. The main difficulty is the necessity to suppress (or to subtract in a reliable way) the strong gamma radiation which is emitted together with the quasimolecular x-rays in a deep inelastic collision.

REFERENCES

* Supported by MINERVA Foundation Munich/Germany.

1. J.R. MacDonald, M.D. Brown, and T. Chiao, Phys. Rev. Lett. 30, 471 (1973).
2. See e.g. the recent review article by R. Anholt, Rev. Modern Phys. 57, 995 (1985).
3. W. Pieper and W. Greiner, Z. Phys. 218, 327 (1969).
   B. Muller, J. Rafelski and W. Greiner, Z. Phys. 257, 62 and 183 (1972).
4. I. Tserruya, R. Schuch, H. Schmidt-Böcking, J. Barrette, Wang Da-Hai, B.M. Johnson, M. Meron, and K.W. Jones, Phys. Rev. Lett. 50, 30 (1983).
5. R. Schuch, H. Schmidt-Böcking, I. Tserruya, B.M. Johnson, K.W. Jones, and M. Meron, Z. Phys. A320, 185 (1985).
6. M. Meron, B.M. Johnson, K.W. Jones, R. Schuch, H. Schmidt-Böcking, and I. Tserruya, Nucl. Instrum. Methods B10/11, 64 (1985).
7. R. Krieg, E. Bozek, U. Gollerthan, E. Kankeleit, G. Klotz-Engmann, M. Krämer, U. Meyer, H. Oeschler, and P. Senger, Phys. Rev. C. Submitted.
8. Ch. Stoller, M. Nessi, E. Morenzoni, W. Wölfli, W.E. Meyerhof, J.D. Molitoris, E. Grosse, and Ch. Michel, Phys. Rev. Lett. 53, 1329 (1984).
9. J. Kirsch, B. Muller, and W. Greiner, UFTP preprint 154/1985.
10. F. Bell, H.D. Betz, H. Panke, E. Spindler, W. Stehling and M. Kleber, Phys. Rev. Lett. 35, 841 (1975).

11. H.D. Betz, F. Bell, H. Panke, W. Stehling, E. Spindler, and M. Kleber, Phys. Rev. Lett. 34, 1256 (1975).
H. Schmidt-Böcking, R. Schuch, I. Tserruya, R. Schule, H.J. Specht, and K. Bethge, Z. Phys. A284, 39 (1978).
P. Vincent and J.S. Greenberg, J. Phys. B12, L641 (1979).

12. I. Tserruya, H. Schmidt-Böcking, R. Schule, K. Bethge, R. Schuch, and H.J. Specht, Phys. Rev. Lett. 36, 1451 (1976).

13. H. Schmidt-Böcking, R. Anholt, R. Schuch, P. Vincent, K. Stiebing, and H.U. Jäger, J. Phys. B15, 3057 (1982).
R. Schuch, G. Gaukler, G. Nolte, K.W. Jones, and B.M. Johnson, Phys. Rev. A22, 2513 (1980).

14. See ref. 2 for other attempts, in particular anisotropy measurements, to obtain spectroscopic information about quasimolecular orbitals.

15. H. Schmidt-Böcking, W. Lichtenberg, R. Schuch, J. Volpp, and I. Tserruya, Phys. Rev. Lett. 41, 859 (1978).

16. W. Lichten, Phys. Rev. A9, 1458 (1974).
B. Müller. In: "Proc. Ninth Int. Conf. on the Physics of Electronic and Atomic Collisions," J.S. Risley and R. Geballe, eds., Univ. of Washington Press, Seattle (1976), p. 481.

17. J.H. Macek and J.S. Briggs, J. Phys. B7, 1312 (1974).

18. P. Thieberger, J. Barrette, B.M. Johnson, K.W. Jones, M. Meron, and H.E. Wegner, IEEE Trans. Nucl. Sci. NS 30, 1431 (1983).

19. G. Gaukler, H. Schmidt-Böcking, R. Schuch, R. Schule, H.J. Specht, and I. Tserruya, Nucl. Instrum. Methods 141, 115 (1977).

20. R. Anholt – private communication.

21. R. Schuch – private communication.

22. See e.g. J.N.L. Connor and R.A. Marcus, J. Chem. Phys. 55, 5636 (1971).

23. K.H. Heinig, H.U. Jäger, H. Richter, H. Woittennek, W. Frank, P. Gippner, K.H. Kaun, and P. Manfrass, J. Phys. B10, 1321 (1977).

24. B. Fricke, W.-D. Sepp, and T. Morovic, Z. Phys. A318, 369 (1984).

25. J.S. Blair, P. Dyer, K.A. Snover, T.A. Trainer, Phys. Rev. Lett. 44, 1712 (1978).

26. R. Anholt, Z. Phys. A288, 257 (1978).

27. J.P. Blocki, H. Feldmeier and W.J. Swiatecki, to be published in Nucl. Phys. A.

28. J.R. Nix, A.J. Sierk, H. Hofmann, F. Scheuter and D. Vautherin, Nucl. Phys. A424, 239 (1984).

29. H. Schmidt-Böcking, private communication.

RELATIVISTIC DENSITY FUNCTIONAL THEORY

R.M. Dreizler, E. Engel and P. Malzacher

Institut für Theoretische Physik
Universität Frankfurt
D-6000 Frankfurt am Main, W. Germany

Density functional methods have become quite a successful tool for the discussion of groundstate properties of nonrelativistic many body systems[1] . The question of a density functional theory for relativistic systems then arises quite naturally. If we are, for instance, dealing with a heavy atom, the simplest starting point is the Hamiltonian commonly used as the basis of the relativistic Dirac-Fock scheme[2] . This Hamiltonian consists of a sum of Dirac Hamiltonians, describing the motion of the electrons in an external four potential, the instantaneous Coulomb interaction between the electrons and the Breit interaction, which takes into account retardation effects as well as the exchange of transverse photons in low order. One may encounter some difficulties if one bases the discussion on this Hamiltonian, due to the fact that stable, normalizable bound state solutions do not exist[3] . One usually avoids these difficulties by restriction to the positive energy sector.

For the development of a relativistic density functional theory it is thus preferable to start directly from a field theoretical basis, that is QED in the case of Coulomb systems. Before we attempt, however, to summarize the status of this development, it is opportune to take a closer look at the nonrelativistic case for some background information.

1. REMARKS ON NONRELATIVISTIC DENSITY FUNCTIONAL THEORY

Our many body system is then characterized by the Hamiltonian

$$H = T + V + W$$

consisting of the nonrelativistic kinetic energy of the particles, the potential of the particles in an external field and the potential due to two body forces. The aim is the discussion of the groundstate of the N-particle system, in particular the calculation of the groundstate energy.

One possible path towards this aim is indicated by the theorem of Hohenberg and Kohn[4] . The theorem states that the groundstate energy can be represented as a unique functional of the groundstate density alone

$$E_o = E_o\,[n_o(\underline{r})] \;.$$

As this theorem constitutes the cornerstone of density functional theory, we will present an outline of the proof. We consider a fixed two body interaction, but subject the system to different one particle potentials, which are all supposed to lead to nondegenerate groundstates

$$H_v |\psi_{vo}\rangle = E_{ov} |\psi_{vo}\rangle .$$

Collecting all the potentials in a set $\mathbf{V}$ and the corresponding groundstates in a set $\boldsymbol{\Psi}$, we have defined a map

$$A : \mathbf{V} \to \boldsymbol{\Psi} .$$

For all the groundstate wavefunctions we calculate the groundstate densities via

$$n_v(\underline{r}) = \langle \psi_{vo}|\hat{n}(\underline{r})|\psi_{vo}\rangle$$

$$= N \sum_{\alpha} \int dx_z \ldots dx_N |\psi_{vo}(\underline{r},\underline{x}_2,\ldots \underline{x}_N)|^2$$

thus establishing a second map

$$B : \boldsymbol{\Psi} \to \mathbf{N} ,$$

where $\mathbf{N}$ is the set of all groundstate densities. The two maps are surjective by construction: Each element of the starting set is mapped onto one element of the final set.

The gist of the proof of the theorem is then: The maps are also injective (one to one) and hence by a standard theorem of the theory of sets, bijective (fully invertible).

In order to demonstrate injectivity of the map A one shows that two potentials V and V', which are elements of the set $\mathbf{V}$ and which differ by more than a mere constant, can not lead to the same wavefunction. In the case of map B one shows that $|\psi_{vo}\rangle \neq |\psi_{v'o}\rangle$ implies $n_{vo}(\underline{r}) \neq n_{v'o}(\underline{r})$. The details of the rather simple proofs of these statements can be gleaned from the original paper of Hohenberg and Kohn.

It is the existence of the map B^{-1} which allows us to state

$$\langle \psi_o|\hat{O}|\psi_o\rangle = \langle \psi_o[n_o]|\hat{O}|\psi_o[n_o]\rangle$$

$$= O[n_o] .$$

The groundstate expectation value of any observable is a unique functional of the exact groundstate density. In particular, we can write

$$E_o[n_o] = T[n_o] + W[n_o] + \int d^3r \, v(\underline{r}) n_o(\underline{r}).$$

As a corollary to the theorem one concludes, as a consequence of the Ritz principle, that

$$E_o[n_o] \le E_o[n] .$$

The Hohenberg-Kohn functional evaluated with the correct groundstate density is smaller than the same functional evaluated with any other density from the set $\mathbf{N}$. This indicates that (given the functional) one may calculate the groundstate density and energy variationally.

$$\frac{\delta}{\delta n}(E_o[n] + \lambda \int d^3r\, n(\underline{r})) = 0$$

The catchis obviously the explicit form of the functional, a point to which we will return shortly.

A second variational approach is provided by the scheme of Kohn and Sham[5] , which introduces the intermediary of an orbital picture. This scheme can be set up by the following argument:

Assume that there exists an auxiliary N-particle problem, characterized by

$$\hat{H}_a = \hat{T} + \hat{V}_a ,$$

which yields the same groundstate density as a specified interacting system

$$n_{ao}(\underline{r}) = n_o(\underline{r}).$$

One can then argue as follows:

(i) Write the Hohenberg-Kohn functional in the form

$$E_o[n_o] = T_s[n_o] + \frac{1}{2}\int d^3r\, d^3r'\, w(\underline{r},\underline{r}')n_o(\underline{r})n_o(\underline{r}')$$

$$+ \int d^3r\, v(\underline{r})n_o(\underline{r}) + E_{xc}[n_o] .$$

The exchange-correlation part is defined as

$$E_{xc}[n_o] = W[n_o] - W_{Hartree}[n_o] + T[n_o] - T_s[n_o] .$$

It involves the difference of the full interaction energy and the "Hartree-term" as well as the difference of the kinetic energies of the interacting system, T , and the auxiliary problem, T_s .

(ii) Expand the density and the quantity T_s in terms of orbitals

$$n(\underline{r}) = 2 \sum_{i=1}^{N/2} \varphi_i^*(\underline{r})\varphi_i(\underline{r})$$

$$T_s[n] = -\frac{h^2}{2m}\int d^3r \sum_{i=1}^{N/2} \varphi_i^*(\underline{r})(\Delta\varphi_i(\underline{r})) .$$

(iii) Replace the variation with respect to the density by variation with respect to the orbitals and obtain (with the subsidiary condition of normalized orbitals)

$$-\frac{h^2}{2m}\Delta\varphi_i(\underline{r}) + v_{eff}(\underline{r})\,\varphi_i(\underline{r}) = \varepsilon_i\varphi_i(\underline{r}) .$$

The effective <u>local</u> potential is given by

$$v_{eff}(\underline{r}) = v(\underline{r}) + \int d^3r'\, w(\underline{r},\underline{r}')n(\underline{r}') + \frac{\delta E_{xc}[n]}{\delta n(\underline{r})} .$$

If the Hohenberg-Kohn functionals $T[n]$, $W[n]$ and $T_s[n]$ are known (at least in some approximation), then the last term can be specified explicitly.

The effective potential depends on the density. The KS-scheme thus amounts to a selfconsistent orbital scheme, which contains as a particular approximation the Hartree-Fock-Slater method

$$\frac{\delta E_{xc}}{\delta n} = \text{const } n^{1/3}$$

and is, in principle, able to transcend the HF limit.

It is (as in HF or HFS) common practice to interpret the orbitals as occupied and virtual orbitals and the corresponding energies as excitation or removal energies. This interpretation is not correct, at least from a purist's point of view: The orbitals merely serve as a vehicle for the representation of the (exact) groundstate density, which can be used to calculate the groundstate energy. One can demonstrate by simple arguments that the Slater determinant constructed from the Kohn-Sham orbitals with lowest energy is not the groundstate wavefunction of the interacting system. The density calculated from these orbitals does nonetheless represent the exact groundstate density, provided $v_{eff}(\underline{r})$ can be specified exactly.

The major task before practical applications of either variational scheme is the derivation of suitable energy functionals. It should be obvious that approximations will have to be accepted. The substantial amount of literature on this topic can be classified under three different headings.

A. THE ELECTRON GAS APPROACH

For this wellknown many body model one distinguishes two cases. The homogeneous electron gas: The interacting electrons move in a neutralizing background of positive charges, which are smeared out uniformly. The inhomogeneous electron gas: An additional point charge is introduced at a specific space point. The homogeneous case is the basis for the derivation of the socalled local-density approximation[6] . The argument goes as follows:

(i) Calculate the groundstate energy of the gas, which can be expressed in the form

$$\frac{E_o}{V} = \varepsilon_o = \varepsilon_o(k_F)$$

The energy density is a function of the Fermi momentum. The density of the system is related to the Fermi momentum by

$$k_F = (3\pi^2 n)^{1/3} \quad .$$

(ii) Replace the constant density of the model system by the density of any inhomogeneous system

$$n \to n(\underline{r}) \quad .$$

The philosophy behind this arguments is the hope that any system behaves locally (at least in an acceptable approximation) like the homogeneous gas. As a matter of principle the resulting functional should only be applied to situations with a slowly varying density. It has, however, been used with astonishing success for the discussion of atoms, molecules and solids[7].

The simplest expression for the groundstate energy density is its Hartree-Fock limit

$$\varepsilon_o^{(HF)} = \frac{3}{10}(3\pi^2)^{2/3}\frac{\hbar^2}{m}n^{5/3} - \frac{3}{4}\left(\frac{3}{\pi}\right)^{1/3} e^2 n^{4/3} .$$

One recognizes the kinetic energy expression of the Thomas-Fermi model, which can also be derived by a variety of other arguments, and an exchange term, which leads to a Slater type exchange. The calculation of the remaining energy contribution, the correlation energy, can be attempted with any current many body method: RPA, (which is exact in the high density limit), the method of local field corrections, cluster expansions ($e^{\hat{s}}$ methods), hypernetted chain methods, Monte Carlo methods. It is probably fair to state, that the results obtained with the last four approaches have converged to an answer, which can for instance be parametrized in the form

$$\varepsilon_o^{(cor)} = \{ - 0.112 + 0.035 \ln [\frac{3}{4\pi}\frac{1}{n}]^{1/3} \} n \quad \text{(in ry units)}$$

with a spread of a few percent in the magnitude of the coefficients.

For simplicity we have only quoted results for the spin-saturated case. Spin-polarised systems can be treated in a similar fashion.

The next step is the calculation of gradient corrections, that is terms involving $\nabla n(\underline{r})$. For this purpose one treats the inhomogeneity introduced into the gas in linear response. One then obtains a correction to the groundstate energy, which corresponds to second order perturbation theory in the inhomogeneous perturbation. We will, however, not attempt to summarize the substantial details, as we will discuss the emergence of gradient terms under the heading.

B. DIRECT GRADIENT EXPANSION (THE KIRZHNITS METHOD).

The starting point[8] is here a representation of the density operator in the form

$$\hat{\rho} = 2\theta(\varepsilon_F - \hat{t} - \hat{v}_{eff}) , \quad \varepsilon_F = \frac{k_F^2}{2m}$$

The discussion will again be restricted to the spin saturated case, hence the factor of 2.

If $\hat{v}_{eff}$ is the local Kohn-Sham potential, this expression yields the exact groundstate density. One can also evaluate the one particle density matrix

$$\gamma(\underline{r}',\underline{r}) = \langle \underline{r}'|\hat{\rho}|\underline{r} \rangle$$

by insertion of a complete set of KS solutions

$$= \sum_i \langle \underline{r}'|\theta(\varepsilon_F - \hat{h}_{KS})|\varphi_i\rangle\langle\varphi_i|\underline{r}\rangle$$

$$= \sum_{i_{occ}} \varphi_i^*(\underline{r}')\,\varphi_i(\underline{r}) .$$

This expression is not the exact one particle density matrix. It can, however, be established, in retrospect, that it is very close to γ_{HF}.

For the arguments that follow[9] we do not rely on a specific knowledge of $\hat{v}_{eff}$ and write

$$\hat{\rho} = 2\,\theta(\hat{\varepsilon}_F - \hat{t})$$

with the local operator

$$\hat{\varepsilon}_F = \varepsilon_F - \hat{v}_{eff} \ , \ < \underline{r}|\hat{\varepsilon}_F|\underline{r}' > \ = \varepsilon_F(\underline{r})\delta^{(3)}(\underline{r}-\underline{r}') \ .$$

One may also evaluate the density matrix by plane wave expansion

$$\gamma(\underline{r},\underline{r}') = 2 \int d^3k \ <\underline{r}'|\theta(\hat{\varepsilon}_F-\hat{t}) \ |\underline{k}><\underline{k}'|\underline{r}' > \ .$$

The technical problem encountered is the calculation of the matrix element

$$< \underline{r}|\theta(\hat{\varepsilon}_F - \hat{t})|\underline{k} > .$$

This would be simple if the two operators in the argument of the step function were to commute

$$\rightarrow \theta(\varepsilon_F(\underline{r}) - \frac{k^2}{2m} \ ) \ .$$

As the operators do not commute, one has to rely on expansion in terms of multiple commutators of the two operators. As we have e.g.

$$< \underline{r}|[\hat{\varepsilon}_F,\hat{t}]|\underline{r}' > \ = \frac{\hbar^2}{2m} \delta^{(3)}(\underline{r}-\underline{r}')(\Delta\varepsilon_F(\underline{r}) + 2\,\underline{\nabla}\varepsilon_F(\underline{r})\cdot\underline{\nabla} \ )$$

this procedure yields, after considerable algebra, an expression for the density matrix in the form

$$\gamma(\underline{r},\underline{r}') \ = \gamma(\varepsilon_F(\underline{r}),\partial_i\varepsilon_F(\underline{r}), \ \ldots,|\underline{r}-\underline{r}'| \ ),$$

that is an expression in terms of the unknown function $\varepsilon_F(\underline{r})$ and its derivatives. With this result one obtains directly the kinetic energy density

$$\tau(r) = [\ -\frac{\hbar^2}{2m}\Delta\gamma(\underline{r},\underline{r}') \ ]_{\underline{r} \, = \, \underline{r}'}$$

$$= \tau(\varepsilon_F,\partial_i\varepsilon_F,\ldots \ ) \ ,$$

the Coulomb exchange energy density

$$\varepsilon_x(\underline{r}) = \frac{e^2}{2} \int d^3r' \ \{ \frac{\gamma(\underline{r},\underline{r}')\gamma(\underline{r}',\underline{r})}{|r-r'|} \ \}$$

$$= \varepsilon_x(\varepsilon_F,\partial_i\varepsilon_F \ , \ \ldots \ )$$

and naturally the density itself

$$n(r) = \gamma(\underline{r},\underline{r}) \ = n(\varepsilon_F,\partial_i\varepsilon_F, \ \ldots) \ .$$

The expressions, obtained at this level, can be classified as a semiclassical expansion, the formal expansion parameter being $\hbar$.

If one inverts the expansion of the density order by order in the gradient terms and inserts the inversion in a consistent manner into the expansion for the kinetic and exchange densities, one obtains e.g. to second order

$$\tau(r) = c_o n(r)^{5/3} + c_2 \frac{(\nabla n)^2}{n}$$

$$c_o = \frac{3}{10}(3\pi^2)^{2/3}\frac{\hbar^2}{m} \qquad c_2 = \frac{1}{72}\frac{\hbar^2}{m}$$

$$\varepsilon_x(\underline{r}) = - d_o\, n(r)^{4/3} - d_2 \frac{(\nabla n)^2}{n^{4/3}}$$

$$d_o = \frac{3}{4}\left(\frac{3}{\pi}\right)^{1/3} e^2 \qquad d_2 = \frac{7}{432\pi(3\pi^2)^{1/3}} \quad e^2 \quad .$$

Higher order terms (for the kinetic energy) can be found in the literature[10].

We note that the lowest order terms are the same as those obtained by the electron gas argument. This statement also applies to the lowest gradient term for the exchange energy density. One should also note that in view of the genesis of this expansion, the kinetic energy calculated here corresponds to the term T_s, the kinetic energy of an equivalent noninteracting system as introduced in the discussion of the KS scheme.

C. THE WIGNER-KIRKWOOD APPROACH

The same results as in the previous section can be obtained if one basis the discussion on the one particle Bloch density matrix

$$C(\underline{r},\underline{r}',\beta) = < \underline{r}|e^{-\beta\hat{h}}|\underline{r}' >$$

rather than the Dirac density matrix γ. This approach is more popular in nuclear physics[11]. In view of the close relationship by Laplace transformation

$$\gamma(\underline{r},\underline{r}') = \frac{1}{2\pi i}\int_{c-i\infty}^{c+i\infty} \frac{d\beta}{\beta} C(\underline{r},\underline{r}',\beta)e^{\varepsilon_F \beta}$$

it is not surprising that the same results are reproduced.

There would remain the task to present an overview over the large body of explicit applications of the functionals obtained. We shall, however, restrict ourselves to a few statements.

1. Solution of the Hohenberg-Kohn variational problem with extended Thomas-Fermi functionals (2nd order gradient terms) reproduces HF groundstate energies to an accuracy of better than 1%, at low cost so to speak. The densities that emerge do not reproduce the shell structure.

2. Solution of the Kohn-Sham equations, including correlation corrections, yields a very consistent description of atomic and molecular groundstates, in particular also a more consistent description of molecular binding properties than obtained in the HF approximation.

2. RELATIVISTIC DENSITY FUNCTIONAL THEORY

As the arguments leading to the Hohenberg-Kohn theorem are largely independent of the form of the Hamiltonian, it should come as no surprise that the extension of the formal arguments to the relativistic domain can be carried through with little effort. This was first demonstrated by Rajagopal and coworkers[12] for the Hamiltonian density describing Dirac particles in interaction with the electromagnetic field and an external four potential. Some differences do occur. The density has to be replaced by a four current

$$n(r) \to j^{\mu}(r) \quad \text{with} \quad \partial_{\mu} j^{\mu} = 0$$

and one has to face (as usual) gauge questions.

The same authors also discuss the relativistic extension of the KS-scheme. Using the units $\hbar = m = c = 1$ the relativistic Kohn Sham equations read

$$[\, - i\underline{\alpha} \cdot \underline{\nabla} + (\beta - 1) + v_{eff}(\underline{r}) \,]\, \psi_i(\underline{r}) = \varepsilon_i \psi_i(\underline{r}) \quad .$$

The energy scale is shifted by the mass of the electron. The effective local potential is now

$$v_{eff}(\underline{r}) = V_{ext} + e^2 \int \frac{n(r')}{|\underline{r}-\underline{r}'|} d^3r' + \frac{\delta E_{xc}}{\delta n(r)}$$

$$+ e\, \underline{\alpha} \cdot (\underline{A}_{ext} + e \int \frac{\underline{j}(r')}{|\underline{r}-\underline{r}'|} d^3r' + \frac{\delta E_{xc}}{\delta \underline{j}(\underline{r})}) \quad .$$

The density is specified by a sum over occupied solutions of the relativistic Kohn Sham equations

$$n(\underline{r}) = \sum_{i_{occ}} \psi_i^{\dagger}(r) \psi_i(\underline{r}) \quad .$$

The exchange correlation energy $E_{xc}[j^{\mu}]$ is defined as the groundstate expectation value of the QED-Hamiltonian minus the external potential contributions and the noninteracting kinetic energy as well as the Hartree-type terms.

We will in the following concentrate on the practical question: The derivation of explicit functionals. There is a history of efforts in this direction going back to 1932, when Vallarta and Rosen[13] first set up the simplest relativistic density functional model. The intervening history can be characterized as a continuous fight against divergence questions. In order to sort out the situation, we will first look at:

2.1 RELATIVISTIC GRADIENT EXPANSIONS ON THE BASIS OF THE DIRAC EQUATION

The first task is to specify the density operator in reasonable analogy to the nonrelativistic case. If one relies on the usual picture employed in the Dirac-Fock scheme and in turn suggested for the KS scheme we would write

$$\hat{\rho}_{DF} = \theta(\varepsilon_F - \hat{h}_{eff}) - \theta(-1 - \hat{h}_{eff})$$

where h_{eff} represents a Dirac Hamiltonian with an effective (scalar) potential. This ensures that one only takes into account the sum over occupied bound orbitals, by cutting out an appropriate window[9] in the complete spectrum of negative continuum, discrete and positive continuum states of the effective Dirac equation at hand.

If one relies, however, on QED arguments in the Furry[14] boundstate interaction picture one would write

$$n(\underline{r}) = \frac{1}{2} \left[\sum_{\varepsilon<\varepsilon_F} \psi_\varepsilon^\dagger (\underline{r}) \psi_\varepsilon(\underline{r}) - \sum_{\varepsilon>\varepsilon_F} \psi_\varepsilon^\dagger (r) \psi_\varepsilon(r) \right]$$

The spinors are the solutions of the effective Dirac problem. This expression can be recasted as

$$= \sum_{-1<\varepsilon<\varepsilon_F} \psi_\varepsilon^\dagger(\underline{r}) \psi_\varepsilon(\underline{r}) + \frac{1}{2} \left[\sum_{\varepsilon<-1} \psi_\varepsilon^\dagger(r) \psi_\varepsilon(\underline{r}) - \sum_{\varepsilon>-1} \psi_\varepsilon^\dagger(\underline{r}) \psi_\varepsilon(\underline{r}) \right] .$$

The first term is the usual density due to the bound electrons, the second term describes, as suggested by Wichmann and Kroll[15], the vacuum polarization density. The corresponding density operator is

$$\hat{\rho}_{QED} = \frac{1}{2} \left[\theta(\varepsilon_F - \hat{h}_{eff}) - \theta(\hat{h}_{eff} - \varepsilon_f) \right]$$

which can also be written as

$$= \theta(\varepsilon_F - \hat{h}_{eff}) - 1/2 \quad .$$

Independent of the form of the density operator adopted, we have the task to carry through the gradient expansion for the standard operator

$$\theta(\varepsilon_F - \hat{h}_{eff}) = \theta(\varepsilon_F - \hat{h}_{oD})$$

where $\hat{h}_{oD}$ represents the free Dirac Hamiltonian.

Again the technique is plane wave expansion

$$n(\underline{r}) = \mathrm{tr} \sum_{s=1}^{4} \int d^3k < \underline{r}|\hat{\rho}|\underline{k}s > < \underline{k}s|\underline{r} >$$

$$h_o(r) = \mathrm{tr} \sum_{s=1}^{4} \int d^3k < \underline{r}|\hat{\rho}|\underline{k}s > < \underline{k}s|\hat{h}_{oD}|\underline{r} > .$$

The differences encountered with respect to the nonrelativistic case are due to the fact that negative and positive energy states of the free Dirac equation are required for completeness.

The details involved in carrying through the gradient expansion are quite substantial[16]. The results at the level of the representation of the quantities in question in terms of $\varepsilon_F(\underline{r})$ and its derivatives to second order in the gradient terms can be written as

$$n(r) = \frac{1}{3\pi^2} \, p_F{}^3 + \frac{1}{3\pi^2} \lim_{x\to\infty} (x^2-1)^{3/2}$$

$$+ f(\varepsilon_F) \, (\underline{\nabla}\varepsilon_F)^2 + g(\varepsilon_F) \, (\Delta\varepsilon_F)$$

with $p_F = (\varepsilon_F{}^2 - 1)^{1/2}$ and

$$f(\varepsilon_F) = \frac{1}{3\pi^2} \left(- \frac{\varepsilon_F{}^2}{8 \, p_F{}^3} + \frac{3}{8 \, p_F} \right)$$

$$g(\varepsilon_F) = \frac{1}{3\pi^2} \left\{ \left[\frac{5}{12} - \frac{1}{2} \lim_{x\to\infty} (\ln 2x) \right] + \frac{1}{2} \operatorname{arsinh} (p_F) + \frac{\varepsilon_F}{4p_F} \right\} ,$$

$$h_o(\underline{r}) = \frac{1}{8\pi^2} \left[p_F(p_F{}^2 + 1)^{1/2} \, (2p_F{}^2 - 1) - \operatorname{arsinh} (p_F) \right]$$

$$- \frac{1}{8\pi^2} \lim_{x\to\infty} [\, ... \,]$$

$$+ \tilde{f} \, (\varepsilon_F)(\underline{\nabla}\varepsilon_F)^2 + \tilde{g}(\varepsilon_F)(\Delta\varepsilon_F)$$

with $\tilde{f}(G_F) = \frac{1}{3\pi^2} \left\{ \left[- \frac{5}{24} + \frac{1}{4} \lim_{x\to\infty} (\ln 2x) \right] \right.$

$$\left. - \frac{1}{4} \operatorname{arsinh}(p_F) + \frac{1}{4} \frac{\varepsilon_F}{p_F} - \frac{1}{8} \frac{\varepsilon_F{}^3}{p_F{}^3} \right.$$

$$\tilde{g}(G_f) = \frac{1}{3\pi^2} \left\{ \frac{\varepsilon_F{}^2}{4 \, p_F} + \frac{p_F}{4} \right\} .$$

We find divergent terms in both the expressions for the density and the energy density. The occurence of divergent contributions is not astonishing, as we have summed over all negative energy states, but the situation is more subtle. To appreciate this point one should have followed the genesis of these terms in detail. The divergent terms in zeroth order are straightforward consequences of the negative energy contributions, the divergencies in second order are, however, due to both negative and positive energy contributions. This excludes the simplest procedure: to neglect these terms without further discussion.

If we adopt the density operator $\hat{\rho}_{DF}$, the density and free energy density are given by[16]

$$n(\underline{r}) = n(\varepsilon_F(\underline{r})) - n(\gamma_F(\underline{r}))$$

with the upper and lower effective Fermi energy

$$\varepsilon_F(\underline{r}) = \varepsilon_F - v_{eff}(\underline{r}) \text{ and } \gamma_F(\underline{r}) = -1 - v_{eff}(\underline{r}) .$$

The divergent terms cancel as we have

$$\underline{\nabla}\varepsilon_F(\underline{r}) = \underline{\nabla}\gamma_F(\underline{r}) .$$

However, further processing of the resulting equations leads to difficulties. The elimination of v_{eff} can not be done directly. Indirect elimination by setting up variational equations with a constraint expressing the relation between the upper and lower local Fermi energy leads to severe troubles. We were not able to solve the variational equations due to additional divergencies (a kind of turning point problem near the position of the nucleus for the case of atoms).

We thus are definitly forced to look at the density operator $\hat{\rho}_{QED}$. In this case one can put forward the following arguments[17]

a) In zeroth order the divergent term in the expression for the density is cancelled exactly. The free energy expression remains unchanged. This fact can be understood directly. For the density a normal ordering procedure and a charge symmetric formulation are equivalent. For the energy they are not. As we used a charge symmetric form in the gradient expansion (rather than a normal ordered form), we obtain zero point energy contributions. The only remedy at this level is to take them out by hand.

b) The divergencies that occur in second order can not be explained by the lack of normal ordering. They are related to UV divergencies, which are encountered in calculating the vacuum polarization in QED. The standard remedy is charge renormalization. We can in the present case perform charge renormalization in a simple fashion.

As the divergent term in the expression for the density is proportional to

$$\Delta\varepsilon_F = -\Delta v_{eff} = 4\pi e^2 \rho_{eff}$$

it can be interpreted as an (infinite) polarization charge. Introducing the polarization vector

$$-\underline{P} = \frac{1}{6\pi^2}\left[\frac{5}{6} - \lim_{x\to\infty}(\ln 2x)\right]\underline{\nabla}\varepsilon_F$$

so that the divergence of $\underline{P}$ is the divergent density contribution

$$-\underline{\nabla}\cdot\underline{P} = \frac{1}{6\pi^2}\left[\frac{5}{6} - \lim_{x\to\infty}(\ln 2x)\right]\Delta\varepsilon_F$$

one finds for the corresponding energy density

$$\frac{1}{2}\underline{P}\cdot\underline{\nabla}\varepsilon_F = \frac{1}{12}\left[-\frac{5}{6} + \lim_{x\to\infty}(\ln 2x)\right](\underline{\nabla}\varepsilon_F)^2 .$$

This is exactly the divergent term in the energy expression. Consistent charge renormalization requires that both terms be dropped, or more explicitly be incorporated in the redefinition

$$v_{eff} \to \varepsilon^{1/2} v_{eff}$$

$$e \to e/\varepsilon^{1/2} \quad ,$$

with ε denoting the dielectric constant.

For the remaining expressions the elimination of the unknown function $\varepsilon_F(\underline{r})$ can be carried through as in the nonrelativistic case. Using the abbreviation

$$x = (3\pi^2 n)^{1/3}$$

one obtains for the relativistic kinetic energy density ($\tau = h_o - n$)

$$\tau(r) = \frac{1}{8\pi^2} [\, x(2x^2+1)(x^2+1)^{1/2} - \text{arsinh}(x) \,] - \frac{x^3}{3\pi^2}$$

$$+ \frac{1}{72} [\, \frac{1}{(x^2+1)^{1/2}} + \frac{2x}{x^2+1} \text{arsinh}(x) \,] \frac{(\nabla n)^2}{n} \quad .$$

This result reduces to the nonrelativistic expression in the nonrelativistic limit ($x \ll 1$ in the units used).

Let us take stock of what we have learned:

i) One encounters two types of divergencies, if one carries through the gradient expansion on the basis of the Dirac formulation. They are both QED effects not properly taken into account: normal ordering questions and charge renormalization. Once the divergencies are identified, they can be dealt with at this level of discussion.

ii) The proper definition of the density operator is the one associated with the Furry picture. It includes vacuum polarization contributions. This implies that vacuum polarization effects can and must be included even in the simplest density functional models. It also implies, on the other hand, that the calculation of the density by a sum over bound states used in the relativistic Dirac-Fock scheme and suggested for the relativistic Kohn-Sham scheme is at least partially deficient. Some modifications are required for these schemes, which should be relevant for regions of high density.

So far we have discussed only the density functional expression for the relativistic kinetic energy and not for the exchange energy. We have actually attempted to calculate this term, but again run into difficulties with divergent terms. The divergencies in the local contribution can be suppressed (even if the arguments are not too clean), little can be done about the divergencies in the gradient term. The reason for this failure is the following: Divergencies in the exchange energy are connected with mass renormalization, which in turn is associated with the fermion self energy insertion of QED. In the effective potential picture, used in the present argument, the question of mass renormalization can not be approached by simple means. We have to tackle the next stage.

2.2 GRADIENT EXPANSION ON THE BASIS OF QED

For brevity we present only a general idea of the arguments involved (for details see Ref. 18).

The starting point is the HF-limit[19] of QED, which is expressed by the following Dyson equation for the fermion propagator

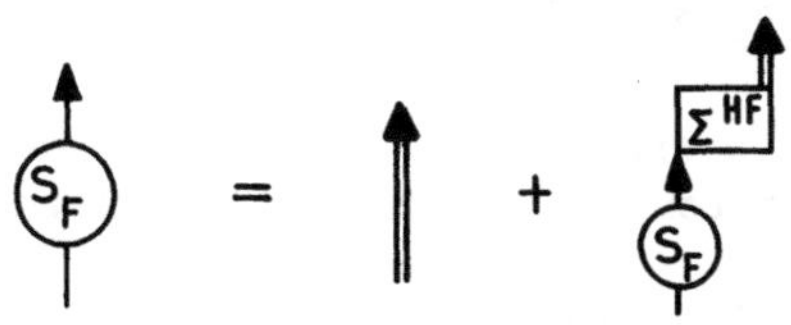

The double line stands for the propagation of the fermions in an external scalar potential. The selfenergy insertion is given by

with

$$S_F(x,y) = \langle g | T\psi(x)\bar{\psi}(y) | g \rangle$$

If one goes back to the full set of structural eqs. of QED one can state the approximations involved: The full photon propagator is replaced by the free photon propagator. All vertex corrections are neglected. The corresponding groundstate energy is

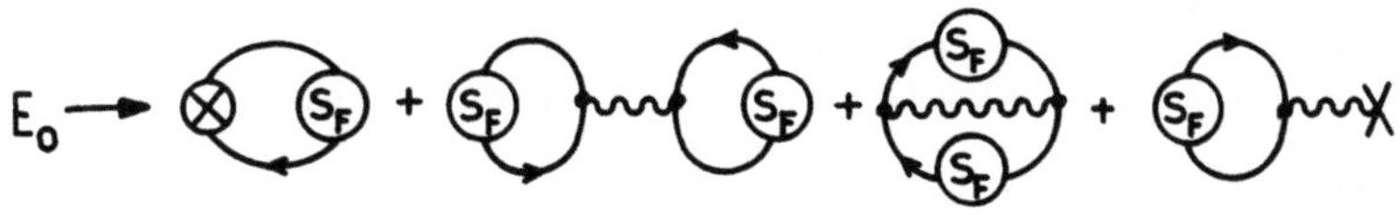

representing kinetic (or free) energy, the direct and exchange terms and the external potential energy.

The scheme could also have been taken from any text on the field theoretical treatment of the nonrelativistic many body problem. In QED there is, however, one difference. The exchange contribution is divergent and needs to be renormalized. This is the mass renormalization indicated above. It can be handled e.g. by the counterterm technique.

The relation between the fermion four current and the propagator is

$$j^{\nu}(x) = - \lim_{y \to x} [\gamma^{\nu} S_F(x,y)] \; .$$

Again one finds that this expression is not well defined due to UV divergencies. The renormalization to be applied here is the charge renormalization.

In order to set up the gradient expansion we imagine (in analogy to the simpler argument) that the HF propagator can be replaced by a propagator of an effective field theory

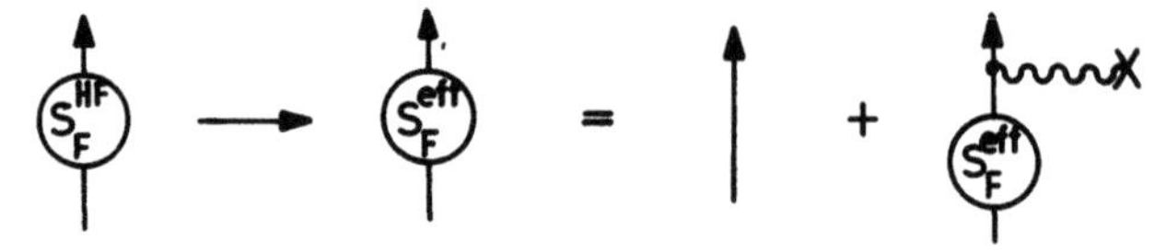

where the simple line stands for the free propagator and the insertion for an effective four potential (as e.g. envisaged in the relativistic KS scheme). The effective propagator can be expressed in terms of the solutions of the effective Dirac eqs., and, by a similar argument as used for the discussion of the density operator, be rephrased in terms of plane wave expansion. The gradient expansion can then be evaluated as usual. Insertion of the gradient expansion of the propagator into the energy expression and the expression for the four current yields the gradient expansion of these quantities. The standard inversion finally gives

$$E_o = E_o [j_{\nu}, \partial^i j_{\nu} , \ldots] \; .$$

The results for the simplest case that the effective potential has only a scalar component, can be stated as follows:

i) The kinetic energy expression is unchanged from the previous result. This then justifies our simpler renormalization argument.

ii) The gradient contribution to the exchange has not yet been calculated. The local part of the relativistic exchange energy density as a function of density is shown in Figs. 1 and 2. We first look at the low density region and compare the full relativistic case (first calculated by Jancovici[20] in the Coulomb gange) with the relativistic Coulomb contribution and the nonrelativistic exchange proportional to $-n^{4/3}$. The curves are relatively close. On a more extended density scale one notices, however, marked differences. The effect of the transverse contributions leads to a markedly different behaviour at higher densities.
For orientation one should be aware of the following maximal density values

$$Z = 50 \qquad \rho_{max} \sim 10^5 \; a_o^{-3}$$

$$Z = 100 \qquad \rho_{max} \sim 10^6 \; a_o^{-3} \; .$$

There remains the task to discuss numerical applications. We are not able to offer a full scale discussion as we are still in the midst of the numerical work. We can, however, offer some first indications. One of the simplest things one can do, is to test the quality of the kinetic energy expression

$$T_{DF} = \int d^3r [\tau_{RTF}(\underline{r}) + B(x) \frac{(\underline{\nabla}\, n)^2}{n}]$$

against the solution of the Dirac equation in a Coulomb field

$$T = \sum_{j=1}^{N} \psi_j^{\dagger}(\underline{r}) [- i\underline{\alpha}\cdot\underline{\nabla} + \beta] \psi_j(\underline{r}) - N$$

by inserting

$$n(r) = \sum_{j=1}^{N} \psi_j^{\dagger}(\underline{r})\psi_j(\underline{r})$$

into the density functional expressions. This is done in Table 1.

The zeroth order term gives an energy accurate within 5 and 10%. The inclusion of the gradient correction improves the energies consistently. The picture looks even more favourable if one compares the free energies rather than the kinetic energies. (As one actually calculates the free energy by gradient expansion, comparison of these quantities does make sense). There are two possibilities to improve the agreement:

i) One can, as is done in the nonrelativistic case, fudge the gradient term by a factor independent of N in order to represent the effect of higher order gradient terms.

(ii) For large values of N = Z one should actually include vacuum polarisation effects.

A reasonable simple density functional model is the RTFW model (no e-e interaction)

$$E_o[n] = \int d^3r [\tau_{RTF} + B(x) \frac{(\underline{\nabla} n)^2}{n}]$$

$$+ \int d^3r\, v(\underline{r}) n(\underline{r})$$

which for spherically symmetric situations leads to the Hohenberg-Kohn variational eqs.

$$(x^2+1)^{1/2} - 1 + \mu - v(\underline{r}) + \frac{1}{24} [(B + x\frac{dB}{dx})(\frac{x'}{x})^2 + 4B\frac{x'}{rx} + 2B\frac{x''}{x}] = 0.$$

The Lagrange multiplier μ, which controlls the particle number, is equivalent to the chemical potential. Naturally this model only constitutes a test case, as one might as well solve the Dirac equation directly.

The solution of the variational eq. (with suitable boundary conditions) for $\mu = 2$ (corresponding to $\varepsilon_F = -1$) gives the number of electrons that have " dived into the negative continuum". Some results for a nucleus with an extended nuclear density (a Fermi profile) and A = 2.5 Z are shown in Table 2. The results are comparable to those obtained with the Dirac equation. Similar calculations including the effects of the e-e interaction should eventually become possible.

In Table 3, we show some energy values for a neutral atom with N = Z = 90 for both the Coulomb potential and a Fermi profile nucleus with A = 2.5 Z.

If one looks at the density profile for this case (Fig. 3) one sees that (as in the nonrelativistic case) the RTFW density does not reproduce the shell structures but averages smoothly between them. One additional feature becomes apparent if one replots the density on a logarithmic scale, in order to emphasize the behaviour at the origin. The behaviour of the density at the origin is difficult to extract and is in error for the case of a point nucleus[17] . It is, however, normalizable even in this case.

Table 1. Comparison of the kinetic energy in the relativistic TF and TFW models for a noninteracting Coulomb-system with exact results (in relativistic units).

Z = N	T	T_{RTF}	T_{RTFW}
10	.01069	.01008	.01073
30	.1507	.1431	.1500
50	.5260	.4942	.5185
70	1.253	1.173	1.239
90	2.509	2.301	2.472
110	4.841	4.294	4.827

Table 2. The number of electrons below the Fermi level $\varepsilon_F = -1$ as a function of Z for neutral atoms

Z	100	120	140	160	180	200	220
N	.0002	.005	.062	.427	1.131	2.089	3.331

Table 3. Groundstate energy values in the (noninteracting) RTFW model for N = Z = 90 in relativistic units

point nucleus exact	$E_o = -2.132$
point nucleus RTFW	$E_o = -2.073$
extended nucleus RTFW	$E_o = -2.201$

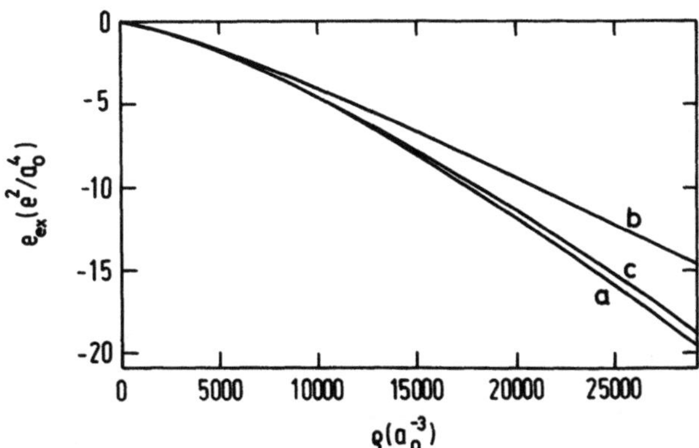

Fig. 1 The local exchange energy density as a function of density for a the nonrelativistic case, b and c the relativistic case with and without transverse contributions.

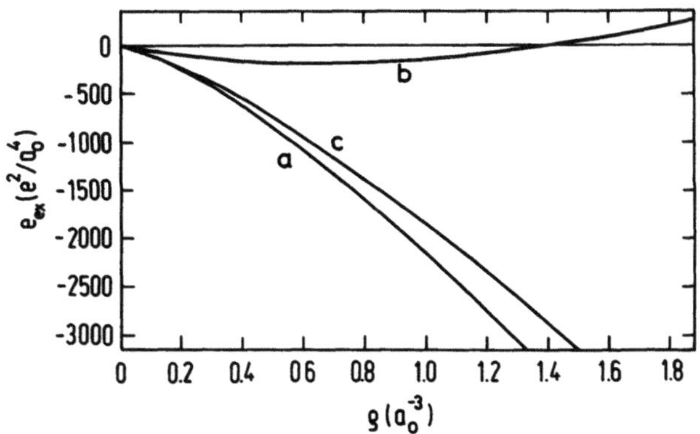

Fig.2 As Figure 1 on an extended density scale.

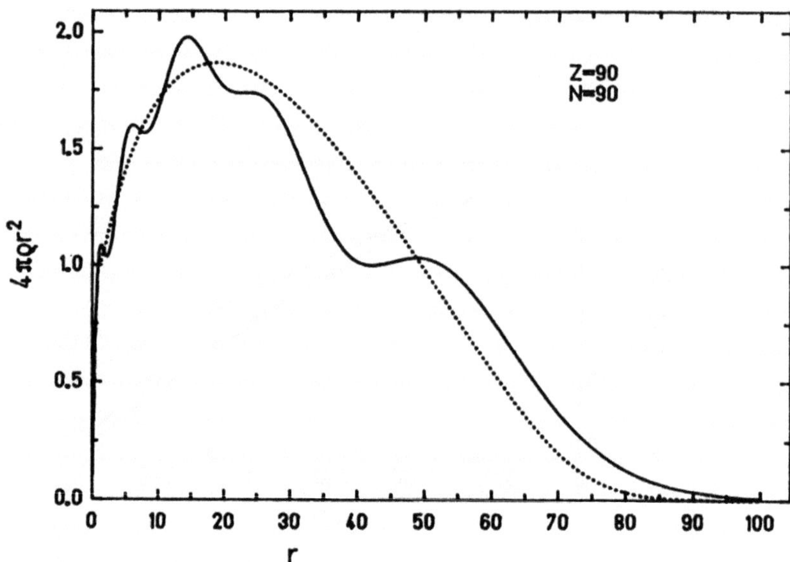

Fig. 3a The groundstate density of an atom with N = Z = 90 (point charge) in the RTFW model as a function of r. (Dirac equation(—), RTFW model (.....))

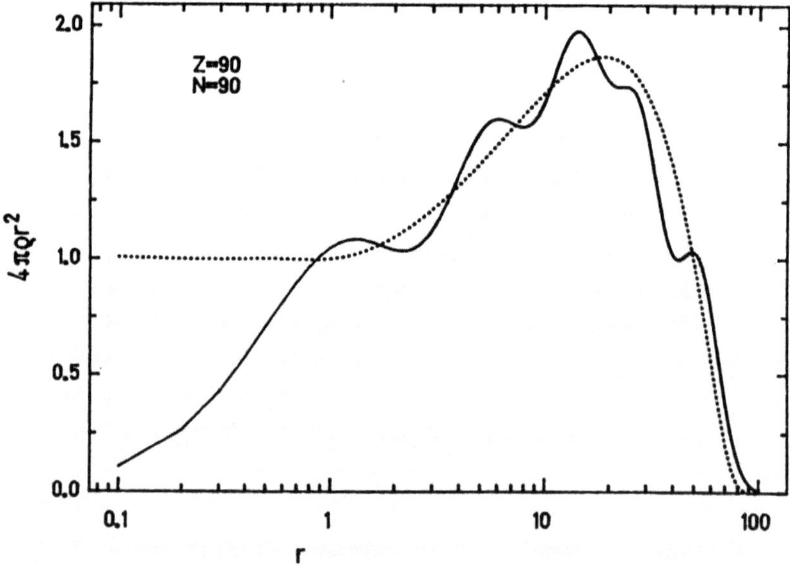

Fig. 3b The groundstate density of an atom with N = Z = 90 (point charge) in the RTFW model as a function of ln r.

References

1. See e.g.
   a) R.M. Dreizler and J. da Providencia, eds.
   " Density Functional Methods in Physics ", NATO ASI Series B 123, Plenum Press, New York (1985)
   and
   b) S. Lundqvist and N.H. March, eds.
   " Theory of the Inhomogeneous Electron Gas "
   Plenum Press, New York (1983)
2. See e.g. J.P. Desclaux, Physica scripta 21:436 (1980) and references given there
3. G.E. Brown and D.G. Ravenhall, Proc. Roy. Soc.(London),A258:552(1951)
4. P. Hohenberg and W. Kohn, Phys. Rev. 136B:864(1964)
5. W. Kohn and L.J. Sham, Phys. Rev. 140A:1133 (1965)
6. See e.g. W. Kohn and P. Vashishta in Ref. 1b, p. 79
7. See e.g. A.R. Williams and U. von Barth in Ref. 1b, p. 189
8. D.A. Kirzhnits, " Field Theoretical Methods in Many Body Theory " Pergamon Press, Oxford (1967)
9. See e.g. R.M. Dreizler and E.K.U. Gross in Ref. 1a, p. 81
10. C.H. Hodges, Can. J. Phys. 51:1428(1973); D.R. Murphy, Phys. Rev. 24A:1682 (1981)
11. See e.g. M. Brack in Ref. 1a, p. 331
12. See e.g. M.V. Ramana and A.K. Rajegopal, Adv. Chem. Phys. 64:231(1983) and references given there
13. M.S. Vallarta and N. Rosen, Phys. Rev. 41:708 (1932)
14. W.H. Furry, Phys. Rev. 81:115 (1951)
15. E.H. Wichmann and N.M. Kroll, Phys. Rev. 101:843 (1956)
16. R.M. Dreizler and E.K.U. Gross in:"QED of Strong Fields", NATO ASI Series B80, W. Greiner, ed., Plenum Press, New York(1983)
17. P. Malzacher and R.M. Dreizler, Z. Phys. D2:37 (1986)
18. E. Engel and R.M. Dreizler, submitted for publication
19. P.G. Reinhardt, W. Greiner and A. Arenhövel, Nucl. Phys.166A:173(1971)
20. B. Jancovici, Nuovo Cim. 25:428 (1962)

NONPERTURBATIVE RADIATIVE AND BOUND-STATE EQUATIONS FOR STRONG QUANTUMELECTRODYNAMICS

A. O. Barut

Department of Physics
The University of Colorado
Boulder, Co. 80309

INTRODUCTION

QED is a good theory. But QED is also an incomplete theory and a theory with limitations. Incomplete, because the renormalization procedure must be mathematically firmly justified and the convergence of the theory must be shown. And there are limitations, because primarily we wish to understand the behaviour of e^+e^--systems, for example, at short distances nonpertubatively. For these reasons I discuss the results of recent work on a reformulation of electromagnetic phenomena in terms of the central concept of "Selfenergy of the electron", rather than in terms of the two second quantized fields A_μ and Ψ. QED is so firmly established that there can only be a reformulation of certain concepts and techniques, not a new theory. It is very useful to look at physical theories from many different points of view and QED should be no exception.

QED

Almost all physical and chemical phenomena, with the exception of nuclear forces and radioactivity perhaps, can be reduced to the interaction of light with the electrons. The quantum theory of these electromagnetic interactions is based on the following postulates:

1) The theory must be relativistic (Poincare invariance of the set of equations).
2) The standard probability interpretation of quantum theory holds (Hermitian total Hamiltonian, Unitarity of the S-matrix).
3) The interaction is local (minimal U(1)-coupling of the Dirac current with the electromagnetic field).
4) Separately existing and second quantized electromagnetic and matter fields.

Unfortunately we do not have a closed finite theory satisfying these principles. Either the theory is infinite, or if one forces it to be finite one violates one of the above postulates, for example, the introduction of a cut-off. What we have however is a perturbative QED. This is a simple and powerful procedure to reduce the amplitude of any process involving electrons and photons into a sum of combinations of elementary processes. There are three elementary pieces in the action integral: The propagation of a photon (plane wave) from a space-time point to another; the propagation of an electron (plane wave) from a point to another, and one basic interaction at a point of strength e - the emission or absorption of light by the electron. One has the rules for combinatorial repetitions of these three actions and an expansion in powers of the interaction e. We then add all possible amplitudes for a process and then sqare it to find the probability of the process. In the lowest order amplitudes everything is fine and one obtains finite results as a function of two paramers only, e and mass of the electron m. In the next order there are three types of infinite integrals. For example, for the scattering of an electron from an external field, the three infinite terms are shown in Fig. 1 (external field is denoted by X).

These infinities come from the lower or upper limit of the integration of the virtual photon in the loop. Thus, as the theory stands, virtual photons of very high (ultraviolet) and very low (infrared) momenta are not correctly accounted by the pertubative QED. The procedure of "Pertubative Renormaliza-

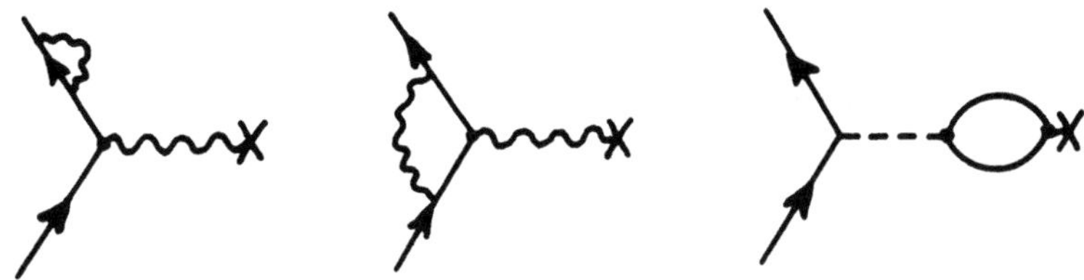

Fig. 1

tion" puts cut-off's to these integrals at λ_{min} and Λ_{max}, separates finite and infinite terms and absorbs the three infinite terms A, B, C into a redefinition of the parameters e and m.

One has however to treat the infinite terms as though they were of order α (i. e. $\Lambda^2 << \Lambda$), and reexpress in the finite terms the bare values of the parameter e_o, m_o by the renormalized values e_r, m_r also to order α. This process is repeated at every order of the perturbation theory. No new infinites arise, and the procedure of absorbing the infinities into e and m can be made at every order, more and more changing e and m. But the convergence of the renormalized series is not known.

The attitudes about the renormalization varies from "the best that one can do", to being called a "bad theory" by Dirac who himself had invented field quantization. In recent years, renormalizability is considered to be the ultimate aim of many new field theories of weak, strong and quantum-gravitational interactions. One goes to great lengths and pays the price of introducing large number of new hypothetical fields just to be able to make the theory renormalizable. Therefore, it is worth to be reminded again and again of the issue of renormalizabili ty, and of the need to have a finite theory to begin with, once thought to be one of the major problems of theoretical physics.

The practical limitations of perturbative QED arise once in the existence of very large number of graphs and the errors associated with them. For example, in (g-2) calculation to order $(\alpha/\pi)^4$ there are "891 graphs, each integrand is composed of 5000-to-/15000 terms, some with 10 variables of integration, with intricate subtractions and overlapping and nested divergences"[1]. A limitation will come when the errors in n^{th} order are bigger than the whole $(n+1)^{th}$-term.

A more immediate limitation comes in the bound-state problems. Here one has to start from a 3-dimensional bound-state equation (Schrödinger- or Dirac-type equation), and bound-state QED is "not at all mechanical, ... it is an art".[2] We shall come back to this point. Finally, more crucially, there is the unknown territory electromagnetic interactions at short distances, where the radiative corrections grow and it is not sufficient to take just a few loop graphs; one needs non-perturbative techniques.

In the next Section I describe an approach to QED which omits the fourth postulate, namely the separate existence of a <u>quantized</u> radiation field A_μ. Since it is much simpler and more direct and makes fewer postulates than QED, the new approach should be compared with experiment. It is simply based on the fact that we can eliminate A_μ between the coupled Maxwell-Dirac equations.

QED BASED ON SELF ENERGY

Electrodynamics, as formulated by Lorentz, is a selfconsistent system of Maxwell equations $F_{\mu\nu}{}^{,\nu}=j_\mu$ coupled to an equation for matter whose current is j_μ. In a particular gauge $A^\mu{}_{,\mu}=0$, we have $\Box A_\mu=j_\mu$, and can solve for A_μ:

$$A_\mu^{\text{self}}(x)=\int dy\, D_{\mu\nu}(x-y)j^\nu(y) \tag{1}$$

with suitable boundary conditions to be specified. We can than insert (1) into the classical, Schrödinger or Dirac (or any other matter) equation of motion:

$$m\ddot{x}_\mu = e\, F^{ext}_{\mu\nu}\dot{x}^\nu + e\, F^{self}_{\mu\nu}\dot{x}^\nu \tag{2}$$

or

$$[\gamma^\mu(i\partial_\mu - eA^{ext}_\mu) - m]\Psi = \gamma^\mu\Psi(x)\int dy\, D_{\mu\nu}(x-y)\bar{\Psi}\gamma^\nu\Psi(y). \tag{3}$$

Here A^{ext}_μ or $F^{ext}_{\mu\nu}$ is a fixed given field whose sources are not dynamical variables.

This seemingly simple step has far reaching consequences on the nature of light:

- All light originates from electrons
- Light carries information about its source (current $j_\mu(x)$) and light emitted by a source depends on the environment (on the Green function D(x-y) of the environment.)
- Quantized properties of light reflect quantized properties of source.
- There are different kinds of lights which we can experimentally distinghish.
- The concept of the mathematical photon looses its individuality.
- Since A_μ has been eliminated, no separate existence of a quantized radiation field needs to be postulated.
- There maybe a danger of double counting the degrees of freedom if we include both the degrees of freedom of the source and of the field independently, for example, in the quantum 2-body problem.

It is generally stated that one can not formulate a number of quantum effects, such as spontaneous emission, vacuum polarization, Casimir effect, without the quantization of the electromagnetic field separately. I shall now discuss how these effects arise in a theory based solely on first quantized Ψ-field with the elimination of A_μ.

First let us consider the classical radiation theory given by eq. (2) after which the quantum theory based on self-energy is modelled. The self-energy term $eF^{self}_{\mu\upsilon}\dot{x}^{\upsilon}$, using the potential (1) can be analyzed, for example by analytic continuation[3], and can be transformed into

$$-\delta m\ddot{x}_{\mu}+\frac{2}{3}e^{2}(\dddot{x}_{\mu}+(\ddot{x})^{2}\dot{x}_{\mu})$$

The first term can be absorbed into the left-handside of eq. (2), it is an inertial term, and we are left with the correct, finite, nonperturbative term.

$$(m+\delta m)\ddot{z}_{\mu}=e\ F^{ext}_{\mu\upsilon}\dot{z}^{\upsilon}+\frac{2}{3}e^{2}(\dddot{x}_{\mu}+(\ddot{x})^{2}\dot{x}_{\mu}). \qquad (4)$$

This procedure is still not completely satisfactory. The self-potential is a generalized function, and one should use the Riesz potential, rather than the Lienard-Wiechert potential. The Riesz potential depends on a complex parameter α and is given by[4]

$$A^{\alpha}_{\mu}(x)=\frac{1}{H(\alpha)}\int_{D^{P}_{\Lambda}} dy(x-y)^{\alpha-4}j_{\mu}(y) \qquad (5)$$

$$H(\alpha)=2^{\alpha-1}\pi\Gamma(\frac{\alpha}{2})\Gamma(\frac{\alpha-2}{2}).$$

Here D^{P}_{Λ} is the inside of the backward light-cone with x at the vertex: For the world-line of a point particle the corresponding self field is

$$F^{\alpha}_{\mu\upsilon}(x)=e\frac{(\alpha-4)}{H(\alpha)}\int_{-\infty}^{1\cdot cs_{o}} (x-y)^{\alpha-6}[(x-y)_{\mu}u_{\upsilon}-(x-y)_{\upsilon}u_{\mu}]dT \qquad (6)$$

where s_{o} is the intersection of the trajectory with D^{P}_{Λ} and $u_{\mu}=\dot{x}_{\mu}$ the velocity of the particle. If we take the limit $\alpha\to 2+0$, we find the correct and finite $F_{\mu\upsilon}$, and eq. (4) without the δm-term. Thus, the classical theory based on self energy is non perturbative and finite. The problem of infinities may very well lie in the proper definition of self energy integrals.

We would like to apply the same procedure now to the Dirac equation with self-field[5], eq. (3). It is more convenient, for this as well as for the 2-body problem, to work with the action integral directly, rather than the equations of motion. The action has the physical interpretation of energy of bound states, and of invariant scattering amplitude for scattering problems, hence can be immediately connected with the observable quantities.

The action $W = \int d^4x\, L$ of a number of DIRAC fields $\Psi_i(x)$ coupled to the EM-field $A_\mu(x)$ can be transformed by using eq. (1) into an action-at-a-distance form where only the sources appear:

$$W=\sum_j \int dx \bar{\Psi}_j(\gamma^\mu i\partial_\mu - m_j)\Psi_j + \frac{1}{2}\sum_{j,k} e_j e_k j_j^\mu(x) D_{\mu\nu}(x-y) j_k^\nu(y)\,dxdy \qquad (7)$$

we have also transformed the field action $-\frac{1}{4}F_{\mu\nu}F^{\mu\nu}$ as well as the interaction $j^\mu(x)A_\mu(x)$. In particular, for a 2-body system we obtain

$$W= \sum_{j=1,2} \int dx \bar{\Psi}_j(\gamma^\mu i\partial_\mu - m_j)\Psi_j + e_1 e_2 \int dxdy\; j_1^\mu(x)D(x-y)j_{2\mu}(y)$$

$$+\frac{1}{2}e_1^2\int dxdy\; j_1^\mu(x)D(x-y)j_{1\mu}(y) + \frac{1}{2}e_2^2\int dxdy\; j_2^\mu(x)D(x-y)j_{2\mu}(y) \qquad (8)$$

As a special case, if the second particle is very heavy, $m_2 \to \infty$, we can omit the kinetic energy, and self-energy of m_2 (they are fixed); the mutual interaction term proportional to $e_1 e_2$ becomes an external potential $A_\mu^{ext}(1)$, hence

$$W \to \int dx \bar{\Psi}_1(\gamma^\mu(i\partial_\mu - e_1 A_\mu^{ext}) - m_1)\Psi_1$$

$$+\frac{1}{2}e_1^2 \int dxdy \bar{\Psi}_1 \gamma^\mu \Psi_1 D(x-y) \bar{\Psi}_1 \gamma_\mu \Psi_1(y). \qquad (9)$$

The variation of W with respect to $\bar{\Psi}_1$ then gives eq. (3).

We remark here that Schrödinger already in 1926 after his four papers on ware mechanics wanted to introduce an additional self energy potential into his equation which he considered to be incomplete[6], essentially from considerations of the energy momentum conservation of the coupled field-matter system.

We shall now study the system (9) first and then return to (8).

A Fourier transform of $\Psi(x,t)$

$$\Psi(x,t)=\sum_n \Psi_n(\vec{x},t)e^{-iE_n t} \tag{10}$$

when inserted into (9) and time integrations performed using the representation

$$D_{\mu\upsilon}(x-y)=g_{\mu\upsilon}\left[(-\frac{1}{(2\pi)4}\right]P\int dk\frac{e^{ik(x-y)}}{k^2} + \frac{i\pi}{2k}(\delta(k^o+k)+\delta(k_o-k)) \tag{11}$$

of the causal Green's function, gives

$$\epsilon=\sum_n\int d\vec{x}\Psi_n^+(\gamma^o E_n-\vec{\gamma}\cdot\vec{p}+e\vec{A}^{ext}-m)\Psi_n$$

$$-\frac{e^2}{2}\sum_n\int d\vec{x}\bar{\Psi}_n(\bar{x})\gamma^\mu\Psi_n(\bar{x})P\int\frac{d\vec{k}}{(2\pi)^3}\ \frac{e^{i\vec{k}\cdot(\vec{x}-\vec{y})}}{k^2}\int d\vec{y}\sum_s\Psi_s(\vec{y})\gamma_\mu\Psi_s(\vec{y})$$

$$+\frac{e^2}{2}\sum_{n,m}\int d\vec{x}\bar{\Psi}_n(\vec{x})\gamma^\mu\Psi_m(\vec{x})\int\frac{d\vec{k}}{(2\pi)^3}e^{i\vec{k}\cdot(\vec{x}-\vec{y})}\frac{i\pi}{2k}[\delta(E_m-E_n+k)+\delta(E_m-E_n-k)]$$

$$\cdot\int d\vec{y}\bar{\Psi}_m(\vec{y})\gamma_\mu\Psi_n(\vec{y})$$

$$+\frac{e^2}{2}\sum_{n,m}\int d\vec{x}\bar{\Psi}_n(\vec{x})\gamma^\mu\Psi_m(\bar{x})P\int\frac{d\vec{k}}{(2\pi)^3}e^{i\vec{k}\cdot(\vec{x}-\vec{y})}\left[\frac{1}{E_m-E_n-k} - \frac{1}{E_m-E_n+k}\right]$$

$$\cdot \int d\vec{y}\bar{\Psi}_m(\vec{y})\gamma_\mu \Psi_n(\vec{y}) \quad (12)$$

The relation between W and ϵ is so that we have cancelled a $\delta(E)$-factor from both sides of (9); E has the dimension of energy, W that of an action. The action W of QED vanishes identically when the equations of motion are inserted in. We see from eq. (12) that, without the self energy terms; i. e. with the first two terms alone, the Ψ_n are the solutions of the Dirac equation in the given external field A_μ^{ext}. We also see the physical meaning of the Fourier components in the expansion (10). In the presence of self energy terms the unperturbed energies and states are not the exact stationary energies or wave functions. The problem is to find a set of E_n and Ψ_n which would make ϵ vanish. These would be then an exact nonperturbative solution of QED for this special problem. In the absence of such an exact solution, we solve (12) by iteration. Anticipating that the correctly evaluated self-energy contributions will be small, we choose Ψ_n to be the solutions $\Psi_n^{(o)}$ of the external field problem, and set $E_n = \Delta E_n^{(o)}$, and immediately obtain from the vanishing of E the following energy shift formula for a level n:

$$\Delta E_n = \frac{e^2}{2}\int d\vec{x}\bar{\Psi}_n(\vec{x})\gamma^\mu \Psi_n(\vec{x}) P\int \frac{d\vec{k}}{(2\pi)^3}\,\frac{e^{i\vec{k}\cdot(\vec{x}-\vec{y})}}{k^2}\sum_s \int d\vec{y}\bar{\Psi}_s(\vec{y})\gamma_\mu \Psi_s(\vec{y})$$

$$-\frac{e^2}{2}\sum_{\substack{m \\ m<n}}\int d\vec{x}\bar{\Psi}_n(\vec{x})\gamma^\mu \Psi_m(\vec{x})\int \frac{d\vec{k}}{(2\pi)^3}\, e^{i\vec{k}\cdot(\vec{x}-\vec{y})}\frac{i\pi}{2k}\delta(E_m - E_n - k)$$

$$\cdot \int d\vec{y}\,\bar{\Psi}_m(\vec{y})\gamma^\mu \Psi_n(\vec{y})$$

$$-\frac{e^2}{2}\sum_m \int d\vec{x}\,\bar{\Psi}_n(\vec{x})\gamma^\mu \Psi_m(\vec{x}) P\int \frac{d\vec{k}}{(2\pi)^3} e^{i\vec{k}\cdot(\vec{x}-\vec{y})}\cdot$$

$$\left[\frac{1}{E_m - E_n - k} - \frac{1}{E_m - E_n + k}\right]\int d\vec{y}\,\bar{\Psi}_m(\vec{y})\gamma_\mu \Psi_n(\vec{y}) \quad (13)$$

The meaning of the three terms in eq. (13) is as follows. The first term represents the analogue of vacuum polarization; it is the observable electrostatic energy of all positive and negative energy levels. The second term gives an imaginary contribution to ΔE and describes the spontaneous emission, rather the width of excited levels. Finally the third term is the real self- energy shift (or the so called Bethe-term). The contribution of the anomalous magnetic moment is included in the third term and can be easily extracted. It is thus remarkable that all radiative effects occur simultaneously in a single formula. It is easy to see that these formulas reduce to their well-known form in the nonrelativistic limit. Only the ground state of the system is stable, all other excited states decay. That means the exact Hamiltonian has only a continuous spectrum.

In the evaluation of the sums and integrals in (13), where the infinities of the standard QED appear, the situation is now as follows. Since we are working with localized wave functions, for example, relativistic Coulomb wave functions, space-integrals over $d\vec{r}$, $d\vec{r}'$ are finite and there are no infrared divergences. The infinities may come a priori from summing over an infinite set of continuum states.

With respect to summing over an infinite set of states our functions are distributions or generalized functions. We have found that such functions are properly defined only by their proper poles and residues in the Mellin transform plane. [Special functions of mathematical physics seem to be defined in this way(Fox functions[7])]. We shall explain this procedure in the example of the vacuum polarization, the first term of eq. (13)[8]. This term is more crucial because it is linearly divergent in QED, whereas the third term (self-energy) is only logarithmically divergent. Vacuum polarization in Coulomb field was treated in a classic paper by Wichman and Kroll[9]. It involves an integral over the positive and negative energy continuum. After performing all the $\vec{k}$, $\vec{r}$, $\vec{r}'$, integrations, we make a Mellin transformation in the energy variable before integrating over the energies. In the Mellin variable s[1.c], we find two kinds of poles on the real axis. A set of fixed spurious poles at s=0,1,2,3,..., and a set of physical poles depending on the quantum number of states. One realizes immediately that the infinities come from the fixed poles, whereas the physical poles give correctly the observable vacuum polarization contributions in powers of $\alpha(z\alpha)^{2\gamma_{n+1}}$, $\alpha(z\alpha)^{2\gamma_{n+2}}$,... Now the spurious poles can easily be avoided by introducing a convergence factor e^{-er} into the integrals, and the final results do not depend on e.

We have thus outlined two novel issues in QED. One is the formulation of radiative effects in terms of self-energy - without the apparatus of field quantization and operator formalism. The other is the proper definition of self-energy integrals in terms of Mellin transformation which results in a finite theory with no renormalization of the two fundamental parameters of the theory, e and m.

TWO BODY QED

We shall now discuss the bound-state problems together with the radiative effects in a nonperturbative way. The goal is to have a covariant two or many-body equation in which not only the mutual interaction between particles but also the self energy effects occur as radiative potentials. This will lead in particular to a formulation of Lamb-shift in a relativistic system like positronium, for which to my knowledge no theory exists so far.

We return to the 2-body action (8). We multiply each kinetic energy term in (8) with the mormalization integral of the other particle; for example the first term with $\int dy\bar{\Psi}_2 n^{\mu}\gamma_{\mu}\Psi_2\delta(y\cdot n-t)=\int d\vec{y}\Psi_2^{+}\Psi_2=1$. Similarily we multiply the self-energy third and 4th-terms of (8) <u>twice</u> with the normalization integrals of the other particle. The purpose is to be able to write (8) entirely in terms of the composed field

$$\Phi(x,y)=\Psi_1(x)\Psi_2(y).$$

Because of the D-functions $\Phi(x,y)$ is defined only for pair of points x and y which can be connected by a light ray. The action (8) becomes

$$W=\int dx d\vec{y}\bar{\Phi}(x,y)[(\gamma^{\mu}p_{1\mu}-m_1)\otimes\gamma\cdot n+\gamma\cdot n\otimes(\gamma^{\mu}p_{2_{\mu}}-m_2)$$

$$+e_1e_2\gamma^{\mu}\otimes\gamma_{\mu}D(x-y)+\frac{1}{2}e_1^2\gamma^{\mu}\otimes\gamma\cdot n\int dz du D(x-z)\bar{\Phi}(z,u)\gamma_{\mu}\otimes\gamma\cdot n\Phi(z,u)$$

$$+\frac{1}{2}e_2^2\gamma\cdot n\otimes\gamma^{\mu}\int dz du D(y-u)\bar{\Phi}(z,u)\gamma\cdot n\otimes\gamma_{\mu}\Phi(z,u)]\Phi(x,y) \qquad (14)$$

where dy and du are actually by construction 3-dimensional integrals over space-like surfaces. We see here the Möller potential $e_1e_2\gamma^{\mu}\otimes\gamma_{\mu}D(x-y)$, as well as the nonlinear and nonlocal potentials proportional to $\frac{1}{2}e_1^2$ and $\frac{1}{2}e_2^2$.

The variation of W with respect to $\bar{\Phi}(x,y)$ gives a covariant 2-body equation[10]. The center of mass and relative coordinates are exactly separable, because W can be written as

$$W=\int dR d\vec{r}\bar{\Phi}(R,\vec{r})[\Gamma^{\mu}P_{\mu}-\vec{k}\cdot\vec{p}-e_1e_2\frac{\gamma^{\mu}\cdot\gamma_{\mu}}{r}-k_m-\gamma^{\mu}\cdot\gamma^{o}A_{\mu}^{self}(1)$$

$$-\gamma^{o}\cdot\gamma^{m}A_{\mu}^{self}(2)]\Phi(R,\vec{r}) \qquad (15)$$

where

$$\Gamma^{\mu}=a\gamma^{\mu}\otimes\gamma\cdot n+(1-a)\gamma\cdot n\otimes\gamma^{\mu}$$

$$\vec{k}=\vec{\gamma}\otimes\gamma\cdot n-\gamma\cdot n\otimes\vec{\gamma}$$

$$k_m=m_1 I\otimes\gamma^{o}+m_2\gamma^{o}\otimes I$$

$$A_{\mu(1)}^{self}=\int dz d\vec{u}D(x-z)\bar{\Phi}(z,u)\gamma_{\mu}\cdot\gamma_{o}\Phi(z,u) \qquad (16)$$

$$A_{\mu(2)}^{self}=\int dz d\vec{u}D(y-u)\bar{\Phi}(z,u)\gamma_{o}\cdot\gamma_{\mu}\Phi(z,u)$$

where the center of mass and relative coordinates are defined as follows (as four vectors)

$$R^{o}=ax_1+(1-a)x_2 \ ; \ a=\frac{m_1}{(m_1+m_2)}$$

$$r=x_1-x_2$$

$$P=p_1+p_2 \tag{17}$$

$$p=(1-a)p_1-ap_2.$$

We expand $\phi(R,\vec{r})$ into a Fourier series

$$\phi(R,\vec{r})=\Sigma u_n(P_n)e^{iP_n\cdot R}\Psi_n(\vec{r}) \tag{18}$$

and obtain after some calculation

$$W=\sum_{nmrs}\delta^4(P_n-P_m+P_r-P_s)\frac{d^3P_n}{(2\pi)^3}\frac{d^3P_m}{(2\pi)^3}\frac{d^3P_r}{(2\pi)^3}\frac{d^3P_s}{(2\pi)^3}$$

$$\Big[\bar{u}_n(P_n)\int d\vec{r}\bar{\Psi}_n(\Gamma^{\mu}P_{\mu}+H_{rel})\Psi_m(\vec{r})u_m(P_m)$$

$$+\frac{1}{2}e_1^2\int\bar{u}_n(P_n)\frac{\overset{(1)}{T}{}_{\mu}^{nm}(\bar{k})u_m\bar{u}_r\overset{(1)}{T}{}_{rs}^{\mu}(-\bar{k})}{(E_n-E_m)^2-\bar{k}_2}u_s(P_s)$$

$$+\frac{1}{2}e_2^2\Big[\bar{u}_n(P_n)\frac{\overset{(2)}{T}{}_{\mu}^{nm}(\bar{k})u_m\bar{u}_r\overset{(2)}{T}{}_{rs}^{\mu}(-\bar{k})}{(E_n-E_m)^2-\bar{k}^2}u_s(P_s)\Big] \tag{19}$$

where we have put $\vec{k}=\vec{P}_m$, $H_{rel}=\vec{k}\cdot\vec{p}+k_m+e_1e_2\frac{\gamma_{\mu}\otimes\gamma^{\mu}}{r}$ and defined the formfactors

$$\overset{(1)}{T}{}_{\mu}^{nm}\equiv\int d\vec{r}\bar{\Psi}_n(\vec{r})\gamma_{\mu}\cdot\gamma^{o}e^{-i\vec{k}\cdot\vec{r}(1-a)}\Psi_m(\vec{r})$$

$$\overset{(2)}{T}{}_{\mu}^{nm}\equiv\int d\vec{r}\bar{\Psi}_n(\vec{r})\gamma_{o}\cdot\gamma_{\mu}\Psi_m(\vec{r})e^{-i\vec{k}\cdot\vec{r}a}$$

Equation (19) has two important limits:

(A) limit to unbound particles and to Feynman graphs: Here we disregard the relative variables $\vec{r}$, $\Psi_n(\vec{r})\to 1$, $U\to U_1(P_1)U_2(P_2)$ (direct product of 2 particles). Then the selfenergy of particle 1, for example, is one of the terms in W

$$W_1^{self}=\frac{1}{2}e_1^2\sum_m \frac{d^3P_m}{(2\pi)^3}\bar{u}_{1n}\gamma^\mu u_{1m}\gamma_\mu u_{1n}\frac{1}{(P_n-P_m)^2} \tag{20}$$

(B) The second limit is to the fixed center problem (H-Atom). Since the center of mass is fixed, we take $U\to 1$, $m_2\to\infty$, $\gamma^\mu\cdot\gamma^o\to\gamma^\mu\cdot 1$, $a\to 0$, $P_n\to(E_n,\vec{o})$, $P_m=(E_m,\vec{k})$ (intermediate state).

$$W=\sum_{nmrs}\delta(E_n-E_m+E_r-E_s)\int\delta^3(-\vec{P}_m-\vec{k})\frac{d^3P_m}{(2\pi)^3}\delta^3(\vec{P}_r-\vec{k})\frac{d^3P_r}{(2\pi)^3}$$

$$\int d\vec{r}\bar{\Psi}_n(\gamma^o E_m+H_{rel})\Psi_m$$

$$+\frac{1}{2}e_1^2\int d\vec{r}\bar{\Psi}_n(\vec{r})\gamma^\mu\Psi_m(\vec{r})d\vec{k}\frac{e^{i\vec{k}\cdot(\vec{r}-\vec{r}\,')}}{(E_n-E_n)^2-\vec{k}_2}\bar{\Psi}_r(\vec{r})\gamma_\mu\Psi_s(\vec{r}\,')d\vec{r}\,' \tag{21}$$

which is precisely our eq. (12). The last term simply is

$$\frac{1}{2}e_1^2\int d\vec{k}\frac{T^\mu_{nm}(\vec{k})\cdot T^{rs}_\mu(-\vec{k})}{(E_n-E_m)^2-\vec{k}^2} .$$

Eq. (19) is the basis of a fully relativistic nonperturbative treatment of a two-body system, like positronium, including the radiative effects, like Lamb-shift, vacuum polarization and spontaneous emission as (non-local) potential terms in the wave equation. We may solve it iteratively, as in eq. (13), by inserting the solutions of the 2-body problem, i. e.

$$(\Gamma^\mu P_\mu+H_{rel})\Psi_m^{(o)}(\vec{r})u_m^{(o)}(P)=0$$

This first iteration is probably sufficient for the accuracy of experiments at the present time.

CONCLUSIONS

I have discussed a straightforward approach to radiative effects in QED in the framework of the first-quantized Dirac fields which is based on the concept of self-energy of the electron. The electromagnetic field A_μ has been eliminated alltogether, and the formalism of quantized field operators is not used.

How does this approach compare with the perturbative QED? We have used localized bound-state wave function everywhere, hence there are no infrared difficulties and individual integrals over bound states converge. We have results to all order in $(Z\alpha)$, the coupling to the external field. In this sense this compares with the Furry picture. We have further used an iterative procedure and calculated energy shifts $\Delta E^{(o)}$ with the external field wave functions $\Psi_n^{(o)}$. Next one must calculate with the shifted self energy potentials the new wave functions $\Psi_n^{(1)}$ and then calculate the new energy shifts $\Delta E^{(1)}$. It is very difficult to see at this time how this procedure corresponds to higher order Feynman graphs. The best strategy would be to calculate the next iteration in a problem like (g-2) where we know the higher order Feynman graphs and compare it directly with experiment.

The non-perturbative 2-body equation that we have derived by the same method as the self energy has a number of remarkable properties: It is relativistic; accounts fully for spin and recoil properties of both particles; the center of mass is exactly separable; the angular and radial parts are exactly separable. The major part of the potential is exactly soluble. When applied to the spectra of positronium, muonium and hydrogen[11], we have obrained agreement with QED up to order α^5. The evalutation of self-energy terms is in progress.

REFERENCES

1. T. Kinoshita, and W.B. Lindquist, Phy. Rev. Lett. 47, 1573 (1981)

2. G.T. Bodwin, D.R. Yennie and M.H. Gregorio, Rev. Mod. Phys. 57, 723 (1985)

3. A.O. Barut, Phys. Rev. D10, 3335 (1974)

4. M. Riesz, Acta Mathematica, 81, 1-223 (1949)

5. A.O. Barut and J. Kraus, Found of Phys. 13, 189 (1983)
A.O. Barut and J.F. van Huele, Phys. Rev. A32, 3187 (1985)

6. E. Schrödinger, Ann d. Physik 82, 265 (1926)

7. G. Fox, Trans. Amer. Math. Soc. 98, 395 (1961) B.L.J. Braaksma, Composito Math. 15, 239 (1964)

8. A.O. Barut and J. Kraus, "Nonperturbative QED without Infinities". Submitted for publication.

9. E. Wichman and N. Kroll, Phys. Rev. 101, 83 (1956)

10. A.O. Barut, in Lectures Notes in Physics, Vol. 180, (Springer 1983), p. 33
A.O. Barut and S, Komy, Fortschritte der Physik 33, 309 (1985)
A.O. Barut and N. Ünal, Fortschritte der Physik 33, 319 (1985)

11. A.O. Barut and N. Ünal, New Approach to Bound-State QED I and II. Spectra of positronium, muonium and Hydrogen; ICTP preprint IC/86/119; Physica (in press) and J. Math. Phys. (in press, Dec. 1986)

MAGNETIC RESONANCES AND THE POSITRON PEAK IN HEAVY-ION COLLISIONS

Asim O. Barut

Department of Physics
The University of Colorado
Boulder, Co. 80309

INTRODUCTION

In this lecture I discuss the possibility of magnetic dipole resonances in the e^+e^--system just above the two-body threshold. Wong and Becker[1] have recently applied this idea in order to explain the positron peak observed in heavy-ion collisions[2]. Can the e^+e^--system form a magnetic resonance with mass around $M \cong 3m$ and a width of about $\tau \sim 10^{-18}$sec.? I shall examine the assumptions underlying this model more precisely, the nature of the neglected terms, a simple derivation and interpretation of the mass formula and the possibility of other higher mass sharp resonances in the (e^+e^-)-system.

MAGNETIC RESONANCES

Magnetic resonances are a direct consequence of the motion of a charge, in the simplest case, in the field of a localized magnetic dipole moment. More generally we may consider two particles both with charges and magnetic moments. The existence of resonances can be easily seen in many models,[3] (classical, semi-classical, relativistic and nonrelativistic quantum models) as long as a point magnetic dipole moment is assumed. The trapping of charged particles in the dipole magnetic field of the earth is a familiar macroscopic example. Intuitively speaking the more singular spin-orbit and spin-spin forces produce an attractive force at short distances followed by a repulsion at still shorter distances, and new resonance states occur in the potential well so produced.

The crucial question is how localized are the magnetic moments in the relativistic case. In the first order Dirac equation there is no singular potential associated with the normal magnetic moment; only in the second order Dirac equation and in nonrelativistic limit the electron appears to have the vector potential $\vec{A} = \mu\frac{\vec{\sigma}\times\vec{r}}{r^3}$ of a localized magnetic dipole. For the two-body problem, the Möller-potential in the first order equation, shows that each particle, in addition to the Coulomb force, sees a vector potential of the form $A_1 = e_1e_2\frac{\vec{\alpha}_2}{r}$, where $\vec{\alpha}_2$ is the Dirac velocity matrix of the other particle. This is a new relativistic magnetic force which vanishes in the nonrelativistic limit, which is however not as singular as a dipole potential. In addition, we have self-energy effects.

It is well known that self-energy gives in the lowest order an anomalous magnetic moment singular potential[4] of the form $\vec{A} = a\frac{\sigma\wedge\vec{r}}{r^3}$ with $a=(\alpha/2\pi)\frac{e}{2m}$.

Unfortunately we do not know the anomalous magnetic moment at higher orders or nonperturbatively. More important is the fact that the anomalous magnetic moment depends on the external field. The general question which includes the above problem is the magnitude of self-energy forces between two particles in addition to Coulomb Breit mutual forces. Self-energy forces are well-known and measured in H-atom or positronium, but they are small at atomic distances. For more tightly bound systems, for example for very large Z-atoms they will become relatively large. In addition to causing energy shifts to the Coulomb levels (e.g. Lamb shift) the self-energy forces may actually dominate at short distances and form new states, namely the magnetic resonances. A simple physical way to observe this is as follows. The self-magnetic field produced by an electron in circular orbit or radius r_o, $B^{self} \sim e/r_o^2$, becomes equal to the cyclotron magnetic field $B^{cyc} \sim \frac{mv}{er_o}$, $(v_o \simeq c)$, necessary to keep a charge going around the circle, at $r \simeq \alpha/m = 2.8$ Fermis. Applied to the e^+e^--system this means that the field produced by e^- at $r=\alpha/m$ is sufficient to keep e^+ in that same orbit[5], and vice versa. I have called this "the magnetic self-focusing principle" The corresponding metastable states are the soliton-like solutions of the nonlinear Dirac equation with self-energy [6]. Such solutions are well-known for the nonlinear Schrödinger equation. At distances of the order of Bohr radius $r_B \sim 1/\alpha m$, the self-field is very small: $B^{self} = \alpha^2 B^{cycl.}$, corresponding to an anomalous magnetic moment correction due to binding of the order of $\alpha^2(\frac{e}{2m})$. Whereas at distances $r_o \sim \alpha/m$, anomalous magnetic monument correction due to binding becomes of the order 1 (in units of e/2m). This means that the anomalous magnetic moment becomes equal to the total magnetic moment and the self-organization solution becomes possible. Note that the normal magnetic moment as given by the minimal coupling in the Dirac equation decreases with energy as $\mu_{Normal} \sim \frac{e}{2E}$. This leads us to conjecture that the total magnetic moment of the bound electron remains about constant.

To conclude, we have showed on physical grounds that the self energy effects between, for example, e^+ and e^- at very short distances ($r \sim$ 1-2 fermis) becomes very much larger than the Coulomb interactions and dominant. Expressed as an effective anomalous magnetic moment these effects can produce new localized metastable states of mass >> 2m, the magnetic resonances.

MODELS

Since the self energy potentials are rather non-local and complicated, we may start with a relativistic (in fact covariant) 2-body equation[7] which can be derived from field theory in which self energy effect are included as electric and anomalous magnetic form factors : e and a. (Dirac and Pauli couplings). In the center of mass frame, the equation is

$$\{\vec{\alpha}_1.(\vec{p}_1-e_1\vec{A}_2)+\vec{\alpha}_2.(\vec{p}_2-e_2\vec{A}_1)+\beta_1 m_1+\beta_2 m_2+$$

$$+V+a_1\sigma^{(1)}_{\mu\nu} F_2^{\mu\nu}+a_2\sigma^{(2)}_{\mu\nu} F_1^{\mu\nu}\}\psi=E\psi. \quad (1)$$

$$(\vec{p}_1 = -\vec{p}_2),\ V = e_1 e_2\frac{1-\vec{\alpha}_1.\vec{\alpha}_2}{r}$$

where the vector potential $\vec{A}$ has two parts, an electric and a magnetic,

$$\vec{A}_i = \frac{1}{2}e_i \frac{\vec{\alpha}_i}{r} + \frac{1}{2}a_i \frac{(\beta\vec{\sigma}^{(i)}) \wedge \vec{r}}{r^3} , \; i = 1,2 \tag{2}$$

and

$$F_i^{\mu\nu} = A_i^{\nu,\mu} - A_i^{\mu,\nu} \tag{3}$$

Note that the components of $\vec{A}$ do not commute. The terms with coefficients e_1e_2, e_ia_j and a_ia_j are the Coulomb, spin-orbit and spin-spin terms, respectively.

Eq. (1) has been analyzed elsewhere into its radial and angular parts and approximately solved in the Coulomb region[8]. This system in general is very complex. The simpler interesting case, considered in Ref. 1, is obtained if we first neglect in eq. (1), the last two terms and keep in the vector potentials $\vec{A}$ only the anomalous magnetic part $a\frac{(\beta\vec{\sigma})\wedge\vec{r}}{r^3}$:

$$\{\vec{\alpha}_1.(\vec{p}_1-e_1\vec{A}_2^a) + \vec{\alpha}_2.(\vec{p}_2-e_2\vec{A}_1^a) + \beta_1m_1+\beta_2m_2 + V\}\psi = E\psi \;. \tag{4}$$

It is further assumed that $a=1(e/2m)$, i.e. anomalous magnetic moment is equal to the full magnetic moment (instead of $a=(\frac{\alpha}{2\pi})(e/2m)$ for a free electron). This last assumption could be justified on the basis of "self-focusing principle" that I have explained above. The neglect of the Breit-electric vector potentials, $e\frac{\alpha}{r}$, in (2) could also be justified since this term is small at relatively low energies. The two other neglected spin-orbit potentials are

$$e_2a_1 \frac{\vec{\gamma}_1.\hat{r}}{r^3}$$

$$e_2e_1 \frac{\vec{\gamma}_2.\hat{r}}{r^3}$$

and the anomalous spin-spin potentials are neglected:

$$\frac{-a_1a_2}{r^3}\left[3(\beta\vec{\sigma})_1.\hat{r}(\beta\vec{\sigma})_2.\hat{r}-(\beta\vec{\sigma}_1).(\beta\vec{\sigma}_2)-3\vec{\gamma}_1.\hat{r}\gamma_2.\hat{r}+\vec{\gamma}_1.\vec{\gamma}_2\right]$$

In spite of neglected terms, eq. (4) is still complicated. We look for a very special symmetric solution in which the two particles move with exactly the same quantum numbers and wave functions:

$$\{\vec{\alpha}.(\vec{p}-eA^a)+\beta m+\frac{V}{2}\}\psi = \frac{E}{2}\psi \tag{5}$$

Where $\vec{\alpha},\beta,\vec{p},m$ refer to particle 1 or 2. For equal mass case and for the state ^3P_o (i.e. $\ell=1$, $s=1$, $j=o$) that one is looking for this is justified. It means that both particles move rigidly connected like a single particle. This reduces the 6 degrees of freedom of the 2-body problem into three.

There is a "dual" equation to (5), namely the motion of a Dirac particle with an anomalous magnetic moment in the Coulomb field which has been analyzed completely[9] and to which we shall later refer. Eq. (5) describes a Dirac particle with charge in the Coulomb and dipole field of another fixed particle.

The second-order equation corresponding to (5) is[10]

$$\left\{ p^2+2\frac{r_e}{r^3}\vec{S}_2.\vec{L}+\frac{r_e}{2r^3}\hat{S}_{12}+\frac{r_e^2}{2r^4}\left(1-4\vec{S}_1.\hat{r}\vec{S}_2.\hat{r}\right) - \frac{V'}{E+2m-\frac{\alpha}{r}}\frac{\partial}{\partial r}+\frac{mr_e}{E+2m-\frac{\alpha}{r}}\left[\frac{2\vec{S}_1.\vec{L}}{r^3}-\frac{r_e^2}{6r^4}\left(\hat{S}_{12}-8\vec{S}_1.\vec{S}_2\right)\right]\right\}\phi = \left[\left(\frac{E-V}{2}\right)^2-m^2\right]\phi \tag{6}$$

This equation agrees with that of Ref. 1 except the term $-4\vec{S}_1.\hat{r}\vec{S}_2.\hat{r}$ in the fourth term. They also further assume $\vec{S}_1 = \frac{1}{2}\vec{S}$, $\vec{S}_2 = \frac{1}{2}\vec{S}$. Here $\hat{S}_{12} = 4(3\vec{S}_1.\vec{S}_2-\vec{S}_1.\vec{S}_2)$ and $r_e = \alpha/m$.

Wong and Becker have solved eq. (6) — without the term $-4\vec{S}_1.\hat{r}\vec{S}_2.\hat{r}$ — numerically and obtained a resonance state with M=1.579MeV. A simple way of getting this result is the following[3]: Since magnetic interactions dominate at short distances, we may first neglect the Coulomb terms in (6). There it is known that (6), as well as the dual problem mentioned above[11], has an exact "zero-energy' solution, i.e. E=2m. Now Coulomb perturbation gives an energy shift $\Delta E=\langle|V|\rangle \cong m$, so that the total mass is $M\cong 3m$=1,533MeV.

More precisely, the radial equations of the 8-component eq. (5) are [10]

$$\begin{aligned}
&\left(m-\frac{E}{2}-\frac{V}{2}\right)u_+ + \frac{d}{dr}v_- + \frac{\sqrt{j(j+1)}}{r}v_+ = o\\
&\left(m-\frac{E}{2}-\frac{V}{2}\right)u_- + \left(\frac{d}{dr}-\frac{1}{r}+\frac{2ea}{r^2}\right)v_+ + \frac{\sqrt{j(j+1)}}{r}v_- = o\\
&\left(-m-\frac{E}{2}-\frac{V}{2}\right)v_+ - \left(\frac{d}{dr}+\frac{1}{r}-\frac{2ea}{r^2}\right)u_- + \frac{\sqrt{j(j+1)}}{r}u_+ = o\\
&\left(-m-\frac{E}{2}-\frac{V}{2}\right)v_- - \frac{d}{dr}u_+ + \frac{\sqrt{j(j+1)}}{r}u_- = o
\end{aligned} \tag{7}$$

There is second set of 4 equations obtained from (7) by a symmetry transformation.

In order to obtain (7) from the radial equations of the full 2-body equation (1) (see Appendix) we have to take not only the limits $m_1=m_2$, $e_1a_2=e_2a_1$; $a_1a_2=o$ $E\to E/2$, $V\to V/2$, but also drop a number of terms which presumably do not contribute to the very symmetric solution (5).

For j=o (^3P_o-state) and leaving the Coulomb part V/2 out as a perturbation we have the simple equations

$$\begin{aligned}
&(m-E/2)u_+ + \frac{d}{dr}v_- = o\\
&(-m-E/2)v_- - \frac{d}{dr}u_+ = o
\end{aligned}$$

$$\begin{aligned}
&(m-E/2)u_- + \left(\frac{d}{dr}-\frac{1}{r}+\frac{2ea}{r^2}\right)v_+ = o\\
&(-m-E/2)v_+ - \left(\frac{d}{dr}+\frac{1}{r}-\frac{2ea}{r^2}\right)u_- = o
\end{aligned} \tag{8}$$

There is a zero-binding energy solution of these equations

$E=2m;\ u_-=c\frac{1}{r}e^{-2ea/r},\ u_+=v_+=v_-=o$.

From (8) we have then

$$\left\{-\frac{d^2}{dr^2}+\left(\frac{2}{r^2}-\frac{4K}{r^3}+\frac{K^2}{r^4}\right)\right\}u_-=\left(\left(\frac{E}{2}\right)^2-m^2\right)u_- \tag{9}$$

$$K = 2ea$$

Eq. (9) belongs to a class of exactly soluble resonance problems[12]. The effective potential $V_{eff}=\frac{2}{r^2}-\frac{4K}{r^3}+\frac{K^2}{r^4}$ has a form very close to the numerical solution of Wong and Becker[1], for $K=e\frac{e}{2m}=\frac{\alpha}{2m}=1.41$, specially at distances, $r<1$ Fermi. Also the wave function agrees with the numerical solution. At longer distances the Coulomb term considerable modifies the potential and the wave function, and we get an energy shift of the order of $\Delta m=\langle|V|\rangle\tilde{=}m$, as was mentioned above. For a more accurate comparison we need to solve the set of equations (7).

Starting from the 2-body equation (1) and specializing to $\Delta m=o$, $\lambda=e_1a_2+e_2a_1=2ea=\tilde{K},\mathcal{I}=e_1a_2-e_2a_1=o$, $V=o$, the more general equation instead of (9) is

$$\left\{-\frac{d^2}{dr^2}+\left(\frac{2}{r^2}-\frac{6\tilde{K}M}{Er^3}+\frac{9\tilde{K}^2}{4r^4}\right)+\left(E-\frac{9\tilde{K}^2}{Er^4}\right)\left[\frac{d}{dr}\left(\frac{1}{E-9\tilde{K}^2/r^4}\right)\right]\frac{d}{dr}\right\}(rv_o)$$
$$=\left(1-\frac{9\tilde{K}^2}{e^2r^4}\right)\left(\left(\frac{E}{2}\right)^2-m^2\right)(rv_o) \tag{10}$$

Compared to (9), $\tilde{K}=\frac{2}{3}K$ and the last terms in the potential are missing.

The dual system, a Dirac particle with charge e, and anomalous magnetic moment a, in the field of a fixed charge e_2 is also exactly soluble if we neglect the Coulomb potential [11,10]. The Hamiltonian is

$$H=\vec{\alpha}_1.\vec{p}_1+\beta_1m_1+ie_2\frac{\hat{r}}{r^2}\cdot\alpha_1(\beta\vec{\alpha})_1 \tag{11}$$

and leads to the coupled equations

$$\begin{pmatrix} E^2-m^2+\frac{d^2}{dr^2}-\frac{j(j+1)}{r^2}-\frac{2e_2a_1}{r^3}-\frac{(e_2a_1)^2}{r^4} & 2\frac{\sqrt{j(j+1)}}{r^2}\left(1+\frac{e_2a_1}{r}\right) \\ \frac{2\sqrt{j(i+1)}}{r}\left(1+\frac{e_2a_1}{r}\right) & E^2-m^2+\frac{d^2}{dr^2}-\frac{j(j+1)+2}{r^2}-\frac{4e_2a_1}{r^3}-\frac{(e_2a_1)^2}{r^4}\end{pmatrix}\begin{pmatrix}u_1\\u_o\end{pmatrix}=o$$

plus two other similar equations. For j=o, we have a zero-binding solution (E=m), for j=1, $\sqrt{E^2-m^2}=-i\frac{1}{ea}$. For j=2 a resonance solution $i\sqrt{E^2-m^2}=\frac{1}{ea}[2\pm\sqrt{2i}]$, ...

SIZE, ENERGY AND UNCERTAINTY RELATIONS

The minimum of the effective potential in (9) is around $\frac{1}{2}$ Fermi. The question therefore arises how we could have such a low energy, $E\simeq3m$, resonance at such short distances. If a state ψ is fairly localized at a distance r_o from the origin and if we put, as usual, $\Delta r \sim r_o$, then the uncertainty relation $\Delta p \Delta r \sim \hbar$ becomes essentially the angular momentum quantization for circular orbits: $pr_o = n\hbar$. Hence $p \sim 1/r_o$ (for $n=1$, $\hbar=1$). This would mean for $r_o = 1/2$ Fermi, a huge energy $E \sim p \sim 300$ MeV! The H-atom and the relativistic $1/2r^2$-potentials give examples of such localized states: In the former case $H = p^2/2m - \alpha/r$ and the minimum of H subject to $pr = n\hbar$ gives $E_n = -\frac{1}{2}\frac{m\alpha^2}{n^2}$. In the latter case $H = 2p - \frac{\alpha}{2m}\frac{1}{r^2}$, and again the minimum of H subject to $pr = n\hbar$ gives $E_n = \frac{2m}{\alpha} n^2$, which for $n=1$ coincides with the pion mass $E_1 = 140$ MeV.

However, the solution discussed in eqs. (6), (7), is not a localized solution. The Coulomb potential pulls, so to speak, the ψ function to larger distances as shown in Fig. 1

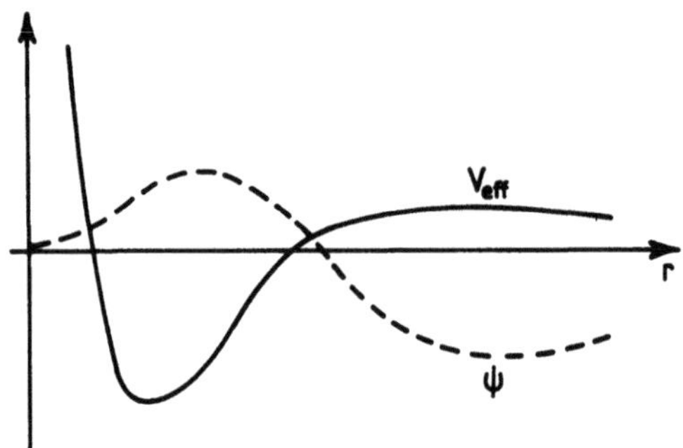

Fig. 1 Schematic form of the effective radial potential and radial nonlocalized wave function.

We have an unsharp r, hence a sharper p; $\Delta p \Delta r \sim h$ still holds, of course, but the energy essentially comes from the Coulomb part, $E \cong 2m + \langle |V| \rangle$ and is very large ($E \sim p \sim 1/\Delta r$), and $\Delta r \gg r_o$.

HIGH j, LOCALIZED SOLUTIONS

For $j > o$ we can have more localized solutions. In fact eq. (11) has a resonance solution for $j \cong 2$ with $E_{res} \sim \frac{m2\sqrt{2}}{\alpha}$, of the order of pion mass.

For the realistic hadron models based on leptons and magnetic interactions [3] the complete form of the interactions must be considered. In fact the full extent of the self energy potentials at short distances is not known[13]. Moreover, the role of the neutrinos must be considered in particle physics. It is possible that pions will be made of a basic (e^+e^-) core with $j=2$, say, and with a $(\nu\bar{\nu})$-cloud such that the total $J=o$. The role of ν is best seen from the dominant decay of pions into $(\mu\bar{\nu})$.

CONCLUSIONS

We have discussed the possible application of magnetic resonances to the formation of (e^+e^-)-resonance states at an effective mass of $M \simeq 3m$ corresponding to the positronium peaks observed in heavy ion collision. The importance of relativistic treatment and of anomalous magnetic moment terms is emphasized. The charge and anomalous magnetic moment coupling "constants" are actually form factors which are difficult to calculate at present nonperturbatively at short distances. We have discussed under what conditions a very simplified model is obtained which shows resonance of the desired type. The assumptions of the model in Ref. 1 are not proven. More numerical calculations are necessary to ascertain these results. The effective potentials derived from the complete equations (A1) are highly complicated energy-dependent functions; it is likely that they can predict at different energies a number of resonances in the range 3m all the way to the pion mass.

APPENDIX

The complete set of two-body radial equations for the (e^+e^-)-system including all the anomalous magnetic moment terms are with $M=2m$, $\Delta m=o$, $\lambda=2ea$; $e_1e_2=-\alpha$, $a_1a_2=a^2$.

$$\left(E+\frac{2\alpha}{r}+\frac{4a^2}{r^3}-\frac{\lambda^2}{Er^4}\right)(rz_1)+\frac{4\sqrt{j(j+1)}}{Mr}\left(\frac{d}{dr}-\frac{1}{r}-\frac{\lambda M}{2Er^2}\right)(ry_{oo})$$

$$-2\left(\frac{d}{dr}-\frac{\lambda M}{2Er^2}\right)(ru_2)-\frac{2\sqrt{j(j+1)}}{Mr}(rv_o)=o$$

$$\left(E+\frac{4\alpha}{r}-\frac{4j(j+1)}{r^2}+\frac{6\lambda}{Mr^2}\left(\frac{d}{dr}-\frac{1}{r}\right)\right)(ry_{oo})-\frac{2\lambda\sqrt{j(j+1)}}{Er^3}(rz_1)$$

$$+2\left(\frac{d}{dr}+\frac{1}{r}-\frac{3\lambda E}{2Mr^2}\right)(rv_o)+\frac{2M\sqrt{j(j+1)}}{Er}(ru_2)=o$$

$$\left(E-\frac{M^2}{E}+\frac{2\alpha}{r}-\frac{4a^2}{r^3}\right)(ru_2)+2\left(\frac{d}{dr}+\frac{\lambda M}{2Er^2}\right)(rz_1)+\frac{2M\sqrt{j(j+1)}}{Er}(ry_{oo})=o$$

$$\left[E\left(E+\frac{2\alpha}{r}+\frac{8a^2}{r^3}\right)-M^2\right](rv_o)-2\left(\left(E+\frac{2\alpha}{r}+\frac{8a^2}{r^3}\right)\left(\frac{d}{dr}-\frac{1}{r}\right)+\frac{3\lambda M}{r^2}\right)(ry_{oo})$$

$$-\frac{2M\sqrt{j(j+1)}}{r}(rz_1)=o \tag{A1}$$

For j=o, in particular, we have two sets of two-coupled equations:

$$\left(E+\frac{2\alpha}{r}+\frac{4a^2}{r^3}-\frac{\lambda^2}{Er^4}\right)(rz_1)-2\left(\frac{d}{dr}-\frac{\lambda M}{2Er^2}\right)(ru_2)=o$$

$$\left(E-\frac{M^2}{E}+\frac{2\alpha}{r}-\frac{4a^2}{r^3}\right)(ru_2)+2\left(\frac{d}{dr}+\frac{\lambda M}{2Er^2}\right)(rz_1)=o$$

$$\left(E+\frac{4\alpha}{r}+\frac{6\lambda}{Mr^2}\left(\frac{d}{dr}-\frac{1}{r}\right)\right)(ry_{oo})+2\left(\frac{d}{dr}+\frac{1}{r}-\frac{3\lambda E}{2Mr^2}\right)(rv_o)=o$$

$$\left[E\left(E+\frac{2\alpha}{r}+\frac{8a^2}{r^3}\right)-M^2\right](rv_o)-2\left[\left(E+\frac{2\alpha}{r}+\frac{8a^2}{r^3}\right)\left(\frac{d}{dr}-\frac{1}{r}\right)+\frac{3\lambda M}{r^2}\right](ry_{oo})=o \tag{A(2)}$$

The numerical solution of these equations should gives us a better justification of the solution of Ref. 1, in particular show the role of the neglected a^2-terms.

REFERENCES

1. C.-Y. Wong and R. L. Becker, Oak Ridge National Laboratory preprint (1986).
2. See contributions to these proceedings.
3. A. O. Barut, Surveys in High Energy Physics 1:113 (1980); Amer. Inst. Physics Conference Proc. 71:73 (1980); Ann. der Physik, 43:83 (1986) and in Quantum Electrodynamics of Strong Fields, Plenum Press, N.Y. (1983), (W. Greiner, ed.), p. 755.
4. H. Bethe and E. Salpeter, Quantum Mechanics of One and Two-Electron Atoms, Plenum Press, N.Y. (1977).
5. A. O. Barut, Suppl. Hadr. Journal 1:314 (1985), Lett. Mathem. Phys. 10:195 (1985).
6. A. O. Barut, to be published.
7. A. O. Barut, in Lecture Notes in Physics, vol. 180:332 (Springer, 1983). A. O. Barut and S. Komy, Fortschritte der Physik, 33:309 (1985).
8. A. O. Barut and N. Ünal, Fortschritte der Physik, 33:319 (1985).
9. A. O. Barut and J. Kraus, J. Math. Phys. 17:1115 (1976).
10. A. O. Barut and N. Ünal, (to be published).
11. A. O. Barut, J. Math. Phys. (in press, 1986) and "New Approach to Bound-state QED, I and II", Physica (in press).
12. A. O. Barut, M. Berrondo and G. Garcia-Calderón, J. Math. Phys. 21:1851 (1980), A. O. Barut, J. Math. Phys. 21:568 (1980).
13. A. O. Barut, see These Proceedings.

ATOMIC PROCESSES IN RELATIVISTIC HEAVY ION COLLISIONS*

Ulrich Becker, Norbert Grün, Klaus Momberger and Werner Scheid

Institut für Theoretische Physik der Universität Giessen
West Germany

ABSTRACT

Theoretical methods for the description of atomic processes in relativistic heavy ion collisions are reviewed. We present probabilities and cross sections for K-shell ionisation and pair creation which are obtained by first order perturbation theory, coupled channel and finite difference methods in semiclassical approximation.

1. INTRODUCTION

Atomic collision experiments with relativistic heavy ions are already feasible today. Nearly all existing experiments have been carried out with the BEVALAC (LBL, Berkeley). Cross sections for K-vacancy production by 4.88 GeV protons on elements between Ni and U were measured by Anholt et al. (1976). Bak et al. (1982) looked for the impact-parameter dependence of K-shell ionization of Ge with relativistic protons. K-vacancy production cross sections for 3 GeV ^{12}C ions on targets ranging from Ti to Pb were reported by Anholt et al. (1977). Data on projectile charge states of low -Z ions (140-2100 MeV/amu C, Ne and Ar ions) were taken by Crawford and coworkers (Greiner et al. 1977, Crawford 1979). Gould et al. (1984) measured electron capture by U^{91+} and U^{92+} and ionisation of U^{90+} and U^{91+} at energies of 962 and 437 MeV/amu. In the last three years Meyerhof and Anholt and coworkers published the results of a series of experiments with relativistic heavy ions: Target K-vacancy and L-X-ray production cross sections in collisions with relativistic (82-670 MeV/amu) heavy ions ranging from Ne to U (Anholt et al. 1984b), X-ray studies in order to observe radiative electron-capture photons with 197 MeV/amu Xe projectiles (Anholt et al. 1984a), projectile 1s ionisation cross sections and cross sections for electron capture by 82-, 140- and 200 MeV/amu Xe projectiles incident on a variety of thin solid targets between Be and Au (Anholt 1985, Meyerhof 1985), charge states of emergent projectiles, Kα X-ray production cross sections, K radiative electron-capture photon cross sections for 82- and 197 MeV/amu Xe projectiles incident on solid targets (Anholt and Meyerhof 1986), radiative processes for Xe to U projectiles at 82 to 422 MeV/amu (Anholt et al.

* Supported by GSI (Darmstadt) and BMFT (Bonn)

1986), and multiple ionisation of Ne, Ar, Kr and I by 420 MeV/amu U^{n+} ($n \approx 91$) (Kelbch et al. 1986). R. Anholt and H. Gould are preparing a report on "Relativistic Heavy-Ion-Atom Collisions" for publication in Advances in Atomic and Molecular Physics.

The importance of application of relativistic heavy ion physics has been outlined by Gould (1984). For example, the pair production could be used as a direct measure of the luminosity for the planned relativistic nuclear collider with ions of 100 GeV/amu. The capture of an electron from the pair by one of the nuclei can influence the beam survival in relativistic nuclear colliders and sets limits on the luminosity.

The theoretical interest in atomic physics with relativistic heavy ion is presently concentrated on the evaluation of cross sections for ionisation, electron capture, pair-creation and radiative processes. The special aspect of these calculations is the relativistically contracted and enhanced electromagnetic field of the projectile charge in the rest frame of the target atom or ion. For relativistic projectile velocities these fields become more and more transversal. This effect is the basis of the equivalent-photon method (Weizsäcker-Williams method, see Heitler 1954), where the relativistic projectile is assumed as the source of real photons and the cross sections are obtained by folding the photon density with the photoreaction cross sections.

In general, approximate methods are appropriate to describe atomic processes with relativistic heavy ions, since the various probabilities for inner shell ionisation, charge exchange and pair-creation are small with exception of collisions with small impact parameters. For example, we mention the completely classical relativistic Monte-Carlo calculation of the K-shell ionisation (Teubner et al. 1980, 1982), the numerical solutions of the time-dependent Dirac equation by the finite difference method (Becker et al. 1983, 1986) and finite element method (Bottcher 1984), the binary encounter method (Garcia 1970, McGuire and Richard 1973) and a relativistic formulation of the eikonal approximation for nonradiative capture (Eichler 1985). The approximate methods most commonly used are the plane wave Born approximation (Davidovic et al. 1978, Anholt 1979) and the first order perturbation theory in semiclassical approximation (Bang and Hansteen 1959, Amundsen and Aashamar 1981, Valluri et al. 1984, Becker et al. 1985, 1986). The latter method has also been applied for relativistic nuclear Coulomb excitation by Winther and Alder (1979).

In the following we review calculations of probabilities and cross sections for K-shell ionisation (Section 2) and for pair creation (Section 3) in relativistic heavy ion collisions. The methods applied are based on the semiclassical approximation. They are the perturbation theory of first order, coupled channel and finite difference methods. In Section 4 we discuss the possibility and limits of channeling experiments with relativistic heavy ions.

2. EXCITATION AND IONISATION

In this Section we discuss probabilities and cross sections for the excitation and ionisation of K-shell electrons of very heavy, highly charged ions by relativistic projectiles. Beside the use of the plane wave Born approximation by Anholt since 1976 and Davidovic et al. (1978), a common method for treating these processes theoretically is based on the semiclassical description of the collision. Within a semiclassical approximation the motion of the projectile and target nuclei is described by classical mechanics, whereas the electrons are treated quantum mechanically by solving the time-dependent Dirac-equation.

The conditions for the classical description of internuclear motion are given by Mott and Massey (1965). The first condition states that the wave length of the projectile nucleus has to be small in comparison with the typical extension of the wave function of the electron at the target nucleus.

$$kd \gg 1 \tag{1}$$

For protons of E_{lab}=1GeV the wave number is k=8/fm. If they are scattered at U^{91+} (d~500fm), we find kd~4000, which fulfils the unequality (1) by far. The second condition originates from the Heisenberg uncertainty principle. In order that the uncertainty in the scattering angle is small, the change of momentum of the projectile nucleus by the field of the target nucleus has to be large compared with $\hbar/b$, where b is the impact parameter.

$$\Delta p \gg \hbar/b \tag{2}$$

With $\Delta p = 2Z_P Z_T e^2/bv_P$ (Jackson 1962), where Z_P and Z_T are the charge numbers of the projectile and target nucleus, respectively, and v_P the projectile velocity, we obtain the condition

$$2Z_P Z_T \alpha \gg 1. \tag{3}$$

Here, we have assumed a relativistic velocity of the projectile ($v_P \sim c$). For an Uranium nucleus as projectile and target the condition (3) is very well fulfilled.

Further approximations, which are used in order to simplify the internuclear motion, are the assumption, that the projectile moves on a straight line with constant velocity, and the neglection of the recoil of the target nucleus. The validity of these approximations can be estimated by calculating the scattering angle of the projectile in the Coulomb field of the target nucleus and the change of the target velocity due to the recoil. One finds the following relations:

$$\theta = \frac{\Delta p}{p_P} = \frac{2Z_P Z_T e^2}{v_P^2 b M_P \gamma}, \tag{4}$$

$$\Delta v_T = \frac{\Delta p}{M_T} = \frac{2Z_P Z_T e^2}{v_P b M_T}, \tag{5}$$

where $\gamma = 1/(1-v_P^2/c^2)^{1/2}$.

For the scattering of U^{92+} on U^{92+} at E_{lab}=1GeV/amu and b=30fm one obtains a scattering angle $\theta=0.13^\circ$ and a small change of the target velocity of $\Delta v_T = 4.2\cdot10^{-3}c$.

For these reasons we can describe the reaction in the rest frame of the target nucleus. In this frame the projectile charge, moving on a straight-line trajectory with constant velocity, generates the following Liénard-Wiechert potential

$$\vec{A} = \phi\, v_P \vec{e}_z / c$$
$$\phi = Z_P e\gamma/((x-b)^2+y^2+\gamma^2(z-v_P t)^2)^{1/2}. \tag{6}$$

The potential has to be inserted as a time-dependent perturbation in the Dirac equation for the electron field:

$$i\hbar\frac{\partial}{\partial t}\psi(\vec{r},t) = (c\vec{\alpha}(\vec{p}+\frac{e}{c}\vec{A}(\vec{r},t))+\beta mc^2-\frac{Z_T e^2}{r}-e\phi(\vec{r},t))\psi(\vec{r},t). \tag{7}$$

In the following we solve this equation by different methods: The first method uses the finite difference method and integrates Eq.(7) numerically in time. The second method treats Eq.(7) in perturbation theory of first order. In this case the wave function $\psi(\vec{r},t)$ is expanded into atomic eigenstates of the target ion. Finally, the third method is a coupled channel procedure and leads to the expansion coefficients of the wave function by solving coupled differential equations.

2.1 THE FINITE DIFFERENCE METHOD

Calculations of differential probabilities carried out in first-order perturbation theory have demonstrated that the K-shell ionisation probabilities exceed the unitary limit for very heavy ions and small impact parameters. These results show that non-perturbative methods are needed for the evaluation of the inner shell excitation and ionisation probabilities at small impact parameters. In the following we present a non-perturbative solution of the time-dependent Dirac equation for relativistic collisions of U^{92+} on U^{91+} applying a finite difference algorithm.

Within this method we solve Eq.(7) using an approximation of the time-development operator

$$\psi(\vec{r},t+\Delta t)=(1+\frac{i\Delta t}{2\hbar}H(\vec{r},t+\frac{\Delta t}{2}))^{-1}(1-\frac{i\Delta t}{2\hbar}H(\vec{r},t+\frac{\Delta t}{2}))\psi(\vec{r},t). \qquad (8)$$

This equation is exact including the terms of the order $(\Delta t)^2$. In order to calculate the effects of these operators on the wave function, we introduce a grid and discretize the wave function in space. A detailed description of the numerical procedure is given by Becker et al.(1983). Other numerical techniques are presented in the review of Bottcher (1984).

In order to keep the numerical procedure manageable on the present computers, we reduced the scattering problem to the case of rotational symmetry about the internuclear axis; i.e. we chose an impact parameter b=0 and used cylindrical coordinates z and $\rho=\sqrt{x^2+y^2}$.

Fig.1 shows the time development of the spatial probability density of the electron in a collision of U^{92+} on U^{91+} at E_{lab}=1GeV/amu and zero impact parameter. At the initial time t=0 the density of the electron is determined by the $1s_{1/2}$-state of hydrogen-like Uranium, whereas the projectile nucleus is assumed to be at z_P=-5300 fm. During the approach of the projectile the wave function of the electron does not change much, which can be deduced by comparing the density for t=0 and $2\cdot10^{-20}$sec. In contrast to non-relativistic collisions a strong localisation of the electron wave function is not observed. After the projectile has passed the point of closest approach, the main excitation and ionisation processes take place. In the densities at $t=2.6\cdot10^{-20}$sec and $4.26\cdot10^{-20}$sec, we observe a large ionisation component which leads to the emission of the electron in forward direction with approximately the velocity of the projectile.

At the last depicted time we projected the numerically evaluated wave function on the set of eigenstates of the target ion. After the collision the initial $1s_{1/2}$-state is only occupied by 13% and the excited bound states by about 10%. Fig.2 shows the differential ionisation probabilities (solid curves) as a function of the total energy E_f of the ionized electron. For the discussion we depict also ionisation probabilities obtained by the first order perturbation theory (dashed curves, see Section 2.2 for details). By comparing the differential ionisation probabilities belonging to $s_{1/2}$-states (κ_f=-1), we find the results of the first order approximation by a factor 2.5 larger than the non-perturbative ones, which reflects the non-unitarity of the first order calculation. In contrast, the finite difference method conserves the norm of the wave function. Nevertheless the slopes and structures of the solid and dashed curves agree roughly for all values of κ_f.

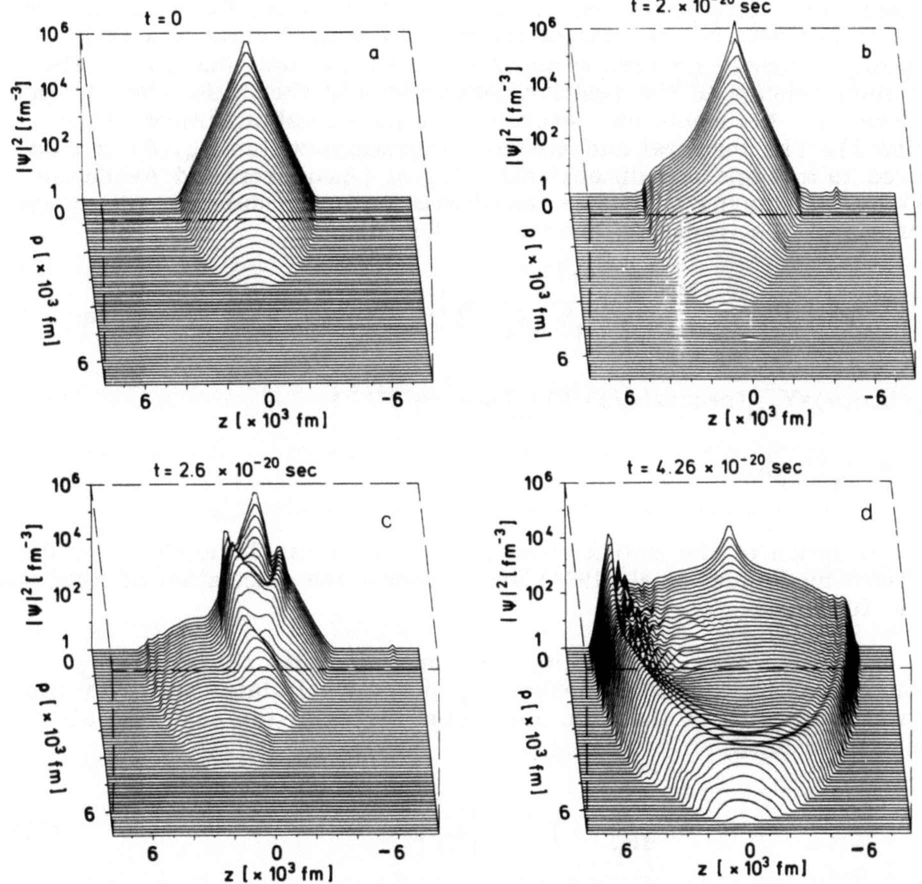

Fig.1 Time development of the electron density distribution in the z-ρ-plane for a collision of U^{92+} on U^{91+} at E_{lab}=1GeV/amu. The position of the target nucleus is fixed at z=0 and ρ=0. The projectile nucleus moves along the z-axis from the negative to the positive z-direction with the position at zp(t).

a) t=0, z_p=-5300 fm, b) $t=2\cdot10^{-20}$sec, z_p=-44fm,
b) $t=2.6\cdot10^{-20}$sec, zp=1533fm, d) $t=4.26\cdot10^{-20}$sec, z_p=5896fm.

2.2 PERTURBATION THEORY OF FIRST ORDER

The perturbation theory of first order is useful for the application to relativistic projectile velocities and impact parameters $b\neq0$. The general formalism has been given by Bang and Hansteen (1959), the special treatment for relativistic ions by Winther and Alder (1979), Amundsen and Aashamar (1981) and Valluri et al. (1984).

The amplitude for a transition of an electron (electrons) from a state ψ_i to a state ψ_f due to the field of the projectile is given in perturbation theory of first order by

$$a_{fi}= (i\gamma Z_p e^2/\hbar) \int_{-\infty}^{\infty} dt \; \exp(i\omega_{fi}t)\langle\psi_f|(1-\beta\alpha_z)/r'|\psi_i\rangle \;, \tag{9}$$

$$\text{where} \quad \omega_{fi}=(E_f-E_i)/\hbar, \quad \beta=v_p/c, \quad r'=((x-b)^2+y^2+\gamma^2(z-v_pt)^2)^{1/2} \;. \tag{10}$$

For initial and final states we use Coulomb-Dirac wave functions centered at the target ion. In the case of ionisation the final state is a state of the positive energy continuum and normalized per unit energy. If the final state belongs to the positive continuum and the initial state to the negative one, the amplitude describes the pair-creation process (see Section 3). The temporal and spatial integrations within Eq.(9) can be reduced to a single one-dimensional integral (Amundsen and Aashamar 1981) by introducing a Fourier transformation and a subsequent multipole expansion of the transition operator in Eq.(9).

$$a_{fi}=(8\pi i Z_p e^2/\hbar v_p)\sum_{\ell,m} i^{\ell-m} \int_q^{\infty} ds \frac{s}{s^2-q^2\beta^2}\langle\psi_f|(1-\beta\alpha_z)j_\ell(sr)Y_{\ell m}(\hat{r})|\psi_i\rangle \cdot B_{\ell m}(b,q,s),$$

$$B_{\ell m}(b,q,s)=Y_{\ell m}(\arccos(q/s),0)J_m(b(s^2-q^2)^{1/2}), \tag{11}$$

$$q = (E_f-E_i)/v_p\hbar.$$

The calculation of the matrix element in Eq.(11) can be analytically done as shown by Valluri et al.(1984) and involves the calculation of hypergeometric functions.

The differential probability for ionisation is obtained by summing over the quantum number κ_f of the final states and over the magnetic substates of the final and initial states in the case of ionisation from a filled shell.

$$\frac{dI_b}{dE_f} = \sum_{\kappa_f,\mu_f,\mu_i} |a_{fi}|^2 \tag{12}$$

The differential ionisation cross section is obtained by an integration of Eq.(12) over the impact parameter:

$$\frac{d\sigma}{dE_f}=8\pi\left(\frac{Z_pe^2}{\hbar v_p}\right)^2 \int_q^{\infty} ds\frac{s}{(s^2-q^2\beta^2)^2} \sum_{\kappa_f,\mu_f,\mu_i} |M_{fi}|^2, \tag{13}$$

where

$$M_{fi}=4\pi \sum_\ell i^\ell Y_{\ell,\mu_f-\mu_i}(\arccos(q/s),0)\langle\psi_f|(1-\beta\alpha_z)j_\ell(sr)Y_{\ell,\mu_f-\mu_i}(\hat{r})|\psi_i\rangle. \tag{14}$$

As pointed out by Amundsen and Aashamar (1981), the differential cross section in semiclassical approximation is identical to the cross section obtained in plane wave Born approximation if the perturbing potential is treated in the Lorentz gauge and the deflection of the projectile is neglected.

Fig.2 shows the differential probabilities (dashed curves) as function of the energy of the emitted electron for an $U^{92+}+U^{91+}$ collision at E_{lab}=1GeV/amu and zero impact parameter. The problem, that these results violate the condition of unitary, has been discussed in Section 2.1.

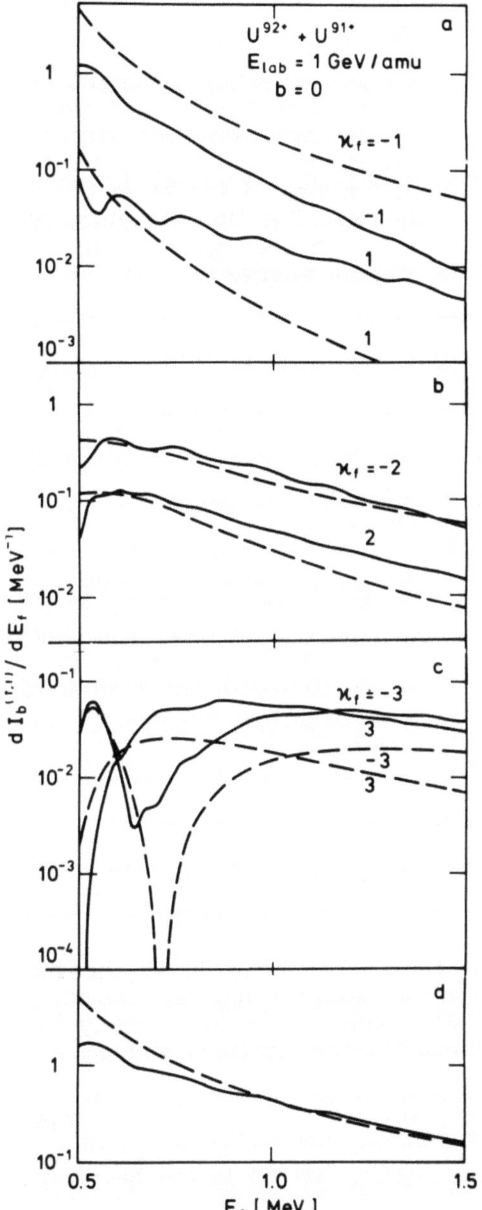

Fig.2

The differential ionisation probabilities for the transition of the $1s_{1/2}$ electron into the positive energy continuum as a function of the final electron energy E_f for a collision of U^{92+} on U^{91+} at E_{lab}=1GeV/amu and zero impact parameter. The final states are a) the $s_{1/2}(\kappa_f=-1)$ and $p_{1/2}(\kappa_f=1)$ states, b) the $p_{3/2}(\kappa_f=-2)$ and $d_{3/2}(\kappa_f=2)$ states, c) the $d_{5/2}(\kappa_f=-3)$ and $f_{5/2}(\kappa_f=3)$ states, and d) the sum over the states of the positive energy continuum. The solid curves are obtained by projecting on the wave function of the electron calculated by the finite difference method at $t=4.26\cdot10^{-20}$sec. The dashed curves are calculated by employing first order perturbation theory within the semiclassical approximation.

In the following we present results for the ionisation of the K-shell of the target ion, which we assume to be filled with two electrons in all considered systems. In Fig.3 we have plotted ionisation probabilities integrated over the electron energy E_f as function of the incident projectile energy for various impact parameters. The curves approach a constant value for $E_{lab} \to \infty$. This saturation effect is due to an analytical property of the transition amplitudes (9), which was studied by Amundsen and Aashamar (1981) for the special case of inner-shell ionisation and generalized by Becker et al. (1986) to all transitions between the eigenstates of the target ion, induced by the perturbing potential of the projectile. In Section 3 we demonstrate that this saturation effect occurs also for pair creation probabilities.

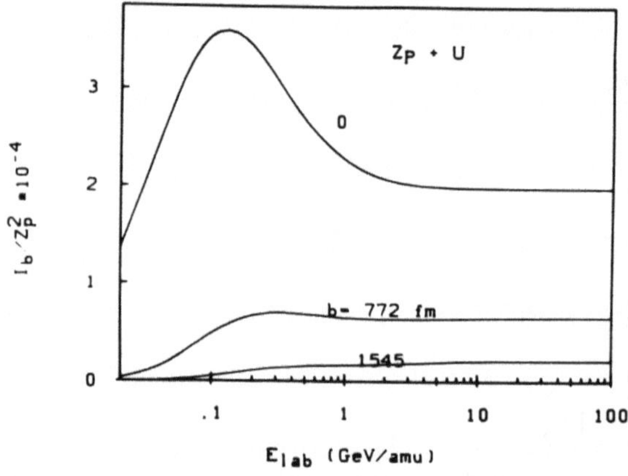

Fig. 3

K-shell ionisation probabilities for systems Zp+U scaled with Z_p^2 for three different impact parameters (b=0,772fm, 1545fm) as function of the projectile energy. The fine structure formula has been used for the binding energies.

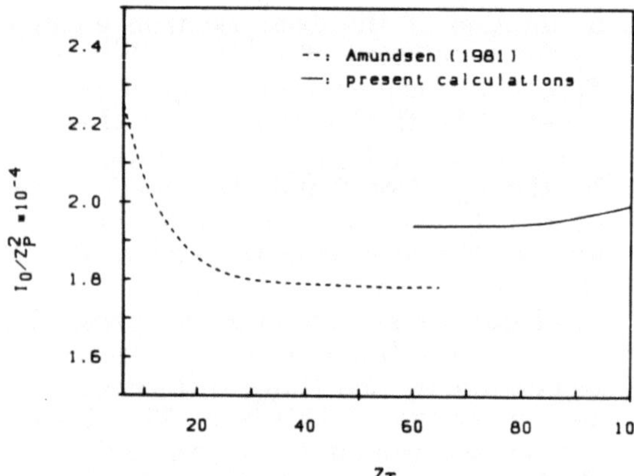

Fig. 4

K-shell ionisation probabilities (solid curve) at b=0 for 49.6 GeV/amu projectiles as a function of the target charge in comparison with the results of Amundsen and Aashamar (1981) (dashed curve).

Fig.4 shows that the probability I_0 for K-shell ionisation is independent of the target charge at high relativistic energies and zero impact parameter. Amundsen and Aashamar (1981) found this result already for $Z_T \leq 65$ using nonrelativistic wave functions for the electron, and gave a scaling law for I_0:

$$I_o = 1.8 \cdot 10^{-4} Z_p^2 \pm 10\%. \tag{15}$$

Our calculations show the validity of the scaling law up to the heaviest target charges.

The ionisation cross section rises with increasing projectile energy due to the relativistic contraction of the projectile potential, which leads to contributions from larger and larger impact parameters. Fig.5 shows the K-shell ionisation cross section as a function of $\gamma = (1-\beta^2)^{-1/2}$. Our results (full curve) are compared with a calculation of Anholt (1979) (broken curve). Both curves show the same qualitative behaviour when the collision energy is increased. For $\gamma > 2$ the absolute difference between the calculations increases slowly as γ increases, which may be caused by the fact that Anholt uses semi-relativistic Darwin wave functions in contrast to the Coulomb-Dirac wave functions used in our calculation. Both calculations take experimental binding energies for the K-shell electrons in order to compare the results with measurements of the K-shell ionisation of atoms.

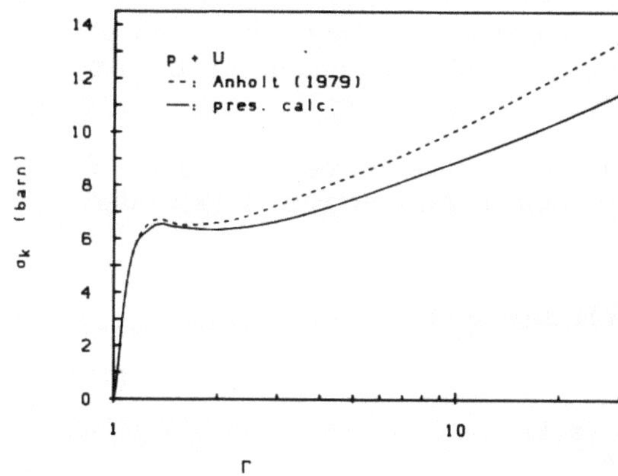

Fig. 5

K-shell ionisation cross section for the system p+U as a function of $\Gamma=(1-v_p^2/c^2)^{-\frac{1}{2}}$ (solid curve). The dashed curve is taken from Anholt (1979). In both curves experimental binding energies have been used.

Finally in Fig.6 we show a comparison of the calculations with experimental data for the K-shell ionisation cross section for 4.88 GeV protons scattered on various targets. The full curve presents our result, the dashed one is the result of a plane wave Born approximation obtained by Davidovic et al. (1978) and Anholt (1979). The experimental values are taken from Anholt et al. (1977). The good agreement of both calculations with experiment is typical for collisional systems with energies E_{lab}>500 MeV/amu (see Anholt et al. 1984). Recent measurements of the ionisation cross section of U^{91+} by Au at 960 MeV/amu (Anholt 1986) also show good agreement with calculations in first-order perturbation theory.

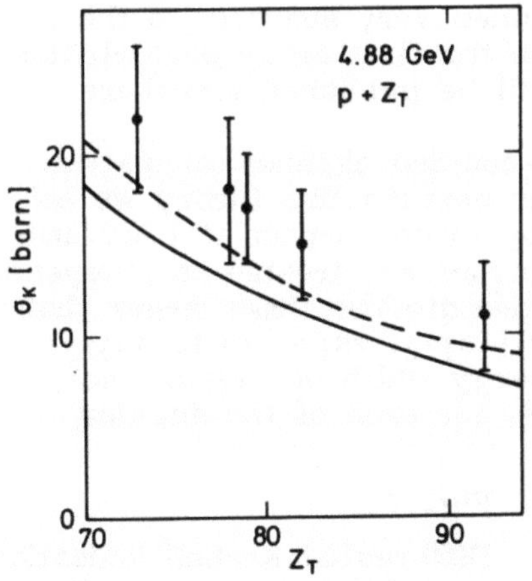

Fig. 6

K-shell ionisation cross sections for 4.88 GeV protons on various targets as a function of the target charge. The experimental data are taken from Anholt et al.(1977). The solid curve gives the present calculations, the dashed curve the PWBA results of Davidović et al.(1978) and Anholt (1979).

2.3 COUPLED CHANNEL CALCULATIONS

Fig.3 indicates the failure of the first order perturbation theory in the case of the K-shell ionisation in U+U collisions. Using the scaling of the first order perturbation theory with Z_P^2, we obtain ionisation probabilities close to unity for small impact parameters, which is in contrast to the basic assumptions of the perturbation theory. Instead of applying perturbation theory of higher order, we expand the time-dependent solution of the Dirac-equation (7) in a finite set of basis functions, which we take as Coulomb-Dirac functions centered at the target ion, and solve coupled time-dependent differential equations of first order for the expansion coefficients.

The main problem in these calculations is the proper treatment of the continuum channels. The wave function of an electron is expanded as (b.s. = bound states)

$$\psi_i(\vec{r},t) = \sum_{b.s.} a_{in}(t)\phi_n(\vec{r})\exp(-i\omega_n t) + \sum_{E_k>mc^2} a_{iE_k}\tilde{\phi}_{E_k}(\vec{r},t). \tag{16}$$

For the continuum states we adopted the following discretisation

$$\tilde{\phi}_{E_k}(\vec{r},t) = \int_{E_k-\Delta E_k/2}^{E_k+\Delta E_k/2} dE\ \phi_E(\vec{r})\exp(-iEt/\hbar). \tag{17}$$

By introducing these functions we approximate the coefficients a_{iE_k} by step functions which have constant values in the intervals $[E_k-\Delta E_k/2,\ E_k+\Delta E_k/2\ ]$. The new continuum states are wave packets with their centers moving with the momentum $p_k=(E_k^2-m^2c^4)^{\frac{1}{2}}/c$. A similar approach is used by Mehler et al. (1985) for nonrelativistic heavy ion collisions, but with the difference that the time-dependent phase factor in Eq.(17) is omitted. Therefore, the wave packets of Mehler et al. (1985) are stationary and centered at the origin during the collision.

The coupling matrix elements are obtained very similarly to the calculation of the transition matrix elements in the first order perturbation theory. The details of these calculations will be published elsewhere.

Fig.7 shows a first application of the coupled channel calculation. As test and comparison with the first order perturbation theory we calculated the ionisation probability of U^{90+} by proton impact at 1GeV/amu and an impact parameter b=0. The two electrons are treated as independent. Therefore, the resulting ionisation probabilities are two times larger than those for U^{91+}. We chose 9 bound states ($1s_{1/2}$, $2s_{1/2}$ up to $3d_{5/2}$) and 10 continuum wave packets with an energy width of $\Delta E_f=0.2\ mc^2$, equally distributed from $E_f=mc^2$ up to $3mc^2$, for each of the angular

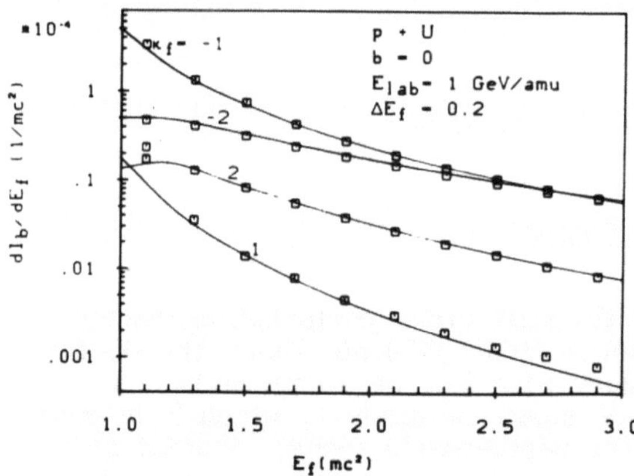

Fig. 7

Differential K-shell ionisation probability in a p+U-collision at b=0 and E_{lab}=1GeV/amu as a function of the electron energy E_f for various angular momenta of the ionised electron denoted by κ_f. The solid lines give the results of the first order perturbation theory, whereas the squares show the results of a coupled channel calculation with a width of the continuum wave packets of $\Delta E_f=0.2mc^2$.

momenta $s_{1/2}, p_{1/2}, p_{3/2}$ and $d_{3/2}$. These channels were found to be the dominant ones for the excitation and ionisation of K-shell electrons according to the finite difference method and the first order perturbation theory. The squares in Fig.7 present the results of the coupled channel calculation and are compared with the results of the first order perturbation theory (solid lines). The agreement between both calculations is very good as one might expect for the weak perturbation of the incoming proton.

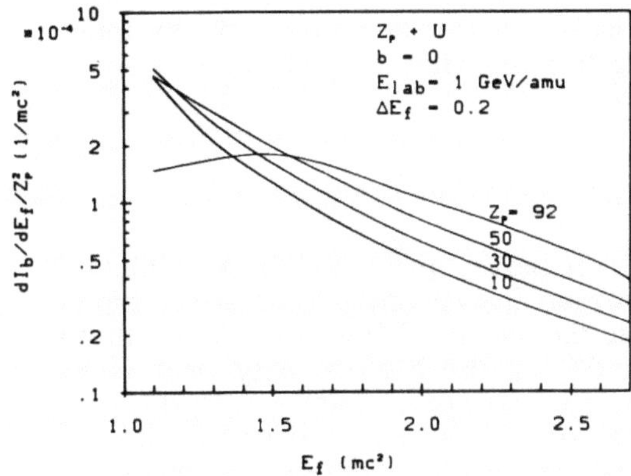

Fig. 8

Differential K-shell ionisation probabilities in Zp+U collision systems at b=0 and E_{lab}=1GeV/amu as a function of the energy of the ionised electron. The curves are obtained from coupled channel calculations by using the same basis set as in Fig.7.

The next step was to increase the charge of the projectile. Fig.8 shows the resulting differential ionisation probabilities summed over all included angular momenta for different projectile charges (Zp=10,30, 50,92). The results are divided by Z_p^2 in order to demonstrate the effects of perturbations of higher order and of unitarity which prevent the ionisation probabilities from rising over unity. Especially in U+U collisions, higher order processes, induced by strong continuum-continuum couplings, shift the probabilities from low energetic continuum states to higher ones. We note, that the actual set of basis functions, cited above, seems to be insufficient to describe the differential ionisation probabilities in the U+U collision system. The weak decrease of the differential probability of the U+U system indicates the necessity to extend the continuum channels to higher energies. Nevertheless this has only a minor influence on the total ionisation probability.

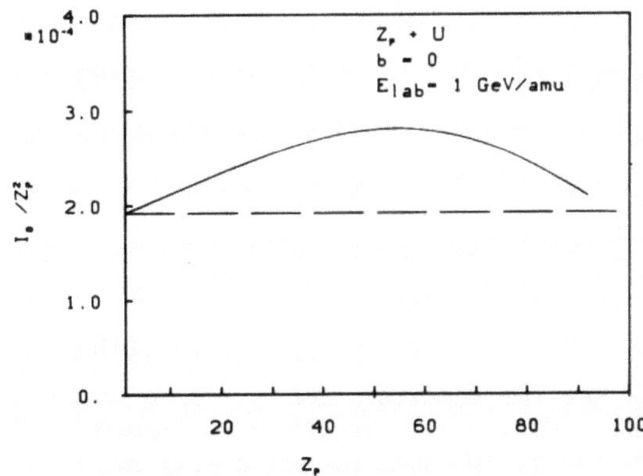

Fig. 9

K-shell ionisation probabilities in Zp+U collision systems at b=0 and E_{lab}=1GeV/amu as a function of Zp. The solid line is obtained from coupled channel calculations whereas the broken line gives the result of first order perturbation theory which scales with Z_p^2.

Fig.9 shows the total ionisation probabilities divided by Z_P^2 (full curve) as a function of Zp for b=0 in comparison with those of first order perturbation theory (broken curve). Higher order perturbations increase the ionisation probabilities above the values of the first order perturbation theory. We find a maximum around Zp=55. Beyond this point the requirements of unitarity prevent the ionisation probability from rising faster than Z_P^2 leading to a constant value in the limit of very strong perturbations. This effect may be already seen in recent experiments of Anholt (1986), measuring K-shell ionisation cross sections of U^{91+} by various projectiles at 960 MeV/amu. The behaviour of the scaled cross section as a function of Zp is very similar to the behaviour of the ionisation probabilities shown in Fig.9, indicating that this effect is still present also at larger impact parameters.

3. PAIR PRODUCTION

Cross sections for electron-positron pairs, produced in a collision of a heavy particle colliding with another heavy particle at rest, have already been derived in the 1930s on the basis of the equivalent-photon method (Weizsäcker and Williams method) for the high-energy limit (for references to the older literature see, for example, Heitler (1954)). In this Section we discuss calculations for pair creation carried out by Becker et al. (1986) using the first order perturbation theory. Since the pair creation probabilities are small, it is not necessary to treat higher orders of the perturbation theory. The pair production is of great interest for relativistic heavy ion colliders (Gould 1984). For example, the electron capture after pair production may impose certain limits on the beam luminosity and lifetime, if the corresponding cross section is high enough.

In Section 3.1 we present results for the direct electron-positron pair production when both particles are emitted into continuum states of the target ion. In Section 3.2 we discuss the creation of a bound electron together with a positron in the continuum. This process is denoted as electron capture from pair production. In this connection we also discuss results for muon capture from muon pair creation.

3.1 DIRECT ELECTRON-POSITRON PAIR PRODUCTION

The double differential probability for the production of an electron-positron pair within the energy intervals dE_p and dE_e of the energies of the positron and electron, respectively, is given in the first order perturbation theory

$$\frac{d^2 I_b}{dE_p dE_e} = \sum_{\kappa_p \mu_p \kappa_e \mu_e} | a_{pe} |^2, \tag{18}$$

$$a_{pe} = \frac{8\pi i Z_p e^2}{\hbar v_p} \int_q^\infty ds \frac{s}{s^2 - q^2 \beta^2} \sum_{\ell m} i^{\ell - m} \langle \psi_e | (1-\beta\alpha_z) j_\ell(sr) Y_{\ell m}(\hat{r}) | \psi_p \rangle B_{\ell m}(b,q,s), \tag{19}$$

where $q=(E_p+E_e)/\hbar v_p$, and ψ_e and ψ_p are the electron and positron wave functions, which are chosen as Coulomb-Dirac wave functions centered at the target ion. The matrix element in Eq.(19) can be calculated analytically by using a new approach for treating Appell functions (Becker et al. 1986).

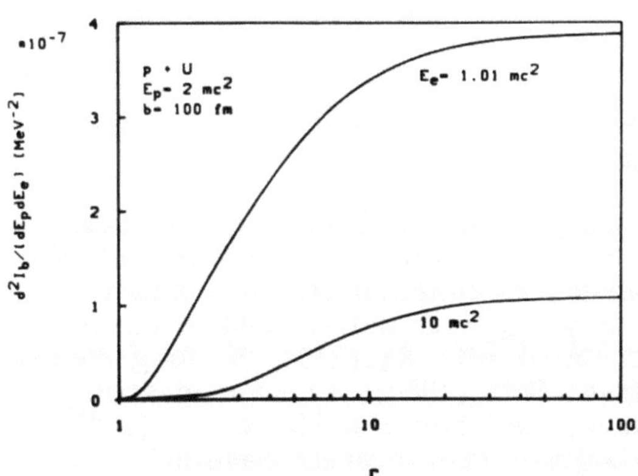

Fig. 10

Double differential probability for direct electron-positron pair creation in a proton-Uranium collision as function of $\gamma = \Gamma=(1-v_p^2/c^2)^{-\frac{1}{2}}$. The curves are calculated at an impact parameter b=100fm and a positron energy $E_p=2mc^2$ for two electron energies E_e=1.01 and 10mc².

Fig.10 shows the double differential probability d^2I_b/dE_pdE_e for a collision of a proton on an Uranium nucleus (unscreened: Z_T=92) as a function of $\gamma=(1-\beta^2)^{-\frac{1}{2}}$. For projectiles with charge numbers $Z_P>1$ the probabilities have to be multiplied by Z_P^2 according to Eq.(18). For $\gamma\to\infty$ the curves of Fig.10 tend to constant values dependent on E_p, E_e and b. This behaviour is the same as that shown in Fig.3 for the probabilities of K-shell ionisation.

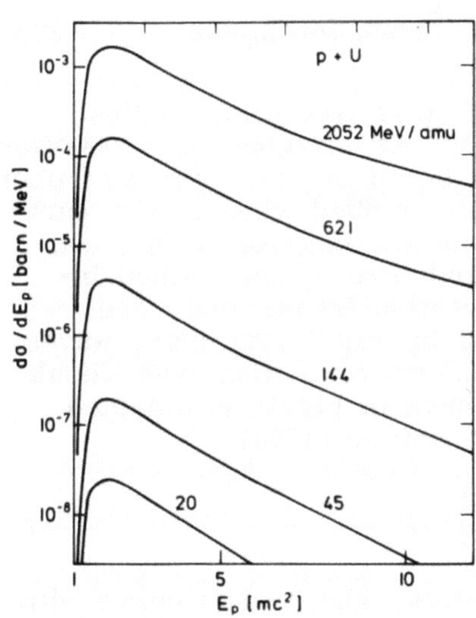

Fig. 11

Single differential cross section for direct electron-positron pair creation in a proton-Uranium collision as a function of the positron energy E_p for incident energies E_{lab}=20,45, 144,621 and 2052 MeV/amu as indicated in the figure.

Fig.11 shows the single-differential pair creation cross section $d\sigma/dE_p$ for various incident energies of E_{lab}=20,45,144,621 and 2052 MeV/amu, corresponding to relative velocities of 0.2, 0.3, 0.5, 0.8 and 0.95 c, respectively. The curves have the typical behaviour of positron spectra for pair creation in a Coulomb field. The steep increase for small positron energies is caused by the repulsion of the wave function of the slow positrons in the Coulomb field of the target nucleus.

In a last step we have integrated $d\sigma/dE_p$ over the positron energy and obtained the total cross section for direct pair production. The resulting cross section for a p+U collision is shown in Fig.12 as a function of the incident energy (full curve). Our values agree with the calculations of Anholt et al. (1983) in the range of $E_{lab}\leq20$ MeV/amu. The method used by these authors is restricted to low incident energies.

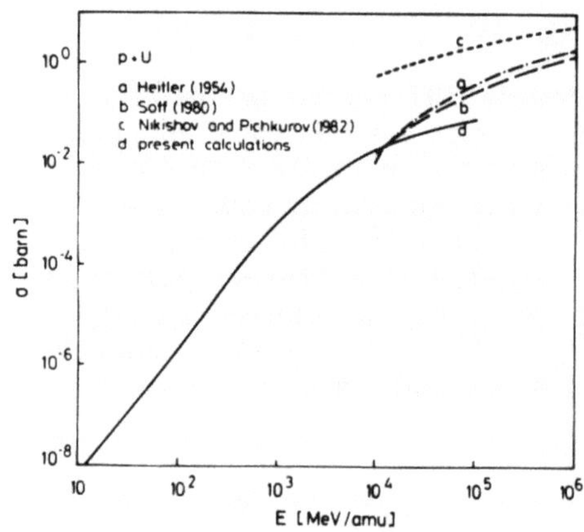

Fig. 12

Cross section for direct electron-positron pair creation in a proton-Uranium collision as function of the incident energy E_{lab}: a) dotted-dashed curve calculated with Eq.(21), b) long-dashed curve: Soff (1980), c) short-dashed curve: Nikishov and Pichkurov (1982), d) solid curve: present results.

The cross section increases approximately linearly up to energies E_{lab}= 1 GeV/amu. This indicates a scaling law for the cross section of pair creation in a collision of a charge Z_P with an Uranium target in the energy range of 20 MeV/amu $\leq E_{lab} \leq$ 1 GeV/amu:

$$\sigma = 1.45\cdot 10^{-11} Z_P^2 (E_{lab}/(\mathrm{MeV/amu}))^{2.6} \mathrm{barn} \text{ for } Z_T=92. \qquad (20)$$

The scaling law (20) is no longer valid for E_{lab} >1 GeV/amu. In this region the cross section increases more slowly. At energies E_{lab}>10GeV/amu we can compare our results with calculations based on the equivalent-photon method (see Heitler 1954). This method can be applied when impact parameter smaller than the Compton wave length of the electron do not contribute significantly to the cross section. Furthermore, the method includes the Bethe-Heitler formula for pair creation by photons. The electrons and positrons are usually described by relativistic plane waves (see Achieser and Berestezki (1962); for Coulomb corrections see Nikishov and Pichkurov (1982)). The dotted-dashed curve in Fig.12 is obtained with the expression given in the textbook of Heitler (1954):

$$\sigma = \frac{28}{27\pi}\left(\frac{Z_P Z_T}{137} r_o\right)^2 \left(\ln\left(\frac{1}{4}\gamma\right)\right)^3 , \qquad (21)$$

where r_o=2.818fm is the classical electron radius. The broken curve with the long dashes is the result of Soff (1980) and that with short dashes the result of Nikishov and Pichkurov (1982). The agreement of our calculations with the result of Soff (1980) and the values calculated from Eq.(21) is fairly good for energies smaller than 30GeV/amu, but they tend to disagree for high relativistic energies. This may be partly due to numerical difficulties in our evaluation of the contributions of the high multipole components with ℓ>10, which may lead to an underestimation of the high-energy tails of the electron-positron spectra.

3.2 PAIR PRODUCTION WITH INNER-SHELL CAPTURE

The calculation of probabilities and cross sections for this type of pair creation is in complete analogy to the calculation of the probabilities for inner-shell ionisation. In the corresponding formulas one only substitutes the state of the positive continuum by a state of the negative continuum.

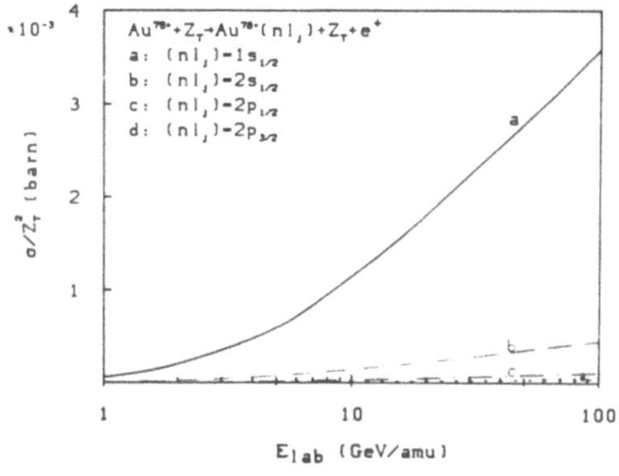

Fig. 13

Cross sections for pair production with inner shell capture in $Au^{79+}+Z_T$ collisions as a function of the incident energy for the capture into: a) the $1s_{1/2}$-state, b) the $2s_{1/2}$-state, c) the $2p_{1/2}$-state and d) the $2p_{3/2}$-state of Au^{78+}. The cross sections scale with Z_T^2.

Fig.13 shows the pair production cross sections with capture of the electron into the four lowest bound states of the Au^{79+}-ion, which is the projectile in the notation of Fig.13. Experiments for measuring these reactions have been proposed by Gould (1986). The cross sections show a similar asymptotic behaviour for $\gamma \to \infty$ as those for K-shell ionisation (see Fig.5), increasing proportionally to $\ln\gamma$. The cross section decreases rapidly with increasing main quantum number and angular momentum of the bound state. For example, we find a decrease of about a factor of 10 when going from the K-shell to the L-shell. This effect is caused by the very high momentum transfer in the pair creation process, which is transferred only for impact parameters smaller than the K-shell radius.

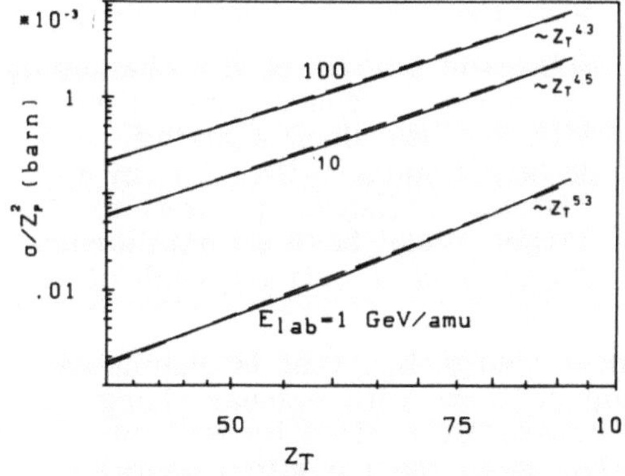

Fig. 14

Cross sections for pair production with capture into the K-shell of the target ion as a function of the target charge for incident energies of 1, 10 and 100 GeV/amu. The dashed lines show linear least-square fits from which the scaling law is obtained.

Fig.14 shows the dependence of the cross section for pair production with capture into the target K-shell as function of the target charge number. The curves can be approximated by straight lines within a range of target charge numbers $40 \le Z_T \le 92$ (dashed lines). This behaviour implies a scaling law for the cross section as $\sigma \sim (Z_T)^a$ with an exponent ranging from a=5.3 at 1 GeV/amu to a=4.3 at 100 GeV/amu. This dependence of the cross section reflects the strong influence of the target charge on the bound state wave functions.

As a further point we want to mention the order of magnitude of the cross section for pair production with capture into the K-shell for $U^{92+}+U^{92+}$ collisions. This cross section is about 70 barn for E_{lab}=100 GeV/amu and 200 barn for E_{lab}=20 TeV/amu. For the latter energy Gould (1984) has discussed a critical limit of roughly 100 barn which would set limits on beam survival or luminosity for future relativistic heavy ion colliders.

In this connection we can give some first results for the creation of pairs of muons in relativistic heavy ion collisions. A first estimate for the cross section is obtained by

$$\sigma_{\mu\pm} = \sigma_{e\pm}(\lambda_\mu/\lambda_e)^2, \tag{22}$$

where λ_e and λ_μ are the Compton wave lengths of the electron and muon, respectively. The formula yields cross sections for muon production which are by a factor $4 \cdot 10^4$ smaller than the ones for the electron-positron production. This estimate does not include the effects of the finite extension of the nuclear charges, which can be neglected in the case of the electron-positron production because the Compton wave length of the electron is large compared to the nuclear radius. In the case of muon pair production the Compton wave length is about 3-4 times smaller than the radius of the nuclear charge distribution with the important consequence that the finite extension of the nuclear charge plays a dominant role. First calculations of muon pair production with capture of the μ^--particle in the K-shell have shown that the finite distribution of the projectile and target nuclear charge leads to an additional decrease of the cross section by a factor of 10^{-5} in a $U^{92+}+U^{92+}$-collision at E_{lab}=100 GeV/amu.

4. COHERENT PROCESSES BY CHANNELING OF RELATIVISTIC IONS

In this Section we suggest experiments on the coherent effect of a string of heavy ions for atomic excitation, ionisation and pair production probabilities. Such experiments can be carried out by channeling of relativistic heavy ions through an oriented crystal.

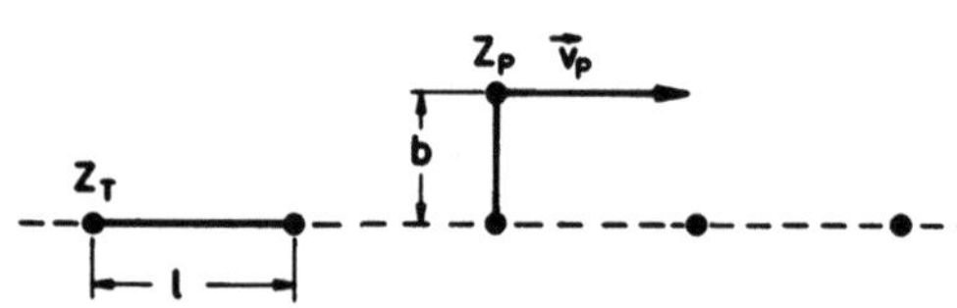

Fig. 15

Schematic picture of the channeling experiment. The channeled ion with Zp flies along a string of target nuclei with Z_T in a distance b (impact parameter). The target nuclei have an equilibrium distance of ℓ between each other.

In Fig.15 we consider a heavy ion of charge Z_P, that is channeled with a relativistic velocity along a string of atoms with nuclear charge Z_T. We assume that the channeled ion gets coherently excited by N unscreened, equidistant charges Z_T. In this case, the transition operator in Eq.(10) has to be replaced by the following expression, where ℓ is the atomic distance:

$$(1-\beta\alpha_z)/r' \rightarrow (1-\beta\alpha_z) \sum_{j=0}^{N-1} 1/((x-b)^2+y^2+\gamma^2(z-v_P t+j\ell)^2)^{1/2}. \tag{23}$$

By using this formula, the coherent transition amplitude a_{fi}^{coh} can be calculated in first order perturbation theory and written as a product of the atomic amplitude and an amplification factor $(q=(E_f-E_i)/\hbar v_P)$:

$$|a_{fi}^{coh}|^2 = |a_{fi}|^2 S^2(q\ell/\gamma), \tag{24}$$

where

$$S(x) = \sin(Nx/2)/\sin(x/2).$$

The coherent transition probability is amplified by the factor N^2 if the momentum transfer q assumes one of the values

$$q = 2\pi n\gamma/\ell, \quad n=0,1,2.... \tag{25}$$

This resonance effect is in competition with the incoherent action of the target charges which contributes a factor N in the transition probability only.

Up to here, our arguments were independent of the special transition under consideration. However, we expect striking effects if one of the states, initial or final, is a continuum state and the other one a bound state. In this case we may observe very sharp peaks in the differential probabilities, since the resonance condition (25) can be fulfilled with the energy of the continuum state. Sharp lines would appear in the electron-spectra for inner-shell ionisation and in the positron-spectra for pair creation with inner-shell capture.

The coherent resonance effect is disturbed by thermal vibrations of the string atoms in beam direction. These perturbations may destroy the coherence effect of the string. Bazylev and Zhevago (1979) have used a simple theory to account for this effect. According to the Debye theory the mean square of the amplitude for thermal vibrations is obtained as

$$u^2 = \frac{3\hbar^2}{M_T k\Theta}\left(\left(\frac{T}{\Theta}\right)^2 \int_0^{\Theta/T} \frac{x\,dx}{\exp(x)-1} + \frac{1}{4}\right) \tag{26}$$

Here, M_T is the mass of the string atoms, T the temperature and Θ the Debye temperature.

The mean square of the vibrational amplitude affects the amplification factor in Eq.(24). Bazylev and Zhevago (1979) have substituted the following averaged expression for S^2:

$$\overline{S^2} = \int \Big| \sum_{j=0}^{N-1} \exp(-i(jq\ell/\gamma + qz_j/\gamma)) \Big|^2 \cdot P(z_o, \dots z_{N-1})\, d^N z, \tag{27}$$

where z_j is the displacement of the j-th string atom out of its equilibrium position. $P(z_o, \dots z_{N-1})$ is the probability of finding the first atom at displacement z_o, the second ion at displacement z_1 and so on. If the thermal vibrations of the atoms in the direction of the string are described by N independent oscillators, one obtains:

$$\overline{S^2} = N\cdot(1-\exp(-u^2q^2/\gamma^2)) + S^2\left(\frac{q\ell}{\gamma}\right)\exp(-u^2q^2/\gamma^2). \tag{28}$$

According to this equation the averaged amplification factor $\overline{S^2}$ consists of an incoherent and coherent part. The possibility of observing the coherent action of the string is dependent on the value of u^2q^2/γ^2.

Let us consider pair creation in a Pb crystal. For pair creation we have $q=2mc/\hbar$. Using the mean square of the vibrational amplitude of the Pb crystal of $10^7\text{fm}^2 \le u^2 \le 10^8\text{fm}^2$ for $0 \le T \le 300^oK$, we obtain for a projectile velocity with $\gamma=10$ the following limits of the exponential factor in Eq.(28): $2\cdot 10^{-12} \le \exp(-u^2q^2/\gamma^2) \le 7\cdot 10^{-2}$. This result means that at least 100 string atoms are necessary and also the target has to be cooled down to the temperature of liquid Helium in order to observe the coherence effects in pair creation. The experimental situation can be improved by looking for reactions with smaller values of q, e.g. K-shell ionisation, and higher projectile velocities.

5. CONCLUSIONS

The experimental and theoretical interest in atomic physics with relativistic heavy ions is directed towards the exploration of the effects arising from the strong electromagnetic field generated by the projectile charge. As shown, the present experimental data can be explained with approximate theories tailored to the relativistic problem.

Future theoretical problems are, for example, corrections of the cross sections due to recoil effects, the cross section for muon pair production by taking the nuclear charge distributions of the projectile and target nuclei into account, and the coupling between the continuum states and its effect on the probability for K-shell ionisation at impact parameters, where the first order perturbation theory fails. The multiple ionisation by relativistic heavy ions is an interesting problem, which is not yet solved for heavy target atoms (see, for example, Kelbch et al. 1986).

Principal problems arise for extremely high relativistic collisions. The question of a possible limitation of the cross sections as function of the incident energy can not yet be answered without detailed investigations of the validity of the semiclassical approximation and the perturbation theory with respect to the normalisation of the probabilities. Last not least, the problem of the linearity of the Maxwell equations (Rafelski et al. 1971) could also be studied in these reactions.

REFERENCES

Achieser, A.I. and Berestezki, W.B., 1962, Quantenelektrodynamik (Frankfurt/M., Harri Deutsch).

Amundsen, P.A. and Aashamar, K.,1981, J.Phys.B 14: 4047.

Anholt, R., Nagamiya, S., Rasmussen, J.O., Bowman, H., Ioannou-Yannou, J.G., and Rauscher, E., 1976, Phys.Rev. A 14: 2103.

Anholt, R., Ioannou-Yannou, J., Bowman, H., Rauscher, E, Nagamiya, S., Rasmussen, J.O., Shibata, T., and Ejiri, H., 1977, Phys.Lett. 59A: 429.

Anholt, R., 1979, Phys.Rev. A 19: 1004.

Anholt, R., Jakubaßa-Amundsen, D.H., Amundsen, P.A., and Aashamar, K., 1983, Phys.Rev. A 27: 680.

Anholt, R., Andriamonje, S.A., Morenzoni, E., Stoller, Ch., Molitoris, J.D., Meyerhof, W.E., Bowman, H., Xu, J.S., Xu, Z.Z., Rasmussen, J.O., and Hoffmann, D.H.H., 1984a, Phys.Rev.Lett. 53: 234.

Anholt, R., Meyerhof, W.E., Stoller, Ch., Morenzoni, E., Andriamonje, S.A., Molitoris, J.D., Baker, O.K., Hoffmann, D.H.H., Bowman, H., Xu, J.S., Xu, Z.Z., Frankel, K., Murphy, D., Crowe, K., and Rasmussen, J.O., 1984b, Phys.Rev. A 30: 2234 .

Anholt, R., Meyerhof, W.E., Gould, H., Munger, Ch., Alonso, J., Thieberger, P., and Wegner, H.E., 1985, Phys.Rev. A 32: 3302.

Anholt, R., and Meyerhof, W.E., 1986, Phys.Rev. A 33: 1556.

Anholt, R., Stoller, Ch., Molitoris, J.D., Spooner, D.W., Morenzoni, E., Andriamonje, S.A., Meyerhof, W.E., Bowman, H., Xu, J.S., Xu, Z.Z., Rasmussen, J.O., and Hoffmann, D.H.H., 1986, Phys.Rev. A 33: 2270.

Anholt, R., 1986, private communication.

Bak, J.F., Melchart, G., Uggerhøj, E., Forster, J.S., Jensen, P.R., Madsbøll, H., Møller, S.P., Petersen, G., Schiøtt, H.E., Regall, R., and Siffert, P., 1982, Phys.Rev. A 25: 1334.

Bang, J., and Hansteen,J.M., 1959, K. Dansk. Vidensk. Selsk. Mat.-Fys. Meddr. 31 No.13.

Bazylev, V.A., and Zhevago, N.K., 1979, Sov.Phys. JETP 50 : 161.

Becker, U., Grün, N., and Scheid, W., 1983, J.Phys.B 16: 1967.

Becker, U., Grün, N., and Scheid, W., 1985, J.Phys.B 18: 4589.
Becker, U., Grün, N., Scheid, W., and Soff, G., 1986, Phys.Rev. Lett. 56: 2016.
Becker, U., Grün, N., and Scheid, W., 1986, J.Phys.B 19: 1347.
Bottcher, C., 1984, in Physics of Electronic and Atomic Collisions, ed.by J. Eichler et al., (North Holland, New York).
Crawford, H.J., 1979, Ph.D. thesis, University of California (LBL Report No.8807, 1979) unpublished.
Davidović, D.M., Moiseiwitsch, B.L., and Norrington, P.H., 1978, J.Phys.B 11: 847.
Eichler, J., 1985, Phys.Rev. A 32: 112.
Garcia, J.D., 1970, Phys.Rev. A 1: 280.
Gould, H., Greiner, D., Lindstrom, P., Symons, T.J.M., and Crawford, H., 1984, Phys.Rev. Lett. 52: 180.
Gould, H., 1984, Lawrence Berkeley Laboratory Technical Information, LBL 18593 UC-28.
Gould, H., 1986, private communication.
Greiner, D.E., Beiser, F.S., and Heckman, H.H., 1977, in Abstracts of the X. International Conference on the Physics of Electronic and Atomic Collisions, Paris 1977, ed.by M. Barat and J. Reinhardt.
Heitler, W., 1954, The Quantum Theory of Radiation (Oxford: Oxford University Press): 264
Jackson, J.D., 1975, Classical Electrodynamics (John Wiley & Sons, Inc., New York).
Kelbch, S., Ullrich, J., Rauch, W., Schmidt-Böcking, H., Horbatsch, M., Dreizler, R.M., Hagmann, S., Anholt, R., Schlachter, A.S., Müller, A., Richard, P., Stoller, Ch., Cocke, C.L., Mann, R., Meyerhof, W.E., and Rasmussen, J.D., 1986, J.Phys.B 19: L47.
McGuire, J.H., and Richard, P., 1973, Phys.Rev. A 8: 1374.
Mehler, G., de Reus, T., Müller, U., Reinhardt, J., Müller, B., Greiner, W., and Soff, G., 1985, Nucl.Instr.Meth. A246: 559.
Meyerhof, W.E., Anholt R., Eichler, J., Gould, H., Munger, Ch., Alonso, J., Thieberger, P., and Wegner, H.E., 1985, Phys.Rev. A 32: 3291.
Mott, N.F., and Massey, H.S.W., 1965, The Theory of Atomic Collisions, (Oxford University Press).
Nikishov, A.I., and Pichkurov, N.V., 1982, Sov.J.Nucl.Phys. 35: 561.
Rafelski, J., Fulcher, L.P., and Greiner, W., 1971, Phys.Rev.Lett. 27: 958.
Soff, G., 1980, in Proceedings of XVIII Winter School in Bielsko-Biala, Poland, ed.by: A. Balanda and Z. Stachura: 201.
Teubner, E., Terlecki, G., Grün, N., and Scheid, W., 1980, J.Phys.B 13: 523.
Teubner, E., Grün, N., and Scheid, W., 1982, J.Phys.B 15: 1269.
Valluri, S.R., Becker, U., Grün, N., and Scheid, W., 1984, J.Phys. B 17: 4359.
Winther, A., and Alder, K., 1979, Nucl.Phys. A319: 518.

This publication presents the essential part of the doctoral thesis of Ulrich Becker, Giessen (D26), 1986.

μ- AND τ-PAIR PRODUCTION FROM RELATIVISTIC HEAVY-ION COLLISIONS*

C. Bottcher and M. R. Strayer

Physics Division
Oak Ridge National Laboratory
Oak Ridge, Tennessee 37831

INTRODUCTION

In these lectures we shall attempt to address the question of μ- and τ-pair production from the motional Coulomb fields available at the new relativistic heavy-ion accelerators. It is well known that electrons and positrons are produced from such collisions in sizeable multiplicities,[1] and Gould[2] has suggested that heavy lepton pair creation may be possible at RHIC.

We shall divide our discussion of these phenomena into two parts. In the first part, a semiclassical field theory is developed which is appropriate for families of leptons which are coupled electromagnetically. The field equations are mapped on to a lattice of collocation points using basis spline methods, and techniques for solving the resulting lattice equations are outlined.

In the second part, we shall examine the properties of the transverse electromagnetic field near the heavy-ion beam and present physical arguments as to the feasibility of pair creation under a variety of circumstances. Using the Dirac-Hartree equations developed in part one, we shall dynamically evolve the vacuum, using the appropriate fields, and compute μ-pair and τ-pair production cross sections.

SEMICLASSICAL FIELD THEORY

In a pedagogical sense, our treatment of pair production in the presence of strong fields is similar to the early version of the adiabatic basis method developed by Greiner and co-workers.[3] However, there are important differences which we shall note in the following discussion. A more complete account of the method is given in Ref. 4. The Dirac-Hartree

*Research sponsored by the U.S. Department of Energy under contract DE-AC05-84OR21400 with Martin Marietta Energy Systems, Inc.

field equations can be obtained from three principal assumptions: i) existence of a semiclassical action and an effective Lagrangian, ii) identification of the initial state and the vacuum state, and iii) unitary time-evolution of these states.

(i) Assuming a minimal electromagnetic coupling of electrons, muons, and tauons through a classical electromagnetic field, the effective Lagrangian density case can be written as[5]

$$\mathscr{L}(x) = \mathscr{L}_e + \mathscr{L}_\mu + \mathscr{L}_\tau - 1/4\, F_{\mu\nu}F^{\mu\nu} + J_\mu A^\mu \tag{1}$$

where, in Eq. (1), J_μ is a conserved classical external current, A^μ and $F^{\mu\nu}$ are respectively the four-vector and field tensor of the classical electromagnetic field, and where the terms $\mathscr{L}_\ell$ are given by

$$\mathscr{L}_\ell(x) = \overline{\psi}_\ell(x)\left[\gamma_\mu\left(i\partial^\mu - A^\mu\right) - m_\ell\right]\psi_\ell(x),\ \ell = e,\mu,\tau. \tag{2}$$

We note that this Lagrangian separately conserves electron, muon, and tauon numbers, as illustrated by the diagram in Fig. 1. Thus, the different terms in Eq. (1) are only coupled through A^μ, and we shall assume, for simplicity, that this coupling can be ignored. Thus, for each species of lepton we take[6]

$$S_\ell = \int d^4x \ {}_\ell\langle\Phi(t)\mid :\mathscr{L}(x): \mid\Phi(t)\rangle_\ell \tag{3}$$

where $|\Phi(t)\rangle$ denotes the many-lepton state at a time t which evolves from the initial state, and where the normal ordering is with respect to a reference state which must be specified. This form of the action has been extensively used in nuclear physics to obtain Hartree-Fock-Bogoliubov equations.[7] In Eq. (3) the dynamical coordinates which are varied to make the action stationary are A^μ and the parameters labeling the wavefunction are $\Phi(t)$, not the lepton field operators. Thus, in what follows, it will be convenient to work in the Schrödinger picture.

(ii) The initial state is assumed to be a single Slater determinant so that

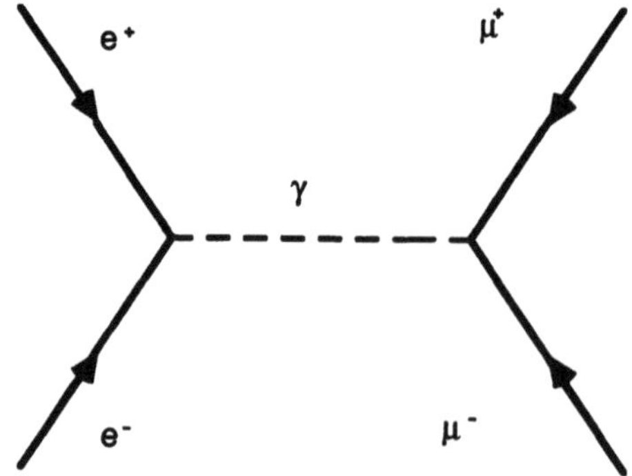

Fig. 1

$$|\Phi(t)\rangle \underset{t\to-\infty}{\to} |\Phi_o\rangle. \tag{4}$$

We shall assume, for initial times, that there is a well-defined Dirac Hamiltonian with a spectrum as shown schematically in Fig. 2. All of the states with energies less than the label 0 are occupied in the vacuum state $|0\rangle$, which we shall identify as the reference state. The single-particle states with labels between 0 and f will comprise the initial state $|\Phi_o\rangle$. By construction, the single-particle states in Fig. 2 are complete and orthonormal

$$1 = \sum_\lambda \{ |\chi_\lambda^{(+)}\rangle\langle\chi_\lambda^{(+)}| + |\chi_\lambda^{(-)}\rangle\langle\chi_\lambda^{(-)}| \},$$

$$\text{and} \quad \langle\chi_\lambda^{(s)}|\chi_{\lambda'}^{(s')}\rangle = \delta_{\lambda\lambda'}\delta_{ss'} \tag{5}$$

With the choice of reference state as given above, we can identify $|\psi_\lambda^{(+)}\rangle$ and $|\chi_\lambda^{(-)}\rangle$ as single-particle and single anti-particle wavefunctions. In the second quantized representation, we have particle and anti-particle annihilation operators a_λ and b_λ respectively so that

$$a_\lambda|0\rangle = b_\lambda|0\rangle = 0$$

and (6)

$$\{a_\lambda, a_{\lambda'}^\dagger\} = \{b_\lambda^\dagger, b_{\lambda'}\} = \delta_{\lambda\lambda'}.$$

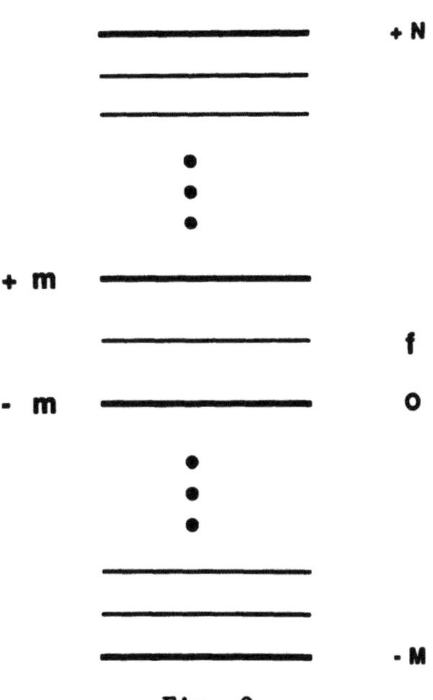

Fig. 2

All anticommutation combinations of the operators a and b not given in Eq. (6) are zero, and the initial state is

$$| \Phi_o> = \prod_{0<\lambda<f} a_\lambda^\dagger | 0>. \tag{7}$$

(iii) We assume that the dynamics governing the time evolution of the wavefunction in Eq. (4) is unitary; that is

$$| \Phi(t)> = S(t) | \Phi_o>, \tag{8}$$

where $S^\dagger S = SS^\dagger = 1$. There are several important consequences of this assumption. Equations (7) and (8) together guarantee that the state $\Phi(t)$ is at all times a single Slater determinant. This may be seen from the following equation

$$| \Phi(t)> = S(t)a_1^\dagger a_2^\dagger \ldots a_f^\dagger | 0> \tag{9}$$

By inserting $S(t)^\dagger S(t)$ between adjacent operators in the above, we can rewrite $\Phi(t)$ as

$$| \Phi(t)> = \alpha_1^\dagger(t)\alpha_2^\dagger(t)\ldots\alpha_f^\dagger(t) | 0(t)> \tag{10}$$

where the operators $\alpha_\lambda^\dagger(t)$ and the state $0(t)$ are given by

$$\alpha_\lambda^\dagger(t) = S(t)a_\lambda^\dagger S^\dagger(t), \qquad \lambda > 0, \tag{11}$$

and

$$| 0(t)> = S(t) | 0>. \tag{12}$$

Equation (10) is a time-dependent Slater determinant, where $0(t)$, as given in Eq. (12), is the vacuum for the operators $\alpha_\lambda(t)$. It is easy to show that $0(t)$ is also the vacuum for operators $\beta_\lambda(t)$, defined as

$$\beta_\lambda^\dagger(t) = S(t)\, b_\lambda^\dagger S^\dagger(t), \qquad \lambda < 0, \tag{13}$$

and thus we can identify a complete and orthonormal set of one-particle states at any time t by

$$\begin{aligned} | \psi_\lambda^{(+)}(t)> &= \alpha_\lambda^\dagger(t) | 0(t)> \qquad \lambda > 0, \\ | \psi_\lambda^{(-)}(t)> &= \beta_\lambda^\dagger(t) | 0(t)> \qquad \lambda < 0. \end{aligned} \tag{14}$$

These states, and the determinant in Eq. (10), contain dynamical excitations of the vacuum through the term $0(t)$. We should like to emphasize that the one-particle states in Eq. (14) cannot be interpreted as physical-particle or anti-particle states, because of these vacuum excitations. Physical lepton or anti-lepton states can only be identified in terms of the projections of these states onto the initial states which define the particle and anti-particle spectrum.

The Dirac-Hartree equations of motion are obtained from the stationary principle for the action in Eq. (3). As previously stated, we shall work in the Schrödinger picture. Because of the normal ordering inherent in the matrix element defining the semiclassical action, we need to expand the field operators in terms of the states in Eqs. (5) and (6)

$$\Psi(x) = \sum_\lambda \psi_\lambda^{(+)}(x)a_\lambda + \psi_\lambda^{(-)}(x)b_\lambda^\dagger. \tag{15}$$

In this representation, Eq. (3) is written as

$$\mathcal{S}_\ell = \int dt' \{ \langle\psi_j^{(+)} | L | \psi_k^{(+)}\rangle\langle\Phi_o | S(t)^\dagger : a_j^\dagger a_k : S(t) | \Phi_o\rangle$$
$$+ \langle\psi_j^{(+)} | L | \psi_k^{(-)}\rangle\langle\Phi_o | S(t)^\dagger : a_j^\dagger b_k^\dagger : S(t) | \Phi_o\rangle$$
$$+ \langle\psi_j^{(-)} | L | \psi_k^{(+)}\rangle\langle\Phi_o | S(t)^\dagger : b_j a_k : S(t) | \Phi_o\rangle \qquad (16)$$
$$+ \langle\psi_j^{(-)} | L | \psi_k^{(-)}\rangle\langle\Phi_o | S(t)^\dagger : b_j b_k^\dagger : S(t) | \Phi_o\rangle\}$$

where

$$L(x) = i\partial_t - h(x), \; h(x) = \vec{\alpha}\cdot(-i\vec{\nabla}-\vec{A}) + \beta m + A_o \qquad (17)$$

This representation of the action is almost normal ordered; only the last term need be changed. In order to evaluate Eq. (16), we need to calculate matrix elements of operators which have a form $S(t)^\dagger a_j^\dagger a_k S(t)$. This is carried out as follows. We expand the operators $\alpha_j(t)$ and $\beta_j^\dagger(t)$ as

$$\alpha_j(t) = U_{jk}(t)a_k + V_{jk}(t)b_k^\dagger$$
$$\beta_j^\dagger(t) = -V_{jk}(t)a_k + U_{jk}(t)b_k^\dagger, \qquad (18)$$

where all of the time dependence is contained in the expansion coefficients U(t) and V(t). Equation (18) represents the most general expansion which satisfies the anticommutation relations, Eq. (6), and the constraints of unitarity. The properties of the transformation (18) are more evident in the finite representation obtained by truncating the positive and negative continuum, as illustrated in Fig. 2. Thus for the operators, α_j and β_j are limited to $0 < j < N$ for α_j, and $-m < j < 0$ for β_j. In this truncated Hilbert space, Eq. (18) becomes an M+N dimensional unitary transformation of the vector comprised of the set of operators α_j and $\beta_j^\dagger$,

$$\begin{pmatrix} \alpha(t) \\ \beta^\dagger(t) \end{pmatrix} = \begin{pmatrix} U & V \\ -V & U \end{pmatrix} \begin{pmatrix} a \\ b^\dagger \end{pmatrix}, \qquad (19)$$

where the elements of the transformation matrix are the expansion coefficients in Eq. (18). It is straightforward to invert Eq. (19) and obtain the matrix elements needed to evaluate the action. In matrix form these are

$$\begin{pmatrix} S^\dagger aS \\ S^\dagger b^\dagger S \end{pmatrix} = \begin{pmatrix} U^\dagger & -V^\dagger \\ V^\dagger & U^\dagger \end{pmatrix} \begin{pmatrix} a \\ b^\dagger \end{pmatrix}. \qquad (20)$$

The norm of the vector in Eqs. (19) and (20) is invariant under the finite rank transformation (19)

$$N \doteq (a^\dagger b) \begin{pmatrix} a \\ b^\dagger \end{pmatrix}$$
$$= a_j^\dagger a_j + b_j b_j^\dagger, \qquad (21)$$

and thus is a constant of the motion. Equation (19) differs from the lepton number only by a time-independent constant, and hence gives lepton number conservation.

In this treatment, the matrix of coefficients U(t) and V(t) are unknown variational parameters which are determined by finding stationary values of the action

$$\delta S/\delta U = \delta S/\delta V = 0. \tag{22}$$

These yield equations of motion

$$h \mid \psi_q^{(s)}(t)\rangle = i\partial_t \mid \psi_q^{(s)}(t)\rangle,$$

for

$$\mid \psi_q^{(+)}(t)\rangle = U_{qk}^*(t) \mid \chi_k^{(+)}\rangle + V_{qk}^*(t) \mid \chi_k^{(-)}\rangle \quad q < f \tag{23}$$

and

$$\mid \psi_q^{(-)}(t)\rangle = -V_{kq}^*(t) \mid \chi_k^{(+)}\rangle + U_{kq}^*(t) \mid \chi_k^{(-)}\rangle \quad q < 0$$

where h is given by Eq. (17). Classical field equations are obtained in a similar way

$$\delta S/\delta A^{\nu} = 0, \tag{24}$$

and result in the usual Maxwell's equations

$$\partial^{\mu} F_{\mu\nu} = J_{\nu} + \langle\Phi(t) \mid : \ \overline{\psi}(x)\gamma_{\nu}\psi(x): \ \mid \Phi(t)\rangle \tag{25}$$

where the current matrix element is evaluated using the methods outlined above. Equations (23) and (25) comprise Dirac-Hartree equations for a set of orbitals, which may be solved without explicit reference to the U and V matrices.

BASIS SPLINE EXPANSION

In this section we shall address a method of solving Eqs. (23) using the basis spline collocation method. Full details of this technique are given in Ref. 4. For simplicity, we shall consider the one-dimensional Dirac equation as given below,

$$h_o \psi(x,t) = i\partial_t \psi(x,t) \tag{26}$$

where the Hamiltonian is spin-degenerate, so that it suffices to specify h in a two-component spinor representation as

$$\begin{aligned} h &= \alpha_x\left(-i\partial_x - A_x\right) + \beta m + A_o \\ &= \begin{bmatrix} A_o + m & -i\partial_x - A_x \\ -i\partial_x - A_x & A_o - m \end{bmatrix} \end{aligned} \tag{27}$$

We shall assume that the field equations, Eq. (25), are integrable and develop numerical methods of solving Eq. (26) which emphasize accuracy, stability, and ease of programming. Our method requires the expansion of the spinor in Eq. (26) on a basis of spline functions[8] of order N,

$$\psi(x) = U_k^N(x)\psi^k, \quad k = 1,\dots,n, \tag{28}$$

where we shall use the convention that repeated indices are summed. Splines of order N are piecewise $(N-1)^{th}$ differentiable polynomials, for which the index k is associated with some space interval. Examples of these functions are shown in Fig. 3. Since the number of B-splines in Eq. (28) is finite, they cover a finite interval. In a completely different context, calculations using B-splines have been given by Dreizler.[9] The set of space points X_k, associated with the spline functions, U_k^N, do not provide an adequate representation for operators of the form Eq. (27). There are three problems which must be addressed: i) finding a local representation that sidesteps the issue of constructing matrix elements of h by numerical integration; ii) representing the derivatives on the spinors so that the boundary conditions on the upper and the lower components are correct; and iii) satisfying current conservation conditions on the space lattice that are implicitly contained in the Dirac Hamiltonian. The latter point is, of course, essential to guarantee lepton number conservation in numerical calculations.

There are a set of space points associated with each spline function which minimizes the error in the expansion, Eq. (28). These points, ξ_α $\alpha = 1,\dots,n$, called collocation points, provide an optimal representation of a function on a finite interval. The set of collocations may be evaluated using several different methods;[8,10] however, for equally spaced points x_k, $k = 1,\dots n$, we may take

$$\xi_\alpha = \left(X_{\alpha+\mu} + X_{\alpha+\mu+1}\right)/2 \quad \mu = [N/2]. \tag{29}$$

These points are shown as the open circles in Fig. 3. Thus, the functions Ψ evaluated at ξ_α are given by the transformation

$$\psi_\alpha = B_{\alpha k}\, \psi^k, \tag{30}$$

with

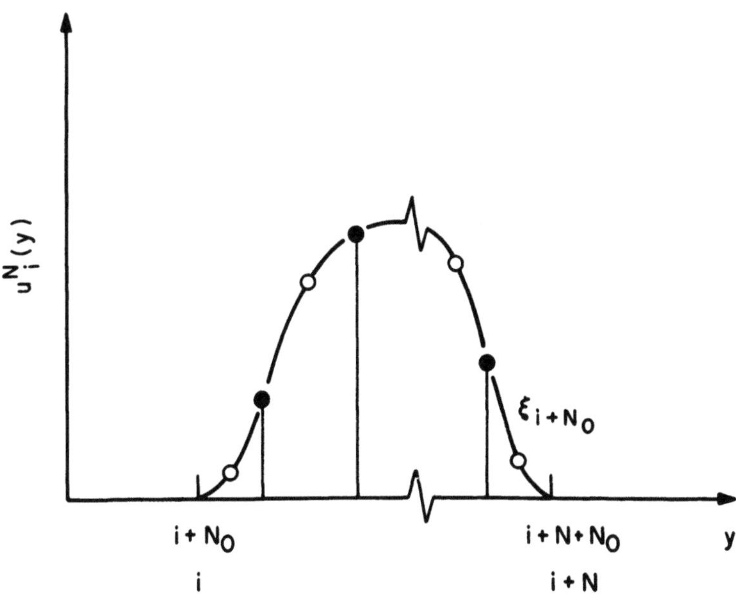

Fig. 3

$$B_{\alpha k} = U_k^N(\xi_\alpha).$$

For B-splines of finite order, the matrix B is banded with a bandwidth of N. Also, the inverse transformation is well behaved, thus giving the coefficient of ψ^k in terms of ψ_α,

$$\psi^k = B^{k\alpha}\ \psi_\alpha, \qquad B^{k\alpha} = [B^{-1}]^{k\alpha}. \tag{31}$$

Representations of differential operators can be easily obtained as matrices in collocation space; for example, $\partial_x^2 = \Delta$ becomes

$$\Delta_\alpha{}^\beta \equiv B''_{k\alpha}\ B^{k\beta}, \qquad B''_{k\alpha} = \partial_x^2\ U_k^N(x)\Big|_{x = \xi_\alpha}, \tag{32}$$

where the matrix $B^{k\beta}$ plays the role of a metric in collocation space. Equation (32) yields a highly accurate representation of the second derivative operator on a lattice, as illustrated in Fig. 4, where we give the result of the matrix Δ acting on the vector f(q)

$$F_2(q) = \Delta f(q), \qquad f(q) = \frac{\cos\left(q\xi_\alpha\right)}{-q^2}. \tag{33}$$

In Fig. 4 $\xi_\alpha = 0$, so that F_2 should take on a constant value, one. We compare the results of the ordinary finite-difference method, dashed curve, with the two B-spline results of order 3 and 11, as a function of q. The B-spline results are clearly superior, and, in general, for B-splines of order N, and for n collocation points, the error in the representation Eq. (32) is

$$\text{error} \sim n^{-N+1}. \tag{34}$$

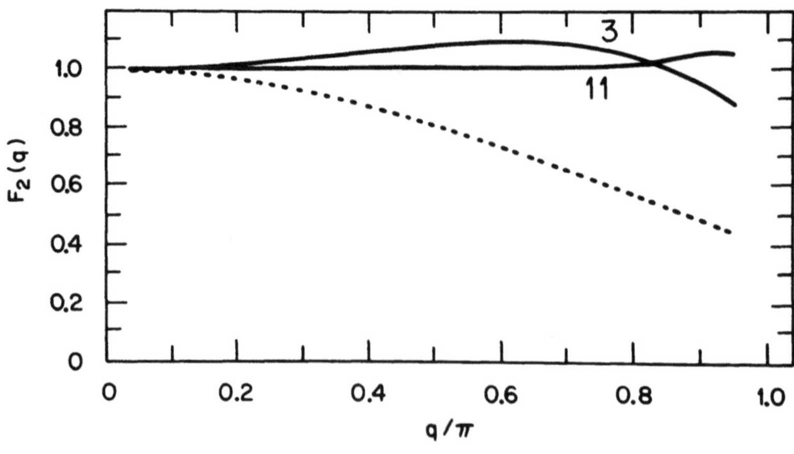

Fig. 4

This form of the second derivative operator has a unique decomposition into upper and lower triangular form using the Cholesky decomposition,[11]

$$\Delta = D^- D^+, \tag{35}$$

where in Eq. (35) D^- is a lower and D^+ is an upper triangular matrix, and where we have imposed the condition

$$D^-_{\alpha\alpha} = \left| D^+_{\alpha\alpha} \right| \tag{36}$$

in order to achieve uniqueness. In this decomposition, we identify two types of first derivatives in which boundary conditions at ξ_1 are contained in D^+ and boundary conditions at ξ_n are contained in D^-. This decomposition has two important consequences for the Hamiltonian, Eq. (27). It resolves the problem of fermion doubling on the lattice,[12,13] and it maintains an exact current conservation on the lattice. Thus, the representation of Eq. (27) on a collocation lattice is

$$h = \begin{bmatrix} A_o + m & -iD^+ - A_x \\ -iD^- - A_x & A_o - m \end{bmatrix}, \tag{37}$$

where the potentials are local functions of space

$$\left(A_\mu\right)_\alpha^{\ \beta} = \delta_{\alpha\beta}\, A_\mu\left(\xi_\alpha\right). \tag{38}$$

Thus, Eq. (26) on the collocation lattice becomes

$$h_\alpha^{\ \beta}\, \psi_\beta(t) = i\partial_t\, \psi_\alpha(t). \tag{39}$$

A more extensive discussion of this method is given in Ref. 4; here we shall employ these techniques to study μ- and τ-pair production from the vacuum.

TRANSVERSE FIELD MODEL FOR PAIR PRODUCTION

In the collisions of two relativistic nuclei, the transverse, near-zone, electromagnetic field becomes very large. For two beams of uranium each at an energy per nucleon of 100 GeV, Gould[2] estimates μ-pair and τ-pair cross sections, respectively

$$\sigma_{\mu^+\mu^-} \sim 1 \text{ mb}$$

$$\sigma_{\tau^+\tau^-} \sim 1\ \mu\text{b}.$$

These estimates are based on a perturbative treatment of the production, which is equivalent to the production out of the field of a time-like virtual photon that subsequently pair decays. However, these considerations suggest that the QED vacuum must undergo large rearrangements near such heavy ions. In the case of real photons coupled to static fields,[14] the dimensionless parameter which sets the scale for pair production, κ, is

$$\kappa = \left(\frac{\omega}{m}\right)\left(E/E_o\right) \tag{40}$$

where ω is the frequency of the photon field, m is the mass of the lepton, E is electric field strength, and E_o is the critical field,

$$E_o = m^2/e.$$

For μ and τ leptons, pair production becomes large whenever $\omega \simeq m$, and $E \simeq E_o$, as given below

	E_o(MV/fm)	ω^{-1} (fm/c)
μ	60	1.85
τ	15×10^3	0.1

Thus, if the transverse fields near the heavy ions have strengths and frequency components similar to these, we expect to observe sizeable amounts of pair production.

If we consider hadronic mechanisms for producing lepton pairs, we can get an idea as to the cross section scales. At these relativistic velocities, the Drell-Yan[15] mechanism sets the scale for the production, as shown in Fig. 5. Here the hard scattering of quark anti-quark pairs annihilate to give a time-like photon which pair decays. The total cross section for μ-pair production is approximately[16]

$$\sigma_{\mu^+\mu^-} \sim \frac{1.3\ A^{5/3}}{4\ M_\mu^{\ 2}}\ (\mathrm{fm}^2),$$

which for uranium collisions at 100 GeV per nucleon, gives

$$\sigma_{\mu^+\mu^-} \sim 1\ \mathrm{mb}$$

and about

$$\sigma_{\tau^+\tau^-} \sim 1\ \mu\mathrm{b}.$$

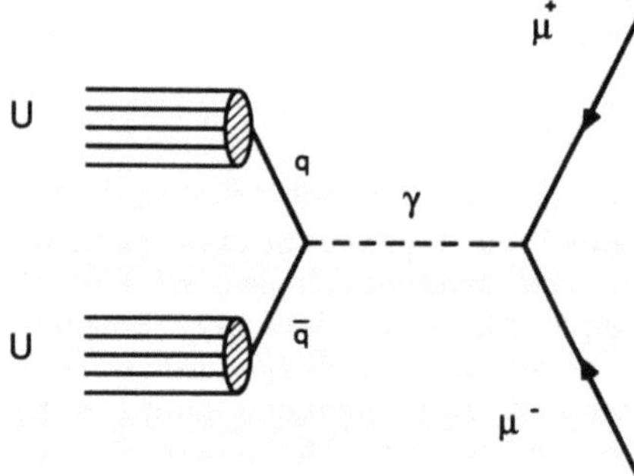

Fig. 5

These cross sections are comparable to those of Gould for the vacuum production of such pairs. The number of electron pairs produced in the heavy-ion collision is approximately 10^2 b; these pairs can convert to μ-pairs via the process shown in Fig. 1, which result in a production cross section of

$$\sigma_{e^+e^-\to\mu^+\mu^-} \sim 10^2 \text{ nb}.$$

Thus from these considerations, we conclude that excitations out of the QED vacuum near such heavy-ion beams are at least as large as other processes. For a heavy-ion collision as shown in Fig. 6, the near zone, transverse field a distance b from the beam axis is

$$E_\perp = \frac{Ze\gamma b}{(b^2+\beta^2t^2)^{3/2}} ,$$

this field is shown schematically in Fig. 6. Note that the maximum field strength is simply $Ze\gamma/b^2$ and that, due to causality, it has an approximate width

$$\Delta t \sim b/\gamma\beta.$$

These parameters are given below for two combinations of heavy-ion beams, and for b = 10 fm.

γ	$E_\perp$ (MV/fm)	Δt (fm/c)
17	25	0.6
10×10^3	30×10^3	5×10^{-4}

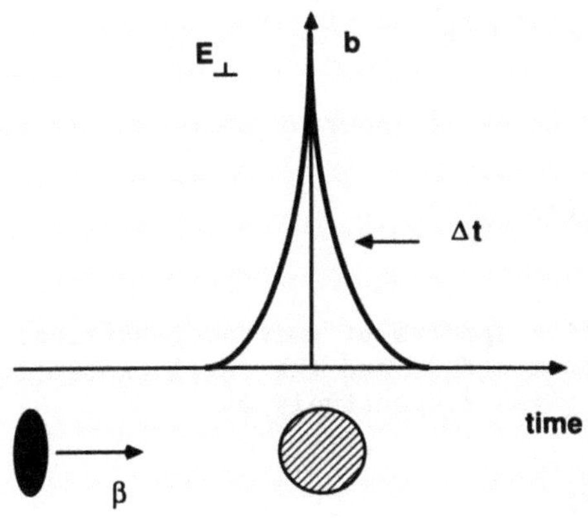

Fig. 6

The case of $\gamma = 17$ corresponds to the type of beams that will be available at the Brookhaven AGS shortly, and the case $\gamma = 20 \times 10^3$ corresponds to the proposed fixed target equivalent energy at RHIC. At AGS energies the field strength is 40% of the critical field for μ-pair production, and at RHIC energies it is 500 times the critical field. Hence, both of these machines should produce sizeable numbers of μ-pairs. At RHIC energies, we see that the field strength is twice the critical field for τ-pair production.

The excitation of such pairs out of the vacuum can be studied with a one-dimensional model as follows. Consider a box of length L in one-dimension, and a time-dependent electric field which is uniform throughout the box,

$$E(x,t) = E_o\, e^{-t^2/\Delta t^2} \tag{41}$$

The size of the box, the values of E_o, and the time history of the field are fixed to reproduce the values of the transverse field as previously discussed, so the E_o and Δt are functions of (γ,b). We shall only consider vacuum states and ignore binding effects so that the states in the box represent equivalent positive and negative energy continuum states. We shall use 74 positive energy and 74 negative energy continuum states and calculate the evolution of the vacuum, Eq. (23), using Eq. (41). Note that (41) is gauge equivalent to the interaction,

$$\begin{aligned} A_x &= 0, \\ A_o &= xE(t), \end{aligned} \tag{42}$$

which are used in the actual calculations. From Eq. (23), the vacuum evolution is obtained by time evolving all of the states $\psi_q^{(-)}$ with $-\nu < q < 0$. The truncation of the states in the vacuum to $q > -\nu$ is under our control and permits us to construct the density of states in the vacuum, evolve it in time, and examine its convergence properties. For example, the lowest energy state in the box has an energy $E_{-M} \sim -250\ mc^2$, whereas the density of states is usually cut off at about $E_\nu \sim -5\ mc^2$. With the wavefunction given by

$$|\,\psi_q^{(-)}(t)\rangle = -V_{kq}^*(t)\,|\,\chi_k^{(+)}\rangle + U_{kq}^*(t)\,|\,\chi_k^{(-)}\rangle,$$

there are three quantities of interest which can be constructed from the projections,

$$P_k^s = \sum_{-\nu<q<0} |\langle\chi_k^{(s)}\,|\,\psi_q^{(-)}\rangle|^2. \tag{43}$$

These are the inclusive spectra of emitted particles, $\overline{dp}/dE_k$, the density of states in the vacuum $\rho(\overline{\varepsilon}_k)$, and the total probability density of producing a pair, $\omega(\gamma,b)$, given respectively as

$$\begin{aligned} \overline{dp}/d\varepsilon_k &= \overline{P}_k^{(+)}/\Delta E_k, \\ \rho(\overline{\varepsilon}_k) &= \overline{P}_k^{(-)}/\Delta E_k, \\ \omega(\gamma,b) &= \sum_k \overline{P}_k^{(+)}, \end{aligned} \tag{44}$$

where

$$\overline{P}_k^{(s)} = (P_{k+1}^{(s)} + P_k^{(s)})/2,$$

and

$$\Delta E_k = | \varepsilon_{k+1} | - | \varepsilon_k | .$$

A typical example of the vacuum evolution is shown in Figs. 7-10 for the case of τ-pair production 15 fm from colliding beams of 50 GeV per nucleon uranium. In natural units, $\hbar = c = m = e = 1$, this would correspond to field strengths of 0.2 and a time width of 1.0. In these calculations, the lepton number is conserved to better than $1:10^{10}$. Figure 7 shows the pair production probability as a function of time. Here the time scale t_o is the Compton time of the tau, approximately

$$t_o \sim 10^{-24} \text{ sec.}$$

Note that the final pair multiplicity is about 10^{-2}, which is moderately large, and the sharp rise and subsequent fall of the probability in time, indicating rearrangement of the vacuum. Associated with this production probability is the inclusive τ^- spectra shown in Fig. 8. Here the differential probability, in units of the τ mass, is shown as a function of the kinetic energy of the τ^-, also in units of the τ mass. The dashed curves denote the contribution to the spectra from the individual states comprising the vacuum. In this example, we are propagating 12 states, and most of the yield occurs at kinetic energies less than the τ-mass. It is instructive to examine the density of states in the vacuum during the initial phase of the evolution, Fig. 9, and at the end of the time evolution, Fig. 10. Note that the ordinate scales for these figures are logarithmic, and hence the initial density of states is approximately $1/\Delta E_k$ for energies less than

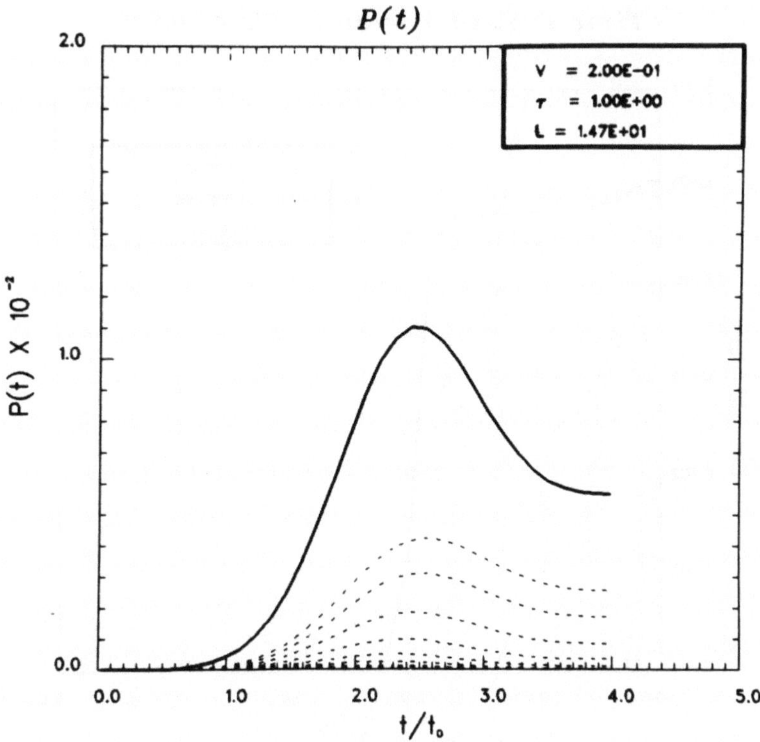

Fig. 7

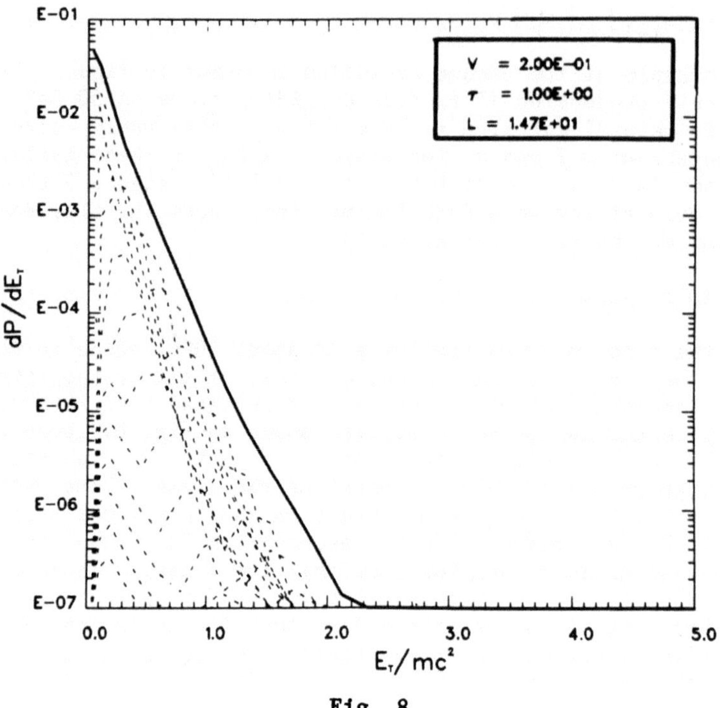
Tauon Pair Spectrum t/t₀ = 3.985
V = 2.00E-01
τ = 1.00E+00
L = 1.47E+01
dP/dE_T
E-01
E-02
E-03
E-04
E-05
E-06
E-07
0.0
1.0
2.0
3.0
4.0
5.0
E_T/mc²

Fig. 8

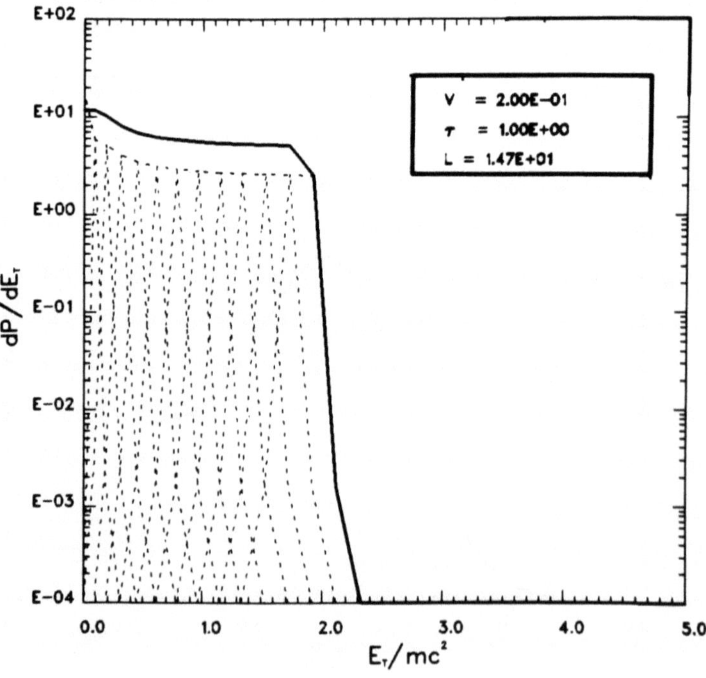
Time Evolved Vacuum t/t₀ = 0.625
V = 2.00E-01
τ = 1.00E+00
L = 1.47E+01
dP/dE_T
E+02
E+01
E+00
E-01
E-02
E-03
E-04
0.0
1.0
2.0
3.0
4.0
5.0
E_T/mc²

Fig. 9

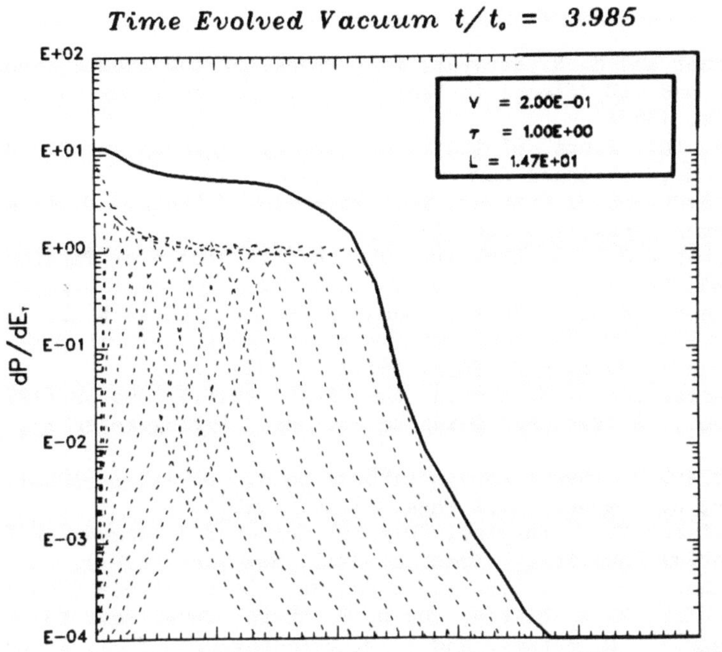

Fig. 10

about 2 mc^2, and zero for energies greater than 2 mc^2, while at the end of the evolution, there is considerable rearrangement of this function with an exponential falloff up to about 4 mc^2.

We can approximately reconstruct production cross sections from the observed probability density. In the above example, $P \sim 3.6 \times 10^{-3}$ yields, $\omega \simeq 4 \times 10^{-3}\ fm^{-1}$ over a region of space of approximately 15 fm. Using a simple geometrical argument, the emission probability transversed to the heavy-ion beam times the transverse area would result in

$$\sigma \sim b^4\ \omega^2 \tag{45}$$

where b is the impact parameter. Thus for 50 GeV per nucleon uranium beams,

$$\sigma_{\tau^+\tau^-} \sim 6\ \text{mb}.$$

Similar calculations for μ-pairs result in

E/A (GeV)	$\omega\ (fm^{-1})$	σ (mb)
21	4×10^{-4}	.08
10^2	5×10^{-3}	12

In summary, we conclude that the near-zone electromagnetic field in relativistic heavy-ion collisions produces large changes in the QED vacuum which have characteristically large Fourier frequency components. This is reflected in the sizeable emission of heavy leptons in a direction transverse to the heavy-ion beam.

REFERENCES

1. C. Bottcher and M. R. Strayer, in: "Procs. of the Atomic Theory Workshop and QED Effects in Heavy Atoms," National Bureau of Standards, 1985.
2. H. Gould, in: "Procs. of the Atomic Theory Workshop and QED Effects in Heavy Atoms," National Bureau of Standards, 1985.
3. J. Reinhardt and W. Greiner, in: "Heavy-Ion Science," D. Allan Bromley, ed., Plenum Press, New York (1985).
4. C. Bottcher and M. R. Strayer, submitted to Annals of Physics (N.Y.), June 1985.
5. C. Itzykson and J. B. Zuber, "Quantum Field," McGraw-Hill, New York, (1980).
6. J. W. Negele, Revs. Mod. Phys. 54:913 (1982).
7. A. K. Kerman and S. E. Koonin, Ann. Phys. (N.Y.) 100:332 (1976).
8. C. de Boor, "A Practical Guide to Splines," Springer-Verlag, New York (1978).
9. R. Dreizler, Z. Physik A309:5 (1982); ibid., A307:211 (1982).
10. U. Asher, J. Computational Phys. 34:401 (1980).
11. V. Vemuri and W. J. Karplus, "Digital Computer Treatment of Partial Differential Equations," Prentice-Hall, New York (1981).
12. C. Bottcher and M. R. Strayer, Phys. Rev. Lett. 54:669 (1985).
13. C. M. Bender, K. A. Milton, and D. H. Sharp, Phys. Rev. D31:383 (1985).
14. V. N. Bair, V. M. Katkov, and V. M. Strakhovenko, Nucl. Instr. and Meth. B16:5 (1986).
15. R. C. Hwa and K. Kajantie, Phys. Rev. D32:1109 (1985).
16. M. Gyulassy, private communication.

EXPERIMENTS ON FEW-ELECTRON VERY HIGH-Z IONS

Harvey Gould and Charles Munger*

Materials and Molecular Research Division,
Lawrence Berkeley Laboratory, University of California,
Berkeley California 94720

INTRODUCTION

The production[1] in 1983 of a beam of bare U^{92+} at the Lawrence Berkeley Laboratory's Bevalac[2], the Bevatron and Super-HILAC operating in tandem, demonstrated the feasibility of experiments using few-electron uranium. Since then, experiments by Anholt and collaborators[3–8] have led to an understanding of the physics of charge changing collisions at relativistic energies. Charge changing collisions at relativistic energies are, in many cases, now better understood than charge changing collisions at non relativistic energies. In 1984 x rays from radiative electron capture into the K shell of uranium was observed[3,6] by Anholt et. al. and x rays from n=2 → n=1 transitions in hydrogenlike uranium (U^{91+}) and heliumlike uranium (U^{90+}) were observed by Munger and Gould[9]. A preliminary value[10] for the Lamb shift in heliumlike uranium was obtained by Munger and Gould in 1986.

This article discusses the measurement of the Lamb shift in heliumlike uranium and outlines future experimental tests of QED using few-electron very high atomic number (Z) ions. We conclude with a discussion of the possibility of using ultrarelativistic atomic collisions to produce very heavy leptons.

PRODUCTION OF FEW-ELECTRON VERY HIGH-Z IONS

Few-electron uranium and other very high-Z ions are produced by stripping beams of relativistic atoms. The experimentally determined charge state distributions for relativistic uranium ions which have passed through equilibrium thickness targets is shown in Fig. 1. Equilibrium thickness—the thickness at which additional material no longer changes the charge state is typically a few ten's of mg/cm^2 for high-Z targets. The processes for electron capture and loss by relativistic heavy ions are well understood and cross sections for ionization, for radiative electron capture and for nonradiative electron capture can be reliably calculated[3–8,11,12].

The Bevalac produces uranium beams at energies up to 960 MeV/amu at intensities of 10^6 ions per pulse (duty cycle is typically 12 pulses per minute). An upgrade presently in progress is designed to increase the beam intensity by a factor of ten. The beam emittence is 30π mm-mR or better.

*Also, Department of Physics, University of California at Berkeley, Berkeley, California, 94720.

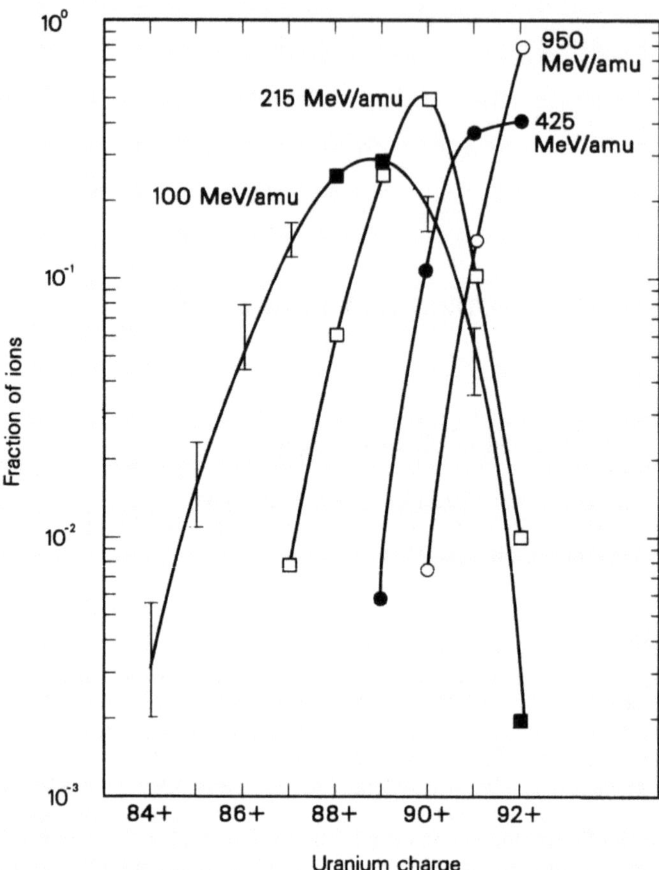

Fig. 1. Charge state distribution of relativistic uranium after passing through an equilibrium thickness target.[39] A Cu (Z=29) target was used for 950 MeV/amu, 425 MeV/amu, and 100 MeV/amu uranium and a Au (Z=79) target was used for 215 MeV/amu uranium.

LAMBSHIFT IN VERY HIGH Z FEW-ELECTRON IONS

The measurement of the Lamb shift in a very high-Z atom is a test of quantum electrodynamics (QED) in a strong Coulomb field. QED is well tested for free particles and in the weak fields of low-Z atoms. But just as Newton's laws work well in weak gravitational fields but fail in strong gravitational fields, experiments in low-Z atoms do not rule out a possible failure of QED in the strong Coulomb field of a very high-Z atom.

At Z=92, the contributions to the Lamb shift in a one-electron atom are the self-energy[13,14] of ≈ -56 eV, the vacuum polarization[15] of $\approx +14$ eV and the finite nuclear size correction[15] of ≈ -33 eV, where a negative value indicates the interaction decreases the binding energy. Vacuum polarization, but not self-energy, is well tested in muonic atom experiments. High-Z Lamb shift measurements primarily test the self-energy in a strong Coulomb field[16].

What is significant about the self energy at Z=92 is it arises almost entirely from terms[17] which are of very high order in $Z\alpha$ (where α is the fine structure constant). Because these terms are large only at very high Z they are not tested in low-Z Lamb shift and fine structure experiments.

The contribution of the higher order terms in the self-energy can be seen by comparing the series expansion of the self energy with an evaluation of the 2 $^2S_{1/2}$ self-energy to all orders[13,14,17–19] in $Z\alpha$. If we write the self energy Σ_n in a power series in α and $Z\alpha$, we have:

$$\Sigma_n = n^{-3} (\alpha/\pi) m_o c^2 \Big[[A_{40} + A_{41} \ln(Z\alpha)^{-2}](Z\alpha)^4 + A_{50}(Z\alpha)^5 + [A_{60} + A_{61}\ln(Z\alpha)^{-2} + A_{62}\ln^2(Z\alpha)^{-2}](Z\alpha)^6 + A_{70}(Z\alpha)^7 + \text{higher order terms} \Big] \quad (1)$$

Where n is the principal quantum number and m_o is the electron rest mass. Values of the coefficients $A_{40} - A_{70}$ can be found in Ref. 19. Fig. 2 shows the ratio of the higher order terms in the self-energy to the total self energy. In neutral hydrogen the higher order terms in the self-energy contribute about 0.1 parts per million to the Lamb shift, nearly 100 times smaller than the uncertainty due to proton structure[20]. At Z=92 however, the higher order terms are essentially the entire self-energy contribution, and make up over half of the total Lamb shift.

MEASUREMENT OF THE LAMB SHIFT IN HELIUMLIKE URANIUM

We have obtained a preliminary value for the Lamb shift in heliumlike U^{90+} of 69.1 (8.0) eV which is in agreement with the theoretical value[13–15] of 75 eV for the one-electron Lamb shift at Z=92.

We choose heliumlike uranium over hydrogenlike uranium for this measurement because both the $2\,^2S_{1/2}$ and $2\,^2P_{1/2}$ states of hydrogenlike uranium decay very rapidly making it very difficult to observe the decays outside of the target where the hydrogenlike uranium is formed. With hydrogenlike uranium there is the risk that interactions with the target will perturb the energy levels of the atom. In heliumlike uranium, however, the $2\,^3P_0$ state is metastable[21] (55 ps lifetime) because angular momentum conservation forbids single photon decays to the $1\,^1S_0$ ground state. Consequently decay of the state can be observed well downstream from the target foil.

Our value for the Lamb shift in heliumlike uranium was obtained from a measurement of the lifetime of the $2\,^3P_0$ state (see Fig. 3). In heliumlike uranium about 70% of the $2\,^3P_0$ state decays by an allowed electric-dipole (E1) transition[22] to the $2\,^3S_1$ state, making the $2\,^3P_0$ lifetime sensitive to the $2\,^3P_0 - 2\,^3S_1$ energy splitting. The remainder decays to the ground state by a two-photon electric-dipole magnetic-dipole (E1M1) transition[21]. The $2\,^3P_0 - 2\,^3S_1$ splitting arises

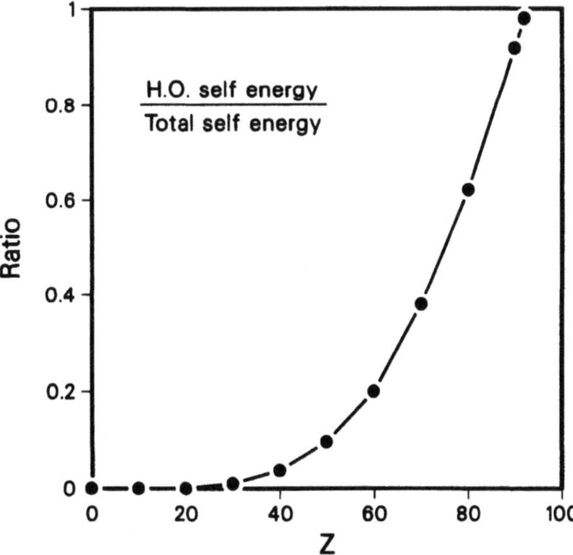

Fig. 2. Ratio of the higher order terms in the self-energy to the total self-energy obtained by comparing the series expansion value through term A_{70} $(Z\alpha)^7$ with a numerical calculation to all orders in $Z\alpha$.

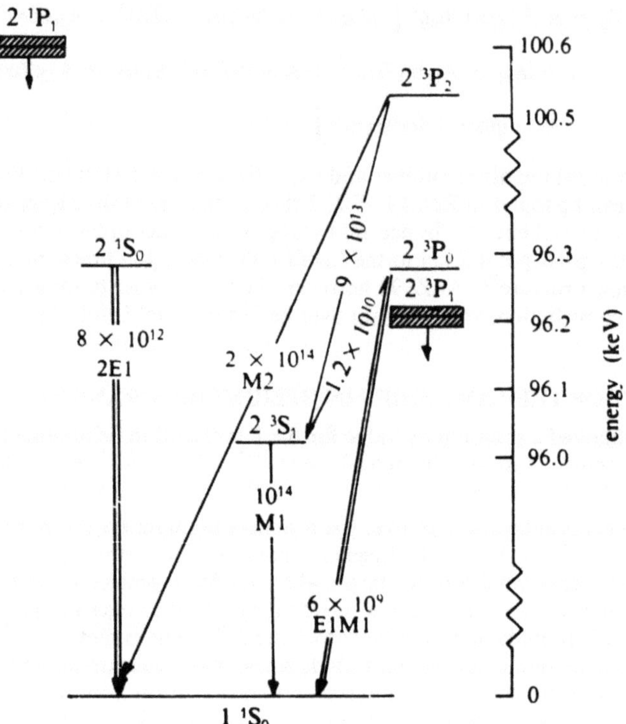

Fig. 3. Energy level diagram of the n=1 and n=2 states of heliumlike uranium. Decays without labels are E1 decays and the cross hatching on the $^1P_1$ and $^3P_1$ states indicates the approximate radiative width. At Z=92 singlet-triplet mixing is close to 100% and singlet-triplet classification (and LS coupling) is used only for the convenience of the authors. The non-QED contributions to the energy levels for heliumlike uranium were obtained from Ref. 23 to which were added the one-electron self energy and vacuum polarization[13–15]. The lowest order correction[16] to the QED terms for the presence of the second electron, of order 1/Z, is neglected. These values for heliumlike uranium are in general agreement with values calculated in Ref.'s. 21,24. Decay rates are taken from Ref.'s 21,23, and where appropriate from hydrogenic matrix elements given in Ref. 22.

from the Lamb shift and from the Coulomb interaction between the two electrons in the 2 $^3P_0$ and in the 2 $^3S_1$ states. The latter is calculated to be 329 eV (Ref. 23,24) and 326 eV (Ref. 21). The measured 2 $^3P_0$ lifetime, the calculated E1M1 decay rate and the calculated E1 matrix element[22] and the non-QED contributions to the 2 $^3P_0$ − 2 $^3S_1$ splitting are combined to determine the Lamb shift. The E1M1 decay rate and the E1 matrix element are insensitive to QED effects at the present experimental accuracy. The effect of the second electron on the Lamb shift[16] is to decrease it by a term of order 1/Z which we neglect for Z=92 in the present experiment.

The indirect method of measuring the 2 $^3P_0$ lifetime rather than a direct measurement of the 2 $^3P_0$ − 2 $^3S_1$ transition energy was chosen because it is easier than measuring the energy of the 0.25 keV photon from this transition. An attractive feature of the lifetime measurement is that the 2 $^3P_0$ − 2 $^3S_1$ transition is followed rapidly by emission of a 96 keV x ray from the 2 $^3S_1$ − 1 $^1S_0$ decay. The 96 keV x ray is easy to detect. The 10^{-14} sec lifetime of the 2 $^3S_1$ state has no effect on the measurement of the 6×10^{-11} sec 2 $^3P_0$ lifetime provided sufficient distance is allowed for the initial 2 $^3S_1$ population to decay. A spectrum taken 0.5 cm downstream from the target is shown in Fig. 4. The 2 $^3P_0$ lifetime is measured by the beam-foil time-of-flight technique.

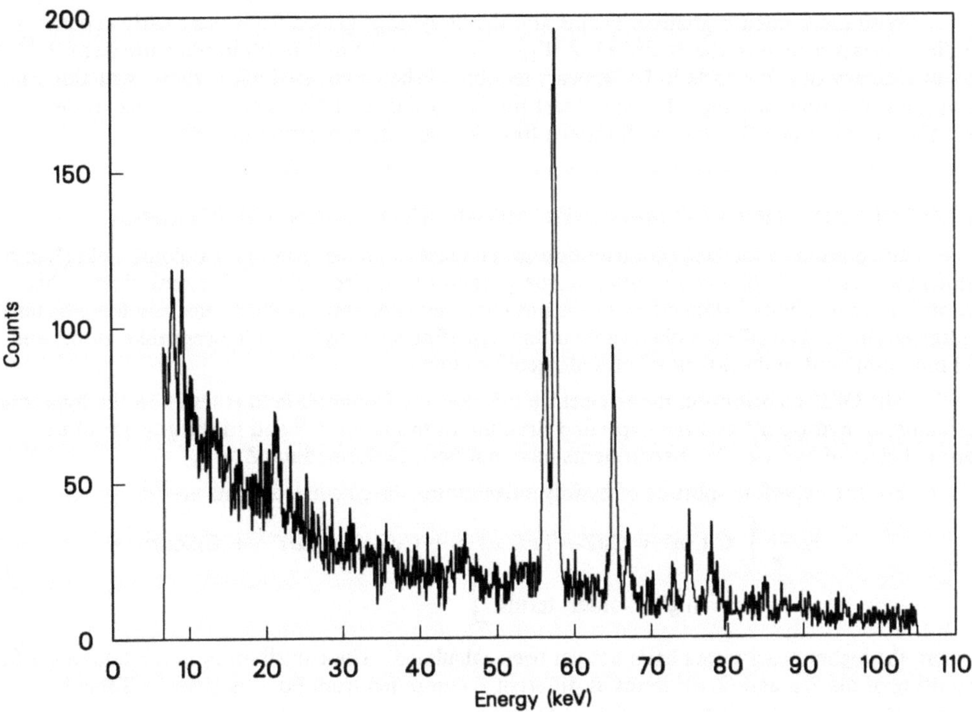

Fig. 4. X-ray spectra 0.5 cm downstream from the target. The Doppler-shifted x-ray peak from the decay of $^2P_0 \rightarrow {^2S_1} \rightarrow {^1S_0}$ is near 78 keV. Peaks at 73 keV, 75, keV and 82 - 86 keV are from fluorescence of Pb. and at those near 57 keV and 65 keV from Ta which are used used for shielding the x-ray detector and for Soller and slits respectively. Cascades from higher excited states, if present, would produce a peak at 81 keV. This spectra represents 135 minutes of counting, or about 10^8 uranium ions. Background is caused by bremsstrahlung of electrons in the target off of the uranium projectile, elastic scattering of the electrons in the target followed by bremsstrahlung of the electrons either in the target, or, if the electrons leave the target, in the walls of the vacuum chamber, and by fragments from nuclear disintegrations in the target and by nuclear reactions by fast neutrons in the Ge detectors. Other sources of background may also exist.

Heliumlike uranium in the 2 $^3P_0$ state is prepared by stripping a beam of 220 MeV/amu uranium 39+ which is obtained from the Lawrence Berkeley Laboratory's Bevalac. An aluminum target produces an equilibrium charge state distribution of roughly 5% U^{92+}, 30% U^{91+}, 60% U^{90+}, 5% U^{89+}. The hydrogenlike U^{91+} fraction is magnetically selected and transported to a 1 mg/cm^2 Pd target. About half of the U^{91+} ions are converted to heliumlike U^{90+}, with about 1% of these being formed in the 2 $^3P_0$ state or in states which rapidly decay to the 2 $^3P_0$ state.

The price for using the 2 $^3S_1$ decay is to make the measurement more sensitive to possible cascade feeding from long-lived states of high principle quantum number (n) and angular momentum (J). Decays of high n,J states cascade down the yrast chain and reach the 2 $^3P_2$ state which has a branching ratio of 1/3 to the 2 $^3S_1$ state by E1 decay and 2/3 to the ground state by electric quadrupole decay. It is the branch to the 2 $^3S_1$ state which is a potential source of systematic error. In heliumlike uranium , however it takes a state of about n=25 and high angular momentum before the cascade time is comparable to the lifetime of the 2 $^3P_0$ state. The populations of excited states fall rapidly with increasing n and we estimate[4,11] that no more than 2% of the population lies above n=25 and with only a small fraction of these in high J states. We therefore expect no observable effect from cascades on our measured 2 $^3P_0$ lifetime with our present statistical accuracy of 6%. Furthermore we designed the apparatus so that the difference in transition energy to the ground state from the 2 $^3S_1$ and 2 $^3P_2$ states is resolved by our germanium x-ray detector. Cascades through the 2 $^3P_2$ state would show up in our spectra as a resolved peak lying just above the decay peak. Figure 2 shows no evidence of a cascade peak nor does any of our data to within background statistics.

With more intense uranium beams and the knowledge gained from these early experiments a direct measurement of the $\approx$ 284 eV 2 $^2P_{1/2}$ − 2 $^2S_{1/2}$ splitting[25] in lithiumlike uranium (U^{89+}) to an accuracy of a few-parts in 10^4 appears feasible. When compared with atomic structure calculations of similar accuracy this would test the Lamb shift to 0.1%. The nuclear size of the uranium nucleus is sufficiently well known from muonic atom measurements[26].

QED CONTRIBUTIONS TO MAGNETIC MOMENTS OF BOUND ELECTRONS

In addition to the QED contribution to the mass of an electron in a Coulomb field (Lamb shift) there is also a QED contribution to the g-factor of the electron in a Coulomb field. This contribution is a bound state effect and is not tested by experiments which measure the g-factor of a free electron. The effect is observable in the hyperfine splitting[27,28] of hydrogenlike atoms and of muonium and in the g-factor[29] of hydrogenlike atoms.

The QED contribution to the electron g-factor in a Coulomb field is tested in the hyperfine structure of hydrogen[28] and the hyperfine structure of muonium[28,30] and in the g-factor of the ground state of hydrogen[31]. Experiments have not been performed for $Z > 1$.

For the hyperfine splitting of hydrogenlike atoms the calculated terms are[27,28],

$$E_F \frac{\alpha}{\pi} \Big[C_1(Z\alpha) + C_2(Z\alpha)^2\ln^2(Z\alpha)^{-2} + C_3(Z\alpha)^2\ln(Z\alpha)^{-2} + C_4(Z\alpha)^2 + \text{higher order terms} \Big] \tag{2}$$

where the higher order terms have not yet been calculated. The contribution to the total hyperfine splitting of the $Z\,\alpha$ and $(Z\,\alpha)^2$ terms at different Z computed from Eq. 2 is given in Table I.

The term of order $(Z\alpha)^2$ contributes about 1% of the hyperfine splitting at Z=81 (the anomalous magnetic moment of the free electron contributes roughly 0.1%). In addition, at Z=81, the $(Z\alpha)^2$ term is larger than the lower order $Z\alpha$ term. At very high Z terms of order $(Z\alpha)^3$ and higher could be larger than the lower order terms. In the calculation of higher order terms it is necessary to consider the energy of the electron bound by both strong Coulomb and magnetic fields[32].

The g_J factor of a bound electron also has QED contributions which are not present for a free electron and which become relatively large at high Z (Ref. 29). The leading term is $\alpha/\pi\,(Z\alpha)^2$ which contributes 3×10^{-8} in hydrogen and 3×10^{-4} in hydrogenlike uranium. The relative contribution to the g_J factor is smaller and of higher order than for the hyperfine splitting.

HYPERFINE STRUCTURE AND G_J EXPERIMENTS

Tests of the QED contribution to the hyperfine splitting of an electron bound in a Coulomb field are limited in hydrogen at a few ppm due to the uncertainty in the proton polarizability and in muonium to a few tenths of a ppm due to uncertainties in the muon mass and fine structure constant[28]. These experiments test the term of order $(Z\alpha)^2$ to about 10% and are probably insensitive to higher order terms. Measurements of the g_J in the ground state of hydrogen[31] achieved a precision of 1×10^{-8} which tests the leading order term to about 30%. Experiments[33] in He^+ are not yet of sufficient sensitivity to see the contribution.

Measurements of the ground state hyperfine structure of hydrogenlike thallium using storage rings have been suggested by Bemis and Gould[34]. The ground state hydrogenlike thallium ($I = 1/2$) F=1 − F=0 transition energy is calculated to be 3800 Å without QED corrections and the magnetic dipole decay (M1) rate for F=1 → F=0 is $\approx 10^3\ s^{-1}$. Confinement of hydrogenlike thallium in a storage ring would then produce a spectrum from the F=1 → F=0 allowed M1 decay and optical spectroscopy would be used to determine the ground state hyperfine interval.

Table I. Bound state QED contributions to hyperfine splitting

Z	$C_1\,(Z\alpha)$	$C_4\,(Z\alpha)^2$
1	1×10^{-4}	2×10^{-6}
19	2×10^{-3}	7×10^{-4}
81	8×10^{-3}	1×10^{-2}

ULTRARELATIVISTIC PAIR PRODUCTION

The cross section for producing electron-positron pairs from the Coulomb field of two colliding (bare) point nuclei is given for the limiting case of kinetic energies much larger than the electron (lepton) rest mass by[35]:

$$\sigma_{pair} = (28/27\pi)\, \alpha^2\, Z_1^2\, Z_2^2\, r_0^2\, \log^3\gamma \tag{3}$$

where $\gamma = (1-\beta^2)^{-1/2}$ with $\beta = v/c$ and Z_1 and Z_2 are the nuclear charges, and r_0 is the classical electron radius. The formula is for point nuclei and impact parameter cut-off's for real nuclei will reduce the cross sections by an estimated one to two orders of magnitude. Ultrarelativistic heavy ion accelerators which have been proposed[36] (RHIC), approved (the booster synchrotron for the Brookhaven Alternating Gradient Synchrotron)[37] or are under construction (The heavy ion facility for the CERN SPS) will produce large quantities of electron-positron pairs. Production of heavier leptons is also possible. If the cross section scales inversely as the square of the mass, a 10^7 ion per second beam of 15 GeV/amu uranium in a 1 g/cm^2 uranium target would produce one tau pair per minute. Such a beam-dump experiment could be used to search for heavier leptons and other very heavy particles.

It is possible for the electron produced in pair production, to be captured into the K-shell of one of the uranium atoms which produced the pair.[8,38] The electrons most likely to be captured after pair production are those having momenta which overlap the momenta of the final bound K electron. This means that electrons with kinetic energies less than the uranium K-shell binding energy may be captured. The energy distribution of the pairs produced by colliding 15 GeV/amu uranium ions with a fixed uranium target extends to about 10 MeV, but is peaked at lower energies. The fraction of electrons within the K shell with kinetic energies of less than 130 keV would then be roughly 130 keV/10 MeV $\approx$ 0.01, hence about 1 percent of the electrons could be captured into the K-shell. For muons, with larger binding energies, the capture fraction could be larger.

ACKNOWLEDGMENTS

We thank Mr. Roy Bossingham, Dr. Benedict Feinberg, Mr. Walter Kehoe, Dr. Richard McDonald Professor Richard Mowat and Dr. Alfred Schlacter for assistance in running the Lamb shift experiment. We are happy to acknowledge the help given by, and to thank, among others: Dr. Curtis Bemis Jr., Professor Gordon Drake, Dr. Peter Mohr, and Professor Jonathan Sapirstein, We especially thank the operators and the staff and the management of the Lawrence Berkeley Laboratory's Bevalac for making experiments with relativistic heavy ions possible. This work was supported by the Director, Office of Energy Research: Office of Basic Energy Sciences, Chemical Sciences Division; and in part by the Office of High Energy and Nuclear Physics, Nuclear Science Division, of the U.S. Department of Energy under Contract No. DE-AC-03-76SF00098.

REFERENCES

1. H. Gould, D. Greiner, P. Lindstrom, T.J.M. Symons, and H. Crawford, Phys. Rev. Lett. 52, 180 (1984) (Errata- Phys. Rev. Lett. 52, 1654 [1984]).
2. J.R. Alonso *et al*, Science 217, 1135 (1982).
3. R. Anholt, W.E. Meyerhof, Ch. Stoller, E. Morenzoni, S.A. Andriamonje, J.D. Molitoris, D.H.H. Hoffmann, H. Bowman, J.S. Xu, Z.Z. Xu and J.O. Rasmussen, Phys. Rev. Lett. 53, 234 (1984).
4. W.E. Meyerhof, R. Anholt, J. Eichler, H. Gould, Ch. Munger, J. Alonso, P. Thieberger and H.E. Wegner, Phys. Rev. A32, 3291 (1985).
5. R. Anholt, W.E. Meyerhof, H. Gould, Ch. Munger, J. Alonso, P. Thieberger and H.E. Wegner, Phys. Rev. A32, 3302 (1985).
6. R. Anholt and W.E. Meyerhof, Phys. Rev. A33, 1556 (1986).
7. R. Anholt, Ch. Stoller, J.D. Molitoris, D.W. Spooner, E. Morenzoni, S.A. Andriamonje, W.E. Meyerhof, H. Bowman, J.S. Xu, Z.Z. Xu, J.O. Rasmussen, and D.H.H. Hoffmann, Phys. Rev. A33, 2270 (1986).
8. R. Anholt and H. Gould, Relativistic Heavy-Ion-Atom Collisions, to be published in: "Advances in Atomic and Molecular Physics," B. Bederson, ed., Academic Press, Orlando FL (1987); Lawrence Berkeley Laboratory Report No. LBL-20661.

9. C. Munger and H. Gould, Bull. Am. Phys. Soc. 30, 860 (1985).
10. C. Munger and H. Gould, submitted to Phys. Rev. Lett. (1986).
11. J. Eichler, Phys. Rev. A32, 112 (1985); R. Anholt, and J. Eichler, Phys. Rev. A31, 3505 (1985).
12. R. Anholt, Phys. Rev. A31, 3579 (1985).
13. P.J. Mohr, Phys. Rev. A26, 2338 (1982).
14. W.R. Johnson, and G. Soff, Atomic Dat. and Nuclear Dat. Tables, 33, 405 (1985).
15. P.J. Mohr, Atomic Dat. and Nuclear Dat. Tables 29, 453 (1983).
16. S.J. Brodsky and P.J. Mohr, Quantum Electrodynamics in Strong and Supercritical Fields in "Topics in Current Physics: Quantum Electrodynamics in Strong and Supercritical Fields," I.A. Sellin, ed., Springer, Berlin, 1978. Vol. 5, p.3.
17. A.M. Desiderio and W.R. Johnson, Phys. Rev. A3, 1267 (1971).
18. G.E. Brown, J.S. Langer and G.W. Schafer, Proc. Roy. Soc. (London) A251, 92 (1959); G.E. Brown and D.F. Mayers, Proc. Roy. Soc. (London) A251, 105 (1959); G.W. Erickson, Phys. Rev. Lett. 47, 780 (1971); K.T. Cheng and W.R. Johnson, Phys. Rev. A14, 1943 (1976).
19. For a discussion ot the series expansion and values for the coefficients see for example, P.J. Mohr, Ann. Phys. (N.Y.) 88, 26 (1974); J. Sapirstein, Phys. Rev. Lett. 47, 1723 (1981).
20. See for example, S.R. Lundeen and F.M. Pipkin, Phys. Rev. Lett. 46, 232 (1981).
21. G.W.F. Drake, Nucl. Instr. Meth. in Phy. Research B9, 465 (1985).
22. M. Hillery and P.J. Mohr, Phys. Rev. A21, 24 (1980); H. Gould, R. Marrus, and P.J. Mohr, Phys. Rev. Lett. 33, 676 (1974); G.W.F. Drake, Astrophys. J. 158, 1199 (1969).
23. C.D. Lin, W.R. Johnson and A. Dalgarno, Phys. Rev. A15, 154 (1977); W.R. Johnson and F. Parpia, private communication.
24. P.J. Mohr, Phys. Rev. A 32, 1949 (1985); P.J. Mohr, Private communication.
25. K.T. Cheng, Y.-K. Kim, and J.P. Desclaux, Atomic Data and Nucl. Data Tables, 24, 111 (1979); Y.-K. Kim and J.P. Desclaux, Phys. Rev. Lett. 36, 139 (1976); see also L. Armstrong, Jr., W.R. Fielder, and D.L. Lin, Phys. Rev. A14, 1114 (1976); C.F. Fisher and T. Brage, private communication.
26. J.D. Zumbro, E.B. Shera, Y. Tanaka, C.E. Bemis, Jr., R.A. Naumann, M.V. Hoehn, W. Reuter, and R.M. Steffen, Phys. Rev. Lett. 20, 1888 (1984).
27. S.J. Brodsky and G.W. Erickson, Phys. Rev. 148, 26 (1966)
28. J.R. Sapirstein, Phys. Rev. Lett. 51, 985 (1983).
29. H. Grotch, Phys. Rev. Lett. 24, 39 (1970); H. Grotch and R. Hegstrom, Phys. Rev. A4, 59 (1971).
30. F.G. Marion et al., Phys. Rev. Lett. 49, 993 (1982).
31. J.S. Tideman and H.G. Robinson, Phys. Rev. Lett. 39, 602 (1977).
32. S.J. Brodsky, and J. Primack, "The Electromagnetic Interaction of Composite Systems" Ann. Phys. (N.Y.) 52, 315 (1969).
33. C.E. Johnson and H.G. Robinson, Phys. Rev. Lett. 45, 250 (1980).
34. C.E. Bemis Jr. and H. Gould (private communication).
35. E.J. Williams, Kgl. Dansk. Vid. Selsk. 13, No. 4 (1935); C.F.V. Weizsäcker, A. Phys. 88, 612 (1934); H.J. Bhabha, Proc. Roy. Soc. A152, 559 (1935); H.J. Bhabha, Proc. Camb. Phil. Soc. 31, 394 (1935); L. Landau and L. Lifshitz, Phys. Zs. Sov. U. 6, 244 (1934); Y. Nishina, S. Tomonaga and M. Kobayasi, Sci. Pap. Ins. Phys. Chem. Research, Japan 27, 137 (1935).
36. RHIC and Quark Matter: "Proposal for a Relativistic Heavy Ion Collider at Brookhaven National Laboratory" Brookhaven National Laboratory Report No. BNL 51801 (UC-28) [Particle Accelerators and High Voltage Machines – TIC-4500], Aug. 1984.

37. "Proposal for a 15A-GeV Heavy Ion Facility at Brookhaven" Brookhaven National Laboratory Report No. BNL-32250 (1983).

38. H. Gould, "Atomic Physics Aspects of a Relativistic Nuclear Collider" Lawrence Berkeley Laboratory Report No. LBL-18593 (UC-28), Nov. 1984.

39. R. Anholt, W.E. Meyerhof, X.-Y. Xu, H. Gould, B. Feinberg, R.M. McDonald, H.E. Wegner and P. Thieberger, submitted to Phys. Rev. A.

A NOVEL APPROACH TO LAMB SHIFT MEASUREMENTS IN HIGH Z HYDROGENIC IONS

J.D. Silver

University of Oxford
Clarendon Laboratory
Parks Road
Oxford
U.K.

We have been studying highly ionised atoms in Oxford since 1974. Why? Well, the main motivation is to test relativistic and quantum electrodynamic (QED) effects. Why do we do that? One reason is that QED has stood up extremely well to all the tests which have been applied to it - if we show it doesn't work, we'll be famous. A brief "historical" introduction indicating how we got set on this track probably can't do any harm, so I will give one. In 1972, I got interested in a wealth of work then appearing in the literature in the general area of beam foil spectroscopy. We had, at that time, several ion beam accelerators in Oxford, as well as a strong tradition in experimental atomic spectroscopy (which originated from the work of D.A. Jackson and H.G. Kuhn in the 1930's). I decided it would be useful to learn the technique of beam foil spectroscopy in one of the centres then active, and then attempt to carry out my own experiments in Oxford. I had been particularly impressed by the very clean application of the fast beam technique to a new measurement of the Lamb shift in atomic hydrogen made by C. Fabjan and F.M. Pipkin at Harvard, so I spent some time working in the group of M. Dufay in Lyon, learning the "art" of beam foil spectroscopy from M-L Gaillard and J. Desesquelles. During my stay, I applied Pipkin's technique to make some measurements of Lamb shifts in He^+ and Li^{++}, and formulated plans to extend that sort of measurement to higher Z. Whilst at Lyon, I also came across a paper by S.O. Kastner which pointed out the possibility of "Lamb shift" measurements in Helium-like ions, in particular by accurate measurement of the wavelengths of the 1sns ^3S - 1snp ^3P transitions.

So much for the "history" - since 1974, I and my students and visitors have studied a wide range of one and two electron ions. The general aim has been to make measurements which are of fundamental interest in atomic physics, and primarily we have been interested in accurate measurement of energy level separations, though with the odd excursion into radiative lifetime measurement. For hydrogen-like ions, the energy of any given bound state (for a spin-zero nucleus) may be expressed as a sum of Dirac contributions, nuclear motion and size corrections, and radiative QED effects. A comparison of measured values of the energy level differences with theoretical values serves as a test of the theory, and may be interpreted as a test of radiative QED effects insofar as the other contributions are well known. For helium-like ions, the situation is rather more complicated, and a proper ab-initio calculation of relativistic and QED effects has not yet been carried out. Despite this, there have been a large number of measurements of selected intervals in helium-like ions. Indeed measurements of the 1s2s ^3S - 1s2p ^3P transition wavelengths have now developed into something of an "industry" - we have been responsible for measurements in N^{5+}, O^{6+}, F^{7+}, Ne^{8+}, Mg^{10+}, Al^{11+}, Si^{14+}, we are currently engaged on more refined experiments on Ne^{8+} and Ar^{16+}, and we are planning improved experiments on systems such as Fe^{24+}, and Xe^{52+}. Other groups are also studying other systems, right up to the very interesting study of U^{90+} described at this meeting by H. Gould and C. Munger. I will not discuss experiments on helium-like ions here, but perhaps I should mention that the approach of most theorists so far to this interval has been to split the energy into non-relativistic, relativsitic, and QED contributions, and then to calculate each of these various contributions by various methods. S.P. Goldman and G.W.F. Drake have published an estimate of the largest uncertainty in these contributions, and this is claimed to lie in the relativistic energy, where the omission of terms of order $\alpha^4 Z^4$ is estimated to lead to uncertainties of $\pm$ 1·2 $(Z/10)^4$ cm^{-1} in the 1s2s ^3S - 1s2p ^3P transition energies. It is heartening to know that an ab-initio QED calculation of the interval for high Z helium-like ions is planned by P.J. Mohr, and it will be very interesting to compare his calculated values with experiment.

If we study the hydrogen-like ions, it seems safe to assume that the relativistic contributions to the energy may be predicted exactly by the Dirac equation. The bound state energy for hydrogen-like ions E_b may be expressed as

$$E_b = E_{Dirac} + E_{QED} + E_{small}$$

The QED contribution to the bound state being studied, E_{QED}, may then be extracted by measuring E_b, and subtracting $(E_{Dirac} + E_{small})$. The contribution E_{small} contains small nasties such as corrections due to nuclear size and motion. For high Z systems, the nuclear sizes of the isotopes are in many cases not yet very well known, so that the accuracy of this form of QED test can be limited by the uncertainty in E_{small} in real cases. It is interesting to apply this approach to an accurate measurement of the Lyman α wavelength of hydrogenic Li^{++}, carried out by Bengt Edlen for his Doctoral thesis in 1933[1]. It appears that this work was the first measurement of a 1s Lamb shift, predating Lamb's work by over a decade, as later pointed out by von Keussler[2]. It is also not, in retrospect, surprising that the 1s Lamb shift was first observed in a hydrogenic ion, since its approximate Z^4 scaling makes it grow as $\sim Z^2$ relative to E_b. Since my own original motivation in starting work with simple highly ionised atoms was to "test QED", the rather muddy situation with the two-electron ion led me to decide, in 1978, that some of our research effort should be concentrated on accurate measurement of energy level separations in hydrogenic ions. The first problem facing the designer of any sensible QED test is - how accurate do I need to make my test in order that it will be interesting or significant? This question is very hard to answer for Lamb shift measurements in high Z hydrogenic ions, which are really QED tests for electrons bound in STRONG FIELDS (which is of course, why I am here in Maratea!). I spent some time thinking about this question during a visit I made to Berkeley in 1979, and I concluded that one possible criterion could be: <u>make the measurements as accurate as the uncertainty in the theory!</u> I looked at the literature, and discovered the beautiful work of R.D. Deslattes, which showed that the best measurements of x ray wavelengths at that time should be accurate to $\sim$ 1 ppm. For high Z hydrogenic ions, the uncertainty in P.J. Mohr's theoretical value of the Lamb shift[3] is $\sim$ ·1% of the shift. If you decide to make your QED test on the 1s level, you can make a (rather optimistic!) guess that you will be able to measure the 1s $^2S_{1/2}$ - 2p $^2P_{1/2,3/2}$ Lyman α wavelengths with an accuracy of 1 ppm, and you then find that you need to study ions where Z is sufficiently high that the 1s Lamb shift is $\sim$ 0·1% of the 1s - 2p separation. This first happens when Z $\sim$ 40. For high Z hydrogenic ions, around Z $\sim$ 30-40, the best laboratory source for spectroscopy appears at present to be the beam foil source, for which fluxes of $\sim 10^{10}$ ions/second in the hydrogenic charge state may be expected. An obvious technique for accurate measurement of transition wavelengths in these ions is shown schematically in figure 1.

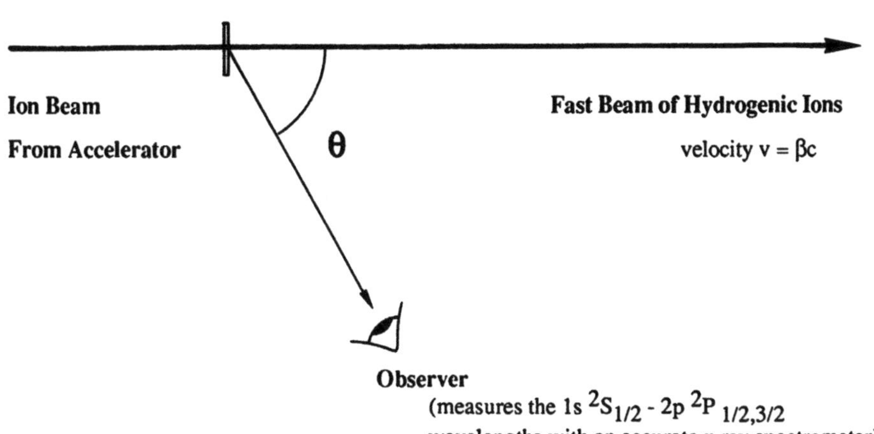

Fig 1: Experimental arrangement for simplest measurement of Lyman α wavelengths.

The main problem with this straightforward approach lies in the appreciable Doppler shifts which arise when viewing a fast beam source, even close to $\theta = 90^o$. If the observer's spectrometer is calibrated with a stationary source, ppm accuracy requires a knowledge of the observation angle θ at the microradian level (!), and $\gamma = (1-\beta^2)^{-1/2}$ also needs to be known to ppm accuracy. This sort of problem, as well as competition for accelerator time from other groups[4], led me to suggest a novel technique in 1981 which is not so sensitive to Doppler effects. The principle of the technique is to use the 4:1 ratio of the wavelengths of Balmer β and Lyman α to make a comparison between the Balmer β transition in first order diffraction, and the Lyman α transition in fourth order[5]. This technique is similar in concept to that used by Wieman and Hansch in a measurement of the 1s $^2S_{1/2}$ Lamb shift in atomic hydrogen, and indeed the method occurred to me whilst reading their work in preparation for a grant application to the UK SERC. A detailed consideration of the Lymanα/Balmerβ spectrum[5] shows that, rather fortuitously, the separation between the components of interest is small enough that the wavelength of Lymanα relative to Balmer β may be accurately determined, yet sufficiently large that only moderate resolution is required to separate the transitions. In this novel method, the hydrogenic transitions studied all arise from the same charge state in the same fast ion beam, so that the relative Doppler shifts $\delta\lambda/\lambda_0$ will be very nearly the same for all components. This is the major advantage of the technique, compared with other work using wavelength calibration from stationary sources. In addition, the first experiment which has actually worked using this method employs a very ingenious double spectrometer[6] which independently provides a further form of Doppler cancellation, by simultaneous observation at two angles, ϕ and $180^0 + \phi$, to the ion beam direction. In this arrangement, the sum of the two observed Doppler shifted wavelengths is $2\gamma\ \lambda_0$, independent of ϕ. (The ion beam velocity $v = \beta c$, c is the speed of light, $\gamma = (1-\beta^2)^{-1/2}$, as before, and the rest frame wavelength of the transition studied is λ_0). The experimental arrangement used in the first successful demonstration of the technique (in April 1986) is shown in figure 2.

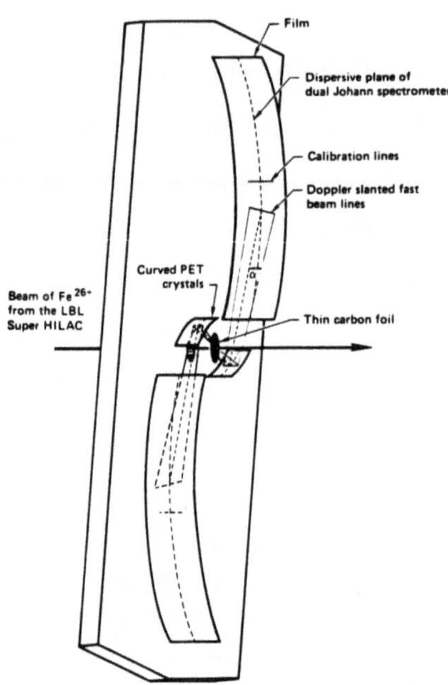

Fig 2: Experimental arrangement used in the present experiment.

A beam of bare iron nuclei (Fe^{26+}) at an energy of 0·489 GeV (8·74 MeV/A, $\beta = 0.135$) from the Lawrence Berkeley Laboratory Super-HILAC was passed through a thin (25 $\mu g/cm^2$) carbon foil. Some of these nuclei capture an electron in the foil, and we found that under these excitation conditions, lines in the Balmer series of hydrogenic Fe^{25+} were clearly seen for upper level principal quantum numbers n up to about 10. In particular, there was a good probability for the production of the n=2 to n=4 hydrogenic Balmer beta transitions.

I and my colleagues in Oxford first tried the novel technique on hydrogenic Kr^{35+} ions at GSI, Darmstadt in 1983 using our own Johannson curved crystal x-ray spectrometer. The reason for starting with hydrogenic krypton is explained above. When I first became aware of the work of D.D. Dietrich using the double spectrometer, it appeared very sensible to collaborate and apply the most suitable instrument to the problem. Our first collaborative efforts to make the technique work were in 1985, with 3 GeV krypton ion beams at GANIL, Caen, and later with 1 GeV germanium ion beams at GSI. These experiments used thick exciting foils ($\sim$ 600 $\mu g/cm^2$) and incident ions in charge states lower than fully stripped. Figures 3 and 4 show the sort of data we obtained. The experiments established that hydrogenic Lyman α transitions were strongly excited and could readily be observed in third, fourth and fifth orders of diffraction with suitable crystals, but we failed to observe the Balmer β transitions. It is now

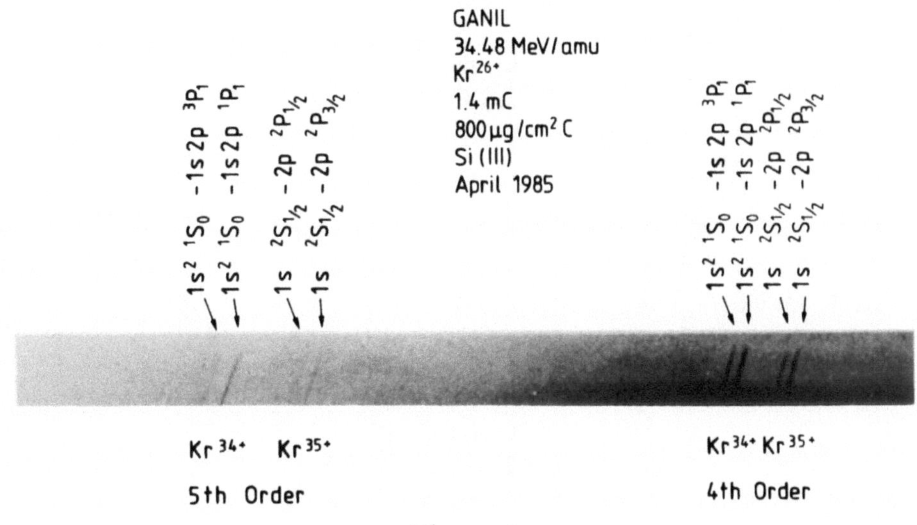

Figure 3

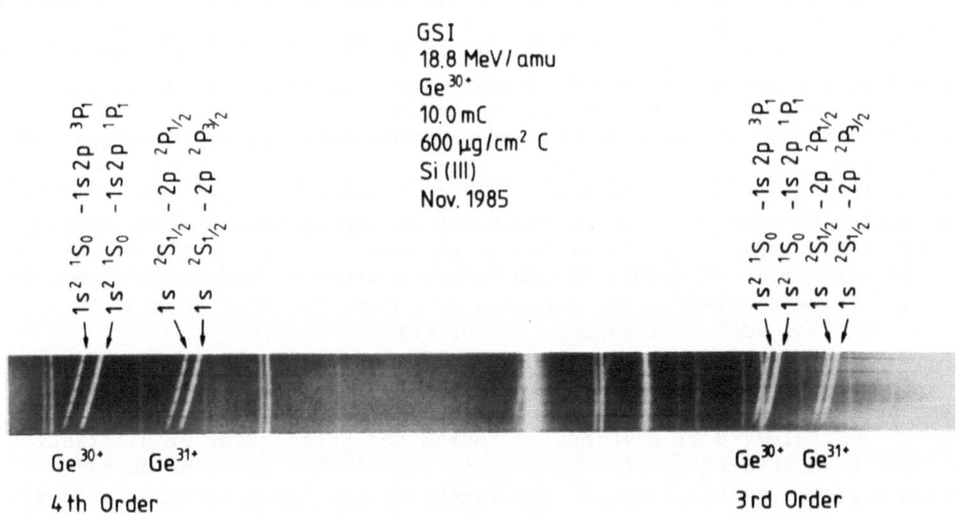

Figure 4

clear from (lower Z and hence somewhat easier) recent experiments with Fe^{25+} that the combination of a fully stripped ion beam with a thin exciter foil gives the best excitation conditions so far realised for the simultaneous observation of Lyman α and Balmer β radiation[7]. The dispersing crystals used for the Fe^{25+} experiment were Quartz $(10\bar{1}0)$ and PET (002). Spectra were subsequently put into digital form (intensity transmissivity as a function of position) using an accurate computer-controlled digital microdensitometer[8]. A print of a microdensitometer scan over the Lymanα/ Balmerβ region from one of the PET films is shown in figure 5. The data is currently under analysis in Oxford, and details of the experiment will be more fully described in A.F. McClelland's Doctoral thesis, and a

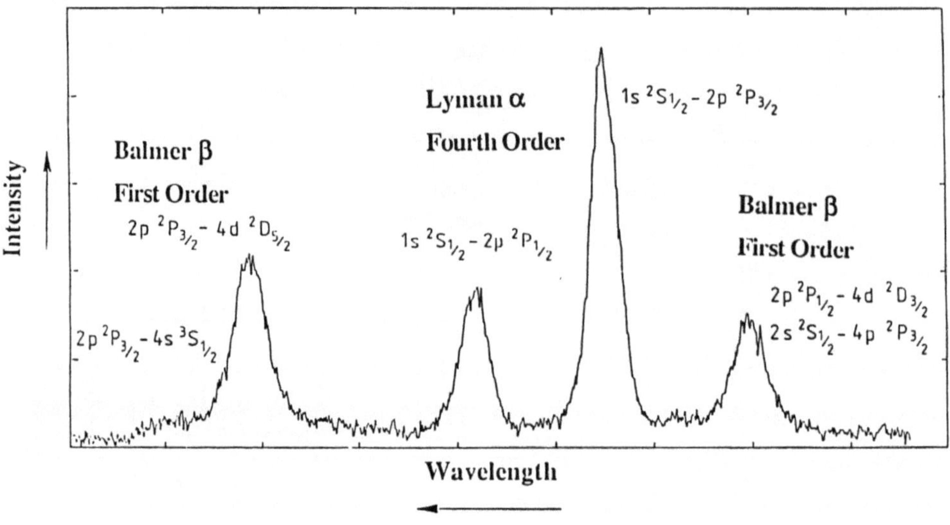

Fig. 5: Microdensitometer scan of photographic spectrum of hydrogenic iron, Fe^{25+}

forthcoming publication. Using wavelengths of the Fe^{25+} n=2 - n=4 Balmer β transitions from G.W. Erickson[9] to calibrate the films leads to preliminary results for the Lyman α wavelengths of: 1s $^2S_{1/2}$ - 2p $^2P_{3/2}$ 1·77815 ± 0·00019 Å, and 1s $^2S_{1/2}$ - 2p $^2P_{1/2}$ 1·78364 ± 0·00019 Å, where the errors arise from statistical errors in finding peak centres, the calibration procedure used, the uncertainty in the refractive index correction in going from first to fourth order, and possible electric field shifts. This last possible source of error arises because the observed radiation is emitted in part from ions moving within the carbon targets, and it is known that fields of order 6×10^8 volts/cm may be experienced by ions moving through solid targets[10], and is not fundamental to the technique, which may also be applied in principle to gas target sources of hydrogenic ions, to fast ion beams excited by gas targets, or indeed to other sources of hydrogenic ions such as Tokamaks, ECR or CRYEBIS sources.

In conclusion, a novel technique for the measurement of the ground state Lamb shifts in hydrogenic ions by simultaneous observation of the Lyman α and Balmer β transitions in fourth and first order of diffraction has been demonstrated for the hydrogenic ion Fe^{25+}. The data so far obtained yields values for the wavelengths of the Lyman α transitions with an accuracy of ~100 ppm, which gives the ground state Lamb shift to an accuracy of 17% and a value for the 2p electron fine-structure splitting with an accuracy of 3%. The accuracy of this first successful experiment is limited by the statistical fluctuations in the signal and resolution obtained in the spectra, the extent to which we understood the spectral profiles, the accuracy to which the refractive index correction is presently known, and the possible effect of electric fields within the targets. Experimental improvements may be envisaged which should result in relative Lyman α Balmer β wavelength accuracies at the ppm level for hydrogenic ions in the range Z = 20 to Z = 40, and the technique should also be applicable to the much higher Z hydrogenic ions (Z = 80 ∿ 90) which will become available in the next few years when the new generation of high energy heavy ion accelerators becomes operational.

ACKNOWLEDGEMENTS

A lot of people have contributed to the success of the experiments I have mentioned here. I would like to acknowledge advice and help from R.D. Deslattes and M. Hart on the techniques of x-ray spectroscopy, and I. Martinson for bringing Edlén's work on Li^{++} to my attention. I have had useful conversations on theoretical aspects of the Lamb shift with G.W. Erickson and P.J. Mohr. My Oxford colleagues B.P. Duval, H.A. Klein, J. Laursen and F. Moscatelli, and more recently J. Takacs, E.C. Finch and J.M. Laming and S.D. Rosner have worked on different stages of the experiments, and it is not clear that there would have been any data to talk about at all without the efforts of A.F. McClelland. We also received encouragement and support from E.W.J. Mitchell. In all the experiments tried since 1985, we have collaborated with D.D. Dietrich and G.C. Chandler from Livermore, and P.O. Egan worked on the two most recent experiments. The Livermore instrument is probably the most suitable spectrometer for demonstrating the feasibility of the technique, so that again it is not clear that there would have been any data to talk about at this time without our collaboration. The experiments would also not have been possible without the skilled operation of the Unilac at GSI, the dual cyclotron at GANIL and the SuperHILAC at LBL, and I am indebted to the accelerator staff at all three institutions for their help.

REFERENCES

1. B. Edlén, Nova Acta Regiae Societatis Scientarium Uppsaliensis, Ser IV, Vol 9 (1934).

2. V. von Keussler, Zeit. f. Naturforsch, 4a, 158 (1949).

3. P.J. Mohr, At. Data and Nucl. Tab., 29, 435 (1983).

4. M. Tavernier, J.P. Briand, P. Indelicato, D. Liesen and P. Richard, J. Phys. B18, L327 (1985) and references therein.

5. A.F. McClelland, J.S. Brown, B.P. Duval, E.C. Finch, H.A. Klein, J. Laursen, D. Lecler, P.H. Mokler and J.D. Silver, Nucl. Inst. and Meths. B9, 682 (1985).

6. D.D. Dietrich, G.A. Chandler, R.J. Fortner, C.J. Hailey and R.E. Stewart, Phys. Rev. Lett. 54, 1008 (1985) and references therein.

7. J.-P. Rozet, Private communication 1985, J. Reed and C.J. Sofield, Private communication 1985.

8. J.S. Brown, C.W. Band, E.C. Finch, R.A. Holt, H.A. Klein, J. Laursen, A.F. McClelland, N.J. Peacock, J.D. Silver, M.F. Stamp and J. Takacs, Nucl. Inst. and Meths. B9, 682 (1985).

9. G.W. Erickson, J. Phys. Chem. Ref. Data, 6, 831 (1977).

10. S. Datz, C.D. Moale, O.H. Crawford, H.F. Krause, P.F. Dittner, J. Comez del Campo, J.A. Biggerstatt, P.D. Miller, P. Hvelplund and H. Knudsen, Phys. Rev. Lett. 40, 843 (1978).

SYMMETRY VIOLATION IN ATOMS

A. Schäfer[*)], B. Müller[+)], G. Soff[+)] and W. Greiner[+)]

*) Gesellschaft für Schwerionenforschung, Darmstadt
+) Institut für Theoretische Physik, Joh. Wolfg. Goethe-Universität, Frankfurt am Main

I. INTRODUCTION

The extremely sophisticated techniques developped in atomic physics especially in combination with laser-optics allows the measurement of very small effects. Such experiments are therefore able to detect e.g. the parity violating interaction due to neutral currents[1-3]. Before we discuss the type of effects one observes and how one calculates the theoretical predictions let us make some general remarks about the role of the discrete symmetries for the wave-function of a bound Dirac-particle (for problems with spherical symmetry).

II. THE DIRAC WAVE FUNCTIONS AND THE DISCRETE SYMMETRIES

The Dirac spinor has four complex components and thus for any spherically symmetric problem it can be characterized by eight real functions of the radial coordinate r.

$$\psi = \begin{bmatrix} i(g_1(r)+ig_2(r))\chi^{\mu}_{\varkappa} + i(g_3(r)+ig_4(r))\chi^{\mu}_{-\varkappa} \\ (f_1(r)+if_2(r))\chi^{\mu}_{-\varkappa} + (f_3(r)+if_4(r)\ \chi^{\mu}_{\varkappa} \end{bmatrix} \tag{1}$$

The usual form with just two real functions can be obtained from this general ansatz with the help of the discrete symmetries. For Dirac fields these can be defined as the operators

$$\hat{P}: \quad \gamma_o\, x\left(\ldots\right)|_{\vec{x}\to-\vec{x}} \tag{2}$$

$$\hat{T}: \quad \gamma_1\gamma_3\, x\left(\ldots\right)^*|_{t\to-t} \tag{3}$$

$$\hat{C}: \quad x\left(\ldots\right)^*|_{x_\mu\to-x_\mu} \tag{4}$$

If e.g. P is conserved one can impose that ψ is a state with definite parity

$$\hat{P}\,\psi = \psi \quad \text{or} \quad \hat{P}\,\psi = -\,\psi\ . \tag{5}$$

Inserting 1 into this equation one finds that either g_3, g_4, f_3 and f_4 or g_1,g_2,f_1 and f_2 have to vanish. Similar constraints are obtained for T and C eigenstates. These results are summarized in Table I.

Table I. The General Form of the Dirac Wave-function for Different Conserved Symmetries.

conserved symmetry	general form of the solution	# coupled equations
-	$\psi = \begin{bmatrix} i(g_1+ig_2)\chi^{\mu}_{\varkappa}+i(g_3+ig_4)\chi^{\mu}_{-\varkappa} \\ (f_1+if_2)\chi^{\mu}_{-\varkappa}+(f_3+if_4)\chi^{\mu}_{\varkappa} \end{bmatrix}$	8
P	$\psi = \begin{bmatrix} i(g_1+ig_2)\chi^{\mu}_{\varkappa} \\ (f_1+if_2)\chi^{\mu}_{-\varkappa} \end{bmatrix}$	4
T	$\psi = \begin{bmatrix} ig_1\chi^{\mu}_{\varkappa}-g_4\chi^{\mu}_{-\varkappa} \\ f_1\chi^{\mu}_{-\varkappa}+if_4\chi^{\mu}_{\varkappa} \end{bmatrix}$	4
C	$\psi = \begin{bmatrix} ig_1\chi^{\mu}_{\varkappa}+ig_3\chi^{\mu}_{-\varkappa} \\ f_1\chi^{\mu}_{-\varkappa}+f_3\chi^{\mu}_{\varkappa} \end{bmatrix}$	4
C,P,T	$\psi = \begin{pmatrix} ig_1\chi^{\mu}_{\varkappa} \\ f_1\chi^{\mu}_{-\varkappa} \end{pmatrix}$	2

Due to the CPT-theorem it is impossible to violate just one symmetry. The various possibilities are therefore most easily classified according to the concerved symmetries. In general the field-theoretical mechanisms leading to the possibility of symmetry violation will still conserve one of the discrete symmetries (e.g. the weak-interaction conserves T). Thus in general one has to use Dirac wave functions with four radial functions in these problems. In the following we will discuss a special P and C-violating effect in atoms, due to the weak neutral currents.

III. PARITY-VIOLATION IN CESIUM[4]

The main effect leading to parity violation in atoms is the following. The electrons in an atom are bound by the exchange of virtual photons. However in addition to photons neutral vector bosons Z^0 are also exchanged (see Fig. 1). As Z^0 may couple either as a vector or as an axial vector, this exchange violates parity. Thus the electron states are no longer eigenstates of the parity operator. Instead e.g. every $s_{1/2}$ state will have a small admixture of $p_{1/2}$ wave functions. As a consequence the ordinary multipole selection rules are violated. In this work we will present the calculation of one specific transition forbidden by the usual selection rules, namely the electric dipole strength for the transition $6s \leftrightarrow 7s$ in cesium. The value of this matrix element has been determined experimentally by M.A. Bouchiat et al.[2]

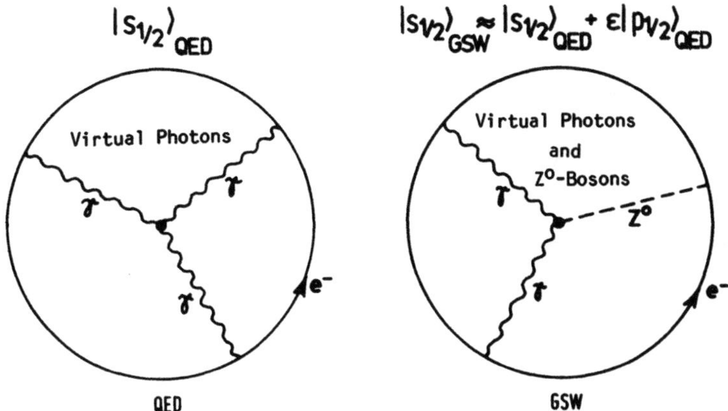

Fig. 1. The exchange of Z bosons between the electron and the nucleus leading to states with mixed parity. QED stands for quantum electrodynamics and GSW for the Glashow-Salam-Weinberg model.

To describe the Z^0-exchange between the electrons and the nucleus the usual relativistic Hartree Fock equations

$$[\vec{\alpha}.\vec{p}+\beta m+V_{cb}+\hat{V}_{hf}]\Psi_i=E_i\Psi_i \quad \underline{6}$$

are modified by the additional term

$$D=\rho(r).\gamma_5 . \quad \underline{7}$$

Here $\rho(r)$ is a combination of the density distributions of protons, $\rho_p(r)$, and neutrons, $\rho_n(r)$, in the nucleus.

$$\rho(r)= - \frac{G}{2}[C_n\rho_n(r)+C_p\rho_p(r)] \quad \underline{8}$$

$$C_p= \frac{1}{2} - 2\sin^2\theta_w; \quad C_n= - \frac{1}{2}; \quad \gamma_5= \begin{pmatrix} 0 & -1 \\ -1 & 0 \end{pmatrix}. \quad \underline{9}$$

The derivation of these equations from the Glashow-Salam-Weinberg lagrangian is not straightforward. The GSW model describes the Z^0-coupling to the quarks. The substitution of the quark-fields by the nucleon-fields can only be done in a kind of lowest order way. Nevertheless these equations become a very good approximation if one inserts the charge-distribution obtained experimentally, e.g. by electron-nucleon scattering. The reason is that the photon couples to the quark in the same way as the Z^0-bosons such that all necessary corrections are already included in the effective charge-distribution. We used the value $\sin^2\theta_w=0.22$ for our calculations[5] and the Fermi distribution given in ref. 6 for $\rho_p(r)$ and $\rho_n(r)$(c=5.67fm,t=2.3fm). G is the Fermi constant. (The results obey approximate scaling relations which makes it easy to correct them for other choices of $\sin^2\theta_w$.) It is straightforward to show that $\hat{D}$ commutes with $\hat{J}$ and $\hat{T}$, but does not commute with $\hat{P}$ and $\hat{C}=\hat{P}\hat{T}$.

$$\left[\vec{J},\rho(r)\gamma_5\right]_- = \gamma_5\left[\vec{r}\times\left(\frac{\vec{r}}{r}\frac{\partial}{\partial r}\rho(r)\right)\right]_- + \frac{1}{2}\rho(r)\left[\vec{\Sigma},\gamma_5\right]_- = 0 \qquad \underline{10}$$

$$\hat{P}\gamma_5\rho(r)\hat{P}^{-1} = \gamma_o\gamma_5\gamma_o\rho(r) = -\gamma_5\rho(r)$$

$$\hat{T}\gamma_5\rho(r)\hat{T}^{-1} = \gamma_1\gamma_3\gamma_5\gamma_3\gamma_1\rho(r) = \gamma_5\rho(r) \qquad \underline{11}$$

$$\hat{C}\gamma_5\rho(r)\hat{C}^{-1} = -\gamma_5\rho(r)$$

We therefore use the ansatz of table I which we write in the form:

$$\Psi(x) = \frac{1}{r}$$

$$\begin{pmatrix} ig_\varkappa(r)X^\mu_\varkappa(\theta,\varphi)-g_{-\varkappa}(r)X^\mu_\varkappa(\theta,\varphi) \\ f_\varkappa(r)X^\mu_{-\varkappa}(\theta,\varphi)-if_{-\varkappa}(r)X^\mu_\varkappa(\theta,\varphi \end{pmatrix} e^{-iEx_o} \qquad \underline{12}$$

and combine the radial functions to one four component vector

$$F':=(g_\varkappa, g_{-\varkappa}, f_\varkappa, f_{-\varkappa}) , \qquad \underline{13}$$

where the superscript t denotes matrix transposition. The components $g_\varkappa$ and $f_\varkappa$ correspond to the two components of the ordinary parity conserving wave function. Inserting this ansatz in (6) one gets:

$$\hat{H}F(r)= \begin{pmatrix} m+V_{cb}+V_{hf} & 0 & \left[\frac{d}{dr}-\frac{\varkappa}{r}\right] & \rho(r) \\ 0 & m+V_{cb}+V_{hf} & \rho(r) & -\left[\frac{d}{dr}+\frac{\varkappa}{r}\right] \\ -\left[\frac{d}{dr}+\frac{\varkappa}{r}\right] & \rho(r) & V_{cb}+V_{hf}-m & 0 \\ \rho(r) & \left[\frac{d}{dr}-\frac{\varkappa}{r}\right] & 0 & V_{cb}+V_{hf}-m \end{pmatrix} F(r). \qquad \underline{14}$$

In the Hartree-Fock potential the contribution from the valence electron of cesium is neglected. Thus one has to sum over closed shells only and the Hartree Fock potential has the rather simple form:

$$\hat{V}_{hf}F_b(r)=e^2\sum_s(2j_s+1)\int dr'\Bigg(\frac{1}{r_>}F^t_s(r').F_s(r').F_s(r')F_b(r)$$

$$-\sum_L \frac{r^L_<}{r^{L+1}_>}\begin{pmatrix} j_b & j_s & L \\ -\frac{1}{2} & \frac{1}{2} & 0\end{pmatrix}^2 [F^1_s(r').F_b(r')$$

$$.\ F_s(r)\pi(l_s+l_b+L)-F^t_s(r').\hat{\Gamma}.F_b(r').\hat{\Gamma}.F_s(r)\pi(l_s+l_b+L+1)]\Bigg)$$

$$\underline{15}$$

with:

$\pi(L)=0$ if L is odd, $\pi(L)=1$ if L is even

$$\hat{\Gamma}=\begin{pmatrix} 0 & -1 & 0 & 0 \\ 1 & 0 & 0 & 0 \\ 0 & 0 & 0 & 1 \\ 0 & 0 & -1 & 0 \end{pmatrix} \qquad \underline{16}$$

(Here the indices b and s characterize the electron states.) $\hat{V}_{hf}$ deviates from the usual expression by the second term in the sum over L. This second term contributes expressions of the form

$$\hat{V}_{hf} g^b_{-\kappa}(r) = e^2 \sum_{L,s} \frac{r^L_<}{r^{L+1}_>}(2j_s+1) \begin{pmatrix} j_b & j_s & L \\ -\frac{1}{2} & \frac{1}{2} & 0 \end{pmatrix}^2 . \pi(l_s+l_b+L+1)$$

$$. \int dr' g^s_{-\kappa}(r') g^b_{\kappa}(r') . g^s_{\kappa}(r) \ . \qquad \underline{17}$$

Thus even for states b for which the effects of the weak electron-nucleus interaction are small (e.g. for states with high angular momentum) $g^b_{-\kappa}$ can become important due to the electromagnetic coupling to the 'wrong' parity components of the other states (see Fig. 2).

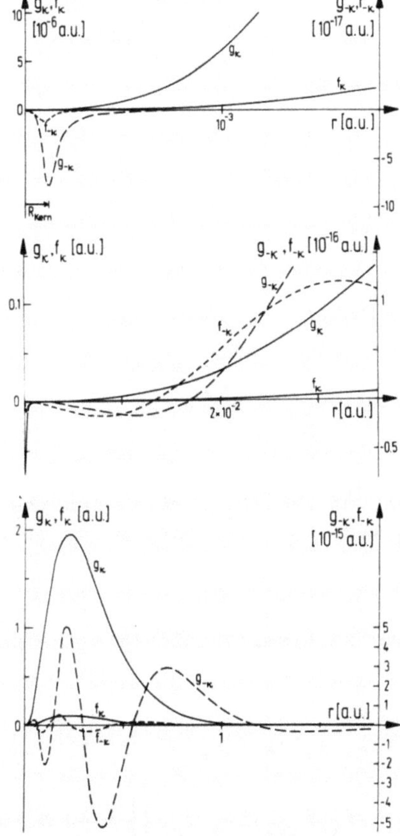

Fig. 2. The four components of the $3d_{5/2}$ wave function in atomic units. In all three parts of the figure r is plotted on a linear scale.

We also included the neutral weak interaction between the electrons, which however turned out to be negligible.

In general the 'wrong' parity components are large, when the overlap of the 'right' parity components with the nucleus is large. They decrease drastically in size if the main quantum-number or the angular-quantum-number is increased. Furthermore for $s_{1/2}$ states the 'wrong' parity components look like a $p_{1/2}$ state and vice versa (see Figure 3).

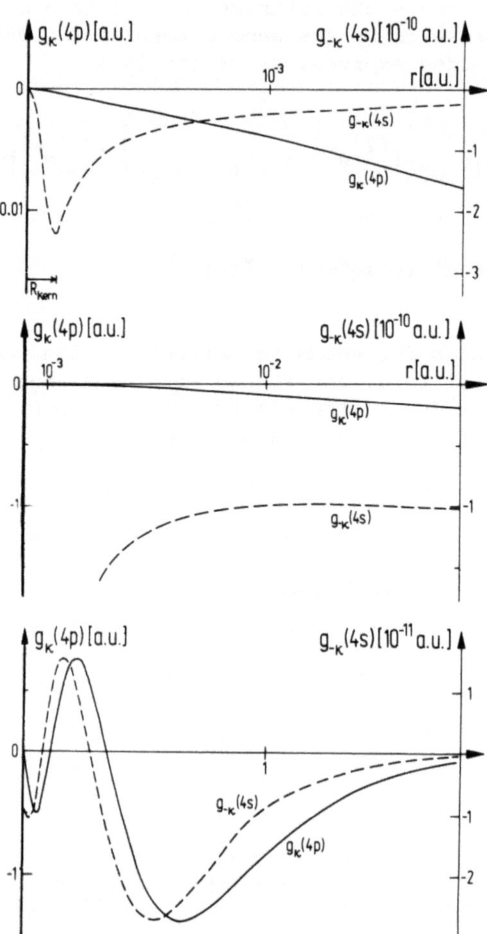

Fig. 3. The $g_\varkappa$ component of the $4p_{1/2}$ state and the $g_{-\varkappa}$ component of the $4s_{1/2}$ state in atomic units. In all three parts of the figure r is plotted on a linear scale.

With these wave functions we finally calculate the parity-violating E1 matrix element 7s⟷6s:

$$E1 = \frac{e}{w}\int dr\Big\{j_2(wr)\Big(g_{-\varkappa_7}f_{\varkappa_6}+f_{\varkappa_7}g_{-\varkappa_6}\Big) + j_1(wr)\Big(g_{-\varkappa_7}g_{\varkappa_6} - g_{\varkappa_7}g_{-\varkappa_6}+f_{\varkappa_7}f_{-\varkappa_6}-f_{-\varkappa_7}f_{\varkappa_6}\Big)\Big\} = -8.4.10^{-12}\, iea_o \quad .$$

18

This result is compared with the experiments in Fig. 4, using the scaling properties of E1. The agreement is excellent.

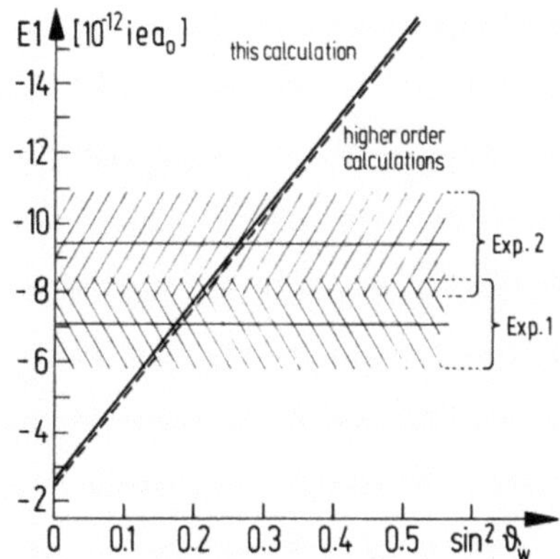

Fig. 4. Comparison between the theoretical result for the E1 matrixelement between the 6s and 7s state in cesium and the experiment values. The higher order calculations have been done by A.-M. Martensson, W. Johnson et al., and V.A. Dzuba et al.(ref. 3).

REFERENCES

1. For a review see in At. Phys. 7 (1981) the articles of: C. Bouchiat, E. D. Commins, pp. 83 and 121; E. N. Fortson, L. Wilets, Adv. At. Mol. Phys. 16:319 (1980).
2. M. A. Bouchiat, J. Guena, L. Hunter, L. Pottier, Phys. Lett. 117B:358 (1982), Phys. Lett. 134B:463 (1984).
3. V. A. Dzuba, V. V. Flambaum, P. G. Silvestrov, P. P. Sushkov, Phys. Lett. 103A:265 (1984); A. M. Martensson-Pendrill, 9th International Conference on Atomic Physics Seattle (1984); W. R. Johnson, D. S. Guo, M. Indrees, J. Sapirstein, Phys. Rev. A. 32:2093 (1985)
4. A. Schäfer, B. Müller, and W. Greiner, Z. Phys. A 322:539 (1985).
5. J. F. Wheater, Phys. Lett. 105B:483 (1981).
6. R. Engfer, H. Schneuwly, J. L. Vuilleumier, H. K. Walter, A. Zehnder, At. Data Nucl. Data Tables 14:509 (1974).

ATOMIC PHYSICS AND THE DIMENSIONALITY OF SPACE

Andreas Schäfer	Berndt Müller
GSI	Inst. f. Theoretische Physik
P.O. Box 110541	Joh.Wolfg.Goethe-Universität
D-6100 Darmstadt 11	D-6000 Frankfurt

INTRODUCTION

The extremely high precision reached in QED experiments provides stringent bounds for any change in the standard framework of low energy physics (i.e. for energies close to the electron mass). In his contribution J. Reinhardt discusses e.g. such bounds for the coupling to a new light particle. In this lecture I want to discuss a rather different possibility, namely to derive from the Lamb-shift a bound for the deviation of the spatial dimension from the integer value three. Before doing so it is obviously necessary to discuss the concept of fractal dimensions and to explain their possible origin.

THE CONCEPT OF FRACTAL DIMENTSIONS

I shall discuss three motivations as to why non-integer dimensions should be considered.
The first one is a purely formal argument. In modern field theory the technique of dimensional continuation was developed to allow for a systematic and consistent renormalization. A crucial fact used by this scheme is that all the logarithmic divergences disappear if the space-time dimension is $d = 4-\varepsilon$, $\varepsilon > o$ with an arbitrarily small ε. As the dimensional continuation allows to formulate any field theory in arbitrary non-integer dimensions one can argue however, that perhaps space-time is really non-integer, that any renormalizable theory is really finite, and that our concept of a three dimensional space is only a (very good) approximation of the reality (Zeilinger and Svozil, 1985).

The second and perhaps most fascinating possibility to allow for the existence of fractional dimensions is based on the properties of 'fractals', mathematical objects introduced in physics mainly by B. Mandelbrot 1977, 1985. To explain what is meant by this term let us discuss first Hausdorff's definition of dimensions. To understand this definition one should remember that the length of a curve, the area of a surface and the volume of a body can be determined by counting the number of rods of length ε, squares of area ε^2 or cubes of volume ε^3 which fit into it. Normally the limit $\varepsilon \to o$ of these sums exists and the resulting number is identified with the length of the curve, the area of the surface and the volume of the body respectively.

length of a curve: $L = \lim_{\varepsilon\to o} \Sigma\, \varepsilon^1$

area of a surface: $A = \lim_{\varepsilon\to o} \Sigma\, \varepsilon^2$ (1)

volume of a body: $V = \lim_{\varepsilon\to o} \Sigma\, \varepsilon^3$

The exponent of ε appearing in equation (1) is obviously equal to the dimension of the object considered. This connection is exploited by Hausdorff's definition of dimensionality, which can be split up into the following steps.

1.) You start with a metric space, i.e. with a point-set and some definition for the separation between two points

$$d(x,y) = \text{separation of the points } x \text{ and } y. \tag{2}$$

Using this metric it is possible to define subsets E_{i,ε_i} as

$$E_{i,\varepsilon_i} = \left\{ x_i : d(x,x_i) < \varepsilon_i \right\} \quad . \tag{3}$$

2.) We now turn to some other point set M the length, surface, volume ...., more generally the 'content' of which we want to measure. To this end we need all combinations of the E_{i,ε_i} which cover M completely. And we define the 'Hausdorff measure' μ_H of M by

$$\mu_H(M) = \lim_{\varepsilon\to o} \inf_{\{E_i\}} \left\{ \sum_i \varepsilon_i^D : M \subset \bigcup_i E_{i,\varepsilon_i};\ \forall_i \varepsilon_i < \varepsilon \right\} , \tag{4}$$

i.e. for all coverings of M by the subsets E_{i,ε_i} with $\varepsilon_i<\varepsilon$ we choose that for which $\sum_i \varepsilon_i^D$ is smallest. μ_H is then the value of this sum in the limit $\varepsilon\to o$. It clearly depends on the value of D whether the limit $\lim_{\varepsilon\to o} \sum_i \varepsilon_i^D$ exists. Let me illustrate this for a simple example.

Let us discuss R^2 with the maximum-metric

$$d(x,y) = \max\left(|x_1-y_1|,\ |x_2-y_2|\right) \tag{5}$$

The subsets E_{i,ε_i} are then squares centered around x_i.

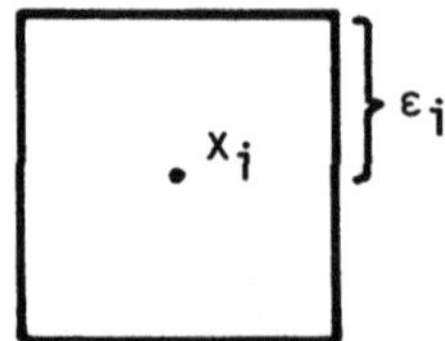

We now choose for M a square of side-length L. Then, for D<2 the sum $\sum_i \varepsilon_i^D$ becomes larger if one goes to a covering with smaller squares. Therefore the infimum $\inf_{\{E_i\}}$ choose the covering where the E_{i}, ε_i are most equal in size. For ε=L/4, L/8, L/2N one gets.

$\varepsilon = L/4$ $\qquad \inf_{\{E_i\}}\{\ldots\} \hat{=}$ $\qquad \sum_i \varepsilon_i^D = 4\left(\frac{L}{4}\right)^D$

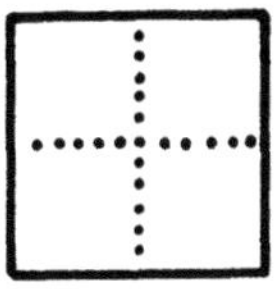

$\varepsilon = L/8$ $\qquad \inf_{\{E_i\}}\{\ldots\} \hat{=}$ $\qquad \sum_i \varepsilon_i^D = 16.\left(\frac{L}{8}\right)^D$

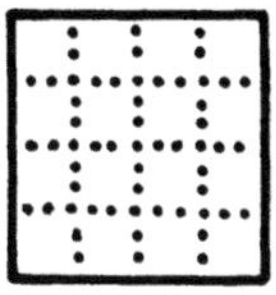

$\varepsilon = L/2N$ $\qquad \sum_i \varepsilon_i^D = N^2.\left(\frac{L}{2N}\right)^D$

It is clear that for any value of D<2 the sum diverges as ε goes to zero. On the other hand for D>2 the value of the sum $\sum_i \varepsilon_i^D$ decreases if one goes to a covering with smaller squares. Thus the $\inf_{\{E_i\}}$ will choose an infinitely fine covering $\{E_i\}$ and the right hand side of equation (4) is zero even for finite ε. The value of D one has to choose is the one where this behaviour changes. Thus in our example D=2.

With this example in mind the general definition of the Hansdorff dimension is now clear:

$$D_H = \sup\left\{D\in \mathbb{R} : D>o \ ; \quad \mu_H(M,D) = \infty\right\} \tag{6}$$

D_H is the upper limit of all values of D for which the expression (4) diverges.

The great advantage of this definition is that it works for any metric, measurable space. Thus one can attribute a dimension to any such space. Doing so one is faced with the rather astonishing result that there are point-sets to which one has to attribute a non-integer dimension. The easiest example is a Koch curve defined as limit of the following graphical series.

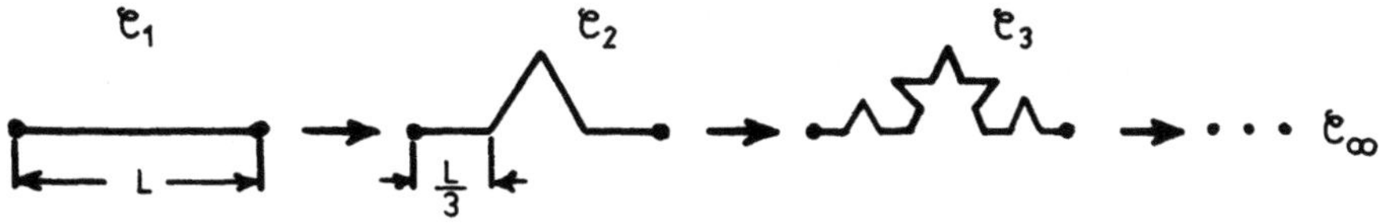

$\mathfrak{C}_1 \rightarrow \mathfrak{C}_2 \rightarrow \mathfrak{C}_3 \rightarrow \cdots \mathfrak{C}_\infty$

For any $\varepsilon \ll L$ one gets for C_1 : $\sum_i \varepsilon_i^1 = L$, C_2 : $\sum_i \varepsilon_i^1 = \frac{4}{3}L$, C_3 : $\sum_i \varepsilon_i^1 = \left(\frac{4}{3}\right)^2 L$ etc. . Thus it is already clear that the sum does not exist for D=1. It is possible however to deduce from this series the value one has to choose for D. The crucial observation is that if one goes from ε to $\varepsilon/3$ one resolves the next level of structuring. Thus one gets

$$\frac{L(\varepsilon/3)}{L(\varepsilon)} = \frac{4}{3} \tag{7}$$

which has the solution $L(\varepsilon) = \text{const.}\ \varepsilon^{1-\frac{\log 4}{\log 3}}$. (8)

From

$$\sum_i \varepsilon_i^1 \sim \varepsilon^{1-\frac{\log 4}{\log 3}} \tag{9}$$

and $\varepsilon_i \sim \varepsilon$ it is now straightforward to conclude

$$1 \sim \sum_i \varepsilon_i^1 \cdot \varepsilon^{-1+\frac{\log 4}{\log 3}} \sim \sum_i \varepsilon_i^{\frac{\log 4}{\log 3}} \tag{10}$$

i.e. one gets a finite, non-zero result only for $D = \frac{\log 4}{\log 3} = 1.2618...$ According to the definition (6) we therefore have to conclude that the dimension of the Koch curve discussed is 1.2618... .

This is the basic idea of fractals: There are point sets with an infinitely fine microscopic structure to which one has to attribute a non-integer dimension.

On the other hand one can (and in fact does) describe space as a discrete lattice. The basic postulate of all lattice gauge calculations is that if the lattice constants are choosen small enough for a given energy-scale such a discrete point set is indistinguishable from the continuous Minkowski space. If one chooses however a fractal point set instead of e.g. a simple cubic lattice the resulting continuum limit could lead to a space with a fractal dimension.

This ideas received an enormous impetus by intensive recent work (Gefen et al., 1983; Bhanot et al., 1984;) e.g. indicating a close relationship between these fractional dimensions and those appearing in the theoretical treatment of phase transitions. We do not want to discuss these results here. Instead let us just note the following properties of these lattices.

i.) The dimension of a 'fractal' point set can take arbitrary noninteger values.

ii.) If one introduces a finite resolution, e.g. a typical scale l_0, into the generalized definition of the dimension of a point set one finds that the space dimension D will in general be a function of this resolution. Within the framework of this interpretation a non-vanishing value of D-3 would indicate a non-trivial microscopic lattice structure of space.

A third possibility to motivate the idea of non-integer dimensions is to interpret the dimension as an effective description of additional, compactified dimensions, postulated e.g. by the extended Kaluza-Klein theories (Wetterich 1985). As all field propagators are affected in the same way by these additional dimensions it could very well be that all the tiny effects of this microscopic structure can be parametrized in first order by just one parameter D_{eff}.

Keeping these motivations in mind we now turn to the calculation of the corrections to the Lamb-shift, due to a non-integer space dimension.

THE LAMB SHIFT IN D=3-ε DIMENSIONS

To analyze the consequences of a non-integer dimension for the Lamb shift in hydrogen it is necessary to include the spin of the electron, as otherwise one would not be able to distinguish between the generalized $2p_{1/2}$ and $2p_{3/2}$ state. It is also advisable to include relativistic effects. Although they are negligible in three space dimensions it is by no means clear that also the changes due to a non-vanishing ε are small as compared to the changes in the non-relativistic contributions. Thus to treat this problem we can in principle start with either a Schroedinger equation including relativistic and fine-structure terms or with the Dirac equation. As the spinor-structure of the latter makes the generalization to arbitrary dimensions difficult (although not impossible), we use the Schroedinger equation. In case that the relativistic effects turned out to be important we would probably have to analyze also the fully relativistic Dirac equation.

We proceed as follows. First we generalize the standard treatment of the Lamb shift as found in Davydov, 1965 to arbitrary integer dimensionality of space D=3,4,5, ... This step requires some group theory to find the higher- dimensional expressions for the fine-structure term. We are then able to calculate the Lamb shift as a function of D : $\Delta E=\Delta E(D)$. This expression can be continued to arbitrary non-integer values of D and we evaluate

$$\Delta E(D=3-\varepsilon) - \Delta E(D=3) \simeq -[\partial \Delta E/\partial D]_{D=3}\, \varepsilon \tag{11}$$

Equation (11) together with the experimental bounds on additional contributions to the Lamb-shift in hydrogen leads to an upper bound for $|\varepsilon|$.

We start with the three-dimensional Hamiltonian

$$(H_o + W_1 + W_2 + W_3)\, \psi = E\, \psi \tag{12}$$

with

$$H_o = -1/(2m)\, \nabla^2 + V \quad ; \quad V= -\alpha/r \tag{13}$$

$$W_1 = \nabla^2 V/(8m^2) \tag{14}$$

$$W_2 = -(E-V)^2/(2m) \tag{15}$$

$$W_3 = 1/(2m^2 r)\, \partial V/\partial r\, \vec{s}.\vec{L} \tag{16}$$

The generalization to D dimensions is straightforward.

$$(H_{0,D} + W_{1,D} + W_{2,D} + W_{3,D}) \phi = E \phi \tag{17}$$

with

$$H_{0,D} = -1/(2m) \; (\partial_r^2 + (D-1)/r \; \partial_r) + c_2[(L)_D]/(2mr^2) + V \tag{18}$$

$$W_{1,D} = e/(8m^2) \; 2\pi^{D/2}/[\Gamma(D/2)] \; (D-2) \; r_o^{D-3} \; \rho(r) \tag{19}$$

$$W_{2,D} = -(E-V)^2/(2m) \tag{20}$$

$$W_{3,D} = \begin{cases} 0 & \text{for } (2s_{1/2})_D \\ (D-1)/(4m^2 r) \; \partial V/\partial r & \text{for } (2p_{1/2})_D \end{cases} \tag{21}$$

Let us now turn to the eigenvalues of the energy which can be expressed as expectation value of the Hamiltonian (17)

$$E_{i,D} = \int d^D r \; \phi^*_{i,D} \; H_D \; \phi_{i,D} \tag{22}$$

with the normalization condition

$$\int d^D r \; \phi^*_{i,D} \; \phi_{i,D} = 1 \tag{23}$$

Due to the rotational symmetry all angular integrals in (22) and (23) can be carried out, leaving only radial integrals to be performed.

$$E_{i,D} = r_o^{3-D} \int dr \; r^{D-1} \; f_{i,D}(r) \; H_{i,D}(r) \; f_{i,D}(r) \tag{24}$$

with

$$r_o^{3-D} \int d^D r \; r^{D-1} \; (f_{i,D})^2 = 1 \qquad D=3,4,5 \ldots \tag{25}$$

The crucial point is that these integrals can be continued to arbitrary dimensions. We indicate this continuation by replacing $E_{i,D}$ by $E_i(D)$ etc.

$$E_i(D) = r_o^{3-D} \int dr \; r^{D-1} \; f_i(D,r) \; H_i(D,r) \; f_i(D,r) \tag{26}$$

with

$$r_o^{3-D} \int d^D r \; r^{D-1} \; (f_i(D,r))^2 = 1 \quad D = \text{real number} \tag{27}$$

As D is very close to 3 it is sufficient to expand equation (26) and (27) around this value. Using a generalized Hellmann-Feynman-theorem (Schäfer and Müller, 1986) we get for the contributions from $H_{0,D}$, $W_{1,D}$, $W_{2,D}$ and $W_{3,D}$ to the Lamb shift.

$$\Delta E(H_{0,D}) = -\varepsilon \int dr\ r^2\ f(2p_{1/2})\ [-1/(2mr)\ \partial_r + 1/(2mr^2)$$

$$- (\alpha/r)\ \log(r_o/r)]\ f(2p_{1/2}) \tag{28}$$

$$+ \varepsilon \int dr\ r^2\ f(2s_{1/2})\ [-1/(2mr)\ \partial_r - (\alpha/r)\ \log(r_o/r)]\ f(2s_{1/2}) = -\varepsilon \frac{\alpha}{12a}$$

$$\Delta E(W_{1,D}),\ \Delta E(W_{2,D}),\ \Delta E(W_{3,D}) \ll \Delta E\ (H_{0,D}) \tag{29}$$

with $a = \frac{1}{m\alpha}$.

Our results thus reads

$$\Delta E = -2.27\text{eV}\ \ \varepsilon \tag{30}$$

As the theoretical and experimental value for the Lamb shift agree within (see the talk given by P. Mohr)

$$|\Delta E(\text{exp}) - \Delta E(\text{theor})| < 0.02\ \text{MHz} = 8.2.10^{-11}\text{eV} \tag{31}$$

we get a rather stringent bound for the deviation of the space dimension from the value three.

$$|\varepsilon| < 3.6.10^{-11} \tag{32}$$

Let us note that we also derived the bound $|\varepsilon|<10^{-9}$ from the perihelion shift of mercury (Müller and Schäfer 1986, Schäfer and Müller 1986) and that similar calculations have been done by de Mesa et al. 1985 and Jarlskog et al. 1985.

We feel that such bounds are important for any microscopic model of space.

REFERENCES

Bhanst C., Neuberger H., and Shapiro J., 1984: "Simulation of a Critical Ising Fractal' Phys. Rev. Lett. 53:2277.

Davydov A. S., 1965: "Quantum Mechanics,' Pergamon, Oxford.

deMesa A. G., Guzman A., and Yndurain F. J., 1985: Determination of the Number of Non-Compact Dimensions,' University of Madrid preprint 17/85.

Gefen, Y., Meir, Y., Mandelbrot, B. B., and Aharony, A., 1983: 'Geometric Implementation of Hypercubic Lattices with Noninteger Dimensionality by Use of Low Lacunarity Fractal Lattices' Phys. Rev. Lett. 50:145.

Jarlskog, C., and Yndurain, F. J., 1985: 'A precision determination of the number of spatial dimensions' CERN preprint TH 4244/85.

Mandelbrot, B. B., 1977: 'Fractals: Form Chance and Dimension', Freeman, San Francisco, 1985: 'The Fractal Geometry of Nature', Freeman, San Francisco.

Müller, B., and Schäfer, A., 1986: 'Improved Bounds on the Dimension of Space-Time', Phys. Rev. Lett. 56:1215.

Schäfer, A., and Müller, B., 1986: 'Bounds for the Fractal Dimension of Space' to be published in J. Phys. A.

Wetterich, C., 1985: 'Kaluza-Klein Cosmology and the Inflationary Universe', Nucl. Phys. B252:309.

Zeilinger, A., and Svozil, K., 1985: 'Measuring the Dimension of Space-Time', Phys. Rev. Lett. 54:2553.

AN INTRODUCTION TO SKYRMIONS AS APPLIED IN NUCLEAR PHYSICS

J.M. Eisenberg*
School of Physics and Astronomy
Raymond and Beverly Sackler Faculty of Exact Sciences
Tel Aviv University, Tel Aviv, Israel

ABSTRACT

After a very brief discussion of the origins of the Skyrme model for hadronic physics, we treat the form of the effective lagrangian proposed by Skyrme, the nature of the equations of motion derived from it in the hedgehog model, projection techniques for the construction of states referring to nucleons or delta baryons, and the physical properties of these particles as derived from the model. We then explore these same properties with a version of the model that achieves stabilization by omega exchange and extend to other possible terms in the lagrangian. The scene is then prepared for applications of the method to nuclear situations. We begin with a derivation of the nucleon-nucleon potential from the skyrmion model using an extension of the approach of Heitler and London. Unfortunately, this method does not yield the well-known attraction in the central part of the NN potential, and we therefore consider various extensions of the model lagrangian and of the generalized Heitler-London approximation scheme in an attempt to correct this situation. We close with a discussion of the extension of the skyrmion approach to models meant to represent hadronic matter in bulk, including cases pertaining to the special conditions under which a quark-gluon plasma might be formed.

I. ORIGINS IN QUANTUM CHROMODYNAMICS AND IN HISTORY

We start with the suppositions that hadronic physics is well described by quantum chromodynamics[1-7], but that the application of this theory in the context of the relatively low energies encountered in nuclear physics is prohibitively difficult. This situation calls for the application of some sort of systematic approximation scheme, however it

----

* The author's work in the area of skyrmions has been heavily influenced by his collaboration with R.R. Silbar and G. Kälbermann, who should be seen as partners in the drafting of this series of lectures but share no responsibility for its shortcomings.

is not immediately obvious what parameter of smallness is available for this purpose. In an answer to this challenge, 't Hooft[8] showed that if one supposes that the number of colors N_C involved in quantum chromodynamics is not 3, but rather some arbitrarily large number, then it is possible greatly to simplify the structure of the theory. In that limit, quantum chromodynamics goes over to an effective field theory of mesons having relatively weak coupling and therefore permitting the use of the tree approximation for the lagrangian. Furthermore[9], baryons may emerge in such a theory as solitons (i.e., nontrivial, self-sustaining, localized solutions of the field equations). At low energies, the theory becomes the nonlinear sigma model for broken chiral symmetry.

Thus, in the situation where we are prepared to think of 3 as a large number (or tolerate intrinsic imprecisions of 33 percent) and to restrict ourselves to low energies, we have a chain leading from quantum chromodynamics to an effective chiral lagrangian. This does not yet tell us with which effective lagrangian we are to work, that is, the precise form of the lagrangian and values of its coupling constants are not yet dictated by these considerations. There have been attempts to derive, with suitable approximations--perhaps the better words are "to motivate"--specific effective lagrangians from theories bearing a close relationship to our presumed starting point, namely quantum chromodynamics. {Thus, for example, one approach[10] takes as its point of deparure an expression for the action that is not equivalent to that of quantum chromodynamics but has the property that the vacuum expectation values of certain combinations of operators are the same for both; while another[11] starts from a lagrangian having quarks in interaction with a sigma field meant to represent the effects of gluon exchange; and yet another[12] attempts to deal more directly with the quantum chromodynamics action while enforcing a particular identification of the combinations of fields that are believed to represent mesons.}

For our present purposes, the story really begins at this stage, that is, we accept the existence of an effective lagrangian, to be used in the sense of a tree approximation, from which we hope to derive the properties of hadronic physics at low energies, though only to an accuracy of about 33 percent insofar as we can anticipate at this stage. The lagrangian in question will refer to meson fields and will contain baryons as soliton solutions. It need not necessarily have quarks as degrees of freedom (unlike bag models, say)--much less gluons, which were lost from the picture when we went over to the effective lagrangian--, thus simplifying the description even more. Obviously, from the point of view of nuclear physics this is a very attractive point of departure and it is very appealing to see to what extent one can derive nuclear properties from such a model. Two caveats are in order, however, from the start: Since the theory is a priori not known to be accurate to better than about 33 percent, it may well emerge that many nuclear properties are intrinsically unaccessible since they involve rather small energies (say on the scale of 10 MeV--or even much less) relative to the baryon masses that must of course also be an intrinsic feature of this same theory. Furthermore, the restriction of the validity of the theory to low energies may also make it unsuitable for many aspects of nuclear physics because the tail of energy distributions for the Fermi motion of nucleons in nuclei may violate this requirement; "low energy" here presumably refers to energies much less than the value of the quantum chromodynamics cutoff parameter, that is, much less than a few hundred MeV.

The way is now clear for us to pick a chiral effective lagrangian and proceed to calculate all that is of interest to us--refining the form of the lagrangian and the values of its coupling constants as we go--seeking agreement with experiment at about the 33 percent level. Clearly, we shall first apply the model to single-baryon properties until we have developed some confidence in it and only then go on to consider systems having more than one nucleon. A remarkable historical peculiarity of this program is that it was well launched by Skyrme[13] years before any of the specialized concepts and vocabulary of the above discussion were known. Already in the early 1960s, Skyrme proposed an effective lagrangian of appropriate character and surmised that its solutions related to baryon properties by reason of the occurrence of a conserved quantity in the theory which to Skyrme suggested the conservation of baryon number. This lagrangian was then developed extensively as the natural candidate for the present program by Adkins, Nappi, and Witten[14].

The present lectures start from this choice of Skyrme's lagrangian and the treatment of it by Adkins, Nappi, and Witten. They then consider how the model may be changed or extended by considering alternate choices of lagrangian terms. Once the single-baryon situation has been explored, applications to two-baryon--and, in particular, two-nucleon--systems are studied in some detail. The limit of very large baryon number is explored a bit at the end, partly as a curiosity, partly for for what it suggests regarding possible further approximations that may allow for a simpler treatment of the model, and partly because it may hint at what this particular model has to say about conditions for the formation of a quark-gluon plasma. Since the current popularity of the Skyrme approach has led to the writing of a goodly number of fine reviews of the general aspects of the theory[15-19] in the past few years, and since an avowed purpose of the present school is to make material of this sort available to nonspecialists in the fields discussed, these lectures do not attempt to be comprehensive, but do try to develop at least some of the details of some of the calculational techniques in the hopes of providing a starting point for those who may wish to join the research in this area.

II. ONE-BARYON PROPERTIES

A. The Skyrme lagrangian

The simplest choice of lagrangian density that possesses the qualities of chiral symmetry and stability that we require is that of the Skyrme model,

$$L = \frac{1}{16} F_\pi^2 \, \mathrm{tr}[\partial^\mu U \, \partial_\mu U^\dagger] + \frac{1}{32e^2} \mathrm{tr}[(\partial_\mu U)U^\dagger, (\partial_\nu U)U^\dagger]^2, \tag{1}$$

where U is a 2 x 2 matrix, whose more detailed properties will be explored in the sequel. The constants appearing in Eq. (1) are F_π, the pion decay constant--here taken in a form such that the experimental value for this quantity is 186 MeV--and a dimensionless parameter $\underline{e}$. The form of the lagrangian taken here is the most minimal one guaranteeing stability, which it does by virtue of its second term, as we shall discuss soon. One other possible term involving four derivatives of U does exist, as we shall see, but leads to destabilization of the solution and also brings about the presence of dynamical terms with more than two time derivatives in the lagrangian, which causes problems in carrying out the usual quantization procedure. Last, we note before plunging into work with this particular structure that one could choose to stabilize the lagrangian with still higher derivatives in U or with mechanisms such as the exchange of a (repulsive) omega meson, and we shall eventually return to an exploration of these options.

The 2 x 2 matrix U transforms under $SU(2)_L \times SU(2)_R$ according to $U \to AUB^{-1}$, where A is a matrix in $SU(2)_L$ and B is in $SU(2)_R$. The matrix U is to be taken to be unitarity, $U^\dagger U = 1$, which yields here the so-called nonlinear sigma model. In seeking solutions to Eq. (1) we shall first restrict ourselves--at least for the moment--to static solutions, so that we may replace the form in (1) with

$$L = \frac{F_\pi^2}{16} \operatorname{tr}(\mathbf{L}\cdot\mathbf{L}) + \frac{1}{32e^2} \operatorname{tr} \sum_{m,n} [[L_m,L_n]\,[L_m,L_n]], \tag{2}$$

where we have defined

$$\mathbf{L} = U^\dagger \nabla U. \tag{3}$$

Our entire study of the solutions of lagrangians in the Skyrme model will be restricted to a form originally introduced by Skyrme in this context and now called the "hedgehog" ansatz. This form allows for great simplifications in the treatment of this (and other) problems involving the pion field in isospin space in one way or another, but has the difficulty that it mixes isospin and configuration spaces in strange and not always desirable ways, and destroys hopes of Lorentz invariance in theories based on it. The ansatz in question consists in defining

$$U_0(\mathbf{r}) = \exp\,[i\, F(r)\, \boldsymbol{\tau}\cdot\hat{\mathbf{r}}], \tag{4}$$

where F(r) is a function of the radial variable only. Substituting Eq. (4) into Eq. (3) leads to a convenient form for $\mathbf{L}$, namely [exercise!]

$$\mathbf{L} = i\,\frac{\sin F}{r} \exp\,(-i\,\boldsymbol{\tau}\cdot\hat{\mathbf{r}})\,\boldsymbol{\tau} + i\,\boldsymbol{\tau}\cdot\hat{\mathbf{r}}\,\hat{\mathbf{r}} \left[F' - \frac{\sin F}{r} \exp\,(-i\,\boldsymbol{\tau}\cdot\hat{\mathbf{r}})\right]. \tag{5}$$

Alternatively, this may be written as

$$L_m = i \sum_n \tau_n\, b_{nm}, \tag{6}$$

where

$$b_{nm} = A\,\delta_{nm} + B\,\hat{r}_l\,\varepsilon_{nlm} + C\,\hat{r}_n\,\hat{r}_m, \tag{7a}$$

with

$$A = \frac{\sin F \cos F}{r}, \qquad B = \frac{\sin^2 F}{r}, \qquad C = F' - \frac{\sin F \cos F}{r}. \tag{7b}$$

Using these expressions, it is easily shown [exercise!] that the lagrangian density may be written

$$L = -\frac{1}{8} F_\pi^2 \left[\left(\frac{dF}{dr}\right)^2 + 2\,\frac{\sin^2 F}{r^2}\right] - \frac{1}{2e^2}\,\frac{\sin^2 F}{r^2} \left[\frac{\sin^2 F}{r^2} + 2\left(\frac{dF}{dr}\right)^2\right]. \tag{8}$$

The hamiltonian of the system for this static case is simply the integral of the negative of the lagrangian density, and thus is given by

$$M = 4\pi \int_0^\infty \left[\frac{1}{8} F_\pi^2 \left[\left(\frac{dF}{dr}\right)^2 + 2\,\frac{\sin^2 F}{r^2}\right] + \frac{1}{2e^2}\,\frac{\sin^2 F}{r^2}\left[\frac{\sin^2 F}{r^2} + 2\left(\frac{dF}{dr}\right)^2\right]\right] r^2\, dr, \tag{9}$$

which also represents the mass of the system since the hadron has been taken at rest. (In order to avoid confusion, it may help to point out that because of the properties of the traces taken here, and the unitarity of U, one could also have written the original lagrangian of Eq. (2) such that one of the Ls in the first term there carries a hermitian conjugate and the sign of that term is reversed.)

B. Equations of motion and their solution

We may now proceed to determine the equations of motion for the skyrmion. Recall that we are now dealing here with a restriced variety of variations since we have limited ourselves to work with the hedgehog ansatz. The only possible variation is thus in the form of the radial function F(r), and the pertinent Euler-Lagrange equation is that derived from

$$\frac{d}{dr}\frac{\delta M}{\delta F'} - \frac{\delta M}{\delta F} = 0, \tag{10}$$

which leads [exercise!] to the nonlinear differential equation for F(r),

$$\left(\frac{1}{4} F_\pi^2 r^2 + \frac{2}{e^2}\sin^2 F\right) F'' + \frac{1}{2} F_\pi^2\, r\, F' + \frac{2}{e^2}\sin F \cos F\, F'^2 - \frac{1}{2} F_\pi^2 \sin F \cos F - \frac{2}{e^2}\,\frac{\sin^3 F \cos F}{r^2} = 0. \tag{11}$$

Simplifying the expression and introducing the dimensionless combination $\tilde{r} = e\, F_\pi\, r$, we have

$$\left(\frac{1}{4}\tilde{r}^2 + 2\sin^2 F\right) F'' + \frac{1}{2}\tilde{r}\, F' + \sin 2F\, F'^2 - \frac{1}{4}\sin 2F - \frac{\sin^2 F \sin 2F}{\tilde{r}^2} = 0. \tag{12}$$

This equation we must now solve numerically, but of course before we can proceed to that stage we must first ask ourselves what the relevant boundary conditions are here. It is immediately clear from Eq. (9) that the system energy will become infinite unless certain conditions are met for F(r): Unless sin F(r) vanishes as $r \to \infty$, the first major term in the integrand for M will blow up because of the part $\sin^2 F(r)$ appearing there. (Indeed, if this term appeared by itself--without the second piece involving four derivatives in U--then one can see that the contribution of the first term will be measured roughly by the radial extent of the system, so that if only that term were present one would have a collapsing solution.) Thus we must expect that $F(r) \to n\pi$, where <u>n</u> is an integer, as $r \to \infty$. On the other hand, as r becomes small the second major term in M tends to increase because of the part in $\sin^4 F(r)/r^2$ there; this in turn requires that $F(r) \to m\pi$, where <u>m</u> is an integer, as $r \to 0$. Thus the boundary conditions we must impose are that F(r) goes to an integer multiple of π at each of its endpoints. This can also be seen from the topological properties of the hedgehog ansatz in Eq. (4), but since we shall not make much use of those features we shall content ourselves with the present line of argumentation. It should already begin to be clear that these boundary conditions for F(r)--which have been forced upon us by the very dynamics of this system--imply a kind of necessary stability in F(r). This is because the requirement of integer values of $F(r)/\pi$ means that it will not be possible to carry out small changes in F(r)--at least at its endpoints. Changes must involve a leap by an entire integer.

The significance of the integers determining the end values of F(r)--or at least of the integer which is the difference N=m-n--becomes clearer if we consider the quantity

$$B^\mu = \frac{\varepsilon^{\mu\nu\alpha\beta}}{24\pi^2} \operatorname{tr}[(U^\dagger\partial_\nu U)(U^\dagger\partial_\alpha U)(U^\dagger\partial_\beta U)]. \tag{13}$$

For the static case this may also be written as

$$B^\mu = \frac{\varepsilon^{\mu\nu\alpha\beta}}{24\pi^2} \operatorname{tr}(L_\nu\, L_\alpha\, L_\beta). \tag{14}$$

It can be shown with no great difficulty [exercise!] that the four-vector B^μ is conserved, that is $\partial_\mu B^\mu = 0$. This conservation makes use only of the general structure of $B^\mu(r)$ and not of the equations of motion [that is to say, $B^\mu(r)$ is not a current based on Noether's theorem]. Thus by the usual procedures in field theory the volume integral of the zeroth component of B^μ is constant in time. This component is

$$B^0(r) = -\,\varepsilon_{ijk} \operatorname{tr}(L_i\, L_j\, L_k) = -\frac{1}{2\pi^2 r^2} \sin^2 F \cdot F'(r), \tag{15}$$

where the last (simple) expression is derived using the form in Eq. (6). It is now easy to verify the conservation of the spatial integral of $B^0(r)$ since that quantity is a complete differential,

$$B^0(r) = -\frac{1}{4\pi^2 r^2}\frac{d}{dr}\left[F(r) - \frac{1}{2}\sin 2\,F(r)\right], \tag{16}$$

whence

$$4\pi\int_0^\infty B^0(r)\, r^2\, dr = \frac{1}{\pi}[\,F(0) - F(\infty)\,] = N. \tag{17}$$

Since this integer $\underline{N}$ relates to a conserved quantity, Skyrme was inspired to identify it with the baryon number in this theory. Thus if we choose to deal with the case N = 1, so that F(r) falls through one step in units of π as r goes from zero to infinity--say from $F(0) = \pi$ to $F(\infty) \to 0$, the corresponding soliton will represent a hadron with baryon number unity. The solution in question is referred to as a soliton because it is self-sustaining in the sense that is has a stable, nonzero energy and is well localized in space: It demands for itself a nontrivial result even with no need incident energy from asymptotic distances (as one has say in scattering problems).

Before turning to the other properties of the single-baryon skyrmion, it may be useful to note briefly numerical methods for its calculation. The most straightforward of these is simply to use the well-known Runge-Kutta method[20], guessing the initial slope F'(0) and then stepping out from the origin. The initial guess must then be improved successively so that the point at which F(r) begins to blow up or go negative is moved out to sufficiently large r to cause no ill consequences for the calculations at hand. A criterion for the accuracy of such a method can be obtained from the fact that the two major terms in the energy M of Eq. (9) contribute equally for a correct solution F(r). {Exercise: Show this to be the case by scaling the radial variable $r \to \lambda r$ and treating λ as a variational parameter. This result is closely related to Derrick's theorem[21], and provides an alternative way to show the stability of the skyrmion--at least under a particular class of otherwise troublesome variations.} Unfortunately for some of the other skyrmion lagrangians we shall consider below, this simple technique is completely inadequate because of instabilities in the numerics--not of course in the physics of the skyrmion model--and more refined techniques must be developed. One possibility is to use the same Runge-Kutta method, but now with a mechanized least-squares procedure to fit the asymptotic form of F(r) and thus to optimize the choice of F'(0). Another approach is to use a relaxation method to solve the nonlinear differential equation; this proves extremely stable and reliable, but tends to use a great deal of computer time.

The solution for a single baryon, N = 1, is shown in Fig. 1. It has a rather featureless fall from π to 0 as the dimensionless variable $\tilde{r}$ increases, and becomes quite negligible by the time $\tilde{r} \approx 10$. We shall explore analytic approximations to this solution in later sections of these lectures, and we shall also consider slightly different dynamics for skyrmions.

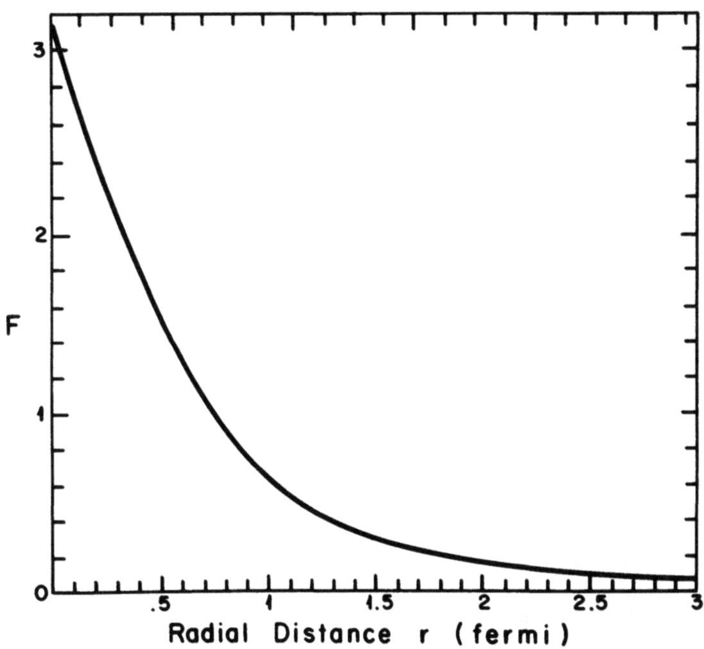

Fig. 1. The numerical result for the single-baryon, N = 1, skyrmion F(r) in the hedgehog ansatz.

C. Projection techniques

As we have already partly anticipated, the hedgehog ansatz has the serious shortcoming that it does not refer to states of well-defined spin and isospin quantum numbers. A way out of this situation has been developed by Adkins, Nappi, and Witten[14]; it is based on techniques that have long been familiar in nuclear physics, where it is often necessary to project states of good quantum numbers from "intrinsic" states not possessing a particular symmetry (see, for example, the discussion of projection techniques in volume 3 of Ref.[22]).

The procedure starts here with the observation that if the soliton form of Eq. (4) is a solution of the equations of motion then so is the transformed form $U = A\, U_0\, A^{-1}$, where A is any constant SU(2) matrix. For fixed A, the transformed solution is also not an eigenstate of spin or isospin. But we now treat A as a collective coordinate, eventually making it into a quantum variable. To do this, we make A time dependent, so that the transformation is now

$$U = A(t)\, U_0\, A^{-1}(t). \tag{18}$$

We use this form to construct a hamiltonian H from our original lagrangian density of Eq. (1) and then carry out the usual procedure of (first) quantization in order to introduce quantum variables for A. In the process of diagonalizing the hamiltonian H we shall be able to classify states according to spin and isospin, in particular identifying the nucleon and delta baryon on the way. Substituting Eq. (18) into Eq. (1) yields for the lagrangian (after integration over volume)

$$\tilde{L} = - M + \lambda\, \mathrm{tr}\left[\partial_0 A\, \partial_0 A^{-1}\right], \tag{19}$$

where M is the static soliton mass given in Eq. (9) and

$$\lambda = \frac{2\pi}{3e^3 F_\pi} \int \sin^2 F \left[1 + 4\left(F'^2 + \frac{\sin^2 F}{\tilde{r}^2}\right)\right] \tilde{r}^2\, d\tilde{r}. \tag{20}$$

The integral in Eq. (20), when evaluated with the soliton solution of Fig. 1, proves to have the numerical value 50.9.

We now parametrize the SU(2) matrix A according to

$$A = a_0 + i\, \mathbf{a}\cdot\boldsymbol{\tau}, \qquad \text{with} \quad a_0^2 + \mathbf{a}^2 = 1, \tag{21}$$

where the side condition guarantees the unitarity of A. In terms of this parametrization, we have

$$\tilde{L} = - M + 2\lambda \sum_{i=0}^{3} \dot{a}_i^2. \tag{22}$$

To carry out the usual quantization procedure we first introduce the conjugate momentum

$$\pi_i = \partial\tilde{L}/\partial\dot{a}_i = 4\,\lambda\, \dot{a}_i, \tag{23}$$

whence the hamiltonian becomes

$$H = \Sigma\, \pi_i\, \dot{a}_i - L = 4\,\lambda\, \Sigma\, \dot{a}_i\, \dot{a}_i - L = M + \frac{1}{8\lambda}\, \Sigma\, \pi_i^2. \tag{24}$$

The usual prescription of quantization is to replace the conjugate momentum with a derivative (so that classical Poisson brackets go over to commutators),

$$\pi_i \to - i\, \partial/\partial a_i, \tag{25}$$

yielding the quantized hamiltonian

$$H = M + \frac{1}{8\lambda} \sum_{i=0}^{3} \left[- \frac{\partial^2}{\partial a_i^2} \right]. \tag{26}$$

The constraint in the second part of Eq. (21) is to be imposed here,

$$\sum_{i=0}^{3} a_i^2 = 1, \tag{27}$$

so that the operator in Eq. (26) has the meaning of a laplacian on a three-dimensional sphere rather than one in four dimensions. The analogous consideration to determine the operation of a three-dimensional laplacian restricted to the two-dimensional surface of a sphere is

$$\begin{aligned} \nabla^2 = \nabla\cdot\nabla &= \frac{1}{r^2} (\mathbf{r}\times\nabla)^2 + \frac{1}{r^3} (\mathbf{r}\cdot\nabla)\ r\ (\mathbf{r}\cdot\nabla) \\ &= \frac{1}{r^2} \frac{\partial}{\partial r} r^2 \frac{\partial}{\partial r} - \frac{1}{r^2} \mathbf{J}^2 \rightarrow - \mathbf{J}^2, \end{aligned} \tag{28}$$

where $\mathbf{J}$ is the angular momentum operator, and the replacement in the last part of the equation represents restriction to a unit sphere.

The generalization of this to our case is immediately obvious. It suggests wave functions that are polynomials in a_i along the lines of the usual spherical harmonics in the three-dimensional case. Thus we have, for example,

$$- \nabla^2 (a_0 + i\ a_1)^l = l\ (l + 2)\ (a_0 + i\ a_1)^l \tag{29}$$

We can construct operators satisfying the required algebra of spin and isospin (distinguishing between the two by requiring the isospin operator to relate correctly to the electromagnetic current we shall construct below):

$$J_k = \frac{i}{2} \left[a_k \frac{\partial}{\partial a_0} - a_0 \frac{\partial}{\partial a_k} - \varepsilon_{klm}\ a_l \frac{\partial}{\partial a_m} \right] \qquad \text{(spin)} \tag{30}$$

$$I_k = \frac{i}{2} \left[a_0 \frac{\partial}{\partial a_k} - a_k \frac{\partial}{\partial a_0} - \varepsilon_{klm}\ a_l \frac{\partial}{\partial a_m} \right]. \qquad \text{(isospin)} \tag{31}$$

[Exercise: Show that the laplacian on the unit sphere in four dimensions is given in terms of the above angular momentum operator by $\nabla^2 = - 4\ \mathbf{J}\cdot\mathbf{J}$.]

Since it is our purpose to build baryons--i.e., fermions--out of the U field, which is essentially mesonic--and thus bosonic--the question arises as to how we shall reconcile the different statistics arising. It emerges that there are two consistent ways to quantize the soliton[23], one of which embodies the requirement that the wave functions satisfy $\psi(-A) = \psi(A)$, and corresponds to bosons, and the other of which has $\psi(-A) = -\psi(-A)$, and represents fermions. Thus for our discussion of baryons we need odd polynomials in $\underline{a}$. Moreover, it is a property of the algebra here [exercise!] that one must have spin equal to isospin, I = J. The resulting wave functions for the nucleon states are

$$\begin{aligned} |p\!\uparrow\rangle &= \frac{1}{\pi} (a_1 + i\ a_2), \qquad & |p\!\downarrow\rangle &= - \frac{i}{\pi} (a_0 - i\ a_3), \\ |n\!\uparrow\rangle &= \frac{i}{\pi} (a_0 + i\ a_3), \qquad & |n\!\downarrow\rangle &= - \frac{1}{\pi} (a_1 - i\ a_2), \end{aligned} \tag{32}$$

and for delta states

$$|\Delta^{++}, s_z = \tfrac{3}{2}\rangle = \frac{\sqrt{2}}{\pi} (a_1 + i\, a_2)^3,$$

$$|\Delta^{+}, s_z = \tfrac{1}{2}\rangle = -\frac{\sqrt{2}}{\pi} (a_1 + i\, a_2) [1 - 3 (a_0^2 + a_3^2)]. \tag{33}$$

We must now answer as to the nature of the wave-function averaging implied by the four-dimensional a-space as restricted to a three-dimensional sphere. It is easy to verify [exercise!--see also the calculation of the surface area of a sphere in n-dimensions in Ref.[24]] that the appropriate averaging is

$$\langle\cdots\rangle = 2 \int da_1\, da_2\, da_3\, da_0\, \delta(1 - a_1^2 - a_2^2 - a_3^2 - a_0^2) \cdots$$

$$= \int da_1\, da_2\, da_3 \frac{1}{|a_0|} \Bigg|_{a_0 = \pm (1 - a_1^2 - a_2^2 - a_3^2)^{1/2}} \cdots . \tag{34}$$

For some purposes (including, for example, formal algebraic manipulation in computers) it is more convenient to work with angular variables on the three-dimensional unit sphere in four-dimensional space. For that purpose, one may assign

$$a_0 = \cos\theta, \quad a_1 = \sin\theta \sin\phi \cos\chi, \quad a_2 = \sin\theta \sin\phi \sin\chi,$$

$$\text{and} \quad a_3 = \sin\theta \cos\phi, \tag{35}$$

and the averaging integral becomes

$$\langle\cdots\rangle = \int_0^\pi \sin^2\theta\, d\theta \int_0^\pi \sin\phi\, d\phi \int_0^{2\pi} d\chi \cdots. \tag{36}$$

It is then very easy to use either of the above forms in order to check that the wave functions of Eqs. (32) and (33) are correctly orthogonal and normalized.

Last, we may now return to the hamiltonian of Eq. (26) and note that the wave functions of the form $(a_0 + i\, a_1)^l$, for example, are, of course, eigenfunctions of it, with eigenvalues

$$E = M + \frac{1}{8\lambda}\, l\, (l + 2), \tag{37}$$

where $l = 2J$; the energies for the nucleon and delta then emerge from Eqs. (32) and (33) as

$$M_N = M + \frac{1}{2\lambda}\frac{3}{4} \quad \text{and} \quad M_\Delta = M + \frac{1}{2\lambda}\frac{15}{4}. \tag{38}$$

Using the (numerical) result $\lambda = (2\pi/3e^3 F_\pi)\ 50.9$ from Eq. (20) together with Eq. (9)--which numerically yields $M = 36.5\ F_\pi/e$--Adkins, Nappi, and Witten[14] treat F_π and e as parameters to be fit such that these masses achieve their experimental values. They then find $F_\pi = 129$ MeV and e = 5.45, the former number being some 30 percent below the experimental 186 MeV.

D. Matrix elements for electroweak processes and hadronic coupling constants

In the work of Adkins, Nappi, and Witten[14] there is given a quite complete discussion of how to calculate expectation values or transition matrix elements for all manner of electromagnetic or weak processes involving a single baryon, including isoscalar and isovector root-mean-square radii, magnetic moments, and the axial-vector coupling constant g_A of weak interactions. There is, as well, a treatment of strong-interaction coupling constants. Since it is not our purpose here to be encyclopedic, we shall not reproduce most of those arguments, instead restricting ourselves to the main lines of determining the electroweak current (along with a few suggested exercises to fill in some detail); a brief discussion of g_A; some exploration of $g_{\pi NN}$, the pion-nucleon coupling constant, and the Goldberger-Treiman relation into which it enters; and the presentation of results for single baryons. These are the main tools we shall require for treatment of nuclear aspects of the problem.

Isoscalar currents in the present model are already given by the baryon density of Eqs. (13)-(16). The isovector currents are generated through the usual procedure of Noether's theorem. That is, one may either use a phase transformation $U \rightarrow e^{i\phi}U$ or the usual gauge replacement $\partial_\mu \rightarrow \partial_\mu - iqA_\mu$ in the lagrangian in order to produce the isovector current. In Ref.[14] the procedure used is to apply the transformation for the V - A current generated by a change in U through $\delta U = iQU$, whence one finds

$$J^\mu_{V-A} = \frac{i}{8}F_\pi^2 \, \mathrm{tr}[(\partial^\mu U)U^\dagger Q] + \frac{i}{8e^2} \, \mathrm{tr}[[(\partial_\nu U)U^\dagger, Q] \, [(\partial^\mu U)U^\dagger, (\partial^\nu U)U^\dagger]]. \quad (39)$$

[Exercises: (i) In the static case and for the hedgehog ansatz, show directly from this expression and the equation of motion (12) that the current is conserved, i.e. $\nabla\cdot J = 0$. Explain why the axial part of the current is also conserved here. (ii) Using the gauge replacement method, find the expression for the current.] It should be noted that this current pertains for the exact solution of the original lagrangian of Eq. (1); when the projection methods of the previous section are used, the conservation properties of this quantity are no longer clear. To obtain the V + A current from the above, one must replace $U \leftrightarrow U^\dagger$. The separate vector or axial vector currents are of course to be obtained by adding or subtracting the two expressions and dividing by 2.

From the time-like part of the vector current one can now extract expressions for the r.m.s. radius with I = 1, while the I = 0 case is taken from $B^0(r)$. (The isovector expressions diverges for massless pions, as taken here, but is convergent for nonzero m_π.) Correspondingly, the I = 0 and 1 magnetic moments arise from the spatial parts of these four-vectors. The axial coupling constant g_A is obtained by (careful and slightly subtle) integration over the spatial part of the axial current $\int dr \, A_i{}^a(r)$, where a refers to the isospin index and i to the spatial one. Alternatively, one may relate g_A directly to the asymptotic behavior of F(r) through the conservation of axial current in this limit of massless pion. We sketch this point here for its bearing on phenomena arising from the one-pion tail that will be of interest to us below:

From the expression for the isovector current in Eq. (39), one has

$$A_i{}^a = \frac{i}{8} F_\pi^2 \, \mathrm{tr} \left[\left[(\partial_i U_0)U_0^\dagger + U_0^\dagger \partial_i U_0\right] A^{-1}\tau^a A\right] + \text{higher derivatives}, \quad (40)$$

and $U_0 = \exp[i\ \boldsymbol{\tau}\cdot\hat{\mathbf{r}}\ F(r)] \to 1 + i\ B\ \boldsymbol{\tau}\cdot\hat{\mathbf{r}}/r^2$ for large r, where we have used the fact [exercise!] that asymptotically $F(r) \to B/r^2$. Thus one easily has a closed form for A at large r, and, converting the volume integral over A that is required for g_A into a surface inetgral at infinity, one finds that $g_A = 2\pi/3\ F_\pi^2\ B$; the quantity B is now to be extracted from the numerical result for F(r) in Fig. 1.

To close the ring by establishing the version, in this context, of the Goldberger-Treiman relation, we first note that in conventional field theory with pion and nucleon fields one has--in the nonrelativistic limit--a lagrangian of the form

$$L = \frac{1}{2}\,(\nabla\vec{\pi})^2 + \frac{g_{\pi NN}}{2M_N}\,\nabla\vec{\pi}\cdot(\overline{\psi}\ \boldsymbol{\sigma}\vec{\tau}\ \psi). \tag{41}$$

From this one can easily derive the usual differential equation for the pion field in terms of an axial-vector, isovector nucleon source. In this case the equation is essentially the Poisson equation with a dipole source, and the solution is well known,

$$\langle N'|\vec{\pi}(\mathbf{r})|N\rangle = -\frac{g_{\pi NN}}{8\pi M_N}\,\frac{\mathbf{r}}{r^3}\cdot\langle N'|\boldsymbol{\sigma}\vec{\tau}|N\rangle. \tag{42}$$

This may now be compared with the expression for the pion field which can be extracted from the expansion of U about its vacuum value*,

$$U = 1 + 2i\,\frac{\boldsymbol{\tau}\cdot\boldsymbol{\pi}}{F_\pi}, \tag{43}$$

together with Eq. (18) and the asymptotic expression for F(r). One finds

$$\begin{aligned}\langle N'|\vec{\pi}|N\rangle &= -\frac{B}{6}\,F_\pi\,\frac{\mathbf{r}}{r^3}\cdot\langle N'|\boldsymbol{\sigma}\vec{\tau}|N\rangle \\ &= -\frac{g_A}{4\pi F_\pi}\,\frac{\mathbf{r}}{r^3}\cdot\langle N'|\boldsymbol{\sigma}\vec{\tau}|N\rangle,\end{aligned} \tag{44}$$

from which, by comparison with Eq. (42), we find

$$g_A = \frac{F_\pi g_{\pi NN}}{2M_N}, \tag{45}$$

the Goldberger-Treiman relation, which here provides a prediction for the πNN coupling constant. We shall show results for all these quantities a little later, after we have developed some alternative skyrmion lagrangians for the sake of comparison (see subsection II.G).

E. Short-range features

The last considerations in the previous subsection imply that the skyrmion model correctly embodies those aspects of baryon behavior that relate to long-range features dominated by the one-pion tail. As we

* Note that the field U may be related to the conventional σ and $\vec{\phi}$ fields of the nonlinear sigma model through $U = \sigma + i\ \vec{\tau}\cdot\vec{\phi}$. The first, kinetic term of the lagrangian of Eq. (1) then embodies the usual chiral invariant form for these fields.

shall see, this is a property that carries over also to the two-baryon situation, where the one-pion exhange potential is automatically obtained correctly. It is natural to ask how well the model may do with short-range issues, and this may still be rather easily answered in the context of the original, unadorned version of the model. In particular, suppose we ask what happens if we seek solutions involving $N = 2, 3, \ldots$, that is, having two or three or more baryons superimposed on each other. This is presumably the condition addressed if we solve the original nonlinear differential equation for $F(r)$ in (12), but with a boundary condition that $F(0) = 2\pi, 3\pi, \ldots$.

This question was addressed numerically (we shall return to a fuller discussion of other aspects of this issue in section IV below) in Ref.[25], where it was found that the sizes and energies of the resulting solutions are as given in Table 1. The main point to be emphasized there is that the energy of the baryonic system rises considerably faster than linearly, so that one must suppose there is a great deal of surplus energy, or repulsion, as more and more baryons are superimposed. Thus, for example, the two-baryon system has nearly three times as much energy as do two noninteracting baryons, and this extra 1,000 MeV is taken as representing the action of the strongly repulsive core known to act when two nucleons are close together.

Table 1
Sizes and energies of superimposed multi-baryon systems

Baryon number	$r_{rms}(N)$ [fm]	E(N) [MeV]	E(N)/E(N=1)
1	0.48	1420	1
2	0.68	4250	2.98
3	0.81	8440	5.93

Thus, before even seriously addressing the problem of the nucleon-nucleon force in the skyrmion model we may expect to find there two of its dominant features: the long-range one-pion tail and the short-range repulsion (usually ascribed to omega exchange).

F. Another possible term in the lagrangian

In addition to the term of fourth order in the derivative of the skyrmion that we have treated in the lagrangian of Eq. (1), there exists another possible form having the general structure $(\partial_\mu U)^4$, namely

$$L_{4s} = \frac{\gamma}{8e^2}\,[\mathrm{tr}(\partial_\mu U\, \partial^\mu U^\dagger)]^2 + \frac{\gamma}{8e^2}\,[\mathrm{tr}(\mathbf{L}\cdot\mathbf{L})]^2, \tag{46}$$

where γ is an additonal constant to be determined from experiment. In the notation introduced in Eq. (46), we are labeling the terms in the lagrangian by the number of derivatives of U contained in them and their symmetry with respect to them, so that the original Skyrme lagrangian of Eq. (1) would now be designated as $L = L_2 + L_{4a}$, and the new, total lagrangian by $L_2 + L_{4a} + L_{4s}$. We have also shown in (46) the form L_{4s} arrived at in the static limit.

The structure in Eq. (46) was avoided in early work with the Skyrme lagrangian because it makes a negative contribution to the energy, thus tending to destabilize the skyrmion, and because it leads to terms of fourth order in the time derivatives of $\underline{a}$, which invalidates the usual quantization procedure. Nonetheless, the existence of such a term is inferred from pion-pion scattering, and a quite remarkable relationship between $\pi\pi$ scattering and the results of the Skyrme approach is thus possible, as emphasized by Donoghue, Golowich, and Holstein[26]. The physical point is that the skyrmion, which relates, on the one hand, to baryon properties in the field U, also refers to the mesons fields through $U = \sigma + i\,\vec{\tau}\cdot\vec{\phi}$, as remarked in the footnote attached before Eq. (43). Thus the fourth-order terms $L_{4a,s}$, in addition to stabilizing the baryon in this model, also represent terms in $\pi\pi$ scattering. [Exercise: Convince yourself that the counting of pion fields entering here is through the derivatives of U, not through the number of appearances of U itself.] The values of the parameters e and γ may be determined from a detailed knowledge of this scattering. Ref.[26] finds

$$F_\pi^2 = \frac{1}{\pi b_0^0}, \qquad \frac{1}{e^2 F_\pi^4} = 40\pi(a_2^0 - a_2^2), \qquad \gamma = \frac{1}{4}\,\frac{a_2^0 + 2a_2^2}{a_2^0 - a_2^2}, \tag{47a}$$

where the partial amplitudes of $\pi\pi$ scattering are parametrized according to

$$\mathrm{Re}\,A_l^1 = q^{21}\,(a_l^1 + b_l^1\,q^2). \tag{47b}$$

Using the (rather poorly known) experimental values for these quantities, one has[26] F_π = 186 MeV (of course), $\gamma = 0.16 \pm 0.04$, and a value for e (≈ 8) such that M = 690 ± 300 MeV, λ^{-1} = 500 ± 200 MeV, and, finally, $\bar{M}_N$ = 880 ± 300 MeV--a quite astounding chain of prediction.

G. Stabilizing the skyrmion through omega exchange

It was early realized[27] that there are other mechanisms for stabilizing the skyrmion apart from the term in four derivatives of U in Eq. (1). Among these are the introduction of terms representing the exchange of an isoscalar, vector meson: an ω-meson. The new lagrangian density now reads

$$\begin{aligned} L = &-\frac{1}{4}\,(\partial_\mu\omega_\nu - \partial_\nu\omega_\mu)(\partial^\mu\omega^\nu - \partial^\nu\omega^\mu) + \frac{1}{2}\,m_\omega^2\omega_\mu\omega^\mu + \beta\omega_\mu B^\mu \\ &+ \frac{1}{16}\,F_\pi^2\,\mathrm{tr}[(\partial_\mu U)(\partial^\mu U^\dagger)] + \frac{1}{8}\,F_\pi^2 m_\pi^2(\mathrm{tr}\,U - 2), \end{aligned} \tag{48}$$

where ω_μ is the field of the ω-meson of mass m_ω, which is coupled through the third term to the Skyrme baryon four-current $B^\mu(r)$. The rest of the quantities and terms in Eq. (48) are the familiar ones we have treated above, though we have here added a pion mass (chiral symmetry breaking) term that could also have been introduced in the original $L_2 + L_{4a}$ (see Ref.[28]). In the static case, this lagrangian density becomes

$$\begin{aligned} L = &\frac{1}{2}\,(\nabla\omega)^2 + \frac{1}{2}\,m_\omega^2\omega^2 - \frac{\beta}{2\pi^2}\,\omega\,\frac{\sin^2 F}{r^2}\,F' - \frac{F_\pi^2}{8}\left[F'^2 + 2\frac{\sin^2 F}{r^2}\right] \\ &- F_\pi^2\frac{m_\pi^2}{4}\,(1 - \cos F), \end{aligned} \tag{49}$$

where, for this static situation, ω is the time-like component of the meson field.

[We note that the appearance of the ω-field in this form gives the illusion of an instability in the solution because the first two terms in L, when translated into the hamiltonian, will have the property of reducing the energy. But this is not to be viewed as a true physical effect, since the ω-field is here to be seen only as an auxiliary field, representing the combination

$$\omega(r) = -\beta \int_0^\infty G(r,r')\, B^0(r')\, r'^2\, dr', \tag{50a}$$

where $G(r,r')$ is the Green function

$$G(r,r') = \frac{1}{2m_\omega rr'} \left[e^{-m_\omega|r-r'|} - e^{-m_\omega(r+r')}\right]. \tag{50b}$$

This is a situation very similar to that of the Coulomb field in noncovarient field theory treatments of the electromagnetic interaction.]

The equations of motion derived from Eq. (49) with the usual hedgehog ansatz are

$$F'' + \frac{2}{r} F' - m_\pi^2 \sin F - \frac{1}{r^2} \sin 2F = -4 \frac{\bar{\beta}}{m_\omega r^2} \bar{\omega}' \sin^2 F, \tag{51a}$$

and

$$\bar{\omega}'' + \frac{2}{r} \bar{\omega}' - m_\omega^2 \bar{\omega} = - \frac{\bar{\beta}}{m_\omega} \frac{\sin^2 F}{r^2} F', \tag{51b}$$

where we have introduced the dimensionless combination

$$\bar{\omega} = \frac{\omega}{F_\pi} \qquad \text{and} \qquad \bar{\beta} = \frac{\beta m_\omega}{2\pi^2 F_\pi}. \tag{51c}$$

The coupling constant $\bar{\beta}$ can be related to the decay $\omega \to 3\pi$ because the three derivatives of U appearing in B(r) can also be viewed as three pion fields, just as in the discussion in subsection II.F regarding $\pi\pi$ scattering. It can, of course, also be seen as pertaining to the ωNN coupling. As with the L_4 term, the constants emerging from these viewpoints are consistent with each other. The numerical solution of the coupled differential equations for the case of omega stabilization is considerably more tedious than for the $L_2 + L_4$ case because the destabilizing effects of the omega are felt in the numerical work. The relaxation method or least-square fitting of asymptotic forms to determine reliable boundary parameters seems to be required almost all of the time. Results for the F(r) and $\omega(r)$ solutions are shown in Fig. 2.

We can now bring a consolidated table of results for the various one-baryon physical quantities calculated from the skyrmion model, both for the lagrangian of Eq. (1)--but with a pion mass term added, as in Eq. (48)--and for the omega-stabilized case. Perhaps somewhat surprisingly the ω-stabilization yields results that agree at least as well with experiment as do those of the original Skyrme form.

Before leaving the matter of omega stabilization, it should be noted that in the limit of very massive ω-mesons the propagation [in Eq. (50b), for example] becomes pointlike, the omega field is no longer necessary even for its auxiliary role, and the stabilizing coupling of the skyrmion is a term involving $B^\mu(r)B_\mu(r)$, that is, the direct coupling of sixth order in derivatives of U,

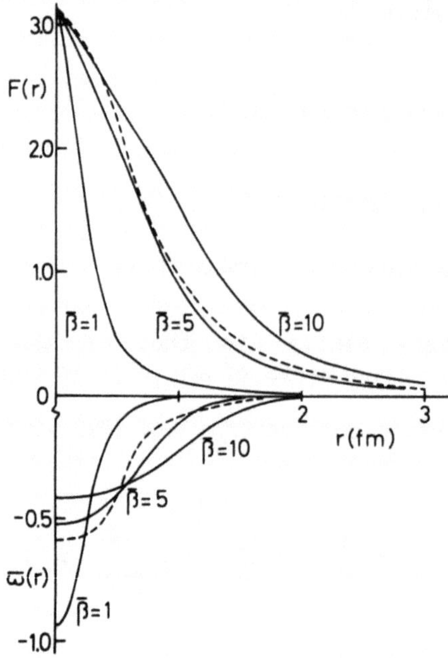

Fig. 2. The radial fields F(r) and ω(r) for the skyrmion stabilized by omega exchange.

Table 2
Physical properties of single baryons in the Skyrme model

Physical quantity	$(\partial U)^4$-stabilized theory	ω-stabilized theory	Experiment
F_π	108 MeV	124 MeV	186 MeV
r_{rms}(ch, I=0)	0.68 fm	0.74 fm	0.72 fm
r_{rms}(ch, I=1)	1.04 fm	1.06 fm	0.88 fm
r_{rms}(mag, I=0)	0.95 fm	0.92 fm	0.81 fm
r_{rms}(mag, I=1)	1.04 fm	1.02 fm	0.81 fm
μ_p	1.97	2.34	2.79
μ_n	-1.24	-1.46	-1.91
g_A	0.65	0.82	1.23
$g_{\pi NN}$	11.9	13.0	13.5
$g_{\pi N\Delta}$	17.8	19.5	20.3
$\mu_{N\Delta}$	2.3	2.7	3.3

$$L_6 = - \frac{\varepsilon^2}{2} B^2(r), \tag{52}$$

in the static case. In point of fact there is no reason not to incorporate all of these orders of ∂U at once, as has indeed been done[29-30] primarily in order to refine the theory for application to the nucleon-nucleon force.

III. USING THE SKYRMION IN TWO-BARYON SYSTEMS

A. Recalling the nucleon-nucleon force and the Heitler-London method

Before we plunge into a discussion of the application of skyrmion approaches to the two-nucleon problem, it would be well briefly to review two well-known topics: the main features of the NN force (so we will know what our target is), and the method of Heitler and London, which they originally applied to molecules, but has served ever since as a means of extracting a static potential from a system where something of the fundamental dynamics is understood but there is little hope of doing a full and precise calculation.

As for the nucleon-nucleon force, it is convenient for our purpose to think of it in terms of one-boson exchange potentials. To capsulize the work of many years (see, for example, chapter 2 of volume 3 of Ref.[22] for a partial review), the NN potential may be thought of as resulting from the exchange, mainly, of four mesons: π, σ, ω, and ρ. Of these, the pion (J^π=0^-, I=1) is responsible--as the lightest meson--for the longest-range part of the force. It is thus the major ingredient in binding the deuteron, which is a very diffuse structure. It is less involved in nuclear binding because its coupling with nucleons requires the presence of significant spin and isospin in the baryonic system and these are largely quenched in nuclei where the nucleons tend to arrange themselves so as to have minimal spin and isospin.

In nuclei the binding is largely supplied by a mechanism that goes under the name of the σ-meson (J^π=0^+, I=0 and 1), with anticipated mass of 500 to 600 MeV in order to match the presence of attraction in the spatial range of 1 to 2 fm. Unfortunately, a meson--or reasonably sharp resonance--with the appropriate quantum numbers and mass has not been found, so that it is not clear precisely what effect is being subsumed within the so-called σ-meson. A
possible candidate has been suggested in the form of dynamic two-pion exchange in which one does not return to the NN state as an intermediate state, as shown in Fig. 3. (The avoidance of the NN intermediate state is

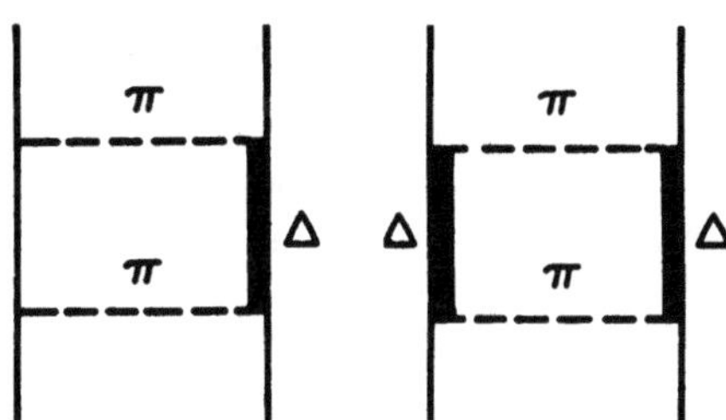

Fig. 3. Box diagrams representing a possible mechanism "equivalent" to σ-meson exchange.

so that the diagram in question will correspond to a new part of the potential, and not merely an iteration of previous pieces--which iteration is to take place in the course of solving the Schrödinger equation for the NN system.)

The ω-meson ($J^{\pi}=1^{-}$, I=0) is the source of the well-known strong repulsion between two nucleons at close separation distances. Thus, it acts to counter the influence of the σ-meson to some degree, though at a different spatial range. In a somewhat similar way, the ρ-meson works against the effect of the pion--again at a shorter range because of its greater mass. Of the four mesons discussed here, we have already noted that the skyrmion should account quite handily for pion-exchange effects because it correctly contains pion tail features. It presumably also deals with ω-exchange through the various repulsive mechanisms that stabilize the skyrmion, and which, we have seen, lead to strong repulsion at short range in the N ≥ 2 systems. The ρ-meson is unlikely to make its presence felt unless specific ρ-exchange features are introduced in the skyrmion (as has been suggested; see, for example, Ref.[31]). This leaves as the major challenge the determination of the origins of the σ-exchange mechanism within the skyrmion picture. As a problem that has never been fully resolved, it would be especially useful if the skyrmion approach could shed some light on this matter. Moreover, as the main agency of nuclear binding it is of particular concern to those of us who study nuclear structure.

A convenient way of attacking the problem of interaction between two systems, each of whose inner dynamics is believed to be understood, was suggested long ago in the context of the hydrogen molecule by Heitler and London, and served as the basis of the understanding of homopolar binding in molecules (see, for example, the lucid and thorough discussion in Ref.[32]). The basic idea there was to assume--as a trial wave function for use in a variational principle calculation--that the two-body system has a wave function made up of a simple combination of products of the wave functions of the one-body systems. One then evaluates the expectation value of the hamiltonian for the total two-body system for a fixed separation between the centers of the individual constituents and subtracts from this energy the value obtained when the two centers are very far apart, identifying the result as the (static) potential acting between the two bodies of the composite system. That is, the potential is seen as the energy change found in the system due to the mutual interaction between the bodies. The equivalent procedure for our present problem will be to take as the description of a two-baryon system a product of one-baryon skyrmions, centered about points a fixed distance R apart. One then evaluates the energy change at various fixed values of R, and identifies this as the static potential in the two-baryon system. The method must of necessity remain a little crude (unless one is prepared truly to work within a variational framework to generate successive refinements to the original ansatz). It is definitely in order to warn the reader at this stage that efforts[33] to calculate solutions for the N = 2 skyrmion on a lattice cast doubts as to the validity of the product ansatz at separation distances as small as, say, 0.7 fm. However for practical purposes one is forced to use this ansatz at least as a point of departure since without it one is at a loss as to how to project out nucleon-nucleon states as required for the NN force. Thus we shall work within this framework and hope for its successive (and successful!) improvement at later stages.

B. The static NN potential obtained from the product ansatz

The basic point of deparutre for treatment of the nucleon-nucleon force is the product assumption,

$$U_{N=2}(\mathbf{r}, \mathbf{R}) = \left[A\, U_0(\mathbf{r} + \tfrac{1}{2}\mathbf{R})\, A^\dagger\right] \left[B\, U_0(\mathbf{r} - \tfrac{1}{2}\mathbf{R})\, B^\dagger\right], \tag{53}$$

where $\mathbf{R}$ is the separation between the skyrmions, and A and B are the rotation matrices for the two separated baryons; these last will eventually be used in order to carry out a "projection" onto the nucleon states that interest here. It is easily seen [Exercise!] that for the calculation of the static potential it suffices to work with a relative rotation between the skyrmions, according to

$$U_{N=2}(\mathbf{r}, \mathbf{R}) = \left[U_0(\mathbf{r} + \tfrac{1}{2}\mathbf{R})\right] \left[C\, U_0(\mathbf{r} - \tfrac{1}{2}\mathbf{R})\, C^\dagger\right], \tag{54}$$

where

$$C = A^\dagger B = c_0 + i\, \vec{\tau}\cdot\vec{c}, \tag{55a}$$

and

$$c_0 = a_0\, b_0 + \vec{a}\cdot\vec{b}, \qquad \vec{c} = a_0\, \vec{b} - b_0\, \vec{a} + \vec{a} \times \vec{b}. \tag{55b}$$

Eq. (54) for $U_{N=2}$ is now substituted into the expression for the hamiltonian--this time without benefit of spherical symmetry, which is, of course, destroyed by the directional dependence fixed by R--and the integration over $\mathbf{r}$ carried out, leading to an expression of the structure

$$V(\mathbf{R}, C) = (\alpha_1 + \beta_1) + (\alpha_2 + \beta_2)\, c_0^{\,2} + (\alpha_3 + \beta_3)\, c_Z^{\,2} + \beta_4\, c_0^{\,4} + \beta_5\, c_0^{\,2}\, c_Z^{\,2} + \beta_6\, c_Z^{\,4}, \tag{56}$$

where we have taken the direction of $\hat{\mathbf{z}}$ to be parallel to R. The quantities α refer to terms arising from the first, kinetic part of the lagrangian in Eq. (1), while those labeled β are from the second, "stabilizing" piece. The last step in the calculation of the potential is to average over $\{c_0, \vec{c}\}$. Doing this for various combinations of NN states in order to generate the full spread of spin and isospin states contained here, one obtains a nucleon-nucleon potential of the structure

$$V(\mathbf{R}) = V_C(R) + \vec{\tau}_1\cdot\vec{\tau}_2\, [\vec{\sigma}_1\cdot\vec{\sigma}_2\, V_{SS}(R) + S_{12}\, V_T(R)], \tag{57}$$

where the $\vec{\tau}$s and $\vec{\sigma}$s refer to nucleon isospin and spin operators, and S_{12} is the usual tensor operator (see, for exapmle, chapter 2 of volume 3 of Ref.[22]). Thus the NN potential emerging from the skyrmion approach--at least under the present approximations of a static lagrangian and the product ansatz--is restricted to a central part V_C, an isovector, spin-spin part V_{SS}, and an isovector, tensor part V_T. These are related to the results of the r-integration defined through Eq. (56) according to

$$V_C = (\alpha_1 + \beta_1) + \tfrac{1}{4}(\alpha_2 + \beta_2 + \alpha_3 + \beta_3) + \tfrac{1}{8}(\beta_4 + \beta_6) + \tfrac{1}{24}\beta_5,$$

$$V_{SS} = \tfrac{1}{36}(\alpha_2 + \beta_2) - \tfrac{1}{108}(\alpha_3 + \beta_3) + \tfrac{1}{48}\beta_4 + \tfrac{1}{432}\beta_5 - \tfrac{1}{144}\beta_6,$$

$$V_T = \tfrac{1}{54}(\alpha_3 + \beta_3) + \tfrac{1}{432}\beta_5 + \tfrac{1}{72}\beta_6. \tag{58a,b,c}$$

Results for the NN potential arising from the lagrangian of Eq. (1) and based on this scheme--or a very similar one--have been obtained by two groups[34,35] and are shown in Fig. 4. As was to be expected, the long-range behavior features a spin-spin part and a tensor part whose strengths are appropriate to that of a one-pion exchange potential with

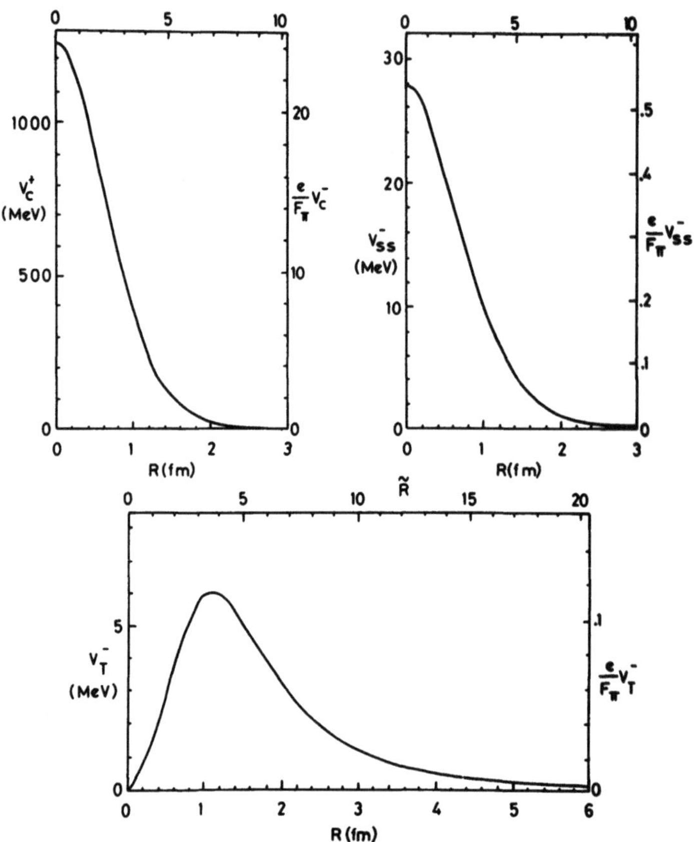

Fig. 4. The central, spin-spin, and tensor potentials obtained from the original Skyrme lagrangian of Eq. (1). (From Ref.[35].)

a πNN coupling constant as given in the Skyrme approach. The central potential shows strong repulsion near the origin, of a magnitude suggested by the excess energy of the N = 2 system over two N = 1 systems as shown in Table 1. What is utterly lacking is the attraction attributed to "σ-meson" exchange. Unfortunately, this situation is not in the least improved by using ω-stabilization[36] in place of the original stabilizing term with four derivatives of Skyrme; indeed that approach leads to very similar results to those of the previous one, as shown in Fig. 5.

C. Whence cometh attraction?

The search is thus on for a possible source of attraction in the NN potential obtained from the Skyrme approach--assuming, of course, that it is not simply lost in the 33 percent uncertainties of the model or in its inability to deal with any other than low-enrgy, long-range effects. One suggestion[37] is to reverse the sign of the term L_{4a} and bring about stabilization through the omega mechanism L_6. This will certainly produce attraction of an appropriate nature, but it violates our expectations with regard to the connection between the Skyrme lagrangian and the known ππ interaction.

A more attractive possibility might be to include <u>all</u> terms through sixth order in the lagrangian,

$$L = L_2 + L_{4a} + L_{4S} + L_6, \tag{59}$$

with the attraction coming from L_{4S}, while L_{4a} and L_6 provide adequate

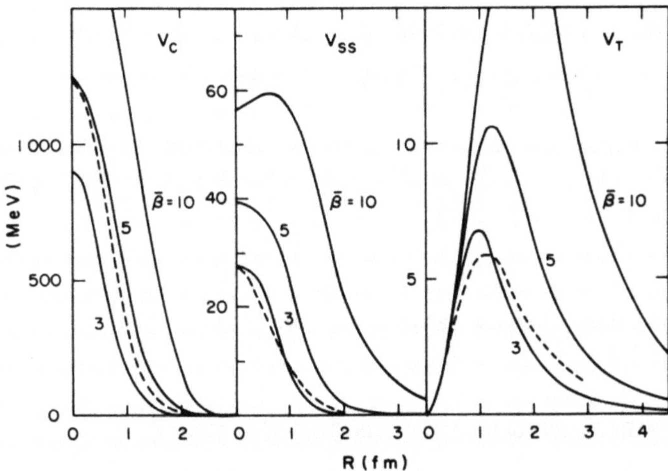

Fig. 5. The central, spin-spin, and tensor potentials obtained from a lagrangian with ω-stabilization. For the sake of comparison, the results of the calculation with conventional stabilization[35] are superimposed here. (From Ref.[36].)

stabilization to counteract a maximal L_{4S}. This has also been tried, with mixed results[38,30]. Though under certain ansätze for the ω-field one may find attraction[38], the particular form for ω(**R**, **r**) taken violates the basic connection of Eq. (50) between ω and the baryon density B(**R**, **r**)--the latter now also calculated by substituting the product ansatz for U and therefore containing terms referring to each baryon individually, but also a mixed product containing both. When this situation is rectified (which is most easily done using the L_6 form rather than a lagrangian with finite ω-mass; see both Ref.[30,38]), the attraction disappears.

The usual argument giving the origins of "σ-exchange" in the box diagrams with intermediate Δ-states, as in Fig. 3, suggests that the answer for the present problem may lie in broadening the Heitler-London product ansatz for the two-baryon system to include an admixture of ΔΔ-states along with the NN part. Such an approach would allow one to see the slow--and presumably slight--conversion of NN to ΔΔ as R gets smaller and the two baryons come closer. This has also been tried[30], but the effect is much too small to induce attraction.

It is very tempting to suppose that the answer may lie in a mutual deformation of the skyrmions as they approach each other, an effect not permitted in the simple product ansatz, but suggested, as we noted above, by lattice calculations[33]. Such calculations[39-41] are difficult to carry out with any degree of completeness, and at the moment seem to have as their main conclusion the need for serious concern about the validity of anything less than an elaborate lattice-computation program.

Last--and with only tangential connection--we note that it is also necessary to take into account the <u>dynamics</u> of the skyrmion-skyrmion interaction[42,43], which up till now we have ignored. This produces nonnegligible effects[43,44], which are, however, not large enough to invalidate the general conclusions concerning the NN potential, and certainly do not lead to attraction. [The general scheme in question again relates closely to techniques developed in nuclear physics in the context of trying to deal with collective motions side-by-side with single-particle features, and is sometimes called in that framework the "dragging" model. One extends the original product ansatz to allow for displacement of one skyrmion relative to the other through the appearance of a time-dependent collective coordinate $\mathbf{R}(t)$. The new form is

$$U_{N=2}(\mathbf{r}, \mathbf{R}) = U_{+}[\mathbf{r} + \tfrac{1}{2}\mathbf{R}(t)]\, U_{-}[\mathbf{r} - \tfrac{1}{2}\mathbf{R}(t)],$$

where + and - refer to the two baryons situated at their respective centers $\mathbf{r} \pm \mathbf{R}/2$. When this is substituted into the time-dependent form of the lagrangian, Eq. (1), one obtains a kinetic term (to order $\dot{R}^3$)

$$\tfrac{1}{2} M_{kl} \dot{R}_k \dot{R}_l,$$

where M_{kl} is a mass tensor which may be calculated in terms of the skyrmion solution.]

IV. THE LARGE-BARYON-NUMBER LIMIT FOR THE SKYRMION

A. Introductory comment

We have applied the skyrmion to N = 1 and N = 2 systems (their use for N = 3 is discussed in Ref.[45]), and we now contemplate a leap to the treatment of skyrmions with very large N. Our purpose here is partly to see what the Skyrme approach might have to say about conditions under which a quark-gluon plasma might be created in nuclear globs with high energy density (see, for example, the reviews on quark-gluon plasmas in Refs.[46-48]). The determination of such conditions is sufficiently difficult and murky at present as to justify even rough phenomenological assessments for that purpose. Further, the study of the $N \to \infty$ limit also serves to offer some small new insights into aspects of the Skyrme problem and its solution.

As a rough measure of conditions under which a quark-gluon plasma may be created, we shall seek to reach energy densities roughly twice those of the single nucleon. It emerges that these are not hard to achieve within the Skyrme model; in fact the energy density rises quite rapidly with baryon number, so that even quite light nuclear clusters might in principle sustain such a transition. Our methods will relate strictly to the spherical skyrmion solution, so that we do not try to study crystal-like or extended forms of multi-baryon systems[49-51] which might permit a fuller description of the transition between this highly dense form and the more normal structure of nuclei.

B. Applying adiabatic invariants to skyrmions

In a prescient paper[52], which, along with an even earlier numerical study[53], preceded the present renewal of interest in the Skyrme model, an analytic method was developed for handling the skyrmion in the $N \to \infty$ limit. We shall here apply and somewhat extend that method, and compare it with numerical results (see[54]). Our point of departure is Eq. (9), the expression for the static energy in units of $F_\pi \overline{e}$ for the skyrmion with baryon number N [which does not, of course, appear explicitly there, but rather in the boundary condition for F(r)]. We now follow Ref.[52] in changing variables to

$$r = 2 \cot t \qquad \text{and} \qquad F(r) = q(t), \tag{60}$$

where in what follows one is to think of $\underline{t}$ as a slowly-varying independent variable playing the role of time, and of $\underline{q}$ as the dependent variable which is the generalized coordinate of our problem and is rapidly varying for large N. In terms of these variables, one has [exercise!]

$$M = 2\pi \int_0^{\pi/2} \left[\tfrac{1}{2} A \dot{q}^2 + B\right] dt, \tag{61}$$

where

$$A = 1 - \cos 2q \sin^2 t \qquad \text{and} \qquad B = \frac{\sin^2 q}{\sin^2 t} + \frac{\sin^4 q}{2 \cos^2 t}. \tag{62}$$

The boundary conditions for these new variables are $q(0) = 0$ and $q(\pi/2) = N\pi$.

The method of adiabatic invariants is now applied[53]. For a system hamiltonian that varies only slowly in time, the adiabatic invariant

$$I = \oint p \, dq, \tag{63}$$

changes exceedingly slowly. Using the usual procedure for constructing the hamiltonian, we find the conjugate momentum and the hamiltonian,

$$p = A \dot{q} \qquad \text{and} \qquad H = \frac{p^2}{2A} - B = e, \tag{64}$$

where $\underline{e}$ is the approximately conserved energy of the system (assuming that A and B are approximately independent of $\underline{t}$). As we shall verify below, for large N the derivative $\dot{q}$ is proportional to N so that B may be neglected in H of Eq. (64) as compared with $1/2\, A \dot{q}^2$. Thus we have

$$e = \frac{p^2}{2A}, \tag{65}$$

and

$$I = \sqrt{2e}\, g(t) \qquad \text{and} \qquad e = \frac{I^2}{2g^2(t)}, \tag{66}$$

where

$$g(t) = \int_0^{\pi} \sqrt{A(q,t)}\, dq = \int_0^{\pi} \sqrt{1 - \cos 2q \sin^2 t}\, dq$$

$$= 2\sqrt{1 + \sin^2 t}\;\; E\left[\sqrt{\frac{2 \sin^2 t}{1 + \sin^2 t}}\right], \tag{67}$$

with E the complete elliptic integral of the second kind. Using the properties of this function, or working directly with the definition above, one can easily show that the extreme values reached by g(t) are

$$r = 0,\ t = \pi/2: \qquad g(\pi/2) = 2\sqrt{2} = 2.83,$$

and

$$r \to \infty,\ t = 0: \qquad g(0) = \pi = 3.14,$$

so that g(t) varies hardly at all over the range of $\underline{t}$, and we can roughly approximate it by

$$g(t) = 3. \tag{68}$$

In Ref.[52] there is now given a rather complete scheme for approximating the skyrmion energy and carrying out successive improvements in the approximation. We choose instead to take a somewhat cruder path that has the advantage for our purpose of being more transparent and allowing for work in configuration space, thus permitting the approximate evaluation of all the interesting skyrmion properties, including, for example, the rms radius of the Skyrme solution.

We consider points t_k such that

$$q(t_k) = k\,\pi, \qquad k = 0, 1, 2, \ldots, N; \qquad t_0 = 0, \ldots, t_N = \pi/2. \tag{69}$$

These are the points at which q(t) completes changes by π. The period of the system we take as

$$T(t_k) = t_{k+1} - t_k. \tag{70}$$

By the usual methods of classical mechanics, this period is related to the adiabatic invariant through

$$T = \frac{dI}{de} = \frac{g}{\sqrt{2e}} = \frac{g^2}{I}. \tag{71}$$

If we now replace the finite differences in Eq. (70) by a continuous variable for a smooth function t(k), we have

$$\frac{dt}{dk} = T = \frac{g^2}{I}, \tag{72}$$

whence

$$t = k\,\frac{g^2}{I}. \tag{73}$$

For $k = N$ we have $t = \pi/2$, which can be inverted to give

$$I = \frac{2Ng^2}{\pi} = \frac{18N}{\pi}, \tag{74}$$

from which we see that the system energy is, from Eqs. (61) and (66),

$$M = 18\,N^2. \tag{75}$$

The corresponding N^2-contribution to the energy when the full functional dependence of g(t) is retained has[52] a coefficient of 18.46 instead of the 18 in Eq. (75).

C. An approximate closed-form expression for F(r) when $N \to \infty$

From Eq. (69), but using the smooth variable t(k), we have

$$q(t_k) = q\left(\frac{kg^2}{I}\right) = q\left(\frac{k\pi}{2N}\right) = k\,\pi, \tag{76}$$

whence

$$q(t) = 2\,N\,t, \tag{77}$$

or

$$F(r) = 2\,N \text{ arc cot } \frac{r}{2}. \tag{78}$$

This very simple form fulfills the boundary condition $F(0) = N\pi$, and yields a linear slope at the origin, as is indeed found. As $r \to \infty$, it falls as r^{-1}, instead of as r^{-2}, as it should. This comes about because of our neglect of the term B in H, and causes problems in evaluating any quantity that depends on the tail of the skyrmion. Nonetheless, Eq. (78) is very useful for finding qualitative features and surprisingly successful even for semiquantitative purposes. A graph of this function as compared with numerical results for N = 1, 5, 18, and 30 is shown in Fig. 6.

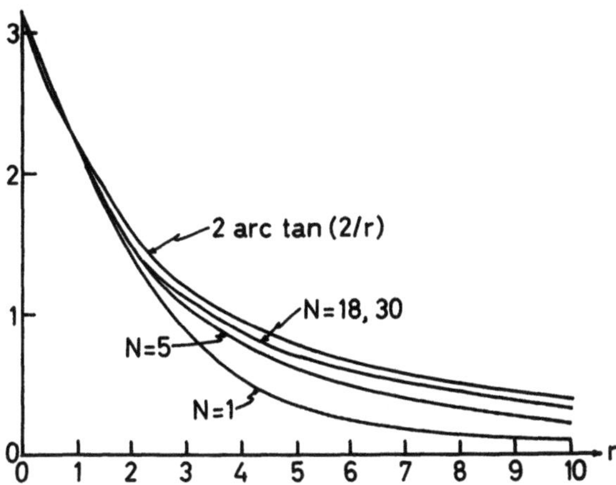

Fig. 6. The function F(r)/N as obtained numerically for N = 1, 5, 18, and 30 and in a closed form in Eq. (78).

Eq. (78) may be substituted into Eq. (9) for the system energy, yielding integrals that may be performed in r-space through the method of residues [exercise!]. The terms with $F'^2(r)$ contribute according to N^2, as anticipated, while those not containing $F'(r)$ are proportional to N. The result for the mass is

$$M = 2\,\pi^2\,N\,(N + 5/4) = 19.74\,N\,(N + 1.25), \tag{79}$$

to be compared with the more accurate result[52]

$$M = 18.46\,N\,(N + 0.8726) \tag{80}$$

in our units. (We shall use the latter form for our comparisons below.)

Again using the closed form for F(r), we may estimate the N-dependence of the skyrmion mean square radius,

$$\langle r^2\rangle = \frac{\int_0^\infty B^0(r)\, r^2\, dr}{\int_0^\infty B^0(r)\, dr} = -\frac{2}{N\pi}\int_0^\infty \sin^2 F\cdot F'\, r^2\, dr, \tag{81}$$

which may also be evaluated using the calculus of residues {Exercise: Do this using the substitution $z = \exp[iF/2N]$} to give

$$\langle r^2\rangle = \frac{2}{N\pi}\int_0^{\frac{\pi}{2}} \left[2\cot\frac{F}{2N}\right]^2 \sin^2 F\, dF = \frac{4}{\pi i}\oint \left[\frac{z+z^{-1}}{z-z^{-1}}\right]^2 \left[\frac{z^{2N}-z^{-2N}}{2i}\right]^2 \frac{d}{}$$

$$= 16\, N - 4, \tag{82}$$

so that the root mean square radius is proportional to $N^{1/2}$, and

$$r_{rms} = 4\, N^{1/2}. \tag{83}$$

Our numerical results for r_{rms} are well fit by the form

$$r_{rms} = 2.17\, N^{1/2}, \tag{84}$$

which suggests that the tail of the expression in (78) is indeed inadequate for this estimate--although it does yield an appropriate N-dependence.

The consequence of all this is to find that the hadronic volume here goes roughly as N^2, for large N, while the volume varies as $N^{3/2}$, so that the energy density is proportional to $N^{1/2}$. In practice, with the slightly better estimates available here (as shown in Table 3), one finds that already for $N > 13$ the energy density is more than twice that for $N = 1$. Since the hadronic density must go as $N^{-1/2}$, this high energy density occurs at the low hadronic density [scaling by $N^{1/2}$ from the experimental $r_{rms}(N=1) = 0.72$ fm] of 0.12 baryons/fm^3. This presumably comes about because of the large number of antiquarks present in the plasma within this description. This is a perfectly reasonable expectation for the quark-gluon plasma, and supports the notion that skyrmions may be able to add to our understanding of the behavior of these intriguing states of nuclear matter.

This work was supported in part by the Israel Academy of Sciences and Humanities--Basic Research Foundation, and by funds from the Yuval Ne'eman Chair in Theoretical Nuclear Physics.

Table 3
Energies and root-mean-square radii for skyrmions with large N

N	r_{rms}	$2.17\ N^{1/2}$	M [Eq. (9)]	M [Eq. (80)]	energy density
1	2.12	2.17	36.45	34.57	0.808
2	2.93	3.07	108.41	106.06	0.876
3	3.58	3.76	217.22	214.46	0.964
6	5.25	5.32	759.86	761.21	1.210
9	6.69	6.51	1645.1	1640.2	1.419
15	8.07	8.40	4395.8	4395.1	1.768
30	11.93	11.89	17071.5	17097.3	2.431

The energy density shown in the last column is based on the closed-form approximations of columns 3 and 5.

REFERENCES

1. F.E. Close, "Introduction to quarks and partons" (Academic, New York, 1979).
2. I.J.R. Aitchison and A.J.G. Hey, "Gauge theories in particle physics" (Adam Hilger, Bristol, 1982).
3. K. Huang, "Quarks leptons and gauge fields" (World Scientific, Singapore, 1982).
4. P. Becher, M. Böhm, and H. Joos, "Eichtheorien der starken and elektroschwachen Wechselwirkung" (Teubner, Stuttgart, 1983).
5. F. Yndurâin, "Quantum chromodynamics" (Springer, New York, 1983).
6. T.-P. Cheng and L.-F. Li, Gauge theory of elementary particle physics (Clarendon, Oxford, 1984).
7. F. Halzen and A.D. Martin, "Quarks and leptons" (Wiley, New York, 1984).
8. G. 't Hooft, Nucl. Phys. B72 (1974) 461; B75 (1974) 461.
9. E. Witten, Nucl. Phys. B160 (1979) 57; B223 (1983) 433.
10. A. Zaks, Nucl. Phys. B260 (1985) 241.
11. L.-H. Chan, Phys. Rev. Lett. 55 (1985) 21.
12. P. Simic, Phys. Rev. Lett. 55 (1985) 40; Rockefeller University report number 85/124.
13. T.H.R. Skyrme, Proc. Roy. Soc. London, Ser. A 260 (1961) 127; 262 (1961) 237; Nucl. Phys. 31 (1962) 556.
14. G.S. Adkins, C.R. Nappi, and E. Witten, Nucl. Phys. B228 (1983) 552.
15. M. Rho, lectures given at the International School of Nuclear Physics, Erice, April, 1983.
16. R.R. Silbar, lectures given at Los Alamos, November and December, 1984, and at the University of Massachusetts, October, 1985.
17. R. Vinh Mau, lectures given at the 24th International Universitätswochen für Kernphysik, Schladming, February, 1985.
18. I. Zahed and G.E. Brown, lectures given at a Los Alamos Summer School, June, 1985.
19. G. Holzwarth and B. Schwesinger, review article commissioned for Repts. Prog. Phys.; preprint, Siegen University, December, 1985.
20. M. Abramowitz and I.A. Stegun, "Handbook of mathematical functions" (Dover, New York, 1965).

21. T.D. Lee, "Particle physics and introduction to field theory" (Harwood, Chur, Switzerland, 1981) p. 138.
22. J.M. Eisenberg and W. Greiner, "Nuclear theory" (North-Holland, Amsterdam, 1970 and 1972).
23. D. Finkelstein and J. Rubinstein, J. Math. Phys. 9 (1968) 1762.
24. K. Huang, "Statistical mechanics" (Wiley, New York, 1963) p. 152.
25. A.D. Jackson and M. Rho, Phys. Rev. Lett. 51 (1983) 751.
26. J.F. Donoghue, E. Golowich, and B.R. Holstein, Phys. Rev. Lett. 53 (1984) 747.
27. G.S. Adkins and C.R. Nappi, Phys. Lett. 137B (1984) 251.
28. G.S. Adkins and C.R. Nappi, Nucl. Phys. B233 (1984) 109.
29. M. Lacombe, B. Loiseau, R. Vinh Mau, and W.N. Cottingham, Phys. Lett. 169B (1986) 121.
30. G. Kälbermann, J.M. Eisenberg, R.R. Silbar, and M.M. Sternheim, submitted.
31. G.S. Adkins, preprint, 1985.
32. L. Pauling and E.B. Wilson, Jr., "Introduction to quantum mechanics" (McGraw-hill, New York, 1935).
33. H.M. Sommerman, H.W. Wyld, and C.J. Pethick, Phys. Rev. Lett. 55 (1985) 476.
34. A. Jackson, A.D. Jackson, and V. Pasquier, Nucl. Phys. A432 (1985) 567.
35. R. Vinh Mau, M. Lacombe, B. Loiseau, W.N. Cottingham, and P. Lisboa, Phys. Lett. 150B (1985) 259.
36. J.M. Eisenberg, A. Erell, and R.R. Silbar, Phys. Rev. C33 (1986) 1531.
37. A. Jackson, A.D. Jackson, A.S. Goldhaber, G.E. Brown, and L.C. Castillejo, Phys. Lett. 154B (1985) 101.
38. M. Lacombe, B. Loiseau, R. Vinh Mau, and W.N. Cottingham, Phys. Lett. 169B (1986) 121.
39. U.B. Kaulfuss and U.-G. Meissner, Phys. Rev. D31 (1985) 3024.
40. E. Sorace and M. Tarlini, preprint, Florence, July, 1985.
41. M. Oka, K.F. Liu, nad H. Yu, preprint, University of Pennsylvania, December, 1985.
42. L. Wilets, private communication.
43. M. Oka, preprint, University of Pennsylvania, February, 1986.
44. J.M. Eisenberg, unpublished.
45. U.-G. Meissner and U.B. Kaulfuss, Phys. Rev. C30 (1984) 2058.
46. G. Baym, in "Common problems in low- and medium-energy nuclear physics", B. Castel, B. Goulard, and F.C. Khanna, eds. (Plenum, New York, 1979) p. 213.
47. B. Müller, "The physics of the quark-gluon plasma", Lecture notes in physics 225 (Springer, Berlin, 1985).
48. M. Gyulassy, "Introduction to QCD thermodynamics and the quark-gluon plasma", to appear in "Progress in Physics", vol. 15 (Pergamon, Oxford, 1986).
49. M. Kutschera, C.J. Pethick, and D.G. Ravenhall, Phys. Rev. Lett. 53 (1984) 1041.
50. I. Klebanov, Nucl. Phys. 262 (1985) 133.
51. N.K. Glendenning, LBL Report-19032, January, 1985.
52. E.B. Bogomol'nyi and V.A. Fateev, Sov. J. Nucl. Phys. 37 (1983) 134 [Yad. Fiz. 37 (1983) 228].
53. V.P. Nisichenko, Dep. VINITI, Ak. Nauk SSSR, No. 2484-80, and Dissertation, Mir, Patrice Lumumba University, 1981, as quoted in Ref.[52].
54. G. Kälbermann and J.M. Eisenberg, preprint, Tel Aviv University, March, 1986.

THE BARYON-BARYON INTERACTION AND THE QUARK MODEL

Amand Faessler

Institut für Theoretische Physik
Universität Tübingen
Auf der Morgenstelle 14
D-7400 Tübingen, West Germany

ABSTRACT

The ^3S, ^1S and P wave phase shifts of the nucleon-nucleon interaction are calculated in the six quark model using the resonating group method. For large distances the model is supplemented by π, σ, ρ and ω-meson exchange. The role of the orbital $[42]_r$ symmetry for the short range repulsion is studied. It is shown that at short distances the orbital $[42]_r$ symmetry plays an important role which is even enlarged by the colour magnetic interaction. The $[42]_r$ symmetry enforces the short range repulsion by a node which it requests at short distances. The mechanism is complicated by the fact, that the orbital $[6]_r$ symmetry is admixed by about the same weight. We show that for meson exchanges which mediate the long range behaviour we can now use the SU_3 flavour ratios of the meson-nucleon coupling constants even for the ω-nucleon coupling. For the ω-meson one had to use in the OBEP's ω-N coupling constant twice to three times as large as predicted by SU_3 flavour to describe the short range repulsion. We also calculated the nucleon-hyperon interaction and describe the NΛ-scattering in agreement with the data.

INTRODUCTION

The nucleon-nucleon (N-N) scattering yields phase shifts[1], which are characteristic for a strong short range repulsion. This short range repulsion has been described in the past by the exchange of the ω-meson between the nucleons[2]. But to obtain the observed repulsion one needed to increase the ω meson-nucleon coupling constant $g^2_{\omega NN}/4\pi$ to twice or even three times the value predicted by SU_3-flavour in relation to the ρ meson-nucleon coupling ($g^2_{\omega NN}/4\pi = 9 \cdot g^2_{\rho NN}/4\pi = 4.5$). All the other meson-nucleon coupling constants follow roughly the SU_3-flavour relations. The discrepancy for the ω meson-nucleon coupling reflects the fact that ω exchange is not the mechanism responsible primarily for the short range repulsion.

With the advent of the quark model and QCD one suggested that the intrinsic quark structure of the two interacting nucleons can explain this short range repulsion[3,4]. These early trials have serious short comings: (i) They used[3,4,5] the Born-Oppenheimer approximation, which would be only justified, if the effective mass of the quarks is small compared to that of the nucleon. This is not the case in the constituent quark model. Even in the current quark model the energy eigenvalue of the quark is about one third of the nucleon mass. (ii) Another serious short coming of these calculations is the neglect of the orbital $[42]_r$ symmetry for the six quarks at distance zero between the two nucleons. The importance of the $[42]_r$ symmetry has first been pointed out by Neudatchin and coworkers[6,7]. The ansatz for the two nucleon wave function as three quarks in an oscillator potential at $\vec{r}/2$ and three quarks in another oscillator potential at $-\vec{r}/2$ suppresses the $[42]_r$ symmetry like[8]

$$F(r) = [1 - f^2 - f^4 + f^6]^{1/2}$$
$$f(r) = \exp(-r^2/4b^2) \tag{1}$$
$$b^2 = \frac{\hbar}{m\omega}$$

at small distances r. Harvey[5] included the $[42]_r$ symmetry even at small distances by allowing the excitation of two quarks in p states in the Born-Oppenheimer approximation. He obtained a large effect. The repulsive core[3,4] disappeared. This opened up again the question for the nature of the short range repulsion.

Oka and Yazaki[9] and Faessler and coworkers[8,10] showed that a non-adiabatic treatment with the help of the resonating group method including the $[42]_r$ symmetry yields hard core phase shifts for the ^1S and ^3S interaction between two nucleons. This short range repulsion is also strongly influenced by the colour magnetic interaction.

In detail we summarize shortly in chapter two the model and discuss a possible mechanism for the short range repulsion. In chapter 3 we give the results and in chapter 4 we summarize the main conclusions.

INTERACTION OF TWO NUCLEONS IN THE QUARK MODEL

At short distances between two nucleons we describe the nucleon-nucleon (NN) interaction by the exchange of gluons and quarks. The gluon exchange is determined by the quark gluon vertex.

$$\mathcal{L}^{QCD}_{int} = i(g/2)\bar{q}_i \, \gamma^{\mu} \, \lambda^{(a)}_{ik} q_k \, G^{(a)}_{\mu} \tag{2}$$

Here $\lambda^{(a)}$ are the eight Gell-Mann colour SU_3 matrices. q_i are the Dirac spinors of the quarks with colour i. Eq. (2) includes a sum over repeated indices $\mu = 0,1,2,3$, $a = 1,2,\ldots 8$ and i,k = red, blue, yellow. The one gluon exchange between two quarks in the non-relativistic reduction is given by[11]:

$$V^{OGEP}_{(i,j)} = \frac{\alpha_s}{4} \, \lambda^{(a)}_i \cdot \lambda^{(a)}_j \, [\frac{1}{r_{ij}} - \frac{\pi}{m_q^2} \delta(\vec{r}_{ij})(1 + \frac{2}{3}\vec{\sigma}_i \cdot \vec{\sigma}_j)] + \ldots \tag{3}$$

m_q is the constituent quark mass and $\alpha_s = g^2/4\pi$ the strong fine structure constant. The dots indicate terms of tensor and two-body spin-orbit nature of the quark-quark interaction. They don't play a role if we restrict us to ^1S and ^3S interaction between nucleons. But they have to be included for the higher partial waves. The quark-quark interaction (3) is the leading term only for large momentum transfer and therefore short distances between nucleons. But even there one should not take (3) literally. It probably gives only the rough dependence. Its quality is improved by fitting the parameters $m_u = m_d, m_s$ and α_s to the nucleon and the Δ mass.

The total six quark Hamiltonian (for different quark masses see eq. (14))

$$\hat{H}_6 = \sum_{i=1}^{6} [m_i + \frac{p_i^2}{2m_q}] + \sum_{i<j=1}^{6} [V^{OGEP}(i,j) - \lambda_i^{(a)} \cdot \lambda_j^{(a)} \; a \; r_{ij}] \tag{4}$$

must include also a colour confinement term. The parameter a [MeV fm^{-1}] is adjusted to the charge root mean square radius of the proton including the pion cloud[12].

For large distances one can not exchange colour objects like quarks and gluons. Thus we go back there to meson exchange. But we have to guarantee asymptotic feedom. This is done by allowing the coupling of the mesons to the quarks only near the surface of the nucleon.

$$g_{qq\mu}(r) = c_\mu \, r^2 \tag{5}$$

The free parameters c_μ will be adjusted for each meson μ to the meson-nucleon coupling constants[13] at zero momentum transfer $g^2_{\pi NN}(q=0)/4\pi = 14.1$, $g^2_{\sigma NN}(q=0)/4\pi = 5.65$ and $g^2_{\rho NN}/4\pi = 0.5$. For the ω-nucleon coupling we shall see that the flavour SU_3 value $g^2_{\omega NN} = 9 \cdot g^2_{\rho NN}$ yields a good agreement for the NN phase shifts. This is opposed to the OBEP's[2,13] where the ω-nucleon coupling constant has to be blown up by a factor two to three to describe the short range repulsion.

In a first step we include only the exchange of the two lightest mesons (m_π = 14/ MeV; m_σ = 520 MeV). The ansatz for the resonating group wave function is

$$\Psi_{6q} = A\{|\overline{NN}\rangle \chi_N(r) + |\overline{\Delta\Delta}\rangle \chi_\Delta(r) + |\overline{CC}\rangle \chi_C(r)\} \tag{6}$$

The Kohn-Hulthen variational principle

$$\langle \delta\psi_{6q} | \hat{H}_6 - E | \psi_{6q} \rangle = 0 \tag{7}$$

yields for ^1S and ^3S channels three coupled integral equations for the relative wave functions $\chi_N(r)$, $\chi_\Delta(r)$ and $\chi_C(r)$. If one describes the six quark wave function by three quarks in an oscillator at $-\vec{r}/2$ and by three quarks in an oscillator at $\vec{r}/2$, the antisymmetrized wave function has the form[8]:

$$A\{|NN,r\rangle\} = \frac{1}{3} D(r) \, | [6]_r \{33\}_{ST} [222]_C \rangle + \frac{2}{3} F(r) [\, | [42]_r \{33\}_{ST} [222]_C \rangle - [42]_r \{51\}_{ST} [222]_C \rangle] \tag{8}$$

with:

$$D(r) = [1 + 9f^2 + 9f^4 + f^6]^{1/2}$$

$$F(r) = [1 - f^2 - f^4 + f^6]^{1/2}$$

$$f = \exp(-r^2/4b^2)$$

Instead of (8) we are using a basis which allows that the variational principle chooses freely the amplitudes of the different orbital symmetries characterized by Young tableaux.

$$\begin{pmatrix} |NN\rangle \\ |\Delta\Delta\rangle \\ |CC\rangle \end{pmatrix} = A \begin{pmatrix} |(\{33\}_{ST}[222]_C)\ [\tilde{6}]\rangle \\ |(\{33\}_{ST}[222]_C)\ [\tilde{42}]\rangle \\ |(\{51\}_{ST}[222]_C)\ [\tilde{42}]\rangle \end{pmatrix} \tag{9}$$

$$A = \begin{pmatrix} \frac{1}{3} & \frac{2}{3} & \frac{2}{3} \\ -\sqrt{\frac{4}{45}} & -\sqrt{\frac{16}{45}} & -\sqrt{\frac{25}{45}} \\ \sqrt{\frac{4}{5}} & -\sqrt{\frac{1}{5}} & 0 \end{pmatrix}$$

The above basis states contain no spatial dependence. ST and colour C are only coupled to the symmetry conjugate to the orbital one. The states $|NN\rangle$, $|\Delta\Delta\rangle$ and $|CC\rangle$ contain also the internal spatial variables for each nucleon. A solution of the resonating group problem (7) yields for each symmetry its own radial dependence. The ansatz (9) includes quark excitations in higher orbital states. If one solves the relative wave function by a coupled system of Hill-Wheeler-Griffin equations

$$\sum_{\beta=N,\Delta,C} \int dr'\ [\langle\alpha,r|\hat{H}_6 + V_\pi + V_\sigma|\beta,r'\rangle$$
$$-E\langle\alpha,r|N|\beta,r'\rangle] \cdot \chi_\beta(r') = 0 \quad , \tag{10}$$

it is useful to look to the wave function in the following form:

$$\tilde{\chi}_\alpha(r) = \sum_\beta \int \langle\alpha,r|N|\beta,r'\rangle^{1/2}\ \chi_\beta(r')dr' \tag{11}$$

According to (9) it can be decomposed into the symmetry basis. Fig. 1 shows such relative wave functions in the ^3S channel. One sees a node in the [42] orbital symmetries. The nodes are reflected in the norm-kernels in oscillator functions, which are diagonal in the symmetry basis.

$$\langle n,l{=}0|N_{[6]_r}|n,l{=}0\rangle = 1 + 9\ (\tfrac{1}{3})^{2n}$$
$$\langle n,l{=}0|N_{[42]_r}|n,l{=}0\rangle = 1 - (\tfrac{1}{3})^{2n} \tag{12}$$

One sees that the Os state is Pauli forbidden for $[42]_r$. This yields the node in the orbital [42] symmetric relative wave functions. If we would have only $[42]_r$ the node would guarantee a hard core phase shift. (One should stress, that the relative wave function is not unique. For the $[42]_r$ symmetry one can add any admixture of Os oscillator functions without changing any observable. This can remove the node in the relative wave function. It depends only on the way how the non-orthogonal basis states $|NN\rangle$, $|\Delta\Delta\rangle$ and $|CC\rangle$ are orthogonalized[24]. The non-orthogonality is here included by the overlap kernel $\langle\alpha,r|N|\beta,r'\rangle$. In discussing the relative wave function in the NN channel it is natural to orthogonalize the $|\Delta\Delta\rangle$ and $|CC\rangle$ channels relative to $|NN\rangle$. But this orthogonalization and the possible removal of the Pauli forbidden Os admixture in the $[42]_r$ symmetry does not change any observable).

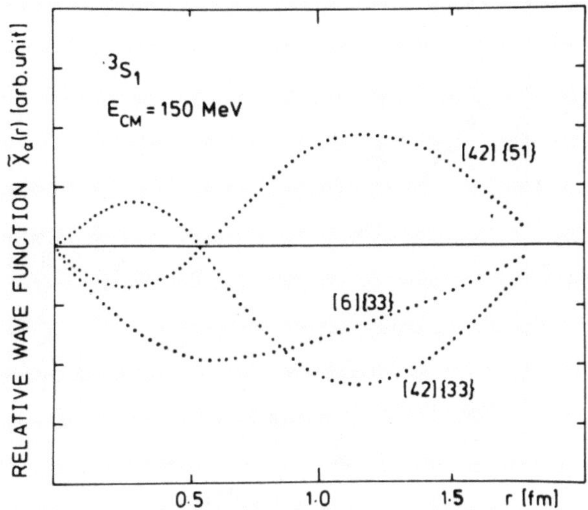

Fig. 1. Relative ^3S wave functions in the symmetry basis at $E_{CM} = 150$ MeV as a function of the distance r between the two nucleons. (m_q = 336 MeV, α_s = 1.3, a_c = 41 MeV fm^{-1}).

The colour magnetic part of the quark-quark wave function is essential in enlarging the [42] orbital symmetry in the relative NN wave function.

At the end of this chapter we are hopefully convinced that the node of the $[42]_r$ symmetry is responsible for the hard core phase shift. But the $[42]_r$ symmetry has only its important position due to the colour magnetic part of the quark-quark force.

Table 1. Decomposition of $\tilde{\chi}_{NN}(r)$ into the spatial symmetries [6] and [42] without and with the colour magnetic interaction (CMI). The numbers given are the amplitudes squared $|\langle ns|\tilde{\chi}_{NN}[f]\rangle|^2$ for the expansion into oscillator functions $|0s\rangle$, $|2s\rangle$, $|2s\rangle$ and $|3s\rangle$. The wave function $\tilde{\chi}_{NN}(r)$ is renormalized from $\chi_{NN}(5)$, so that the overlap kernel is the δ-function.

	[f]	$\|0s\rangle$	$\|1s\rangle$	$\|2s\rangle$	$\|3s\rangle$
without	[6]	0.450	0.045	0.005	0
CMI	[42]	0	0.160	0.039	0.001
without	[6]	0.132	0.062	0.026	0.011
CMI	[42]	0	0.219	0.188	0.088

This result is also supported by table 1. It shows the decomposition of the relative wave function between two nucleons $\chi_{NN}(r)$ into the spatial symmetries [6] and [42] and their expansion into oscillator wave functions $|n,l=0\rangle$ calculated in the NN channel only. The results with the colour magnetic force (with CMI) shows the $[42]_O|1s\rangle$ as the strongest admixture. While without CMI this role is played by $[6]_O|0s\rangle$.

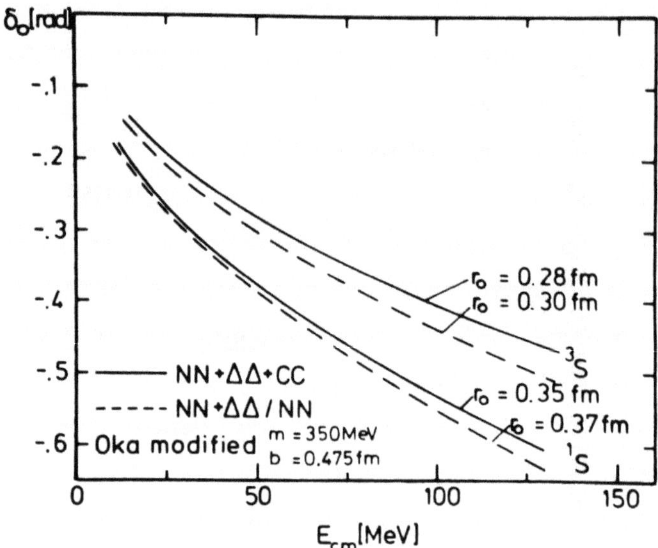

Fig. 2. ^1S and ^3S phase shifts calculated[14] without π and σ exchange in the three channel case (NN, ΔΔ, CC; solid line) and in the two channel case (NN, ΔΔ; dashed line). One obtains a typical hard core phase shift $\delta = -r_O k = -r_O \times [3\ E_{CM}\ m_q]^{1/2}/\hbar c$ with the equivalent hard core radius r_O. (m_q = 336 MeV, α_s = 1.3 and a_c = 41 MeV fm^{-1}).

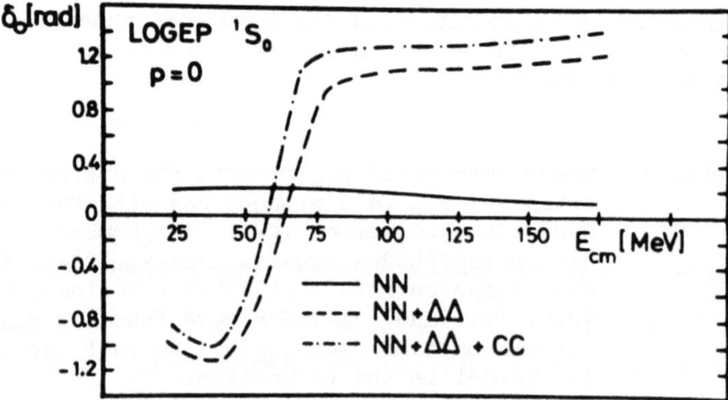

Fig. 3. ^1S phase shifts calculated[14] without π and σ exchange and with putting the colour magnetic interaction to zero. Now one sees no hard core behaviour for the NN phase shifts given here. This result is independent if one takes only one (NN) or two (NN+ΔΔ) or three (NN+ΔΔ) or three (NN+ΔΔ+CC) channels.

RESULTS

The solution of the coupled system of resonating group equations (10) yields the relative wave functions $\chi_N(r)$, $\chi_\Delta(r)$ and $\chi_C(r)$ for the channels with two asymptotic nucleons, two asymptotic Δ's and the six quark hidden colour state, respectively. The asymptotic form of the relative wave function in the NN-channel gives the phase shift for the nucleon-nucleon scattering. This calculation is performed including quark and gluon exchange $\hat{H}_6$ and the exchange of π- and σ-mesons. The parameters are adjusted as described in chapter 2. The results are given for the ^1S and ^3S in figures 4 and 5, respectively.

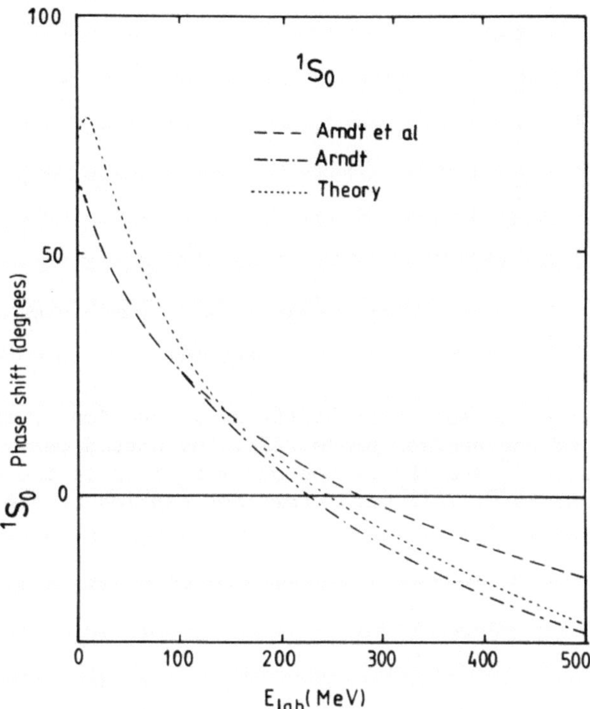

Fig. 4. *^1S nucleon-nucleon phase shifts as a function of the laboratory energy. The dotted curve gives the results of the theory presented[12]. The dashed and the dashed-dotted line are two different sets of experimental phase shifts[15]. (m_q = 336 MeV, α_s = 1.3 and a_c = 41 MeV fm^{-1}).*

In the results presented in figures 4 and 5 we included only the exchange of the two lightest mesons, the π- and σ-meson. It is interesting to see what happens if we include also in addition the ρ- and the ω-mesons. This question is especially exciting since in the one meson exchange potentials the ω-meson is solely responsible for the short range repulsion. Out of these reasons one adjusts in the OBEP's the ω-nucleon coupling constant to a value which is by a factor two to three times larger than the flavour SU_3 value derived from the ρ-nucleon coupling. If we now have the correct nature of the short range repulsion, we should get a satisfactory fit to the phase shifts, if we use the flavour SU_3 value $g^2_{\omega NN}/4\pi = 4.5$. Figure 6 shows the ^3S phase shifts for different ω-NN coupling constants in radians as a function of the nucleon bombarding energy. The factor g_O is defined as the deviation from the flavour SU_3 value.

The figure shows that for a reasonably good nuclear radius one gets a good agreement for the phaseshifts for a value of the parameter $g_O = 1$ in agreement with the flavour SU_3 symmetry. These results support strongly our conviction that we found indeed the correct nature of the short range nuc-

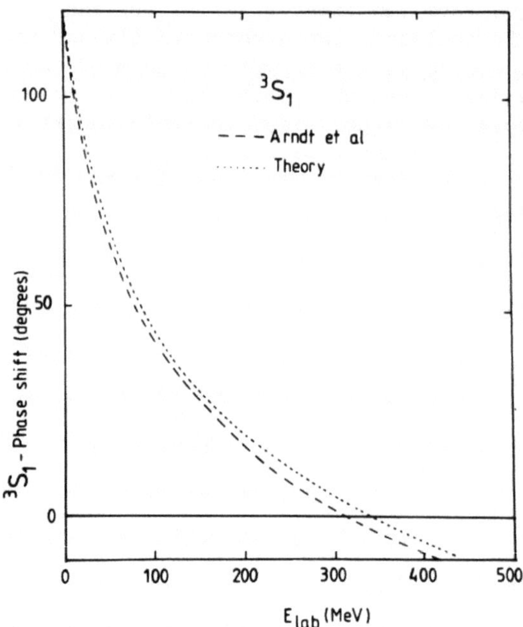

Fig. 5. ^3S nucleon-nucleon phase shifts as a function of the laboratory energy of the nucleon projectile. The dotted curve is the present calculation[12] and the dashed curve represents the experimental phase shifts[15]. (For parameters see caption of Fig. 4).

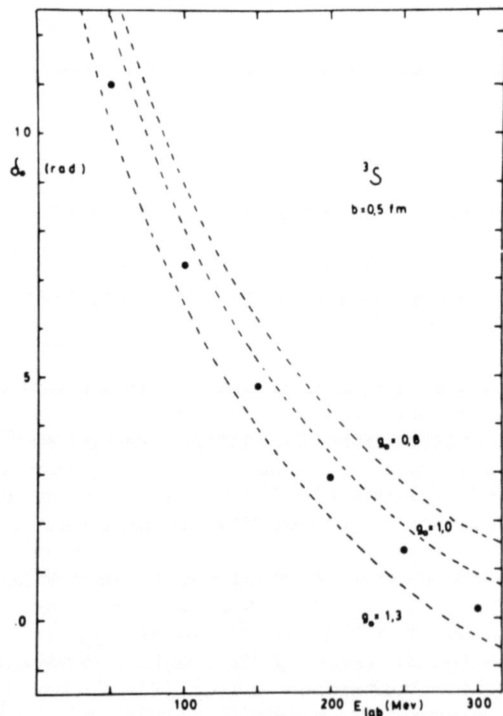

Fig. 6. ^3S nucleon-nucleon phase shifts in radians as a function of the laboratory energy for the oscillator length $b = 0.5$ fm. The ω-nucleon coupling constant is chosen as $g^2_{\omega NN}/4\pi = 4.5\ g_O$. The data are taken from ref. 15.

leon-nucleon repulsion. Recently we have extended this model to include also the finite size of the pions. This has the advantage that one can take into account the pion cloud in determining the energy of the proton and still keep the nucleon stable at finite radius[16]. For a point pion the pion self energy exerts on the nucleon such a pressure that it is compressed to the radius zero. In addition we calculated also the $^1P_1$ and the averaged ^3P_J partial waves.

$$\delta(\overline{^3P_J}) = \frac{1}{9}\,[\delta(^3P_0) + 3\ (^3P_1) + 5\delta(^2P_2)] \qquad (13)$$

Figures 7 and 8 give the $^1P_1$ and the averaged ^3P_J phase shifts. The solid line includes full antisymmetrization while the dashed line emits the antisymmetrization between quarks in nucleon one and quarks in nucleon two for the contribution to the Hamilton overlap from the exchange of the pions.

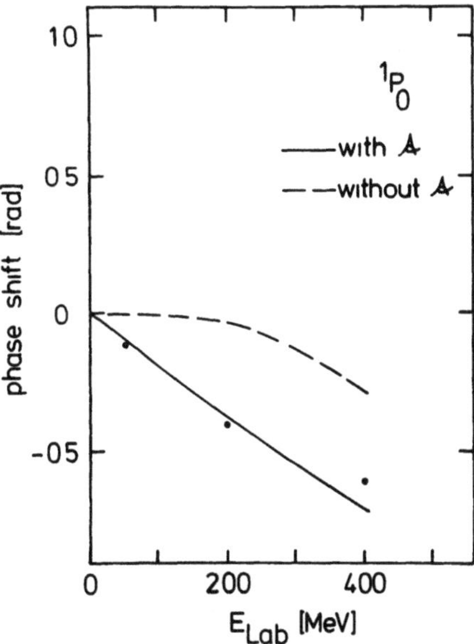

Fig. 7: $^1P_1$ phase shift in radians as a function of the laboratory nucleon bombarding energy. The solid line includes the full antisymmetrization of the quarks in nucleon one and two while the dashed line neglects for the pion exchange contribution this antisymmetrization. The dots indicate the experimental values.

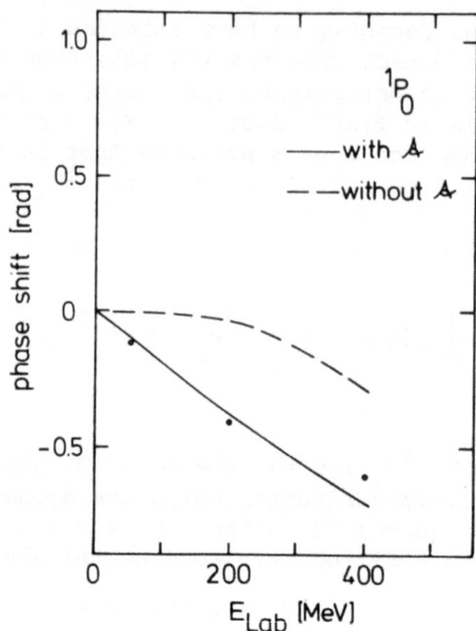

Fig. 8. ^3P_J *averaged according to equation (13) phase shifts in radians as a function of the laboratory bombarding energy of the nucleon. The solid line includes the full antisymmetrization between quarks in nucleon one and quarks in nucleon two. The dashed line neglects this antisymmetrization for the contribution of the pion exchange.*

THE LAMBDA-NUCLEON INTERACTION

The lambda-nucleon (Λ-N) can be calculated in the same way as the nucleon-nucleon interaction. The one-gluon exchange (3) must now include the different masses for the up and down quarks and for the strange quark. We choose this mass by 150 MeV larger than the up and down quark masses[18].

$$V^{OGEP}_{(i,j)} = \frac{\alpha_s}{4} \lambda_i^{(a)} \cdot \lambda_j^{(a)} \left[\frac{1}{r_{ij}} - \frac{\pi}{2} \delta(r_{ij}) \cdot \left(\frac{1}{m_i^2} + \frac{1}{m_j^2} + \frac{4}{3} \frac{\vec{\sigma}_i \cdot \vec{\sigma}_j}{m_i m_j}\right)\right] \tag{14}$$

The quark and gluon exchange between the hyperon and the nucleon yields the short range part of the interaction. One has to add the confinement potential (4) and for large distances meson exchange. The meson exchange potential is a generalization of the pion potential of ref. 19 to include all pseudoscalar mesons and also mixing of the octet η_8 with the singlet η_1. A mixing angle of $\theta = -23^o$ is assumed.

$$V^{PSM}_{(i,j)} = \frac{1}{3}\,\frac{g^2_{qqM}}{4\pi}\,\frac{\mu^2}{4M_A M_B}\,\frac{\Lambda^2}{\Lambda^2-\mu^2}\; 1^{-\mu^2 b^2/3}$$

$$\cdot\mu\;\left[\frac{e^{-\mu+r_{ij}}}{\mu r_{ij}} - \frac{\Lambda^3}{\mu^3}\cdot\frac{e^{-\Lambda r_{ij}}}{\Lambda r_{ij}}\right]\vec{\sigma}_i\cdot\vec{\sigma}_j\cdot O^F_{ij} \tag{15}$$

with: $g_{qq8} = \frac{3}{5}\, g_{NN\pi} = \frac{3}{5}\, 14.2$

$g_{qq1} = g_{NN\eta_1} = \frac{\sqrt{5}}{6}\, g_{NN\pi}$

here μ is the mass of the pseudoscalar meson in question, Λ a size parameter describing the finite size of the meson, M_A and M_B are the masses of the baryons A, B between which the mesons are exchanged and O^F_{ij} is an operator in flavour space according to table 2.

Table 2. Meson parameters

	μ(MeV)	Λ(MeV)	O^F_{ij}
π	138	832	$\sum_{K=1}^{3} \lambda^F_K(i)\cdot\lambda^F_K(j)$
K	495	832	$\sum_{K=y}^{7} \lambda^F_K(i)\cdot\lambda^F_K(j)$
η	549	4000	$\cos\theta\cdot\lambda^F_8(i)\cdot\lambda^F_8(j) - \sin\theta$
η'	958	4000	$\sin\theta\cdot\lambda^F_8(i)\cdot\lambda^F_8(j) + \cos\theta$

There are two independent coupling constants g_{qq8} and g_{qq1} describing the coupling of the octet or singlet mesons with the quarks. Because the coupling constants are assumed to be independent of flavour, they can be determined by looking to the long range part of (15) and comparing it with the one meson exchange potential for the NN interaction. In this way one finds the relations given in eq. (15).

In addition to these potentials acting on the quark level we add a phenomenological σ meson potential on the baryon level with a microscopic calculated form factor[20] representing the contributions of two pion exchange.

$$V_\sigma(r) = -\frac{g^2_\sigma}{4\pi}\,\frac{1}{2m^2R^2r}\begin{cases} 1-e^{-mr}-e^{-2mR}\sinh(mr) & \text{for } r<2R \\ [\cosh(2mR)-1]e^{-mr} & \text{for } r<2R \end{cases} \tag{16}$$

Here m=520 MeV is the mass of the σ-meson and R=0.72 fm is a size parameter fitted to the form factor.

In the kinetic energy we use an average quark mass. These approximations to use a mean mass in the kinetic energy is unavoidable if one uses the same symmetry space as for the nucleon-nucleon interaction. The parameters of the short range part of the interaction are fixed by requiring that the mass differences $M_\Lambda - M_N = 177$ MeV, $M_\Sigma - M_N = 253$ MeV are reproduced and that the stability condition $dM_N/db = 0$ is satisfied. The parameters obtained by this procedure are shown in table 3 for two values of the oscillator length b without and with the meson cloud included. One sees that the meson cloud contributes to about one half of the mass splittings.

Table 3. Parameter sets for OGE

b(fm)	0.50	0.55	0.50	0.55
$m_u = m_d$ (MeV)	313	313	313	313
m_s (MeV)	535.2	535.2	425.4	443.2
α_s	0.82	1.09	0.26	0.48
a_c (MeV/fm^2)	85.2	47.0	58.3	33.2
$(M_\Lambda - M_N)_{Meson}$ (MeV)	0	0	74.7	61.9
$(M_\Sigma - M_N)_{Meson}$ (MeV)	0	0	135.3	114.2
$g_\sigma^2/4\pi$	0	0	2.3	2.6

Fig. 9 shows the total elastic cross section for Λp-scattering as a function of the incident Λ-momentum in the laboratory frame. The theoretical results are compared with the experimental data given in the references 21 to 23.

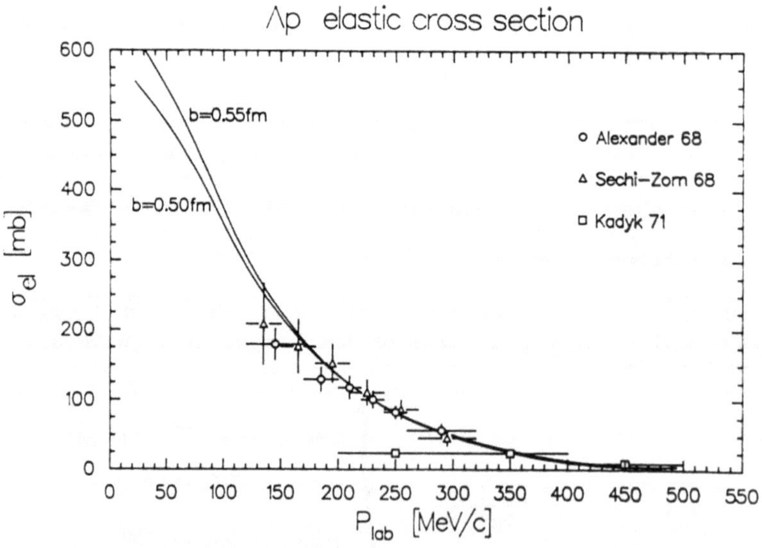

Fig. 9. Total elastic cross section for Λp-scattering as a function of the incident Λ-momentum in the laboratory frame. Two theoretical curves are compared with the experimental data from refs. 21-23. The two theoretical curves are distinguished by the two sets of parameters adjusted for the two different oscillator lengths b=0.5 fm and b=0.55 fm.

The comparison between theory and experiment for the Λ-proton scattering shows a surprisingly good agreement between theory and experiment.

CONCLUSIONS

The NN phase shifts are calculated using the quark model with a QCD inspired quark-quark force. The short range part of the NN force is given by quark and gluon exchange. The long range part is described by π and σ-meson exchange. The data fitted in the model are five values connected with three quarks only: The nucleon mass, the Δ mass, the root mean square radius of the charge distribution of the proton including the pion cloud, the π-N and the σ-N coupling constant at zero momentum transfer. The ^1S and ^3S phase shifts are nicely reproduced. The short range repulsion is decisively influenced by the node in the $[42]_r$ relative wave function. Very important is the colour magnetic quark-quark force which enlarges the $[42]_r$ admixture. We are also able to describe the higher phase shifts.

In the OBEP's the short range repulsion is connected with the exchange of the ω-meson. But to reproduce the short range repulsion one had to blow up the ω-N coupling constant by a factor 2 to 3 compared to flavour SU_3. With quark and gluon exchange the best fit to the ω-N coupling constant lies close to the SU_3 flavour value. This fact strongly supports the notion that we have found the real nature of the short range repulsion of the NN interaction.

In chapter 4 we have applied the same theoretical description to the Λ-nucleon scattering including for the short range interaction quark-gluon exchange and for the long range the exchange of the whole pseudo scalar octet and singlet mesons (π, K, η, η'). For the elastic Λ-proton cross section we obtain a good agreement with the data.

I would like to thank Dr.'s Bräuer, Fernandez, Shimizu and Dipl.-Phys. Straub with whom I was working on the results reported above.

REFERENCES

1. R. Arndt et al, Phys. Rev. C15, 1002 (1977) and R. Arndt, "Nucleon-Nucleon Interaction," 1977 Vancouver Conference AIP Conference Proceedings, No. 41, 117 (1977).
2. K. Holinde, Phys. Rep. 68, 191 (1981).
3. D.A. Liberman, Phys. Rev. D16, 1542 (1977).
4. C. E. De Tar, Phys. Rev. D17, 323 (1978).
5. M. Harvey, Nucl. Phys. A352, 301 and 326 (1980).
6. V. G. Neudatchin, I. T. Obukhovsky, V. I. Kukulin, N. F. Golanova, Phys. Rev. C11, 128 (1975).
7. V. G. Neudatchin, Yu. F. Smirnov, R. Tamagaki, Progr. Theor. Phys. 58, 1072 (1977).
8. A. Faessler, F. Fernandez, G. Lübeck, K. Shimizu, Phys. Lett. B112, 555 (1983).
9. M. Oka, K. Yazaki, Progr. Theor. Phys. 66, 551 and 572 (1981).
10. A. Faessler, G. Fernandez, G. Lübeck, K. Shimizu, Phys. Lett. B112, 201 (1982).
11. A. De Rujula, H. Georgi, S. L. Glashow, Phys. Rev. D12, 147 (1975).
12. A. Faessler, F.Fernandez, Phys. Lett. B124, 145 (1983).
13. K. Holinde, R. Machleidt, Nucl. Phys. A256, 479 (1976).
14. E. G. Lübeck, Diplomarbeit, Universität Tübingen, February 1982.
15. See reference 1, p. 117.

16. G. Strobel, K. Bräuer, A. Faessler, Nucl. Phys. A347, 605 (1985).
S. Furui, S.B. Khadkikar, A. Faessler, Nucl. Phys. A437, 619 (1985).
17. K. Bräuer, A. Faessler, F. Fernandez, K. Shimizu, Z. Phys. A320, 609 (1985).
18. A. Faessler, U. Straub, Tübingen preprint 1986 (to be published).
19. F. Fernandez, E. Oset, Salamanca preprint 04/85, to be published in Nucl. Phys.
20 A. Faessler, F. Fernandez, K. Bräuer, S. Kuyucak, K. Shimizu, Tübingen preprint, 1985.
21. G. Alexander et al., Phys. Rev. 173, 1452 (1968).
22. B. Sechi-Zorn et al., Phys. Rev. 175, 1735 (1968).
23. J.A. Kadyk et al., Nucl. Phys. B27, 13 (1971).
24. G. Spitz, E.W. Schmid, Few body systems 1, 37 (1986).

NUCLEAR MATTER AT HIGH DENSITIES AND TEMPERATURE

Amand Faessler

Institut für Theoretische Physik
Universität Tübingen
Auf der Morgenstelle 14
D-7400 Tübingen, West Germany

ABSTRACT

The properties of nuclear matter for high densities and high temperatures are discussed. We concentrate especially on the possibility of pion condensation, Lee-Wick condensation and quark matter. We show that the same diagrams which lead to pion condensation provide a shielding aginst pion condensation if the diagrams coming from antisymmetrization are included up to the same order. While the RPA bubble series leads to pion condensation the bubbles inside the bubbles screen against it. The same is true for the Lee-Wick condensation which is a polarization of nuclear matter with the quantum numbers (scalar and isoscalar) of the σ-meson. Finally we discuss quark matter at high densities and high temperatures and explore experimental possibilities to detect it in relativistic heavy-ion collisions.

INTRODUCTION

Nuclear matter is defined as infinitely extended having at each position in three-dimensional space the same density for the protons and the neutrons. In momentum space it is a Fermi sphere with the radius of the Fermi momentum k_F. All the states inside the sphere are occupied and the states outside are empty. Such an agglomeration of matter does naturally nowhere exist. Nevertheless nuclear matter plays a decisive role in testing different nucleon-nucleon interactions and especially in studying different many-body approaches for solving nuclear structure problems. The reason for this important role of nuclear matter is the fact that we believe that we have quasi-experimental data on nuclear matter although it does not exist in the form described above.

The first experimental information comes from the Weizsäcker mass formula.

$$B = a_v A - a_s A^{2/3} - a_c Z^2/A^{1/3} - a_A (Z-N)^2/(4A) + \text{Pairing} \tag{1}$$

The binding energy B has a volume term a_V proportional to the mass number

A of the nucleus. This binding energy is then reduced by the fact that some nucleons are sitting at the surface and have fewer neighbours than the nucleons inside the nucleus (a_S). In addition one obtains a reduction of the binding energy from the Coulomb repulsion and the asymmetry energy. Pairing effects distinguish between even-even, even-odd and odd-odd mass nuclei. In infinite nuclear matter one expects only a contribution from the volume term since one is neglecting the Coulomb repulsion between the protons. This gives a total energy per nucleon in infinite nuclear matter.

$$E/A = -a_v = -16\pm 1 \text{ [MeV]} \tag{2}$$

We obtain a sedond information on infinite nuclear matter by assuming that the roughly constant denstiy which we find in the centre of heavy nuclei is also the saturation density in nuclear matter. From the equivalent radius of a homogeneous charge distribution $R=1.12\ A^{1/3}$ fm one derives for the saturation density in nuclear matter

$$\rho_0 = 0.17 \quad \text{nucleons/fm}^3 \tag{3}$$

The third "experimental" information on infinite nuclear matter can be obtained by the energy of the breathing mode 0^+ state in nuclei[1,2]. The extraction of the incompressibility of infinite nuclear matter from the breathing mode in a finite nucleus yields a value which was thought to be surprisingly low[1,2].

$$K = 9\ \rho_0^2 \left[\frac{1}{A}\frac{\partial^2 E(\rho)}{\partial\rho^2}\right]_{\rho_0} = 220\pm 30 \text{ MeV} \tag{4}$$

Calculations of supernova explosions seem to suggest even lower values although these indications have to be taken with care[3].

Fig. 1 shows the total energy per nucleon in infinite nuclear matter qualitatively as a function of the density ρ in units of the saturation density ρ_0 and as a function of the temperature T. The three quasi-experimental information listed in eq. (2), (3) and (4) are reflected in Fig. 1 by the fact that the minimum is reached at $E/A = -16$[MeV] for the value $\rho/\rho_0=1$. The incompressibility given in eq. (4) requests a special curvature in the minimum of the total energy per nucleon E/A.

The purpose of this lecture is not to study nuclear matter under regular conditions as we have it in the centre of heavy nuclei but to study it under extreme conditions that means, at high densities and at high temperatures. Fig. 2 shows the qualitative phase diagram of nuclear matter as a function of the density and the temperature. In this lecture we shall especially discuss the possibility of (i) pion condensation, (ii) Lee-Wick (σ)-condensation and (iii) quark matter.

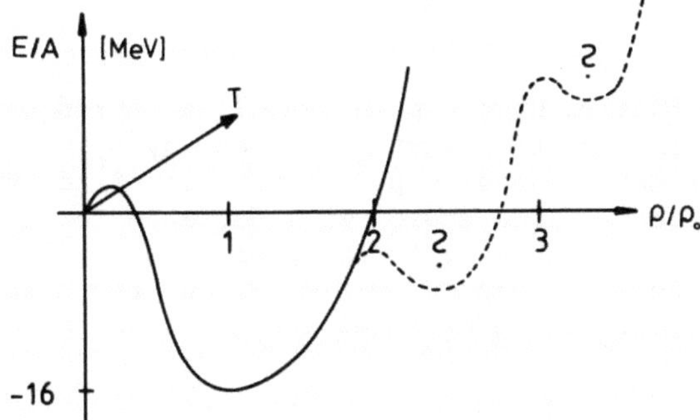

Fig. 1. The total energy per nucleon E/A [MeV] qualitatively indicated as a function of the density ρ in units of the saturation density $\rho_O = 0.17\ fm^{-3}$. A third axis is indicated for the temperature T to remember that the total energy per nucleon E/A is changing with the temperature T.

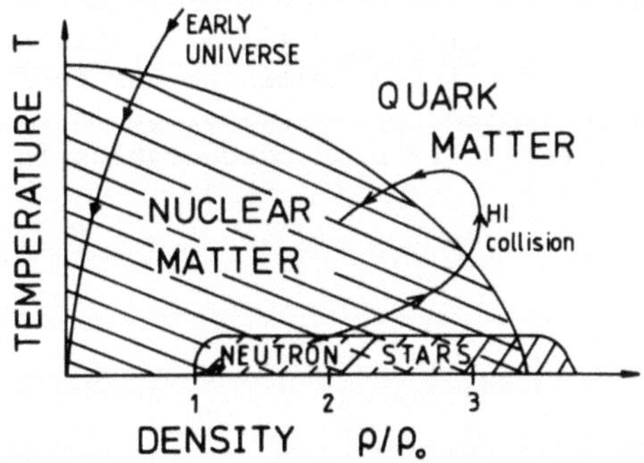

Fig. 2. Qualitative phase diagram of nuclear matter as a function of the density ρ in units of $\rho_O = 0.17\ fm^{-3}$ and the temperature T. The trajectory coming from the upper right part shows qualitatively the development of the early universe. The large hatched area indicates the region of nuclear matter. At high temperatures and high densities one expects a phase transition to quark matter. This phase transition could be reached at relatively low temperatures but high densities in large neutron stars or in the laboratory by a relativistic collision between two heavy nuclei (HI collision).

PION CONDENSATION

Let us assume that we have a pion travelling through space with momentum k_π. We now ask the question what is the energy E_π of this pion as a function of the nuclear matter density ρ (see Fig. 3). Since the nucleon-pion interaction is attractive, the energy will decrease for a given pion momentum k_π with increasing density ρ. If the pion gets more bound than its rest mass at a critical density (ρ/ρ_0=2 in Fig. 3) one can create as many $\pi^+\pi^-$ and π^0's without needing any additional energy. This phenomenon is called pion condensation.

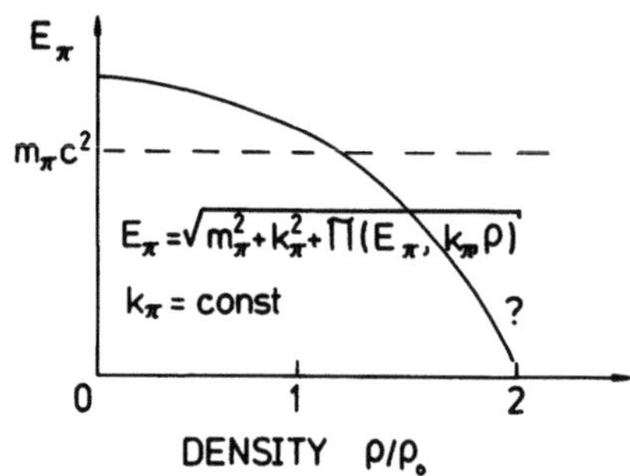

Fig. 3. Pion energy as a function of the nuclear matter density ρ in units of the saturation density ρ_0=0.17 fm^{-3}. This qualitative sketch is given for a constant pion momentum and indicates pion condensation for twice nuclear matter density. In this lecture we investigate the question if pion condensation is indeed a realistic possibility.

In the non-relativistic reduction the pion-nucleon interaction has the form:

$$H_{\pi NN} = \ldots \pi \ldots = \ldots \pi$$

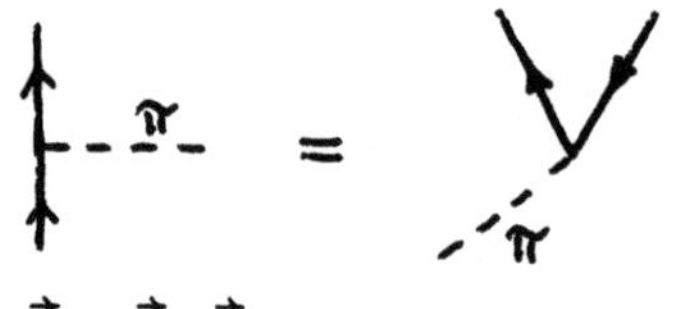

$$\propto \psi^+_{N'} \, (\vec{k}_\pi \cdot \vec{\sigma}_N)(\vec{\tau}_N \cdot \vec{\phi}_\pi)\psi_N \qquad (5)$$

$$\Longrightarrow \psi^+_{N'} \, \sigma_z \tau_z \frac{\partial}{\partial z} \phi_{\pi^0} \, \psi_N$$

if we imagine a standing wave of a π^0 in nuclear matter one has a sinus or cosinus in nuclear matter. The interaction (5) shows that nucleons and pions interact near the nodes of the pion wave function where the derivative $(\partial/\partial z)\phi_{\pi^0}$ is largest. If one assumes that protons with spin-up are attracted to one node one realizes immediately that at the same node also neutrons with spin-down are attracted due to the $\sigma_z\tau_z$ structure of the interaction. At the next node the sign of the derivative of the pion wave function has changed and protons with spin-down and neutrons with spin-up are attracted. Thus pion condensation would lead to a spin-isospin polarization of nuclear matter in form of a regular lattice.

The possible behaviour of the pion energy in nuclear matter now for a constant density ρ=const. but as a function of the pion momentum k_π is indicated in Fig. 4.

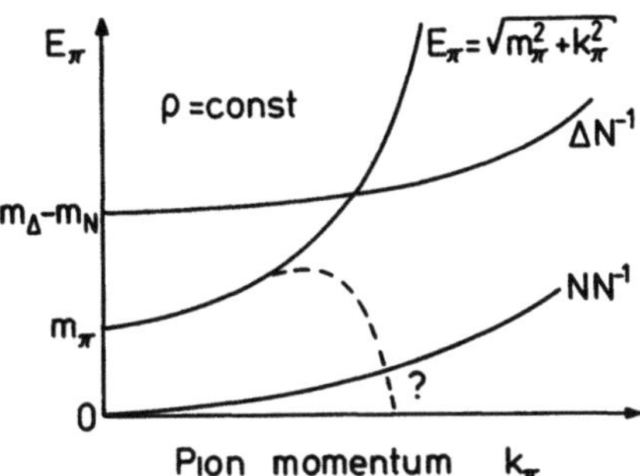

Fig. 4. Pion energy in nuclear matter for the constant density ρ as a function of the pion momentum k_π. The solid line E_π indicates the free pion energy without interaction with nuclear matter. The line labelled by NN^{-1} shows qualitatively the centre of the ridge for the nucleon-particle - nucleon-hole excitations. The curve described with ΔN^{-1} corresponds to the excitation of nuclear matter by a Δ-particle and a nucleon-hole. At zero momentum k_π=0 the excitation energy of the Δ-particle - nucleon-hole excitation is m_Δ-m_N=1232-938=294 MeV. The dashed curve indicates qualitatively a possible change of the pion energy E_π due to the interaction with the nucleon-particle - nucleon-hole NN^{-1} and ΔN^{-1} excitations of nuclear matter. If this dashed curve reaches at some density ρ and at some momentum k_π the value E_π=0 one speaks of pion condensation.

The interaction of the pion with nuclear matter is described by diagrams given in Fig. 5. Fig. 5 shows on the left side a typical RPA diagram for the propagation of a pion in nuclear matter. They lead to pion condensation[6] at about twice nuclear matter density and the Fermi momentum k_F=280 MeV/c. If one includes the antisymmetrization in the same order one finds the same collective RPA diagrams inside bubbles. These exchange diagrams shield with the same strength against pion condensation as the diagram of the left hand side of Fig. 5 favours pion condensation. We succeeded[9] in summing up all diagrams with bubbles in bubbles in bubbles up to infinite order. The calculation shows no indication for pion condensation. In these calculations we included also the self energy of the particle and the hole lines which yield modifications of the single-particle energies. One should stress that this calculation is not performed in the Landau limit where one can not expect anyway to get pion condensation since in the Landau limit one has $k_{\pi}=0$ in some of the diagrams and in that limit the pion-nucleon interaction in the p-wave is anyway zero.

A review with more technical details about the screening of pion condensation has been published by Müther[10] in Progress in Particle and Nuclear Physics.

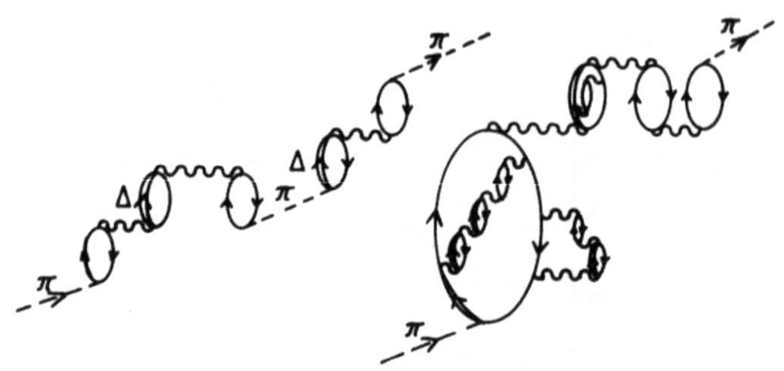

Fig. 5. The left diagram shows the propagation of a pion in nuclear matter allowing for RPA NN^{-1} and ΔN^{-1} nuclear matter excitations. Such diagrams lead to pion condensation[4,5] near the Fermi momentum[6] k_F=280 MeV/c. But if one includes in the same order also the antisymmetrized diagrams shown on the right[7,8,9] one obtains no pion condensation.

LEE-WICK CONDENSATION

Lee and Wick have proposed that the corresponding effect in nuclear matter as pion condensation can happen if a σ-meson of scalar and isoscalar character travels through nuclear matter. Indeed a calculation including only RPA type diagrams (see Fig. 6) shows that this condensation already happens at densities below normal nuclear matter densities. If one includes the antisymmetrized diagrams of the same order which yield bubbles in bubbles as shown in Fig. 7, the Lee-Wick condensation disappears completely. Thus we can say that Lee-Wick condensation is shielding against Lee-Wick condensation.

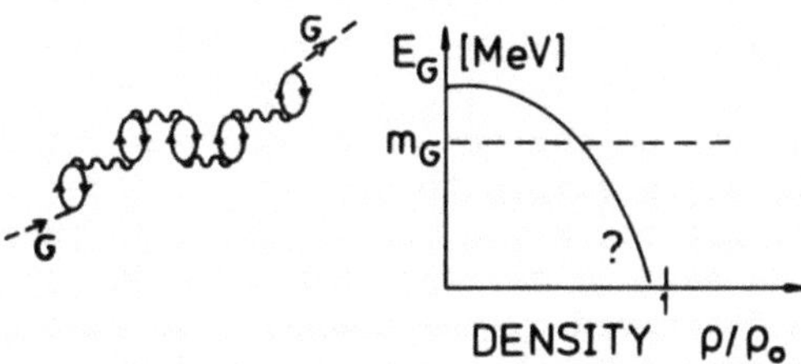

Fig. 6. The left side shows a RPA diagram for the propagation of a σ-meson in nuclear matter. Due to the isoscalar nature of the σ-meson only nucleon-particle - nucleon-hole excitations are possible. The right hand side shows the energy of the σ-meson as a function of the nuclear density ρ in units of the saturation density ρ_0=0.17 fm^{-3}. One finds already a total σ-meson energy E_σ=0 for densities below normal nuclear matter saturation density. The situation is totally changed if one includes the exchange diagrams (see Fig. 7).

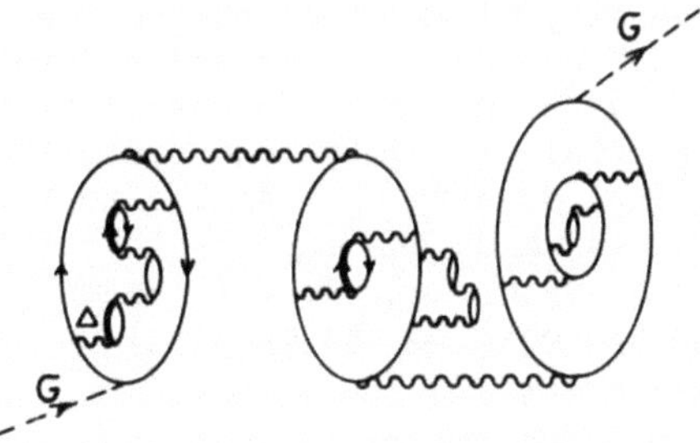

Fig. 7. Antisymmetrization of the RPA diagram shown in Fig. 6 and the inclusion of the single-particle self energy leads to diagrams as the one shown in this figure. In the bubbles inside bubbles we can have again Δ-resonances as for the diagrams in pion condensation. These bubbles in bubbles shield against the Lee-Wick condensation (σ)-condensation. We included also in our calculations[9] the single-particle self energy in the Landau approximation. The non-linear system of integral equations[9] which has to be solved to describe the propagation of a σ-meson in nuclear matter has been calculated without recourse to the Landau approximation[10].

QUARK MATTER

If one takes a piece of nuclear matter and presses it together so that the vacuum between the nucleons disappears one expects that the bags of the nucleon start to overlap and the quarks and the gluons inside the bags can move freely. In this way one obtains a phase transition between nuclear matter and the quark-gluon plasma (quark matter). The critical density for the phase transition must be the one which one has inside a nucleon. Thus on the first sight it seems simple to estimate the density where one expects the phase transition. But one would expect that the quark content of the nucleon wave functions have to overlap. Since we do not know for sure which piece of the nucleon described by the charge root mean square radius $\langle r^2\rangle^{1/2}=0.83$ fm is occupied by the quark bag and which is occupied by the surrounding meson cloud one has difficulties to estimate the critical density ρ_c.

$$\rho_c = \frac{3}{4\pi} R_{Nq}^{-3} = 0.24\, R_{Nq}^{-3} \tag{6}$$

Here R_{Nq} is the bag radius of the quark content of the nucleon wave function. In the MIT bag model this value is about 1 fm and thus one calculates easily for the critical density of nucleons per fm^3 a value of $\rho_c=1.4\ \rho_0$ ($\rho_0=0.17\ fm^{-3}$ saturation density of normal nuclear matter). In the Stony Brook bag the bag radius has a value of the order of 0.4 fm. This corresponds to a critical density of 22 times the nuclear saturation density ρ_0. In the quark model which we use in Tübingen[12,13] in which we obtain a stable nucleon radius by including the pressure of the pion cloud with finite size pions, we obtain an equivalent radius R_{Nq} of about 0.7 fm. This corresponds to a critical density $\rho_c = 4\rho_0$. 4 to 5 times the normal nuclear matter density are values which one expects to obtain in the relativistic collision between two heavy nuclei. Thus it seems possible to study the phase transition from hadronic to the quark-gluon plasma in relativistic heavy ion collisions.

In heavy ion collisions one is not only compressing nuclear matter but one is also heating it up. If the temperature of nuclear matter reaches values which are comparable to the rest mass of the pion one creates these lightest mesons. Using the Boltzmann distribution one easily calculates a critical temperature at which the mesons fill the whole space and where one expects a phase transition to quark matter. At zero baryonic density this yields a value $T_c=2\ m_\pi=280$ MeV. Therefore we obtain easily for the phase diagram shown in Fig. 2 estimates for the critical density at zero temperature ρ_c and for the critical temperature at zero density T_c. This figure shows also qualitatively a trajectory which one could obtain in heavy ion collisions.

Now let us assume that we can reach in relativistic heavy ion collisions with about 50 GeV per nucleon in the laboratory such high densities and temperatures that we have a phase transition to the quark gluon plasma (quark matter). How could one then detect that one indeed had as a transient intermediate state the quark gluon plasma? Figure 8 shows[14] an artist view of an idea what can happen if two nuclei are colliding. This picture gives the extreme view of the participant-spectator model. One forms a hot fire ball which is expanding and emits particles which can be measured. The data are momenta, energies and compositions of these emitted particles. How could one extract fingerprints from these data for quark matter?

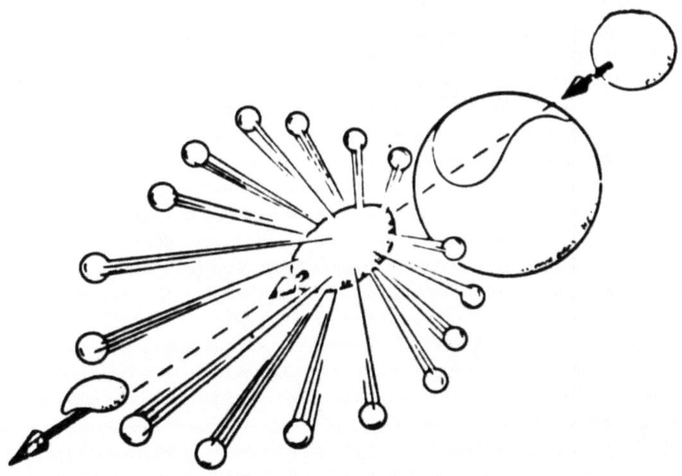

Fig. 8. Pictorial sketch of the extreme participant spectator model. The projectile is coming from the upper right corner. The target and the projectile spectator material is shown. The participants form a hot fireball from which particles are emitted. The fireball travels with an intermediate speed.

According to the extreme participant-spectator model the number of particles contained in the fireball should vary with the size of the two nuclei involved. Fig. 9 shows that the data are different. There the charged particle multiplicity is shown as a function of the total kinetic energy for the collision of different nuclei with a Uranium target. One sees there that the collision of ^{20}Ne and ^{40}Ar with the total kinetic energy of 40 GeV have the same charged particle multiplicity. Thus the correct picture must be more a fireball expanding into the spectator material while it is cooling down and giving its energy to more and more participants, thereby producing a higher multiplicity with increasing energy.

The picture evolving from the data shown in Fig. 9 is also supported by data of Nagamiya[15]. Fig. 10. shows the "Temperature" of K^+,p and π's after a Ne on Pb sollision with 2.1 GeV/A. The three temperatures are quite different. How can different thermometers using the exponential slope of the emitted particle spectra yield different temperatures? The pion has a very small mean-free path in nuclear matter (λ_π=0.5 fm). The mean-free path of a proton is larger ($\approx$ 2 fm). The K^+, which is the "weak" interacting particle of strong interactions, has a mean free path of $\approx$ 7 fm. The K^+, which has the weakest interaction, decouples from the thermal equilibrium first. Therefore it shows us the temperature of the earliest stage, when the fireball was still small and very hot. After the fireball has further expanded and cooled down the protons decouple from the thermal equilibrium. After further expansion and cooling the pions decouple. So the pions reflect the temperature at the latest stage of the fireball.

This interpretation is supported by measurements of the size of the fireball by the Hanbury-Brown-Twiss effect[15]. From $\pi\pi$-correlations one obtains the largest radius (R_π=3.5 fm); while the interaction radius from the proton-proton correlations is $R_p \approx 2.4$ fm in agreement with the above interpretation (see Fig. 10).

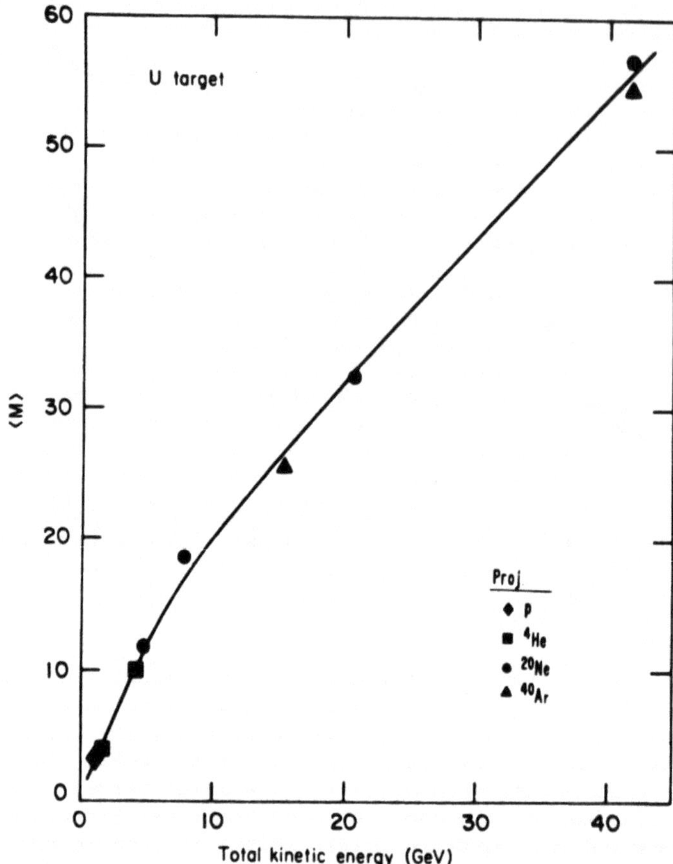

Fig. 9. Particle multiplicity as a function of the total kinetic energy of the projectile for reactions of different nuclei on a uranium target. Contrary to the extreme participant spectator model the charged particle number multiplicity is increasing for example for ^{20}Ne on U if the energy is raised from 5 to 40 GeV.

Let us go to higher energies, towards the quark-gluon plasma and let's discuss the boiling of hadronic matter into a quark-gluon plasma. Fig. 11 of Stöcker et al. shows the "temperature" for different heavy ion reactions as a function of the bombarding energy per nucleon. The different reactions yield a wide range of temperatures. If one focusses on the expected temperature for only protons and neutrons in the interaction region, then temperature has to go up as the energy input is increased (see dashed curve in Fig. 11). But it is obvious from the data, that nature is not following this line. The measured values cluster around a line which is much lower. Is there a limiting temperature as predicted by Hagedorn, due to producing heavier particles in the interaction region? When he assumes an exponentially growing spectrum of the elementary particles as a function of their mass and fits it to the hadron density between 1 and 2 GeV, he finds a limiting temperature of the order of 134 MeV. The data points in Fig. 11 are spread out and one cannot conclude from them if there is really a limiting temperature or if the temperature is increasing up to infinity.

But one anyway expects if one increases the energy input in heavy ion collisions more and more that nuclear matter boils at the limiting temperature into the quark-gluon plasma. Thus the limiting temperature should be identical with the phase transition from hadronic into quark matter[17,18].

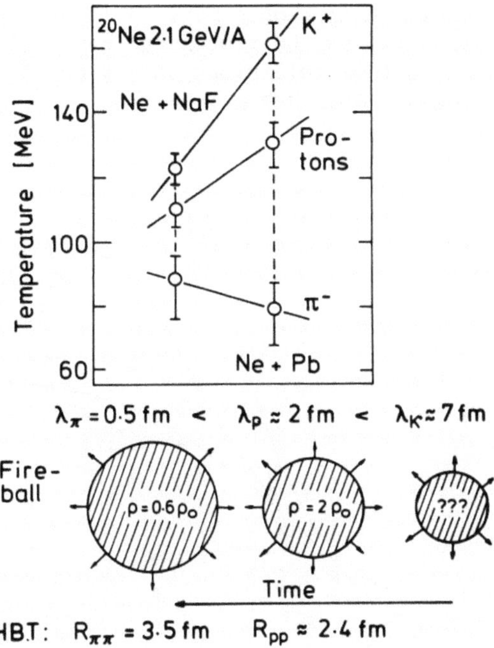

Fig. 10. Temperature determined by the particle-spectrum of K^+, protons and π^- emitted from the interaction area in a collision of ^{20}Ne with 2.1 GeV/A on NaF and Pb. One sèes that the pions have the lowest, the protons an intermediate and K^+-mesons the highest temperature. This can be explained by taking into account that pions have the smallest mean-free path (λ_π=0.5 fm) in nuclear matter while protons have an intermediate value (λ_p=2 fm) and K^+ are interacting most weakly with nuclear matter (λ_K=7 fm). Thus the kaons decouple earliest from the thermal equilibrium in the fireball and they show the early high temperature. After some expansion and cooling of the fireball the protons decouple and show an intermediate temperature. The pions decouple last and show the temperature in a state of the fireball. This interpretation is supported by the measurement of the size of the interaction area with the Hanbury-Brown-Twiss-effect. The interaction radius for the protons is about 2.4 and for the pions 3.5 fm.

Monte-Carlo calculations on the lattice[18] show that the phase transi tion between hadronic matter and the quark-gluon plasma should lie around T_C=200 MeV.

Another question is what are the experimental fingerprints of this quark-gluon plasma. These have been discussed in refs. 17 and 19. One idea might be that due to the large number of up and down quarks after the deconfinement of hadronic matter into quark matter makes it very costly to produce anti-up and anti-down quarks. Due to the low Fermi surface for strange matter it seems cheaper to produce pairs of a strange and an anti-strange quarks. Thus one expects to form in a relativistic nucleus-nucleus collision more anti-Ξ = $\bar{s}\bar{s}\bar{d}$ than anti-Λ = $\bar{s}\bar{u}\bar{d}$ and fewer antiprotons = $\bar{u}\bar{u}\bar{d}$. So the ratio of anti-Λ's to antiproton's should be larger than 1; the ratio anti-Ξ's to antiprotons should be even larger. Experimentally one has measured the $\bar{\Lambda}/\bar{p}$-ratio in the proton-nucleus collisions. It is smaller than 1. So in proton-nucleus collisions one is not forming a quark-gluon plasma. One needs relativistic heavy ions.

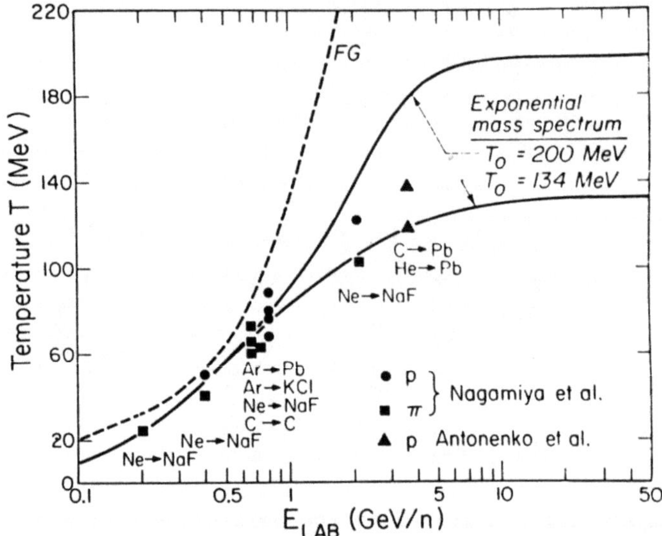

Fig. 11. The dashed line shows the temperature dependence expected from a free proton-neutron gas. The solid lines represent the temperature expected according to the Hagedorn model with an exponential mass spectrum. The limiting temperature of 134 MeV could be expected if one fits the exponential form of the hadronic state density to the known hadrons between 1 and 2 GeV. The measured data do not yet indicate a limiting temperature which should also be identical with the critical temperature for the phase transition from hadron to quark matter.

Another fingerprint for quark-gluon matter has been proposed by Gyulassy[17]. In an extreme relativistic heavy-ion collision a quark-gluon plasma tube is formed along the collision channel of the two heavy ions. The spatial position along the tube is proportional to the time, since one is on the light cone. If the quark-gluon plasma-tube starts to cool down it begins to condense into hadronic matter, as water-vapor in a rain cloud condenses into water drops. These drops of hadronic matter, into which the quark-gluon plasma condenses, hadronize into different particles which are emitted into the same angular direction. Thus one expects a peaked structure of the particles per angle or also per rapidity intervals since pseudo rapidity is connected with the angle in the laboratory frame.

$$y = \text{rapidity} = -\ln\left(\text{tg}\,\vartheta_{LAB}/2\right) \tag{7}$$

Such a peaked structure has been found in cosmic ray events[20]. If this highly speculative interpretation of the data would be correct we have already seen the fingerprints of the quark-gluon plasma. But one has definitely to be careful especially as long as one has no laboratory data. At the moment at CERN three different groups under the spokesmen of H.J. Specht, R. Stock and H. Gutbrod and at Brookhaven under the leadership of O. Hansen are preparing relativistic heavy ion collision experiments to look for the quark-gluon plasma. But even after the completion of these experiments one should not immediately expect a clear cut answer of "yes" or "no" but one has to be prepared that it will take some time to be sure if one indeed has seen the quark-gluon plasma.

CONCLUSIONS

In these lectures we were studying nuclear matter under extreme conditions. Extreme conditions mean high density ($\rho/\rho_O>2$) and at high temperatures (T>160 MeV). We were discussing (i) the possibility of pion and (ii) Lee-Wick condensation and of (iii) quark matter.

We saw that pion-condensation which one finds theoretically at twice nuclear matter density if one neglects antisymmetrization[6] is shielded off by pion-condensation if one includes the antisymmetrization[7,8,9]. The same thing happens for the Lee-Wick condensation.

The bigger prospect for finding it in the data has the quark-gluon plasma. If one compresses nuclear matter so that the vacuum between the nucleons disappears and the quark sectors of the nucleon start to overlap one expects that the quarks and the gluons inside the nucleons are deconfined and one obtains a free quark-gluon matter. Before one reaches the quark-gluon matter one should approach a limiting temperature due to the fact that most of the energy put into hadronic matter is then used to create heavier particles and not to heat up the nucleons. Monte-Carlo calculations of QCD on the lattice[18] show that this phase transition into the quark-gluon plasma should happen between a temperature of T_C=160 to 240 MeV.

At the end we discussed possible fingerprints for the quark-gluon plasma. It seems that none of the fingerprints considered alone is convincing. Thus one has to be prepared that one needs many data of relativistic heavy ion collisions to convince ourselves that indeed we have seen the quark-gluon plasma.

REFERENCES

1. A. Faessler, J. E. Galonska, K. Galonska, K. Goeke, S. A. Moszkowski, Nucl. Phys. A239, 477 (1975).
2. S. Krewald, J. E. Galonska, A. Faessler, Phys. Lett. 55B, 267 (1975).
3. S. Kahana, Proceedings of the 5. International School of Intermediate Nuclear Physics 1985 in Verona/Italy; see this Volume.
4. A. B. Migdal, Zh. Eksp. Teor. Fiz. 61, 2210 (1971). (Sov. Phys. - JETP 34, 1184 (1972)).
5. R. F. Sawyer, D. J. Scalapino, Phys. Rev. Lett. 29, 382 (1972). R. F. Sawyer, D. J. Scalapino, Phys. Rev. D7, 953 (1973).
6. G. E. Brown, W. Weise, Phys. Rep. Rep. 27C, 1 (1976).
7. W. H. Dickhoff, A. Faessler, J. Meyer-ter-Vehn, H. Müther, Phys. Rev. C23, 1154 (1981).
8. W. H. Dickhoff, A. Faessler, J. Meyer-ter-Vehn, H. Müther, Nucl. Phys. A368, 445 (1981).
9. W. H. Dickhoff, A. Faessler, H. Müther, S. S. Wu, Nucl. Phys. , A405, 534 (1983).
10. H. Müther, Progress in Particle and Nuclear Physics, 14, 123 (1985).
11. T. D. Lee, G. C. Wick, Phys. Rev. D9, 2291 (1974).
12. A. Faessler, F. Fernandez, Phys. Lett. 124B, 145 (1983).
13. G. Strobel, K. Bräuer, A. Faessler, F. Fernandez, Nucl. Phys. A437, 605 (1985).
14. A. Faessler, Nuc. Phys. A400, 565c (1983).
15. S. Nagamiya, Nucl. Phys. A400, 399c (1983).

16. H. Stöcker, A. A. Ogloblin, W. Greiner, Z. Phys. A303, 259 (1981).
17. M. Gyulassy, Nucl. Phys. A400, 31c (1983) and Progress in Particle and Nuclear Physics 15, 403 (1985).
18. H. Satz, Nucl. Phys. A400, 541c (1983).
19. H.J. Specht, Nucl. Phys. A400, 43c (1983).
20. J. Iwai et al., Nuovo Cimento 69A, 295 (1982).

THE NON-TOPOLOGICAL SOLITON BAG MODEL

Lawrence Wilets

Institute for Nuclear Theory
Department of Physics, FM-15
University of Washington
Seattle, Washington 98195, USA

1. MODELING QCD

Static, boundary condition bag models[1] have had considerable success in reproducing the spectra and properties of low-lying hadronic states involving light quarks. This is especially true of hybrid models, such as the Cloudy Bag Model [2] (CBM), which incorporate pions as an elementary field in order to restore PCAC.

A difficulty which all such models encounter, however, has been the handling of the *dynamics* of the confinement mechanism. The Friedberg-Lee[3,4,5] soliton model, which I discuss here, has the important feature that confinement is effected by a quantal scalar field. The model has been extended beyond the original classical interpretation to permit quantum-dynamical calculations. The effective Lagrangian contains the field and its time derivative, so that a Hamiltonian can be constructed which contains the field and its conjugate momentum. Methods familiar from nuclear theory can therefore be used to construct fully quantal states of the system. In doing so, one can employ the coherent (or, more generally, the single mode) state approximation for the scalar field part of the state vector. This is related to the mean field approximation, but is quantal.

The Lagrangian for the non-topological soliton model is the usual QCD Lagrangian supplemented by a non-linear scalar sigma field. The sigma field may be interpreted as representing the gluon condensate arising from the non-linear interactions of the color fields. Since the gluons are also represented in the Lagrangian, there is clearly double counting, in principle at least. This does not arise to the order of one gluon exchange, since the sigma field is color-singlet, and one gluon always carries color; two gluon exchange could involve two gluons in a color singlet state and hence could also be contained in the exchange of a sigma quantum. The Lagrangian also contains a color-dielectric function which is a function of the sigma field. The form of the dielectric function assures color confinement.

Parameters associated with the sigma field are adjusted to yield physical results for calculated hadronic properties. If we could solve the model Lagrangian exactly, we would obtain the exact QCD results when the model parameters are adjusted to decouple the sigma field from the system. Since the parameters of the sigma field are to be readjusted at every level of the calculation to fit key data, the results of calculations should converge to the exact QCD values. Of course, no one has yet been clever enough to calculate hadronic properties exactly in QCD; so long as the calculations based on the soliton model remain at a relatively simple level, the model should be regarded as phenomenological.

2. THE MODEL

The non-topological soliton model[4] is described by a covariant, gauge-invariant Lagrangian density

$$\mathcal{L} = \mathcal{L}_q + \mathcal{L}_\sigma + \mathcal{L}_{q,\sigma} + \mathcal{L}_G, \tag{2.1}$$

where the various components have the following meaning:

The Dirac term is

$$\mathcal{L}_q = \sum_f \overline{\psi}_f (i\gamma_\mu \partial^\mu - m_f)\psi_f, \tag{2.2a}$$

with f denoting flavor (the flavor index will be suppressed in what follows, the field operator ψ is taken to incorporate flavor componenets); the scalar field term is

$$\mathcal{L}_\sigma = \frac{1}{2}\partial_\mu \sigma\, \partial^\mu \sigma - U(\sigma); \tag{2.2b}$$

the fermion-scalar interaction term is

$$\mathcal{L}_{q,\sigma} = -g\overline{\psi}\sigma\psi; \tag{2.2c}$$

and the term involving gluons is

$$\mathcal{L}_G = -\tfrac{1}{4}\kappa(\sigma) F^c_{\mu\nu} F^{\mu\nu}_c - g_S \overline{\psi}\gamma_\mu \tfrac{1}{2}\lambda^c A^\mu_c \psi. \tag{2.2d}$$

The gauge field tensor is given by

$$F^a_{\mu\nu} = \partial_\mu A^a_\nu - \partial_\nu A^a_\mu + g_S f^a_{\ bc} A^b_\mu A^c_\nu, \tag{2.3}$$

and the quartic sigma self interaction is

$$U(\sigma) = \frac{a}{2}\sigma^2 + \frac{b}{3!}\sigma^3 + \frac{c}{4!}\sigma^4 + B. \tag{2.4}$$

The constants a, b and c are fixed within a range so that $U(\sigma)$ has two minima, one at $\sigma = 0$ and a lower minimum at $\sigma = \sigma_v$, the vacuum value. Then $U'(0) = U'(\sigma_v) = 0$. The constant B is determined to make $U(\sigma_v) = 0$ and has the same meaning as the MIT bag constant: namely, the energy density of the cavity.

The dielectric function $\kappa(\sigma)$ satisfies the following conditions in order to guarantee absolute color confinement:

$$\kappa(0) = 1, \qquad \kappa(\sigma_v) = 0$$

and must also satisfy

$$\kappa'(\sigma_v) = 0.$$

It is important to distinguish statements about the *model* and statements about the *approximations*. The model yields absolute color confinement and absolute confinement of quarks in color singlet clusters. It is free of the color Van der Waals problem.

Except for $\kappa(\sigma)$ in (2.2d), the model is renormalizable. Since it is already an effective model, it could be argued that there is no need for renormalizeability. This would be true if we were to limit calculations to the classical level, or to the level of tree graphs, for the sigma field. However, we are interested in dynamical calculations which require a quantal treatment of the sigma field. It is therefore convenient to employ a renormalizeable model, so that the quantum corrections are controlable. At the level of calculations performed. the non-renormalizeability arising from $\kappa(\sigma)$ has offered no difficulties. When such do arise, one could always introduce a cut-off to regularize any divergences.

For appropriate limits of the parameters, the model is sufficiently general so that the mean field solutions can be made to approach either the MIT or SLAC bag model solutions.

Note that if we set $g = 0$ and $\kappa = 1$, the sigma field decouples and Eqs. (2.2 a & d) yield the exact QCD Lagrangian. Since at every level of approximation the model parameters are to be renormalized to yield physical results. the model has the possibility of converging to the exact theory (as more exact solutions are obtained) with the decoupling of the σ field, and the recession of the cutoff in κ to infinity.

3. PARAMETERS OF THE MODEL

The five input parameters are:

$U(\sigma)$ contains a, b, c;

the σ-quark coupling constant g;

and the strong coupling constant $\alpha_S = g_S^2/4\pi$.

Some of the key data to be fit are:

(1) The mean baryon mass $\overline{m} = \frac{1}{2}(m_N + m_\Delta) = 1087$ MeV;

(2) The proton root-mean-square charge radius $< r_p^2 >^{\frac{1}{2}} = 0.83$ fm;

(3) The neutron mean-square charge radius $< r_n^2 > = -0.12$ fm^2;

(4) The proton magnetic moment $\mu_p = 2.7928456$ nm;

(5) The neutron magnetic moment $\mu_n = -1.9130418$;

(6) The ratio of the axial to vector coupling constants $g_A/g_V \simeq 1.26$;

(7) The delta-nucleon mass splitting $m_\Delta - m_N = 297$ MeV;

(8) The coefficient of the linear term in the heavy $q\overline{q}$ potential (from charmonium, bottomonium, *etc.*), called the string tension, $t \approx 925$ MeV/fm;[6]

(9) The chiral condensate.[7] $< \overline{\psi}\psi > = (225 \pm 25 \text{ MeV})^3$.

There are also very interesting quantities which have not yet been measured. We look to both experiment and lattice gauge calculations to determine their values. These include:

(10) The bag constant B, which is the volume energy of the cavity. In the MIT model it has a value of about 57 MeV/fm^3. It is also related to the deconfinement transition temperature, although the connection is clearly model-dependent.

(11) We identify the 0^{++} glueball with excitations of the sigma field according to $m_{GB}^2(0^{++}) = U''(\sigma_v)$.

4. THE MEAN FIELD APPROXIMATION[5]

As a preliminary to dynamical calculations, we consider first static solutions to the field equations. The simplest of these is the mean field approximation (MFA). If we neglect gluonic interactions, the Hamiltonian can be written

$$H = \int d^3r \mathcal{H}(\vec{r}),$$

$$\mathcal{H}(\vec{r}) = \psi^\dagger(\vec{r})\left(-i\vec{\alpha}\cdot\vec{\nabla} + g\beta\sigma(\vec{r})\right)\psi(\vec{r}) + \tfrac{1}{2}\pi(\vec{r})^2 + \tfrac{1}{2}|\vec{\nabla}\sigma(\vec{r})|^2 + U(\sigma) \tag{4.1}$$

We can always separate the sigma field according to

$$\sigma = \sigma_0(\vec{r}) + \sigma_1, \quad \dot{\sigma} = \pi = \pi_0(\vec{r}) + \pi_1, \tag{4.2}$$

where $\sigma_0(\vec{r})$ is a time-independent c-number and σ_1 is the quantum fluctuation operator. Because σ_0 is static, $\pi_0 = 0$. The operators satisfy the equal-time Bose commutation relations

$$[\pi(\vec{r},t),\sigma(\vec{r}',t)] = [\pi_1(\vec{r},t),\sigma_1(\vec{r}',t)] = -i\delta^3(\vec{r}-\vec{r}'). \tag{4.3}$$

Similarly, we can represent the quark field operators by

$$\psi = \sum_k c_k\psi_k(\vec{r}), \tag{4.4}$$

where the c_k satisfy the equal-time Fermi anti-commutation relations

$$[c_k(t), c^\dagger_{k'}(t)]_+ = \delta_{kk'}, \tag{4.5}$$

and the $\psi_k(\vec{r})$ are any complete and orthonormal set of spinor-color-flavor functions.

In the MFA, we consider a fixed occupation number of valence quarks (3 quarks for nucleons, a quark-antiquark pair for mesons). The Hamiltonian density can be written

$$\mathcal{H} = \mathcal{H}(\sigma_0) + \mathcal{H}'(\sigma_0)\sigma_1 + \tfrac{1}{2}\mathcal{H}''(\sigma_0)\sigma_1^2 + \tfrac{1}{6}U'''(\sigma_0)\sigma_1^3 + \tfrac{1}{24}c\sigma_1^4, \tag{4.7}$$

where $U''' = \mathcal{H}'''$ and $c = \mathcal{H}''''$. Extremization of the expectation value of H with respect to $\psi^\dagger_k$ (subject to the normalization constraint) and with respect to σ_0 leads to the coupled set of equations

$$(\vec{\alpha}\cdot\vec{p} + \beta g\sigma_0)\psi_k = \epsilon_k\psi_k\ , \tag{4.8a}$$

$$-\nabla^2\sigma_0 + U'(\sigma_0) + g\sum_{k(valence)}\psi^\dagger_k\beta\psi_k = 0. \tag{4.8b}$$

The first is a linear eigenvalue problem (if σ_0 is given): the second is a non-linear inhomogeneous equation (if the ψ_k are given). Satisfaction of the extremization conditions also eliminates terms proportional to σ_1 in the energy. In addition to employing the mean field approximation, the set of equations (4.8) neglects the negative energy sea of quarks, i.e. vacuum polarization. We will return to this matter later.

Various techniques are available for solving Eqs. (4.8)[5,8] or their non-spherical generalizations.[9] An example of a solution for the baryon ($s^3_{\frac{1}{2}}$) is shown in Fig. 1. Although the MFA implies single quark excitations with a continuum at $g\sigma_v$, the inclusion of OGE in the chromodielectric medium leads to a continuum only for three quark excitations (for baryons) beginning at $3g\sigma_v$, or about 2 GeV. Even that continuum is an artifice of the approximations.

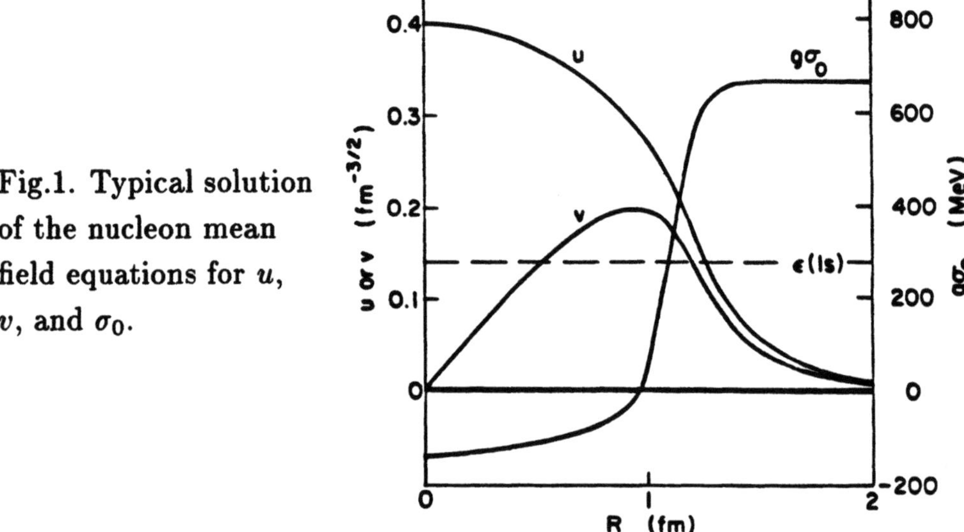

Fig.1. Typical solution of the nucleon mean field equations for u, v, and σ_0.

5. SMALL AMPLITUDE OSCILLATIONS.[5]

In terms of the quark states obtained from solving Eqs. (4.8), and the Hamiltonian density (4.7), the Hamiltonian can be written

$$H = E_0 + \sum_k{}' \epsilon_k \, c_k^\dagger c_k + \int d^3r \Big\{ \tfrac{1}{2}[\pi_1^2 + |\vec{\nabla}\sigma_1|^2 + U''(\sigma_0(\vec{r}))\sigma_1^2]$$

$$+ \tfrac{1}{6}U'''(\sigma_0(\vec{r}))\sigma_1^3 + \tfrac{1}{24}c\sigma_1^4 + g \sum_{k\ell}{}' \, \overline{\psi}_k(\vec{r})\sigma_1\psi_\ell(\vec{r})c_k^\dagger c_\ell \Big\}. \qquad (5.1)$$

where $E_0 = < \int d^3r \; \mathcal{H}(\sigma_0(\vec{r})) >$. The primes on the sums denote subtraction of the expectation of those same terms in the ground state.

Note the vanishing of terms linear in σ_1 in the ground state. The quantum part of the soliton field can be expanded in an orthonormal set:

$$\sigma_1 = \sum_n \left(\frac{1}{2\omega_n}\right)^{1/2} \left(a_n^\dagger s_n^* + a_n s_n\right) , \tag{5.2a}$$

$$\pi_1 = i \sum_n \left(\frac{\omega_n}{2}\right)^{1/2} \left(a_n^\dagger s_n^* - a_n s_n\right) . \tag{5.2b}$$

We now work in the Schrödinger picture. The commutation relations (4.3) are satisfied independently of the choice of the ω_m. However the Hamiltonian (5.1) simplifies if we *choose* $s_n(r)$ and ω_n to satisfy

$$\left(-\nabla^2 + U''(\sigma_0) - \omega_n^2\right) s_n(\vec{r}) = 0 ; \tag{5.3}$$

then

$$H = E_0 + \sum_k{}' \epsilon_k\, c_k^\dagger\, c_k + \sum_n \omega_n \left(a_n^\dagger a_n + \tfrac{1}{2}\right) + H_1 + H_3 + H_4 , \tag{5.4}$$

where $\mathcal{E}_{\sigma_0}$ is the integral of $\mathcal{H}_{si}(\sigma_0)$ and the other corresponding Hamiltonian densities are given by

$$\mathcal{H}_1 = g \sum_{k\ell n}{}' \overline{\psi}_k\, s_n\, \psi_\ell \left(\frac{1}{2\omega_n}\right)^{1/2} a_n + h.c. , \tag{5.5a}$$

$$\mathcal{H}_3 = \tfrac{1}{6}\,(b + c\sigma_0)\sigma_1^3 , \tag{5.5b}$$

$$\mathcal{H}_4 = \tfrac{1}{24} c\sigma_1^4 , \tag{5.5c}$$

where σ_1 in $\mathcal{H}_3$ and $\mathcal{H}_4$ is represented by (5.2a).

Eq. (5.3) defines normal modes for oscillations about the mean field solution. Since we consider $\sigma_0(r)$ to be spherically symmetric, we can set

$$s_n(\vec{r}) \equiv s_{\ell mn}(\vec{r}) = r^{-1}\, u_{\ell n}(r)\, Y_{\ell m}(\theta, \phi) , \tag{5.6}$$

whence (5.3) becomes

$$\left(-\frac{d^2}{dr^2} + \frac{\ell(\ell+1)}{r^2} + U''(\sigma_0(\vec{r})) - \omega_{\ell n}^2\right) u_{\ell n} = 0 . \tag{5.7}$$

$U''(\sigma_0(r))$ has a sharp dip in the vicinity of the bag surface, so that the lowest normal modes are surface modes. For the lowest state we find ω_{00} ranges between 300 and 600 MeV. depending on the parameter set. The quark (ϵ) and soliton (ω) spectra are displayed *schematically* in Fig. 2.

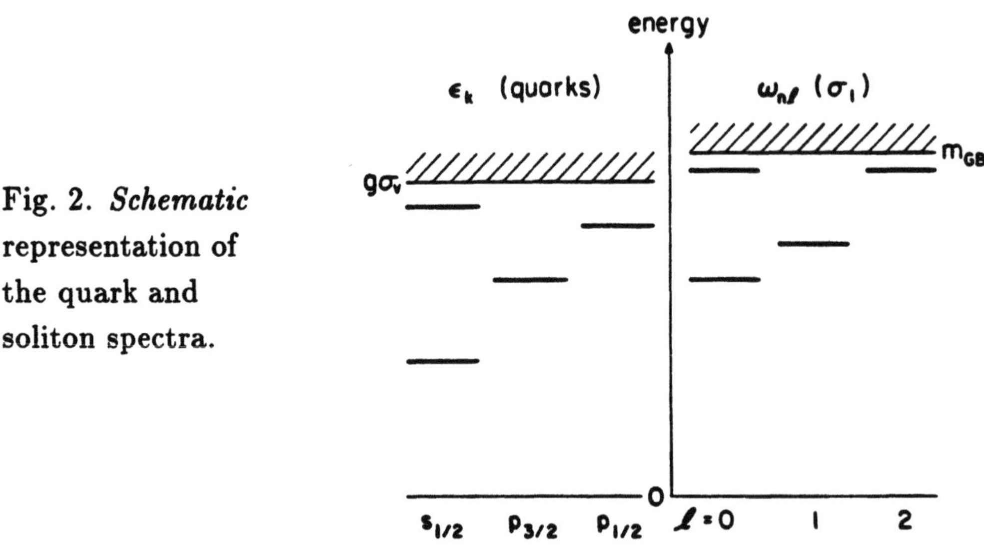

Fig. 2. *Schematic* representation of the quark and soliton spectra.

The soliton-quark interaction term, H_1, couples the σ_1 excitations to quark particle-hole pairs. $q\bar{q}$ virtual excitations are interpreted as giving rise to a "meson" cloud surrounding the bag or, more specifically, the nucleon. I will return to the meson later.

Although we began with a covariant Lagrangian, the MFA destroys covariance by the selection of a preferred frame. Inclusion of the σ_1 part of the soliton field can restore covariance. For example, the localized MFA bag has $< \vec{P}^2 > \neq 0$. A state with $\vec{P} = 0$ is spread out over all space. Thus $< r^2 >$ will increase (ultimately to ∞) as the approximations are improved, and we are alerted to the fact that $< r^2 >$ is not a measure of nucleon size. I return to this in Section 7.

The terms H_3 and H_4 involve only σ_1 (not quark) operators and lead to a restructuring of the soliton spectra. For present purposes, we will neglect these terms and assume that their effects can be absorbed into the effective parameters. although there may be important physical effects in these terms.

6. QUANTUM ALTERNATIVES TO THE MEAN FIELD APPROXIMATION[10]

Although there is no loss in generality in the separation of the sigma operator into a c-number part and a quantum fluctuation, there are sometimes difficulties with the procedure: the operator σ_1 is defined with respect to $\sigma_0(\vec{r})$, and so are the a_n and $a_n^\dagger$ of Eq. (5.2). Matrix elements connecting states of different σ_0 are thus not easily calculated. For this reason, we consider an approach which avoids introduction of a spatially-dependent mean field. The closest quantum analogue of the mean field is the coherent state. A simple generalization of this is the general single mode.

6.1 The Coherent State[10,11]

We expand σ and π in a full set of modes,

$$\sigma = \sigma_v + \sum_n \left(\frac{1}{2\omega_n}\right)^{1/2} \left(a_n^\dagger s_n^* + a_n s_n\right) ,$$

$$\pi = i \sum_n \left(\frac{\omega_n}{2}\right)^{1/2} \left(a_n^\dagger s_n^* - a_n s_n\right) , \tag{6.1}$$

where the ω_n are as yet undetermined and the $\{s_n\}$ are a complete, orthonormal set. For example, they can be chosen to be plane wave. $\sigma(\vec{r},t)$ and $\pi(\vec{r},t)$ satisfy the usual equal time commutation relations if

$$\left[a_n(t),\ a_{n'}^\dagger(t)\right] = \delta_{nn'} ,$$

$$\left[a_n(t),\ a_{n'}(t)\right] = \left[a_n^\dagger(t),\ a_{n'}^\dagger(t)\right] = 0 \tag{6.2}$$

A "coherent" state (CS) in one mode, say $n = 0$, is obtained by the construction

$$|\lambda >= e^{\lambda a_0^\dagger}|0> , \tag{6.3}$$

where $a_n|0>= 0$, for all n. The coherent state is an eigenstate of the annihilation operator a_0:

$$a_0|\lambda >= \lambda|\lambda > \tag{6.4}$$

and

$$< \lambda|\lambda >= e^{|\lambda|^2} . \tag{6.5}$$

Multiple-mode coherent state vectors can be constructed by taking a product of exponentiated operators:

$$|\lambda_1\lambda_2 \cdots >= \prod_n e^{\lambda_n a_n^\dagger}|0>\equiv e^{\Sigma_n \lambda_n a_n^\dagger}|0>\equiv e^{\Lambda A_0^\dagger}|0> \tag{6.6}$$

Again one has

$$a_m|\lambda_1\lambda_2\cdots>=\lambda_m|\lambda_1\lambda_2\cdots> \tag{6.7}$$

However, (6.6) can be regarded as a single mode state by identifying

$$A_0^\dagger = \Lambda^{-1}\sum_n \lambda_n a_n^\dagger\,, \tag{6.8}$$

where

$$\Lambda = \sqrt{\sum_n |\lambda_n|^2}\,, \tag{6.9}$$

with one member of a transformed set of operators $A_k^\dagger$.

Consider now the normalized state

$$|f;cs>= e^{\Sigma_n\sqrt{\omega_n/2}\,f_n a_n^\dagger}|0>e^{-\frac{1}{2}\Sigma_n\omega_n|f_n|^2} \tag{6.10}$$

where the factor $(\omega_n/2)^{1/2}$ is chosen for convenience. Then

$$\begin{aligned}
<f;cs|\sigma|f;cs> &= \sigma_v + \frac{1}{2}\sum_n [f_n s_n(\vec{r}) + f_n^* s_n^*(\vec{r})] \equiv \sigma_0(\vec{r})\\
<f;cs|:\sigma^n:|f;cs> &= <f;cs|\sigma|f;cs>^n = \sigma_0^n(\vec{r})\\
<f;cs|\pi|f;cs> &= -\frac{i}{2}\sum_n \omega_n [f_n s_n(\vec{r}) - f_n^* s_n^*(\vec{r})] \equiv \pi_0(\vec{r})\\
<f;cs|:\pi^n:|f;cs> &= <f;cs|\pi|f;cs>^n = \pi_0^n(\vec{r})
\end{aligned} \tag{6.11}$$

Equations (6.11) look classical, but that is because of the normal ordering operation. In fact, there are quantum fluctuations, and, for example.

$$<f;cs|\sigma^2|f;cs>\neq<f;cs|\sigma|f;cs>^2, \tag{6.12}$$

since σ^2 is not normal ordered.

The Hamiltonian can be reorganized so that it is expressed in terms of normal-ordered operators only. If we choose the $s_n(\vec{r})$ to be plane waves, say $e^{i\vec{k}_n\cdot\vec{r}}/L^{3/2}$, then with suitable counter terms the new $H \equiv: H:$ is of the form (4.1) except that all operators are normal ordered. The normal ordering does depend upon the choice of ω_n. *In what follows we will take H to be normal ordered with respect to a chosen set of* ω_n.

In the spirit of the MFA, we write the total state vector

$$|\Psi>= \prod_{k-valence} c_k^\dagger\, e^{\frac{1}{2}\Sigma_n\sqrt{\omega_n}f_n a_n^\dagger}|0>e^{-\frac{1}{2}\Sigma_n\omega_n|f|^2}\,. \tag{6.13}$$

For the static bag case, it is reasonable to take $\sum_n f_n s_n$ real whence $< f; cs|\pi|f; cs >= 0$. Then

$$< \Psi|H|\Psi >= H(\sigma_0) \tag{6.14}$$

where $H(\sigma_0)$ is just Eq. (4.1) with $\sigma \to \sigma_0(\vec{r})$, a c-number. We are again led to the MFA equations. However, the state vector contains fluctuations of the σ-field, and this has consequences for recoil and center-of-mass corrections. Note that in this case the mode frequencies ω_n do not even appear.

6.2 The General Single Mode Approximation.[11]

One generalization of the coherent state is the general single mode (SM) approximation, which is essentially the Tamm-Dancoff or Tomanaga approximation. Rather than the exponential functional form for the creation operators given in (6.10), one may assume a general function of the single mode:

$$|f; sm >= F\left(\Lambda^{-1}\sum_n \sqrt{\omega_n/2} f_n a_n^\dagger\right)|0> \ . \tag{6.15}$$

This has been studied by Lübeck *et al.*[11] where the function F was expanded in a power series:

$$F(A^\dagger) = N\sum_m \frac{F_m}{m!}(A^\dagger)^m \ . \tag{6.16}$$

Here $A^\dagger$ is the normalized argument of F appearing in (6.15). For the cases tested, there were some differences from the coherent state, but the F_m varied slowly over a fair range of m-values and the series converged reasonably rapidly. The algebra for evaluating matrix elements is quite tractable.

7. RECOIL: PROJECTION AND BOOST

A composite structure localized in a particular reference frame is a wave packet containing a distribution of momentum components. This is the case for a bag described in the MFA. Denoting the total linear momentum operator by

$$\vec{P} = -\frac{1}{2}\int d^3r \ \left[i\psi^\dagger \overleftrightarrow{\nabla} \psi + \{\pi, \vec{\nabla}\sigma\}\right] \tag{7.1}$$

and the state vector for a localized quark-soliton bag state by $|B>$, then

$$<B|\vec{P}|B>= 0, \text{ but } <B|P^2|B> \; > \; 0.$$

Thus localized states contain spurious center-of-mass energy and center-of-mass fluctuational motion. The underlying translational invariance shows up as spurious states in the excitation spectrum.

In a non-relativistic theory, it is straightforward to construct the center-of-mass coordinate, which is just the mean of the quark positions if only (equal-mass) quarks are present, although it is complex in practice to isolate the center-of-mass motion in a many-body system. In a relativistic field theory, the corresponding center-of-energy *operator* is not a tractable object.[12,13] An *approximate* center-of-energy operator is[12,14]

$$\vec{R} = \int d^3r \, \vec{r} \, \mathcal{H}(\vec{r}) / < H > , \tag{7.2}$$

which can be shown to satisfy the commutation relations

$$[R_i, P_j] = i\delta_{ij} \; H / < H > \tag{7.3}$$

and

$$\frac{d\vec{R}}{dt} = i[H, \vec{R}] = \vec{P} / < H > ; \tag{7.4}$$

unfortunately, the components of R do not commute among themselves, but give rather

$$[R_i, R_j] = \epsilon_{ijk} M_k / < H >^2 , \tag{7.5}$$

where M_k is the angular momentum operator. (7.3) is the condition that $\vec{R}$ be conjugate to $\vec{P}$, which is satisfied when acting on an eigenstate or in its expectation value. (7.4) is a "very pleasant" and non-trivial result. Recall that for the Dirac position operator $\vec{r}_D = \int d^3r \psi^\dagger \vec{r} \psi$ we have (for a local potential)

$$\frac{d\vec{r}_D}{dt} = \int d^3r \; \psi^\dagger \vec{\alpha} \psi , \tag{7.6}$$

which has eigenvalues equal in magnitude to the speed of light. Wave packets, however, satisfy (7.4), and so does the position operator in the Foldy-Wouthuysen representation.[15] The operator $\vec{R}$ is closely related to the F-W position operator.

7.1 Momentum Projection[11]

The soliton coherent state described in the previous section is localized and so has no definite momentum. To construct an eigenstate of zero momentum from it, we use the projection method of Peierls and Yoccoz[16]. The zero-momentum projected state is

$$|\vec{P}=0> = \int d^3X\,|\vec{X}>, \tag{7.7}$$

where $|\vec{X}>$ is a bag state localized about the point $\vec{X}$, and has the form

$$|\vec{X}> = \exp\left[\sum \sqrt{\tfrac{1}{2}\omega_k}\, f_{\vec{k}}(\vec{X})\, a^{\dagger}_{\vec{k}}\right]\, c^{\dagger}_1(\vec{X})\, c^{\dagger}_2(\vec{X})\, c^{\dagger}_3(\vec{X})|0>, \tag{7.8}$$

where $f_{\vec{k}}(\vec{X})$ is the Fourier transform of the σ field distribution centered at $\vec{X}$,

$$f_{\vec{k}}(\vec{X}) = e^{-i\vec{k}\cdot\vec{X}} f_{\vec{k}}(0), \tag{7.9}$$

and the operators $b^{\dagger}_i(\vec{X})$ create quark states also centered at $\vec{X}$,

$$\psi_i(\vec{r},\vec{X}) = \psi_i(\vec{r}-\vec{X}). \tag{7.10}$$

The generalization to the single mode approximation for the sigma part of the state vector is straightforward.

Since the sigma field is expanded in terms of plane waves, the operators are translationally invariant. The quark basis we have used is not translationally invariant but we neglect differences from unity of the overlaps of quark vacua centered on different points. This procedure cannot be exact but no obvious problems have arisen as yet. If it does cause trouble, one can always go to a plane-wave basis and work with a Dirac Hamiltonian projected onto positive-energy plane waves. A better procedure would be to calculate the distortion of the Dirac sea explicitly, see Sec. 13.

7.2 Boost[11]

To calculate other nucleon properties (e.g. electro-magnetic form factors) we need states with non-zero momentum. These could be constructed using finite-momentum projection. However, such a procedure has well-known difficulties: namely, the various states of good momentum are not related to each other by the appropriate Lorentz transformation.[16,17] Instead we operate on the zero-momentum state with the appropriate Lorentz boost operator to produce an approximate four-momentum eigenstate. The boosted state is defined by

$$|\vec{y}> = e^{i\vec{y}\cdot\vec{K}}|\vec{P}=0>, \tag{7.11}$$

$$\vec{K} = \int d^3r\, \vec{r}\, \mathcal{H}(\vec{r}), \tag{7.12}$$

and $\mathcal{H}(\vec{r})$ is the Hamiltonian density. The quantity $\vec{y}$ is the rapidity in the direction of the velocity:

$$\vec{y} = \tfrac{1}{2}\hat{v} \ln\left(\frac{1+v}{1-v}\right). \tag{7.13}$$

Unless the state $|\vec{P} = 0>$ is an exact energy eigenstate, the state $|\vec{y}>$ is not an exact momentum eigenstate but it does lead to expectation values of energy and momentum with the correct Lorentz transformation properties. This can be seen as follows:

The operator $\vec{K}$ obeys the following commutation rules [*cf.* Eqs. (7.3) and (7.4)],

$$\begin{aligned} [\vec{K}, H] &= i\vec{P}, \\ [K_i, P_j] &= i\delta_{ij} H. \end{aligned} \tag{7.14}$$

If we define

$$\begin{aligned} E(y) &= <\vec{y}|H|\vec{y}>, \\ P_{\parallel}(y) &= <\vec{y}|\hat{v}\cdot\vec{P}|\vec{y}>, \end{aligned} \tag{7.15}$$

then from (7.14) it follows that

$$\begin{aligned} \frac{dE}{dy} &= P_{\parallel}(y), \\ \frac{dP_{\parallel}}{dy} &= E(y). \end{aligned} \tag{7.16}$$

These differential equations have solutions (for the boundary conditions $E(0) = M$, $P_{\parallel}(0) = 0$):

$$\begin{aligned} E(y) &= M\cosh(y), \\ P_{\parallel}(y) &= M\sinh(y), \end{aligned} \tag{7.17}$$

where $M = E(0)$ is the expectation value of H in the $\vec{P} = 0$ state. The boosted state is more conveniently labelled by the expectation value of the momentum, $\vec{P} = \hat{v}M\sinh(y)$. In particular, we note that

$$E^2 = M^2 + P^2.$$

The use of the boost to calculate magnetic moments and form factors has been described by Lübeck[30].

7.3 Variation After Projection

It is well known in nuclear physics that energy variation before projection can be a dangerous procedure. In the present case, the expectation value of the energy depends upon the functions

$$u(r),\quad v(r),\quad \sigma_0(r)\ [\text{or } f_k],\quad F_m \text{ and } \omega_k,$$

where u and v are the upper and lower components of $\psi_0(\vec{r})$. The coefficients F_m appear only in the case of the general single mode. A full variation with respect to all of these functions appears to be prohibitive at present. Instead, the *forms* obtained from solving the mean field equations, now denoted by a tilde, were utilized, and scaling was introduced as follows:

$$\sigma_0(r) - \sigma_{vac} = \xi[\tilde{\sigma}_0(r/\lambda) - \sigma_{vac}] \tag{7.18a}$$

$$u(r) = \tilde{u}(r/\delta), \qquad v(r) = \gamma\tilde{v}(r/\delta). \tag{7.18b}$$

The normalization of ψ_0 must, of course, be readjusted. We have also tried an alternative parametrization of σ_0, using a Fermi function form,

$$\sigma_0(r) - \sigma_V = -\frac{\xi\,\sigma_V}{1+\exp\left[(r-r_0)/\mu\right]}. \tag{7.19}$$

This allows independent variations of the surface thickness and radius of σ_0.

As already noted the commutation relations (4.3) are satisfied for any set of ω_k. Another choice, say Ω_k, defines operators $A_{\vec{k}}$ which are linear combinations of $a_{\vec{k}}$ and $a^{\dagger}_{\vec{k}}$, and can be regarded as a Bogoliubov transformation on the creation and annihilation operators:

$$A_{\vec{k}} = \tfrac{1}{2}\left(\sqrt{\frac{\Omega_k}{\omega_k}}+\sqrt{\frac{\omega_k}{\Omega_k}}\right)a_{\vec{k}} + \tfrac{1}{2}\left(\sqrt{\frac{\Omega_k}{\omega_k}}-\sqrt{\frac{\omega_k}{\Omega_k}}\right)a^{\dagger}_{-\vec{k}}\,. \tag{7.20}$$

Such a transformation is well-defined as long as no Ω_k vanishes.

The corresponding "vacuum" state $|\Omega>$, defined by

$$A_{\vec{k}}|\Omega> = 0. \qquad \text{for all } \vec{k}. \tag{7.21}$$

is a Gaussian wave packet in functional space. Its principal axes are given by the plane-wave expansion functions, and its widths are $\Omega_{\vec{k}}^{-\frac{1}{2}}$. The special choice

$$\omega_k^2 = m^2 + k^2 \tag{7.22}$$

with

$$m^2 = m_{GB}^2 = U''(\sigma_v) \tag{7.23}$$

diagonalizes the Hamiltonian for small oscillations about the mean σ field in the absence of quarks and minimizes the energy of the corresponding vacuum.

The coherent state used to describe the nucleon is a displaced Gaussian wave packet of the same form as $|\Omega>$, but centered on the classical field configuration $\sigma_0(r)$. The general single-mode state allows for replacement of this Gaussian by a more general function for the chosen single mode.

A better starting point for the nucleon state would be to expand the σ field in distorted waves (see Sec.13). However, we have here committed ourselves to a plane-wave basis in order to facilitate projection. In choosing a coherent state with $\Omega_k \neq \omega_k$ to descibe the nucleon, we also change the vacuum for the nucleon, and produce an infinite shift in the vacuum contribution to the energy. Our procedure is to calculate only the (finite) differences in energies, and other properties, between the nucleon and the (unobservable) vacuum state, $|\Omega>$. In order to focus on the quantum fluctuations in the vicinity of the nucleon, we invoke the variational principle for the difference in energy between the nucleon and the vacuum. Since the energy of each eigenstate is stationary with respect to arbitrary variations, so also is the energy difference, at least for independent variations of the parameters used to describe both states; we constrain the Ω_k to be the same in both states — a somewhat dangerous procedure. To test whether this procedure is meaningful we utilize certain virial theorems, described next. As noted below, optimization with respect to the Ω_k leads to significant improvement in the satisfaction of the virial theorems.

7.4. Virial Theorems

The time derivatives of the expectation value of any time-independent operator vanish in a stationary eigenstate of the Hamiltonian:

$$\frac{d}{dt} < O > = i < [H, O] > = 0. \tag{7.24}$$

We have considered the following operators:

$$\int \psi^\dagger(\vec{r})\, \vec{r} \cdot \vec{p}\, \psi(\vec{r})\, d^3r, \quad \int \pi(\vec{r})\, d^3r, \quad \text{and} \quad \int :\sigma(\vec{r})\, \pi(\vec{r}): d^3r.$$

The commutators of these operators with respect to the (normal ordered) Hamiltonian lead to corresponding "virial theorems", which require the vanishing of

$$V_1 \equiv < \int d^3r\, \psi^\dagger(\vec{r})[\,\vec{\alpha}\cdot\vec{p} - g\,\beta\,\vec{r}\cdot\vec{\nabla}\sigma(\vec{r})\,]\psi(\vec{r}) > / <H>, \qquad (7.25a)$$

$$V_2 \equiv < \int d^3r\, [-\nabla^2\sigma(\vec{r}) + U'(\sigma(\vec{r})) + g\bar{\psi}(\vec{r})\psi(\vec{r})] >, \qquad (7.25b)$$

$$V_3 \equiv < \int d^3r : \left[-\pi(\vec{r})^2 + \sigma(\vec{r})\frac{\partial H}{\partial\sigma(\vec{r})}\right] :> / <H>. \qquad (7.25c)$$

(The factors $< H >^{-1}$ are introduced to render the V_i dimensionless.) The first is a generalization of the familiar virial theorem of non-relativistic quantum mechanics. In that case, it is equivalent to energy minimization with respect to the scale of length for the wave function. The second is the integral of the field equation for $-i\dot{\pi}$. The third is the σ-field analogue of the first. In the first two cases, commutation does not destroy normal ordering while in the last case it does. Eq.(7.25c) contains a normal-ordered piece plus divergent terms. However, with $\omega_k = \sqrt{m^2 + k^2}$, the divergent terms cancel leaving only the finite part.

Each of these quantities vanishes in the mean field approximation if one has obtained self consistent solutions. None of these quantities vanishes automatically in the projected state. The satisfaction of these quantities after variation is a necessary, but obviously not sufficient, condition for a good solution to the field equations.

7.5 Numerical results

Projection of a bag state removes spurious center-of-mass energy. In Fig. 3 is a comparison of the energy of a projected bag state with the energy otained by subtraction of the mean-square momentum, *i.e.* $(<H>^2 - <P^2>)^{1/2}$, for one family of parameters and varying c. Note that the two curves cross at $c \simeq 3,000$.

The best test we have for the quality of a wave function is the satisfaction of the virial theorems. For this, it was essential to include variation of the ω_k, which was effected by varying the m in Eq. (7.22) and including an exponential cut-off parameter. As can be seen in Fig. 4, the three virials cross zero *near* the same point and near the (very flat) minimum in $<H>$.

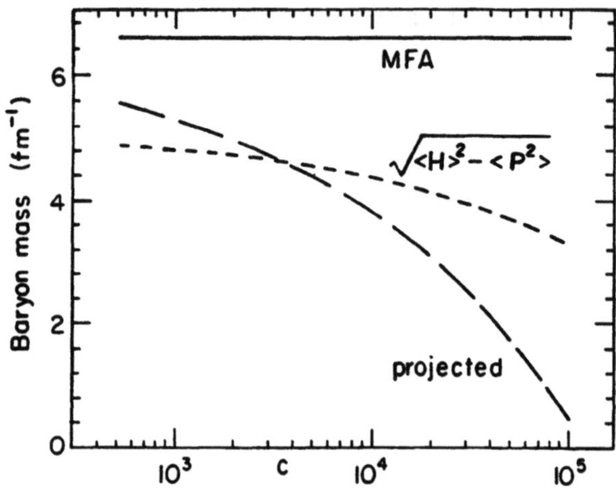

Fig. 3. Energy of a projected nucleon bag as a function of the model parameter c, compared with subtraction of the spurious center-of-mass momentum. [Ref. 11)]

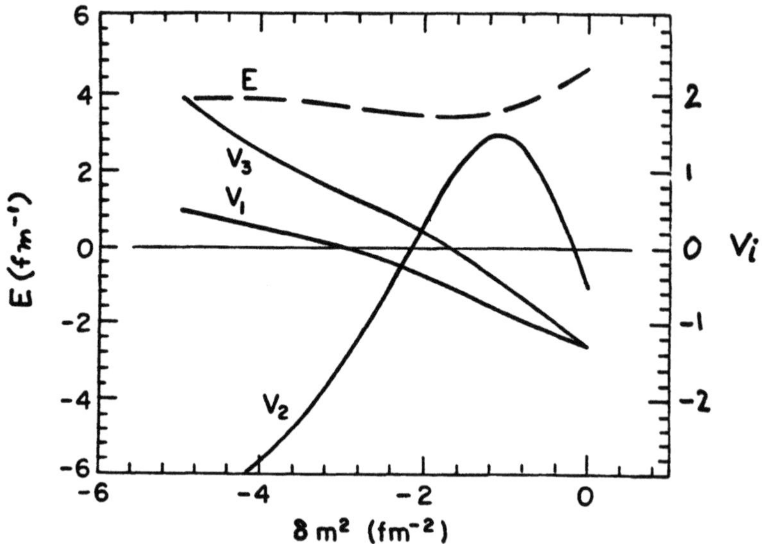

Fig. 4. Dependence of the nucleon projected energy and virial theorems on the parameter δm^2. [Ref. 11]

The actual change in energy due to projection and variation is large but not very interesting since the parameters are readjusted to fit key data, such as masses and sizes. What is interesting is the quantitive fit to other data compared with the mean field approximation. Lübeck *et al.*[11] found significant improvement in reproducing the values of the nucleon magnetic moments. the axial coupling g_A, and the pionic mass.

There are (at least) two methods for calculating the nucleon magnetic moment. One is the expectation value of the magnetic moment operator in the $\vec{P} = 0$ state,

$$\mu_s = \left\langle \int d^3r (\vec{r} \times \vec{j}(\vec{r}))_z \right\rangle_{m=\frac{1}{2}}. \tag{7.26a}$$

The other is obtained by boosting the rest state to a momentum $\pm\vec{q}/2$, and calculating the matrix element of the current linear in $\vec{q}$:

$$\mu_b = \frac{1}{2M} \left\langle \left(\vec{K} \times \vec{j}(0) \right)_{z,+} \right\rangle_{m=\frac{1}{2}}, \tag{7.26b}$$

where the subscript "+" indicates an averaging over both orderings of $\vec{K}$ and $\vec{j}(0)$. The values so obtained would agree if the state vectors were energy eigenstates. The degree of agreement is another measure of the quality of the wave functions. Lübeck[30] finds agreement to a few percent.

8. THE GENERATOR COORDINATE METHOD

The method of generator coordinates[18,17] (GCM) has been applied to large amplitude bag dynamics,[10] with particular application to N-N scattering,[8] π-N coupling,[19] and $N\overline{N}$ reactions.[20]

Consider a parameter or set of parameters $\{\alpha\}$ which describe the static configuration of a system of quarks and the soliton field and let $|\alpha. n>$ denote a set of basis states which is complete for any α. A method of obtaining these basis states is described briefly later. The GCM state vector is written

$$|\Psi> = \sum_n \int \varphi_n(\alpha) |\alpha, n> d\alpha. \tag{8.1}$$

Since the set is complete for each α, the expansion is overcomplete. In practice, this causes no problem since the sum is truncated to a small number of terms. In what follows, we consider only a single term and suppress n. In practical cases, several configurations may be required. The generalization is straightforward.

The weight function $\phi(\alpha)$ is obtained by extremizing the expectation value of the Hamiltonian

$$\int d\alpha \phi^*(\alpha) < \alpha|H|\alpha' > \phi(\alpha')\, d\alpha'$$

subject to the normalization constraint. Then

$$\int d\alpha' < \alpha|H - E|\alpha' > \phi(\alpha') = 0. \tag{8.2}$$

This is the basic Hill-Wheeler GCM integral equation for $\phi(\alpha)$. Depending upon whether the spectrum is discrete or continuous, it is either an eigenvalue or a scattering equation. Although we can work with the integral equation (which does, however, require regularization), it is instructive to consider the *approximate* differential equation which it satisfies. For a system which has well developed collective motion, we expect $< \alpha|H - E|\alpha' >$ to fall off rapidly as a function of $\alpha - \alpha'$. To utilize that property, it is convenient to introduce the mean and relative parameters

$$\overline{\alpha} = \tfrac{1}{2}(\alpha + \alpha'), \qquad \delta = \alpha - \alpha'. \tag{8.3}$$

Then

$$\begin{aligned} &< \Psi|H - E|\Psi > \\ &= \int d\overline{\alpha} \int d\delta \phi^*(\overline{\alpha} + \tfrac{1}{2}\delta) < \overline{\alpha} + \tfrac{1}{2}\delta|H - E|\overline{\alpha} - \tfrac{1}{2}\delta > \phi(\overline{\alpha} - \tfrac{1}{2}\delta) \\ &= \int d\overline{\alpha} \int d\delta \left[\phi^*(\overline{\alpha}) + \tfrac{1}{2}\delta\, \phi'^*(\overline{\alpha}) + \tfrac{1}{8}\delta^2 \phi''^*(\overline{\alpha}) + \cdots\right] \\ &\qquad \times < \overline{\alpha} + \tfrac{1}{2}\delta|H - E|\overline{\alpha} - \tfrac{1}{2}\delta > \\ &\qquad \times \left[\phi(\overline{\alpha}) - \tfrac{1}{2}\delta\phi'(\overline{\alpha}) + \tfrac{1}{8}\delta^2\phi''(\overline{\alpha}) + \cdots\right]. \end{aligned} \tag{8.4}$$

Since $< \overline{\alpha} + \frac{1}{2}\delta|H - E|\overline{\alpha} - \frac{1}{2}\delta >$ is even in δ, the only integrals which survive are of the form

$$\mathcal{O}_n(\overline{\alpha}) = \int < \overline{\alpha} + \tfrac{1}{2}\delta|\mathcal{O}|\overline{\alpha} - \tfrac{1}{2}\delta > \delta^n d\delta \tag{8.5}$$

for n even. Here the operator $\mathcal{O}$ is either H or $N \equiv 1$. Through order ϕ'' and $|\phi'|^2$, we have

$$\begin{aligned} &< \Psi|H - E|\Psi > \\ &= \int d\overline{\alpha} \left[\phi^*(H_0 - E\, N_0)\phi + \tfrac{1}{4}(H_2 - E\, N_2)(-\phi'^*\phi' + \tfrac{1}{2}\phi''^*\phi + \tfrac{1}{2}\phi^*\phi'')\right]. \end{aligned} \tag{8.6}$$

This may be cast into the more familiar form

$$\int \tilde{\phi}^* \left(-\frac{d}{d\overline{\alpha}} \frac{1}{2B(\overline{\alpha})} \frac{d}{d\overline{\alpha}} + V(\overline{\alpha}) - E\right) \tilde{\phi}\, d\overline{\alpha} \tag{8.7}$$

with

$$V(\overline{\alpha}) = \frac{H_0}{N_0} + N_0^{-\frac{1}{2}} \frac{d}{d\overline{\alpha}}(H_2 - E\,N_2)\frac{d}{d\overline{\alpha}}N_0^{-\frac{1}{2}} + \frac{1}{8\,N_0}\frac{d^2}{d\overline{\alpha}^2}(H_2 - E\,N_2), \tag{8.8a}$$

$$B(\overline{\alpha}) = -\frac{N_0}{H_2 - E\,N_2}, \tag{8.8b}$$

and

$$\tilde{\phi} = N_0^{\frac{1}{2}}\phi. \tag{8.8c}$$

We choose to define α such that $\alpha \to r$ as $r \to \infty$, where r is (say) the separation of two bags. However, $V(\overline{\alpha})$ has no simple interpretation as a potential until $B(\overline{\alpha})$ is determined!

Following Fujiwara[21], we introduce a change of variable from $\overline{\alpha}$ to x (which is the same as r for a head-on collision) such that

$$x = X(\overline{\alpha}) = \int^{\overline{\alpha}} [B(\alpha')/\mu]^{\frac{1}{2}} d\alpha' = -\int_{\overline{\alpha}}^{\infty} \left[\left(\frac{B(\alpha')}{\mu}\right)^{\frac{1}{2}} - 1\right] d\alpha' + \overline{\alpha}, \tag{8.9a}$$

$$\tilde{\phi}(\overline{\alpha}) = \left(\frac{B(\overline{\alpha})}{\mu}\right)^{\frac{1}{4}} \psi(x). \tag{8.9b}$$

where $\mu = M/2$ is the reduced nucleon mass. Then variation of Eq.(8.7) with respect to ψ^* yields

$$\left(-\frac{1}{2\mu}\frac{d^2}{dx^2} + V + V_1 - E\right)\psi(x) = 0 \tag{8.10}$$

with

$$V_1 = -\frac{B''}{8\,B^2} + \frac{7(B')^2}{32\,B^3}. \tag{8.11}$$

This is consistent with $B \to \mu$ as $\alpha \to x$ (or r). The generalization to three dimensions is straightforward.

9. THE N-N INTERACTION

The pioneering calculations on the N-N interaction in the context of the MIT bag were performed by DeTar.[22] These were static calculations of quarks interacting through one gluon exchange in a deformed cavity. While instructive, one cannot interpret the results in terms of an N-N potential until one has well-defined dynamics.

The equation for the dynamical collision of two bags has been calculated by Schuh *et al.*[8] without (as yet) gluon exchange for zero impact parameter. The state vector $|\alpha>$ is determined by extremizing the expectation value of the total Hamiltonian, $<\alpha|H|\alpha>$ with respect to a variational mean field wave function for the quarks and a coherent state wave function for the soliton field subject to a constraint

$$<\alpha|Q|\alpha> = Q_0 \tag{9.1}$$

where Q is some moment of the quark distribution

$$Q = \int \overline{\psi}\, q(\vec{r})\, \psi\, d^3r. \tag{9.2}$$

The constrained mean field equations now assume the form

$$\{\vec{\alpha} + \beta[g\sigma_0(\vec{r}) - \lambda\, q(\vec{r})] - \epsilon_k\}\psi_k = 0, \tag{9.3a}$$

$$-\nabla^2\sigma_0 + U'(\sigma_0) + g \sum_{k(valence)} \overline{\psi}_k\psi_k = 0. \tag{9.3b}$$

where λ is a Lagrange multiplier. Instead of specifying the constraint function $q(\vec{r})$ explicitly and solving the pair of equations (9.3 a & b) simultaneously, it is more physical to specify the function in square brackets

$$\mathcal{V}(\vec{r}) \equiv g\sigma_0(\vec{r}) - \lambda\, q(\vec{r}). \tag{9.4}$$

This plays the role of a generating potential for the quarks. $\mathcal{V}(\vec{r})$ is determined by folding the volume formed by the union (intersection) of two spheres whose centers are separated by a distance $|\alpha|$ with a Yukawa form factor. Prolate (oblate) deformations are obtained for α positive (negative).

The mass parameter $B(\overline{\alpha})$ is strongly $\overline{\alpha}$-dependent, as is shown in Fig. 5. The resulting $x(\overline{\alpha})$ is shown in Fig. 6, and the effective potential $V_{eff} = V + V_1$ is shown is Fig. 7. For $x \to -\infty$, $V_{eff} \to 6g\sigma_v \approx 4$ GeV. For the zero-impact parameter trajectory, the boundary condition on $\psi(x)$ is $\psi(-\infty) = 0$. V_{eff} is significantly weaker than the "static" energy $<\alpha|H|\alpha>$, by roughly 40-50%. The results clearly indicate the importance of dynamics in the scattering process: static energies do not tell the story.

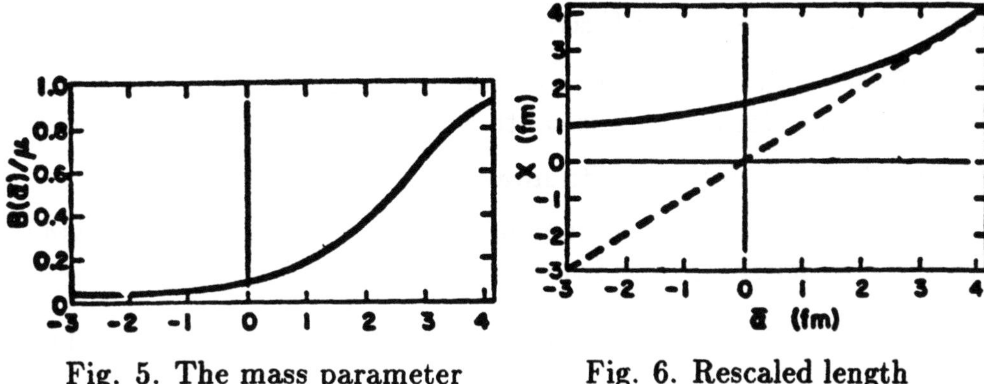

Fig. 5. The mass parameter $B(\overline{\alpha})$.

Fig. 6. Rescaled length parameter $x(\overline{\alpha})$. [Ref. 9]

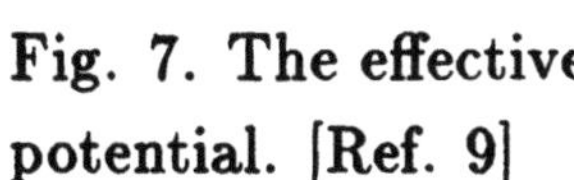

Fig. 7. The effective potential. [Ref. 9]

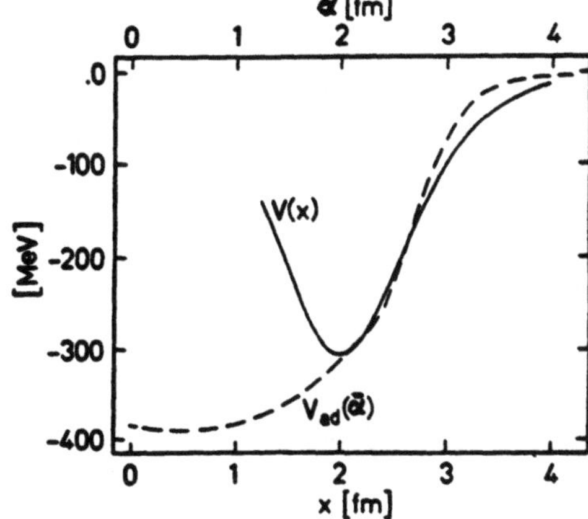

Crawford[38] has approached the nucleon-nucleon scattering problem by dividing space into an inner soliton bag region and an outer N-N potential region, with the appropriate continuity conditions at the boundary. The inner region is described by six quarks in a deformed sigma field. The quark wave functions are approximated by ("sudden") displaced three quark clusters, in contrast to the adiabatic quark states of Schuh *et al.* The quarks interact through an effective contact potential to sumulate one gluon exchange. The strength of the interaction and the matching radius are adjustable parameters. The outer region is described through the use of a Reid potential. Quite satisfactory phase shifts are obtained.

10. ONE GLUON EXCHANGE

To the order of one-gluon exchange, the gluon field equations linearize and are formally identical to Maxwell's equations *in media.*

10.1 Absolute color confinement

In order to effect color confinement, the dielectric function $\kappa(\sigma)$ is required to satisfy the conditions

$$\kappa(0) = 1, \qquad \kappa(\sigma_v) = 0\,. \tag{10.1}$$

This can be seen by considering, for example, a spherically symmetric system with static charge density, $Q\,\delta^3(\vec{r})$ inside a cavity with $\kappa = 1$; outside the cavity, $\kappa \to 0$ as $r \to \infty$. The chromo-electrostatic field equations are then the familiar

$$\begin{aligned} \vec{\nabla}\cdot\vec{D} &= \rho & \qquad \vec{\nabla}\times\vec{E} &= 0 \\ \vec{D} &= \kappa\vec{E} & \qquad \vec{E} &= -\vec{\nabla}A_0\,, \end{aligned} \tag{10.2}$$

where here, however, all quantities are operators in color space. From Gauss's law we find immediately that

$$D_r = \frac{Q}{4\pi r^2}$$

and the energy is

$$\mathcal{E} = \frac{1}{2}\int d^3r\,\vec{D}\cdot\vec{E} = \frac{1}{2}\int d^3r\,\frac{D_r^2}{\kappa(r)} = \frac{Q^2}{8\pi}\int_0^\infty \frac{dr}{r^2}\,\frac{1}{\kappa(r)} \tag{10.3}$$

The divergence at the lower limit of integration is associated with the usual self-energy. However, outside the cavity the integral diverges because $\kappa \to 0$ (exponentially in the model). The energy is infinite unless the charge vanishes. The argument does not depend on spherical symmetry. Any net charge gives rise to a $\vec{D}$-field over all space. Therefore the charge within the cavity must be in a color-singlet state. One or two quark excitations, which means charge separation, lead to a strong restoring force and hence cannot cannot yield continuum states. This is the basis of the argument that within the OGE-MFA approximation the continuum begins at $3\,g\,\sigma_v$.

10.2 The string constant

The spectra of heavy mesons, e.g. charmonium and the Υ, have been fit with a non-relativistic potential of the form

$$V(r) = -\frac{4}{3}\,\frac{\alpha_s}{r} + \frac{r}{a^2}\,, \tag{10.4}$$

where the coefficient of the linear term, $1/a^2 \equiv t$, is the string constant, numerically equal to about 925 MeV/fm.

The calculation of the string constant has been made in the MIT model by Johnson and Thorn[23] and in the soliton model by Bickeböller *et al.*[24]

10.3 There is no color Van der Waals problem!

The vanishing of κ in the vacuum guarantees that there is no (long-range) interaction between isolated bags. The problem of the r^{-6} [or rather r^{-7}, as Prof. Casimir has taught us] Van der Waals potential plagues most quark models. Subsequent to this School, Danos[25] clarified to me that because the stress-energy tensor is bilinear in E and D, and in B and H, and therefore vanishes in the vacuum, no gluonic forces can be transmitted through the vacuum, in the Abelian approximation. (E and B can be finite.) Although T. D. Lee[35] seems to have appreciated this, he nevertheless proposed a gluon mass term to prevent penetration of the vacuum by E and B. This has the undesirable features of destroying gauge invariance and color confinement. At the level of the present calculations, the form for κ alone is sufficient to remove the Van der Waals problem.

10.4 The linearized gluon propagator[26]

In a scalar medium, the gluon propagator is diagonal in the color indices,

$$G^{cc'}_{\mu\mu'} = \delta^{cc'} G_{\mu\mu'}. \tag{10.5}$$

The potentials are then given by

$$A^c_\mu(\vec{r}, t) = \int d^3r'\, dt'\, G_{\mu\nu}(\vec{r}, t; \vec{r}', t') J^c_\nu(\vec{r}', t'). \tag{10.6}$$

In the linearized (one gluon exchange) approximation, the gluon propagator satisfies equations which are identical to those for a Maxwell propagator in medium. We work in the transverse gauge, so that

$$\partial^i\, \kappa\, G_{ij} = 0. \tag{10.7}$$

The scalar Green's function $G_{00} = G$ satisfies the time-independent differential equation

$$\vec{\nabla} \cdot \kappa(\vec{r}) \vec{\nabla} G(\vec{r}, \vec{r}') = -\delta^3(\vec{r} - \vec{r}'). \tag{10.8}$$

To simplify this equation and to avoid infinities for $\kappa \to 0$ one can define

$$\overline{G}(\vec{r}, \vec{r}') \equiv \sqrt{\kappa(\vec{r})}\, G(\vec{r}, \vec{r}') \sqrt{\kappa(\vec{r}')}, \tag{10.9}$$

where $\overline{G}$ satisfies

$$(\nabla^2 - W(\vec{r}))\overline{G}(\vec{r},\vec{r}\,') = -\delta^3(\vec{r} - \vec{r}\,'), \tag{10.10}$$

with the "potential" $W(\vec{r})$ given by

$$W(\vec{r}) \equiv \tfrac{1}{4}|\vec{\nabla} \ln \kappa(\vec{r})|^2 + \tfrac{1}{2}\nabla^2 \ln \kappa(\vec{r}). \tag{10.11}$$

The differential equation for the vector potential is time-dependent. After a Fourier transformation in the time and defining the transverse dyadic Green's function as

$$\overline{G}^{ii'}(\vec{r},\vec{r}\,'\omega) \equiv \kappa(\vec{r})G^{ii'}(\vec{r},\vec{r}\,',\omega), \tag{10.12}$$

one obtains

$$\begin{aligned}(\omega^2 + \nabla^2)\overline{G}^{ii'}(\vec{r},\vec{r}\,',\omega) - \epsilon_{ikl}\partial^k \left(\epsilon_{lmn}\overline{G}^{mi'}(\vec{r},\vec{r}\,'\omega)\partial^n \ln \kappa(\vec{r})\right) \\ = -\delta_t^{ii'}(\vec{r} - \vec{r}\,'),\end{aligned} \tag{10.13}$$

with the restriction

$$\partial^i\overline{G}^{ii'}(\vec{r},\vec{r}\,',\omega) = 0. \tag{10.14}$$

The transverse gauge condition requires the asymmetric definition of $\overline{G}$ (10.12) . The subscript "t" on the delta-function in (10.13) denotes that it is a transverse delta-function.

It is surprising to note that the Maxwell propagator in terms of spherical harmonics for a homogeneous medium ($\kappa = 1$) had not been formulated correctly until 1979, when it was presented by Johnson, Howard and Dudley.[27] The essential stumbling block had been the failure to realize that the transverse delta function in Eq. (10.13) is non-local!

Bickeböller[24] has formulated the solution for an arbitrary $\kappa(\vec{r})$ and has developed a computer code for the case of axial symmetry. He has also calculated OGE effects in hadrons, both perturbatively and self-consistently.[28]

It is of interest to compare the results of using a confined and a free gluon propagator. The resulting energy of interaction is approximately twice as large for a confined propagator as for a free one. This can be understood qualitatively by noting that the $\vec{D}$ and $\vec{H}$ fields are confined to the bag, and are hence stronger than for the free propagator. For many purposes, such as spherical. static bag states, it is often sufficient to use a free propagator with an effective α_S. However, when dealing with deformed bags. such as in NN collisions, the effect of confinement depends upon shape, as noted by DeTar.[22]

11. THE PION AND DRESSING OF THE BARYONS[19]

Dethier has described the pion as a $q\bar{q}$ bag, and this structure has been coupled to nucleons and deltas using the coupled-state generator coordinate method, including one gluon exchange. As first noted by DeGrand *et al.*[29], OGE splits the pion from the rho and omega mesons, bringing it to within a factor of two of the physical pion mass for the same parameters that fit the nucleon and the delta. This in itself was an encouraging result. Removal of spurious center-of-mass energy by projection lead to a further reduction in the pion mass. Lübeck[30] finds that full variation of the projected pion state vector came close to yielding the physical pion mass.

In Dethier's calculation parameters of the model are adjusted so that the composite structure reproduces the nucleon mass, the proton charge radius and the N-Δ mass-splitting. The neutron charge radius is calculated to be $< r^2 >_n = -0.08$ fm^2 compared with the experimental value of -0.12 fm^2. The resulting description is similar to the Cloudy Bag Model. The probability of a "pion" in a nucleon is calculated to be 47%, indicating that two pion channels may be important. An interesting feature of the model is the partial restoration of PCAC in an integral sense: The radial integral of the divergence of the axial current is -0.12 fm^{-1} compared with 0.35 for the q^3 nucleon core and 1.25 for the MIT model. The effective "pion" tail is too short; this is characteristic of Hartree-type calculations. since the "eigenvalue" associated with the "pion" is lower than the separation energy (which should be the pion mass); the difference is the "rearrangement energy." The nucleon-pion coupling constant is *calculated* to be 0.19 compared with the experimental value of 0.28.

12. SUMMARY OF THE HADRON STATES CALCULATIONS

The nucleon, delta, pi, rho and omega have now been calculated including recoil corrections (through projection and variation) and one gluon exchange; the pion bag has been coupled to the bare nucleon and delta bags. The various calculations are each quite complex, and no individual calculation encompasses the entire program. The picture which emerges, however, is that the soliton model is capable of giving a very satisfactory picture of hadronic structure. This is a necessary condition for the credibility of the dynamic calculations. The pieces of the program also show that it is not always necessary to do the complete calculation, depending upon what properties one wishes to describe.

In all of the results quoted, the nucleon mass and proton radius have been fixed. The projected state calculations using a free gluon propagator[30] yield very good results for the nucleon magnetic moments, g_A/g_V, and the meson (pi, rho and omega) masses. The delta-nucleon splitting is about 2/3 of the experimental value. The confined propagator is consistent with these results, requiring, however, a smaller value of α_S. The use of a self-consistent propagator did not change the results appreciably.[28]

In all of these calculations, the chromo-electric self energy was included throughout the size of the Hilbert space of the valence quarks. To this order, the chromo-electric energy contributions vanish. The magnetic self energy, however, causes difficulties in fitting the hadronic spectrum. It is strongly attractive and could not easily be compensated for by readjustment of the soliton parameters. Therefore, the magnetic self energy was omitted in the calculations reported here. The problem has been addressed in depth for the MIT bag by Hansson and Jaffe[31].

The coupling of the pion bag to the nucleon and delta permits a fitting of the delta-nucleon splitting, and is responsible for roughly 40% of that splitting. The resultant picture is thus complete and satisfactory. It is of interest to note that the self-consistent treatment of gluons results leads to a somewhat smaller fitted coupling between the quarks and the sigma field, consistent with the expectation (hope) that higher order calculations would lead to a withering away of the sigma field.

13. ONE LOOP QUANTUM CORRECTIONS

Although the soliton model Lagrangian may be regarded as an effective Lagrangian, we employ it for dynamic calculations and as such treat it as a complete quantum mechanical relativistic field theory. Questions of consistency and renormalization must be addressed. For example, the valence and sea quarks, in the above discussions, have not been treated on an equal footing; indeed the sea quarks have not been distorted at all and have been assumed to be orthogonal to the valence quarks. Similarly, the zero-point fluctuations of the sigma field have not been included.

One consistent approach to the classification of higher order corrections is through an expansion in powers of $\hbar$. To lowest order in $\hbar$ we have the level of tree graphs, which is essentially classical. To next order is the one loop approximation. This involves distortion of the Dirac sea and distortion of the sigma modes.

13.1 Dirac sea corrections

The distortion of the Dirac sea in the presence of the sigma field is what we call in QED *vacuum polarization.* A uniform sigma field is equivalent to the introduction of quark mass $g\sigma$ and yields an energy density which is quartic in σ plus a term proportional to $\sigma^4 \ln \sigma$. These terms lead to a renormalization of $U(\sigma)$. The $\ln\sigma$ term can be approximated and regrouped with the polynomial[32]. For the case of a non-uniform system, there are inhomogeneous corrections to the Dirac sea. The leading order term is proportional to $|\nabla\sigma|^2$ and incorporated as a sigma field renormalization. Thus this part of the quantum corrections is already contained in the "effective Lagrangian." There are, however, further non-local corrections and the gradient expansion may not be convergent. These have been studied in the case of the linear sigma model by Kahana and Ripka[33], and are currently under investigation in the soliton model by Perry, Li and Wilets.[34] What is required is the spectrum of negative energy states. This can be calculated either by generating the eigenfunctions and eigenvalues of these states or by constructing the Green's function for imaginary frequencies.

13.2 Soliton field quantum corrections

The normal modes of oscillation of the sigma field in vacuum are plane waves of frequency

$$\omega_k^2 = U''(\sigma_v) + k^2.$$

In the presence of a quark cluster, the eigenfunctions and eigenfrequencies for the soliton states satisfy Eq. (5.3). The explicit inclusion of the energies contributed by the soltion zero-point oscillations, as for the Dirac sea, gives rise not only to a renormalization of the Lagrangian parameters but also to nonlocality.

14. WORK IN PROGRESS AND FUTURE DIRECTIONS

14.1 N-$\overline{N}$ Annihilation

An extension of the NN interaction problem is the nucleon-antinucleon reaction leading to annihilation and rare processes.[20] This is being described by the coupled-channel Hill-Wheeler equation. In the case of $N\overline{N}$ annihilation, for example, rearrangement is assumed to proceed *via* a $q\overline{q}$,$qq\overline{q}\overline{q}$ doorway state with the internal mechanism provided by the quark-sigma interaction. Previous studies had indicated that one gluon exchange was too small to describe annihilation.

14.2 The many bag problem

The nuclear many body problem is being approached as a many bag problem. Achtzehnter *et al.*[36] have solved the self-consistent mean field equations for a periodic (crystalline) structure of the sigma and quark fields. By varying the lattice spacing, the energy and wavefunctions were studied as a function of the density; at moderate density the system popped into the uniform plasma phase.

An alternative approach currently being studied[37] is that of solving the equations in a spherical Wigner-Seitz cell. The first block band is filled to the top with a dilution factor to accomodate three quarks per cell. The sigma field has zero gradient at the origin and at the cell radius. The quark functions satisfy the appropriate boundary conditions at the bottom and at the top of the band; the scalar quark density is interpolated linearly in between. The equations are solved self-consistently. One gluon exchange energies are added perturbatively by use of the confined, linearized gluon propagator in the presence of the dielectric function $\kappa(\sigma)$. The sigma field used in $\kappa(\sigma)$ is that obtained self-consistently for r inside the cell; beyond the cell boundary, σ is taken to be an approximate angular average of the distributed cells.

14.3 A Pauli equation for the nucleon

Although the Dirac equation has been used with considerable success to describe nucleon scattering, the procedure has been criticized because the nucleon has a structure large compared with its Compton wavelength. Achtzehnter[39] is currently working on the derivation of two-component Pauli equation to describe the nucleon in external vector and scalar potentials. The results are to be compared with the Foldy-Wouthuysen transformed Dirac equation for a point particle in external fields.

15. ACKNOWLEDGMENTS

I wish to express my gratitude to my many collaborators who have contributed so much to these projects, and from whom I have learned so much. I would like to mention, in particular, M. Bickeböller, M. C. Birse, K. Bräuer, G. Crawford, J.-L. Dethier, R. Goldflam, E. M. Henley, Ming Li, E. G. Lübeck, R. Perry, J. J. Rehr and A. Schuh.

This work was supported in part by the U. S. Department of Energy.

REFERENCES

1. A. Chodos, R. L. Jaffe, K. Johnson, C. B. Thorn and V. F. Weisskopf, Phys. Rev. **D9** 3471, (1974).
2. G. A. Miller, A. W. Thomas and S. Théberge, Phys. Lett. **91B**, 192 (1980); S. Théberge, A. W. Thomas and G. A. Miller, Phys. Rev. **D22**, 2838 (1980); A. W. Thomas, S. Théberge and G. A. Miller, *ibid* **24**, 216 (1981).
3. K. Huang and D. R. Stump, Phys. Rev. **D 14**, 223 (1976).
4. R. Friedberg and T. D. Lee, Phys. Rev. **D15**, 1694 (1977); **D16**, 1096 (1977); **D18**, 2623 (1978); T. D. Lee, *Particle Physics and Introduction to Field Theory* (Harwood Academic, New York, 1981).
5. R. Goldflam and L. Wilets, Phys. Rev. **D25**, 1951 (1982);
6. E. Eichten, K. Gottfried, T. Kinoshita, K. D. Lane and T. M. Yan, Phys. Tev. **D 21**, 203 (1980); W. Buchmüller, Phys. Lett. **112 B**, 479 (1982); S. R. Gupta, S. F. Radford and W. W. Repko, Phys. Rev. **D 26**, 3305 (1982); P. Moxhay and J. L. Rosner, *ibid.* **28**, 1132 (1983).
7. J. Gasser and H. Leutwyler, Phys. Rep. **87 C**, 78 (1982).
8. R. Saly, Comput. Phys. Commun. **30**, 411 (1983); R. Saly and M. K. Sundaresan, Phys. Rev. **D29**, 525 (1984);

   Th. Köppel and M. Harvey, Phys. Rev. **D 31**, 171 (1985).
9. A. Schuh, 'Nukleon-Nukleon-Wechselwirkung im Soliton Bag Modell,' Doctoral Dissertation, University of Heidelberg (1985); "Dynamics of 6 Quark Systems in the Soliton Bag Model," A. Schuh, H. J. Pirner and L. Wilets, Int. Workshop XIII on Gross Properties of Nuclei and Nuclear Excitations, Hirschegg, Austria (1985);

   A. Schuh, H. J. Pirner and L. Wilets, submitted to Physics Letters B.
10. L. Wilets, in *Hadrons and Heavy Ions*, Lecture Notes in Physics **231**, 317 (Springer, Berlin, 1985).
11. L. Wilets, M. C. Birse, E. G. Lübeck and E. M. Henley, Nuc. Phys. **A 434**, 129c (1984);

    E. G. Lübeck, M. C. Birse, E. M. Henley and L. Wilets, Phys. Rev. **D 33**, 234 (1986).
12. M. H. L. Pryce, Proc. R. Soc. London **A195** (1948);
13. R. A. Krajcik and L. L. Foldy, Phys. Rev. **D10**, 1777 (1974).

14. J.-L. Dethier, R. Goldflam, E. M. Henley and L. Wilets, Phys. Rev. **D27**, 2193 (1983);
A. D. Fokker, *Relativiteitstheorie* (Groningen: P. Noordhoff, 1929).

15. L. L. Foldy and S. A. Wouthuysen, Phys. Rev. **78**, 29 (1949).

16. R. E. Peierls and J. Yoccoz, Proc. Phys. Soc. London **A70**, 381 (1957).

17. J. J. Griffin and J. A. Wheeler, Phys. Rev. **108**, 311 (1957).

18. D. L. Hill and J. A. Wheeler, Phys. Rev. **89**, 1106 (1953).

19. J.-L. Dethier, "A Soliton Bag Model of the Nucleon and the Delta Dressed by a Quark-Antiquark Pion," Doctoral Dissertation, University of Washington (1985);
J.-L. Dethier and L. Wilets, Phys. Rev. **D**, to be published.

20. G. Crawford, J-L. Dethier, L. Wilets and M. Alberg, Proc. Int. Conf. on Antinucleon- and Nucleon-Nucleus Interactions, Telluride, Colorado (1985).

21. I. Fujiwara, Prog. Theor. Phys. **21**, 902 (1959).

22. C. DeTar, Phys. Rev. **D 17**, 302 and 323 (1978).

23. K. Johnson and C. B. Thorn, Phys. Rev **D13**, 1934 (1976).

24. M. Bickeböller, M. C. Birse, H. Marschall and L. Wilets, Phys. Rev. **D 31**, 2892 (1985).

25. M. Danos, private communication.

26. M. Bickeböller, 'Der Gluon-Propagator im Soliton-Bag-Modell,' Diplom Thesis, University of Bonn, 1984; M. Bickeböller, R. Goldflam and L. Wilets, J. Math. Phys. **26**. 1810 (1985).

27. W. A. Johnson, A. Q. Howard, and D. G. Dudley, Radio Science **14**, 961 (1979).

28. M. Bickeböller, "The Gluon Propagator in the Soliton Bag Model and its Applications," Doctoral Dissertation, University of Washington (1986).

29. T. DeGrand, R. L. Jaffe, K. Johnson and J. Kiskis, Phys. Rev. **D 12**, 2060 (1975).

30. E. G. Lübeck, "Momentum Projection and Revativistic Boost of Solitons," Doctoral Dissertation, University of Washington (1986).

31. T. H. Hansson and R. L. Jaffe, Phys. Rev. **D 28**, 882 (1983); Annals of Physics (N.Y.) **151**, 204 (1983).

32. M. Li. M. C. Birse and L. Wilets, "Phase Transition in the Soliton Bag Model," Journal of Physics **G**, to be published.

33. S. Kahana and G. Ripka, Nucl. Phys. **A 429**, 462 (1984).

34. R. Perry, M. Li, and L. Wilets, private communication.

35. T. D. Lee, Phys. Rev. **D 19**, 1802 (1979).

36. J. Achtzehnter, W. Scheid and L. Wilets, Phys. Rev. **D 32**, 2414 (1985).

37. M. Bickeböller, J. J. Rehr and L. Wilets, private communication.

38. G. Crawford, "Quarks and the Saturation Properties of Nuclear Matter", Doctoral Thesis, University of Washington (1986).

39. J. Achtzehnter, private communication.

PAIR PRODUCTION AND QUANTUM TRANSPORT IN STRONG COLOR FIELDS

M. Gyulassy[1]

Institut für Theoretische Physik
Johann-Wolfgang-Goethe Universität
Frankfurt/M., West Germany

H.-Th. Elze, A. Iwazaki, and D. Vasak[2]

Nuclear Science Division
Lawrence Berkeley Lab
Berkeley, CA 94720

INTRODUCTION

In the first lecture the neutralization rate of covariant constant SU(N) fields due to quark and gluon pair creation is computed following Ref.[1]. In the second lecture recent work in Refs.[2,3,4] on the quantum transport theory of quarks and gluons in self-consistent SU(N) fields is reviewed. The generalized quantum Vlasov and constraint equations are obtained in the Abelian Dominance Approximation. Phenomenological consequences for quark-gluon plasma production in ultra-relativistic nuclear collisions are discussed.

1. QUARK AND GLUON PAIR PRODUCTION

Pair production in strong external fields is one of the most basic and interesting phenomena associated with strong fields [5,6,7] Most work on this problem thus far has been in connection with Abelian (QED) fields although the pair production rate of gluons in covariant constant non-Abelian SU(2) fields is also known [8,9]. Recently, the rates for both quark and gluon pair production in covariant constant SU(N) fields were computed in [1]. Phenomenologically, those rates are interesting in connection with estimates for the formation time of quark-gluon plasmas in ultra-relativistic nuclear collisions [11].

---

[1]Permanent address: Nuclear Science Division, Lawrence Berkeley Lab, Berkeley, CA. Work supported by the U.S. Department of Energy under Contract DE-AC03-76SF00098 and an Alexander von Humboldt Foundation Senior U.S. Scientist Award

[2]H.T.E. and D.V. acknowledge support by DAAD-NATO Postdoctoral Fellowships

Abelian Case

In string models for multiparticle production [12]-[16], one imagines that multiple soft gluon exchange leaves both projectile and target in color non-singlet states. Because of confinement though, the resulting color electric fields extending between the receeding projectile and target fragments are assumed to be confined to narrow color flux tubes of some finite transverse area $A_\perp \sim 1\text{fm}^2$. The field is furthermore assumed to be a constant along the beam direction with a strength $E \sim gQ/A_\perp$, where Q is the net "charge" on the projectile. In *Abelian* models [12,13,15,16], Q is treated simply as an integer and the pair production rate per unit volume, $w_{1/2}(\sigma, m)$, for $q\bar{q}$ pairs of mass, m, is taken from the well known Schwinger formula [5,7,12]

$$w_{\frac{1}{2}}(\sigma, m) = \theta(\sigma)\frac{\sigma}{4\pi^2}\sum_{n=1}^{\infty}\frac{1}{n}\int_{m^2}^{\infty} dE_\perp^2 \exp(-n\pi E_\perp^2/\sigma) \approx \theta(\sigma)\frac{\sigma^2}{24\pi} \quad , \tag{1}$$

with $\theta(x) = 0(1)$ for $x < (>)0$, and where the approximation holds for $\pi m^2/\sigma \ll 1$. In the above expression $\sigma = gE$ is the effective string tension, which phenomenologically is taken to be on the order of 1 GeV/fm.

With (1) the rate with which the external "color" field is neutralized due to $q\bar{q}$ production can be estimated as in Ref.[16]. First we recognize that conservation of energy requires that in order to produce a pair, each with transverse energy $m_\perp$ and zero longitudinal momentum, that pair must separate by at least a distance $r_c \sim 2m_\perp/\sigma$. The minimum volume for pair production is thus $\delta V \sim A_\perp 2m_\perp/\sigma$. Once a pair has been produced in this finite volume element the field between them is reduced by $\delta E = g/A_\perp$ if we assume that all flux lines are confined to the original flux tube. With this crucial model assumption we can then estimate [16]

$$\frac{dE(t)}{dt} \approx -\frac{g}{A_\perp} w_{\frac{1}{2}}\delta V \propto \sigma\langle m_\perp\rangle = -aE(t)^{3/2} \quad , \tag{2}$$

where a is a positive constant. The power 3/2 follows from dimensional considerations since locally $E(t)$ is the only dimensioned quantity that can set the time scale (also $\langle m_\perp\rangle \propto E^{1/2}$ for the same reason). Writing $E(t) = gQ(t)/A_\perp$, the solution of (2) is simply

$$Q(t) = Q_0/(1 + t/\tau_{1/4}(Q_0))^2 \quad , \tag{3}$$

with

$$\tau_{1/4} = \tau_1/Q_0^{\frac{1}{2}} \quad , \tag{4}$$

in terms of the characteristic time $\tau_1 \sim 1$ fm/c required to neutralize 3/4 of the field strength in an elementary string with unit charge. The most important feature of the above solution is that it shows that the neutralization time of strong ($Q_0 \gg 1$) Abelian fields *decreases* as $Q_0^{-1/2}$. This is good news from the point of view of creating a quark-gluon plasma. The faster the fields neutralize the more time will there be for the quarks and gluons to come into chemical and thermal equilibrium and thus to produce interesting signatures [11] of this new state of matter.

Non-Abelian SU(N) Case

Encouraged by the above Abelian analysis we turn next to investigate non-Abelian aspects of the problem. In particular, how do we treat the non-Abelian character of the charges of both quarks and gluons? How do quarks and gluons compete in the neutralization process? How close is the resulting quark-gluon plasma to local equilibrium conditions at time $\tau_{1/4}$? To answer these questions we compute the pair production rates for SU(N) in the one-loop $O(\hbar)$ approximation [1].

In conventional notation the generators, $t_a, a = 1, \cdots, N^2-1$, of SU(N) in the fundamental representation satisfy $[t_a, t_b] = i f_{abc} t_c$, $tr(t_a t_b) = \delta_{a,b}/2$. The gluon field matrix is denoted by $A_\mu \equiv A^a_\mu t_a$, the covariant derivative matrix by $D_\mu = \partial_\mu + igA_\mu$, and the field tensor by $F_{\mu\nu} = [D_\mu, D_\nu]/(ig)$. The Heisenberg field equations for quarks and gluons are

$$(i\gamma^\mu D_\mu - m_f)\psi_f = 0 \ , \tag{5}$$

$$[D^\mu, F_{\mu\nu}] = gJ_\nu \ , \tag{6}$$

where $J_\nu = \sum_f \bar{\psi}_f \gamma_\nu t_a \psi_f t_a$ and f labels the quark flavors.

A covariant-constant field [8], which satisfies Eq.(6) in the source free region, is of the form

$$F_{\mu\nu} = \langle F_{\mu\nu}\rangle U(x) n_a t_a U^\dagger(x) \ , \tag{7}$$

where $\langle F_{\mu\nu}\rangle$ is independent of x_μ, and n_a is an N^2-1 dimensional color vector. The unitary matrix $U(x) = \exp(i\theta_a(x)t_a)$ implements arbitrary local gauge transformations. The covariant derivative is given by $D_\mu = U(x)(\partial_\mu - \frac{1}{2}ig\langle F_{\mu\nu}\rangle x^\nu n_a t_a)U^\dagger(x)$.

Since $n_a t_a$ is Hermitian, there exists an x_μ *independent* matrix $V \in$ SU(N) such that $V n_a t_a V^\dagger$ is diagonal. We can thus transform the external field into diagonal form by making a gauge transformation $G = V^\dagger U^\dagger(x)$ under which $D_\mu \to G D_\mu G^\dagger$. Since all real $N \times N$ traceless diagonal matrices can be expanded in terms of the $N-1$ diagonal matrices, h_i, representing the Cartan subgroup of SU(N), it is then most convenient to expand A_μ in that gauge in terms of the Cartan-Weyl basis of SU(N). That basis consists [10] of $N-1$ Abelian generators, h_i, and the $N(N-1)$ non-Abelian generators, $\{e_{ij},\ i,j = 1,\cdots,N;\ i \neq j\}$, which satisfy

$$[h_i, h_j] = 0 \ , \ [h_i, e_{jk}] = (\vec{\eta}_{jk})_i e_{jk} \ , \ [e_{ij}, e_{jk}] = \frac{1}{\sqrt{2}} e_{ik} \quad \text{for } i \neq j \neq k \ . \tag{8}$$

The h_j are the Gell-Mann matrices $h_j = (2j(j+1))^{-\frac{1}{2}} diag(1,\cdots,1,-j,0,\cdots,0)$, with $-j$ appearing in the $j+1$ column. The $\vec{\eta}_{ij} = \vec{\epsilon}_i - \vec{\epsilon}_j$ are the root vectors of SU(N) as expressed in terms of the elementary weight vectors $\vec{\epsilon}_i = (\vec{h})_{ii} = ((h_1)_{ii}, \cdots, (h_{N-1})_{ii})$. (Note that $e_{ij} = \hat{e}_i \hat{e}_j^\dagger/\sqrt{2}$ in terms of the N orthonormal unit vectors $\hat{e}_i$.)

In the gauge which diagonalizes $\langle A_\mu\rangle$, the external field can be expressed in terms of $N-1$ Abelian components, $\vec{H}^\mu = (H^\mu_1, \cdots, H^\mu_{N-1})$, as

$$\langle A^\mu\rangle = \sum_{i=1}^{N} H^\mu_i h_i \equiv \vec{H}^\mu \cdot \vec{h} \ , \tag{9}$$

where $\vec{h} \equiv (h_1, \cdots, h_{N-1})$. To take quantum fluctuations around this external field into account, the gluon field operator can then be written as

$$A_\mu = \langle A_\mu\rangle + B_\mu = \vec{H}_\mu \cdot \vec{h} + B_\mu \ , \tag{10}$$

where B_μ represents the quantum fluctuations.

The physical significance of $\vec{\epsilon}_i$ can be seen from Eq.(5) by considering the equation of motion for the tranformed quark field, $\psi' \equiv G^\dagger \psi$. Eq.(5) then reduces to the set of equations

$$(\gamma_\mu(i\partial^\mu - g\vec{\epsilon}_c \cdot \vec{H}^\mu) - m_f)\psi'_c = O(B\psi') \ . \tag{11}$$

The approximation of neglecting higher order quantum fluctuations beyond the one loop order is equivalent to neglecting the $O(B\psi')$ terms on the right hand side of (11). We therefore see that the equations for the N quarks (of each flavor) in the prime basis decouple and reduce to Abelian type equations in the one loop approximation with $\vec{H}^\mu$ playing the role of an effective electromagnetic potential that couples to quarks with effective "charges" $g\vec{\epsilon}_c$. Since we already know (1) the pair creation rate, $w_{\frac{1}{2}}(eF^{\mu\nu}; m)$, of fermions in an external Abelian field $F^{\mu\nu}$, we can immediately write down the pair creation rate per unit volume of ψ'_c quarks of flavor f as

$$w_{q_{c,f}} = w_{\frac{1}{2}}(g\vec{\epsilon}_c \cdot \vec{F}^{\mu\nu}; m_f) \ , \tag{12}$$

where $\vec{F}^{\mu\nu} = \partial^\mu \vec{H}^\nu - \partial^\nu \vec{H}^\mu$ is a constant $N-1$ dimensional vector.

Turning next to gluons, the equations of motion for B^μ in the one loop approximation are obtained by linearizing Eq.(6) in B^μ. It is most convenient to expand B^μ in the Cartan-Weyl basis as

$$B^\mu \equiv B_a^\mu t_a = \vec{C}^\mu \cdot \vec{h} + \sum_{i \neq j = 1}^{N} W_{ij}^\mu e_{ij} \ . \tag{13}$$

Inserting (10) into Eq.(6) and using the Cartan-Weyl expansion (13) for B^μ together with the algebra (8) leads to the following equations of motion for the fluctuations $\vec{C}^\mu$ and W_{mn}^μ in the (linearized) one loop approximation to

$$\partial_\mu(\partial^\mu \vec{C}^\nu - \partial^\nu \vec{C}^\mu) = 0 \ , \tag{14}$$

$$(D_{mn})_\mu (D_{mn}^\mu W_{mn}^\nu - D_{mn}^\nu W_{mn}^\mu) - (W_{mn})_\mu [D_{mn}^\mu, D_{mn}^\nu] = 0 \ , \tag{15}$$

where the effective covariant derivative D_{mn}^μ is given by

$$D_{mn}^\mu = \partial^\mu + ig\vec{\eta}_{mn} \cdot \vec{H}^\mu \ . \tag{16}$$

We therefore see that the Abelian fluctuations, $\vec{C}^\mu$, obey free field equations whereas the non-Abelian fluctuations, W_{mn}^μ obey Abelian vector field equations in the external field, $\vec{H}^\mu$, with an anomalous magnetic moment coupling [9]. Note that $[D_{mn}^\mu, D_{mn}^\nu] = ig\vec{\eta}_{mn} \cdot \vec{F}^{\mu\nu}$. Obviously these equations decouple in this approximation. The effective "charge" of the W_{mn}^μ gluon is given by $g\vec{\eta}_{mn}$. Pair production in SU(N) covariant constant fields is thus equivalent to $N(N-1)/2$ different SU(2) problems. Therefore, the pair creation rate per unit volume of $W_{mn}W_{nm}$ gluon pairs can be calculated from the known [8,9] rate, $w_1(gF^{\mu\nu})$, of vector mesons for SU(2) covariant constant fields as

$$w_{g_{mn}} = w_1(g\vec{\eta}_{mn} \cdot \vec{F}^{\mu\nu}) \ , \tag{17}$$

where $\vec{F}^{\mu\nu}$ is the same external covariant constant SU(N) field as in Eq.(12) and the spin 1 rate is given by [17]

$$w_1(\sigma) = \theta(\sigma) \frac{\sigma}{4\pi^2} \sum_{n=1}^{\infty} \frac{(-1)^{n+1}}{n} \int_0^\infty dp_\perp^2 \exp(-n\pi p_\perp^2/\sigma) = \frac{1}{2} w_{\frac{1}{2}}(\sigma; 0) \ . \tag{18}$$

The case of particular interest in phenomenological applications [12,14] corresponds to constant color electric fields created between interacting partons in high energy collisions. For that case $\vec{F}^{30} = -\vec{F}^{03} = \vec{E} = \vec{Q}E_0$, where $E_0 = g/A_\perp$ for a flux tube of transverse area $A_\perp$ and $\pm\vec{Q}$ are the effective color charges of the projectile and target. The elementary $q\bar{q}$ string corresponds in this picture to $\vec{Q} = \vec{\epsilon}_c$. The adjoint $g\bar{g}$ string corresponds to $\vec{Q} = \vec{\eta}_{ij}$. Note that $\vec{\eta}\cdot\vec{\eta}/(\vec{\epsilon}\cdot\vec{\epsilon}) = 2N/(N-1)$.

In high energy nuclear collisions [14] or very high energy hadronic collisions multiple soft gluon exchange may lead to large effective charges. If $\mathcal{N}$ gluons are exchanged with random charges, then $\langle Q^2\rangle = \mathcal{N}(1+N^{-1})^{-1}$ since only $N(N-1)$ of the N^2-1 gluons are "charged" with $\vec{\eta}_{ij}\cdot\vec{\eta}_{ij} = 1$. Of course such color electric fields are unstable against pair production as we saw above. The pair production rate per unit volume of massless quark-antiquark and gluon pairs is given from Eqs.(12,17) by

$$w_q = \frac{N_f g^2}{24\pi}\sum_{c=1}^{N}(\vec{\epsilon}_c\cdot\vec{E})^2 = \frac{Q^2}{2}N_f w_0 \ , \ w_g = \frac{g^2}{48\pi}\sum_{i>j=1}^{N}(\vec{\eta}_{ij}\cdot\vec{E})^2 = \frac{Q^2}{4}N w_0 \ , \tag{19}$$

where $w_0 = (gE_0)^2/(24\pi)$ and N_f is the number of quark flavors such that $\pi m_f^2/\sigma \ll 1$.

We can now answer one of our initial questions. The ratio of quarks to gluons just after the color field is neutralized is $(q/g)_{neut} \approx 2N_f/2N = 1$ for $N = N_f = 3$. In comparison, local equilibrium at zero chemical potential is characterized by $(q/g)_{equil} = (\frac{3}{4}NN_f)/(N^2-1) = 0.84$. Thus, this ratio is surprisingly close to equilibrium in spite of the fact that non-equilibrium tunneling was involved. Note however that no "neutral" gluons are produced via pair production.

Generalizing the discussion in the Abelian case, we can estimate the rate of neutralization for strong covariant constant fields $(Q^2 \gg \eta^2 = 1)$ by

$$\frac{d\langle\vec{E}\rangle}{dt} \approx -\sum_f\sum_{\vec{\epsilon}_i}\frac{g\vec{\epsilon}_i}{A_\perp}w_{\vec{\epsilon}_i}\delta V_{\vec{\epsilon}_i} - \sum_{\vec{\eta}_{ij}}\frac{g\vec{\eta}_{ij}}{A_\perp}w_{\vec{\eta}_{ij}}\delta V_{\vec{\eta}_{ij}} \ , \tag{20}$$

where the pair production rates per unit volume and effective volume elements, $\delta V_{\vec{q}}$, depend on the particular charges, $\vec{q}$, as well as on the mean external charge, $\langle\vec{Q}(t)\rangle = \langle\vec{E}(t)\rangle A_\perp/g$. The minimum volume elements for pair production are again constrained by energy conservation [16] with now $r_c = 2m_\perp/\sigma(\vec{q},\vec{Q})$ and where the effective string tension is given by

$$\sigma(\vec{q},\vec{Q}) = g\vec{q}\cdot\vec{E} = 2\vec{q}\cdot\vec{Q}\sigma_A \ . \tag{21}$$

Here $\sigma_A = g^2/(2A_\perp)$ is the effective string tension for adjoint strings. The factor of two arises due to interactions between the pairs as discussed in Refs.[1,13]. The average volume required for pair production is therefore

$$\delta V_{\vec{q}} \approx r_c A_\perp = \frac{2\langle m_\perp\rangle}{\sigma(\vec{q},\vec{Q})}A_\perp \ . \tag{22}$$

The neutralization rate is dominated by production of pairs with small mass $(\pi m^2/\sigma \ll 1)$, for which $\langle m_\perp\rangle_{s=\frac{1}{2}} = (2-2^{-1/2})^{-1}\langle m_\perp\rangle_{s=1}$. Therefore, Eq.(20) reduces to

$$\frac{d\langle\vec{Q}(t)\rangle}{dt} \approx -\frac{2^{3/2}}{\tau_1}\left[N_f{\sum_{\pm\vec{\epsilon}_i}}'\vec{\epsilon}_i(\vec{\epsilon}_i\cdot\langle\vec{Q}\rangle)^{3/2} + (1-2^{-3/2}){\sum_{\vec{\eta}_{ij}}}'\vec{\eta}_{ij}(\vec{\eta}_{ij}\cdot\langle\vec{Q}\rangle)^{3/2}\right] \ , \tag{23}$$

where $\tau_1 \propto (g^2\sqrt{\sigma_A})^{-1}$ is independent of Q and where the sums are restricted to charges with $\vec{q}\cdot\vec{Q} > 0$.

Eq.(23) controls the rate of color neutralization of strong covariant constant SU(N) fields and is the natural generalization of the equation derived for the Abelian case in Ref.[16]. The SU(N $\geq$ 3) case is only complicated by the fact that Eq.(23) is a vector equation. The power 3/2 again follows from dimensional considerations under the assumption that the only dimensional quantity in the problem is $\vec{E}$.

For $\vec{Q}(0)$ pointing along one of the weight or root vectors Eq.(23) reduces to an Abelian equation

$$dQ/dt = -Q^{3/2}/\tau(\hat{Q}) \ , \tag{24}$$

where the relaxation time is given in the two special cases by

$$\begin{aligned} \tau(\hat{e}) &= \tau_1((N-1)/2N)^{1/4}\left[N_f((1-N^{-1})^{3/2}+N^{-3/2})+(1-2^{-3/2})N\right]^{-1} \\ \tau(\hat{\eta}) &= \tau_1\left[N_f+(1-2^{-3/2})(N+2^{3/2}-2)\right]^{-1} \end{aligned} \ . \tag{25}$$

The terms proportional to N_f are those due to $q\bar{q}$ pair production. For SU(2) the two times are of course identical. Amazingly, for SU(3) they only differ in the fourth decimal place for $N_f = 3$.

The numerical coincidence of $\tau(\hat{e})$ and $\tau(\hat{\eta})$ for SU(3) has the pleasant consequence that $\tau(\hat{n}) \approx 0.18\tau_1$ is independent of the orientation of the color charge in the Cartan subspace to a very high accuracy. Therefore for SU(3), the vector nature of Eq.(23) is irrelevant, and the solution is accurately given by (3) with Q replaced by the $N-1$ dimensional Cartan charge vector, $\vec{Q}$ and with the characteristic time required to neutralize 3/4 of the initial color field replaced by

$$\tau_{1/4}(Q_0) = 0.36\ \tau_1\ Q_0^{-1/2} \ . \tag{26}$$

The factor 0.36 represents the combined effect of both $q\bar{q}$ and $g\bar{g}$ production and significantly shortens the neutralization time over the Abelian case where only $q\bar{q}$ contribute.

There are several interesting phenomenological consequences of the above results in connection with ultra-relativistic nuclear collisions. First, most of the quarks and gluons produced in the neutralization of strong color fields may appear at proper times an order of magnitude smaller than in elementary pp or e^+e^- collisions. Obviously shorter times imply that plasmas with higher initial energy densities can be produced [16] and the chances that local equilibrium can be reached are greater. On the other hand, the equilibration time due to ordinary kinetic phenomena could be long compared to the neutralization time [18]. Fortunately, the color neutralization mechanism leads to initial conditions that are not far from local equilibrium. First, as we noted, in strong fields the neutralization mechanism leads to production of quarks and gluons at comparable rates (for SU(3)). Furthermore, u,d,s quarks are produced with nearly the same abundance since their masses become irrelevant. Thus the chemical composition of the non-equilibrium plasma produced through neutralization is not far from that in local equilibrium. Second, the distributions of transverse momenta from (1,18) are approximately exponential as in local equilibrium although gluons materialize with ~30% larger transverse momentum than quarks due to their larger effective charge. Thus we conclude that the non-equilibrium quantum tunneling dynamics and the specific non-Abelian features of gluon pair production may play an important role in creating plasma initial conditions close to local equilibrium at very early times in the collision.

2. QUANTUM CHROMO TRANSPORT THEORY

Once the quarks and gluons are produced, they are subject to accelerations in the neutralizing external field and scatter with each other as the plasma evolves closer toward equilibrium. To follow that evolution in detail we must turn to transport theory. First, we recall the familiar classical transport theory leading to the Vlasov-Boltzmann equation. Then the Wigner function and its equation of motion are derived in non relativistic quantum mechanics. The formulation of gauge covariant quantum transport theory is then introduced. A generalized quantum Vlasov and associated constraint equation are derived in connection with Abelian QED. Finally, features of non-Abelian transport theory are analyzed in the Abelian Dominance Approximation.

Classical Transport Theory

A classical plasma is characterized by the phase space density, $f(x,p,t) = \sum_i \delta(x - x_i(t))\delta(p - p_i(t))$, where the classical trajectories $(x_i(t), p_i(t))$ obey

$$\dot{x}_i = p_i/m \ , \ \dot{p}_i = F_{ext}(x_i) + F_2(x_i) \ , \tag{27}$$

where F_{ext} is the external force and $F_2(y) = \int d^3x d^3p \nabla_x V(y-x) f(x,p,t)$ is the self-consistent force that is a functional of the phase space density. From (27) we see that

$$\dot{f} = \sum_i (\dot{x}_i \nabla_{x_i} + \dot{p}_i \nabla_{p_i})\delta(x - x_i(t))\delta(p - p_i(t)) = -(\frac{p}{m}\cdot\nabla_x + F(x)\cdot\nabla_p) f(x,p,t) \ . \tag{28}$$

Taking the ensemble average of this (Klimontovich) equation leads to the Vlasov-Boltzmann equation:

$$(md/dt + p\cdot\nabla_x + m\langle F(x)\rangle\cdot\nabla_p)\langle f(x,p,t)\rangle = C(\langle f\rangle) \ , \tag{29}$$

where the mean force is given by $F_{ext} + F_2$ with f replaced by its ensemble average, $\langle f\rangle$, and where the collision term can be extracted [19] from the correlation term

$$C(f) = -m\int d^3x' d^3p' \nabla_{x'} V(x-x')\nabla_{p'}(\langle f(x,p,t) f(x',p',t)\rangle - \langle f(x,p,t)\rangle\langle f(x',p',t)\rangle) \ . \tag{30}$$

Neglecting two-body correlations in the plasma is equivalent to neglecting the collision term. In that case (29) reduces to the Vlasov equation which applies only to "collisionless" plasmas. What makes the collision term particularly difficult to calculate in QED is the infinite Coulomb cross section. Medium polarization effects that screen long range fields must then be taken into account. In addition a physical assumption must be made regarding different relaxation time scales for correlation functions (the BBGKY hierarchy). This procedure leads in the quasi-linear approximation to the Balescu-Lenard form for the collision term involving the dielectric function.

Quantum Transport

The quantum mechanical analogue of the phase space density is the Wigner function [20]. We know that $\psi^\dagger(x)\psi(x)$ is the density of particles in coordinate space. Also

$\psi^\dagger(p)\psi(p)$ is the density in momentum space. A natural candidate for the density in (x,p) phase space is then

$$W(x,p) = \psi^\dagger(x)\delta(p-\hat{p})\psi(x) \ , \tag{31}$$

where $\hat{p}$ is the momentum operator. The delta function acts as a projector onto momentum space and is defined by the Fourier transform as

$$W(x,p) = \int \frac{d^3y}{(2\pi\hbar)^3} e^{-ipy/\hbar}\psi^\dagger(x)e^{iy\hat{p}/\hbar}\psi(x) \ . \tag{32}$$

Recall that $\exp(iy\hat{p}/\hbar)$ generates a translation by y when acting to the right or $-y$ when acting to the left. In order to make W Hermitian we will make it act halfway to the left and halfway to the right by representing $\hat{p}$ as $\frac{1}{2}i\hbar(\partial_x - \partial_x^\dagger)$. In this way we recover the familiar expression for the Wigner function

$$W(x,p) = \int \frac{d^3y}{(2\pi\hbar)^3} e^{ipy/\hbar}\psi^\dagger(x+\tfrac{1}{2}y)\psi(x-\tfrac{1}{2}y) \ , \tag{33}$$

in terms of a mixed Fourier transform. The second quantized Wigner operator, $\hat{W}$, is then obtained by replacing $\psi(x)$ by the Heisenberg field operator $\hat{\psi}(x)$. The advantage of second quantization is of course that it allows us to pass from a single particle to many body quantum theory. The ensemble average of $\hat{W}$ is what corresponds to the classical phase space density.

The equation of motion for $\hat{W}$ obviously follows from the Schrödinger equation, $(i\hbar\partial_t - \hbar^2\nabla^2/2m + \hat{U}(x))\hat{\psi}(x) = 0$. To proceed we compute $p\cdot\nabla\hat{W}(x,p)$. Within the integration, p can be replaced by $-i\hbar\nabla_y$ after integrating by parts. Then, since $\nabla_y\cdot\nabla_x = \frac{1}{2}(\nabla^2_{x+y/2} - \nabla^2_{x-y/2})$, we obtain using the Schrödinger equation

$$p\cdot\nabla_x\hat{W} = m\int \frac{d^3y}{(2\pi\hbar)^3} e^{ipy/\hbar}(-\partial_t + \frac{i}{\hbar}[\hat{U}(x+\tfrac{1}{2}y) - \hat{U}(x-\tfrac{1}{2}y)])\hat{\psi}^\dagger(x+\tfrac{1}{2}y)\hat{\psi}(x-\tfrac{1}{2}y) \ . \tag{34}$$

Expanding in Taylor series, $U(x+\frac{1}{2}y) = \exp(\frac{1}{2}y\nabla_x)U(x)$, we can pull the exponential out of the integral by replacing y by $-i\hbar\nabla_p$ term by term in the series. With this trick, the right hand side of (34) can be written formally as

$$m(-\partial_t + \frac{2}{\hbar}[\sin(\tfrac{1}{2}\hbar\triangle)\hat{U}])\hat{W} \ ,$$

where we introduced [2] the "triangle" operator, $\triangle \equiv \nabla_x\cdot\nabla_p$ with ∇_x only acting on the potential and ∇_p only acting on the Wigner operator. In this way we have derived the non-relativistic quantum transport equation

$$(m\partial_t + p\cdot\nabla_x - \nabla_x\hat{U}\cdot\nabla_p)\hat{W}(x,p) = Q(\hbar\triangle)\hat{U}(x)\hat{W}(x,p) \ , \tag{35}$$

where the "quantum" operator is given by

$$Q(\hbar\triangle) = \frac{2}{\hbar}\sin(\tfrac{1}{2}\hbar\triangle) - \triangle = -\frac{\hbar^2}{24}\triangle^3 + O(\hbar^4) \ . \tag{36}$$

In the $\hbar \to 0$ limit we recover the Vlasov-Boltzmann equation. The Vlasov part follows from approximating the ensemble average of $\hat{U}\hat{W}$ on the l.h.s. by $\langle\hat{U}\rangle\langle\hat{W}\rangle$. The collision term is contained in the correlation function $C(x,p) = \triangle(\langle\hat{U}\hat{W}\rangle - \langle\hat{U}\rangle\langle\hat{W}\rangle)$ just as in the classical case. The terms on the right hand side correspond to genuine quantum corrections. It is however important to note that quantum corrections also occur in the collision term, C, since quantum fluctuations affect correlation functions [21].

The importance of quantum corrections in the collisionless domain depends on the magnitude of the dimensionless ratio $\hbar\Delta \sim \hbar/(\Delta R_U \Delta P_W)$, where ΔR_U is the characteristic *spatial* scale of variation of the potential and where ΔP_W is the characteristic *momentum* scale over which $W(x,p)$ varies appreciably. Classical Vlasov transport theory thus applies only for relatively slowly varying potentials and phase space densities. In Ref. [21] it was shown that quantum corrections modify the collision terms if the density is so high that the two body scattering rate approaches $\hbar/\Omega$, where Ω is the characteristic single-particle energy in the plasma.

Gauge Covariant Wigner Operator

Turning next to gauge theories we are faced with the new complication of formulating transport theory in a gauge invariant way. Recall that under a general non-Abelian gauge transformation, $G(x) = \exp(i\theta_a(x)t_a)$, the fields, $\psi(x)$ and $A_\mu(x)$, in the fundamental and adjoint representations transform as

$$\psi(x) \to G(x)\psi(x) \quad , \quad A_\mu(x) \to G(x)A_\mu(x)G^\dagger(x) + G(x)\partial_\mu G^\dagger(x)/(ig) \ . \tag{37}$$

Thus, the covariant derivative and field tensor transform covariantly, i.e., $D_\mu(x) \to G(x)D_\mu(x)G^\dagger(x)$. The relativistic generalization [22,23] of Wigner operator (31) as given by

$$\hat{W}_{\alpha\beta}(x,p) = \hat{\bar{\psi}}_\beta \delta^4(p-\hat{p})\hat{\psi}_\alpha(x) \tag{38}$$

is not satisfactory for gauge theories because under a gauge transformation:

$$\hat{W} \to \hat{\bar{\psi}}(x)G^\dagger(x) \otimes \delta^4(p-\hat{p})G(x)\hat{\psi}(x) \neq G(x)\hat{W}G^\dagger(x) \ .$$

We must demand that $\hat{W}$ transforms covariantly to insure for example that the momentum distribution, $\int d^4x \langle Tr\hat{W}(x,p)\rangle$, is gauge invariant. The problem is that ordinary derivatives do not commute with $G(x)$. Thus the ordinary derivative of a field, $\partial\psi(x) = \lim(\psi(x+\epsilon) - \psi(x))/\epsilon$ is not well defined in a gauge theory because under a gauge transformation it transforms inhomogeneously, $\partial\psi \to G\{\partial\psi + (G^{-1}\partial G)\psi\}$. The field at neighboring points can be compared only with the *covariant* derivative, $D_\mu(x)$. Under a gauge transformation $D_\mu(x)\psi(x) \to G(x)D_\mu(x)\psi(x)$. This is of course why covariant derivatives were invented in the first place. In addition, the covariant derivative is special because it represents the *kinetic* in contrast to the *canonical* momentum. Recall that in classical electrodynamics the kinetic momentum is given by $\pi^\mu = p^\mu - eA^\mu$, where p^μ is the *canonical* momentum conjugate to the coordinate x^μ. In quantum mechanics p^μ can be represented when sandwiched between $\hat{\psi}^\dagger$ and $\hat{\psi}$ by the operator $\hat{p}^\mu = \frac{1}{2}i(\partial^\mu - \partial^{\dagger\mu})$ where $\partial^\dagger$ acts to the left in bilinear forms. Thus, the operator representing the kinetic momentum can be represented by $\hat{\pi}_\mu = \hat{p}_\mu - eA_\mu(x) = \frac{1}{2}i(D - D^\dagger)$ in terms of the *gauge covariant* derivatives. Note that up to spin corrections it is π^μ and not p^μ which is on shell since $\hat{\pi}^2 = m^2 + O(\hbar)$ follows from the Dirac equation. We are thus led as emphasized in [2,3,4] to define the gauge covariant Wigner operator by substituting the gauge covariant derivative D^μ and its adjoint in place of ∂^μ and its adjoint in the naive gauge dependent definition.

Applying this minimal substitution rule gives

$$\begin{aligned}\hat{W}(x,p) &= \bar{\psi}(x)\,\delta^4(p-\hat{\pi}(x))\,\psi(x)\,,\\ &= \int \frac{d^4y}{(2\pi)^4}e^{-ip\cdot y}\,\bar{\psi}(x)e^{\frac{1}{2}y\cdot D^\dagger(x)}\otimes e^{-\frac{1}{2}y\cdot D(x)}\psi(x)\,. \end{aligned} \tag{39}$$

The symbol $\otimes$ indicates that $\hat{W}$ is a 4×4 matrix in spinor indices and a $N\times N$ matrix in color indices. Because $\hat{\pi}_\mu$ transforms covariantly, so does $\hat{W}$ Not only have we achieved gauge covariance but also it is now obvious that $\langle Tr\hat{W}(x,p)\rangle$ has the desired interpretation of being the Lorentz scalar density of particles at x with kinetic momentum p.

Next we need to know how $\exp(-y\cdot D(x))$ acts on $\psi(x)$. We note [2] that

$$\begin{aligned}\exp(-y\cdot D_x) &= \lim_{n\to\infty}\,(\,1-\Delta y(\partial_x+igA(x))\,)^n\\ &= \lim_{n\to\infty}\,\Big(\,(1-ig\Delta y\cdot A(x))\,e^{-\Delta y\cdot\partial_x}\,(1+O(\Delta y^2))\,\Big)^n\\ &= \lim_{n\to\infty}\,(1-ig\Delta y\cdot A(x))\cdots(1-ig\Delta y\cdot A(x-(n-1)\Delta y)\,\exp(-y\cdot\partial_x)\\ &= U(x,x-y)e^{-y\cdot\partial_x}\,, \end{aligned} \tag{40}$$

where $\Delta y\equiv y/n$. Thus the effect of the covariant translation operator is to translate the field and keep track of the phase via the path ordered phase factor (or link operator)

$$U(b,a)=Pexp\left\{-ig\int_0^1 ds\frac{dz^\mu}{ds}A_\mu(z(s))\right\}=P\prod_{s=0}^{1}(1-igA(z(s))\cdot dz(s))\,, \tag{41}$$

Note that the path of integration is uniquely determined to be a *straight line* between the end points, $z(s)\equiv z(b,a,s)=a+(b-a)s$ with $0\le s\le 1$.

Finally, we can express the gauge covariant Wigner operator in the form proposed in Ref.[24],

$$\hat{W}(x,p)=\int\frac{d^4y}{(2\pi)^4}e^{-ip\cdot y}\,\bar{\psi}(x+\tfrac{1}{2}y)U(x+\tfrac{1}{2}y,x)\otimes U(x,x-\tfrac{1}{2}y)\psi(x-\frac{1}{2}y)\,. \tag{42}$$

The advantage of our definition (39), is that there is no path ambiguity on account of the physical requirement that the momentum variable corresponds to the kinetic momentum. Of course, $U(b,a)$ depends on the path except in the trivial case, when $A^\mu\propto S\partial^\mu S^{-1}$ is a pure gauge field with $F^{\mu\nu}=0$.

Gauge Covariant Operator Equation

To derive the equation of motion for $\hat{W}$ we need the following relation derived in [2] for how the link operator varies when the endpoints are varied:

$$\begin{aligned}\delta U(b,a) &= U(b+db,a+da)-U(b,a)\\ &= -igA(b)\cdot db\,U(b,a)+igU(b,a)A(a)\cdot da\\ &\quad +ig\int_0^1 ds\,U(b,z(s))F_{\mu\nu}(z(s))U(z(s),a)(b-a)^\mu(da+(db-da)s)^\nu\,, \end{aligned} \tag{43}$$

with $z(s)\equiv a+(b-a)s$.

Consider first the equation obeyed by the kernel of the Wigner operator

$$\hat{\rho}(x_2,x_1) = \bar{\psi}(x_2)U(x_2,x) \otimes U(x,x_1)\psi(x_1) = 0 \ . \tag{44}$$

where $x_1 \equiv x - \frac{1}{2}y$ and $x_2 \equiv x + \frac{1}{2}y$ and thus $x = \frac{1}{2}(x_1 + x_2)$. From the Dirac equation

$$\bar{\psi}(x_2)U(x_2,x) \otimes U(x,x_1)\left(i\gamma^\mu D_\mu(x_1) - m\right)\psi(x_1) = 0 \ . \tag{45}$$

Taking the derivative to the left we get

$$\begin{aligned}(i\gamma_\mu D^\mu(x_1) - m)\hat{\rho}(x_2,x_1) \;&=\; A^\mu(x_1)\psi(x_1) \\ &\quad + i\gamma_\mu\bar{\psi}(x_2)U(x_2,x) \otimes [\partial^\mu_{x_1}U(x,x_1)]\psi(x_1) \\ &\quad + i\gamma_\mu\bar{\psi}(x_2)[\partial^\mu_{x_1}U(x_2,x)] \otimes U(x,x_1)\psi(x_1) \ , \end{aligned} \tag{46}$$

where we have defined the product of any operator $\hat{O}$ times $\hat{\rho}$ such that

$$\begin{aligned}\hat{O}\hat{\rho}(x_2,x_1) \;&\equiv\; \bar{\psi}(x_2)U(x_2,x) \otimes \hat{O}U(x,x_1)\psi(x_1) \ , \\ \hat{\rho}(x_2,x_1)\hat{O} \;&\equiv\; \bar{\psi}(x_2)U(x_2,x)\hat{O} \otimes U(x,x_1)\psi(x_1) \ . \end{aligned} \tag{47}$$

Using (43) we find that

$$\begin{aligned}\partial^\mu_{x_1}U(\tfrac{1}{2}(x_1+x_2),x_1) \;&=\; -\tfrac{1}{2}igA^\mu(x)U(x,x_1) + igU(x,x_1)A^\mu(x_1) \\ &+ig\int_0^1 ds\, U(x,z)F_\nu{}^\mu(x;z;x_1)U(z,x_1)\cdot\tfrac{1}{2}(x_2-x_1)^\nu\cdot(1-\tfrac{1}{2}s) \ , \end{aligned} \tag{48}$$

and,

$$\begin{aligned}\partial^\mu_{x_1}U(x_2,\tfrac{1}{2}(x_1+x_2)) \;&=\; +\tfrac{1}{2}igU(x_2,x)A^\mu(x) \\ &+ig\int_0^1 ds\, U(x_2,z)F_\nu{}^\mu(x_2;z;x)U(z,x)\cdot\tfrac{1}{2}(x_2-x_1)^\nu\cdot\tfrac{1}{2}(1-s) \ , \end{aligned} \tag{49}$$

where $F_{\mu\nu}(b;z;a) = F_{\mu\nu}(z)$ for z along the straight line path between a and b. The Wigner kernel thus obeys

$$\begin{aligned}&i\gamma_\mu\left(\partial^\mu_{x_1}\hat{\rho}(x_2,x_1) + \tfrac{1}{2}ig[A^\mu(x),\hat{\rho}(x_2,x_1)]\right) - m\hat{\rho}(x_2,x_1) \\ &= -\tfrac{1}{2}g(x_2-x_1)^\nu\gamma^\mu\left(\int_0^1 ds\,(1-\tfrac{1}{2}s)U(x,z)F_{\nu\mu}(x;z;x_1)U(z,x_1)U(x_1,x)\hat{\rho}(x_2,x_1)\right. \\ &\left.+\hat{\rho}(x_2,x_1)\int_0^1 ds\,\tfrac{1}{2}(1-s)U(x,x_2)U(x_2,z)F_{\nu\mu}(x_2;z;x)U(z,x)\right) \ . \end{aligned} \tag{50}$$

To simplify this expression further, we define the covariant derivative of a second-rank tensor $\mathcal{T}$ by

$$\mathcal{D}(x)\mathcal{T}(x) \equiv \partial_x\mathcal{T}(x) + ig[A(x),\mathcal{T}(x)] \ . \tag{51}$$

One may show by a similar proof as in eq.(40) that

$$e^{-y\cdot\mathcal{D}(x)}\,\mathcal{T}(x) \;=\; U(x,x-y)\mathcal{T}(x-y)U(x-y,x) \ . \tag{52}$$

Changing variables from x_2, x_1 to the midpoint and relative distance variables $x \equiv \frac{1}{2}(x_1 + x_2)$ and $y \equiv x_2 - x_1$, and using eqs.(51, 52), we can write (50) as

$$-i\gamma_\mu \left(\partial_y^\mu - \tfrac{1}{2}D^\mu(x)\right) \hat{\rho}(x + \tfrac{1}{2}y, x - \tfrac{1}{2}y) - m\hat{\rho}(x + \tfrac{1}{2}y, x - \tfrac{1}{2}y)$$

$$= -\tfrac{1}{2}gy^\nu\gamma^\mu \left(\int_0^1 ds\,(1 - \tfrac{1}{2}s)\,[e^{-\frac{1}{2}(1-s)y\cdot D(x)}\,F_{\nu\mu}(x)]\,\hat{\rho}(x + \tfrac{1}{2}y, x - \tfrac{1}{2}y) \right.$$

$$\left. +\hat{\rho}(x + \tfrac{1}{2}y, x - \tfrac{1}{2}y) \int_0^1 ds\,\tfrac{1}{2}(1 - s)\,[e^{\frac{1}{2}sy\cdot D(x)}\,F_{\nu\mu}(x)] \right) . \qquad (53)$$

Taking the relative Fourier transform of (53) and reinstating $\hbar$, we obtain finally the following equation for the Wigner operator:

$$\left(\gamma^\mu p_\mu - m + \tfrac{1}{2}i\hbar\gamma^\mu D_\mu(x)\right)\hat{W}(x,p) = \tfrac{1}{4}\hbar g\gamma_\mu \left(\{Q_1(\tfrac{1}{2}\hbar\triangle)\,F^{\mu\nu}(x), \partial_\nu^p W(x,p)\} \right.$$

$$\left. + [Q_2(\tfrac{1}{2}\hbar\triangle)\,F^{\mu\nu}(x), \partial_\nu^p W(x,p)]\right) , \qquad (54)$$

where

$$Q_1(x) = i\frac{\sin x}{x} + \frac{\sin x - x\cos x}{x^2} \quad , \quad Q_2(x) = i\frac{x\sin x + \cos x - 1}{x^2} + \frac{1 - \cos x}{x} \; . \qquad (55)$$

and the *triangle operator*, $\triangle$, is given in this case by

$$\triangle = \partial_p \cdot D = \partial_p \cdot \partial_x + \frac{1}{\hbar}ig\partial_p \cdot [A(x),\ ] \, , \qquad (56)$$

We emphasize that $D(x)$ on the right hand side only acts on $F_{\nu\mu}$. To obtain this final expression we used the same trick as in the non-relativistic case of replacing y by $i\partial_p$ using $\int d^4y e^{-ip\cdot y} f(y)g(y) = f(i\partial_p) \int d^4y e^{-ip\cdot y} g(y)$ and performing the s integral explicitly. Because D_μ, $F_{\mu\nu}$, $\triangle$ and $\hat{W}(x,p)$ transform covariantly under a gauge transformation, eq.(54) is the sought after *gauge covariant operator equation* of motion. The ordering of the field operators follows the rules (47).

Equation (54) is completely equivalent to the original Dirac's equation for the Heisenberg field $\hat{\psi}$. However, it is in a form that allows us to extract the quantum transport equation. As we shall see in the next section it in fact contains two equations. One is the generalized quantum transport equation. The second is a generalized mass shell constraint equation that arises on account of the Heisenberg uncertainty principle.

Quantum Vlasov and Constraint Equations in the Abelian Limit

The physical content of (54) is especially clear in the Abelian limit when in addition $F_{\mu\nu}$ is replaced as the self-consistent c-number field [4]. In that case the operator ordering is irrelevant. Specifically we consider the Hartree approximation in which $F_{\mu\nu}$ is approximated by the mean (c-number) field $\bar{F}_{\mu\nu}$ satisfying Maxwell's equations

$$\partial_\mu \bar{F}^{\mu\nu}(x) = \langle J^\mu(x)\rangle = e\,tr \int d^4p\,\gamma^\mu W^H(x,p) \, . \qquad (57)$$

Of course, an arbitrary external current could be added to the r.h.s. of (57). The trace is over spinor indices of $W^H(x,p) = \langle\hat{W}(x,p)\rangle$ corresponding to the ensemble average of the Wigner operator in the Hartree approximation. In this case the *triangle operator* is simply the mixed derivative

$$\triangle \equiv \partial^p \cdot \partial_x \ . \tag{58}$$

Notice that the triangle operator has the dimension of inverse action so that in units where $\hbar \neq 1$ it always appears multiplied by $\hbar$. Therefore, a power series expansion of the Bessel functions coincides with an expansion in terms of the ratio of $\hbar$ to a characteristic angular momentum, L, of the plasma. That characteristic angular momentum measures the product of the space-time scale, ΔR_F, over which the field tensor, $\bar{F}^{\mu\nu}(x)$, varies appreciably and the momentum scale, ΔP_W, over which the Wigner function varies appreciably. Therefore, a necessary condition for the validity of a power series expansion of the Bessel functions is that the mean field is slowly varying in the sense

$$\hbar \ll \Delta R_F \Delta P_W \ . \tag{59}$$

In the Abelian case the linear equation (54) can be written in the following suggestive form:

$$(\gamma \cdot K - m)\, W^H(x,p) = 0\,, \tag{60}$$

in terms of the operator $K^\mu \equiv \Pi^\mu + \frac{1}{2} i\hbar \nabla^\mu$, where

$$\nabla^\mu \equiv \partial_x^\mu - e j_0(\tfrac{1}{2}\hbar\triangle)\, \bar{F}^{\mu\nu}\, \partial_\nu^p\,, \tag{61}$$

$$\Pi^\mu \equiv p^\mu - \tfrac{1}{2}\hbar e j_1(\tfrac{1}{2}\hbar\triangle)\, \bar{F}^{\mu\nu}\, \partial_\nu^p\,, \tag{62}$$

and where $j_l(x)$ are conventional spherical Bessel functions. We have reinstated $\hbar$ explicitly to show clearly the quantum character of this equation.

To second order in $\hbar$ these operators are given by

$$\begin{aligned} \nabla^\mu &= \partial_x^\mu - e\bar{F}^{\mu\nu}\partial_\nu^p + \tfrac{1}{24} e\hbar^2 \triangle^2 \bar{F}^{\mu\nu}\partial_\nu^p \ + \ O(\hbar^4)\ , \\ \Pi^\mu &= p^\mu - \tfrac{1}{12} e\hbar^2\, \triangle \bar{F}^{\mu\nu}\partial_\nu^p \ + \ O(\hbar^4)\,. \end{aligned} \tag{63}$$

The generalized constraint and transport equations can be extracted by first multiplying (60) by $(\gamma \cdot K + m)$. Noting that $\gamma^\mu\gamma^\nu = g^{\mu\nu} - i\sigma^{\mu\nu}$, the following quadratic quantum equation of motion follows:

$$(K^2 - m^2 - \tfrac{1}{2} i\sigma^{\mu\nu}[K_\mu, K_\nu])\, W^H(x,p) = 0 \ . \tag{64}$$

Adding and subtracting the adjoint of this equation then leads to

$$(\Pi^2 - m^2 - \tfrac{1}{4}\hbar^2\nabla^2)\, W^H(x,p) \ = \ -\tfrac{1}{4}\hbar I^{\mu\nu}\{\sigma_{\mu\nu}, W^H(x,p)\} + \tfrac{1}{4} i\hbar R^{\mu\nu}[\sigma_{\mu\nu}, W^H(x,p)]\ , \tag{65}$$

$$\Pi^\mu \nabla_\mu\, W^H(x,p) \ = \ \tfrac{1}{4} i I^{\mu\nu}[\sigma_{\mu\nu}, W^H(x,p)] + \tfrac{1}{4} R^{\mu\nu}\{\sigma_{\mu\nu}, W^H(x,p)\}\ , \tag{66}$$

where the l.h.s. of the quantum transport equation (66) acquires the remarkably simple form on account of the relation $[\nabla_\mu, \Pi^\mu] = 0$ that holds to all orders in $\hbar$ because of the special form of Π^μ and ∇^μ and the relation $dj_0(z)/d_z = -j_1(z)$.

The real operators R and I associated with the spin terms on the r.h.s. are given by

$$\begin{aligned} R^{\mu\nu} + iI^{\mu\nu} &\equiv \hbar^{-1}[K^\mu, K^\nu] = -\tfrac{1}{2}\hbar\triangle j_0(\tfrac{1}{2}\hbar\triangle)\, e\bar{F}^{\mu\nu} - i[j_0(\tfrac{1}{2}\hbar\triangle) - \tfrac{1}{2}\hbar\triangle j_1(\tfrac{1}{2}\hbar\triangle)]\, e\bar{F}^{\mu\nu} \\ &= -ie\bar{F}^{\mu\nu} - \tfrac{1}{2}\hbar e\triangle\bar{F}^{\mu\nu} + O(\hbar^2 e\triangle^2\bar{F}^{\mu\nu})\ . \end{aligned} \tag{67}$$

The generalized quantum constraint and quantum Vlasov equations [4] (65) and (66), specify the dynamics of Abelian plasmas in the "collisionless" regime. They generalize the non-relativistic quantum transport equation (35) to QED of spin-$\frac{1}{2}$ particles. Unlike previous work [26]-[29] based on gauge dependent formulations and/or with assumptions about "slowly varying" fields, the above Lorentz covariant, gauge invariant equations apply for arbitrarily strong and varying c-number fields. Quantum corrections in the colisionless domain may be computed systematically to any order in $\hbar$ by expanding the Bessel functions in powers of the triangle operator. Together with eq. (57) they form a coupled set of equations to determine simultaneously the fermion Wigner function and the field $\bar{F}^{\mu\nu}(x)$ in a self-consistent way. They form a closed set of equations because the expectation value of the current is directly related to the Wigner function in the Hartree approximation via (57)

The "collision" terms neglected in this approximation follow from the operator equation (54) only when correlations such as $\langle F^{\mu\nu}\hat{W}\rangle - \bar{F}^{\mu\nu}W$ are not neglected, i.e., only if the operator character of the gauge field is explicitly taken into account. Using the methods outlined in Ref.[22] for non-gauge theories, it is possible to compute explicitly such collisions terms. However, for gauge theories with long range forces, such as QED, the infrared divergences arising in perturbation theory need special care [27]. We note again that quantum corrections to the collision terms are small only in relatively dilute plasmas [21].

To second order in $\hbar$, the quantum Vlasov equation including spin corrections is thus given by

$$
\begin{aligned}
&(p\cdot\partial_x - ep_\mu\bar{F}^{\mu\nu}\partial_\nu^p)\, W^H(x,p) + \tfrac{1}{4}ie\bar{F}^{\mu\nu}\left[\sigma_{\mu\nu}, W^H(x,p)\right] \\
&= -\tfrac{1}{12}\hbar^2 e\triangle\bar{F}_{\mu\nu}[\partial_x^\nu - e\bar{F}^{\nu\lambda}\partial_\lambda^p]\partial_p^\mu\, W^H(x,p) - \tfrac{1}{8}\hbar e\triangle\bar{F}^{\mu\nu}\left\{\sigma_{\mu\nu}, W^H(x,p)\right\}\ .
\end{aligned} \tag{68}
$$

Equation (68) reduces to the equation derived in [25] for scalar QED when the spin terms are neglected.

It is important to appreciate that the above quantum Vlasov equation is not complete without the quantum constraint equation (65). As soon as quantum corrections become important in the transport equation, quantum corrections to the constraint equation must also be considered as emphasized in [4]. To second order in $\hbar$, the constraint equation in the Hartree approximation is given by

$$
\begin{aligned}
(p^2 - m^2)\, W^H(x,p) = {}& \tfrac{1}{4}\hbar e\bar{F}^{\mu\nu}\{\sigma_{\mu\nu}, W^H(x,p)\} - \tfrac{1}{8}i\hbar^2 e\triangle\bar{F}^{\mu\nu}[\sigma_{\mu\nu}, W^H(x,p)] \\
& +\hbar^2(\tfrac{1}{6}ep\cdot\triangle\bar{F}\cdot\partial_p + \tfrac{1}{12}e(\partial_x^\mu\bar{F}_{\mu\nu})\partial_\nu^p + \tfrac{1}{4}(\partial_x - e\bar{F}\cdot\partial_p)^2)W^H(x,p)
\end{aligned} \tag{69}
$$

From eqs. (68) and (69) we see that quantum transport theory reduces to classical transport theory only if several conditions are satisfied simultaneously. In addition to the condition (59) necessary for the validity of the triangle operator expansion, the field strength must be small compared to the typical energy scale of particles in the plasma, i.e.,

$$
\hbar e\|\bar{F}^{\mu\nu}\| \ll m^2 + |\vec{p}|^2\ . \tag{70}
$$

Thus the field strength has to be relatively weak and slowly varying for classical theory to hold. In addition, spatial and momentum gradients of the Wigner function need to be sufficiently small so that $\Delta m^2 = \frac{1}{4}\hbar^2(\partial_x - e\bar{F}\partial_p)^2 \ll m^2$. These off mass shell corrections obviously arise as a result of the uncertainty principle.

We remark that in the Hartree approximation the spinor structure of the above equations has been resolved in Ref.[4]. The scalar and axial vector components of W^H were shown to specify W^H completely to any order in $\hbar$. Furthermore the classical BMT equation for spinning particles was derived from (68) in the semiclassical limit.

Quark Transport and the Abelian Dominance Approximation

We turn next to the non Abelian features of (54). There is a very important difference between the triangle operator expansion in the Abelian and non-Abelian cases. The Abelian expansion is equivalent to the semiclassical expansion in powers of $\hbar$. In the non-Abelian case, there appears a commutator in (56) that is of *zeroth* order in $\hbar$!! We can still write the linear operator equation in the form (60). However, the operators Π^μ and ∇^μ are now given by

$$\begin{aligned}
\nabla^\mu &\equiv \mathcal{D}_x^\mu - \tfrac{1}{2}g\{ImQ_1(\tfrac{1}{2}\hbar\Delta)\,F^{\mu\nu},\ \}\,\partial_\nu^p - \tfrac{1}{2}g[ImQ_2(\tfrac{1}{2}\hbar\Delta)\,F^{\mu\nu},\ ] \\
&= \frac{1}{\hbar}ig[A^\mu(x),\ ] + \partial_x^\mu - \tfrac{1}{2}g\{F^{\mu\nu},\ \}\partial_\nu^p - \tfrac{1}{24}g^3\{[A^\alpha,[A^\beta,F^{\mu\nu}]],\ \}\partial_\nu^p\partial_\alpha^p\partial_\beta^p + \cdots\,, \\
\Pi^\mu &\equiv p^\mu - \tfrac{1}{4}\hbar g\left(\{ReQ_1(\tfrac{1}{2}\hbar\Delta)\,F^{\mu\nu},\ \} + [ReQ_2(\tfrac{1}{2}\hbar\Delta)\,F^{\mu\nu},\ ]\right)\partial_\nu^p\,.
\end{aligned} \tag{71}$$

We see that something terrible has happened. The non-Abelian commutator has changed the leading order in $\hbar$ for the operator ∇^μ to be $O(1/\hbar)$, and there now arise an infinite number of commutators and anti-commutators to zeroth order in $\hbar$. Thus the simple classical limit we found in the Abelian case becomes hopelessly complicated in the non-Abelian case.

The only sensible semiclassical limit of a non-Abelian theory is one where the *covariant* derivative of $F^{\mu\nu}$ is small. Only for covariantly slowly varying fields, in the sense that

$$\langle F\hat{W}\rangle \gg \langle \mathcal{D}F\ \partial_p\hat{W}\rangle\ , \tag{72}$$

is it possible to carry out an expansion in powers of the Δ operator. The covariant constant field case considered in lecture 1 trivially satisfies this condition. In particular for covariant constant fields, the linear operator equation for $\hat{W}$ reduces to

$$\left(\gamma_\mu(p^\mu + \tfrac{1}{2}i\hbar(\mathcal{D}^\mu - \tfrac{1}{2}g\{F^{\mu\nu},\ \}\partial_\nu^p - \tfrac{1}{4}g[F^{\mu\nu},\ ]\partial_\nu^p) - m\right)\hat{W}(x,p) = 0\ . \tag{73}$$

In the general case for strong or rapidly varying fields the full quantum equation, eq.(54), must be solved. This would be equivalent to solving the original field equations, i.e. hopeless at this time. Only under the rather restrictive conditions, (72), can we expect that the transport theory for quarks reduces to a simpler, more manageable form.

Proceeding as in the Abelian case, we can isolate the transport equation for (73) by converting it into a quadratic equation and isolating the anti-Hermitian part. This leads to the *semiclassical transport equation for the QCD Wigner operator* [2]:

$$\begin{aligned}
&p\cdot\mathcal{D}(x)\,\hat{W}(x,p)\ +\ \tfrac{1}{2}gp^\mu\partial_p^\nu\left\{\,F_{\nu\mu}(x),\hat{W}(x,p)\,\right\}\ +\ \tfrac{1}{4}ig\left[\,\sigma^{\mu\nu}F_{\mu\nu}(x),\hat{W}(x,p)\,\right] \\
&= -\ \tfrac{1}{8}ig\partial_p^\mu\left[\,F_{\mu\nu}(x),(\mathcal{D}^\nu(x)\,\hat{W}(x,p))\,\right]\ -\ \tfrac{1}{16}ig^2\partial_p^\mu\partial_p^\nu\left[\,F_\mu{}^\lambda(x)F_{\nu\lambda}(x),\hat{W}(x,p)\,\right] \\
&\quad + O(\Delta F)\ .
\end{aligned} \tag{74}$$

Equation (74) is the non Abelian generalization of Vlasov's equation in the following sense: i) it is still an operator equation; ii) the first and second term on the l.h.s. of eq.(74) present the usual combination of phase space variables and derivatives, however, modified by (anti)commutators; iii) explicit spin-dependent corrections arise; iv) quantum corrections can be systematically calculated via expansion in powers of the Δ operator. The general form for the transport equations valid to all orders in Δ was also derived in Ref.[2]. Neglecting all but the first two terms on the l.h.s. recovers the classical transport equation for spinless quarks first derived in Ref.[24]. Furthermore, we have shown that the $O(\Delta F)$ corrections not involving $\sigma^{\mu\nu}$ that follow from (54) reproduce the result for scalar quarks obtained in Ref.[30] modulo a sign error therein.

For covariant constant or slowly varying fields, the color structure of the above equation is particularly transparent. We consider essentially Abelian fields such that all components of $F^{\mu\nu}$ can be simultaneously diagonalized in the same gauge:

$$\langle F^{\mu\nu}(x)\rangle \equiv S(x)\vec{F}^{\mu\nu}\cdot\vec{h}S^{-1}(x) \ , \tag{75}$$

where h_i the $N-1$ diagonal Gell-Mann matrices and $S(x)$ is a particular gauge transformation. We now make the bold model dependent *ansatz* that the ensemble of quarks is such that $\hat{W}$ is diagonal in the *same* gauge where $\langle F_{\mu\nu}(x)\rangle$ is diagonal. By assumption then we can express

$$\langle\hat{W}\rangle \equiv S(x)\left(\sum_{j=1}^{N-1} W^j h_j + W^0\mathbf{1}\right) S^{-1}(x) \ , \tag{76}$$

in terms of N Wigner *functions* depending on (x,p). The decomposition in spinor space proceeds as in the Abelian case [4]. Note that the property $\hat{W}^\dagger = \gamma^0\hat{W}\gamma^0$ insures that the N diagonal elements in color space $\langle\hat{W}\rangle^{ii}$ are real numbers after taking appropriate traces in spinor space. Equation (76) is a strong model assumption and we refer to it as the Abelian Dominance Approximation [10] for reasons that become clear below. There is no guaranty that such ensembles of quarks exist in nature or how well they approximate the conditions of the quark-gluon plasma formed in nuclear collisions. We proceed in the spirit of the MIT bag and string models and adopt this assumption for phenomenological purposes.

When (76) holds, it is most convenient to work in the $S(x)$ gauge, where

$$\langle\hat{W}\rangle_{ij} = (\vec{W}\cdot\vec{h} + W^0_{ij}\mathbf{1}) = \delta_{ij}(\vec{W}\cdot\vec{\epsilon}_j + W^0) \equiv \delta_{ij}f_j \ , \tag{77}$$

where the "charges" $\vec{\epsilon}_j$ are just the elementary weight vectors of SU(N) discussed in lecture 1.

The semiclassical transport equations for the $f_j(x,p)$ in this approximation are obtained by taking the expectation value of eq.(74) in this ensemble. We see that the color structure of those equations simplifies considerably:

$$(p\cdot\partial_x + g\vec{\epsilon}_j\cdot\vec{F}_{\mu\nu}p^\nu\partial_p^\mu)\, f_j(x,p) = -\tfrac{1}{4}ig\vec{\epsilon}_j\cdot\vec{F}_{\mu\nu}\,[\,\sigma^{\mu\nu}, f_j(x,p)\,] + \tilde{C}_j(x,p) \ , \tag{78}$$

where $\tilde{C}_j$ represents correlation terms of the form

$$\tilde{C}_j(x,p) = -\tfrac{1}{2}gp^\mu\partial_p^\nu\left(\langle\{F_{\nu\mu},\hat{W}(x,p)\}\rangle - 2\langle F_{\nu\mu}\rangle\langle\hat{W}(x,p)\rangle\right)_j + \cdots \ . \tag{79}$$

As usual, the collision terms are contained in such correlations.

Note that all the non-Abelian commutator terms dropped out for the model *ansatz* (76) and the transport theory for quarks has reduced to an effective multicomponent Abelian plasma theory. The $f_i(x,p)$ just correspond to the phase space densities for quarks with "charge" $g\vec{e}_i$. Just as the pair production rate for $q_i\bar{q}_i$ in the semiclassical limit could be computed as if the $i^t h$ colored quarks couple with an effective Abelian field with $e_i F^{\mu\nu}_{eff} = g\vec{e}_i \cdot \vec{F}^{\mu\nu}$, we see that those quarks obey an effective Abelian Vlasov equation with the same $e_i F^{\mu\nu}_{eff}$ appearing in the force term. Therefore, the quark plasma dynamics is very simple and intuitive in this Abelian Dominance Approximation.

Gluon Transport and the Abelian Dominance Approximation

To develop a gauge covariant transport theory for gluons we start with the following definition of the gauge covariant gluon Wigner operator proposed in Ref. [3].

$$\hat{\Gamma}_{\mu\nu}(x,p) \equiv \int \frac{d^4y}{(2\pi)^4} e^{-ip\cdot y}[e^{\frac{1}{2}y\cdot D(x)} F_\mu{}^\lambda(x)] \otimes [e^{-\frac{1}{2}y\cdot D(x)} F_{\lambda\nu}(x)] \ , \tag{80}$$

where the covariant derivative of a second-rank tensor $\mathcal{T}$ is defined as in (51). The connection between gauge dependent and independent Wigner functions for Abelian photons is discussed in detail in Ref. [4]. In eq.(80) we suppressed four color indices of $\hat{\Gamma}$ but explicitly indicated its tensor structure. $\hat{\Gamma}$ is closely related to the energy-momentum tensor of the field,

$$\hat{T}_{\mu\nu}(x) \equiv Tr\left(F_\mu{}^\lambda F_{\lambda\nu} + \tfrac{1}{4}g_{\mu\nu}F_{\lambda\tau}F^{\lambda\tau}\right) = Tr\int d^4p\left(\hat{\Gamma}_{\mu\nu}(x,p) - \tfrac{1}{4}g_{\mu\nu}\hat{\Gamma}_\lambda{}^\lambda(x,p)\right) \ , \tag{81}$$

where the trace refers to color indices, $Tr A \otimes B \equiv A_{ab}B_{ba}$. Eq.(81) provides the connection between $\hat{\Gamma}$ and observables of the gauge field. Formally,

$$Tr\hat{\Gamma}_{\mu\nu}(x,p) = F_\mu{}^\lambda(x)\delta^4(p - iD(x))F_{\lambda\nu}(x) \ , \tag{82}$$

which shows that $Tr\hat{\Gamma}(x,p)$ measures the energy-momentum flux of gluons at x with kinetic momentum p.

The derivation of the quantum transport equation for $\hat{\Gamma}$ proceeds as for $\hat{W}$ [3]. We want to calculate $p \cdot \tilde{D}\hat{\Gamma}_{\mu\nu}$, where the generalized covariant derivative acting on tensors of the structure of $\hat{\Gamma}$ is defined by

$$\tilde{D} A \otimes B \equiv [DA] \otimes B + A \otimes [DB] \ . \tag{83}$$

Expressing the action of the covariant translation operators in (80) in terms of link operators, we carry $p \cdot \tilde{D}$ into the integrand converting $p \to -i\partial_y$ (by partial integration) and $y \to i\partial_p$. We calculate all necessary derivatives of link operators which occur in the integrand by applying eq.(43). These straightforward though tedious manipulations lead to a very complex equation for $\hat{\Gamma}_{\mu\nu}(x,p)$ recorded in [3]. That equation becomes manageable only in the semiclassical limit obtained by expanding it to lowest order in powers of $\triangle \sim \hbar\partial_p D$.

However, there is an additional complication for gluons not encountered for quarks. The $F^{\mu\nu}$ field contains in general *coherent* as well as *incoherent* parts. The coherent part corresponds to the external or self-consistent mean field $\bar{F}$. The incoherent part describes the interesting fluctuation part that most closely corresponds to a random gas of gluons for which $\langle F^{\mu\nu}\rangle = 0$. We separate these two parts by denoting the incoherent part of $\hat{\Gamma}$ by

$$\hat{G}_{\mu\nu} \equiv \hat{\Gamma}_{\mu\nu} - \bar{\Gamma}_{\mu\nu} \tag{84}$$

where $\bar{\Gamma}_{\mu\nu}$ is defined by eq.(80) with $F \to \bar{F} \equiv \langle F\rangle$. Barred quantities henceforth refer to $\bar{A}$. It is important to realize that $\langle\hat{\Gamma}\rangle \neq \bar{\Gamma}$ and therefore $G \equiv \langle\hat{G}\rangle \neq 0$. The classical mean field is determined from

$$[\bar{D}_\mu, \bar{F}^{\mu\nu}] = g(\langle J^\nu\rangle + \langle j^\nu\rangle) \ , \tag{85}$$

where $\langle J^\nu\rangle$ is the mean color current due to quarks and $\langle j^\nu\rangle$ is the mean induced color current due to the incoherent gas of gluons. We determine that induced current so as to simplify the transport equation for G.

We found that current source terms can be eliminated from in the transport equation by choosing the induced gluon current to satisfy

$$\bar{D}_\mu(x)\langle j_\nu(x)\rangle + iTr'\int d^4p\,\{G_{\mu\nu}(x,p) - G_{\nu\mu}(x,p)\} = 0 \ , \tag{86}$$

with $(Tr'G)_{ac} = G_{abbc}$. In this case the gauge covariant analogue of Vlasov's equation for gluon fluctuations was found [3] to be

$$\begin{aligned} p\cdot\bar{\bar{D}}\,G_{\mu\nu} = {} & \tfrac{1}{2}gp^\sigma\partial_p^\tau \left\{\left[\bar{F}_{\tau\sigma}, G_{\mu\nu}\right]_R + \left[G_{\mu\nu}, \bar{F}_{\tau\sigma}\right]_L\right\} \\ & +\tfrac{1}{8}ig\partial_p^\sigma \left\{\left[\bar{F}^\tau_{\ \sigma}, \bar{\bar{D}}_\tau G_{\mu\nu}\right]_R - \left[\bar{\bar{D}}_\tau G_{\mu\nu}, \bar{F}^\tau_{\ \sigma}\right]_L\right\} \\ & +\tfrac{1}{16}ig^2\partial_p^\sigma\partial_p^\eta \left\{\left[\,\bar{F}_{\tau\sigma}\,,\,\left[\bar{F}^\tau_{\ \eta}, G_{\mu\nu}\right]_R + \left[G_{\mu\nu}, \bar{F}^\tau_{\ \eta}\right]_L\,\right]_R\right. \\ & \left. -\left[\,\left[\bar{F}^\tau_{\ \eta}, G_{\mu\nu}\right]_R + \left[G_{\mu\nu}, \bar{F}^\tau_{\ \eta}\right]_L\,,\,\bar{F}_{\tau\sigma}\,\right]_L\right\} \ , \end{aligned} \tag{87}$$

where the commutators are defined by

$$\left[\bar{F}, A\otimes B\right]_R \equiv \left[\bar{F}, A\right]\otimes B \ , \quad \left[A\otimes B, \bar{F}\right]_L \equiv A\otimes\left[\bar{F}, B\right] \ . \tag{88}$$

Eq.(87) describes a gluon plasma in the so-called collisionless regime to zeroth order in the triangle operator expansion. Eqs.(85,86,87) together form a consistent set of gauge covariant equations for the study of gluon fluctuations under the influence of a classical color field. Note that there is no need to solve for $\bar{\Gamma}$ since eq.(85) determines the classical field. Corrections to the above equations in powers of $\triangle \sim \hbar\partial_p D$ can be calculated systematically from the exact transport equation in [3].

The color structure of the above equations simplifies greatly in the Abelian Dominance Approximation. If $\bar{F}^{\mu\nu}$ is to have the simple form in (75) then so must the mean currents. In particular $Tr'G$ must be diagonal in the same gauge as $\bar{F}$ is. This means that it must be possible to decompose G as

$$G_{\mu\nu}(x,p) \equiv G^{ij}_{\mu\nu}(x,p)e_{ij}\otimes e_{ji} + G^i_{\mu\nu}(x,p)h_i\otimes h_i \equiv G^{\vec{\eta}}_{\mu\nu}(x,p) + G^h_{\mu\nu}(x,p) \ , \tag{89}$$

with $G^{ij}_{\mu\nu} \equiv G^{ji}_{\nu\mu}$ *and* $G^i_{\mu\nu} \equiv G^i_{\nu\mu}$. In this case

$$\left[\vec{h}, G\right]_R = \vec{\eta}_{ij}G^{ij}e_{ij}\otimes e_{ji} = \left[G, \vec{h}\right]_L \ . \tag{90}$$

The induced gluon current satisfying (86) is

$$\langle j_\nu(x)\rangle = -\int d^4p \int d^4q \, e^{iq\cdot x} \frac{q^\mu}{q^2} \, diag(\dots, \tfrac{1}{2}\sum_{j,j\neq i}\{G^{ij}_{\mu\nu}(q,p) - G^{ij}_{\nu\mu}(q,p)\}, \dots) \ . \quad (91)$$

The $N-1$ Abelian mean fields, $F^{\mu\nu}_i$, related to $\bar{F}$ by $\bar{F}^{\mu\nu} = F^{\mu\nu}_i h_i$ satisfy in this approximation

$$\partial_\mu \vec{F}^{\mu\nu} = g(\vec{J}^\nu + \vec{\jmath}^\nu) \ . \quad (92)$$

Inserting eq.(89) into the transport eq.(87), we can isolate the transport equations for the G^{ij}, G^i-components by multiplying with e_{ij}, h_i respectively and taking traces. Thus we find that the $N-1$ diagonal components of G defined in eq.(89) obey Abelian transport equations,

$$p \cdot \partial_x \, G^i_{\mu\nu}(x,p) = 0 \ . \quad (93)$$

Since G^i do not contribute to $\langle j\rangle$ in (91), and do not interact with the mean field $\vec{F}$, the $N-1$ *neutral gluons* described by its components G^i completely decouple from the dynamics of the system in this approximation. Using eq.(90), the Vlasov type equations for the $N(N-1)$ *charged gluons* described by the components G^{ij} are found to be

$$\left(p\cdot\partial_x - gp^\sigma \partial^\tau_p \vec{F}_{\tau\sigma}\cdot\vec{\eta}_{ij}\right) G^{ij}_{\mu\nu}(x,p) = 0 \ . \quad (94)$$

Notice the appearance of the effective coupling $g\vec{\eta}$ for charged gluons just as for quarks we found the effective coupling to be $g\vec{\epsilon}$.

In conclusion, we see that the transport theory for quark-gluon plasmas reduces in the Abelian Dominance Approximation to a multicomponent effectively Abelian plasma theory. In SU(N) there are N quarks with effective charges $g\vec{\epsilon}_i$, $N-1$ neutral gluons, and $N(N-1)$ charged gluons with effective charges $g\vec{\eta}_{ij}$. These particles are produced in pairs and accelerate in an effective $N-1$ component mean field, $\vec{F}^{\mu\nu}$. This suggests a simple extension of Abelian string models and transport theories that could be applied in the phenomenology of nuclear collisions [2]. Obviously, we have only scratched the surface of the topic of Quantum Chromo Transport theory and many fascinating theoretical challenges remain: the structure of collision terms with attention to color electric and magnetic screening, the inclusion of source terms in the phase space formulation due to color neutralization, and a deeper understanding of whether there exists a classical plasma transport limit for quark-gluon plasmas in the first place.

Acknowledgments: We especially thank U. Heinz for many stimulating discussions.

References

[1] M. Gyulassy, A. Iwazaki, Phys. Lett. 165B (1985) 157.

[2] H.-Th. Elze, M. Gyulassy, D. Vasak, Nucl. Phys. B276 (1986) 706.

[3] H.-Th. Elze, M. Gyulassy, D. Vasak, Transport Equations for the QCD Gluon Wigner Operator, LBL-21652, 1986, Phys. Lett. B in press.

[4] D. Vasak, M. Gyulassy, H.-Th. Elze, Quantum Transport Theory for Abelian Plasmas, LBL-21632, 1986, submitted to Ann. Phys.

[5] J. Schwinger, Phys.Rev. 82 (1951) 664.

[6] C. Itzykson and J.-B. Zuber: Quantum Field Theory, McGraw-Hill, New York, 1980.

[7] W. Greiner, B. Müller, J. Rafelski, Quantum Electrodynamics of Strong Fields, Springer-Verlag, Berlin 1985.

[8] I.A.Batalin, S.G.Matinyan, G.K.Savvidi, Sov.J.Nucl. Phys. 26 (1977) 214.

[9] J.Ambjorn, R.J.Hughes, Ann.Phys. 145 (1983) 340.

[10] Z.F. Ezawa, A. Iwazaki, Phys.Rev. D25 (1982) 2681; D26 (1982) 631.

[11] B. Müller: The Physics of the Quark-Gluon Plasma, Springer Lecture Notes in Physics, Springer, Berlin, 1985; J. Cleymans, R. V. Gavai, and E. Suhonen, Phys. Rep. 130 (1986), 217; L. Schroeder, M. Gyulassy, eds., Proc. of the 5th International Conference on Ultra-Relativistic Nucleus-Nucleus Collisions, Asilomar, USA, 1986, to be published in Nucl. Phys.

[12] A.Casher, H.Neuberger, S.Nussinov, Phys.Rev. D20 (1979) 179; E.G.Gurvich, Phys.Lett. 87B (1979) 386.

[13] N.K.Glendenning, T.Matsui, Phys.Rev. D28 (1983) 2890.

[14] T.S.Biro, H.B.Nielsen, and J.Knoll, Nucl.Phys. B245 (1984) 449.

[15] A. Bialas, W. Czyz, Phys. Rev. D31 (1985) 198.

[16] K.Kajantie and T.Matsui, Phys. Lett. 164B (1985) 373.

[17] S.G.Matinyan, G.K. Savvidi, Sov.J.Nucl. Phys. 25 (1977) 118.

[18] P. Danielewicz and M. Gyulassy, Phys. Rev. D31 (1985) 53.

[19] S. Ichimaru: Basic Principles of Plasma Physics, W.A. Benjamin Inc., London, 1973.

[20] E. Wigner, Phys. Rev. 40 (1932), 749.

[21] P. Danielewicz, Ann. Phys. 152 (1984) 239, 305.

[22] S. R. de Groot, W. A. van Leeuwen and Ch. G. van Weert: Relativistic Kinetic Theory, North-Holland, Amsterdam, 1980.

[23] P. Carruthers and F. Zachariasen, Rev. Mod. Phys. 55 (1983), 245.

[24] U. Heinz, Phys. Lett. 144B (1984) 228, Ann. Phys. (N.Y.) 161 (1985), 48.

[25] E. A. Remler, Phys. Rev. D 16 (1977), 3464.

[26] R. Hakim, Riv. Nuovo Cim. 1 no. 6 (1978); R. Hakim and H. Sivak, Ann. Phys. (N.Y.) 139 (1982), 230; H. D. Sivak, Ann. Phys. 159 (1985) 351, Phys. Rev. C34 (1986), 653.

[27] B. Bezzerides, D. F. Dubois, Ann. Phys. 70 (1972) 10.

[28] M. Ploszajczak, M. J. Rhoades-Brown: Non-Equilibrium Truncation Schemes for Statistical Mechanics of Relativistic Matter, Stony Brook Preprint (1986) to be published.

[29] O. T. Serimaa, J. Javanainen and S. Varró, Phys. Rev. A 33 (1986), 2913.

[30] J. Winter, J. de Phys. 45 C6 (1984) 53.

ANTIMATTER CLUSTERS FROM HADRONIZING QUARK-GLUON PLASMA

Ulrich Heinz

Physics Department
Brookhaven National Laboratory
Upton, NY 11973

ABSTRACT

We employ a realistic model for the phase transition between the color deconfining quark-gluon plasma phase and the color confining hadronic gas phase of nuclear matter to discuss the question how quarks and antiquarks hadronize in an expanding quark-gluon plasma. We pay particular attention to the problem associated with the latent heat and latent entropy set free in the hadronization process. Assuming a specific space-time scenario for the phase transition, we compute relative abundances of different hadronic particles and resonances produced during hadronization and show that in particular antinucleons and light antinuclei are enhanced above their equilibrium abundances in a hadron gas of similar density and temperature. We interpret this enhancement as a possible signature for the existence of a transient quark-gluon plasma phase in relativistic heavy-ion collisions which can complement the widely discussed strangeness signal. We point out, however, that detailed dynamical studies of the hadronization process are necessary in order to definitively settle questions about the quantitative yields.

INTRODUCTION

After several years of intensive discussions whether in high energy collisions ($E_{lab} \gtrsim 10$ GeV/n) between heavy nuclei (preferrably A $\gtrsim$ 200) a new state of nuclear matter, the quark-gluon plasma, would be formed and how one would experimentally detect it[1,2], the chase for the quark-gluon plasma is finally on. First experiments are scheduled for the Alternating Gradient Synchrotron (AGS) at Brookhaven National Laboratory and the SPS at CERN towards the end of this year and in early 1987, with projectile ions up to ^{32}S at 15 GeV/n at Brookhaven and up to ^{40}Ca at 60 and 225 GeV/n at CERN. Although the chances to produce a quark-gluon plasma in these first experiments are somewhat limited due to the small size of the projectile ions, we expect an abundance of valuable information on hadronic dynamics in relativistic nuclear collisions, thus filling in the big holes in our theoretical models for collisions in which this hadronic dynamics is modified by a transient quark-gluon plasma phase.

One of the biggest problems in identifying a quark-gluon plasma possibly formed in such a collision is the fact that the bulk of the emitted particles are color-singlet hadrons which are subject to the strong interaction and have very limited memory of the initial hot quark-gluon plasma phase. Rather, their momentum distribution etc. is determined mostly by the hadronization process in which the more or less thermally distributed quarks and gluons coalesce to form the mesons and (anti-) baryons which will finally be detected. Due to its non-perturbative nature, this hadronization process is the most difficult step in a dynamical description of quark matter formation during heavy-ion collisions. On the other hand, its theoretical understanding may be essential in order to extract the necessary information from the spectrum of emitted hadrons that will allow us to prove that a quark-gluon plasma had been formed.

An important piece of information of that kind may be the chemical composition of the emitted hadrons, i.e. the abundance ratios between different species of mesons and (anti)baryons. It has been suggested that the abundance of strange particles[3] and antibaryons (including light antinuclei[4]) should be higher in collisions with a transient plasma phase than in purely hadronic collisions. The argument is based on a fast timescale for chemical equilibration of light antiquark and strage quark and antiquark densities at relatively high levels in the quark-gluon plasma phase (due to the large density of gluons and the small quark masses in this phase[5]), and on a destruction of chemical equilibrium by the hadronization process which allows to transfer the information on temperature and baryon density contained in those quark and antiquark abundances into the final hadron phase. Since the time before freeze-out of these hadrons into non-interacting, free-streaming particles is too short to reestablish chemical equilibrium on a hadronic level, the chemical composition of the detected hadrons should to some extent reflect the chemical composition in the transient quark-gluon plasma phase and should show striking deviations from what we expect in purely hadronic collisions by extrapolation from our experience from high energy p-p and p-A data.

I will concentrate in this lecture on antinucleon and antinucleus formation from hadronizing quark matter; the strange particle aspects will be discussed in a later lecture by J. Rafelski. Particles containing only antiquarks are in both cases selected because they are initially absent and therefore provide the best signal-to-noise ratio. Furthermore (this is particularly relevant for strange particles) once they are produced, the most dangerous thing that can happen to them is annihilation in a collision with a baryon; rescattering between pairs of antibaryons or antibaryons and mesons with flavor exchange are negligibly rare[3], whereas similar processes affect the baryons to a considerable extent leading to deviations in the final flavor distribution from the one originally produced during hadronization. Non-strange antibaryons and antinuclei have the advantage of being absolutely stable in vacuum and (except for the antineutron) negatively charged and are therefore easily detected. Furthermore, the light quark equilibration time in the original plasma phase is so short that saturation of the thermodynamic phase space limit can be reliably assumed for $\bar{q}$; for s and $\bar{s}$ this is generally not true unless the initial plasma temperature is larger than the strange quark mass of ~ 150 MeV, a condition which is hardly achievable in the experiments planned for the next few years. [Since for the available beam energies more or less complete stopping of the colliding nuclei within each other's volume is likely, a large fraction of the beam energy will be transformed into compression energy, leading to large baryon densities at moderate temperatures[6] ($\mu_q/T > 1$)]. At the higher beam energies planned for the future the temperature will increase, without appreciable increase in baryon density due to the onset of nuclear transparency[1,2]. Therefore, the ratio between the quark baryonic chemical potential and the temperature μ_q/T will decrease ($\mu_q/T \lesssim 1$) and the main argument for strange antibaryons over nonstrange antibaryons, namely that due to strangeness conservation $\bar{s}$ is not suppressed like $\bar{q}$ by a factor $e^{-\mu_q/T}$, will disappear.

However, there are also disadvantages of non-strange antibaryons: entropy conservation requires the production of many pions[7] in the hadronization process which drain a large fraction of $\bar{q}$'s from the interesting antibaryon channels. The same is only true to a much lesser degree for strange particles because the lightest strange mesons, K and $\bar{K}$, are four times heavier than pions. Additionally, the annihilation cross section for antinucleons is considerably higher than for $\bar{\Lambda},\bar{\Sigma}$ etc., making annihilation during the final hadron phase more likely for non-strange antibaryons. Finally, the vacuum itself outside the hadrons contains a condensate of light $q\bar{q}$ pairs which is responsible for chiral symmetry breakdown in the hadron phase[2]; the possible coupling of the light quarks from the plasma to this condensate during the hadronization process introduces a so far unavoidable uncertainty into all existing models for hadron production from a quark-gluon plasma.

THE MODEL EQUATION OF STATE

Before we can follow the quark and antiquarks through the phase transition defined by the hadronization process, we have to locate that phase transition and discuss its thermodynamic properties. We will here present a model in which the quark-gluon plasma is described as a weakly interacting gas of quarks and gluons subject to an external vacuum pressure given by the bag constant, while the hadron phase consists of a noninteracting gas mixture of baryonic and mesonic resonances with a finite proper volume. This is a highly sophisticated version of the so-called bag-model equation of state (EOS)[2] and yields, as we will see, results which are in good qualitative agreement with other models and with Monte Carlo results for numerical simulation of QCD on a lattice.

Calculations of the grand canonical potential for weakly interacting massless quarks and gluons from QCD in the 1-loop approximation yield[8]

$$-\Omega_Q = P_Q = \frac{8\pi^2}{45}T^4\left(1-\frac{15\alpha_s}{4\pi}\right) + \sum_f \left\{\frac{7\pi^2}{60}T^4\left(1-\frac{50\alpha_s}{21\pi}\right)+\left(\frac{\mu_f^2T^2}{2}+\frac{\mu_f^4}{4\pi^2}\right)\left(1-\frac{2\alpha_s}{\pi}\right)\right\} - B \quad (1)$$

where the sum is over quark flavors, and an external "bag" pressure from the outside true vacuum has been added. The quark chemical potentials μ_f contain a baryonic contribution μ_q and, in the case of strange quarks, an additional contribution $\tilde{\mu}_s$ due to the strangeness of these particles. Hence $\mu_u = \mu_d = \mu_q$ and $\mu_s = \mu_q + \tilde{\mu}_s$; for a system of zero net strangeness $\tilde{\mu}_s = -\mu_q$, and $\mu_s = 0$ in the quark-gluon plasma. From (1) all other thermodynamic variables, baryon density ρ_b, entropy density s, and energy density ε, are derived by straightforward differentiation:

$$\rho_{b,Q} = -\frac{1}{3}\frac{\partial\Omega_Q}{\partial\mu_q}\,; \quad (2)$$

$$s_Q = -\frac{\partial\Omega_Q}{\partial T}\,; \quad (3)$$

$$\varepsilon_Q = 3\mu_q\rho_{b,Q} + T\,s_Q - P_Q = 3P_Q + 4B\,. \quad (4)$$

In working out (4) one realizes that the outside vacuum pressure B also plays the role of a vacuum energy density in the quark-gluon plasma which is by an amount B higher than the energy density of the outside vacuum.

Although perturbation theory in QCD is problematic because a perturbative expansion of Ω_Q breaks down at order α_s^3 we include the 1-loop corrections to get a qualitative feeling of the effect of color interactions. We parametrize α_s by a QCD scale parameter Λ[4,6] which we vary over a range of values from 0 to 400 MeV, as stated when we show results.

For the hadron resonance gas we start from the expressions for a mixture of noninteracting fermions and bosons:

$$P_H = \sum_i P_i = \sum_i \frac{d_i}{6\pi^2} \int_{m_i}^{\infty} \frac{(E^2-m_i^2)^{3/2}\, dE}{e^{(E-\mu_i)/T} \pm 1} ; \tag{5}$$

$$\rho_{b,H} = \sum_i b_i \rho_i = \sum_i \frac{b_i d_i}{2\pi^2} \int_{m_i}^{\infty} \frac{E(E^2-m_i^2)^{1/2}\, dE}{e^{(E-\mu_i)/T} \pm 1} ; \tag{6}$$

$$\varepsilon_H = \sum_i \varepsilon_i = \sum_i \frac{d_i}{2\pi^2} \int_{m_i}^{\infty} \frac{E^2(E^2-m_i^2)^{1/2}\, dE}{e^{(E-\mu_i)/T} \pm 1} ; \tag{7}$$

$$s_H = (\varepsilon + P - \sum_i \mu_i \rho_i)_H / T \ . \tag{8}$$

The sum goes over the spectrum of measured meson and baryon resonances, i.e., π, η, η', ρ, ω, K, K*, ϕ, ..., N, $\bar{N}$, Δ, $\bar{\Delta}$, N*, $\bar{N}^*$, ... d_i is the spin-isospin degeneracy factor for species i. Initially we concentrate on non-strange particles, and we will discuss the effects of strangeness towards the end of this lecture. For non-strange particles only baryon number is conserved, and their chemical potential is proportional to the baryon number b_i carried by each particle of species i:

$$\mu_i = b_i \mu_b \qquad \sum_i \mu_i \rho_i = \mu_b \rho_b \ . \tag{9}$$

THE HADRONIZATION PHASE TRANSITION

The phase coexistence line is determined by the three conditions

$$T_H = T_Q \ \text{(thermal equilibrium)} ; \tag{10a}$$

$$P_H = P_Q \ \text{(mechanical equilibrium)} ; \tag{10b}$$

$$\mu_b = 3\mu_q \ \text{(chemical equilibrium)} . \tag{10c}$$

The factor 3 in the last condition comes from the fact that quarks carry baryon number 1/3.

For small T,μ the pressure in the quark-gluon phase is smaller than in the hadronic phase, mainly because of the negative vacuum pressure -B. There the hadronic phase is stable. However, in general the quark-gluon plasma pressure increases with T and/or μ_q faster than the hadronic pressure. The reason is that many of the hadronic degrees of freedom are suppressed in the relevant temperature and density regime by their large masses, such that the quark-gluon phase has more effective degrees of freedom. These lead to a faster increase in pressure and, for a given μ_q, to a critical temperature $T_{cr}(\mu_q)$ at which $P_H = P_Q$, and the transition to the plasma phase takes place.

Note, however, that at extremely high T or μ where all the hadronic masses can be neglected, the hadronic phase as defined in (5)-(8) has the higher number of degrees of freedom and, if taken seriously, would again have the larger pressure and be stable (Fig. 1). This is due to the huge number of contributing resonances and their large spin-isospin degeneracy factors. Even for identical number of degrees of freedom, the pressure

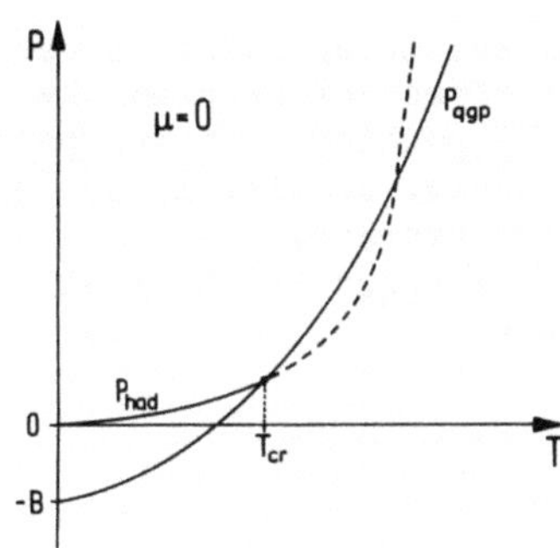

Fig. 1. Schematic behavior of the pressure in the hadronic resonance gas and in the quark-gluon plasma as a function of T at $\mu = 0$.

rises faster with increasing μ in the hadronic phase than in the quark-gluon phase, since in the plasma each quantum state can be filled by three times as many particles as in the hadronic gas, thus reducing the Pauli pressure. From the qualitative nature of Figure 1, it is conceivable that there exists a range of values for B and μ_q such that the two curves for P_H and P_Q never cross, and the hadronic phase never decays into the quark-gluon plasma. With the model (5)-(8) of pointlike, non-interacting hadronic resonances, this occurs actually for large values of μ_q and B (see Fig. 2).

This pathological behaviour is related to the one noted by Hagedorn[9]: in his statistical bootstrap model, he assumed that the spectrum of hadronic resonances continues towards higher masses with an exponentially increasing number of degrees of freedom, and this led to an increase of the pressure beyond limits at a finite, so-called 'limiting temperature' T_0 $\left(\sim O(m_\pi)\right)$. The origin for this singular behavior lies in the assumption of non-interacting pointlike hadrons which allows for the thermal production of arbitrarily many hadrons in a given volume and eventually leads to arbitarily high energy densities and pressures.

A simple remedy is provided by an inclusion of a finite proper volume for the hadrons[10-12]. This allows for a thermal production of more hadrons only until the fireball volume is completely filled with particles, and quenches the production of heavier hadrons which have larger proper volumes[10,11]. (Typical assumptions, which can be based on the MIT bag model for hadrons, take the proper volume of a hadron with mass m as $V_{had}=m/4B$[11]; or, in a system with energy density ε, a proper volume $V_{excl} = \varepsilon/4B$ is excluded from the available phase space for the thermally distributed hadrons[10]. In either case, B is a parameter which is usually taken to be identical with the QCD vacuum pressure discussed above.)

Within the bootstrap model[9,10] this correction for the proper volume is the only consistent one since, in the philosophy of the model, all other interactions between the hadrons are completely taken into account by choosing the correct hadronic mass spectrum. In our case where the hadronic mass spectrum is limited by our selection of resonances to be taken into account, further interactions should, in principle, be considered. The inclusion of interactions into a thermodynamic description is easy if the interactions are assumed to be local, i.e., they depend only on the particle densities[13].

A strong short-range repulsive interaction between the hadrons might be expected to have a similar effect of keeping the hadrons from getting too close to each other, thereby pushing even for pointlike hadrons the point, where particle densities, energy densities, and pressure become too big, to larger values for the temperature. We shall now discuss both these corrections to (5)-(8) with respect to their effect on the phase transition towards a quark-gluon plasma.

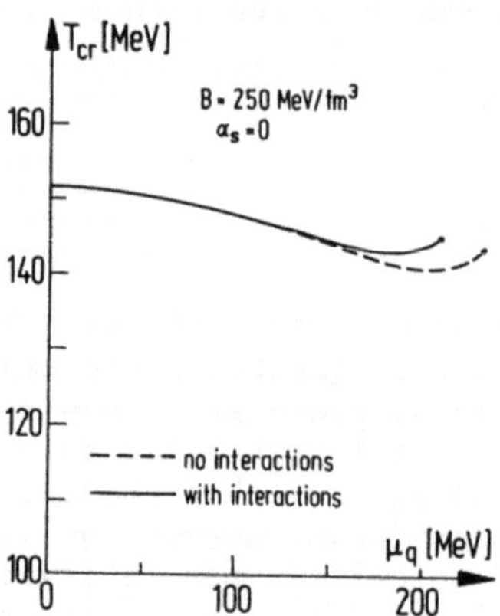

Fig. 2. The critical line for the transition from a gas of hadron resonances into a quark-gluon plasma. Dashed line: free, pointlike hadrons. Solid line: including a local interaction between the pointlike hadrons which is repulsive at large baryon densities.

Density Dependent Interactions

If we simulate the short-range repulsion between hadrons by a vector exchange interaction and take the zero range limit such that the strength of the interaction depends only on the (baryon) density, its effect on the thermodynamics is easily included[4,13] by substituting in the thermal distribution functions:

$$\mu_i \longrightarrow \tilde{\mu}_i = \mu_i - W_i(\rho) \tag{11}$$

where $W_i(\rho)$ is the interaction felt by particles of species i. Working out the appropriate derivatives of the grand canonical potential Ω with the modification we find

$$\rho_b = \sum_i b_i \rho_i(\tilde{\mu}_i, T) ; \tag{12}$$

$$\varepsilon = \sum_i \rho_i W_i + \sum_i \varepsilon_i(\tilde{\mu}_i, T) \equiv \varepsilon_{int} + \varepsilon_{therm} ; \tag{13}$$

$$P = \sum_{i,j} \rho_i \rho_j \frac{\partial W_i}{\partial \rho_j} + \sum_i P_i(\tilde{\mu}_i, T) \equiv P_{comp} + P_{therm} ; \tag{14}$$

$$s = [\varepsilon - \sum_i \mu_i \rho_i + (P - P_{comp})]/T . \tag{15}$$

$\rho_i(\tilde{\mu}_i, T)$ etc. are given by integrals (5)-(8) with the substitution (11). The expressions for the interaction energy ε_{int} and P_{comp} simplify if $W_i(\rho)$ is assumed to be proportional to the baryon number b_i of species i:

$$W_i(\rho) = b_i W(\rho_b) \quad \text{with} \quad W(\rho_b) = -W(-\rho_b) . \tag{16}$$

Then $\sum_i \rho_i W_i = \rho_b W(\rho_b)$ and $\sum_{i,j} \rho_i \rho_j \; \partial W_i/\partial \rho_j = \rho_b^2 \; \partial W/\partial \rho_b$.

We tested the effect of this modification on the hadron gas equation of state with a form for $W(\rho_b)$ which reproduces saturation of nuclear matter at $\rho_0 = 0.17$ fm^{-3} with a binding energy of -16 MeV and a compressibility of 200 MeV, and which increases for large ρ_b linearly with the density. Such a repulsive interaction at high densities effectively raises the single particle energy levels, thereby reducing their occupation probability at a given temperature and reducing the thermal pressure. In view of Figure 1 and our attempts to force a crossing point with the quark-gluon plasma pressure curve also at high densities, this is a desired effect. However, the increase of $W(\rho_b)$ leads also to a strong compression pressure P_{comp} which is an undesirable effect because it more or less happens to cancel the reduction in P_{therm}. As shown in Figure 2, the net effect is negligible, and the system still avoids a phase transition at large enough baryon densities. Apparently we were not able to improve on this unphysical interaction because we still allowed the particles to get arbitrarily close to each other (even though at a tremendous cost in energy). Thus we did not actually reduce the allowed phase space to the extent that is required by the finite extension of the hadronic particles. As we will now see, such a van-der-Waals type correction for the hadronic proper volume is the essential ingredient into our thermodynamic description to obtain a transition to quark matter at low temperature and high baryon density.

Proper Volume Connection for Hadrons

There exist several suggestions in the literature as to how to take into account the finite proper volume of the particles within a thermodynamic framework. Without theoretical prejudice, we will use the prescription derived by Hagedorn[14] within the so-called pressure ensemble because it is extremely easy to implement: in the first step all thermodynamic quantities are calculated for pointlike hadrons using (5)-(8). The physical values for these quantities are obtained by finally applying a correction factor $1/(1+\varepsilon_H^{pt}/4B)$:

$$P_H = P_H^{pt} / \{1 + \varepsilon_H^{pt}/4B\} , \tag{17}$$

$$\varepsilon_H = \varepsilon_H^{pt} / \{1 + \varepsilon_H^{pt}/4B\} , \tag{18}$$

$$\rho_{b,H} = \rho_{b,H}^{pt} / \{1 + \varepsilon_H^{pt}/4B\} , \tag{19}$$

$$s_H = s_H^{pt} / \{1 + \varepsilon_H^{pt}/4B\} . \tag{20}$$

The parameter B here is taken to be identical with the bag pressure in the quark-gluon plasma EOS; 4B is the energy density inside an MIT bag, and from (18) it is seen to form an upper limit to the energy density in the hadron gas. Before this limit is reached, we expect the phase transition to quark matter to have occurred. This is indeed borne out in the calculations[4], and the prescription (17)-(20) leads to a reasonable phase diagram for all baryon densities, shown in Figure 3a.

In Figure 3a-f we show for a specific value of B (B = 250 MeV/fm^3) the phase coexistence line and the behavior of the thermodynamic quantities

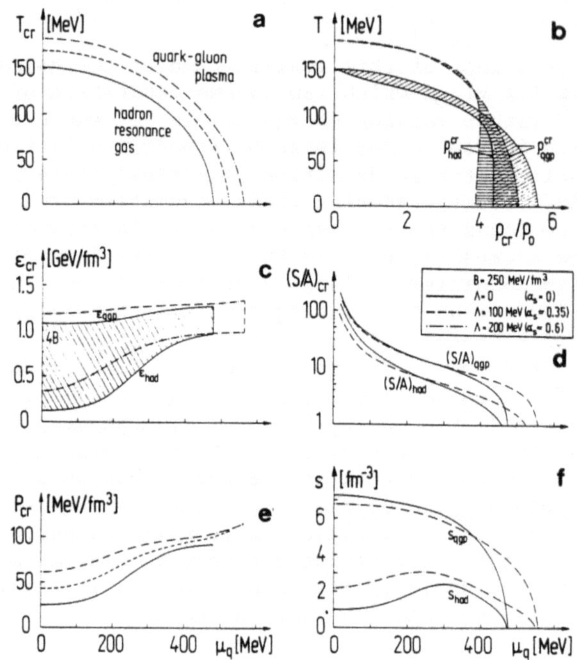

Fig. 3. a) The critical line (T,μ_q) for the hadronization transition, taking into account the proper volume of the hadrons in the resonance gas; b) baryon density; c) energy density; d) entropy per baryon; e) pressure; and f) entropy density along the critical line. Except in (b), all abscissae denote the chemical potential μ_q. Different style curves correspond to different values of α_s. The shaded region in (c) denotes the amount of latent heat. From Ref. 4.

along this critical line. Switching on the interactions in the quark-gluon plasma leads to an increase in the critical values for T and μ_q and to larger critical pressures and energy densities; this is due to the attractive nature of these interactions and the resulting reduction of the pressure in the quark-gluon plasma phase. Increasing the value of B has the same effect, and in both cases the point of pressure equilibrium (see Fig. 1) is shifted to higher values of T and μ_q.

Figure 3 shows that the phase transition in our model for hadronic matter is of first order, with large discontinuities in the energy density, entropy density, baryon density, and entropy per baryon. The latent heat (shaded region in Fig. 3c) is sizeable, namely of order 1 GeV/fm^3; the energy density increases by roughly an order of magnitude before the phase transition is completed at $\varepsilon \sim$ 1-2 GeV/fm^3 (the exact value depends on the choice of B and Λ, but is more or less constant along the critical line). Similarly, there is a latent entropy of about a factor 3-5, i.e., the entropy density is considerably higher in the quark matter phase than in the hadron resonance gas (although not by a factor 12-15, as obtained with simpler and coarser models for the hadron gas[15]).

The first order nature of this phase transition is common to all existing models and basically due to the procedure of matching two different EOS, each of which is well motivated only safely away from the critical point. However, the amount of latent heat and latent entropy is similar to other models[1,2], and in particular Monte Carlo simulations of exact QCD on a lattice are also compatible with these numbers: although there a first order phase transition is only seen for a purely gluonic theory, and the inclusion of dynamical quarks tends to smear out the transition to the point where it may not be a phase transition in the mathematical sense any more[16], a similar increase in energy density and entropy density over a very small temperature interval is seen in these calculations[16,17]. Thus, for practical purposes our model will presumably be a very reasonable approximation to reality.

THE ENTROPY PROBLEM

Since the entropy discontinuity will play an important role in the dynamics of the hadronization process, let us study it in a little more detail. In Figure 4 we show the variation of the entropy density with the temperature for two values of the baryon chemical potential, μ_b = 0 and μ_b = 750 MeV. We plot $s(\hbar c/T)^3$ because at μ_b = 0 for a gas of massless particles this would be a constant counting the number of degrees of freedom (as exemplified by the solid line labelled "free quark-gluon gas"). On the plasma side we included color interactions with $\alpha_s \approx 0.4$; note the strong decrease in s relative to the free gas limit. At $T = T_{cr}$ we see the entropy discontinuity across the phase transition. Below T_{cr} we split s into its contributions from different hadron resonances contained in the mixture. One sees that near T_{cr} the pions account for only about 30% of the entropy density, and that quantitatively the η, ρ, and ω mesons lead to a 100% correction, another 50% on top of that coming from the higher lying resonances. This trend starts already near T = 100 MeV, i.e., well below the phase transition, showing that a pure pion gas is quantitatively a poor approximation for the hadronic phase even at μ_b = 0 and overestimates the entropy discontinuity by a factor of 3. Qualitatively the same conclusions hold at finite μ_b; the curve for the free quark-gluon gas increases in this plot at small T like 1/T due to the term $\sim\mu^2 T$ in the entropy density. Again the interactions in the plasma reduce s by nearly 50%. The bump below the phase transition is an artifact of the way we plot s (we use units of $(T/\hbar c)^3$) and just means that between T $\approx$ 100 MeV and T_{cr} the entropy density does not increase as fast as T^3; this turns out to be mainly due to the van der Waals correction for the proper hadron volume which begins to suppress

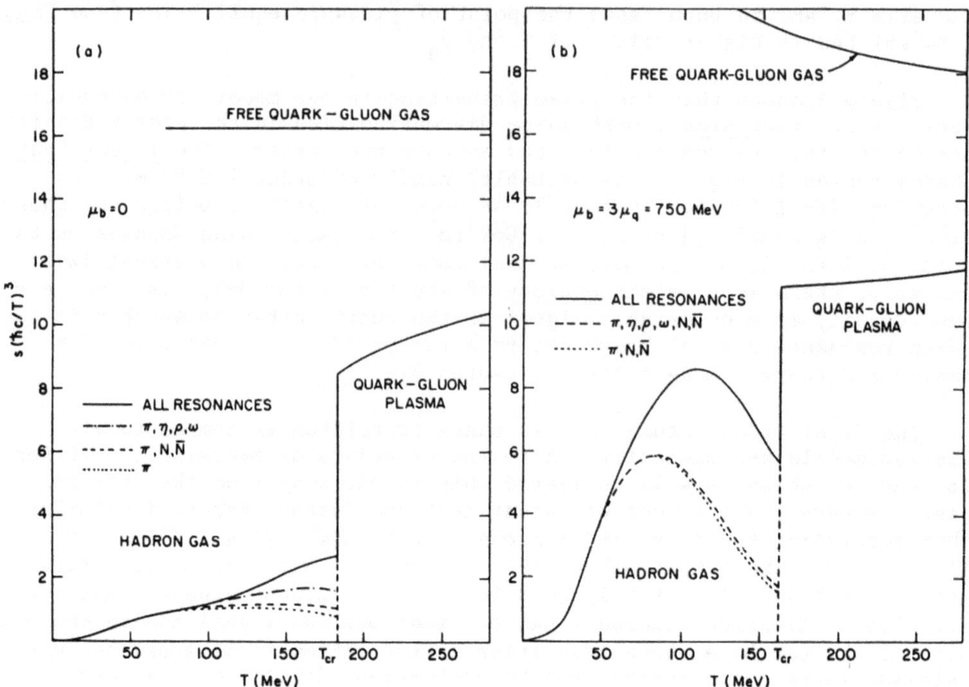

Fig. 4. The entropy density s in units of $(T/\hbar c)^3$ as a function of temperature. On the hadron gas side of the phase transition, contributions from different sets of hadronic resonances have been singled out in order to demonstrate their relative importance. On the plasma side, the limit for a non-interacting quark-gluon gas is indicated for comparison. The QCD scale parameter was chosen as $\Lambda = 200$ MeV, leading to $\alpha_s \approx 0.4$ just above the phase transition. The parameters are: $B = 250\ \mathrm{MeV/fm^3}$; $\mu_q = 0$ (Fig. 3a) and $\mu_q = 250$ MeV (Fig. 3b).

the effective degrees of freedom. Still it is obvious that the presence of all the high lying resonances leads to a big contribution to s already well below the phase transition. For a quantitative description it is therefore essential to take them into account.

Since the phase transition in our model is of first order, hadronization (if it happens locally on a fast time scale) will proceed through a mixed phase where hadron gas and plasma coexist inside the collision zone. It has been argued[15] that the lifetime of this mixed phase is determined by the entropy discontinuity because the entropy flux through the surface separating plasma from hadron gas is limited. Model calculations with a massless pion gas approximating the hadronic phase give rise to very long mixed phase lifetimes[15] ($\gtrsim 20$ fm/c); however, our calculations show that the presence of hadronic resonances reduces the entropy discontinuity by a factor of 3 or 4, thereby also reducing the above predictions for the lifetime of the mixed hadronization phase. This conclusion is not easily avoided by invoking a deflagration shock wave[18] from which the hadronic matter emerges at a temperature $T < T_{cr}$, where the pion gas approximation becomes increasingly better; for it to be sufficient, the temperature would need to jump across the shock front by about 70 MeV.

While the discontinuity of the entropy density thus affects the dynamics of the hadronization process, there is also a problem with conservation of overall entropy[3,7,19]. Obviously the total entropy must not decrease during hadronization. To illustrate the implications of this constraint, let us consider the case $\mu_q = 0$. The total entropy of the quark-gluon plasma is then given by

$$S \approx 4\left(N_q + N_{\bar{q}} + N_g\right) , \tag{21}$$

since for massless fermions S/N = 4.202 and for massless bosons S/N = 3.602. At the critical temperature $N_g \sim 2N_q = 2N_{\bar{q}}$, hence

$$S_Q \approx 8\left(N_q + N_{\bar{q}}\right) . \tag{22}$$

If we approximate the hadron phase by a massless pion gas (only for illustration), we find

$$S_H \approx 4N_\pi = 2\left(\tilde{N}_q + \tilde{N}_{\bar{q}}\right) \tag{23}$$

where $\tilde{N}_q$ and $\tilde{N}_{\bar{q}}$ are the total number of valence quarks and valence antiquarks in the hadron gas. Obviously, the condition $S_H \geq S_Q$ requires the number of valence quarks in the hadronic phase to be considerable larger than the number of quarks in the plasma; otherwise it is not possible to absorb all the entropy present in the gluons that disappear during hadronization from the excitation spectrum. Koch, Müller, and Rafelski therefore suggested[3,19] that during hadronization gluons fragment into additional quark-antiquark pairs, thereby saving the entropy balance. Both the thermally excited quarks and those from gluon fragmentation then recombine into mesons and baryons and determine the chemical composition of the hadron gas.

THE HADRONIZATION PROCESS - ANTINUCLEON AND ANTINUCLEI ABUNDANCES

In Figure 5 we show the light quark and antiquark densities along the critical line of phase coexistence. The hadron gas curve is obtained by counting the valence quarks and antiquarks contained in the hadrons. It is seen that the quark and antiquark density is about a factor 3 or so larger when the phase transition to quark matter is completed than what it was when the first hadrons started to dissolve. On the way back, i.e., during hadronization, additional $q\bar{q}$ production from gluon fragmentation will tend to increase that factor. On the other hand, the hadronization process will also involve some increase in volume of the system, tending to reduce the quark densities. The precise evolution governed by the counterplay of these two effects contributes the problem of hadronization dynamics and will determine the chemical composition of the hadron resonance mixture emerging from the hadronization process. This question can by no means be considered solved at the present time, but there exist several theoretical attempts, one of which we are now going to discuss in some detail. [Note that in Figure 5 the curves for the s and $\bar{s}$ densities would start at $\mu_q = 0$ about a factor 3 below the curves shown for light quarks, and due to the vanishing chemical potential of the strange quarks in the plasma phase and its relative smallness in the hadron phase[20], their values will stay nearly constant until the phase transition temperature begins to drop drastically near $\mu_q \approx 300$ MeV. On the other hand, the curves in Figure 5 for the light antiquark densities do not drop by a factor of 3 before $\mu_q \gtrsim 150$-200 MeV. Only then (i.e., for $\mu_q/T \gtrsim 1$) strange antiquarks become more abundant than light antiquarks and there will be an appreciably higher chance to form, say, strange rather than non-strange antibaryons. It is not clear without a complete dynamical calculation under which conditions regions with $\mu_q/T > 1$ will be formed in a heavy-ion collision. To get some idea one may look at Figure 4 of Ref. 6; even in the hydrodynamic approximation, i.e., assuming complete stopping of the nuclei, μ_q/T does not appreciably increase with the collision energy but appears to be limited by $\mu_q/T \sim 2$-3.]

Let us start by assuming that the plasma, once formed, has expanded and cooled down again to a point on the critical line $T(\mu_q)$ where hadronization begins. At this point we can compute the quark and antiquark densities ρ_q and $\rho_{\bar{q}}$ (see Fig. 5). After hadronization these quarks and antiquarks have to end up in hadrons; this will occur by recombination, e.g., of 3 quarks going into a baryon, a quark and an antiquark forming a meson, or 12 antiquarks forming an $\bar{\alpha}$-nucleus. Let us assume that these processes happen

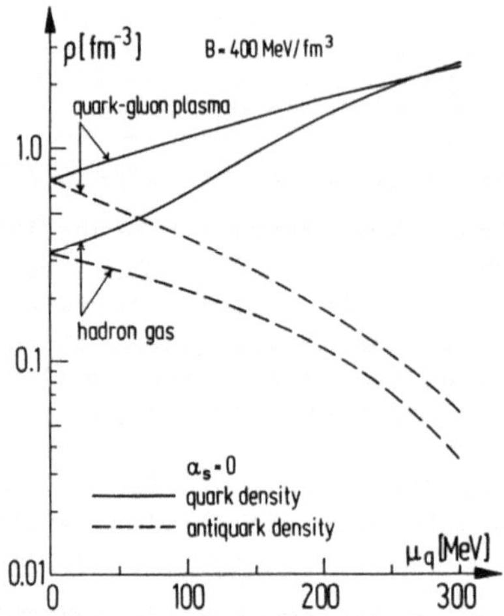

Fig. 5. The critical values for the quark and antiquark densities along the phase transition line shown in Fig. 3. In the hadron gas the (anti)quark density is computed by counting valence (anti)quarks. B = 400 MeV/fm^3.

fast, and that at each point there is chemical equilibrium between the quarks and antiquarks on the one side and the hadrons forming from them on the other side. Then, if we define Lagrange multipliers ("chemical potentials") ν, $\bar{\nu}$ to keep count of the average quark and antiquark numbers (in the grand canonical sense), the chemical potential of the hadrons will be related to ν and $\bar{\nu}$ by their valence quark contents:

$$\begin{aligned}
&\mu_\pi = \nu + \bar{\nu} = \mu_\rho = \mu_\omega = \mu_\eta = \dots && \text{(mesons)} \\
&\mu_i = 3\nu = \mu_\Delta = \mu_{N^*} = \dots && \text{(baryons)} \\
&\mu_d = 6\nu && \text{(deuteron)} \qquad (24)\\
&\mu_{\bar{N}} = 3\nu = \mu_{\bar{\Delta}} = \mu_{\bar{N}^*} = \dots && \text{(antibaryons)} \\
&\mu_{\bar{d}} = 6\bar{\nu} && \text{(antideuteron)}
\end{aligned}$$

etc. These equations are easily generalized, if strange quarks are to be included. Note that although all baryons, all mesons, etc. have the same chemical potential, their production rates during hadronization will decrease with increasing resonance mass due to the mass term in the thermal distribution functions. This is in contrast to the recombination + fragmentation model by Koch, Müller, and Rafelski[3,19] where the number of produced hadrons of a given species is given only by a combinatorial factor depending on its valence quark content, and the mass difference, say between an N and a Λ or a Λ and a Σ, does not enter.

Given the relations (24), all hadron densities $\rho_i(T,\mu_i(\nu,\bar{\nu}))$ can be determined as a function of ν and $\bar{\nu}$, and hence the hadronic valence quark content

$$\tilde{\rho}_q(\nu,\bar{\nu}) = \sum_i n_i \rho_i(T,\mu_i(\nu,\bar{\nu})) \; ;$$
$$\tilde{\rho}_{\bar{q}}(\nu,\bar{\nu}) = \sum_i \bar{n}_i \rho_i(T,\mu_i(\nu,\bar{\nu})) \; . \tag{25}$$

Here n_i ($\bar{n}_i$) counts the number of valence quarks (antiquarks) in hadron i.

To determine ν and $\bar{\nu}$ we have to solve a set of matching conditions

$$\tilde{\rho}_q = \lambda\rho_q = (\rho_q + \Delta\rho_q) \cdot V_Q/V_H \; ;$$
$$\tilde{\rho}_{\bar{q}} = \bar{\lambda}\rho_{\bar{q}} = (\rho_{\bar{q}} + \Delta\rho_{\bar{q}}) \cdot V_Q/V_H \; . \tag{26}$$

Here $\Delta\rho_q = \Delta\rho_{\bar{q}}$ is the number of additional quark-antiquark pairs per unit volume formed by gluon fragmentation in the hadronization process. This number is in principle determined by entropy conservation, but in practice one has to know the volume expansion factor V_Q/V_H and the (T,μ) trajectory the system takes during hadronization in order to evaluate it. The reason for this complication is[4] that the latent heat is set free during hadronization and, depending on how much of it is converted into collective hydrodynamic flow energy, leads to reheating. Subramanian et al.[21] worked out such a scenerio, assuming conversion at constant entropy per baryon, however without including the effects of gluon fragmentation. Considerable reheating is observed[21]. No complete dynamical hadronization calculation to account for <u>all</u> these effects has been performed until now.

Given these uncertainties and no way to resolve them without a complete dynamical model, we will here show results of a calculation where we assumed in (26) <u>$\lambda = \bar{\lambda} = 1$</u>. From entropy conservation one estimates that this corresponds to a volume expansion factor $V_H/V_Q \sim 2$ if the phase transition occurs at constant temperature. Possible reheating effects are neglected and have to be studied in the future.

In Figure 6 we show the results for the hadron densities emerging from hadronization at a value (T,μ_q) as they are obtained after solving the matching conditions (26) with $\lambda = \bar{\lambda} = 1$ for ν and $\bar{\nu}$ and inserting into (24) (solid lines). One sees that the solid lines (hadron densities from plasma hadronization) consistently lie above the dashed lines (equilibrium hadron densities), due to the need to absorb the larger density of quarks and antiquarks initially present in the plasma. The gain factors increase with the size of the hadronic cluster formed, because the gain factors per quark essentially multiply. They are of order 3 or so for antibaryons, and for anti-alpha nuclei they reach 2 orders of magnitude. Although the absolute yield drops exponentially with the cluster mass, the value for $\rho_{\bar{\alpha}}$ in Figure 6 at $\mu = 0$ for a fireball volume corresponding to, say, half a uranium nucleus translates into about 1 $\bar{\alpha}$-events per hour in the planned Relativistic Heavy Ion Collider if a luminosity of 10^{27} cm^{-2} sec^{-1} is assumed; without plasma formation one would have to wait for several days.

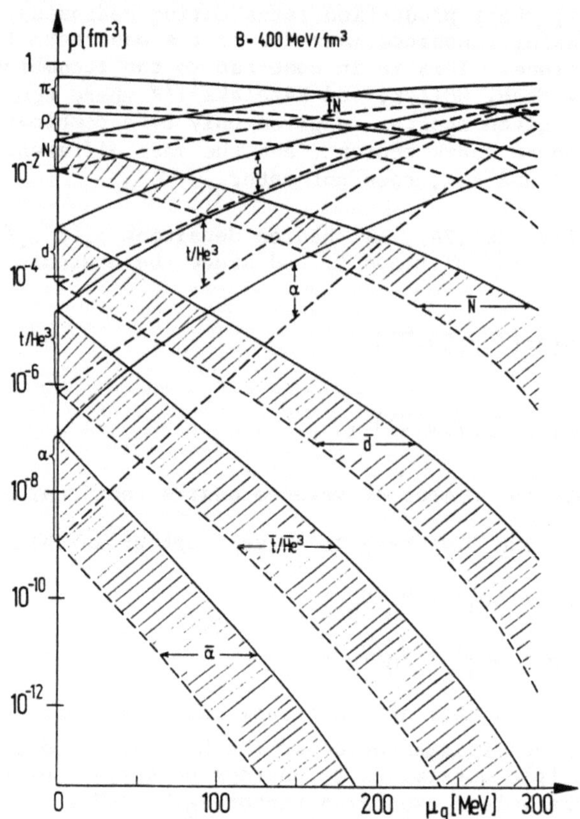

Fig. 6. Hadronic cluster densities as a function of the quark baryonic chemical potential in the initial quark-gluon plasma phase. Dashed curves denote the densities in an equilibrium hadron gas at the critical temperature corresponding to the given μ_q. Solid curves are the results obtained from a hadronization calculation which keeps the density of quarks and antiquarks fixed at the level of the original quark-gluon plasma (for details see text). $B = 400\ \mathrm{MeV/fm^3}$.

THE INFLUENCE OF STRANGE PARTICLES ON THE HADRONIZATION PROCESS

Before closing, we will now briefly look at the modifications that we will have to expect from the inclusion of strange particles. For a detailed account of strange hadrons from a quark-gluon plasma see Rafelski's talk in this volume. We will here concentrate on modifications to the critical quantities along the phase transition; a hadronization calculation for strange hadrons incorporating these changes has not yet been completed.

The grand canonical potential of the quark-gluon plasma now takes the form $(q,\bar{q}$ denotes light (massless) quarks$)$:

$$-\Omega_Q = P_Q = P_{q\bar{q}g} + \frac{1}{\pi^2}\int_{m_s}^{\infty} dE(E^2-m_s^2)^{3/2}\left\{\frac{1}{e^{\beta(E+\mu_q+\tilde{\mu}_s)}+1} + \frac{1}{e^{\beta(E-\mu_q-\tilde{\mu}_s)}+1}\right\}. \quad (27)$$

The sum over hadrons in the hadron resonance gas expressions is extended to include $K^{\pm}$, K^0, $\bar{K}^0$, Λ, $\bar{\Lambda}$, ϕ ... . The strange hadron chemical potentials in chemical equilibrium are determined as before; e.g. $\mu_{K^-} = \mu_q+\tilde{\mu}_s+\mu_{\bar{q}} = \tilde{\mu}_s$, $\mu_\Lambda = 3\mu_q+\tilde{\mu}_s$, $\mu_\phi = 0$, etc. Imposing the conservation of total strangeness at its zero initial value $S = 0$ leads to $\tilde{\mu}_s = -\mu_q$ in the plasma phase; however, in the hadron gas this relationship has to be violated[20] for nonvanishing μ_q: due to the suppression of light antiquarks at finite μ_q the dominant strange particles are Λ's, whereas the strange antiquarks are mostly contained in K^+ and K^0. It is easily seen that the assumption $\tilde{\mu}_s = -\mu_q$ for the valence quarks does not lead to a system with zero strangeness, due to the chemical potential of the additional light quarks in the Λ's. Therefore,

$$\tilde{\mu}_{sH} \neq \tilde{\mu}_{sQ} = -\mu_q \ . \tag{28}$$

Hence, we now have two non-identical critical lines (T_{cr}, μ_q) and $(T_{cr}, \tilde{\mu}_{sH})$ shown in Figure 7. The difference between the two curves is a measure for the discontinuity of $\tilde{\mu}_s$ across the phase transition. They can be considered as different projections of one critical line in the 3-dimensional space spanned by $(T, \mu_q, \tilde{\mu}_{sH})$. It is seen that the strange particles reduce the critical temperature by a few percent, due to a relative increase of the pressure in the plasma phase over the hadron phase (where strange quarks have less influence owing to their larger effective mass).

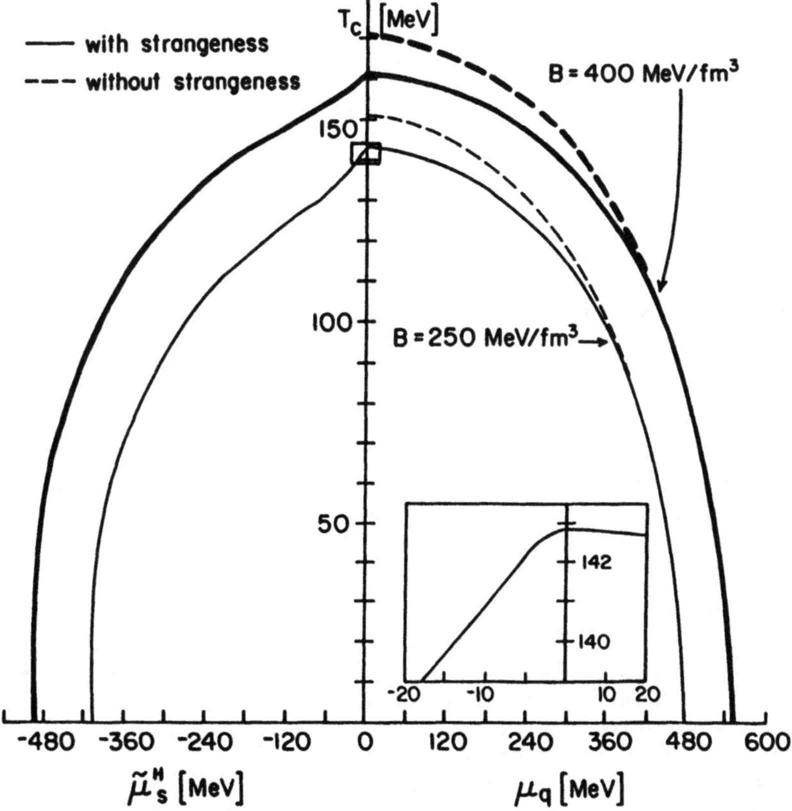

Fig 7. Projections of the critical line $(T, \mu_q, \tilde{\mu}_{sH})$ on the (T, μ_q) and $(T, \tilde{\mu}_{sH})$ planes. The dashed lines are for comparison and show the critical line (T, μ_q) in absence of strange particles. Two different bag constants have been chosen as indicated. Note the reduction in the critical temperature by about 5%. The inset shows that both curves are smooth near $\mu = 0$. From Ref. 20.

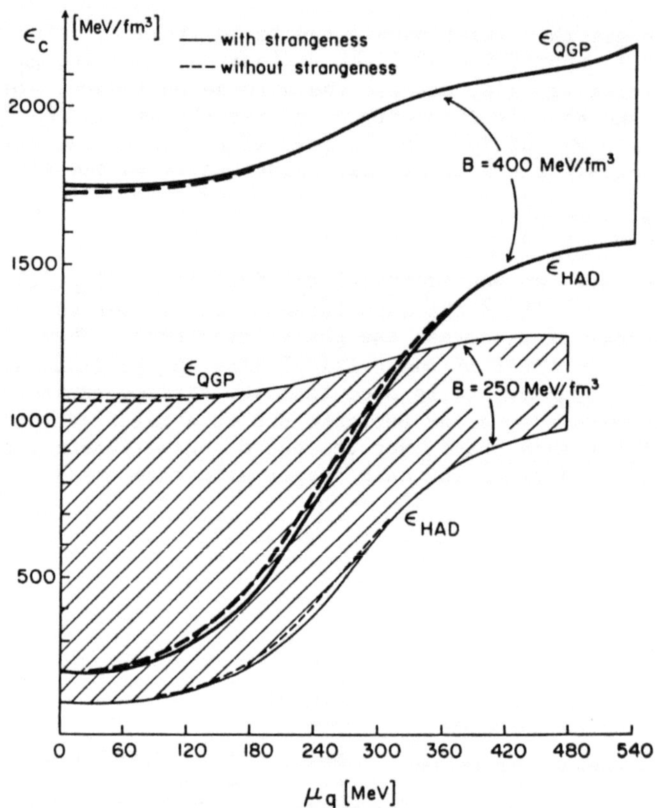

Fig. 8. The energy density along the critical line of Fig. 7, parametrized by μ_q. The curve labelled ε_{had} (ε_{QGP}) denotes its value on the hadron gas (quark-gluon plasma) side of the phase transition. The shaded area demonstrates the large amount of latent heat set free in the hadronization process. Solid (dashed) lines incorporate (omit) strange particles. Plotted against μ_q, the influence of strangeness on the critical energy densities is minimal. From Ref. 20.

It is important to note that this change in T_{cr} has virtually no effect on the values for the critical pressure and critical energy density (Fig. 8). The influence of strange particles just leads to a shift of the crossing point between the two pressure curves in Figure 1 to a smaller value of T_{cr}, but at a nearly unchanged value for P_{cr}.

CLOSING REMARKS

In this lecture I presented a model for the hadronization phase transition which is sophisticated enough to allow an analysis of the chemical composition of the hadron gas emerging from a hadronizing quark-gluon plasma. The phase transition was obtained by matching the equation of state for a weakly interacting quark-gluon plasma with a gas of hadronic resonances taken from the particle data tables. We have seen that accounting for the proper volume of the hadrons in the hadron gas is an essential ingredient to obtain a reasonable phase diagram also at large baryon densities, whereas the inclusion of longer range interactions between the hadrons does not qualitatively affect the phase transition.

In our model the phase transition is of first order and exhibits a large latent heat and latent entropy. Although the actual transition in nature may not be of first order, we discussed that even there the energy and entropy densities show strong variations over a small temperature interval, and that our model will be a good qualitative approximation. We reviewed the argument by Koch, Müller, and Rafelski that during hadronization the total entropy balance can only be satisfied if additional quark-antiquark pairs are created in the hadronization process.

The hadronization itself is a difficult problem for which, so far, no microscopy model exists. We discussed here a statistical model which is based on chemical equilibrium between quarks and hadrons during the clustering process. Rafelski in his lecture discusses a clustering model mainly based on combinatorial considerations. I believe that these two models are complementary, and reality has to lie somewhere in between. Both models are, so far, incomplete as no full self-consistent dynamical study has been performed in which the respective hadronization model is coupled to a hydrodynamical evolution which correctly treats the energy <u>and</u> entropy balance. In particular, the volume expansion factors during hadronization are uncertain and can only be bounded between extremes. Still, both models for the clustering process agree qualitatively in their prediction that antibaryons and light antinuclei should be <u>enhanced</u> in hadronization of a quark-gluon plasma over their hadronic equilibrium values. Rafelski will argue that this enhancement works even better for strange antibaryons ($\Lambda,\Sigma,...$), particularly at large μ_b, but I have given arguments that looking for nonstrange antibaryons may be easier experimentally and nearly as promising quantitatively. For both strange and nonstrange particles, generally the gain factors increase while the yields decrease with increasing size of the observed cluster.

Still, the quantitative predictions are uncertain, and it cannot be stressed often enough that the questions about actual yields to be expected in experiment cannot be reliably settled without a full dynamical calculation. This should be (and is for me) the highest priority project in the context of using strange and antimatter as a signature for quark-gluon plasma formation.

ACKNOWLEDGEMENT

I would like to thank my collaborators W. Greiner, K.S. Lee, M. Rhoades-Brown, H. Stöcker, and P.R. Subramanian, who had an essential part in obtaining the results I presented in this lecture. Interesting and clarifying discussions with P. Koch and B. Müller are gratefully acknowledged.

This work was supported by the U. S. Department of Energy under contract DE-AC02-76CH00016.

REFERENCES

1. "Statistical Mechanics of Quarks and Hadrons", H. Satz (ed.), North Holland, Amsterdam (1981);
   "Quark Matter '83", T.W. Ludlam and H.E. Wegner (eds.), <u>Nucl.Phys.A</u> 418 (1984);
   "Quark Matter '84", K. Kajantie (ed.), <u>Lecture Notes in Physics</u> 221, Springer, Heidelberg (1985);
   "Quark Matter '86", M. Gyulassy and L.S. Schroeder (eds.), to appear in <u>Nucl.Phys.A</u> (1986).
2. E.V. Shuryak, <u>Phys.Rep.</u> 61:71 (1980) and 115:151 (1984);
   B. Müller, "The Physics of the Quark-Gluon Plasma", <u>Lecture Notes in Physics</u> 225, Springer, Heidelberg (1985);
   J. Cleymans, R.V. Gavai, and E. Suhonen, <u>Phys.Rep.</u> 130:217 (1986).
3. J. Rafelski, <u>Nucl.Phys.A</u> 418:215c (1984);
   P. Koch, B. Müller, and J. Rafelski, <u>Phys.Rep.</u> (1986), in print.

4. U. Heinz, P.R. Subramanian, and W. Greiner, Z.Phys.A 318:247 (1984);
U. Heinz, P.R. Subramanian, H. Stöcker, and W. Greiner, J.Phys.G (1986), in print.
5. J. Rafelski and B. Müller, Phys.Rev.Lett. 48:1066 (1982).
6. H. Stöcker, Nucl.Phys.A 418:587c (1984).
7. N.K. Glendenning and J. Rafelski, Phys.Rev.C 31:823 (1985).
8. J. Kapusta, Nucl.Phys.B 148:461 (1979).
9. R. Hagedorn, Suppl. Nuovo. Cimento 3:147 (1965); 6:311 (1968).
10. R. Hagedorn and J. Rafelski, in Ref. 1, p. 237 and 253.
11. J. Kapusta, Phys. Rev. D 23:2444 (1981).
12. P.R. Subramanian, L.P. Csernai, H. Stöcker, J.A. Maruhn, W. Greiner, and H. Kruse, J. Phys. G 7:L241 (1981).
13. U. Heinz, W. Greiner, and W. Scheid, J. Phys. G 5:1383 (1979).
14. R. Hagedorn, Z. Phys. C 17:265 (1983).
15. B. Friman, K. Kajantie, and P.V. Ruuskanen, Nucl. Phys. B 266:468 (1986);
H. von Gersdorff, L. McLerran, M. Kataja, and P.V. Ruuskanen, Phys. Rev. D (1986) in print.
16. R.V. Gavai and F. Karsch, Nucl. Phys. B 261:273 (1985).
17. H. Satz, Ann. Rev. Nucl. Part. Sci. 35:245 (1985).
18. L. van Hove, Z. Phys. C 21:93 (1983).
19. B. Müller, in "Quark Matter '86", Ref. 1.
20. K.S. Lee, M.J. Rhoades-Brown, and U. Heinz, Phys. Lett. B 174:123 (1986).
21. P.R. Subramanian, H. Stöcker, and W. Greiner, Phys. Lett. B 173:468 (1986).

COLOUR INTERACTIONS IN GIANT QUARK BAGS

Matthias Grabiak, Stefan Schramm, and Walter Greiner

Institut für Theoretische Physik
Johann Wolfgang Goethe-Universität, Postfach 111932
D-6000 Frankfurt/Main, West Germany

INTRODUCTION

In the lecture presented by L. Neise[1] about giant quark bags it was assumed that the quarks move freely inside the bag. In the following we want to examine the effects of the interaction between the quarks due to the exchange of gluons within the MIT-bag-model[2], i.e. we shall adopt the point of view that the surface of the bag is a sharp boundary which separates the nonperturbative vacuum outside the bag from the perturbative phase inside (Fig. 1). No gluon field can exist outside the bag, whereas inside the bag the coupling constant is assumed to be small so that we can describe the colour interaction perturbatively. Only the interactions between quarks and gluons are taken into account because the gluonic self-interactions lead to higher order terms. We have no real justification that this approximation can be applied in the case of giant quark bags.

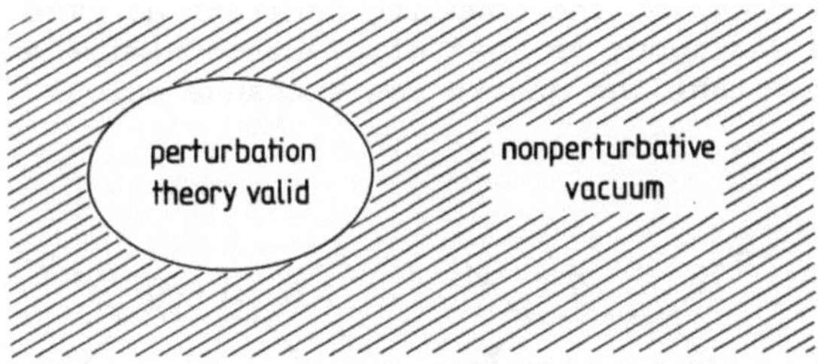

Fig. 1. nonperturbative vacuum: No colour field can exist, quarks cannot penetrate into this region.

inside the bag: Quarks move around, colour interaction mediated by gluons, weak qG, GG coupling.

For a nucleon one can argue that the separation of the quarks is small, and due to asymptotic freedom this will lead to a relatively small effective coupling constant to be used within this model. For a giant quark bag this argument does not hold because the radius of the bag is rather large in this case. On the other hand the quark density in a giant quark bag is a bit larger than in a light hadron[1] and this could help to keep the coupling constant small[3]. Even if this is not the case the perturbative calculation will at least give some qualitative insight into the colour interactions of a giant quark bag provided that these objects really exist.

THE MIT-BAG MODEL[2]

According to the MIT-bag model quarks and gluons can only exist inside the bag volume. Here they are described by the Lagrangian[3]

$$L = -\frac{1}{4}F^a_{\mu\nu}\, F^{\mu\nu}_a + \bar{\psi}(i\not{\partial} + g\, T_a\, \not{A}_a)\psi \tag{1}$$

with a fixed coupling constant g. The quark field ψ has three colour components ψ_r, ψ_b and ψ_g describing red, blue and green quarks. T_a are the generators of the colour gauge group SU(3), which for quarks are given by $T_a=\lambda_a/2$, where λ_a are the Gell-Mann matrices. A^a_μ are the eight gauge potentials and the electromagnetic field $F^a_{\mu\nu}$ is defined as

$$F^a_{\mu\nu} = \partial_\mu A_\nu - \partial_\nu A_\mu + g\, f_{abc}\, A^b_\mu\, A^c_\nu \quad . \tag{2}$$

We now want to expand the colour interactions in powers of the coupling constant g. To zeroth order quarks and gluons decouple and the quarks move freely inside the bag. Consequently the energies of hadrons and giant quark bags depend only on the kinetic energy of the quarks but not on the way the quark spins are coupled together. When the masses of the three lightest quarks are neglected this leads to the SU(6) symmetry with the fundamental sextett (u↑, u↓, d↑, d↓, s↑, s↓). On the other hand the proton with spin 1/2 has a mass of 938 MeV, whereas the mass of the Δ^+-resonance with spin 3/2 is 1232 MeV, even though both particles have the same quark content and differ only in the spin coupling. The mass splitting between p and Δ^+ can be attributed to the colour interactions between the quarks[4]. To see this let us examine the colour interactions up to the second order of the coupling constant g. The interaction term $g\bar{\psi}T_a\not{A}_a\psi$ is quite analogous to that of electrodynamics and leads to similar graphs, except that the quarks can exchange colour (Fig. 2).

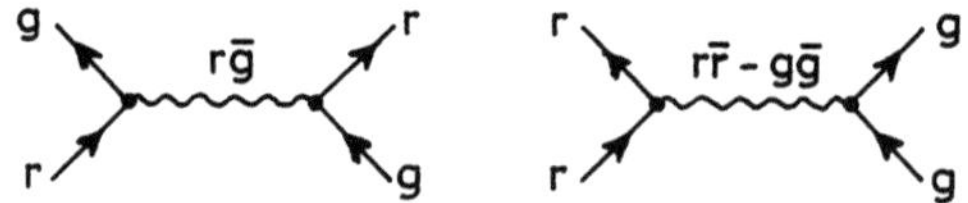

Fig. 2. Colour interactions between quarks.

The graphs of Fig. 2 are already of second order in the coupling constant g. Taking into account the self-interactions of the gluons due to the nonlinear nature of $F^a_{\mu\nu}$ leads to higher order terms which are dropped in the second order calculation. This means that $F^a_{\mu\nu}$ is replaced by $\partial_\mu A^a_\nu - \nabla_\nu A^a_\mu$ and we have essentially eight noninteracting Maxwell fields. This is usually called the Abelian approximation.

The colour electric charge density is independent of the quark spin.

This means the colour electric energy is the same for p and Δ^+, because in both particles all quarks sit in the lowest s-state. Therefore we are mainly interested in the magnetic colour interactions which depend also on the quark spins. We have also another reason to restrict ourselves to magnetism. An electric field will always lead to an increase of the energy, but we can always put three quarks of different colour into the same spatial wave function so that they do not produce an electric field. On the other hand the presence of a magnetic field corresponds to a decrease of the total energy. Similar to electrodynamics we get the following equations:

$$\vec{B}_a = \vec{\nabla}\times\vec{A}_a \quad ,$$

$$\vec{\nabla}\times\vec{B}_a = \vec{\nabla}\times(\vec{\nabla}\times\vec{A}_a) = -\Delta\vec{A}_a + \vec{\nabla}(\vec{\nabla}\cdot\vec{A}_a) = \vec{j}_a \tag{3}$$

with the quark current $\vec{j}_a$. The magnetic energy is given by

$$E_{magn} = \int d^3r \left[\frac{1}{2}\vec{B}_a^2 - \vec{j}_a\, \vec{A}_a \right] = -\frac{1}{2}\int d^3r\ \vec{B}_a^2 \tag{4}$$

where we have inserted the solution (3) in the last step. This shows that a nonvanishing magnetic field will lower the energy. Similar to electrodynamics the colour currents try to align in order to produce a large magnetic field. If we qualitatively understand different colours as different charges then we have the following situation: The quartz combination r↑b↑g↑ found in the Δ-resonance is unfavourable because the quark currents cancel each other whereas in the combination r↑b↑g↓ found in the proton r↑b↑ is again repulsive whereas r↑g↓ and b↑g↓ are attractive. Thus we can understand why the energy of the proton is lower than that of the Δ-resonance.

MAGNETIC INTERACTIONS IN GIANT QUARK BAGS

The energy levels in a giant quark bag exhibit a typical shell structure where a shell consists of all degenerate energy levels which share the same value of κ and n^1. In the following we shall assume that each shell is occupied by a fixed number of quarks. We take into account processes where a quark changes its colour and makes a transition from one magnetic substate to another, but we neglect processes where a quark is excited into a higher orbit with different n and κ. A closed shell does not produce a magnetic field for the following reason. If three quarks occupy a substate with the same quantum numbers their wave function has to be ântisymmetric with respect to colour. This means that they are automatically coupled to a colour singlet in which the quark currents cancel each other. This is however not the whole story because the magnetic self-energy of the quarks has to be treated carefully. Emitting a gluon a quark can be scattered to any orbit, reabsorb the gluon and get back into its original state. This process will modify the quark propagator. Following reference 4 we shall neglect this modification but consequently omit the self-energy of a quark. This will increase the energy of the quark bag, because there remain only the interactions of different quarks in the same substate which is repulsive because the quarks have the same spin but different colour.

In our calculation of the colour magnetic interaction of a giant quark bag we use the following approximations. We assume that the bag is spherical and that the quarks are massless. We omit the self-energy of the gluons and do not take into account processes where quarks are excited to other shells. Now the quark i produces the magnetic field $\vec{B}_a^{(i)}$, which obeys[4]

$$\vec{\nabla}\times\vec{B}_a^{(i)} = \vec{j}_a^{(i)} \; , \quad \vec{j}_a^{(i)} = g\psi_{(i)}^{+}\,\vec{\alpha}\,T_a\,\psi_{(i)} \; ,$$

$$\vec{\nabla}\cdot\vec{B}_a^{(i)} = 0 \quad . \tag{5}$$

In addition, the magnetic field has to fulfil a boundary condition at the surface of the bag. The nonperturbative vacuum behaves like a perfect dielectric medium. This means that the colour electric field perpendicular to the bag surface vanishes so that due to Gauss' law the total colour charge inside the bag must be zero so that we have colour confinement within the MIT-bag model. On the other hand the magnetic field along the bag surface also vanishes and we get the boundary condition

$$\vec{r}\times\vec{B}_a^{(i)} = 0 \tag{6}$$

for a spherical bag. The solution of (5) is

$$\vec{A}_a(\vec{r}) = \int d^3r' \,\frac{1}{4\pi|\vec{r}-\vec{r}\,'|}\;\vec{j}_a(\vec{r}\,') + \vec{A}_a^h \tag{7}$$

where the homogeneous part obeys $\Delta\vec{A}_a^h = 0$ and is chosen in such a way that the boundary condition (6) is fulfilled.

Now we have to insert the wave functions of the quarks into the definition of the current. In the following we shall omit the colour part of the quark wave function because it is irrelevant for the calculation of the shape of the magnetic field. The wave function of the quark reads[2]

$$\psi_{\varkappa n\nu} = A_{\varkappa n}\begin{pmatrix} j_{\ell_\varkappa} & X_\varkappa^\nu \\ -ij_{\ell_{-\varkappa}} & X_{-\varkappa}^\nu \end{pmatrix} \tag{8}$$

where $A_{\varkappa n}$ is a normalization factor and the functions j are spherical Bessel functions. $X_\varkappa^\nu$ are spinor functions where $Y_{\ell m}$ and the spinor $X_{\pm 1/2}$ are coupled to angular momentum $\ell=|\varkappa|-1/2$ and the magnetic projection ν. The matrix element of the current between the magnetic substates μ and ν is found to be

$$\vec{j}_{\mu\nu} = i|A_{\varkappa n}|^2\, j_{\ell_\varkappa}\, j_{\ell_{-\varkappa}}\left(X_{-\varkappa}^{\mu *}\,\vec{\sigma}\,X_\varkappa^\nu - X_\varkappa^{\mu *}\,\vec{\sigma}\,X_{-\varkappa}^\nu\right) \tag{9}$$

In order to calculate the magnetic field we use the identity

$$\frac{1}{4\pi|\vec{r}-\vec{r}\,'|} = \sum_{\ell=0}^{\infty}\sum_{m=-\ell}^{\ell}\frac{4\pi}{2\ell+1}\,Y_{\ell m}(\Omega)Y_{\ell m}^{*}(\Omega)\frac{r_<^{\ell}}{r_>^{\ell+1}} \tag{10}$$

and for simplicity we restrict ourselves to the dipole part with $\ell=1$. The calculation is straightforward but tedious. The final result for the magnetic energy of a given shell is

$$E_{\varkappa n} = -\alpha_s\frac{\varepsilon_{\varkappa n}}{R_0}\sum_{i>j}\left\langle \vec{J}_{(i)}\cdot\vec{J}_{(j)}\,\lambda_a^{(i)}\,\lambda_a^{(j)}\right\rangle \tag{11}$$

where $\vec{J}_{(i)}$ is the angular momentum operator of the quark i and $\lambda_a^{(i)}$ is the pertaining Gell-Mann matrix. α_s is the strong coupling constant and R_o is the radius of the bag. Equation (11) confirms our expectation that the energy is lowered if the quark currents are aligned because it tells us that it is favourable to have the quantity $\vec{J}\lambda_a$ of different quarks parallel.

For a closed shell the magnetic field vanishes. In this case the magnetic energy is given by the negative self-energy of the quarks which is

$$+\frac{1}{2}\alpha_s \frac{\varepsilon_{\varkappa n}}{R_o} \sum_i \left\langle \vec{J}^2_{(i)} \quad \lambda_a^{(i)\,2} \right\rangle =$$

$$= \frac{1}{2}\alpha_s \frac{\varepsilon_{\varkappa n}}{R_o} \,.\, 2 \,.\, 3 \,.\, (2j+1)\; j\; (j+1) \,.\, \frac{16}{3} =$$

$$= 16\;\alpha_s \frac{\varepsilon_{\varkappa n}}{R_o}\; j\; (j+1)\; (2j+1) \quad . \qquad (12)$$

We have taken into account that there are two flavours, three colour and 2j+1 magnetic substates and that the expectation value of λ_a^2 is 16/3 in the fundamental representation of the gauge group. We have calculated the magnetic self-energy for different values of $\varkappa$ and n. The results are summarized in table 1. The energy is divided by 2j+1 in order to get the

Table 1

$\frac{ER_o}{2j+1}$ in MeV-fm $\qquad (\alpha_s = 2.2)$

$\varkappa$ \ n	1	2	3
-1	923	442	250
-2	872	488	346
-3	912	561	412
-4	955	621	
-5	986	671	
1	611	354	218
2	630	406	304
3	696	478	
4	751	528	
5	794		

The radius increases for large quark nuclei, so that the energy per particle decreases.

The magnetic self-interaction is less important for larger nuclei.

energy per quark in a given shell and multiplied by the bag radius. We see that the energy slightly increases for higher values of $\varkappa$ because this corresponds to higher values of the angular momentum. On the other hand the energy decreases for higher values of n. Finally the radius of the bag increases for large quark nuclei, so that the energy per particle decreases. We conclude that the magnetic self-energy becomes less important for larger quark bags.

A TWO-DIMENSIONAL MODEL

In half closed shells the energy can be lowered if the quarks are coupled in a suitable way. Therefore one might ask whether it is favourable to have many half-closed shells with all quarks coupled in such a way that they produce a large colour-magnetic field. To answer this question we have performed a two-dimensional model calculation with one flavour and a U(1) interaction between the quarks. We assume that there is a constant magnetic field B inside the bag. The magnitude of this field is determined self-consistently. We have to solve the Dirac equation

$$[E\gamma_o - \vec{\gamma}\cdot(-i\vec{\nabla}-e\vec{A})]\psi = o \quad ,$$
$$\vec{A} = \frac{1}{2} B \vec{e}_\varphi \tag{13}$$

in two dimensions together with the linear boundary condition

$$(-i\vec{\gamma}\cdot\vec{e}_r\psi - \psi)|_{surface} = o \quad . \tag{14}$$

The total energy is given by

$$E\cdot R = \frac{1}{2}\pi B^2 + n \sum_i \varepsilon_i \tag{15}$$

where n is a degeneracy factor which takes into account colour and flavour, R is the bag radius and ε_i is the energy of the quark i. The energy (15) is minimized with respect to the magnetic field B. This yields the equation

$$\pi B + n \sum_i \frac{d\varepsilon_i}{dB} = 0 \quad . \tag{16}$$

For positive B this equation has a solution only if the total susceptibility $\sum_i d\varepsilon_i/dB$ of all quarks is negative. The larger this susceptibility the larger the magnetic field will be. For a vanishing magnetic field each quark state is degenerate with the corresponding state of opposite spin and angular momentum. For an increasing magnetic field one of these states goes down to zero while the energy of the other state increases (Fig. 3). The increase of the latter state overwhelms the decrease of

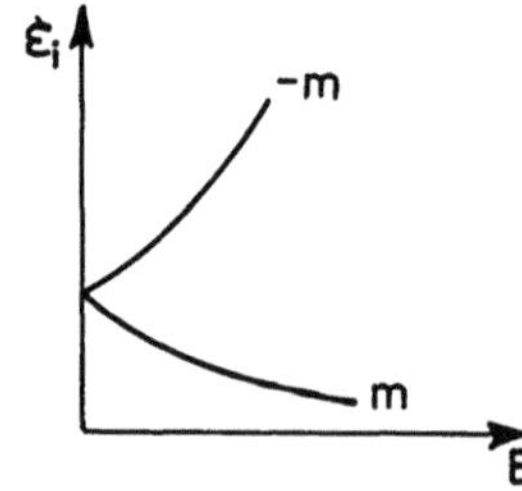

Fig. 3. The splitting of two degenerate levels when the magnetic field is turned on.

the first, so that the total susceptibility of both states is positive. Therefore these states cannot give rise to a magnetic field if both are occupied by quarks. This is related to the fact that in closed shells the net current vanishes so that the corresponding magnetic field vanishes. For weak coupling constant there is only an effect if one has a half-closed shell. On the other hand for a large coupling constant and a large magnetic field there is the possibility of level crossings (Fig. 4).

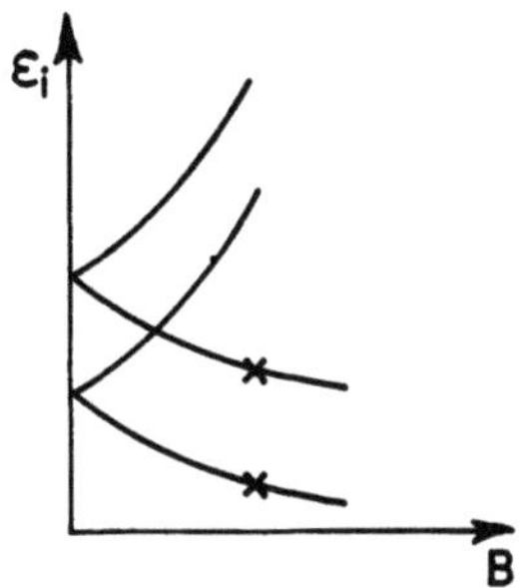

Fig. 4. Rearrangement due to level crossing.

Instead of occupying a rising level and a falling one as well it may be favourable to occupy two decreasing levels as indicated by the crosses in Fig. 4. The total contribution of these two levels to the susceptibility is then negative and a magnetic field may be produced. Examining the energy levels at large magnetic fields we found the following result. At low energies there are some decreasing levels. The slope of the lowest levels is rather small so that they do not contribute much to the susceptibility. For a magnetic field with gB=10 there are just six of them. Above the region where there are only decreasing levels there are nearly as many increasing levels as there are decreasing ones, and up to statistical fluctuations the net effect is rather small. This means that only a small number of quark current namely those corresponding to the lowest energy levels, are aligned. For a larger number of quarks the magnetic energy per quark becomes smaller and smaller. For gB=10 the total energy is 942,2 in arbitrary units if 100 levels are occupied. The decrease of the energy due to the magnetic field is only 3.3, and a coupling constant of g=7.9 was needed in order to produce this field.

CONCLUSIONS

The colour magnetic interaction between the quarks produces the splitting between the proton and the Δ^+-resonance. For larger bags the magnetic energy becomes less important and a magnetic field is only produced if there are half-closed shells. On the other hand in a model calculation no solution was found where many quarks are coupled together in order to produce a huge magnetic field.

REFERENCES

1. Talk of L. Neise, presented at this conference; D. Vasak and W. Greiner Do Nucleons Dissolve in Giant Nuclei, UFTP Preprint 163 (1985).

2. A. Chodos, R. L. Jaffe, K. Johnson, C. B. Thorn, and V. F. Weisskopf, Phys. Rev. D9:3471 (1974).
3. J. J. Kapusta, Phys. Rev. D2∅:989 (1979); P. Moreley and L. Kislinger, Phys. Rep. 51:63 (1979).
4. T. De Grand, R. L. Jaffe and J. Riskis, Phys. Rev. D12:2060 (1975).

THE QCD VACUUM*

Michael Danos

National Bureau of Standards
Gaithersburg, Maryland 20899

QCD is inherently a strong-field system in the infrared. Hence the physical vacuum itself without external charges is already non-perturbatively polarized; it has an energy density lower than that of the QED-type so-called perturbative vacuum by the amount of the "bag energy density." It will be shown that the physical vacuum can be described by an analogue of the BCS-state of superconductivity. This vacuum exhibits color confinement.

As time passes, more and more data becomes available to be compared with QCD predictions. The agreement between predictions and experiment is surprisingly good — dramatically so. One therefore by now is forced to conclude that QCD with very high probability is the correct description of hadron physics, in the same way as QED is the discription of atomic physics.

In fact, QCD is very similar to QED.[1] The only essential difference is that in QED the Boson field is neutral, in QCD it carries charge. Being charged, the "photons" of QCD can themselves emit "photons;" the Lagrangian (and the Hamiltonian) contain 3- and 4-gluon field vertices. Consequently the "renormalized charge" in QCD has a different behavior than that in QED. This results from the form of the vacuum polarization which yields

$$g_s = \frac{g_0}{1 - \eta \log(k^2/\mu^2)} \tag{1}$$

where for $k^2 = \vec{k}^2 - \omega^2 >> \mu^2$

$$\eta = \begin{cases} \dfrac{g_0}{3\pi} , \; \mu^2 = m_e^2 & \text{QED} \; , \\[2ex] \dfrac{g_0}{3\pi} \left(n_q - \dfrac{33}{2}\right) & \text{QCD} \; ; \end{cases} \tag{2}$$

n_q is the number of quark flavors (presumably $n_q = 6$). For QCD this is usually re-written as

$$g_s = \frac{6\pi}{33 - 2n_q} \frac{1}{\log(k^2/\Lambda^2)} \quad . \tag{3}$$

Here Λ is a parameter; from experimental data Λ is found to have a value somewhere between 100 and 500 MeV. Supposedly (3) is valid for all k^2.

The essential difference between QED and QCD is now evident. For QED g_s grows with increasing k^2, while in QCD it decreaes with increasing k^2. This is denoted by "QCD in an asymptotically free theory." The perturbative treatment should become ever better as one goes to higher energy processes. On the other hand, g_s has a pole at $k^2/\Lambda^2 = 1$. A perturbative treatment there is impossible: any field there is in effect an "infinitely strong" field. The low-energy regime of QCD therefore can be treated only by non-perturbative methods.

I shall now describe a method which takes advantage of these properties of QCD in that it uses a non-perturbative treatment for the low-momentum, i.e., the infrared, parts of the theory while permitting a perturbative treatment of the high-momentum parts.

We begin with the Lagrangian (a, b: color indices; μ,ν: Minkowski indices)

$$L = -\frac{1}{4} F^a_{\mu\nu} F^a_{\mu\nu} - \bar{\psi} \gamma_\mu D_\mu \psi \tag{4}$$

$$D_\mu = \partial_\mu - ig \lambda^b A^b_\mu \tag{5a}$$

$$F^a_{\mu\nu} = D_\mu A^a_\nu - D_\nu A^a_\mu \tag{5b}$$

From it (after fixing the QCD gauge — we chose the Coulomb gauge) one obtains the Hamiltonian. This allows one to write a Schrödinger equation, which for stationary states has the familiar form

$$H|S\rangle = E_s|S\rangle \quad . \tag{6}$$

At this point we provide the step which will allow us to use different methods for solving the low-momentum and the high-momentum sectors.[2] We split the Hilbert space into two parts, and write for our state vector

$$|S\rangle = \begin{pmatrix} F \\ G \end{pmatrix} \quad . \tag{7}$$

Herewith we have

$$H \begin{pmatrix} F \\ G \end{pmatrix} \equiv \begin{pmatrix} X & Y \\ Y^+ & Z \end{pmatrix} \begin{pmatrix} F \\ G \end{pmatrix} = E \begin{pmatrix} F \\ G \end{pmatrix} \quad . \tag{8}$$

Eliminating $|G\rangle$ we obtain

$$X|F\rangle + Y \frac{1}{E-Z} Y^{+}|F\rangle \equiv H_{Eff} \ |F\rangle = E|F\rangle \quad . \tag{9}$$

Now we specify the space $|F\rangle$ to contain only low-momentum components. This is clearly a very small part of the full Hilbert space. Hence the non-perturbative treatment can be limited to a small part of the Hilbert space. This then allows the application of a number of well-known powerful techniques[3] which, for technical reasons, could not be employed in the full Hilbert space. At the same time the ultraviolet renormalization can be taken care of in the space G by perturbative methods, totally independently from the treatment of the infrared sector, i.e., the space F.

For the gluons we use a plane wave expansion for the (Schrödinger picture) fields

$$A = \frac{1}{(2\pi)^{3/2}} \sum\int \frac{d^3k}{\sqrt{2\omega_k}} \left[a_{\underset{\sim}{k}ia}\, \varepsilon_i \ell^{i\underset{\sim}{k}\cdot\underset{\sim}{r}} + a^{+}_{\underset{\sim}{k}ia}\, \varepsilon_i \ell^{-i\underset{\sim}{k}\cdot\underset{\sim}{r}}\right] \quad . \tag{10}$$

Here $\underset{\sim}{k}$ specifies the momentum, i the polarization, a the color, and ε_i are the real Cartesian components of the polarization vectors. The field operators are defined to obey

$$[a_{\underset{\sim}{k}ia}, a^{+}_{\underset{\sim}{k}'i'a'}] = \delta_{\underset{\sim}{kk}'}\ \delta_{aa'}\ (\delta_{ii'} - \frac{k_i k_{i'}}{k^2}) \quad . \tag{11}$$

We now specify the space F as containing all states which contain arbitrary numbers of transversal gluons with $|k| < M$. The states in G then contain at least one quark (in our case, $q\bar{q}$ pair), or at least one gluon with $|k| > M$, in addition to an arbitrary number of soft gluons.

We now are ready to specify the model of the vacuum. We consider the particular coherent pairing state,[4]

$$|V'\rangle = e^{\frac{1}{2}\int_0^M d^3k\theta_k\, (a^{+}_{\underset{\sim}{k}}\cdot a^{+}_{\underset{\sim}{\underline{k}}})}\ |0\rangle \tag{12}$$

where the dot indicates formation of a singlet in both Minkowski and color spaces, and $\underset{\sim}{\underline{k}} = -\underset{\sim}{k}$. Note that θ_k depends only on $k = |\underset{\sim}{k}|$. In eq (12) we have omitted the time dependence; it will have to be added to achieve the complete Schrödinger picture state vector. Also, the state $|V'\rangle$ is not normalized. In equation (12) the separation between the spaces F and G is given as the upper limit of the integral over the momenta, denoted by M.

The state (12) contains an unspecified number of pairs. It cannot be achieved in a perturbative treatment as it is connected with the perturbative vacuum $|0\rangle$ only by an infinite number of applications of the Hamiltonian. The parameter θ_k is a variational parameter which will be used to find a (local) minimum of the Hamiltonian. It is advantageous to replace eq (12) by

$$|V\rangle = e^{\frac{1}{2}\int d^3k\theta_k\, B_{\underset{\sim}{k}}}\ |0\rangle \quad , \tag{13}$$

$$B_{\underset{\sim}{k}} = \left(a^{+}_{\underset{\sim}{k}} \cdot a^{\pm}_{\underset{\sim}{k}} - a_{\underset{\sim}{k}} \cdot a_{-\underset{\sim}{k}}\right) \quad ; \tag{14}$$

this state, the operator (14) being anti-Hermitian, respects the normalization of $|0\rangle$.

Note that it would be misleading to call the state of the system described by $|V\rangle$ a "gluon condensate." Namely, even though only the gluons are treated explicitly in (13), (14), the actual state contains also an unspecified number of qq-pairs. However, being in the part G of the Hilbert space they are hidden in the effective force, which we now discuss. To that end we rewrite (9) as a variational problem

$$\langle F|X|F\rangle + \langle F|Y \frac{1}{E-Z} Y^{+}|F\rangle = E \tag{15}$$

$$\delta\{\langle F|X|F\rangle + \langle F|Y \frac{1}{E-Z} Y^{+}|F\rangle\} = 0 \ . \tag{16}$$

The kinetic energy part of the first term of (15) is simply,

$$H_0 = \int^{M} d^3k \, \langle |k| \rangle \, a^{+}_{\underset{\sim}{k}} \cdot a_{\underset{\sim}{k}} \equiv \int^{M} d^3k \, \omega_k \, a^{+}_{\underset{\sim}{k}} \cdot a_{\underset{\sim}{k}} \quad , \tag{17}$$

It is the second term of (15), i.e., in the effective force, where the essential model assumptions have to be made since an exact inversion of the operator (E-Z) is not possible, even though it concerns the space G and the inversion could be treated perturbatively.

We will write the effective interaction as a series in powers of the field operators of the space F, i.e., in essence an expansion in powers of the gluon density. Taking over the results of Appendix A, eqs. (A.10) and (A.20) we have for a typical lowest order term

$$X' + Y \frac{1}{E-Z} Y^{+} + H_I' = \alpha_0 \int d^3x \; :(A\cdot A)^2: \quad . \tag{18}$$

We here have introduced the dimensionless strength parameter α_0, which depends implicitly on the energy M which separates the spaces F and G. In higher order it also contains the QCD running coupling constant g_s, eq. (4).

From the results of Appendix A where a number of lowest order graphs are computed one may conclude that in QCD the quantity α_0 seems to be negative, which is the sign needed for the possibility of the existence of a condensate. However, at the same time the resulting Hamiltonian is not positive definite, i.e., the Hamiltonian $H_0 + H_I'$ is not bounded from below; the particle density would tend to infinity. Therefore, with this Hamiltonian one cannot expect the vacuum state to be stable. This, of course, is also true if one takes for H_I' an attractive QED-type Coulomb force. Since the original Hamiltonian representing a renormalizable theory presumably is bounded from below the effective Hamiltonian (18) certainly is an insufficient representation of the effective force H_{eff} of (9). To achieve a bounded Hamiltonian one has to continue the expansion begun in (18). To that end we add the sixth-order terms, e.g.,

$$H_{II}' = \frac{\beta_0}{M^2} \int d^3x \; :(A\cdot A)^3: \tag{19}$$

where we have extracted the scale M^2 in order to define the dimensionless interaction strength parameter β_0. Analogously to the case of eq. (18), the numerical value of β_0 can be computed perturbatively in the space G; from Appendix A we see that to the same level of assurance of the validity of the results as for α_0, the constant β_0 seems to be positive. Herewith we have reproduced the essential characteristics of the effective interaction: attraction at low density, stabilized at higher density.

In order to solve our model we introduce the Boson analogue of the Bogoliubov quasi-particle transformation[4] which we write as

$$\begin{pmatrix} b^+_{\underset{\sim}{k}ia} \\ b_{-\underset{\sim}{k}ia} \end{pmatrix} = \begin{pmatrix} u_k & -v_k \\ -v_k & u_k \end{pmatrix} \begin{pmatrix} a^+_{\underset{\sim}{k}ia} \\ a_{-\underset{\sim}{k}ia} \end{pmatrix} \tag{20}$$

with

$$u_k^2 - v_k^2 = 1 \tag{21}$$

for each $k = |\underset{\sim}{k}|$. Hence the quasi-particle operators b and b^+ also obey the commutation relations (11). Furthermore we demand

$$b|V\rangle = 0 \tag{22}$$

where $|V\rangle$ is taken to be of the form (13). In order to fulfill (21) identically we introduce the Bogoliubov angle θ_k and write

$$u_k = \cosh\theta_k \tag{23a}$$

$$v_k = \sinh\theta_k \quad . \tag{23b}$$

In our calculation we shall use a more complete interaction, viz.,

$$H_I = \alpha_0 \int d^3x:(A\cdot A)^2: + \frac{\gamma_0}{M^2} \int d^3x:(A\cdot A)(\Pi\cdot\Pi): + \frac{\delta_0}{M^4} \int d^3x:(\Pi\cdot\Pi)^2 \quad . \tag{18a}$$

$$H_{II} = \frac{\beta_0}{M^2} \int d^3x:(A\cdot A)^3: + \frac{\eta_0}{M^8} \int d^3x:(\Pi\cdot\Pi)^3 \quad . \tag{19a}$$

Forming the expectation value of (18a) and (19a) together with (17) in the physical vacuum $|V\rangle$, we find

$$\frac{E}{\Omega} \equiv \zeta = \int_0^M d^3k \frac{(x_k-1)^2}{4x_k} \omega_k + \frac{\alpha}{8} Y^2 + \frac{1}{M^2} \frac{\beta}{12} Y^3 + \frac{\gamma}{4M^2} YZ + \frac{\delta}{8M^4} Z^2 + \frac{\eta}{12M^8} Z^3 \tag{24}$$

$$\Omega = 2(N^2 - 1)\tilde{\Omega} \tag{24a}$$

Here $(2\pi)^3 \tilde{\Omega}$ is the volume of the quantization box and $2(N^2-1) = 16$ for SU(3). Also, all strength constants of (24) are dimensionless. In detail,

$$x_k = e^{2\theta_k} = (u_k + v_k)^2 \tag{25}$$

$$\alpha = 8 \frac{\alpha_0}{4(2\pi)^3} \left(2(N^2-1) + \frac{4}{3}\right) \tag{26a}$$

$$\beta = 12 \frac{\beta_0}{8(2\pi)^6} \left[2(N^2-1) + \frac{4}{3}\right]\left[2(N^2-1) + \frac{8}{3}\right] \tag{26b}$$

$$\gamma = 4 \frac{\gamma_0}{4(2\pi)^3} 2(N^2-1) \tag{26c}$$

$$\delta = 8 \frac{\delta_0}{4(2\pi)^3} \left[2(N^2-1) + \frac{4}{3}\right] \tag{26d}$$

$$\eta = 12 \frac{\eta_0}{8(2\pi)^6} \left[2(N^2-1) + \frac{4}{3}\right]\left[2(N^2-1) + \frac{8}{3}\right] \tag{26e}$$

$$Y = \int_0^M d^3k \frac{x_k - 1}{\omega_k} \tag{27a}$$

$$Z = \int_0^M d^3k \left(\frac{1}{x_k} - 1\right) \omega_k \quad . \tag{27b}$$

Except for the factor $2(N^2-1)/(2\pi)^3$ the expression in (24) thus is directly the energy density. The quantities (26) are simply numerical constants, while the quantities (27) are also numerical constants, which, however, being functionals, depend on the actual form of the solution.

To obtain the minimum of (24) by varying θ_k it is most convenient to perform the variation independently with respect to u_k and v_k. To that end we add the condition (21) to the minimization, and search for

$$\begin{pmatrix} \frac{\partial}{\partial u_k} \\ \frac{\partial}{\partial v_k} \end{pmatrix} \left(\zeta - \frac{1}{2} \int d^3k \; \varepsilon_k \; (u_k^2 - v_k^2 - 1)\right) = 0 \tag{28}$$

where ε_k is the Lagrange parameter. This leads to the following condition

$$\varepsilon_k \begin{pmatrix} u_k \\ -v_k \end{pmatrix} = \begin{pmatrix} \omega_k + \Gamma_k & \Delta_k \\ \Delta_k & \omega_k + \Gamma_k \end{pmatrix} \begin{pmatrix} u_k \\ v_k \end{pmatrix} \quad , \tag{29}$$

where

$$\Gamma_k = \frac{1}{2}\frac{G}{\omega_k} + \frac{1}{2} F\omega_k \tag{30a}$$

$$\Delta_k = \frac{1}{2}\frac{G}{\omega_k} - \frac{1}{2} F\omega_k \tag{30b}$$

$$G = \alpha Y + \frac{\beta}{M^2} Y^2 + \frac{\gamma}{M^2} Z \tag{30c}$$

$$F = \frac{\gamma}{M^2} Y + \frac{\delta}{M^4} Z + \frac{\eta}{M^8} Z^2 \quad . \tag{30d}$$

The solution of (29) is

$$\varepsilon_k = \sqrt{\omega_k^2 + G} \quad \sqrt{1 + F} \quad . \tag{31}$$

As seen from the eqs. (25) through (27), the equations are too involved to allow an analytical treatment. The actual solution must be found by numerical methods. We will show the results of such a calculation below in Fig. 1.

In order to proceed with the discussion of the spectrum we need the expression of the Hamiltonian in terms of the quasi-particle operators, The Wick decomposition of eq. (28) yields

$$\frac{H}{\Omega} = \zeta + \frac{1}{2}\int d^3k(v_k,u_k)\begin{pmatrix}\omega_k + \Gamma_k & \Delta_k \\ \Delta_k & \omega_k + \Gamma_k\end{pmatrix}\begin{pmatrix}u_k \\ v_k\end{pmatrix}(b^+_{\underset{\sim}{k}}\cdot b^+_{\underset{\sim}{k}} - b_{\underset{\sim}{k}}\cdot b_{\underset{\sim}{k}})$$

$$+ \int d^3k(u_k,v_k)\begin{pmatrix}\omega_k + \Gamma_k & \Delta_k \\ \Delta_k & \omega_k + \Gamma_k\end{pmatrix}\begin{pmatrix}u_k \\ v_k\end{pmatrix}(b^+_{\underset{\sim}{k}}\cdot b_{\underset{\sim}{k}}) + v_{residual} \quad . \tag{32}$$

Here $v_{residual}$ represents also the higher order terms of the perturbation expansion of H_{eff} beyond H_I and H_{II}. Upon insertion of the solution, i.e., using the results (28)-(29), eq. (32) acquires the form

$$H = E + \int d^3k \ \varepsilon_k \ b^+_{\underset{\sim}{k}} \cdot b_{\underset{\sim}{k}} + V_{residual} \tag{33}$$

This shows that for the case where $V_{residual}$ can be neglected, ε_k indeed is the quasi-particle energy and b^+ creates quasi-particles. From (31) we see immediately that the existence of a gap requires $G > 0$. Namely, for $\alpha = \beta = \gamma = 0$ we have $\varepsilon_k = \omega_k \sqrt{1 + \delta Z}$ which does not exhibit a gap, only a dilatation of the spectrum. On the other hand, if $G > 0$ then indeed $\sqrt{G}$ is the gap energy, i.e., the quasi-particle mass; the spectrum again is modified by the dilatation factor.

We now are in the position to discuss the character of the solutions. In view of the fact that we have not performed a full calculation of H_{eff} we shall do this freely, i.e., without prejudicing the sign and magnitude of the interaction constants of (24). That means, we will study the behavior of the solutions for a selection of cases when only some of the constants at a time are not zero. This way we will learn the essential characteristics of the solutions. For the general case the solution must and can be obtained numerically.

We begin with the case $\alpha, \beta \neq 0$, all other constants $= 0$. Then $F = 0$. In view of the form (27a), G may have either sign. We thus investigate ζ as function of G. The extrema of this curve give the possible states of the system.

The branch $G > 0$ poses no difficulty. For $\alpha < 0$ and $Y > 0$, $\beta > 0$ is required to stabilize the system. Then the curve has one minimum. Thus, if $G > 0$ one may obtain a gap. Note that Γ_k and Δ_k, i.e., the mean field and the pairing strength, are singular for $k \to 0$, as a result of the relativistic measure contained in (10). At the same time the Bogoliubov angle, $\theta = \frac{1}{2} \log (k/\sqrt{k^2 + G})$ has there a singularity — confirming that the condensation is an infrared phenomenon. Still, of course, $|V\rangle$ remains normalized.

The branch $G < 0$ requires a more careful analysis. To wit: ε_k, eq (31), has a branchpoint at $k^2 = |G|$ and is imaginary for $k^2 < |G|$. Since the Hamiltonian is hermitean this indicates that the space of the variational functions is inadequate. That this is indeed the case one sees by investigating the following Ansatz for the Bogoliubov angle:

$$\theta_k = 0 \quad \text{for} \quad 0 \leq k < m \tag{34a}$$

$$\theta_k = \phi_m \, \Delta_m(k) + \tilde{\theta}_k \tag{34b}$$

where the distribution $\Delta_m(k)$ is defined by

$$\Delta_m(k) = \int \frac{d^2\hat{k}}{4\pi \, \tilde{\Omega}} \, \delta^3(k - m) \tag{34c}$$

and where $\tilde{\theta}_k$ has support in the interval (m,M]. The function $\tilde{\theta}_k$ and the constants ϕ_m and m are the variational parameters. The distribution Δ is built to be idempotent so that the decomposition (34b) holds for all analytical functions of θ_k. Consequently (27a) is replaced by

$$\tilde{Y} = \int_m^M d^3k \, \frac{\tilde{x}_k - 1}{\omega_k} + \frac{x_m - 1}{\omega_m \tilde{\Omega}} \tag{35}$$

where

$$\tilde{x}_k = e^{2\tilde{\theta}_k} \quad ; \quad x_m = e^{2\phi_m} \quad . \tag{36}$$

Herewith the energy density becomes

$$\zeta = \int_m^M d^3k\, \omega_k \frac{(x_k - 1)^2}{4x_k} + \omega_k \frac{(x_m - 1)^2}{4x_m\tilde{\Omega}} + \frac{\alpha}{8}\tilde{Y}^2 + \frac{\beta}{12M^2}\tilde{Y}^3 \quad . \tag{37}$$

Now we find that the minimum arrises at the solution

$$m = 0 \quad , \quad \tilde{\theta}_k = 0 \quad , \quad G = 0 \quad , \quad \zeta_{min} = \frac{\alpha^3}{24\beta^2} M^4 \quad , \tag{38a}$$

with

$$x_m = - \frac{\alpha M^2}{\beta} m\, M^2\, \tilde{\Omega} \quad . \tag{38b}$$

For a quantization box of size L^3 this gives for $m \to 0$,

$$x_m \to - \frac{\alpha}{\beta} \frac{1}{8\pi^2} M^2 L^2 \quad . \tag{38c}$$

In fact, this singular case of the Bogoliubov transformation turns out to be precisely the Bose-Einstein condensation with a spectrum $\varepsilon_k = k$ and with a population density ϕ_0 at $k = 0$. This way we have the important result:

If the minimum at $G < 0$ of (37) is lower than the minimum at $G > 0$, no gap ensues and we have Bose-Einstein condensation with a vacuum state $e^{\frac{1}{2}\theta_0 a_0^+ a_0^+}|0\rangle$ (case A of Fig. 1); only if the minimum at $G > 0$ is lower, we have a superconducting vacuum $|V\rangle$ with a non-vanishing gap (case B). The dashed parts of the curves in Fig. 1 show the energy for the variational functions without the distribution (34), the full lines show the complete result.

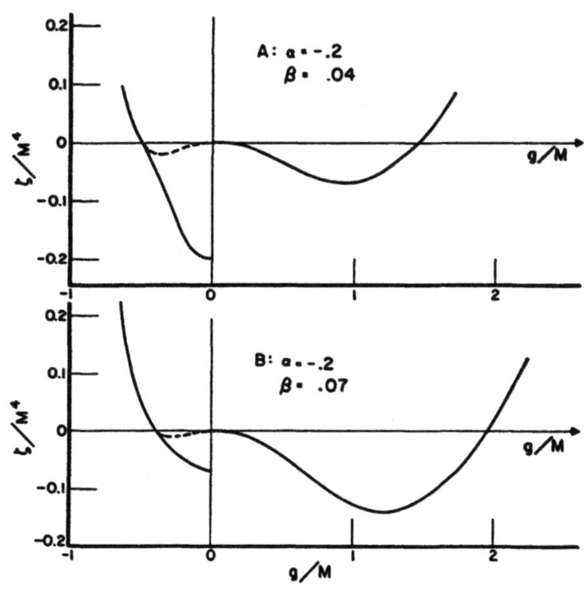

Fig. 1

Of particular importance is the vertex $A^2\Pi^2$, i.e., $\gamma \neq 0$. A solution which exhibits a gap exists for $G > 0$; $-1 < F < 0$. In that case the Π^2 of H_0 and the Π^2 of the vertex play against each other; they have the same dependence on the density. We find for this solution that the energy of the condensate is finite only if $\gamma < \gamma_c = \frac{1}{2}\pi$. In this regime saturation is provided by the term Π^2 in H_0. Then we find for $\alpha = 0$

$$\lim_{\gamma\to\gamma_c} \zeta = -\frac{\pi}{36} M^4 \quad ,$$

$$\lim_{\substack{\gamma\to\gamma_c \\ k\to o}} \varepsilon_k = \frac{2}{3} M \quad , \tag{39}$$

while at the same time $F \to -1$ and $G \to +\infty$. It therefore is possible to have a gap arise from $A^2\Pi^2$, however, then $\gamma < \gamma_c$ is necessary. This is not the case for a form factor $1/\Delta$, i.e., for the QED form of the Coulomb term. In that case H is not bounded from below. By the way, this shows that the replacement of the QCD Coulomb term by the QED form is not justifiable for the infrared; such a replacement would destabilize the solutions into the $\gamma > \gamma_c$ regime.

One of the principal consequences of the vacuum condensate is the confinement of color. However, a full treatment of the confinement demands an understanding of the transition region $|V\rangle \leftrightarrow |0\rangle$, i.e., of the structure of the vacuum-"bag" interface, which would require the breaking of the translational invariance[5] - which our vacuum state possesses. Hence we here can give only a qualitative description of the penetration of a color field from the perturbative into the condensed vacuum.

Consider a "weak" field, A_e, in the region of the perturbative vacuum, say region W. In that case the higher order terms are unimportant, and the system in region W is well described by

$$H_W = \frac{1}{2} \int_W d^3x \;:(\mathring{A}\cdot\mathring{A} - A\cdot\nabla^2 A): \quad . \tag{40}$$

In the physical vacuum, say region Ω, however, the presence of the condensate does not allow the neglect of the higher terms, and hence

$$H_\Omega = \frac{1}{2} \int_\Omega d^3x \;:(\mathring{A}\cdot\mathring{A} - A\cdot\nabla^2 A): + \alpha\int_\Omega d^3x \;:(A\cdot A)^2: + \beta\int_\Omega d^3x \;:(A\cdot A)^3: \;. \tag{41}$$

Denoting the condensate field by A_c and keeping in (41) only the terms quadratic in A_e, we have, in the mean-field approximation,

$$H = \frac{1}{2} \int d^3x \;:(\mathring{A}_e\cdot\mathring{A}_e - A_e\cdot\nabla^2 \Delta A_e): + m \int_\Omega d^3x \;:(A_e\cdot A_e): \tag{42}$$

where

$$m = \langle V| \;:\alpha(A_c\cdot A_c) + \beta(A_c\cdot A_c)^2: \;|V\rangle \quad . \tag{43}$$

Herewith we have for the equations of motion for $A_e|0\rangle$

$$\ddot{A}_e = \nabla^2 A_e - mA_e \qquad x \text{ in } \Omega \tag{44}$$

$$\ddot{A}_e = \nabla^2 A_e \qquad x \text{ in } W \quad . \tag{45}$$

Recalling (30c) we see from (43), (44) that in the physical vacuum $\sqrt{G}$ plays the role of a gluon mass. As long as $m > 0$, in the limit of small energy of A_e, for $\nu^2 < m$ we have from eq (47)

$$q^2 = \nu^2 - m < 0 \tag{46}$$

i.e., an exponential damping of A_e in Ω, i.e., in the physical vacuum, while according to eq (45) the field can freely propagate in the perturbative vacuum with $k^2 = \nu^2$.

Since the magnitude of m is related to the gap size, a high-energy gluon, i.e., for which $\nu^2 > m$, could penetrate the physical vacuum; however it will destroy that vacuum by emitting quasi-particles, and in the process lose energy until $\nu^2 < m$, and thus in the end will be turned around. In other words, both low and high energy gluons are totally reflected at the $|0\rangle$ - $|V\rangle$ interface.

Next, consider the energy-momentum character of $|V\rangle$. Namely, since our treatment is not manifestly covariant a boost of the solution must be carried out in detail. Still, since our treatment does not break translational invariance, we at least, in contrast to the bag model, have no difficulties associated with the center-of-mass problem. We begin by discussing the energy of the solution $|V(t)\rangle$, which here we write in full, i.e., (13) augmented by its time dependence. The expectation value (24),

$$\langle V(t)|\ H\ |V(t)\rangle = E_V < 0 \tag{47}$$

cannot actually be the physical eigenvalue of the vacuum; in the utilized quantized form all energies must be non-negative. The result (47) simply implies the need for a kind of gauge transformation. This is equivalent to the case of classical electrodynamics where one can shift the energy scale arbitrarily up or down by the addition of a constant scalar potential, i.e., by a global gauge transformation. The same can be done here by a redefinition of the phase of the state vector $|V\rangle$:

$$|V_0\rangle = e^{+iE_V t}\ |V(t)\rangle\ ; \tag{48}$$

with this phase the new state vector obeys

$$\frac{\partial}{\partial t}\ |V_0\rangle = 0\quad . \tag{49}$$

Remembering that we work in the Schrödinger picture this then yields for the vacuum energy

$$E_0 = 0\quad . \tag{50}$$

This way the vacuum $|V_0\rangle$ does not contribute to the cosmological term of gravity. On the other hand, now the perturbative vacuum acquires the energy $|E_V|$, which is the energy needed to replace $|V_0\rangle$ by $|0\rangle$. Since in our model system the state $|V_0\rangle$ occupies a volume $\tilde{\Omega}$ in position space, the number $|E_V|$ actually represents an energy density

$$\tilde{\zeta} = \frac{|E_V|}{\tilde{\Omega}}\quad . \tag{51}$$

With this normalization the state $|V_0\rangle$ has the momentum four-vector

$$P = (E,\vec{k}) = (0,0) \tag{52}$$

which indeed remains the same in all frames. However, a formal boost of $|V\rangle$ would be wrong since the separation F - G of the Hilbert space in eqs (8), (12) is not boost-invariant. Therefore, in order to describe the vacuum in a boosted system one must perform the complete calculation in that boosted system. Then, of course, the form of $|V\rangle$ in that system is exactly the same as in the original system. Of course, this formal boost non-invariance does not invalidate the accuracy of the solutions for a given frame of reference, say, the lab system.

APPENDIX A: EFFECTIVE INTERACTION

We here give a justification for the form of the effective interaction eqs. (18), (19). We will limit the discussion to the lowest order graphs. The following terms with four external fields are possible:

$$I = \,:(A\cdot A)(A\cdot A):\, C_1 \tag{A.1}$$

$$II = \,:(A\cdot A)(\Pi\cdot\Pi):\, C_2 \tag{A.2}$$

$$III = \,:(\Pi\cdot\Pi)(\Pi\cdot\Pi):\, C_3 \tag{A.3}$$

Equations (A.1), (A.2), (A.3) are written in an obvious symbolic fashion. In terms of the gluon operators, the operators A and Π are

$$A_{ki\beta} = a^+_{ki\beta} + a_{\bar{k}i\beta} \tag{A.4}$$

$$\Pi_{ki\beta} = a^+_{ki\beta} - a_{\bar{k}i\beta} \quad . \tag{A.5}$$

Here the indices denote the momentum, the polarization and the color respectively. The following elementary four-field vertices exist:

$V_4 =$  A, A, A, A (A.6a)

$V_c =$  A, A, Π, Π . (A.7a)

We have, explicitly

$$V_4 = \frac{1}{4} g^2 \int d^3x\, A^\alpha_i A^\beta_j A^\gamma_i A^\delta_j\, f^{a\alpha\beta} f^{a\gamma\delta}$$

$$= \frac{1}{2}\frac{g^2}{(4\pi)^3}\int dp\,dq\,dk\,d\ell\; A^\alpha_{pi} A^\beta_{qj} A^\gamma_{ki} A^\delta_{\ell j}\, f^{a\alpha\beta} f^{a\gamma\delta}\, \delta^3(p+q+k+\ell)/\sqrt{\omega_p\omega_q\omega_k\omega_\ell} \tag{A.6b}$$

$$V_c = -\frac{1}{2} g^2 \int d^3x\, d^3y\, A^\alpha_{xi}\Pi^\beta_{xi}\, \frac{1}{\Delta}(x-y)\, A^\gamma_{yj}\Pi^\delta_{yj}\, f^{a\alpha\beta} f^{a\gamma\delta}$$

$$= -\frac{g^2}{(4\pi)^3}\int dp\,dq\,dk\,d\ell\; A^\alpha_{pi}\,\Pi^\beta_{qj}\,\frac{4}{|p+q|^2}\,A^\beta_{kj}\,\Pi^\delta_{\ell j}\, f^{a\alpha\beta} f^{a\gamma\delta}\, \delta^3(p+q+k+\ell) \ . \tag{A.7b}$$

Note that V_c involves the Coulomb propagator. In the loop the integration will be over high momenta; hence the replacement of the covariant derivative by the simple derivative, i.e., the QCD Coulomb propagator by the QED propagator in (A.7b). We see that in second order, the operator I can be achieved by iterating either (A.6a) or (A.7a); the operator II by either iterating (A.7a) or by a product of (A.6a) and (A.6b); and the operator III by iterating (A.7a).

We now sketch the evaluation in second order of the operator I. This involves the contraction of the internal lines of the loop, i.e., the integration in the space G. To that end we consider the propagator $(E - Z)^{-1}$ of (4b). Since E is the desired eigenvalue, and we are interested in the lowest eigenstate, E is either zero, if no condensation takes place, or negative and equal the condensate energy, in case of condensation. Still, E drops out also in that case. Namely, since the space G contains also low momentum gluons, Z is very complicated and looks like H_{eff}. Indeed, Z can be written as $Z = E + \int d^3k\, \omega_k\, b_k^+ \cdot b_k + Z_{residual}$. In principle, $Z_{residual}$ could be evaluated in an iterative manner; to the approximation of a small $Z_{residual}$, the propagator becomes simply the reciprocal of the sum of the one-particle energies ω_k.

We now have for the interaction Hamiltonian, omitting for the time being the color couplings,

$$g^4 \int \frac{\delta^3(p+q+p'+q')}{\sqrt{\omega_p\omega_q\omega_{p'}\omega_{q'}}} \cdot \frac{\delta^3(k+\ell+k'+\ell')}{\sqrt{\omega_k\omega_\ell\omega_{k'}\omega_{\ell'}}} A_p A_q \Big[A_{p'} A_{q'} \frac{-1}{H_0} A_{k'} A_{\ell'} + \frac{\omega_{p'}\omega_{q'}}{|p'+p|^2} \Pi_{p'} \Pi_{q'} \frac{-1}{H_0} \Pi_{k'} \Pi_{\ell'} \frac{\omega_{k'}\omega_{\ell'}}{|k'+k|^2} \Big] A_k A_\ell \tag{A.8}$$

The form factor in this expression is

$$-g^4 \int_M^U d^3p' \left[\frac{1}{\omega_{p'}\omega_{p'+p+q}} + \frac{\omega_{p'}\omega_{p'+p+q}}{|p'+p|^2} \right] \frac{1}{\omega_{p'}+\omega_{p'+p+q}} \tag{A.9a}$$

which for small external momenta becomes

$$\sim -g^4 \int dp\,dq\,dk\,d\ell\, \frac{\delta^3(p+q+k+\ell)}{\sqrt{\omega_p\omega_q\omega_k\omega_\ell}} (A_p \cdot A_q)(A_k \cdot A_\ell) \int_M^U d^3r/\omega_r^3 \quad . \tag{A.9b}$$

Here U is an upper limit associated with the ultraviolet renormalization which we will discuss below; and M is the mass introduced in (12) which specifies the boundary between the spaces F and G. Note that V_4 and V_c yield the same dependence of the effective interaction on M. Adding the contribution of the elementary vertex (A.6) we have, in position space,

$$H_{eff}(A^4) \sim \left(g^2 - g^4 \log \frac{U}{M}\right) \int d^3x\, \left(A(x) \cdot A(x)\right)^2 \tag{A.10}$$

The force II, eq. (A.2) requires no ultraviolet renormalization; it has the small momentum limit

$$\sim - g^4 \int d^3p\, d^3q\, d^3k\, d^3\ell\, \delta^3(p+q+k+\ell) \frac{\sqrt{\omega_k\omega_\ell}}{\sqrt{\omega_p\omega_q}} (A_p \cdot \Pi_k)\, (A_q \cdot \Pi_\ell) \int_M^U d^3r/\omega_r^5 \;, \tag{A.11}$$

and thus we find

$$H_{eff}(A^2\Pi^2) \sim - g^4/M^2 \int d^3x\, \left(A(x) \cdot \Pi(x)\right)^2 \tag{A.12}$$

to which, of course, must be added the Coulomb term (A.7b).

The operator III, eq. (A.3) has the position space form

$$H_{eff}(\Pi^4) \sim -g^4/M^2 \int d^3x \, \left(\Pi(x)\cdot\Pi(x)\right)^2 \quad . \tag{A.13}$$

Again no renormalization is required.

We now proceed to the higher terms which are supposed to stabilize the Hamiltonian. They result from continuing the graph expansion of the effective interaction (9). The lowest order graphs arise by replacement of the two- point loop by a three-point loop, i.e., a triangle. Thus for the term in (A.6) we have

$$g^6 \int \frac{\delta^3(p+q+p'+q')}{\sqrt{\omega_p\omega_q\omega_{p'}\omega_{q'}}} \times A_p A_q A_{p'} A_{q'} \frac{-1}{H_0} \frac{\delta^3(k+\ell+k'+\ell')}{\sqrt{\omega_k\omega_\ell\omega_{k'}\omega_{\ell'}}} A_k A_\ell A_{k'} A_{\ell'} \frac{-1}{H_0} \frac{\delta^3(r+s+r'+s')}{\sqrt{\omega_r\omega_s\omega_{r'}\omega_{s'}}} A_r A_s A_{r'} A_{s'} \tag{A.14}$$

Here again no renormalization is required and we have

$$H_{eff}(A^6) \sim g^6/M^2 \int d^3x \, \left(A(x)\cdot A(x)\right)^3 \tag{A.15}$$

Indeed, they have the opposite sign of the fourth-power interaction owing to the presence of two (negative) energy denominators. Similarly, using the Coulomb vertex (A.7a) we find

$$H_{eff}(\Pi^6) \sim g^6/M^8 \int d^3x \, \left(\Pi(x)\cdot\Pi(x)\right)^3 \quad . \tag{A.16}$$

We now consider the renormalization which is required for the term $\sim A^4$ (A.10). Since QCD is a renormalizable theory the renormalization can be performed by using counterterms in the Lagrangian. Developing the coupling constant up to second order in the Goldstone series yields

$$g = g_0(1 + \tfrac{1}{2}\, c\, g_0^2 \log U/\mu) \tag{A.17}$$

where μ defines the renormalization point. At the same time, the sum of all graphs up to second order which describe soft gluon scattering has the form factor

$$g^2 - g^4 c \log U/t \tag{A.18}$$

when t is the momentum transfer. Using (A.17) the ultra-violet momentum U drops out and (A.18) becomes instead

$$g_0^2 - g_0^4 c \log M/t \quad . \tag{A.19}$$

We now return to our computation of the effective force. For each graph of the Goldstone series there exists an equivalent graph for the coupling constant. The difference between the two graphs resides in the difference between the energy denominators: The propagators for the effective force contain the perturbed energies, as the Goldstone series is a Brilloin-Wigner expansion. The difference between the two series thus vanishes in the ultraviolet limit, and hence we find for the effecive interaction

$$H_{eff}(A^4) \sim \left(g^2 - g^4 c \log \frac{\mu}{M}\right) = \alpha_s = g_0^2\left(1 - cg_0^2 \log \frac{\mu}{M}\right) \quad (A.20)$$

Finally, comparing with the form (4) we find

$$\mu = e^{\frac{1}{2}cg_0^2} \Lambda \gg \Lambda \quad (A.21)$$

for our case.

APPENDIX B: COULOMB INTERACTION

The usual geometric series expansion for the Coulomb gauge Coulomb propagator,

$$\frac{1}{\Delta - L} = \frac{1}{\Delta} + \frac{1}{\Delta} L \frac{1}{\Delta} + \dots \quad (B.1)$$

where

$$L = g\, A \cdot \nabla \times \quad (B.2)$$

(here $\times$ = vector product in color space) cannot be utilized for the infrared, owing to the destabilization of the solutions by the term $1/\Delta$, as discussed earlier. On the other hand, in the space-axial gauge, $\vec{A}\cdot\vec{n} = 0$, with $\vec{n}$ a space-like unit vector, the Coulomb propagator is simply $1/\Delta$. The use of this gauge is however prohibitively complicated in cases where the system, e.g., a hadron, has a definite angular momentum. One must therefore work in a gauge which respects angular momentum. A possible way to avoid the operator $1/\Delta$ when using the Coulomb gauge is to "measure" the Coulomb interaction against a simpler shift operator, say M, for which one can construct the inverse. Then one has

$$\frac{1}{\Delta - L} = \frac{1}{\Delta - M} + \left[\frac{1}{\Delta - L} - \frac{1}{\Delta - M}\right]$$

$$= \frac{1}{\Delta - M} + \frac{1}{\Delta}(L - M)\frac{1}{\Delta} + \dots \quad (B.3)$$

REFERENCES

*This lecture represents work (to be published) done in collaboration with D. Gogny and D. Irakane, Centre d'Etudes de Bruyeres-le-Chatel, France.

1. For a general introduction to QCD see, for example, T. D. Lee, "Particle Physics and the Introduction to Field Theory," Harwood Acad. Publ., New York (1981).
2. M. Danos, D. Gogny, and D. Irakane, NBSIR 83-2759, Washington, D.C. (1983).
3. M. Danos, V. Gillet, and M. Cauvin, "Methods in Relativistic Nuclear Physics," North Holland Publishing Co., Amsterdam, New York, Oxford (1984).
4. See, for example, P. Ring and P. Schuck, Appendix E in "The Nuclear Many- Body Problem," Springer Verlag, New York, Heidelberg, Berlin (1980).
5. This has been discussed in ref. 2.

GLUON CONDENSATION IN QUARK-GLUON PLASMA

I. Lovas

Central Research Institute for Physics

H-1525 Budapest, Hungary

INTRODUCTION

According to the generally accepted view, at high enough temperature or/and density a transition may take place from the hadronic phase of the matter into the quark-gluon plasma phase.[1,2] We hope that in high energy heavy-ion collisions there is a chance to reach that region of physical conditions where this phase transition is possible. How the plasma is produced and what kind of phase transitions takes place are questions of extreme interest. A number of approaches has been developed to estimate the temperature and the density of the transition.[3,4,5] In the quark gluon plasma another phase transition, associated with the restoration of the chiral invariance is also expected.[6] We will assume that the transition from hadronic phase into the quark-gluon plasma is already accomplished and we will search for further possibilities of phase transitions of the plasma, applying the mean field approximation.

In field theoretical studies the mean field approximation is one of the easiest methods. It is based upon the assumption that the expectation values of some boson fields may be different from zero. These expectation values can be considered as classical fields and the motion of fermions is determined by these classical fields generated selfconsistently by the averaged currents of the quantised fermion fields.

The application of the mean field approximation in the QCD seemed at first to be rather successful.[7,8] A transition into the deconfined quark phase has been found and it was concluded that this phase transition is associated with the spontaneous break down of the local gauge symmetry, since the appearence of a non-vanishing expectation value of the non-gauge invariant potential $\langle A_\mu^c(x)\rangle$ is conceivable only if the local gauge invariance is broken. This conclusion, however, turned out to be false, as it was proved by Elitzur,[9] already in 1975. According to Elitzur's theorem the local gauge invariance can not be broken spontaneously. The appearence of a nonvanishing expectation value $\langle A_\mu^c(x)\rangle$ has nothing to do with a spontaneous break down of the local gauge symmetry, it is the approximation scheme which is responsible for the symmetry violation.

This failure of the mean field approximation practically switched off the interest in its application to the QCD. The mean field approximation, however, may be considered as a useful guide line in the exploration of the properties of the true, symmetric theory, as it was emphasised by Elitzur.

In this paper, starting from the field equations of the QCD we will derive equations in the framework of the mean field approximation. Searching for solutions of these equations we will show that non-vanishing values for the mean fields can be obtained. It should be kept in mind, however, that the exact values of $\langle A_\mu^c(x)\rangle$ in the true, symmetric theory must vanish necessarily, and the non-vanishing values of $\bar{A}_\mu^c(x)$ obtained in the framework of the mean field approximation may play the role only of an approximate order parameter.

We will show that in the gluon condensed phase, characterised by the non-vanishing values of the approximate order parameters $\bar{A}_\mu^c(x)$, a static, periodic chromomagnetic field may be present and this induces in a self-consistant way a similar periodic behaviour of the color-spin density of the quarks and antiquarks.[10] We assume that such a transition leading to the gluon condensed phase can be found also in the true, symmetric theory, using a more appropriate order parameter instead of $\bar{A}_\mu^c(x)$. Other type of gluon-condensations, characterized by color singlet, scalar order parameters, has been investigated in a number of papers.[11,12,13,14] Recently a somewhat similar approach has been developed by Celenza and Shakin for the treatment of the hadronic structure.[15]

MEAN FIELD APPROXIMATION

The field equations of the QCD in conventional notations are given by

$$i\gamma^\mu\partial_\mu\psi + gA_\mu^a\gamma^\mu T^a\psi - m\psi = 0 \ , \tag{1}$$

$$\partial_\nu F^{\nu\mu a} = J^{\mu a} \ , \tag{2}$$

where the field strength $F^{\mu\nu a}(x)$ and the vector current $J^{\mu a}(x)$ are defined as follows:

$$F^{\mu\nu a}(x) = \partial^\mu A^{\nu a} - \partial^\nu A^{\mu a} + gf^{abc}A^{\mu b}A^{\nu c} \ , \tag{3}$$

$$J^{\mu a}(x) = gf^{abc}A_\nu^b F^{\mu\nu c} + g\bar{\psi}\gamma^\mu T^a\psi \ . \tag{4}$$

The generators and the structure constants of the gauge group are denoted by T^a and f^{abc}, respectively. As a consequence of eq.(2), the divergence of the color vector current vanishes, so that the color charge

$$Q_c = \int J^{o\ a=3}(x)\ d^3x \tag{5}$$

is a conserved quantity. The field equations (1) and (2) are symmetric under the local gauge transformations.

We assume that the field operators carrying both color and Lorentz indeces may have non-vanishing expectation values:

$$\langle A^{\mu a}\rangle = \bar{A}^{\mu a} \neq 0 \ .$$

Let us introduce the following decomposition

$$A^{\mu a}(x) = \bar{A}^{\mu a}(x) + \alpha^{\mu a}(x) \ , \tag{6}$$

where both $A^{\mu a}(x)$ and $\alpha^{\mu a}(x)$ are field operators, while $\bar{A}^{\mu a}(x)$ is a c-number and the expectation value of $\alpha^{\mu a}(x)$ vanishes by definition:

$$<\alpha^{\mu a}> = 0 \ .$$

Substituting the decomposition (6) into eq.(2) and taking expectation value on both sides, the following equation is obtained for the mean field $\bar{A}^{\mu a}(x)$:

$$\partial_\nu \bar{F}^{\nu\mu a} = gf^{abc}\bar{A}^b_\nu \bar{F}^{\mu\nu c} + <g\bar{\psi}\gamma^\mu T^a\psi> - M^2\bar{A}^{\mu a} \ . \tag{7}$$

Deriving this equation, it is assumed that in addition to $\alpha^{\mu a}(x)$ all of its derivatives have vanishing expectation values and all the expectation values of products of independently fluctuating quantities are also zero. In this way all terms containing $\alpha^{\mu a}(x)$ or its derivatives drop out except for the products of non-independently fluctuating quantities, which may have non-vanishing expectation values:

$$M^2(\nu) = -g^2 \sum_{\substack{\mu\neq\nu \\ c}} <\alpha_{\mu c}\alpha^{\mu c}> \ .$$

The parameter $M^2(\nu)$ formally plays the role of the mass associated with the excitations of the mean gluon field. For the sake of simplicity $M^2(\nu) = M^2$ is assumed.

In order to have a logically consistent description, the gluon operators $A^{\mu a}(x)$ in eq.(1) are substituted by their expectation values $\bar{A}^{\mu a}(x)$ and the coupling constant g by a renormalized one.

It is worthwhile to point out that the field equations obtained by the mean field approximation can be derived from the lagrangian given by

$$L = \frac{i}{2}(\bar{\psi}\gamma^\mu\partial_\mu\psi - \partial_\mu\bar{\psi}\gamma^\mu\psi) - g\bar{\psi}\gamma^\mu\bar{A}^a_\mu T^a\psi - \frac{1}{4}\bar{F}^a_{\mu\nu}\bar{F}^{\mu\nu a} - m\bar{\psi}\psi + \frac{1}{2}M^2\bar{A}^{\mu a}\bar{A}_{\mu a} \ . \tag{8}$$

Having this lagrangian we have a unique way to introduce the energy-momentum tensor $T^{\mu\nu}$ of the system using the standard definition.

SELF-CONSISTENT EQUATIONS

One of the possible strategies for the solution of the coupled, non-linear set of field equations is to introduce an ansatz for the solution. Substituting this ansatz into the field equations one must check if the resulting equations can be solved for the parameters of the ansatz.

For the mean gluon field we introduce the following ansatz

$$\bar{A}^{\mu a}(x) = a^\mu\theta^a(kx)/g \ , \tag{9}$$

which is separable in the Lorentz and color indeces and for the case of SU(2) color symmetry the spacetime dependence is given by

$$\theta^1(kx) = \sin kx, \ \theta^2(kx) = \cos kx, \ \theta^3(kx) = 0 \ , \tag{10}$$

$$a_\mu k^\mu = 0, \qquad a_o = 0, \qquad k_o = 0.$$

The choice of this ansatz is motivated by the remarkable fact, that the Dirac-equation can be solved exactly.[16,17] It is easy to show[10] that the energy eigenvalues for quarks ($E_\pm$) and antiquarks ($\bar{E}_\pm$) are given by

$$E_\pm = \bar{E}_\pm = [\vec{p}^2 + m^2 + \frac{1}{4}(a^2+k^2) \pm ((\vec{a}\,\vec{p})^2 + (\vec{k}\,\vec{p})^2 + \frac{1}{4}a^2k^2)^{1/2}]^{1/2} \tag{11}$$

Having the fermion single-particle solutions, the vector current $J^{\mu a}$ can be calculated in a straightforward way.

Substituting the ansatz (9) into eq.(7) the following set of self-consistent equations can be derived

$$<j_v^{\mu 1}> = 0\ , \tag{12}$$

$$<j_v^{\mu 2}> = \frac{1}{g}\{M^2 + \vec{k}^2\}\, a^\mu\ , \tag{13}$$

$$<j_v^{\mu 3}> = \frac{1}{g}\,\vec{a}^2\, k^\mu\ , \tag{14}$$

where

$$<j_v^{\mu a}> = <g\bar{\psi}_v\gamma^\mu T^\alpha\psi_v>$$

and the quasi-particle field

$$\psi_v(x) = R^+\psi(x) \tag{15}$$

is defined by the help of the operator

$$R = \exp(-ikxT^3)\ . \tag{16}$$

Our considerations are confined to the study of the case of SU(2), the generalization for SU(3) is straightforward.

The solution of the self-consistent set of equations provide us the amplitude a^μ and the wave vector k^μ. It is worthwhile to mention that in the mean field approximation the color vector current is conserved only if $a_\mu k^\mu = 0$. This condition can be satisfied if $\vec{a}\vec{k} = 0$, $a_o = 0$ and $k_o = 0$. In this case the chromoelectric field vanishes and a static, periodic chromomagnetic field, orthogonal both to $\vec{a}$ and $\vec{k}$ vectors is present; the color density of the gluons and that of the quarks and antiquarks vanish; the energy eigenvalues for quarks and antiquarks are identical. Without loss of generality the system of coordinates can be chosen in such a way that $k_1 = k_2 = 0$ and $a_2 = a_3 = 0$ are fulfilled. The only non-vanishing parameters are $a = a_1$ and $k = k_3$.

THERMODYNAMICAL EQUILIBRIUM

We assume that the quark-gluon matter under consideration has a volume V and it is in thermodynamical equilibrium with the surrounding world. Neither the energy nor the number of baryons is fixed. The system is characterized by the temperature T and the baryonic chemical potential μ. Thus the thermodynamical properties of the system can be described by the following grand canonical density matrix:

$$\rho = Z^{-1} \exp \{-(H-\mu N)/T\} , \qquad (17)$$

where the partition function Z is given by

$$Z = \mathrm{Tr} \exp \{-(H-\mu N)/T\} . \qquad (18)$$

Here $H = \int \bar{T}^{oo}\, dV$ is the Hamiltonian and N denotes the number of baryons. We introduce also the thermodynamical potential defined by

$$\Omega = -T \, \ell n \, Z . \qquad (19)$$

In the mean field approximation the quarks behave as independent quasiparticles. In this approximation the Dirac-equation can be solved without any further restriction, so one is able to calculate explicitely the thermodynamical potential density

$$\omega = \frac{\Omega}{V} = \in - T\sum_{\lambda} \int \frac{d^3p}{(2\pi)^3} \, \ell n \, \{(1+e^{-(E_\lambda-\mu)/T})(1+e^{-(\bar{E}_\lambda+\mu)/T})\} , \qquad (20)$$

where $\in$ is given by

$$\in = \frac{1}{2g^2} (M^2 + \vec{k}^2)\vec{a}^2 . \qquad (21)$$

The energy eigenvalues of the quarks and antiquarks are denoted by E_λ and $\bar{E}_\lambda$, respectively. The 2x2 independent spin and color states are labelled by the index λ.

In addition to the temperature T and chemical potential μ the generalized thermodynamical potential density is the function also of the parameters a and k:

$$\omega(T, \mu; a, k) .$$

The thermodynamical equilibrium is characterized by the minimum of the generalized thermodynamical potential density. This means that at prescribed values of T and μ the conditions

$$\frac{\partial\omega}{\partial a} = 0 , \qquad \frac{\partial\omega}{\partial k} = 0 \qquad (22)$$

should be satisfied. It is not difficult to prove, that eqs.(22), which are the necessary conditions of the thermodynamical equilibrium are the independent ones among the self-consistent set of eqs.(12) and (13). The solution of eqs.(22) define the dependence of the parameters a and k on the temperature and chemical potential in the thermodynamical equilibrium:

$$a(\mu, T) , \qquad k(\mu, T) .$$

Using these functions the generalized thermodynamical potential density for the gluon-condensed phase can be obtained as a function of the temperature and of the baryon chemical potential:

$$\omega_c(\mu, T) = \omega(\mu, T; a(\mu,T), k(\mu,T)) . \qquad (23)$$

The thermodynamical potential density of the normal phase is given by

$$\omega_n(\mu, T) = \omega(\mu, T; a = 0, k = 0) . \qquad (24)$$

The other thermodynamical characteristics of the system can be easily derived from functions (23) and (24), e.g. the baryon density ρ, the entropy density s and the specific heat can be obtained as

$$\rho_B = -\frac{\partial\omega}{\partial\mu}, \quad s = -\frac{\partial\omega}{\partial T} \quad \text{and} \quad C = T\frac{\partial s}{\partial T}, \text{ respectively.}$$

GLUON-CONDENSED PHASE

In the normal phase the mean gluon field is equal to zero and quarks and antiquarks form a non-interacting Fermigas. In the other phase gluons are condensed forming a static, periodic chromomagnetic field. The gluon-condensed phase has a layered structure. The 3-vector component of the color vector current is proportional to the vector potential, $J^{\alpha a}(x) \sim \bar{A}^{\alpha a}(x)$ ($\alpha = 1, 2, 3$). So in the layers of the thickness π/k a flow of the color charges appears, the direction of which is opposite in the neighbouring layers. The quarks and the antiquarks form a lattice like structure in the periodic mean gluon field, which can be characterized by the spin-color density given as

$$\rho_{sc} = \langle\bar{\psi}\gamma^o(\frac{1}{2} + i\gamma^3\gamma^1T^1)\psi\rangle . \tag{25}$$

In our case the spin-color density has the form

$$\rho_{sc} = \frac{1}{2}\rho_B + \rho_o \cos kx , \tag{26}$$

where

$$\rho_o = \langle\bar{\psi}_v\gamma^o i\gamma^3\gamma^1T^1\psi_v\rangle . \tag{27}$$

The baryon density ρ_B and the amplitude of the oscillations of the spin-color density ρ_o can be divided into two terms, one coming from quarks and the other from the antiquarks:

$$\rho_B = \rho_B^{(q)} - \rho_B^{(\bar{q})} , \quad \rho_o = \rho_o^{(q)} - \rho_o^{(\bar{q})} .$$

Let us call the color charges blue and red. Then, according to eq.(26) the density of the spin-up red and spin-down blue quarks and similarly the density of the spin-up blue and spin-down red quarks vary periodically around the half value of the quark density. The same can be told about the antiquarks. The spatial spin-color density oscillations of quarks and antiquarks have the same phase, but in general are of different amplitudes. This seems to be the analogue of the Overhauser effect.[18,19]

PHASE TRANSITIONS

The minimum of the generalized thermodynamical potential density has been found numerically and the thermodynamical state functions as the baryon density ρ_B, the entropy density s and the specific heat C have been computed at the thermodynamical equilibrium.

According to Gibbs' law, in phase equilibrium the temperatures, the pressures ($p = -\omega$) and the chemical potentials must be equal for the two phases. A typical phase diagram is shown on Fig. 1.

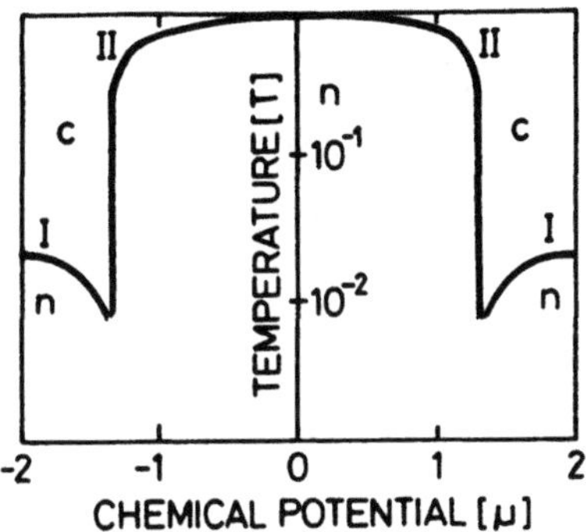

Fig. 1. Phase diagram for the quark-gluon matter ($M^2 = 0.1$, $g = 15.0$, $m = 1.0$)

In the region above the curves labelled by I and II, the set of eqs. (11)-(13) has nontrivial solution. Thus, at a given value of the temperature there is a density interval in which the gluon-condensed phase can exist and it is thermodynamically more favoured than the normal phase.

The condensed phase is characterized by the nonvanishing mean value of the gluonic vector potential. Therefore in the condensed phase a number of symmetries are broken. Since the normal and the condensed phases have different symmetries, the curve of phase equilibrium can not be terminated by a single point.

It was established that at the formation of the gluon-condensed phase from the low density normal phase the amplitude of the mean gluon field a increases continuously from zero along an isotherm. The square of the amplitude a^2 was found proportional to the temperature difference $t = T - T_{cr}$ at a given value of the baryion chemical potential, where T_{cr} denotes the temperature at which the phase transition takes place. On the other hand, the wave vector k being different from zero has a finite jump k_{cr} as a result of the phase transition. At a given value of the chemical potential the difference $k - k_{cr}$ was found proportional to the temperature difference t too. The only exception is the $M^2 = 0$ case, when $a = k$, as a consequence of the symmetry of the self-consistent equations. In that case the wave vector has no jump ($k_{cr} = 0$) and k is proportional to t.

If the temperature is low enough, a phase equilibrium can be reached by increasing the chemical potential. Then the condensed phase disappears by a phase transition of the first order (branch I), accompanied by an absorption of the latent heat $Q = T_{cr}\,(s_c - s_n) < 0$ (given to the system by the surrounding world). At the same time the specific volume of a baryon increases, since the inequality $1/\rho_c < 1/\rho_n$ holds.

Through branch II a second order phase transition takes place. Indeed, no change in the entropy and in the baryon density appears. Let us now examine the specific heat. The difference of the specific heats of the condensed and normal phases can be written as follows:

$$\Delta C = C_c - C_n = -T_c\left(\frac{\partial^2\omega}{\partial T\partial a}\frac{\partial a}{\partial T} + \frac{\partial^2\omega}{\partial T\partial k}\frac{\partial k}{\partial T}\right) . \tag{28}$$

It was established numerically, that the second derivatives $\partial^2\omega/\partial T\partial a$ and $\partial^2\omega/\partial T\partial k$ have non-zero values at branch II. As the temperature difference goes to zero $t \to 0$, the wave vector goes also $k \sim t \to 0$, $\partial k/\partial t = \partial k/\partial T$ remaining constant. Since $a^2 \sim t$, $\partial a/\partial T \sim t^{-1/2}$ goes to infinity as $t \to 0$. Thus the first term on the r.h.s. of eq.(27) causes an anomaly in the specific heat. In the case of $M^2 = 0$, the specific heat has only a finite jump, since $a = k \sim t$. According to the general convention, the anomalies of the system can be characterized by the following critical exponents:

$$a \sim t^{\beta}, \quad C \sim t^{-\alpha}, \quad \frac{\partial a}{\partial k} \sim t^{-\gamma}.$$

In our case they have the following values

$$\alpha = 1/2(0), \quad \beta = 1/2(1) \quad \text{and} \quad \gamma = 1/2(0)$$

for $M^2 \neq 0$ ($M^2 = 0$), and satisfy the general relation

$$\alpha + 2\beta + \gamma = 2 .$$

Microscopically the phase transition is connected with the redistribution of the fermions among the single-particle energy levels. On Fig. 2 the single-particle energy levels $E_\pm$ and $E_\pm^{(n)}$ are shown for the condensed and the normal states, respectively, as functions of the longitudinal momentum p_3 (at $p_1 = p_2 = 0$) in a typical case. When phase transition through branch II takes place, a condensate with $k = k_{cr} >> a \approx 0$ is formed. In this case the energy level E_- has a minimum at $p_3 = \pm \frac{1}{2}\sqrt{k^2-a^2} \neq 0$ (Fig. 2).

Fermions, originally occupying the twice degenerated levels in the Fermi sphere of the radius μ in the momentum space, can reorder and occupy two spheres of the same radii with centers removed from each other by the "distance" $\sqrt{k^2-a^2}$. Since this reordering appears at infinitesimally small value of the gluonic amplitude a, no variation of the baryon density takes place.

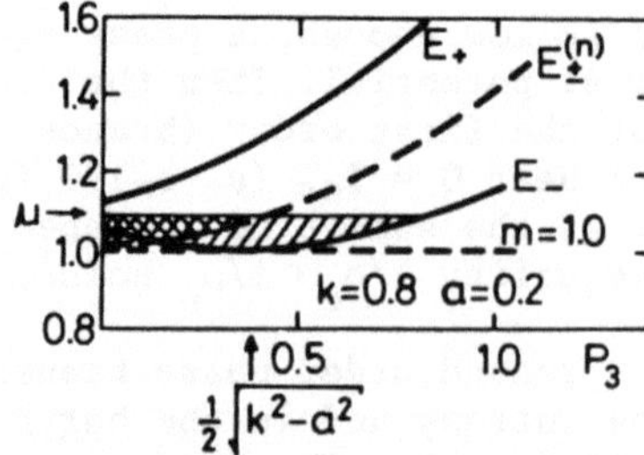

Fig. 2. Fermion single-particle energy levels $E_\pm$ and $E_\pm^{(n)}$ in the condensed and normal phases, respectively, in the case of $M^2 \neq 0$. The hatched and the double hatched areas denote the single-particle states occupied in the condensed and the normal phases, respectively.

The dependence of the phase diagram on the coupling constant g has also been examined. The phase transition at given value of the chemical potential takes place at higher temperatures if the coupling constant decreases. It has to be noted that the gluon-condensed phase may exist only above some minimal value of g.

SUMMARY

Let us suppose, the quark-gluon matter of given volume is cold and dilute enough to form a free quark-antiquark Fermi-gas. Putting it into a bath to keep the temperature constant the following happens if the baryon chemical potential is increased, i.e. more and more quarks are pushed into the system. At first, the baryon density reaches a value at which the formation of the gluon-condensed phase becomes energetically favoured. In the gluon-condensed phase with broken color symmetry, a mean gluon field is continuously building up having a spatial periodicity of the length $2\pi/k$. This phase transition is a second order one, connected with the redistribution of the fermions among the single-particle levels in the momentum space without the variation of the Fermi-momentum. With further increase of the chemical potential the amplitude of the mean gluon field grows and a strong gluon field of order of magnitude of the chemical potential builds up. In the meantime the periodicity length $2\pi/k$ has a minimum. At low enough temperatures a phase transition of first order takes place when the chemical potential is further increased. Then the fermions redistribute among the single-particle states with sudden decrease of the Fermi-momentum and symmetries broken in the condensed phase are restored.

From the phase diagram depicted on Fig. 1 one may expect that the gluon-condensed phase exists at any high enough temperatures. This is the case only if one uses a really constant coupling constant g, as we did. However if one would use the running coupling constant of QCD the gluon-condensed phase ceased to exist at some high enough temperature, since the effective coupling constant would decrease.

If $M^2 = 0$ then the local gauge invariance is restored and the periodic field $\bar{A}^{o}_{\mu}(x)$ is gauge equivalent with a constant field leading to a constant chromomagnetic field.

REFERENCES

1. A.M. Polyakov, Phys. Lett. 72B, 477 (1978).
2. L. Susskind, Phys. Rev. D20, 2610 (1979).
3. J. Kuti, B. Lukács, J. Polonyi and K. Szlachányi, Phys. Lett. 95B, 75 (1980).
4. J. Kuti, J. Polonyi and K. Szlachányi, Phys. Lett. 98B, 199 (1981).
5. I. Montvay and E. Pietarinen, Phys. Lett. 110B, 148 (1982), Phys. Lett. 115B, 151 (1982).
6. R.D. Pisarski, Phys. Lett. 110B, 155 (1982).
7. R. Balian, J.M. Drouffe and C. Itzykson, Phys. Rev. D10, 3376 (1974).
8. J. Kogut and L. Susskind, Phys. Rev. D11, 995 (1975).
9. S. Elitzur, Phys. Rev. D12, 3979 (1975).
10. I. Lovas, W. Greiner, P. Hraskó, E. Lovas and K. Sailer, Phys. Lett. 156B, 255 (1985).
    K. Sailer, W. Greiner and I. Lovas, Phys. Rev. (in press)
11. G.K. Saviddy, Phys. Lett. 71B, 173 (1977).
12. N.K. Nielsen and P. Olesen, Nucl. Phys. B144, 376 (1978).
13. H.B. Nielsen and M. Ninomiye, Nucl. Phys. B156, 1 (1979).
14. J. Balog and A. Patkós, Phys. Lett. 98B, 205 (1981).

15. L.S. Celenza and C.M. Shakin, Phys. Rev. D32, 1807 (1985).
16. B. Banerjee, N.K. Glendenning, and M. Gyulassy, Nucl. Phys. A461, 326 (1981).
17. I. Lovas, J. Németh and K. Sailer, Phys. Lett. 135B, 258 (1984).
18. A.W. Overhauser, Phys. Rev. 128, 1438 (1962).
19. M. Noga, Nucl. Phys. B220, 185 (1983).

HOW TOPOLOGICAL CONCEPTS LEAD TO QUANTUM NUMBERS FOR BARYONS*

L. C. Biedenharn**

Institut für Theoretische Kernphysik
der Universität Bonn
Nussallee 14-16, D-5300 Bonn
West Germany

INTRODUCTION

Strong interaction physics is currently believed to be determined by quantum chromodynamics. If this is correct, then it follows that all of nuclear physics must, somehow, be a consequence. Unravelling these consequences will be a daunting task, for nuclear physics is the regime of long-distances (on the scale of Λ_{QCD}) and it is precisely here that the enormously strong confinement forces will be operative. A direct approach, starting from QCD, is not presently known, but strong hints as to the nature of the confinement regime have come from 't Hooft's idea of taking the number of colors (N_c) in QCD to be large[1]. Assuming confinement, the result[2] is an effective field theory of colorless low-lying mesons, without explicit quarks. Baryons become extended objects and involve topological concepts applied to non-linear self-interacting meson fields.

These ideas are new to nuclear physics, and being unfamiliar, may appear as paradoxical. The purpose of the present paper is to discuss these concepts at an introductory level with emphasis on the structure, and the logic, omitting inessential detail. (More detail, and references to the original literature, may be found in the recent reviews[3,4].) See Figure 1 (next page).

The accepted wisdom has long been that to describe spinorial particles in a Lagrangian field theory one must include at the outset at least one primitive spinor field (as for example in the Heisenberg-Pauli non-linear spinor field theory). The theory sketched above purports to accomplish exactly the opposite: *to make baryons out of mesons, that is fermions (spinors) from bosons!* This is surprising, even startling, but actually it is as old as 1931 when Dirac discussed magnetic monopoles. Consider a (spinless) point charge moving under the influence of a fixed (spinless) point magnetic monopole g. This quantal system for $eg/c = \hbar/2$ (the Dirac quantization condition for N=1) is a *spinor* (probability amplitude changes sign under a rotation about any axis by 2π), yet none of the constituents has spin 1/2. The resolution of this "paradox" (Fierz 1943) is that the

---

* Supported in part by the National Science Foundation

** Alexander von Humboldt Stiftung Re-Invitation Program on leave from Duke University, Durham, NC 27706 USA.

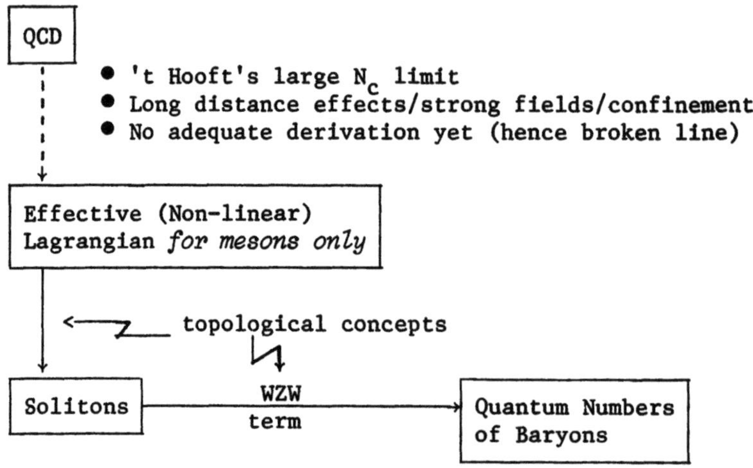

Figure 1

electromagnetic field in the system contains angular momentum $\hbar/2$. This helps one to accept the result, yet some of the paradoxical flavor remains if we recall that the EM field is itself bosonic.

We shall demonstrate below that the baryons (fermions) are analogous structures built out of mesons (bosons) in an $SU3_{flavor}$ world of "hidden" quarks with qualitative, topological, concepts determining their quantum numbers.

THE DIRAC MAGNETIC MONOPOLE EXAMPLE

Topological ideas first entered physics in the magnetic monopole problem--the motion of a (spinless) point particle, of mass m, carrying an electric charge e and moving under the influence of a fixed (spinless) magnetic pole g. The radial motion separates from the angular motion, and we can simplify greatly by considering the angular motion alone. This motion is on the two-sphere, S^2, which is topologically the same as the manifold SU(2)/U(1).

The *classical* motion (Poincaré 1896) presents no problem: one works directly with the magnetic field in the equations of motion: $m d\vec{v}/dt = eg\vec{v} \times \vec{r}/r^3$ and finds four integrals of the motion, $\vec{J}_{c\ell} = m\vec{r} \times \vec{v} - (eg/c)\hat{r}$ and $H = 1/2\, m\vec{v}^2$. (Note the radial angular momentum.)

The *quantal* motion, by contrast, presents fundamental difficulties since a Hamiltonian formulation requires a vector potential, and, although there are many such potentials (differing by gauge rotations), each is singular somewhere. For example, the monopole vector potential: $\vec{A} = g\frac{(1-\cos\theta)}{\sin\theta}\,\hat{e}_\phi$, is singular at $\theta = \pi$.

The way out of this difficulty is simple, but far-reaching[5]: we divide the sphere into two overlapping hemispheres, such that motion in one hemisphere has the singularity always in the opposite hemisphere. The 'wave function' is now a pair of functions $\{\psi_a, \psi_b\}$ each defined only in the appropriate hemisphere (a,b). In the overlap region the two functions are related by a gauge transformation: $\psi_a = S_{ab}\psi_b$.

This pair of functions for angular motion in the space parametized by $(\phi,\theta) \in S^2 \approx SU(2)/U(1)$, can be treated as eigenfunctions of the (dimensionless) angular momentum operators: $\vec{J} = \hbar^{-1}\vec{r}\times(\vec{p} - \frac{e\vec{A}}{c}) - \mu\hat{r}$, $\mu = eg/\hbar c$. Since the vector potential occurs in this expression for $\vec{J}$, we see that the angular momentum operators themselves must be pairs of operators, properly adapted to be non-singular in the appropriate hemisphere. In each region the operators obey angular momentum commutation relations and join smoothly in the overlap region.

To construct the monopolar harmonics (angular eigen-sections), we recall that the (complex-conjugated) rotation matrices $\underset{\sim}{D}^{*(j)}(\phi\theta\psi)$ are eigenfunctions of the three commuting operators J^2, J_3 and J'_3 with eigenvalues $(j(j+1), m, m'$ respectively. (Here the J'_i are the angular momentum operators projected on the intrinsic frame.) One realizes J_3 and J'_3 in the standard way as differential operators acting on the Euler angles parametrizing the three-sphere (S^3, the manifold SU2).

The monopolar harmonics are, however, pairs of functions defined on the angular variables ϕ,θ parametrizing the two-sphere S^2. The key to understanding this situation is to recognize that on a given hemisphere the eigen-section for that region maps a point in S^2 into a point in S^3. This map preserves the angular momentum eigenvalues, that is, the eigen-section (the monopolar harmonic) and the rotation matrix in S^3 both have the same eigenvalues for the operators J^2, J_3 and J'_3.

In other words, instead of the concept of a wave function (a map from configuration space into the complex numbers) we now have an intermediate step: configuration space (the two-sphere S^2, with distinct regions) is first mapped into a larger structure (the three-sphere S^3) and then mapped from S^3 to the complex numbers.

It remains only to make explicit the mapping, σ, from S^2 into S^3:

$$\text{Region a)}\quad \sigma_a:\ (\phi,\theta,-\mu),$$
$$\text{Region b)}\quad \sigma_b:\ (\phi,\theta,\mu)\ .$$

Once we have points in S^3 we can use the D^*-matrices as the appropriate eigenfunctions. We have thereby arrived at the monopolar harmonics, in agreement with the results determined by more detailed calculations :

$$\psi_{j,m,-\mu}(\phi\theta) = \begin{cases} D^*_{m,-\mu}(\phi,\theta,-\phi), & 0 \le \theta \le \frac{\pi}{2}+\varepsilon \\ D^*_{m,-\mu}(\phi,\theta,\phi), & \frac{\pi}{2} - \varepsilon \le \theta \le \pi. \end{cases}$$

To calculate a matrix element of an operator using the monopolar harmonics as a basis, one splits the integration règion into two parts using the monopolar harmonics and operators appropriate to each region. (It is interesting to note that with this definition, the canonical momentum operator $\vec{p} = (\hbar/i)\vec{\nabla}$ is *not* Hermitian, in addition to not being gauge-invariant, and is accordingly *not* observable.)

The famous charge quantization condition found by Dirac is obtained, heuristically, by arguing that all angular momenta are quantized in units of $\hbar/2$, so that the intrinsic angular momentum--the radial angular momentum in the monopole problem--obeys: $eg/c = N\cdot\hbar/2$, where N is an integer.

These results have counterparts for the baryons, as discussed below.

QCD can be formulated for any number of colors, N_c, independently of the number of quark-flavors. The resulting theory involves $(N^2{}_c - 1)$ gauge gluons interacting in an $SU(N_c)$ gauge-invariant way with N_c quarks (coupling constant g). 't Hooft investigated[1] the limit $N_c >> 1$, with $g^2 N_c$ fixed, and found the dominant Feynman diagrams to have exactly the structure characterizing the quantized dual string with (confined) quarks at the ends. All mesons constructed out of a quark and antiquark become stable non-interacting particles (lifetimes varying as $N^{-1}{}_c$) with meson-meson scattering cross-sections behaving as $N^{-2}{}_c$. Moreover in this limit meson masses behave as $(N_c)^0$, Zweig's rule recomes exact, and there are no exotic states.

These results bear some resemblance to the physical world with $N_c = 3$, and suggest that the low-energy, long-distance, regime may be described for large, but finite, N_c as a weakly interacting effective field theory of colorless low-lying mesons. Baryons (as originally suggested by Skyrme[6]) should appear as topological excitations of this non-linear effective field theory.

Lacking a valid *a priori* derivation from QCD, one constructs, heuristically, the effective field theory on the basis of symmetry principles; the (meson field) action chosen is that for phenomenological non-linear chiral-dynamics. The dynamical degrees of freedom are SU(2) matrices $U(\vec{x},t)$ which may be related (in a parametrization dependent way) to the pion field:

$$U(\vec{x},t) = \exp\left(\frac{2i}{F_\pi}\vec{\pi}(\vec{x},t)\cdot\vec{\tau}\right),$$ where F_π (the decay constant for the pion) $\cong$ 186 MeV.

Defining the matrix vector: $R_\mu = -i\, U^+ \partial_\mu U$, one has the action[6]:

$$S_{Skyrme} = \frac{F_\pi{}^2}{16}\int d^4x\ \mathrm{tr}(R_\mu R^\mu) - \frac{1}{32e^2}\int d^4x\ \mathrm{tr}(R_{\mu\nu}R^{\mu\nu}), \qquad (1)$$

where: $R_{\mu\nu} = (-i)[R_\mu, R_\nu] = \partial_\mu R_\nu - \partial_\nu R_\mu$, and e is a dimensionless constant.

The first term in eq. (1) is taken from phenomenological chiral-dynamics; the second term (due to Skyrme) is *ad hoc* and chosen to ensure stability. (Later work shows that this term may be viewed as arising from vector mesons.)

Restricting attention to configurations $U(\vec{x},t)$ with finite energy, we see that for the energy functional to be convergent, all derivatives of U must vanish as $|\vec{x}|$ tends to infinity at fixed t. Thus $U(\vec{x},t)$ must tend to a spacetime independent matrix U_0.

Requiring finite energy therefore implies that the point at spatial infinity maps to a fixed element in the field manifold, U_0. Expressed topologically, this shows that configuration space $\mathbb{R}^3$ has been compactified to the three sphere: $S^3 \cong \mathbb{R}^3 + \{\infty\}$.

Topological concepts now enter the problem in a decisive way. Recall that the matrices U belong to SU2, which, as a manifold, is the three sphere S^3. Thus a finite energy configuration $U_{fin.en.}(\vec{x},t)$ may be considered as a mapping from S^3 (compactified space) into S^3 (SU2 group manifold). All such maps are characterized by integers, the homotopy class of the map. (Expressed topologically, the homotopy classes of the maps $S^3 \to S^3$ are elements of the Abelian group $\pi_3(S^3) = \mathbb{Z}$.) Since a Hamiltonian time development of a given initial configuration U(x,t) is continuous the homotopy class does not change with time and is therefore a constant of the motion. This constant is given by:

$$B[U] \equiv \frac{-i}{24\pi^2} e_{ijk} \int d^3x \; tr(R_i R_j R_k). \tag{2}$$

B[U] is the charge of the current density:

$$j_\mu^{(B)}(\vec{x},t) = \frac{\varepsilon_{\mu\nu\lambda\sigma}}{24\pi^2} \; tr(R_\nu R_\lambda R_\sigma) \; . \tag{3}$$

The current $j_\mu^{(B)}$ is conserved independently of the equations of motion for $U(\vec{x},t)$. (Such conservation laws are characterized as topological conservation laws.)

Skyrme[6] identified B[U] as the *baryon number*. We will show that indeed B[U] is the *same* baryon number that is defined in the quark model (or in QCD).

To find an energy minimizing solution in the $B=1$ sector, it pays to make the solution as slowly varying as possible. The relevant concept here is *equivariance*: given a mapping f from a manifold M to a manifold M' one says that it is equivariant with respect to an abstract group G--which acts on M and M'--if $(G\circ f)(x) = f(G\circ x)$, $x \in M$, $f(x) \in M'$. To find the equivariance group one considers the degeneracy manifold of the state. For static solutions, time translations make no change, while space translations change the center of mass location. Lorentz boosts change the energy so that only the spatial rotation group SO(3) does not change the energy, the center of mass location or the (vanishing) momentum. Out of the $SU(2) \times SU(2)$ chiral isospin group only the diagonal subgroup SU(2) does not change the boundary conditions at spatial infinity. We are thus left with the group $SU(2)_I \times SO(3)_J$. It is easy to see that a solution invariant under $SU(2)_I$ must have $B=0$ and similarly invariance under $SO(3)_J$ forces $B=0$. Thus the best one can do is to demand invariance under diagonal subgroup $SU(2)_K$ formed by the pairs (A,R(A)) where R(A) is the SO(3) matrix representing the SU(2) matrix A.

In other words $U(\vec{x})$ should be equivariant under G(=SU(2)) which acts on space through rotations: $\vec{x} \to R(A)\vec{x}$ and simultaneously on the SU(2) isospin manifold by conjugation: $U \to AUA^{-1}$.

Equivariance implies that the combined generator: $\vec{K} = \vec{I} + \vec{J}$ commutes with an equivariant $U(\vec{x})$. For the Skyrme model the equivariant U(x) can be written as: $U(\vec{x}) = e^{if(r)\hat{r}\cdot\tau}$, where the function f(r) is to be determined by solving the Euler-Lagrange equation. (Numerical results are in [3],[4] with citations of the original literature.)

This linking together of two very disparate symmetries is curious, but not unexpected. Pauli in treating pseudoscalar meson field theory by strong coupling techniques (in 1943) was led to introduce exactly the same composition of spin and isospin! Nuclear physicists are familiar with similar examples of 'broken symmetry' in treating deformed nuclei (rotationally asymmetric intrinsic states) and in applications of the Hartree-Fock approximation (broken time-reversal symmetry). *The quantum states are projections from these intrinsic states and do not themselves manifest the broken symmetry.*

To obtain the quantum states, one needs next to quantize the fluctuations about an equivariant (classical) solution. Starting from the equivariant $B=1$ static solution, one can generate any other solution on the energy degeneracy manifold $(SU(2)_I \times SU(2)_J)/SU(2)_K$ by conjugation: $U(\vec{x}) \to A\,U(\vec{x})\,A^+$, and the infinitesimal conjugations give rise to zero modes. The customany approximation is to quantize only these zero modes.

To this end one elevates the matrices A to the status of dynamical variables A(t) (that is to say, collective coordinates, familiar in nuclear rotational models). Substituting: $U_{B=1}(\vec{x},t) = A(t)\, U_{B=1}(\vec{x})A(t)^{-1}$, into the action, one can obtain the quantal Hamiltonian for the dynamical variables A(t).

This quantization procedure is immediate as it is formally equivalent to the problem of a free particle constrained to move on the energy degeneracy manifold S^3. In particular, the generators of $SU(2)_I \times SU(2)_J$ are now operators and in view of the invariance of the classical soliton under $SU(2)_K$ fulfill the operator relation $\vec{I}^2 = \vec{J}^2$.

The quantum problem thus has $SU(2)_I \times SU(2)_J$ invariance and one can diagonalize the effective Hamiltonian simultaneously with the three commuting operators $\vec{I}^2 = \vec{J}^2$, I_3, J_3. The Hamiltonian has a discrete rotational spectrum with eigenvalues $E_1 + J(J+1)/2I$, where the allowed values of J are 0, 1/2, 1, 3/2, 2, and the corresponding eigenfunctions are the rotation matrices: $D^{(I)*}_{I_3,J_3}(A)$.

For the B = 1 system, (choosing half-integer spins) the result is just as in the old strong coupling model: the nucleon corresponds to the lowest state (1/2,1/2); the delta to the isobar (3/2,3/2), with the state (5/2,5/2) presumably an artifact of the model.

This model gives a surprisingly good account of the properties of the nucleon and delta, and can be extended to meson-nucleon scattering and the nucleon-nucleon interaction (see the reviews [3],[4]).

THE ANOMALY TERM

The Skyrme model, described above, has two basic shortcomings:

(1) the Lagrangian, eq. (1), has an unwanted discrete symmetry not found in QCD, and,

(2) the quantization allows both fermionic and bosonic states.

Witten[7] showed that both difficulties are removed by adding an anomaly term, Γ, to the action:

$$S_{\text{Skyrme-Witten}} = S_{\text{Skyrme}} + N_c \Gamma \,, \tag{4}$$

where Γ is the "anomaly" and N_c can be shown to be an integer. The anomaly cannot be written as an integral over space-time, but rather appears in the form:

$$\Gamma = \frac{1}{240\pi^2} \int d\,\Sigma^{ijklm}\, \mathrm{Tr}(R_i R_j R_k R_l R_m) \,, \tag{5}$$

with $d\,\Sigma^{ijklm}$ a volume element in an extended five-dimensional space whose boundary is compactified space-time.

The anomaly term Γ is very complicated in appearance, technically difficult to manipulate, and cannot be discussed here. (Details are given in ref.(8).)

Conceptually, however, the anomaly term plays a rôle analogous to the interaction term in the magnetic monopole example. Just as that interaction resulted in quantized radial angular momentum, so does the anomaly term contribute quantized 'intrinsic hypercharge', which, as we shall see, largely determines the baryonic quantum numbers.

The anomaly term Γ vanishes for $SU(2)_f$ (since the integral is five dimensional and $SU(2)_f$ has but three dimensions). *Thus one is forced to generalize to three flavors,* $SU(3)_f$.

At first glance the extension to three flavors is completely straightforward: $\vec{\pi}(\vec{x},t) \to \phi_a(\vec{x},t)$ (the pseudoscalar octet, $a = 1 \ldots 8$) and $\tau \to \lambda_a$ (the Gell-Mann matrices), so that $U(\vec{x},t) = \exp\left(\frac{2i}{F_\pi} \phi_a(\vec{x},t) \cdot \lambda_a\right)$ in the action (4). However when we try to determine the $B = 1$ (classical) ground state, there is a problem: How shall we achieve equivariance? The resolution is to choose a three-dimensional sub-algebra of SU(3) and define equivariance under this sub-group of SU(3) and spatial rotations. Using the isospin subgroup of SU3, one finds for the static $B = 1$ minimal energy (classical) solution:

$$U^{SU(3)}(\vec{x}) = \left(\begin{array}{cc|c} \multicolumn{2}{c|}{U^{SU(2)}(\vec{x})} & 0 \\ & & 0 \\ \hline 0 & 0 & 1 \end{array}\right), \tag{6}$$

where $U^{SU(2)}(\vec{x})$ is the 2×2 matrix of the minimal energy $B = 1$ Skyrme ansatz.

When we proceed to the quantization of this minimal energy solution, we find a surprise. By construction, $U^{SU(3)}$ in eq. (6), *commutes with all transformations generated by* λ_8. It follows that in the quantization:

$$U(\vec{x}) \to U(\vec{x},t) = A(t)\, U^{SU(3)}(\vec{x})\, A^{-1}(t),$$

the collective coordinates $A(t)$ *are invariant to all right translations generated by* λ_8, that is:

$$A(t) \longrightarrow A(t)\, e^{i\phi\lambda_8} \cong A(t).$$

Thus we conclude: *the manifold for the collective coordinates* $A(t)$ *is not* SU(3) *but* SU(3)/U(1).

One sees from this result that the probability amplitudes for the baryons ($B = 1$ sector) are the *monopolar harmonics of* SU(3)/U(1) *in precise analogy to the monopolar harmonics* SU(2)/U(1) *of the Dirac monopole example.*

It remains only to avail ourselves of the technical evaluation of the anomaly in [8], where it is shown that the "intrinsic hypercharge", Y_{int}, (the analog of the intrinsic (radial) angular momentum) is given by:

$$Y_{int} = N_c\, B/3\ ,$$

where N_c is the integer coefficient of the anomaly, and B is the baryon number given by Skyrme, eq. (2).

DETERMINATION OF THE QUANTUM NUMBERS

With the information accumulated in the preceding sections it is now straightforward to determine the quantum numbers of the baryons in the $B = 1$ sector.

Consider the information supplied by the anomaly. If we take--from experiment--the lowest energy baryon to be the nucleon belonging (experimentally) to an $SU(3)_f$ octet, *then* it follows that $N_c B/3$ = integer. Thus for $B = 1$, experiment determines the number of colors, N_c, to be a multiple of 3. Moreover if we go to the intrinsic frame, we see that the (experimental) hypercharge of the nucleon forces the intrinsic hypercharge to be 1. This implies that N_c is *exactly* 3. (In the intrinsic frame the spin (1/2) and isospin (1/2) also agree.) One sees, too, (taking now $N_c = 3$) that in the intrinsic frame (since the nucleon has strangeness zero) the Skyrme baryon number B *necessarily agrees with the baryon number of the quark model.*

Using now $N_c = 3$, we conclude (from the anomaly determination of the intrinsic hypercharge), that the baryon $SU(3)_f$ quantum numbers all have *triality zero.*

By equivariance, the spin (J) of the baryons is determined by the intrinsic isospin. Using the determination above that $N_c = 3$, the intrinsic hypercharge becomes B, and in consequence, one finds the Michel-Lurçat rule: $2J \equiv B \mod 2$.

Using the Hamiltonian to determine the energy levels shows that the level ordering for $B = 1$ is: (8,1/2), (10,3/2), ($\overline{10}$,1/2), ...

CONCLUSION

It is quite remarkable, we feel, that the topological and structural properties of the Skyrme-Witten model can result in such impressively quark-like qualitative results for the baryons, all from a model with no explicit quarks!

Acknowledgements

Thanks are due to my collaborators Professor Yossef Dothan and Dr. Marco Tarlini for their contributions to the work presented here, and to Professor Walter Greiner for organizing this Advanced Study Institute.

REFERENCES:

1. G. 't Hooft, Nucl. Phys. B72 (1974); ibid. B75; (1974) 461.

2. E. Witten, Nucl. Phys. B160 (1979) 57.

3. G. Holzwarth and B. Schwesinger, "Baryons in the Skyrme Model", Reports on Progress in Physics, (to be published).

4. Yossef Dothan and L.C. Biedenharn, "Old Models Never Die: The Revival of the Skyrme Model", Comments on Nuclear and Particle Physics, (to be published, Fall 1986).

5. L.C. Biedenharn, J.D. Louck, Encl. for Math. and Appl., Vol. 9: "The Racah-Wigner Algebra in Quantum Theory", Addison-Wesley (Reading, MA) 1981.

6. T.H.R. Skyrme, Proc. Roy. Soc. A260 (1961); J. Math. Phys. 12 (1971) 1735.

7. E. Witten, Nucl. Phys. B223, (1983) 422; ibid. 433.

8. L.C. Biedenharn and Yossef Dothan, "Monopolar Harmonics in $SU(3)_f$ as Eigenstates of the Skyrme-Witten Model for Baryons", pp. 19-34 in "From SU3 to Gravity" (Ne'eman Festschrift), Cambridge University Press, Cambridge UK, 1986; see also,

E. Guadagnini, Nucl. Phys. B236, (1984) 35.

PRESSURE ENSEMBLE AND DENSE NUCLEAR MATTER WITH FINITE SIZE NUCLEONS AT ZERO TEMPERATURE

Allard Schnabel

Institute of Theoretical Physics and Astrophysics
University of Cape Town
Rondebosch 7700
Cape, R.S.A

INTRODUCTION

In this seminar I will describe nuclear matter in the context of the pressure ensemble. My aim is to find the equation of state of dense nuclear matter near the phase transition to quark-gluon-plasma. There are already strong indications that the finite nucleon volume is probably the key physical property of dense nuclear matter[1,2,3,4]. I present a thermodynamical model which takes into account the finite size of the particles (nuclei). This is possible if the following assumptions are made:

a) There are numerous particles, such that there are many physical states of the system.

b) The system is in thermal, chemical and static equilibrium which means that we can speak of a local temperature, T and can define a chemical potential, μ.

Up to now only numerical calculations for T=0 have been done whereas the presented theory is not restricted to zero temperature. The thermodynamical picture on which this approach is based is shown in fig. 1. We are looking at a

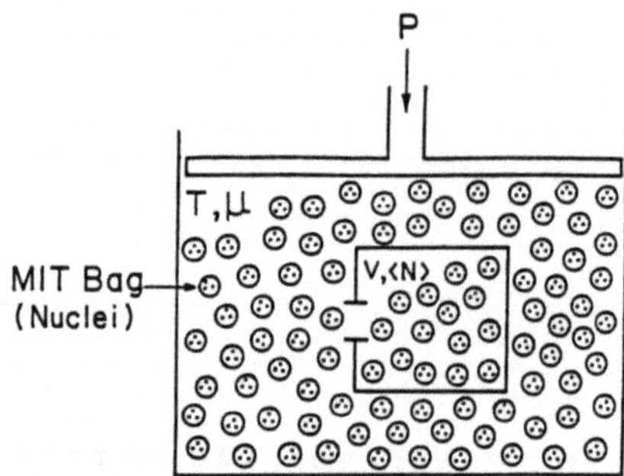

Fig. 1. Used thermodynamical picture of a nucleon gas.

system contained in a temperature and particle bath which has the pressure P. The particles of our "gas" are MIT-Bags with a volume v. Our description has to be grand canonical to allow particle creation and annihilation.

DEFINITIONS

Normally thermodynamic theories are based on the assumption of pointlike particles. In this case we can write the grand partition function as:

$$\Xi(\mu,T,V) = e^{\ln \Xi} = e^{V*F(T,\lambda)} =: \sum_{N=0}^{\infty} \frac{\lambda^N}{N!} \left[\frac{d^N}{dy^N} e^{V*F(T,y)} \right]_{y=0} \tag{1}$$

where we have done a Taylor expansion of $e^{V*F(\mu,T)}$ in μ with:

$$F(T,y) = \frac{\pm g}{(2\pi)^3} \int \ln (1 \pm y\ e^{-E/T})\ d^3p \tag{1a}$$

+ Fermi particle
- Bose particle

For particles of volume v we have to take two different effects into account:

1. The sum over all possible particle numbers can no longer be infinite. We have to exclude configurations where the particles occupy more volume than there is available in our system of volume V. Therefore we terminate the sum when the "box" is full of particles.

2. We have to correct the co-ordinate space. When we count all states the system can choose we will end up with a volume $\tilde{V}$ which is smaller than the volume of the whole box V. The reduction is proportional to the number of particles in the system and to their volume v.

$$\tilde{V} = \int_{\text{possible states}} d^3q = V - \text{factor} * Nv \tag{2}$$

The factor depends on the particle density in the system. The factor will increase from 1 in a low density system to 1.35 for the most densely packed crystal structure. In this case we have only one possible space configuration left in our system. Because we hope to be far away from a crystal structure we choose the factor to be 1. This is a good assumption as long as the particles can move.

$$\tilde{V} = V-Nv \tag{3}$$

With these corrections the grand partition function for equal particles with a finite volume becomes:

$$\Xi(\mu,T,V,v) = \sum_{N=0}^{V/v} \frac{\lambda^N}{N!} \left[\frac{d^N}{dy^N} e^{\tilde{V}*F(T,y)} \right]_{y=0} \tag{4}$$

OUR APPROACH

R. Hagedorn[5,6,7] pointed out that a nice way to solve the problem is to do another Laplace transformation.

$$\underline{\text{def.:}} \quad \pi(P,T,\mu) := \int_0^\infty e^{-V\xi} \; \Xi(V,T,\mu) \; dV \tag{5}$$

with $\xi = P/T$

I like to call the function π the pressure function. This function is related to the pressure partition function[8,9] by a dimensional normalisation constant k.

$$\Xi(P,T,\mu) = k\ \pi(P,T,\mu) \tag{6}$$

The pressure partition function obeys the relation:

$$\Omega(P,T,\mu) = -T\ \ln\ \Xi(P,T,\mu) \tag{7}$$

to a "grand pressure potential" $\Omega(P,T,\mu)$. The missing normalisation constant k doesn't matter because when we calculate the mean values <N>, <V>, <E> the constant always drops out. e.g:

$$\langle N\rangle = -\left.\frac{\partial\Omega(P,T,\mu)}{\partial\mu}\right|_{P,T} = \left.\frac{T\ \partial\ \ln\ \Xi(P,T,\mu)}{\partial\mu}\right|_{P,T} \tag{8}$$

$$= \frac{T\left.\frac{\partial\Xi(P,T,\mu)}{\partial\mu}\right|_{P,T}}{\Xi(P,T,\mu)} = \frac{Tk\left.\frac{\partial}{\partial\mu}\int_0^\infty e^{-V\xi}\ \Xi(V,T,\mu)\ dV\right|_{P,T}}{k\int_0^\infty e^{-V\xi}\ \Xi(V,T,\mu)\ dV}$$

$$= \frac{T\left.\frac{\partial\pi(P,T,\mu)}{\partial\mu}\right|_{P,T}}{\pi(P,T,\mu)}$$

We see that it is possible to get the mean values from the pressure function if we evaluate:

$$\langle N\rangle = \frac{T\left.\frac{\partial\pi(P,T,\mu)}{\partial\mu}\right|_{P,T}}{\pi(P,T,\mu)} \tag{9a}$$

$$\langle E\rangle = -\frac{T\left.\frac{\partial\pi(P,T,\mu)}{\partial\beta}\right|_{P,\mu}}{\pi(P,T,\mu)} \tag{9b}$$

$$\langle V\rangle = -\frac{T\left.\frac{\partial\pi(P,T,\mu)}{\partial\xi}\right|_{T,\mu}}{\pi(P,T,\mu)}\ . \tag{9c}$$

with $\xi = P/T$ and $\beta = 1/T$

Because π is not dimentionless it is not permitted to take the ln of π, as is often done. From the equations above we find that the energy density and particle density is:

$$\varepsilon(P,T,\mu) := \frac{\langle E\rangle}{\langle V\rangle} = \frac{\left.\frac{\partial\pi(P,T,\mu)}{\partial\beta}\right|_{P,T}}{\left.\frac{\partial\pi(P,T,\mu)}{\partial\xi}\right|_{T,\mu}} \tag{10a}$$

$$\nu(P,T,\mu) := \frac{\langle N\rangle}{\langle V\rangle} = -\frac{\left.\frac{\partial\pi(P,T,\mu)}{\partial\mu}\right|_{P,T}}{\left.\frac{\partial\pi(P,T,\mu)}{\partial\xi}\right|_{T,\mu}} \tag{10b}$$

We will now use pressure function in a simple example:

Example: Gas of equal pointlike particles with mass m in Boltzmann limit.

In this case the grand partition function is:

$$\Xi(V,T,\mu) = \sum_{N=0}^{\infty} \lambda^N Z_N(V,T) = \sum_{N=0}^{\infty} \frac{\lambda^N}{N!} Z_1(V,T)^N \tag{11}$$

$$Z_1 = V \int e^{-E/T} \frac{g\, d^3p}{(2\pi)^3} \qquad E = \sqrt{p^2 + m^2}$$

$$Z_1 = V\, f(T) \qquad f(T) = \int e^{-E/T} \frac{g\, d^3p}{(2\pi)^3}$$

Inserting $\Xi(V,T,\mu)$ into the definition of the pressure function we get:

$$\pi(P,T,\mu) = \int_0^{\infty} e^{-\xi V} \sum_{N=0}^{\infty} V^N \frac{\lambda^N}{N!} f(T)^N \, dV \tag{12}$$

Note that we can change the order of sum and integration. This is the step which improves the calculation in other applications.

$$\pi(P,T,\mu) = \sum_0^{\infty} \frac{[\lambda f(T)]^N}{N!} \int_0^{\infty} V^N e^{-\xi V} \, dV$$

After doing the integration we have:

$$\pi(P,T,\mu) = \sum_0^{\infty} \frac{[\lambda f(T)]^N}{N!} \frac{N!}{\xi^{N+1}}$$

and the geometrical series finally gives us:

$$\pi(P,T,\mu) = \frac{1}{\xi} \sum_{N=0}^{\infty} \left[\frac{\lambda f(T)}{\xi} \right]^N = \frac{1}{\xi} \frac{1}{1-\frac{\lambda f(T)}{\xi}}$$

$$\pi(P,T,\mu) = \frac{1}{\xi - \lambda f(T)} \tag{13}$$

This function has a pole at $\xi_0 = \lambda * f$. To see what the meaning of this pole is we compare with what we know from the grand canonical description. There we get:

$$\Omega(T,V,\lambda) = -PV = -T \ln \Xi(T,V,\lambda) \tag{14}$$

and in the thermodynamic limit:

$$P := \lim_{V\to\infty} -\frac{\Omega}{V} = \lim_{V\to\infty} \frac{T}{V} \ln \Xi(T,V,\lambda) \tag{15}$$

For our example we get the wellknown relation:

$$P = T\lambda f(T) \tag{16}$$

If we compare this with the pressure ensemble we see that the pole of the pressure function $\xi_0 T$ is P in the thermodynamic limit. Indeed the pole ξ_0 always represents the thermodynamic limit. To see this we rewrite the pressure function, π.

$$\pi(P,T,\mu) = \int_0^\infty e^{-V\xi} \; \Xi(V,T,\mu) \; dV$$

$$= \int_0^\infty e^{-V(\xi - \frac{1}{V} \ln \Xi(V,T,\mu))} \; dV \tag{17}$$

The pressure function, π can only have a pole if the expression $(\xi - \frac{1}{V} \ln \Xi(T,V,\mu)$ goes to 0 in the limit $V \to \infty$.

$$0 = \lim_{V\to\infty} (\xi_0 - \frac{1}{V} \ln \Xi(T,V,\mu))$$

$$\xi_0 = \lim_{V\to\infty} \frac{1}{V} \ln \Xi(T,V,\mu)$$

Which is with $\xi = P/T$ the pressure in the thermodynamic limit.

$$P = \lim_{V\to\infty} \frac{T}{V} \ln \Xi(T,V,\lambda)$$

q.e.d.

Since we are only interested in the thermodynamic limit we will always search for the pole of the pressure function because only this point is relevant. Now we are ready to introduce the particle volume. The pressure function in the Boltzmann limit becomes:

$$\pi(P,T,\mu,v) = \int_0^\infty \varepsilon^{-\xi V} \sum_{N=0}^{\infty} \frac{\lambda^N}{N!} \tilde{V}^N f(T)^N \Theta(\tilde{V})\, dV \tag{18}$$

with $\tilde{V} = V - N*v$.

After doing similar steps as before (see appendix 1), we get:

$$\pi(P,T,\mu,v) = \frac{1}{\xi - \lambda f e^{-\lambda v}} \tag{19}$$

If we use the proper quantum distribution function the calculation is more complicated (see appendix 2) and yields:

$$\pi(P,T,\mu,v) = \frac{1}{\xi - F(\lambda'=\lambda e^{-\xi v})} \tag{20}$$

$$F(T,\lambda') = \pm \int \ln\,(1 \pm \lambda' e^{-E/T})\, \frac{g\, d^3p}{(2\pi)^3}$$

+ Fermi particle
- Bose particle

Boltzmann limit: $F= \lambda f$

The temperature zero case

Let us consider a Fermi gas of finite size particle ("nucleons") at zero temperature. In this case we have to fill all the states up to the Fermi level. We find that the fermi level is shifted. It is not where $E=\mu$. The Fermi energy is now (see also figure 2):

$$E_f = \mu_f = \mu - \xi_0 Tv = \mu - Pv \tag{21}$$

because $\lambda' \, e^{-E/T} = e^{\frac{\mu - \xi_0 Tv - E}{T}} = e^{\frac{\mu_f - E}{T}}$

Although we have not incorporated an explicit interaction, we see that the finite volume effect acts like a repelling force (potential) which increases with density.

Hard sphere gas of finite size particle at T=0

If we start calculating with a reasonable volume of our particle (nuclei) e.g. $v = 2.1\ fm^3$ and mass e.g. $m = 939$ Mev, a phase transition to a quark-gluon-plasma will always be somewhere close to the point where the box is full of particles. There the pressure and the energy per particle is increasing rapidly (see fig. 9 below).The "hard-ball" assumption for the nuclei is unsatisfactory for dense nuclear matter close to the phase transition. Therefore it is better to describe the nuclei as a quark bag, i.e. permit that the volume of the bag will depend on ambient conditions.

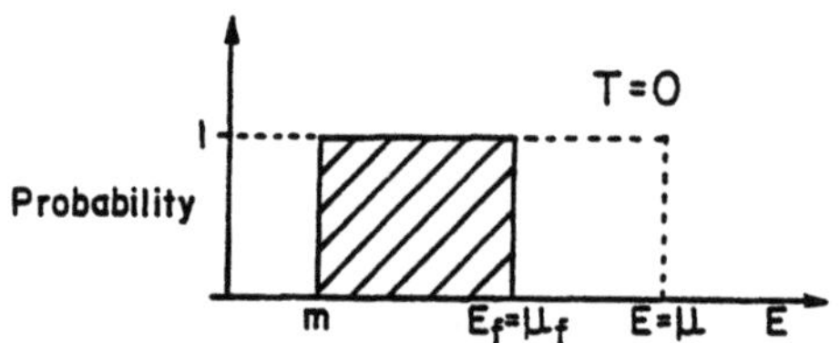

Fig. 2. Filled energy levels for T=0

Gas of MIT-Bag particle at zero temperature

In the MIT-Bag model[10] the particles (Bags) are described by:

$$m(v) = \frac{c'}{r} + Bv = \frac{c}{v^{1/3}} + Bv \qquad B = \text{Bag-constant} \quad (22)$$

The constant c is fitted so that the minimum mass of eq. 22 is the nucleon mass. The behaviour of the function is also shown in figure 3. This is definitely a good assumption if one looks at only one particle.

There is nothing against the assumption that the particle will shrink and become heavier in respect to eq. 22 if we compress a box filled up with MIT-bags.

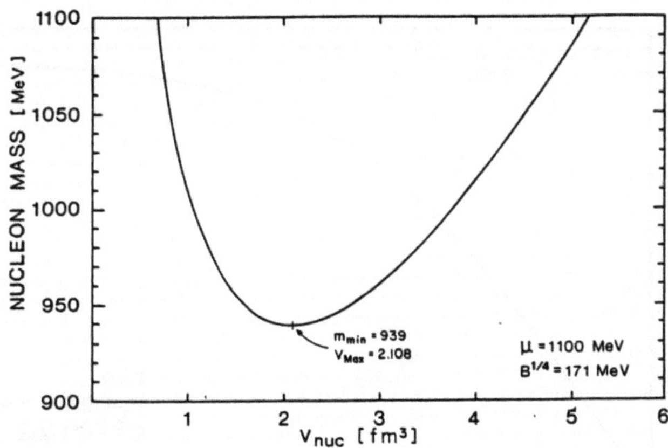

Fig. 3. MIT-Bag relation between mass and volume of the particles.

In this sense the particle volume v is a new independent parameter. However, the thermodynamical principle (R.Peierls[11]) tells us which particle volume the system will choose. The system goes to the point where:

$$\left.\frac{\partial\Omega(P,T,\mu,v)}{\partial v}\right|_{\mu,T,V} = 0 \qquad (23)$$

It goes to the minimum of the grand potential. Figure 4 shows the grand potential per box volume as a function of the particle volume. The function always has a minimum and is only defined in an area where μ is greater than the particle mass. If we take the minimum, calculate the density and all the other properties at this point and do the same for different μ we can study the behaviour of our model gas at different densities.

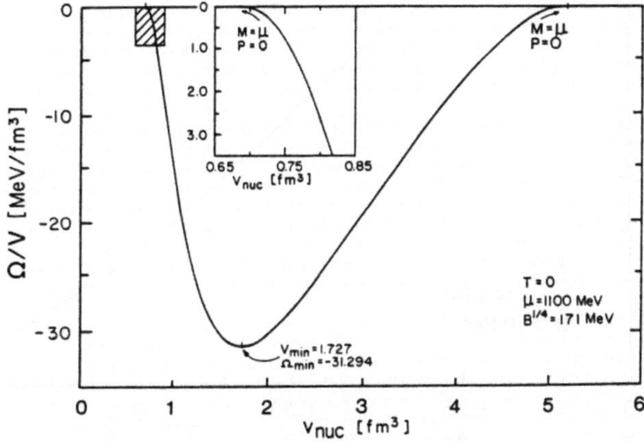

Fig. 4. Grand potential per box volume in dependance of the nuclei volume.

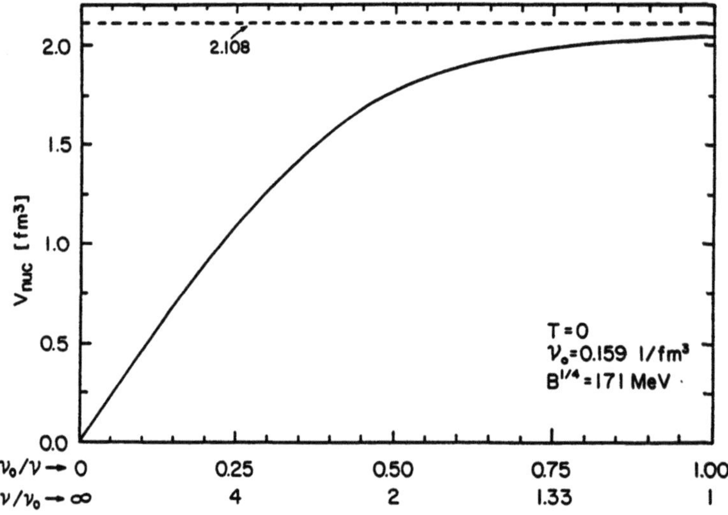

Fig. 5. Density dependence of the nuclei volume.

Figure 5 shows how the particle volume changes if we compress the gas. The dashed line shows the limit that the volume has for infinite dilution. At normal nuclear density the nucleon is already smaller. The situation where the box is completely full is never reached. For very high densities the particles reduce their own volume faster than the box is compressed. This is shown in figure 6 showing the percentage of space which is taken up by the volume of particles. Interesting is the point where the curve touches the y-axis. The densest crystal structure would occupy 74.0 % of the box volume. It is not yet understood why our curve comes so close to that point.

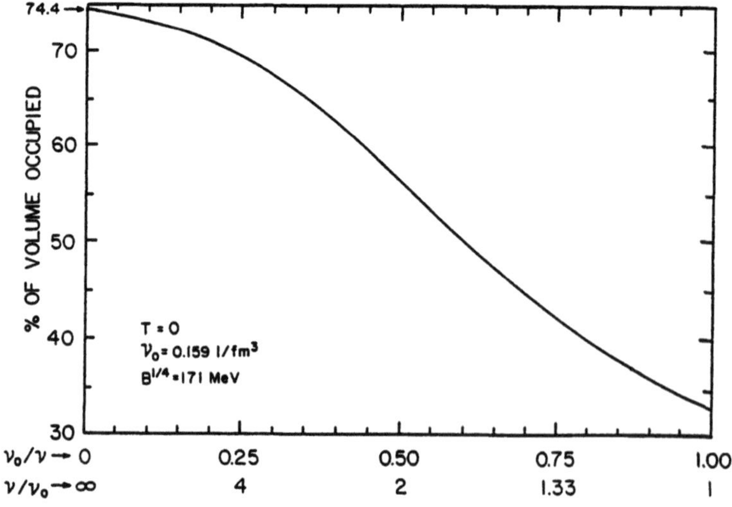

Fig. 6. Percentage of space occupied by the particles in the system

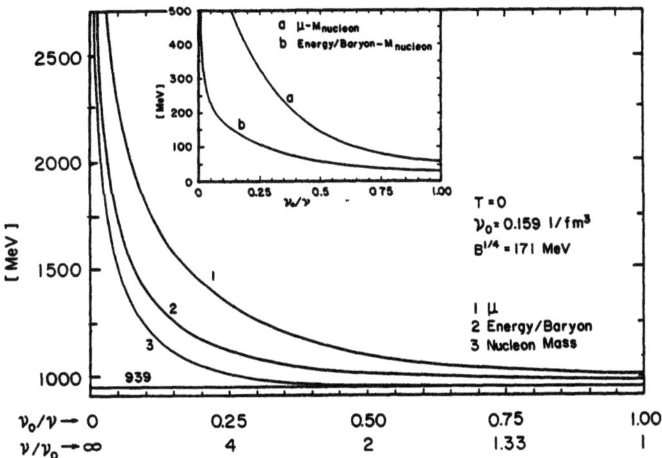

Fig. 7. Density dependence of chemical potential, nucleon mass and energy per baryon for our model nucleon gas.

Figure 7 shows several curves which have the same units. The first curve marked 1 is the chemical potential μ which is always greater than the rest mass of the particle, in the same figure curve marked 3. The difference between both is shown as curve (a) in the inset graph. We see that the difference is increasing at higher densities, so that we expect that we could neglect the mass, even if it is increasing a lot, against the faster increasing chemical potential in the very dense region. Curve 2 in that figure shows the energy per particle (baryon). Here we deviate from the normal representation which does not include the rest mass, because our rest mass changes. The difference between the energy per particle and the changing rest mass is shown as curve b in the inset. The difference is the kinetic energy of the particle. It grows as we reach higher densities. As we can neglect the rest mass against the kinetic energy and also against the chemical potential we expect to have something like a relativistic gas at higher densities. Figure 8 shows that the gas behaves in this way because at high density P/ε approaches the relativistic Fermi gas result 1/3. In figure 9 our results are compared with other theories[12,13]. We see that our model is not in contradiction with the other models for higher densities where the particle volume is important. At normal densities it is clear that the binding energy is missing because we did not consider any attractive force. In the curve marked quark gas, the properties of a gas of point-like up and down quarks without interaction is presented.

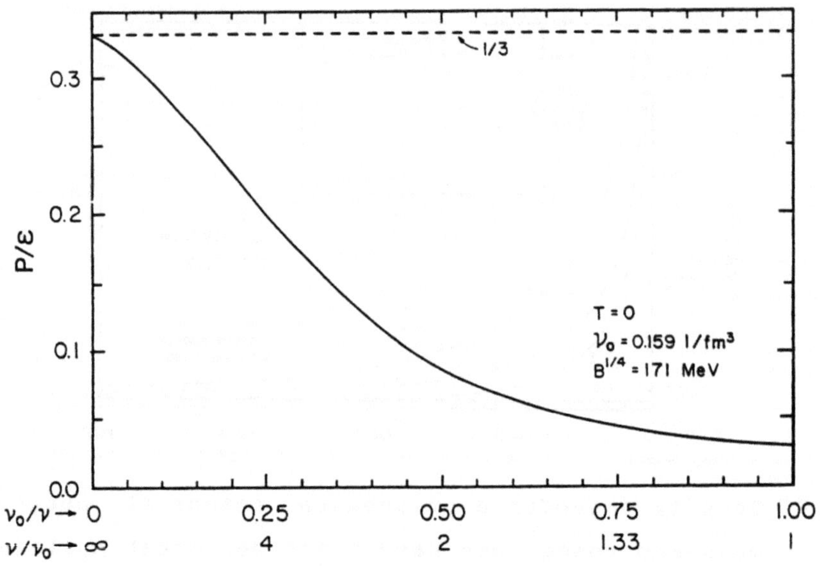

Fig. 8. Ratio of pressure to energy density as function of the particle density

Comparing this curve with our MIT-bag gas results we are led to expect a phase transition to the quark-matter at about four times the normal nuclear density if T=0. We note that since our nuclear gas is stiffer than conventional models, this phase transition occurs somewhat earlier than in other calculations.

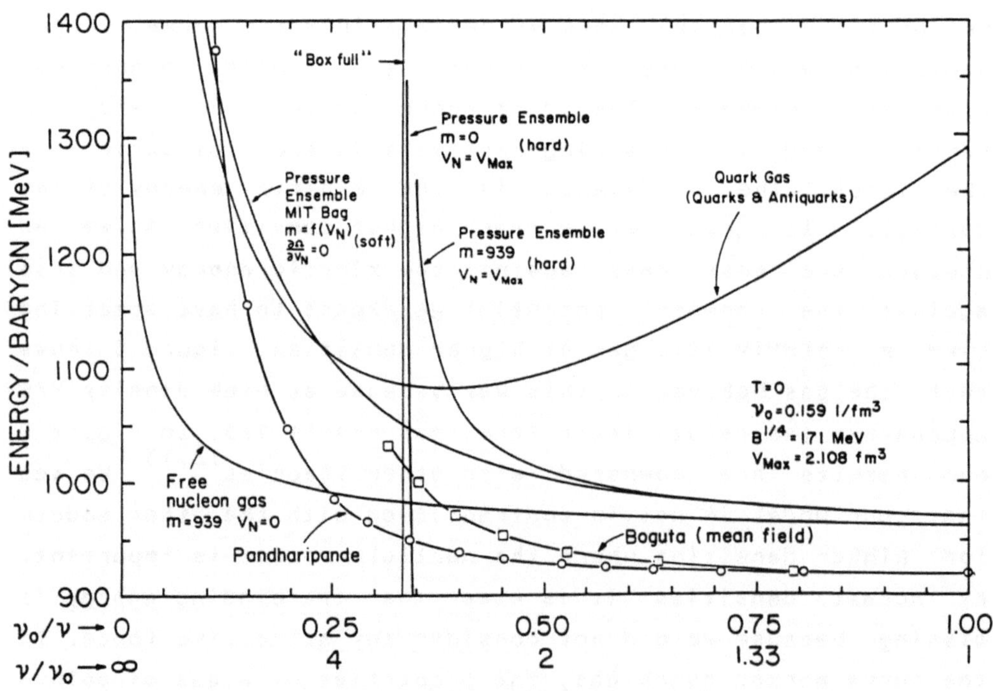

Fig. 9. Comparison of our results with other theories.

OPEN QUESTIONS, FUTURE WORK, PROBLEMS

1. How can we include the low density nuclear attraction?

2. All presented calculations are done at T=0. To do the calculations for T≠0 is more complicated for two reasons:

(a) For T≠0 we have to deal with a gas consisting of different particles e.g. $N, \bar{N}, \Delta, \bar{\Delta}, \pi^0, \pi^+, \pi^-, \varrho, \ldots$. All these particles have different volumes. The physical state is now where the thermodynamic potential has a minimum with respect to these volumes. This requires a search for an absolute minimum in more dimensions, (technical problem) if it exists at all, (theoretical problem). It is plausible that several minima could exist, irdicating several phases of hadronic matter, (physical problem).

(b) A special problem are the bosons. Their description in the context of the bag model is bad (especially for the pion).

3. What is the value of the bag constant? Its value is still rather uncertain. Also, it is not clear if it is really a constant if we go to higher densities and temperatures. If it is affected we will get an additional dependence of v in eq. 22 on ambient conditions.

4. Can we do a calculation with a better expression of the factor used in equation 3 to describe the reduction of the available co-ordinate space?.

The presented investigations are done in collaboration with J. Rafelski, R. Hagedorn, J. Cleymans and H. Eggers and are part of the work done towards my Ph.D. thesis.

Appendix 1: Pressure function for equal particles of volume v

In the Boltzmann limit we write the grand partition function as:

$$\Xi(V,T,\mu,v) = \sum_{N=0}^{V/v} \frac{\lambda^N}{N!} Z_1(\tilde{V},T)^N = \sum_{N=0}^{V/v} \frac{\lambda^N}{N!} \tilde{V}^N f(T)^N$$

$$= \sum_{N=0}^{\infty} \frac{(\lambda \tilde{V} f(T))^N}{N!} \Theta(\frac{V}{v} - N)$$

$$Z_1 = \tilde{V} \int e^{-E/T} \frac{g\, d^3p}{(2\pi)^3} \qquad E = \sqrt{p^2 + m^2}$$

$$Z_1 = \tilde{V} f(T) \qquad \tilde{V} = V - Nv$$

Inserting the grand partition function into the definition of the pressure function yields:

$$\pi(P,T,\mu,v) = \int_0^{\infty} e^{-V\xi} \sum_{N=0}^{\infty} \frac{(\lambda \tilde{V} f(T))^N}{N!} \Theta(\frac{V}{v} - N) \quad dV$$

Now we can interchange sum and integration:

$$\pi(P,T,\mu,v) = \sum_{N=0}^{\infty} e^{-V\xi} \int_0^{\infty} \frac{(\lambda \tilde{V} f(T))^N}{N!} \Theta(\frac{V}{v} - N) \quad dV$$

If we substitute:

$$x = \frac{V}{v} - N$$

we get with:

$$\tilde{V} = V - Nv = v\,(\frac{V}{v} - N) = xv$$

$$V = vx + Nv \qquad dV = v\, dx \qquad \Theta(\frac{V}{v} - N) = \Theta(x)$$

and the integration boundaries become:

for: $V = 0 \implies x = -N \implies x = 0$
because of the $\Theta(x)$ function

for: $V = \infty \implies x = \infty$

Everything inserted yields:

$$\pi(P,T,\mu,v) = \sum_{N=0}^{\infty} e^{-\xi N v} \frac{(\lambda\ v\ f(T))^N}{N!}\ v \int_0^{\infty} x^N\ e^{-\xi v x}\ dx$$

After doing the integration we have:

$$\pi(P,T,\mu,v) = \sum_{N=0}^{\infty} \frac{1}{\xi} \left[\frac{(\lambda\ f(T)}{\xi\ e^{\xi v}} \right]^N$$

and after summation we get:

$$\pi(P,T,\mu,v) = \frac{1}{1 - \lambda\ f(T)\ e^{-\xi v}}$$

q.e.d.

<u>Appendix 2:</u> Pressure partition function for equal particles with volume v and mass m.

If we insert our grand partition function for finite size particles in the definition of the pressure function we get:

$$\pi(P,T,\mu,v) = \int_0^{\infty} e^{-V\xi} \sum_{N=0}^{\infty} \frac{\lambda^N}{N!} \left[\frac{d^N}{dy^N} e^{\tilde{V}*F(T,y)} \right]_{y=0} \Theta(\frac{V}{v} - N)\ dV$$

$$\text{with:}\quad F(T,y) = \frac{\pm g}{(2\pi)^3} \int \ln\ (1\pm\ \lambda e^{-\frac{E}{T}})\ dp$$

$$\tilde{V} = V - Nv \qquad E = \sqrt{p^2 + m^2}$$

+ Fermi distribution
− Bose distribution

Here we can change the sum and the integration:

$$\pi(P,T,\mu,v) = \sum_{N=0}^{\infty} \int_0^{\infty} e^{-V\xi} \frac{\lambda^N}{N!} \left[\frac{d^n}{dy^n} e^{\tilde{V}*F(T,y)} \right]_{y=0} \Theta(\frac{V}{v} - N)\ dV$$

If we substitute:

$$x = V - Nv = \tilde{V}$$

we get with:

$$V = x + Nv \qquad dV = dx \qquad \Theta(\frac{V}{v} - N) = \Theta(V - Nv) = \Theta(x)$$

and the integration boundaries become:

for: $V = 0 \implies x = -Nv \implies x = 0$ because of the $\Theta(x)$ function

for: $V = \infty \implies x = \infty$

Everything inserted yields:

$$\pi(P,T,\mu,v) = \sum_{N=0}^{\infty} \frac{[\lambda e^{-\varepsilon v}]^N}{N!} \left[\frac{d^n}{dy^n} \int_0^{\infty} e^{-x(\varepsilon - F(T,y))} dx \right]_{y=0}$$

Now we can do the integration and get:

$$\pi(P,T,\mu,v) = \sum_{N=0}^{\infty} \frac{[\lambda e^{-\varepsilon v}]^N}{N!} \left[\frac{d^n}{dy^n} \frac{1}{\varepsilon - F(T,y)} \right]_{y=0}$$

This expression is the Taylor expansion of $(\varepsilon - F(T,y))]^{-1}$ at the point $y=\lambda' = \lambda\, e^{-\varepsilon v}$ so that we can write:

$$\pi(P,T,\mu,v) = \frac{1}{\varepsilon - F(T,\lambda' = \lambda e^{-\varepsilon v})}$$

$$F(T,\lambda') = \pm \int \ln\,(1 \pm \lambda' e^{-E/T})\, \frac{g\, d^3p}{(2\pi)^3}$$

+ Fermi distribution
- Bose distribution

q.e.d.

REFERENCES

1 U. Heinz, P.R. Subramanian & W. Greiner
"Can Antibaryons signal the formation of a Quark-Gluon Plasma?"
Contributed paper for the "Workshop on Experiments for RHIC"
held at Brookhaven National Laboratory, April 15-19, 1985

2 Kang S. Lee, M.J. Rhoades-Brown & U. Heinz
"The deconfining phase transition - influence of hadron resonances and strangeness"
Phys. Let. 174b 2 (1986) 123

3 J. Cleymans, K. Redlich, H. Satz & E. Suhonen
"On the phenomenology of deconfinement and chiral symmetry restoration"
Bielefeld (Germany) Preprint BI-TP 86/19 June 1986

4 F. Karsch & H. Satz
"Thermodynamics of extended hadrons"
Phys. Rev. D 21 4 (1980) 1168

5 R. Hagedorn
"Miscellaneous elementary remarks about the phase transition from hadron gas to a quark-gluon-plasma
Cern TH 4100/85

6 R. Hagedorn
"The pressure ensemble as a tool for describing the hadron-quark phase transition"
Z. Phys. C Particles and fields 17. 265-281 (1983)

7 R. Hagedorn, J. Rafelski
"Hot hadronic matter and nuclear collisions"
Phys. Let. 97b 1 (1980) 136

8 E.A. Guggenheim
"Grand Partition Functions and so-called 'Thermodynamic Probability'"
J. Chem. Phys. 7 (1939) 103

9 A. Muenster "Statistical Thermodynamics"
Volume 1, Springer Verlag Berlin (1969) 168ff

10 A. Chodos, R.L. Jaffe, K. Johnson, C.B. Thorn & V.F Weisskopf
"New extended model of hadrons"
Phys. Rev. D9 (1974) 3471

11 R. Peierls
"On a minimum property of the free energy"
Phys Rev 54 (1938) 918

12 V.R. Pandharipande
"Hyperonic matter"
Nucl. Phys. A178 (1971) 123

13 J. Boguta & H. Stocker
"Systematics of nuclear matter properties in a non-linear relativistic field theory"
Phys. Let. 120b 4,5,6 (1983) 286

BOUNDARY CONDITIONS AND THE STRUCTURE OF THE VACUUM

Corinne A. Manogue

School of Natural Sciences
Institute for Advanced Study
Princeton, NJ 08540
USA

INTRODUCTION

The idea of the ether, a background against which all motion in the vacuum takes place, was abandoned in 1905 with the advent of special relativity, which showed that there is no preferred inertial reference frame in which the supposed ether could be at rest. But in quantum mechanics, the need for a zero-point energy of the vacuum raises the question of whether the vacuum does indeed have some structure. With Casimir's prediction in 1948 [1] that two perfectly conducting parallel plates should experience, solely due to this zero-point energy, a mutual attractive force inversely proportional to the fourth power of the distance between the plates, and with the experimental confirmation of this prediction by Sparnaay in 1958, [2] it became clear that the concept of vacuum structure would have to be taken seriously. Since then, many calculations have been made in an attempt to elucidate this structure. Two types of calculations in particular have shed light on this issue. The first of these is calculations which expand on Casimir's work by examining how the vacuum is affected by the presence of various boundaries. The second of these is calculations of the vacuum production of particles due to the presence of supercritical potentials, both electromagnetic and gravitational.

This paper will discuss qualitatively a few of these results in an attempt to show that:

1) The vacuum does have structure, almost as if it were a (Lorentz invariant!) ether.

2) Many of the properties of the vacuum can be derived from the imposition of "boundary conditions" on solutions of the relevant field equations. (These "boundary conditions" may take the form of an external potential background which distorts the solutions of the relevant field equations in much the same way as traditional boundary conditions.)

No attempt will be made to provide an exhaustive review of the literature. In addition to the references cited, the interested reader is referred to the excellent books by Greiner, Müller, and Rafelski [3] and by Birrell and Davies, [4] both for their clear and detailed treatment of most of the topics mentioned here and for their comprehensive bibliographies.

In Section I, several extensions of the Casimir effect are calculated, demonstrating some aspects of vacuum structure. In particular, the effect of moving boundaries (accelerating and rotating) is examined. In Section II, particle production in the presence of supercritical electromagnetic and gravitational potentials is discussed, highlighting similarities between the two situations and stressing the need for the careful choice of a basis of solutions appropriate to the "boundary conditions".

I. EXTENSIONS OF THE CASIMIR EFFECT

The original version of the Casimir effect [1] determined the change in total vacuum energy due to the presence of boundaries (two parallel plates). The vacuum energy for the situation with no boundaries was calculated by summing up $\frac{1}{2}\hbar\omega$ for each mode of a basis of solutions of the electromagnetic field equations, where ω is the frequency of that solution. The boundaries require a new basis of solutions with different frequencies and therefore a different sum. The difference in the total energy in the two situations is physically observable [2] as a force between the plates.

More recent calculations focus on the vacuum expectation value of the stress-energy tensor. This is a local rather than a global quantity and hence gives more detailed information about the structure of the vacuum. For the parallel plate case, the integral over the region between the plates of the time-time component of the stress-energy tensor reproduces the original Casimir result for the total energy. Due to divergences of the stress-energy tensor near boundaries [5] (an effect which will be discussed in detail below), this is not always the case unless the renormalization procedures are consistent.

The vacuum expectation value of the stress-energy tensor is given by the coincidence limit of a certain differential operator (determined from the Lagrangian) acting on the Green function for the field in question with the given boundary conditions. (See Ref. 4 for details.) Therefore it is the Green function which is of interest. The Green function for the massless scalar field (the only field to be discussed in the rest of this section) in Minkowski space is

$$G(x,x') = \frac{i}{4\pi^2}\,\frac{1}{-(t-t')^2+(x-x')^2+(y-y')^2+(z-z')^2+i\varepsilon} \tag{1}$$

where x,x' on the left-hand side refer to the spacetime points (t,x,y,z), (t',x',y',z') in cartesian coordinates. The Green function for many problems with boundaries can be found using the method of images. For example, in the case of two parallel plates at $x=\pm a/2$, the Green function is an infinite sum of terms like (1) above with x' replaced by its image formed by reflection in one of the boundaries or replaced by the image of an image (see Figure 1), i.e.

$$G(x,x')_{d,n} = \frac{-i}{4\pi^2}\sum_{n=-\infty}^{\infty}\frac{(\pm 1)^n}{(t-t')^2+\left(x-(-1)^n x'+\frac{na}{2}\right)^2+(y-y')^2+(z-z')^2+i\varepsilon} \tag{2}$$

Cancellations between the terms are such that with the top sign the total Green function is zero on the plates (Dirichlet boundary conditions, d), and with the bottom sign the first derivative of the total Green function is zero on the plates (Neumann boundary conditions, n).

It is clear from (1) that the coincidence limit (i.e. the limit as x' approaches x) of the Green function is infinite. This leads to divergences in the stress-energy tensor. The divergences from (1) are just due to the

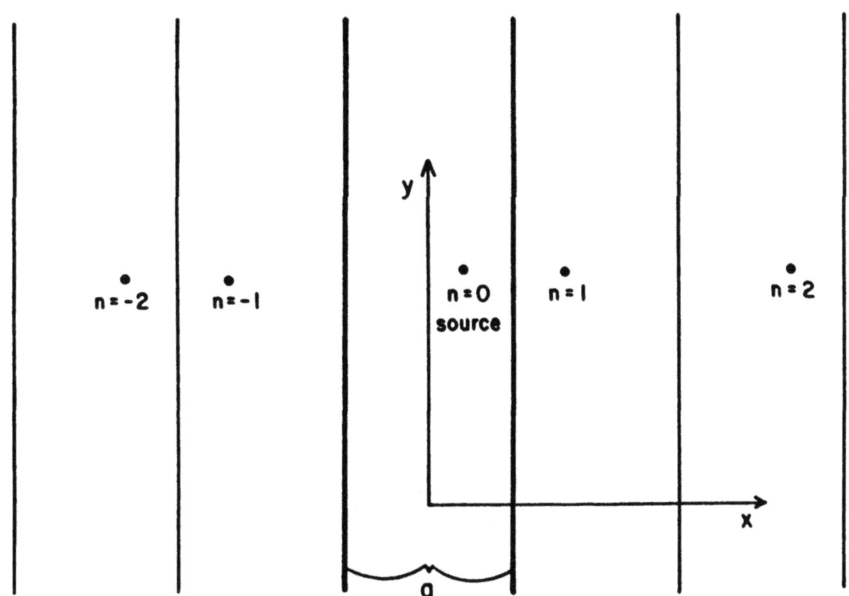

Fig. 1. Images for two infinite parallel plates.

zero-point fluctuations in Minkowski space. Since one is always interested in the difference between unbounded and bounded spacetimes, the Green function is always renormalized by throwing away this term (e.g. the n=0 term in (2)) and the divergences do not appear in the final answer. However, in the coincidence limit, as x approaches a boundary so does the image of x' formed by a single reflection in that boundary. This causes other divergences in the Green function as it approaches the boundary and usually causes divergences in the stress-energy tensor. In the special case of the conformal stress-energy tensor and a straight boundary the high degree of symmetry causes delicate cancellations to occur and there is no contribution from the images formed by an odd number of reflections. Thus, the conformal stress-energy tensor for a massless scalar field confined by two parallel plates is

$$\langle 0|T^{\mu\nu}|0\rangle_{d,n} = \frac{\pi^2}{1440a^2}\,\mathrm{diag}(-1,-3,1,1) \tag{3}$$

in agreement with the Casimir result except for a factor of one-half due to the difference in number of degrees of freedom between the electromagnetic and scalar fields. In the generic case of curved boundaries, without the high degree of symmetry, it has been shown that divergences near the boundary do occur. [5] These are ultraviolet divergences which arise because of the imposition of perfect conductor boundary conditions. Boundaries made of any ordinary matter will become transparent to modes with wavelengths shorter than the interatomic spacing in the boundary, thus imposing a cut-off. However, if the interatomic spacing is small compared to other distance scales in the problem, such as the spacing between the plates, then components of the stress-energy tensor and other observables can become quite large near the boundary, so that the divergences should not be thrown away entirely, only cut off appropriately. In cases where the boundary is not ordinary matter (e.g. in bags), the question of what cut-off, if any, can be imposed is a crucial one.

After stationary boundaries, one is naturally led to ask what happens if the boundary moves. Lorentz invariance implies that there can be no physically measurable differences between a boundary moving rigidly with constant (linear) velocity and the same boundary at rest.

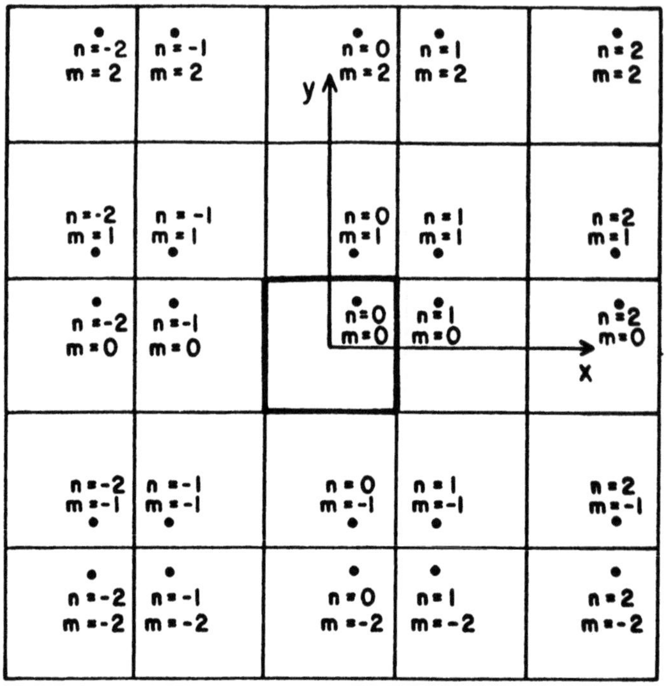

Fig. 2. Images for an infinitely long square box.

A more interesting motion is a plane boundary moving with constant acceleration normal to itself.[6] The stress-energy tensor far from the boundary has a black body spectrum corresponding to radiation with a temperature proportional to the acceleration. The dependence on distance from the plane is just what one would expect from the equivalence principle for a gas of massless scalar particles in thermal equilibrium in the gravitational field of an infinitely large flat "earth".

Another interesting motion is rotation. The simplest rotating boundary is an infinitely long circular cylinder rotating around its long central axis. Since the boundary is smooth and parallel to the direction of motion there is no way for the vacuum to detect that the cylinder is moving. Detailed calculation[7] does indeed show that there is no difference between the stress-energy tensors inside rotating and non-rotating cylinders. (Of course there <u>is</u> a stationary Casimir effect due to the mere presence of a cylindrical boundary.)

In order for the vacuum to sense the rotation, there must be some kind of bump on the boundary. Calculationally, the simplest kind of bump is to give the cylinder a square cross-section.[7] The Green function for a non-rotating box can be found using the method of images. (See Figure 2.) The Green function for the rotating box can be found, to first order in the angular velocity, using a variational principle, from the solution for the non-rotating box. Only the momentum components receive a first order correction. The results for $<0|T^{tx}|0>$ and $<0|T^{ty}|0>$ are plotted as a vector in Figure 3 for the conformal stress-energy tensor with Dirichlet and Neumann boundary conditions. These results show not only the expected divergence near the boundary (the boundary is curved in a spacetime sense), but also complicated motion (even vortices in the Neumann case!) inside the box where the motion of the corners does not directly push the vacuum. One is tempted to think of a viscosity for the vacuum but this has not been either verified or quantified.

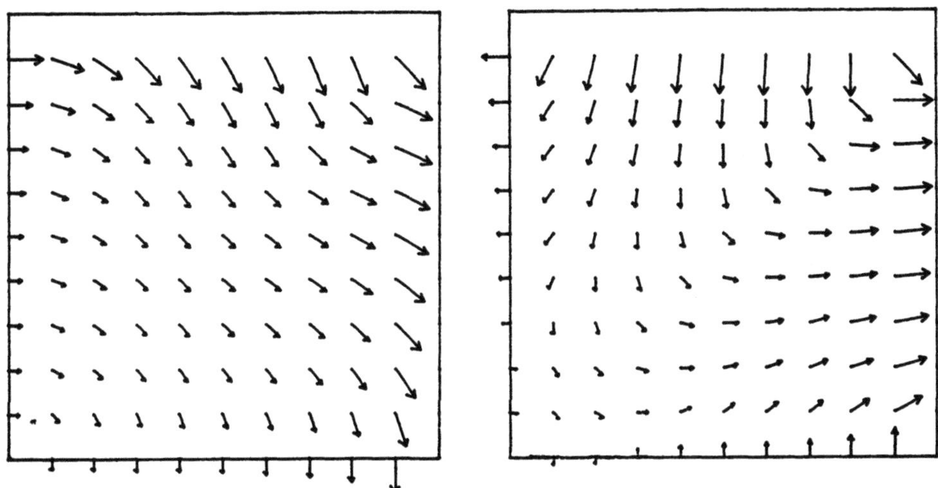

Fig. 3. $\langle 0|T^{tx}|0\rangle$ and $\langle 0|T^{ty}|0\rangle$ inside a rotating box, plotted together as a vector. Only the upper right quadrant of a cross-section of the box is shown, i.e. the center of of the box is the lower left corner of the drawing. The lengths of the vectors have been scaled by taking the eighth root in order to show the structure near the center. The figure on the left is for Dirichlet boundary conditions while the right shows the Neumann case.

II. PARTICLE CREATION IN SUPERCRITICAL ELECTROMAGNETIC AND GRAVITATIONAL POTENTIALS

The simplest example of particle creation by a supercritical potential is the classic Klein paradox [8] where a massive spin-½ fermion field or a massive spin-0 boson field is minimally coupled to an external electric potential which goes asymptotically to the constant values $\pm V/2$ as x goes to $\pm\infty$ (hereafter called to the left and to the right). (See Figure 4.) Considerable physical information can be obtained for this example merely by looking at the solutions ϕ of the field equations, before second quantization. In either the fermion or boson case, the solutions go asymptotically like

$$\begin{aligned} \phi &\sim e^{i(qx-\omega t)} e^{i\underline{k}\cdot\underline{y}} \qquad \text{(to the left)} \\ \phi &\sim e^{i(rx-\omega t)} e^{i\underline{k}\cdot\underline{y}} \qquad \text{(to the right)} \end{aligned} \tag{4}$$

where $\underline{y}$ stands for the two-vector (y,z) and q, r and $\underline{k}$ are related to the mass m and the frequency ω by the dispersion relations

$$\begin{aligned} q^2 &= \left(\omega + \frac{eV}{2}\right)^2 - (\underline{k}^2 + m^2) \\ r^2 &= \left(\omega - \frac{eV}{2}\right)^2 - (\underline{k}^2 + m^2) \end{aligned} \tag{5}$$

In the fermion case these asymptotic solutions multiply constant spinors.

If V is greater than twice the mass of the field, the frequency range divides naturally into five regions depending on the nature of the mode solutions as determined from the dispersion relations. (See Figure 4.)

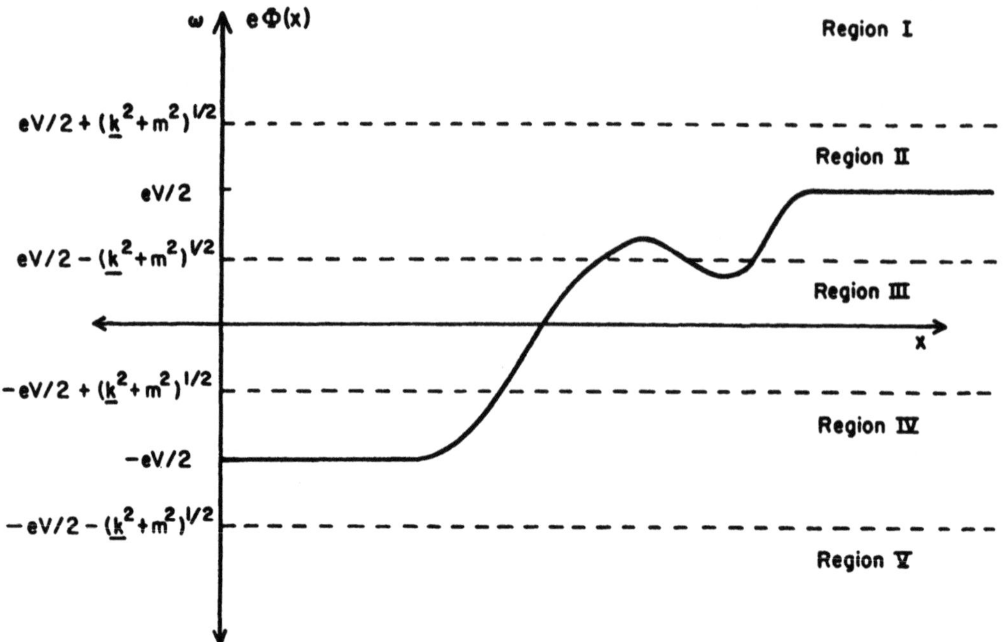

Fig. 4. An example of the type of potential considered, showing the five regions for the frequency.

Modes with frequency in regions I, III, V are oscillatory in both asymptotic regions. Modes with frequency in region II are oscillatory to the left but exponentially damped to the right. Modes with frequency in region IV are exponentially damped to the left but oscillatory to the right. If the potential is not supercritical, i.e. if V is not greater than twice the mass of the field, then region III does not exist. Notice that for a mode in region III, the frequency is greater than the asymptotic value of the potential to the left. The mode looks like a particle there. But on the right, the frequency is greater than the asymptotic value of the potential and the mode looks like an antiparticle. This situation is the Klein paradox.

A mode of a single frequency can be considered to be the limit of an infinitely broad wave packet. Then one can ask which direction these waves are moving. For example, on the left the packet will travel to the right if its group velocity is positive, i.e. if

$$\frac{\partial\omega}{\partial q} = \frac{q}{\omega + eV} > 0 \tag{6}$$

This means that for frequencies greater than the asymptotic value of the potential, the packet will travel to the right only if its wave number is positive, but for frequencies less than the asymptotic value of the potential, the packet will travel to the right only if its wave number is <u>negative</u>.

A complete basis of solutions to the field equations, called in-modes, is given by the following two types of modes: those that travel in from the left at early times and are scattered off the barrier, with a piece reflected back to the left and a piece transmitted (perhaps exponentially damped) toward the right at late times, and those which behave similarly with left and right interchanged.

For the boson case, if the basis mode has frequency in region III, then it can be shown that the reflected current is greater than the incident current. [9] This is a phenomenon known as superradiance, In the fermion case the reflected current is always less than the incident current. Superradiance does not occur. Notice that if one misidentifies which way the transmitted part of the mode is travelling, then it can incorrectly appear as if fermion modes are superradiant. One is then dealing with a solution which has pieces incoming from both sides instead of just one (an out-mode as defined below). The reflected current is not greater than the sum of the two incoming currents. An interesting historical note is that although many modern textbooks make exactly this error, Klein did not. In fact, in his original paper, [8] he thanks Bohr for pointing out the group velocity argument to him.

The frequencies for which the boson modes are superradiant (i.e. region III) show up in the second quantized picture for both bosons and fermions as follows. A second complete basis of solutions can be defined, the out-modes. These solutions consist of waves which travel in from both sides at early times in just such a way that they go out at late times in only one direction. Either the in-basis or the out-basis can be used to define a vacuum state by defining creation $a^{\dagger}_{in}, a^{\dagger}_{out}$ and annihilation a_{in}, a_{out} operators for the two bases. The in-vacuum $|in\rangle$ has no particles coming in from either side at early times, the out-vacuum $|out\rangle$ has no particles escaping to either side at late times. If the system is in the in-vacuum at early times, one can ask what number of out-type particles will be present at late times (i.e. $\langle in|a^{\dagger}_{out}a_{out}|in\rangle$). The actual number of particles depends upon details of the shape of the potential between the asymptotic regions and on the type of field, but the frequencies of the created particles all lie in region III.

In the supercritical gravitational potential of a Kerr black hole an in-basis of modes can be defined [10] analogously to the Klein paradox example. (See Figure 5.) This basis consists of two types of solutions:

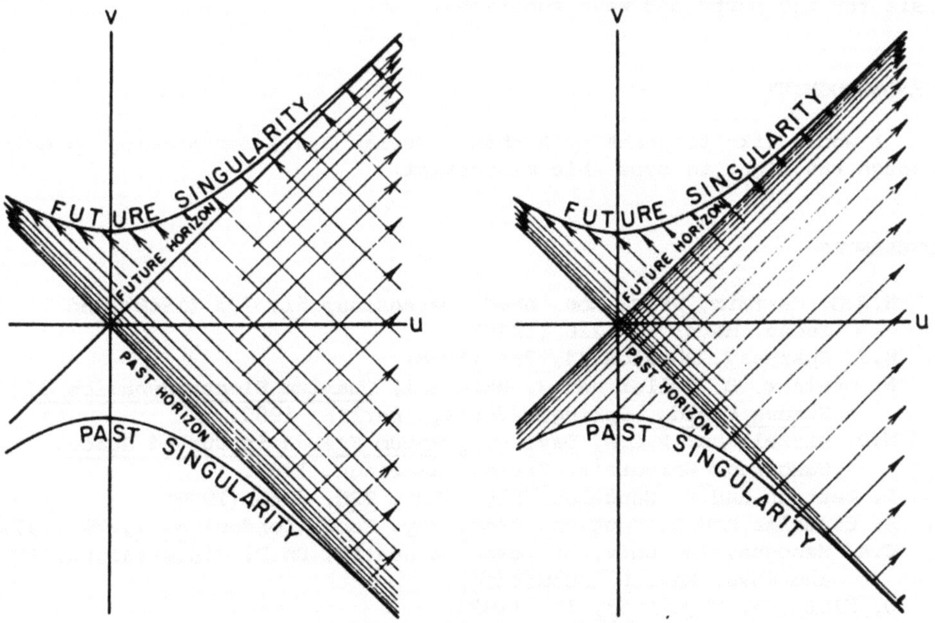

Fig. 5. In-modes for the Kerr black hole. This figure was taken from Ref. 10.

those which come in from infinity at early times and scatter off the potential, with a piece reflected back towards infinity and a piece which falls through the future horizon to the future singularity, and those which emerge through the past horizon at early times and scatter off the potential, with a piece reflected back through the future horizon and a piece which escapes to infinity. Again there is a specific region of frequencies (depending on the angular velocity and mass of the black hole) for which boson modes are superradiant. In agreement with the Klein paradox, fermion modes are never superradiant.

CONCLUSION

We have seen from various extensions of the Casimir effect that the vacuum certainly does have structure. This is most evident in the case of moving boundaries. The vacuum can be pushed along by an accelerating plane (or equivalently supported by a flat "earth") and stirred by a rotating box. Whether or not this implies that the vacuum has a viscosity is still unclear.

We have also seen that "boundary conditions" can have quite dramatic effects both on the solutions of the wave equation and on the physically observable quantities calculated from these solutions. These "boundary conditions" can be in the form of real boundaries (as in the extensions of the Casimir effect) or in the form of background potentials which distort the form of the solutions, sometimes to the extent of adding whole new types of solutions (as in the superradiant regions of the Klein paradox and Kerr black hole examples). It should be noted that what constitutes a complete basis for a Hilbert space (and indeed what constitutes the Hilbert space itself) depends upon the "boundary conditions". One often looks for a basis of solutions subject to one set of "boundary conditions" by perturbing around a complete basis of solutions subject to another set of "boundary conditions". This works very successfully in some examples. However, if the "boundary conditions" for the two bases are sufficiently different (e.g. possibly when a potential becomes supercritical), one might completely miss some important modes because the original solutions do not form a complete basis for the perturbed wave functions.

ACKNOWLEDGMENT

I would like to thank my husband, Tevian Dray, for staying up halfway through the night to type this manuscript.

REFERENCES

1. H.B.G. Casimir, Proc. Kon. Akad. Wetenschap 51, 793 (1948) and Indag. Math. 10, 226 (1948).
2. M.J. Sparnaay, Physica 24, 751 (1958).
3. W. Greiner, B. Müller and J. Rafelski, Quantum Electrodynamics of Strong Fields, Springer-Verlag, Berlin, 1985.
4. N.D. Birrell and P.C.W. Davies, Quantum Fields in Curved Space, Cambridge University Press, Cambridge, 1982.
5. D. Deutsch and P. Candelas, Phys. Rev. D20, 3063 (1979).
6. P. Candelas and D. Deutsch, Proc. Roy. Soc. (London) A354, 79 (1977).
7. C.A. Manogue, The Univ. of Texas at Austin, Ph.D. dissertation, 1984 and Phys. Rev. D (submitted).
8. O. Klein, Z. Physik 53, 157 (1929).
9. C.A. Manogue, The Univ. of Texas at Austin, Ph.D. dissertation, 1984 and Ann. Phys. (submitted).
10. B.S. DeWitt, Phys. Rep. 19C, 295 (1975).

QUANTUM EFFECTS IN STRONG GRAVITATIONAL FIELDS

P.C.W. Davies

Department of Physics, School of Physics
The University
Newcastle upon Tyne NE1 7RU, England

I. INTRODUCTION

A consistent quantum theory of gravitation remains an elusive goal, and one of the great outstanding challenges to fundamental theoretical physics. The most promising current set of ideas stems from the concepts of supersymmetry and supergravity, and goes under the name of superstring theory. This theory seeks to provide a unification of all the forces and particles of physics, and would thus provide a quantum description of gravitation as part of the package.[1]

Quantum gravitational effects are not expected under laboratory conditions. The relevant scales, on dimensional grounds, are the Planck length $(G\hbar/c^5)^{1/2} \approx 10^{-43}$s and the Planck energy $\approx 10^{19}$ GeV. By comparison the unification energy for the strong and electroweak interactions is about 10^{14} GeV, while direct laboratory simulation can currently reach only about 10^2 GeV.

On the other hand, quantum effects involving non-gravitational fields occur on atomic energy and length scales, at least twenty powers of ten from the Planck regime. This implies a huge range of sizes and energies over which quantum effects are important, but for which the gravitational field may be treated as a classical background. We may call this the <u>semi-classical approximation</u>[2]. Semi-classical quantum gravity - or the theory of quantum fields propagating in a classical background gravitational field (i.e. curved spacetime) has been very actively developed over the last decade or so. (For a detailed treatment see reference 2).

To give an example of this approximation, a black hole of mass 10^{15} gm has a size comparable to an atomic nucleus, and the behaviour of, say, the electromagnetic field in its vicinity is certainly subject to important quantum effects. Given that the Newtonian gravitational field near the surface of a 10^{15} gm black hole is about 10^{30} g, we can legitimately call this "quantum physics in strong gravitational fields".

Before proceeding a word must be said about the consistency of the semi-classical approximation[3]. The well-known principle of equivalence, upon which all theories of gravitation, and especially the general theory of relativity, are founded, states that all particles and fields couple equally strongly to gravity. One of the novel features of gravitation, therefore, is that the gravitational field itself couples to gravity; in short, gravity gravitates (gravitational energy is a source of gravitation). This means that if a strong gravitational field produces, say, photon effects it must, more or less equally strongly, produce similar "graviton" effects. This tells us that one cannot really ignore true quantum gravitational effects if one wants to discuss the action of gravity on other quantized fields.

This problem can be accommodated easily enough, however, by separating the gravitational field into a strong background component and a linearized perturbation. The former remains classical, and the latter is quantized and regarded as simply another quantum field propagating on the given classical background. This approximation is consistent so long as one does not start considering higher-order graviton-graviton processes.

What quantum effects, then, might be expected from intense background gravitational fields? These broadly divide into two categories : particle creation and vacuum stress-energy production. I shall consider each of these in turn.

II. PARTICLE CREATION

The creation of particles in strong external fields can be understood heuristically as follows. The quantum vacuum contains pairs of virtual particles and antiparticles that are continually forming and annihilating. According to the Heisenberg uncertainty principle, the rest mass, m, of a particle, and that of its antiparticle, can be "borrowed" for a time t C $\hbar/mc^2$, corresponding to a distance $\hbar/mc$ equal to a Compton wavelength. If during this brief interval the virtual pair interacts with an external field, energy will be exhanged. It may happen, if the field is strong enough, that sufficient energy will be imparted to the pair to pay for their rest mass, $2mc^2$. In this case, the Heisenberg "loan" need not be repaid, and the virtual pair will be promoted to a real particle pair. This can lead to the "decay of the vacuum", a phenomenon that is well known in quantum electrodynamics[4].

An external electric field will tend to drag a charged virtual pair apart, thus preventing annihilation and leading naturally to pair creation from the vacuum. For a Coulomb field Ze^2/r to provide mc^2 of energy over a Compton wavelength $r \sim \hbar/mc$ requires

$$\frac{Ze^2}{\hbar/mc} \sim mc^2 \tag{2.1}$$

or

$$Ze^2/\hbar c \sim 1. \tag{2.2}$$

Turning to the gravitational case, an external gravitational field accelerates both particles and antiparticles equally in the same direction (principle of equivalence), so at first sight it might appear as if no creation occurs. There will, however, be tidal forces due to the variation of gravitational field between the temporarily separated pair, causing them to fall at different rates and thereby to be pulled apart. (This can be most directly seen in the case of the expanding universe. The pair are literally drawn apart by the expansion of space.) As a result, intense gravitational fields can cause the quantum vacuum to become unstable and decay by pair creation, of both charged and neutral particles.

In the gravitational case, the Coulomb energy Ze^2/r is replaced by Newtonian gravitational energy GMm/r, and (2.2) becomes

$$GMm/\hbar c \sim 1 \tag{2.3}$$

or

$$Mm \sim m^2_{Planck}. \tag{2.4}$$

Again, requiring that the gravitational energy be supplied over about a Compton wavelength, we must demand that the gravitating mass M be confined to a Compton region

$$\text{size of mass} \lesssim \frac{\hbar}{mc} \tag{2.5}$$

which from (2.3) requires

$$\text{size of mass} \lesssim \frac{GM}{c^2} \approx \text{gravitational radius} \tag{2.6}$$

The requirement that the vacuum decays gravitationally is thus that the source of the field be confined within its gravitational radius. In popular parlance this means that the object is a black hole. The emission of particles by black holes was discovered by Hawking[5] (see section 6), and is the most famous example of the spontaneous decay of the quantum vacuum by gravitational effects.

Plausible though the above heuristic ideas may be, the rigorous discussion of particle creation by external gravitational fields involves some subtleties, chiefly because the very concept of a quantum particle in an external gravitational field is problematical.

Fundamentally, this is because particles are defined in quantum field theory in terms of field <u>modes</u>, which inhabit the whole spacetime and are therefore global in nature. The presence of a gravitational field introduces spacetime curvature, and possibly a non-trivial topology, which result in ambiguities in the definition of the vacuum state.

The problem is intimately associated with the <u>localizability</u> of particles. To fix ideas, suppose we have a photon of wavelength λ, and we try to localize it (in one dimension) within a length L. This confinement introduces a momentum uncertainty of order $\hbar/L$. As $L \rightarrow \lambda$, the uncertainty is comparable to the momentum of the photon itself, and the identity of the photon is lost. We thus expect important effects when geometrical features, such as the curvature of the spacetime, are comparable in size to the particle's wavelength. For most particles the Compton wavelength is very small by the standards of gravitational physics, and we conclude that such effects only occur for very large spacetime curvatures, or very intense gravitational fields.

III. PARTICLE CONCEPT TIED TO FLAT SPACETIME

In this section the particle concept will be examined more precisely. Consider a massive scalar field in Minkowski space. The field equation is (in units $\hbar = c = 1$)

$$(\Box + m^2)\phi(\underline{x}, t) = 0 \tag{3.1}$$

which may be solved as a superposition of positive and negative frequency plane wave modes

$$\phi = \sum_{\underline{k}} \left(a_{\underline{k}}\, e^{i\underline{k}.\underline{x} - i\omega t} + a_{\underline{k}}^{*}\, e^{-i\underline{k}.\underline{x} + i\omega t} \right) \tag{3.2}$$

where $a_{\underline{k}}$ and $a_{\underline{k}}^*$ are complex amplitudes (including a normalization factor). The field is quantized in the Heisenberg picture by letting these amplitudes become annihilation and creation operators $\hat{a}_{\underline{k}}$ and $\hat{a}_{\underline{k}}^\dagger$, and defining a vacuum (no-particle) state by

$$a_{\underline{k}}|0\rangle = 0 \qquad \forall\, \underline{k}. \tag{3.3}$$

The mode decomposition used is intimately connected with the symmetries of Minkowski space. In particular, the vacuum state $|0\rangle$ is invariant under the action of the Poincaré group, as are the set of modes. In curved spacetime, however, there are generally no such symmetries, and hence no privileged set of modes in terms of which the field may be decomposed.

Suppose we have two alternative sets of modes, labelled u_i and v_j. Then

$$\phi = \sum_i \left(\hat{a}_i u_i + \hat{a}_i^\dagger u_i^*\right) \tag{3.4}$$

and also

$$\phi = \sum_j \left(\hat{b}_j v_j + \hat{b}_j^\dagger v_j^*\right). \tag{3.5}$$

As both sets of modes are complete we may expand each v_j in terms of the set u_i,

$$v_j = \sum_i \left(\alpha_{ji} u_i + \beta_{ji} u_i^*\right) \tag{3.6}$$

and vice versa, where α and β are expansion coefficients. Equation (3.6) is an example of a so-called Bogoliubov transformation[6].

Each set of operators may be used to define a vacuum state :

$$\hat{a}_i |0_a\rangle = 0 \tag{3.7}$$

and

$$\hat{b}_i |0_b\rangle = 0. \tag{3.8}$$

In general, however, $|0_a\rangle \neq |0_b\rangle$. Indeed, if we evaluate the particle number operator, $N_i = a_i^\dagger ai$, for "a-type" particles in the "b-type" vacuum state we get

$$\langle 0_b | \hat{a}_i^\dagger \hat{a}_i | 0_b\rangle = \sum_j |\beta_{ji}|^2. \tag{3.9}$$

The right hand side will be non-vanishing whenever $\beta_{ji} \neq 0$, which occurs whenever the Bogoliubov transformation (3.6) mixes the positive and negative frequencies of one set of modes in defining another. Physically, (3.9) implies the $|0_b\rangle$ vacuum contains a-type particles and vice versa. But now the question arises of which particles are physically "real". Which vacuum state are we to choose to correspond to "true emptiness" when there is no privileged set of field modes as there was in (3.3)?

The question can only be answered by appealling to a particle detector. Simple model detectors introduced by Unruh[7] and DeWitt[8] have been extensively studied[9]. The detector is treated as a classical point object with quantized internal states. It is set initially in its ground state, and then allowed to follow a specified world line in the spacetime of interest, with the internal dynamics coupled to the given quantum state (assumed here to be a vacuum state). It is then possible to calculate the excitation rate, i.e. the probability per unit time of the detector going "click".

The general form of this rate for a scalar detector is a factor depending on the structure of the detector multiplied by a structurally-independent response function

$$\int_{-\infty}^{\infty} d\tau \int_{-\infty}^{\infty} d\tau' \; e^{-iE(\tau-\tau')} \; G^{+}(x(\tau), x(\tau')) \tag{3.10}$$

where E is the energy, τ the proper time along the detector's world line $x^{\mu} = x^{\mu}(\tau)$ and G^{+} is the Wightman function $\langle 0|\phi(x)\phi(x')|0\rangle$ evaluated in the state $|0\rangle$, with x, x' taken to lie on the world line. For Minkowski space and the standard vacuum state (3.3) one finds that (3.10) vanishes if the detector is static or in uniform motion. This is expected. It merely confirms our intuition that the vacuum state of conventional quantum field theory in the absence of gravitation is perceived as devoid of particles by all inertial observers.

In curved spacetime, however, (3.10) will generally be non-zero, indicating that particles are detected even in a so-called vacuum state. For example, in an expanding Friedmann cosmological model, no choice of field modes or detector world line produces a zero detector response.

A striking case is for de Sitter space. This may be described by the space-time metric

$$ds^2 = dt^2 - e^{2t/\alpha}\left[dx^2 + dy^2 + dz^2\right] \tag{3.11}$$

(α = constant). The scalar wave equation is very easy to solve in this spacetime if we set the mass m to zero, and include a so-called conformal coupling term R/6, where R is the Riemann curvature scalar, which for de Sitter space is a constant ($12\alpha^{-2}$). Thus

$$(\Box + \tfrac{1}{6}R)\phi = 0. \tag{3.12}$$

Equation (3.12) separates in the coordinate system (3.11) and mode solutions can be written down explicitly in terms of exponentials. A de Sitter-invariant vacuum state $|0\rangle$ can then be defined in the way discussed in the previous section. The Wightman function is readily evaluated for such a state. One finds

$$G^+(x, x') = - \frac{\eta\eta'}{4\pi^2\alpha^2[(\Delta\eta - i\varepsilon)^2 - |\Delta \underline{x}|^2]} \tag{3.13}$$

where η is the so-called conformal time, defined by

$$\eta = \int e^{-t/\alpha}\, dt = -\alpha e^{-t/\alpha} \tag{3.14}$$

$\Delta\eta = \eta - \eta'$, etc. and ε is an infinitesimal quantity.

Substitution of (3.13) in (3.10) yields the factor

$$\frac{E}{2\pi(e^{2\pi\alpha E} - 1)} \tag{3.15}$$

indicating a uniform response rate with a <u>Planck</u> (thermal) spectrum corresponding to a temperature $T = 1/(2\pi\alpha k)$, where k is Boltzmann's constant. This curious result means that a detector at rest in de Sitter space responds as though immersed in a bath of thermal radiation at a temperature inversely proportional to the radius of the de Sitter horizon[10] (i.e. small de Sitter spaces are hot).

It is worth remarking that de Sitter space forms the basis of the recent inflationary universe scenario models, and that the thermal quality indicated above plays a key role in the behaviour of the inflationary phase[11]. In this respect it could prove to be a fundamental feature in explaining the large-scale structure of the cosmos.

In the calculation above the detector was supposed to be "at rest" in de Sitter space, i.e. freely falling. If the detector is accelerated, its response changes. For a uniform acceleration a the response remains thermal, with an equivalent temperature[12]

$$T = \frac{1}{2\pi k}\left(\alpha^{-2} + a^2\right)^{1/2}. \tag{3.16}$$

Notice that we can put $\alpha = \infty$ and recover the Minkowski space limit. The temperature remains finite, and equal to

$$T = \frac{a}{2\pi k}. \tag{3.17}$$

This result tells us that <u>even in flat spacetime, our accelerated particle detector responds as though immersed in a bath of thermal radiation</u>[7,13]. Thus, one man's vacuum state is another man's thermal bath! It is even possible that something like this "acceleration radiation" has been detected[14]. This example provides a graphic demonstration of the nebulousness of the particle concept once one has departed from conventional flat space quantum field theory for inertial observers, tied as the concept is to the special features of the Poincaré symmetry group.

IV. PARTICLE CREATION IN COSMOLOGY

It has long been known that an expanding universe causes quantum particle creation[15]. Heuristically this can be understood as due to the disturbance of the vacuum field modes by the background spacetime. But given the nebulous character of particles in curved spacetime, can we be sure that cosmological expansion creates "real" particles?

Consider the idealized case of a homogeneous, isotropic universe with a scale factor (i.e. distance between typical "galaxies") $a(t)$ that approaches Minkowski space at $t \to \infty$ and $t \to -\infty$, with a period of smooth expansion in between.

One can define conventional plane wave field modes in the far past (the "in" region) and in the far future (the "out" region), but because of the period of expansion the in-modes, when propagated to the out region, do not coincide with the out-modes. They are, in fact, related by a Bugoliubov transformation that mixes positive and negative frequencies. We may conclude, then, that the in-vacuum state (i.e. the vacuum state defined with respect to the in-modes) contains out-particles, or, in more physical language, that there are particles in the out region which were not present in the in region. It is then natural to say that these particles have been <u>created</u> during the expansion phase.

That this interpretation is correct may be verified with a model particle detector. By direct calculation one can confirm that an inertial detector in the in region has zero response, but that such a detector when slowly switched on in the out region, detects particles. Some explicit models exist that give the particle spectrum[2]. It is not possible, however, to say exactly <u>when</u> the particles are created by leaving the detector on during the period of expansion, for in this region of curved spacetime the above-mentioned ambiguities in the particle concept intrude. Typically the frequency of the created particles is comparable to the expansion <u>rate</u>, and hence the particle creation events cannot be localized within the expansion region, in accordance with the remarks at the end of section 2.

With these interpretational cautions in mind, one may calculate particle creation rates in a wide range of cosmological models[2]. Although the technical details are beyond the scope of this presentation, good insight into the general principles can be gained by making use of the concept of vacuum viscosity.

V. VACUUM VISCOSITY: REVIVAL OF THE AETHER

It is well known that the quantum vacuum is a ferment of energetic activity involving virtual particles being continually created and destroyed. Less well known is the fact that this vacuum texture can display viscous properties[16], and is in many respects reminiscent of the aether[17].

A striking demonstration of vacuum viscosity has been given by Manogue[18], who considered a massless scalar field vacuum state inside a rotating box with reflecting walls, using various boundary conditions.

As the box rotates along its axis the moving boundaries disturb the field modes within the box and "stir up" the vacuum energy. There is a non-zero vacuum energy density even in the absence of rotation due to the discretization of the field modes that are constrained to satisfy the boundary conditions (this is an example of the famous Casimir effect[19]). The rotating box, however, does more than simply rigidly push this vacuum energy around. In fact, Manogue's results show that the momentum flux near the centre of the box is greatly in excess of this, and can be thought of as due to the viscous acceleration of the vacuum energy inside the box produced by the moving walls. The pattern of flow is shown in reference 18.

We may conclude that if the box were left spinning, vacuum viscosity would cause irreversible damping of the motion, so that the box would eventually come to <u>rest</u>, and be found to contain entropy. This entropy would be in the form of particles created by the rotation of the mirrors. The fact that particles are created by moving reflecting boundaries has been known for some time[20] but it is intriguing to interpret this phenomenon as due to the frictional effects of vacuum viscosity.

Applying these ideas to gravitational fields, one can ask about entropy production, i.e. particle creation, in the expanding universe. Prolific production is expected in anisotropically expanding universes, as the vacuum is continually being sheared. Explicit calculations verify this expectation[21]. Generally speaking, shear viscosity of the vacuum is the most important source of created particles.

In the case of isotropic expansion there is no shear, but particle creation can still occur. To see how, we may take a cue from classical cosmology. In a homogeneous, isotropic universe (a Friedmann model universe), filled with heat radiation (i.e. a relativistic fluid), the expansion produces no entropy. Instead, the radiation remains thermal, its temperature merely scaling with the cosmological scale factor a(t) :

$$T \propto 1/a(t). \tag{5.1}$$

Thus the entropy density, which is proportional to T^3 for Planck radiation, scales like a^{-3}, i.e. volume^{-1}. It follows that the total entropy in an expanding volume remains unchanged.

There is a similar result for a universe filled with a <u>non-relativistic</u> fluid at temperature T. This time $T \propto a^{-2}$. The entropy, however, still remains unchanged. By contrast, a <u>partially</u> relativistic fluid, when isotropically expanded, <u>does</u> create entropy, because its spectrum distorts under the expansion and entropy is created when it relaxes back to equilibrium. This is therefore an example of <u>bulk</u> viscosity[22].

These results have a direct analogue in the case of the quantum vacuum in a Friedmann universe. A relativistic field has no vacuum viscosity in this universe and so there is no creation of massless particles. Likewise, no very massive particles are created. On the other hand, particles with a mass $m \sim \dot{a}/a$ will be produced in proliferation.

To take a specific example, consider the two-dimensional expanding model universe

$$ds^2 = C(\eta)(d\eta^2 - dx^2) \tag{5.2}$$

where

$$C(\eta) = A + B \tanh \rho\eta \tag{5.3}$$

and A, B, ρ are constants. The expansion scale factor has the general form shown in Fig. 1. An exact calculation[23] yields a created particle number spectrum in the out-region of

$$N_k = \frac{\sinh^2(\pi\omega_-/\rho)}{\sinh(\pi\omega_{in}/\rho)\ \sinh(\pi\omega_{out}/\rho)} \tag{5.4}$$

where

$$\omega_{in} = \left[k^2 + m^2(A-B)\right]^{1/2} \tag{5.5}$$

$$\omega_{out} = \left[k^2 + m^2(A+B)\right]^{1/2} \tag{5.6}$$

$$\omega_\pm = \tfrac{1}{2}(\omega_{out} \pm \omega_{in}). \tag{5.7}$$

Now ρ is essentially an expansion rate parameter, and as $\rho \to 0$ the expansion slows to a halt. In this slow expansion limit (5.4) yields

$$N_k \approx e^{-2\pi\omega_{in}/\rho}. \tag{5.8}$$

For m $\gg \rho$, $\omega_{in}/\rho \gg 1$ and the particle number is exponentially suppressed. For the massless limit, $\omega_{out} = \omega_{in}$ and $\omega_- = 0$ so (5.4) vanishes identically, and there is strictly no particle production. Only when m $\sim \rho^{-1}$ will significant particle production occur.

VI. BLACK HOLES

Perhaps the most famous example of particle creation by strong gravitational fields concerns black holes. First consider the case of a rotating hole. There is a well-known effect in general relativity called "the dragging of inertial frames". It means, picturesquely speaking, that a free-falling particle orbiting a rotating body is "dragged around" by the rotation. Close to a black hole the effect is so pronounced that it is actually impossible for a body to remain at rest relative to an inertial observer at infinity[24]. A rotating massive body will also drag the quantum vacuum around, and because the effect depends on distance from the body there will be differential vacuum rotation rates. Vacuum viscosity will therefore produce a drag, which will act to decelerate the rotating body. The friction generated appears as heat radiation produced in the vacuum region around the body. This radiation flows away, taking angular momentum with it, and eventually the body would be reduced to

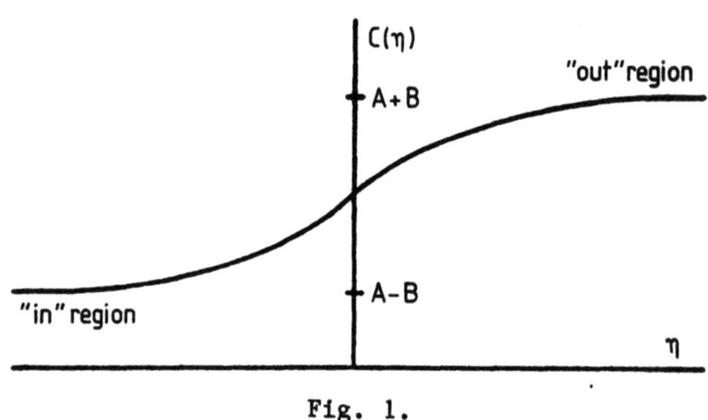

Fig. 1.

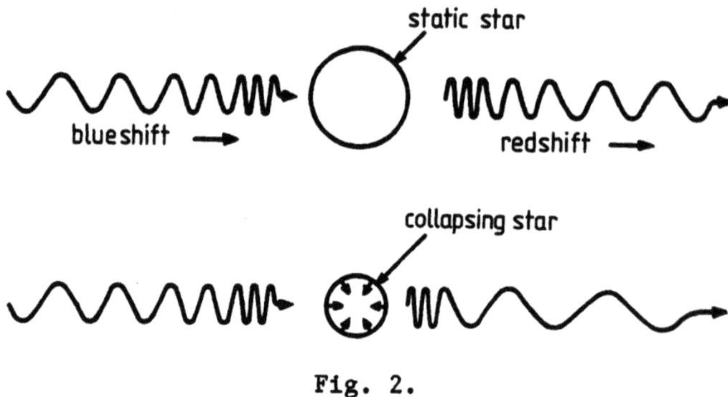

Fig. 2.

rest. The effect was originally predicted for black holes by Zeld'ovich[25], and confirmed in detailed calculations by Unruh[26] and Starobinski[27].

In addition to the effect of its rotation, a black hole produces heat radiation by another, more fundamental, mechanism - the so-called Hawking effect[28]. Black holes swallow everything in their vicinity, including the quantum vacuum, and as the vacuum streams down towards the hole, vacuum viscosity causes it to heat up and glow, emitting thermal radiation.

Hawking's original treatment of this effect was along the following lines. Consider a quantum field (e.g. the electromagnetic field) in the vicinity of a static spherical gravitating ball. The field may be decomposed into ingoing and outgoing spherically symmetric wave modes, and a vacuum for "incoming" modes defined in the usual way.

As the converging waves fall towards the body (see Fig. 2) they are blue shifted. The waves then travel through the centre of the body (assumed transparent) and out again, to become outgoing spherical modes. On the outward journey they are red shifted. Far from the hole the red shift just compensates for the blue shift, and there is no change in mode frequency or form. Thus the vacuum state associated with the outgoing modes is the same as that of the ingoing modes. There is no particle creation.

Suppose, now, that the ball collapses to a black hole. The collapse time is comparable to the light travel time across the interior of the ball. This means that by the time a given ingoing mode has traversed the interior the ball will have shrunk appreciably. Because the red shift escalates as the radius of the ball approaches its critical Schwarzschild (gravitational) radius at which it becomes a black hole, the blue shift on the inward journey is not sufficient to pay for the red shift on the outward journey, and so there is a net red shift. A calculation shows, in fact, that the modes are exponentially red shifted. Thus an ingoing wave of the form

$$\frac{1}{r} \exp[-i\omega(t-r)] \tag{6.1}$$

emerges from the collapsing ball having the form

$$\frac{1}{r} \exp\left[i\omega\left(A e^{-\kappa(t-r)} + B\right)\right] \tag{6.2}$$

where A and B are constants and κ is the so-called surface gravity of the hole, which for the Schwarzschild (spherically symmetric) case is 1/4M (in units with G = c = 1).

The violently modified modes (6.2) may be compared with the standard outgoing modes (1/r) exp $[i\omega(t - r)]$ associated with the conventional vacuum state. Thus the vacuum state corresponding to the in modes, which yields no particles in the in region (i.e. prior to collapse) is no longer a vacuum state in the out region (i.e. far from the hole after collapse). A straightforward computation of the Bogoliubov transformation between these two sets of modes shows that the in vacuum state contains, in the out region, particles with a thermal spectrum corresponding to a temperature

$$T = \frac{\kappa}{2\pi k} = \frac{1}{8\pi k M} \tag{6.3}$$

for a Schwarzschild black hole. This is Hawking's famous result. It implies that black holes are not black but emit heat radiation at a characteristic temperature inversely proportional to their mass. They get hotter as they lose energy, implying a negative specific heat. The process therefore escalates, leading to the evaporation and eventual explosive disappearance of the hole. This dramatic process could provide observational support for quantum effects in strong gravitational fields[29].

Black hole evaporation raises two questions. Where are the particles created? How does the hole lose energy?

The first problem concerns the fact that nothing is supposed to escape from a black hole. However, as we have seen, quantum particles cannot be fully localized to within about a wavelength. Thermal radiation at temperature $1/8\pi Mk$ implies a typical wavelength of order M, which is comparable to the size of the hole. Thus there is a position uncertainty over a region comparable with the black hole's horizon size, and it is wrong to think of the particles as "coming from" a particular location within, on, or outside the surface of the hole.

The second question, concerning the flow of energy from the hole, will be dealt with in the final section of this paper.

Before leaving the subject of black holes, I should like to make an interesting new suggestion. There has been a recent flurry of excitement over the possibility of a "fifth force" of nature, a weak component of gravitation that is baryon-number dependent, causing nuclei to fall at different rates from electrons[30]. If it exists this force would lead to the rate of fall being dependent on the chemical nature of the test body, in violation of the principle of equivalence and the time-honoured tradition, going back to Galileo, that all bodies fall equally fast.

If a fifth force of this sort exists, it means gravitation can discriminate between matter and antimatter. Neutrons and antineutrons, for example would fall at different rates. This presumably implies that there would be particle creation from the vacuum even in a <u>uniform</u> gravitational field, though I have not yet checked by calculation.

A more exciting possibility concerns black holes. It might be expected that, when matter collapses to form a black hole, the "fifth force field" is swallowed by the hole as well; in colloquial parlance, a black hole would have no fifth force hair. In that case, the effective value of G for a black hole will be G_∞, i.e. the astronomical long-range value, rather than the effective value at the Earth's surface. This question cannot, however, be properly investigated until more is known about the nature of the fifth force field. But even if the force is swallowed, its relatively long range implies that it will take a relatively long time to fade away. For the claimed range $\lambda \approx 200$ m one might expect a fade-out time $\lambda/c \approx 10^{-6}$s. Although this is negligible for a macroscopic black hole, the situation for microscopic holes is dramatically different.

It has been suggested that black holes of mass $\lesssim 10^{15}$ gm might have formed from the ultra-dense cosmological material during the very early universe. According to Hawking's formula, primordial black holes with masses $\lesssim 10^7$ gm would have evaporated away again before the end of the first microsecond. It is then possible that such holes form and evaporate before they get around to swallowing the fifth force "hair". In which case the evaporation characteristics of these holes will be slightly, but very significantly, modified by the presence of the fifth force.

A hole of mass $\lesssim 10^7$ gm is hot enough to emit baryons prolifically. Because the fifth force is baryon-dependent, it will emit baryons and antibaryons at different rates. The effect would be cumulative, and result in a final baryon-antibaryon, or matter-antimatter, asymmetry once the hole had disappeared. We thus have an interesting model for the matter-antimatter asymmetry in the universe. The magnitude of the observed cosmic asymmetry (only about one part in 10^9) is perhaps, small enough to be explained by a weak baryon-dependent force.

VII. VACUUM STRESS

Energy is a source of gravitation. The quantum vacuum contains "zero point" energy. Formally this is an infinite quantity :

$$\sum \tfrac{1}{2}\hbar\omega = \infty. \tag{7.1}$$

To make sense of the theory, this infinite vacuum energy must be "renormalized away". In flat space-time a simple infinite subtraction will do, but in curved space-time a much more elaborate renormalization procedure is necessary[2].

The source of gravitation in Einstein's theory is not energy as such, but the stress-energy-momentum tensor $T_{\mu\nu}$. Thus one finds $\langle T_{\mu\nu}\rangle$ for all quantum states, including the vacuum. Renormalization consists of absorbing the divergent terms in $\langle T_{\mu\nu}\rangle$ into geometrical terms on the left hand side of the gravitational field equations, by infinite re-scaling of coupling constants such as the Newtonian G. The technical details are very involved and will not be discussed here.

Suffice it to say that when a gravitational field is present, i.e. spacetime is curved, the quantum vacuum activity is disturbed. The effect of curving up a space both produces particles by vacuum viscosity, and induces a non-zero stress-energy. Thus although $\langle T_{\mu\nu}\rangle$ is formally divergent, <u>changes</u> in $\langle T_{\mu\nu}\rangle$ due to changes in geometry can be finite and explicitly calculable.

I shall limit myself here to a single example - the black hole. A quite general result of renormalization is that a classical field for which it so happens that

$$T^{\mu}{}_{\mu} = 0 \tag{7.2}$$

(e.g. the electromagnetic field) before quantization acquires a non-zero stress tensor trace on quantization

$$\langle T^{\mu}{}_{\mu}\rangle \neq 0. \tag{7.3}$$

This is known as a <u>conformal anomaly</u> because the condition (7.2) is an expression of the invariance of the classical field under conformal transformations.

A massless scalar field in two spacetime dimensions has the conformal anomaly[31]

$$\langle T^{\mu}{}_{\mu} \rangle = -R/24\pi . \tag{7.4}$$

Consider such a field in the vicinity of a two dimensional model black hole described by the quite general metric

$$ds^2 = C(r)\,dt^2 - C^{-1}(r)\,dr^2 \tag{7.5}$$

$$= C(r)\left[dt^2 - dr^{*2}\right] \tag{7.6}$$

where

$$r^* = \int C^{-1}\,dr . \tag{7.7}$$

The horizon of the hole occurs where C = 0 (e.g. in the Schwarzschild case C = 1 - 2M/r) and I assume $C \to 1$ as $r \to \infty$ so that the spacetime is asymptotically flat.

Again, quite generally, the quantum stress-energy-momentum tensor <u>must</u> satisfy the covariant conservation law

$$\langle T^{\mu\nu} \rangle_{;\nu} = 0 . \tag{7.8}$$

Assuming that the situation is stationary, we obtain from (7.8)

$$\langle T^{r}{}_{t} \rangle = \text{constant} \tag{7.9}$$

$$\frac{\partial}{\partial r}\left[C \langle T^{r}{}_{r} \rangle\right] = \frac{1}{2} C' \langle T^{\mu}{}_{\mu} \rangle . \tag{7.10}$$

Making use of the general result (7.4), and the fact that R = C"/4 for this spacetime, (7.10) integrates to give

$$\langle T^{r}{}_{r} \rangle = \frac{1}{96\pi C} \int C' C''\,dr$$

$$= \frac{C'^2}{192\pi C} + \text{constant} \tag{7.11}$$

Finally, I require that $\langle T_r{}^r \rangle$ remain finite at the event horizon C = 0. This must be so for any physically acceptable state as the horizon is, locally, an unexceptionable region of spacetime. This last condition suffices to fix the constant in (7.11) :

$$\langle T^{r}{}_{r} \rangle = \frac{C_h'^2 - C'^2}{192\pi C} \tag{7.12}$$

where C_h' is the value of C' at the horizon, C = 0. Now the definition of the surface gravity of the black hole is, in this case

$$\kappa = \frac{1}{2}\left(\frac{\partial C}{\partial r}\right)_{C=0} \tag{7.13}$$

so (7.12) becomes

$$\langle T^r{}_r \rangle = \frac{4\kappa^2 - C'^2}{192\pi C}. \tag{7.14}$$

Far from the black hole, spacetime is flat, and we may put $C = 1$, $C' = 0$, so (7.14) yields

$$\langle T^r{}_r \rangle \underset{r\to\infty}{\longrightarrow} \kappa^2/48\pi. \tag{7.15}$$

Physically, this is precisely the vacuum stress of a flux of thermal radiation at a temperature $T = \kappa/2\pi k$, the Hawking temperature (6.3). Thus, using only the covariant conservation law, the conformal anomaly (both general results), finiteness on the horizon and stationarity, we obtain the black hole radiance effect at the correct temperature. Notice that this did not depend on a specific choice of gravitational field.

One can go on to investigate the value of $\langle T_{\mu\nu} \rangle$ at the horizon of the hole. It turns out that there is an ingoing negative energy flux down the hole which diminishes its mass at precisely the rate required to pay for the Hawking flux at infinity. Thus the law of mass-energy conservation is respected, and we can answer the question, raised at the end of the previous section, of how a black hole can radiate energy. It does this not by positive energy getting out (which seems to violate the law that nothing can escape from a black hole), but by negative energy flowing in. The existence of negative energy densities and fluxes is a curious, but rather general feature of quantum field theory in the presence of external gravitational fields.[32]

As a final example of the use of vacuum energies, I mention briefly the recent interest in higher-dimensional unified theories, such as Kaluza-Klein and the superstring theories. These theories require the additional space dimensions to be "compactified" (i.e. rolled up very small). In the case of eleven-dimensional Kaluza-Klein theory, for example, the extra seven dimensions form a seven-sphere. There has been much investigation of the vacuum stress-energy induced within these compactified higher dimensions to see if it would act as a stabilizing factor, preventing them from undergoing gravitational collapse.[33] If quantum vacuum energy turns out to be the key to a achieving realistic compactified configurations, then the physics of quantum fields in curved background spacetimes could be an absolutely crucial element in the unification of nature.

REFERENCES

1. For a recent review see M. Green, Nature 314 : 409 (1985).
2. N.D. Birrell and P.C.W. Davies, "Quantum Fields in Curved Space", Cambridge University Press, Cambridge (1982).
3. M.J. Duff, Inconsistency of quantum field theory in curved space-time, in : "Quantum Gravity 2 : A Second Oxford Symposium", C.J. Isham, R. Penrose, and D.W. Sciama, eds., Clarendon Press, Oxford (1981).
4. For a heuristic introduction see J. Rafelski and B. Muller, "The Structured Vacuum : Thinking About Nothing", Verlag Harri Deutsch, Frankfurt am Main (1985).
5. S.W. Hawking, Commun. Math. Phys. 43 : 199 (1975).
6. N.N. Bogoliubov, Zh. Eksp. Teor. Fiz. 34 : 58 (1958).
7. W.G. Unruh, Phys. Rev. D 14 : 870 (1976).
8. B.S. DeWitt, Quantum Gravity : the new synthesis, in : "General Relativity : an Einstein Centenary Survey", S.W. Hawking and W. Israel, eds., Cambridge University Press, Cambridge (1979).
9. See, for example, K.J. Hinton, PhD thesis, University of Newcastle upon Tyne (1985).
10. G.W. Gibbons and S.W. Hawking, Phys. Rev. D 15 : 2738 (1977).
11. See, for example, "The Very Early Universe", G.W. Gibbons, S.W. Hawking and S.T.C. Siklos, eds., Cambridge University Press, Cambridge (1983) and A.D. Linde, Rep. Prog. Phys. 47 : 925 (1984).
12. P.C.W. Davies, Phys. Rev. D. 30 : 737 (1984).
13. P.C.W. Davies, J. Phys. A : Gen. Phys. 8 : 365 (1975).
14. J.S. Bell, R.J. Hughes and J.M. Leinaas, Z. Phys. C 28 : 75 (1985).
15. L. Parker, Phys. Rev. 183 : 1057 (1969).
16. Ya. B. Zeldovich and I.D. Novikov, "Relativistic Astrophysics Volume 2 : The Structure and Evolution of the Universe", University of Chicago Press, Chicago (1983), page 592.
17. B.S. DeWitt, The quantum aether, section 14.2 in reference 8.
18. C. Manogue, see article in this volume.
19. H.B.G. Casimir, Proc. Kon. Ned. Akad. Wet. 51 : 793 (1948).
20. P.C.W. Davies and S.A. Fulling, Proc. Roy. Soc. London A 348 : 393 (1976); 356 : 337 (1977).

21. Ya. B. Zeldovich, Pis'ma Zh. Eksp. Teor. Fiz. 12 : 443 (1970); Ya. B. Zeldovich and A.A. Starobinski, Pis'ma Zh. Eksp. Teor. Fiz. 26 : 373 (1977).

22. See, for example, C.W. Misner, K.S. Thorne and J.A. Wheeler, "Gravitation", Freeman, San Francisco (1973), section 40.7.

23. C. Bernard and A. Duncan, Ann. Phys. (NY) 107 : 201 (1977).

24. C.W. Misner, K.S. Thorne and J.A. Wheeler, "Gravitation", Freeman, San Francisco (1973) section

25. Ya. B. Zeldovich. JETP Letts. 14 : 180 (1971).

26 W.G. Unruh, Phys. Rev. D 10 : 3194 (1974).

27 A.A. Starobinski, Sov. Phys. JETP 37 : 28 (1973).

28. See reference 5 and section 8.1 of reference 2.

29. M.J. Rees, Nature 266 : 333 (1977).

30. E. Fishback, D. Sudarsky, A. Szafer, C. Talmadge, Phys. Rev. Letts. 56 : 3 (1986).

31. P.C.W. Davies, S.A. Fulling and W.G. Unruh, Phys. Rev. D 13 : 2720 (1976).

32. P.C.W. Davies and S.A. Fulling, Proc. Roy. Soc. London A 356 : 237 (1977).

33. D.J. Toms, The importance of quantum effects in Kaluza-Klein theory, Can. J. Phys. 64 : to appear (1986).

TEMPERATURE CORRECTIONS TO THE CASIMIR EFFECT

Günter Plunien, Berndt Müller and Walter Greiner

Institut für Theoretische Physik
Johann Wolfgang Goethe-Universität, Postfach 111932
D-6000 Frankfurt/Main, West Germany

INTRODUCTION

The existence of the Casimir effect[1], i.e. the attraction between neutral, perfectly conducting parallel plates, has demonstrated that the change of the zero-point energy of the electromagnetic field due to the presence of external constraints represents a physical and measurable quantity[2-5]. Stimulated by Casimir's early investigations[1,6-8] various approaches have been made on calculating Casimir energies of other quantum fields under constraints (for a review see e.g.[9]).

Accordingly the physical vacuum energy of a quantum field is defined as a difference between the zero-point energies in the presence and absence of external boundaries. Treated as a function of a suitable chosen parameter λ specifying the external constraints this implies the definition:

$$E_C(\lambda) = \left[E_o(\lambda) - E_o(\lambda_o)\right]_{reg} \quad , \tag{1}$$

where the subscript "reg" indicates that an appropriate regularization is required to obtain a finite result for the Casimir energy.

Under normal conditions fields are far from being in the vacuum state and, therefore, beside the virtual quantum fluctuations also real quanta are excited forming a quantum statistical ensemble interacting with external constraints. The free energy of a given ensemble "quantum field plus external constraints", which is assumed to be in thermodynamic equilibrium, allows for a consideration of finite temperature corrections to the Casimir effect:

$$F_C(\lambda,T) = \left[F(\lambda,T) - F(\lambda_o,T)\right]_{reg} \quad . \tag{2}$$

Since the total partition function Z of the constrained, but otherwise non-interacting system of particles factorizes into a part Z_o containing the zero-point oscillations and $\tilde{Z}$ referring to real occupied states forming the ensemble, i.e. $Z = Z_o\,\tilde{Z}$, the Casimir free energy appears as sum of the Casimir energy at zero temperature and a thermal part $\tilde{F}_C$ which is the difference between normal free energies:

$$F_C(\lambda,T) = E_C(\lambda) + \tilde{F}_C(\lambda,T) \ . \tag{3}$$

The following treatment of Casimir free energies is based on the representation in terms of infinite mode summations. According to this method one considers a given quantum field in the presence of boundaries enclosed by a large but finite quantization volume. Correspondingly the complete eigenmode spectrum, i.e. all eigenfrequencies $\omega_\alpha(\lambda)$ of the constrained respectively those $\omega_\alpha(\lambda_o)$ of the free system must be known. In general their analytical dependence on the geometrical parameter λ is unknown and has to be calculated numerically.

INTEGRAL REPRESENTATIONS FOR CASIMIR FREE ENERGY

We have derived the integral representations for Casimir free energies for ensembles of constrained non-interacting bosons and fermions at thermodynamic equilibrium. As the convenient quantity we introduce a so-called regularized mode sum function of the constrained ensemble depending exclusively on the geometry of the boundary.

Massless bosons

In thermodynamic equilibrium a system of non-interacting bosons may be characterized by means of the canonical partition function

$$Z = \mathrm{Tr}\left[e^{-\beta\hat{H}}\right] , \qquad \beta = \frac{1}{T} , \tag{4}$$

which leads to the explicit factorized form when evaluating the trace:

$$Z(\beta,\lambda) = \left[\prod_\alpha e^{-\beta\omega_\alpha/2}\right]\left[\prod_\alpha (1-e^{-\beta\omega_\alpha})^{-1}\right] \equiv Z_o(\beta,\lambda)\tilde{Z}(\beta,\lambda). \tag{5}$$

The free energy thus is the sum of the zero-point energy E_o and a thermal contribution $\tilde{F}$:

$$\begin{aligned} F(\lambda,T) &= E_o(\lambda) + \tilde{F}(\lambda,T) \\ &= \sum_\alpha \frac{\omega_\alpha}{2} + T \sum_\alpha \ell n\left(1-e^{-\omega_\alpha/T}\right) \ . \end{aligned} \tag{6}$$

This yields the following parameter integral (the parameter s is the euclidian proper time [10,11])for the free energy:

$$F(\lambda,T) = -\frac{1}{2}\int_0^\infty \frac{ds}{s}(4\pi s)^{-\frac{1}{2}} \vartheta_3\left(0\Big|\frac{i}{\pi s(2T)^2}\right)\Omega(s,\lambda) \quad , \tag{7}$$

where we have introduced the abbreviation

$$\Omega(s,\lambda) = \sum_\alpha e^{-s\omega_\alpha^2} \quad , \tag{8}$$

which we call <u>mode sum function</u> together with the Jacobi theta function[14]

$$\vartheta_3(0|it) = \sum_{m=-\infty}^{\infty} e^{-\pi m^2 t} \ . \tag{9}$$

Expressions of this type have been used in connection with thermal effects in external gauge fields[13,14]. Equation (7) greatly facilitates the definition of the Casimir free energy because the regularization procedure in the sense of eqs. (1) and (2) need only be applied to the mode sum function Ω. The rest of the integrand does not depend on the boundary conditions. Substituting $s = \lambda^2/\tau$ we may further write:

$$F_C(\lambda,T) = -\frac{1}{2\lambda}\int_0^\infty d\tau(4\pi\tau)^{-\frac{1}{2}}\vartheta_3\left(o\left|\frac{i\tau}{\pi(2\lambda T)^2}\right.\right)\Omega_{reg}(\frac{\lambda^2}{\tau},\lambda) \quad , \tag{10a}$$

together with the regularization mode sum function

$$\Omega_{reg}(\frac{\lambda^2}{\tau},\lambda) = \left[\Omega(\frac{\lambda^2}{\tau},\lambda) - \Omega(\frac{\lambda_o^2}{\tau},\lambda_o)\right]_{reg} \quad . \tag{10b}$$

One may assume that regularization can be performed in such a way that the representation of the Casimir energy at T = o, i.e.:

$$E_C(\lambda) = -\frac{1}{2\lambda}\int_0^\infty d\tau(4\pi\tau)^{-\frac{1}{2}}\Omega_{reg}(\frac{\lambda^2}{\tau},\lambda) \tag{11}$$

leads to a finite result. The representation (10a) expresses that temperature corrections can be derived by considering terms $|m|\geq 1$ of the series expansion of the Jacobi theta function. It is the convenient starting point for the derivation of the low-temperature expansion of the Casimir free energy. In order to evaluate the high-temperature limit one can use a general property of the Jacobi theta function[12] which relates reciprocal arguments

$$\vartheta_3(o|it) = \frac{1}{\sqrt{t}}\,\vartheta_3(o|\frac{i}{t}) \quad . \tag{12}$$

Applying this transformation to eq. (10a) one obtains the Casimir free energy in a form which allows an evaluation for the limiting case $\lambda T>>1$:

$$F_C(\lambda,T) = -\frac{T}{2}\int_0^\infty \frac{d\tau}{\tau}\,\vartheta_3\left(o\left|i\frac{\pi(2\lambda T)^2}{\tau}\right.\right)\Omega_{reg}(\frac{\lambda^2}{\tau},\lambda) \quad . \tag{13}$$

We have thus found general integral representations for the Casimir free energy of constrained boson fields. The formal advantage is that temperature and the functional dependence on the boundaries are carried separately. A particular ensemble is completely defined by the regularized mode sum function which if known allows the evaluation of the temperature corrections.

Constrained fermions

Considering a constrained system of non-interacting fermions in thermodynamic equilibrium at a fixed energy and total charge the corresponding partition function reads

$$Z = \mathrm{Tr}\ e^{-\beta(\hat{H}-\mu'\hat{Q})} = \mathrm{Tr}\ e^{-\beta(\hat{H}-\mu\hat{N})} \quad , \tag{14}$$

with the chemical potential $\mu = \mu' q_o$ where q_o denotes the elementary charge of the fermion. By means of standard representations for the Hamiltonian $\hat{H}$

and the particle number operator $\hat{N}$ in terms of single particle creation- and annihilation operators the trace in eq. (14) can be evaluated leading to the total partition function.

$$Z = \left[\prod_p e^{\frac{\beta}{2}(\omega_p-\mu)}\right]\left[\prod_p (1+e^{-\beta(\omega_p-\mu)})\right]\left[\prod_{\bar{p}} e^{\frac{\beta}{2}(\omega_{\bar{p}}+\mu)}\right]\left[\prod_{\bar{p}}(1+e^{-\beta(\omega_{\bar{p}}+\mu)})\right]$$

$$\equiv Z_o^p \, \tilde{Z}^p \, Z_o^{\bar{p}} \, \tilde{Z}^{\bar{p}} \; . \tag{15}$$

The partition function factorizes again into parts $\tilde{Z}_p^P$ and $\tilde{Z}_{\bar{p}}^{\bar{P}}$ referring to real particles and the corresponding contributions Z_o^P and $Z_o^{\bar{P}}$ due to the quantum fluctuations. Performing similar manipulations as in the boson case one can derive an integral representation for the free energy. For particles one obtains the contribution

$$F^P(\lambda,T) = \frac{1}{2}\int_0^\infty \frac{ds}{s}(4\pi s)^{-\frac{1}{2}} \, \vartheta_4\left(0\left|\frac{i}{\pi s(2T)^2}\right.\right) \Omega^P(s,\lambda,\mu) \quad , \tag{16}$$

where the corresponding mode sum function is defined as

$$\Omega^P(s,\lambda,\mu) = \sum_p e^{-s(\omega_p-\mu)^2} \; . \tag{17}$$

The temperature dependence is carried by the Jacobi theta function

$$\vartheta_4(0|it) = \sum_{m=-\infty}^{\infty} (-1)^m \, e^{-\pi m^2 t} = \vartheta_3(\tfrac{\pi}{2}|it) \quad . \tag{18}$$

The free energy contribution $F^{\bar{P}}$ referring to antiparticle modes takes the same form as (16) only with the slight change to $(\omega_{\bar{p}}+\mu)$ in the mode sum function. Assuming that regularization is achieved by introducing appropriate regularized mode sum functions the Casimir free energy of constrained fermions follows as:

$$F_C(\lambda,T) = \frac{1}{2}\int_0^\infty \frac{ds}{s}(4\pi s)^{-\frac{1}{2}}\vartheta_3\left(\frac{\pi}{2}\left|\frac{i}{\pi s(2T)^2}\right.\right)\left[\Omega^P_{reg}(s,\lambda,\mu)+\Omega^{\bar{P}}_{reg}(s,\lambda,\mu)\right] \; . \tag{19}$$

This result reveals that the Casimir free energy of constrained bosons and fermions can be treated in a unique manner. Again the separation of temperature and geometry dependence is possible.

Notice that the difference between bosons and fermions due to their statistics is reflected in a modified temperature dependence of the Casimir free energy. In view of general property of the Jacobi theta function, $\vartheta_3(u|q) = \vartheta_3(u+\pi|q)$, the thermal factor is given by $\vartheta_3\left(\sigma\pi\left|\frac{i}{\pi s(2T)^2}\right.\right)$ for particles with arbitrary spin σ.

EXAMPLE: TEMPERATURE CORRECTIONS TO THE CASIMIR EFFECT IN d DIMENSIONS

As an example for a successful application of the integral representation we now calculate the Casimir free energy for the d dimensional version of the conducting parallel plate configuration . The continuation in dimensions is <u>not</u> used as regularization method.

Consider the electromagnetic field inside a large but finite cubic cavity of volume L^d bounded by perfectly conducting walls in which

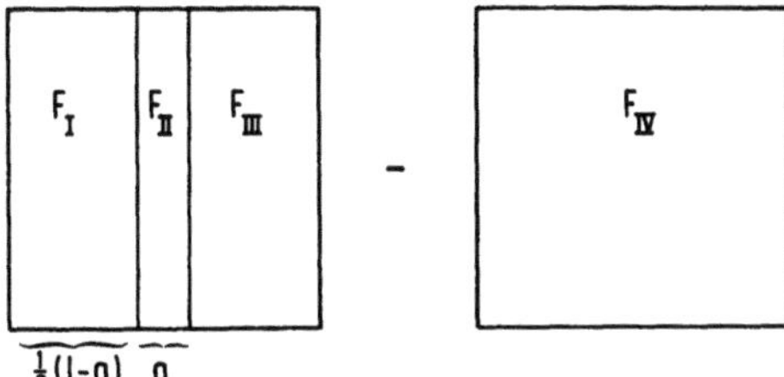

Fig. 5.1. Subtraction of the free energies in the case of the d dimensional conducting parallel plate configuration.

two conducting square planes of length L are placed parallel to each other at an adjustable distance a (see fig. 5.1). According to the method of mode summation the Casimir free energy can be defined by the following energy subtraction:

$$F_C(a,T,d) = 2F_I\left((L-a)/2,T,d\right) + F_{II}(a,T,d)-F_{IV}(L,T,d) \quad , \tag{20}$$

assuming L>>a. The eigenfrequencies of the constrained electromagnetic field are determined by the boundary conditions on perfect conductors. In region II one has:

$$\omega = \left(\vec{k}_{\parallel}^2 + \vec{k}_{\perp}^2\right)^{\frac{1}{2}} , \quad \vec{k}_{\parallel} = (k_1,\ldots,k_{d-1},0) , \quad \vec{k}_{\perp} = (0,\ldots,0,k_d=\frac{n\pi}{a}) \quad , \tag{21}$$

where the components of the parallel projection $\vec{k}_{\parallel}$ are treated as continuous variables while the component k_d perpendicular to the plates takes discrete values only. In the other regions the eigenfrequencies are obtained from eq. (21) by replacing the separation a by (L-a)/2 or L respectively, and replacing summation over n by integration. Writing down explicitly the Casimir free energy in accordance to eq. (20) one identifies the regularized mode sum function:

$$\Omega_{reg}(s,a,d) = 2\left(\frac{L}{2\pi}\right)^{d-1} \int d^{d-1}k_{\parallel} \left\{ \sum_{n=0}^{\infty}{}' \exp\left\{-s[k_{\parallel}^2 + (\frac{n\pi}{a})^2]\right\} \right.$$

$$\left. - \int_0^{\infty} dn \exp\left\{-s[k_{\parallel}^2 + (\frac{n\pi}{a})^2]\right\}\right\} . \tag{22}$$

The prime on the summation sign indicates that the term n=o must be multiplied by 1/2 (only one polarization). This expression can be analytically calculated by means of Poisson's sum formula. Substituting $s = a^2/_\tau$ the result is found to be

$$\Omega_{reg}\left(\frac{a^2}{\tau},a,d\right) = -\left(\frac{L}{2}\right)^{d-1}\left(\frac{\tau}{\pi}\right)^{\frac{d}{2}}\frac{2}{a^{d-1}}\sum_{n=1}^{\infty} e^{-n^2\tau} \quad . \tag{23}$$

Now we are in a position to calculate the Casimir energy by means of the representations eq. (10a) or (13) in terms of the regularized mode sum function. Inserting (23) into eq. (10a) one obtains, after performing the parameter integration and summations:

$$F_C(a,T,d) = -\frac{L^{d-1}\Gamma\left(\frac{d+1}{2}\right)}{2^d\pi^{(d+1)/2}a^d}\left[\zeta(d+1)+S_{(d+1)/2}\left(\frac{1}{2aT}\right)\right] , \tag{24a}$$

where the dependence on temperature is carried by the function

$$S_{(d+1)/2}\left(\frac{1}{x}\right) = 2\sum_{m=1}^{\infty}\sum_{n=1}^{\infty}\left[\left(\frac{m}{x}\right)^2+n^2\right]^{-(d+1)/2} , \quad x = 2aT$$

$$= \frac{\pi^{1/2}\Gamma\left(\frac{d}{2}\right)}{\Gamma\left(\frac{d+1}{2}\right)}\zeta(d)\,x^d - \zeta(d+1)x^{d+1}$$

$$+ \frac{4\pi^{(d+1)/2}}{\Gamma\left(\frac{d+1}{2}\right)}\sum_{m=1}^{\infty}\sum_{n=1}^{\infty}\left(\frac{nx}{m}\right)^{\frac{d}{2}} K_{d/2}(2\pi nm/x) \quad . \tag{24b}$$

The first term in eq. (24a) is the Casimir energy at zero temperature. It coincides with the result found by Ambjørn and Wolfram[15] obtained by dimensional regularization. If one is interested in the low-temperature limit, i.e. (aT)<<1, then it is legitimate to consider only the term $n = m = 1$, and the modified Bessel function $K_{d/2}$ can be replaced by its asymptotic expansion. In this limiting case the Casimir free energy reads:

$$F_C(a,T,d) = -\frac{L^{d-1}\Gamma\left(\frac{d+1}{2}\right)\zeta(d+1)}{2^d\pi^{(d+1)/2}\,a^d} - \frac{L^{d-1}\Gamma\left(\frac{d}{2}\right)T^d}{\pi^{d/2}}$$

$$+ 2\,\frac{L^{d-1}a}{\pi^{(d+1)/2}}\,\Gamma\left(\frac{d+1}{2}\right)\zeta(d+1)T^{d+1}$$

$$- \frac{L^{d-1}T^{(d+1)/2}}{2^{(d+3)/2}a^{(d-1)/2}}\,e^{-\frac{\pi}{aT}}\left[1+\sum_{k=1}^{\infty}\frac{\Gamma\left(\frac{d+1}{2}+k\right)\left(\frac{aT}{2\pi}\right)^k}{\Gamma\left(\frac{d+1}{2}-k\right)\Gamma(k+1)}\right]+O\left(e^{-\frac{2\pi}{aT}}\right) . \tag{25}$$

As low-temperature corrections to the usual Casimir energy beside a term which is independent from the separation (it will not contribute to the Casimir force $\mathcal{F}_c = -L^{-d+1}(\partial F_c/\partial a)$) there occurs a generalized Stefan Boltzmann term. The next terms vanish exponentially. Notice, that for an odd number of dimension d the asymptotic expansion for the modified Bessel function terminates if $k = (d-1)/2$. For d=3 the result for the force given by Mehra[16] is recovered.

Let us now turn to the high-temperature expansion of the Casimir free energy. The most direct and easiest way to derive it is to start from the result eq. (24a). Making use of the obvious property of the function $S_{(d+1)/2}$, i.e.:

$$S_{(d+1)/2}\left(\frac{1}{x}\right) = x^{d+1}\, S_{(d+1)/2}(x) \quad , \quad x = 2aT \quad , \tag{26}$$

we obtain directly the expression

$$F_C(a,T,d) = -\left(\frac{L}{2}\right)^{d-1} \frac{\Gamma\left(\frac{d}{2}\right)\zeta(d)}{\pi^{d/2}a^{d-1}}\, T$$

$$- \left(\frac{L}{2}\right)^{d-1} \frac{2T}{\pi^{d/2}\, a^{d-1}} \sum_{m=1}^{\infty} \sum_{n=1}^{\infty} 2\left(2\pi\frac{m}{n}aT\right)^{\frac{d}{2}} K_{d/2}(4\pi mnaT) \quad .(27)$$

The corresponding asymptotic expansion for the limiting case $aT \gg 1$ is found to be

$$F_C(a,T,d) = -\left(\frac{L}{2}\right)^{d-1} \frac{T}{\pi^{d/2}\, a^{d-1}} \left\{\Gamma\left(\frac{d}{2}\right)\zeta(d) + 2\pi^{\frac{d}{2}}(2aT)^{(d-1)/2}\, e^{-4\pi aT} \cdot\right.$$

$$\left.\left[1 + \sum_{k=1}^{\infty} \frac{\Gamma\left(\frac{d+1}{2} + k\right)(8\pi aT)^{-k}}{\Gamma\left(\frac{d+1}{2} - k\right)\Gamma(k+1)}\right]\right\} + O\left(e^{-8\pi aT}\right) \quad .(28)$$

The leading term depends linearly on T while all other corrections are exponentially small. For d = 3 the corresponding Casimir force coincides again with the result found by Mehra[16].

CONCLUSIONS

The Casimir free energy of constrained bosons and fermions can be calculated in a unique manner by means of an integral representation. It has the formal advantage that the dependence on the boundary and on temperature enter separately. A constrained system is characterized exclusively by the corresponding regularized mode sum function. As we have illustrated for the Casimir free energy of the electromagnetic field between perfectly conducting parallel plates in an arbitrary number of dimensions, the calculation is straightforward and not plagued with further divergencies.

REFERENCES

1. H. B. G. Casimir, Proc. Kon. Ned. Akad. Wet. 51:793 (1948).
2. B. V. Deriagin and I. I. Abrikosova, Sov. Phys. JEPT 3:819 (1957).
3. M. J. Sparnaay, Physica 24:751 (1958).

4. A. van Silfhout, Proc. Kon. Ned. Akad. Wet. B69:501 (1966).
5. W. Arnold, S. Hunklinger, and K. Dransfeld, Phys. Rev. B19:6049 (1979).
6. H. B. G. Casimir, J. Chim. Phys. 46:407 (1949).
7. H. B. G. Casimir, Philips Rev. Rept. 6:167 (1951).
8. H. B. G. Casimir, Physica 19:846 (1953).
9. G. Plunien, B. Müller, and W. Greiner, Phys. Rep. 134:87 (1986).
10. V. Fock, Physik. Z. Sowjetunion 12:404 (1937).
11. J. Schwinger, Phys. Rev. 82:664 (1951).
12. I. S. Gradshteyn and I. M. Ryzhik, Table of Integrals, Series and Products (Academic Press, New York, 1980).
13. W. Dittrich, Phys. Lett. B83:67 (1979).
14. B. Müller and J. Rafelski, Phys. Lett. B101:111 (1981).
15. J. Ambjorn and S. Wolfram, Ann. Phys. 147:1 (1983).
16. J. Mehra, Physica 37:145 (1967).

THE FUTURE OF NUCLEAR PHYSICS

D. Allan Bromley

Henry Ford II Professor of Physics
Yale University
New Haven, Connecticut 06511 USA

ABSTRACT

Nuclear physics is on the threshold of new and exciting developments stemming both from new understanding of the nuclear many-body system and new higher-energy probes—electrons, kaons and heavy-ions—that are becoming available or on the horizon. The discovery of new dynamic symmetries and the parallel establishment of an algebraic, rather than a geometric, approach has revolutionized our understanding of nuclear structure and promises a similar inpact on nuclear dynamics. The response of the nuclear system to ever increasing angular momentum and temperature is being mapped out with characteristic gross changes in shape and behavior. Higher-energy electron beams are probing the nuclear interior—long an ignored domain—and establishing the roles of meson currents and the validity of shell model concepts deep in the nuclear interior. They will soon be able to map the transition from nucleonic to quark matter within the nucleus—the appearance of nucleon isobars and their melting with the beginning of quark deconfinement. Current lattice gauge calculations predict the deconfinement phase transition at a temperature of about 200 MeV. Relativistic heavy-ion studies have already demonstrated that densities several times normal are achieved in central collisions; the next generation of experiments in this field holds promise of probing the quark-gluon plasma and of giving access to the equation of state of nuclear matter. Quantum chromodynamics may provide the key to understanding of the strong nuclear interaction at a fundamental level—perhaps equivalent to what has already been achieved in quantum electrodynamics—as new phenomena emerge at relativistic energies. But at the same time, it bears emphasis that radioactivity--double beta decay, searches for finite neutrino mass and for neutron-antineutron oscillations, for example—and studies with low-energy Van de Graaff and cyclotron accelerators continue to provide data that are critical to our understanding of nuclear phenomena. As our microscopic knowledge of nuclear structure improves, the nucleus becomes an ever more powerful microscopic laboratory wherein some of the most fundamental symmetries and interactions in nature can be subjected to quantitative tests. All contribute to the fabric of nuclear science and to the rich interfaces between nuclear and its sister sciences.

INTRODUCTION

In any field of activity, it is essential from time to time to stand back for a longer look, for better perspective. I have been asked to address that task with respect to nuclear physics. Where is it going? What does its future look like? Does it, indeed, have a future?

To this latter question, my answer must be an emphatic "Yes!" Indeed, as I shall try to illustrate in this paper, we are at a particularly interesting and promising time in the history of our science. In a very real sense, we are now in the 55th year of nuclear physics, dating back to 1932, that remarkable year with its discovery of the neutron, the deuteron, the positron, isospin, the invention of the cyclotron and the Van de Graaff; and the experimental proof of general relativity. In these 55 years, we have made incredible discoveries, and our understanding has increased at least exponentially.

But despite this progress, we tend to forget that our studies thus far have been limited: to cold nuclei near their ground states; to low-spin nuclei where centrifugal effects are negligible; to species relatively near to stability; to the surface of nuclei to particle and collective degrees of freedom involving these surfaces; and to consideration of nuclei as nucleon complexes only.

But all of this is changing rapidly: as measurements to ever-higher precision make first-order theories and understanding unacceptable; as new dynamic symmetries and perhaps supersymmetries are identified in both nuclear structure and nuclear dynamics; as new, higher-energy electron accelerators illuminate the nuclear interior, mesonic effects, nucleon isobars, and take us at least to the threshold of quark deconfinement; as new, higher-energy, heavy-ion machines illuminate high spin phenomena, take us to and beyond the limits of nuclear stability, explore the nuclear equation of state, deconfine quarks and restore chiral symmetry, recreate the earliest moments of creation, and probe the final moments of gravitational collapse; as higher-energy proton machines are used increasingly as pion and particularly kaon factories to study both nuclear structure and dynamics and the physics of hypernuclei; and as the relationship of quantum chromodynamics to the nuclear many-body system is more fully understood.

In what follows, I shall attempt to illustrate all these statements; inasmuch as I shall be spanning the entire field of nuclear physics, it will simply not be possible for me to give detailed references. The choice of illustrations will inevitably be a very personal one, and I make no claim whatever to completeness; and my examples—particularly of new facilities—will be taken primarily from U.S. experience, since I happen to be particularly familiar with it.

HEATING THE NUCLEAR SYSTEM

One of the problems of longest standing in nuclear physics is that of understanding what happens as energy and angular momentum are added to the nuclear many-body system. This general behavior is illustrated in Fig. 1 where I take as an example the 56-body ^{56}Ni system and illustrate schematically the gross changes that are expected with increasing temperature and spin. Close to the ground state, the valence nucleons and surface collective modes absorb the added energy, but with further heating, the nucleus becomes an oblate spheriod and spins collectively; at still higher temperatures the nuclear shape changes abruptly to that of a prolate spheroid, and in the case of ^{56}Ni, we now know that even up to excitation energies of some 70 MeV, very sharp quantum states are superimposed on a dense continuum. Some of these states have been shown to correspond to a spontaneous separation of the 56 nucleons into two identical ^{28}Si subnuclei which engage in simple molecular motion relative to one another. Other states, only a few hundred keV different in total energy, have been shown to correspond to a spontaneous separation into ^{16}O and ^{40}Ca clusters, and although they have not been specifically identified, it seems obvious that there are many other binary and more complex cluster structures that emerge spontaneously. No one knows how far these sharp cluster or molecular states extend into the nuclear continuum; all that can be said at this point is that wherever probes have been available with adequate resolution to see these sharp states, the states have appeared.

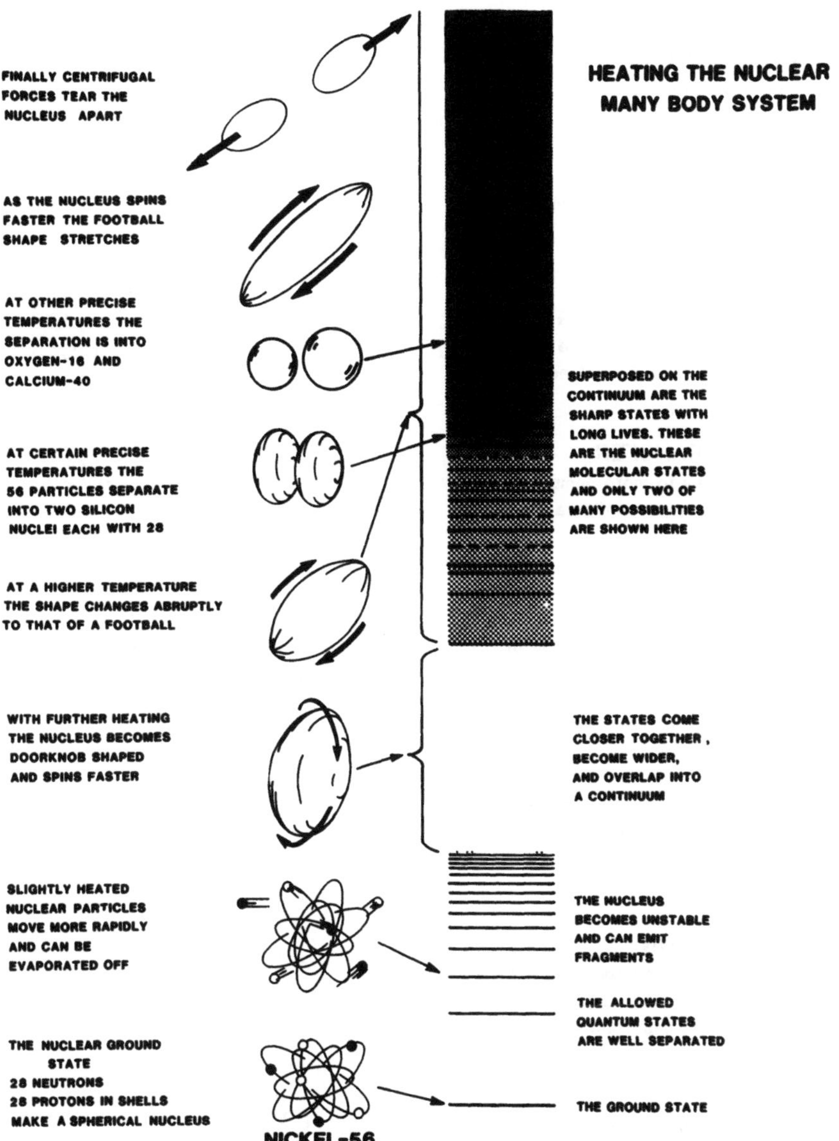

Figure 1: A schematic illustration of the change in nuclear structure with increasing excitation energy and angular momentum for the 56 body system ^{56}Ni. The increased shading with excitation energy is intended to represent the exponentially increasing level density with excitation, and the sketches to the left of the level diagram are highly schematic representations of gross structural features.

As the nucleus spins faster, the prolate spheroid stretches and eventually, at some critical angular momentum, the nucleus spontaneously disintegrates under centrifugal forces. In the case of ^{56}Ni, simple calculations suggest that this happens at a total angular momentum of roughly 50$\hbar$ which is only some 6 units beyond spin values that have now been identified unambiguously in this system.

While Fig. 1 focuses on the structure of nuclei, Fig. 2 provides a corresponding overview of nuclear interactions and dynamics ranging from the most simple to the most complex. Elastic scattering, governed by the Coulomb interaction, is of course the simplest possible interaction in that it involves no nuclear degrees of freedom at all. This is followed by nucleon transfer, inelastic scattering and cluster transfer. More complex is quasifusion and deep inelastic

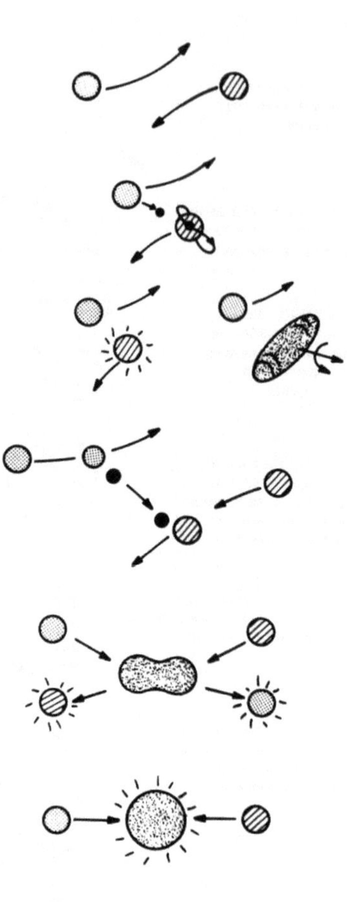

ELASTIC SCATTERING
no internal excitation

NUCLEON TRANSFER
one or at most a few degrees of freedom involved

INELASTIC SCATTERING
rotations, vibrations and surface collective modes involved

CLUSTER TRANSFER
opens way to more complex molecular type configurations

QUASIFUSION DEEP INELASTIC
involves massive exchange of mass and energy but far from thermodynamic equilibrium. Friction enters the quantum domain

FUSION COMPOUND NUCLEUS
approaches thermodynamic equilibrium, memory of formation lost

FUSION–FISSION
fission following equalibration but increasing evidence for fission before full fusion , important intermediate processes

HIGH ENERGY FIREBALL
projectile gouges a hot section out of an otherwise spectator remnant

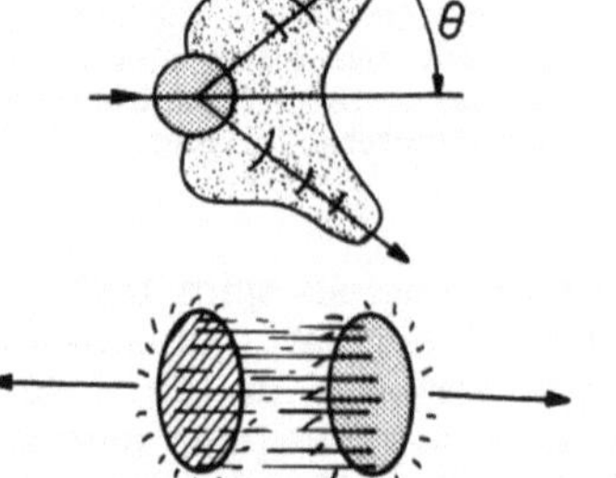

HIGHER ENERGY COMPRESSION
shock phenomena lead to characteristic sidesplash or Mach angle emission and to densities substantially greater than normal . Possible new forms of matter

ULTRA HIGH ENERGY QUARK–GLUON PLASMA
projectile and target pass completely through one another leaving baryon free central firetube:
quark deconfinement, quark gluon plasma

Figure 2: An overview of nuclear interactions ranging from the simplest—elastic scattering—to the most complex at ultrahigh energies. Again the figures are very highly schematic representations of the interactions involved.

scattering where substantial exchange of mass and energy takes place; but the system still remains far from thermodynamic equilibrium. This, too, is the regime where the last of the classical parameters—friction—enters the quantum domain. Next comes complete fusion and the formation of a compound nucleus where it is still assumed that the famous Bohr hypothesis of complete thermodynamic equilibration holds. There is, however, increasing evidence for significant reaction flux via a mechanism which approaches fusion but does not complete it, and wherein, after significant exchange of energy and matter, the projectile and target separate again in a quasifission process.

In moving to substantially higher energies, there is convincing evidence that, particularly in peripheral interactions, the projectile shears a sector out of the target, heating it to very high temperatures in the process with subsequent particle evaporation and emission from the hot fragment while the remainder of the target plays a passive, spectator role. At even higher energies, recent measurements which have separated out central collisions from the much more probable peripheral ones demonstrate clear evidence for hydrodynamic shock formation with characteristic shock flow angles and compression leading to densities several times that normal in nuclear matter. Finally, at ultrahigh energy, nuclei become transparent to one another, and in collisions at these energies, projectiles and target, can pass completely through one another, heating both to very high temperatures and leaving a central **fire tube** having essentially zero baryon number and deconfined quarks in an effective quark-gluon plasma—an entirely new form of matter.

Figure 3 represents a different and still instructive gross characterization of nuclear phenomena in terms of the nuclear size (proportional to $A^{1/3}$) and the

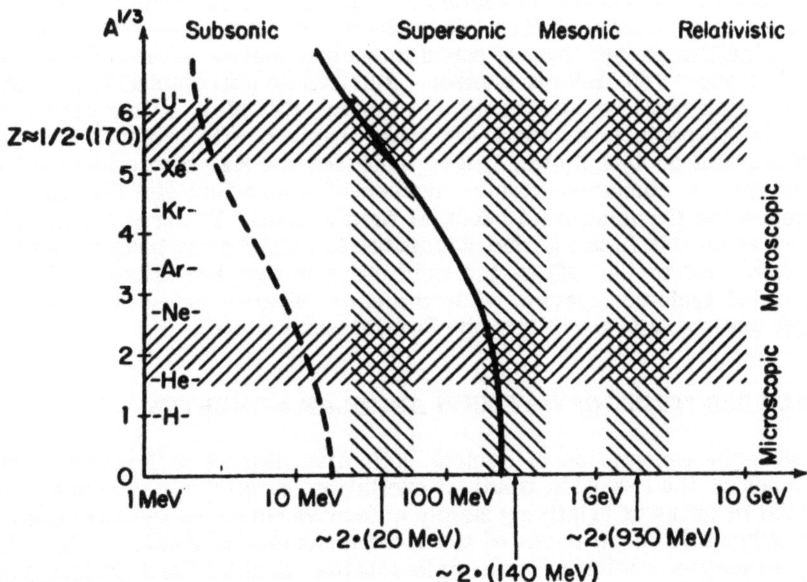

Figure 3: The nuclear energy regime with size plotted against available energy. The shaded bands are very roughly defined and intended only to separate crudely different behavioral regimes. All the area to the left of the dashed line is accessible to an Emperor tandem accelerator, while all that to the left of the solid black line is accessible to the Michigan State cyclotrons and the entire diagram below some 2 GeV per nucleon, is accessible to the Berkeley Bevalac. The upper horizontal shaded band simply indicates the size at which in identical particle collisions the Coulomb field would be expected to become supercritical.

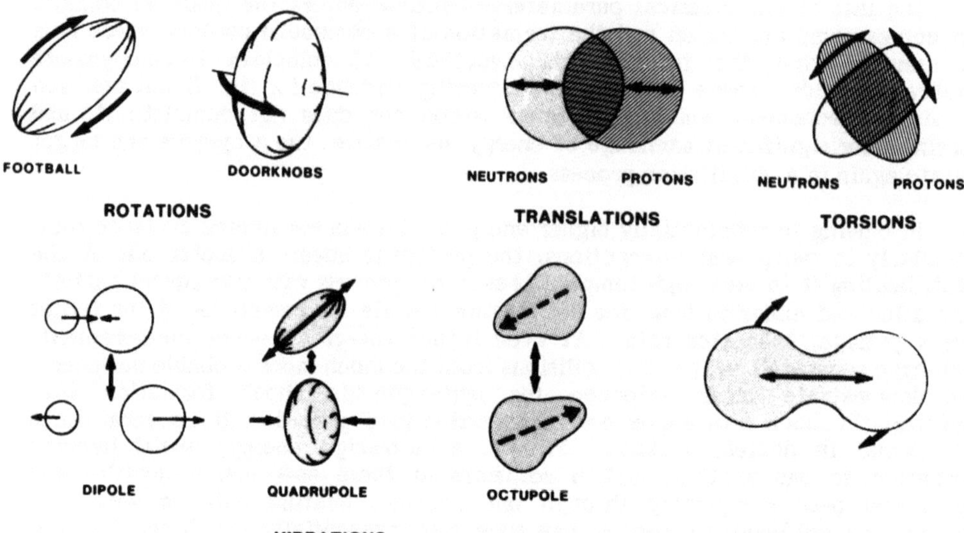

Figure 4: A schematic illustration of the primary varieties of nuclear collectivity. Rotations and vibrations are those familiar in the nuclear liquid drop model, the translational oscillation gives rise to the giant nuclear resonance while the torsional oscillation is the so-called scissors mode recently discovered in rare-earth nuclei. In molecular collectivity the participant nuclei retain their individual characteristics in large measure and undergo both vibrations and rotation much as in the atomic case.

available energy. The subsonic regime is that familiar in classical nuclear physics and extends up to roughly 20 MeV per nucleon, while the supersonic range in which nucleon velocities exceed that of sound in nuclear matter extends up to the meson threshold at about 140 MeV per nucleon. Relativistic phenomena become important above the rest mass of the nucleon at about 930 MeV. Considering the size variable, microscopic phenomena are those involved with projectiles lighter than say lithium, and on this figure, I have indicated the size where identical particle collisions give a combined atomic number of approximately 170 and where in consequence the Coulomb field becomes supercritical. The area to the left of the dashed curve on this figure is that accessible to an MP class tandem Van de Graaff accelerator; that to the left of the solid curve is that accessible to the Michigan State coupled cyclotron system, while the entire diagram below energies of roughly 2 GeV per nucleon is accessible to the Bevalac accelerator.

NUCLEAR SPECTROSCOPY AT HIGH ANGULAR MOMENTUM

With the recognition of nuclear collective degrees of freedom in the early 1950s, much of the low-lying quantum excitation of relatively cold nuclei has been understood in terms of relatively simple collective rotations and vibrations. Figure 4 shows schematic illustrations of the most important of these. Perhaps the most striking collective motion is the simple rotation of oblate and prolate quadrupole deformed nuclei. The giant resonance in nuclei has been recognized for an even longer time as a simple collective excitation wherein the neutron and proton fluids oscillate coherently in translation against one another, leading to separation of the centers of mass and charge, and therefore, to an electric dipole moment. Only within the past few years has it been recognized that in the case of deformed nuclei, the neutron and proton fluids can execute a torsional oscillation with respect to one another in the so-called **scissors** mode. While the giant resonance is found typically from 14 to 20 MeV in excitation, the scissors mode is found at about 3 MeV in the rare earth nuclei, for example. Molecular collective oscillations are those wherein the participant nuclei retain their intrinsic characteristics while

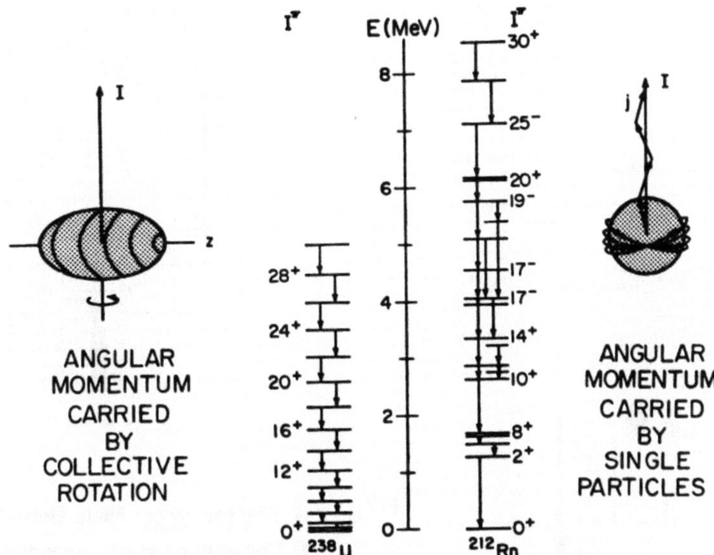

Figure 5a: A schematic illustration of the way in which the nuclear many-body system responds to increased angular momentum. Two quite distinct mechanisms are involved. On the left is the case where increased angular momentum is accommodated through increased angular velocity of the nucleus, rotating effectively as a rigid body, whereas on the right. increased angular momentum is accommodated by sequential realignment of individual particle orbits, so that their angular momentum can add coherently to the total of the system. The characteristic excitation spectra are illustrated for the two cases with those of ^{238}U and ^{212}Rn respectively.

executing standard molecular rotations and vibrations. Finally, vibrations of many multiple orders have been recognized with the quadrupole one being of greatest importance and longest standing; only within the past few years has it been discovered that collective dipole vibrations occur in the vicinity of closed shells where valence alpha particles and perhaps heavier fragments oscillate with respect to the remaining core nucleus. During this same period, there has also been growing evidence for the existence of both dynamic and static octupole deformations, the former—with the dipole collective configurations—corresponding to reflection asymmetric intrinsic shapes.

All of these collective excitations are now relatively well understood, both macroscopically and microscopically.

Particularly with the availability of precision heavy-ion accelerators, it has become possible to add very large amounts of angular momentum to the nuclear system without simultaneously adding so much energy that the system disintegrates spontaneously; this has made possible spectroscopy at very much higher angular momentum than was previously accessible. How then does the nuclear many-body system accommodate this increased angular momentum? As illustrated in Fig. 5a, there are two quite distinct mechanisms: in the first, the angular momentum is carried by collective rotation of the entire nucleus, resulting in a very regular and characteristic rotor spectrum as shown for the case of ^{238}U; the second is more complex in that the increasing angular momentum is carried by the sequential alignment of single particle orbitals so that their angular momentum adds to the nuclear total. This latter mechanism results in a very irregular excitation spectrum as illustrated here for the case of ^{212}Ra.

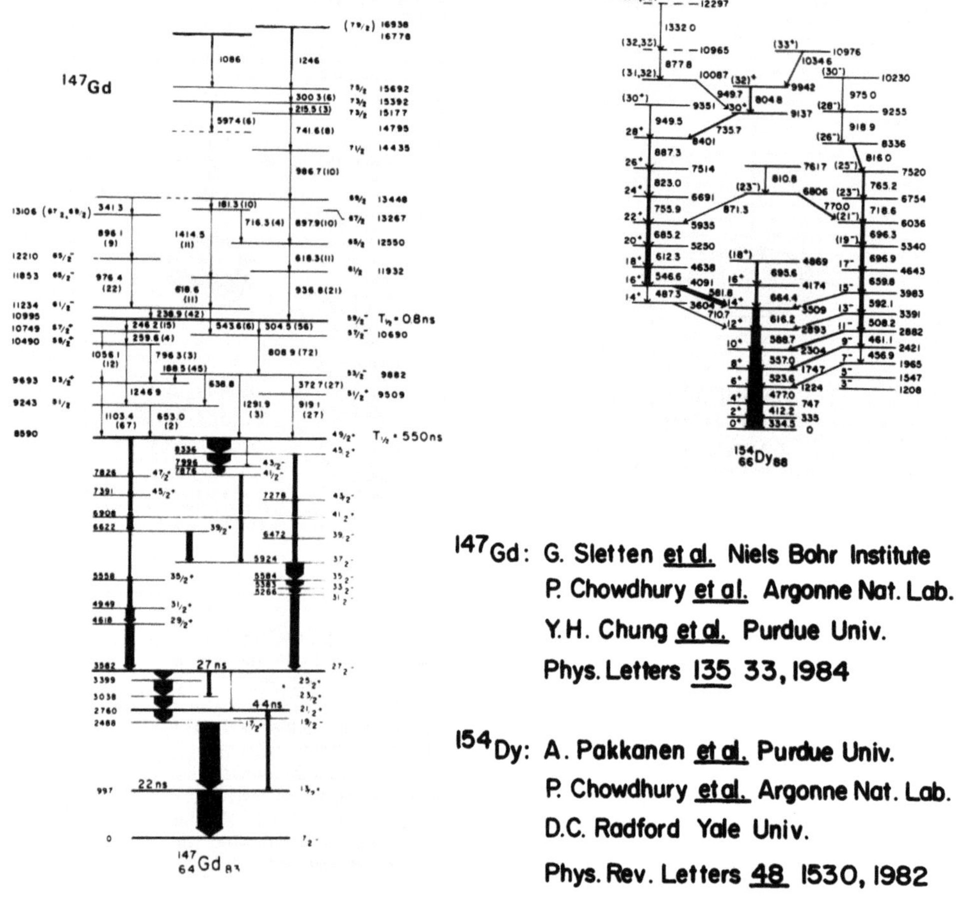

Figure 5b: An illustration of collective and single-particle structure in typical rare-earth nuclei. In the case of ^{147}Gd, the structure below the isomeric state at 8.5 MeV is purely collective, while that at higher excitation is typical of that resulting from alignment of single particle orbitals. In the case of ^{154}Dy, all of the indicated excitation spectrum is of collective origin excepting the states at 10.1 and 11 MeV deexcited by electric dipole transitions. These and the states above them are of aligned particle origin.

Figure 5b illustrates these two types of behavior in two well-studied rare-earth nuclei, ^{147}Gd and ^{154}Dy. In the case of ^{147}Gd, the structure is clearly collective up to the isomeric state at 8.5 MeV and dominated by a single-particle phenomenon at higher excitation energies. In the case of ^{154}Dy, the entire low-energy spectrum is clearly collective up to the states at 10.0 and 10.9 MeV where single-particle phenomena are first identified.

From measurements such as these, it has become possible to map out the general behavior expected as both neutrons and angular momenta are added to closed shell nuclei. In Fig. 6, this is illustrated for the case of the dysprosium isotopes; the shaded region is that which has thus far been studied in detail, and it is clear that a very large unknown region is still open for exploration. At low angular momentum, on adding neutrons to the magic N = 82 system, a range of vibrators is encountered, followed by deformed prolate rotors. As the angular momentum is increased, at low neutron numbers the higher angular momentum is encompassed by aligning valence particle orbitals with respect to a deformed oblate core. With larger neutron number, the particle orbits are aligned with respect to a static prolate deformation, and although there is no concrete experimental

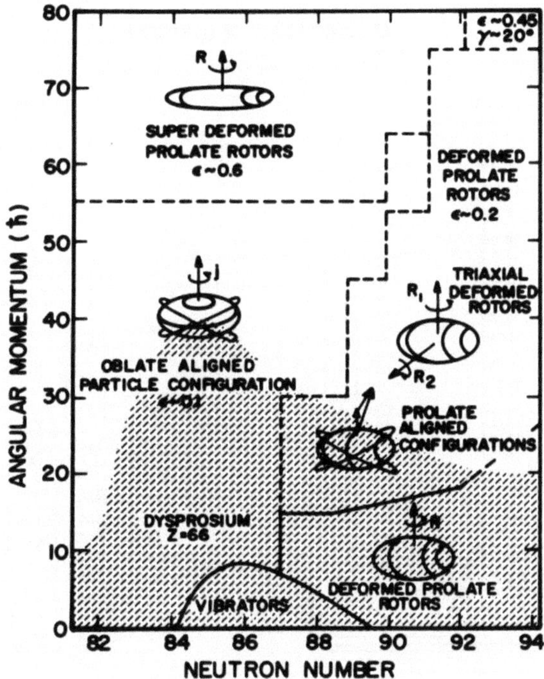

Figure 6: A schematic representation of the shapes of dysprosium isotopes as a function of angular momentum and neutron number. The shaded region is that studied experimentally thus far, while the unshaded region remains a matter of conjecture. The indicated dividing lines are extremely crude and based on simple model predictions.

evidence as yet, it is anticipated that, with increasing angular momentum and large neutron number, triaxial rotor configurations will be involved. Whatever the neutron number, at sufficiently high angular momentum centrifugal forces can be expected to lead to superdeformed prolate rotors, in general showing axial symmetry, except in the case of very large neutron number where there is the possibility of sustained triaxiality. An enormous amount of work remains to be done in mapping out this domain across the periodic table.

The availability of new generations of gamma-ray detectors, including the shielded bismuth germinate (BGO) units has in the recent past led to a renaissance in the study of discreet gamma-ray spectroscopy at high excitation and at high angular momentum. Figure 7 is typical of recent measurements on odd- and even-mass dysprosium—in this case, work from the Copenhagen group—and again illustrating striking collective behavior.

Figure 8, showing very recent work of Twin and his collaborators at the Daresbury Laboratory in England, is a magnificient technological **tour de force** in that for the first time it demonstrates very clearly discreet gamma-ray lines from states in ^{152}Dy at angular momenta up to 60ℏ, far beyond any previously observed. These spectra demonstrate the power of the BGO units when coupled with heavy-ion reactions induced by projectiles such as ^{48}Ca. Figure 9 plots the yrast states as determined from the spectra of Fig. 8 showing the crossing of the low deformation band by a superdeformed rotational band of much higher moment of inertia. These measurements open up an entirely new domain of study of nuclear structure where centrifugal forces more nearly balance the nuclear binding and make possible much more sensitive probing of the microscopic structure at high angular momentum than has previously been possible.

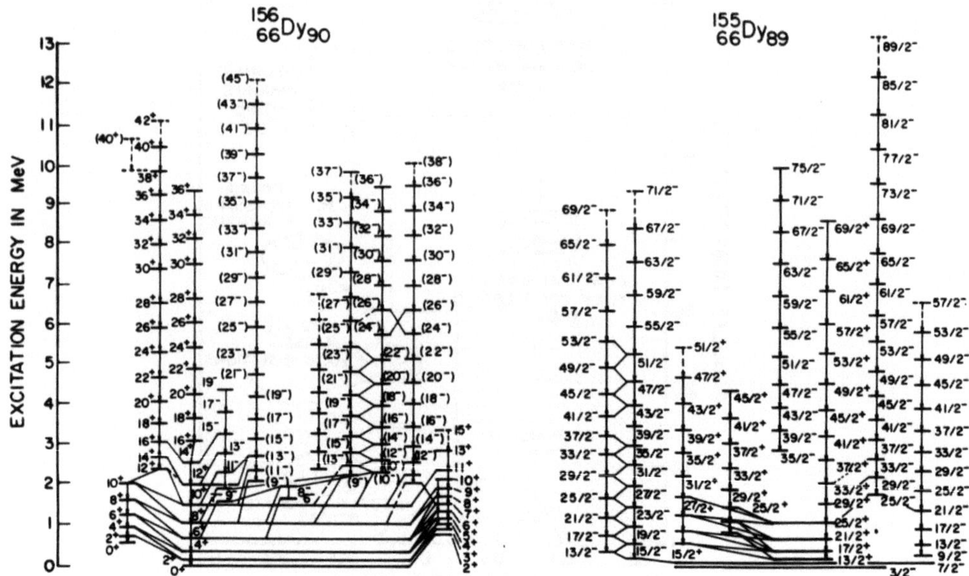

Figure 7: Typical modern spectroscopy of rare-earth nuclei. These data were obtained by a Copenhagen group working at the Daresbury National Laboratory in the United Kingdom and are typical of modern studies on even- and odd-mass, rare-earth nuclei, respectively.

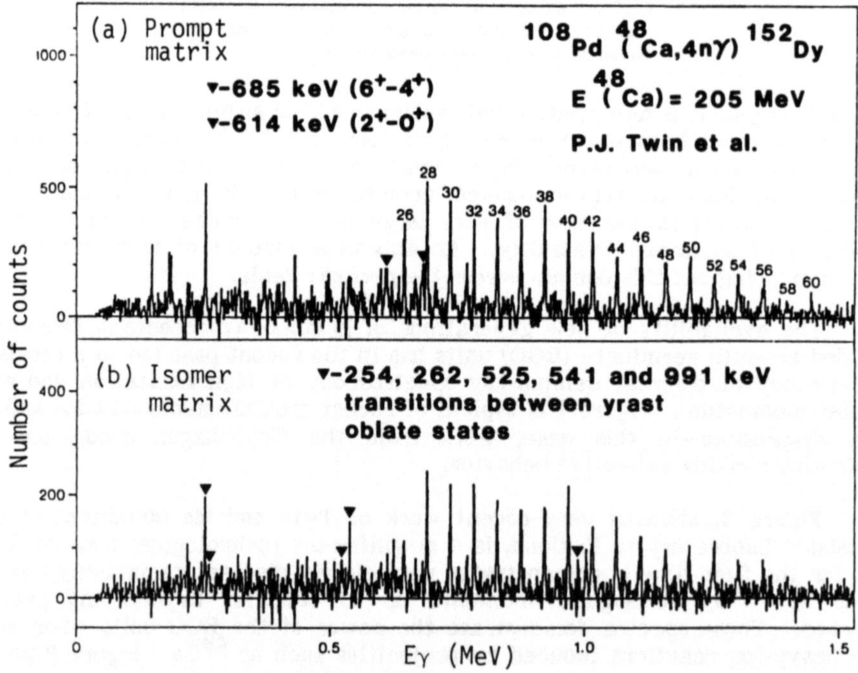

Figure 8: Very recent data from the Daresbury National Laboratory obtained using a large BGO gamma detection array and 205 MeV ^{48}Ca projectiles on ^{108}Pd targets. The numbers in the upper panel indicate the angular momentum, in units of ℏ, of the initial state in the transition. The transitions indicated by the inverted triangles are identified in the partial spectrum shown in Figure 9.

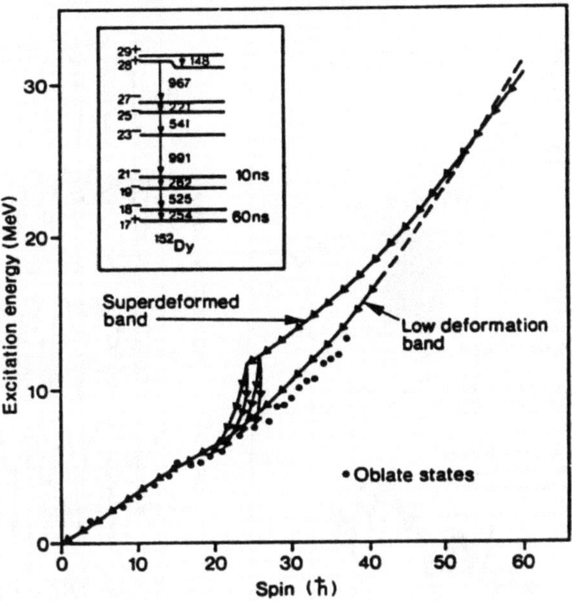

Figure 9: Yrast spectroscopy in ^{152}Dy. Both the normal low deformation band and the superdeformed (see Fig. 6) are shown, as are the oblate states along the yrast line. The transitions shown in the inset in this figure are identified by the inverted triangles in Fig. 8. These data represent, by a substantial margin, the highest angular momenta that have yet been identified unambiguously in nuclear physics.

Of particular importance here is the fact that recent theoretical calculations, using modern quantum chaos and information theory techniques, by Alhassid and his collaborators at Yale have predicted the occcurrence of phase transitions and tricritical points in the high spin spectrum of deformed nuclei. Such phenomenà, if they are indeed present, should be clearly discernible in discreet spectra similar to those shown in Fig. 8.

Up until these recent data of Twin **et al.** shown in Fig. 8, the record in nuclear physics for the highest unambiguously identified angular momenta was established in measurements on the elastic scattering of silicon on silicon, as mentioned above in conjunction with Fig. 1. Figure 10 shows, at the upper left, data of Betts **et al.** at Yale and at lower left a much higher resolution expansion of the shaded region taken with adequate resolution to show the sharp molecular states mentioned previously in the ^{56}Ni compound nucleus. The spins of these compound states were established from the elastic scattering angular distributions shown on the right of Fig. 10 where as each new partial wave becomes active the angular distribution can be represented almost precisely by the square of the corresponding Legendre polynomial. Interest here centers on the fact that spontaneous centrifugal disruption of the compound system would be expected at a total spin of 50 units which will be readily attainable with the larger electrostatic accelerators now just coming into operation. Fig. 11 again shows the sharp structure in the ^{56}Ni as well as in the ^{48}Cr compound system reached in the scattering of ^{24}Mg on ^{24}Mg as well as the apparent absence of such structure in both the ^{58}Ni and ^{60}Ni compound systems. On the right of this figure are shown the potential energy contours, as calculated by Leander **et al.**, for the L = 40 partial wave for these four compound systems, and it is at least suggestive that in the case

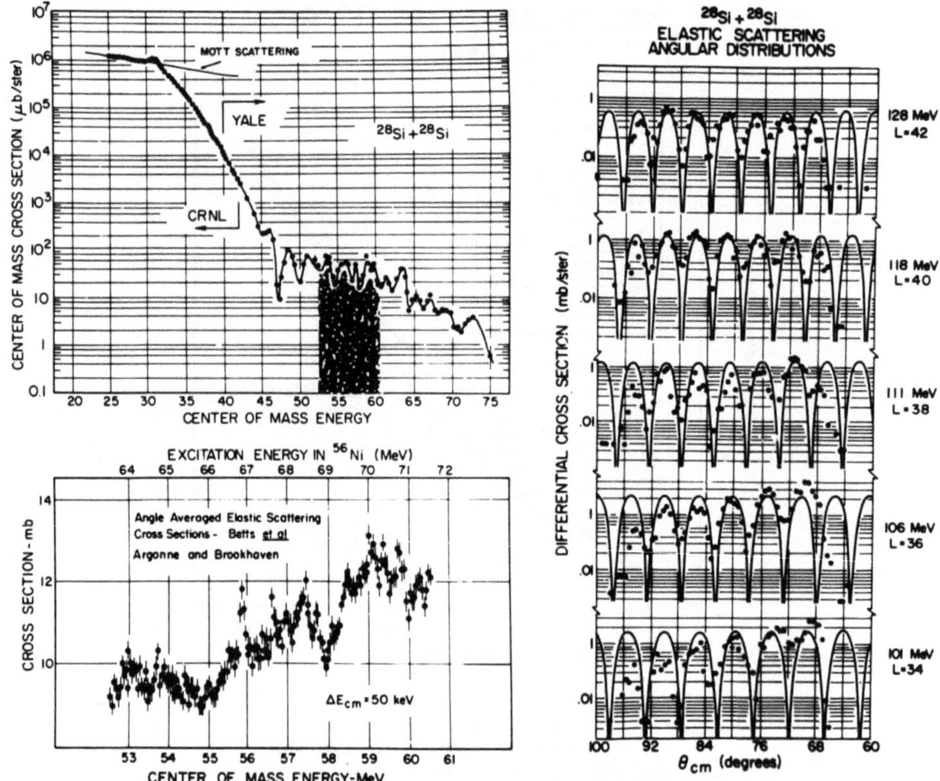

Figure 10: Aspects of the elastic scattering of $^{28}Si + ^{28}Si$. The upper left panel shows combined data from Yale and Chalk River from center-of-mass energies of 25 to 75 MeV. The shaded region is expanded in the lower left figure to show the fine structure corresponding to sharp states in the compound-nucleus ^{56}Ni. The panel on the right shows typical angular distributions measured at the indicated energies. When a particular partial wave becomes active in the scattering, the angular distribution can be reproduced with the square of the appropriate Legendre polynomial. The situation is particularly simple in the case of identical particles scattering because symmetry precludes the appearance of odd partial waves.

of ^{48}Cr and of ^{56}Ni secondary potential minima develop at very large deformation; the obvious suggestion—although it is not yet proven—is that the resonances correspond to states in these very highly deformed potential minima. What are required to resolve these speculations are data on much heavier nuclear systems than have been studied thus far and, in these systems, at angular momentum much higher than have been studied in the lighter systems.

That molecular interactions of this general type may persist even to the heaviest nuclear systems is illustrated in Fig. 12 showing work of Greenberg and his collaborators at GSI in West Germany. Originally these studies were stimulated by the suggestion of Greiner and his collaborators that, in the supercritical Coulomb fields attained during nuclear collisions with a combined atomic number in excess of 170, spontaneous ionization of the Dirac vacuum becomes possible. What was unexpected in the early measurements was the appearance in many of these supercritical systems of sharp positron lines superimposed on the anticipated positron spectrum, as shown in the upper left panel of this figure. In order to obtain an adequately long supercritical lifetime to correspond to the width of this positron peak, it has been suggested that, in a very small fraction of the collisions, long-lived molecular states may be formed, and the calculations thus far suggest that molecular binding is more probable in the nose-to-nose configuration labeled "Yukawa" than in the perhaps more obvious configuration labeled "proximity."

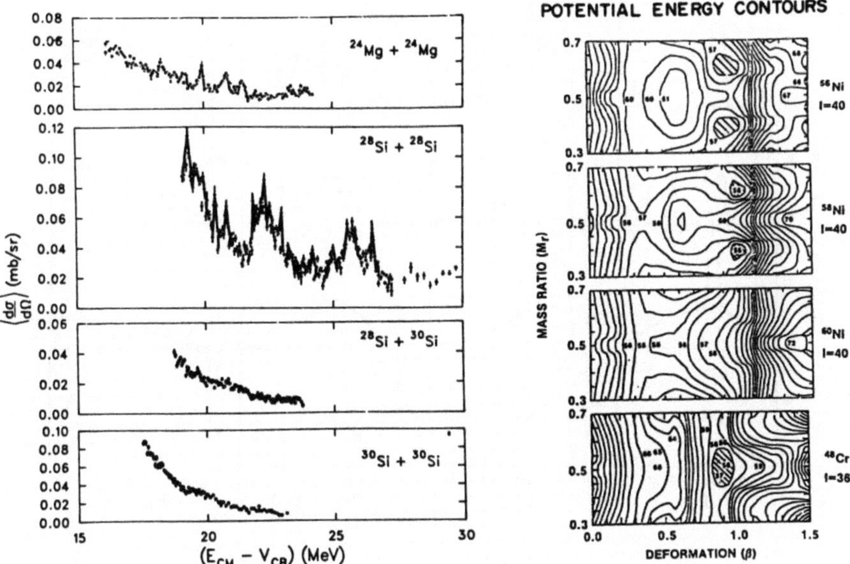

Figure 11: The left panel shows excitation functions obtained by Betts and his collaborators at the Argonne National Laboratory, while the right panel shows potential energy contours calculated by Leander **et al.** It is striking to note that in the case of ^{48}Cr and of ^{56}Ni, where the excitation functions show sharp structure, the potential energy contours show the existence of large deformation secondary minima in the potential surfaces, and the suggestion is that the resonance states correspond to quantum excitations in these large deformation minima.

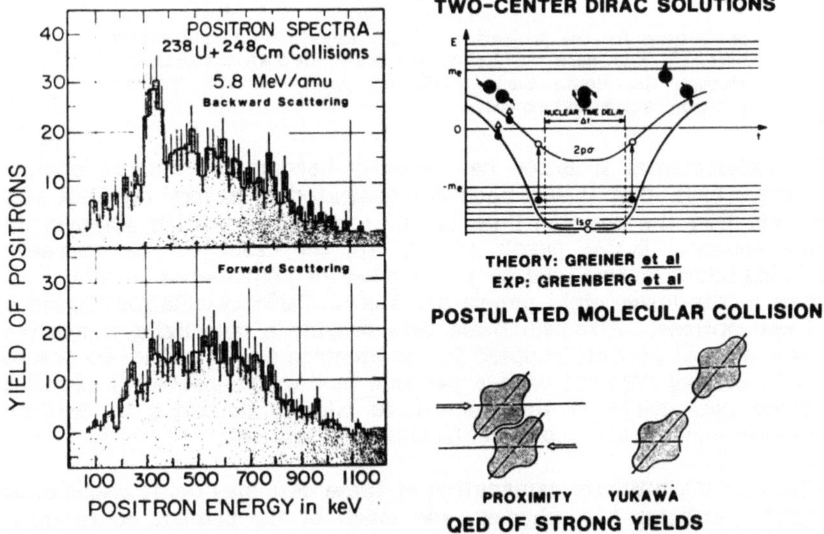

Figure 12: On the left are shown positron spectra taken in the upper panel in coincidence with backward scattered uranium ions and in the lower panel in coincidence with forward scattered uranium ions. The sharp positron lines superimposed on the continuous spectrum is quite obvious in the upper panel. On the upper right is shown the Dirac diagram for a supercritical nuclear scattering system and the nuclear time delay, corresponding to the sharp positron spike, is assumed to result from the formation of a transient molecular complex in the nuclear scattering of U on Cm. Preliminary calculations have demonstrated that the probability of molecular binding in the collision is greater in the case of a nose-to-nose collision, labeled here as Yukawa, than in the nestled case labeled here as proximity.

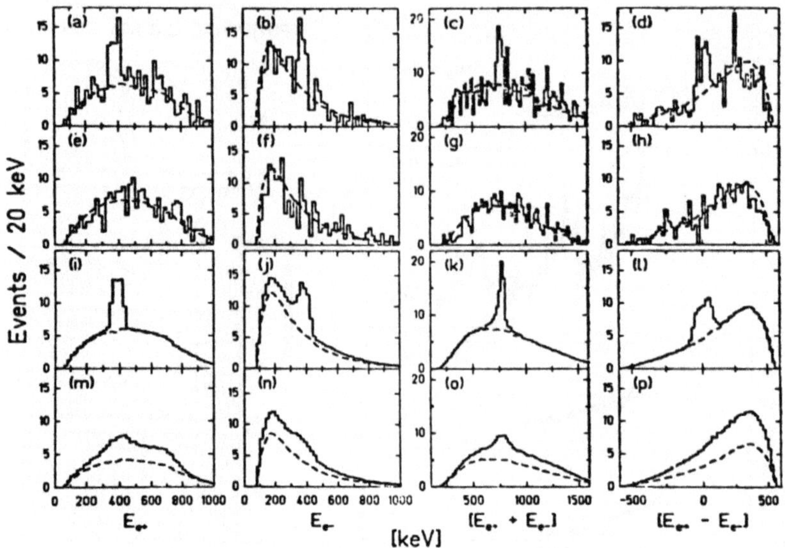

(a)-(h) : Projections of the experimental data

(i)-(l) Monte Carlo calculations assuming two-body decay of a neutral particle

(m)-(p) Monte Carlo calculations assuming internal pair conversion of a nuclear state

DATA OF COWAN et al. P.R.L. 56 444 (1986)

Figure 13: Electron positron coincidence events measured in U + Cm collisions together with Monte Carlo predictions for the projections of the experimental data shown here. In each case the dashed line drawn through the data and through the Monte Carlo predictions represents the dynamic positron background expected.

The experimental situation has recently become much more complex, as shown in Fig. 13, in that it has been demonstrated in a very difficult series of measurements that the positron lines occur in coincidence with electron lines of comparable energy. In the panels of Fig. 13, the dashed curves represent the dynamic background, normalized to the number of coincidence events, while the solid curves in the lower eight panels are Monte Carlo calculations based on the indicated assumptions. Although these data provide a tantalizing indication of a possible new neutral particle coupled to the electromagnetic field, no one has yet succeeded in coming forward with a particle having the necessary properties to explain these data while at the same time missing detection in earlier high-precision measurements on simple electromagnetic systems.

Whatever the ultimate explanation of these data may be, it seems clear that it will involve exciting new physics, and much of the present conference is of course devoted to attempting to determine what that physics might be.

SYMMETRIES IN NUCLEI

Symmetries in nuclei fall into three broad classes, as shown in Fig. 14. The geometric symmetries were first enunciated explicitly by A. Bohr in 1952. The kinematic symmetries are the fundamental ones usually characterized by P, C and T as well as Lorentz invariance. All of these date from the 1930s as indeed does the first of the dynamic symmetries, that of isospin, represented mathematically by the group U(2) and the supermultiplet theory of Wigner basic to alpha-particle models of the nucleus and governed by the group U(4). The isobaric mass

SYMMETRIES IN NUCLEI

A. Geometric Symmetries

(i) Axis of rotational symmetry in deformed nuclei	Bohr(1952)
(ii) Planes of reflection symmetry in deformed nuclei	Bohr(1952)

B. Kinematic Symmetries

(i) Parity (P)
(ii) Charge Conjugation (C)
(iii) Time Reversal Invariance (T)
(iv) Lorentz Invariance

C. Dynamic Symmetries

(i) Isospin U (2)	Heisenberg (1932)
(ii) Supermultiplets U (4)	Wigner(1936)
(iii) Isobaric Mass Formulae	Garvey and Kelson (1969)
(iv) Harmonic Oscillator Symmetries U (3)	Elliott (1958)
(v) Interacting Boson Structure Model	

$$U(6) \begin{cases} \nearrow U(5)\supset O(5)\supset O(3)\supset O(2) \\ \rightarrow U(3)\supset O(3)\supset O(2) \\ \searrow O(6)\supset O(5)\supset O(3)\supset O(2) \end{cases}$$

Iachello, Arima, Talmi (1974)

(vi) Algebraic Model For Scattering — Iachello (1981)

$$U(4) \begin{cases} \nearrow O(4)\supset O(3)\supset O(2) \\ \searrow U(3)\supset O(3)\supset O(2) \end{cases}$$

(vii) Supersymmetries

(a) Kinematic	Weiss-Zumino (1974)
(b) Dynamic	Bars-Iachello-Nambu (1980)

Figure 14: A compilation of the geometric kinematic and dynamic symmetries found in the nuclear many-body system.

expressions of Garvey and Kelson also correspond to a dynamic symmetry but one based on algebraic identities involving the nucleon-nucleon interactions. The U(3) harmonic oscillator symmetries introduced by Elliott in 1958 arose from the observation by Paul that shell and collective models of nuclei such as ^{19}F agreed with one another more closely than did either with experiment. More recently, the interacting boson model of Iachello, Arima and Talmi, dating from 1974, is governed by the group U(6) describing the behavior of postulated s and d bosons, whereas the algebraic model for scattering and indeed for binary cluster configurations, introduced by Iachello in 1981, is based on the U(4) group.

Finally, as indicated in Fig. 14, we have the two supersymmetries, the kinematic one introduced by Weiss and Zumino in 1974 in their work on supergravity and the dynamic one introduced by Bars, Iachello and Nambu in 1980 in their work on nuclear phenomena.

All these symmetries have played central roles in making possible the correlation and understanding of vast masses of nuclear data and indeed data in a great many physical systems. Within the past few years, however, the introduction of dynamic symmetries into the study of nuclear structure, with its characteristic replacement of the geometric techniques represented by the insertion of a model potential into a Schrodinger equation with subsequent solution for the model eigenvalues and eigenfunctions with an algebraic approach based on the above mentioned groups, has had revolutionary consequences. As illustrated in Fig. 15,

SYMMETRIES IN NUCLEI - I

Group Decomposition

$$
\begin{array}{lll}
 & \nearrow U(5) \supset O(5) \supset O(3) \supset O(2), & \text{(I)} \\
U(6) & \rightarrow SU(3) \supset O(3) \supset O(2), & \text{(II)} \\
 & \searrow O(6) \supset O(5) \supset O(3) \supset O(2). & \text{(III)}
\end{array}
$$

Quantum Numbers

$$
\begin{array}{ccccccccc}
U(6) & \supset & O(6) & \supset & O(5) & \supset & O(3) & \supset & O(2) \\
\downarrow & & \downarrow & & \downarrow & & \downarrow & & \downarrow \\
N & & \sigma & & \tau & (\nu_\Delta) & L & & M_L
\end{array}
$$

Mass Formulae

$$E^{(I)}(N, n_d, v, n_\Delta, L, M_L) = E_0^{(I)} + \epsilon n_d + \alpha n_d(n_d + 4) + \beta v(v + 3) + \gamma L(L + 1),$$

$$E^{(II)}(N, \lambda, \mu, K, L, M_L) = E_0^{(II)} + \kappa(\lambda^2 + \mu^2 + \lambda\mu + 3\lambda + 3\mu) + \kappa' L(L + 1),$$

$$E^{(III)}(N, \sigma, \tau, \nu_\Delta, L, M_L) = E_0^{(III)} + A\sigma(\sigma + 4) + B\tau(\tau + 3) + CL(L + 1).$$

Vibrator Spectra

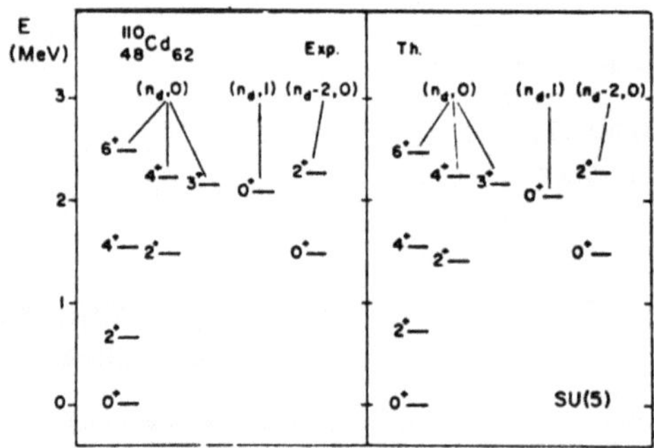

Rotor Spectra

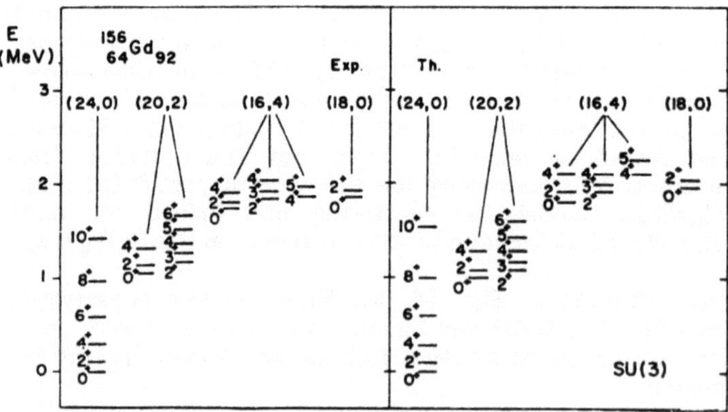

Figure 15: The upper part of this figure shows the decomposition of the U(6) group governing the interacting boson model for nuclear structure. Also shown are the quantum numbers appropriate to the chain and the mass formulae which result from the Casimir operators corresponding to the chain member. In the two lower panels, the excitation spectra corresponding to cases one and two are compared with experimental data on ^{110}Cd and ^{156}Gd respectively.

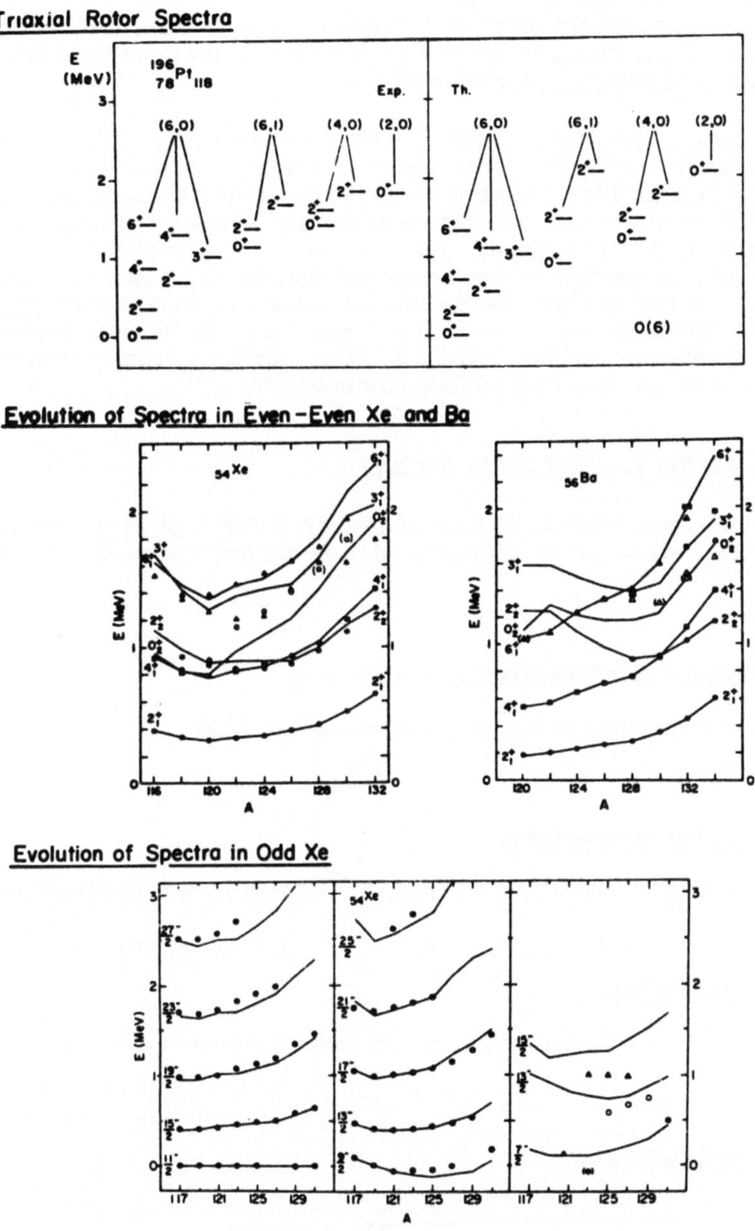

Figure 16: The upper panel compares theoretical predictions with experiment for the case of ^{196}Pt, the first nucleus to be identified as having the triaxial rotor configuration (case III of Fig. 15). The center panel compares the predictions of an interacting boson model with the excitation spectrum of the xenon and barium isotopes respectively, while the lower panel shows the evolution of spectra in odd xenon isotopes and compares the IBA predictions with the experimental data where available.

the U(6) group has only three decomposition chains which contain the O(3) group, a necessary condition for the resulting eigenfunctions to have angular momentum as a good quantum number. The quantum numbers appropriate to the various groups are as shown in this figure, and inserting the Casimir operators for the appropriate groups gives rise to the mass formulae shown for these three group chains. From the structure of these mass formulae, it is quickly obvious that the chains correspond, respectively, to an harmonic oscillator, to an axial rotor and to a triaxial rotor, respectively, each with characteristic spectra, and the lower two panels in Fig. 15 and the upper one in Fig. 16 show the now typical success that this model enjoys in reproducing experimental data.

Much more important, however, than the reproduction of these spectra for nuclei in regions where the three above mentioned models have long enjoyed considerable success, the interacting boson model has the tremendous advantage of being equally applicable in the transition regions where previously none of the models were valid. The central panel in Fig. 16 for example shows the IBA calculation of the spectra of the xenon and barium isotopes, respectively, as a function of neutron number, illustrating the transition from one of the limiting cases noted above to the other. The lower panel shows calculations on the evolution of spectra in the odd-mass xenon isotopes, again comparing IBA calculations with the experimental data now available.

DYNAMIC SUPERSYMMETRIES IN NUCLEI

To the extent that the success of the IBA makes it physically reasonable to consider even-even nuclei as consisting of well-defined bosons and thus as bosons

POSSIBLE SUPERSYMMETRY IN NUCLEI - I

A j = 3/2 VALENCE NUCLEON COUPLED TO AN O(6) TRIAXIAL ROTOR CORE

SUPERGROUP DECOMPOSITION

$$\begin{array}{cccccccc} U(6/4) \supset & U^{(B)}(6) \otimes U^{(F)}(4) \supset & & O^{(B)}(6) \otimes SU^{(F)}(4) \supset & Spin(6) \supset & Spin(5) \supset & Spin(3) \supset & Spin(2) \\ \downarrow & \downarrow & \downarrow & \downarrow & \downarrow & \downarrow & \downarrow & \downarrow \\ \mathcal{N} & N & M & \Sigma & \sigma_1\,\sigma_2,\sigma_3 & \tau_1,\tau_2(\nu_\Delta) & J & M_J \end{array}$$

MASS FORMULA

$$\begin{aligned} E(\mathcal{N}, N, M, \Sigma, \sigma_1, \sigma_2, \sigma_3, \tau_1, \tau_2, \nu_\Delta, J, M_J) = \mathcal{E}_0 + \mathcal{E}_1 N + \mathcal{E}_2 N^2 + A\Sigma(\Sigma + 4) \\ + A'\left[\sigma_1(\sigma_1 + 4) + \sigma_2(\sigma_2 + 2) + \sigma_3^2\right] \\ + B\left[\tau_1(\tau_1 + 3) + \tau_2(\tau_2 + 1)\right] + CJ(J + 1) \end{aligned}$$

SUPERMULTIPLETS

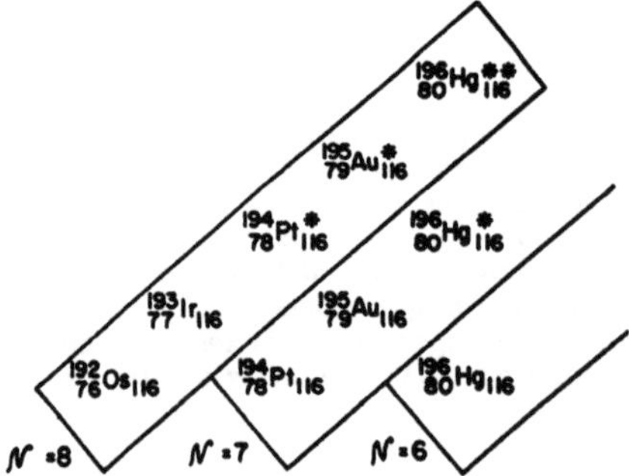

Figure 17: The super group decomposition corresponding to a j = 3/2 valence nucleon coupled to a O(6) triaxial core leads to the indicated mass formula, and the lower panel shows the region around ^{193}Ir where supersymmetric behavior has been most clearly identified.

themselves, odd-mass nuclei with a single valence nucleon should be describable by a dynamic supersymmetry treating fermions and bosons on an equal footing. Figure 17 shows the case of a j = 3/2 valence nucleon coupled to an O(6) triaxial rotor core, illustrating the decomposition of the relevant supergroup U(6/4) and the corresponding mass formula.

Whereas dynamic symmetries lead to families of quantum states within a single nucleus, dynamic supersymmetries lead to supermultiplets of states in adjacent nuclei, as well as to relations between the ground state masses of these nuclei, electromagnetic transitions within them and spectroscopic factors for transfer reactions linking them. The supersymmetries are characterized by the primary quantum number n, and in the lower panel of this figure, I show parts of three of the supermultiplets. In Fig. 18, taken from work of Cizewski et al., I show the predicted deexcitation behavior of even- and odd-mass nuclei in the supermultiplet which reproduces, with high accuracy, the experimental observations, and in the lower panel I show that the model calculations for single proton pickup reactions also reproduce the experimental data quite well.

POSSIBLE SUPERSYMMETRY IN NUCLEI - II

A j=3/2 VALENCE NUCLEON COUPLED TO AN O(6) TRIAXIAL ROTOR CORE

EXCITATION SPECTRA

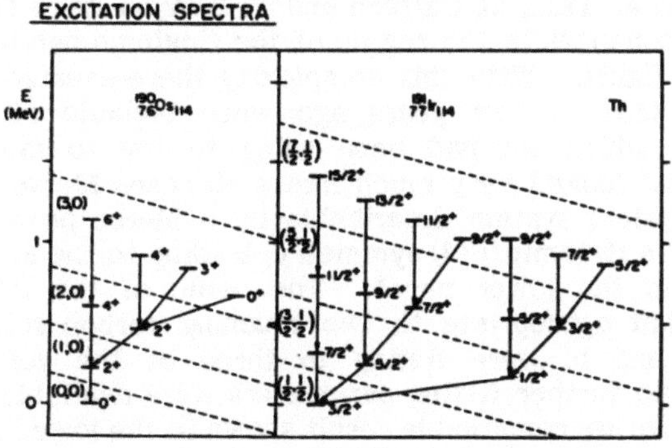

SPECTROSCOPIC FACTORS

Pt(t,α)Ir

$\frac{3}{2}^+$, $\sigma_1 = N - \frac{1}{2}$

$\frac{3}{2}^+$, $\sigma_1 = N + \frac{1}{2}$ ← 0^+, $\sigma_1 = N$

RATIO OF STRENGTHS, S, FOR STATES $(\sigma_1 \tau_1)$

$$\frac{S(N-\frac{1}{2},\frac{1}{2})}{S(N+\frac{1}{2},\frac{1}{2})} \approx \frac{N}{N+4} \sim 0.70$$

ONLY TWO STATES, BOTH $J = \frac{3}{2}$, WILL BE SEEN

	E_x	J^π	S_{rel}	S_{thy}
^{193}Ir	0	$\frac{3}{2}^+$	≡1.00	≡1.00
N=7	180	$\frac{3}{2}^+$	0.07	0
	460	$\frac{3}{2}^+$	0.69	0.64
	73	$\frac{1}{2}^+$		

Figure 18: The upper panel shows predicted supersymmetric deexcitation patterns for even- and odd-mass nuclei in the iridium region. These predicted patterns are in excellent agreement with the experimental data. The lower panel shows a comparison of the spectroscopic factors obtained experimentally and predicted from a supersymmetric model for the case of proton pickup on platinum targets.

It has been established that the supersymmetry model works remarkably well in the vicinity of ^{190}Ir. Furthermore, it has been possible to show that as the supersymmetry is broken in moving away from ^{193}Ir, reproduction of the experimental observations can be regained by varying only a single parameter—namely, the strength of the coupling of the valence fermion to the bosonic core. Obviously one of the major challenges now facing those studying nuclear structure is that of finding other areas of the periodic table where supersymmetric effects are equally apparent. Although the case has not yet been firmly made, there is growing evidence that such an area is to be found in the neutron deficient barium region.

DYNAMIC SYMMETRY IN NUCLEAR INTERACTIONS

Despite the success of the interacting boson model in reproducing the structure of nuclei i.e. the nuclear bound states, there has always been considerable question as to whether the group theoretical and algebraic approaches could be used to describe the unbounded wave functions encountered in nuclear scattering and other interactions. The first evidence that this might, in fact, be possible came from studies on the molecular resonances in the C + C system, as shown in Fig. 19. The first three of these resonances were discovered in the first measurements with the first tandem accelerator at the Chalk River Laboratories in 1959. Subsequent, more detailed studies at Yale, at Caltech and at Munster have finally given rise to some 39 discreet resonances in the region of the Coulomb barrier, as shown in the upper panel of this figure. With this complexity there seemed little hope of any simple explanation until a few years ago when Iachello recognized that the effective potential, which we had been using to try to model these nuclear molecular resonances, looked very much like a standard Morse potential and that furthermore any physical system describable by a Morse potential could equally well be described by a dynamic U(4) symmetry, leading to the analytic form for the eigenvalues shown in the lower panel. The value of the fitting parameter c appearing here is that appropriate to two touching carbon spheres, whereas the other three, D, a and b, were fitting to three of the very well-determined resonances, so that no further fitting parameters were available for the remaining 36 resonances. The rather remarkable result shown in the lower panel suggests that a U(4) description of these states is an entirely reasonble one and led to substantially increased insight into the mechanisms whereby these resonances occur.

Within the past year it has been recognized more generally that noncompact groups of the form SO(m,n) may have much broader applicability in the understanding of nuclear interactions. As shown in Fig. 20, Pauli had recognized in 1926 that nonrelativistic Coulomb scattering has a dynamic symmetry SO(3,1). More generally, in 1986, Alhassid **et al.** have demonstrated that once the appropriate dynamic symmetry can be established, it is possible to write down the pertinent S matrix elements analytically and further that the parameter m is the dimensionality of the problem, while n is the number of parameters required to establish the interaction under study.

Figure 21 illustrates the success of this approach in some preliminary applications that we have undertaken at Yale. The data here are those of Korotky **et al.** on the ^{13}C + ^{13}C system, and the dashed curve is the best fit that we were able to obtain after substantial optical model searches with six parameters available. The solid curve, on the other hand, is that obtained using the scattering matrix elements appropriate to an SO(3,2) symmetry with the complex parameter ν chosen as indicated (i.e. two-fitting parameters). Equivalent success has been obtained in reproducing heavy-ion elastic angular distributions.

As indicated in Fig. 20, the S matrix elements used here quite naturally factor, and we are hoping that it may be possible to apply this formalism to more complex interactions such as inelastic scattering, transfer reactions and so on by

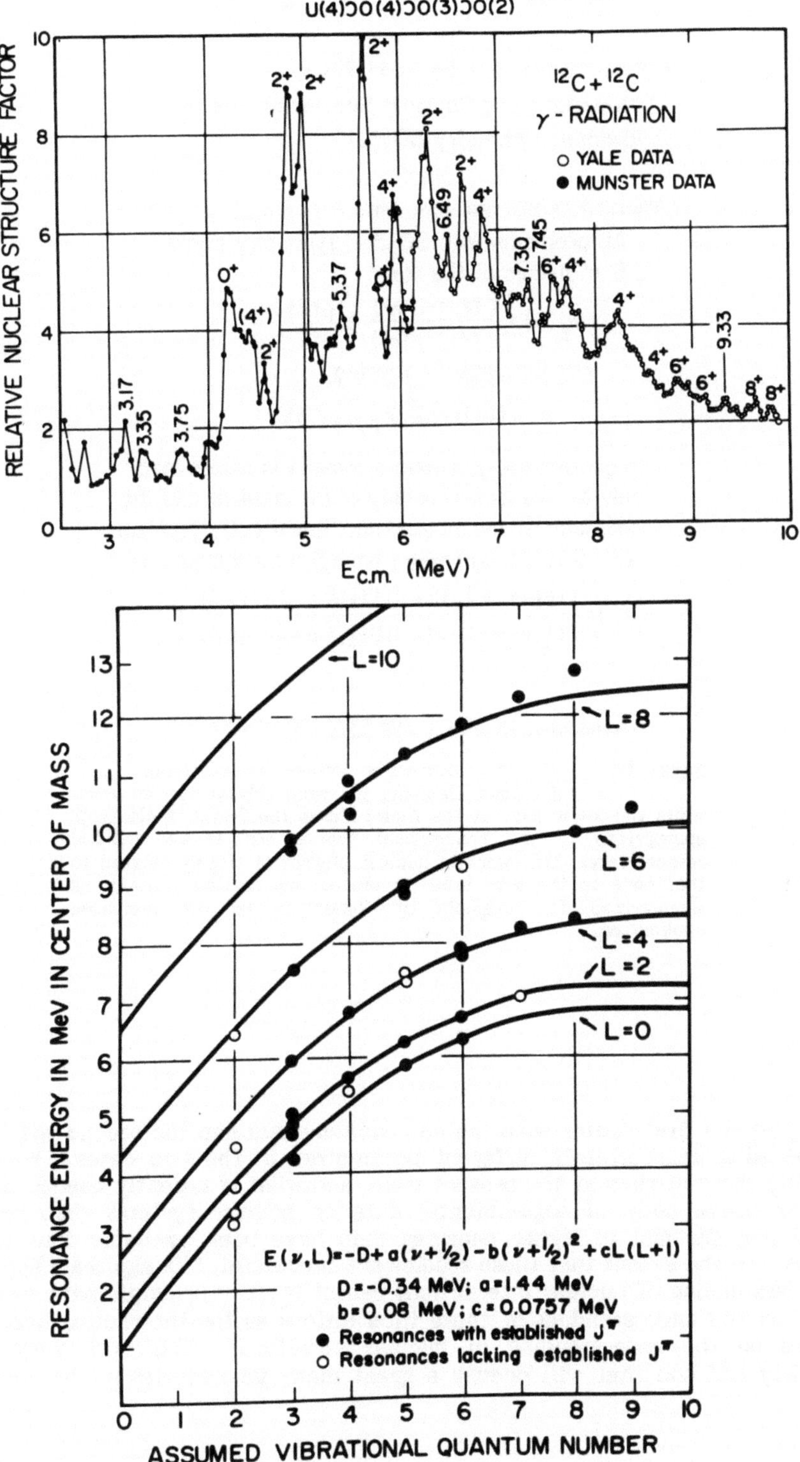

Figure 19: The upper panel shows the molecular resonances in the $^{12}C + ^{12}C$ scattering system in the region of the Coulomb barrier. The lower panel shows a fit to the indicated eigenvalue expression.

DYNAMIC SYMMETRIES IN SCATTERING

$$\frac{d\sigma}{d\Omega} = \left| \frac{1}{2ik} \Sigma(2\ell+1)(1-S_\ell(k))\, P_\ell(\cos\theta) \right|^2$$

W. Pauli, Zeits f. Phys. 36 336 (1926)

"Nonrelativistic Coulomb Scattering has the dynamic symmetry SO(3,1)"

Y. Alhassid, F. Iachello and J Wu. Phys. Rev. Lett. 56 271 (1986)

"All problems with SO(3,1) symmetry imply S matrices of the form

$$S_\ell(k) = \frac{\Gamma(\ell+1+if(k))}{\Gamma(\ell+1-if(k))}\, e^{i\phi(k)}"$$

For Coulomb Scattering

$$f(k) = \alpha Z_1 Z_2 \mu c/\hbar k$$

In general, the governing symmetry is determined only by the dimensionality of the problem and the character of the interaction. Specifically problems with SO(3,2) symmetry have S matrix elements

$$S_\ell(k) = \frac{\Gamma(\frac{1}{2}[\ell+\nu+\frac{3}{2}+if(k)])\,\Gamma(\frac{1}{2}[\ell-\nu+\frac{1}{2}+if(k)])}{\Gamma(\frac{1}{2}[\ell+\nu+\frac{3}{2}-if(k)])\,\Gamma(\frac{1}{2}[\ell-\nu-\frac{1}{2}-if(k)])}\, e^{i\Delta(k)}$$

G Veneziano, Nuovo Cimento 57A 190 (1968)

Figure 20: The upper expression defines the scattering S-matrix elements for each orbital partial wave, while the lower part of the figure shows the S-matrix elements appropriate to the noncompact groups SO(3,1) and SO(3,2) respectively. This work in nuclear physics is closely related to that done in the late 1960s in elementary particle physics, as summarized for example by Veneziano in the indicated publication.

identifying the first factor with the entrance channel and the second with the exit channel using appropriately different parameters in the two cases. We are just beginning these studies at the present time, and what is urgently needed is a much more extensive body of experimental data on heavier systems than have been studied thus far and to higher energies than have been available thus far in all systems. To the extent that these studies are successful, this algebraic approach to the understanding of nuclear interactions can at least in principle have as great an impact on the understanding of these interactions as the interacting boson model has had on the understanding of nuclear structure. This is a very exciting possibility and one that will occupy a great many laboratories in the coming few years.

NEW FACILITIES IN LOW-ENERGY NUCLEAR PHYSICS

All of the topics that I have discussed thus far are within the realm of what is now called "classical nuclear physics" i.e. that accessible to tandem Van de

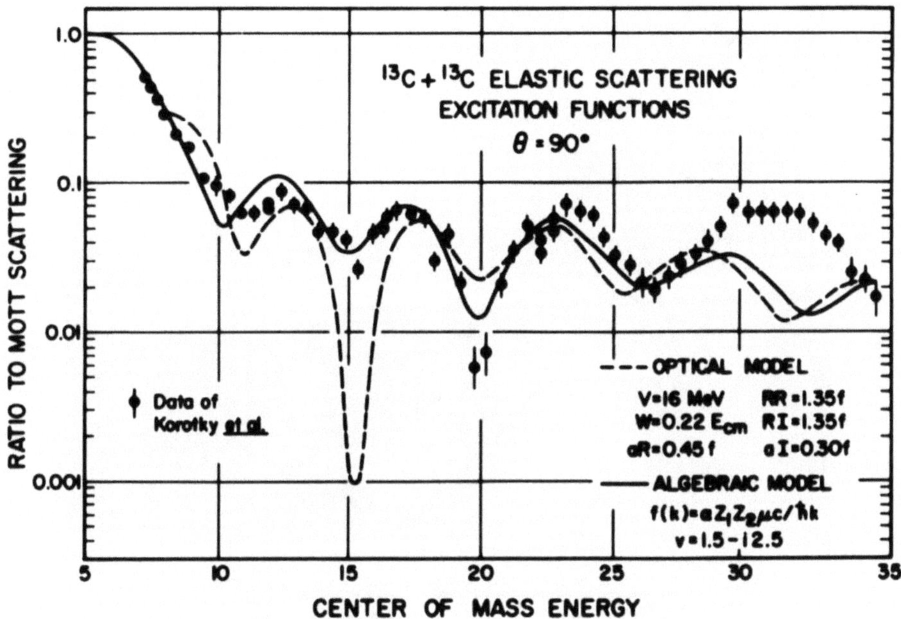

Figure 21: A comparison of the fits attained with an optical model and with the algebraic model of Figure 20 to data of Korotky et al. on the $^{13}C + ^{13}C$ 90° elastic scattering excitation function.

Graaffs, cyclotrons, linear accelerators and appropriate combinations of all three, accelerating essentially all particles from protons to uranium nuclei. As is obvious, there remains an enormous open frontier available for study with these facilities, and fortunately, a series of new and very powerful machines are just now either coming into operation or will come into operation within the next year. Included here are the tandem-linear accelerator combinations typified by the ATLAS unit at the Argonne National Laboratories, the double superconducting cyclotron system at the Michigan State University, the large tandem electrostatic accelerators at the Oak Ridge and Daresbury National Laboratories and appropriate combinations in a great many laboratories around the world. In Fig. 22, I illustrate the conversion project that is something more than half completed in my own Laboratory at Yale where we are replacing the first of the MP class tandem accelerators—which operated originally at 10 MV on terminal—with the first of the much larger ESTU tandems which we anticipate to operate somewhere between 22 and 25 MV on terminal.

STUDIES WITH ELECTRON ACCELERATORS

Until very recently, work with electron accelerators characteristically suffered from low energy resolution and serious background problems, so that it was extremely difficult if not impossible to isolate individual quantum states in the nuclei under study. That this is no longer the case is illustrated in striking fashion by Fig. 23 taken from recent work from the Netherlands. Here, inelastic electron scattering, with the removal of a proton from ^{208}Pb to leave a residual ^{207}Tl nucleus, is shown at three different momentum transfers. Clearly, the resolution is more than adequate to separate the proton groups coming from specific shell-model states in the lead nucleus.

Although highly complementary to work done with the lower-energy tandem accelerators and cyclotrons, these data are still restricted to phenomena in the nuclear surface. There has long been considerable question as to whether the simple shell model concepts apply deep in the nuclear interior i.e. do nucleons

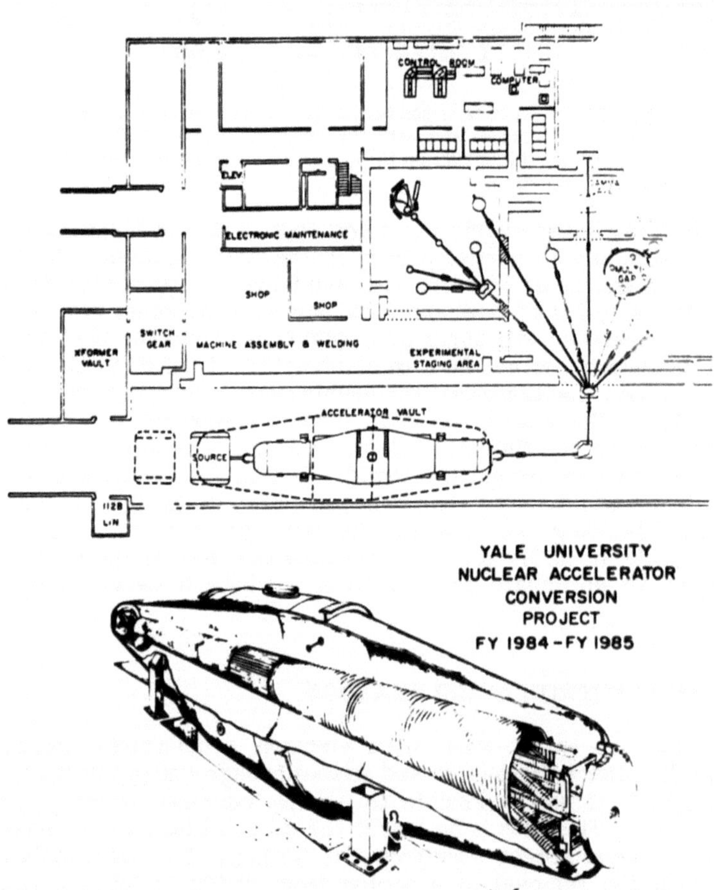

Figure 22: The photograph (top) shows the first of the MP tandem accelerators in the accelerator vault of the A.W. Wright Nuclear Structure Laboratory at Yale. The magnet in the foreground is the momentum reference for the entire system and also serves to direct the accelerated beams through the heavy shielded wall on the right into the target areas. A schematic diagram of the ground floor layout of the Laboratory is given (below), and the new ESTU accelerator now under construction, is shown in dashed outline superimposed on the MP accelerator. A cutaway view of the ESTU machine is shown at the lower right.

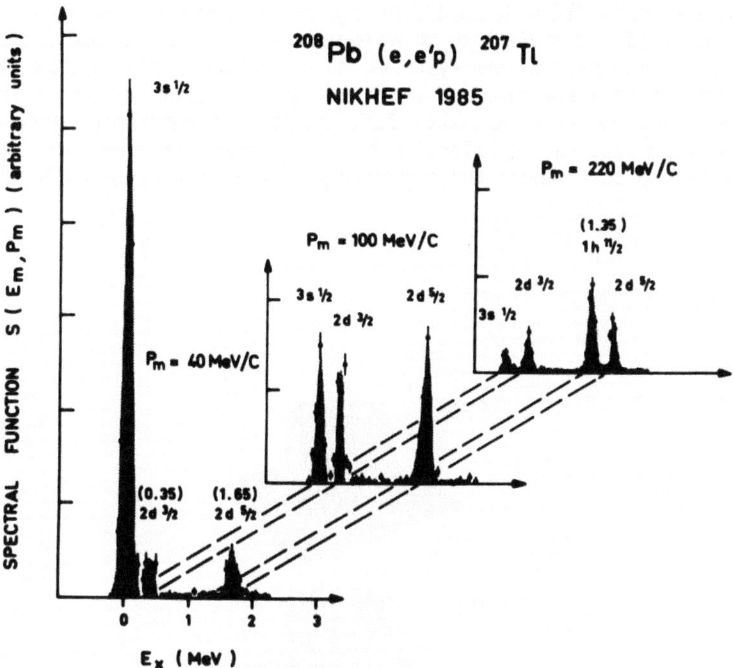

Figure 23: Inelastic electron scattering on ^{208}Pb with removal of protons from the indicated shell-model states. The experimental results are shown for three different momentum transfers showing the characteristic behavior of the different shell-model orbitals as a function of momentum transfer.

behave when they are in the central region of a heavy nucleus as they would in a free state or as they do on the surface of nuclei. Figure 24, taken from work of Frois and his collaborators at Saclay answers this question in rather dramatic fashion. Scattering high-energy electrons on both ^{206}Pb and ^{205}Tl, the difference between the two situations can be attributed to the presence of the additional 3s proton in ^{206}Pb, and subtraction of the pertinent data leads to the 3s wave function shown at the bottom of this figure. As is clear, this is a classic 3s wave function, demonstrating once and for all that at least as probed by high-energy electrons, the shell model **does** provide an excellent representation of nucleon behavior, even in the center of heavy nuclei.

Figure 25 is an outline plan of the 750 MeV electron linac facility at the Bates Laboratory at MIT; the MIT group currently has a proposal under consideration to extend the energy of this machine to 1 GeV while retaining the energy resolution typified by the data of Figs. 24 and 23.

There has, in fact, been substantial and growing pressure from the international nuclear science community for electron accelerator facilities with higher duty factor and higher energy while still retaining excellent energy resolution. One of the most striking data leading to this increased interest is the so-called EMC effect first reported by the European Muon Collaboration group from studies on the inelastic scattering of high-energy muons on nuclei. Plotted in Fig. 26 are their results, together with older data on the similar inelastic scattering of high-energy electrons measured at SLAC—data which had not been analyzed until publication of the EMC results. What is shown here is the ratio of the form factors for high-energy muon and electron scattering on nucleons in an iron target and in a deuterium target. Conventional wisdom had always been that the specific target would make no difference in these studies any more than would the chemical composition of a standard target used in conventional nuclear physics. The obvious departure of this ratio from 1, as a function of the Bjorken scaling parameter X, has been interpreted in a variety of ways to indicate at least partial deconfinement of the quarks in the heavy nuclear systems. I shall return to this matter later, since there is considerable disagreement as to the proper explanation of these data. Whatever that may be, however, they have stimulated substantial interest in the possibility of using high-energy electron beams to scan the transition from normal hadronic matter to that where the underlying quarks are beginning to act as free agents.

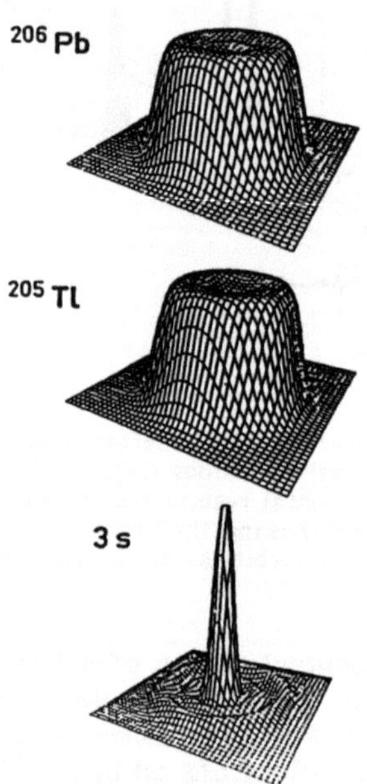

Figure 24: Inelastic scattering measurements on ^{206}Pb and ^{205}Tl from Saclay. The subtraction of the upper two figures gives rise to the 3s proton wave function in the lower panel. This is a classic 3s wave function shape and demonstrates conclusively that the shell model concepts hold deep in the nuclear interior.

Figure 27, illustrating the scattering of electrons from protons, is useful in illustrating this transition from hadronic to quark matter and in demonstrating what energy electrons are required to study the transition. What is plotted here is the scattering form factor as a function of energy loss and of scaling parameter. At low energies, the scattering is simply from the nucleons in their ground state; but with higher energy, the nucleons themselves are excited to the delta resonance and then to the various N* resonances, so that the nuclear system being studied changes from one comprised of nucleons to one where a number of the nucleons are themselves in excited configurations. Moving higher in energy, the current evidence suggests that the nucleons begin to leak quarks and on occasion to form six quark entities, so that in effect the nucleons gradually melt into a quark-gluon

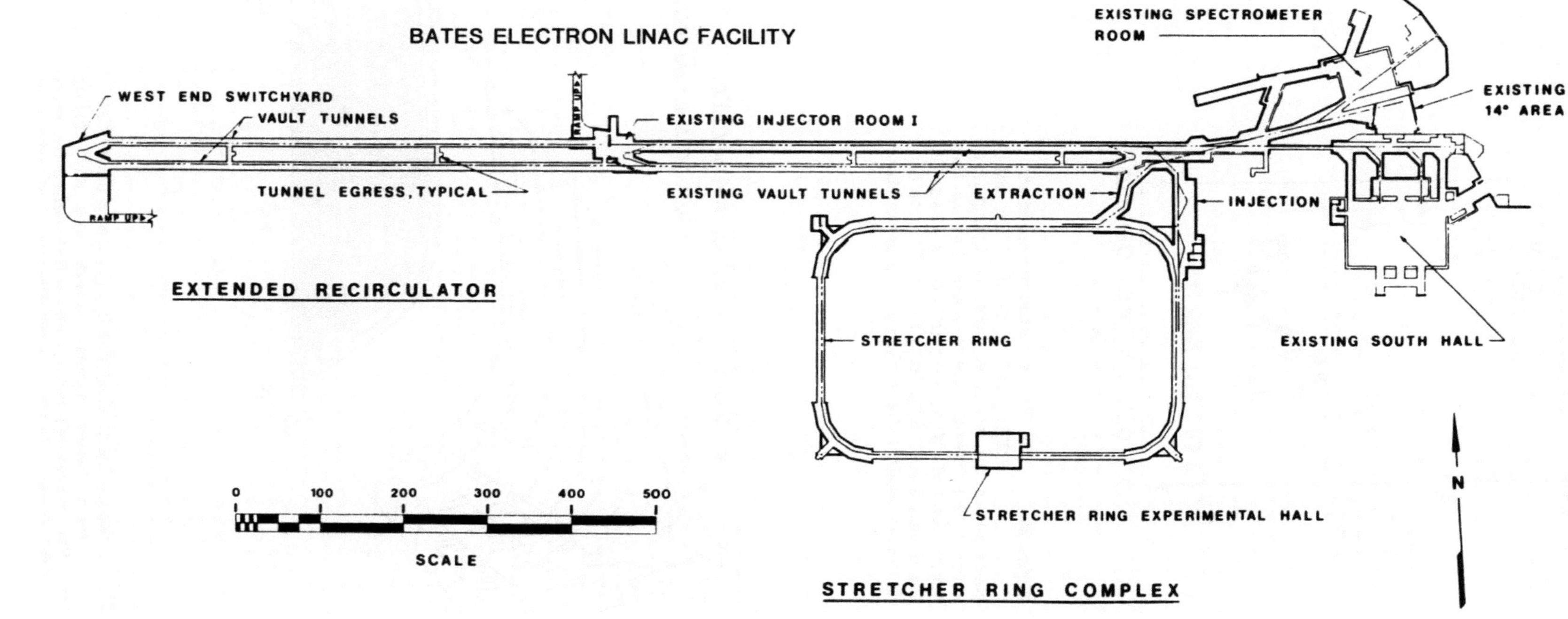

Figure 25: The Bates Electron Linac Facility at the Massachusetts Institute of Technology. Currently this facility operates at 750 MeV without either the extended recirculator or the stretcher ring complex. These are both parts of a proposed upgrading of the facility to provide 1 GeV operation with higher duty cycle.

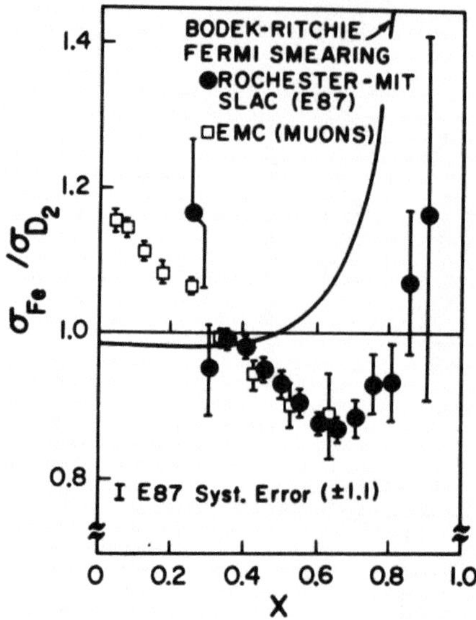

Figure 26: The EMC effect as measured at CERN and at SLAC. The solid curve shows the predicted behavior taking into account the internal Fermi momentum in the nuclei involved. What is plotted here is a ratio of the inelastic scattering form factors for iron and deuterium against the Bjorken scaling parameter.

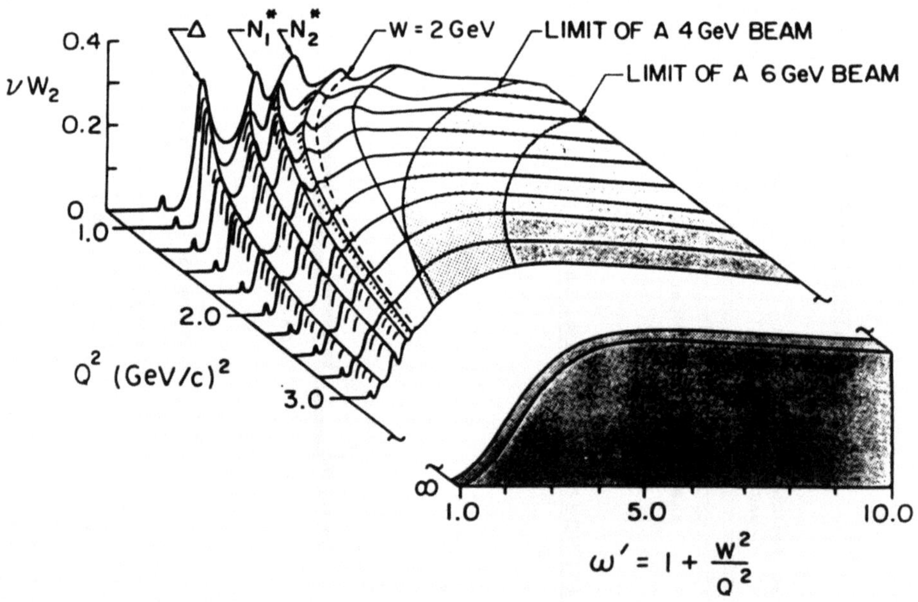

Figure 27: Scattering of electrons from nucleons plotted as form factor against energy loss and scaling parameter. The curves labeled 2,4 and 6 GeV define the right boundary of the regime available to experimental study with such a facility.

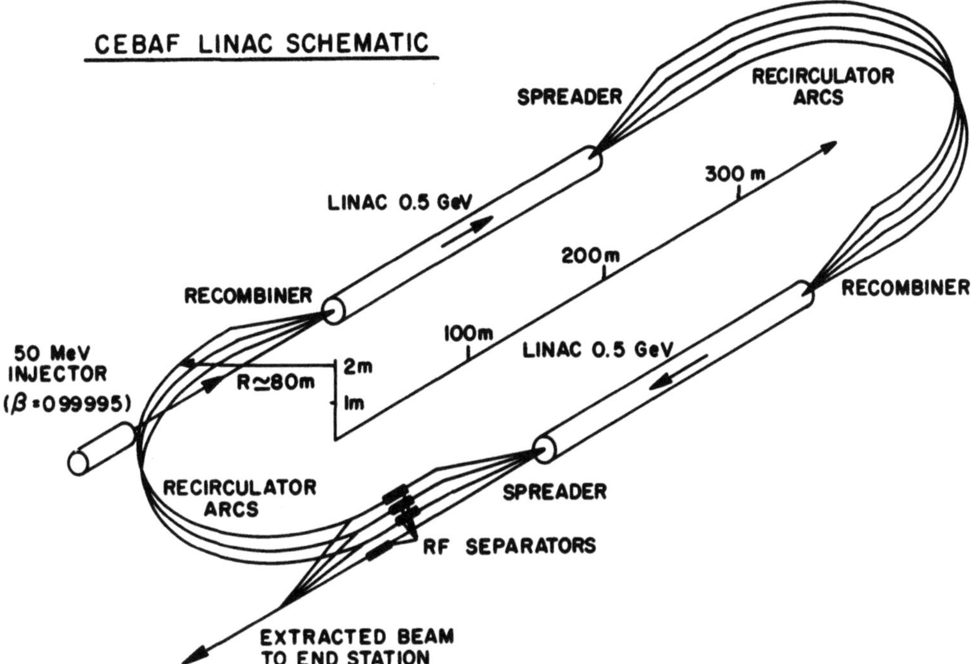

Figure 28: A schematic layout of the planned CEBAF electron accelerator facility. Two 0.5 GeV superconducting linacs will be arranged as shown, and spreaders make it possible to extract the beam at several different energies below the maximum value. The recirculator arcs have been designed so that they can encompass additional magnetic dipoles to increase the maximum energy to 16 GeV at some future time.

plasma. It must be emphasized however that complete conversion into the quark-gluon plasma cannot be expected at energies less than many tens of GeV. As shown in this figure, with a 2 GeV accelerator it will not be possible to get much beyond the situation where the nucleons are excited. With a 4 GeV beam, it becomes possible to at least begin to probe the quark domain, and with a 6 GeV beam a much more solid foothold can be expected in the scaling regime, although again it must be emphasized that even with 6 GeV, one would be nowhere near a regime where quantum chromodynamics, for example, could be used in perturbative fashion.

Obviously, this transition region from hadronic to quark matter will be an exceedingly complex one and one that will be very difficult to analyze from a theoretical point of view. Compensating for this difficulty, however, is the fact that the physics will be correspondingly very interesting.

In the U.S., the decision has already been made to proceed with the construction of a 100% duty cycle accelerator with an initial guaranteed maximum electron energy of 4 GeV. This so-called Continuous Electron Beam Acclerator Facility (CEBAF) is shown schematically in Fig. 28. Recent progress in the design and fabrication of superconducting linear accelerator cavities has been such that the design of this facility is now based on their use. The recirculator arcs have been designed so that if, and when, the research program makes it desirable the maximum energy of the machine can be increased from 4 to 16 GeV through the addition of further sections to the two linacs and additional dipole magnets to the recirculator arcs.

The CEBAF facility is now approved for construction beginning during 1987. Figure 29, which plots the duty factor versus the maximum beam energy for a number of the world's electron accelerator facilities demonstrates the unique position that it will be occupy. A very active scientific program has already been

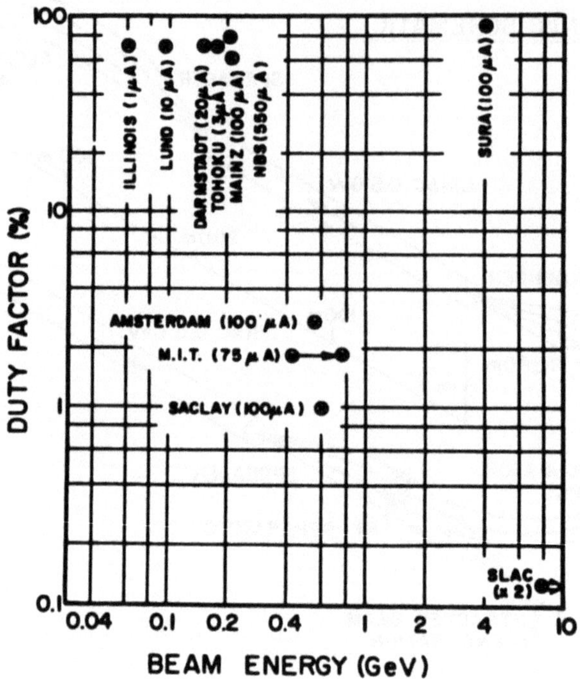

Figure 29: A comparison of the world's major electron accelerator facilities plotting duty factor against maximum beam energy.

developed for this facility as soon as it becomes functional in 1991. Included in this is the study of the role of mesons, of excited baryons and of quarks in nuclei; of the strong force, including nonperturbative features of quantum chromodynamics; of the effect of the nuclear medium on nucleon structure and on quark interactions. Among the advantages that electrons have in these studies are first, that the electromagnetic interaction is extremely well-understood (perturbative QED); the electrons probe the entire nuclear volume; and the energy loss and momentum transfer can be varied independently.

Before leaving electrons, however, it is also important to consider recent Saclay data, as shown in Fig. 30. These studies on the electrodisintegration of the deuteron should provide a strong cautionary note for those interested in probing quark phenomena in the nuclear domain. As is obvious from this figure, calculations which assume the deuteron to consist of only a neutron and protron (the impulse approximation) bear no resemblance to the experimental data whatever. What is truly remarkable is the degree to which the calculations reproduce the experimental results as soon as they include the presence of soft pion exchange in addition to the impulse approximation. Indeed, these calculations give an excellent reproduction of the experimental data down to distances of 0.2 Fm without invocation of any quark effects whatever. The fact that the solid curve here, which includes the effect of both rho meson exchange and nucleon excitation to the delta resonance makes so little change in the predicted cross section reflects the fact that the rho and delta contributions are essentially equal, of opposite sign, hence cancel.

It is essential to bear these results in mind, and not leap too quickly to the invocation of quark phenomena. This is a point that has been emphasized repeatedly by G.E. Brown and is a point well-taken. There has already been considerable literature pollution in attempts to see quark signatures in data which were entirely explicable in terms of more traditional nuclear physics.

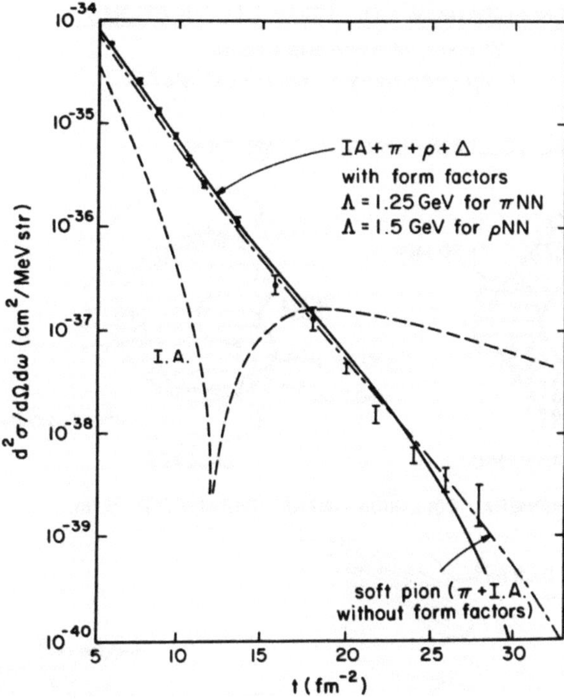

Figure 30: The cross section for electric disintegration of the deuteron, as measured at Saclay. The dashed curve labeled IA is that obtained from an impulse approximation calculation assuming the presence of only a neutron and a proton. The dot-dashed curve is that obtained in an impulse approximation calculation which also includes soft exchange pions; it provides a remarkable fit to the experimental data down to radii of 0.2 Fm. The solid curve includes additionally the effects of ρ meson exchange and excitation of the nucleons to the Δ resonance. The fact that this makes essentially no change in the prediction results from the fact that the ρ and Δ contributions are of essentially identical magnitude and opposite sign.

HEAVY-ION INTERACTIONS AT HIGH ENERGY

There is general agreement that if we are to see quark effects in unambiguous fashion and really probe the nuclear equation of state, we will do so in the simplest fashion using very high-energy heavy-ions.

Until this past year, however, there has been considerable uncertainty and ambiguity as to the character of these very high-energy heavy-ion interactions. As shown in Fig. 31, it was not at all clear whether the collisions were better described with hydrodynamic models wherein the nuclear matter was assumed to behave like an Euler fluid or by cascade models wherein the collision was assumed to proceed through a time ordered sequence of nucleon-nucleon collisions. The problem was primarily an experimental one in that it was extremely difficult to separate out the central collisions of greatest interest from the much more numerous peripheral ones, each having their characteristic observables as indicated in this figure. During the past year, studies by the joint LBL-GSI group, using the plastic ball detector at the Bevalac accelerator to unambiguously separate out central collisions, have obtained the data shown in Fig. 32. The striking feature of these

NEON + URANIUM COLLISION CALCULATIONS

Stocker, Maruhn and Greiner

Neon Laboratory Energy - 400 MeV/A

A. CENTRAL COLLISION - IMPACT PARAMETER = 0

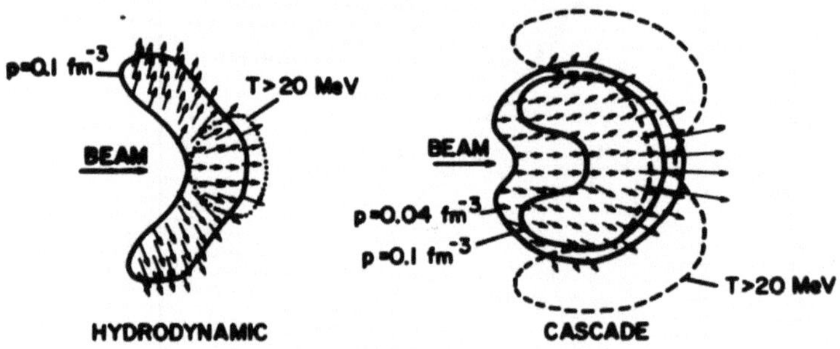

B. SEMIPERIPHERAL COLLISION - IMPACT PARAMETER = 6 fm

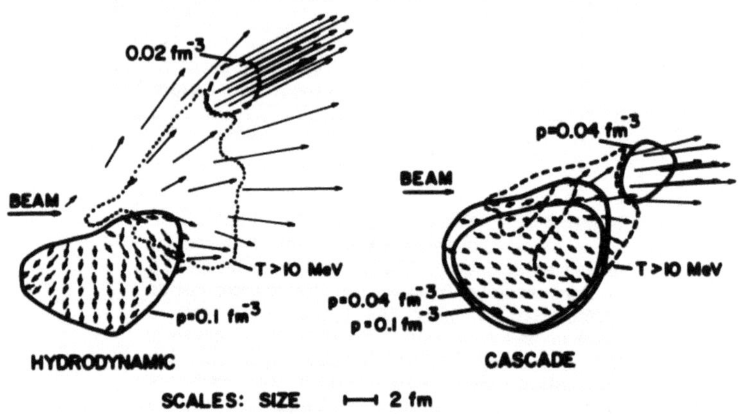

Figure 31: Calculations of Stocker, Maruhn and Greiner on Ne + U collisions. The left figure is appropriate to a hydrodynamic model, whereas the right is based on a cascade model. The upper panel treats central collisions, whereas the lower panel treats peripheral collisions having an impact parameter of 6 fm. The velocity field resulting from the collision is shown by the magnitude and orientation of the arrows with the scale as shown. The zones of maximum temperature are indicated by the dashed and dotted outlines. Density contours are also indicated.

data is the unambiguous appearance of hydrodynamic shock angles, particularly in moving from the Ca + Ca up to the much heavier Au + Au studies, thus demonstrating the physical reality of the hydrodynamic picture of heavy-ion collisions. This has the very important consequence that densities substantially higher than normal can be anticipated in the heavy-ion collision, whereas had a cascade model been the appropriate one, there would have been negligible compression.

Up to now essentially all our information on the behavior of heavy-ion collisions at very high energy has been obtained at the Dubna Synchrophasotron and the Berkeley Bevalac; the latter is shown in Fig. 33 with the beam path indicated from the Superhilac injector down the hillside to the former Bevatron proton synchrotron now used jointly as the Bevalac to obtain something over 2 GeV per nucleon for all ions up to and including uranium. A modern copy of this facility is currently under construction at GSI in Darmstadt, West Germany, and the Lawrence

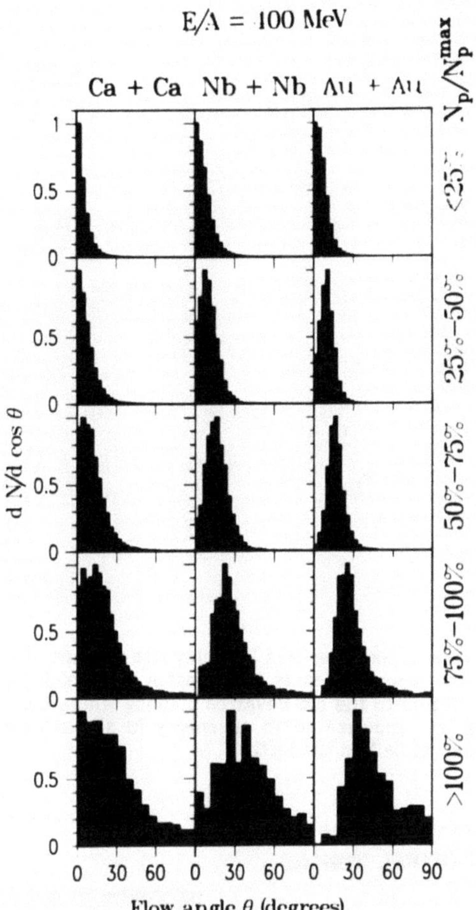

Figure 32: Results from identical particle collisions at 100 MeV per nucleon measured at the Berkeley Bevalac for Ca, Nb and Au. What is plotted here is the angular distribution, and the multiplicity of charged particles from the collision increases in moving from the top to the bottom of the figure. The hydrodynamic flow angle is clearly evident, particularly in the heavier systems with high multiplicity.

Berekeley Laboratory has proposed a major upgrading of the Bevalac for use in the medium energy (i.e. less than or equal to 5 GeV per nucleon) heavy-ion energy regime; a competing but somewhat lower energy proposal has also been put forward by the Oak Ridge National Laboratory in the U.S.

THE NUCLEAR EQUATION OF STATE

Much of the interest in high-energy heavy ions can be associated with attempts to understand the equation of state for nuclear matter. Figure 34 is a somewhat reworked version of the much more familiar PV diagram familiar from classical thermodynamics. Here I plot temperature against the density, and the goal again is that of going from hadronic matter to quark-gluon plasma where the

Figure 33: An aerial view of the Lawrence Berkeley site showing at the top the Superhilac injector. The arrow indicates the transfer line which passes across the Superhilac parking lot, then down a hillside to the old Bevatron building where the heavy-ion beam is injected into the synchrotron ring and accelerated to an energy in access of 2 GeV per nucleon before extraction into the large target hall on the left.

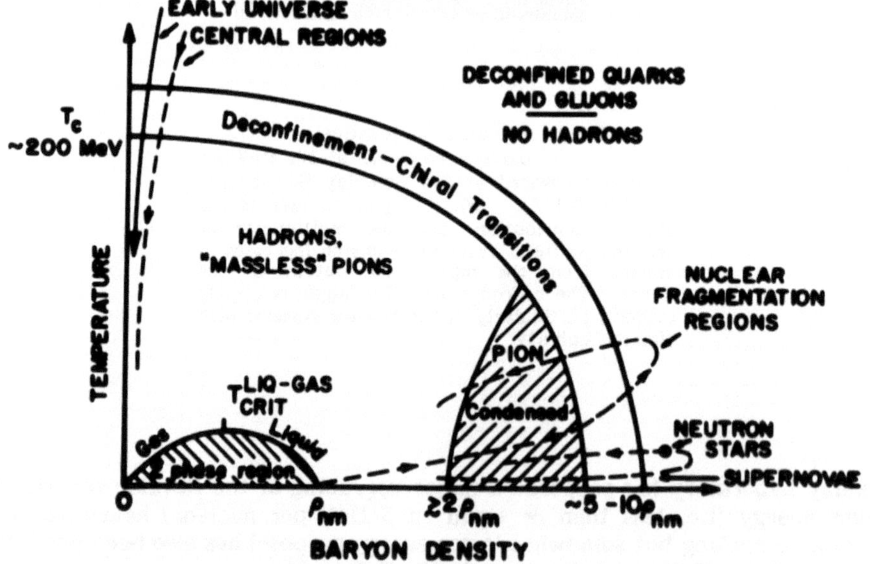

Figure 34: A schematic phase diagram for nuclear matter. The transition between hadronic and quark matter would be sharp and first-order in infinite nuclear matter, but is smeared into the band shown because of the finite size of nuclei. The trajectories followed in the early universe and in supernovae creation of neutron stars are indicated as is the path anticipated in ultrarelativistic heavy-ion collisions leading to nuclear fragmentation. As indicated at density more than twice normal, it is anticipated that pion condensation should occur, whereas at densities less than normal liquid-gas phase transitions should be discernable.

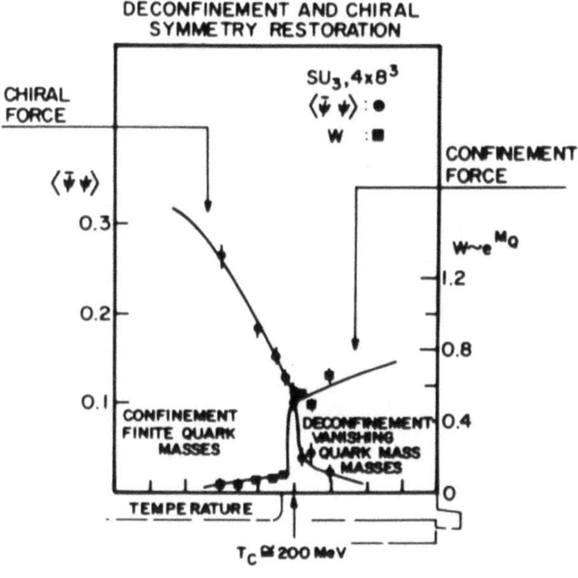

Figure 35: The results of coupled first-order phase transition calculations in lattice gauge theory. Shown here, as functions of temperature are the chiral and the confinement forces with an indicated critical temperature of 200 MeV.

quarks are entirely deconfined. If nuclei were of infinite extent, it would be assumed that the transition between hadronic and quark matter would be a first-order one at a precisely defined temperature for example; the fact that nuclei are of finite size, however, smears this transition region into the band shown.

There remains considerable ambiguity and uncertainty as to the actual parameters that should be applied to this diagram. The belief that the transition temperature should be something like 200 MeV comes from modern lattice gauge calculations with results such as those shown in Fig. 35. Plotted here as functions of temperature are the chiral force and the confinement force, the latter plotted as the Wilson parameter W. It appears clear from these calculations that at very low density, one would expect critical temprature of about 200 MeV. It is obvious from simple physical considerations that as the density increases, the critical temperature will decrease, as the bags containing the quarks in individual nucleons are pushed into closer and closer proximity. It is still, however, not clear at very low temperatures what density would be required to give deconfinement, although all calculations place this critical density somewhere between five and ten times normal nuclear matter density.

This figure shows a number of other interesting phenomena for which we can search. If indeed there is substantial compression in heavy-ion collisions—as the data of Fig. 32 suggest—then so also should there be rarefaction of the nuclear matter, and as shown in Fig. 34, a liquid-gas phase transition should occur. This should be observable in the characteristics of the fragments emitted from the collisions giving rise to the rarefactions. Considerable effort has already been devoted to searching for the pion condensation predicted in almost all models of heavy-ion collisions. The pion condensate, if it exists, would be an extremely interesting physical system in that a coherence phenomenon quite analogous to that of photons in a laser should be observable in the condensate. Unhappily, none of the measurements thus far have given any evidence suggesting its presence.

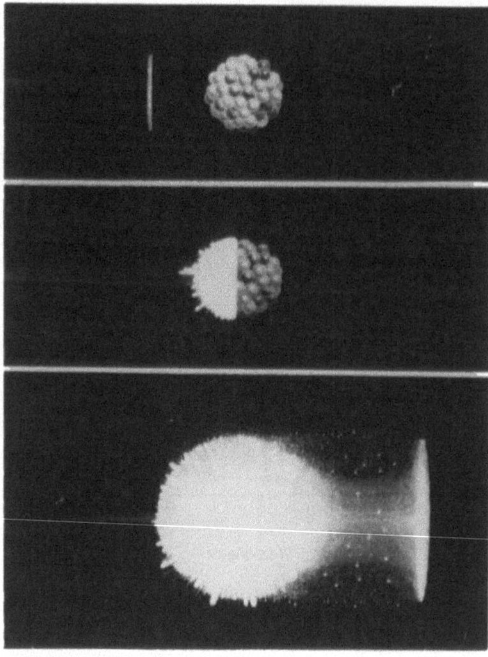

Figure 36: An artist's conception of the collision of a uranium projectile with a uranium target at ultrarelativistic energies. The upper panel shows the situation prior to the collision, the middle one during the collision and the bottom one immediately after the collision when both target and projectile have been raised to very high temperature and the so-called fire tube containing the quark gluon plasma has been formed between them.

As is clear from this figure, nature has already probed its behavior along both the temperature and density axes. During the expansion and cooling of the early universe, matter came down along the temperature axis, and in neutron stars and in supernovae, it is anticipated that densities in excess of the critical one are obtained. In accelerator induced collisions, the hoped-for phase path is indicated by the dashed line passing outward into the nuclear fragmentation region. Whether this is actually attainable or not remains to be determined experimentally.

Figure 36 is an artist's conception of the collision between an ultrarelativistic uranium projectile and a uranium target. David Shirley of LBL has commented that while this figure is long on relativity, it is very short on quantum mechanics; although this is unquestionably true, it is an instructive figure. The upper panel shows the incident uranium relativistically foreshortened into a disk just prior to its incidence on the target. As indicated earlier, at these energies, nuclei are essentially transparent to one another, so the projectile passes through the target but in doing so heats both the target and itself to extremely high temperatures. After passing through the target as shown in the lower panel, we have two regions, formerly the projectile and the target, now at very high temperature and still with high baryonic number, while in between is a region which we assume to have essentially zero baryon number but a high density of high temperature quarks and gluons--in other words, a quark-gluon plasma, a totally new state of matter. If we are to study this quark-gluon plasma, we obviously require

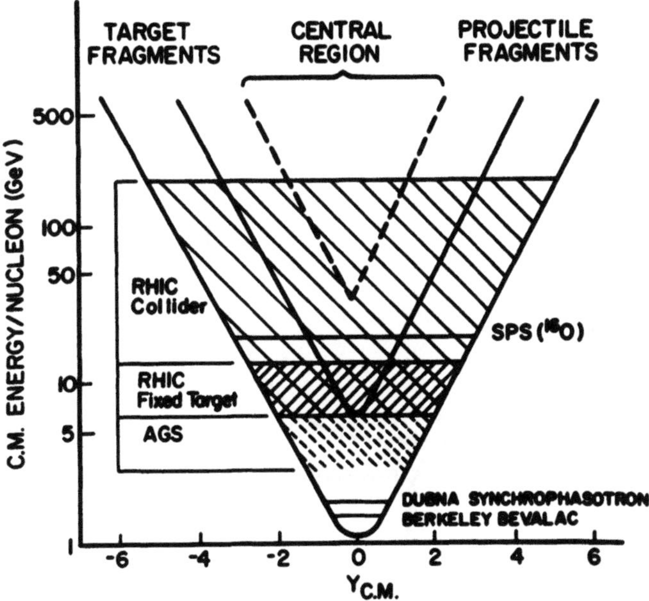

Figure 37: The kinematics of relativistic heavy-ion collisions plotted as a function of the center-of-mass energy per nucleon against the center-of-mass rapidity. The regions occupied by target and projectile fragments are shown, as is the central region where the quark gluon plasma—the fire tube—would be expected. Also shown on this diagram are the regions accessible to a variety of existing and proposed heavy-ion accelerators.

characteristic signatures for its presence. I shall return to this question below. We also however need to have the target and projectile-like regions sufficiently separated so that we can gain experimental access to the so-called fire tube containing the quark-gluon plasma. This raises the question of how high the projectile energies must be to achieve this separation.

This matter is addressed in Fig. 37 where I plot the center of mass energy per nucleon in GeV against the rapidity (effectively a measure of the momentum transfer). We are interested in the so-called central region well separated from the region containing either the target or the projectile fragments. I have shown on this figure the regions accessible to various present and proposed accelerators. Obviously, neither the Dubna Synchrophasotron nor the Berkeley Bevalac can give access to the fire tube, nor indeed can the AGS at Brookhaven when it becomes operational later this year. The SPS at CERN, which will accelerate ^{16}O beams later this year and ^{32}S beams next year, comes the closest of anything to be available in the near future. However, even it, as shown here, will be somewhat marginal in terms of giving adequate separation of the central region from the target and projectile regions with high baryon density. The only facility that truly makes available the central region is the Relativistic Heavy-Ion Collider (RHIC) now proposed by the Brookhaven National Laboratory and that operating in its collider rather than its fixed target mode. Figure 38 shows a schematic diagram of the proposed RHIC accelerator with the beam from the tandem Van de Graaff source being transferred to the alternating gradient synchrotron for acceleration up to 15.2 GeV per nucleon. This part of the facility will be functional in autumn of

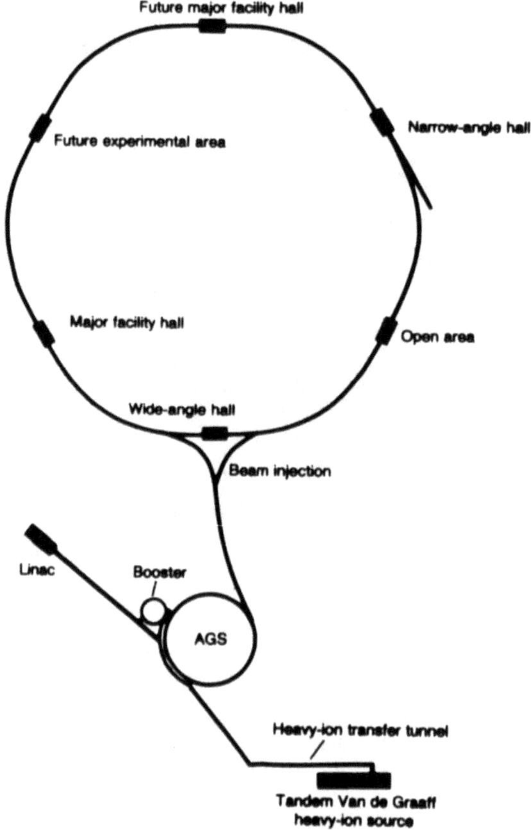

Figure 38: A schematic diagram of the proposed relativistic heavy-ion accelerator at the Brookhaven National Laboratory. The tandem Van de Graaff and the alternating gradient synchrotron with its linac injector are already in existence, as is the heavy-ion transfer tunnel. The booster synchrotron is approved for construction, and the tunnel and much of the associated hardware for the large ring are in existence following their construction as part of the ISABELLE high-energy physics project.

1986. Also approved for construction is the indicated booster which will increase the range of heavy ions from sulfur to gold and at the same time increase the proton beam intensity—and, therefore, the produced kaon flux—from the AGS by a factor of 4, making it into a mini-kaon factory.

The RHIC proposal will take the acclerated beam from the AGS, inject it into the existing tunnel constructed originally for the proposed high-energy physics ISABELLE accelerator and accelerate it further to 100 GeV per nucleon. With injection and acceleration in both directions in the main ring, it will then be possible to examine head-on collisions of these beams at the six indicated collision regions. Figure 39 is an aerial view of the Brookhaven site with the components of Fig. 38 shown in place. This proposal for RHIC has received the highest priority rating of the U.S. Nuclear Science Advisory Committee and approval in principle by the U.S. Department of Energy; now the question is one of phasing it into the national program in nuclear physics. The fact that the RHIC proposal uses essentially the entire prior investment in the ISABELLE accelerator makes it a remarkably cost-effective proposal and one that will give clear access to the

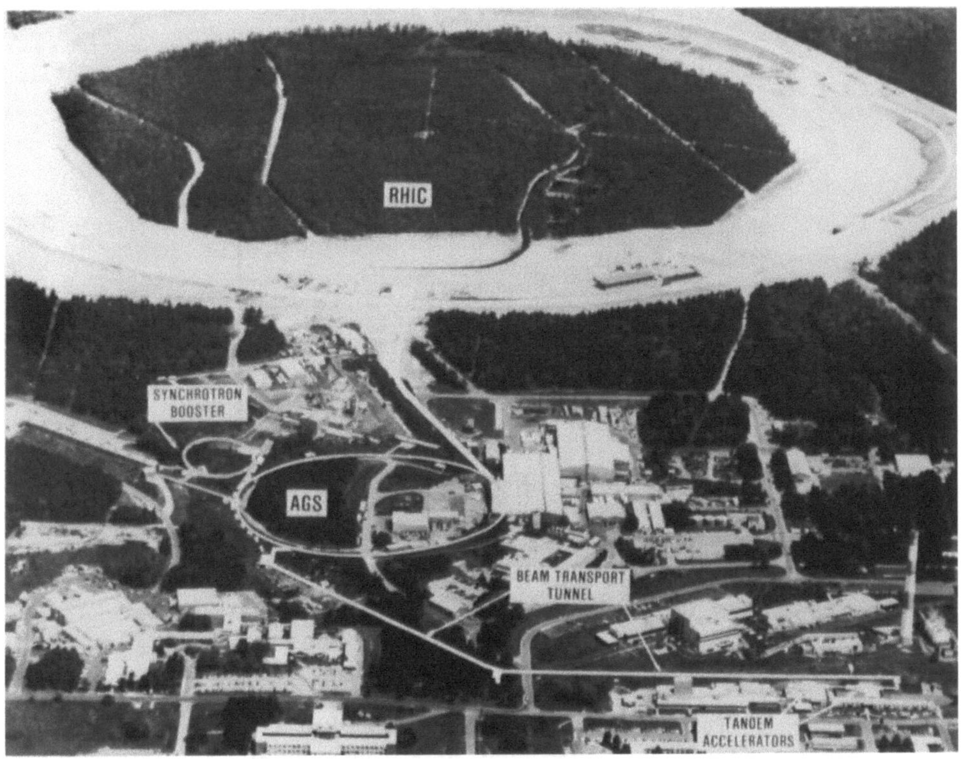

Figure 39: An aerial view of the Brookhaven National Laboratory's central site showing the various components of the proposed RHIC accelerator.

central rapidity region and, on the basis of current knowledge, to the quark-gluon plasma.

The proposal has received the highest priority rating from NSAC because it addresses some exceedingly fundamental problems in both nuclear and particle physics as well as in astrophysics. Among these are the deconfinement of quarks in nuclei, the creation of the quark-gluon plasma, the recreation of the early moments of creation, the physics of gravitational collapse of massive stars, the equation of state of nuclear matter, the possibility of entirely new kinds of matter, and the strong force including perturbative QCD. It bears emphasis that these studies are highly complementary to those now being carried out on the high-energy frontiers of particle physics. In particle physics, the emphasis is on delivering ever increasing energy to smaller and smaller volumes in the hope that this energy will be materialized into new particles; in nuclear physics the emphasis is on delivering ever increasing energy to volumes large enough to contain many nucleons (quarks) and thus large enough to permit the possibility of collective phenomena. The availability of a wide range of projectile and target species in these heavy-ion interactions also will be of great assistance in the isolation of specific interaction features and phenomena.

Figure 40 summarizes the status of heavy-ion accelerators in the U.S. and in Europe when RHIC is completed. The German project for an upgraded Bevalac (SIS) covers the same domain as indicated here for the Bevalac.

I have noted that signatures for the quark-gluon plasma are essential. Several have been suggested, including strange particle, antiparticles and lepton

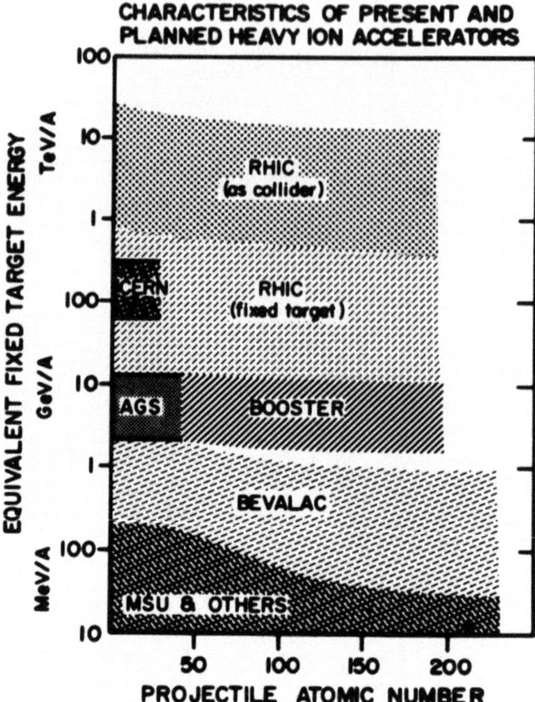

Figure 40: A Blann diagram showing the equivalent fixed target energy available from current and proposed U.S. and European heavy-ion accelerators as a function of projectile atomic number. The SIS accelerator under construction at GSI will have essentially the same profile as that shown here for the Bevalac.

pairs; but there is no clear concensus regarding the most sensitive or regarding the establishment of the natural background production rates of these signatures through a simple condensation of the plasma. Only experimental studies can shed light on these questions. And it is of particular importance that too much attention and expectation not be focused on the preliminary and limited studies that will become available later this year from Brookhaven and CERN. It may well be that these studies will indeed provide unambiguous evidence for the quark-gluon plasma--but if they do not, we must not lose heart and must remember that only very light projectiles will be involved, perhaps too light to generate an adquate fire tube.

THE IMPORTANCE OF HIGH-ENERGY PROTON MACHINES

Although not yet included in the national plan for U.S. nuclear physics, there is growing interest, worldwide, in the availability of more intense sources of kaons and higher mesons that will make possible the more commonplace introduction of strangeness into the nuclear domain. Substantial progress has already been made in this area using time on elementary particle physics machines at CERN, at Brookhaven and at Fermilab. Much of the emphasis thus far has been on the study of hypernuclei wherein reactions involving an incident kaon and outgoing pion convert one of the neutrons in the target nucleus into a lambda and thus create a hypernucleus. Povh and his collaborators at CERN as well as Barnes and others in the U.S. have already made substantial progress toward understanding the lambda-nucleon interaction, and as shown in Fig. 41, a substantial number of hypernuclei have already been studied.

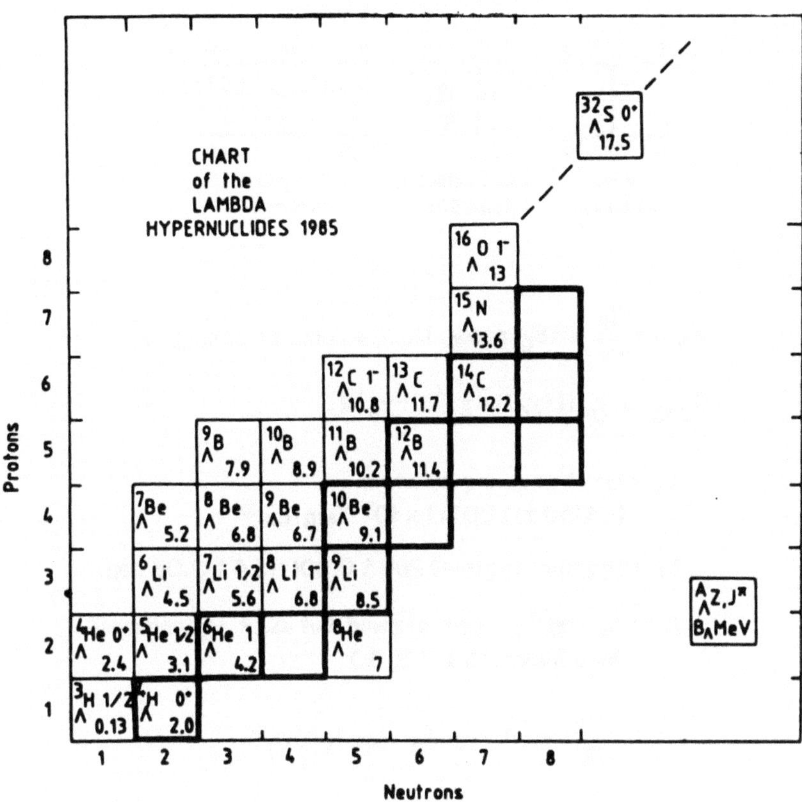

Figure 41: A Z vs. N plot of the lambda hypernuclei that have been identified and studied up to 1985.

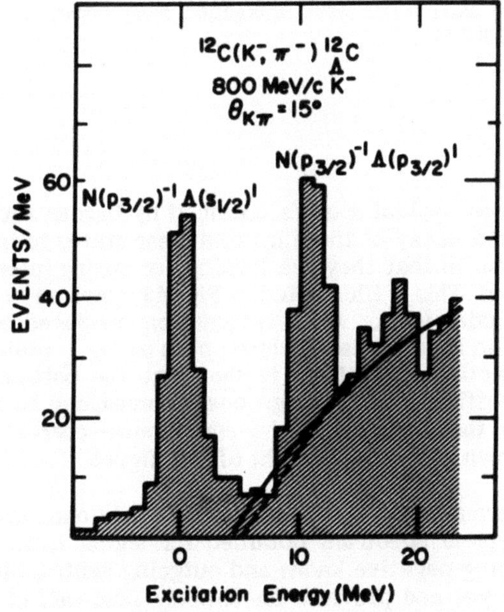

Figure 42: Typical spectra obtained in the (K^-, π^-) reaction on ^{12}C with 800 MeV/c kaons leading to the indicated lambda hypernuclear states.

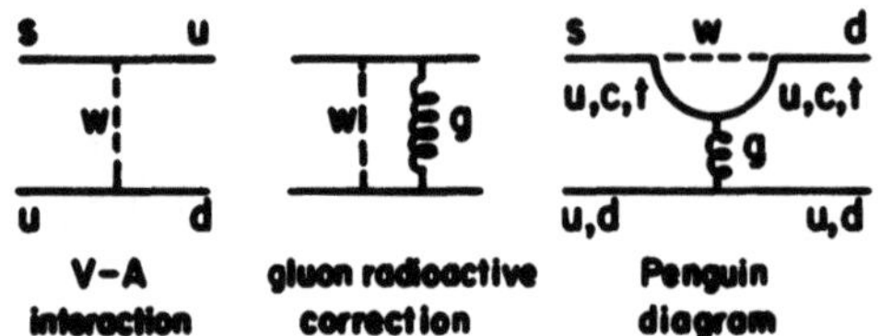

$$H_{V-A} = \frac{G_F}{\sqrt{2}} \sin\theta_C \cos\theta_C \, O_{V-A} + \text{complex conjugate}$$

$$O_{V-A} = \bar{u}\gamma_\mu(1-\gamma_5)s\bar{d}\gamma^\mu(1-\gamma_5)u$$

G_F = Fermi coupling constant
$(1.4150 \pm 0.0011) \times 10^{-49}$ erg·cm^2

θ_C = Cabibbo angle = $0.269 \pm 0.001 = 15.4 \pm 0.06$ deg.

$\Delta I = 1/2$ and $3/2$ both allowed but $\Delta I = 1/2$ enhanced by a factor $15 \leq F \leq 30$

$$-\frac{G_F}{\sqrt{2}} \sin\theta_1 \cdot \cos\theta_2 \cdot \cos\theta_3 \sum_j C_j O_j$$

Wilson coefficients: $C_1 = 1.51$; $C_2 = -0.87$; $C_3 < 10^{-2}$; $C_4 < 10^{-2}$

$$O_1 = O_{V-A}; O_2 = \bar{d}\gamma_\mu(1-\gamma_5)S\bar{u}\gamma^\mu(1-\gamma_5)u$$

Figure 43: Feynman diagrams and the formalism for establishing strong interaction corrections to the vector axial-vector weak interaction decay occurring in lambda hypernuclei.

Figure 42 shows typical spectra obtained by Barnes and coworkers who have used their data on the decay of these hypernuclear states to investigate the inverse of the usual situation, in that they are looking for strong interaction corrections to the weak interaction. This is illustrated in Fig. 43 which shows these corrections to the normal vector-axial vector weak interaction mediated by the W boson. The pure weak interaction Hamiltonian is shown here as h_{V-A} while that in the presence of the strong interaction corrections is shown at the bottom of this figure. The first two Wilson coefficients—the large ones—correspond to the gluon radioactive correction whereas the latter two—the small ones—correspond to the so-called penguin diagram shown at the upper right of this figure.

Lambda hypernuclei are of course not the only ones available for study, and Fig. 44 shows a typical spectrum obtained for sigma hypernuclei produced in a reaction with incoming negative kaons and outgoing neutral pions. This entire field is in its infancy as yet and provides an entirely new way of probing not only the nuclear interior—because the Pauli principle permits the addition of strange hyperons into orbits that are already filled with nucleons—but also, as illustrated in Fig. 43, fundamental studies on the elementary interactions themselves.

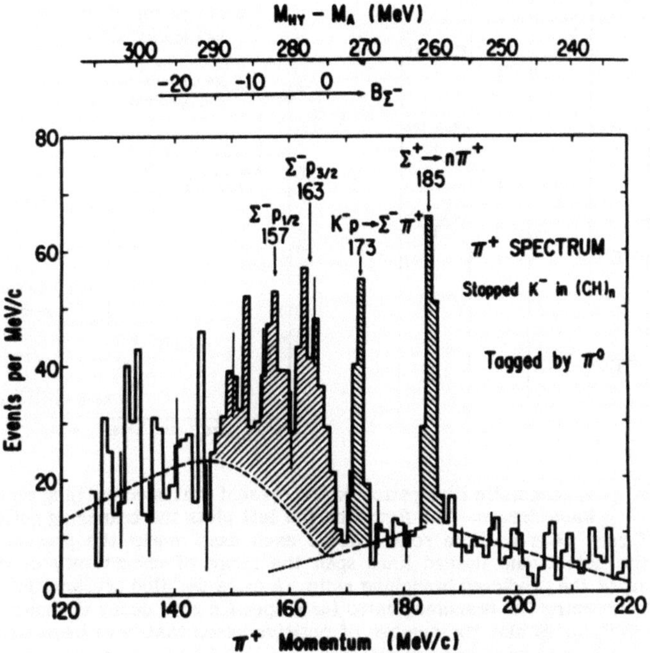

Figure 44: A typical positive pion spectrum from the bombardment of ^{12}C with negative kaons showing sigma hypernuclear states.

RARE KAON DECAYS

Since the discovery of the kaon, emphasis has been focused on the so-called rare decay modes because they provide stringent tests for the standard models. What is normally considered is the branching ratio defined as the ratio of the forbidden decay rate, as for example those shown at the top of the left panel of Fig. 45, to the allowed decay rate, that is, the decay of the kaon into a muon plus neutrino. The standard model suggests, for example, that the K^+ decay to $\pi^+\nu\bar{\nu}$ should have a branching ratio of about 10^{-11}; but new interactions which have at least been proposed could raise this to between 10^{-9} and 10^{-10}, as shown here, while the presence of massive neutrinos could lower it to 10^{-12} as shown by the dashed line. Currently, the experimental limit is, as shown by the solid line, about 10^{-7}. This is shown in greater detail on the right of this figure where I show the branching ratio limits as a function of time, together with the number of positive kaons which were studied in the different experiments. It is clear that with the proposed kaon factories it would be possible to scan the regions of branching ratio predicted by the standard model, including variations with massive neutrinos and new interactions. Here, the relationship with elementary particle physics is a close one, of course, and this is a new area into which nuclear physicists are moving with enthusiasm.

Earlier I mentioned the EMC effect and the current uncertainty as to its explanation. In Fig. 46, I again show the experimental data and list four quite distinct and contradictory conclusions that have been published based on analyses of these results. As shown in the lower part of this figure, if we are to resolve this question and establish which of these four competing explanations is the valid one, it will be necessary to go beyond the deep inelastic scattering of electrons and muons. As shown here for electron and muon scattering, both quarks and antiquarks enter equally into the determination of the form factor, and the form factor is

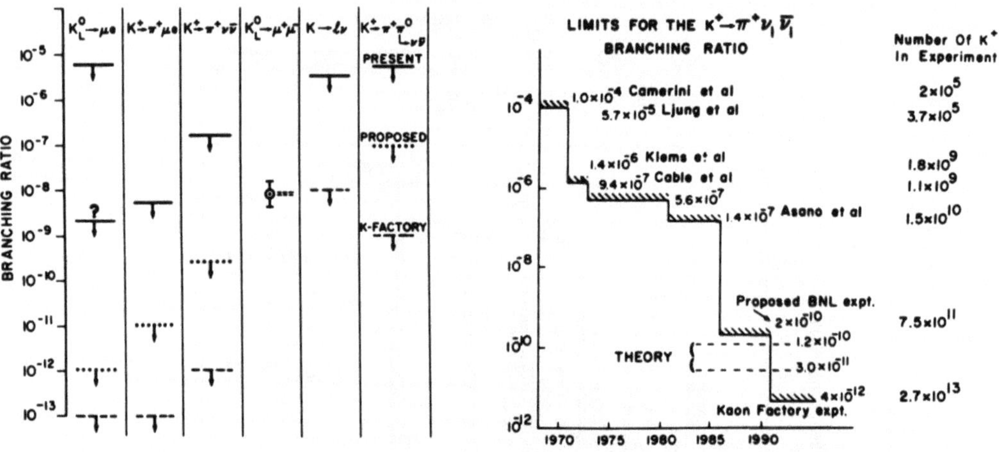

Figure 45: A schematic illustration of the present status concerning rare kaon decays. The figure on the left plots the branching ratio for a variety of rare decays. The solid line in each case marks the present limit, whereas the dotted and dashed lines span the range of uncertainty of current calculations of the predicted branching ratio. A more detailed presentation of the situation concerning the branching ratio for a specific rare decay is shown on the right, which includes also the number of positive kaons that have been studied in each of the indicated proposed experiments.

insensitive to flavor since it is summed over all three flavors. If instead one uses pions or kaons as the probe particles, one takes advantage of the Drell-Yan process which involves the production of virtual, very heavy photons. Now the ratio of form factors has two terms: pion scattering emphasizes the first term and therefore gives results that are similar to those obtained with electrons and muons; kaons, on the other hand, emphasize the second term and in consequence are sensitive to the presence of sea, as opposed to valence, quarks and are also sensitive to flavor. Thus, by comparing deep inelastic scattering of kaons with that of electrons, muons and pions, it should be possible to establish unambiguously which of the four alternative explanations listed in Fig. 46 is the physically valid one.

Kaon factories have been proposed in Canada, in Japan, in Switzerland and in the United States. In the U.S. case, the proposal is shown in outline in Fig. 47 and would make use of the Los Alamos 800 MeV proton linac as an injector first to a 6 GeV synchrotron ring and then to a 45 GeV ring. There is a wide range of fascinating physics that awaits the availability of facilities on this high-intensity frontier.

THE UNITY OF NUCLEAR PHYSICS

But let me emphasize that the physics does not separate neatly as a function of projectile energy. As I illustrate in Fig. 48, it is well recognized that the effect of the Z^0 boson is highly suppressed for in lepton-lepton and in lepton-hadron collisions. Where it would be expected to show up strongly is in the weak interaction between nucleons; but there, its detection becomes exceedingly difficult because of the presence of the strong interaction. In order to be able to detect the weak interaction effects at all, it is essential to have a signature which is unambiguously attributable to weak phenomena. Fortunately, this is available in parity nonconservation. What I plot in Fig. 48 is the asymmetry in the scattering of polarized protons from nucleons—in this case, from protons in hydrogen and water targets. The low-energy point was taken with the Los Alamos tandem accelerator, the next with the injector to the SIN cyclotron in Zurich, that at 800 MeV at the Los Alamos Meson Physics Facility and that at 6 GeV as the last experiment on the ZGS before its demise at the Argonne National Laboratory. What is important here

QUARKS IN NUCLEI

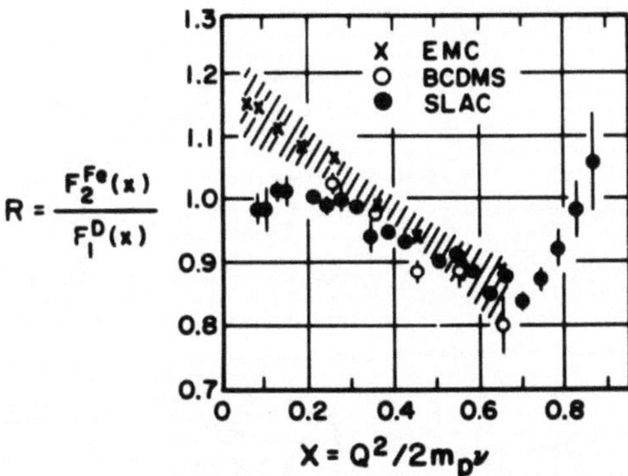

1. Valence quarks in nucleons deconfined but held in nuclear volume
2. Nucleon interactions produce multibaryon clusters
3. Nucleon size increases in the nuclear environment
4. Density effects on pion currents enhance quark sea

DEEP INELASTIC SCATTERING OF ELECTRONS AND MUONS

$$F_2(x) = x \sum_i e_i^2 \left[q_i(x) + \bar{q}_i(x) \right]$$

Both quarks and antiquarks enter equally: insensitive to flavor

DRELL-YAN PROCESS

$$R_{DY}^{NA}(x_1, x_2) = \frac{\sum_f e_f^2 \left[\bar{q}_f^{h}(x_1) q_f^{A}(x_2) + q_f^{h}(x_1) \bar{q}_f^{A}(x_2) \right]}{\sum_f e_f^2 \left[\bar{q}_f^{h}(x_1) q_f^{N}(x_2) + q_f^{h}(x_1) \bar{q}_f^{N}(x_2) \right]}$$

Pion scattering emphasizes the first term—quarks
Kaon scattering (or proton at large x_1) emphasizes second term — sea quarks
Latter also sensitive to flavor

Figure 46: The upper panel shows the EMC data as a ratio of the iron to the deuterium form factors plotted as a function of the Bjorken scaling variable. Listed under the figure are the four quite distinct explanations that have been proposed in the literature for these data, and below that is the formalism for deep inelastic scattering of electrons and muons and of pions and kaons respectively, the latter invoking the Drell-Yan process and the creation of a virtual superheavy photon.

is to note first of all that the entire energy range from 10 MeV to 100 GeV is very important to understanding the phenomena involved. A precise series of measurements even at tandem accelerator energies could, for example, serve to distinguish between meson exchange and quark bag models, and a new series of high precision measurements under way at the TRIUMF facility in Canada will provide critical data between the second and third points on this figure. This provides an excellent illustration of the fact that the boundaries between nuclear and particle physics are completely artificial ones, and their crumbling in the past few years is a very healthy sign.

NONACCELERATOR NUCLEAR PHYSICS

After all this discussion focusing on accelerators, it is absolutely essential to emphasize that there is much exciting and challenging work to be done in nuclear

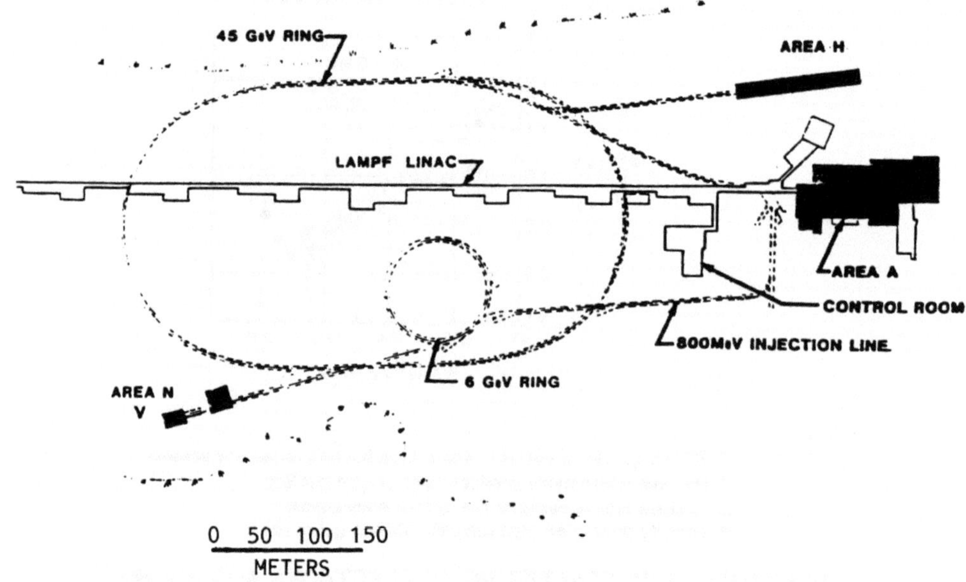

Figure 47: The proposed LAMPF II facility at Los Alamos. The existing 800 MeV proton linac will be used as an injector, first to a 6 GeV booster ring and then to a 45 GeV ring. The outline of the Los Alamos MESA is also sketched showing that the proposed facility occupies essentially all the available real estate at Los Alamos.

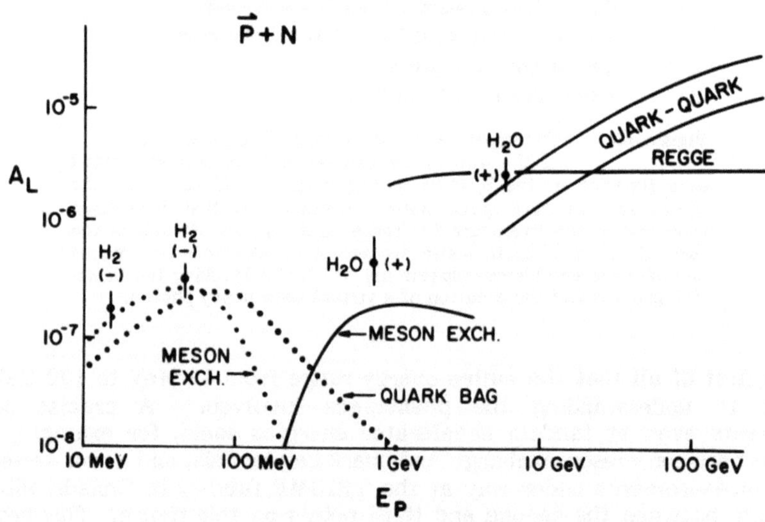

Figure 48: The asymmetry in the scattering of polarized protons from protons is here plotted against the energy of the incident polarized protons. The point at 15 MeV was obtained with the Los Alamos tandem accelerator that had 50 MeV, with the injector to the SIN cyclotron in Zurich, that at 800 MeV at Los Alamos and that at 6 GeV from the ZGS accelerator at the Argonne National Laboratory. Also shown here are the predictions of a variety of different models. It is clear that much more extensive and more precise data are required before any firm conclusions can be drawn.

SUPERALLOWED BETA DECAY STUDIES

Koslowsky *et al.* – Chalk River Nuclear Laboratories

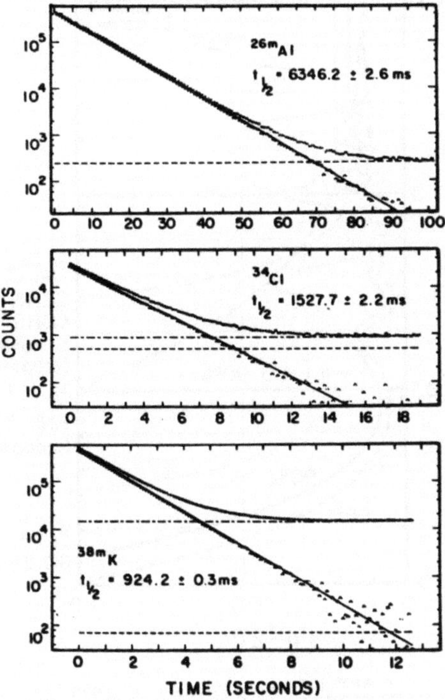

Figure 49: Typical decay curves in the study of superallowed beta decay at the Chalk River National Laboratories.

$$\begin{bmatrix} d' \\ s' \\ b' \end{bmatrix} = \begin{bmatrix} V_{ud} & V_{us} & V_{ub} \\ V_{cd} & V_{cs} & V_{cb} \\ V_{td} & V_{ts} & V_{tb} \end{bmatrix} \begin{bmatrix} d \\ s \\ b \end{bmatrix}$$

weak eigenstates — mass eigenstates

By convention, the charge $2/3\,e$ quarks are unmixed

Superallowed beta decays yield $|V_{ud}| = 0.9752 \pm 0.0003$

Hyperon and $K_{\ell 3}$ decays yield $|V_{us}| = 0.231 \pm 0.003$

Neutrino production of charm yields $|V_{cd}| = 0.24 \pm 0.03$

Assuming unitarity and only 3 generations yields the following values.

$\lvert V_{ub}\rvert = 0.007 \pm 0.007$	$\lvert V_{td}\rvert = 0.012 \pm 0.012$
$\lvert V_{cs}\rvert = 0.972 \pm 0.001$	$\lvert V_{ts}\rvert = 0.052 \pm 0.017$
$\lvert V_{cb}\rvert = 0.053 \pm 0.017$	$\lvert V_{tb}\rvert = 0.998 \pm 0.001$

Figure 50: The Kobayashi-Maskawa quark mixing matrix. As indicated, the superallowed beta decay data yields a determination of the appropriate matrix element with an order of magnitude greater accuracy than is obtained from the study of hyperon or kaon decays and two orders of magnitude greater than that obtained in the neutrino production of charm.

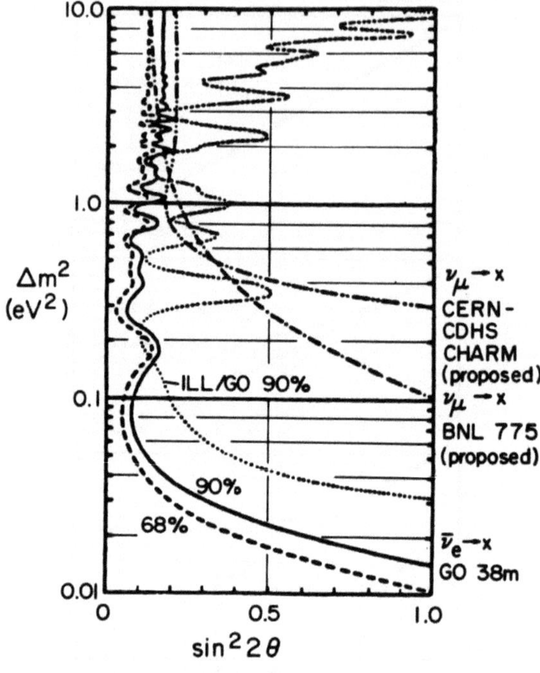

$$P(\nu_{\ell 1} \to \nu_{\ell 2}) \sim \frac{\sin^2 2\theta}{2} \left[1 - \cos(2.53\, \Delta m^2 (L/E_\nu))\right]$$

Figure 51: The current status of measurements setting limits on the mass of different neutrino flavors. The probability of oscillation between neutrino flavors is shown at the bottom of this figure and is a function both of the angle θ and of Δm^2, the square of the mass difference between the neutrino flavors. What is plotted here is Δm^2 vs. the $\sin^2 (2\theta)$, and the region to the right of each of the curves has been excluded by that experiment. As is obvious the only nonexcluded regions now correspond to very small values of θ or to very small values of Δm^2.

physics that requires no accelerator at all. Beta decay studies since the earliest days of our field have continued to provide vital new information. In Fig. 49, for example, I show measurements on superallowed beta decay from the Chalk River Laboratories and, in Fig. 50, how the results of these superallowed decays provide by far the most precise input we have to establishing the matrix elements in the Kobayashi-Maskawa quark mixing matrix. Figure 51 illustrates yet another very topical area where beta decay and neutron oscillation searches are of critical importance in attempting to establish limits on the mass of different neutrino flavors. Additional studies on double beta decay, for example, continue to provide critical information on lepton number conservation and other important questions in nuclear and particle physics. The list is a long one and one that shows no sign of dwindling.

THE FRONTIERS OF MODERN NUCLEAR PHYSICS

Together, all of these areas make up the frontier of nuclear physics, as I illustrate in very highly schematic fashion in Fig. 52. Nuclear physics is by its very

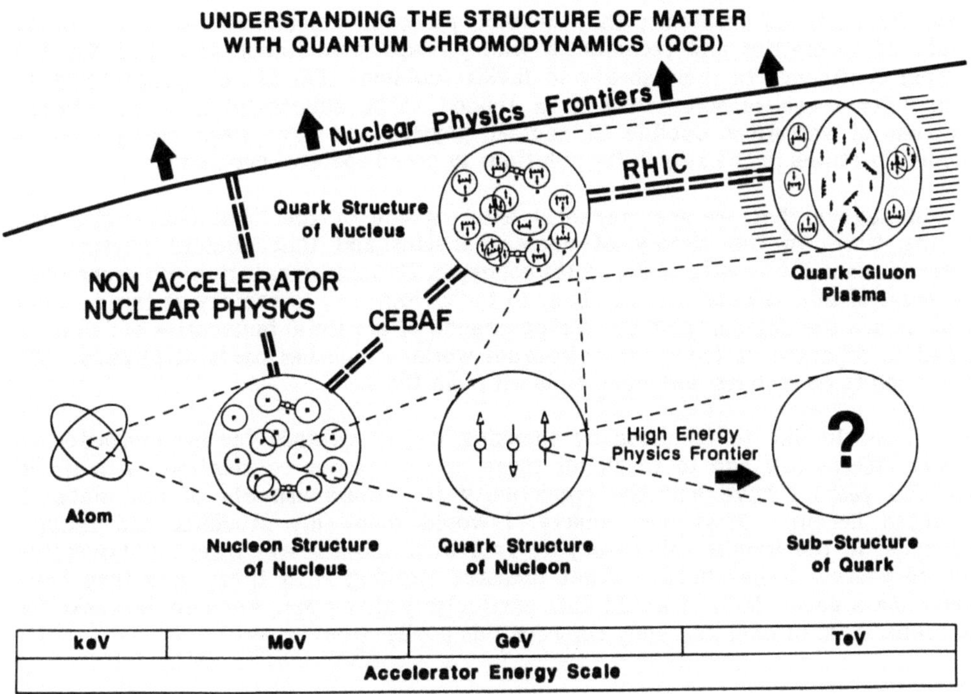

Figure 52: A very schematic illustration of the frontiers of modern particle physics and of modern nuclear physics. Shown here is the full range of activities beginning with the atom with its characteristic keV energies moving to the nucleus considered as a complex of nucleons with its characteristic MeV energy range and on to higher energies. At the GeV level, the quark structure of the nucleon and, therefore, of the nucleus becomes apparent while it moving to TeV energies in the nuclear physics domain, the emphasis is on the quark-gluon plasma and on collective phenomena that might be involved in this new state of matter. In particle physics, in the TeV domain, the emphasis is on the possible substructure of the quark itself and the testing of the predictions of the now standard model and variants of it. In particle physics, the frontier is toward ever higher energies, whereas in nuclear physics the frontier is a broader one ranging all the way from nonaccelerator nuclear physics to that in the TeV range on the quark-gluon plasma. The role of CEBAF in moving from hadronic to quark structure of the nucleus is indicated here, as is that of RHIC in moving from quark structure of the nucleus to the anticipated quark-gluon plasma.

nature a very broad ranging science, and information from beta decay can be of critical importance in the understanding of phenomena measured at the highest energies available from our largest accelerators. It is essential that we keep this fabric intact and that we recognize the interdependence of the different areas represented within our field and on its boundaries with other sciences.

Although this is not the place to develop it, it bears emphasis too that in a major way the future of nuclear physics is strongly coupled to its many applications in other sciences, in technology and throughout society. Nuclear physics in many ways is unique in its scope, both intrinsic and extrinsic, in the range of phenomena it covers within science and in the range of problems to which it can make a contribution in contemporary society.

It also bears mentioning that those trained in modern nuclear physics have, not through any intrinsic merit of the field but rather through a rather happy coincidence, had to develop a systems approach to their work in order to make progress. Students in our field must learn to mobilize the resources and talents of a wide range of technologies and support personnel if they are to make progress; they develop confidence in attacking what at the beginning frequently appear to be broad, unstructured problems; they develop personal expertise in a wide ranging variety of specialties and techniques; and, perhaps most important, they develop personal confidence in their ability to tackle problems that are completely new to them. For these reasons, they have tended to be successful in a very broad spectrum of activities outside of nuclear physics and have been recognized as welcome partners in the search for solutions to broad societal problems.

I believe that we are currently entering one of the most challenging and exciting times in the history of nuclear physics and that nuclear physics has important contributions to make in its concepts, its instrumentation, its techniques and, most of all, its people to science, to technology and to society at large. I am happy to see the beginning of the disappearance of artificial boundaries which have tended to separate us from our colleagues working in other parts of physics. We have much to contribute and much to learn from them.

I should like to conclude by thanking Walter Greiner for his invitation to prepare this review and to thank all those, too numerous to mention, with whom over the years I have had the opportunity to discuss aspects of the material presented herein. Most particularly, I would thank my students and faculty colleagues in the Wright Laboratory at Yale with whom I have interacted over the past 20 years. It has been a great pleasure working with them, and they have taught me a great deal. I would also particularly thank Ms. Annalee Jacunski for the preparation of camera-ready copy for this publication.

VAN DER WAALS FORCES AND ZERO POINT ENERGY

H. B. G. Casimir

De Zegge 7

NL-5591 TT Heeze, The Netherlands

In the theory of the so-called Casimir effect two lines of approach are coming together. The first one is concerned with Van der Waals forces, the second one with zero-point energy. Two recent reviews show that there has been considerable activity along both lines. In 1984 Barash and Ginzburg[1] published a review paper "Some problems in the theory of Van der Waals forces" that was reprinted in an English translation in 1985. It contains 327 references. In March 1986 there appeared in Physics Reports a survey "The Casimir effect" by Plunien, Müller and Greiner[2] that deals primarily with zero point energy. It contains 156 references, only partly identical with those in the Barash-Ginzburg paper. Clearly there is no shortage of material to study. However, much of the more recent work is rather above my rapidly aging head and in any case I am utterly unable to add something new. So I shall go back 35 or 40 years, to the beginning of this development, when things were still simple. Perhaps this can help to understand the basic principles involved. Also, this offers a good example of the relation that can exist between fairly fundamental work and industrial research.

Let me begin with Van der Waals forces. Johannes Diderik van der Waals (1837-1923) is famous for his equation of state (most of us learnt about it at school)

$$(v-b)\left(p + \frac{a}{v^2}\right) = RT \quad ,$$

in which b accounts for the volume of the molecules and a/v^2 for the attraction between them. This equation played an important role in late 19th century and early 20th century physics and, although it is by no means rigorously valid, it provided valuable guidance for the liquefaction of gases, especially of hydrogen and helium,so in a way it became a starting point for low temperature physics.

To assume an attractive force between neutral atoms or molecules was a bold step and when one began to understand more about the structure of the atom it became even more surprising that atoms like helium that have no permanent dipole moment nor any higher moment should attract each other.

Fritz London was the first to work out the quantum mechanical theory.

For two atoms, A and B, the total Hamiltonian will be given by

$$H = H_A + H_B + S_{AB} \quad ,$$

where S_{AB} is the electrostatic interaction. If we restrict ourselves to dipole interactions we can write

$$S_{AB} = \frac{\vec{q}_A \cdot \vec{q}_B}{R^3} - 3 \frac{(\vec{q}_A \cdot \vec{R})(\vec{q}_B \cdot \vec{R})}{R^5} \quad , \tag{1}$$

where q_A and q_B are dipole moments and R is the distance between the centres of the atoms. The first order perturbation vanishes - the expectancy value of S_{AB} is zero - but the second approximation gives a non-vanishing result. It can be written in a number of ways. For reasons that will become clear later on I prefer the following formulation. The static polarizability of an atom is the sum of partial polarizabilities, each one corresponding to a transition from the ground state to a specific excited state:

$$\alpha(o) = \sum_k \alpha_k \quad , \tag{2}$$

where α_k is given by

$$\alpha_k = \frac{2}{3} \frac{|q_{ok}|^2}{\hbar\omega_k}$$

with

$$\hbar\omega_k = E_k - E_o \quad .$$

The second order interaction energy can now be written as

$$U = -\frac{3}{2} \frac{hc}{R^6} \sum_{k,l} \frac{\alpha_k \alpha_l}{\lambda\!\!\!^{-}_k + \lambda\!\!\!^{-}_l} \quad , \tag{3}$$

where the index l refers to the second atom and where $\lambda\!\!\!^{-}_k = c/\omega_k$.

These London-Van der Waals forces do not only figure in the equation of state, they also play an important role in surface phenomena and contribute to the cohesion of solids. It is an interesting feature of these forces that they are additive and hence give rise to attraction between macroscopic objects. The simplest case is the attraction between two thick flat plates at a distance a. If we write $- A/R^6$ for the attractive potential between two atoms and if the elements of volume dv_1 and dv_2 contain $\rho_1 dv_1$ and $\rho_2 dv_2$ atoms, then the total interaction energy is given by the sixfold integral

$$U = -\iint \frac{\rho_1 \rho_2 A}{R^6} \, dv_1 dv_2$$

which leads to

$$U = -\frac{2\pi}{24}\rho_1\rho_2 \cdot \frac{A}{a^2} \cdot S \quad , \tag{4}$$

where S is the total area of the plates (edge corrections being neglected). Slightly more complicated is the interaction between two spherical particles. The integrals are elementary but tedious. However, it is easily shown that if the shortest distance between the surface of the spheres - this distance we denote by a - is small compared with either radius,we may apply Eq. (4) with an effective area

$$S_{eff} = \frac{2\pi R_1 R_2}{R_1+R_2} \cdot a \quad .$$

The notion of Van der Waals attraction between spherical particles was turned to good account by E.J.W. Verwey and J. Th. G. Overbeek at the Philips Research Laboratories who during World War II and shortly thereafter developed a theory of the stability of colloids in which these attractive forces are counteracted by repulsive forces between double layers[3]. Essentially the same theory was worked out by Derjagin and Landau - as was discovered after the war. Varwey, who was my co-director at the Philips Research Laboratories, had written a thesis on colloids and it was natural that he was interested in completing his work, but was this of any interest to the Philips organization? As a matter of fact it was. Colloidal solutions and suspensions are frequently used in the electronic industries when surfaces have to be covered with fine powders, as in cathode-ray tubes and in luminescent lamps and it was an accepted policy of the research laboratories to try to really understand empirical procedures and not to be satisfied with a recipe that worked well in practice but was not understood.

It was Overbeek who took the next step. It was found experimentally that suspensions of coarse particles were rather more stable than was predicted by theory and so great was his confidence in the main ideas of that theory that he concluded that it must be the $1/R^6$ law that is at fault. He then put forward the following argument. If we want to visualize the London force we may imagine that in Atom A an electron is going round and hence a dipole is rotating. It will produce a rotating field at atom B that will induce by polarization a rotating dipole. This in turn gives rise to an electric field at A that is synchronous with the rotating dipole and hence gives rise to a non-vanishing interaction energy. But if the distance is comparable with the wavelength corresponding to the frequency of rotation this induced field will show a phase-lag and the interaction energy will diminish. The argument is suggestive; it is also misleading. An atom in its fundamental state is not surrounded by a periodically varying and propagating field: if this were so there would be an outward pointing Poynting vector and the fundamental state would lose energy! All the same Overbeek's idea prompted Polder and myself to look for radiative corrections to the London theory[4]. It occurred to us that we might first study the interaction of an atom with a perfectly conducting wall, taking into account the interaction with the radiative field. In this case the Hamiltonian is

$$H = H_o + S + G_r \quad ,$$

where H_o is the Hamiltonian for the atom in free space, and S the electrostatic interaction of the atomic dipole with its electrostatic image,

$$S = -\frac{q_x^2+q_y^2+2q_z^2}{16a^3} ,$$

a is the distance of the atom to the wall. G_r represents the interaction with the quantized radiation field and can be written either as

$$G_r = \sum_n \left\{- \frac{e}{mc} (\vec{p}_n\cdot\vec{A}) + \frac{e^2}{2mc^2} A^2\right\} ,$$

the form used by Polder and myself, or as

$$G_r = \sum_n (\vec{q}_n\cdot\vec{E}) .$$

The first order in S, which is proportional to e^2, gives now a non-vanishing results, namely,

$$U = \bar{S} = -\frac{4}{3} \sum_k |q_{ok}|^2/a^3 ; \tag{5}$$

To find radiative corrections we have to calculate the second order perturbation caused by G_r. The result is surprising. Each term in Eq.(5) has to be multiplied by a correction factor that depends only on the ratio of a and λ_k and that is a monotonically decreasing function of that ratio. In the limit of large a one finds for the energy of interaction the surprisingly simple formula

$$U = -\frac{3\hbar c}{8\pi R^4}\cdot \alpha(o) , \tag{6}$$

where $\alpha(o)$ is the static polarizability(Cf. Eq. (2)).

This elegant result gave us the courage to tackle in a similar way the influence of the radiation field on the interaction between two atoms. The Hamiltonian is

$$H = H_A + H_B + S_{AB} + G_r^A + G_r^B ,$$

and we have to calculate all terms that are proportional to e^4 eliminating those that are part of the selfenergy. The work was done in the autumn of 1946, the year before the Lambshift was discovered and before Schwinger and others began to formulate QED and by modern standards the calculations may look rather clumsy. However, the results have later been confirmed in many different ways. They can be summarized as follows. Each term in Eq.(3) is multiplied by a correction factor that decreases monotonically with increasing distance. For large distances the interaction is given by

$$U = -\frac{23}{4\pi} \cdot \frac{\hbar c}{R^7} \cdot \alpha(o)^A \cdot\alpha(o)^B , \tag{7}$$

where $\alpha(o)^A$, $\alpha(o)^B$ are the static polarizabilities.

But how did zero-point energy get into this picture? During a visit to Copenhagen I told Niels Bohr about our results concerning Van der Waals forces. He reacted by saying "that is nice, that is something new". When

I remarked that I was still looking for a simpler derivation of the asymtotic formulae he muttered "it must be a manifestation of zero-point energy". As far as I remember that was the whole of our conservation on this problem, but it led into a new direction[5].

We are all of us familiar with zero-point energy of simple mechanical systems. The energy levels of a harmonic oscillator for instance are given by

$$E_n = (n + \frac{1}{2}) \hbar\omega \quad ,$$

and the lowest energy is

$$E_o = \frac{1}{2} \hbar\omega \quad .$$

This is connected with the uncertainty principle. If we know that the kinetic energy of the oscillator is very small we do not know its potential energy: it may be very large. If we know the oscillator is close to its equlibrium state then its kinetic energy may be large. The fundamental state is the best possible compromise. Of course the existence of zero-point energy has been amply confirmed, not only for vibrations of molecules but also for lattice vibrations in solids. They lead to a broadening of X-ray diffraction lines, but even more striking is in my opinion the vapour pressure difference between neon isotopes. For vapour pressures at low temperatures we have the approximate formula

$$\ln p = C - \frac{\lambda_o}{RT} \quad ,$$

where λ_o is the heat of evaporation at absolute zero, and if two isotopes have a slightly different λ_o we have

$$\ln p_1/p_2 = - \frac{\delta\lambda_o}{RT} \quad .$$

Now the main difference in heat of evaporation between two isotopes is the difference in zero-point energy of the acoustical waves. For a monoatomic substance like neon all frequencies are proportional to $1/A^{1/2}$ where A is the atomic mass and the zero-point energy of the lattice vibrations will be of the form $B/A^{1/2}$ where B can be calculated with fair accuracy from Debye's theory. So we find

$$\delta\lambda_o \approx \frac{1}{2} B \frac{\delta A}{A^{3/2}} = \frac{1}{2} \lambda_o \frac{\delta A}{A} \quad .$$

The vapour pressure difference is sufficiently large to make possible a partial separation of neon isotopes by destillation. Incidentally, J. Haantjes, who did this work for his thesis at Leiden, later became the foremost expert on TV circuitry at Philips Research Laboratories.

There is a close analogy between the electromagnetic waves in a cavity and the acoustical waves in a solid. To a microwave expert the standing waves in a cavity - the modes of vibration - appear as very tangible things. From microwave theory we can also take a perturbation formula. If

$$\vec{E} = \vec{E}_o (x,y,z) \sin \omega t$$

is a mode of vibration and if at a point x_1, y_1, z_1 we insert a particle small compared with the wavelength and having a polarizability α then the change of frequency is to a first approximation given by

$$\frac{\delta\omega}{\omega} = -\,2\pi\,\frac{\alpha|E_0(x_1,y_1,z_1)|^2}{\iiint|E_0\ (x,y,z)|^2 dv} \quad .$$

The total zero-point energy of an empty cavity would be

$$U_0 = \sum_n \frac{1}{2}\hbar\,\omega_n \quad ,$$

where we have to sum over all modes. This sum is very divergent and we have to get rid of it either by just dropping it or by a less naive approach. Introduce a particle with polarizability α . The sum

$$\delta U_0 = \sum_n \frac{1}{2}\hbar\,\delta\omega_n \quad ,$$

is still divergent although it diverges less rapidly. Now consider this expression in two different situations I and II, and write down

$$\delta_I\,U_0 - \delta_{II}\,U_0 = \sum_n \left(\frac{1}{2}\hbar\,\delta_I\omega_n - \frac{1}{2}\hbar\,\delta_{II}\,\omega_n\right) \quad .$$

From the point of view of a physicist this sum can often be considered to be convergent, a statement that can be made more precise by defining

$$U = \lim_{\gamma\to 0}\ \sum_n \left(\frac{1}{2}\hbar\,\delta_I\,\omega_n - \frac{1}{2}\hbar\,\delta_{II}\,\omega_n\right) e^{-\gamma\omega_n} \quad .$$

In order to calculate the interaction between an atom and a perfectly conducting wall we consider a large cubic cavity and choose for II the situation when the atom is far from all walls; in situations I the atom is at a distance a from one of the walls. The calculation of U is now perfectly straightforward: we need only classical electrodynamics and simple integrations. I sometimes refer to this procedure as "Poor Man's Q.E.D.". It can also be used to calculate the interaction of two atoms. The unperturbed modes are then the modes of a cavity with one particle. We study the difference in zero-point enerty due to a second particle when it is far away from the first particle (situation II) and when it is at a distance R (situation I). The calculations are slightly more complicated but again, classical electrodynamics and elementary integrations are all that is required. And in this way Eq.(6) and Eq. (7) are confirmed.

There exists a certain analogy with the neon case: also there the difference between two situations leads to a measurable effect. But we have no isotopes of vacuum, we have to insert particles. It occurred to me that one might also change the shape of the vacuum and that idea led me to consider the interaction between two perfectly conducting planes[6]. We consider a cubic cavity with a dividing partition that can be either in the middle or close to one of the walls. This leads to the well-known formula for the interaction energy:

$$U = -\frac{\pi^2}{720}\,\frac{\hbar c}{a^3}\cdot S \quad , \qquad (8)$$

where S is the total area of the plate. If we denote by a_μ the distance measured in microns we find for the attractive force

$$F = 0.013 \frac{1}{a_\mu^4} \text{ dyne/cm}^2 \quad .$$

This force is small but measurable.

An interaction energy proportional to $1/a^3$ is also found if we integrate the interaction energy given by Eq.(7). One finds

$$U = - \frac{23}{120} \hbar c \cdot \frac{\alpha(o)^2}{a^3} \quad . \tag{9}$$

If we relate α to the dielectric constant ε by the formula

$$\frac{4\pi\alpha}{3} = \frac{\varepsilon-1}{\varepsilon+2} \quad ,$$

we find

$$U = - \frac{23}{120} \cdot \left(\frac{3}{4\pi}\right)^2 \cdot \left(\frac{\varepsilon-1}{\varepsilon+2}\right)^2 \cdot \hbar c \cdot S/a^3 \quad . \tag{10}$$

Although the derivation of Eq.(10) is only valid for $(\varepsilon-1) \ll 1$ it is interesting to remark that for $\varepsilon \to \infty$ this leads to

$$U = 0.0109 \hbar c \cdot S/a^3 \quad ,$$

whereas the correct value is

$$U = - 0.0137 \hbar c \; S/a^3 \quad .$$

Lifshitz[7] has made a rigorous theory for the attraction between dielectric plates. His formula reduces to Eq. (10) for $(\varepsilon-1) \ll 1$ and to Eq. (8)for $\varepsilon \to \infty$ and he has also treated the case of real imperfect conductors. His method is different from mine, but it has later been shown that his results can also be obtained by "poor man's Q.E.D."

This formula for the attraction between conducting plates has been confirmed with reasonable accuracy by several experimentalists. The forces are small but at distances of one micron or less they can become important. And technology moves more and more into the sub-micron region. As an amusing example I mention some observations that were made in the Philips Laboratories[8]. For establishing contacts with an aluminium strip on a microcircuit one often uses a gold-bonding technique. A thin wire carries at its end a little sphere of gold. This is pressed onto the substrate by a "micro soldering iron". Now if R is the radius of the sphere and δ the deviation of the sphere from its equilibrium position and h the distance of the centre of the sphere in its equilibrium position to the substrate, then the distance of the surface of the sphere to the substrate is given by $a = (h-R-\delta)$ and the attractive force is of the form

$$F_a = b/a^3 = b/(h-R-\delta)^3$$

for in this case the effective area is given by $S = \pi Ra$. On the other hand the elastic force acting on the sphere is

$$F_{el} = k \cdot \delta \; ;$$

k can be determined by measuring the frequency of oscillation of the sphere on its wire (of course away from the substrate). Now if we approach the wire to the substrate the wire will bend but at a certain distance the equilibrium becomes unstable and the sphere flips against the substrate. This was observed for many wires and although this is not a precision method for determining hc the agreement with theory was satisfactory. I think it is a fascinating thought that contacts with microcircuits can be put into place by the zero-point pressure of empty space (see Figure 1 below).

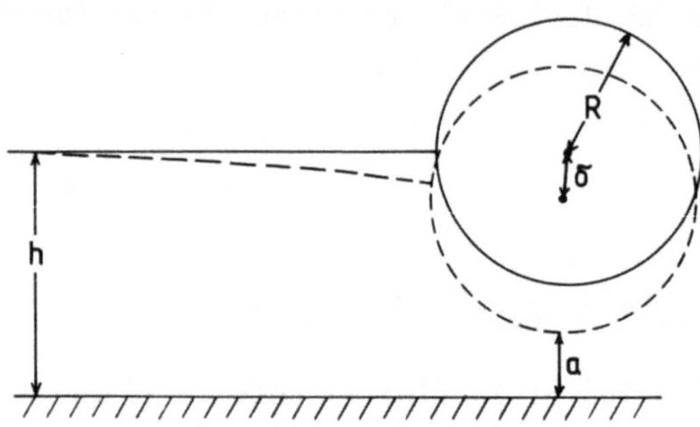

Figure 1

REFERENCES

1. Yu. S. Barash and V. L. Ginzburg, Usp. Fiz. Nauk 143:345-389 (1984) (Russian); Sov. Phys. Usp. 143:467-491 (English)
2. G. Plunien, B. Müller, and W. Greiner, Physics Reports 134:87-193 (1986)
3. E.J.W. Verwey and J. Th. G. Overbeek; Theory of the Stability of Lyophobic Colloids, Elsevier, Amsterdam (1948)
4. H.B.G. Casimir and D. Polder, Nature 158:787-788 (1946); Phys. Rev. 73:360-372 (1948)
5. H.B.G. Casimir, Journ. Chimie Physique 46:407-410 (1949)
6. H.B.G. Casimir, Proc. Kon. Ned. Akad. Wet 51:793-795 (1948)
7. E.M.Lifshitz, Zh. Eks. Teor. Fiz, 29:94 (1955)
8. M.J.Sparnaay and P.W.J. Jochems; Third Int. Congress of Surface Activity, Cologne 1960; Vol. 2 Sec. B/III/1, no. 56, p. 375

SUMMARY TALK: THEORETICAL

Abraham Klein

Physics Department
University of Pennsylvania
Philadelphia, Pa. 19104-6396 USA

I. INTRODUCTION

My very weak link to strong field physics began in June, 1971 when I spent a month in Frankfurt at the kind invitation of Walter Greiner, who proved, as always, an excellent teacher as well as an excellent host. I still recall the pervasive attitude which was that diving into the negative energy continuum was somehow not a respectable activity, at least for a physical phenomenon, and all reasonable efforts should be made to avoid it. However, all such efforts proved fruitless and within a couple of years the basic physics of spontaneous positron emission was elucidated by Müller, Rafelski, and Greiner and independently by Lewis Fulcher and myself (only thirteen years after the phenomenon was recognized in a paper by Voronko and Kolesnikov, a paper which was unearthed a few years after the work cited)(see W. Greiner's introduction to the Lahnstein proceedings).

Since then and up to the time of the Lahnstein Conference, I held two conflicting attitudes toward these ideas. The first was that one was not, after all, dealing with anything so fundamental from the point of view of elementary particles, but rather with an amusing detail of atomic structure in the extreme relativistic domain. This point of view was, however, modified by the extreme interest of audiences whenever I gave a colloquium on this subject, which I did over a period which was much longer than was really conscionable in view of my overall tenuous connection with the subject.

My present attitude is sharply different. On the one hand, if I may dare to comment, the experiments have uncovered surprises which we do not yet understand, but which at least promise an important discovery. On the other hand, the Frankfurt group, in addition to carrying its full and tedious responsibility to do the calculations necessary to analyze the experiments, has also broadened and deepened its interests to include strong field physics of all types, of which there are now a number of interesting examples. Thus my review responsibility is very broad, as will be evident in the account which follows.

During the two weeks of the "school" I attended over four dozen lectures in theoretical physics. The breadth of the subject matter is proven by the fact that I have been able to identify eight different subjects (Secs. II-IX) on which more than one lecture was given, and have included a Sec. X for interesting odds and ends.

Toward the end of the first week of the conference I arrived at a new concept of the role of the summary speaker (at least new in my experience): I would actually try to summarize the lectures. This seemed an obvious solution to me because I am not actively involved in any of the research topics covered. Therefore my best hope of surviving my ordeal was to comment, perforce superficially in most cases, on as wide a variety of topics as possible, since I couldn't go very deeply into any of them. This has not stopped me from rendering judgments in some instances, for which I hope I will be forgiven.

In the following I quote only the names of the lecturers whose detailed accounts appear in the preceding pages. I depend on them for the identities of their collaborators and predecessors. The order of presentation of the subject matter is partly determined by the temporal sequence at the conference.

II. PRESENT STATUS OF QED

By QED we mean a system of electrons, positrons and photons in "external fields". For precision experiments ((g-2, e.g.) the dynamical structure of the external fields becomes relevant.

A. Precision Tests. (Mohr)

We may distinguish two sets of experiments: (i) A single electron or muon in a weak external field, of which examples are the (g-2) experiments and the Lamb shift in H. (ii) Two "bound" particles with no external field, such as positronium. The upshot of a thorough review is that there is no convincing evidence of any discrepancy between theory and experiment. Only one instance was cited, the orthopositronium decay rate for which the experimental rate definitely exceeds the theoretical one. It was pointed out from the audience, however, that experimental deficiencies nesessarily yield a deviation of the observed sign, so that the onus is currently on the experimental side.

B. Lamb Shift of One-Electron Atoms, High Z. (Mohr)

It is now experimentally "routine" to strip all or almost all electrons from an atom. For one-electron atoms, it is far from routine to measure the transition $2p_{\frac{1}{2}} \rightarrow 1s_{\frac{1}{2}}$ as has been done for a number of elements. Theoretically, the difficulty is that expansion in powers of the Coulomd field of the nucleus cannot be tolerated. The natural tool is the Furry Interaction Representation (1949). In practice, this means that self-energy and vacuum polarization effects must be treated with the exact Coulomb Green's function. In diagrammatic terms, for instance, the self energy is written

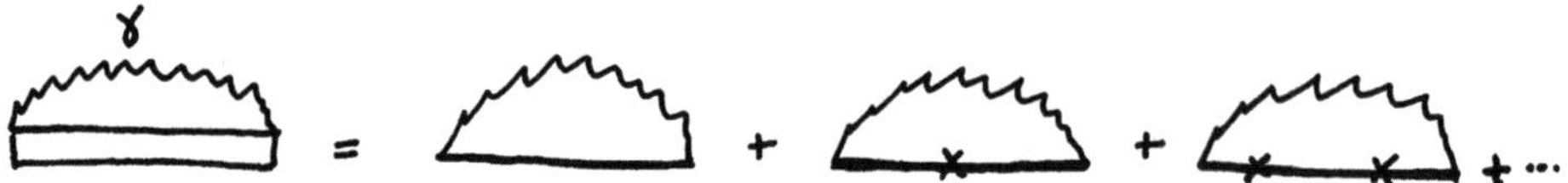

the doubled line representing the Coulomb propagator and the right hand side its formal expansion in αZ, valid only for sufficiently small Z.

Currently the LS at high Z can be measured to a percent. Theory and experiment are in concord to this level of accuracy.

C. Many Electron Theory, High Z. (Mohr)

This problem represents the current frontier of LS theory. Up to now treatments of the effects of screening have been rudimentary (effective charges) and only partially successful in bringing theory and experiment into agreement. A systematic theory is in the developmental stage. One is seeking a modified Furry picture in which the external field acting on a given electron is replaced by an effective field

$$A_\mu^{ext} \to A_\mu = A_\mu^{ext} + \delta A_\mu \quad ,$$

where δA_μ is the potential due to all other electrons. One would think first of choosing δA_μ to be the Hartree-Fock field of the electron distribution, but technically, this is too difficult to implement. How best to choose this potential so that it is both practical and accurate is an open problem.

D. QED for Quarks in a Cavity (Mohr)

The observation that the neutron-proton mass difference, measured in units of the quark mass is of the order of the fine structure constant,

$$\frac{m_n - m_p}{m_q} = 0.59 \frac{\alpha}{\pi}$$

suggests that this difference can be understood as one virtual photon self-energy effects of quarks. Such a calculation, interesting technically, was carried out within the framework of the static limit of the MIT bag model (fixed radius R). It yields the wrong sign for all possible values of the quark mass, $m = m_u = m_d \equiv (m_q)$.

E. Suggested Revision of QED (Barut)

Barut has challenged the present form of QED and proposed a revised theory with the following differences from QED:

(i) No real photons enter.
(ii) Fermions are treated in "first quantized" form.
(iii) For problems without real photons, the new theory agrees with conventional QED to order α and to all orders in the externally applied electromagnetic field.

There appear to be differences to order α^2, however, and since terms of this order have been confirmed by precision QED experiments ((g-2) and LS in H), doubt is cast on the correctness of this theory until this matter is settled.

III. (RELATIVISTIC) THOMAS FERMI THEORY. (Dreizler)

The aim of Thomas-Fermi (TF) theory, which has its genesis in almost the earliest days of quantum mechanics, is to construct a ground state energy functional of the density, $E_0[n(\underline{r})]$, such that the variational principle ($\delta E_0/\delta n(\underline{r}) = 0$) subject to suitable constraints (usually number conservation) determines the density $n(\underline{r})$ and the energy E_0. The application of this method to atomic, nuclear, and solid state physics has been stimulated by a now famous theorem of Kohn and Hohenberg which asserts the existence of a unique functional $E_0[n(\underline{r})]$. The proof is non-constructive, however, and the real task is to construct an accurate functional with the available tools of many body theory.

For non-relativistic nearly homogeneous systems such an accurate functional is known in the form of a sum,

$$E_o = E_T + E_x + E_{correl},$$

of kinetic energy, exchange, and correlation terms calculated up to the first corrections (gradient corrections) depending on ∇n. The great advantage of ease of application of this theory is somewhat balanced by the loss of any shell structure detail. The latter can be regained, however, by one relatively simple change known as the Kohn-Sham (KS) alternative. It is to replace the customary Fermi gas approximation to E_T by the kinetic energy of Hartree-Fock theory,

$$E_T = \sum_o \int \phi_o^*(-h^2\nabla^2/2m)\phi_o \ ,$$

where

$$n(\underline{r}) = \sum_o \phi_o(\underline{r})\phi_o^*(\underline{r}) \quad ,$$

and then vary with respect to ϕ_o^*. Both the original TF theory and the KS alternative have been remarkably successful in application.

The forefront of this subject lies in the effort to construct the TF approximation to QED (including gradient corrections). This effort has a long history of failures. In this regard Dreizler reported an important new result. The TF approximation with gradient corrections has been calculated for QED in the Hartree-Fock approximation which includes, as is well known, both self-energy and vacuum polarization corrections. Applications are just beginning.

IV. WHAT IS THE ORIGIN OF THE POSITRON PEAKS?

This is the core problem of the conference. I wish I knew the answer (this may be my last chance for fame). At this point I can do little more then recount the suggestions which we have heard during the past days. Though some of the speakers have been living with the exciting experimental facts for almost a year, none of the theoretical suggestions is so far very convincing. We shall deal with the heart of the matter first and then turn back to summarize the important auxiliary analyses which appear to rule out conservative interpretations.

A. The Data

For details we refer to the beautiful summary talks of Backe and Kienle, and for a theorist's point of view to the talk of B. Müller upon whom we shall lean heavily in this part of our summary. We call attention to the following observations:

(i) The decay of several quasi-bound (unstable) objects (≤ 4) with masses (1.2 MeV $\lesssim M \lesssim$ 1.8 MeV) into positron-electron (e_+e_-) pairs correlated back to back in the CM system.

(ii) The decaying object has a small velocity in the CM system of the colliding ions.

(iii) From the observed width we have a lifetime $\tau_s > 3 \times 10^{-20}$ sec.

(iv) The cross-section is "large", $\frac{d\sigma}{d\Omega} \sim 10\,\mu b/sr$.

(v) the cross-section grows like a high power of the united charge Z_u of the colliding ions, and is seen for subcritical systems.

(vi) The energy dependence for fixed charge is not yet completely clear.

To explain this data we have so far only incomplete conjectures or already discredited ones. The favorite conjecture at the moment is that we are observing the decay of a poly positronium complex, and we shall deal with this "idea" first.

B. Remarks on Polypositronium Complex (Müller, Rafelski)

For this suggestion, one must first explain the composition of the object. Is it a quasi-bound system of a few e_+ and e_- as suggested by Müller or is it a truly many-body system as advocated by Rafelski? The possible self-binding of such systems by purely electromagnetic forces seems unlikely, but perhaps merits more attention than it has so far received. Müller has conjectured a new many-particle coupling $\sim(\bar{\psi}\psi)^n$, where n is an integer ≥ 3; of course no one is thrilled with such an ad hoc hypothesis.

Granted the existence of such a quasi-bosonic object, one can suggest an explanation (Rafelski) for relative ease of production. Here the electric and magnetic fields of the colliding ions may provide nearly supercritical fields acting on the bosons, thus enhancing production.

As long as we are dealing with such vague statements, let us add to the catalogue. The fact that positron peaks are observed for subcritical systems, $Z<Z_c$, leads one to remember (Klein,Wong) that $Z>150$ is already interesting, in that for $150 \lesssim Z \lesssim Z_c$, positrons can be bound to the heavy ion system and one can think of bound e_+e_- systems. But these are bound in the first instance to the ion field, rather than to each other, so that the binding mechanism we are seeking is still absent.

Two intriguing related questions were raised which may provide a clue if indeed they make any sense: (i) Can the properties of the vacuum be effected by strong time-dependent external electromagnetic fields in ways other than those already explored and understood?

(ii) Can poly positronium be produced by "compression" of the e_+e_- wavefunctions in strong electric fields?

C. Other Esoteric Mechanisms Including Production of a New Neutral Particle (Schäfer, J. Reinhardt, Balantenkin)

In a seminar, A. Schäfer reviewed all the speculations of the Frankfurt group concerning the origin of the positron lines. There are two major categories:

Particle families	:	Massive gluon
		Higgs Particles
Composites	:	Magnetic resonances (Barut)
		Polypositronium

We refer to Schäffer's report for a discussion of the ad hoc particle hypothesis and to the proceedings of the Lahnstein conference for a detailed account of Barut's ideas (also this volume). As above, we consider these suggestions speculative and unconvincing.

Reinhardt studied the production of a new neutral particle under the assumption that it could be described by a simple effective (e.g. trilinear $= \bar{\psi}\psi\phi$) couplings. If such couplings exist, they contribute to QED, e.g. to the value of (g-2). Such contributions, bounded as they are by experiment, put stringent limits on the size of the coupling constants for such processes. If we now turn around and calculate production rates, we find them too low by orders of magnitude. Other objections can also be raised, such as much too broad momentum distributions of the decaying $e_\pm$.

B. Balantenkin repeated and extended many of Reinhardt's arguments and conclusions, especially investigating whether the "new particle" could be a version of the long-sought-after axion. The current data leaves little room for this possibility.

D. Has Spontaneous Positron Production Been Seen? (Müller, Heinz, Schramm)

Let us recall that in the first generation of experiments, one observed only a broad e_+ energy spectrum produced by "shake-off of the vacuum polarization", the width being associated naturally with the collision time $\tau_{coll} \sim 10^{-21}$sec. When the first narrow positron lines were observed for energies in the neighborhood of or slightly above the Coulomb barrier, there followed the Frankfurt hypothesis that the heavy ion (HI) potential could develop an attractive pocket in which the system could be trapped and delayed some fraction of the collision events. A suitable delay time in the well could possibly account for the narrow positron line, which is the spontaneous emission line. This hypothesis is not supported by the available experimental and theoretical evidence, but nevertheless constitutes an interesting line of inquiry upon which we report.

The theoretical model upon which the analysis is based was developed in detail in talks by Heinz (semi-classical approximation) and Schramm (quantum mechanical method) and summarized in the second lecture of Müller. We consider here only the semi-classical approximation. Here one assumes that initially the two ions follow a classical trajectory until they reach a separation R (at time t = 0). From t = 0 to t = T they suffer an elastic scattering described by a unitary matrix S, and reach once more the radius R, where the classical description of the ions resumes. The amplitude, a_{ij} for the electronic system to make a transition from an initial state i of the two-center Dirac equation (TCD) to a final state f is written as

$$a_{if} = \sum_k a_{ik}^{(-)} S(E-\varepsilon_k) a_{kf}^{(+)} ,$$

where (∓) refer as usual to incoming and outgoing waves, E is the total energy and ε_k is an eigenvalue of the TCD equation. The total probability P_{if} can be put into the form

$$P_{if} = \int_0^\infty dT\, f(T) |a_{if}(T)|^2 ,$$

as an integral of probabilities weighted by a delay time distribution f(T) which is a Fourier transform

$$f(T) = \int_{-\infty}^{\infty} d\omega\, e^{i\omega T} <S^*(E) S(E-\omega)>_{av} .$$

Referring for details and figures to the lecture of Müller, it turns out to be easy to fit the positron peak for any given choice of colliders, for instance U + U, but then, in contradiction to experiment, the peak will shift in energy (and height) for other collision partners and, of course, disappear when $Z_u < Z_c$. This outcome can only be avoided by a most unprobable set of ad hoc assumptions.

V. COUPLED CHANNELS; NUCLEAR PHYSICS

A. Atomic Processes as a Clock for the Measurement of Nuclear Reaction Times (Soff)

The use of inelastic nuclear scattering (in fact "deep inelastic scattering") to achieve increased collision times was originally suggested in order to enhance the possibility of observing spontaneous positron production. Though the original hope has not yet been realized, instead there appears to be developing a new method of studying nuclear interaction times (delay times) on the basis of atomic measurements.

As Soff demonstrated in his lecture, all of the following processes are promising in this respect:

(i) δ ray (continuum electron) energy distribution.
(ii) positron spectra.
(iii) K-vacancy production.
(iv) Molecular orbital x-rays.

In the case of (i) positive results have already been obtained (delay times of order 10^{-21}sec.) In addition a scaling law for intermediate energy collisions has been developed. The analysis in this section was all based on classical motion, including friction, for the ions.

B. Quantum Theory of Positron Production (Heinz, Schramm)

This work extends the analysis of Soff, as has already been partly explained in our discussion of Müller's talk. The matter involved here is technical, and we therefore defer to the lectures in question for such detail.

C. Conversion Processes as a Possible Source for Narrow Positron Lines (Soff)

In this lecture we were first treated to a review of the standard theory of internal conversion coefficients in order to establish that the theoretical machinery was under control. It was then applied to the observed supercritical system. The conclusion is once more that artificial and unbelievable physical conditions must be postulated to explain the singles peak. No possible explanation was forthcoming for the correlated events.

D. Do Giant Nuclear Molecules Occur, i.e. Is There a Pocket in the Interatomic Potential? (Oberacker, Pinkston)

Oberacker gave a systematic review of existing methods for computing HI potentials. He then described an improved microscopic approach which utilizes the adiabatic approximation: For any fixed separation, R, the nucleons of the approaching ions are taken to move in a simple (axially symmetric) two center shell potential with a few adjustable parameters. The systems are assumed to remain in the lowest energy Slater determinant $|\phi_0(R)\rangle$ of that potential. Let H be a many-body Hamiltonian consisting of a sum of the kinetic energy and a two-body force of Skyrme type with parameters fitted to a series of ground state energies. The interatomic potential is then given by the "Heitler-London" formula:

$$V(R) = \langle\phi_0(R)|H|\phi_0(R)\rangle - \langle\phi_0(\infty)|H|\phi_0(\infty) \quad .$$

The figure reproduced in Oberacker's lecture shows no pocket in the most favorable of circumstances. This agrees with recent independent findings.

During the discussion period, Greiner objected that the dynamics depended not only on the potential energy function, but also on the collective mass. The existing literature on the microscopic theory of low energy fission and fusion reactions supports the contention that the matter is not fully settled by the potential energy.

In a related talk, Pinkston employed simple models based on standard nuclear reaction theory to extend the theory of Heinz (Sec. B) to include the effect of inelastic channels on the formation of pockets. He concludes that the changes appear to go in the wrong direction.

He also discussed the interpretation of HI one-nucleon transfer data and concluded that a satisfactory interpretation of the data does not require nuclear pockets.

VII. DEVELOPMENTS IN ATOMIC THEORY

A. Atomic Physics with Relativistic Heavy Ions (Scheid, Becker)

These authors described a new and comprehensive program for computing ionization and pair production processes in highly relativistic kinematic regimes, $0.2 \geq (v_p/c) < 1$, where v_p is the projectile velocity. Extreme approximations were made for the ion motion: (i) Projectile moves in a straight line. (ii) Target recoil is neglected. For the electrons a number of approximations of increasing complexity were investigated including classical approximations, semi-classical approximations, finite difference methods, coupled channel calculations, and perturbation theory. These were applied to K-shell ionization, direct pair production and pair production with electron capture.

The results found are of essential interest both for the design and use of heavy ion colliders. For pictorial interest, we call the attention of the reader particularly to the time development after impact of a U K-shell electronic wave function, reproduced in the lecture of Becker.

B μ and τ Pair Production in Relativistic Heavy Ion Collisions (Strayer)

Strayer described a new Dirac-Hartree approach to the description of atomic processes in HI collisions. The state vector is at all times a Slater determinant $|\Phi(t)\rangle$ propagated in time from an initial Slater determinant $|\Phi(0)\rangle$ according to the equation

$$|\Phi(t)\rangle = S(t)|\Phi(0)\rangle \quad ,$$

which can be considered to be a time-dependent Thouless or Bogoliubov-Valatin transformation. $|\Phi(t)\rangle$ is part of a complete set $|\Phi_m\rangle$, which is needed to calculate the probabilities

$$P_m(t) = |\langle\Phi_m|\Phi(t)\rangle|^2 \quad .$$

The state $|\Phi(t)\rangle$ is constructed from fermion mode operators for the "orbitals" which are solutions of the Dirac-Hartree equations (in turn derivable from a variational principle)

$$i\,\dot{\psi}_n = \{\vec{\alpha}\cdot(\vec{p}-\vec{A}^{eff}) + \beta m + A_o^{eff}\}\psi_n,$$

$$\partial^{\mu}(F_{\mu\nu})^{eff} = J_{\mu} + \langle\Phi(t)|:j_{\mu}:|\Phi(t)\rangle \quad ,$$

where J_{μ} is the current of the external field of the ions and j_{μ} that of the electron-positron field.

To solve these equations, a method of splines, which is a generalized finite difference technique, was described.

Applications are at a very early stage. A one dimensional model of close collisions for production of μ and τ mesons was described. Production rates found are large compared to those estimated by Scheid and Becker using the Weizacker-Williams approximation. This discrepancy remains to be resolved, and the calculations are to be extended to realistic models.

C. Photoproduction of Positrons for Nearly Critical Electric Fields (Lemmer)

Calculations were reported based on the relativistic random phase approximation with full treatment of the continuum using coordinate space methods. For $Z \to Z_c$, enhanced photoproduction was predicted.

D. Symmetry Violation in Atoms (Schäfer)

In the 7s→6s transition in the Ce atom, one measures interference between the parity conserving M1 and a parity violating E1. The latter is presumed to have its origin in short-range parity violating forces derived from the "standard" (Glashow, Salam, Weinberg) model. In his lecture Schäfer first reviewed the modifications which are implied for the solution of the Dirac equation in external potentials which violate one or more of the discrete symmetries P,C,T. He then discussed the form of the parity-violating short range electron-nucleon interaction responsible for the E1 mixing. The main new contribution to the subject is the calculation of Dirac-Hartree-Fock wave functions for use in the evaluation of the required electromagnetic transition.

The result depends on only one unknown parameter, the Weinberg angle. Comparison with experiment yields a result completely in accord with the value obtained from decay experiments.

VII. MODELS OF QCD.

In this section, we describe those lectures which involved elements of QCD or simplified models based on or suggested by QCD.

A. Non-Topological Soliton Bag (Wilets)

We deal here with a field theoretical model which contains as a limiting case either the well-known MIT bag model or the (not so well-known) SLAC bag model. It is defined by a Lagrangian density

$$\mathcal{L} = \underset{1}{\mathcal{L}_q(\text{quark})} + \underset{3}{\mathcal{L}_\sigma(\text{sigma})} + \underset{1}{\mathcal{L}_{q,\sigma}} + \underset{1+\kappa(\sigma)}{\mathcal{L}_G(\text{gluon})},$$

where below each term we indicate the number of parameters associated with that term, and where every term is "standard" except for the gluon term which is written

$$\mathcal{L}_G = -\frac{1}{4}\kappa(\sigma)\, F_{\mu\nu}^{\ \ c} F_c^{\mu\nu} - g_s(\bar{\psi}\gamma_\mu \tfrac{1}{2}\lambda A_c^{\ \mu}\psi) \quad ,$$

containing a susceptibility function $\kappa(\sigma)$. In this theory, quark confinement is imposed by a static σ field and is not absolute. On the other hand, absolute color confinement can be achieved by a suitable choice of the function $\kappa(\sigma)$, which, however, destroys the renormalizability of the model.

Applications proceed initially in a standard way. One starts with the mean field approximation which breaks translational and Lorentz invariance. The techniques used to restore the symmetry and improve the accuracy of the theory are borrowed in part from nuclear physics - among them generator coordinates and variation after projection. The generally good agreement with experiment is encouraging for further development of the model.

In the writer's view this model is a serious one for its apparent ability to correlate experiment, but lacks aesthetic appeal.

B. Skyrmions (Biedenharn, Eisenberg)

We heard two lectures on this model, which is beautiful but not yet serious.

A basic feature of the Skyrmion model, which we do not try to summarize technically, is the occurrence of a soliton solution, called the "hedgehog". Biedenharn combined topological and group-theoretical arguments in an impressive way in order to explain, among other results, how the hedgehog solution describes baryons.

Eisenberg emphasized the problems of finding the practical implications of Skyrmion models. From its theoretical origin as the limit of QCD as the number of colors increases without limit, he emphasized that we should not expect an accuracy greater than 30%. He gave a lucid explanation of how the single hedgehog solution is actually calculated in practice. He discussed the non-uniqueness of the Lagrangian used in this theory. There was considerable interest in the fact that an approximate analytic calculation of a phase transition to a quark-gluon plasma can be carried out. In the end, however, the model is not in good practical shape because of its "failure" thus far in the two-nucleon problem. A "Heitler-London" calculation of the two nucleon interaction (based on the single hedgehog solution) fails to reveal the main attractive part of the central force (identified these days with π-meson exchange). No clear way out of this shortcoming was offered.

C. Baryon-Baryon Interaction and Quarks (Faessler)

Faessler described his "successful" version of a two baryon interaction derived from an ersatz version of QCD. The model, which contains relatively few parameters, has, in its simplest form, the following elementary interactions among quarks:

(i) A one-gluon exchange potential.
(ii) A linear confining potential which is supposed to have its origin in multi-gluon interactions.
(iii) A one pion exchange potential.
(iv) A one-sigma exchange potential.

(i) and (ii) are supposed to account for the short range part of the force and (iii) and (iv) for the long range part. As in all such studies, a six quark wave function is studied by means of the resonating group approximation. Individual quarks are found in a simple harmonic oscillator well. When we reduce the direct product of the wave functions of three quarks in each of the wells, there occurs, among others, a mixed symmetry basic function, omitted in previous calculations, which Faessler showed, is mainly responsible for the effective hard core behavior.

The calculations show good agreement with the data; the agreement is improved by further refinement of the model.

VIII. QUARK-GLUON PLASMA

This is a subject of widespread interest (which grew up around the writer when he wasn't looking), and in which, at the moment, the ratio of data to theory is an infinitesimal of higher order, so that nobody's theory is yet definitively wrong.

A. Formation of Anti-Matter Clusters During Hadronization of a Quark-Gluon Plasma (Heinz)

We refer to Heinz's lecture for the striking pictorial representations of the formation of a quark-gluon plasma and the subsequent transformation back into hadrons - hadronization. These representations are also relevant for discussion of several subsequent lectures.

Heinz argues that in the hadronization process following formation of a quark-gluon plasma in high energy nucleus-nucleus collisions, one has "excessive production of antihadrons" compared to the equivalent antihadron content (in terms of q and $\bar{q}$) of the plasma. This result is required by overall entropy conservation. The study of antihardon production may therefore be a suitable signal for the formation of the plasma.

B. Color Neutralization of SU(N) Strings (Gyulassy)

One picture of a high energy nucleus-nucleus collisions postulates two objects strongly excited in passing through each other and sharing between them a strong color electromagnetic field. The purpose of this contribution is to develop a theory of how to neutralize this color field by hadronization. The theory described is claimed to be "applicable over the last few units of rapidity."

Gyulassy first reviewed the theory of pair production in strong uniform fields in QED. He pointed out how the famous results of Schwinger could be rederived by a semi-classical method which could be applied to QCD. It turns out that the results of this generalization - to SU(N) - can be described by an effective Abelian field and is thus similar to the QED result.

C. Quantum Transport Theory of Quark-Gluon Plasmas (Gyulassy)

The quark gluon plasma when (if?) produced is not a system in equilibrium. It is therefore necessary to describe the equilibrium processes by transport theory. Gyulassy first reviewed the derivation of classical transport theory, starting from the non-relativistic Schrodinger equation, the basic tool being the Wigner transform. The extension to QED requires a careful consideration of gauge invariance and the choice of an invariant density as the basic quantity to study. The further extension to QCD is then formally straightforward, though technically more laborious. The machinery thus developed remains to be applied.

D. Gluon Condensation in Quark-Gluon Plasma (Lovas)

Lovas described a solution of the QCD equations for the quark-gluon plasma in which the normal plasma could "condense" into a phase in which both the gluon field and the quark density could have a periodic structure, provided the gluon field was given a mass. In the limit that this mass was sent to zero, the condensate persisted but became spatially homogeneous. The phase transition could be either first or second order and has the unusual feature that the condensate is established at high temperature.

E. Missing Pieces in the Description of the Quark Gluon Plasma (Rafelski)

Rafelski pinpoints gluons as the essential elements in the ultimate experimental verification of the production of a quark gluon plasma. The argument is essentially: (i) The gluon content of the plasma is "known". (ii) The gluons "like" to produce $s\bar{s}$ (strange pairs). In fact the $s\bar{s}$ productions may be up to several orders of magnitude greater than what would be predicted if the production in hadron-hadron collisions took place in ordinary hadron matter without the intervention of the plasma.

F. Do Nucleons Dissolve in Giant Nuclei? (Neise, Grabiak, Schnabel)

According to the first two authors, nucleons can dissolve into a giant (MIT) bag provided the bag constant is small enough. The consequences of this dissolution were discussed.

Schnabel discussed the phase transition of a bag of MIT bags, in particular emphasizing the importance of allowing the ingredient bags to be compressible.

IX. VACUUM STATES IN QED, QCD ...

A. Vacuum States in Strong Fields (Müller)

Müller began with a pedagogical discussion of the vacuum of QED in strong external electromagnetic fields. He then gave an introduction to recent developments which have disclosed models with fractionally charged vacua. This is definitely a phenomenon of supercriticality associated as well with models with suitable topological properties. A "realistic" model in current research is the hybrid model of quarks and Skyrmions. In this model, part of the charge can reside on the quark and part on the soliton (hedgehog).

B. QCD Vacuum (Danos)

Danos described a theory of the vacuum of the gluon field, which can be extended to a theory of the vacuum of QCD. The structure of the vacuum state is sought as a pair condensate of "low energy" transverse gluons. (Such pairs naturally carry momentum zero.) The first step of the calculation is to derive by projection methods an effective Hamiltonian in the low energy space and then apply a Bogoliubov-Valatin transformation to find the quasi-particles of the new vacuum.

Detailed study shows that depending on the parameters of the theory, the solution can be either a Bose-Einstein condensate or the pair condensate as originally proposed. Problems that remain to be solved in this theory are: (i) Confinement has been only partially established. (ii) The calculation does not maintain relativistic covariance.

C. Quantum Effects in Strong Gravitational Fields (Davies)

The discussion focussed initially on the difficulties associated with defining particles in curved spaces. One situation in which unambiguous physics can be extracted is that in which asymptotically flat spaces (in and out spaces) can be defined. The concept of model particle detectors in curved spaces was introduced and illustrated by examples. Quoting work of Manogue (see below), it was pointed out that not only can a suitably defined vacuum create particles, but it can be endowed with the property of viscosity. The highlight of a second lecture was a discussion of the Hawking radiation from black holes.

D. The Casimir Effect (Casimir, Manogue, Plunien)

We were privileged to hear an historical account of the development of what is now known as the Casimir effect. Manogue described results for the Casimir effect in gravitational fields, and Plunien described the mathematical tools currently available for studying the Casimir effect both at zero and at finite temperatures.

X. ODDS AND ENDS

A. Model Studies of Atoms and Molecules in Strong External Fields (Kleber)

Kleber described several simplified but non-trivial models: (i) A one-dimensional model of an atom in a strong external field. (ii) A model of laser assisted quantum tunneling.

B. Polarization Correlation in Two-Particle States (Merzbacher).

The speaker first reviewed the Einstein-Podolsky-Rosen "paradox". He then described a recent test of Bell's inequality, which, within limited accuracy is violated as required by quantum mechanics. The measurement is of the polarization correlation of two photons emitted back-to-back in the 2s-1s transition in Hydrogen.

C. Dimensionality of Space (Schäfer)

What are the observable consequences of attribiting to space-time dimensions other than four? There are several motivations for considering such modifications: Perhaps the most physically cogent these days are the multi-dimensional string theories of elementary particles. The difference between theory and experiment for the Lamb Shift (within the experimental error) is analyzed as a possible manifestation of higher dimensions. The value three for the space dimension is verified to better than one part in 10^{10}.

D. Nuclear Matter under Extreme Conditions (Faessler).

Faessler has extended the previous theories of pion condensation by arguing that previous authors have neglected exchange effects which are the same order of magnitude as the direct terms included heretofore. He finds enough effective cancellation as to preclude pion condensation.

A similar calculation precludes the Lee-Wick phase transition.

ACKNOWLEDGEMENT

The preparation of this summary would have been impossible without the cooperation of most of the speakers in providing me with at least temporary access to their lecture materials and in clarifying a number of issues. I am deeply indebted to Walter Greiner for giving me this opportunity to renew my interest in the area of strong field physics. This work was supported in part under U. S. Department of Energy Grant Number 40264-5-20441.

EXPERIMENTAL SUMMARY

P. Kienle

Gesellschaft für Schwerionenforschung mbH
D 6100 Darmstadt

First, I like to apologize for not giving an entertaining summary talk. I got so scared by the presentations of the positron peak story that I decided to concentrate on describing you what is really going on in order to relieve us all a bit.

In doing so I have to apologize a second time. You all realize that we have heard about other exciting experimental developments, which I will mention only in few sentences.

Saenger reported about δ-ray measurements, which represent an impressive progress in the application of atomic clocks to probe nuclear reaction times. Observables like the energy transfer, which are strongly correlated with the reaction times were fixed using kinematically complete measurements of the outgoing particles in deep inelastic collisions. Tserruya reported about experiments to probe the energies of quasi molecular states as function of the internuclear distance. Interference patterns of the Mo X-rays, observed when holes are introduced externally in a collision, are a sensitive tool for probing quasi molecules. Gould reported about cross section measurements for various atomic processes and precision spectroscopy of few electron system studied at medium energies using Bevalac beams. Silver presented a "selfcalibrating" and "Dopplerfree" X-ray spectroscopy method for measuring QED shifts in high Z-systems.

Five years after the Lahnstein meeting on the physics of strong fields,[1] during which first results on positron spectroscopy in heavy ion collisions have been reported, the question on the existence and origin of the mysterious positron peaks kept us excited and frustrated at the same time. We are excited because of the tremendous experimental progress of the field, demonstrated in the talks of Oeschler, Kozhuhàrov, Bokemeyer, Krämer, Cowan, Koenig and Stiebing. Backe has summarized the problems and gave an outlook for the future aspects of positron spectroscopy. We are also deeply frustrated because we have seen no solution of the problem and even worse, there may none in sight. Backe talked about hidden variables which seem to influence the results of the experiments. This sounded to me as if there really is something wrong with the experiments. I should like to address my summary to this question, but like to rephrase it as follows: "What do we think we know, what do we not know and what might we learn to know about e^+-production in heavy collisions?"

We have learnt during the last five years of studying e^+-production in heavy ion collisions at bombarding energies close to the Coulomb barrier, that it is a rare process with creation probabilities of the order of 10^{-7} - 10^{-8} per collision and keV positron energy. I think we have also learnt how to measure its main charactistics, at least we have identified the difficulties like target deterioration and have worked very hard to improve our very complex measuring methods constantly. I am the first time optimistic that we will soon have no hidden variables any more in the experiments. Whether this helps to find a clear solution of the problem I don't know. There are several complex sources of positrons in collisions of 5.9 MeV/u U with uranium, which we have to separate, measure their characteristics, and identify their origin.

Let me first talk about sources which emit continuous positron spectra, which we understand well. One is connected with nuclear excitations with energies E larger than $2m_0c^2$, which decay with small branches ($\sim 10^{-4}$) by internal pair conversion. The resulting positron spectrum of a transition in a heavy nuclei is sawtooth shaped with a maximum at $E_{e+} = E-2m_0c^2$. In actinides and other heavy nuclei, the level density at excitation energies above $2m_0c^2$ is high enough that the resulting positron spectrum becomes more or less continuous. It can be determined by measuring the other possible decay modes; γ-ray and internal conversion electron spectra. Using theoretical or empirical conversion coefficients one is able to reconstruct from measured γ- and conversion electron spectra the associated positron spectra.

Fig. 1 shows γ-spectra above $2m_0c^2$ energy in coincidence with scattered projectiles and/or recoils, detected in the angular range indicated for collisions of U+Pd, U+U, U+Th and Pb+Th at 5.9 MeV/u bombarding energy.[2] One observes in all cases smooth γ-spectra with exponential fall offs towards higher energies with the exception of the Pb-Th collisions, where the 2.61 MeV (3^--O^+) transition of ^{208}Pb can be recognized as a Doppler broadened weak line. Most of the high energy γ-rays result from the statistical E1 decay cascade of excited nuclei with a small E2 contribution at the low energy end. This leads to a positron spectrum which is determined dominantly by E1 pair conversion.

Fig. 2 shows a summary of early results[3],[4] for the ratio of observed positrons relative to those expected from γ-ray spectra for collisions of ^{238}U with various targets at 5.9 and 5.2 MeV/u bombarding energies. The positrons and γ-rays were measured in coincidence with heavy ions scattered at $(45 \pm 10)^{\circ}$ in the laboratory system. E1-conversion is assumed for all systems. The factor f introduced for taking into account a small E2-conversion can be taken as unity for most cases. Note that for all collision systems with united charges $Z_u = Z_1 + Z_2$ smaller than 160, all positrons observed originate from nuclear excitations. For $Z_u > 160$ a steep increase of the positron production above the nuclear contribution has been observed. A new source of positrons with the characteristics that its strength rises steeply with increasing charge Z_u of the collision system has been discovered.[3] This special feature of the new positron source, which made its detection above nuclear background possible is shown quantitativly in Fig. 3 where the positron production probability P_{e+} for the positron energy intervals indicated is plotted as function of Z_u after subtraction of the nuclear background.[2] A fit of the Z_u-dependence of P_{e+} using a power law, $P_{e+} = cZ_u^n$, gives exponents n_{exp} between 17 and 16 depending on the positron energy. From the strong Z_u-dependence it is clear that this source of positrons is of electromagnetic origin. Before we explain this exceptional Z_u-dependence, we look at the impact parameter dependence of the positron production, which shows also a characteristic scaling behaviour.[5]

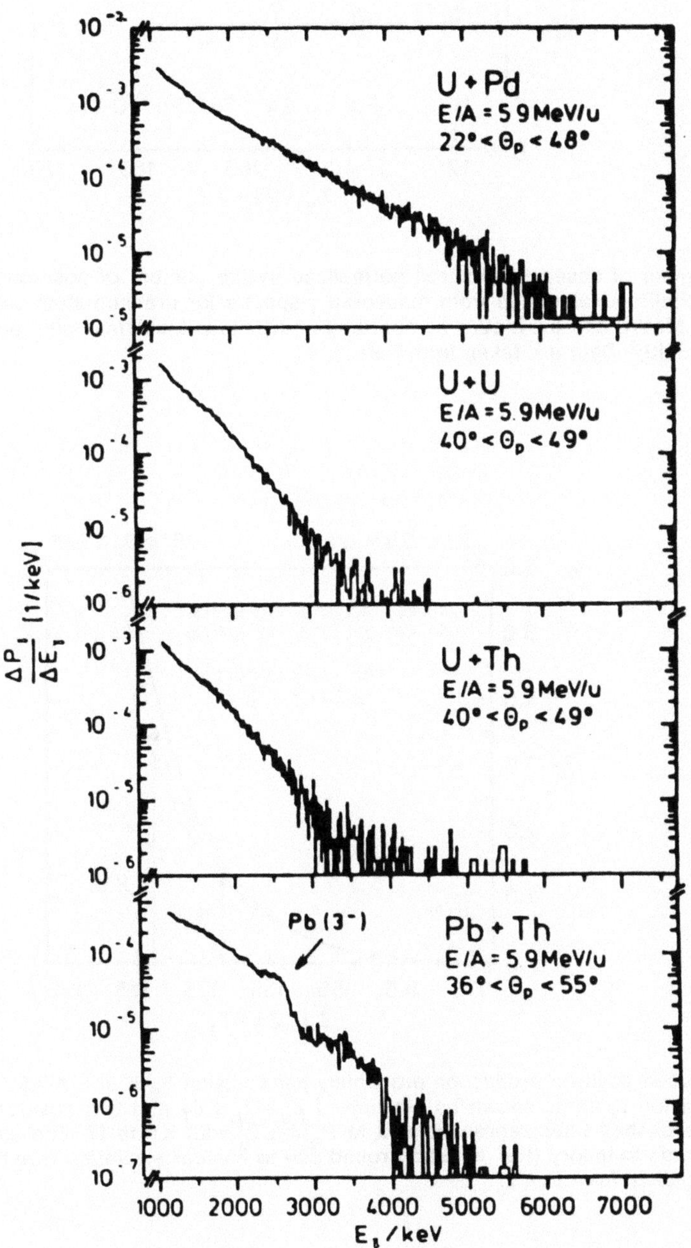

Fig. 1 Typical γ-spectra from collisions as marked. The γ-rays were measured in coincidence with particles scattered in angular ranges Θ_p as indicated (Ref. 2).

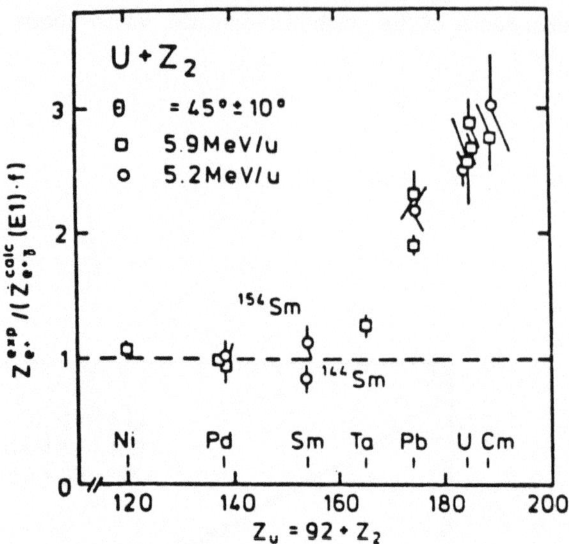

Fig. 2 Number of observed positron normalized by the number of positrons from nuclear excitations calculated from measured γ-spectra for uranium-atom collisions (Z_2) at 5.9 MeV/u and 5.2 MeV/u bombarding energy in coincidence with ions scattered at $(45 \pm 10)^o$. Data are taken from Refs. 3, 4.

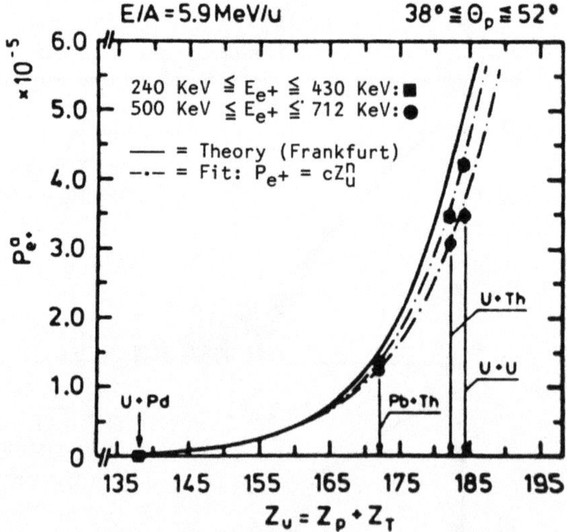

Fig. 3 Induced positron production probability per collision P_e+ at E/A = 5.9 MeV/u for the collision systems shown as function of $Z_u = Z_p + Z_T$ and two positron energy bins. The dashed lines represent a fit to $P_e^a+ = cZ_u^n$ with n = 16-17. The solid line corresponds to theory (Ref. 8). Background due to nuclear excitation is subtracted. Data are taken from Ref. 2.

Fig. 4 shows[4] the total positron production probability per collision for $^{238}U + ^{208}Pb$ encounters at 5.9 MeV/u as function of the distance of closest approach R_o. The distance of closest approach R_o was determined by measuring the projectile scattering angle θ_p and assuming Rutherford trajectories. For a unique measurement of θ_p, a kinematic coincidence between the projectile and recoil was required with an angular resolution sufficient to separate U- and Pb-nuclei. After subtraction of the nuclear contribution from the total e^+-production probability one notes that the additional positron source (atomic e^+) scales like $P_{e^+} \sim \exp(-2q_o R_o)$ with $q_o = \Delta E/\hbar v$, the minimum momentum transfer necessary for an energy transfer ΔE in a collision with relative velocity v of the participants.[6] It has been shown that the observed[7] scaling law for e^+-production is expected to hold for a mechanism by which the positrons are produced by the monopole part of the time changing Coulomb potential $V_c(Z_1, Z_2, R(t))$ of two colliding nuclei with charge Z_1, Z_2 and time changing internuclear distance R(t). In the framework of this model the strongly time changing Coulomb field induces electron transitions from bound states into the positive energy continuum as well as transitions from the negative to the positive energy continuum which lead to electron-positron pairs. The continuum continuum transitions may occur directly or via intermediate bound states, a complicated process which has to be calculated with coupled channel techniques. This has been done[8] by assuming classical trajectories for the ions, and calculating the sum of the coupled channels transition amplitudes along the ion trajectories induced by the time changing monopole part of the Coulomb potential. Adiabatic basis states of the strong Coulomb fields $V_c(Z_1, Z_2, R)$ were used. The strong fields of high

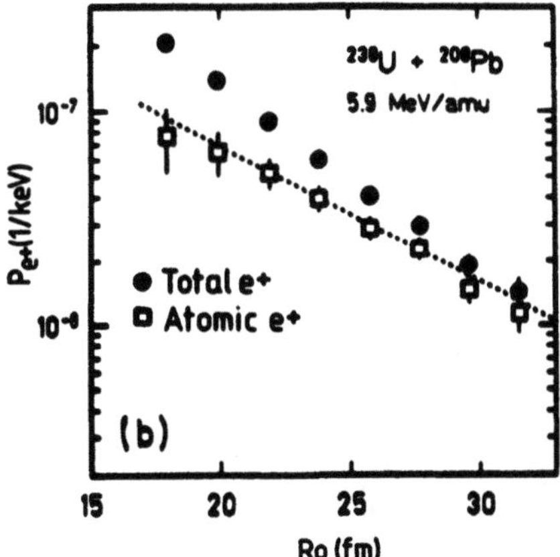

Fig. 4 Energy-averaged positron production probabilities P_{e^+} at the distance of closest approach R_o for the U-Pb ($Z_u \cong 174$) system at 5.9 MeV/u bombarding energy. Full symbols: measured P_{e^+}-values; open symbols: nuclear background subtracted; dashed line: atomic theory for elastic collisions (U. Müller, private communication). Data are taken from Ref. 4

Z_u-systems at small R, produce a relativistic enhancement of all relevant matrix elements. This can be directly observed in the absolute transition rates, but even more directly in the strong increase of the transition probabilities, when the interaction potential is changed, for example by changing Z_u or R_0, as shown in Fig. 3 and Fig. 4 respectively.

An even more detailed comparison of the theoretical predictions for induced positron production by a strong time changing Coulomb field has been done by measuring positron spectra as function of Z_u and R_0. Fig. 5 shows that theoretical calculations[8] reproduce the e^+-spectra for Pb+Th collisions in the positron energy interval from 200 keV to 1 MeV and for scattering angles between $19^0$and 67^0 quantitatively. For U-Th and U-U collisions shown in Fig. 6 the calculated induced positron spectra (solid line a) reproduce the continuous parts of the spectra again very well; when a common normalization factor of 0.8 is used for the theory. This agreement using the same normalization factor holds for both collison systems in a scattering angular range from 12^0 to 49^0.[9]

We have also seen data for U-Ta-collisions[10] (Fig. 7) with $Z_u = 165$, where the small contribution for induced positron production is well reproduced by theory[8] normalized with a factor 0.8. There are more data on the U-Au and Pb-Pb-collision systems ready for analysis. But I think we are safe to conclude at this point that we understand induced positron production within a normalization uncertainty of about 20 % for a large range of conditions, which vary the interaction strength of the time changing Coulomb field considerably.

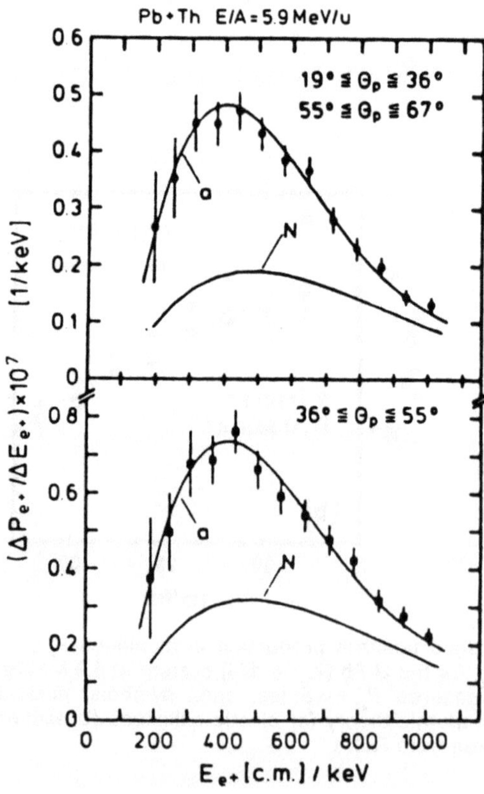

Fig. 5 Positron spectra from Pb-Th-collisions at E/A=5.9 MeV/u in coincidence with heavy ions detected in the scattering angular ranges indicated. The nuclear background (N) is already subtracted. Curve (a) is a theoretical calculation for induced positron creation, without fitting any scale (U. Müller, private communication). Data are taken from Ref. 2.

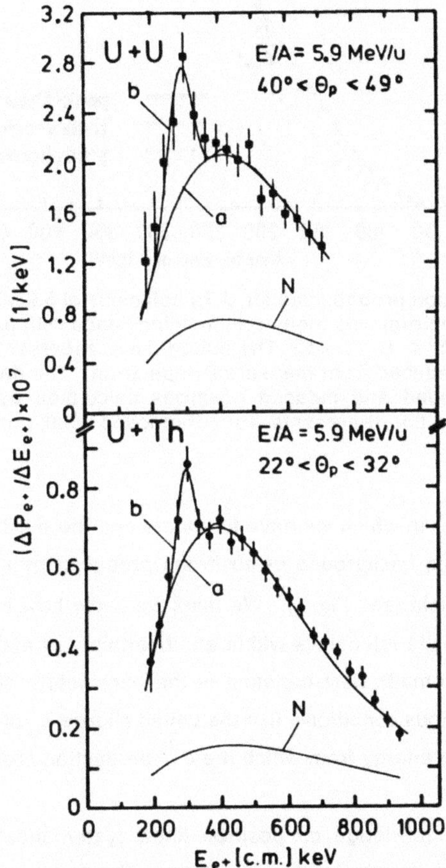

Fig. 6 Positron spectra of U-Th and U-U-collisions at E/A=5.9 MeV/u in coincidence with heavy ions detected in the scattering angular ranges indicated. The nuclear background (N) is subtracted. Curve (a) represents a calculation for induced positrons (U. Müller, private communication), fitted with a normalization constant (~ 0.80). Curve (b) is a superimposed fit of a line function. Data are taken from Ref. 2, 12.

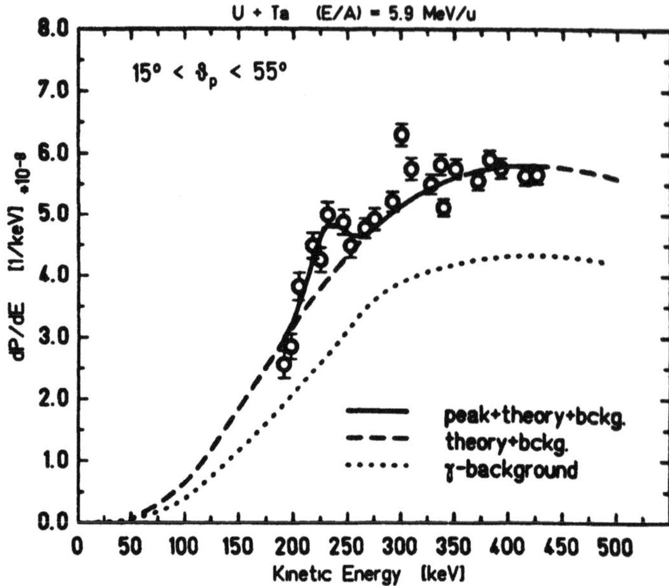

Fig. 7 Positron production probabilities for U-Ta-collisions at 5.9 MeV/u bombarding energy. The positron spectrum was measured in coincidence with particles scattered into the angular range $15^0 < \theta \quad p < 55^0$ The dotted curve represents positrons from nuclear excitations as deduced from measured γ-spectrum. The dashed curve is the sum of nuclear background and induced positrons calculated by a coupled channel approach. The solid line represents a superimposed weak positron line. Data are taken from Ref. 10.

We have set the stage now in which we have to understand the positron lines. They are superimposed on the continuous background of positrons produced by nuclear excitation and the time changing Coulomb field (see Fig. 6). We think we know how to determine their spectral forms and also their absolute intensities within an uncertainty of maximum 20 %. This knowledge is useful for a systematic determination of the parameters of the e^+-lines, especially their intensities under various conditions, like the united charge Z_u of the system, the scattering angle and the bombarding energy from which the e^+-production cross section depends.

In order to discuss our knowledge on positron lines systematically we like to ask a few questions, which we try to answer in view of all data. The first question may be: Do positron lines occur at all in heavy ion collisions? The answer could be that in all heavy systems with $Z_u > 180$ lines were observed,[11,12,13,9] reproduceably even with different devices (Fig. 8). There are problems related to serious target deterioration during heavy ion bombardment. This has the effect that the projectile energy is lowered during the bombardment in a uncontrollable fashion, demonstrated by the energy spectra of scattered particles taken with a monitor detector at different times of an experiment in which a uranium metal target of 800 $\mu g/cm^2$ thickness was bombarded with 5.9 MeV/u ^{238}U-ions (Fig. 9). One notes a shift of the elastic scattering line towards lower energies and more serious a strong low energy tailing. Although care was taken by exchanging targets after certain bombarding times or when some criteria were not met anymore in the monitor spectra, the effective target thickness and bombarding energy was hard to control. This could be one experimental reason why one had to apply a normalization factor of less than one in the comparison of theoretical and experimental induced positron production; the bombarding energy might have been effectively lower than assumed. This experimental

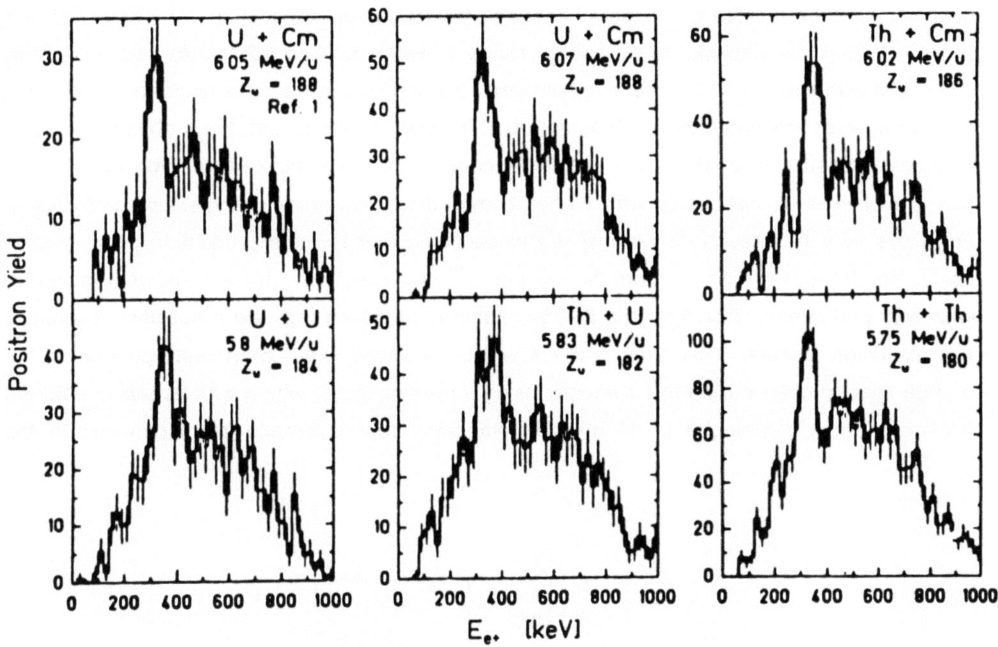

Fig. 8 Positron spectra for five collision systems as indicated (Ref. 13). The first panel (upper left corner) repeats data from Schweppe et al. (Ref. 11).

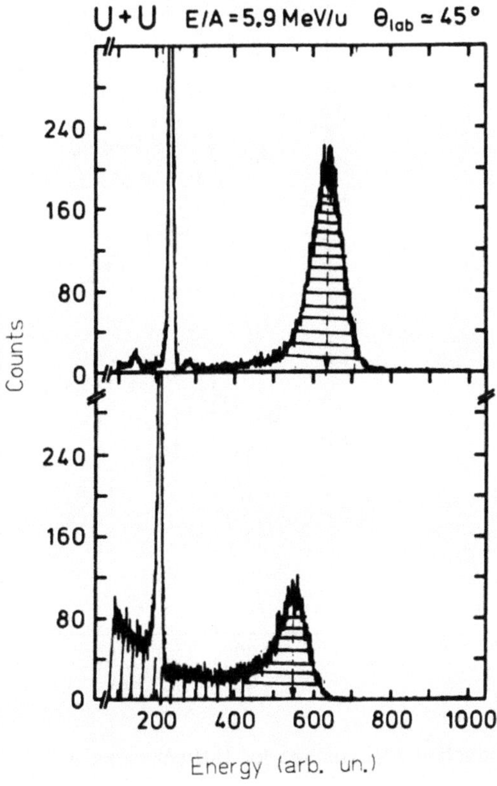

Fig. 9 Monitor spectrum of U-ions scattered by an uranium metal target of 800 μg/cm^2 thickness, taken at different times of the experiment.

deficiency prevented also a measurement of a reliable excitation function. The data[2,9] shown in Fig. 10 for U-U-collisions in the energy range between 5.7 and 6.2 MeV/u were carefully controlled with respect to the effective bombarding energy using the criteria that the shift of the monitor spectra amounted to less than a certain fraction for one target, but it obviously suffers from low statistics what the line intensity is concerned. The induced positron contribution (curve a) increases noticeably with higher bombarding energy as is expected theoretically. From this very first result one is inclined to conclude that the line production cross section scales not drastically different from the induced e^+-cross section. This is somewhat in conflict with very preliminary data from the EPOS-collaboration, which indicate a possible resonance like excitation function. This open problem will be resolved with experiments planned at the orange spectrometer during this summer. Using a rotating target wheel with targets of different thickness being exchanged in 10 ms time intervals, the occurence of resonances in the

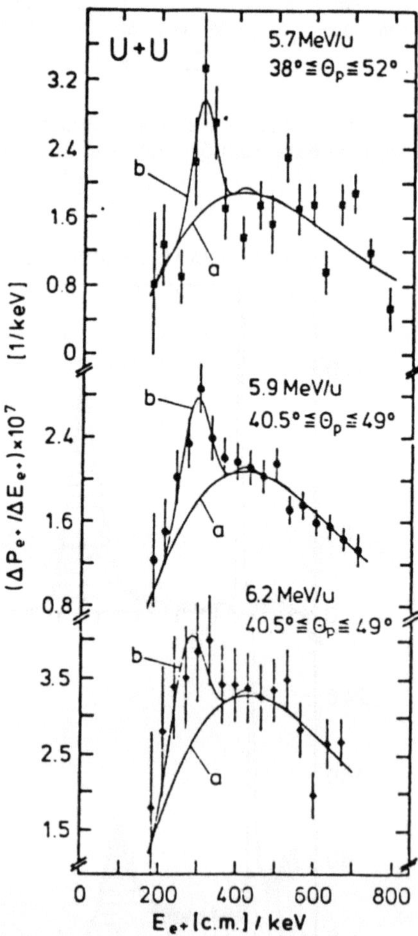

Fig. 10 Positron production probabilities for U-U-collisions at 5.7, 5.9 and 6.2 MeV/u bombarding energies and scattering angular ranges as indicated. The nuclear background is subtracted in all spectra. The curves marked a and b are theoretical curves for induced and spontaneous positron emission, respectively. Data are taken from Ref. 2, 9.

excitation function should be reliably detected. We also hope to measure reproduceable excitation functions in a large bombarding energy range, because target deterioration effects seem to be no severe problem any more with the target wheel. Excitation functions in symmetric collision systems may also help to solve the question on the dependence of the e^+-line production probability from the distance of closest approach R_0. In symmetric systems R_0 can only be determined uniquely at the c.m. scattering angle 90^0. Thus in order to study the R_0-dependence of the e^+ production in a transparent way in symmetric systems one has to resort to the measurement of excitation functions. We come back to this problem in detail when we discuss the measurements of the scattering angle dependence of e^+-production.

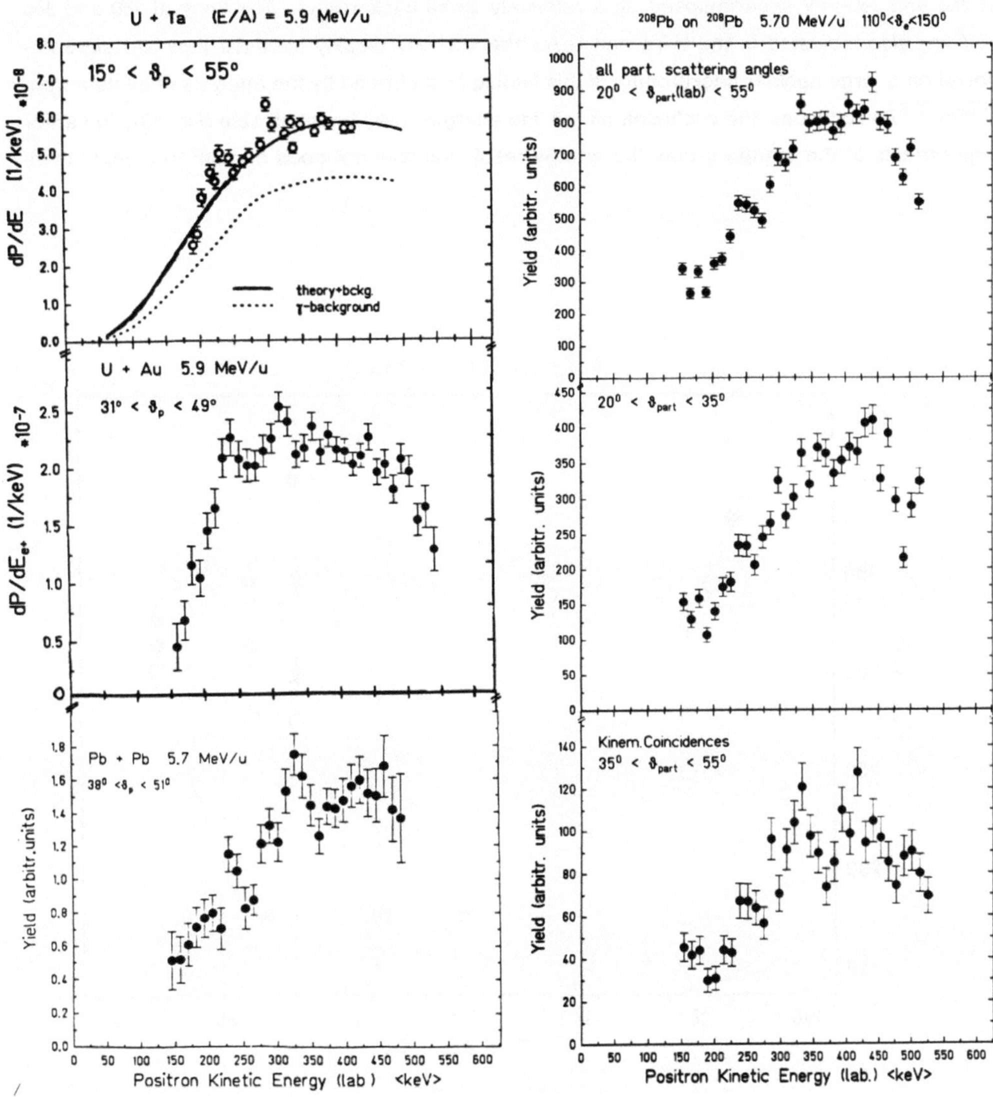

Fig. 11

Positron spectra of the undercritical collision system U+Ta (Z_u=165), U+Au (Z_u=171) and Pb+Pb (Z_u=164) at bombarding energies as indicated.

Fig. 12

Preliminary positron spectra for the Pb+Pb collision systems at 5.70 MeV/u bombarding energy for particle scattering as indicated.

We like to address the next problem with the question: Do we observe lines in undercritical systems? Very recent high statistics and high resolution data[14] presented by Koenig from the new orange spectrometer set up (Fig. 11) show indications of lines in Pb-Pb, U-Ta and U-Au collision with intensities which are at least one order of magnitude smaller than in U-U-collisions. Especially a complete analysis of the data from Pb-Pb-collisions (Fig. 12) which could be only gained with the rotating target wheel may give even a new suprising touch to the story.

In ^{208}Pb-^{208}Pb collisions the nuclear background is very low. It is mainly due to the excitation of the 2.61 (3^-)-state which causes small background that can easily be determined. The preliminary spectra from the orange spectrometer shown by Koenig indicate possibly two e^+-lines at 250 and 340 keV superimposed on a relatively small background. The lines at 250 and 340 keV are also indicated in the U-Ta and U-Au spectra very weakly because they are superimposed on a large nuclear background. If this finding is confirmed by the analysis of all data from ^{208}Pb-^{208}Pb collisions, the confusion on the line energies may be resolvable (Fig. 13). In earlier experiments of the orange group, the energy resolution was not good enough to identify them

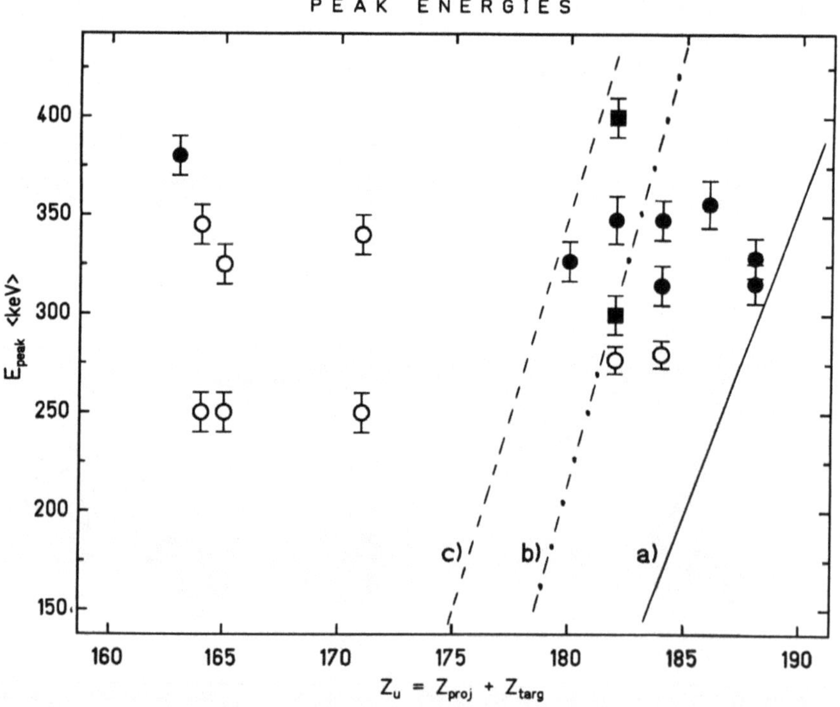

Fig. 13 Summary of all positron line energies observed in various systems. Solid points are data from the EOPOS collaboration, open circles represent measurements by the orange spectrometer. The lines represent calculations of peak energies for spontaneous positron emission from nuclear systems with either deformed, touching configurations with the centers separated by 17 fm (a) or spherical nuclei with normal densities (b), both with 50 fold ionization. Bare nuclei with spherical shape are presented by line (c).

all, in view of large backgrounds. A similar statement may also apply to the e^+- single spectra of the EPOS-spectrometer, which suffer in principle from a large Doppler broadening, if the source moves with the c.m. velocity of the collision system. "Kinematic cuts", which seem to lower the background, were successfully used to dig out e^+-line structures around 320 keV (Fig. 8), which may have contributions from several lines. This is supported by very recent e^+-e^--coincidence sum coincidence spectra,[15] which show at least two lines one around 310 keV and the other at 380 keV. If the source is moving with c.m. velocity, the sum coincidence spectra, with the e^+ and e^- detected preferentially at 180°, show less Doppler broadening. This leads to a better energy resolution and also background suppression. The detection of the 250 keV line in the present coincidence studies is difficult due to the small efficiency of EPOS at low energies. The orange spectrometer data have lower efficiency at energies of 380 keV, so that no definite statement can be made yet on the existence of such a line.

Next I like to discuss the Z_u-dependence of the e^+-line production. The data from the orange spectrometer indicates a strong increase of the e^+-line production with Z_u(Fig. 14) similar to the induced positron creation, which is well understood as a high field effect. More information will be available from Pb-Pb-collisions. There seems to be a strong controversary of this data set with the cross section for the e^+-line production in Th-Ta-collisions, which was determined by EPOS to be in the order of 10 μb/sr, indicating together with similar values[13] for the heavy collision systems an independence of the e^+ production probability from Z_u. It is most important to resolve this inconsistency because a strong Z_u-dependence suggested by the orange spectrometer measurements would strongly favour a pair creation process by strong electromagnetic fields.

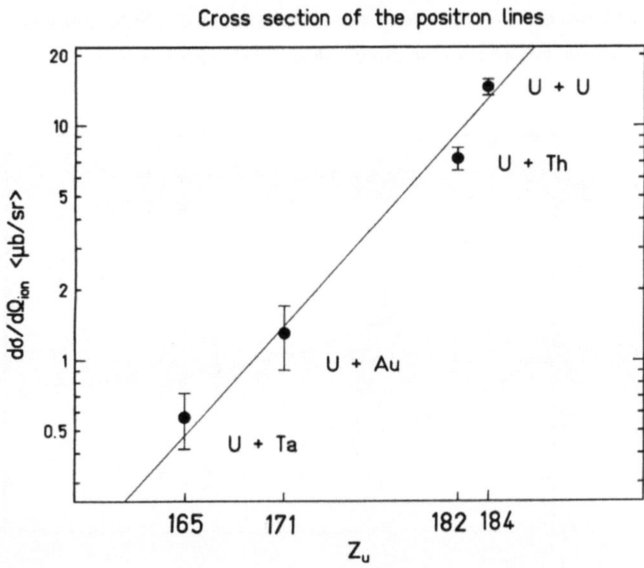

Fig. 14 Cross section of the positron line at ~ 260 keV for four different collision systems (undercritical (U+Ta, U+Au) as well as overcritical (U+Th, U+U)). The fitted line shows a Z_u^{30}-dependence of the cross section.

Most important in this context are also the results on the scattering angle dependence of the e^+-production. Fig. 15 shows results for the symmetric U-U and the slightly asymmetric U-Th collisions.[9] Both data are difficult to interprete because they contain contributions from the scattering angles Θ and $(\pi\text{-}\Theta)$, which cannot be seperated by principle for symmetric systems like U+U. In the case of U-Th the resolution of the measuring device did not allow an identification of the collision partner. In view of this interpretation difficulty, the fit of a 1/sin Θ-distribution, which would be expected for a quasimolecular reaction, is probable only fortuitously reproducing the data so well.[9] More likely the data should be interpreted as presenting a cross section which rises very strongly for backward scattering angles, which are contained in the small angles. The data actually have a strong indication for such a behaviour, because the cross section seems to rise for angles smaller than 90°. If this behaviour could be substantiated by a measurement in a asymmetric system with complete particle resolution, it would be consistent with the interpretation of the strong $Z_u$dependence that the process is electromagnetic in nature. In this case the cross section would scale with the strength of the Coulomb interaction which depends on Z_u and $R_o$the distance of closest approach. R_o can be decreased in Rutherford scattering by increasing the scattering angle or increasing the bombarding energy. Experiments planned at the orange spectrometer should resolve this important issue very soon.

Finally I will discuss the implication of the results of the e^+-e^--coincidence experiments pioneered by the EPOS- collaboration[16] and presented so nicely by T. Cowan.[15] Their main result is that in coincidence with the positron lines, electron lines with the same energies are emitted. It is indicated that the coincidence spectra show sum lines which appear to be sharper than the single lines. After a complete evaluation of their data, they should be able to make a statement on the degree of back to back correlation in the e^+-e^--emission. For this they have to understand the very complicated response function of their apparatus on emission angle and e^+/e^--energy quantitatively. We have the impression that this is indeed the case using powerful Monte Carlo simulations based on efficiency calibrations with sources.

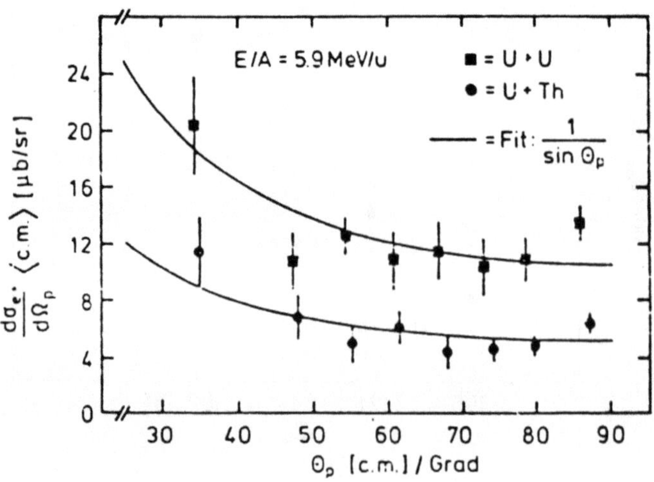

Fig. 15 Differential production cross section of the positron line as function of the scattering angle Θ_P for U-U and U-Th collisions. $d\sigma_{e^+}/d\Omega_P$is fitted by a $(1/\sin\Theta_P)$ angular distribution (Ref. 9)

Fig. 16 shows the exciting first results of e^+-e^--coincidence measurements in U-Th collisions at 5.83 MeV/u bombarding energy. The pairs are measured in coincidence with ions scattered between 20° and 70° in the laboratory system. The coincidence spectra gated on certain electron (a), positron (b), sum (c) and difference (d) energies show lines, which indicate back to back decay into an electron positron pair with an energy of (380 ± 15) keV for the positron and (375 ± 15) keV for the electron. The appearance of a line at zero difference energy is the strongest evidence that the pair does not originate from the decay of an excited nuclear state for which one expects a variable partition of the available energy on the electron and positron (three body decay). The width of the line in the sum spectrum is a measure of the degree of 180^ocorrelation of the e^+-e^--pair decay, because in such a case one expects first order cancellation of the Dopplershifts introduced by a moving c.m. system.

Preliminary results of new coincidence measurements were reported by Cowan, Bokemeyer and Stiebing during this conference for U-Th collisions at 5.87 MeV/u. The pair events with a very narrow sum energy of about 800 keV were cleanly seen again in the difference as well as the coincident e^+ and e^--spectra (Fig. 17, upper panel). Surprisingly one observes for different kinematic conditions pair events with a sum energy of about 600 keV and corresponding lines in the difference and single spectra (lower panel). Obviously better statistics are needed to get a clear cut answer concerning the conditions which determine the appearance of correlated e^+-e^--pairs. Nevertheless it seems to be indicated by the coincidence results shown that

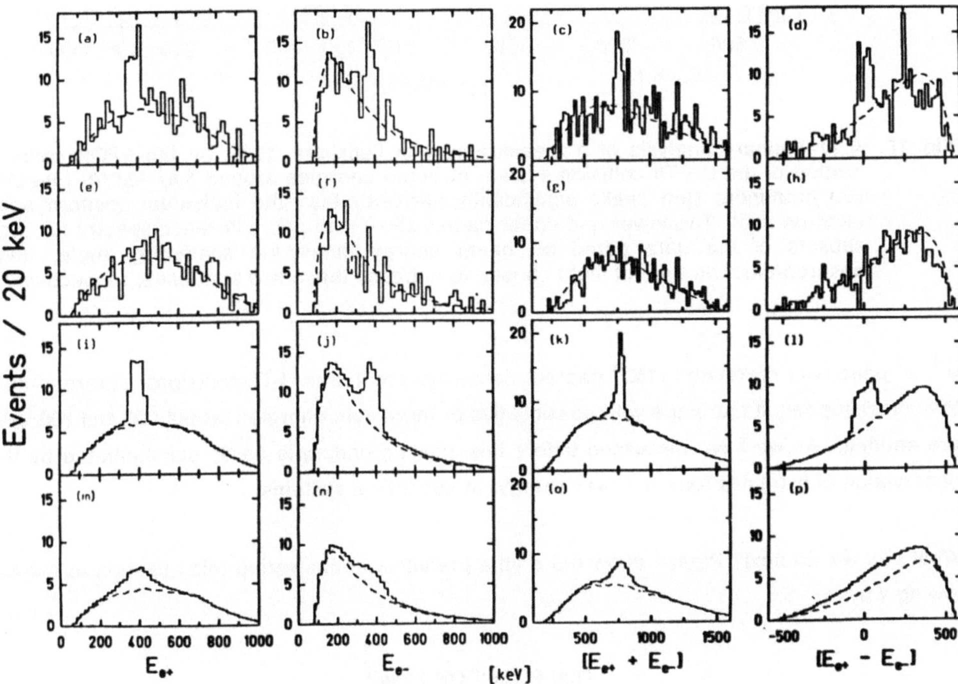

Fig. 16 Parts (a)-(h) show projections of the measured intensity distribution of Fig. 31 (a); parts (i)-(p) represent Monte Carlo simulations: (i)-(l) for the two body decay of a neutral particle (scaled by $6x10^{-4}$), and (m)-(p) for internal pair conversion of a nuclear state (scaled by $6x10^{-3}$). The columns correspond to projections onto the E_e+, E_e-, (E_e+ + E_e-), and (E_e+ - E_e-) axes, respectively. Parts (a) - (d), as well as (i) - (l) and (m) - (p), correspond to the gates labelled A-D in Fig. 31 (a), respectively. Parts (e) to (h) are the average of similar gates adjacent to either side of gates A-D, respectively. Data are taken from Ref. 16.

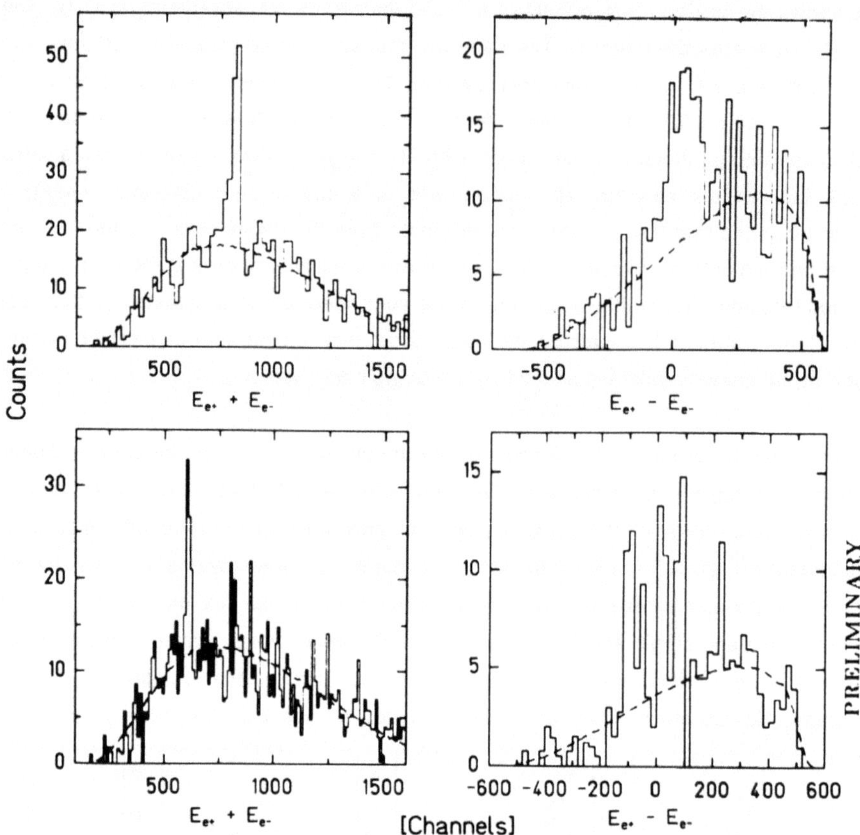

Fig. 17 A preliminary analysis of a measurement in February, 1986, by the EPOS collaboration of the U+Th collision system at beam energies around 5.87 MeV/u reveals two prominent sum peaks significantly narrower than the individual positron and electron lines. The lower and upper panels show sum and difference spectra for two subsets of the data gated on beam energy, heavy-ion scattering angle, and positron/electron time of flight chosen to enhance these two sum lines, respectively.

e^+-e^- pairs with correlated (180° degree) decay are created in U-Th-collisions. There is also the surprising result that pairs with possibly two or more sum energies (about 600 and 800 keV) are emitted. As we have discussed before this exciting finding is partly substantiated by the observation of two lines (one at lower energy) in subcritical systems.

What can we do next? Please allow me a little prejudice in answering this question in the following way:

"Let POSEIDON blow!"

Poseidon (*P*ositron-*E*lectron- *I*dentification by *D*ouble- *O*ra*n*gespectrometry) a double orange spectrometer for coincidence spectroscopy of e^+-e^--pairs (Fig. 18) will be installed in the frame of an extension program by adding a second spectrometer in the direction of the beam. The target is placed at the object point of both spectrometers, which will be equipped with identical existing detectors.

The backward spectrometer may be used for detecting positrons while the forward one may focus coincident electrons. The spectrometers may be also used as two independent e^+-spectrometers for doubling the detection efficiency and simultaneous measurements of e^+-spectra in and opposite to the beam direction. Positron sensitive heavy ion detectors may be placed in the forward cone of the forward spectrometer or outside of the current coils. This

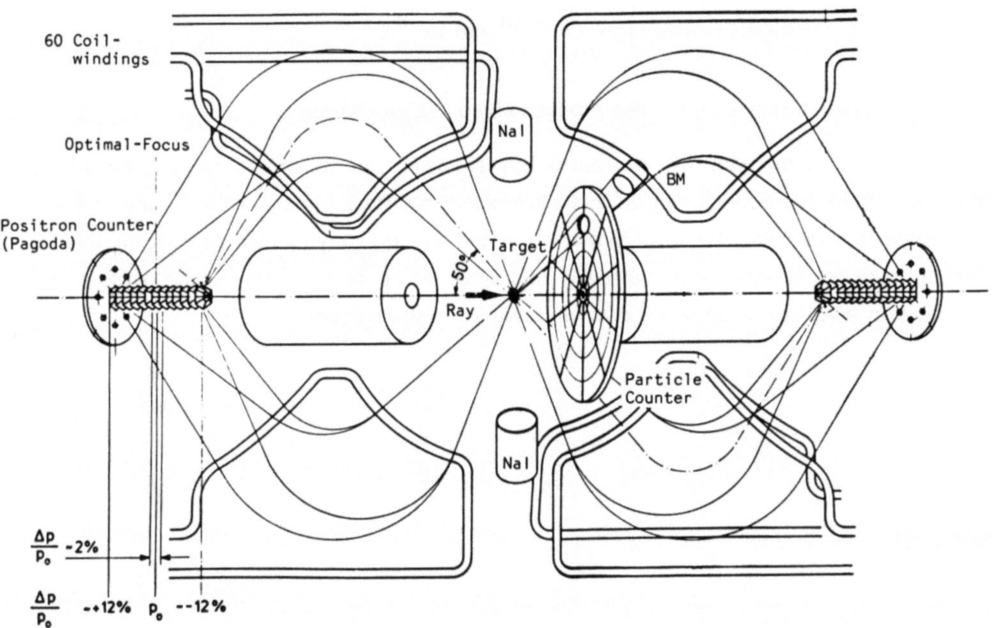

Fig. 18 Double orange spectrometer POSEIDON (Positron-Electron-Identification Double-Orange-Spectrometer)

instrument would have a 100 % coincidence detection efficiency for back to back emitted e^+-e^--pairs. With the existing detection devices the angular acceptance *and* definition in polar direction is $180^\circ \pm 25^\circ$ and in azimuthal direction $180^\circ \pm 60^\circ$. The polar angular definition could be improved by measuring e^+-e^--time of flight differences, whereas an improvement of the azimuthal angular resolution would require a new detection system.

I should like to conclude my summary with my best *guess* concerning the origin of the positron lines:

- The e^+-lines may originate from objects, which decay to e^+-e^--pairs (two body decay)

- The objects weigh roughly 3 m_e.

- The energies of the objects seem to show fine structures ($\sim$ 80 keV), which points towards a spectroscpy of composite systems.

- The objects are produced by strong electromagnetic fields.

- The life time of the object ranges between 10^{-10} and 10^{-19}s.

- I heard rumors that they may be also produced by e^+-e^--scattering.

It is up to you to name with my "Warrant of apprehension" the delinquent tracked for a long time. I am afraid that nobody can do that now except our dear friend Walter Greiner, who is responsible for everything. We want to thank him and his staff very much for arranging this interesting meeting in this most charming and quiet place of Italy. So, gracia Walter and arrivederci in ACQUA Fredda as soon as we have found the solution.

References

1. W. Greiner, Quantum Electrodynamics of Strong Fields, in: NATO Advanced Study Institutes, Series B, Physics Volume 80 (1981) Plenum Press, New York and London.

2. H. Tsertos, Thesis, TU München (1985). C. Kozhuharov, P. Kienle, E. Berdermann, H. Bokemeyer, J. S. Greenberg, Y. Nahayama, P. Vincent, H. Backe, L. Handschug, E. Kankeleit, Phys. rev. Lett. 42:376 (1979).

3. H. Backe and C. Kozhuharov, Progress in Atomic Spectroscopy Part C, ed. by H. J. Beyer and H. Kleinpoppen, Plenum Press 459 (1984).

4. H. Bokemeyer, GSI preprint, 84-43 (1984).

5. J. Bang and J. M. Hansteen, Phys. Lett. 72A:218 (1979).

6. P. Armbruster and P. Kienle, Z. Physik A291:399 (1979).

7. D. Liesen, P. Armbruster, F. Bosch, S. Hagmann, P.H. Mokler et al. Phys. Rev. Lett. 44:983 (1980).

8. U. Müller, G. Soff, T. de Reus, J. Reinhardt, B. Müller, W. Greiner, Z. Physik A313:263 (1983).

9. H. Tsertos, E. Berdermann, F. Bosch, M. Clemente, P. Kienle, W. Koenig, C. Kozhuharov and W. Wagner, Phys. Lett. B162:273 (1985).

10. E. Berdermann, F. Bosch, M. Franosch, S. Huchler, J. Kemmer, P. Kienle, C. Kozhuharov, H. Tsertos, W. Wagner, GSI Scientific Report 1985 p. 179

11. J. Schweppe, A. Gruppe, K. Bethge, H. Bokemeyer, T. Cowan, H. Folger, J. S. Greenberg, H. Grein, S. Ito, R. Schule, D. Schwalm, K. E. Stiebing, N. Trautmann, P. Vincent, M. Waldschmidt, Phys. Rev. Lett. 51:2261 (1983).

12. M. Clemente, E. Berdermann, P. Kienle, H. Tsertos, W. Wagner, C. Kozhuharov, F. Bosch, W. Koenig, Phys. Lett. B137:41 (1984).

13. T. Cowan, H. Backe, M. Begemann, K. Bethge, H. Bokemeyer, H. Folger, J. S. Greenberg, H. Grein, A. Gruppe, Y. Kido, M. Klüver, D. Schwalm, J. Schweppe, K. E. Stiebing, N. Trautmann, P. Vincent, Phys. Rev. Lett. 54:1761 (1985).

14. W. Koenig, Contribution to this conference

15. T. Cowan, Contribution to this conference

16. T. Cowan, H. Backe, K. Bethge, H. Bokemeyer, H. Folger, J. S. Greenberg, K. Sakaguchi, D. Schwalm, J. Schweppe, K. E. Stiebing, P. Vincent, Phys. Rev. Lett. 56:444 (1986).

INDEX

MIX
Papier aus verantwortungsvollen Quellen
Paper from responsible sources
FSC® C105338

If you have any concerns about our products,
you can contact us on
ProductSafety@springernature.com

In case Publisher is established outside the EU,
the EU authorized representative is:
Springer Nature Customer Service Center GmbH
Europaplatz 3, 69115 Heidelberg, Germany

Printed by Libri Plureos GmbH
in Hamburg, Germany